D1068611

Springer Texts in Statistics

Springer Texts in Statistics

Alfred: Elements of Statistics for the Life and Social Sciences
Berger: An Introduction to Probability and Stochastic Processes, Second Edition
Bilodeau and Brenner: Theory of Multivariate Statistics
Blom: Probability and Statistics: Theory and Applications
Brockwell and Davis: Introduction to Times Series and Forecasting, Second Edition
Carmona: Statistical Analysis of Financial Data in S-Plus
Chow and Teicher: Probability Theory: Independence, Interchangeability, Martingales, Third Edition
Christensen: Advanced Linear Modeling: Multivariate, Time Series, and Spatial Data; Nonparametric Regression and Response Surface Maximization, Second Edition
Christensen: Log-Linear Models and Logistic Regression, Second Edition
Christensen: Plane Answers to Complex Questions: The Theory of Linear Models, Third Edition
Creighton: A First Course in Probability Models and Statistical Inference
Davis: Statistical Methods for the Analysis of Repeated Measurements
Dean and Voss: Design and Analysis of Experiments
du Toit, Steyn, and Stumpf: Graphical Exploratory Data Analysis
Durrett: Essentials of Stochastic Processes
Edwards: Introduction to Graphical Modelling, Second Edition
Everitt: An R and S-Plus® Companion to Multivariate Analysis
Finkelstein and Levin: Statistics for Lawyers
Flury: A First Course in Multivariate Statistics
Gut: Probability: A Graduate Course
Heiberger and Holland: Statistical Analysis and Data Display: An Intermediate Course with Examples in S-PLUS, R, and SAS
Jobson: Applied Multivariate Data Analysis, Volume I: Regression and Experimental Design
Jobson: Applied Multivariate Data Analysis, Volume II: Categorical and Multivariate Methods
Kalbfleisch: Probability and Statistical Inference, Volume I: Probability, Second Edition
Kalbfleisch: Probability and Statistical Inference, Volume II: Statistical Inference, Second Edition
Karr: Probability
Keyfitz: Applied Mathematical Demography, Second Edition
Kiefer: Introduction to Statistical Inference
Kokoska and Nevison: Statistical Tables and Formulae
Kulkarni: Modeling, Analysis, Design, and Control of Stochastic Systems
Lange: Applied Probability
Lange: Optimization
Lehmann: Elements of Large-Sample Theory

(continued after index)

E.L. Lehmann Joseph P. Romano

Testing Statistical Hypotheses

Third Edition

E.L. Lehmann
Professor of Statistics Emeritus
Department of Statistics
University of California, Berkeley
Berkeley, CA 94720
USA

Joseph P. Romano
Department of Statistics
Stanford University
Sequoia Hall
Stanford, CA 94305
USA
romano@stanford.edu

Library of Congress Cataloging-in-Publication Data
A catalog record for this book is available from the Library of Congress.

ISBN 0-387-98864-5 Printed on acid-free paper.

Printed in the United States of America.

9 8 7 6 5 4 3

springer.com

Dedicated to the Memory of

Lucien Le Cam (1924-2000) and John W. Tukey (1915-2000)

Preface to the Third Edition

The Third Edition of *Testing Statistical Hypotheses* brings it into consonance with the Second Edition of its companion volume on point estimation (Lehmann and Casella, 1998) to which we shall refer as *TPE*2. We won't here comment on the long history of the book which is recounted in Lehmann (1997) but shall use this Preface to indicate the principal changes from the 2nd Edition.

The present volume is divided into two parts. Part I (Chapters 1–10) treats small-sample theory, while Part II (Chapters 11–15) treats large-sample theory. The preface to the 2nd Edition stated that "the most important omission is an adequate treatment of optimality paralleling that given for estimation in *TPE*." We shall here remedy this failure by treating the difficult topic of asymptotic optimality (in Chapter 13) together with the large-sample tools needed for this purpose (in Chapters 11 and 12). Having developed these tools, we use them in Chapter 14 to give a much fuller treatment of tests of goodness of fit than was possible in the 2nd Edition, and in Chapter 15 to provide an introduction to the bootstrap and related techniques. Various large-sample considerations that in the Second Edition were discussed in earlier chapters now have been moved to Chapter 11.

Another major addition is a more comprehensive treatment of multiple testing including some recent optimality results. This topic is now presented in Chapter 9. In order to make room for these extensive additions, we had to eliminate some material found in the Second Edition, primarily the coverage of the multivariate linear hypothesis.

Except for some of the basic results from Part I, a detailed knowledge of small-sample theory is not required for Part II. In particular, the necessary background should include: Chapter 3, Sections 3.1–3.5, 3.8–3.9; Chapter 4: Sections 4.1–4.4; Chapter 5, Sections 5.1–5.3; Chapter 6, Sections 6.1–6.2; Chapter 7, Sections 7.1–7.2; Chapter 8, Sections 8.1–8.2, 8.4–8.5.

Of the two principal additions to the Third Edition, multiple comparisons and asymptotic optimality, each has a godfather. The development of multiple comparisons owes much to the 1953 volume on the subject by John Tukey, a mimeographed version which was widely distributed at the time. It was officially published only in 1994 as Volume VIII in *The Collected Works of John W. Tukey.*

Many of the basic ideas on asymptotic optimality are due to the work of Le Cam between 1955 and 1980. It culminated in his 1986 book, *Asymptotic Methods in Statistical Decision Theory.*

The work of these two authors, both of whom died in 2000, spans the achievements of statistics in the second half of the 20th century, from model-free data analysis to the most abstract and mathematical asymptotic theory. In acknowledgment of their great accomplishments, this volume is dedicated to their memory.

Special thanks to George Chang, Noureddine El Karoui, Matt Finkelman, Brit Katzen, Mee Young Park, Elizabeth Purdom, Armin Schwartzman, Azeem Shaikh and the many students at Stanford University who proofread several versions of the new chapters and worked through many of the over 300 new problems. The support and suggestions of our colleagues is greatly appreciated, especially Persi Diaconis, Brad Efron, Susan Holmes, Balasubramanian Narasimhan, Dimitris Politis, Julie Shaffer, Guenther Walther and Michael Wolf. Finally, heartfelt thanks go to friends and family who provided continual encouragement, especially Joe Chavez, Ann Marie and Mark Hodges, David Fogle, Scott Madover, Tom Neville, David Olachea, Janis and Jon Squire, Lucy, and Ron Susek.

E. L. Lehmann
Joseph P. Romano

January, 2005

Contents

Part I

Small-Sample Theory

1
The General Decision Problem

1.1 Statistical Inference and Statistical Decisions

The raw material of a statistical investigation is a set of observations; these are the values taken on by random variables X whose distribution P_θ is at least partly unknown. Of the parameter θ, which labels the distribution, it is assumed known only that it lies in a certain set Ω, the *parameter space. Statistical inference* is concerned with methods of using this observational material to obtain information concerning the distribution of X or the parameter θ with which it is labeled. To arrive at a more precise formulation of the problem we shall consider the purpose of the inference.

The need for statistical analysis stems from the fact that the distribution of X, and hence some aspect of the situation underlying the mathematical model, is not known. The consequence of such a lack of knowledge is uncertainty as to the best mode of behavior. To formalize this, suppose that a choice has to be made between a number of alternative actions. The observations, by providing information about the distribution from which they came, also provide guidance as to the best decision. The problem is to determine a rule which, for each set of values of the observations, specifies what decision should be taken. Mathematically such a rule is a function δ, which to each possible value x of the random variables assigns a decision $d = \delta(x)$, that is, a function whose domain is the set of values of X and whose range is the set of possible decisions.

In order to see how δ should be chosen, one must compare the consequences of using different rules. To this end suppose that the consequence of taking decision d when the distribution of X is P_θ is a *loss*, which can be expressed as a nonnegative real number $L(\theta, d)$. Then the long-term average loss that would result from the use of δ in a number of repetitions of the experiment is the expectation

$E[L(\theta, \delta(X))]$ evaluated under the assumption that P_θ is the true distribution of X. This expectation, which depends on the decision rule δ and the distribution P_θ, is called the *risk function* of δ and will be denoted by $R(\theta, \delta)$. By basing the decision on the observations, the original problem of choosing a decision d with loss function $L(\theta, d)$ is thus replaced by that of choosing δ, where the loss is now $R(\theta, \delta)$.

The above discussion suggests that the aim of statistics is the selection of a decision function which minimizes the resulting risk. As will be seen later, this statement of aims is not sufficiently precise to be meaningful; its proper interpretation is in fact one of the basic problems of the theory.

1.2 Specification of a Decision Problem

The methods required for the solution of a specific statistical problem depend quite strongly on the three elements that define it: the class $\mathcal{P} = \{P_\theta, \theta \in \Omega\}$ to which the distribution of X is assumed to belong; the structure of the space D of possible decisions d; and the form of the loss function L. In order to obtain concrete results it is therefore necessary to make specific assumptions about these elements. On the other hand, if the theory is to be more than a collection of isolated results, the assumptions must be broad enough either to be of wide applicability or to define classes of problems for which a unified treatment is possible.

Consider first the specification of the class $\mathcal{P}$. Precise numerical assumptions concerning probabilities or probability distributions are usually not warranted. However, it is frequently possible to assume that certain events have equal probabilities and that certain other are statistically independent. Another type of assumption concerns the relative order of certain infinitesimal probabilities, for example the probability of occurrences in an interval of time or space as the length of the internal tends to zero. The following classes of distributions are derived on the basis of only such assumptions, and are therefore applicable in a great variety of situations.

The *binomial* distribution $b(p, n)$ with

$$P(X = x) = \binom{n}{x} p^x (1-p)^{n-x}, \qquad x = 0, \ldots, n. \quad 0 \le p \le 1. \tag{1.1}$$

This is the distribution of the total number of successes in n independent trials when the probability of success for each trial is p.

The *Poisson* distribution $P(\tau)$ with

$$P(X = x) = \frac{\tau^x}{x!} e^{-\tau}, \qquad x = 0, 1, \ldots, \quad 0 < \tau. \tag{1.2}$$

This is the distribution of the number of events occurring in a fixed interval of time or space if the probability of more than one occurrence in a very short interval is of smaller order of magnitude than that of a single occurrence, and if the numbers of events in nonoverlapping intervals are statistically independent. Under these assumptions, the process generating the events is called a *Poisson process*. Such processes are discussed, for example, in the books by Feller (1968), Ross (1996), and Taylor and Karlin (1998).

The *normal* distribution $N(\xi, \sigma^2)$ with probability density

$$p(x) = \frac{1}{\sqrt{2\pi}\sigma} \exp\left[-\frac{1}{2\sigma^2}(x-\xi)^2\right], \qquad -\infty < x, \xi < \infty, \quad 0 < \sigma. \tag{1.3}$$

Under very general conditions, which are made precise by the central limit theorem, this is the approximate distribution of the sum of a large number of independent random variables when the relative contribution of each term to the sum is small.

We consider next the structure of the decision space D. The great variety of possibilities is indicated by the following examples.

Example 1.2.1 Let $X_1, \ldots, X_n$ be a *sample* from one of the distributions (1.1)–(1.3), that is let the X's be distributed independently and identically according to one of these distributions. Let θ be p, τ, or the pair (ξ, σ) respectively, and let $\gamma = \gamma(\theta)$ be a real-valued function of θ.

(i) If one wishes to decide whether or not γ exceeds some specified value γ_0, the choice lies between the two decisions $d_0 : \gamma > \gamma_0$ and $d_1 : \gamma \le \gamma_0$. In specific applications these decisions might correspond to the acceptance or rejection of a lot of manufactured goods, of an experimental airplane as ready for flight testing, of a new treatment as an improvement over a standard one, and so on. The loss function of course depends on the application to be made. Typically, the loss is 0 if the correct decision is chosen, while for an incorrect decision the losses $L(\gamma, d_0)$ and $L(\gamma, d_1)$ are increasing functions of $|\gamma - \gamma_0|$.

(ii) At the other end of the scale is the much more detailed problem of obtaining a numerical estimate of γ. Here a decision d of the statistician is a real number, the estimate of γ, and the losses might be $L(\gamma, d) = v(\gamma)w(|d - \gamma|)$, where w is a strictly increasing function of the error $|d - \gamma|$.

(iii) An intermediate case is the choice between the three alternatives $d_0 : \gamma < \gamma_0$, $d_1 : \gamma > \gamma_1$, $d_2 : \gamma_0 \le \gamma \le \gamma_1$, for example accepting a new treatment, rejecting it, or recommending it for further study. ■

The distinction illustrated by this example is the basis for one of the principal classifications of statistical methods. Two-decision problems such as (i) are usually formulated in terms of *testing a hypothesis* which is to be accepted or rejected (see Chapter 3). It is the theory of this class of problems with which we shall be mainly concerned here. The other principal branch of statistics is the theory of *point estimation* dealing with problems such as (ii). This is the subject of *TPE2*. The intermediate problem (iii) is a special case of a *multiple decision procedure*. Some problems of this kind are treated in Ferguson (1967, Chapter 6); a discussion of some others is given in Chapter 9.

Example 1.2.2 Suppose that the data consist of samples $X_{ij}, j = 1, \ldots, n_i$, from normal populations $N(\xi_i, \sigma^2), i = 1, \ldots, s$.

(i) Consider first the case $s = 2$ and the question of whether or not there is a material difference between the two populations. This has the same structure as problem (iii) of the previous example. Here the choice lies between the three decisions $d_0 : |\xi_2 - \xi_1| \leq \Delta$, $d_1 : \xi_2 > \xi_1 + \Delta$, $d_2 : \xi_2 < \xi_1 - \Delta$, where Δ is preassigned. An analogous problem, involving $k+1$ possible decisions, occurs in the general case of k populations. In this case one must choose between the decision that the k distributions do not differ materially, $d_0 : \max |\xi_j - \xi_i| \leq \Delta$, and the decisions $d_k : \max |\xi_j - \xi_i| > \Delta$ and ξ_k is the largest of the means.

(ii) A related problem is that of ranking the distributions in increasing order of their mean ξ.

(iii) Alternatively, a standard ξ_0 may be given and the problem is to decide which, if any, of the population means exceed the standard. ■

Example 1.2.3 Consider two distributions—to be specific, two Poisson distributions $P(\tau_1)$, $P(\tau_2)$—and suppose that τ_1 is known to be less than τ_2 but that otherwise the τ's are unknown. Let $Z_1, \ldots, Z_n$ be independently distributed, each according to either $P(\tau_1)$ or $P(\tau_2)$. Then each Z is to be classified as to which of the two distributions it comes from. Here the loss might be the number of Z's that are incorrectly classified, multiplied by a suitable function of τ_1 and τ_2. An example of the complexity that such problems can attain and the conceptual as well as mathematical difficulties that they may involve is provided by the efforts of anthropologists to classify the human population into a number of homogeneous races by studying the frequencies of the various blood groups and of other genetic characters. ■

All the problems considered so far could be termed *action problems*. It was assumed in all of them that if θ were known a unique correct decision would be available, that is, given any θ, there exists a unique d for which $L(\theta, d) = 0$. However, not all statistical problems are so clear-cut. Frequently it is a question of providing a convenient summary of the data or indicating what information is available concerning the unknown parameter or distribution. This information will be used for guidance in various considerations but will not provide the sole basis for any specific decisions. In such cases the emphasis is on the inference rather than on the decision aspect of the problem. Although formally it can still be considered a decision problem if the inferential statement itself is interpreted as the decision to be taken, the distinction is of conceptual and practical significance despite the fact that frequently it is ignored.[1] An important class of such problems, estimation by interval, is illustrated by the following example. (For the more usual formulation in terms of confidence intervals, see Sections 3.5, 5.4 and 5.5.)

Example 1.2.4 Let $X = (X_1, \ldots, X_n)$ be a sample from $N(\xi, \sigma^2)$ and let a decision consist in selecting an interval $[\underline{L}, \bar{L}]$ and stating that it contains ξ. Suppose that decision procedures are restricted to intervals $[\underline{L}(X), \bar{L}(X)]$ whose expected length for all ξ and σ does not exceed $k\sigma$ where k is some preassigned constant. An appropriate loss function would be 0 if the decision is correct and would otherwise depend on the relative position of the interval to the true value of ξ. In this case there are many correct decisions corresponding to a given distribution $N(\xi, \sigma^2)$. ■

[1] For a more detailed discussion of this distinction see, for example, Cox (1958), Blyth (1970), and Barnett (1999).

It remains to discuss the choice of loss function, and of the three elements defining the problem this is perhaps the most difficult to specify. Even in the simplest case, where all losses eventually reduce to financial ones, it can hardly be expected that one will be able to evaluate all the short- and long-term consequences of an action. Frequently it is possible to simplify the formulation by taking into account only certain aspects of the loss function. As an illustration consider Example 1.2.1(i) and let $L(\theta, d_0) = a$ for $\gamma(\theta) \leq \gamma_0$ and $L(\theta, d_1) = b$ for $\gamma(\theta) > \gamma_0$. The risk function becomes

$$R(\theta, \delta) = \begin{cases} aP_\theta\{\delta(X) = d_0\} & \text{if} \quad \gamma \leq \gamma_0, \\ bP_\theta\{\delta(X) = d_1\} & \text{if} \quad \gamma > \gamma_0, \end{cases} \tag{1.4}$$

and is seen to involve only the two probabilities of error, with weights which can be adjusted according to the relative importance of these errors. Similarly, in Example 1.2.3 one may wish to restrict attention to the number of misclassifications.

Unfortunately, such a natural simplification is not always available, and in the absence of specific knowledge it becomes necessary to select the loss function in some conventional way, with mathematical simplicity usually an important consideration. In point estimation problems such as that considered in Example 1.2.1(ii), if one is interested in estimating a real-valued function $\gamma = \gamma(\theta)$, it is customary to take the square of the error, or somewhat more generally to put

$$L(\theta, d) = v(\theta)(d - \gamma)^2. \tag{1.5}$$

Besides being particularly simple mathematically, this can be considered as an approximation to the true loss function L provided that for each fixed θ, $L(\theta, d)$ is twice differentiable in d, that $L(\theta, \gamma(\theta)) = 0$ for all θ, and that the error is not large.

It is frequently found that, within one problem, quite different types of losses may occur, which are difficult to measure on a common scale. Consider once more Example 1.2.1(i) and suppose that γ_0 is the value of γ when a standard treatment is applied to a situation in medicine, agriculture, or industry. The problem is that of comparing some new process with unknown γ to the standard one. Turning down the new method when it is actually superior, or adopting it when it is not, clearly entails quite different consequences. In such cases it is sometimes convenient to treat the various loss components, say $L_1, L_2, \ldots, L_r$, separately. Suppose in particular that $r = 2$ and the L_1 represents the more serious possibility. One can then assign a bound to this risk component, that is, impose the condition

$$EL_1(\theta, \delta(X)) \leq \alpha, \tag{1.6}$$

and subject to this condition minimize the other component of the risk. Example 1.2.4 provides an illustration of this procedure. The length of the interval $[\underline{L}, \bar{L}]$ (measured in σ-units) is one component of the loss function, the other being the loss that results if the interval does not cover the true ξ.

1.3 Randomization; Choice of Experiment

The description of the general decision problem given so far is still too narrow in certain respects. It has been assumed that for each possible value of the random variables a definite decision must be chosen. Instead, it is convenient to permit the selection of one out of a number of decisions according to stated probabilities, or more generally the selection of a decision according to a probability distribution defined over the decision space; which distribution depends of course on what x is observed. One way to describe such a randomized procedure is in terms of a nonrandomized procedure depending on X and a random variable Y whose values lie in the decision space and whose conditional distribution given x is independent of θ.

Although it may run counter to one's intuition that such extra randomization should have any value, there is no harm in permitting this greater freedom of choice. If the intuitive misgivings are correct, it will turn out that the optimum procedures always are of the simple nonrandomized kind. Actually, the introduction of randomized procedures leads to an important mathematical simplification by enlarging the class of risk functions so that it becomes convex. In addition, there are problems in which some features of the risk function such as its maximum can be improved by using a randomized procedure.

Another assumption that tacitly has been made so far is that a definite experiment has already been decided upon so that it is known what observations will be taken. However, the statistical considerations involved in designing an experiment are no less important than those concerning its analysis. One question in particular that must be decided before an investigation is undertaken is how many observations should be taken so that the risk resulting from wrong decisions will not be excessive. Frequently it turns out that the required sample size depends on the unknown distribution and therefore cannot be determined in advance as a fixed number. Instead it is then specified as a function of the observations and the decision whether or not to continue experimentation is made *sequentially* at each stage of the experiment on the basis of the observations taken up to that point.

Example 1.3.1 On the basis of a sample $X_1, \ldots, X_n$ from a normal distribution $N(\xi, \sigma^2)$ one wishes to estimate ξ. Here the risk function of an estimate, for example its expected squared error, depends on σ. For large σ the sample contains only little information in the sense that two distributions $N(\xi_1, \sigma^2)$ and $N(\xi_2, \sigma^2)$ with fixed difference $\xi_2 - \xi_1$ become indistinguishable as $\sigma \to \infty$, with the result that the risk tends to infinity. Conversely, the risk approaches zero as $\sigma \to 0$, since then effectively the mean becomes known. Thus the number of observations needed to control the risk at a given level is unknown. However, as soon as some observations have been taken, it is possible to estimate σ^2 and hence to determine the additional number of observations required. ■

Example 1.3.2 In a sequence of trials with constant probability p of success, one wishes to decide whether $p \leq \frac{1}{2}$ or $p > \frac{1}{2}$. It will usually be possible to reach a decision at an early stage if p is close to 0 or 1 so that practically all observations are of one kind, while a larger sample will be needed for intermediate values of p. This difference may be partially balanced by the fact that for intermediate values a loss resulting from a wrong decision is presumably less serious than for the more extreme values. ■

Example 1.3.3 The possibility of determining the sample size sequentially is important not only because the distributions P_θ can be more or less informative but also because the same is true of the observations themselves. Consider, for example, observations from the uniform distribution over the interval $(\theta - \frac{1}{2}, \theta + \frac{1}{2})$ and the problem of estimating θ. Here there is no difference in the amount of information provided by the different distributions P_θ. However, a sample $X_1, X_2, \ldots, X_n$ can practically pinpoint θ if $\max |X_j - X_i|$ is sufficiently close to 1, or it can give essentially no more information then a single observation if $\max |X_j - X_i|$ is close to 0. Again the required sample size should be determined sequentially. ■

Except in the simplest situations, the determination of the appropriate sample size is only one aspect of the design problem. In general, one must decide not only how many but also what kind of observations to take. In clinical trials, for example, when a new treatment is being compared with a standard procedure, a protocol is required which specifies to which of the two treatments each of the successive incoming patients is to be assigned. Formally, such questions can be subsumed under the general decision problem described at the beginning of the chapter, by interpreting X as the set of all available variables, by introducing the decisions whether or not to stop experimentation at the various stages, by specifying in case of continuance which type of variable to observe next, and by including the cost of observation in the loss function.

The determination of optimum sequential stopping rules and experimental designs is outside the scope of this book. An introduction to this subject is provided, for example, by Siegmund (1985).

1.4 Optimum Procedures

At the end of Section 1.1 the aim of statistical theory was stated to be the determination of a decision function δ which minimizes the risk function

$$R(\theta, \delta) = E_\theta[L(\theta, \delta(X))]. \tag{1.7}$$

Unfortunately, in general the minimizing δ depends on θ, which is unknown. Consider, for example, some particular decision d_0, and the decision procedure $\delta(x) \equiv d_0$ according to which decision d_0 is taken regardless of the outcome of the experiment. Suppose that d_0 is the correct decision for some θ_0, so that $L(\theta_0, d_0) = 0$. Then δ minimizes the risk at θ_0 since $R(\theta_0, \delta) = 0$, but presumably at the cost of a high risk for other values of θ.

In the absence of a decision function that minimizes the risk for all θ, the mathematical problem is still not defined, since it is not clear what is meant by a best procedure. Although it does not seem possible to give a definition of optimality that will be appropriate in all situations, the following two methods of approach frequently are satisfactory.

The nonexistence of an optimum decision rule is a consequence of the possibility that a procedure devotes too much of its attention to a single parameter value at the cost of neglecting the various other values that might arise. This suggests the restriction to decision procedures which possess a certain degree of impartiality, and the possibility that within such a restricted class there may exist a procedure with uniformly smallest risk. Two conditions of this kind, invariance and unbiasedness, will be discussed in the next section.

Instead of restricting the class of procedures, one can approach the problem somewhat differently. Consider the risk functions corresponding to two different decision rules δ_1 and δ_2. If $R(\theta, \delta_1) < R(\theta, \delta_2)$ for all θ, then δ_1 is clearly preferable to δ_2, since its use will lead to a smaller risk no matter what the true value of θ is. However, the situation is not clear when the two risk functions intersect as in Figure 1.1. What is needed is a principle which in such cases establishes a preference of one of the two risk functions over the other, that is, which introduces an ordering into the set of all risk functions. A procedure will then be optimum if its risk function is best according to this ordering. Some criteria that have been suggested for ordering risk functions will be discussed in Section 1.6.

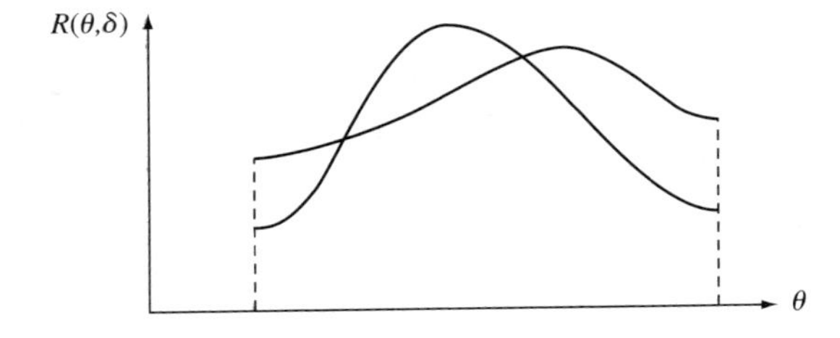

Figure 1.1.

A weakness of the theory of optimum procedures sketched above is its dependence on an extraneous restricting or ordering principle, and on knowledge concerning the loss function and the distributions of the observable random variables which in applications is frequently unavailable or unreliable. These difficulties, which may raise doubt concerning the value of an optimum theory resting on such shaky foundations, are in principle no different from those arising in any application of mathematics to reality. Mathematical formulations always involve simplification and approximation, so that solutions obtained through their use cannot be relied upon without additional checking. In the present case a check consists in an overall evaluation of the performance of the procedure that the theory produces, and an investigation of its sensitivity to departure from the assumptions under which it was derived.

The optimum theory discussed in this book should therefore not be understood to be prescriptive. The fact that a procedure δ is optimal according to some optimality criterion does not necessarily mean that it is the right procedure to use, or even a satisfactory procedure. It does show how well one can do in this particular direction and how much is lost when other aspects have to be taken into account.

The aspect of the formulation that typically has the greatest influence on the solution of the optimality problem is the family $\mathcal{P}$ to which the distribution of the observations is assumed to belong. The investigation of the *robustness* of a proposed procedure to departures from the specified model is an indispensable feature of a suitable statistical procedure, and although optimality (exact or asymptotic) may provide a good starting point, modifications are often necessary before an acceptable solution is found. It is possible to extend the decision-theoretic framework to include robustness as well as optimality. Suppose robustness is desired against some class $\mathcal{P}'$ of distributions which is larger (possibly much larger) than the give $\mathcal{P}$. Then one may assign a bound M to the risk to be tolerated over $\mathcal{P}'$. Within the class of procedures satisfying this restriction, one can then optimize the risk over $\mathcal{P}$ as before. Such an approach has been proposed and applied to a number of specific problems by Bickel (1984) and Kempthorne (1988).

Another possible extension concerns the actual choice of the family $\mathcal{P}$, the model used to represent the actual physical situation. The problem of choosing a model which provides an adequate description of the situation without being unnecessarily complex can be treated within the decision-theoretic formulation of Section 1.1 by adding to the loss function a component representing the complexity of the proposed model. Such approaches to *model selection* are discussed in Stone (1981), de Leeuw (1992) and Rao and Wu (2001).

1.5 Invariance and Unbiasedness[2]

A natural definition of impartiality suggests itself in situations which are symmetric with respect to the various parameter values of interest: *The procedure is then required to act symmetrically with respect to these values.*

Example 1.5.1 Suppose two treatments are to be compared and that each is applied n times. The resulting observations $X_{11}, \ldots, X_{1n}$ and $X_{21}, \ldots, X_{2n}$ are samples from $N(\xi_1, \sigma^2)$ and $N(\xi_2, \sigma^2)$ respectively. The three available decisions are $d_0 : |\xi_2 - \xi_1| \leq \Delta$, $d_1 : \xi_2 > \xi_1 + \Delta$, $d_2 : \xi_2 < \xi_1 - \Delta$, and the loss is w_{ij} if decision d_j is taken when d_i would have been correct. If the treatments are to be compared solely in terms of the ξ's and no outside considerations are involved, the losses are symmetric with respect to the two treatments so that $w_{01} = w_{02}$, $w_{10} = w_{20}$, $w_{12} = w_{21}$. Suppose now that the labeling of the two treatments as 1 and 2 is reversed, and correspondingly also the labeling of the X's, the ξ's, and the decisions d_1 and d_2. This changes the meaning of the symbols, but the formal decision problem, because of its symmetry, remains unaltered. It is then natural to require the corresponding symmetry from the procedure δ and ask that $\delta(x_{11}, \ldots, x_{1n}, x_{21}, \ldots, x_{2n}) = d_0$, d_1, or d_2 as $\delta(x_{21}, \ldots, x_{2n}, x_{11}, \ldots, x_{1n}) = d_0$, d_2, or d_1 respectively. If this condition were not satisfied, the decision as to which population has the greater mean would depend on the presumably quite

[2]The concepts discussed here for general decision theory will be developed in more specialized form in later chapters. The present section may therefore be omitted at first reading.

accidental and irrelevant labeling of the samples. Similar remarks apply to a number of further symmetries that are present in this problem. ■

Example 1.5.2 Consider a sample $X_1, \ldots, X_n$ from a distribution with density $\sigma^{-1} f[(x-\xi)/\sigma]$ and the problem of estimating the location parameter ξ, say the mean of the X's, when the loss is $(d-\xi)^2/\sigma^2$, the square of the error expressed in σ-units. Suppose that the observations are originally expressed in feet, and let $X_i' = aX$ with $a = 12$ be the corresponding observations in inches. In the transformed problem the density is $\sigma'^{-1} f[(x'-\xi')/\sigma']$ with $\xi' = a\xi$, $\sigma' = a\sigma$. Since $(d'-\xi')^2/\sigma'^2 = (d-\xi)^2/\sigma^2$, the problem is formally unchanged. The same estimation procedure that is used for the original observations is therefore appropriate after the transformation and leads to $\delta(aX_1, \ldots, aX_n)$ as an estimate of $\xi' = a\xi$, the parameter ξ expressed in inches. On reconverting the estimate into feet one finds that if the result is to be independent of the scale of measurements, δ must satisfy the condition of scale invariance

$$\frac{\delta(aX_1, \ldots, aX_n)}{a} = \delta(X_1, \ldots, X_n)\ . \ ■$$

The general mathematical expression of symmetry is invariance under a suitable group of transformations. A group G of transformations g of the sample space is said to leave a statistical decision problem invariant if it satisfies the following conditions:

(i) It leaves invariant the family of distributions $\mathcal{P} = \{P_\theta, \theta \in \Omega\}$, that is, for any possible distribution P_θ of X the distribution of gX, say $P_{\theta'}$, is also in $\mathcal{P}$. The resulting mapping $\theta' = \bar{g}\theta$ of Ω is assumed to be onto[3] Ω and 1:1.

(ii) To each $g \in G$, there corresponds a transformation $g^* = h(g)$ of the decision space D onto itself such that h is a homomorphism, that is, satisfies the relation $h(g_1 g_2) = h(g_1)h(g_2)$, and the loss function L is unchanged under the transformation, so that

$$L(\bar{g}\theta, g^* d) = L(\theta, d).$$

Under these assumptions the transformed problem, in terms of $X' = gX$, $\theta' = \bar{g}\theta$, and $d' = g^* d$, is formally identical with the original problem in terms of X, θ, and d. Given a decision procedure δ for the latter, this is therefore still appropriate after the transformation. Interpreting the transformation as a change of coordinate system and hence of the names of the elements, one would, on observing x', select the decision which in the new system has the name $\delta(x')$, so that its old name is $g^{*-1}\delta(x')$. If the decision taken is to be independent of the particular coordinate system adopted, this should coincide with the original decision $\delta(x)$, that is, the procedure must satisfy the *invariance* condition

$$\delta(gx) = g^* \delta(x) \qquad \text{for all} \quad x \in X, \quad g \in G. \tag{1.8}$$

Example 1.5.3 The model described in Example 1.5.1 is invariant also under the transformations $X_{ij}' = X_{ij} + c$, $\xi_i' = \xi_i + c$. Since the decisions d_0, d_1, and d_2

[3]The term *onto* is used in indicate that $\bar{g}\Omega$ is not only contained in but actually equals Ω; that is, given any θ' in Ω, there exists θ in Ω such that $\bar{g}\theta = \theta'$.

concern only the differences $\xi_2 - \xi_1$, they should remain unchanged under these transformations, so that one would expect to have $g^* d_i = d_i$ for $i = 0, 1, 2$. It is in fact easily seen that the loss function does satisfy $L(\bar{g}\theta, d) = L(\theta, d)$, and hence that $g^* d = d$. A decision procedure therefore remains invariant in the present case if it satisfies $\delta(gx) = \delta(x)$ for all $g \in G$, $x \in X$. ■

It is helpful to make a terminological distinction between situations like that of Example 1.5.3 in which $g^* d = d$ for all d, and those like Examples 1.5.1 and 1.5.2 where invariance considerations require $\delta(gx)$ to vary with g. In the former case the decision procedure remains unchanged under the transformations $X' = gX$ and is thus truly invariant; in the latter, the procedure varies with g and may then more appropriately be called *equivariant* rather than invariant. Typically, hypothesis testing leads to procedures that are invariant in this sense; estimation problems (whether by point or interval estimation), to equivariant ones. Invariant tests and equivariant confidence sets will be discussed in Chapter 6. For a brief discussion of equivariant point estimation, see Bondessen (1983); a fuller treatment is given in *TPE*2, Chapter 3.

Invariance considerations are applicable only when a problem exhibits certain symmetries. An alternative impartiality restriction which is applicable to other types of problems is the following condition of unbiasedness. Suppose the problem is such that for each θ there exists a unique correct decision and that each decision is correct for some θ. Assume further that $L(\theta_1, d) = L(\theta_2, d)$ for all d whenever the same decision is correct for both θ_1 and θ_2. Then the loss $L(\theta, d')$ depends only on the actual decision taken, say d', and the correct decision d. The loss can thus be denoted by $L(d, d')$ and this function measures how far apart d and d' are. Under these assumptions a decision function δ is said to be unbiased with respect to the loss function L, or L-unbiased, if for all θ and d'

$$E_\theta L(d', \delta(X)) \geq E_\theta L(d, \delta(X))$$

where the subscript θ indicates the distribution with respect to which the expectation is taken and where d is the decision that is correct for θ. Thus δ is unbiased if on the average $\delta(X)$ comes closer to the correct decision than to any wrong one. Extending this definition, δ is said to be *L-unbiased* for an arbitrary decision problem if for all θ and θ'

$$E_\theta L(\theta', \delta(X)) \geq E_\theta L(\theta, \delta(X)). \tag{1.9}$$

Example 1.5.4 Suppose that in the problem of estimating a real-valued parameter θ by confidence intervals, as in Example 1.2.4, the loss is 0 or 1 as the interval $[\underline{L}, \bar{L}]$ does or does not cover the true θ. Then the set of intervals $[\underline{L}(X), \bar{L}(X)]$ is unbiased if the probability of covering the true value is greater than or equal to the probability of covering any false value. ■

Example 1.5.5 In a two-decision problem such as that of Example 1.2.1(i), let ω_0 and ω_1 be the sets of θ-values for which d_0 and d_1 are the correct decisions. Assume that the loss is 0 when the correct decision is taken, and otherwise is given by $L(\theta, d_0) = a$ for $\theta \in \omega_1$, and $L(\theta, d_1) = b$ for $\theta \in \omega_0$. Then

$$E_\theta L(\theta', \delta(X)) = \begin{cases} aP_\theta\{\delta(X) = d_0\} & \text{if} \quad \theta' \in \omega_1, \\ bP_\theta\{\delta(X) = d_1\} & \text{if} \quad \theta' \in \omega_0, \end{cases}$$

so that (1.9) reduces to

$$aP_\theta\{\delta(X) = d_0\} \geq bP_\theta\{\delta(X) = d_1\} \quad \text{for} \quad \theta' \in \omega_0,$$

with the reverse inequality holding for $\theta \in \omega_1$. Since $P_\theta\{\delta(X) = d_0\} + P_\theta\{\delta(X) = d_1\} = 1$, the unbiasedness condition (1.9) becomes

$$\begin{array}{lll} P_\theta\{\delta(X) = d_1\} \leq \frac{a}{a+b} & \text{for} & \theta \in \omega_0, \\ P_\theta\{\delta(X) = d_1\} \geq \frac{a}{a+b} & \text{for} & \theta \in \omega_1 \ . \ \blacksquare \end{array} \tag{1.10}$$

Example 1.5.6 In the problem of estimating a real-valued function $\gamma(\theta)$ with the square of the error as loss, the condition of unbiasedness becomes

$$E_\theta[\delta(X) - \gamma(\theta')]^2 \geq E_\theta[\delta(X) - \gamma(\theta)]^2 \quad \text{for all } \theta, \theta'.$$

On adding and subtracting $h(\theta) = E_\theta\delta(X)$ inside the brackets on both sides, this reduces to

$$[h(\theta) - \gamma(\theta')]^2 \geq [h(\theta) - \gamma(\theta)]^2 \quad \text{for all } \theta, \theta'.$$

If $h(\theta)$ is one of the possible values of the function γ, this condition holds if and only if

$$E_\theta\delta(X) = \gamma(\theta) \ . \ \blacksquare \tag{1.11}$$

In the theory of point estimation, (1.11) is customarily taken as the definition of unbiasedness. Except under rather pathological conditions, it is both a necessary and sufficient condition for δ to satisfy (1.9). (See Problem 1.2.)

1.6 Bayes and Minimax Procedures

We now turn to a discussion of some preference orderings of decision procedures and their risk functions. One such ordering is obtained by assuming that in repeated experiments the parameter itself is a random variable Θ, the distribution of which is known. If for the sake of simplicity one supposes that this distribution has a probability density $\rho(\theta)$, the overall average loss resulting from the use of a decision procedure δ is

$$r(\rho, \delta) = \int E_\theta L(\theta, \delta(X))\rho(\theta)\, d\theta = \int R(\theta, \delta)\rho(\theta)\, d\theta \tag{1.12}$$

and the smaller $r(\rho, \delta)$, the better is δ. An optimum procedure is one that minimizes $r(\rho, \delta)$, and is called a *Bayes solution* of the given decision problem corresponding to a priori density ρ. The resulting minimum of $r(\rho, \delta)$ is called the *Bayes risk* of δ.

Unfortunately, in order to apply this principle it is necessary to assume not only that θ is a random variable but also that its distribution is known. This assumption is usually not warranted in applications. Alternatively, the right-hand side of (1.12) can be considered as a weighted average of the risks; for $\rho(\theta) \equiv 1$ in particular, it is then the area under the risk curve. With this interpretation the choice of a weight function ρ expresses the importance the experimenter attaches to the various values of θ. A systematic Bayes theory has been developed which interprets ρ as describing the state of mind of the investigator towards θ. For an account of this approach see, for example, Berger (1985a) and Robert (1994).

If no prior information regarding θ is available, one might consider the maximum of the risk function its most important feature. Of two risk functions the one with the smaller maximum is then preferable, and the optimum procedures are those with the *minimax* property of minimizing the maximum risk. Since this maximum represents the worst (average) loss that can result from the use of a given procedure, a minimax solution is one that gives the greatest possible protection against large losses. That such a principle may sometimes be quite unreasonable is indicated in Figure 1.2, where under most circumstances one would prefer δ_1 to δ_2 although its risk function has the larger maximum.

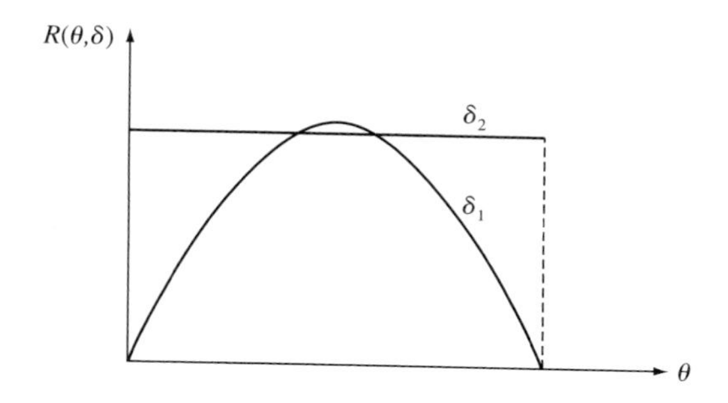

Figure 1.2.

Perhaps the most common situation is one intermediate to the two just described. On the one hand, past experience with the same or similar kind of experiment is available and provides an indication of what values of θ to expect; on the other, this information is neither sufficiently precise nor sufficiently reliable to warrant the assumptions that the Bayes approach requires. In such circumstances it seems desirable to make use of the available information without trusting it to such an extent that catastrophically high risks might result if it is inaccurate or misleading. To achieve this one can place a bound on the risk and restrict consideration to decision procedures δ for which

$$R(\theta, \delta) \leq C \qquad \text{for all} \quad \theta. \tag{1.13}$$

[Here the constant C will have to be larger than the maximum risk C_0 of the minimax procedure, since otherwise there will exist no procedures satisfying (1.13).] Having thus assured that the risk can under no circumstances get out of hand, the experimenter can now safely exploit his knowledge of the situation, which may be based on theoretical considerations as well as on past experience; he can follow his hunches and guess at a distribution ρ for θ. This leads to the selection of a procedure δ (a *restricted Bayes solution*), which minimizes the average risk (1.12) for this a priori distribution subject to (1.13). The more certain one is of ρ, the larger one will select C, thereby running a greater risk in case of a poor guess but improving the risk if the guess is good.

Instead of specifying an ordering directly, one can postulate conditions that the ordering should satisfy. Various systems of such conditions have been investigated and have generally led to the conclusion that the only orderings satisfying these systems are those which order the procedures according to their Bayes risk with respect to some prior distribution of θ. For details, see for example Blackwell and Girshick (1954), Ferguson (1967), Savage (1972), Berger (1985a), and Bernardo and Smith (1994).

1.7 Maximum Likelihood

Another approach, which is based on considerations somewhat different from those of the preceding sections, is the method of maximum likelihood. It has led to reasonable procedures in a great variety of problems, and is still playing a dominant role in the development of new tests and estimates. Suppose for a moment that X can take on only a countable set of values $x_1, x_2, \ldots$, with $P_\theta(x) = P_\theta\{X = x\}$, and that one wishes to determine the correct value of θ, that is, the value that produced the observed x. This suggests considering for each possible θ how probable the observed x would be if θ were the true value. The higher this probability, the more one is attracted to the explanation that the θ in question produced x, and the more likely the value of θ appears. Therefore, the expression $P_\theta(x)$ considered for fixed x as a function of θ has been called the *likelihood* of θ. To indicate the change in point of view, let it be denoted by $L_x(\theta)$. Suppose now that one is concerned with an action problem involving a countable number of decisions, and that it is formulated in terms of a gain function (instead of the usual loss function), which is 0 if the decision taken is incorrect and is $a(\theta) > 0$ if the decision taken is correct and θ is the true value. Then it seems natural to weight the likelihood $L_x(\theta)$ by the amount that can be gained if θ is true, to determine the value of θ that maximizes $a(\theta)L_x(\theta)$ and to select the decision that would be correct if this were the true value of θ. Essentially the same remarks apply in the case in which $P_\theta(x)$ is a probability density rather than a discrete probability.

In problems of point estimation, one usually assumes that $a(\theta)$ is independent of θ. This leads to estimating θ by the value that maximizes the likelihood $L_x(\theta)$, the *maximum-likelihood estimate* of θ. Another case of interest is the class of two-decision problems illustrated by Example 1.2.1(i). Let ω_0 and ω_1 denote the sets of θ-values for which d_0 and d_1 are the correct decisions, and assume that $a(\theta) = a_0$ or a_1 as θ belongs to ω_0 or ω_1 respectively. Then decision d_0 or d_1 is taken as $a_1 \sup_{\theta\in\omega_1} L_x(\theta) <$ or $> a_0 \sup_{\theta\in\omega_0} L_x(\theta)$, that is as

$$\frac{\sup\limits_{\theta\in\omega_0} L_x(\theta)}{\sup\limits_{\theta\in\omega_1} L_x(\theta)} > \quad \text{or} \quad < \frac{a_1}{a_0}. \tag{1.14}$$

This is known as a *likelihood ratio procedure*.[4]

[4]This definition differs slightly from the usual one where in the denominator on the left-hand side of (1.14) the supremum is taken over the set $\omega_0 \cup \omega_1$. The two definitions agree whenever the left-hand side of (1.14) is ≤ 1, and the procedures therefore agree is $a_1 < a_0$.

Although the maximum likelihood principle is not based on any clearly defined optimum considerations, it has been very successful in leading to satisfactory procedures in many specific problems. For wide classes of problems, maximum likelihood procedures will be shown in Chapter 13 to possess various asymptotic optimum properties as the sample size tends to infinity; also see *TPE*2, Chapter 6. On the other hand, there exist examples for which the maximum-likelihood procedure is worse than useless; where it is, in fact, so bad that one can do better without making any use of the observations (see Problem 6.28).

1.8 Complete Classes

None of the approaches described so far is reliable in the sense that the resulting procedure is necessarily satisfactory. There are problems in which a decision procedure δ_0 exists with uniformly minimum risk among all unbiased or invariant procedures, but where there exists a procedure δ_1 not possessing this particular impartiality property and preferable to δ_0. (Cf. Problems 1.14 and 1.16.) As was seen earlier, minimax procedures can also be quite undesirable, while the success of Bayes and restricted Bayes solutions depends on a priori information which is usually not very reliable if it is available at all. In fact, it seems that in the absence of reliable a priori information no principle leading to a unique solution can be entirely satisfactory.

This suggests the possibility, at least as a first step, of not insisting on a unique solution but asking only how far a decision problem can be reduced without loss of relevant information. It has already been seen that a decision procedure δ can sometimes be eliminated from consideration because there exists a procedure δ' *dominating* it in the sense that

$$\begin{aligned} R(\theta, \delta') &\leq R(\theta, \delta) \qquad \text{for all } \theta \\ R(\theta, \delta') &< R(\theta, \delta) \qquad \text{for some } \theta. \end{aligned} \tag{1.15}$$

In this case δ is said to be *inadmissible*; δ is called *admissible* if no such dominating δ' exists. A class $\mathcal{C}$ of decision procedures is said to be *complete* if for any δ not in $\mathcal{C}$ there exists δ' in $\mathcal{C}$ dominating it. A complete class is *minimal* if it does not contain a complete subclass. If a minimal complete class exists, as is typically the case, it consists exactly of the totality of admissible procedures.

It is convenient to define also the following variant of the complete class notion. A class $\mathcal{C}$ is said to be *essentially complete* if for any procedure δ there exists δ' in $\mathcal{C}$ such that $R(\theta, \delta') \leq R(\theta, \delta)$ for all θ. Clearly, any complete class is also essentially complete. In fact, the two definitions differ only in their treatment of equivalent decision rules, that is, decision rules with identical risk function. If δ belongs to the minimal complete class $\mathcal{C}$, any equivalent decision rule must also belong to $\mathcal{C}$. On the other hand, a minimal essentially complete class need contain only one member from such a set of equivalent procedures.

In a certain sense a minimal essentially complete class provides the maximum possible reduction of a decision problem. On the one hand, there is no reason to consider any of the procedures that have been weeded out. For each of them, there is included one in $\mathcal{C}$ that is as good or better. On the other hand, it is not possible to reduce the class further. Given any two procedures in $\mathcal{C}$, each of them is better in places than the other, so that without additional information it is not known which of the two is preferable.

The primary concern in statistics has been with the explicit determination of procedures, or classes of procedures, for various specific decision problems. Those studied most extensively have been estimation problems, and problems involving a choice between only two decisions (hypothesis testing), the theory of which constitutes the subject of the present volume. However, certain conclusions are possible without such specialization. In particular, two results concerning the structure of complete classes and minimax procedures have been proved to hold under very general assumptions.[5]

(i) The totality of Bayes solutions and limits of Bayes solutions constitute a complete class.

(ii) Minimax procedures are Bayes solutions with respect to a *least favorable* a priori distribution, that is, an a priori distribution that maximizes the associated Bayes risk, and the minimax risk equals this maximum Bayes risk. Somewhat more generally, if there exists no least favorable a priori distribution but only a sequence for which the Bayes risk tends to the maximum, the minimax procedures are limits of the associated sequence of Bayes solutions.

1.9 Sufficient Statistics

A minimal complete class was seen in the preceding section to provide the maximum possible reduction of a decision problem without loss of information. Frequently it is possible to obtain a less extensive reduction of the data, which applies simultaneously to all problems relating to a given class $\mathcal{P} = \{P_\theta,\ \theta \in \Omega\}$ of distributions of the given random variable X. It consists essentially in discarding that part of the data which contains no information regarding the unknown distribution P_θ, and which is therefore of no value for any decision problem concerning θ.

Example 1.9.1 Trials are performed with constant unknown probability p of success. If X_i is 1 or 0 as the ith trial is a success or failure, the sample $(X_1, \ldots, X_n)$ shows how many successes there were and in which trials they occurred. The second of these pieces of information contains no evidence as to the value of p. Once the total number of successes $\sum X_i$ is known to be equal to t, each of the $\binom{n}{t}$ possible positions of these successes is equally likely regardless of p. It follows that knowing $\sum X_i$ but neither the individual X_i nor p, one can, from a table of random numbers, construct a set of random variables $X_1', \ldots, X_n'$ whose joint distribution is the same as that of $X_1, \ldots, X_n$. Therefore, the information contained in the X_i is the same as that contained in $\sum X_i$ and a table of random numbers. ■

[5] Precise statements and proofs of these results are given in the book by Wald (1950). See also Ferguson (1967) and Berger (1985a). Additional results and references are given in Brown and Marden (1989) and Kowalski (1995).

Example 1.9.2 If $X_1, \ldots, X_n$ are independently normally distributed with zero mean and variance σ^2, the conditional distribution of the sample point over each of the spheres, $\sum X_i^2 =$ constant, is uniform irrespective of σ^2. One can therefore construct an equivalent sample $X_1', \ldots, X_n'$ from a knowledge of $\sum X_i^2$ and a mechanism that can produce a point randomly distributed over a sphere. ■

More generally, a statistic T is said to be *sufficient* for the family $\mathcal{P} = \{P_\theta, \theta \in \Omega\}$ (or sufficient for θ, if it is clear from the context what set Ω is being considered) if the conditional distribution of X given $T = t$ is independent of θ. As in the two examples it then follows under mild assumptions[6] that it is not necessary to utilize the original observations X. If one is permitted to observe only T instead of X, this does not restrict the class of available decision procedures. For any value t of T let X_t be a random variable possessing the conditional distribution of X given t. Such a variable can, at least theoretically, be constructed by means of a suitable random mechanism. If one then observes T to be t and X_t to be x', the random variable X' defined through this two-stage process has the same distribution as X. Thus, given any procedure based on X, it is possible to construct an equivalent one based on X' which can be viewed as a randomized procedure based solely on T. Hence if randomization is permitted (and we shall assume throughout that this is the case), there is no loss of generality in restricting consideration to a sufficient statistic.

It is inconvenient to have to compute the conditional distribution of X given t in order to determine whether or not T is sufficient. A simple check is provided by the following *factorization criterion.*

Consider first the case that X is discrete, and let $P_\theta(x) = P_\theta\{X = x\}$. Then a necessary and sufficient condition for T to be sufficient for θ is that there exists a factorization

$$P_\theta(x) = g_\theta[T(x)]h(x), \tag{1.16}$$

where the first factor may depend on θ but depends on x only through $T(x)$, while the second factor is independent of θ.

Suppose that (1.16) holds, and let $T(x) = t$. Then $P_\theta\{T = t\} = \sum P_0(x')$ summed over all points x' with $T(x') = t$, and the conditional probability

$$P_\theta\{X = x \mid T = t\} = \frac{P_\theta(x)}{P_\theta\{T = t\}} = \frac{h(x)}{\sum h(x')}$$

is independent of θ. Conversely, if this conditional distribution does not depend on θ and is equal to, say $k(x, t)$, then $P_\theta(x) = P_\theta\{T = t\}k(x, t)$, so that (1.16) holds.

Example 1.9.3 Let $X_1, \ldots, X_n$ be independently and identically distributed according to the Poisson distribution (1.2). Then

$$P_\tau(x_1, \ldots, x_n) = \frac{\tau^{\sum x_i} e^{-n\tau}}{\prod_{j=1}^{n} x_j!},$$

[6]These are connected with difficulties concerning the behavior of conditional probabilities. For a discussion of these difficulties see Sections 2.3–2.5.

and it follows that $\sum X_i$ is a sufficient statistic for τ. ■

In the case that the distribution of X is continuous and has probability density $p_\theta^X(x)$, let X and T be vector-valued, $X = (X_1, \ldots, X_n)$ and $T = (T_1, \ldots T_t)$ say. Suppose that there exist functions $Y = (Y_1, \ldots, Y_{n-r})$ on the sample space such that the transformation

$$(x_1, \ldots, x_n) \leftrightarrow (T_1(x), \ldots, T_r(x), Y_1(x), \ldots, Y_{n-r}(x)) \tag{1.17}$$

is 1:1 on a suitable domain, and that the joint density of T and Y exists and is related to that of X by the usual formula

$$p_\theta^X(x) = p_\theta^{T,Y}(T(x), Y(x)) \cdot |J|, \tag{1.18}$$

where J is the Jacobian of $(T_1, \ldots, T_r, Y_1, \ldots, Y_{n-r})$ with respect to $(x_1, \ldots, x_n)$. Thus in Example 1.9.2, $T = \sqrt{\sum X_i^2}, Y_1, \ldots, Y_{n-1}$ can be taken to be the polar coordinates of the sample point. From the joint density $p_\theta^{T,Y}(t, y)$ of T and Y, the conditional density of Y given $T = t$ is obtained as

$$p_\theta^{Y|t}(y) = \frac{p_\theta^{T,Y}(t, y)}{\int p_\theta^{T,Y}(t, y')\, dy'} \tag{1.19}$$

provided the denominator is different from zero. Regularity conditions for the validity of (1.18) are given by Tukey (1958b).

Since in the conditional distribution given t only the Y's vary, T is sufficient for θ if the conditional distribution of Y given t is independent of θ. Suppose that T satisfies (1.19). Then analogously to the discrete case, a necessary and sufficient condition for T to be sufficient is a factorization of the density of the form

$$p_\theta^X(x) = g_\theta[T(x)]h(x). \tag{1.20}$$

(See Problem 1.19.) The following two examples illustrate the application of the criterion in this case. In both examples the existence of functions Y satisfying (1.17)–(1.19) will be assumed but not proved. As will be shown later (Section 2.6), this assumption is actually not needed for the validity of the factorization criterion.

Example 1.9.4 Let $X_1, \ldots, X_n$ be independently distributed with normal probability density

$$p_{\xi,\sigma}(x) = (2\pi\sigma^2)^{-n/2} \exp\left(-\frac{1}{2\sigma^2}\sum x_i^2 + \frac{\xi}{\sigma^2}\sum x_i - \frac{n}{2\sigma^2}\xi^2\right).$$

Then the factorization criterion shows $(\sum X_i, \sum X_i^2)$ to be sufficient for (ξ, σ). ■

Example 1.9.5 Let $X_1, \ldots, X_n$ be independently distributed according to the uniform distribution $U(0, \theta)$ over the interval $(0, \theta)$. Then $p_\theta(x) = \theta^{-n}(\max x_i, \theta)$, where $u(a, b)$ is 1 or 0 as $a \le b$ or $a > b$, and hence $\max X_i$ is sufficient for θ. ■

An alternative criterion of *Bayes sufficiency*, due to Kolmogorov (1942), provides a direct connection between this concept and some of the basic notions of decision theory. As in the theory of Bayes solutions, consider the unknown parameter θ as a random variable Θ with an a priori distribution, and assume for simplicity that it has a density $\rho(\theta)$. Then if T is sufficient, the conditional distribution of Θ given $X = x$ depends only on $T(x)$. Conversely, if $\rho(\theta) \neq 0$ for all θ and if the conditional distribution of Θ given x depends only on $T(x)$, then T is sufficient for θ.

In fact, under the assumptions made, the joint density of X and Θ is $p_\theta(x)\rho(\theta)$. If T is sufficient, it follows from (1.20) that the conditional density of Θ given x depends only on $T(x)$. Suppose, on the other hand, that for some a priori distribution for which $\rho(\theta) \neq 0$ for all θ the conditional distribution of Θ given x depends only on $T(x)$. Then

$$\frac{p_\theta(x)\rho(\theta)}{\int p_{\theta'}(x)\rho(\theta')\,d\theta'} = f_\theta[T(x)]$$

and by solving for $p_\theta(x)$ it is seen that T is sufficient.

Any Bayes solution depends only on the conditional distribution of Θ given x (see Problem 1.8) and hence on $T(x)$. Since typically Bayes solutions together with their limits form an essentially complete class, it follows that this is also true of the decision procedures based on T. The same conclusion had already been reached more directly at the beginning of the section.

For a discussion of the relation of these different aspects of sufficiency in more general circumstances and references to the literature see Le Cam (1964), Roy and Ramamoorthi (1979) and Yamada and Morimoto (1992). An example of a statistic which is Bayes sufficient in the Kolmogorov sense but not according to the definition given at the beginning of this section is provided by Blackwell and Ramamoorthi (1982).

By restricting attention to a sufficient statistic, one obtains a reduction of the data, and it is then desirable to carry this reduction as far as possible. To illustrate the different possibilities, consider once more the binomial Example 1.9.1. If m is any integer less than n and $T_1 = \sum_{i=1}^m X_i$, $T_2 = \sum_{i=m+1}^n X_i$, then (T_1, T_2) constitutes a sufficient statistic, since the conditional distribution of $X_1, \ldots, X_n$ given $T_1 = t_1$, $T_2 = t_2$ is independent of p. For the same reason, the full sample $(X_1, \ldots, X_n)$ itself is also a sufficient statistic. However, $T = \sum_{i=1}^n X_i$ provides a more thorough reduction than either of these and than various others that can be constructed. A sufficient statistic T is said to be *minimal sufficient* if the data cannot be reduced beyond T without losing sufficiency. For the binomial example in particular, $\sum_{i=1}^n X_i$ can be shown to be minimal (Problem 1.17). This illustrates the fact that in specific examples the sufficient statistic determined by inspection through the factorization criterion usually turns out to be minimal. Explicit procedures for constructing minimal sufficient statistics are discussed in Section 1.5 of *TPE2*.

1.10 Problems

Section 1.2

Problem 1.1 The following distributions arise on the basis of assumptions similar to those leading to (1.1)–(1.3).

(i) Independent trials with constant probability p of success are carried out until a preassigned number m of successes has been obtained. If the number of trials required is $X+m$, then X has the *negative binomial* distribution $Nb(p,m)$:

$$P\{X=x\}=\binom{m+x-1}{x}p^m(1-p)^x, \qquad x=0,1,2\ldots.$$

(ii) In a sequence of random events, the number of events occurring in any time interval of length τ has the Poisson distribution $P(\lambda\tau)$, and the numbers of events in nonoverlapping time intervals are independent. Then the "waiting time" T, which elapses from the starting point, say $t=0$, until the first event occurs, has the *exponential* probability density

$$p(t)=\lambda e^{-\lambda\tau}, \qquad t\geq 0.$$

Let T_i, $i\geq 2$, be the time elapsing from the occurrence of the $(i-1)$st event to that of the ith event. Then it is also true, although more difficult to prove, that $T_1,T_2,\ldots$ are identically and independently distributed. A proof is given, for example, in Karlin and Taylor (1975).
(iii) A point X is selected "at random" in the interval (a,b), that is, the probability of X falling in any subinterval of (a,b) depends only on the length of the subinterval, not on its position. Then X has the *uniform* distribution $U(a,b)$ with probability density

$$p(x)=1/(b-a), \qquad a<x<b.$$

Section 1.5

Problem 1.2 *Unbiasedness in point estimation.* Suppose that γ is a continuous real-valued function defined over Ω which is not constant in any open subset of Ω, and that the expectation $h(\theta)=E_\theta\delta(X)$ is a continuous function of θ for every estimate $\delta(X)$ of $\gamma(\theta)$. Then (1.11) is a necessary and sufficient condition for $\delta(X)$ to be unbiased when the loss function is the square of the error.
[Unbiasedness implies that $\gamma^2(\theta')-\gamma^2(\theta)\geq 2h(\theta)[\gamma(\theta')-\gamma(\theta)]$ for all θ, θ'. If θ is neither a relative minimum nor maximum of γ, it follows that there exist points θ' arbitrarily close to θ both such that $\gamma(\theta)+\gamma(\theta')\geq$ and $\leq 2h(\theta)$, and hence that $\gamma(\theta)=h(\theta)$. That this equality also holds for an extremum of γ follows by continuity, since γ is not constant in any open set.]

Problem 1.3 *Median unbiasedness.*
(i) A real number m is a median for the random variable Y if $P\{Y\geq m\}\geq\frac{1}{2}$, $P\{Y\leq m\}\geq\frac{1}{2}$. Then all real a_1,a_2 such that $m\leq a_1\leq a_2$ or $m\geq a_1\geq a_2$ satisfy $E|Y-a_1|\leq E|Y-a_2|$.
(ii) For any estimate $\delta(X)$ of $\gamma(\theta)$, let $m^-(\theta)$ and $m^+(\theta)$ denote the infimum and supremum of the medians of $\delta(X)$, and suppose that they are continuous functions of θ. Let $\gamma(\theta)$ be continuous and not constant in any open subset of Ω. Then the estimate $\delta(X)$ of $\gamma(\theta)$ is unbiased with respect to the loss function $L(\theta,d)=|\gamma(\theta)-d|$ if and only if $\gamma(\theta)$ is a median of $\delta(X)$ for each θ. An estimate with this property is said to be *median-unbiased*.

Problem 1.4 *Nonexistence of unbiased procedures.* Let $X_1, \ldots, X_n$ be independently distributed with density $(1/a)f((x-\xi)/a)$, and let $\theta = (\xi, a)$. Then no estimator of ξ exists which is unbiased with respect to the loss function $(d-\xi)^k/a^k$. *Note.* For more general results concerning the nonexistence of unbiased procedures see Rojo (1983).

Problem 1.5 Let $\mathcal{C}$ be any class of procedures that is closed under the transformations of a group G in the sense that $\delta \in \mathcal{C}$ implies $g^*\delta g^{-1} \in \mathcal{C}$ for all $g \in G$. If there exists a unique procedure δ_0 that uniformly minimizes the risk within the class $\mathcal{C}$, then δ_0 is invariant.[7] If δ_0 is unique only up to sets of measure zero, then it is *almost invariant*, that is, for each g it satisfies the equation $\delta(gx) = g^*\delta(x)$ except on a set N_g of measure 0.

Problem 1.6 *Relation of unbiasedness and invariance.*
(i) If δ_0 is the unique (up to sets of measure 0) unbiased procedure with uniformly minimum risk, it is almost invariant.
(ii) If $\bar{G}$ is transitive and G^* commutative, and if among all invariant (almost invariant) procedures there exists a procedure δ_0 with uniformly minimum risk, then it is unbiased.
(iii) That conclusion (ii) need not hold without the assumptions concerning G^* and $\bar{G}$ is shown by the problem of estimating the mean ξ of a normal distribution $N(\xi, \sigma^2)$ with loss function $(\xi - d)^2/\sigma^2$. This remains invariant under the groups $G_1 : gx = x + b$, $-\infty < b < \infty$ and $G_2 : gx = ax + b$, $0 < a < \infty$, $-\infty < b < \infty$. The best invariant estimate relative to both groups is X, but there does not exist an estimate which is unbiased with respect to the given loss function.
[(i): This follows from the preceding problem and the fact that when δ is unbiased so is $g^*\delta g^{-1}$.
(ii): It is the defining property of transitivity that given θ, θ' there exists $\bar{g}$ such that $\theta' = \bar{g}\theta$. Hence for any θ, θ'

$$E_\theta L(\theta', \delta_0(X)) = E_\theta L(\bar{g}\theta, \delta_0(X)) = E_\theta L(\theta, g^{*-1}\delta_0(X)).$$

Since G^* is commutative, $g^{*-1}\delta_0$ is invariant, so that

$$R(\theta, g^{*-1}\delta_0) \geq R(\theta, \delta_0) = E_\theta L(\theta, \delta_0(X)).]$$

Section 1.6

Problem 1.7 *Unbiasedness in interval estimation.* Confidence intervals $I = (\underline{L}, \bar{L})$ are unbiased for estimating θ with loss function $L(\theta, I) = (\theta - \underline{L})^2 + (\bar{L} - \theta)^2$ provided $E[\frac{1}{2}(\underline{L} + \bar{L})] = \theta$ for all θ, that is, provided the midpoint of I is an unbiased estimate of θ in the sense of (1.11).

Problem 1.8 *Structure of Bayes solutions.*
(i) Let Θ be an unobservable random quantity with probability density $\rho(\theta)$, and let the probability density of X be $p_\theta(x)$ when $\Theta = \theta$. Then δ is a Bayes solution

[7] Here and in Problems 1.6, 1.7, 1.11, 1.15, and 1.16 the term "invariant" is used in the general sense (1.8) of "invariant or equivalent."

of a given decision problem if for each x the decision $\delta(x)$ is chosen so as to minimize $\int L(\theta, \delta(x))\pi(\theta \mid x)\, d\theta$, where $\pi(\theta \mid x) = \rho(\theta)p_\theta(x)/\int \rho(\theta')p_{\theta'}(x)\, d\theta'$ is the conditional (a posteriori) probability density of Θ given x.
(i) Let the problem be a two-decision problem with the losses as given in Example 1.5.5. Then the Bayes solution consists in choosing decision d_0 if

$$aP\{\Theta \in \omega_1 \mid x\} < bP\{\Theta \in \omega_0 \mid x\}$$

and decision d_1 if the reverse inequality holds. The choice of decision is immaterial in case of equality.
(iii) In the case of point estimation of a real-valued function $g(\theta)$ with loss function $L(\theta, d) = (g(\theta) - d)^2$, the Bayes solution becomes $\delta(x) = E[g(\Theta) \mid x]$. When instead the loss function is $L(\theta, d) = |g(\theta) - d|$, the Bayes estimate $\delta(x)$ is any median of the conditional distribution of $g(\Theta)$ given x.
[(i): The Bayes risk $r(\rho, \delta)$ can be written as $\int[\int L(\theta, \delta(x))\pi(\theta \mid x)\, d\theta] \times p(x)\, dx$, where $p(x) = \int \rho(\theta')p_{\theta'}(x)\, d\theta'$.
(ii): The conditional expectation $\int L(\theta, d_0)\pi(\theta \mid x)\, d\theta$ reduces to $aP\{\Theta \in \omega_1 \mid x\}$, and similarly for d_1.]

Problem 1.9 (i) As an example in which randomization reduces the maximum risk, suppose that a coin is known to be either standard (HT) or to have heads on both sides (HH). The nature of the coin is to be decided on the basis of a single toss, the loss being 1 for an incorrect decision and 0 for a correct one. Let the decision be HT when T is observed, whereas in the contrary case the decision is made at random, with probability ρ for HT and $1-\rho$ for HH. Then the maximum risk is minimized for $\rho = \frac{1}{3}$.
(ii) A genetic setting in which such a problem might arise is that of a couple, of which the husband is either dominant homozygous (AA) or heterozygous (Aa) with respect to a certain characteristic, and the wife is homozygous recessive (aa). Their child is heterozygous, and it is of importance to determine to which genetic type the husband belongs. However, in such cases an a priori probability is usually available for the two possibilities. One is then dealing with a Bayes problem, and randomization is no longer required. In fact, if the a priori probability is p that the husband is dominant, then the Bayes procedure classifies him as such if $p > \frac{1}{3}$ and takes the contrary decision if $p < \frac{1}{3}$.

Problem 1.10 *Unbiasedness and minimax.* Let $\Omega = \Omega_0 \cup \Omega_1$ where Ω_0, Ω_1 are mutually exclusive, and consider a two-decision problem with loss function $L(\theta, d_i) = a_i$ for $\theta \in \Omega_j (j \neq i)$ and $L(\theta, d_i) = 0$ for $\theta \in \Omega_i (i = 0, 1)$.
(i) Any minimax procedure is unbiased. (ii) The converse of (i) holds provided $P_\theta(A)$ is a continuous function of θ for all A, and if the sets Ω_0 and Ω_1 have at least one common boundary point.
[(i): The condition of unbiasedness in this case is equivalent to $\sup R_\delta(\theta) \leq a_0a_1/(a_0 + a_1)$. That this is satisfied by any minimax procedure is seen by comparison with the procedure $\delta(x) = d_0$ or $= d_1$ with probabilities $a_1/(a_0 + a_1)$ and $a_0/(a_0 + a_1)$ respectively.
(ii): If θ_0, is a common boundary point, continuity of the risk function implies that any unbiased procedure satisfies $R_\delta(\theta_0) = a_0a_1/(a_0 + a_1)$ and hence sup $R_\delta(\theta_0) = a_0a_1/(a_0 + a_1)$.]

Problem 1.11 *Invariance and minimax.* Let a problem remain invariant relative to the groups G, $\bar{G}$, and G^* over the spaces $\mathcal{X}$, Ω, and D respectively. Then a randomized procedure Y_x is defined to be invariant if for all x and g the conditional distribution of Y_x given x is the same as that of $g^{*-1}Y_{gx}$.
(i) Consider a decision procedure which remains invariant under a finite group $G = \{g_1, \ldots, g_N\}$. If a minimax procedure exists, then there exists one that is invariant. (ii) This conclusion does not necessarily hold for infinite groups, as is shown by the following example. Let the parameter space Ω consist of all elements θ of the free group with two generators, that is, the totality of formal products $\pi_1 \ldots \pi_n$ ($n = 0, 1, 2, \ldots$) where each π_i is one of the elements a, a^{-1}, b, b^{-1} and in which all products $aa^{-1}, a^{-1}a, bb^{-1}$, and $b^{-1}b$ have been canceled. The empty product ($n = 0$) is denoted by e. The sample point X is obtained by multiplying θ on the right by one of the four elements a, a^{-1}, b, b^{-1} with probability $\frac{1}{4}$ each, and canceling if necessary, that is, if the random factor equals π_n^{-1}. The problem of estimating θ with $L(\theta, d)$ equal to 0 if $d = \theta$ and equal to 1 otherwise remains invariant under multiplication of X, θ, and d on the left by an arbitrary sequence $\pi_{-m} \ldots \pi_{-2}\pi_{-1}$($m = 0, 1, \ldots$). The invariant procedure that minimizes the maximum risk has risk function $R(\theta, \delta) \equiv \frac{3}{4}$. However, there exists a noninvariant procedure with maximum risk $\frac{1}{4}$.
[(i): If Y_x is a (possibly randomized) minimax procedure, an invariant minimax procedure Y'_x is defined by $P(Y'_x = d) = \sum_{i=1}^{N} P(Y_{g_i x} = g_i^* d)/N$.
(ii): The better procedure consists in estimating θ to be $\pi_1 \ldots \pi_{k-1}$ when $\pi_1 \ldots \pi_k$ is observed ($k \geq 1$), and estimating θ to be a, a^{-1}, b, b^{-1} with probability $\frac{1}{4}$ each in case the identity is observed. The estimate will be correct unless the last element of X was canceled, and hence will be correct with probability $\geq \frac{3}{4}$.]

Section 1.7

Problem 1.12 (i) Let X have probability density $p_\theta(x)$ with θ one of the values $\theta_1, \ldots, \theta_n$, and consider the problem of determining the correct value of θ, so that the choice lies between the n decisions $d_1 = \theta_1, \ldots, d_n = \theta_n$ with gain $a(\theta_i)$ if $d_i = \theta_i$ and 0 otherwise. Then the Bayes solution (which maximizes the average gain) when θ is a random variable taking on each of the n values with probability $1/n$ coincides with the maximum-likelihood procedure. (ii) Let X have probability density $p_\theta(x)$ with $0 \leq \theta \leq 1$. Then the maximum-likelihood estimate is the mode (maximum value) of the a posteriori density of Θ given x when Θ is uniformly distributed over $(0, 1)$.

Problem 1.13 (i) Let $X_1, \ldots, X_n$ be a sample from $N(\xi, \sigma^2)$, and consider the problem of deciding between $\omega_0 : \xi < 0$ and $\omega_1 : \xi \geq 0$. If $\bar{x} = \sum x_i/n$ and $C = (a_1/a_0)^{2/n}$, the likelihood-ratio procedure takes decision d_0 or d, as

$$\frac{\sqrt{n}\bar{x}}{\sqrt{\sum(x_i - \bar{x})^2}} < k \quad \text{or} \quad > k,$$

where $k = \sqrt{C - 1}$ if $C > 1$ and $k = \sqrt{(1 - C)/C}$ if $C < 1$.

(ii) For the problem of deciding between $\omega_0 : \sigma < \sigma_0$ and $\omega_1 : \sigma \geq \sigma_0$ the likelihood ratio procedure takes decision d_0 or d, as

$$\frac{\sum(x_i - \bar{x})^2}{n\sigma_0^2} < \quad \text{or} \quad > k,$$

where k is the smaller root of the equation $Cx = e^{x-1}$ if $C > 1$, and the larger root of $x = Ce^{x-1}$ if $C < 1$, where C is defined as in (i).

Section 1.8

Problem 1.14 *Admissibility of unbiased procedures.*
(i) Under the assumptions of Problem 1.10, if among the unbiased procedures there exists one with uniformly minimum risk, it is admissible. (ii) That in general an unbiased procedure with uniformly minimum risk need not be admissible is seen by the following example. Let X have a Poisson distribution truncated at 0, so that $P_\theta\{X = x\} = \theta^x e^{-\theta}/[x!(1 - e^{-\theta})]$ for $x = 1, 2, \ldots$. For estimating $\gamma(\theta) = e^{-\theta}$ with loss function $L(\theta, d) = (d - e^{-\theta})^2$, there exists a unique unbiased estimate, and it is not admissible.
[(ii): The unique unbiased estimate $\delta_0(x) = (-1)^{x+1}$ is dominated by $\delta_1(x) = 0$ or 1 as x is even or odd.]

Problem 1.15 *Admissibility of invariant procedures.* If a decision problem remains invariant under a finite group, and if there exists a procedure δ_0 that uniformly minimizes the risk among all invariant procedures, then δ_0 is admissible.
[This follows from the identity $R(\theta, \delta) = R(\bar{g}\theta, g^*\delta g^{-1})$ and the hint given in Problem 1.11(i).]

Problem 1.16 (i) Let X take on the values $\theta - 1$ and $\theta + 1$ with probability $\frac{1}{2}$ each. The problem of estimating θ with loss function $L(\theta, d) = \min(|\theta - d|, 1)$ remains invariant under the transformation $gX = X + c$, $\bar{g}\theta = \theta + c$, $g^*d = d + c$. Among invariant estimates, those taking on the values $X - 1$ and $X + 1$ with probabilities p and q (independent of X) uniformly minimize the risk. (ii) That the conclusion of Problem 1.15 need not hold when G is infinite follows by comparing the best invariant estimates of (i) with the estimate $\delta_1(x)$ which is $X + 1$ when $X < 0$ and $X - 1$ when $X \geq 0$.

Section 1.9

Problem 1.17 In n independent trials with constant probability p of success, let $X_i = 1$ or 0 as the ith trial is a success or not. Then $\sum_{i=1}^n X_i$ is minimal sufficient.
[Let $T = \sum X_i$ and suppose that $U = f(T)$ is sufficient and that $f(k_1) = \cdots = f(k_r) = u$. Then $P\{T = t \mid U = u\}$ depends on p.]

Problem 1.18 (i) Let $X_1, \ldots, X_n$ be a sample from the uniform distribution $U(0, \theta)$, $0 < \theta < \infty$, and let $T = \max(X_1, \ldots, X_n)$. Show that T is sufficient,

once by using the definition of sufficiency and once by using the factorization criterion and assuming the existence of statistics Y_i satisfying (1.17)–(1.19).
(ii) Let $X_1, \ldots, X_n$ be a sample from the exponential distribution $E(a, b)$ with density $(1/b)e^{-(x-a)/b}$ when $x \geq a$ $(-\infty < a < \infty, 0 < b)$. Use the factorization criterion to prove that $(\min(X_1, \ldots, X_n), \sum_{i=1}^n X_i)$ is sufficient for a, b, assuming the existence of statistics Y_i satisfying (1.17)–(1.19).

Problem 1.19 A statistic T satisfying (1.17)–(1.19) is sufficient if and only if it satisfies (1.20).

1.11 Notes

Some of the basic concepts of statistical theory were initiated during the first quarter of the 19th century by Laplace in his fundamental *Théorie Analytique des Probabilités* (1812), and by Gauss in his papers on the method of least squares. Loss and risk functions are mentioned in their discussions of the problem of point estimation, for which Gauss also introduced the condition of unbiasedness.

A period of intensive development of statistical methods began toward the end of the century with the work of Karl Pearson. In particular, two areas were explored in the researches of R. A. Fisher, J. Neyman, and many others: estimation and the testing of hypotheses. The work of Fisher can be found in his books (1925, 1935, 1956) and in the five volumes of his collected papers (1971–1973). An interesting review of Fisher's contributions is provided by Savage (1976), and his life and work are recounted in the biography by his daughter Joan Fisher Box (1978). Many of Neyman's principal ideas are summarized in his Lectures and Conferences (1938b). Collections of his early papers and of his joint papers with E. S. Pearson have been published [Neyman (1967) and Neyman and Pearson (1967)], and Constance Reid (1982) has written his biography. An influential synthesis of the work of this period by Cramér appeared in 1946. Further concepts were introduced in Lehmann (1950, 1951ab). More recent surveys of the modern theories of estimation and testing are contained, for example, in the books by Strasser (1985), Stuart and Ord (1991, 1999), Schervish (1995), Shao (1999) and Bickel and Doksum (2001).

A formal unification of the theories of estimation and hypothesis testing, which also contains the possibility of many other specializations, was achieved by Wald in his general theory of decision procedures. An account of this theory, which is closely related to von Neumann's theory of games, is found in Wald's book (1950) and in those of Blackwell and Girshick (1954), Ferguson (1967), and Berger (1985b).

2
The Probability Background

2.1 Probability and Measure

The mathematical framework for statistical decision theory is provided by the theory of probability, which in turn has its foundations in the theory of measure and integration. The present chapter serves to define some of the basic concepts of these theories, to establish some notation, and to state without proof some of the principal results which will be used throughout Chapters 3–9. In the remainder of this chapter, certain special topics are treated in more detail. Basic notions of convergence in probability theory which will be needed for large sample statistical theory are deferred to Section 11.2.

Probability theory is concerned with situations which may result in different outcomes. The totality of these possible outcomes is represented abstractly by the totality of points in a space $\mathcal{Z}$. Since the events to be studied are aggregates of such outcomes, they are represented by subsets of $\mathcal{Z}$. The union of two sets C_1, C_2 will be denoted by $C_1 \cup C_2$, their intersection by $C_1 \cap C_2$, the complement of C by $C^c = \mathcal{Z} - C$, and the empty set by 0. The probability $P(C)$ of an event C is a real number between 0 and 1; in particular

$$P(0) = 0 \quad \text{and} \quad P(\mathcal{Z}) = 1 \tag{2.1}$$

Probabilities have the property of *countable additivity*,

$$P\left(\bigcup C_i\right) = \sum P(C_i) \qquad \text{if} \quad C_i \cap C_j = 0 \quad \text{for all} \quad i \neq j. \tag{2.2}$$

Unfortunately it turns out that the set functions with which we shall be concerned usually cannot be defined in a reasonable manner for all subsets of $\mathcal{Z}$ if they are to satisfy (2.2). It is, for example, not possible to give a reasonable definition of "area" for all subsets of a unit square in the plane.

The sets for which the probability function P will be defined are said to be "measurable." The domain of definition of P should include with any set C its complement C^c, and with any countable number of events their union. By (2.1), it should also include $\mathcal{Z}$. A class of sets that contains $\mathcal{Z}$ and is closed under complementation and countable unions is a σ-field. Such a class is automatically also closed under countable intersections.

The starting point of any probabilistic considerations is therefore a space $\mathcal{Z}$, representing the possible outcomes, and a σ-field $\mathcal{C}$ of subsets of $\mathcal{Z}$, representing the events whose probability is to be defined. Such a couple $(\mathcal{Z}, \mathcal{C})$ is called a *measurable space*, and the elements of $\mathcal{C}$ constitute the *measurable sets*. A countably additive nonnegative (not necessarily finite) set function μ defined over $\mathcal{C}$ and such that $\mu(0) = 0$ is called a *measure*. If it assigns the value 1 to $\mathcal{Z}$, it is a *probability measure*. More generally, μ is *finite* if $\mu(\mathcal{Z}) < \infty$ and σ-finite if there exist $C_1, C_2, \ldots$ in $\mathcal{C}$ (which may always be taken to be mutually exclusive) such that $\cup C_i = \mathcal{Z}$ and $\mu(C_i) < \infty$ for $i = 1, 2, \ldots$. Important special cases are provided by the following examples.

Example 2.1.1 (Lebesgue measure) Let $\mathcal{Z}$ be the n-dimensional Euclidean space E_n, and $\mathcal{C}$ the smallest σ-field containing all rectangles[1]

$$R = \{(z_1, \ldots, z_n) : a_i < z_i \le b_i, i = 1, \ldots, n\}.$$

The elements of $\mathcal{C}$ are called the *Borel sets* of E_n. Over $\mathcal{C}$ a unique measure μ can be defined, which to any rectangle R assigns as its measure the volume of R,

$$\mu(R) = \prod_{i=1}^{n} (b_i - a_i).$$

The measure μ can be *completed* by adjoining to $\mathcal{C}$ all subsets of sets of measure zero. The domain of μ is thereby enlarged to a σ-field $\mathcal{C}'$, the class of *Lebesgue-measurable* sets. The term *Lebesgue-measure* is used for μ both when it is defined over the Borel sets and when it is defined over the Lebesgue-measurable sets. ■

This example can be generalized to any nonnegative set function ν, which is defined and countably additive over the class of rectangles R. There exists then, as before, a unique measure μ over $(\mathcal{Z}, \mathcal{C})$ that agrees with ν for all R. This measure can again be completed; however, the resulting σ-field depends on μ and need not agree with the σ-field $\mathcal{C}'$ obtained above.

Example 2.1.2 (Counting measure) Suppose the $\mathcal{Z}$ is countable, and let $\mathcal{C}$ be the class of all subsets of $\mathcal{Z}$. For any set C, define $\mu(C)$ as the number of elements of C if that number is finite, and otherwise as $+\infty$. This measure is sometimes called *counting measure*. ■

In applications, the probabilities over $(\mathcal{Z}, \mathcal{C})$ refer to random experiments or observations, the possible outcomes of which are the points $z \in \mathcal{Z}$. When recording the results of an experiment, one is usually interested only in certain of its

---

[1]If $\pi(z)$ is a statement concerning certain objects z, then $\{z : \pi(z)\}$ denotes the set of all those z for which $\pi(z)$ is true.

aspects, typically some counts or measurements. These may be represented by a function T taking values in some space $\mathcal{T}$.

Such a function generates in $\mathcal{T}$ the σ-field $\mathcal{B}'$ of sets B whose inverse image

$$C = T^{-1}(B) = \{z : z \in \mathcal{Z}, T(z) \in B\}$$

is in $\mathcal{C}$, and for any given probability measure P over $(\mathcal{Z}, \mathcal{C})$ a probability measure Q over $(\mathcal{T}, \mathcal{B}')$ defined by

$$Q(B) = P(T^{-1}(B)). \tag{2.3}$$

Frequently, there is given a σ-field $\mathcal{B}$ of sets in $\mathcal{T}$ such that the probability of B should be defined if and only if $B \in \mathcal{B}$. This requires that $T^{-1}(B) \in \mathcal{C}$ for all $B \in \mathcal{B}$, and the function (or transformation) T from $(\mathcal{Z}, \mathcal{C})$ into[2] $(\mathcal{T}, \mathcal{B})$ is then said to be $\mathcal{C}$-measurable. Another implication is the sometimes convenient restriction of probability statements to the sets $B \in \mathcal{B}$ even though there may exist sets $B \notin \mathcal{B}$ for which $T^{-1}(B) \in \mathcal{C}$ and whose probability therefore could be defined.

Of particular interest is the case of a single measurement in which the function of T is real-valued. Let us denote it by X, and let $\mathcal{A}$ be the class of Borel sets on the real line $\mathcal{X}$. Such a measurable real-valued X is called a *random variable*, and the probability measure it generates over $(\mathcal{X}, \mathcal{A})$ will be denoted by P^X and called the probability distribution of X. The value this measure assigns to a set $A \in \mathcal{A}$ will be denoted interchangeably by $P^X(A)$ and $P(X \in A)$. Since the intervals $\{x : x \leq a\}$ are in $\mathcal{A}$, the probabilities $F(a) = P(X \leq a)$ are defined for all a. The function F, the *cumulative distribution function* (cdf) of X, is nondecreasing and continuous on the right, and $F(-\infty) = 0$, $F(+\infty) = 1$. Conversely, if F is any function with these properties, a measure can be defined over the intervals by $P\{a < X \leq b\} = F(b) - F(a)$. It follows from Example 2.1.1 that this measure uniquely determines a probability distribution over the Borel sets. Thus the probability distribution P^X and the cumulative distribution function F uniquely determine each other. These remarks extend to probability distributions over n-dimensional Euclidean space, where the cumulative distribution function is defined by

$$F(a_1, \ldots, a_n) = P\{X_1 \leq a_1, \ldots, X_n \leq a_n\}.$$

In concrete problems, the space $(\mathcal{Z}, \mathcal{C})$, corresponding to the totality of possible outcomes, is usually not specified and remains in the background. The real starting point is the set X of observations (typically vector-valued) that are being recorded and which constitute the *data*, and the associated measurable space $(\mathcal{X}, \mathcal{A})$, the *sample space*. Random variables or vectors that are measurable transformations T from $(\mathcal{X}, \mathcal{A})$ into some $(\mathcal{T}, \mathcal{B})$ are called *statistics*. The distribution of T is then given by (2.3) applied to all $B \in \mathcal{B}$. With this definition, a statistic is specified by the function T and the σ-field $\mathcal{B}$. We shall, however, adopt the convention that when a function T takes on its values in a Euclidean space, unless otherwise stated the σ-field $\mathcal{B}$ of measurable sets will be taken to be the class of

---

[2]The term *into* indicates that the range of T is in $\mathcal{T}$; if $T(\mathcal{Z}) = \mathcal{T}$, the transformation is said to be from $\mathcal{Z}$ *onto* $\mathcal{T}$.

Borel sets. It then becomes unnecessary to mention it explicitly or to indicate it in the notation.

The distinction between statistics and random variables as defined here is slight. The term statistic is used to indicate that the quantity is a function of more basic observations; all statistics in a given problem are functions defined over the same sample space $(\mathcal{X}, \mathcal{A})$. On the other hand, any real-valued statistic T is a random variable, since it has a distribution over $(\mathcal{T}, \mathcal{B})$, and it will be referred to as a random variable when its origin is irrelevant. Which term is used therefore depends on the point of view and to some extent is arbitrary.

2.2 Integration

According to the convention of the preceding section, a real-valued function f defined over $(\mathcal{X}, \mathcal{A})$ is measurable if $f^{-1}(B) \in \mathcal{A}$ for every Borel set B on the real line. Such a function f is said to be *simple* if it takes on only a finite number of values. Let μ be a measure defined over $(\mathcal{X}, \mathcal{A})$, and let f be a simple function taking on the distinct values $a_1, \ldots, a_m$ on the sets $A_1, \ldots, A_m$, which are in $\mathcal{A}$, since f is measurable. If $\mu(A_i) < \infty$ when $a_i \neq 0$, the integral of f with respect to μ is defined by

$$\int f \, d\mu = \sum a_i \mu(A_i). \tag{2.4}$$

Given any nonnegative measurable function f, there exists a nondecreasing sequence of simple functions f_n converging to f. Then the integral of f is defined as

$$\int f \, d\mu = \lim_{n \to \infty} \int f_n \, d\mu, \tag{2.5}$$

which can be shown to be independent of the particular sequence of f_n's chosen. For any measurable function f its positive and negative parts

$$f^+(x) = \max[f(x), 0] \quad \text{and} \quad f^-(x) = \max[-f(x), 0] \tag{2.6}$$

are also measurable, and

$$f(x) = f^+(x) - f^-(x).$$

If the integrals of f^+ and f^- are both finite, then f is said to be *integrable*, and its integral is defined as

$$\int f \, d\mu = \int f^+ \, d\mu - \int f^- \, d\mu.$$

If of the two integrals one is finite and one infinite, then the integral of f is defined to be the appropriate infinite value; if both are infinite, the integral is not defined.

Example 2.2.1 Let $\mathcal{X}$ be the closed interval $[a, b]$, $\mathcal{A}$ be the class of Borel sets or of Lebesgue measurable sets in $\mathcal{X}$, and μ be Lebesgue measure. Then the integral of f with respect to μ is written as $\int_a^b f(x) \, dx$, and is called the Lebesgue integral of f. This integral generalizes the Riemann integral in that it exists and agrees with the Riemann integral of f whenever the latter exists. ■

Example 2.2.2 Let $\mathcal{X}$ be countable and consist of the points $x_1, x_2, \ldots$; let $\mathcal{A}$ be the class of all subsets of $\mathcal{X}$, and let μ assign measure b_i to the point x_i. Then f is integrable provided $\sum f(x_i)b_i$ converges absolutely, and $\int f\,d\mu$ is given by this sum. ■

Let P^X be the probability distribution of a random variable X, and let T be a real-valued statistic. If the function $T(x)$ is integrable, its *expectation* is defined by

$$E(T) = \int T(x)\,dP^X(x). \tag{2.7}$$

It will be seen from Lemma 2.3.2 in Section 2.3 below that the integration can be carried out alternatively in t-space with respect to the distribution of T defined by (2.3), so that also

$$E(T) = \int t\,dP^T(t). \tag{2.8}$$

The definition (2.5) of the integral permits the basic convergence theorems.

Theorem 2.2.1 Fatou's Lemma *Let f_n be a sequence of measurable functions such that $f_n(x) \geq 0$ and $f_n(x) \to f(x)$, except possibly on a set of x values having μ measure 0. Then,*

$$\int f d\mu \leq \liminf \int f_n d\mu\ .$$

Theorem 2.2.2 *Let f_n be a sequence of measurable functions, and let $f_n(x) \to f(x)$, except possibly on a set of x values having μ measure 0. Then*

$$\int f_n\,d\mu \to \int f\,d\mu$$

if any one of the following conditions holds:

(i) **Lebesgue Monotone Convergence Theorem:** *the f_n's are nonnegative and the sequence is nondecreasing;*

or

(ii) **Lebesgue Dominated Convergence Theorem:** *there exists an integrable function g such that $|f_n(x)| \leq g(x)$ for n and x.*

or

(iii) **General Form:** *there exist g_n and g with $|f_n| \leq g_n$, $g_n(x) \to g(x)$ except possibly on a μ null set, and $\int g_n d\mu \to \int g d\mu$.*

Corollary 2.2.1 Vitali's Theorem *Suppose f_n and f are real-valued measurable functions with $f_n(x) \to f(x)$, except possibly on a set having μ measure 0. Assume*

$$\limsup_n \int f_n^2(x) d\mu(x) \leq \int f^2(x) d\mu(x) < \infty\ .$$

Then,

$$\int |f_n(x) - f(x)|^2 d\mu(x) \to 0 \quad .$$

For a proof of this result, see Theorem 6.1.3 of Hájek, Sidák, and Sen (1999).

For any set $A \in \mathcal{A}$, let I_A be its *indicator function* defined by

$$I_A(x) = 1 \text{ or } 0 \qquad as \quad x \in A \text{ or } x \in A^c, \tag{2.9}$$

and let

$$\int_A f\, d\mu = \int f I_A\, d\mu. \tag{2.10}$$

If μ is a measure and f a nonnegative measurable function over $(\mathcal{X}, \mathcal{A})$, then

$$\nu(A) = \int_A f\, d\mu \tag{2.11}$$

defines a new measure over $(\mathcal{X}, \mathcal{A})$. The fact that (2.11) holds for all $A \in \mathcal{A}$ is expressed by writing

$$d\nu = f\, d\mu \quad \text{ or } \quad f = \frac{d\nu}{d\mu}. \tag{2.12}$$

Let μ and ν be two given σ-finite measures over $(\mathcal{X}, \mathcal{A})$. If there exists a function f satisfying (2.12), it is determined through this relation up to sets of measure zero, since

$$\int_A f\, d\mu = \int_A g\, d\mu \qquad \text{for all} \quad A \in \mathcal{A}$$

implies that $f = g$ a.e. μ.[3] Such an f is called the *Radon–Nikodym derivative* of ν with respect to μ, and in the particular case that ν is a probability measure, the *probability density* of ν with respect to μ.

The question of existence of a function f satisfying (2.12) for given measures μ and ν is answered in terms of the following definition. A measure ν is *absolutely continuous* with respect to μ if

$$\mu(A) = 0 \quad \text{implies} \quad \nu(A) = 0.$$

Theorem 2.2.3 (Radon–Nikodym) *If μ and ν are σ-finite measures over $(\mathcal{X}, \mathcal{A})$, then there exists a measurable function f satisfying* (2.12) *if and only if ν is absolutely continuous with respect to μ.*

The *direct* (or *Cartesian*) *product* $A \times B$ of two sets A and B is the set of all pairs (x, y) with $x \in A$, $y \in B$. Let $(\mathcal{X}, \mathcal{A})$ and $(\mathcal{Y}, \mathcal{B})$ be two measurable spaces, and let $\mathcal{A} \times \mathcal{B}$ be the smallest σ-field containing all sets $A \times B$ with $A \in \mathcal{A}$ and $B \in \mathcal{B}$. If μ and ν are two σ-finite measures over $(\mathcal{X}, \mathcal{A})$ and $(\mathcal{Y}, \mathcal{B})$ respectively,

---

[3] A statement that holds for all points x except possibly on a set of μ-measure zero is said to hold *almost everywhere* μ, abbreviated a.e. μ; or to hold a.e. $(\mathcal{A}, \mu)$ if it is desirable to indicate the σ-field over which μ is defined.

then there exists a unique measure $\lambda = \mu \times \nu$ over $(\mathcal{X} \times \mathcal{Y}, \mathcal{A} \times \mathcal{B})$, the *product* of μ and ν, such that for any $A \in \mathcal{A}$, $B \in \mathcal{B}$,

$$\lambda(A \times B) = \mu(A)\nu(B). \tag{2.13}$$

Example 2.2.3 Let $\mathcal{X}, \mathcal{Y}$ be Euclidean spaces of m and n dimensions, and let $\mathcal{A}, \mathcal{B}$ be the σ-fields of Borel sets in these spaces. Then $\mathcal{X} \times \mathcal{Y}$ is an $(m+n)$-dimensional Euclidean space, and $\mathcal{A} \times \mathcal{B}$ the class of its Borel sets. ■

Example 2.2.4 Let $Z = (X, Y)$ be a random variable defined over $(\mathcal{X} \times \mathcal{Y}, \mathcal{A} \times \mathcal{B})$, and suppose that the random variables X and Y have distributions P^X, P^Y over $(\mathcal{X}, \mathcal{A})$ and $(\mathcal{Y}, \mathcal{B})$. Then X and Y are said to be *independent* if the probability distribution P^Z of Z is the product $P^X \times P^Y$. ■

In terms of these concepts the reduction of a double integral to a repeated one is given by the following theorem.

Theorem 2.2.4 (Fubini) *Let μ and ν be σ-finite measures over $(\mathcal{X}, \mathcal{A})$ and $(\mathcal{Y}, \mathcal{B})$ respectively, and let $\lambda = \mu \times \nu$. If $f(x, y)$ is integrable with respect to λ, then*

(i) *for almost all (ν) fixed y, the function $f(x, y)$ is integrable with respect to μ,*

(ii) *the function $\int f(x, y)\, d\mu(x)$ is integrable with respect to ν, and*

$$\int f(x, y)\, d\lambda(x, y) = \int \left[\int f(x, y)\, d\mu(x) \right] d\nu(y). \tag{2.14}$$

2.3 Statistics and Subfields

According to the definition of Section 2.1, a statistic is a measurable transformation T from the sample space $(\mathcal{X}, \mathcal{A})$ into a measurable space $(\mathcal{T}, \mathcal{B})$. Such a transformation induces in the original sample space the subfield[4]

$$\mathcal{A}_0 = T^{-1}(\mathcal{B}) = \left\{ T^{-1}(B) : B \in \mathcal{B} \right\}. \tag{2.15}$$

Since the set $T^{-1}[T(A)]$ contains A but is not necessarily equal to A, the σ-field $\mathcal{A}_0$ need not coincide with $\mathcal{A}$ and hence can be a proper subfield of $\mathcal{A}$. On the other hand, suppose for a moment that $\mathcal{T} = T(\mathcal{X})$, that is, that the transformation T is onto rather than into $\mathcal{T}$. Then

$$T\left[T^{-1}(B)\right] = B \qquad \text{for all} \qquad B \in \mathcal{B}, \tag{2.16}$$

so that the relationship $A_0 = T^{-1}(B)$ establishes a 1:1 correspondence between the sets of $\mathcal{A}_0$ and $\mathcal{B}$, which is an isomorphism—that is, which preserves the set operations of intersection, union, and complementation. For most purposes it is therefore immaterial whether one works in the space $(\mathcal{X}, \mathcal{A}_0)$ or in $(\mathcal{T}, \mathcal{B})$. These generate two equivalent classes of events, and therefore of measurable functions, possible decision procedures, etc. If the transformation T is only into $\mathcal{T}$, the above

[4]We shall use this term in place of the more cumbersome "sub-σ-field."

1:1 correspondence applies to the class $\mathcal{B}'$ of subsets of $\mathcal{T}' = T(\mathcal{X})$ which belong to $\mathcal{B}$, rather than to $\mathcal{B}$ itself. However, any set $B \in \mathcal{B}$ is equivalent to $B' = B \cap \mathcal{T}'$ in the sense that any measure over $(\mathcal{X}, \mathcal{A})$ assigns the same measure to B' as to B. Considered as classes of events, $\mathcal{A}_0$ and $\mathcal{B}$ therefore continue to be equivalent, with the only difference that $\mathcal{B}$ contains several (equivalent) representations of the same event.

As an example, let $\mathcal{X}$ be the real line and $\mathcal{A}$ the class of Borel sets, and let $T(x) = x^2$. Let $\mathcal{T}$ be either the positive real axis or the whole real axis, and let $\mathcal{B}$ be the class of Borel subsets of $\mathcal{T}$. Then $\mathcal{A}_0$ is the class of Borel sets that are symmetric with respect to the origin. When considering, for example, real-valued measurable functions, one would, when working in $\mathcal{T}$-space, restrict attention to measurable function of x^2. Instead, one could remain in the original space, where the restriction would be to the class of even measurable functions of x. The equivalence is clear. Which representation is more convenient depends on the situation.

That the correspondence between the sets $A_0 = T^{-1}(B) \in \mathcal{A}_0$ and $B \in \mathcal{B}$ establishes an analogous correspondence between measurable functions defined over $(\mathcal{X}, \mathcal{A}_0)$ and $(\mathcal{T}, \mathcal{B})$ is shown by the following lemma.

Lemma 2.3.1 *Let the statistic T from $(\mathcal{X}, \mathcal{A})$ into $(\mathcal{T}, \mathcal{B})$ induce the subfield $\mathcal{A}_0$. Then a real-valued $\mathcal{A}$-measurable function f is $\mathcal{A}_0$-measurable if and only if there exists a $\mathcal{B}$-measurable function g such that*

$$f(x) = g[T(x)]$$

for all x.

Proof. Suppose first that such a function g exists. Then the set

$$\{x : f(x) < r\} = T^{-1}(\{t : g(t) < r\})$$

is in $\mathcal{A}_0$, and f is $\mathcal{A}_0$-measurable. Conversely, if f is $\mathcal{A}_0$-measurable, then the sets

$$A_{in} = \left\{x : \frac{i}{2^n} < f(x) \le \frac{i+1}{2^n}\right\}, \qquad i = 0, \pm 1, \pm 2, \ldots,$$

are (for fixed n) disjoint sets in $\mathcal{A}_0$ whose union is $\mathcal{X}$, and there exist $B_{in} \in \mathcal{B}$ such that $A_{in} = T^{-1}(B_{in})$. Let

$$B_{in}^* = B_{in} \cap \{\bigcup_{j \ne i} B_{jn}\}^c .$$

Since A_{in} and A_{jn} are mutually exclusive for $i \ne j$, the set $T^{-1}(B_{in} \cap B_{jn})$ is empty and so is the set $T^{-1}(B_{in} \cap \{B_{in}^*\}^c)$. Hence, for fixed n, the sets B_{in}^* are disjoint, and still satisfy $A_{in} = T^{-1}(B_{in}^*)$. Defining

$$f_n(x) = \frac{i}{2^n} \quad \text{if} \quad x \in A_{in}, \qquad i = 0 \pm 1, \pm 2, \ldots,$$

one can write

$$f_n(x) = g_n[T(x)],$$

where

$$g_n(t) = \begin{cases} \frac{i}{2^n} & \text{for } t \in B^*_{in}, \quad i = 0 \pm 1, \pm 2, \dots, \\ 0 & \text{otherwise.} \end{cases}$$

Since the functions g_n are $\mathcal{B}$-measurable, the set B on which $g_n(t)$ converges to a finite limit is in $\mathcal{B}$. Let $R = T(\mathcal{X})$ be the range of T. Then for $t \in R$,

$$\lim g_n[T(x)] = \lim f_n(x) = f(x)$$

for all $x \in \mathcal{X}$ so that R is contained in B. Therefore, the function g defined by $g(t) = \lim g_n(t)$ for $t \in B$ and $g(t) = 0$ otherwise possesses the required properties. ■

The relationship between integrals of the functions f and g above is given by the following lemma.

Lemma 2.3.2 *Let T be a measurable transformation from $(\mathcal{X}, \mathcal{A})$ into $(\mathcal{T}, \mathcal{B})$, μ a σ-finite measure over $(\mathcal{X}, \mathcal{A})$, and g a real-valued measurable function of t. If μ^* is the measure defined over $(\mathcal{T}, \mathcal{B})$ by*

$$\mu^*(B) = \mu\left[T^{-1}(B)\right] \qquad \text{for all} \quad B \in \mathcal{B}, \tag{2.17}$$

then for any $B \in \mathcal{B}$,

$$\int_{T^{-1}(B)} g[T(x)]\, d\mu(x) = \int_B g(t)\, d\mu^*(t) \tag{2.18}$$

in the sense that if either integral exists, so does the other and the two are equal.

Proof. Without loss of generality let B be the whole space $\mathcal{T}$. If g is the indicator of a set $B_0 \in \mathcal{B}$, the lemma holds, since the left- and right-hand sides of (2.18) reduce respectively to $\mu[T^{-1}(B_0)]$ and $\mu^*(B_0)$, which are equal by the definition of μ^*. If follows that (2.18) holds successively for all simple functions, for all nonnegative measurable functions, and hence finally for all integrable functions. ■

2.4 Conditional Expectation and Probability

If two statistics induce the same subfield $\mathcal{A}_0$, they are equivalent in the sense of leading to equivalent classes of measurable events. This equivalence is particularly relevant to considerations of conditional probability. Thus if X is normally distributed with zero mean, the information carried by the statistics $|X|$, X^2, e^{-X^2}, and so on, is the same. Given that $|X| = t$, $X^2 = t^2$, $e^{-X^2} = e^{-t^2}$, it follows that X is $\pm t$, and any reasonable definition of conditional probability will assign probability $\frac{1}{2}$ to each of these values. The general definition of conditional probability to be given below will in fact involve essentially only $\mathcal{A}_0$ and not the range space $\mathcal{T}$ of T. However, when referred to $\mathcal{A}_0$ alone the concept loses much of its intuitive meaning, and the gap between the elementary definition and that of the general case becomes unnecessarily wide. For these reasons it is frequently more convenient to work with a particular representation of a statistic, involving a definite range space $(\mathcal{T}, \mathcal{B})$.

Let P be a probability measure over $(\mathcal{X}, \mathcal{A})$, T a statistic with range space $(\mathcal{T}, \mathcal{B})$, and $\mathcal{A}_0$ the subfield it induces. Consider a nonnegative function f which is integrable $(\mathcal{A}, P)$, that is $\mathcal{A}$-measurable and P-integrable. Then $\int_A f\,dP$ is defined for all $A \in \mathcal{A}$ and therefore for all $A_0 \in \mathcal{A}_0$. If follows from the Radon–Nikodym theorem (Theorem 2.2.3) that there exists a function f_0 which is integrable $(\mathcal{A}_0, P)$ and such that

$$\int_{A_0} f\,dP = \int_{A_0} f_0\,dP \qquad \text{for all} \quad A_0 \in \mathcal{A}_0, \tag{2.19}$$

and that f_0 is unique $(\mathcal{A}_0, P)$. By Lemma 2.3.1, f_0 depends on x only through $T(x)$. In the example of a normally distributed variable X with zero mean, and $T = X^2$, the function f_0 is determined by (2.19) holding for all sets A_0 that are symmetric with respect to the origin, so that $f_0(x) = \frac{1}{2}[f(x) + f(-x)]$.

The function f_0 defined through (2.19) is determined by two properties:

(i) Its average value over any set A_0 with respect to P is the same as that of f;

(ii) It depends on x only through $T(x)$ and hence is constant on the sets D_x over which T is constant.

Intuitively, what one attempts to do in order to construct such a function is to define $f_0(x)$ as the conditional P-average of f over the set D_x. One would thereby replace the single averaging process of integrating f represented by the left-hand side with a two-stage averaging process such as an iterated integral. Such a construction can actually be carried out when X is a discrete variable and in the regular case considered in Section 1.9; $f_0(x)$ is then just the conditional expectation of $f(X)$ given $T(x)$. In general, it is not clear how to define this conditional expectation directly. Since it should, however, possess properties (i) and (ii), and since these through (2.19) determine f_0 uniquely $(\mathcal{A}_0, P)$, we shall take $f_0(x)$ of (2.19) as the general definition of the *conditional expectation* $E[f(X) \mid T(x)]$. Equivalently, if $f_0(x) = g[T(x)]$, one can write

$$E[f(X) \mid t] = E[f(X) \mid T = t] = g(t),$$

so that $E[f(X) \mid t]$ is a $\mathcal{B}$-measurable function defined up to equivalence $(\mathcal{B}, P^T)$. In the relationship of integrals given in Lemma 2.3.2, if $\mu = P^X$, then $\mu^* = P^T$, and it is seen that the function g can be defined directly in terms of f through

$$\int_{T^{-1}(B)} f(x)\,dP^X(x) = \int_B g(t)\,dP^T(t) \qquad \text{for all} \quad B \in \mathcal{B}, \tag{2.20}$$

which is equivalent to (2.19).

So far, f has been assumed to be nonnegative. In the general case, the conditional expectation of f is defined as

$$E[f(X) \mid t] = E[f^+(X) \mid t] - E[f^-(X) \mid t].$$

Example 2.4.1 (Order statistics) Let $X_1, \ldots, X_n$ be identically and independently distributed random variables with continuous distribution function, and let

$$T(x_1, \ldots, x_n) = (x_{(1)}, \ldots, x_{(n)})$$

where $x_{(1)} \le \cdots \le x_{(n)}$ denote the ordered x's. Without loss of generality one can restrict attention to the points with $x_{(1)} < \cdots < x_{(n)}$, since the probability of two coordinates being equal is 0. Then $\mathcal{X}$ is the set of all n-tuples with distinct coordinates, $\mathcal{T}$ the set of all ordered n-tuples, and $\mathcal{A}$ and $\mathcal{B}$ are the classes of Borel subsets of $\mathcal{X}$ and $\mathcal{T}$. Under T^{-1} the set consisting of the single point $a = (a_1, \ldots, a_n)$ is transformed into the set consisting of the $n!$ points $(a_{i_1}, \ldots, a_{i_n})$ that are obtained from a by permuting the coordinates in all possible ways. It follows that $\mathcal{A}_0$ is the class of all sets that are symmetric in the sense that if A_0 contains a point $x = (x_1, \ldots, x_n)$, then it also contains all points $(x_{i_1}, \ldots, x_{i_n})$.

For any integrable function f, let

$$f_0(x) = \frac{1}{n!} \sum f(x_{i_1}, \ldots, x_{i_n}),$$

where the summation extends over the $n!$ permutations of $(x_1, \ldots, x_n)$. Then f_0 is $\mathcal{A}_0$-measurable, since it is symmetric in its n arguments. Also

$$\int_{A_0} f(x_1, \ldots, x_n)\, dP(x_1) \ldots dP(x_n) = \int_{A_0} f(x_{i_1}, \ldots, x_{i_n})\, dP(x_1) \ldots dP(x_n),$$

so that f_0 satisfies (2.19). It follows that $f_0(x)$ is the conditional expectation of $f(X)$ given $T(x)$.

The conditional expectation of $f(X)$ given the above statistic $T(x)$ can also be found without assuming the X's to be identically and independently distributed. Suppose that X has a density $h(x)$ with respect to a measure μ (such as Lebesgue measure), which is symmetric in the variables $x_1, \ldots, x_n$ in the sense that for any $A \in \mathcal{A}$ it assigns to the set $\{x : (x_{i_1}, \ldots, x_{i_n}) \in A\}$ the same measure for all permutations $(i_1, \ldots, i_n)$. Let

$$f_0(x_1, \ldots, x_n) = \frac{\sum f(x_{i_1}, \ldots, x_{i_n}) h(x_{i_1}, \ldots, x_{i_n})}{\sum h(x_{i_1}, \ldots, x_{i_n})};$$

here and in the sums below the summation extends over the $n!$ permutations of $(x_1, \ldots, x_n)$. The function f_0 is symmetric in its n arguments and hence $\mathcal{A}_0$-measurable. For any symmetric set A_0, the integral

$$\int_{A_0} f_0(x_1, \ldots, x_n) h(x_{j_1}, \ldots, x_{j_n})\, d\mu(x_1, \ldots, x_n)$$

has the same value for each permutation $(x_{j_1}, \ldots, x_{j_n})$, and therefore

$$\begin{aligned}
&\int_{A_0} f_0(x_1, \ldots, x_n) h(x_1, \ldots, x_n)\, d\mu(x_1, \ldots, x_n) \\
&\quad = \int_{A_0} f_0(x_1, \ldots, x_n) \frac{1}{n!} \sum h(x_{i_1}, \ldots, x_{i_n})\, d\mu(x_1, \ldots, x_n) \\
&\quad = \int_{A_0} f(x_1, \ldots, x_n) h(x_1, \ldots, x_n)\, d\mu(x_1, \ldots, x_n).
\end{aligned}$$

It follows that $f_0(x) = E[f(X) \mid T(x)]$.

Equivalent to the statistic $T(x) = (x_{(1)}, \ldots, x_{(n)})$, the set of *order statistics*, is $U(x) = \left(\sum x_i, \sum x_i^2, \ldots, \sum x_i^n\right)$. This is an immediate consequence of the fact, to be shown below, that if $T(x^0) = t^0$ and $U(x^0) = u^0$, then

$$T^{-1}\left(\{t^0\}\right) = U^{-1}\left(\{u^0\}\right) = S$$

where $\{t^0\}$ and $\{u^0\}$ denote the sets consisting of the single point t^0 and u^0 respectively, and where S consists of the totality of points $x = (x_1, \ldots, x_n)$ obtained by permuting the coordinates of $x^0 = (x_1^0, \ldots, x_n^0)$ in all possible ways.

That $T^{-1}(\{t^0\}) = S$ is obvious. To see the corresponding fact for U^{-1}, let

$$V(x) = \left(\sum_i x_i, \sum_{i<j} x_i x_j, \sum_{i<j<k} x_i x_j x_k, \ldots, x_1 x_2 \cdots x_n\right),$$

so that the components of $V(x)$ are the elementary symmetric functions $v_1 = \sum x_i, \ldots, v_n = x_1 \ldots x_n$ of the n arguments $x_1, \ldots, x_n$. Then

$$(x - x_1) \ldots (x - x_n) = x^n - v_1 x^{n-1} + v_2 x^{n-2} - \cdots + (-1)^n v_n.$$

Hence $V(x^0) = v^0 = (v_1^0, \ldots, v_n^0)$ implies that $V^{-1}(\{v^0\}) = S$. That then also $U^{-1}(\{u^0\}) = S$ follows from the 1:1 correspondence between u and v established by the relations (known as Newton's identities):[5]

$$u_k - v_1 u_{k-1} + v_2 u_{k-2} - \cdots + (-1)^{k-1} v_{k-1} u_1 + (-1)^k k v_k = 0$$

for $1 \leq k \leq n$. ■

It is easily verified from the above definition that conditional expectation possesses most of the usual properties of expectation. It follows of course from the nonuniqueness of the definition that these properties can hold only $(\mathcal{B}, P^T)$. We state this formally in the following lemma.

Lemma 2.4.1 *If T is a statistic and the functions $f, g, \ldots$ are integrable $(\mathcal{A}, P)$, then a.e. $(\mathcal{B}, P^T)$*

(i) $E[af(X) + bg(X) \mid t] = aE[f(X) \mid t] + bE[g(X) \mid t]$;

(ii) $E[h(T)f(X) \mid t] = h(t)E[f(X) \mid t]$;

(iii) $a \leq f(x) \leq b$ $(\mathcal{A}, P)$ *implies* $a \leq E[f(X) \mid t] \leq b$;

(iv) $|f_n| \leq g$, $f_n(x) \to f(x)$ $(\mathcal{A}, P)$ *implies* $E[f_n(X) \mid t] \to E[f(X) \mid t]$.

A further useful result is obtained by specializing (2.20) to the case that B is the whole space $\mathcal{T}$. One then has

Lemma 2.4.2 *If $E[|f(X)|] < \infty$, and if $g(t) = E[f(X) \mid t]$, then*

$$E[f(X)] = E[g(T)] \; ; \tag{2.21}$$

that is, the expectation can be obtained as the expected value of the conditional expectation.

Since $P\{X \in A\} = E[I_A(X)]$, where I_A denotes the indicator of the set A, it is natural to define the *conditional probability* of A given $T = t$ by

$$P(A \mid t) = E[I_A(X) \mid t]. \tag{2.22}$$

[5] For a proof of these relations see for example Turnbull (1952), Section 32.

In view of (2.20) the defining equation for $P(A \mid t)$ can therefore be written as

$$\begin{aligned} P^X\left(A \cap T^{-1}(B)\right) &= \int_{A \cap T^{-1}(B)} dP^X(x) \qquad (2.23) \\ &= \int_B P(A \mid t)\, dP^T(t) \qquad \text{for all} \quad B \in \mathcal{B}. \end{aligned}$$

It is an immediate consequence of Lemma 2.4.1 that subject to the appropriate null-set[6] qualifications, $P(A \mid t)$ possesses the usual properties of probabilities, as summarized in the following lemma.

Lemma 2.4.3 *If T is a statistic with range space $(\mathcal{T}, \mathcal{B})$, and $A, B, A_1, A_2, \ldots$ are sets belonging to $\mathcal{A}$, then a.e. $(\mathcal{B}, P^T)$*

(i) $0 \le P(A \mid t) \le 1$;

(ii) *if the sets $A_1, A_2, \ldots$ are mutually exclusive,*

$$P\left(\bigcup A_i \mid t\right) = \sum P(A_i \mid t);$$

(iii) $A \subset B$ *implies* $P(A \mid t) \le P(B \mid t)$.

According to the definition (2.22), the conditional probability $P(A \mid t)$ must be considered for fixed A as a $\mathcal{B}$-measurable function of t. This is in contrast to the elementary definition in which one takes t as fixed and considers $P(A \mid t)$ for varying A as a set function over $\mathcal{A}$. Lemma 2.4.3 suggests the possibility that the interpretation of $P(A \mid t)$ for fixed t as a probability distribution over $\mathcal{A}$ may be valid also in the general case. However, the equality $P(A_1 \cup A_2 \mid t) = P(A_1 \mid t) + P(A_2 \mid t)$, for example, can break down on a null set that may vary with A_1 and A_2, and the union of all these null sets need no longer have measure zero.

For an important class of cases, this difficulty can be overcome through the nonuniqueness of the functions $P(A \mid t)$, which for each fixed A are determined only up to sets of measure zero in t. Since all determinations of these functions are equivalent, it is enough to find a specific determination for each A so that for each fixed t these determinations jointly constitute a probability distribution over $\mathcal{A}$. This possibility is illustrated by Example 2.4.1, in which the conditional probability distribution given $T(x) = t$ can be taken to assign probability $1/n!$ to each of the $n!$ points satisfying $T(x) = t$. Sufficient conditions for the existence of such conditional distributions will be given in the next section. For counterexamples see Blackwell and Dubins (1975).

[6]This term is used as an alternative to the more cumbersome "set of measure zero."

2.5 Conditional Probability Distributions[7]

We shall now investigate the existence of conditional probability distributions under the assumption, satisfied in most statistical applications, that $\mathcal{X}$ is a Borel set in a Euclidean space. We shall then say for short that $\mathcal{X}$ is Euclidean and assume that, unless otherwise stated, $\mathcal{A}$ is the class of Borel subsets of $\mathcal{X}$.

Theorem 2.5.1 *If $\mathcal{X}$ is Euclidean, there exist determinations of the functions $P(A \mid t)$ such that for each t, $P(A \mid t)$ is a probability measure over $\mathcal{A}$.*

Proof. By setting equal to 0 the probability of any Borel set in the complement of $\mathcal{X}$, one can extend the given probability measure to the class of all Borel sets and can therefore assume without loss of generality that $\mathcal{X}$ is the full Euclidean space. For simplicity we shall give the proof only in the one-dimensional case. For each real x put $F(x,t) = P((-\infty, x] \mid t)$ for some version of this conditional probability function, and let $r_1, r_2, \dots$ denote the set of all rational numbers in some order. Then $r_i < r_j$ implies that $F(r_i, t) \le F(r_j, t)$ for all t except those in a null set N_{ij}, and hence that $F(x,t)$ is nondecreasing in x over the rationals for all t outside of the null set $N' = \bigcup N_{ij}$. Similarly, it follows from Lemma 2.4.1(iv) that for all t not in a null set N'', as n tends to infinity $\lim F(r_i + 1/n, t) = F(r_i, t)$ for $i = 1, 2, \dots$, $\lim F(n,t) = 1$, and $\lim F(-n,t) = 0$. Therefore, for all t outside of the null set $N' \cup N''$, $F(x,t)$ considered as a function of x is properly normalized, monotone, and continuous on the right over the rationals. For t not in $N' \cup N''$ let $F^*(x,t)$ be the unique function that is continuous on the right in x and agrees with $F(x,t)$ for all rational x. Then $F^*(x,t)$ is a cumulative distribution function and therefore determines a probability measure $P^*(A \mid t)$ over $\mathcal{A}$. We shall now show that $P^*(A \mid t)$ is a conditional probability of A given t, by showing that for each fixed A it is a $\mathcal{B}$-measurable function of t satisfying (2.23). This will be accomplished by proving that for each fixed $A \in \mathcal{A}$

$$P^*(A \mid t) = P(A \mid t) \qquad (\mathcal{B}, P^T).$$

By definition of P^* this is true whenever A is one of the sets $(-\infty, x]$ with x rational. It holds next when A is an interval $(a,b] = (-\infty, b] - (-\infty, a]$ with a, b rational, since P^* is a measure and P satisfies Lemma 2.4.3(ii). Therefore, the desired equation holds for the field $\mathcal{F}$ of all sets A which are finite unions of intervals $(a_i, b_i]$ with rational end points. Finally, the class of sets for which the equation holds is a monotone class (see Problem 2.1) and hence contains the smallest σ-field containing $\mathcal{F}$, which is $\mathcal{A}$. The measure $P^*(A \mid t)$ over $\mathcal{A}$ was defined above for all t not in $N' \cup N''$. However, since neither the measurability of a function nor the values of its integrals are affected by its values on a null set, one can take arbitrary probability measures over $\mathcal{A}$ for t in $N' \cup N''$ and thereby complete the determination.

If X is a vector-valued random variable with probability distribution P^X and T is a statistic defined over $(\mathcal{X}, \mathcal{A})$, let $P^{X|t}$ denote any version of the family

[7]This section may be omitted at first reading. Its principal application is in the proof of Lemma 2.7.2(ii) in Section 2.7, which in turn is used only in the proof of Theorem 4.4.1

of conditional distributions $P(A \mid t)$ over $\mathcal{A}$ guaranteed by Theorem 2.5.1. The connection with conditional expectation is given by the following theorem. ■

Theorem 2.5.2 *If X is a vector-valued random variable and $E|f(X)| < \infty$, then*

$$E[f(X) \mid t] = \int f(x)\, dP^{X|t}(x) \qquad (\mathcal{B}, P^T). \tag{2.24}$$

Proof. Equation (2.24) holds if f is the indicator of any set $A \in \mathcal{A}$. It then follows from Lemma 2.4.1 that it also holds for any simple function and hence for any integrable function.

The determination of the conditional expectation $E[f(X) \mid t]$ given by the right-hand side of (2.24) possesses for each t the usual properties of an expectation, (i), (iii), and (iv) of Lemma 2.4.1, which previously could be asserted only up to sets of measure zero depending on the functions $f, g, \dots$ involved. Under the assumptions of Theorem 2.5.1 a similar strengthening is possible with respect to (ii) of Lemma 2.4.1, which can be shown to hold except possibly on a null set N not depending on the function h. It will be sufficient for the present purpose to prove this under the additional assumption that the range space of the statistic T is also Euclidean. For a proof without this restriction see for example Billingsley (1995). ■

Theorem 2.5.3 *If T is a statistic with Euclidean domain and range spaces $(\mathcal{X}, \mathcal{A})$ and $(\mathcal{T}, \mathcal{B})$, there exists a determination $P^{X|t}$ of the conditional probability distribution and a null set N such that the conditional expectation computed by*

$$E[f(X) \mid t] = \int f(x)\, dP^{X|t}(x)$$

satisfies for all $t \notin N$.

$$E[h(T)f(X) \mid t] = h(t)E[f(X) \mid t]. \tag{2.25}$$

Proof. For the sake of simplicity and without essential loss of generality suppose that T is real-valued. Let $P^{X|t}(A)$ be a probability distribution over $\mathcal{A}$ for each t, the existence of which is guaranteed by Theorem 2.5.1. For $B \in \mathcal{B}$, the indicator function $I_B(t)$ is $\mathcal{B}$-measurable and

$$\int_{B'} I_B(t)\, dP^T(t) = P^T(B' \cap B) = P^X(T^{-1}B' \cap T^{-1}B) \qquad \text{for all} \quad B' \in \mathcal{B}.$$

Thus by (2.20)

$$I_B(t) = P^{X|t}\left(T^{-1}B\right) \qquad \text{a.e. } P^T.$$

Let $B_n, n = 1, 2, \dots$, be the intervals of $\mathcal{T}$ with rational end points. Then there exists a P-null set $N = \cup N_n$ such that for $t \notin N$

$$I_{B_n}(t) = P^{X|t}\left(T^{-1}B_n\right)$$

for all n. For fixed $t \notin N$, the two set functions $P^{X|t}\left(T^{-1}B\right)$ and $I_B(t)$ are probability distributions over $\mathcal{B}$, the latter assigning probability 1 or 0 to a set as it does or does not contain the point t. Since these distributions agree over the rational intervals B_n, they agree for all $B \in \mathcal{B}$. In particular, for $t \notin N$, the set consisting of the single point t is in $\mathcal{B}$, and if

$$A^{(t)} = \{x : T(x) = t\},$$

it follows that for all $t \notin N$

$$P^{X|t}\left(A^{(t)}\right) = 1. \tag{2.26}$$

Thus

$$\begin{aligned} \int h[T(x)]f(x)\,dP^{X|t}(x) &= \int_{A^{(t)}} h[T(x)]f(x)\,dP^{X|t}(x) \\ &= h(t)\int f(x)\,dP^{X|t}(x) \end{aligned}$$

for $t \notin N$, as was to be proved. ■

It is a consequence of Theorem 2.5.3 that for all $t \notin N$, $E[h(T) \mid t] = h(t)$ and hence in particular $P(T \in B \mid t) = 1$ or 0 as $t \in B$ or $t \notin B$.

The conditional distributions $P^{X|t}$ still differ from those of the elementary case considered in Section 1.9, in being defined over $(\mathcal{X}, \mathcal{A})$ rather than over the set $A^{(t)}$ and the σ-field $\mathcal{A}^{(t)}$ of its Borel subsets. However, (2.26) implies that for $t \notin N$

$$P^{X|t}(A) = P^{X|t}(A \cap A^{(t)}).$$

The calculations of conditional probabilities and expectations are therefore unchanged if for $t \notin N$, $P^{X|t}$ is replaced by the distribution $\bar{P}^{X|t}$, which is defined over $(A^{(t)}, \mathcal{A}^{(t)})$ and which assigns to any subset of $A^{(t)}$ the same probability as $P^{X|t}$.

Theorem 2.5.3 establishes for all $t \notin N$ the existence of conditional probability distributions $\bar{P}^{X|t}$, which are defined over $(A^{(t)}, \mathcal{A}^{(t)})$ and which by Lemma 2.4.2 satisfy

$$E[f(X)] = \int_{\mathcal{T}-N}\left[\int_{A^{(t)}} f(x)\,dP^{(X|t)}(x)\right]\,dP^T(t) \tag{2.27}$$

for all integrable functions f. Conversely, consider any family of distributions satisfying (2.27), and the experiment of observing first T, and then, if $T = t$, a random quantity with distribution $\bar{P}^{X|t}$. The result of this two-stage procedure is a point distributed over $(\mathcal{X}, \mathcal{A})$ with the same distribution as the original X. Thus $\bar{P}^{X|t}$ satisfies this "functional" definition of conditional probability.

If $(\mathcal{X}, \mathcal{A})$ is a product space $(\mathcal{T}\times\mathcal{Y}, \mathcal{B}\times\mathcal{C})$, then $A^{(t)}$ is the product of $\mathcal{Y}$ with the set consisting of the single point t. For $t \notin N$, the conditional distribution $\bar{P}^{X|t}$ then induces a distribution over $(\mathcal{Y}, \mathcal{C})$, which in analogy with the elementary case will be denoted by $P^{Y|t}$. In this case the definition can be extended to all of $\mathcal{T}$ by letting $P^{Y|t}$ assign probability 1 to a common specified point y_0 for all $t \in N$. With this definition, (2.27) becomes

$$Ef(T, Y) = \int_{\mathcal{T}}\left[\int_{\mathcal{Y}} f(t, y)\,dP^{Y|t}(y)\right]\,dP^T(t). \tag{2.28}$$

As an application, we shall prove the following lemma, which will be used in Section 2.7.

Lemma 2.5.1 *Let $(\mathcal{T}, \mathcal{B})$ and $(\mathcal{Y}, \mathcal{C})$ be Euclidean spaces, and let $P_0^{T,Y}$ be a distribution over the product space $(\mathcal{X}, \mathcal{A}) = (\mathcal{T} \times \mathcal{Y}, \mathcal{B} \times \mathcal{C})$. Suppose that another distribution P_1 over $(\mathcal{X}, \mathcal{A})$ is such that*

$$dP_1(t, y) = a(y)b(t)\, dP_0(t, y),$$

with $a(y) > 0$ for all y. Then under P_1 the marginal distribution of T and a version of the conditional distribution of Y given t are given by

$$dP_1^T(t) = b(t)\left[\int a(y)\, dP_0^{Y|t}(y)\right] dP_0^T(t)$$

and

$$dP_1^{Y|t}(y) = \frac{a(y)\, dP_0^{Y|t}(y)}{\int_{\mathcal{Y}} a(y')\, dP_0^{Y|t}(y')}.$$

Proof. The first statement of the lemma follows from the equation

$$\begin{aligned} P_1\{T \in B\} = E_1\left[I_B(T)\right] &= E_0\left[I_B(T)a(Y)b(T)\right] \\ &= \int_B b(T)\left[\int_{\mathcal{Y}} a(y)\, dP_0^{Y|t}(y)\right] dP_0^T(t). \end{aligned}$$

To check the second statement, one need only show that for any integrable f the expectation $E_1 f(Y, T)$ satisfies (2.28), which is immediate. The denominator of $dP_1^{Y|t}$ is positive, since $a(y) > 0$ for all y. ■

2.6 Characterization of Sufficiency

We can now generalize the definition of sufficiency given in Section 1.9. If $\mathcal{P} = \{P_\theta, \theta \in \Omega\}$ is any family of distributions defined over a common sample space $(\mathcal{X}, \mathcal{A})$, a statistic T is *sufficient* for $\mathcal{P}$ (or for θ) if for each A in $\mathcal{A}$ there exists a determination of the conditional probability function $P_\theta(A \mid t)$ that is independent of θ. As an example suppose that $X_1, \ldots, X_n$ are identically and independently distributed with continuous distribution function $F_\theta, \theta \in \Omega$. Then it follows from Example 2.4.1 that the set of order statistics $T(X) = (X_{(1)}, \ldots, X_{(n)})$ is sufficient for θ.

Theorem 2.6.1 *If $\mathcal{X}$ is Euclidean, and if the statistic T is sufficient for $\mathcal{P}$, then there exist determinations of the conditional probability distributions $P_\theta(A \mid t)$ which are independent of θ and such that for each fixed t, $P_\theta(A \mid t)$ is a probability measure over $\mathcal{A}$.*

Proof. This is seen from the proof of Theorem 2.5.1. By the definition of sufficiency one can, for each rational number r, take the functions $F(r, t)$ to be independent of θ, and the resulting conditional distributions will then also not depend on θ. ■

In Chapter 1 the definition of sufficiency was justified by showing that in a certain sense a sufficient statistic contains all the available information. In view of Theorem 2.6.1 the same justification applies quite generally when the sample space is Euclidean. With the help of a random mechanism one can then construct from a sufficient statistic T a random vector X' having the same distribution as the original sample vector X. Another generalization of the earlier result, not involving the restriction to a Euclidean sample space, is given in Problem 2.13.

The factorization criterion of sufficiency, derived in Chapter 1, can be extended to any *dominated* family of distributions, that is, any family $\mathcal{P} = \{P_\theta, \theta \in \Omega\}$ possessing probability densities p_θ with respect to some σ-finite measure μ over $(\mathcal{X}, \mathcal{A})$. The proof of this statement is based on the existence of a probability distribution $\lambda = \sum c_i P_{\theta_i}$ (Theorem 2.2.3 of the Appendix), which is *equivalent* to $\mathcal{P}$ in the sense that for any $A \in \mathcal{A}$

$$\lambda(A) = 0 \quad \text{if and only if} \quad P_\theta = 0 \quad \text{for all } \theta \in \Omega. \tag{2.29}$$

Theorem 2.6.2 *Let $\mathcal{P} = \{P_\theta, \theta \in \Omega\}$ be a dominated family of probability distributions over $(\mathcal{X}, \mathcal{A})$, and let $\lambda = \sum c_i P_{\theta_i}$ satisfy (2.29). Then a statistic T with range space $(\mathcal{T}, \mathcal{B})$ is sufficient for $\mathcal{P}$ if and only if there exist nonnegative $\mathcal{B}$-measurable functions $g_\theta(t)$ such that*

$$dP_\theta(x) = g_\theta[T(x)]\, d\lambda(x) \tag{2.30}$$

for all $\theta \in \Omega$.

Proof. Let $\mathcal{A}_0$ be the subfield induced by T, and suppose that T is sufficient for θ. Then for all $\theta \in \Omega$, $A_0 \in \mathcal{A}_0$, and $A \in \mathcal{A}$

$$\int_{A_0} P(A \mid T(x))\, dP_\theta(x) = P_\theta(A \cap A_0);$$

and since $\lambda = \sum c_i P_{\theta_i}$,

$$\int_{A_0} P(A \mid T(x))\, d\lambda(x) = \lambda(A \cap A_0),$$

so that $P(A \mid T(x))$ serves as conditional probability function also for λ. Let $g_\theta(T(x))$ be the Radon–Nikodym derivative $dP_\theta(x)/d\lambda(x)$ for $(\mathcal{A}_0, \lambda)$. To prove (2.30) it is necessary to show that $g_\theta(T(x))$ is also the derivative of P_θ for $(\mathcal{A}, \lambda)$. If A_0 is put equal to $\mathcal{X}$ in the first displayed equation, this follows from the relation

$$\begin{aligned} P_\theta(A) &= \int P(A \mid T(x))\, dP_\theta(x) = \int E_\lambda[I_A(x) \mid T(x)]\, dP_\theta(x) \\ &= \int E_\lambda[I_A(x) \mid T(x)]\, g_\theta(T(x))\, d\lambda(x) \\ &= \int E_\lambda[g_\theta(T(x)) I_A(x) \mid T(x)]\, d\lambda(x) \\ &= \int g_\theta(T(x)) I_A(x)\, d\lambda(x) = \int_A g_\theta(T(x))\, d\lambda(x). \end{aligned}$$

Here the second equality uses the fact, established at the beginning of the proof, that $P(A \mid T(x))$ is also the conditional probability for λ; the third equality holds because the function being integrated is $\mathcal{A}_0$-measurable and because $dP_\theta = g_\theta\, d\lambda$ for $(\mathcal{A}_0, \lambda)$; the fourth is an application of Lemma 2.4.1(ii); and the fifth employs the defining property of conditional expectation.

Suppose conversely that (2.30) holds. We shall then prove that the conditional probability function $P_\lambda(A \mid t)$ serves as a conditional probability function for all $P \in \mathcal{P}$. Let $g_\theta(T(x)) = dP_\theta(x)/\, d\lambda(x)$ on $\mathcal{A}$ and for fixed A and θ define a measure ν over $\mathcal{A}$ by the equation $d\nu = I_A\, dP_\theta$. Then over $\mathcal{A}_0$, $d\nu(x)/\, dP_\theta(x) = E_\theta[I_A(X) \mid T(x)]$, and therefore

$$\frac{d\nu(x)}{d\lambda(x)} = P_\theta[A \mid T(x)] g_\theta(T(x)) \qquad \text{over } \mathcal{A}_0.$$

On the other hand, $d\nu(x)/d\lambda(x) = I_A(x) g_\theta(T(x))$ over $\mathcal{A}$, and hence

$$\begin{aligned} \frac{d\nu(x)}{d\lambda(x)} &= E_\lambda[I_A(X) g_\theta(T(X)) \mid T(x)] \\ &= P_\lambda[A \mid T(x)] g_\theta(T(x)) \qquad \text{over } \mathcal{A}_0. \end{aligned}$$

It follows that $P_\lambda(A \mid T(x)) g_\theta(T(x)) = P_\theta(A \mid T(x)) g_\theta(T(x))$ $(\mathcal{A}_0, \lambda)$ and hence $(\mathcal{A}_0, P_\theta)$. Since $g_\theta(T(x)) \neq 0$ $(\mathcal{A}_0, P_\theta)$, this shows that $P_\theta(A \mid T(x)) = P_\lambda(A \mid T(x))$ $(\mathcal{A}_0, P_\theta)$, and hence that $P_\lambda(A \mid T(x))$ is a determination of $P_\theta(A \mid T(x))$. ■

Instead of the above formulation, which explicitly involves the distribution λ, it is sometimes more convenient to state the result with respect to a given dominating measure μ.

Corollary 2.6.1 (Factorization theorem) *If the distributions P_θ of $\mathcal{P}$ have probability densities $p_\theta = dP_\theta/d\mu$ with respect to a σ-finite measure μ, then T is sufficient for $\mathcal{P}$ if and only if there exist nonnegative $\mathcal{B}$-measurable functions g_θ on T and a nonnegative $\mathcal{A}$-measurable function h on $\mathcal{X}$ such that*

$$p_\theta(x) = g_\theta[T(x)] h(x) \qquad (\mathcal{A}, \mu). \tag{2.31}$$

PROOF. Let $\lambda = \sum c_i P_{\theta_i}$ satisfy (2.29). Then if T is sufficient, (2.31) follows from (2.30) with $h = d\lambda/d\mu$. Conversely, if (2.31) holds,

$$d\lambda(x) = \sum c_i g_{\theta_i}[T(x)] h(x)\, d\mu(x) = k[T(x)] h(x)\, d\mu(x)$$

and therefore $dP_\theta(x) = g^*_\theta(T(x))\, d\lambda(x)$ where $g^*_\theta(t) = g_\theta(t)/k(t)$ when $k(t) > 0$ and may be defined arbitrarily when $k(t) = 0$. ■

For extensions of the factorizations theorem to undominated families, see Ghosh, Morimoto, and Yamada (1981) and the literature cited there.

2.7 Exponential Families

An important family of distributions which admits a reduction by means of sufficient statistics is the *exponential family*, defined by probability densities of the form

$$p_\theta(x) = C(\theta) \exp\left[\sum_{j=1}^{k} Q_j(\theta) T_j(x)\right] h(x) \tag{2.32}$$

with respect to a σ-finite measure μ over a Euclidean sample space $(\mathcal{X}, \mathcal{A})$. Particular cases are the distributions of a sample $X = (X_1, \ldots, X_n)$ from a binomial, Poisson, or normal distribution. In the binomial case, for example, the density (with respect to counting measure) is

$$\binom{n}{x} p^x (1-p)^{n-x} = (1-p)^n \exp\left[x \log\left(\frac{p}{1-p}\right)\right] \binom{n}{x}.$$

Example 2.7.1 If $Y_1, \ldots, Y_n$ are independently distributed, each with density (with respect to Lebesgue measure)

$$p_\sigma(y) = \frac{y^{[(f/2)-1]} \exp\left[-y/\left(2\sigma^2\right)\right]}{(2\sigma^2)^{f/2}\,\Gamma(f/2)}, \qquad y > 0, \tag{2.33}$$

then the joint distribution of the Y's constitutes an exponential family. For $\sigma = 1$, (2.33) is the density of the χ^2-distribution with f degrees of freedom; in particular for f an integer this is the density of $\sum_{j=1}^f X_j^2$, where the X's are a sample from the normal distribution $N(0,1)$. ■

Example 2.7.2 Consider n independent trials, each of them resulting in one of the s outcomes $E_1, \ldots, E_s$ with probabilities $p_1, \ldots, p_s$ respectively. If X_{ij} is 1 when the outcome of the ith trial is E_j and 0 otherwise, the joint distribution of the X's is

$$P\{X_{11} = x_{11}, \ldots, X_{ns}\} = p_1^{\sum x_{i1}} p_2^{\sum x_{i2}} \cdots p_s^{\sum x_{is}},$$

where all $x_{ij} = 0$ or 1 and $\sum_j x_{ij} = 1$. this forms an exponential family with $T_j(x) = \sum_{i=1}^n x_{ij}$ $(j = 1, \ldots, s-1)$. The joint distribution of the T's is the multinomial distribution $M(n; p_1, \ldots, p_s)$ given by

$$P\{T_1 = t_1, \ldots, T_{s-1} = t_{s-1}\} \tag{2.34}$$

$$= \frac{n!}{t_1! \ldots t_{s-1}!(n - t_1 - \cdots - t_{s-1})!}$$

$$\times p_1^{t_1} \ldots p_{s-1}^{t_{s-1}} (1 - p_1 - \cdots - p_{s-1})^{n - t_1 - \cdots - t_{s-1}}. \ ■$$

If $X_1, \ldots, X_n$ is a sample from a distribution with density (2.32), the joint distribution of the X's constitutes an exponential family with the sufficient statistics $\sum_{i=1}^n T_j(X_i)$, $j = 1, \ldots, k$. Thus there exists a k-dimensional sufficient statistic for $(X_1, \ldots, X_n)$ regardless of the sample size. Suppose conversely that $X_1, \ldots, X_n$ is a sample from a distribution with some density $p_\theta(x)$ and that the set over which this density is positive is independent of θ. Then under regularity assumptions which make the concept of dimensionality meaningful, if there exists a k-dimensional sufficient statistic with $k < n$, the densities $p_\theta(x)$ constitute an exponential family. For proof of this result, see Darmois (1935), Koopman (1936) and Pitman (1937). Regularity conditions of the result are discussed in Barankin and Maitra (1963), Brown (1964), Barndorff–Nielsen and Pedersen (1968), and Hipp (1974).

Employing a more natural parametrization and absorbing the factor $h(x)$ into μ, we shall write an exponential family in the form $dP_\theta(x) = p_\theta(x)\,d\mu(x)$ with

$$p_\theta(x) = C(\theta)\exp\left[\sum_{j=1}^{k}\theta_j T_j(x)\right]. \tag{2.35}$$

For suitable choice of the constant $C(\theta)$, the right-hand side of (2.35) is a probability density provided its integral is finite. The set Ω of parameter points $\theta = (\theta_1, \ldots, \theta_k)$ for which this is the case is the *natural parameter space* of the exponential family (2.35).

Optimum tests of certain hypotheses concerning any θ_j are obtained in Chapter 4. We shall now consider some properties of exponential families required for this purpose.

Lemma 2.7.1 *The natural parameter space of an exponential family is convex.*

PROOF. Let $(\theta_1, \ldots, \theta_k)$ and $(\theta_1', \ldots, \theta_k')$ be two parameter points for which the integral of (2.35) is finite. Then by Hölder's inequality,

$$\begin{aligned}&\int \exp\left[\sum\left[\alpha\theta_j + (1-\alpha)\theta_j'\right]T_j(x)\right]\,d\mu(x)\\ &\quad\le \left[\int \exp\left[\sum \theta_j T_j(x)\right]\,d\mu(x)\right]^{\alpha}\left[\int \exp\left[\sum \theta_j' T_j(x)\right]\,d\mu(x)\right]^{1-\alpha} < \infty\end{aligned}$$

for any $0 < \alpha < 1$.

If the convex set Ω lies in a linear space of dimension $< k$, then (2.35) can be rewritten in a form involving fewer than k components of T. We shall therefore, without loss of generality, assume Ω to be k-dimensional.

It follows from the factorization theorem that $T(x) = (T_1(x), \ldots, T_k(x))$ is sufficient for $\mathcal{P} = \{P_\theta, \theta \in \Omega\}$. ■

Lemma 2.7.2 *Let X be distributed according to the exponential family*

$$dP^T_{\theta,\vartheta}(x) = C(\theta,\vartheta)\exp\left[\sum_{i=1}^{r}\theta_i U_i(x) + \sum_{j=1}^{s}\vartheta_j T_j(x)\right]\,d\mu(x).$$

Then there exist measures λ_θ and ν_t over s- and r-dimensional Euclidean space respectively such that

(i) *the distribution of $T = (T_1, \ldots, T_s)$ is an exponential family of the form*

$$dP^T_{\theta,\vartheta}(t) = C(\theta,\vartheta)\exp\left(\sum_{j=1}^{s}\vartheta_j t_j\right)\,d\lambda_\theta(t), \tag{2.36}$$

(ii) *the conditional distribution of $U = (U_1, \ldots, U_r)$ given $T = t$ is an exponential family of the form*

$$dP^{U|t}_{\theta}(u) = C(\theta)\exp\left(\sum_{i=1}^{r}\theta_i u_i\right)\,d\nu_t(u), \tag{2.37}$$

and hence in particular is independent of ϑ.

PROOF. Let (θ^0, ϑ^0) be a point of the natural parameter space, and let $\mu^* = P^X_{\theta^0,\vartheta^0}$. Then

$$dP^X_{\theta^0,\vartheta^0}(x) = \frac{C(\theta,\vartheta)}{C(\theta^0,\vartheta^0)} \times \exp\left[\sum_{i=1}^r (\theta_i - \theta_i^0) U_i(x) + \sum_{j=1}^s (\vartheta_j - \vartheta_j^0) T_j(x)\right] d\mu^*(x),$$

and the result follows from Lemma 2.5.1, with

$$d\lambda_\theta(t) = \exp\left(-\sum \vartheta_i^0 t_i\right)\left[\int \exp\left[\sum_{i=1}^r (\theta_i - \theta_i^0) u_i\right] dP^{U|t}_{\theta^0,\vartheta^0}(u)\right] dP^T_{\theta^0,\vartheta^0}(t)$$

and

$$d\nu_t(u) = \exp\left(-\sum \theta_i^0 u_i\right) dP^{U|t}_{\theta^0,\vartheta^0}(u). \blacksquare$$

Theorem 2.7.1 *Let ϕ be any function on $(\mathcal{X}, \mathcal{A})$ for which the integral*

$$\int \phi(x) \exp\left[\sum_{j=1}^k \theta_j T_j(x)\right] d\mu(x) \tag{2.38}$$

considered as a function of the complex variables $\theta_j = \xi_j + i\eta_j$ $(j = 1, \ldots, k)$ exists for all $(\xi_1, \ldots, \xi_k) \in \Omega$ and is finite. Then

(i) *the integral is an analytic function of each of the θ's in the region R of parameter points for which $(\xi_1, \ldots, \xi_k)$ is an interior point of the natural parameter space Ω;*

(ii) *the derivatives of all orders with respect to the θ's of the integral (2.38) can be computed under the integral sign.*

PROOF. Let $(\xi_1, \ldots, \xi_k)$ be any fixed point in the interior of Ω, and consider one of the variables in question, say θ_1. Breaking up the factor

$$\phi(x) \exp\left[\left(\xi_2^0 + i\eta_2^0\right) T_2(x) + \cdots + \left(\xi_k^0 + i\eta_k^0\right) T_k(x)\right]$$

into its real and complex part and each of these into its positive and negative part, and absorbing this factor in each of the four terms thus obtained into the measure μ, one sees that as a function of θ_1 the integral (2.38) can be written as

$$\int \exp\left[\theta_1 T_1(x)\right] d\mu_1(x) - \int \exp\left[\theta_1 T_1(x)\right] d\mu_2(x) + i \int \exp\left[\theta_1 T_1(x)\right] d\mu_3(x) - i \int \exp\left[\theta_1 T_1(x)\right] d\mu_4(x).$$

It is therefore sufficient to prove the result for integrals of the form

$$\psi(\theta_1) = \int \exp\left[\theta_1 T_1(x)\right] d\mu(x).$$

Since $(\xi_1^0, \ldots, \xi_k^0)$ is in the interior of Ω, there exists $\delta > 0$ such that $\psi(\theta_1)$ exists and is finite for all θ_1 with $|\xi_1 - \xi_1^0| \le \delta$. Consider the difference

$$\frac{\psi(\theta_1) - \psi(\theta_1^0)}{\theta_1 - \theta_1^0} = \int \frac{\exp\left[\theta_1 T_1(x)\right] - \exp\left[\theta_1^0 T_1(x)\right]}{\theta_1 - \theta_1^0} d\mu(x).$$

The integrand can be written as

$$\exp\left[\theta_1^0 T_1(x)\right] \left[\frac{\exp\left[(\theta_1 - \theta_1^0)T_1(x)\right] - 1}{\theta_1 - \theta_1^0}\right].$$

Applying to the second factor the inequality

$$\left|\frac{\exp(az) - 1}{z}\right| \leq \frac{\exp(\delta|a|)}{\delta} \qquad \text{for} \quad |z| \leq \delta,$$

the integrand is seen to be bounded above in absolute value by

$$\frac{1}{\delta}\left|\exp\left(\theta_1^0 T_1 + \delta|T_1|\right)\right| \leq \frac{1}{\delta}\left|\exp\left[\left(\theta_1^0 + \delta\right) T_1\right] + \exp\left[\left(\theta_1^0 - \delta\right) T_1\right]\right|$$

for $|\theta_1 - \theta_1^0| \leq \delta$. Since the right-hand side integrable, it follows from the Lebesgue dominated-convergence theorem [Theorem 2.2.2(ii)] that for any sequence of points $\theta_1^{(n)}$ tending to θ_1^0, the difference quotient of ψ tends to

$$\int T_1(x) \exp\left[\theta_1^0 T_1(x)\right]\, d\mu(x).$$

This completes the proof of (i), and proves (ii) for the first derivative. The proof for the higher derivatives is by induction and is completely analogous. ■

2.8 Problems

Section 2.1

Problem 2.1 *Monotone class.* A class $\mathcal{F}$ of subsets of a space is a *field* if it contains the whole space and is closed under complementation and under finite unions; a class $\mathcal{M}$ is *monotone* if the union and intersection of every increasing and decreasing sequence of sets of $\mathcal{M}$ is again in $\mathcal{M}$. The smallest monotone class $\mathcal{M}_0$ containing a given field $\mathcal{F}$ coincides with the smallest σ-field $\mathcal{A}$ containing $\mathcal{F}$. [One proves first that $\mathcal{M}_0$ is a field. To show, for example, that $A \cap B \in \mathcal{M}_0$ when A and B are in $\mathcal{M}_0$, consider, for a fixed set $A \in \mathcal{F}$, the class $\mathcal{M}_A$ of all B in $\mathcal{M}_0$ for which $A \cap B \in \mathcal{M}_0$. Then $\mathcal{M}_A$ is a monotone class containing $\mathcal{F}$, and hence $\mathcal{M}_A = \mathcal{M}_0$. Thus $A \cap B \in \mathcal{M}_A$ for all B. The argument can now be repeated with a fixed set $B \in \mathcal{M}_0$ and the class $\mathcal{M}_B$ of sets A in $\mathcal{M}_0$ for which $A \cap B \in \mathcal{M}_0$. Since $\mathcal{M}_0$ is a field and monotone, it is a σ-field containing $\mathcal{F}$ and hence contains $\mathcal{A}$. But any σ-field is a monotone class so that also $\mathcal{M}_0$ is contained in $\mathcal{A}$.]

Section 2.2

Problem 2.2 Prove Corollary 2.2.1 using Theorems 2.2.1 and 2.2.2.

Problem 2.3 *Radon–Nikodym derivatives.*

[We have

$$p_1^Y(y) = \int_{\mathcal{T}} h(y,t)\, d\nu(t) = f(y) \int_{\mathcal{T}} \frac{h(y,t)}{f(y)g(t)} g(t)\, d\nu(t).]$$

Section 2.6

Problem 2.9 *Symmetric distributions.*

(i) Let $\mathcal{P}$ be any family of distributions of $X = (X_1, \ldots, X_n)$ which are symmetric in the sense that

$$P\{(X_{i_1}, \ldots, X_{i_n}) \in A\} = P\{(X_1, \ldots, X_n) \in A\}$$

for all Borel sets A and all permutations $(i_1, \ldots, i_n)$ of $(1, \ldots, n)$. Then the statistic T of Example 2.4.1 is sufficient for $\mathcal{P}$, and the formula given in the first part of the example for the conditional expectation $E[f(X) \mid T(x)]$ is valid.

(ii) The statistic Y of Problem 2.6 is sufficient.

(iii) Let $X_1, \ldots, X_n$ be identically and independently distributed according to a continuous distribution $P \in \mathcal{P}$, and suppose that the distributions of $\mathcal{P}$ are symmetric with respect to the origin. Let $V_i = |X_i|$ and $W_i = V_{(i)}$. Then $(W_1, \ldots, W_n)$ is sufficient for $\mathcal{P}$.

Problem 2.10 *Sufficiency of likelihood ratios.* Let P_0, P_1 be two distributions with densities p_0, p_1. Then $T(x) = p_1(x)/p_0(x)$ is sufficient for $\mathcal{P} = \{P_0, P_1\}$. [This follows from the factorization criterion by writing $p_1 = T \cdot p_0, p_0 = 1 \cdot p_0$.]

Problem 2.11 *Pairwise sufficiency.* A statistic T is pairwise sufficient for $\mathcal{P}$ if it is sufficient for every pair of distributions in $\mathcal{P}$.

(i) If $\mathcal{P}$ is countable and T is pairwise sufficient for $\mathcal{P}$, then T is sufficient for $\mathcal{P}$.

(ii) If $\mathcal{P}$ is a dominated family and T is pairwise sufficient for $\mathcal{P}$, then T is sufficient for $\mathcal{P}$.

[(i): Let $\mathcal{P} = \{P_0, P_1, \ldots\}$, and let $\mathcal{A}_0$ be the sufficient subfield induced by T. Let $\lambda = \sum c_i P_i$ $(c_i > 0)$ be equivalent to $\mathcal{P}$. For each $j = 1, 2, \ldots$ the probability measure λ_j that is proportional to $(c_0/n)P_0 + c_j P_j$ is equivalent to $\{P_0, P_j\}$. Thus by pairwise sufficiency, the derivative $f_j = dP_0/[(c_0/n)\, dP_0 + c_j\, dP_j]$ is $\mathcal{A}_0$-measurable. Let $S_j = \{x : f_j(x) = 0\}$ and $S = \bigcup_{j=1}^n S_j$. Then $S \in \mathcal{A}_0$, $P_0(S) = 0$, and on $\mathcal{X} - S$ the derivative $dP_0/d\sum_{j=1}^n c_j P_j$ equals $(\sum_{j=1}^n 1/f_j)^{-1}$ which is $\mathcal{A}_0$-measurable. It then follows from Problem 2.3 that

$$\frac{dP_0}{d\lambda} = \frac{dP_0}{d\sum\limits_{j=0}^{n} c_j P_j} \frac{d\sum\limits_{j=0}^{n} c_j P_j}{d\lambda}$$

is also $\mathcal{A}_0$-measurable. (ii): Let $\lambda = \sum_{j=1}^\infty c_j P_{\theta_j}$ be equivalent to $\mathcal{P}$. Then pairwise sufficiency of T implies for any θ_0 that $dP_{\theta_0}/(dP_{\theta_0} + d\lambda)$ and hence $dP_{\theta_0}/d\lambda$ is a measurable function of T.]

Problem 2.12 If a statistic T is sufficient for $\mathcal{P}$, then for every function f which is $(\mathcal{A}, P_\theta)$-integrable for all $\theta \in \Omega$ there exists a determination of the conditional expectation function $E_\theta[f(X) \mid t]$ that is independent of θ. [If $\mathcal{X}$ is Euclidean, this follows from Theorems 2.5.2 and 2.6.1. In general, if f is nonnegative there exists a nondecreasing sequence of simple nonnegative functions f_n tending to f. Since the conditional expectation of a simple function can be taken to be independent of θ by Lemma 2.4.1(i), the desired result follows from Lemma 2.4.1(iv).]

Problem 2.13 For a decision problem with a finite number of decisions, the class of procedures depending on a sufficient statistic T only is essentially complete. [For Euclidean sample spaces this follows from Theorem 2.5.1 without any restriction on the decision space. For the present case, let a decision procedure be given by $\delta(x) = (\delta^{(1)}(x), \dots, \delta^{(m)}(x))$ where $\delta^{(i)}(x)$ is the probability with which decision d_i is taken when x is observed. If T is sufficient and $\eta^{(i)}(t) = E[\delta^{(i)}(X) \mid t]$, the procedures δ and η have identical risk functions.] [More general versions of this result are discussed, for example, by Elfving (1952), Bahadur (1955), Burkholder (1961), LeCam (1964), and Roy and Ramamoorthi (1979).]

Section 2.7

Problem 2.14 Let X_i $(i = 1, \dots, s)$ be independently distributed with Poisson distribution $P(\lambda_i)$, and let $T_0 = \sum X_j$, $T_i = X_i$, $\lambda = \sum \lambda_j$. Then T_0 has the Poisson distribution $P(\lambda)$, and the conditional distribution of $T_1, \dots, T_{s-1}$ given $T_0 = t_0$ is the multinomial distribution (2.34) with $n = t_0$ and $p_i = \lambda_i/\lambda$.

Problem 2.15 *Life testing.* Let $X_1, \dots, X_n$ be independently distributed with exponential density $(2\theta)^{-1}e^{-x/2\theta}$ for $x \geq 0$, and let the ordered X's be denoted by $Y_1 \leq Y_2 \leq \cdots \leq Y_n$. It is assumed that Y_1 becomes available first, then Y_2, and so on, and that observation is continued until Y_r has been observed. This might arise, for example, in life testing where each X measures the length of life of, say, an electron tube, and n tubes are being tested simultaneously. Another application is to the disintegration of radioactive material, where n is the number of atoms, and observation is continued until r α-particles have been emitted.

(i) The joint distribution of $Y_1, \dots, Y_r$ is an exponential family with density

$$\frac{1}{(2\theta)^r} \frac{n!}{(n-r)!} \exp\left[-\frac{\sum_{i=1}^{r} y_i + (n-r)y_r}{2\theta}\right], \qquad 0 \leq y_1 \leq \cdots \leq y_r.$$

(ii) The distribution of $[\sum_{i=1}^r Y_i + (n-r)Y_r]/\theta$ is χ^2 with $2r$ degrees of freedom.

(iii) Let $Y_1, Y_2, \dots$ denote the time required until the first, second, ... event occurs in a Poisson process with parameter $1/2\theta'$ (see Problem 1.1). Then $Z_1 = Y_1/\theta'$, $Z_2 = (Y_2 - Y_1)/\theta'$, $Z_3 = (Y_3 - Y_2)/\theta', \dots$ are independently distributed as χ^2 with 2 degrees of freedom, and the joint density $Y_1, \dots, Y_r$ is an exponential family with density

$$\frac{1}{(2\theta')^r} \exp\left(-\frac{y_r}{2\theta'}\right), \qquad 0 \leq y_1 \leq \cdots \leq y_r.$$

The distribution of Y_r/θ' is again χ^2 with $2r$ degrees of freedom.

(iv) The same model arises in the application to life testing if the number n of tubes is held constant by replacing each burned-out tube with a new one, and if Y_1 denotes the time at which the first tube burns out, Y_2 the time at which the second tube burns out, and so on, measured from some fixed time.

[(ii): The random variables $Z_i = (n-i+1)(Y_i - Y_{i-1})/\theta$ $(i = 1, 2, \ldots, r)$ are independently distributed as χ^2 with 2 degrees of freedom, and $[\sum_{i=1}^r Y_i + (n-r)Y_r/\theta = \sum_{i=1}^r Z_i.]$

Problem 2.16 For any θ which is an interior point of the natural parameter space, the expectations and covariances of the statistics T_j in the exponential family (2.35) are given by

$$\begin{aligned} E\left[T_j(X)\right] &= -\frac{\partial \log C(\theta)}{\partial \theta_j} \qquad (j = 1, \ldots, k), \\ E\left[T_i(X)T_j(X)\right] - \left[ET_i(X)ET_j(X)\right] &= -\frac{\partial^2 \log C(\theta)}{\partial \theta_i \partial \theta_j} \qquad (i, j = 1, \ldots, k). \end{aligned}$$

Problem 2.17 Let Ω be the natural parameter space of the exponential family (2.35), and for any fixed $t_{r+1}, \ldots, t_k$ $(r < k)$ let $\Omega'_{\theta_1 \ldots \theta_r}$ be the natural parameter space of the family of conditional distributions given $T_{r+1} = t_{r+1}, \ldots, T_k = t_k$.

(i) Then $\Omega'_{\theta_1, \ldots, \theta_r}$ contains the projection $\Omega_{\theta_1, \ldots, \theta_r}$ of Ω onto $\theta_1, \ldots, \theta_r$.

(ii) An example in which $\Omega_{\theta_1, \ldots, \theta_r}$ is a proper subset of $\Omega'_{\theta_1, \ldots, \theta_r}$ is the family of densities

$$p_{\theta_1 \theta_2}(x, y) = C(\theta_1, \theta_2) \exp(\theta_1 x + \theta_2 y - xy), \qquad x, y > 0.$$

2.9 Notes

The theory of measure and integration in abstract spaces and its application to probability theory, including in particular conditional probability and expectation, is treated in a number of books, among them Dudley (1989), Williams (1991) and Billingsley (1995). The material on sufficient statistics and exponential families is complemented by the corresponding sections in *TPE*2. Much fuller treatments of exponential families (as well as sufficiency) are provided by Barndorff–Nielsen (1978) and Brown (1986).

3
Uniformly Most Powerful Tests

3.1 Stating The Problem

We now begin the study of the statistical problem that forms the principal subject of this book, the problem of hypothesis testing. As the term suggests, one wishes to decide whether or not some hypothesis that has been formulated is correct. The choice here lies between only two decisions: accepting or rejecting the hypothesis. A decision procedure for such a problem is called a *test* of the hypothesis in question.

The decision is to be based on the value of a certain random variable X, the distribution P_θ of which is known to belong to a class $\mathcal{P} = \{P_\theta, \theta \in \Omega\}$. We shall assume that if θ were known, one would also know whether or not the hypothesis is true. The distributions of $\mathcal{P}$ can then be classified into those for which the hypothesis is true and those for which it is false. The resulting two mutually exclusive classes are denoted by H and K, and the corresponding subsets of Ω by Ω_H and Ω_K respectively, so that $H \cup K = \mathcal{P}$ and $\Omega_H \cup \Omega_K = \Omega$. Mathematically, the hypothesis is equivalent to the statement that P_θ is an element of H. It is therefore convenient to identify the hypothesis with this statement and to use the letter H also to denote the hypothesis. Analogously we call the distributions in K the alternatives to H, so that K is the *class of alternatives*.

Let the decisions of accepting or rejecting H be denoted by d_0 and d_1 respectively. A nonrandomized test procedure assigns to each possible value x of X one of these two decisions and thereby divides the sample space into two complementary regions S_0 and S_1. If X falls into S_0, the hypothesis is accepted; otherwise it is rejected. The set S_0 is called the region of acceptance, and the set S_1 the region of rejection or *critical* region.

When performing a test one may arrive at the correct decision, or one may commit one of two errors: rejecting the hypothesis when it is true (error of the first kind) or accepting it when it is false (error of the second kind). The consequences of these are often quite different. For example, if one tests for the presence of some disease, incorrectly deciding on the necessity of treatment may cause the patient discomfort and financial loss. On the other hand, failure to diagnose the presence of the ailment may lead to the patient's death.

It is desirable to carry out the test in a manner which keeps the probabilities of the two types of error to a minimum. Unfortunately, when the number of observations is given, both probabilities cannot be controlled simultaneously. It is customary therefore to assign a bound to the probability of incorrectly rejecting H when it is true and to attempt to minimize the other probability subject to this condition. Thus one selects a number α between 0 and 1, called the *level of significance*, and imposes the condition that

$$P_\theta\{\delta(X) = d_1\} = P_\theta\{X \in S_1\} \leq \alpha \qquad \text{for all} \quad \theta \in \Omega_H. \tag{3.1}$$

Subject to this condition, it is desired to minimize $P_\theta\{\delta(X) = d_0\}$ for θ in Ω_K or, equivalently, to maximize

$$P_\theta\{\delta(X) = d_1\} = P_\theta\{X \in S_1\} \qquad \text{for all} \quad \theta \in \Omega_K. \tag{3.2}$$

Although usually (3.2) implies that

$$\sup_{\Omega_H} P_\theta\{X \in S_1\} = \alpha, \tag{3.3}$$

it is convenient to introduce a term for the left-hand side of (3.3): it is called the *size* of the test or critical region S_1. The condition (3.1) therefore restricts consideration to test whose size does not exceed the given level of significance. The probability of rejection (3.2) evaluated for a given θ in Ω_K is called the *power* of the test against the alternative θ. Considered as a function of θ for all $\theta \in \Omega$, the probability (3.2) is called the *power function* of the test and is denoted by $\beta(\theta)$.

The choice of a level of significance α is usually somewhat arbitrary, since in most situations there is no precise limit to the probability of an error of the first kind that can be tolerated.[1] Standard values, such as .01 or .05, were originally chosen to effect a reduction in the tables needed for carrying out various test. By habit, and because of the convenience of standardization in providing a common frame of reference, these values gradually became entrenched as the conventional levels to use. This is unfortunate, since the choice of significance level should also take into consideration the power that the test will achieve against the alternatives of interest. There is little point in carrying out an experiment which has only a small chance of detecting the effect being sought when it exists. Surveys by Cohen (1962) and Freiman et al. (1978) suggest that this is in fact the case for many studies. Ideally, the sample size should then be increased to permit adequate values for both significance level and power. If that is not feasible one may wish to use higher values of α than the customary ones. The opposite possibility,

[1]The standard way to remove the arbitrary choice of α is to report the p-value of the test, defined as the smallest level of significance leading to rejection of the null hypothesis. This approach will discussed toward the end of Section 3.3.

that one would like to decrease α, arises when the latter is so close to 1 that α can be lowered appreciably without a significant loss of power (cf. Problem 3.11). Rules for choosing α in relation to the attainable power are discussed by Lehmann (1958), Arrow (1960), and Sanathanan (1974), and from a Bayesian point of view by Savage (1962, pp. 64–66). See also Rosenthal and Rubin (1985).

Another consideration that may enter into the specification of a significance level is the attitude toward the hypothesis before the experiment is performed. If one firmly believes the hypothesis to be true, extremely convincing evidence will be required before one is willing to give up this belief, and the significance level will accordingly be set very low. (A low significance level results in the hypothesis being rejected only for a set of values of the observations whose total probability under hypothesis is small, so that such values would be most unlikely to occur if H were true.)

Let us next consider the structure of a randomized test. For any values x, such a test chooses between the two decisions, rejection or acceptance, with certain probabilities that depend on x and will be denoted by $\phi(x)$ and $1 - \phi(x)$ respectively. If the value of X is x, a random experiment is performed with two possible outcomes R and $\bar{R}$, the probabilities of which are $\phi(x)$ and $1-\phi(x)$. If in this experiment R occurs, the hypothesis is rejected, otherwise it is accepted. A randomized test is therefore completely characterized by a function ϕ, the *critical function*, with $0 \le \phi(x) \le 1$ for all x. If ϕ takes on only the values 1 and 0, one is back in the case of a nonrandomized test. The set of points x for which $\phi(x) = 1$ is then just the region of rejection, so that in a nonrandomized test ϕ is simply the indicator function of the critical region.

If the distribution of X is P_θ, and the critical function ϕ is used, the probability of rejection is

$$E_\theta \phi(X) = \int \phi(x)\, dP_\theta(x),$$

the conditional probability $\phi(x)$ of rejection given x, integrated with respect to the probability distribution of X. The problem is to select ϕ so as to maximize the power

$$\beta_\phi(\theta) = E_\theta \phi(X) \qquad \text{for all} \quad \theta \in \Omega_K \tag{3.4}$$

subject to the condition

$$E_\theta \phi(X) \le \alpha \qquad \text{for all} \quad \theta \in \Omega_H. \tag{3.5}$$

The same difficulty now arises that presented itself in the general discussion of Chapter 1. Typically, the test that maximized the power against a particular alternative in K depends on this alternative, so that some additional principal has to be introduced to define what is meant by an optimum test. There is one important exception: if K contains only one distribution, that is, if one is concerned with a single alternative, the problem is completely specified by (3.4) and (3.5). It then reduces to the mathematical problem of maximizing an integral subject to certain side conditions. The theory of this problem, and its statistical applications, constitutes the principle subject of the present chapter. In special cases it may of course turn out that the same test maximizes the power of all alternatives in K even when there is more than one. Examples of such *uniformly most powerful* (UMP) tests will be given in Section 3.4 and 3.7.

In the above formulation the problem can be considered as special case of the general decision problem with two types of losses. Corresponding to the two kinds of error, one can introduce the two component loss functions,

$$\begin{array}{lll} L_1(\theta, d_1) = 1 \text{ or } 0 & \text{as} & \theta \in \Omega_H \text{ or } \theta \in \Omega_K, \\ L_1(\theta, d_0) = 0 & \text{for all } \theta & \end{array}$$

and

$$\begin{array}{lll} L_2(\theta, d_0) = 0 \text{ or } 1 & \text{as} & \theta \in \Omega_H \text{ or } \theta \in \Omega_K, \\ L_2(\theta, d_1) = 0 & \text{for all } \theta . & \end{array}$$

With this definition the minimization of $EL_2(\theta, \delta(X))$ subject to the restriction $EL_1(\theta, \delta(X)) \leq \alpha$ is exactly equivalent to the problem of hypothesis testing as given above.

The formal loss functions L_1 and L_2 clearly do not represent in general the true losses. The loss resulting from an incorrect acceptance of the hypothesis, for example, will not be the same for all alternatives. The more the alternative differs from the hypothesis, the more serious are the consequences of such an error. As was discussed earlier, we have purposely foregone the more detailed approach implied by this criticism. Rather than working with a loss function which in practice one does not know, it seems preferable to base the theory on the simpler and intuitively appealing notion of error. It will be seen later that at least some of the results can be justified also in the more elaborate formulation.

3.2 The Neyman–Pearson Fundamental Lemma

A class of distributions is called *simple* if it contains a single distribution, and otherwise it is said to be *composite*. The problem of hypothesis testing is completely specified by (3.4) and (3.5) if K is simple. Its solution is easiest and can be given explicitly when the same is true of H. Let the distributions under a simple hypothesis H and alternative K be P_0 and P_1, and suppose for a moment that these distributions are discrete with $P_i\{X = x\} = P_i(x)$ for $i = 0, 1$. If at first one restricts attention to nonrandomized tests, the optimum test is defined as the critical region S satisfying

$$\sum_{x \in S} P_0(x) \leq \alpha \tag{3.6}$$

and

$$\sum_{x \in S} P_1(x) = \text{maximum} .$$

It is easy to see which points should be included in S. To each point are attached two values, its probability under P_0 and under P_1. The selected points are to have a total value not exceeding α on the one scale, and as large as possible on the other. This is a situation that occurs in many contexts. A buyer with a limited budget who wants to get "the most for his money" will rate the items according to their *value per dollar*. In order to travel a given distance in the shortest possible time, one must choose the quickest mode of transportation, that is, the one that yields the largest number of *miles per hour*. Analogously in the present problem the most valuable points x are those with the highest value of

$$r(x) = \frac{P_1(x)}{P_0(x)}.$$

The points are therefore rated according to the value of this ratio and selected for S in this order, as many as one can afford under restriction (3.6). Formally this means that S is the set of all points x for which $r(x) > c$, where c is determined by the condition

$$P_0\{X \in S\} = \sum_{x:r(x)>c} P_0(x) = \alpha .$$

Here a difficulty is seen to arise. It may happen that when a certain point is included, the value α has not yet been reached but that it would be exceeded if the point were also included. The exact value α can then either not be achieved at all, or it can be attained only by breaking the preference order established by $r(x)$. The resulting optimization problem has no explicit solution. (Algorithms for obtaining the maximizing set S are given by the theory of linear programming.) The difficulty can be avoided, however, by a modification which does not require violation of the r-order and which does lead to a simple explicit solution, namely by permitting randomization.[2] This makes it possible to split the next point, including only a portion of it, and thereby to obtain the exact value α without breaking the order of preference that has been established for inclusion of the various sample points. These considerations are formalized in the following theorem, the *fundamental lemma of Neyman and Pearson.*

Theorem 3.2.1 *Let P_0 and P_1 be probability distributions possessing densities p_0 and p_1 respectively with respect to a measure μ.*[3]

(i) Existence. *For testing $H : p_0$ against the alternative $K : p_1$ there exists a test ϕ and a constant k such that*

$$E_0\phi(X) = \alpha \tag{3.7}$$

and

$$\phi(x) = \begin{cases} 1 & \textit{when} \quad p_1(x) > kp_0(x), \\ 0 & \textit{when} \quad p_1(x) < kp_0(x). \end{cases} \tag{3.8}$$

(ii) Sufficient condition for a most powerful test. *If a test satisfies* (3.7) *and* (3.8) *for some k, then it is most powerful for testing p_0 against p_1 at level α.*

(iii) Necessary condition for a most powerful test. *If ϕ is most powerful at level α for testing p_0 against p_1, then for some k it satisfies* (3.8) a.e. μ. *It also satisfies* (3.7) *unless there exists a test of size $< \alpha$ and with power* 1.

PROOF. For $\alpha = 0$ and $\alpha = 1$ the theorem is easily seen to be true provided the value $k = +\infty$ is admitted in (3.8) and $0 \cdot \infty$ is interpreted as 0. Throughout the proof we shall therefore assume $0 < \alpha < 1$.

---

[2] In practice, typically neither the breaking of the r-order nor randomization is considered acceptable. The common solution, instead, is to adopt a value of α that can be attained exactly and therefore does not present this problem.

[3] There is no loss of generality in this assumption, since one can take $\mu = P_0 + P_1$.

(i): Let $\alpha(c) = P_0\{p_1(X) > cp_0(X)\}$. Since the probability is computed under P_0, the inequality need be considered only for the set where $p_0(x) > 0$, so that $\alpha(c)$ is the probability that the random variable $p_1(X)/p_0(X)$ exceeds c. Thus $1 - \alpha(c)$ is a cumulative distribution function, and $\alpha(c)$ is nonincreasing and continuous on the right, $\alpha(c-0) - \alpha(c) = P_0\{p_1(X)/p_0(X) = c\}, \alpha(-\infty) = 1$, and $\alpha(\infty) = 0$. Given any $0 < \alpha < 1$, let c_0 be such that $\alpha(c_0) \le \alpha \le \alpha(c_0 - 0)$, and consider the test ϕ defined by

$$\phi(x) = \begin{cases} 1 & \text{when} \quad p_1(x) > c_0 p_0(x), \\ \frac{\alpha - \alpha(c_0)}{\alpha(c_0 - 0) - \alpha(c_0)} & \text{when} \quad p_1(x) = c_0 p_0(x), \\ 0 & \text{when} \quad p_1(x) < c_0 p_0(x). \end{cases}$$

Here the middle expression is meaningful unless $\alpha(c_0) = \alpha(c_0 - 0)$; since then $P_0\{p_1(X) = c_0 p_0(X)\} = 0$, ϕ is defined a.e. The size of ϕ is

$$E_0\phi(X) = P_0\left\{\frac{p_1(X)}{p_0(X)} > c_0\right\} + \frac{\alpha - \alpha(c_0)}{\alpha(c_0 - 0) - \alpha(c_0)} P_0\left\{\frac{p_1(X)}{p_0(X)} = c_0\right\} = \alpha,$$

so that c_0 can be taken as the k of the theorem.

(ii): Suppose that ϕ is a test satisfying (3.7) and (3.8) and that ϕ^* is any other test with $E_0\phi^*(X) \le \alpha$. Denote by S^+ and S^- the sets in the sample space where $\phi(x) - \phi^*(x) > 0$ and < 0 respectively. If x is in $S^+, \phi(x)$ must be > 0 and $p_1(x) \ge kp_0(x)$. In the same way $p_1(x) \le kp_0(x)$ for all x in S^-, and hence

$$\int (\phi - \phi^*)(p_1 - kp_0)\, d\mu = \int_{S^+ \cup S^-} (\phi - \phi^*)(p_1 - kp_0)\, d\mu \ge 0.$$

The difference in power between ϕ and $\phi*$ therefore satisfies

$$\int (\phi - \phi^*)p_1\, d\mu \ge k \int (\phi - \phi^*)p_0\, d\mu \ge 0,$$

as was to be proved.

(iii): Let ϕ^* be most powerful at level α for testing p_0 against p_1, and let ϕ satisfy (3.7) and (3.8). Let S be the intersection of the set $S^+ \cup S^-$, on which ϕ and ϕ^* differ, with the set $\{x : p_1(x) \ne kp_0(x)\}$, and suppose that $\mu(S) > 0$. Since $(\phi - \phi^*)(p_1 - kp_0)$ is positive on S, it follows from Problem 2.4 that

$$\int_{S^+ \cup S^-} (\phi - \phi^*)(p_1 - kp_0)\, d\mu = \int_S (\phi - \phi^*)(p_1 - kp_0)\, d\mu > 0$$

and hence that ϕ is more powerful against p_1 than ϕ^*. This is a contradiction, and therefore $\mu(S) = 0$, as was to be proved.

If ϕ^* were of size $< \alpha$ and power < 1, it would be possible to include in the rejection region additional points or portions of points and thereby to increase the power until either the power is 1 or the size is α. Thus either $E_0\phi^*(X) = \alpha$ or $E_1\phi^*(X) = 1$. ■

The proof of part (iii) shows that the most powerful test is uniquely determined by (3.7) and (3.8) except on the set on which $p_1(x) = kp_0(x)$. On this set, ϕ can be defined arbitrarily provided the resulting test has size α. Actually, we have shown that it is always to define ϕ to be constant over this boundary set. In the trivial case that there exists a test of power 1, the constant k of (3.8) is 0, and one will accept H for all points for which $p_1(x) = kp_0(x)$ even though the test may then have size $< \alpha$.

It follows from these remarks that the most powerful test is determined uniquely (up to sets of measure zero) by (3.7) and (3.8) whenever the set on which $p_1(x) = kp_0(x)$ has μ-measure zero. This unique test is then clearly nonrandomized. More generally, it is seen that randomization is not required except possibly on the boundary set, where it may be necessary to randomize in order to get the size equal to α. When there exists a test of power 1, (3.7) and (3.8) will determine a most powerful test, but it may not be unique in that there may exist a test also most powerful and satisfying (3.7) and (3.8) for some $\alpha' < \alpha$.

Corollary 3.2.1 *Let β denote the power of the most powerful level-α test* ($0 < \alpha < 1$) *for testing P_0 against P_1. Then $\alpha < \beta$ unless $P_0 = P_1$.*

PROOF. Since the level-α test given by $\phi(x) \equiv \alpha$ has power α, it is seen that $\alpha \leq \beta$. If $\alpha = \beta < 1$, the test $\phi(x) \equiv \alpha$ is most powerful and by Theorem 3.2.1(iii) must satisfy (3.8). Then $p_0(x) = p_1(x)$ a.e. μ and hence $P_0 = P_1$. ■

An alternative method for proving some of the results of this section is based on the following geometric representation of the problem of testing a simple hypothesis against a simple alternative. Let N be the set of all points (α, β) for which there exists a test ϕ such that

$$\alpha = E_0\phi(X), \qquad \beta = E_1\phi(X).$$

This set is convex, contains the points (0,0) and (1,1), and is symmetric with respect to the point $(\frac{1}{2}, \frac{1}{2})$ in the sense that with any point (α, β) it also contains the point $(1-\alpha, 1-\beta)$. In addition, the set N is closed. [This follows from the weak compactness theorem for critical functions, Theorem A.5.1 of the Appendix; the argument is the same as that in the proof of Theorem 3.6.1(i).]

For each value $0 < \alpha_0 < 1$, the level-α_0 tests are represented by the points whose abscissa is $\leq \alpha_o$. The most powerful of these tests (whose existence follows from the fact that N is closed) corresponds to the point on the upper boundary of N with abscissa α_0. This is the only point corresponding to a most powerful level-α_0 test unless there exists a point $(\alpha, 1)$ in N with $\alpha < \alpha_0$ (Figure 3.1*b*).

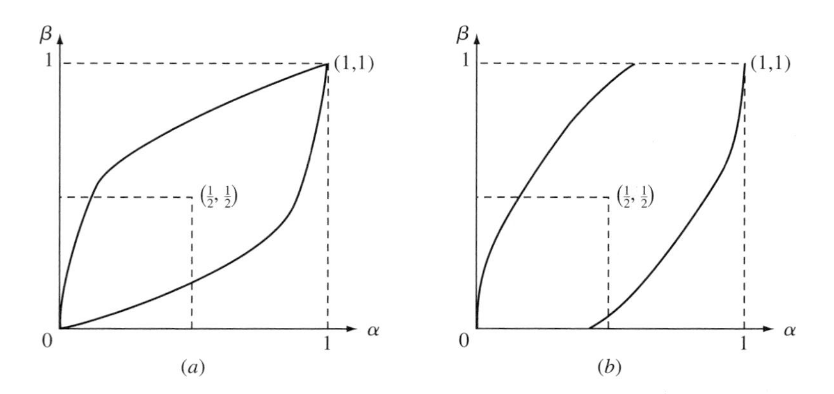

Figure 3.1.

As a example of this geometric approach, consider the following alternative proof of Corollary 3.2.1. Suppose that for some $0 < \alpha_0 < 1$ the power of the most powerful level-α_0 test is α_0. Then it follows from the convexity of N that $(\alpha, \beta) \in N$ implies $\beta \leq \alpha$, and hence from the symmetry of N that N consists exactly of the line segment connecting the points (0,0) and (1,1). This means that $\int \phi p_o \, d\mu = \int \phi p_1 \, d\mu$ for all ϕ and hence that $p_0 = p_1$ (a.e.μ), as was to be proved. A proof of Theorem 3.2.1 along these lines is given in a more general setting in the proof of Theorem 3.6.1.

Example 3.2.1 Suppose X is an observation from $N(\xi, \sigma^2)$, with σ^2 known. The null hypothesis specifies $\xi = 0$ and the alternative specifies $\xi = \xi_1$ for some $\xi_1 > 0$. Then, the likelihood ratio is given by

$$\frac{p_1(x)}{p_0(x)} = \frac{\exp[-\frac{1}{2\sigma^2}(x - \xi_1)^2]}{\exp[-\frac{1}{2\sigma^2}x^2]} = \exp[\frac{\xi_1 x}{\sigma^2} - \frac{\xi_1^2}{2\sigma^2}] . \tag{3.9}$$

Since the exponential function is strictly increasing and $\xi_1 > 0$, the set of x where $p_1(x)/p_0(x) > k$ is equivalent to the set of x where $x > k'$. In order to determine k', the level constraint

$$P_0\{X > k'\} = \alpha$$

must be satisfied, and so $k' = \sigma z_{1-\alpha}$, where $z_{1-\alpha}$ is the $1 - \alpha$ quantile of the standard normal distribution. Therefore, the most powerful level α test rejects if $X > \sigma z_{1-\alpha}$. ■

3.3 p-values

Testing at a fixed level α as described in Sections 3.1 and 3.2 is one of two standard (non-Bayesian) approaches to the evaluation of hypotheses. To explain the other, suppose that, under P_0, the distribution of $p_1(X)/p_0(X)$ is continuous. Then, the most powerful level α test is nonrandomized and rejects if $p_1(X)/p_0(X) > k$, where $k = k(\alpha)$ is determined by (3.7). For varying α, the resulting tests provide an example of the typical situation in which the rejection regions S_α are nested in the sense that

$$S_\alpha \subset S_{\alpha'} \quad \text{if } \alpha < \alpha' . \tag{3.10}$$

When this is the case,[4] it is good practice to determine not only whether the hypothesis is accepted or rejected at the given significance level, but also to determine the smallest significance level, or more formally

$$\hat{p} = \hat{p}(X) = \inf\{\alpha : \; X \in S_\alpha\} , \tag{3.11}$$

at which the hypothesis would be rejected for the given observation. This number, the so-called *p-value* gives an idea of how strongly the data contradict the

---

[4]See Problems 3.17 and 3.58 for examples where optimal nonrandomized tests need not be nested.

hypothesis.[5] It also enables others to reach a verdict based on the significance level of their choice.

Example 3.3.1 (Continuation of Example 3.2.1) Let Φ denote the standard normal c.d.f. Then, the rejection region can be written as

$$S_\alpha = \{X : X > \sigma z_{1-\alpha}\} = \{X : \Phi(\frac{X}{\sigma}) > 1-\alpha\} = \{X : 1 - \Phi(\frac{X}{\sigma}) < \alpha\} .$$

For a given observed value of X, the inf over all α where the last inequality holds is

$$\hat{p} = 1 - \Phi(\frac{X}{\sigma}) .$$

Alternatively, the p-value is $P_0\{X \geq x\}$, where x is the observed value of X. Note that, under $\xi = 0$, the distribution of $\hat{p}$ is given by

$$P_0\{\hat{p} \leq u\} = P_0\{1 - \Phi(\frac{X}{\sigma}) \leq u\} = P_0\{\Phi(\frac{X}{\sigma}) \geq 1-u\} = u ,$$

because $\Phi(X/\sigma)$ is uniformly distributed on (0,1) (see Problem 3.22); therefore, $\hat{p}$ is uniformly distributed on (0,1). ■

A general property of p-values is given in the following lemma, which applies to both simple and composite null hypotheses.

Lemma 3.3.1 *Suppose X has distribution P_θ for some $\theta \subset \Omega$, and the null hypothesis H specifies $\theta \in \Omega_H$. Assume the rejection regions satisfy (3.10).*
(i) If

$$\sup_{\theta \in \Omega_H} P_\theta\{X \in S_\alpha\} \leq \alpha \quad \text{for all } 0 < \alpha < 1, \tag{3.12}$$

then the distribution of $\hat{p}$ under $\theta \in \Omega_H$ satisfies

$$P_\theta\{\hat{p} \leq u\} \leq u \quad \text{for all } 0 \leq u \leq 1 . \tag{3.13}$$

(ii) If, for $\theta \in \Omega_H$,

$$P_\theta\{X \in S_\alpha\} = \alpha \quad \text{for all } 0 < \alpha < 1 , \tag{3.14}$$

then

$$P_\theta\{\hat{p} \leq u\} = u \quad \text{for all } 0 \leq u \leq 1 ;$$

i.e. $\hat{p}$ is uniformly distributed over $(0, 1)$.

PROOF. (i) If $\theta \in \Omega_H$, then the event $\{\hat{p} \leq u\}$ implies $\{X \in S_v\}$ for all $u < v$. The result follows by letting $v \to u$.

(ii) Since the event $\{X \in S_u\}$ implies $\{\hat{p} \leq u\}$, it follows that

$$P_\theta\{\hat{p} \leq u\} \geq P_\theta\{X \in S_u\} .$$

Therefore, if (3.14) holds, then $P_\theta\{\hat{p} \leq u\} \geq u$, and the result follows from (i). ■

---

[5] One could generalize the definition of p-value to include randomized level α tests ϕ_α assuming that they are nested in the sense that $\phi_\alpha(x) \leq \phi_{\alpha'}(x)$ for all x and $\alpha < \alpha'$. Simply define $\hat{p} = \inf\{\alpha : \phi_\alpha(X) = 1\}$; in words, $\hat{p}$ is the smallest level of significance where the hypothesis is rejected with probability one.

Example 3.3.2 Suppose X takes values $1, 2, \dots, 10$. Under H, the distribution is uniform, i.e., $p_0(j) = \frac{1}{10}$ for $j = 1, \dots, 10$. Under K, suppose $p_1(j) = j/55$. The MP level $\alpha = i/10$ test rejects if $X \geq 11 - i$. However, unless α is a multiple of $1/10$, the MP level α test is randomized. If we want to restrict attention to nonrandomized procedures, consider the conservative approach by defining

$$S_\alpha = \{X \geq 11 - i\} \quad \text{if } \frac{i}{10} \leq \alpha < \frac{i+1}{10} .$$

If the observed value of X is x, then the p-value is given by $(11 - x)/10$. Then, the distribution of $\hat{p}$ under H is given by

$$P\{\hat{p} \leq u\} = P\{\frac{11 - X}{10} \leq u\} = P\{X \geq 11 - 10u\} \leq u , \tag{3.15}$$

and the last inequality is an equality if and only if u is of the form $i/10$ for some integer $i = 0, 1, \dots, 10$, i.e. the levels for which the MP test is nonrandomized (Problem 3.21). ■

P-values, with the additional information they provide, are typically more appropriate than fixed levels in scientific problems, whereas a fixed predetermined α is unavoidable when acceptance or rejection of H implies an imminent concrete decision. A review of some of the issues arising in this context, with references to the literature, is given in Kruskal (1978).

3.4 Distributions with Monotone Likelihood Ratio

The case that both the hypothesis and the class of alternatives are simple is mainly of theoretical interest, since problems arising in applications typically involve a parametric family of distributions depending on one or more parameters. In the simplest situation of this kind the distributions depend on a single real-valued parameter θ, and the hypothesis is one-sided, say $H : \theta \leq \theta_0$. In general, the most powerful test of H against an alternative $\theta_1 > \theta_0$ depends on θ_1 and is then not UMP. However, a UMP test does exist if an additional assumption is satisfied. The real-parameter family of densities $p_\theta(x)$ is said to have *monotone likelihood ratio*[6] if there exists a real-valued function $T(x)$ such that for any $\theta < \theta'$ the distributions P_θ and $P_{\theta'}$ are distinct, and the ratio $p_{\theta'}(x)/p_\theta(x)$ is a nondecreasing function of $T(x)$.

Theorem 3.4.1 *Let θ be a real parameter, and let the random variable X have probability density $p_\theta(x)$ with monotone likelihood ratio in $T(x)$.*

(i) *For testing $H : \theta \leq \theta_0$ against $K : \theta > \theta_0$, there exists a UMP test, which is given by*

$$\phi(x) = \begin{cases} 1 & \text{when } T(x) > C, \\ \gamma & \text{when } T(x) = C, \\ 0 & \text{when } T(x) < C, \end{cases} \tag{3.16}$$

[6] This definition is in terms of specific versions of the densities p_θ. If instead the definition is to be given in terms of the distribution P_θ, various null-set considerations enter which are discussed in Pfanzagl (1967).

where C and γ are determined by

$$E_{\theta_0}\phi(X) = \alpha. \tag{3.17}$$

(ii) *The power function*

$$\beta(\theta) = E_\theta \phi(X)$$

of this test is strictly increasing for all points θ for which $0 < \beta(\theta) < 1$.

(iii) *For all θ', the test determined by (3.16) and (3.17) is UMP for testing $H' : \theta \leq \theta'$ against $K' : \theta > \theta'$ at level $\alpha' = \beta(\theta')$.*

(iv) *For any $\theta < \theta_0$ the test minimizes $\beta(\theta)$ (the probability of an error of the first kind) among all tests satisfying (3.17).*

PROOF. (i) and (ii): Consider first the hypothesis $H_0 : \theta = \theta_0$ and some simple alternative $\theta_1 > \theta_0$. The most desirable points for rejection are those for which $r(x) = p_{\theta_1}(x)/p_{\theta_0}(x) = g[T(x)]$ is sufficiently large. If $T(x) < T(x')$, then $r(x) \leq r(x')$ and x' is at least as desirable as x. Thus the test which rejects for large values of $T(x)$ is most powerful. As in the proof of Theorem 3.2.1(i), it is seen that there exist C and γ such that (3.16) and (3.17) hold. By Theorem 3.2.1(ii), the resulting test is also most powerful for testing $P_{\theta'}$ against $P_{\theta''}$ at level $\alpha' = \beta(\theta')$ provided $\theta' < \theta''$. Part (ii) of the present theorem now follows from Corollary 3.2.1. Since $\beta(\theta)$ is therefore nondecreasing the test satisfies

$$E_\theta \phi(X) \leq \alpha \qquad \text{for} \quad \theta \leq \theta_0. \tag{3.18}$$

The class of tests satisfying (3.18) is contained in the class satisfying $E_{\theta_0}\phi(X) \leq \alpha$. Since the given test maximizes $\beta(\theta_1)$ within this wider class, it also maximizes $\beta(\theta_1)$ subject to (3.18); since it is independent of the particular alternative $\theta_1 > \theta_0$ chosen, it is UMP against K.

(iii) is proved by an analogous argument.

(iv) follows from the fact that the test which minimizes the power for testing a simple hypothesis against a simple alternative is obtained by applying the fundamental lemma (Theorem 3.2.1) with all inequalities reversed.

By interchanging inequalities throughout, one obtains in an obvious manner the solution of the dual problem, $H : \theta \geq \theta_0, K : \theta < \theta_0$. ■

The proof of (i) and (ii) exhibits the basic property of families with monotone likelihood ratio: every pair of parameter values $\theta_0 < \theta_1$ establishes essentially the same preference order of the sample points (in the sense of the preceding section). A few examples of such families, and hence of UMP one-sided tests, will be given below. However, the main applications of Theorem 3.4.1 will come later, when such families appear as the set of conditional distributions given a sufficient statistic (Chapters 4 and 5) and as distributions of a maximal invariant (Chapters 6 and 7).

Example 3.4.1 (Hypergeometric) From a lot containing N items of a manufactured product, a sample of size n is selected at random, and each item in the sample is inspected. If the total number of defective items in the lot is D, the number X of defectives found in the sample has the *hypergeometric* distribution

$$P\{X = x\} = P_D(x) = \frac{\binom{D}{x}\binom{N-D}{n-x}}{\binom{N}{n}}, \quad \max(0, n + D - N) \leq x \leq \min(n, D).$$

Interpreting $P_D(x)$ as a density with respect to the measure μ that assigns to any set on the real line as measure the number of integers $0, 1, 2, \dots$ that it contains, and nothing that for values of x within its range

$$\frac{P_{D+1}(x)}{P_D(x)} = \begin{cases} \frac{D+1}{N-D}\frac{N-D-n+x}{D+1-x} & \text{if} \quad n+D+1-N \le x \le D, \\ 0 \text{ or } \infty & \text{if} \quad x = n+D-N \text{ or } D+1, \end{cases}$$

it is seen that the distributions satisfy the assumption of monotone likelihood ratios with $T(x) = x$. Therefore there exists a UMP test for testing the hypothesis $H : D \le D_0$ against $K : D > D_0$, which rejects H when X is too large, and an analogous test for testing $H' : D \ge D_0$. ■

An important class of families of distributions that satisfy the assumptions of Theorem 3.4.1 are the *one-parameter exponential families.*

Corollary 3.4.1 *Let θ be a real parameter, and let X have probability density (with respect to some measure μ)*

$$p_\theta(x) = C(\theta)e^{Q(\theta)T(x)}h(x), \tag{3.19}$$

where Q is strictly monotone. Then there exists a UMP test ϕ for testing $H : \theta \le \theta_0$ against $K : \theta > \theta_0$. If Q is increasing,

$$\phi(x) = 1, \gamma, 0 \qquad \textit{as} \quad T(x) >, =, < C,$$

where C and γ are determined by $E_{\theta_0}\phi(X) = \alpha$. If Q is decreasing, the inequalities are reversed.

A converse of Corollary 3.4.1 is given by Pfanzagl (1968), who shows under weak regularity conditions that the existence of UMP tests against one-sided alternatives for all sample sizes and one value of α implies an exponential family.

As in Example 3.4.1, we shall denote the right-hand side of (3.19) by $P_\theta(x)$ instead of $p_\theta(x)$ when it is a probability, that is, when X is discrete and μ is counting measure.

Example 3.4.2 (Binomial) The binomial distributions $b(p, n)$ with

$$P_p(x) = \binom{n}{x} p^x(1-p)^{n-x}$$

satisfy (3.19) with $T(x) = x, \theta = p, Q(p) = \log[p/(1-p)]$. The problem of testing $H : p \ge p_0$ arises, for instance, in the situation of Example 3.4.1 if one supposes that the production process is in statistical control, so that the various items constitute independent trials with constant probability p of being defective. The number of defectives X in a sample of size n is then sufficient statistic for the distribution of the variables X_i $(i = 1, \dots, n)$, where X_i is 1 or 0 as the ith item drawn is defective or not, and X is distributed as $b(p, n)$. There exists therefore a UMP test of H, which rejects H when X is too small.

An alternative sampling plan which is sometimes used in binomial situations is *inverse binomial sampling.* Here the experiment is continued until a specified number m of successes—for example, cures effected by some new medical treatment—have been obtained. If Y_i denotes the number of trials after the

$(i-1)$st success up to but not including the ith success, the probability that $Y_i = y$ is pq^y for $y = 0, 1, \ldots$, so that the joint distribution of $Y_1, \ldots, Y_m$ is

$$P_p(y_1, \ldots, y_m) = p^m q^{\sum y_i}, \qquad y_k = 0, 1, \ldots, \quad k = 1, \ldots, m.$$

This is an exponential family with $T(y) = \sum y_i$ and $Q(p) = \log(1-p)$. Since $Q(p)$ is a decreasing function of p, the UMP test of $H : p \leq p_0$ rejects H when T is too small. This is what one would expect, since the realization of m successes in only a few more than m trials indicates a high value of p. The test statistic T, which is the number of trials required in excess of m to get m successes, has the negative binomial distribution [Problem 1.1(i)]

$$P(t) = \binom{m+t-1}{m-1} p^m q^t, \qquad t = 0, 1, \ldots. \blacksquare$$

Example 3.4.3 (Poisson) If $X_1, \ldots, X_n$ are independent Poisson variables with $E(X_i) = \lambda$, their joint distribution is

$$P_\lambda(x_1, \ldots, x_n) = \frac{\lambda^{x_1 + \cdots + x_n}}{x_1! \cdots x_n!} e^{-n\lambda}.$$

This constitutes an exponential family with $T(x) = \sum x_i$, and $Q(\lambda) = \log \lambda$. One-sided hypotheses concerning λ might arise if λ is a bacterial density and the X's are a number of bacterial counts, or if the X's denote the number of α-particles produced in equal time intervals by a radioactive substance, etc. The UMP test of the hypothesis $\lambda \leq \lambda_0$ rejects when $\sum X_i$ is too large. Here the test statistic $\sum X_i$ has itself a Poisson distribution with parameter $n\lambda$.

Instead of observing the radioactive material for given time periods or counting the number of bacteria in given areas of a slide, one can adopt an inverse sampling method. The experiment is then continued, or the area over which the bacteria are counted is enlarged, until a count of m has been obtained. The observations consist of the times $T_1, \ldots, T_m$ that it takes for the first occurrence, from the first to the second, and so on. If one is dealing with a Poisson process and the number of occurrences in a time or space interval τ has the distribution

$$P(x) = \frac{(\lambda\tau)^x}{x!} e^{-\lambda\tau}, \qquad x = 0, 1, \ldots,$$

then the observed times are independently distributed, each with the exponential density $\lambda e^{-\lambda t}$ for $t \geq 0$ [Problem 1.1(ii)]. The joint densities

$$p_\lambda(t_1, \ldots, t_m) = \lambda^m \exp\left(-\lambda \sum_{i=1}^m t_i\right), \qquad t_1, \ldots, t_m \geq 0,$$

form an exponential family with $T(t_1, \ldots, t_m) = \sum t_i$ and $Q(\lambda) = -\lambda$. The UMP test of $H : \lambda \leq \lambda_0$ rejects when $T = \sum T_i$ is too small. Since $2\lambda T_i$ has density $\frac{1}{2} e^{-u/2}$ for $u \geq 0$, which is the density of a χ^2-distribution with 2 degrees of freedom, $2\lambda T$ has a χ^2-distribution with $2m$ degrees of freedom. The boundary of the rejection region can therefore be determined from a table of χ^2. $\blacksquare$

The formulation of the problem of hypothesis testing given at the beginning of the chapter takes account of the losses resulting from wrong decisions only in terms of the two types of error. To obtain a more detailed description of the problem of testing $H : \theta \le \theta_0$ against the alternatives $\theta > \theta_0$, one can consider it as a decision problem with the decisions d_0 and d_1 of accepting and rejecting H and a loss function $L(\theta, d_i) = L_i(\theta)$. Typically, $L_0(\theta)$ will be 0 for $\theta \le \theta_0$ and strictly increasing for $\theta \ge \theta_0$, and $L_1(\theta)$ will be strictly decreasing for $\theta \le \theta_0$ and equal to 0 for $\theta \ge \theta_0$. The difference then satisfies

$$L_1(\theta) - L_0(\theta) \gtrless 0 \qquad \text{as} \quad \theta \lessgtr \theta_0. \tag{3.20}$$

The following theorem is a special case of complete class results of Karlin and Rubin (1956) and Brown, Cohen, and Strawderman (1976).

Theorem 3.4.2 (i) *Under the assumptions of Theorem 3.4.1, the family of tests given by* (3.16) *and* (3.17) *with* $0 \le \alpha \le 1$ *is essentially complete provided the loss function satisfies* (3.20).

(ii) *This family is also minimal essentially complete if the set of points* x *for which* $p_\theta(x) > 0$ *is independent of* θ.

PROOF. (i): The risk function of any test ϕ is

$$\begin{aligned} R(\theta, \phi) &= \int p_\theta(x)\{\phi(x)L_1(\theta) + [1 - \phi(x)]L_0(\theta)\}\, d\mu(x) \\ &= \int p_\theta(x)\{L_0(\theta) + [L_1(\theta) - L_0(\theta)]\phi(x)\}\, d\mu(x), \end{aligned}$$

and hence the difference of two risk functions is

$$R(\theta, \phi') - R(\theta, \phi) = [L_1(\theta) - L_0(\theta)] \int (\phi' - \phi) p_\theta \, d\mu.$$

This is ≤ 0 for all θ if

$$\beta_{\phi'}(\theta) - \beta_\phi(\theta) = \int (\phi' - \phi) p_\theta \, d\mu \gtreqless 0 \qquad \text{for} \quad \theta \gtreqless \theta_0.$$

Given any test ϕ, let $E_{\theta_0}\phi(X) = \alpha$. It follows from Theorem 3.4.1(i) that there exists a UMP level-α test ϕ' for testing $\theta = \theta_0$ against $\theta > \theta_0$, which satisfies (3.16) and (3.17). By Theorem 3.4.1(iv), ϕ' also minimizes the power for $\theta < \theta_0$. Thus the two risk functions satisfy $R(\theta, \phi') \le R(\theta, \phi)$ for all θ, as was to be proved.

(ii): Let ϕ_α and $\phi_{\alpha'}$ be of sizes $\alpha < \alpha'$ and UMP for testing θ_0 against $\theta > \theta_0$. Then $\beta_{\phi_\alpha}(\theta) < \beta_{\phi_{\alpha'}}(\theta)$ for all $\theta > \theta_0$ unless $\beta_{\phi_\alpha}(\theta) = 1$. By considering the problem of testing $\theta = \theta_0$ against $\theta < \theta_0$ it is seen analogously that this inequality also holds for all $\theta < \theta_0$ unless $\beta_{\phi_{\alpha'}}(\theta) = 0$. Since the exceptional possibilities are excluded by the assumptions, it follows that $R(\theta, \phi') \lessgtr R(\theta, \phi)$ as $\theta \gtrless \theta_0$. Hence each of the two risk functions is better than the other for some values of θ.

The class of tests previously derived as UMP at the various significance levels α is now seen to constitute an essentially complete class for a much more general decision problem, in which the loss function is only required to satisfy certain broad qualitative conditions. From this point of view, the formulation involving the specification of a level of significance can be considered a simple way of selecting a particular procedure from an essentially complete family.

The property of monotone likelihood ratio defines a very strong ordering of a family of distributions. For later use, we consider also the following somewhat weaker definition. A family of cumulative distribution functions F_θ on the real line is said to be *stochastically increasing* (and the same term is applied to random variables possessing these distributions) if the distributions are distinct and if $\theta < \theta'$ implies $F_\theta(x) \geq F_{\theta'}(x)$ for all x. If then X and X' have distributions F_θ and F'_θ respectively, it follows that $P\{X > x\} \leq P\{X' > x\}$ for all x, so that X' tends to have larger values than X. In this case the variable X' is said to be *stochastically larger* than X. This relationship is made more intuitive by the following characterization of the stochastic ordering of two distributions. ■

Lemma 3.4.1 *Let F_0 and F_1 be two cumulative distribution functions on the real line. Then $F_1(x) \leq F_0(x)$ for all x if and only if there exist two nondecreasing functions f_0 and f_1, and a random variable V, such that* (a) *$f_0(v) \leq f_1(v)$ for all v, and* (b) *the distributions of $f_0(V)$ and $f_1(V)$ are F_0 and F_1 respectively.*

PROOF. Suppose first that the required f_0, f_1 and V exist. Then

$$F_1(x) = P\{f_1(V) \leq x\} \leq P\{f_0(V) \leq x\} = F_0(x)$$

for all x. Conversely, suppose that $F_1(x) \leq F_0(x)$ for all x, and let $f_i(y) = \inf\{x : F_i(x-0) \leq y \leq F_1(x)\}$, $i = 0, 1$. These functions are nondecreasing and for $f_i = f, F_i = F$ satisfy

$$f[F(x)] \leq x \text{ and } F[f(y)] \geq y \qquad \text{for all } x \text{ and } y.$$

It follows that $y \leq F(x_0)$ implies $f(y) \leq f[F(x_0)] \leq x_0$ and that conversely $f(y) \leq x_0$, implies $F[f(y)] \leq F(x_0)]$ and hence $y \leq F(x_0)$, so that the two inequalities $f(y) \leq x_0$ and $y \leq F(x_0)$ are equivalent. Let V be uniformly distributed on (0,1). Then $P\{f_i(V) \leq x\} = P\{V \leq F_i(x)\} = F_i(x)$. Since $F_i(x) \leq F_0(x)$ for all x implies $f_0(y) \leq f_1(y)$ for all y, this completes the proof. ■

One of the simplest examples of a stochastically ordered family is a location parameter family, that is, a family satisfying

$$F_\theta(x) = F(x - \theta).$$

To see that this is stochastically increasing, let X be a random variable with distribution $F(x)$. Then $\theta < \theta'$ implies

$$F(x-\theta) = P\{x \leq x - \theta\} \geq P\{X \leq x - \theta'\} = F(x - \theta'),$$

as was to be shown.

Another example is finished by families with monotone likelihood ratio. This is seen from the following lemma, which establishes some basic properties of these families.

Lemma 3.4.2 *Let $p_\theta(x)$ be a family of densities on the real line with monotone likelihood ratio in x.*

(i) *If ψ is a nondecreasing function of x, then $E_\theta\psi(X)$ is a nondecreasing function of θ; if $X_1, \ldots, X_n$ are independently distributed with density p_θ and ψ' is a function of $x_1, \ldots, x_n$ which is nondecreasing in each of its arguments, then $E_\theta\psi'(X_1, \ldots, X_n)$ is a nondecreasing function of θ.*

(ii) *For any $\theta < \theta'$, the cumulative distribution functions of X under θ and θ' satisfy*

$$F_{\theta'}(x) \leq F_\theta(x) \qquad \text{for all } x.$$

(iii) *Let ψ be a function with a single change of sign. More specifically, suppose there exists a value x_0 such that $\psi(x) \leq 0$ for $x < x_0$ and $\psi(x) \geq 0$ for $x \geq x_0$. Then there exists θ_0 such that $E_\theta\psi(X) \leq 0$ for $\theta < \theta_0$ and $E_\theta\psi(X) \geq 0$ for $\theta > \theta_0$, unless $E_\theta\psi(X)$ is either positive for all θ or negative for all θ.*

(iv) *Suppose that $p_\theta(x)$ is positive for all θ and all x, that $p_{\theta'}(x)/p_\theta(x)$ is strictly increasing in x for $\theta < \theta'$, and that $\psi(x)$ is as in* (iii) *and is $\neq 0$ with positive probability. If $E_{\theta_0}\psi(X) = 0$, then $E_\theta\psi(X) < 0$ for $\theta < \theta_0$ and > 0 for $\theta > \theta_0$.*

Proof. (i): Let $\theta < \theta'$, and let A and B be the sets for which $p_{\theta'}(x) < p_\theta(x)$ and $p_{\theta'}(x) > p_\theta(x)$ respectively. If $a = \sup_A \psi(x)$ and $b = \inf_B \psi(x)$, then $b - a \geq 0$ and

$$\begin{aligned}\int \psi(p_{\theta'} - p_\theta)\, d\mu &\geq a\int_A (p_{\theta'} - p_\theta)\, d\mu + b\int_B (p_{\theta'} - p_\theta)\, d\mu \\ &= (b-a)\int_B (p_{\theta'} - p_\theta)\, d\mu \geq 0,\end{aligned}$$

which proves the first assertion. The result for general n follows by induction.

(ii): This follows from (i) by letting $\psi(x) = 1$ for $x > x_0$ and $\psi(x) = 0$ otherwise.

(iii): We shall show first that for any $\theta' < \theta''$, $E_{\theta'}\psi(X) > 0$ implies $E_{\theta''}\psi(X) \geq 0$. If $p_{\theta''}(x_0)/p_{\theta'}(x_0) = \infty$, then $p_{\theta'}(x) = 0$ for $x \geq x_0$ and hence $E_{\theta'}\psi(X) \leq 0$. Suppose therefore that $p_{\theta''}(x_0)/p_{\theta'}(x_0) = c < \infty$. Then $\psi(x) \geq 0$ on the set $S = \{x : p_{\theta'}(x) = 0 \text{ and } p_{\theta''}(x) > 0\}$, and

$$\begin{aligned}E_{\theta''}\psi(X) &\geq \int_{\tilde{S}} \psi \frac{p_{\theta''}}{p_{\theta'}} p_{\theta'}\, d\mu \\ &\geq \int_{-\infty}^{x_0-} c\psi p_{\theta'}\, d\mu + \int_{x_0}^{\infty} c\psi p_{\theta'}\, d\mu = cE_{\theta'}\psi(X) \geq 0.\end{aligned}$$

The result now follows by letting $\theta_0 = \inf\{\theta : E_\theta\psi(X) > 0\}$.

(iv): The proof is analogous to that of (iii). ■

Part (ii) of the lemma shows that any family of distributions with monotone likelihood ratio in x is stochastically increasing. That the converse does not hold is shown for example by the Cauchy densities

$$\frac{1}{\pi}\frac{1}{1 + (x-\theta)^2}.$$

The family is stochastically increasing, since θ is a location parameter; however, the likelihood ratio is not monotone. Conditions under which a location parameter family possesses monotone likelihood ratio are given in Example 8.2.1.

Lemma 3.4.2 is a special case of a theorem of Karlin (1957, 1968) relating the number of sign changes of $E_\theta\psi(X)$ to those of $\psi(x)$ when the densities $p_\theta(x)$ are *totally positive* (defined in Problem 3.50). The application of totally positive–or equivalently, variation diminishing–distributions to statistics is discussed by Brown, Johnstone, and MacGibbon (1981); see also Problem 3.53.

3.5 Confidence Bounds

The theory of UMP one-sided tests can be applied to the problem of obtaining a lower or upper bound for a real-valued parameter θ. The problem of setting a lower bound arises, for example, when θ is the breaking strength of a new alloy; that of setting an upper bound, when θ is the toxicity of drug or the probability of an undesirable event. The discussion of lower and upper bounds completely parallel, and it is therefore enough to consider the case of a lower bound, say $\underline{\theta}$.

Since $\underline{\theta} = \underline{\theta}(X)$ will be a function of the observations, it cannot be required to fall below θ with certainty, but only with specified high probability. One selects a number $1 - \alpha$, the *confidence level*, and restricts attention to bounds θ satisfying

$$P_\theta\{\underline{\theta}(X) \leq \theta\} \geq 1 - \alpha \qquad \text{for all } \theta. \tag{3.21}$$

The function $\underline{\theta}$ is called a lower *confidence bound* for θ at confidence level $1 - \alpha$; the infimum of the left-hand side of (3.21), which in practice will be equal to $1 - \alpha$, is called the *confidence coefficient* of $\underline{\theta}$.

Subject to (3.21), $\underline{\theta}$ should underestimate θ by as little as possible. One can ask, for example, that the probability of $\underline{\theta}$ falling below any $\theta' < \theta$ should be a minimum. A function $\underline{\theta}$ for which

$$P_\theta\{\underline{\theta}(X) \leq \theta'\} = \text{ minimum} \tag{3.22}$$

for all $\theta' < \theta$ subject to (3.21) is a uniformly *most accurate* lower confidence bound for θ at confidence level $1 - \alpha$.

Let $L(\theta, \underline{\theta})$ be a measure of the loss resulting from underestimating θ, so that for each fixed θ the function $L(\theta, \underline{\theta})$ is defined and nonnegative for $\underline{\theta} < \theta$, and is nonincreasing in this second argument. One would then wish to minimize

$$E_\theta L(\theta, \underline{\theta}) \tag{3.23}$$

subject to (3.21). It can be shown that a uniformly most accurate lower confidence bound $\underline{\theta}$ minimizes (3.23) subject to (3.21) for every such loss function L. (See Problem 3.44.)

The derivation of uniformly most accurate confidence bounds is facilitated by introducing the following more general concept, which will be considered in more detail in Chapter 5. A family of subsets $S(x)$ of the parameter space Ω is said to constitute a family of *confidence sets* at confidence level $1 - \alpha$ if

$$P_\theta\{\theta \in S(X)\} \geq 1 - \alpha \qquad \text{for all} \quad \theta \in \Omega, \tag{3.24}$$

that is, if the random sets $S(X)$ covers the true parameter point with probability $\geq 1 - \alpha$. A lower confidence bound corresponds to the special case that $S(x)$ is a one-sided interval

$$S(x) = \{\theta : \underline{\theta}(x) \leq \theta < \infty\}.$$

Theorem 3.5.1 (i) *For each $\theta_0 \in \Omega$ let $A(\theta_0)$ be the acceptance region of a level-α test for testing $H(\theta_0) : \theta = \theta_0$, and for each sample point x let $S(x)$ denote the set of parameter values*

$$S(x) = \{\theta : x \in A(\theta), \theta \in \Omega\}.$$

Then $S(x)$ is a family of confidence sets for θ at confidence level $1 - \alpha$.

(ii) *If for all* $\theta_0, A(\theta_0)$ *is UMP for testing* $H(\theta_0)$ *at level* α *against the alternatives* $K(\theta_0)$, *then for each* $\theta_0 \notin \Omega, S(X)$ *minimizes probability*

$$P_\theta\{\theta_0 \in S(X)\} \qquad \text{for all} \quad \theta \in K(\theta_0)$$

among all level $1-\alpha$ *families of confidence sets for* θ.

PROOF. (i): By definition of $S(x)$,

$$\theta \in S(x) \qquad \text{if and only if} \quad x \in A(\theta), \tag{3.25}$$

and hence

$$P_\theta\{\theta \in S(X)\} = P_\theta\{X \in A(\theta)\} \geq 1 - \alpha.$$

(ii): If $S^*(x)$ is any other family of confidence sets at level $1-\alpha$, and if $A^*(\theta) = \{x : \theta \in S^*(x)\}$, then

$$P_\theta\{X \in A^*(\theta)\} = P_\theta\{\theta \in S^*(X)\} \geq 1 - \alpha,$$

so that $A^*(\theta_0)$ is the acceptance region of a level-α test of $H(\theta_0)$. It follows from the assumed property of $A(\theta_0)$ that for any $\theta \in K(\theta_0)$

$$P_\theta\{X \in A^*(\theta_0)\} \geq P_\theta\{X \in A(\theta_0)\}$$

and hence that

$$P_\theta\{\theta_0 \in S^*(X)\} \geq P_\theta\{\theta_0 \in S(X)\},$$

as was to be proved. ■

The equivalence (3.25) shows the structure of the confidence sets $S(x)$ as the totality of parameter values θ for which the hypothesis $H(\theta)$ is accepted when x is observed. A confidence set can therefore be viewed as a combined statement regarding the tests of the various hypotheses $H(\theta)$, which exhibits the values for which the hypothesis is accepted $[\theta \in S(x)]$ and those for which it is rejected $[\theta \in \bar{S}(x)]$.

Corollary 3.5.1 *Let the family of densities* $p_\theta(x), \theta \in \Omega$, *have monotone likelihood ratio in* $T(x)$, *and suppose that the cumulative distribution function* $F_\theta(t)$ *of* $T = T(X)$ *is a continuous function in each of the variables* t *and* θ *when the other is fixed.*

(i) *There exists a uniformly most accurate confidence bound* $\underline{\theta}$ *for* θ *at each confidence level* $1-\alpha$.

(ii) *If* x *denotes the observed values of* X *and* $t = T(x)$, *and if the equation*

$$F_\theta(t) = 1 - \alpha \tag{3.26}$$

has a solution $\theta = \hat{\theta}$ *in* Ω *then this solution is unique and* $\underline{\theta}(x) = \hat{\theta}$.

PROOF. (i): There exists for each θ_0 a constant $C(\theta_0)$ such that

$$P_{\theta_0}\{T > C(\theta_0)\} = \alpha,$$

and by Theorem 3.4.1, $T > C(\theta_0)$ is a UMP level-α rejection region for testing $\theta = \theta_0$ against $\theta > \theta_0$. By Corollary 3.2.1, the power of this test against any alternative $\theta_1 > \theta_0$ exceeds α, and hence $C(\theta_0) < C(\theta_1)$ so that the function C is strictly increasing; it is also continuous. Let $A(\theta_0)$ denote the acceptance region

$T \leq C(\theta_0)$, and let $S(x)$ be defined by (3.25). If follows from the monotonicity of the function C that $S(x)$ consists of those values $\theta \in \Omega$ which satisfy $\underline{\theta} \leq \theta$, where

$$\underline{\theta} = \inf\{\theta : T(x) \leq C(\theta)\}.$$

By Theorem 3.5.1, the sets $\{\theta : \underline{\theta}(x) \leq \theta\}$, restricted to possible values of the parameter, constitute a family of confidence sets at level $1 - \alpha$, which minimize $P_\theta\{\underline{\theta} \leq \theta'\}$ for all $\theta \in K(\theta')$, that is, for all $\theta > \theta'$. This shows $\underline{\theta}$ to be a uniformly most accurate confidence bound for θ.

(ii): It follows from Corollary 3.2.1 that $F_\theta(t)$ is a strictly decreasing function of θ at any point t for which $0 < F_\theta(t) < 1$, and hence that (3.26) can have at most one solution. Suppose now that t is the observed value of T and that the equation $F_\theta(t) = 1 - \alpha$ has the solution $\hat{\theta} \in \Omega$. Then $F_{\hat{\theta}}(t) = 1 - \alpha$, and by definition of the function $C, C(\hat{\theta}) = t$. The inequality $t \leq C(\theta)$ is then equivalent to $C(\hat{\theta}) \leq C(\theta)$ and hence to $\hat{\theta} \leq \theta$. It follows that $\underline{\theta} = \hat{\theta}$, as was to be proved.

Under the same assumptions, the corresponding upper confidence bound with confidence coefficient $1 - \alpha$ is the solution $\bar{\theta}$ of the equation $P_\theta\{T \geq t\} = 1 - \alpha$ or equivalently of $F_\theta(t) = \alpha$. ■

Example 3.5.1 (Exponential waiting times) To determine an upper bound for the degree of radioactivity λ of a radioactive substance, the substance is observed until a count of m has been obtained on a Geiger counter. Under the assumptions of Example 3.4.3, the joint probability density of the times $T_i (i = 1, \ldots, m)$ elapsing between the $(i-1)$st count and the ith one is

$$p(t_1, \ldots, t_m) = \lambda^m e^{-\lambda \sum t_i}, \qquad t_1, \ldots, t_m \geq 0.$$

If $T = \sum T_i$ denotes the total time of observation, then $2\lambda T$ has a χ^2-distribution with $2m$ degrees of freedom, and, as was shown in Example 3.4.3, the acceptance region of the most powerful test of $H(\lambda_0) : \lambda = \lambda_0$ against $\lambda < \lambda_0$ is $2\lambda_0 T \leq C$, where C is determined by the equation

$$\int_0^C \chi^2_{2m} = 1 - \alpha \,.$$

The set $S(t_1, \ldots, t_m)$ defined by (3.25) is then the set of values λ such that $\lambda \leq C/2T$, and it follows from Theorem 3.5.1 that $\bar{\lambda} = C/2T$ is a uniformly most accurate upper confidence bound for λ. This result can also be obtained through Corollary 3.5.1. ■

If the variables X or T are discrete, Corollary 3.5.1 cannot be applied directly, since the distribution functions $F_\theta(t)$ are not continuous, and for most values θ_0 the optimum test of $H : \theta = \theta_0$ are randomized. However, any randomized test based on X has the following representation as a nonrandomized test depending on X and an independent variable U distributed uniformly over $(0, 1)$. Given a critical function ϕ, consider the rejection region

$$R = \{(x, u) : u \leq \phi(x)\}.$$

Then

$$P\{(X, U) \in R\} = P\{U \leq \phi(X)\} = E_\phi(X),$$

whatever the distribution of X, so that R has the same power function as ϕ and the two tests are equivalent. The pair of variables (X, U) has a particularly simple representation when X is integer-valued. In this case the statistic

$$T = X + U$$

is equivalent to the pair (X, U), since with probability 1

$$X = [T], \qquad U = T - [T],$$

where $[T]$ denotes the largest integer $\leq T$. The distribution of T is continuous, and confidence bounds can be based on this statistic.

Example 3.5.2 (Binomial) An upper bound is required for a binomial probability p—for example, the probability that a batch of polio vaccine manufactured according to a certain procedure contains any live virus. Let $X_1, \ldots, X_n$ denote the outcome of n trials, X_i being 1 or 0 with probabilities p and q respectively, and let $X = \sum X_i$. Then $T = X + U$ has probability density

$$\binom{n}{[t]} p^{[t]} q^{n-[t]}, \qquad 0 \leq t < n + 1.$$

This satisfies the conditions of Corollary 3.5.1, and the upper confidence bound $\bar{p}$ is therefore the solution, if it exists, of the equation

$$P_p\{T < t\} = \alpha,$$

where t is the observed value of T. A solution does exist for all values $\alpha \leq t \leq n + \alpha$. For $n + \alpha < t$, the hypothesis $H(p_0) : p = p_0$ is accepted against the alternative $p < p_0$ for all values of p_0 and hence $\bar{p} = 1$. For $t < \alpha, H(p_0)$ is rejected for all values of p_0 and the confidence set $S(t)$ is therefore empty. Consider instead the sets $S^*(t)$ which are equal to $S(t)$ for $t \geq \alpha$ and which for $t < \alpha$ consist of the single point $p = 0$. They are also confidence sets at level $1 - \alpha$, since for all p,

$$P_p\{p \in S^*(T)\} \geq P_p\{p \in S(T)\} = 1 - \alpha.$$

On the other hand, $P_p\{p' \in S^*(T)\} = P_p\{p' \in S(T)\}$ for all $p' > 0$ and hence

$$P_p\{p' \in S^*(T)\} = P_p\{p' \in S(T)\} \qquad \text{for all} \quad p' > p.$$

Thus the family of sets $S^*(t)$ minimizes the probability of covering p' for all $p' > p$ at confidence level $1 - \alpha$. The associated confidence bound $\bar{p}^*(t) = \bar{p}(t)$ for $t \geq \alpha$ and $\bar{p}^*(t) = 0$ for $t < \alpha$ is therefore a uniformly most accurate upper confidence bound for p at level $1 - \alpha$.

In practice, so as to avoid randomization and obtain a bound not dependent on the extraneous variable U, one usually replaces T by $X + 1 = [T] + 1$. Since $\bar{p}^*(t)$ is a nondecreasing function of t, the resulting upper confidence bound $\bar{p}^*([t] + 1)$ is then somewhat larger than necessary; as a compensation it also gives a correspondingly higher probability of not falling below the true p.

References to tables for the confidence bounds and a careful discussion of various approximations can be found in Hall (1982) and Blyth (1984). Large sample approaches will be discussed in Example 11.2.7. ■

Let $\underline{\theta}$ and $\bar{\theta}$ be lower and upper bounds for θ with confidence coefficients $1-\alpha_1$ and $1-\alpha_2$, and suppose that $\underline{\theta}(x) < \bar{\theta}(x)$ for all x. This will be the case under the assumptions of Corollary 3.5.1 if $\alpha_1 + \alpha_2 < 1$. The intervals $(\underline{\theta}, \bar{\theta})$ are then *confidence intervals* for θ with confidence coefficient $1 - \alpha_1 - \alpha_2$; that is, they contain the true parameter value with probability $1 - \alpha_1 - \alpha_2$, since

$$P_\theta\{\underline{\theta} \le \theta \le \bar{\theta}\} = 1 - \alpha_1 - \alpha_2 \qquad \text{for all } \theta.$$

If $\underline{\theta}$ and $\bar{\theta}$ are uniformly most accurate, they minimize $E_\theta L_1(\theta, \underline{\theta})$ and $E_\theta L_2(\theta, \bar{\theta})$ at their respective levels for any function L_1 that is nonincreasing in $\underline{\theta}$ for $\underline{\theta} < \theta$ and 0 for $\underline{\theta} \ge \theta$ and any L_2 that is nondecreasing in $\bar{\theta}$ for $\bar{\theta} > \theta$ and 0 for $\bar{\theta} \le \theta$. Letting

$$L(\theta; \underline{\theta}, \bar{\theta}) = L_1(\theta, \underline{\theta}) + L_2(\theta, \bar{\theta}),$$

the intervals $(\underline{\theta}, \bar{\theta})$ therefore minimize $E_\theta L(\theta; \underline{\theta}, \bar{\theta})$ subject to

$$P_\theta\{\underline{\theta} > \theta\} \le \alpha_1, \qquad P_\theta\{\bar{\theta} < \theta\} \le \alpha_2.$$

An example of such a loss function is

$$L(\theta; \underline{\theta}, \bar{\theta}) = \begin{cases} \bar{\theta} - \underline{\theta} & \text{if} \quad \underline{\theta} \le \theta \le \bar{\theta}, \\ \bar{\theta} - \theta & \text{if} \quad \theta < \underline{\theta}, \\ \theta - \underline{\theta} & \text{if} \quad \bar{\theta} < \theta, \end{cases}$$

which provides a natural measure of the accuracy of the intervals. Other possible measures are the actual length $\bar{\theta} - \underline{\theta}$ of the intervals, or, for example, $a(\theta - \underline{\theta})^2 + b(\bar{\theta} - \theta)^2$, which gives an indication of the distance of the two end points form the true value.[7]

An important limiting case corresponds to the levels $\alpha_1 = \alpha_2 = \frac{1}{2}$. Under the assumptions of Corollary 3.5.1 and if the region of positive density is independent of θ so that tests of power 1 are impossible when $\alpha < 1$, the upper and lower confidence bounds $\bar{\theta}$ and $\underline{\theta}$ coincide in this case. The common bound satisfies

$$P_\theta\{\underline{\theta} \le \theta\} = P_\theta\{\underline{\theta} \ge \theta\} = \frac{1}{2},$$

and the estimate $\underline{\theta}$ of θ is therefore as likely to underestimate as to overestimate the true value. An estimate with this property is said to be *median unbiased.* (For the relation of this to other concepts of unbiasedness, see Problem 1.3.) It follows from the above result for arbitrary α_1 and α_2 that among all median unbiased estimates, $\underline{\theta}$ minimizes $EL(\theta, \underline{\theta})$ for any *monotone* loss function, that is, any loss function which for fixed θ has a minimum of 0 at $\underline{\theta} = \theta$ and is nondecreasing as $\underline{\theta}$ moves away from θ in either direction. By taking in particular $L(\theta, \underline{\theta}) = 0$ when $|\theta - \underline{\theta}| \le \triangle$ and $= 1$ otherwise, it is seen that among all median unbiased estimates, $\underline{\theta}$ minimizes the probability of differing from θ by more than any given amount; more generally it maximizes the probability

$$P_\theta\{-\triangle_1 \le \theta - \underline{\theta} < \triangle_2\}$$

for any $\triangle_1, \triangle_2 \ge 0$.

A more detailed assessment of the position of θ than that provided by confidence bounds or intervals corresponding to a fixed level $\gamma = 1 - \alpha$ is obtained by

[7]Proposed by Wolfowitz (1950).

stating confidence bounds for a number of levels, for example upper confidence bounds corresponding to values such as $\gamma = .05, .1, .25, .5, .75, .9, .95$. These constitute a set of *standard confidence bounds*,[8] from which different specific intervals or bounds can be obtained in the obvious manner.

3.6 A Generalization of the Fundamental Lemma

The following is useful extension of Theorem 3.2.1 to the case of more than one side condition.

Theorem 3.6.1 *Let $f_1, \dots, f_{m+1}$ be real-valued functions defined on a Euclidean space $\mathcal{X}$ and integrable μ, and suppose that for given constants $c_1, \dots, c_m$ there exists a critical function ϕ satisfying*

$$\int \phi f_i \, d\mu = c_i, \qquad i = 1, \dots, m. \tag{3.27}$$

Let $\mathcal{C}$ be the class of critical functions ϕ for which (3.27) holds.

(i) *Among all members of $\mathcal{C}$ there exists one that maximizes*

$$\int \phi f_{m+1} \, d\mu.$$

(ii) *A sufficient condition for a member of $\mathcal{C}$ to maximize*

$$\int \phi f_{m+1} \, d\mu$$

is the existence of constants $k_1, \dots, k_m$ such that

$$\begin{aligned} \phi(x) &= 1 \quad \text{when} \quad f_{m+1}(x) > \sum_{i=1}^{m} k_i f_i(x), \\ \phi(x) &= 0 \quad \text{when} \quad f_{m+1}(x) < \sum_{i=1}^{m} k_i f_i(x). \end{aligned} \tag{3.28}$$

(iii) *If a member of $\mathcal{C}$ satisfies (3.28) with $k_1, \dots, k_m \geq 0$, then it maximizes*

$$\int \phi f_{m+1} \, d\mu$$

among all critical functions satisfying

$$\int \phi f_i \, d\mu \leq c_i, \qquad i = 1, \dots, m. \tag{3.29}$$

(iv) *The set M of points in m-dimensional space whose coordinates are*

$$\left(\int \phi f_1 \, d\mu, \dots, \int \phi f_m \, d\mu \right)$$

[8] Suggested by Tukey (1949b).

for some critical function ϕ is convex and closed. If $(c_1, \ldots, c_m)$ is an inner point[9] *of M, then there exist constants $k_1, \ldots, k_m$ and a test ϕ satisfying* (3.27) *and* (3.28)*, and a necessary condition for a member of $\mathcal{C}$ to maximize*

$$\int \phi f_{m+1}\, d\mu$$

is that (3.28) *holds a.e. μ.*

Here the term "inner point of M" in statement (iv) can be interpreted as meaning a point interior to M relative to m-space or relative to the smallest linear space (of dimension $\leq m$) containing M. The theorem is correct with both interpretations but is stronger with respect to the latter, for which it will be proved.

We also note that exactly analogous results hold for the minimization of $\int \phi f_{m+1}\, d\mu$.

Proof. (i): Let $\{\phi_n\}$ be a sequence of functions in $\mathcal{C}$ such that $\int \phi_n f_{m+1}\, d\mu$ tends to $\sup_\phi \int \phi f_{m+1}\, d\mu$. By the weak compactness theorem for critical functions (Theorem 3.4.2 of the Appendix), there exists a subsequence $\{\phi_{n_i}\}$ and a critical function ϕ such that

$$\int \phi_{n_i} f_k\, d\mu \to \int \phi f_k\, d\mu \qquad \text{for} \quad k = 1, \cdots, m+1.$$

It follows that ϕ is in $\mathcal{C}$ and maximizes the integral with respect to $f_{m+1}\, d\mu$ within $\mathcal{C}$.

(ii) and (iii) are proved exactly as was part (ii) of Theorem 3.2.1.

(iv): That M is closed follows again from the weak compactness theorem, and its convexity is a consequence of the fact that if ϕ_1 and ϕ_2 are critical functions, so is $\alpha\phi_1 + (1-\alpha)\phi_2$ for any $0 \leq \alpha \leq 1$. If N (see Figure 3.2) is the totality of points in $(m+1)$-dimensional space with coordinates

$$\left(\int \phi f_1\, d\mu, \ldots, \int \phi f_{m+1}\, d\mu\right),$$

where ϕ ranges over the class of all critical functions, then N is convex and closed by the same argument. Denote the coordinates of a general point in M and N by $(u_1, \ldots, u_m)$ and $(u_1, \ldots, u_{m+1})$ respectively. The points of N, the first m coordinates of which are $c_1, \ldots, c_m$, form a closed interval $[c^*, c^{**}]$.

Assume first that $c^* < c^{**}$. Since $(c_1, \ldots, c_m, c^{**})$ is a boundary point of N, there exists a hyperplane $\prod$ through it such that every point on N lies below or on $\prod$. Let the equation of $\prod$ be

$$\sum_{i=1}^{m+1} k_i u_i = \sum_{i=1}^{m} k_i c_i + k_{m+1} c^{**}.$$

Since $(c_1, \ldots, c_m)$ is an inner point of M, the coefficient $k_{m+1} \neq 0$. To see this, let $c^* < c < c^{**}$, so that $(c_1, \ldots c_m, c)$ is an inner point of N. Then there exists a sphere with this point as center lying entirely in N and hence below $\prod$. It follows

[9] A discussion of the problem when this assumption is not satisfied is given by Dantzig and Wald (1951).

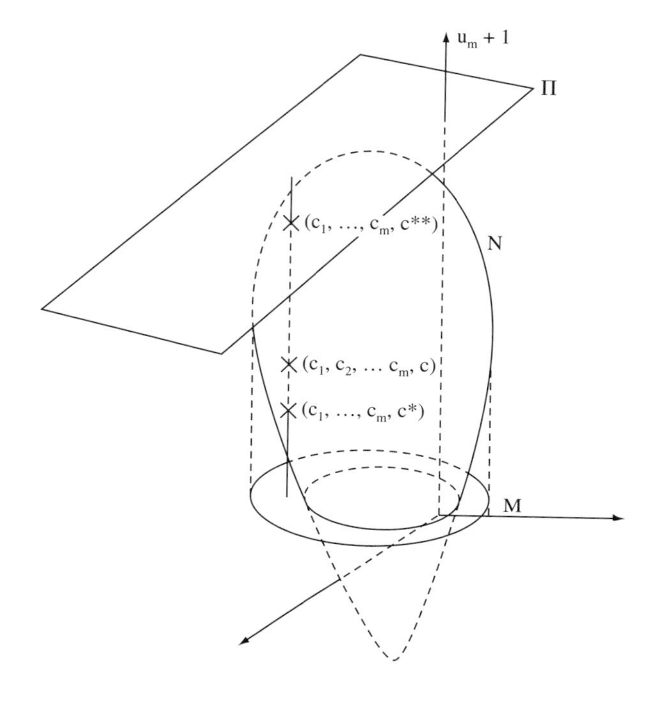

Figure 3.2.

that the point $(c_1, \ldots c_m, c)$ does not lie on $\prod$ and hence that $k_{m+1} \neq 0$. We may therefore take $k_{m+1} = -1$ and see that for any point of N

$$u_{m+1} - \sum_{i=1}^{m} k_i u_i \leq c_{m+1}^{**} - \sum_{i=1}^{m} k_i c_i.$$

That is, all critical functions ϕ satisfy

$$\int \phi \left(f_{m+1} - \sum_{i=1}^{m} k_i f_i \right) d\mu \leq \int \phi^{**} \left(f_{m+1} - \sum_{i=1}^{m} k_i f_i \right) d\mu,$$

where ϕ^{**} is the test giving rise to the point $(c_1, \ldots, c_m, c^{**})$. Thus ϕ^{**} is the critical function that maximizes the left-hand side of this inequality. Since the integral in question is maximized by putting ϕ equal to 1 when the integrand is positive and equal to 0 when it is negative, ϕ^{**} satisfies (3.28) a.e. μ.

If $c^* = c^{**}$, let $(c_1', \ldots, c_m')$ be any point of M other than $(c_1, \ldots, c_m)$. We shall show now that there exists exactly one real number c' such that $(c_1', \ldots, c_m', c')$ is in N. Suppose to the contrary that $(c_1', \ldots, c_m', \underline{c}')$and $(c_1', \ldots, c_m', \bar{c}')$ are both in N, and consider any point $(c_1'', \ldots, c_m'', c'')$ of N such that $(c_1, \ldots, c_m)$ is an interior point of the line segment joining $(c_1', \ldots, c_m')$ and $(c_1'', \ldots, c_m'')$. Such a point exists since $(c_1, \ldots, c_m)$ is an inner point of M. Then the convex set spanned by the three points $(c'_1, \ldots, c'_m, \underline{c}')$, $(c'_1, \ldots, c'_m, \bar{c}')$, and $(c''_1, \ldots, c''_m, c'')$ is contained in N and contains points $(c_1, \ldots, c_m, \underline{c})$ and $(c_1, \ldots, c_m, \bar{c})$ with $\underline{c} < \bar{c}$, which is a contradiction. Since N is convex, contains the origin, and has at most one point on any vertical line $u_1 = c'_1, \ldots, \quad u_m = c'_m$, it is contained in a hyperplane, which passes through the origin and is not parallel to the u_{m+1}-axis. It follows that

$$\int \phi f_{m+1}\, d\mu = \sum_{i=1}^{m} k_i \int \phi f_i\, d\mu$$

for all ϕ. This arises of course only in the trivial case that

$$f_{m+1} = \sum_{i=1}^{m} k_i f_i \qquad \text{a.e. } \mu,$$

and (3.28) is satisfied vacuously. ■

Corollary 3.6.1 *Let $p_1, \ldots, p_m, p_{m+1}$ be probability densities with respect to a measure μ, and let $0 < \alpha < 1$. Then there exists a test ϕ such that $E_i\phi(X) = \alpha$ $(i = 1, \ldots, m)$ and $E_{m+1}\phi(X) > \alpha$, unless $p_{m+1} = \sum_{i=1}^{m} k_i p_i$, a.e. μ.*

Proof. The proof will be by induction over m. For $m = 1$ the result reduces to Corollary 3.2.1. Assume now that it has been proved for any set of m distributions, and consider the case of $m + 1$ densities $p_1, \ldots, p_{m+1}$. If $p_1, \ldots, p_m$ are linearly dependent, the number of p_i can be reduced and the result follows from the induction hypothesis. Assume therefore that $p_1, \ldots, p_m$ are linearly independent. Then for each $j = 1, \ldots, m$ there exist by the induction hypothesis tests ϕ_j and ϕ'_j such that $E_i\phi_j(X) = E_i\phi'_j(X) = \alpha$ for all $i = 1, \ldots, j-1, j+1, \ldots, m$ and $E_j\phi_j(X) < \alpha < E_j\phi'_j(X)$. It follows that the point of m-space for which all m coordinates are equal to α is an inner point of M, so that Theorem 3.6.1(iv) is applicable. The test $\phi(x) \equiv \alpha$ is such that $E_i\phi(X) = \alpha$ for $i = 1, \ldots, m$. If among all tests satisfying the side conditions this one is most powerful, it has to satisfy (3.28). Since $0 < \alpha < 1$, this implies

$$p_{m+1} = \sum_{i=1}^{m} k_i p_i \qquad \text{a.e.}\mu,$$

as was to be proved. ■

The most useful parts of Theorems 3.2.1 and 3.6.1 are the parts (ii), which give sufficient conditions for a critical function to maximize an integral subject to certain side conditions. These results can be derived very easily as follows by the method of undetermined multipliers.

Lemma 3.6.1 *Let $F_1, \ldots, F_{m+1}$ be real-valued functions defined over a space U, and consider the problem of maximizing $F_{m+1}(u)$ subject to $F_i(u) = c_i$ $(i = 1, \ldots, m)$. A sufficient condition for a point u^0 satisfying the side conditions to be a solution of the given problem is that among all points of U it maximizes*

$$F_{m+1}(u) - \sum_{i=1}^{m} k_i F_i(u)$$

for some $k_1, \ldots, k_m$.

When applying the lemma one usually carries out the maximization for arbitrary k's, and then determines the constants so as to satisfy the side conditions.

PROOF. If u is any point satisfying the side conditions, then

$$F_{m+1}(u) - \sum_{i=1}^{m} k_i F_i(u) \le F_{m+1}(u^0) - \sum_{i=1}^{m} k_i F_i(u^0),$$

and hence $F_{m+1}(u) \le F_{m+1}(u^0)$.

As an application consider the problem treated in Theorem 3.6.1. Let U be the space of critical functions ϕ, and let $F_i(\phi) = \int \phi f_i \, d\mu$. Then a sufficient condition for ϕ to maximize $F_{m+1}(\phi)$, subject to $F_i(\phi) = c_i$, is that it maximizes $F_{m+1}(\phi) - \sum k_i F_i(\phi) = \int (f_{m+1} - \sum k_i f_i)\phi \, d\mu$. This is achieved by setting $\phi(x) = 1$ or 0 as $f_{m+1}(x) >$ or $< \sum k_i f_i(x)$. ■

3.7 Two-Sided Hypotheses

UMP tests exist not only for one-sided but also for certain two-sided hypotheses of the form

$$H : \theta \le \theta_1 \text{ or } \theta \ge \theta_2 \qquad (\theta_1 < \theta_2). \tag{3.30}$$

This problem arises when trying to demonstrate equivalence (or sometimes called bioequivalence) of treatments; for example, a new drug may be declared equivalent to the current standard drug if the difference in therapeutic effect is small, meaning θ is a small interval about 0. Such testing problems also occur when one wishes to determine whether given specifications have been met concerning the proportion of an ingredient in a drug or some other compound, or whether a measuring instrument, for example a scale, is properly balanced. One then sets up the hypothesis that θ does not lie within the required limits, so that an error of the first kind consists in declaring θ to be satisfactory when in fact it is not. In practice, the decision to accept H will typically be accompanied by a statement of whether θ is believed to be $\le \theta_1$ or $\ge \theta_2$. The implications of H are, however, frequently sufficiently important so that acceptance will in any case be followed by a more detailed investigation. If a manufacturer tests each precision instrument before releasing it and the test indicates an instrument to be out of balance, further work will be done to get it properly adjusted. If in a scientific investigation the inequalities $\theta \le \theta_1$ and $\theta \ge \theta_2$ contradict some assumptions that have been formulated, a more complex theory may be needed and further experimentation will be required. In such situations there may be only two basic choices, to act as if $\theta_1 < \theta < \theta_2$ or to carry out some further investigation, and the formulation of the problem as that of testing the hypothesis H may be appropriate. In the present section, the existence of a UMP test of H will be proved for one-parameter exponential families.

Theorem 3.7.1 (i) *For testing the hypothesis $H : \theta \le \theta_1$ or $\theta \ge \theta_2$ $(\theta_1 < \theta_2)$ against the alternatives $K : \theta_1 < \theta < \theta_2$ in the one-parameter exponential family*

(3.19) *there exists a UMP test given by*

$$\phi(x) = \begin{cases} 1 & \text{when} \quad C_1 < T(x) < C_2 \quad (C_1 < C_2), \\ \gamma_i & \text{when} \quad T(x) = C_i, \quad i = 1, 2, \\ 0 & \text{when} \quad T(x) < C_1 \text{ or} > C_2, \end{cases} \tag{3.31}$$

where the C's and γ's are determined by

$$E_{\theta_1}\phi(X) = E_{\theta_2}\phi(X) = \alpha. \tag{3.32}$$

(ii) *This test minimizes $E_\theta\phi(X)$ subject to (3.32) for all $\theta < \theta_1$ and $> \theta_2$.*

(iii) *For $0 < \alpha < 1$ the power function of this test has a maximum at a point θ_0 between θ_1 and θ_2 and decreases strictly as θ tends away from θ_0 in either direction, unless there exist two values t_1, t_2 such that $P_\theta\{T(X) = t_1\} + P_\theta\{T(X) = t_2\} = 1$ for all θ.*

PROOF. (i): One can restrict attention to the sufficient statistic $T = T(X)$, the distribution of which by Lemma 2.7.2 is

$$dP_\theta(t) = C(\theta)e^{Q(\theta)t}d\nu(t),$$

where $Q(\theta)$ is assumed to be strictly increasing. Let $\theta_1 < \theta' < \theta_2$, and consider first the problem of maximizing $E_{\theta'}\psi(T)$ subject to (3.32) with $\phi(x) = \psi[T(x)]$. If M denotes the set of all points $E_{\theta_1}\psi(T), E_{\theta_2}\psi(T))$ as ψ ranges over the totality of critical functions, then the point (α, α) is an inner point of M. This follows from the fact that by Corollary 3.2.1 the set M contains points (α, u_1) and (α, u_2) with $u_1 < \alpha < u_2$ and that it contains all points (u, u) with $0 < u < 1$. Hence by part (iv) of Theorem 3.6.1 there exist constants k_1, k_2 and test $\psi_0(t)$ and that $\phi_0(x) = \psi_0[T(x)]$ satisfies (3.32) and that $\psi_0(t) = 1$ when

$$k_1C(\theta_1)e^{Q(\theta_1)t} + k_2C(\theta_2)e^{Q(\theta_2)t} < C(\theta')e^{Q(\theta')t}$$

and therefore when

$$a_1e^{b_1t} + a_2e^{b_2t} < 1 \qquad (b_1 < 0 < b_2),$$

and $\psi_0(t) = 0$ when the left-hand side is > 1. Here the a's cannot both be ≤ 0, since then the test would always reject. If one of the a's is ≤ 0 and the other one is > 0, then the left-hand side is strictly monotone, and the test is of the one-sided type considered in Corollary 3.4.1, which has a strictly monotone power function and hence cannot satisfy (3.32). Since therefore both a's are positive, the test satisfies (3.31). It follows from Lemma 3.7.1 below that the C's and γ's are uniquely determined by (3.31) and (3.32), and hence from Theorem 3.6.1(iii) that the test is UMP subject to the weaker restriction $E_{\theta_i}\psi(T) \leq \alpha$ $(i = 1, 2)$. To complete the proof that this test is UMP for testing H, it is necessary to show that it satisfies $E_\theta\psi(T) \leq \alpha$ for $\theta \leq \theta_1$ and $\theta \geq \theta_2$. This follows from (ii) by comparison with the test $\psi(t) \equiv \alpha$.

(ii): Let $\theta' < \theta_1$, and apply Theorem 3.6.1(iv) to minimize $E_{\theta'}\phi(X)$ subject to (3.32). Dividing through by $e^{Q(\theta_1)t}$, the desired test is seen to have a rejection region of the form

$$a_1e^{b_1t} + a_2e^{b_2t} < 1 \qquad (b_1 < 0 < b_2).$$

Thus it coincides with the test $\psi_0(t)$ obtained in (i). By Theorem 3.6.1(iv) the first and third conditions of (3.31) are also necessary, and the optimum test is therefore unique provided $P\{T = C_i\} = 0$.

(iii): Without loss of generality let $Q(\theta) = \theta$. It follows from (i) and the continuity of $\beta(\theta) = E_\theta\phi(X)$ that either $\beta(\theta)$ satisfies (iii) or there exist three points $\theta' < \theta'' < \theta'''$ such that $\beta(\theta'') \le \beta(\theta') = \beta(\theta''') = c$, say. Then $0 < c < 1$, since $\beta(\theta') = 0$ (or 1) implies $\phi(t) = 0$ (or 1) a.e. ν and this is excluded by (3.32). As is seen by the proof of (i), the test minimizes $E_{\theta''}\phi(X)$ subject to $E_{\theta'}\phi(X) = E_{\theta'''}\phi(X) = c$ for all $\theta' < \theta'' < \theta'''$. However, unless T takes on at most two values with probability 1 or all $\theta, p_{\theta'}, p_{\theta''}, p_{\theta'''}$ are linearly independent, which by Corollary 3.6.1 implies $\beta(\theta'') > c$. ■

In order to determine the C's and γ's, one will in practice start with some trial values C_1^*, γ_1^*, find C_2^*, γ_2^* such that $\beta^*(\theta_1) = \alpha$, and compute $\beta^*(\theta_2)$, which will usually be either too large or too small. For the selection of the next trial values it is then helpful to note that if $\beta^*(\theta_2) < \alpha$, the correct acceptance region is to the right of the one chosen, that is, it satisfies either $C_1 > C_1^*$ or $C_1 = C_1^*$ and $\gamma_1 < \gamma_1^*$, and that the converse holds if $\beta^*(\theta_2) > \alpha$. This is a consequence of the following lemma.

Lemma 3.7.1 *Let $p_\theta(x)$ satisfy the assumptions of Lemma 3.4.2(iv).*

(i) *If ϕ and ϕ^* are two tests satisfying* (3.31) *and $E_{\theta_1}\phi(T) = E_{\theta_1}\phi^*(T)$, and if ϕ^* is to the right of ϕ, then $\beta(\theta) <$ or $> \beta^*(\theta)$ as $\theta > \theta_1$ or $< \theta_1$.*

(ii) *If ϕ and ϕ^* satisfy* (3.31) *and* (3.32), *then $\phi = \phi^*$ with probability one .*

PROOF. (i): The result follows from Lemma 3.4.2(iv) with $\psi = \phi^* - \phi$. (ii): Since $E_{\theta_1}\phi(T) = E_{\theta_1}\phi^*(T)$, ϕ^* lies either to the left or the right of ϕ, and application of (i) completes the proof.

Although a UMP test exists for testing that $\theta \le \theta_1$ or $\ge \theta_2$ in an exponential family, the same is not true for the dual hypothesis $H : \theta_1 \le \theta \le \theta_2$ or for testing $\theta = \theta_0$ (Problem 3.54). There do, however, exist UMP unbiased tests of these hypotheses, as will be shown in Chapter 4. ■

3.8 Least Favorable Distributions

It is a consequence of Theorem 3.2.1 that there always exists a most powerful test for testing a simple hypothesis against a simple alternative. More generally, consider the case of a Euclidean sample space; probability densities $f_\theta, \theta \in \omega$, and g with respect to a measure μ; and the problem of testing $H : f_\theta, \theta \in \omega$, against the simple alternative $K : g$. The existence of a most powerful level α test then follows from the weak compactness theorem for critical functions (Theorem A.5.1 of the Appendix) as in Theorem 3.6.1(i).

Theorem 3.2.1 also provides an explicit construction for the most powerful test in the case of a simple hypothesis. We shall now extend this theorem to composite hypotheses in the direction of Theorem 3.6.1 by the method of undetermined multipliers. However, in the process of extension the result becomes much less explicit. Essentially it leaves open the determination of the multipliers, which now take the form of an arbitrary distribution. In specific problems this usually still involves considerable difficulty.

From another point of view the method of attack, as throughout the theory of hypothesis testing, is to reduce the composite hypothesis to a simple one. This is achieved by considering weighted averages of the distributions of H. The composite hypothesis H is replaced by the simple hypothesis H_Λ that the probability density of X is given by

$$h_\Lambda(x) = \int_\omega f_\theta(x)\, d\Lambda(\theta),$$

where Λ is a probability distribution over ω. The problem of finding a suitable Λ is frequently made easier by the following consideration. Since H provides no information concerning θ and since H_Λ is to be equivalent to H for the purpose of testing against g, knowledge of the distribution Λ should provide as little help for this task as possible. To make this precise suppose that θ is known to have a distribution Λ. Then the maximum power β_Λ that can be attained against g is that of the most powerful test ϕ_Λ for testing H_Λ against g. The distribution Λ is said to be *least favorable* (at level α) if for all Λ' the inequality $\beta_\Lambda \le \beta_{\Lambda'}$ holds.

Theorem 3.8.1 *Let a σ-field be defined over ω such that the densities $f_\theta(x)$ are jointly measurable in θ and x. Suppose that over this σ-field there exist a probability distribution Λ such that the most powerful level-α test ϕ_Λ for testing H_Λ against g is of size $\le \alpha$ also with respect to the original hypothesis H.*

(i) *The test ϕ_Λ is most powerful for testing H against g.*

(ii) *If ϕ_Λ is the unique most powerful level-α for testing H_Λ against g, it is also the unique most powerful test of H against g.*

(iii) *The distribution Λ is least favorable.*

Proof. We note first that h_Λ is again a density with respect to μ, since by Fubini's theorem (Theorem 2.2.4)

$$\int h_\Lambda(x)\, d\mu(x) = \int_\omega d\Lambda(\theta) \int f_\theta(x)\, d\mu(x) = \int_\omega d\Lambda(\theta) = 1.$$

Suppose that ϕ_Λ is a level-α test for testing H, and let ϕ^* be any other level-α test. Then since $E_\theta\phi^*(X) \le \alpha$ for all $\theta \in \omega$, we have

$$\int \phi^*(x) h_\Lambda(x)\, d\mu(x) = \int_\omega E_\theta\phi^*(X) d\Lambda(\theta) \le \alpha.$$

Therefore $\phi*$ is a level-α test also for testing H_Λ and its power cannot exceed that of ϕ_Λ. This proves (i) and (ii). If Λ' is any distribution, it follows further that ϕ_Λ is a level-α test also for testing $H_{\Lambda'}$, and hence that its power against g cannot exceed that of the most powerful test, which by definition is $\beta_{\Lambda'}$. ■

The conditions of this theorem can be given a somewhat different form by noting that ϕ_Λ can satisfy $\int_\omega E_\theta\phi_\Lambda(X)\, d\Lambda(\theta) = \alpha$ and $E_\theta\phi_\Lambda(X) \le \alpha$ for all $\theta \in \omega$ only if the set of θ's with $E_\theta\phi_\Lambda(X) = \alpha$ has Λ-measure one.

Corollary 3.8.1 *Suppose that Λ is a probability distribution over ω and that ω' is a subset of ω with $\Lambda(\omega') = 1$. Let ϕ_Λ be a test such that*

$$\phi_\Lambda(x) = \begin{cases} 1 & \text{if} \quad g(x) > k \int f_\theta(x)\, d\Lambda(\theta), \\ 0 & \text{if} \quad g(x) < k \int f_\theta(x)\, d\Lambda(\theta). \end{cases} \tag{3.33}$$

Then ϕ_Λ is a most powerful level-α for testing H against g provided

$$E_{\theta'}\phi_\Lambda(X) = \sup_{\theta\in\omega} E_\theta\phi_\Lambda(X) = \alpha \qquad \text{for} \quad \theta' \in \omega'. \tag{3.34}$$

Theorems 3.4.1 and 3.7.1 constitute two simple applications of Theorem 3.8.1. The set ω' over which the least favorable distribution Λ is concentrated consists of the single point θ_0 in the first of these examples and of the two points θ_1 and θ_2 in the second. This is what one might expect, since in both cases these are the distributions of H that appear to be "closest" to K. Another example in which the least favorable distribution is concentrated is at a single point is the following.

Example 3.8.1 (Sign test) The quality of items produced by a manufacturing process is measured by a characteristic X such as the tensile strength of a piece of material, or the length of life or brightness of a light bulb. For an item to be satisfactory X must exceed a given constant u, and one wishes to test the hypothesis $H : p \geq p_0$, where

$$p = P\{X \leq u\}$$

is the probability of an item being defective. Let $X_1, \ldots, X_n$ be the measurements of n sample items, so that the X's are independently distributed with common distribution about which no knowledge is assumed. Any distribution on the real line can be characterized by the probability p together with the conditional probability distributions P_- and P_+ of X given $X \leq u$ and $X > u$ respectively. If the distributions P_- and P_+ have probability densities p_- and $p+$, for example with respect to $\mu = P_- + P_+$, then the joint density of $X_1, \ldots, X_n$ at a sample point $x_1, \ldots, x_n$ satisfying

$$x_{i_1}, \ldots, x_{i_m} \leq u < x_{j_1}, \ldots, x_{j_{n-m}}$$

is

$$p^m(1-p)^{n-m} p_-(x_{i_1}) \cdots p_-(x_{i_m}) p_+(x_{j_1}) \cdots p_+(x_{j_{n-m}}).$$

Consider now a fixed alternative to H, say (p_1, P_-, P_+), with $p_1 < p_0$. One would then expect the least favorable distribution Λ over H to assign probability 1 to the distribution (p_0, P_-, P_+) since this appears to be closest to the selected alternative. With this choice of Λ, the test (3.33) becomes

$$\phi_\Lambda(x) = 1 \text{ or } 0 \qquad \text{as} \quad \left(\frac{p_1}{p_0}\right)^m \left(\frac{q_1}{q_0}\right)^{n-m} > \text{ or } < C,$$

and hence as $m <$ or $> C$. The test therefore rejects when the number M of defectives is sufficiently small, or more precisely, when $M < C$ and with probability γ when $M = C$, where

$$P\{M < C\} + \gamma P\{M = C\} = \alpha \qquad \text{for} \quad p = p_0. \tag{3.35}$$

The distribution of M is the binomial distribution $b(p, n)$, and does not depend on P_+ and P_-. As a consequence, the power function of the test depends only on p and is a decreasing function of p, so that under H it takes on its maximum for $p = p_0$. This proves Λ to be least favorable and ϕ_Λ to be most powerful. Since the test is independent of the particular alternative chosen, it is UMP.

Expressed in terms of the variables $Z_i = X_i - u$, the test statistic M is the number of variables ≤ 0, and the test is the so-called *sign test* (cf. Section 4.9). It is an example of a *nonparametric* test, since it is derived without assuming a given functional form for the distribution of the X's such as the normal, uniform, or Poisson, in which only certain parameters are unknown .

The above argument applies, with only the obvious modifications, to the case that an item satisfactory if X lies within certain limits: $u < X < v$. This occurs, for example, if X is the length of a metal part or the proportion of an ingredient in a chemical compound, for which certain tolerances have been specified. More generally the argument applies also to the situation in which X is vector-valued. Suppose that an item is satisfactory only when X lies in a certain set S, for example, if all the dimensions of a metal part or the proportions of several ingredients lie within specified limits. The probability of a defective is then

$$p = P\{X \in S^c\},$$

and P_- and P_+ denote the conditional distributions of X given $X \in S$ and $X \in S^c$ respectively. As before, there exists a UMP test of $H : p \geq p_0$, and it rejects H when the number M of defectives is sufficiently small, with the boundary of the test being determined by (3.35). ■

A distribution Λ satisfying the conditions of Theorem 3.8.1 exists in most of the usual statistical problems, and in particular under the following assumptions. Let the sample space be Euclidean, let ω be a closed Borel set in s-dimensional Euclidean space, and suppose that $f_\theta(x)$ is a continuous function of θ for almost all x. Then given any g there exists a distribution Λ satisfying the conditions of Theorem 3.8.1 provided

$$\lim_{n\to\infty} \int_S f_{\theta_n}(x)\, d\mu(x) = 0$$

for every bounded set S in the sample space and for every sequence of vectors θ_n whose distance from the origin tends to infinity.

From this it follows as did Corollaries 1 and 4 from Theorems 3.2.1 and 3.6.1, that if the above conditions hold and if $0 < \alpha < 1$, there exists a test of power $\beta > \alpha$ for testing $H : f_\theta, \theta \in \omega$, against g unless $g = \int f_\theta\, d\Lambda(\theta)$ for some Λ. An example of the latter possibility is obtained by letting f_θ and g be the normal densities $N(\theta, \sigma_0^2)$ and $N(0, \sigma_1^2)$ respectively with $\sigma_0^2 < \sigma_1^2$. (See the following section.)

The above and related results concerning the existence and structure of least favorable distributions are given in Lehmann (1952b) (with the requirement that ω be closed mistakenly omitted), in Reinhardt (1961), and in Krafft and Witting (1967), where the relation to linear programming is explored.

3.9 Applications to Normal Distributions

3.9.1 Univariate Normal Models

Because of their wide applicability, the problems of testing the mean ξ and variance σ^2 of a normal distribution are of particular importance. Here and in similar problems later, the parameter not being tested is assumed to be unknown, but will not be shown explicitly in a statement of the hypothesis. We shall write, for example, $\sigma \leq \sigma_0$ instead of the more complete statement $\sigma \leq \sigma_0, -\infty < \xi < \infty$.

The standard (likelihood-ratio) tests of the two hypotheses $\sigma \le \sigma_0$ and $\xi \le \xi_0$ are given by the rejection regions

$$\sum (x_i - \bar{x})^2 \ge C \tag{3.36}$$

and

$$\frac{\sqrt{n}(\bar{x} - \xi_0)}{\sqrt{\frac{1}{n-1}\sum (x_i - \bar{x})^2}} \ge C. \tag{3.37}$$

The corresponding tests for the hypotheses $\sigma \ge \sigma_0$ and $\xi \ge \xi_o$ are obtained from the rejection regions (3.36) and (3.37) by reversing the inequalities. As will be shown in later chapters, these four tests are UMP both within the class of unbiased and within the class of invariant test (but see Section 11.3 for problems arising when the assumption of normality does not hold exactly). However, at the usual significance levels only the first of them is actually UMP.

Example 3.9.1 (One-sided tests of variance.) Let $X_1, \ldots, X_n$ be a sample from $N(\xi, \sigma^2)$, and consider first the hypotheses $H_1 : \sigma \ge \sigma_0$ and $H_2 : \sigma \le \sigma_0$, and a simple alternative $K : \xi = \xi_1, \sigma = \sigma_1$. It seems reasonable to suppose that the least favorable distribution Λ in the (ξ, σ)-plane is concentrated on the line $\sigma = \sigma_0$. Since $Y = \sum X_i/n = \bar{X}$ and $U = \sum (X_i - \bar{X})^2$ are sufficient statistics for the parameters (ξ, σ), attention can be restricted to these variables. Their joint density under H_Λ is

$$C_o u^{(n-3)/2} \exp\left(-\frac{u}{2\sigma_0^2}\right) \int \exp\left[-\frac{n}{2\sigma_o^2}(y - \xi)^2\right] \, d\Lambda(\xi),$$

while under K it is

$$C_1 u^{(n-3)/2} \exp\left(-\frac{u}{2\sigma_1^2}\right) \exp\left[-\frac{n}{2\sigma_1^2}(y - \xi_1)^2\right].$$

The choice of Λ is seen to affect only the distribution of Y. A least favorable Λ should therefore have the property that the density of Y under H_Λ,

$$\int \frac{\sqrt{n}}{\sqrt{2\pi\sigma_0^2}} \exp\left[-\frac{n}{2\sigma_0^2}(y - \xi)^2\right] \, d\Lambda(\xi),$$

comes as close as possible to the alternative density,

$$\frac{\sqrt{n}}{\sqrt{2\pi\sigma_1^2}} \exp\left[-\frac{n}{2\sigma_1^2}(y - \xi_1)^2\right].$$

At this point one must distinguish between H_1 and H_2. In the first case $\sigma_1 < \sigma_0$. By suitable choice of Λ the mean of Y can be made equal to ξ_1, but the variance will if anything be increased over its initial value σ_0^2. This suggests that the least favorable distribution assigns probability 1 to the point $\xi = \xi_1$, since in this way the distribution of Y is normal both under H and K with the same mean in both cases and the smallest possible difference between the variances. The situation is somewhat different for H_2, for which $\sigma_0 < \sigma_1$. If the least favorable distribution Λ has a density, say Λ', the density of Y under H_Λ becomes

$$\int_{-\infty}^{\infty} \frac{\sqrt{n}}{\sqrt{2\pi}\sigma_0} \exp\left[-\frac{n}{2\sigma_0^2}(y - \xi)^2\right] \Lambda'(\xi) \, d\xi.$$

This is the probability density of the sum of two independent random variables, one distributed as $N(0, \sigma_0^2/n)$ and the other with density $\Lambda'(\xi)$. If Λ is taken to be $N(\xi_1, (\sigma_1^2 - \sigma_0^2)/n)$, the distribution of Y under H_Λ becomes $N(\xi_1, \sigma_1^2/n)$, the same as under K.

We now apply Corollary 3.8.1 with the distributions Λ suggested above. For H_1 it is more convenient to work with the original variables than with Y and U. Substitution in (3.33) gives $\phi(x) = 1$ when

$$\frac{(2\pi\sigma_1^2)^{-n/2} \exp\left[-\frac{1}{2\sigma_1^2}\sum(x_i - \xi_1)^2\right]}{(2\pi\sigma_0^2)^{-n/2} \exp\left[-\frac{1}{2\sigma_0^2}\sum(x_i - \xi_1)^2\right]} > C,$$

that is, when

$$\sum(x_i - \xi_1)^2 \le C. \tag{3.38}$$

To justify the choice of Λ, one must show that

$$P\left\{\sum(X_i - \xi_1)^2 \le C | \xi, \sigma\right\}$$

takes on its maximum over the half plane $\sigma \ge \sigma_0$ at the point $\xi = \xi_1$, $\sigma = \sigma_0$. For any fixed σ, the above is the probability of the sample point falling in a sphere radius, computed under the assumption that the X's are independently distributed as $N(\xi, \sigma^2)$. This probability is maximized when the center of the sphere coincides with that of the distribution that is, when $\xi = \xi_1$. (This follows for example from Problem 7.15.) The probability then becomes

$$P\left\{\sum\left(\frac{x_i - \xi_1}{\sigma}\right)^2 \le \frac{C}{\sigma^2} \middle| \xi_1, \sigma\right\} = P\left\{\sum V_i^2 \le \frac{C}{\sigma^2}\right\},$$

where $V_1, \ldots, V_n$ are independently distributed as $N(0,1)$. This is a decreasing function of σ and therefore takes on its maximum when $\sigma = \sigma_0$.

In the case of H_2, application of Corollary 3.8.1 to the sufficient statistics (Y, U) gives $\phi(y, u) = 1$ when

$$\begin{aligned} &\frac{C_1 u^{(n-3)/2} \exp\left(-\frac{u}{2\sigma_1^2}\right) \exp\left[-\frac{n}{2\sigma_1^2}(y - \xi_1)^2\right]}{C_0 u^{(n-3)/2} \exp\left(-\frac{u}{2\sigma_0^2}\right) \int \exp\left[-\frac{n}{2\sigma_0^2}(y - \xi)^2\right] \Lambda'(\xi)\, d\xi} \\ &\qquad = C' \exp\left[-\frac{u}{2}\left(\frac{1}{\sigma_1^2} - \frac{1}{\sigma_0^2}\right)\right] \ge C, \end{aligned}$$

that is, when

$$u = \sum(x_i - \bar{x})^2 \ge C. \tag{3.39}$$

Since the distribution of $\sum(X_i - \bar{X})^2/\sigma^2$ does not depend on ξ or σ, the probability $P\{\sum(X_i - \bar{X})^2 \ge C \mid \xi, \sigma\}$ is independent of ξ and increases with σ, so that the conditions of Corollary 3.8.1 are satisfied. The test (3.39), being independent of ξ_1 and σ_1, is UMP for testing $\sigma \le \sigma_0$ against $\sigma > \sigma_0$. It is also seen to coincide with the likelihood-ratio test (3.36). On the other hand, the most powerful test (3.38) for testing $\sigma \ge \sigma_0$ against $\sigma < \sigma_0$ does depend on the value ξ_1 of ξ under the alternative.

It has been tacitly assumed so far that $n > 1$. If $n = 1$, the argument applies without change with respect to H_1, leading to (3.38) with $n = 1$. However, in the discussion of H_2 the statistic U now drops out, and Y coincides with the single observation X. Using the same Λ as before, one sees that X has the same distribution under H_Λ as under K, and the test ϕ_Λ therefore becomes $\phi_\Lambda(x) \equiv \alpha$. This satisfies the conditions of Corollary 3.8.1 and is therefore the most powerful test for the given problem. It follows that a single observation is of no value for testing the hypothesis H_2, as seems intuitively obvious, but that it could be used to test H_1 if the class of alternatives were sufficiently restricted. ∎

The corresponding derivation for the hypothesis $\xi \leq \xi_0$ is less straightforward. It turns out[10] that Student's test given by (3.37) is most powerful if the level of significance α is $\geq \frac{1}{2}$, regardless of the alternative $\xi_1 > \xi_0$, σ_1. This test is therefore UMP for $\alpha \geq \frac{1}{2}$. On the other hand, when $\alpha < \frac{1}{2}$ the most powerful test of H rejects when $\sum(x_i - a)^2 \leq b$, where the constants a and b depend on the alternative (ξ_1, σ_1) and on α. Thus for the significance levels that are of interest, a UMP test of H does not exist. No new problem arises for the hypothesis $\xi \geq \xi_0$, since this reduces to the case just considered through the transformation $Y_i = \xi_0 - (X_i - \xi_0)$.

3.9.2 Multivariate Normal Models

Let X denote a $k \times 1$ random vector whose ith component, X_i, is a real-valued random variable. The mean of X, denoted $E(X)$, is a vector with ith component $E(X_i)$ (assuming it exists). The covariance matrix of X, denoted Σ, is the $k \times k$ matrix with (i, j) entry $Cov(X_i, X_j)$. Σ is well-defined iff $E(|X|^2) < \infty$, where $|\cdot|$ denotes the Euclidean norm. Note that, if A is an $m \times k$ matrix, then the $m \times 1$ vector $Y = AX$ has mean (vector) $AE(X)$ and covariance matrix $A\Sigma A^T$, where A^T is the transpose of A (Problem 3.63).

The multivariate generalization of a real-valued normally distributed random variable is a random vector $X = (X_1, \ldots, X_k)^T$ with the *multivariate normal probability density*

$$\frac{\sqrt{|A|}}{(2\pi)^{\frac{1}{2}k}} \exp\left[-\tfrac{1}{2}\sum\sum a_{ij}(x_i - \xi_i)(x_j - \xi_j)\right], \tag{3.40}$$

where the matrix $A = (a_{ij})$ is positive definite, and $|A|$ denotes its determinant. The means and covariance matrix of the X's are given by

$$E(X_i) = \xi_i, \qquad E(X_i - \xi_i)(X_j - \xi_j) = \sigma_{ij}, \quad (\sigma_{ij}) = A^{-1}. \tag{3.41}$$

The column vector $\xi = (\xi_1, \ldots, \xi_k)^T$ is the mean vector and $\Sigma = A^{-1}$ is the covariance matrix of X.

Such a definition only applies when A is nonsingular, in which case we say that X has a nonsingular multivariate normal distribution. More generally, we say that Y has a multivariate normal distribution if $Y = BX + \mu$ for some $m \times k$ matrix of constants B and $m \times 1$ constant vector μ, where X has some nonsingular multivariate normal distribution. Then, Y is multivariate normal if and only if

[10] See Lehmann and Stein (1948)

$\sum_{i=1}^m c_i Y_i$ is univariate normal, if we interpret $N(\xi, \sigma^2)$ with $\sigma = 0$ to be the distribution that is point mass at ξ. Basic properties of the multivariate normal distribution are given in Anderson (2003).

Example 3.9.2 (One-sided tests of a combination of means.) Assume X is multivariate normal with unknown mean $\xi = (\xi_1, \ldots, \xi_k)^T$ and known covariance matrix Σ. Assume $a = (a_1, \ldots, a_k)^T$ is a fixed vector with $a^T \Sigma a > 0$. The problem is to test

$$H: \sum_{i=1}^k a_i \xi_i \le \delta \quad \text{vs.} \quad K: \sum_{i=1}^k a_k \xi_i > \delta .$$

We will show that a UMP level α test exists, which rejects when $\sum_i a_i X_i > \sigma z_{1-\alpha}$, where $\sigma^2 = a^T \Sigma a$. To see why,[11] we will consider four cases of increasing generality.

Case 1. If $k = 1$ and the problem is to test the mean of X_1, the result follows by Problem 3.1.

Case 2. Consider now general k, so that $(X_1, \ldots, X_k)$ has mean $(\xi_1, \ldots, \xi_k)$ and covariance matrix Σ. However, consider the special case $(a_1, \ldots, a_k) = (1, 0, \ldots, 0)$. Also, assume X_1 and $(X_2, \ldots, X_k)$ are independent. Then, for any fixed alternative $(\xi_1', \ldots, \xi_k')$ with $\xi_1' > \delta$, the least favorable distribution concentrates on the single point $(\delta, \xi_2', \ldots, \xi_k')$ (Problem 3.65).

Case 3. As in case 2, consider $a_1 = 1$ and $a_i = 0$ if $i > 1$, but now allow Σ to be an arbitrary covariance matrix. We can reduce the problem to case 2 by an appropriate linear transformation. Simply let $Y_1 = X_1$ and, for $i > 1$, let

$$Y_i = X_i - \frac{Cov(X_1, X_i)}{Var(X_1)} X_1 .$$

Then, it is easily checked that $Cov(Y_1, Y_i) = 0$ if $i > 1$. Moreover, Y is just a 1:1 transformation of X. But, the problem of testing $E(Y_1) = E(X_1)$ based on $Y = (Y_1, \ldots, Y_k)$ is in the form already studied in case 2, and the UMP test rejects for large values of $Y_1 = X_1$.

Case 4. Now, consider arbitrary $(a_1, \ldots, a_k)$ satisfying $a^T \Sigma a > 0$. Let $Z = OX$, where O is any orthogonal matrix with first row $(a_1, \ldots, a_k)$. Then, $E(Z_1) = \sum_{i=1}^k a_i \xi_i$, and the problem of testing $E(Z_1) \le \delta$ versus $E(Z_1) > \delta$ reduces to case 3. Hence, the UMP test rejects for large values of $Z_1 = \sum_{i=1}^k a_i X_i$. ■

Example 3.9.3 (Equivalence tests of a combination of means.) As in Example 3.9.2, assume X is multivariate normal $N(\xi, \Sigma)$ with unknown mean vector ξ and known covariance matrix Σ. Fix $\delta > 0$ and any vector $a = (a_1, \ldots, a_k)^T$ satisfying $a^T \Sigma a > 0$. Consider testing

$$H: |\sum_{i=1}^k a_i \xi_i| \ge \delta \quad vs \quad K: |\sum_{i=1}^k a_i \xi_i| < \delta .$$

[11]Proposition 15.2 of van der Vaart (1998) provides an alternative proof in the case Σ is invertible.

Then, a UMP level α test also exists and it rejects H if

$$|\sum_{i=1}^{k} a_i X_i| < C ,$$

where $C = C(\alpha, \delta, \sigma)$ satisfies

$$\Phi\left(\frac{C-\delta}{\sigma}\right) - \Phi\left(\frac{-C-\delta}{\sigma}\right) = \alpha \tag{3.42}$$

and $\sigma^2 = a^T \Sigma a$. Hence, the power of this test against an alternative $(\xi_1, \ldots, \xi_k)$ with $|\sum_i a_i \xi_i| = \delta' < \delta$ is

$$\Phi\left(\frac{C-\delta'}{\sigma}\right) - \Phi\left(\frac{-C-\delta'}{\sigma}\right) .$$

To see why, we again consider four cases of increasing generality.

Case 1. Suppose $k = 1$, so that $X_1 = X$ is $N(\xi, \sigma^2)$ and we are testing $|\xi| \geq \delta$ versus $|\xi| < \delta$. (This case follows by Theorem 3.7.1, but we argue independently so that the argument applies to the other cases as well.) Fix an alternative $\xi = m$ with $|m| < \delta$. Reduce the composite null hypothesis to a simple one via a least favorable distribution that places mass p on $N(\delta, \sigma^2)$ and mass $1-p$ on $N(-\delta, \sigma^2)$. The value of p will be chosen shortly so that such a distribution is least favorable (and will be seen to depend on m, α, σ and δ). By the Neyman Pearson Lemma, the MP test of

$$pN(\delta, \sigma^2) + (1-p)N(-\delta, \sigma^2) \qquad vs \qquad N(m, \sigma^2)$$

rejects for small values of

$$\frac{p \exp\left[-\frac{1}{2\sigma^2}(X-\delta)^2\right] + (1-p)\exp\left[-\frac{1}{2\sigma^2}(X+\delta)^2\right]}{\exp\left[-\frac{1}{2\sigma^2}(X-m)^2\right]} , \tag{3.43}$$

or equivalently for small values of $f(X)$, where

$$f(x) = p\exp[(\delta - m)X/\sigma^2] + (1-p)\exp[-(\delta+m)X/\sigma^2] .$$

We can now choose p so that $f(C) = f(-C)$, so that p must satisfy

$$\frac{p}{1-p} = \frac{\exp[(\delta+m)C/\sigma^2] - \exp[-(\delta+m)C/\sigma^2]}{\exp[(\delta-m)C/\sigma^2] - \exp[-(\delta-m)C/\sigma^2]} . \tag{3.44}$$

Since $\delta - m > 0$ and $\delta + m > 0$, both the numerator and denominator of the right side of (3.44) are positive, so the right side is a positive number; but, $p/(1-p)$ is a nondecreasing function of p with range $[0, \infty)$ as p varies from 0 to 1. Thus, p is well-defined. Also, observe $f''(x) \geq 0$ for all x. It follows that (for this special choice of C)

$$\{X : \ f(X) \leq f(C)\} = \{X : \ |X| \leq C\}$$

is the rejection region of the MP test. Such a test is easily seen to be level α for the original composite null hypothesis because its power function is symmetric and decreases away from zero. Thus, the result follows by Theorem 3.8.1.

Case 2. Consider now general k, so that $(X_1, \ldots, X_k)$ has mean $(\xi_1, \ldots, \xi_k)$ and covariance matrix Σ. However, consider the special case $(a_1, \ldots, a_k) =$

$(1, 0, \dots, 0)$, so we are testing $|\xi_1| \geq \delta$ versus $|\xi_1| < \delta$. Also, assume X_1 and $(X_2, \dots, X_k)$ are independent, so that the first row and first column of Σ are zero except the first entry, which is σ^2 (assumed positive). Using the same reasoning as case 1, fix an alternative $m = (m_1, \dots, m_k)$ with $|m_1| < \delta$ and consider testing

$$pN\left((\delta, m_2, \dots, m_k), \Sigma\right) + (1-p)N\left((-\delta, m_2, \dots, m_k), \Sigma\right)$$

versus $N\left((m_1, \dots, m_k), \Sigma\right)$. The likelihood ratio is in fact the same as (3.43) because each term is now multiplied by the density of $(X_2, \dots, X_k)$ (by independence), and these densities cancel. The UMP test from case 1, which rejects when $|X_1| \leq C$, is UMP in this situation as well.

Case 3. As in case 2, consider $a_1 = 1$ and $a_i = 0$ if $i > 1$, but now allow Σ to be an arbitrary covariance matrix. By transforming X to Y as in Case 3 of Example 3.9.2, the result follows (Problem 3.66).

Case 4. Now, consider arbitrary $(a_1, \dots, a_k)$ satisfying $a^T \Sigma a > 0$. As in case 4 of Example 3.9.2, transform X to Z and the result follows (Problem 3.66).

3.10 Problems

Section 3.2

Problem 3.1 Let $X_1, \dots, X_n$ be a sample from the normal distribution $N(\xi, \sigma^2)$.

(i) If $\sigma = \sigma_0$ (known), there exists a UMP test for testing $H : \xi \leq \xi_0$ against $\xi > \xi_0$, which rejects when $\sum(X_i - \xi_0)$ is too large.

(ii) If $\xi = \xi_0$ (known), there exists a UMP test for testing $H : \sigma \leq \sigma_0$ against $K : \sigma > \sigma_0$, which rejects when $\sum(X_i - \xi_0)^2$ is too large.

Problem 3.2 *UMP test for $U(0, \theta)$.* Let $X = (X_1, \dots, X_n)$ be a sample from the uniform distribution on $(0, \theta)$.

(i) For testing $H : \theta \leq \theta_0$ against $K : \theta > \theta_0$ any test is UMP at level α for which $E_{\theta_0}\phi(X) = \alpha$, $E_\theta \phi(X) \leq \alpha$ for $\theta \leq \theta_0$, and $\phi(x) = 1$ when $\max(x_1, \dots, x_n) > \theta_0$.

(ii) For testing $H : \theta = \theta_0$ against $K : \theta \neq \theta_0$ a unique UMP test exists, and is given by $\phi(x) = 1$ when $\max(x_1, \dots, x_n) > \theta_0$ or $\max(x_1, \dots, x_n) \leq \theta_0 \sqrt[n]{\alpha}$, and $\phi(x) = 0$ otherwise.

[(i): For each $\theta > \theta_0$ determine the ordering established by $r(x) = p_\theta(x)/p_{\theta_0}(x)$ and use the fact that many points are equivalent under this ordering.

(ii): Determine the UMP tests for testing $\theta = \theta_0$ against $\theta < \theta_0$ and combine this result with that of part (i).]

Problem 3.3 Suppose N i.i.d. random variables are generated from the same known strictly increasing absolutely continuous cdf $F(\cdot)$. We are told only X, the maximum of these random variables. Is there a UMP size α test of

$$H_0 : N \leq 5 \text{ versus } H_1 : N > 5?$$

If so, find it.

Problem 3.4 *UMP test for exponential densities.* Let $X_1, \ldots, X_n$ be a sample from the exponential distribution $E(a, b)$ of Problem 1.18, and let $X_{(1)} = \min(X_1, \ldots, X_n)$.

(i) Determine the UMP test for testing $H : a = a_0$ against $K : a \neq a_0$ when b is assumed known.

(ii) The power of any MP level-α test of $H : a = a_0$ against $K : a = a_1 < a_0$ is given by

$$\beta^*(a_1) = 1 - (1 - \alpha)e^{-n(a_0 - a_1)/b}.$$

(iii) For the problem of part (i), when b is unknown, the power of any level α test which rejects when

$$\frac{X_{(1)} - a_0}{\sum[X_i - X_{(1)}]} \leq C_1 \text{ or } \geq C_2$$

against any alternative (a_1, b) with $a_1 < a_0$ is equal to $\beta^*(a_1)$ of part (ii) (independent of the particular choice of C_1 and C_2).

(iv) The test of part (iii) is a UMP level-α test of $H : a = a_0$ against $K : a \neq a_0$ (b unknown).

(v) Determine the UMP test for testing $H : a = a_0, b = b_0$ against the alternatives $a < a_0, b < b_0$.

(vi) Explain the (very unusual) existence in this case of a UMP test in the presence of a nuisance parameter [part(iv)] and for a hypothesis specifying two parameters [part(v)].

[(i) The variables $Y_i = e^{-X_i/b}$ are a sample from the uniform distribution on $(0, e^{-a/b})$.]

Note. For more general versions of parts (ii)–(iv) see Takeuchi (1969) and Kabe and Laurent (1981).

Problem 3.5 In the proof of Theorem 3.2.1(i), consider the set of c satisfying $\alpha(c) \leq \alpha \leq \alpha(c - 0)$. If there is only one such c, c is unique; otherwise, there is an interval of such values $[c_1, c_2]$. Argue that, in this case, if $\alpha(c)$ is continuous at c_2, then $P_i(C) = 0$ for $i = 0, 1$, where

$$C = \left\{ x : p_0(x) > 0 \text{ and } c_1 < \frac{p_1(x)}{p_0(x)} \leq c_2 \right\}.$$

If $\alpha(c)$ is not continuous at c_2, then the result is false.

Problem 3.6 Let P_0, P_1, P_2 be the probability distributions assigning to the integers $1, \ldots, 6$ the following probabilities:

	1	2	3	4	5	6
P_0	.03	.02	.02	.01	0	.92
P_1	.06	.05	.08	.02	.01	.78
P_2	.09	.05	.12	0	.02	.72

Determine whether there exists a level-α test of $H : P = P_0$ which is UMP against the alternatives P_1 and P_2 when (i) $\alpha = .01$; (ii) $\alpha = .05$; (iii) $\alpha = .07$.

Problem 3.7 Let the distribution of X be given by

x	0	1	2	3
$P_\theta(X = x)$	θ	2θ	$.9 - 2\theta$	$.1 - \theta$

where $0 < \theta < .1$. For testing $H : \theta = .05$ against $\theta > .05$ at level $\alpha = .05$, determine which of the following tests (if any) is UMP:

(i) $\phi(0) = 1, \phi(1) = \phi(2) = \phi(3) = 0$;

(ii) $\phi(1) = .5, \phi(0) = \phi(2) = \phi(3) = 0$;

(iii) $\phi(3) = 1, \phi(0) = \phi(1) = \phi(2) = 0$.

Problem 3.8 A random variable X has the *Pareto distribution* $P(c, \tau)$ if its density is $c\tau^c/x^{c+1}, 0 < \tau < x, 0 < C$.

(i) Show that this defines a probability density.

(ii) If X has distribution $P(c, \tau)$, then $Y = \log X$ has exponential distribution $E(\xi, b)$ with $\xi = \log \tau$, $b = 1/c$.

(iii) If $X_1, \ldots, X_n$ is a sample from $P(c, \tau)$, use (ii) and Problem 3.4 to obtain UMP tests of (a) $H : \tau = \tau_0$ against $\tau \neq \tau_0$ when b is known; (b) $H : c = c_0$, $\tau = \tau$ against $c > c_0$, $\tau < \tau_0$.

Problem 3.9 Let X be distributed according to $P_\theta, \theta \in \Omega$, and let T be sufficient for θ. If $\varphi(X)$ is any test of a hypothesis concerning θ, then $\psi(T)$ given by $\psi(t) = E[\varphi(X) \mid t]$ is a test depending on T only, an its power function is identical with that of $\varphi(X)$.

Problem 3.10 In the notation of Section 3.2, consider the problem of testing $H_0 : P = P_0$ against $H_1 : P = P_1$, and suppose that known probabilities $\pi_0 = \pi$ and $\pi_1 = 1 - \pi$ can be assigned to H_0 and H_1 prior to the experiment.

(i) The overall probability of an error resulting from the use of a test φ is

$$\pi E_0\varphi(X) + (1 - \pi)E_1[1 - \varphi(X)].$$

(ii) The *Bayes test* minimizing this probability is given by (3.8) with $k = \pi_0/\pi_1$.

(iii) The conditional probability of H_i given $X = x$, the *posterior probability* of H_i is

$$\frac{\pi_i p_i(x)}{\pi_0 p_0(x) + \pi_1 p_1(x)},$$

and the Bayes test therefore decides in favor of the hypothesis with the larger posterior probability

Problem 3.11 (i) For testing $H_0 : \theta = 0$ against $H_1 : \theta = \theta_1$ when X is $N(\theta, 1)$, given any $0 < \alpha < 1$ and any $0 < \pi < 1$ (in the notation of the preceding problem), there exists θ_1 and x such that (a) H_0 is rejected when $X = x$ but (b) $P(H_0 \mid x)$ is arbitrarily close to 1.

(ii) The paradox of part (i) is due to the fact that α is held constant while the power against θ_1 is permitted to get arbitrarily close to 1. The paradox disappears if α is determined so that the probabilities of type I and type II error are equal [but see Berger and Sellke (1987)].

[For a discussion of such paradoxes, see Lindley (1957), Bartlett (1957), Schafer (1982, 1988) and Robert (1993).]

Problem 3.12 Let $X_1, \ldots, X_n$ be independently distributed, each uniformly over the integers $1, 2, \ldots, \theta$. Determine whether there exists a UMP test for testing $H : \theta = \theta_0$, at level $1/\theta_0^n$ against the alternatives (i) $\theta > \theta_0$; (ii) $\theta < \theta_0$; (iii) $\theta \neq \theta_0$.

Problem 3.13 The following example shows that the power of a test can sometimes be increased by selecting a random rather than a fixed sample size even when the randomization does not depend on the observations. Let $X_1, \ldots, X_n$ be independently distributed as $N(\theta, 1)$, and consider the problem of testing $H : \theta = 0$ against $K : \theta = \theta_1 > 0$.

(i) The power of the most powerful test as a function of the sample size n is not necessarily concave.

(ii) In particular for $\alpha = .005, \theta_1 = \frac{1}{2}$, better power is obtained by taking 2 or 16 observations with probability $\frac{1}{2}$ each than by taking a fixed sample of 9 observations.

(iii) The power can be increased further if the test is permitted to have different significance levels α_1 and α_2 for the two sample sizes and it is required only that the expected significance level be equal to $\alpha = .005$. Examples are: (a) with probability $\frac{1}{2}$ take $n_1 = 2$ observations and perform the test of significance at level $\alpha_1 = .001$, or take $n_2 = 16$ observations and perform the test at level $\alpha_2 = .009$; (b) with probability $\frac{1}{2}$ take $n_1 = 0$ or $n_2 = 18$ observations and let the respective significance levels be $\alpha_1 = 0, \alpha_2 = .01$.

Note. This and related examples were discussed by Kruskal in a seminar held at Columbia University in 1954. A more detailed investigation of the phenomenon has been undertaken by Cohen (1958).

Problem 3.14 If the sample space $\mathcal{X}$ is Euclidean and P_0, P_1 have densities with respect to Lebesgue measure, there exists a nonrandomized most powerful test for testing P_0 against P_1 at every significance level α.[12] [This is a consequence of Theorem 3.2.1 and the following lemma.[13] Let $f \geq 0$ and $\int_A f(x)\,dx = a$. Given any $0 \leq b \leq a$, there exists a subset B of A such that $\int_B f(x)\,dx = b$.]

---

[12]For more general results concerning the possibility of dispensing with randomized procedures, see Dvoretzky, Wald, and Wolfowitz (1951).

[13]For a proof of this lemma see Halmos (1974, p. 174.) The lemma is a special case of a theorem of Lyapounov (1940); see Blackwell(1951).

Problem 3.15 *Fully informative statistics.* A statistic T is *fully informative* if for every decision problem the decision procedures based only on T form an essentially complete class. If $\mathcal{P}$ is dominated and T is fully informative, then T is sufficient. [Consider any pair of distributions P_0, $P_1 \in \mathcal{P}$ with densities p_0, p_1, and let $g_i = p_i/(p_0 + p_1)$. Suppose that T is fully informative, and let $\mathcal{A}_{\prime}$ be the subfield induced by T. Then $\mathcal{A}_{\prime}$ contains the subfield induced by (g_0, g_1) since it contains every rejection which is unique most powerful for testing P_0 against P_1 (or P_1 against P_0) at some level α. Therefore, T is sufficient for every pair of distributions (P_0, P_1), and hence by Problem 2.11 it is sufficient for $\mathcal{P}$.]

Problem 3.16 Based on X with distribution indexed by $\theta \in \Omega$, the problem is to test $\theta \in \omega$ versus $\theta \in \omega'$. Suppose there exists a test ϕ such that $E_\theta[\phi(X)] \leq \beta$ for all θ in ω, where $\beta < \alpha$. Show there exists a level α test $\phi^*(X)$ such that

$$E_\theta[\phi(X)] \leq E_\theta[\phi^*(X)] \ ,$$

for all θ in ω' and this inequality is strict if $E_\theta[\phi(X)] < 1$.

Problem 3.17 *A counterexample.* Typically, as α varies the most powerful level α tests for testing a hypothesis H against a simple alternative are nested in the sense that the associated rejection regions, say R_α, satisfy $R_\alpha \subset R_{\alpha'}$, for any $\alpha < \alpha'$. Even if the most powerful tests are nonrandomized, this may be false. Suppose X takes values 1, 2, and 3 with probabilities 0.85, 0.1, and 0.05 under H and probabilities 0.7, 0.2, and 0.1 under K.
(i) At any level $< .15$, the MP test is not unique.
(ii) At $\alpha = .05$ and $\alpha' = .1$, there exist unique nonrandomized MP tests and they are not nested.
(iii) At these levels there exist MP tests ϕ and ϕ' that are nested in the sense that $\phi(x) \leq \phi'(x)$ for all x. [This example appears as Example 10.16 in Romano and Siegel (1986).]

Problem 3.18 Under the setup of Theorem 3.2.1, show there always exists MP tests that are nested in the sense of Problem 3.17(iii).

Problem 3.19 Suppose $X_1, \ldots, X_n$ are i.i.d. $N(\xi, \sigma^2)$ with σ known. For testing $\xi = 0$ versus $\xi \neq 0$, the average power of a test $\phi = \phi(X_1, \ldots, X_n)$ is given by

$$\int_{-\infty}^{\infty} E_\xi(\phi) d\Lambda(\xi) \ ,$$

where Λ is a probability distribution on the real line. Suppose that Λ is symmetric about 0; that is, $\Lambda\{E\} = \Lambda\{-E\}$ for all Borel sets E. Show that, among α level tests, the one maximizing average power rejects for large values of $|\sum_i X_i|$. Show that this test need not maximize average power if Λ is not symmetric.

Problem 3.20 Let f_θ, $\theta \in \Omega$, denote a family of densities with respect to a measure μ. (We assume Ω is endowed with a σ-field so that the densities $f_\theta(x)$ are jointly measurable in θ and x.) Consider the problem of testing a simple null hypothesis $\theta = \theta_0$ against the composite alternatives $\Omega_K = \{\theta : \ \theta \neq \theta_0\}$. Let Λ be a probability distribution on Ω_K.

(i) As explicitly as possibly, find a test ϕ that maximizes $\int_{\Omega_K} E_\theta(\phi) d\Lambda(\theta)$, subject to it being level α.
(ii) Let $h(x) = \int f_\theta(x) d\Lambda(\theta)$. Consider the nonrandomized ϕ test that rejects if and only if $h(x)/f_{\theta_0}(x) > k$, and suppose $\mu\{x : h(x) = kf_\theta(x)\} = 0$. Then, ϕ is admissible at level $\alpha = E_{\theta_0}(\phi)$ in the sense that it is impossible that there exists another level α test ϕ' such that $E_\theta(\phi') \geq E_\theta(\phi)$ for all θ.
(iii) Show that the test of Problem 3.19 is admissible.

Section 3.3

Problem 3.21 In Example 3.21, show that p-value is indeed given by $\hat{p} = \hat{p}(X) = (11 - X)/10$. Also, graph the c.d.f. of $\hat{p}$ under H and show that the last inequality in (3.15) is an equality if and only u is of the form $0, \ldots, 10$.

Problem 3.22 Suppose X has a continuous distribution function F. Show that $F(X)$ is uniformly distributed on $(0, 1)$. [The transformation from X to $F(X)$ is known as the *probability integral transformation.*]

Problem 3.23 Under the setup of Lemma 3.3.1, suppose the rejection regions are defined by

$$S_\alpha = \{X : T(X) \geq k(\alpha)\} \tag{3.45}$$

for some real-valued statistic $T(X)$ and $k(\alpha)$ satisfying

$$\sup_{\theta \in \Omega_H} P_\theta\{T(X) \geq k(\alpha)\} \leq \alpha \ .$$

Then, show

$$\hat{p} = \sup_{\theta \in \Omega_H} P\{T(X) \geq t\} \ ,$$

where t is the observed value of $T(X)$.

Problem 3.24 Under the setup of Lemma 3.3.1, show that there exists a real-valued statistic $T(X)$ so that the rejection region is necessarily of the form (3.45). [*Hint*: Let $T(X) = -\hat{p}$.]

Problem 3.25 (i) If $\hat{p}$ is uniform on $(0, 1)$, show that $-2\log(\hat{p})$ has the Chi-squared distribution with 2 degrees of freedom.
(ii) Suppose $\hat{p}_1, \ldots, \hat{p}_s$ are i.i.d. uniform on $(0, 1)$. Let $F = -2\log(\hat{p}_1 \cdots \hat{p}_s)$. Argue that F has the Chi-squared distribution with $2s$ degrees of freedom. What can you say about F if the $\hat{p}_i$ are independent and satisfy $P\{\hat{p}_i \leq u\} \leq u$ for all $0 \leq u \leq 1$? [Fisher (1934a) proposed F as a means of combining p-values from independent experiments.]

Section 3.4

Problem 3.26 Let X be the number of successes in a n independent trials with probability p of success, and let $\phi(x)$ be the UMP test (3.16) for testing $p \leq p_0$ against $p > p_0$ at level of significance α.

(i) For $n = 6$, $p_0 = .25$ and the levels $\alpha = .05, .1, .2$ determine C and γ, and the power of the test against $p_1 = .3, .4, .5, .6, .7$.

(ii) If $p_0 = .2$ and $\alpha = .05$, and it is desired to have power $\beta \geq .9$ against $p_1 = .4$, determine the necessary sample size (a) by using tables of the binomial distribution, (b) by using the normal approximation.[14]

(iii) Use the normal approximation to determine the sample size required when $\alpha = .05$, $\beta = .9$, $p_0 = .01$, $p_1 = .02$.

Problem 3.27 (i) A necessary and sufficient condition for densities $p_\theta(x)$ to have monotone likelihood ratio in x, if the mixed second derivative $\partial^2 \log p_\theta(x)/\partial\theta\,\partial x$ exists, is that this derivative is ≥ 0 for all θ and x.

(ii) An equivalent condition is that

$$p_\theta(x)\frac{\partial^2 p_\theta(x)}{\partial\theta\,\partial x} \geq \frac{\partial p_\theta(x)}{\partial\theta}\,\frac{\partial p_\theta(x)}{\partial x} \qquad \text{for all } \theta \text{ and } x.$$

Problem 3.28 Let the probability density p_θ of X have monotone likelihood ratio in $T(x)$, and consider the problem of testing $H : \theta \leq \theta_0$ against $\theta > \theta_0$. If the distribution of T is continuous, the p-value $\hat{p}$ of the UMP test is given by $\hat{p} = P_{\theta_0}\{T \geq t\}$, where t is the observed value of T. This holds also without the assumption of continuity if for randomized tests $\hat{p}$ is defined as the smallest significance level at which the hypothesis is rejected with probability 1. Show that, for any $\theta \leq \theta_0$, $P_\theta\{\hat{p} \leq u\} \leq u$ for any $0 \leq u \leq 1$.

Problem 3.29 Let $X_1, \ldots, X_n$ be independently distributed with density $(2\theta)^{-1}e^{-x/2\theta}, x \geq 0$, and let $Y_1 \leq \cdots \leq Y_n$ be the ordered X's. Assume that Y_1 becomes available first, then Y_2, and so on, and that observation is continued until Y_r has been observed. On the basis of $Y_1, \ldots, Y_r$ it is desired to test $H : \theta \geq \theta_0 = 1000$ at level $\alpha = .05$ against $\theta < \theta_0$.

(i) Determine the rejection region when $r = 4$, and find the power of the test against $\theta_1 = 500$.

(ii) Find the value of r required to get power $\beta \geq .95$ against the alternative.

[In Problem 2.15, the distribution of $[\sum_{i=1}^r Y_i + (n-r)Y_r]/\theta$ was found to be χ^2 with $2r$ degrees of freedom.]

Problem 3.30 When a Poisson process with rate λ is observed for a time interval of length τ, the number X of events occurring has the Poisson distribution $P(\lambda\tau)$. Under an alternative scheme, the process is observed until r events have occurred, and the time T of observation is then a random variable such that $2\lambda T$ has a χ^2-distribution with $2r$ degrees of freedom. For testing $H : \lambda \leq \lambda_0$ at level α one can, under either design, obtain a specified power β against an alternative λ_1 by choosing τ and r sufficiently large.

[14] Tables and approximations are discussed, for example, in Chapter 3 of Johnson and Kotz (1969).

(i) The ratio of the time of observation required for this purpose under the first design to the expected time required under the second is $\lambda\tau/r$.

(ii) Determine for which values of λ each of the two designs is preferable when $\lambda_0 = 1, \lambda_1 = 2, \alpha = .05, \beta = 9$.

Problem 3.31 Let $X = (X_1, \ldots, X_n)$ be a sample from the uniform distribution $U(\theta, \theta + 1)$.

(i) For testing $H : \theta \leq \theta_0$ against $K : \theta > \theta_0$ at level α there exists a UMP test which rejects when $\min(X_1, \ldots, X_n) > \theta_0 + C(\alpha)$ or $\max(X_1, \ldots, X_n > \theta_0 + 1$ for suitable $C(\alpha)$.

(ii) The family $U(\theta, \theta+1)$ does not have monotone likelihood ratio. [Additional results for this family are given in Birnbaum (1954b) and Pratt (1958).]

[(ii) By Theorem 3.4.1, monotone likelihood ratio implies that the family of UMP test of $H : \theta \leq \theta_0$ against $K : \theta > \theta_0$ generated as α varies from 0 to 1 is independent of θ_0].

Problem 3.32 Let X be a single observation from the Cauchy density given at the end of Section 3.4.

(i) Show that no UMP test exists for testing $\theta = 0$ against $\theta > 0$.

(ii) Determine the totality of different shapes the MP level-α rejection region for testing $\theta = \theta_0$ against $\theta = \theta_1$ can take on for varying α and $\theta_1 - \theta_0$.

Problem 3.33 Let X_i be independently distributed as $N(i\Delta, 1)$, $i = 1, \ldots, n$. Show that there exists a UMP test of $H : \Delta \leq 0$ against $K : \Delta > 0$, and determine it as explicitly as possible.

Note. The following problems (and some in later chapters) refer to the gamma, Pareto, Weibull, and inverse Gaussian distributions. For more information about these distributions, see Chapters 17, 19, 20, and 25 respectively of Johnson and Kotz (1970).

Problem 3.34 Let $X_1, \ldots, X_n$ be a sample from the *gamma distribution* $\Gamma(g, b)$ with density

$$\frac{1}{\Gamma(g)b^g} x^{g-1} e^{-x/b}, \qquad 0 < x, \quad 0 < b, g.$$

Show that there exist a UMP test for testing

(i) $H : b \leq b_0$ against $b > b_0$ when g is known;

(ii) $H : g \leq g_0$ against $g > g_0$ when b is known.

In each case give the form of the rejection region.

Problem 3.35 A random variable X has the *Weibull distribution* $W(b, c)$ if its density is

$$\frac{c}{b}\left(\frac{x}{b}\right)^{c-1} e^{-(x/b)^c}, \qquad x > 0, b, c > 0.$$

Show that this defines a probability density. If $X_1, \ldots, X_n$ is a sample from $W(b, c)$, with the shape parameter c known, show that there exists a UMP test of $H : b \leq b_0$ against $b > b_0$ and give its form.

Problem 3.36 Consider a single observation X from $W(1, c)$.

(i) The family of distributions does not have monotone likelihood ratio in x.

(ii) The most powerful test of $H : c = 1$ against $c = 2$ rejects when $X < k_1$ and when $X > k_2$. Show how to determine k_1 and k_2.

(iii) Generalize (ii) to arbitrary alternatives $c_1 > 1$, and show that a UMP test of $H : c = 1$ against $c > 1$ does not exist.

(iv) For any $c_1 > 1$, the power function of the MP test of $H : c = 1$ against $c = c_1$ is an increasing function of c.

Problem 3.37 Let $X_1, \ldots, X_n$ be a sample from the *inverse Gaussian* distribution $I(\mu, \tau)$ with density

$$\sqrt{\frac{\tau}{2\pi x^3}} \exp\left(-\frac{\tau}{2x\mu^2}(x-\mu)^2\right), \qquad x > 0, \quad \tau, \mu > 0.$$

Show that there exists a UMP test for testing

(i) $H : \mu \leq \mu_0$ against $\mu > \mu_0$ when τ is known;

(ii) $H : \tau \leq \tau_0$ against $\tau > \tau_0$ when μ is known.
In each case give the form of the rejection region.

(iii) The distribution of $V = r(X_i - \mu)^2/X_i\mu^2$ is χ_1^2 and hence that of $\tau \sum[(X_i - \mu)^2/X_i\mu^2]$ is χ_n^2.

[Let $Y = \min(X_i, \mu^2/X_i)$, $Z = \tau(Y - \mu)^2/\mu^2 Y$. Then $Z = V$ and Z is χ_1^2 [Shuster (1968)].] *Note.* The UMP test for (ii) is discussed in Chhikara and Folks (1976).

Problem 3.38 Let $X_1, \cdots, X_n$ be a sample from a location family with common density $f(x - \theta)$, where the location parameter $\theta \in \mathbf{R}$ and $f(\cdot)$ is known. Consider testing the null hypothesis that $\theta = \theta_0$ versus an alternative $\theta = \theta_1$ for some $\theta_1 > \theta_0$. Suppose there exists a most powerful level α test of the form: reject the null hypothesis iff $T = T(X_1, \cdots, X_n) > C$, where C is a constant and $T(X_1, \ldots, X_n)$ is location equivariant, i.e. $T(X_1 + c, \ldots, X_n + c) = T(X_1, \ldots, X_n) + c$ for all constants c. Is the test also most powerful level α for testing the null hypothesis $\theta \leq \theta_0$ against the alternative $\theta = \theta_1$. Prove or give a counterexample.

Problem 3.39 *Extension of Lemma* 3.4.2. Let P_0 and P_1 be two distributions with densities p_0, p_1 such that $p_1(x)/p_0(x)$ is a nondecreasing function of a real-valued statistic $T(x)$.

(i) If T has probability density p_i' when the original distribution of P_i, then $p_1'(t)/p_0'(t)$ is nondecreasing in t.

(ii) $E_0\psi(T) \leq E_1\psi(T)$ for any nondecreasing function ψ.

(iii) If $p_1(x)/p_0(x)$ is a strictly increasing function of $t = T(x)$, so is $p_1'(t)/p_0'(t)$, and $E_0\psi(T) < E_1\psi(T)$ unless $\psi[T(x)]$ is constant a.e. $(P_0 + P_1)$ or $E_0\psi(T) = E_1\psi(T) = \pm\infty$.

(iv) For any distinct distributions with densities p_0, p_1,

$$-\infty \leq E_0 \log\left[\frac{p_1(X)}{p_0(X)}\right] < E_1 \log\left[\frac{p_1(X)}{p_0(X)}\right] \leq \infty.$$

[(i): Without loss of generality suppose that $p_1(x)/p_0(x) = T(x)$. Then for any integrable ϕ,

$$\int \phi(t)p_1'(t)\,dv(t) = \int \phi[T(x)]T(x)p_0(x)\,d\mu(x) = \int \phi(t)tp_0'(t)\,dv(t),$$

and hence $p_1'(t)/p_0'(t) = t$ a.e.
(iv): The possibility $E_0\log[p_1(X)/p_0(X)] = \infty$ is excluded, since by the convexity of the function log,

$$E_0 \log\left[\frac{p_1(X)}{p_0(X)}\right] < \log E_0\left[\frac{p_1(X)}{p_0(X)}\right] = 0.$$

Similarly for E_1. The strict inequality now follows from (iii) with $T(x) = p_1(x)/p_0(x)$.]

Problem 3.40 F_0, F_1 are two cumulative distribution functions on the real line, then $F_i(x) \leq F_0(x)$ for all x if and only if $E_0\psi(X) \leq E_1\psi(X)$ for any nondecreasing function ψ.

Problem 3.41 Let F and G be two continuous, strictly increasing c.d.f.s, and let $k(u) = G[F^{-1}(u)]$, $0 < u < 1$.
(i) Show F and G are stochastically ordered, say $F(x) \leq G(x)$ for all x, if and only if $k(u) \leq u$ for all $0 < u < 1$.
(ii) If F and G have densities f and g, then show they are monotone likelihood ratio ordered, say g/f nondecreasing, if and only if k is convex.
(iii) Use (i) and (ii) to give an alternative proof of the fact that MLR implies stochastic ordering.

Problem 3.42 Let $f(x)/[1 - F(x)]$ be the "mortality" of a subject at time x given that it has survived to this time. A c.d.f. F is said to be smaller than G in the hazard ordering if

$$\frac{g(x)}{1 - G(x)} \leq \frac{f(x)}{1 - F(x)} \quad \text{for all } x\ . \tag{3.46}$$

(i) Show that (3.46) is equivalent to

$$\frac{1 - F(x)}{1 - G(x)} \quad \text{is nonincreasing.} \tag{3.47}$$

(ii) Show that (3.46) holds if and only if k is starshaped. [A function k defined on an interval $I \subset [0,\infty)$ is starshaped on I if $k(\lambda x) \leq \lambda k(x)$ whenever $x \in I$, $\lambda x \in I$, $0 \leq \lambda \leq 1$. Problems 3.41 and 3.42 are based on Lehmann and Rojo (1992).]

Section 3.5

Problem 3.43 (i) For $n = 5, 10$ and $1 - \alpha = .95$, graph the upper confidence limits $\bar{p}$ and $\bar{p}^*$ of Example 3.5.2 as functions of $t = x + u$.

(ii) For the same values of n and $\alpha_1 = \alpha_2 = .05$, graph the lower and upper confidence limits $\underline{p}$ and $\bar{p}$.

Problem 3.44 *Confidence bounds with minimum risk.* Let $L(\theta, \underline{\theta})$ be nonnegative and nonincreasing in its second argument for $\underline{\theta} < \theta$, and equal to 0 for $\underline{\theta} \geq \theta$. If $\underline{\theta}$ and $\underline{\theta}^*$ are two lower confidence bounds for θ such that

$$P_0\{\underline{\theta} \leq \theta'\} \leq P_\theta\{\underline{\theta}^* \leq \theta'\} \qquad \text{for all} \quad \theta' \leq \theta,$$

then

$$E_\theta L(\theta, \underline{\theta}) \leq E_\theta L(\theta, \underline{\theta}^*).$$

[Define two cumulative distribution functions F and F^* by $F(u) = P_\theta\{\theta \leq u\}/P_\theta\{\underline{\theta}^* \leq \theta\}$, $F^*(u) = P_\theta\{\underline{\theta}^* \leq u\}/P_\theta\{\underline{\theta}^* \leq \theta\}$ for $u < \theta$, $F(u) = F^*(u) = 1$ for $u \geq \theta$. Then $F(u) \leq F^*(u)$ for all u, and it follows from Problem 3.40 that

$$\begin{aligned} E_\theta[L(\theta, \underline{\theta})] &= P_\theta\{\theta^* \leq \theta\} \int L(\theta, u) dF(u) \\ &\leq P_\theta\{\underline{\theta}^* \leq \theta\} \int L(\theta, u) dF^*(u) = E_\theta[L(\theta, \underline{\theta}^*)].] \end{aligned}$$

Section 3.6

Problem 3.45 If $\beta(\theta)$ denotes the power function of the UMP test of Corollary 3.4.1, and if the function Q of (3.19) is differentiable, then $\beta'(\theta) > 0$ for all θ for which $Q'(\theta) > 0$.

[To show that $\beta'(\theta_0) > 0$, consider the problem of maximizing, subject to $E_{\theta_0}\phi(X) = \alpha$, the derivative $\beta'(\theta_0)$ or equivalently the quantity $E_{\theta_0}[T(X)\,\phi(X)]$.]

Problem 3.46 *Optimum selection procedures.* On each member of a population n measurements $(X_1, \ldots, X_n) = X$ are taken, for example the scores of n aptitude tests which are administered to judge the qualifications of candidates for a certain training program. A future measurement Y such as the score in a final test at the end of the program is of interest but unavailable. The joint distribution of X and Y is assumed known.

(i) One wishes to select a given proportion α of the candidates in such a way as to maximize the expectation of Y for the selected group. This is achieved by selecting the candidates for which $E(Y|x) \geq C$, where C is determined by the condition that the probability of a member being selected is α. When $E(Y|x) = C$, it may be necessary to randomized in order to get the exact value α.

(ii) If instead the problem is to maximize the probability with which in the selected population Y is greater than or equal to some preassigned score y_0, one selects the candidates for which the conditional probability $P\{Y \geq y_0|x\}$ is sufficiently large.

[(i): Let $\phi(x)$ denote the probability with which a candidate with measurements x is to be selected. Then the problem is that of maximizing

$$\int \left[\int y p^{Y|x}(y)\ \phi(x) dy \right] p^x(x) dx$$

subject to

$$\int \phi(x) p^x(x) dx = \alpha.]$$

Problem 3.47 The following example shows that Corollary 3.6.1 does not extend to a countably infinite family of distributions. Let p_n be the uniform probability density on $[0, 1 + 1/n]$, and p_0 the uniform density on $(0, 1)$.

(i) Then p_0 is linearly independent of $(p_1, p_2, \ldots)$, that is, there do not exist constants $c_1, c_2, \ldots$ such that $p_0 = \sum c_n p_n$.

(ii) There does not exist a test ϕ such that $\int \phi p_n = \alpha$ for $n = 1, 2, \ldots$ but $\int \phi p_0 > \alpha$.

Problem 3.48 Let $F_1, \ldots, F_{m+1}$ be real-valued functions defined over a space U. A sufficient condition for u_0 to maximize F_{m+1} subject to $F_i(u) \leq c_i (i = 1, \ldots, m)$ is that it satisfies these side conditions, that it maximizes $F_{m+1}(u) - \sum k_i F_i(u)$ for some constants $k_i \geq 0$, and that $F_i(u_o) = c_i$ for those values i for which $k_i > 0$.

Section 3.7

Problem 3.49 For a random variable X with binomial distribution $b(p, n)$, determine the constants $C_i, \gamma (i = 1, 2)$ in the UMP test (3.31) for testing $H : p \leq .2$ or $\leq .7$ when $\alpha = .1$ and $n = 15$. Find the power of the test against the alternative $p = .4$.

Problem 3.50 *Totally positive families.* A family of distributions with probability densities $p_\theta(x)$, θ and x real-valued and varying over Ω and $\mathcal{X}$ respectively, is said to be totally positive of order $r(\text{TP}_r)$ if for all $x_1 < \cdots < x_n$ and $\theta_1 < \cdots < \theta_n$

$$\triangle_n = \begin{vmatrix} p_{\theta_1}(x_1) & \cdots & p_{\theta_1}(x_n) \\ p_{\theta_n}(x_1) & \cdots & p_{\theta_n}(x_n) \end{vmatrix} \geq 0 \qquad \text{for all} \quad n = 1, 2, \ldots, r. \tag{3.48}$$

It is said to be strictly totally positive of order r (STP_r) if strict inequality holds in (3.48). The family is said to be (strictly) totally positive of infinity if (3.48) holds for all $n = 1, 2, \ldots$. These definitions apply not only to probability densities but to any real-valued functions $p_\theta(x)$ of two real variables.

(i) For $r = 1$, (3.48) states that $p_\theta(x) \geq 0$; for $r = 2$, that $p_\theta(x)$ has monotone likelihood ratio in x.

(ii) If $a(\theta) > 0, b(x) > 0$, and $p_\theta(x)$ is STP_r then so is $a(\theta)b(x)p_\theta(x)$.

(iii) If a and b are real-valued functions mapping Ω and $\mathcal{X}$ onto Ω' and $\mathcal{X}'$ and are strictly monotone in the same direction, and if $p_\theta(x)$ is $(\text{STP}_r$, then $p_{\theta'}(x')$ with $\theta' = a^{-1}(\theta)$ and $x' = b^{-1}(x)$ is $(STP)_r$ over $(\Omega', \mathcal{X}')$.

Problem 3.51 *Exponential families.* The exponential family (3.19) with $T(x) = x$ and $Q(\theta) = \theta$ is STP_∞, with Ω the natural parameter space and $\mathcal{X} = (-\infty, \infty)$.

[That the determinant $|e^{\theta_i x_j}|, i, j = 1, \ldots, n$, is positive can be proved by induction. Divide the ith column by $e^{\theta_1 x_i}, i = 1, \ldots, n$; subtract in the resulting determinant the $(n-1)$st column from the nth, the $(n-2)$nd from the $(n-1)$st, $\ldots$, the 1st from the 2nd; and expand the determinant obtained in this way by the first row. Then $\triangle_n$ is seen to have the same sign as

$$\triangle_n' = |e^{\eta_i x_j} - e^{\eta_i x_{j-1}}|, \qquad i, j = 2, \ldots, n,$$

where $\eta_i = \theta_i - \theta_1$. If this determinant is expanded by the first column one obtains a sum of the form

$$\begin{aligned} a_2(e^{\eta_2 x_2} - e^{\eta_2 x_1}) + \cdots + a_n(e^{\eta_n x_2} - e^{\eta_n x_1}) &= h(x_2) - h(x_1) \\ &= (x_2 - x_1)h'(y_2), \end{aligned}$$

where $x_1 < y_2 < x_2$. Rewriting $h'(y_2)$ as a determinant of which all columns but the first coincide with those of $\triangle_n'$ and proceeding in the same manner with the columns, one reduces the determinant to $|e^{\eta_i y_j}|, i, j = 2, \ldots, n$, which is positive by the induction hypothesis.]

Problem 3.52 STP_3. Let θ and x be real-valued, and suppose that the probability densities $p_\theta(x)$ are such that $p_{\theta'}(x)/p_\theta(x)$ is strictly increasing in x for $\theta < \theta'$. Then the following two conditions are equivalent: (a) For $\theta_1 < \theta_2 < \theta_3$ and $k_1, k_2, k_3 > 0$, let

$$g(x) = k_1 p_{\theta_1}(x) - k_2 p_{\theta_2}(x) + k_3 p_{\theta_3}(x).$$

If $g(x_1) - g(x_3) = 0$, then the function g is positive outside the interval (x_1, x_3) and negative inside. (b) The determinant $\triangle_3$ given by (3.48) is positive for all $\theta_1 < \theta_2 < \theta_3$, $x_1 < x_2 < x_3$. [It follows from (a) that the equation $g(x) = 0$ has at most two solutions.]

[That (b) implies (a) can be seen for $x_1, < x_2 < x_3$ by considering the determinant

$$\begin{vmatrix} g(x_1) & g(x_2) & g(x_3) \\ p_{\theta_2}(x_1) & p_{\theta_2}(x_2) & p_{\theta_2}(x_3) \\ p_{\theta_3}(x_1) & p_{\theta_3}(x_2) & p_{\theta_3}(x_3) \end{vmatrix}$$

Suppose conversely that (a) holds. Monotonicity of the likelihood ratios implies that the rank of $\triangle_3$ is at least two, so that there exist constants k_1, k_2, k_3 such that $g(x_1) = g(x_3) = 0$. That the k's are positive follows again from the monotonicity of the likelihood ratios.]

Problem 3.53 *Extension of Theorem 3.7.1.* The conclusions of Theorem 3.7.1 remain valid if the density of a sufficient statistic T (which without loss of generality will be taken to be X), say $p_\theta(x)$, is STP_3 and is continuous in x for each θ.

[The two properties of exponential families that are used in the proof of Theorem 3.7.1 are continuity in x and (a) of the preceding problem.]

Problem 3.54 For testing the hypothesis $H' : \theta_1 \leq \theta \leq \theta_2 (\theta_1 \leq \theta_2)$ against the alternatives $\theta < \theta_1$ or $\theta > \theta_2$, or the hypothesis $\theta = \theta_0$ against the alternatives

$\theta \neq \theta_0$, in an exponential family or more generally in a family of distributions satisfying the assumptions of Problem 3.53, a UMP test does not exist.

[This follows from a consideration of the UMP tests for the one-sided hypotheses $H_1 : \theta \geq \theta_1$ and $H_2 : \theta \leq \theta_2$.]

Problem 3.55 Let f, g be two probability densities with respect to μ. For testing the hypothesis $H : \theta \leq \theta_0$ or $\theta \geq \theta_1 (0 < \theta_0 < \theta_1 < 1)$ against the alternatives $\theta_0 < \theta < \theta_1$, in the family $\mathcal{P} = \{\theta f(x) + (1-\theta) g(x), 0 \leq \theta \leq 1\}$, the test $\varphi(x) \equiv \alpha$ is UMP at level α.

Section 3.8

Problem 3.56 Let the variables $X_i (i = 1, \ldots, s)$ be independently distributed with Poisson distribution $P(\lambda_i)$. For testing the hypothesis $H : \sum \lambda_j \leq a$ (for example, that the combined radioactivity of a number of pieces of radioactive material does not exceed a), there exists a UMP test, which rejects when $\sum X_j > C$.

[If the joint distribution of the X's is factored into the marginal distribution of $\sum X_j$ (Poisson with mean $\sum \lambda_j$) times the conditional distribution of the variables $Y_i = X_j / \sum X_j$ given $\sum X_j$ (multinomial with probabilities $p_i = \lambda_i / \sum \lambda_j$), the argument is analogous to that given in Example 3.8.1.]

Problem 3.57 *Confidence bounds for a median.* Let $X_1, \ldots, X_n$ be a sample from a continuous cumulative distribution functions F. Let ξ be the unique median of F if it exists, or more generally let $\xi = \inf\{\xi' : F(\xi') = \frac{1}{2}\}$.

(i) If the ordered X's are $X_{(1)} < \cdots < X_{(n)}$, a uniformly most accurate lower confidence bound for ξ is $\underline{\xi} = X_{(k)}$ with probability $\rho, \underline{\xi} = X_{(k+1)}$ with probability $1 - \rho$, where k and ρ are determined by

$$\rho \sum_{j=k}^{n} \binom{n}{j} \frac{1}{2^n} + (1-\rho) \sum_{j=k+1}^{n} \binom{n}{j} \frac{1}{2^n} = 1 - \alpha.$$

(ii) This bound has confidence coefficient $1 - \alpha$ for any median of F.

(iii) Determine most accurate lower confidence bounds for the $100p$-percentile ξ of F defined by $\xi = \inf\{\xi' : F(\xi') = p\}$.

[For fixed to the problem of testing $H : \xi = \xi_0$ to against $K : \xi > \xi_0$ is equivalent to testing $H' : p = \frac{1}{2}$ against $K' : p < \frac{1}{2}$.]

Problem 3.58 A *counterexample*. Typically, as α varies the most powerful level α tests for testing a hypothesis H against a simple alternative are nested in the sense that the associated rejection regions, say R_α, satisfy $R_\alpha \subset R_{\alpha'}$, for any $\alpha < \alpha'$. The following example shows that this need not be satisfied for composite H. Let X take on the values $1, 2, 3, 4$ with probabilities under distributions P_0, P_1, Q:

	1	2	3	4
P_0	$\frac{2}{13}$	$\frac{4}{13}$	$\frac{3}{13}$	$\frac{4}{13}$
P_1	$\frac{4}{13}$	$\frac{2}{13}$	$\frac{1}{13}$	$\frac{6}{13}$
Q	$\frac{4}{13}$	$\frac{3}{13}$	$\frac{2}{13}$	$\frac{4}{13}$

Then the most powerful test for testing the hypothesis that the distribution of X is P_0 or P_1 against the alternative that it is Q rejects at level $\alpha = \frac{5}{13}$ when $X = 1$ or 3, and at level $\alpha = \frac{6}{13}$ when $X = 1$ or 2.

Problem 3.59 Let X and Y be the number of successes in two sets of n binomial trials with probabilities p_1 and p_2 of success.

(i) The most powerful test of the hypothesis $H : p_2 \leq p_1$ against an alternative (p_1', p_2') with $p_1' < p_2'$ and $p_1' + p_2' = 1$ at level $\alpha < \frac{1}{2}$ rejects when $Y - X > C$ and with probability γ when $Y - X = C$.

(ii) This test is not UMP against the alternatives $p_1 < p_2$.

[(i): Take the distribution Λ assigning probability 1 to the point $p_1 = p_2 = \frac{1}{2}$ as an a priori distribution over H. The most powerful test against (p_1', p_2') is then the one proposed above. To see that Λ is least favorable, consider the probability of rejection $\beta(p_1, p_2)$ for $p_1 = p_2 = p$. By symmetry this is given by

$$2\beta(p,p) = P\{|Y - X| > C\} + \gamma P\{|Y - X| = C\}.$$

Let X_i be 1 or 0 as the ith trial in the first series is a success or failure, and let Y_1, be defined analogously with respect to the second series. Then $Y - X = \sum_{i-1}^{n}(Y_i - X_i)$, and the fact that $2\beta(p,p)$ attains its maximum for $p = \frac{1}{2}$ can be proved by induction over n.

(ii): Since $\beta(p,p) < \alpha$ for $p \neq 1$, the power $\beta(p_1, p_2)$ is $< \alpha$ for alternatives $p_1 < p_2$ sufficiently close to the line $p_1 = p_2$. That the test is not UMP now follows from a comparison with $\phi(x,y) \equiv \alpha$.]

Problem 3.60 *Sufficient statistics with nuisance parameters.*

(i) A statistic T is said to be *partially sufficient* for θ in the presence of a nuisance parameter η if the parameter space is the direct product of the set of possible θ- and η-values, and if the following two conditions hold: (a) the conditional distribution given $T = t$ depends only on η; (b) the marginal distribution of T depends only on θ. If these conditions are satisfied, there exists a UMP test for testing the composite hypothesis $H : \theta = \theta_0$ against the composite class of alternatives $\theta = \theta_1$, which depends only on T.

(ii) Part (i) provides an alternative proof that the test of Example 3.8.1 is UMP.

[Let $\psi_0(t)$ be the most powerful level α test for testing θ_0 against θ_1 that depends only on t, let $\phi(x)$ be any level-α test, and let $\psi(t) = E_{\eta_1}[\phi(X) \mid t]$. Since $E_{\theta_i}\psi(T) = E_{\theta_i,\eta_1}\phi(X)$, it follows that ψ is a level-α test of H and its power, and therefore the power of ϕ, does not exceed the power of ψ_0.]

Note. For further discussion of this and related concepts of partial sufficiency see Fraser (1956), Dawid (1975), Sprott (1975), Basu (1978), and Barndorff-Nielsen (1978).

Section 3.9

Problem 3.61 Let $X_1, \ldots, X$ and $Y_1, \ldots, Y_n$ be independent samples from $N(\xi, 1)$ and $N(\eta, 1)$, and consider the hypothesis $H : \eta \leq \xi$ against $K : \eta > \xi$. There exists a UMP test, and it rejects the hypothesis when $\bar{Y} - \bar{X}$ is too large.

[If $\xi_1 < \eta_1$, is a particular alternative, the distribution assigning probability 1 to the point $\eta = \xi = (m\xi_1 + n\eta_1)/(m + n)$ is least favorable.]

Problem 3.62 Let $X_1, \ldots, X_m; Y_1, \ldots, Y_n$ be independently, normally distributed with means ξ and η, and variances a σ^2 and τ^2 respectively, and consider the hypothesis $H : \tau \leq \sigma$ a against $K : \sigma < \tau$.

(i) If ξ and η are known, there exists a UMP test given by the rejection region $\sum(Y_j - \eta)^2 / \sum(X_i - \xi)^2 \geq C$.

(ii) No UMP test exists when ξ and η are unknown.

Problem 3.63 Suppose X is a $k \times 1$ random vector with $E(|X|^2) < \infty$ and covariance matrix Σ. Let A be an $m \times k$ (nonrandom) matrix and let $Y = AX$. Show Y has mean vector $AE(X)$ and covariance matrix $A \Sigma A^T$.

Problem 3.64 Suppose $(X_1, \ldots, X_k)$ has the multivariate normal distribution with unknown mean vector $\xi = (\xi_1, \ldots, \xi_k)$ and known covariance matrix Σ. Suppose X_1 is independent of $(X_2, \ldots, X_k)$. Show that X_1 is partially sufficient for ξ_1 in the sense of Problem 3.60. Provide an alternative argument for Case 2 of Example 3.9.2.

Problem 3.65 In Example 3.9.2, Case 2, verify the claim for the least favorable distribution.

Problem 3.66 In Example 3.9.3, provide the details for Cases 3 and 4.

3.11 Notes

Hypothesis testing developed gradually, with early instances frequently being rather vague statements of the significance or nonsignificance of a set of observations. Isolated applications are found in the 18th century [Arbuthnot (1710), Daniel Bernoulli (1734), and Laplace (1773), for example] and centuries earlier in the Royal Mint's Trial of the Pyx [discussed by Stigler (1977)]. They became more frequent in the 19th century in the writings of such authors as Gavarret (1840), Lexis (1875, 1877), and Edgeworth (1885). A new stage began with the work of Karl Pearson, particularly his χ^2 paper of 1900, followed in the decade 1915–1925 by Fisher's normal theory and χ^2 tests. Fisher presented this work systematically in his enormously influential book *Statistical Methods for Research Workers* (1925b).

The first authors to recognize that the rational choice of a test must involve consideration not only of the hypothesis but also of the alternatives against which it is being tested were Neyman and F. S. Pearson (1928). They introduced the distinction between errors of the first and second kind, and thereby motivated their proposal of the likelihood-ratio criterion as a general method of test construction. These considerations were carried to their logical conclusion by Neyman and Pearson in their paper of 1933. in which they developed the theory of UMP tests. Accounts of their collaboration can be found in Pearson's recollections (1966), and in the biography of Neyman by Reid (1982).

The Neyman–Pearson lemma has been generalized in many directions, including the results in Sections 3.6, 3.8 and 3.9. Dantzig and Wald (1951) give necessary conditions including those of Theorem 3.6.1, for a critical function which maximizes an integral subject to a number of integral side conditions, to satisfy (3.28). The role of the Neyman–Pearson lemma in hypothesis testing is surveyed in Lehmann (1985a).

An extension to a selection problem, proposed by Birnbaum and Chapman (1950), is sketched in Problem 3.46. Further developments in this area are reviewed in Gibbons (1986, 1988). Grenander (1981) applies the fundamental lemma to problems in stochastic processes.

Lemmas 3.4.1, 3.4.2, and 3.7.1 are due to Lehmann (1961).

Complete class results for simple null hypothesis testing problems are obtained in Brown and Marden (1989).

The earliest example of confidence intervals appears to occur in the work of Laplace (1812). who points out how an (approximate) probability statement concerning the difference between an observed frequency and a binomial probability p can be inverted to obtain an associated interval for p. Other examples can be found in the work of Gauss (1816), Fourier (1826), and Lexis (1875). However, in all these cases, although the statements made are formally correct, the authors appear to consider the parameter as the variable which with the stated probability falls in the fixed confidence interval. The proper interpretation seems to have been pointed out for the first time by E. B. Wilson (1927). About the same time two examples of exact confidence statements were given by Working and Hotelling (1929) and Hotelling (1931).

A general method for obtaining exact confidence bounds for a real-valued parameter in a continuous distribution was proposed by Fisher (1930), who however later disavowed this interpretation of his work. For a discussion of Fisher's controversial concept of fiducial probability, see Section 5.7. At about the same time,[15] a completely general theory of confidence statements was developed by Neyman and shown by him to be intimately related to the theory of hypothesis testing. A detailed account of this work, which underlies the treatment given here, was published by Neyman in his papers of 1937 and 1938.

The calculation of p-values was the standard approach to hypothesis testing throughout the 19th century and continues to be widely used today. For various questions of interpretation, extensions, and critiques, see Cox (1977), Berger and Sellke (1987), Marden (1991), Hwang, Casella, Robert, Wells and Farrell (1992), Lehmann (1993), Robert (1994), Berger, Brown and Wolpert (1994), Meng (1994), Blyth and Staudte (1995, 1997), Liu and Singh (1997), Sackrowitz and Samuel-Cahn (1999), Marden (2000), Sellke et al. (2001), and Berger (2003).

Extensions of p-values to hypotheses with nuisance parameters is discussed by Berger and Boos (1994) and Bayarri and Berger (2000), and the large-sample

[15]Cf. Neyman (1941b).

behavior of p-values in Lambert and Hall (1982) and Robins et al. (2000). An optimality theory in terms of p-values is sketched by Schweder (1988), and p-values for the simultaneous testing of several hypotheses is treated by Schweder and Spjøtvoll (1982), Westfall and Young (1993), and by Dudoit et al. (2003).

An important use of p-values occurs in meta-analysis when one is dealing with the combination of results from independent experiments. The early literature on this topic is reviewed in Hedges and Olkin (1985, Chapter 3). Additional references are Marden (1982b, 1985), Scholz (1982) and a review article by Becker (1997). Associated confidence intervals are proposed by Littell and Louv (1981).

4
Unbiasedness: Theory and First Applications

4.1 Unbiasedness For Hypothesis Testing

A simple condition that one may wish to impose on tests of the hypothesis $H : \theta \in \Omega_H$ against the composite class of alternatives $K : \theta \in \Omega_K$ is that for no alternative in K should the probability of rejection be less than the size of the test. Unless this condition is satisfied, there will exist alternatives under which acceptance of the hypothesis is more likely than in some cases in which the hypothesis is true. A test ϕ for which the above condition holds, that is, for which the power function $\beta_\phi(\theta) = E_\theta \phi(X)$ satisfies

$$\begin{aligned} \beta_\phi(\theta) &\leq \alpha \quad \text{if} \quad \theta \in \Omega_H, \\ \beta_\phi(\theta) &\geq \alpha \quad \text{if} \quad \theta \in \Omega_K, \end{aligned} \tag{4.1}$$

is said to be *unbiased.* For an appropriate loss function this was seen in Chapter 1 to be a particular case of the general definition of unbiasedness given there. Whenever a UMP test exists, it is unbiased, since its power cannot fall below that of the test $\phi(x) \equiv \alpha$.

For a large class of problems for which a UMP test does not exist, there does exist a UMP unbiased test. This is the case in particular for certain hypotheses of the form $\theta \leq \theta_0$ or $\theta = \theta_0$, where the distribution of the random observables depends on other parameters besides θ.

When $\beta_\phi(\theta)$ is a continuous function of θ, unbiasedness implies

$$\beta_\phi(\theta) = \alpha \quad \text{for all} \quad \theta \text{ in } \omega, \tag{4.2}$$

where ω is the common boundary of Ω_H and Ω_K that is, the set of points θ that are points or limit points of both Ω_H and Ω_K. Tests satisfying this condition are said to be similar *on the boundary* (of H and K). Since it is more convenient to work with (4.2) than with (4.1), the following lemma plays an important role in the determination of UMP unbiased tests.

Lemma 4.1.1 *If the distributions P_θ are such that the power function of every test is continuous, and if ϕ_0 is UMP among all tests satisfying* (4.2) *and is a level-α test of H then ϕ_0 is UMP unbiased.*

PROOF. The class of tests satisfying (4.2) contains the class of unbiased tests, and hence ϕ_0 is uniformly at least as powerful as any unbiased test. On the other hand, ϕ_0 is unbiased, since it is uniformly at least as powerful as $\phi(x) \equiv \alpha$. ■

4.2 One-Parameter Exponential Families

Let θ be a real parameter, and $X = (X_1, \ldots, X_n)$ a random vector with probability density (with respect to some measure μ)

$$p_\theta(x) = C(\theta)e^{\theta T(x)}h(x).$$

It was seen in Chapter 3 that a UMP test exists when the hypothesis H and the class K of alternatives are given by (i) $H : \theta \leq \theta_0$, $K : \theta > \theta_0$ (Corollary 3.4.1) and (ii) $H : \theta \leq \theta_1$ or $\theta \geq \theta_2$ $(\theta_1 < \theta_2)$, $K : \theta_1 < \theta < \theta_2$ (Theorem 3.7.1), but not for (iii) $H : \theta_1 \leq \theta \leq \theta_2$, $K : \theta < \theta_1$ or $\theta > \theta_2$. We shall now show that in case (iii) there does exist a UMP unbiased test given by

$$\phi(x) = \begin{cases} 1 & \text{when} \quad T(x) < C_1 \text{ or } > C_2, \\ \gamma_i & \text{when} \quad T(x) = C_i, \quad i = 1, 2, \\ 0 & \text{when} \quad C_1 < T(x) < C_2, \end{cases} \tag{4.3}$$

where the C's and γ's are determined by

$$E_{\theta_1}\phi(X) = E_{\theta_2}\phi(X) = \alpha. \tag{4.4}$$

The power function $E_\theta\phi(X)$ is continuous by Theorem 2.7.1, so that Lemma 4.1.1 is applicable. The set ω consists of the two points θ_1 and θ_2, and we therefore consider first the problem of maximizing $E_{\theta'}\phi(X)$ for some θ' outside the interval $[\theta_1, \theta_2]$, subject to (4.4). If this problem is restated in terms of $1 - \phi(x)$, it follows from part (ii) of Theorem 3.7.1 that its solution is given by (4.3) and (4.4). This test is therefore UMP among those satisfying (4.4), and hence UMP unbiased by Lemma 4.1.1. It further follows from part (iii) of the theorem that the power function of the test has a minimum at a point between θ_1 and θ_2, and is strictly increasing as θ tends away from this minimum in either direction.

A closely related problem is that of testing (iv) $H : \theta = \theta_0$ against the alternatives $\theta \neq \theta_0$. For this there also exists a UMP unbiased test given by (4.3), but the constants are now determined by

$$E_{\theta_0}[\phi(X)] = \alpha \tag{4.5}$$

and

$$E_{\theta_0}[T(X)\phi(X)] = E_{\theta_0}[T(X)]\alpha. \tag{4.6}$$

To see this, let θ' be any particular alternative, and restrict attention to the sufficient statistic T, the distribution of which by Lemma 2.7.2, is of the form

$$dP_\theta(t) = C(\theta)e^{\theta t}\,d\nu(t).$$

Unbiasedness of a test $\psi(t)$ implies (4.5) with $\phi(x) = \psi[T(x)]$; also that the power function $\beta(\theta) = E_\theta[\psi(T)]$ must have a minimum at $\theta = \theta_0$. By Theorem 2.7.1, the function $\beta(\theta)$ is differentiable, and the derivative can be computed by differentiating $E_\theta\psi(T)$ under the expectation sign, so that for all tests $\psi(t)$

$$\beta'(\theta) = E_\theta[T\psi(T)] + \frac{C'(\theta)}{C(\theta)}E_\theta[\psi(T)].$$

For $\psi(t) \equiv \alpha$, this equation becomes

$$0 = E_\theta(T) + \frac{C'(\theta)}{C(\theta)}.$$

Substituting this in the expression for $\beta'(\theta)$ gives

$$\beta'(\theta) = E_\theta[T\psi(T)] - E_\theta(T)E_\theta[\psi(T)],$$

and hence unbiasedness implies (4.6) in addition to (4.5).

Let M be the set of points $(E_{\theta_0}[\psi(T)], E_{\theta_0}[T\psi(T)])$ as ψ ranges over the totality of critical functions. Then M is convex and contains all points $(u, uE_{\theta_0}(T))$ with $0 < u < 1$. It also contains points (α, u_2) with $u_2 > \alpha E_{\theta_0}(T)$. This follows from the fact that there exist tests with $E_{\theta_0}[\psi(T)] = \alpha$ and $\beta'(\theta_0) > 0$ (see Problem 3.45). Since similarly M contains points (α, u_1) with $u_1 < \alpha E_{\theta_0}(T)$, the point $(\alpha, \alpha E_{\theta_0}(T))$ is an inner point of M. Therefore, by Theorem 3.6.1(iv), there exist constants k_1, k_2 and a test $\psi(t)$ satisfying (4.5) and (4.6) with $\phi(x) = \psi[T(x)]$, such that $\psi(t) = 1$ when

$$C(\theta_0)(k_1 + k_2 t)e^{\theta_0 t} < C(\theta')e^{\theta' t}$$

and therefore when

$$a_1 + a_2 t < e^{bt}.$$

This region is either one-sided or the outside of an interval. By Theorem 3.4.1, a one-sided test has a strictly monotone power function and therefore cannot satisfy (4.6). Thus $\psi(t)$ is 1 when $t < C_1$ or $> C_2$, and the most powerful test subject to (4.5) and (4.6) is given by (4.3). This test is unbiased, as is seen by comparing it with $\phi(x) \equiv \alpha$. It is then also UMP unbiased, since the class of tests satisfying (4.5) and (4.6) includes the class of unbiased tests.

A simplification of this test is possible if for $\theta = \theta_0$ the distribution of T is symmetric about some point a, that is, if $P_{\theta_0}\{T < a - u\} = P_{\theta_0}\{T > a + u\}$ for all real u. Any test which is symmetric about a and satisfies (4.5) must also satisfy (4.6), since $E_{\theta_0}[T\psi(T)] = E_{\theta_0}[(T-a)\psi(T)] + aE_{\theta_0}\psi(T) = a\alpha = E_{\theta_0}(T)\alpha$. The C's and γ's are therefore determined by

$$P_{\theta_0}\{T < C_1\} + \gamma_1 P_{\theta_0}\{T = C_1\} = \tfrac{\alpha}{2},$$
$$C_2 = 2a - C_1, \quad \gamma_2 = \gamma_1.$$

The above tests of the hypotheses $\theta_1 \le \theta \le \theta_2$ and $\theta = \theta_0$ are *strictly unbiased* in the sense that the power is $> \alpha$ for all alternatives θ. For the first of these tests, given by (4.3) and (4.4), strict unbiasedness is an immediate consequence of Theorem 3.7.1(iii). This states in fact that the power of the test has a minimum at a point θ_0 between θ_1 and θ_2 and increases strictly as θ tends away from θ_0 in either direction. The second of the tests, determined by (4.3), (4.5), and (4.6), has a continuous power function with a minimum of α at $\theta = \theta_0$. Thus there exist $\theta_1 < \theta_0 < \theta_2$ such that $\beta(\theta_1) = \beta(\theta_2) = c$ where $\alpha \le c < 1$. The test therefore coincides with the UMP unbiased level-c test of the hypothesis $\theta_1 \le \theta \le \theta_2$, and the power increases strictly as θ moves away from θ_0 in either direction. This proves the desired result.

Example 4.2.1 (Binomial) Let X be the number of successes in n binomial trials with probability p of success. A theory to be tested assigns to p the value p_0, so that one wishes to test the hypothesis $H : p = p_0$. When rejecting H one will usually wish to state also whether p appears to be less or greater than p_0. If, however, the conclusion that $p \ne p_0$ in any case requires further investigation, the preliminary decision is essentially between the two possibilities that the data do or do not contradict the hypothesis $p = p_0$. The formulation of the problem as one of hypothesis testing may then be appropriate.

The UMP unbiased test of H is given by (4.3) with $T(X) = X$. The condition (4.5) becomes

$$\sum_{x=C_1+1}^{C_2-1} \binom{n}{x} p_0^x q_0^{n-x} + \sum_{i=1}^{2} (1-\gamma_i) \binom{n}{C_i} p_0^{C_i} q_0^{n-C_i} = 1 - \alpha,$$

and the left-hand side of this can be obtained from tables of the individual probabilities and cumulative distribution function of X. The condition (4.6), with the help of the identity

$$x \binom{n}{x} p_0^x q_0^{n-x} = np_0 \binom{n-1}{x-1} p_0^{x-1} q_0^{(n-1)-(x-1)}$$

reduces to

$$\sum_{x=C_1+1}^{C_2-1} \binom{n-1}{x-1} p_0^{x-1} q_0^{(n-1)-(x-1)} + \sum_{i=1}^{2} (1-\gamma_i) \binom{n-1}{C_i-1} p_0^{C_i-1} q_0^{(n-1)-(C_i-1)} = 1 - \alpha$$

the left-hand side of which can be computed from the binomial tables.

For sample sizes which are not too small, and values of p_0 which are not too close to 0 or 1, the distribution of X is therefore approximately symmetric. In this case, the much simpler "equal tails" test, for which the C's and γ's are determined by

$$\sum_{x=0}^{C_1-1} \binom{n}{x} p_0^x q_0^{(n-x)} + \gamma_1 \binom{n}{C_1} p_0^{C_1} q_0^{n-C_1} = \gamma_2 \binom{n}{C_2} p_0^{C_2} q_0^{n-C_2} + \sum_{x=C_2+1}^{n} \binom{n}{x} p_0^x q_0^{n-x} = \frac{\alpha}{2},$$

is approximately unbiased, and constitutes a reasonable approximation to the unbiased test. Note, however, that this approximation requires large sample sizes when p_0 is close to 0 or 1; in this connection, see Example 5.7.2 which discusses the corresponding problem of confidence intervals for p. The literature on this and other approximations to the binomial distribution is reviewed in Johnson, Kotz and Kemp (1992). See also the related discussion in Example 5.7.2. ■

Example 4.2.2 (Normal variance) Let $X = (X_1, \ldots, X_n)$ be a sample from a normal distribution with mean 0 and variance σ^2, so that the density of the X's is

$$\left(\frac{1}{\sqrt{2\pi}\sigma}\right) \exp\left(-\frac{1}{2\pi\sigma^2}\sum x_i^2\right).$$

Then $T(X) = \sum X_i^2$ is sufficient for σ^2, and has probability density $(1/\sigma^2)f_n(y/\sigma^2)$, where

$$f_n(y) = \frac{1}{2^{n/2}\Gamma(n/2)} y^{(n/2)-1} e^{(y/2)}, \qquad y > 0,$$

is the density of a χ^2-distribution with n degrees of freedom. For varying σ, these distributions form an exponential family, which arises also in problems of life testing (see Problem 2.15), and concerning normally distributed variables with unknown mean and variance (Section 5.3). The acceptance region of the UMP unbiased test of the hypothesis $H : \sigma = \sigma_0$ is

$$C_1 \le \sum \frac{x_i^2}{\sigma_0^2} \le C_2$$

with

$$\int_{C_1}^{C_2} f_n(y)\, dy = 1 - \alpha$$

and

$$\int_{C_1}^{C_2} y f_n(y)\, dy = \frac{(1-\alpha)E_{\sigma_0}(\sum X_i^2)}{\sigma_0^2} = n(1-\alpha).$$

For the determination of the constants from tables of the χ^2-distribution, it is convenient to use the identity

$$y f_n(y) = n f_{n+2}(y),$$

to rewrite the second condition as

$$\int_{C_1}^{C_2} f_{n+2}(y)\, dy = 1 - \alpha.$$

Alternatively, one can integrate $\int_{C_1}^{C_2} f_n(y)\, dy$ by parts to reduce the second condition to

$$C_1^{n/2} e^{-C_1/2} = C_2^{n/2} e^{-C_2/2}.$$

[For tables giving C_1 and C_2 see Pachares (1961).] Actually, unless n is very small or σ_0 very close to 0 or ∞, the equal-tails test given by

$$\int_0^{C_1} f_n(y)\, dy = \int_{C_2}^{\infty} f_n(y)\, dy = \frac{\alpha}{2}$$

is a good approximation to the unbiased test. This follows from the fact that T, suitably normalized, tends to be normally and hence symmetrically distributed for large n. ■

UMP unbiased tests of the hypotheses (iii) $H : \theta_1 \leq \theta \leq \theta_2$ and (iv) $H : \theta = \theta_0$ against two-sided alternatives exist not only when the family $p_\theta(x)$ is exponential but also more generally when it is strictly totally positive (STP_∞). A proof of (iv) in this case is given in Brown, Johnstone, and MacGibbon (1981); the proof of (iii) follows from Problem 3.53.

4.3 Similarity and Completeness

In many important testing problems, the hypothesis concerns a single real-valued parameter, but the distribution of the observable random variables depends in addition on certain nuisance parameters. For a large class of such problems a UMP unbiased test exists and can be found through the method indicated by Lemma 4.1.1. This requires the characterization of the tests ϕ, which satisfy

$$E_\theta \phi(X) = \alpha$$

for all distributions of X belonging to a given family $\mathcal{P}^X = \{P_\theta, \theta \in \omega\}$. Such tests are called *similar* with respect to $\mathcal{P}^X$ or ω, since if ϕ is nonrandomized with critical region S, the latter is "similar to the sample space" $\mathcal{X}$ in that both the probability $P_\theta\{X \in S\}$ and $P_\theta\{X \in \mathcal{X}\}$ are independent of $\theta \in \omega$.

Let T be a sufficient statistic for $\mathcal{P}^X$, and let $\mathcal{P}^T$ denote the family $\{P_\theta^T, \theta \in \omega\}$ of distributions of T as θ ranges over ω. Then any test satisfying[1]

$$E[\phi(X)|t] = \alpha \qquad \text{a.e. } \mathcal{P}^T \tag{4.7}$$

is similar with respect to $\mathcal{P}^X$, since then

$$E_\theta[\phi(X)] = E_\theta\{E[\phi(X)|T]\} = \alpha \qquad \text{for all} \quad \theta \in \omega.$$

A test satisfying (4.7) is said to have *Neyman structure* with respect to T. It is characterized by the fact that the conditional probability of rejection is α on each of the surfaces $T = t$. Since the distribution on each such surface is independent of θ for $\theta \in \omega$, the condition (4.7) essentially reduces the problem to that of testing a simple hypothesis for each value of t. It is frequently easy to obtain a most powerful test among those having Neyman structure, by solving the optimum problem on each surface separately. The resulting test is then most powerful among all similar tests provided every similar test has Neyman structure. A condition for this to be the case can be given in terms of the following definition.

A family $\mathcal{P}$ of probability distributions P is *complete* if

$$E_P[f(X)] = 0 \qquad \text{for all} \quad P \in \mathcal{P} \tag{4.8}$$

implies

$$f(x) = 0 \qquad \text{a.e. } \mathcal{P}. \tag{4.9}$$

[1]A statement is said to hold a.e. $\mathcal{P}$ if it holds except on a set N with $P(N) = 0$ for all $P \in \mathcal{P}$.

In applications, $\mathcal{P}$ will be the family of distributions of a sufficient statistic.

Example 4.3.1 Consider n independent trials with probability p of success, and let X_i be 1 or 0 as the ith trial is a success or failure. Then $T = X_1 + \cdots + X_n$ is a sufficient statistic for p, and the family of its possible distributions is $\mathcal{P} = \{b(p,n), 0 < p \le 1\}$. For this family (4.8) implies that

$$\sum_{t=0}^{n} f(t) \binom{n}{t} \rho^t = 0 \qquad \text{for all} \quad 0 < \rho < \infty,$$

where $\rho = p/(1-p)$. The left-hand side is a polynomial in ρ, all the coefficients of which must be zero. Hence $f(t) = 0$ for $t = 0, \ldots, n$ and the binomial family of distributions of T is complete. ■

Example 4.3.2 Let $X_1, \ldots, X_n$ be a sample from the uniform distribution $U(0,\theta)$, $0 < \theta < \infty$. Then $T = \max(X_1, \ldots, X_n)$ is a sufficient statistic for θ, and (4.8) becomes

$$\int f(t)\, dP_\theta^T(t) = n\theta^{-n} \int_0^\theta f(t) \cdot t^{n-1}\, dt = 0 \qquad \text{for all} \quad \theta.$$

Let $f(t) = f^+(t) - f^-(t)$ where f^+ and f^- denote the positive and negative parts of f respectively. Then

$$v^+(A) = \int_A f^+(t) t^{n-1}\, dt \text{ and } v^-(A) = \int_A f^-(t) t^{n-1}\, dt$$

are two measures over the Borel sets on $(0, \infty)$, which agree for all intervals and hence for all A. This implies $f^+(t) = f^-(t)$ except possibly on a set of Lebesgue measure zero, and hence $f(t) = 0$ a.e. $\mathcal{P}^T$. ■

Example 4.3.3 Let $X_1, \ldots, X_m$; $Y_1, \ldots, Y_n$ be independently normally distributed as $N(\xi, \sigma^2)$ and $N(\xi, \tau^2)$ respectively. Then the joint density of the variables is

$$C(\xi, \sigma, \tau) \exp\left(-\frac{1}{2\sigma^2}\sum x_i^2 + \frac{\xi}{\sigma^2}\sum x_i - \frac{1}{2\tau^2}\sum y_j^2 + \frac{\xi}{\tau^2}\sum y_j\right).$$

The statistic

$$T = \left(\sum X_i, \sum X_i^2, \sum Y_j, \sum Y_j^2\right)$$

is sufficient; it is, however, not complete, since $E(\sum Y_j/n - \sum X_i/m)$ is identically zero. If the Y's are instead distributed with a mean $E(Y) = \eta$ which varies independently of ξ, the set of possible values of the parameters $\theta_1 = -1/2\sigma^2, \theta_2 = \xi/\sigma^2, \theta_3 = -1/2\tau^2, \theta_4 = \eta/\tau^2$ contains a four-dimensional rectangle, and it follows from Theorem 4.3.1 below that $\mathcal{P}^T$ is complete. ■

Completeness of a large class of families of distributions including that of Example 4.3.1 is covered by the following theorem.

Theorem 4.3.1 *Let X be a random vector with probability distribution*

$$dP_\theta(x) = C(\theta) \exp\left[\sum_{j=1}^{k} \theta_j T_j(x)\right] d\mu(x),$$

and let $\mathcal{P}^T$ be the family of distributions of $T = (T_1(X), \dots, T_k(X))$ as θ ranges over the set ω. Then $\mathcal{P}^T$ is complete provided ω contains a k-dimensional rectangle.

PROOF. By making a translation of the parameter space one can assume without loss of generality that ω contains the rectangle

$$I = \{(\theta_1, \dots, \theta_k) : -a \leq \theta_j \leq a, j = 1, \dots, k\}$$

Let $f(t) = f^+(t) - f^-(t)$ be such that

$$E_\theta f(T) = 0 \qquad \text{for all} \quad \theta \in \omega.$$

Then for all $\theta \in I$, if ν denotes the measure induced in T-space by the measure μ,

$$\int e^{\sum \theta_j t_j} f^+(t)\, d\nu(t) = \int e^{\sum \theta_j t_j} f^-(t)\, d\nu(t)$$

and hence in particular

$$\int f^+(t)\, d\nu(t) = \int f^-(t)\, d\nu(t).$$

Dividing f by a constant, one can take the common value of these two integrals to be 1, so that

$$dP^+(t) = f^+(t)\, d\nu(t) \quad \text{and} \quad dP^-(t) = f^-(t)\, d\nu(t)$$

are probability measures, and

$$\int e^{\sum \theta_j t_j}\, dP^+(t) = \int e^{\sum \theta_j t_j}\, dP^-(t)$$

for all θ in I. Changing the point of view, consider these integrals now as functions of the complex variables $\theta_j = \xi_j + i\eta_j, j = 1, \dots, k$. For any fixed $\theta_1, \dots, \theta_{j-1}, \theta_{j+1}, \dots, \theta_k$ with real parts strictly between $-a$ and $+a$, they are by Theorem 2.7.1 analytic functions of θ_j in the strip $R_j : -a < \xi_j < a, -\infty < \eta_j < \infty$ of the complex plane. For $\theta_2, \dots, \theta_k$ fixed, real, and between $-a$ and a, equality of the integrals holds on the line segment $\{(\xi_1, \eta_1) : -a < \xi_1 < a, \eta_1 = 0\}$ and can therefore be extended to the strip R_1, in which the integrals are analytic. By induction the equality can be extended to the complex region $\{(\theta_1, \dots, \theta_k) : (\xi_j, \eta_j) \in R_j \text{ for } j = 1, \dots, k\}$. It follows in particular that for all real $(\eta_1, \dots, \eta_k)$

$$\int e^{i \sum \eta_j t_j}\, dP^+(t) = \int e^{i \sum \eta_j t_j}\, dP^-(t).$$

These integrals are the characteristic functions of the distributions P^+ and P^- respectively, and by the uniqueness theorem for characteristic functions,[2] the two distributions P^+ and P^- coincide. From the definition of these distributions it then follows that $f^+(t) = f^-(t)$ a.e. ν, and hence that $f(t) = 0$ a.e. $\mathcal{P}^T$, as was to be proved. ■

[2] See for example Section 26 of Billingsley (1995).

Example 4.3.4 (Nonparametric completeness.) Let $X_1, \dots, X_N$ be independently and identically distributed with cumulative distribution function $F \in \mathcal{F}$, where $\mathcal{F}$ is the family of all absolutely continuous distributions. Then the set of order statistics $T(X) = (X_{(1)}, \dots, X_{(N)})$ was shown to be sufficient for $\mathcal{F}$ in Section 2.6. We shall now prove it to be complete. Since, by Example 2.4.1, $T'(X) = (\sum X_i, \sum X_i^2, \dots, \sum X_i^N)$ is equivalent to $T(X)$ in the sense that both induce the same subfield of the sample space, $T'(X)$ is also sufficient and is complete if and only if $T(X)$ is complete. To prove the completeness of $T'(X)$ and thereby that of $T(X)$, consider the family of densities

$$f(X) = C(\theta_1, \dots, \theta_N) \exp(-x^{2N} + \theta_1 x + \cdots + \theta_N x^N),$$

where C is a normalizing constant. These densities are defined for all values of the θ's since the integral of the exponential is finite, and their distributions belong to $\mathcal{F}$. The density of a sample of size N is

$$C^N \exp\left(-\sum x_j^{2N} + \theta_1 \sum x_j + \ldots + \theta_N \sum x_j^N\right)$$

and these densities constitute an exponential family $\mathcal{F}_0$. By Theorem 4.3.1, $T'(X)$ is complete for $\mathcal{F}_0$ and hence also for $\mathcal{F}$, as was to be proved.

The same method of proof establishes also the following more general result. Let X_{ij}, $j = 1, \dots, N_i$, $i = 1, \dots, c$, be independently distributed with absolutely continuous distributions F_i, and let $X_i^{(1)} < \cdots < X_i^{(N_i)}$ denote the N_i observations $X_{i1}, \dots, X_{iN_i}$ arranged in increasing order. Then the set of order statistics

$$(X_1^{(1)}, \dots, X_1^{(N_1)}, \dots, X_c^{(1)}, \dots, X_c^{(N_c)})$$

is sufficient and complete for the family of distributions obtained by letting $F_1, \dots, F_c$ range over all distributions of $\mathcal{F}$. Here completeness is proved by considering the subfamily $\mathcal{F}_0$ of $\mathcal{F}$ in which the distributions F_i have densities of the form

$$f_i(x) = C_i(\theta_{i1}, \dots, \theta_{iN_i}) \exp\left(-x^{2N_i} + \theta_{i1} x + \ldots + \theta_{iN_i} x^{N_i}\right).$$

The result remains true if $\mathcal{F}$ is replaced by the family F_1 of continuous distributions. For a proof see Problem 4.13 or Bell, Blackwell, and Breiman (1960). For related results, see Mandelbaum and Rüschendorf (1987) and Mattner (1996). ■

For the present purpose the slightly weaker property of bounded completeness is appropriate, a family $\mathcal{P}$ of probability distributions being *boundedly complete* if for all bounded functions f, (4.8) implies (4.9). If $\mathcal{P}$ is complete it is a fortiori boundedly complete. An example if which $\mathcal{P}$ is boundedly complete but not complete is given in Problem 4.12. For additional examples, see Hoeffding (1977), Bar-Lev and Plachky (1989) and Mattner (1993).

Theorem 4.3.2 *Let X be a random variable with distribution $P \in \mathcal{P}$, and let T be a sufficient statistic for $\mathcal{P}$. Then a necessary and sufficient condition for all similar tests to have Neyman structure with respect to T is that the family $\mathcal{P}^T$ of distributions of T is boundedly complete.*

PROOF. Suppose first that $\mathcal{P}^T$ is boundedly complete, and let $\phi(X)$ be similar with respect to $\mathcal{P}$. Then

$$E[\phi(X) - \alpha] = 0 \qquad \text{for all} \quad P \in \mathcal{P}$$

and hence, if $\psi(t)$ denotes the conditional expectation of $\phi(X) - \alpha$ given t,

$$E\psi(T) = 0 \qquad \text{for all} \quad P^T \in \mathcal{P}^T.$$

Since $\psi(t)$ can be taken to be bounded by Lemma 2.4.1, it follows from the bounded completeness of $\mathcal{P}^T$ that $\psi(t) = 0$ and hence $E[\phi(X)|t] = \alpha$ a.e. $\mathcal{P}^T$, as was to be proved.

Conversely suppose that $\mathcal{P}^T$ is not boundedly complete. Then there exists a function f such that $|f(t)| \leq M$ for some M, that $Ef(T) = 0$ for all $P^T \in \mathcal{P}^T$ and $f(T) \neq 0$ with positive probability for some $P^T \in \mathcal{P}^T$. Let $\phi(t) = cf(t) + \alpha$, where $c = \min(\alpha, 1 - \alpha)/M$. Then ϕ is a critical function, since $0 \leq \phi(t) \leq 1$, and it is a similar test, since $E\phi(T) = \alpha$ for all $P^T \in \mathcal{P}^T$. But ϕ does not have Neyman structure, since $\phi(T) \neq \alpha$ with positive probability for at least some distribution in $\mathcal{P}^T$. ■

4.4 UMP Unbiased Tests for Multiparameter Exponential Families

An important class of hypotheses concerns a real-valued parameter in an exponential family, with the remaining parameters occurring as unspecified nuisance parameters. In many of these cases, UMP unbiased tests exist and can be constructed by means of the theory of the preceding section.

Let X be distributed according to

$$dP^X_{\theta,\vartheta}(x) = C(\theta, \vartheta) \exp\left[\theta U(X) + \sum_{i=1}^{k} \vartheta_i T_i(x)\right] d\mu(x), \qquad (\theta, \vartheta) \in \Omega, \qquad (4.10)$$

and let $\vartheta = (\vartheta_1, \dots, \vartheta_k)$ and $T = (T_1, \dots, T_k)$. We shall consider the problems[3] of testing the following hypotheses H_j against the alternatives K_j, $j = 1, \dots, 4$:

$$\begin{array}{ll} H_1 : \theta \leq \theta_0 & K_1 : \theta > \theta_0 \\ H_2 : \theta \leq \theta_1 \text{ or } \theta \geq \theta_2 & K_2 : \theta_1 < \theta < \theta_2 \\ H_3 : \theta_1 \leq \theta \leq \theta_2 & K_3 : \theta < \theta_1 \text{ or } \theta > \theta_2 \\ H_4 : \theta = \theta_0 & K_4 : \theta \neq \theta_0. \end{array}$$

We shall assume that the parameter space Ω is convex, and that it is not contained in a linear space of dimension $< k + 1$. This is the case in particular when Ω is the natural parameter space of the exponential family. We shall also assume that there are points in Ω with θ both $<$ and $>$ θ_0, θ_1, and θ_2 respectively.

[3]Such problems are also treated in Johansen (1979), which in addition discusses large sample tests of hypotheses specifying more than one parameter.

Attention can be restricted to the sufficient statistics (U, T) which have the joint distribution

$$dP_{\theta,\vartheta}^{U,T}(u,t) = C(\theta,\vartheta)\exp\left(\theta U + \sum_{i=1}^{k}\vartheta_i t_i\right) d\nu(u,t), \qquad (\theta,\vartheta) \in \Omega. \tag{4.11}$$

When $T = t$ is given, U is the only remaining variable and, by Lemma 2.7.2, the conditional distribution of U given t constitutes an exponential family

$$dP_\theta^{U|t}(u) = C_t(\theta)e^{\theta u}\, d\nu_t(u).$$

In this conditional situation there exists by Corollary 3.4.1 a UMP test for testing H_1, with critical function ϕ_1, satisfying

$$\phi(u,t) = \begin{cases} 1 & \text{when } u > C_0(t), \\ \gamma_0(t) & \text{when } u = C_0(t), \\ 0 & \text{when } u < C_0(t), \end{cases} \tag{4.12}$$

where the functions C_0 and γ_0 are determined by

$$E_{\theta_0}[\phi_1(U,T)|t] = \alpha \qquad \text{for all } t. \tag{4.13}$$

For testing H_2 in the conditional family there exists by Theorem 3.7.1 a UMP test with critical function

$$\phi(u,t) = \begin{cases} 1 & \text{when } C_1(t) < u < C_2(t), \\ \gamma_i(t) & \text{when } u = C_i(t), \quad i = 1,2, \\ 0 & \text{when } u < C_1(t) \text{ or } > C_2(t), \end{cases} \tag{4.14}$$

where the C's and γ's are determined by

$$E_{\theta_1}[\phi_2(U,T)|t] = E_{\theta_2}[\phi_2(U,T)|t] = \alpha. \tag{4.15}$$

Consider next the test ϕ_3 satisfying

$$\phi(u,t) = \begin{cases} 1 & \text{when } u < C_1(t) \text{ or } > C_2(t), \\ \gamma_i(t) & \text{when } u = C_i(t), \quad i = 1,2, \\ 0 & \text{when } C_1(t) < u < C_2(t), \end{cases} \tag{4.16}$$

with the C's and γ's determined by

$$E_{\theta_1}[\phi_3(U,T)|t] = E_{\theta_2}[\phi_3(U,T)|t] = \alpha. \tag{4.17}$$

When $T = t$ is given, this is (by Section 4.2 of the present chapter) UMP unbiased for testing H_3 and UMP among all tests satisfying (4.17).

Finally, let ϕ_4 be a critical function satisfying (4.16) with the C's and γ's determined by

$$E_{\theta_0}[\phi_4(U,T)|t] = \alpha \tag{4.18}$$

and

$$E_{\theta_0}[U\phi_4(U,T)|t] = \alpha E_{\theta_0}[U|t]. \tag{4.19}$$

Then given $T = t$, it follows again from the results of Section 4.2 that ϕ_4 is UMP unbiased for testing H_4 and UMP among all tests satisfying (4.18) and (4.19).

So far, the critical functions ϕ_j have been considered as conditional tests given $T = t$. Reinterpreting them now as tests depending on U and T for the hypotheses concerning the distribution of X (or the joint distribution of U and T) as originally stated, we have the following main theorem.[4]

Theorem 4.4.1 *Define the critical functions ϕ_1 by (4.12) and (4.13); ϕ_2 by (4.14) and (4.15); ϕ_3 by (4.16) and (4.17); ϕ_4 by (4.16), (4.18), and (4.19). These constitute UMP unbiased level-α tests for testing the hypotheses $H_1, \ldots, H_4$ respectively when the joint distribution of U and T is given by (4.11).*

PROOF. The statistic T is sufficient for ϑ if θ has any fixed value, and hence T is sufficient for each

$$\omega_j = \{(\theta, \vartheta) : (\theta, \vartheta) \in \Omega, \theta = \theta_j\}, \qquad j = 0, 1, 2.$$

By Lemma 2.7.2, the associated family of distributions of T is given by

$$dP^T_{\theta_j,\vartheta}(t) = C(\theta_j, \vartheta) \exp\left(\sum_{i=1}^{k} \vartheta_i t_i\right) d\nu_{\theta_j}(t), \quad (\theta_j, \vartheta) \in \omega_j \qquad j = 0, 1, 2.$$

Since by assumption Ω is convex and of dimension $k + 1$ and contains points on both sides of $\theta = \theta_j$, it follows that ω_j is convex and of dimension k. Thus ω_j contains a k-dimensional rectangle; by Theorem 4.3.1 the family

$$\mathcal{P}^T_j = \left\{P^T_{\theta_j,\vartheta} : (\theta, \vartheta) \in \omega_j\right\}$$

is complete; and similarity of a test ϕ on ω_j implies

$$E_{\theta_j}[\phi(U, T)|t] = \alpha.$$

(1) Consider first H_1. By Theorem 2.7.1, the power function of all tests is continuous for an exponential family. It is therefore enough to prove ϕ_1 to be UMP among all tests that are similar on ω_0 (Lemma 4.1.1), and hence among those satisfying (4.13). On the other hand, the overall power of a test ϕ against an alternative (θ, ϑ) is

$$E_{\theta,\vartheta}[\phi(U, T)] = \int \left[\int \phi(u, t)\, dP^{U|t}_\theta(u)\right] dP^T_{\theta,\vartheta}(t). \tag{4.20}$$

One therefore maximizes the overall power by maximizing the power of the conditional test, given by the expression in brackets, separately for each t. Since ϕ_1 has the property of maximizing the conditional power against any $\theta > \theta_0$ subject to (4.13), this establishes the desired result.
(2) The proof for H_2 and H_3 is completely analogous. By Lemma 4.1.1, it is enough to prove ϕ_2 and ϕ_3 to be UMP among all tests that are similar on both ω_1 and ω_2, and hence among all tests satisfying (4.15). For each t, ϕ_2 and ϕ_3 maximize the conditional power for their respective problems subject to this condition and therefore also the unconditional power.

---

[4] A somewhat different asymptotic optimality property of these tests is established by Michel (1979).

(3) Unbiasedness of a test of H_4 implies similarity on ω_0 and

$$\frac{\partial}{\partial\theta}\left[E_{\theta,\vartheta}\phi(U,T)\right] = 0 \qquad \text{on } \omega_0.$$

The differentiation on the left-hand side of this equation can be carried out under the expectation sign, and by the computation which earlier led to (4.6), the equation is seen to be equivalent to

$$E_{\theta,\vartheta}[U\phi(U,T) - \alpha U] = 0 \qquad \text{on } \omega_0.$$

Therefore, since $\mathcal{P}_0^T$ is complete, unbiasedness implies (4.18) and (4.19). As in the preceding cases, the test, which in addition satisfies (4.16), is UMP among all tests satisfying these two conditions. That it is UMP unbiased now follows, as in the proof of Lemma 4.1.1, by comparison with the test $\phi(u,t) \equiv \alpha$.

(4) The functions $\phi_1, \ldots, \phi_4$ were obtained above for each fixed t as a function of u. To complete the proof it is necessary to show that they are jointly measurable in u and t, so that the expectation (4.20) exists. We shall prove this here for the case of ϕ_1; the proof for the other cases is sketched in Problems 4.21 and 4.22. To establish the measurability of ϕ_1, one needs to show that the functions $C_0(t)$ and $\gamma_0(t)$ defined by (4.12) and (4.13) are t-measurable. Omitting the subscript 0, and denoting the conditional distribution function of U given $T = t$ and for $\theta = \theta_0$ by

$$F_t(u) = P_{\theta_0}\{U \leq u | t\},$$

one can rewrite (4.13) as

$$F_t(C) - \gamma[F_t(C) - F_t(C-0)] = 1 - \alpha.$$

Here $C = C(t)$ is such that $F_t(C-0) \leq 1-\alpha \leq F_t(C)$, and hence

$$C(t) = F_t^{-1}(1-\alpha)$$

where $F_t^{-1}(y) = \inf\{u : F_t(u) \geq y\}$. It follows that $C(t)$ and $\gamma(t)$ will both be measurable provided $F_t(u)$ and $F_t(u-0)$ are jointly measurable in u and t and $F_t^{-1}(1-\alpha)$ is measurable in t.

For each fixed u the function $F_t(u)$ is a measurable function of t, and for each fixed t it is a cumulative distribution function and therefore in particular nondecreasing and continuous on the right. From the second property it follows that $F_t(u) \geq c$ if and only if for each n there exists a rational number r such that $u \leq r < u + 1/n$ and $F_t(r) \geq c$. Therefore, if the rationals are denoted by $r_1, r_2, \ldots,$

$$\{(u,t) : F_t(u) \geq c\} = \bigcap_n \bigcup_i \left\{(u,t) : 0 \leq r_i - u < \frac{1}{n}, F_t(r_i) \geq c\right\}$$

This shows that $F_t(u)$ is jointly measurable in u and t. The proof for $F_t(u-0)$ is completely analogous. Since $F_t^{-1}(y) \leq u$ if and only if $F_t(u) \geq y$, $F_t^{-1}(y)$ is t-measurable for any fixed y and this completes the proof. ■

The test ϕ_1 of the above theorem is also UMP unbiased if Ω is replaced by the set $\Omega' = \Omega \cap \{(\theta,\vartheta) : \theta \geq \theta_0\}$, and hence for testing $H' : \theta = \theta_0$ against $\theta > \theta_0$. The assumption that Ω should contain points with $\theta < \theta_0$ was in fact used only to prove that the boundary set ω_0 contains a k-dimensional rectangle, and this remains valid if Ω is replaced by Ω'.

The remainder of this chapter as well as the next chapter will be concerned mainly with applications of the preceding theorem to various statistical problems. While this provides the most expeditious proof that the tests in all these cases are UMP unbiased, there is available also a variation of the approach, which is more elementary. The proof of Theorem 4.4.1 is quite elementary except for the following points: (i) the fact that the conditional distributions of U given $T = t$ constitute an exponential family, (ii) that the family of distributions of T is complete, (iii) that the derivative of $E_{\theta,\vartheta}\phi(U,T)$ exists and can be computed by differentiating under the expectation sign, (iv) that the functions $\phi_1, \ldots, \phi_4$ are measurable. Instead of verifying (i) through (iv) in general, as was done in the above proof, it is possible in applications of the theorem to check these conditions directly for each specific problem, which in some cases is quite easy.

Through a transformation of parameters, Theorem 4.4.1 can be extended to cover hypotheses concerning parameters of the form

$$\theta^* = a_0\theta + \sum_{i=1}^{k} a_i\vartheta_i, \qquad a_0 \neq 0.$$

This transformation is formally given by the following lemma, the proof of which is immediate.

Lemma 4.4.1 *The exponential family of distributions (4.10) can also be written as*

$$dP_{\theta,\vartheta}^X = K(\theta^*, \vartheta) \exp\left[\theta^* U^*(x) + \sum \vartheta_i T_i^*(x)\right] d\mu(x)$$

where

$$U^* = \frac{U}{a_0}, \qquad T_i^* = T_i - \frac{a_i}{a_0}U.$$

Application of Theorem 4.4.1 to the form of the distributions given in the lemma leads to UMP unbiased tests of the hypothesis $H_1^* : \theta^* \leq \theta_0$ and the analogously defined hypotheses H_2^*, H_3^*, H_4^*.

When testing one of the hypotheses H_j one is frequently interested in the power $\beta(\theta', \vartheta)$ of ϕ_j against some alternative θ'. As is indicated by the notation and is seen from (4.20), this power will usually depend on the unknown nuisance parameters ϑ. On the other hand, the power of the conditional test given $T = t$,

$$\beta(\theta'|t) = E_{\theta'}[\phi(U,T)|t],$$

is independent of ϑ and therefore has a known value.

The quantity $\beta(\theta'|t)$ can be interpreted in two ways: (i) It is the probability of rejecting H when $T = t$. Once T has been observed to have the value t, it may be felt, at least in certain problems, that this is a more appropriate expression of the power in the given situation than $\beta(\theta', \vartheta)$, which is obtained by averaging $\beta(\theta'|t)$ with respect to other values of t not relevant to the situation at hand. This argument leads to difficulties, since in many cases the conditioning could be carried even further and it is not clear where the process should stop. (ii) A more clear-cut interpretation is obtained by considering $\beta(\theta'|t)$ as an estimate of $\beta(\theta', \vartheta)$. Since

$$E_{\theta',\vartheta}[\beta(\theta'|T)] = \beta(\theta', \vartheta),$$

this estimate is unbiased in the sense of equation (1.11). It follows further from the theory of unbiased estimation and the completeness of the exponential family that among all unbiased estimates of $\beta(\theta', \vartheta)$ the present one has the smallest variance. (See *TPE*2, Chapter 2.)

Regardless of the interpretation, $\beta(\theta'|t)$ has the disadvantage compared with an unconditional power that it becomes available only after the observations have been taken. It therefore cannot be used to plan the experiment and in particular to determine the sample size, if this must be done prior to the experiment. On the other hand, a simple sequential procedure guaranteeing a specified power β against the alternatives $\theta = \theta'$ is obtained by continuing taking observations until the conditional power $\beta(\theta'|t)$ is $\geq \beta$.

4.5 Comparing Two Poisson or Binomial Populations

A problem arising in many different contexts is the comparison of two treatments or of one treatment with a control situation in which no treatment is applied. If the observations consist of the number of successes in a sequence of trials for each treatment, for example the number of cures of a certain disease, the problem becomes that of testing the equality of two binomial probabilities. If the basic distributions are Poisson, for example in a comparison of the radioactivity of two substances, one will be testing the equality of two Poisson distributions.

When testing whether a treatment has a beneficial effect by comparing it with the control situation of no treatment, the problem is of the one-sided type. If ξ_2 and ξ_1 denote the parameter values when the treatment is or is not applied, the class of alternatives is $K : \xi_2 > \xi_1$. The hypothesis is $\xi_2 = \xi_1$ if it is known a priori that there is either no effect or a beneficial one; it is $\xi_2 \leq \xi_1$ if the possibility is admitted that the treatment may actually be harmful. Since the test is the same for the two hypotheses, the second somewhat safer hypothesis would seem preferable in most cases.

A one-sided formulation is sometimes appropriate also when a new treatment or process is being compared with a standard one, where the new treatment is of interest only if it presents an improvement. On the other hand, if the two treatments are on an equal footing, the hypothesis $\xi_2 = \xi_1$ of equality of two treatments is tested against the two-sided alternatives $\xi_2 \neq \xi_1$. The formulation of this problem as one of hypothesis testing is usually quite artificial, since in case of rejection of the hypothesis one will obviously wish to know which of the treatments is better.[5] Such two-sided tests do, however, have important applications to the problem of obtaining confidence limits for the extent by which one treatment is better than the other. They also arise when the parameter ξ does not measure a treatment effect but refers to an auxiliary variable which one hopes can be ignored. For example, ξ_1 and ξ_2 may refer to the effect of two

[5]The comparison of two treatments as a three-decision problem or as the simultaneous testing of two one-sided hypotheses is discussed and the literature reviewed in Shaffer (2002).

different hospitals in a medical investigation in which one would like to combine the patients into a single study group. (In this connection, see also Section 7.3.)

To apply Theorem 4.4.1 to this comparison problem it is necessary to express the distributions in an exponential form with $\theta = f(\xi_1, \xi_2)$, for example $\theta = \xi_2 - \xi_1$ or ξ_2/ξ_1, such that the hypotheses of interest become equivalent to those of Theorem 4.4.1. In the present section the problem will be considered for Poisson and binomial distributions; the case of normal distributions will be taken up in Chapter 5.

We consider first the Poisson problem in which X and Y are independently distributed according to $P(\lambda)$ and $P(\mu)$, so that their joint distribution can be written as

$$P\{X = x, Y = y\} = \frac{e^{-(\lambda+\mu)}}{x!y!} \exp\left[y \log \frac{\mu}{\lambda} + (x+y)\log\lambda\right].$$

By Theorem 4.4.1 there exist UMP unbiased tests of the four hypotheses $H_1, \ldots, H_4$ concerning the parameter $\theta = \log(\mu/\lambda)$ or equivalently concerning the ratio $\rho = \mu/\lambda$. This includes in particular the hypotheses $\mu \leq \lambda$ (or $\mu = \lambda$) against the alternatives $\mu > \lambda$, and $\mu = \lambda$ against $\mu \neq \lambda$. Comparing the distribution of (X, Y) with (4.10), one has $U = Y$ and $T = X + Y$, and by Theorem 4.4.1 the tests are performed conditionally on the integer points of the line segment $X + Y = t$ in the positive quadrant of the (x, y) plane. The conditional distribution of Y given $X + Y = t$ is (Problem 2.14)

$$P\{Y = y | X + Y = t\} = \binom{t}{y}\left(\frac{\mu}{\lambda+\mu}\right)^y \left(\frac{\lambda}{\lambda+\mu}\right)^{t-y}, \qquad y = 0, 1, \ldots, t,$$

the binomial distribution corresponding to t trials and probability $p = \mu/(\lambda+\mu)$ of success. The original hypotheses therefore reduce to the corresponding ones about the parameter p of a binomial distribution. The hypothesis $H : \mu \leq a\lambda$, for example, becomes $H : p \leq a/(a+1)$, which is rejected when Y is too large. The cutoff point depends of course, in addition to a, also on t. It can be determined from tables of the binomial, and for large t approximately from tables of the normal distribution.

In many applications the ratio $\rho = \mu/\lambda$ is a reasonable measure of the extent to which the two Poisson populations differ, since the parameters λ and μ measure the rates (in time or space) at which two Poisson processes produce the events in question. One might therefore hope that the power of the above tests depends only on this ratio, but this is not the case. On the contrary, for each fixed value of ρ corresponding to an alternative to the hypothesis being tested, the power $\beta(\lambda, \mu) = \beta(\lambda, \rho\lambda)$ is an increasing function of λ, which tends to 1 as $\lambda \to \infty$ and to α as $\lambda \to 0$. To see this consider the power $\beta(\rho|t)$ of the conditional test given t. This is an increasing function of t, since it is the power of the optimum test based on t binomial trials. The conditioning variable T has a Poisson distribution with parameter $\lambda(1+\rho)$, and its distribution for varying λ forms an exponential family. It follows Lemma 3.4.2 that the overall power $E[\beta(\rho|T)]$ is an increasing function of λ. As $\lambda \to 0$ or ∞, T tends in probability to 0 or ∞, and the power against a fixed alternative ρ tends to α or 1.

The above test is also applicable to samples $X_1, \ldots, X_m$ and $Y_1, \ldots, Y_n$ from two Poisson distributions. The statistics $X = \sum_{i=1}^m X_i$ and $Y = \sum_{j=1}^n Y_j$ are then sufficient for λ and μ, and have Poisson distributions with parameters $m\lambda$

and $n\mu$ respectively. In planning an experiment one might wish to determine $m = n$ so large that the test of, say, $H : \rho \leq \rho_0$ has power against a specified alternative ρ_1 greater than or equal to some preassigned β. However, it follows from the discussion of the power function for $n = 1$, which applies equally to any other n, that this cannot be achieved for any fixed n, no matter how large. This is seen more directly by noting that as $\lambda \to 0$, for both $\rho = \rho_0$ and $\rho = \rho_1$, the probability of the event $X = Y = 0$ tends to 1. Therefore, the power of any level-α test against $\rho = \rho_1$ and for varying λ cannot be bounded away from α. This difficulty can be overcome only by permitting observations to be taken sequentially. One can for example determine t_0 so large that the test of the hypothesis $p_1 \leq \rho_0/(1+\rho_0)$ on the basis of t_0 binomial trials has power $\geq \beta$ against the alternative $p_1 = \rho_1/(1+\rho_1)$. By observing $(X_1, Y_1), (X_2, Y_2), \ldots$ and continuing until $\sum(X_i + Y_i) \geq t_0$, one obtains a test with power $\geq \beta$ against all alternatives with $\rho \geq \rho_1$.[6]

The corresponding comparison of two binomial probabilities is quite similar. Let X and Y be independent binomial variables with joint distribution

$$
\begin{aligned}
P\{X = x, Y = y\} &= \binom{m}{x} p_1^x q_1^{m-x} \binom{n}{y} p_2^y q_2^{n-y} \\
&= \binom{m}{x}\binom{n}{y} q_1^m q_2^n \exp\left[y\left(\log\frac{p_2}{q_2} - \log\frac{p_1}{q_1}\right)\right. \\
&\qquad\qquad \left. +(x+y)\log\frac{p_1}{q_1}\right].
\end{aligned}
$$

The four hypotheses $H_1, \ldots, H_4$, can then be tested concerning the parameter

$$\theta = \log\left(\frac{p_2}{q_2} \Big/ \frac{p_1}{q_1}\right),$$

or equivalently concerning the *odds ratio* (also called *cross-product ratio*)

$$\rho = \frac{p_2}{q_2} \Big/ \frac{p_1}{q_1}$$

This includes in particular the problems of testing $H_1' : p_2 \leq p_1$ against $p_2 > p_1$ and $H_4' : p_2 = p_1$ against $p_2 \neq p_1$. As in the Poisson case, $U = Y$ and $T = X + Y$, and the test is carried out in terms of the conditional distribution of Y on the line segment $X + Y = t$. This distribution is given by

$$P\{Y = y | X + Y = t\} = C_t(\rho)\binom{m}{t-y}\binom{n}{y}\rho^y, \qquad y = 0, 1, \ldots, t, \tag{4.21}$$

where

$$C_t(\rho) = \frac{1}{\sum_{y'=0}^{t}\binom{m}{t-y'}\binom{n}{y'}\rho^{y'}}.$$

[6]A discussion of this and alternative procedures for achieving the same aim is given by Birnbaum (1954a).

In the particular case of the hypotheses H'_1 and H'_4, the boundary value θ_0 of (4.13), (4.18), and (4.19) is 0, and the corresponding value of ρ is $\rho_0 = 1$. The conditional distribution then reduces to

$$P\{Y = y | X + Y = t\} = \frac{\binom{m}{t-y}\binom{n}{y}}{\binom{m+n}{t}},$$

which is the hypergeometric distribution.

Tables of critical values by Finney (1948) are reprinted in *Biometrika Tables for Statisticians*, Vol. 1, Table 38 and are extended in Finney, Latscha, Bennett, Hsu, and Horst (1963, 1966). Somewhat different ranges are covered in Armsen (1955), and related charts are provided by Bross and Kasten (1957). Extensive tables of the hypergeometric distributions have been computed by Lieberman and Owen (1961). Various approximations are discussed in Johnson, Kotz and Kemp (1992, Section 6.5). Critical values can also be easily computed with built-in functions of statistical packages such as `R`.[7]

The UMP unbiased test of $\rho_1 = \rho_2$, which is based on the (conditional) hypergeometric distribution, requires randomization to obtain an exact conditional level α for each t of the sufficient statistic T. Since in practice randomization is usually unacceptable, the one-sided test is frequently performed by rejecting when $Y \geq C(T)$, where $C(t)$ is the smallest integer for which $P\{Y \geq C(T) | T = t\} \leq \alpha$. This conservative test is called *Fisher's exact test* [after the treatment given in Fisher (1934a)], since the probabilities are calculated from the exact hypergeometric rather than an approximate normal distribution. The resulting conditional levels (and hence the unconditional level) are often considerably smaller than α, and this results in a substantial loss of power. An approximate test whose overall level tends to be closer to α is obtained by using the normal approximation to the hypergeometric distribution *without* continuity correction. [For a comparison of this test with some competitors, see e.g. Garside and Mack (1976).] A nonrandomized test that provides a conservative overall level, but that is less conservative than the "exact" test, is described by Boschloo (1970) and by McDonald, Davis, and Milliken (1977). For surveys of the extensive literature on these and related aspects of 2 × 2 and more generally $r \times c$ tables, see Agresti (1992, 2002), Sahai and Khurshid (1995) and Martín and Tapia (1998).

4.6 Testing for Independence in a 2 × 2 Table

Two characteristics A and B, which each member of a population may or may not possess, are to be tested for independence. The probabilities or proportion of individuals possessing properties A and B are denoted $P(A)$ and $P(B)$.

If $P(A)$ and $P(B)$ are unknown, a sample from one of the categories such as A does not provide a basis for distinguishing between the hypothesis and the alternatives. This follows from the fact that the number in the sample possessing characteristic B then constitutes a binomial variable with probability $p(B|A)$, which is completely unknown both when the hypothesis is true and when it is

[7]This package can be downloaded for free from `http://cran.r-project.org/`.

false. The hypothesis can, however, be tested if samples are taken both from categories A and A^c, the complement of A, or both from B and B^c. In the latter case, for example, if the sample sizes are m and n, the numbers of cases possessing characteristic A in the two samples constitute independent variables with binomial distributions $b(p_1, m)$ and $b(p_2, n)$ respectively, where $p_1 = P(A|B)$ and $p_2 = P(A|B^c)$. The hypothesis of independence of the two characteristics, $P(A|B) = p(A)$, is then equivalent to the hypothesis $p_1 = p_2$ and the problem reduces to that treated in the preceding section.

Instead of selecting samples from two of the categories, it is frequently more convenient to take the sample at random from the population as a whole. The results of such a sample can be summarized in the following 2×2 contingency table, the entries of which give the numbers in the various categories:

	A	A^c	
B	X	X'	M
B^c	Y	Y'	N
	T	T'	s

The joint distribution of the variables X, X', Y, and Y' is multinomial, and is given by

$$\begin{aligned} P\{X &= x, X' = x', Y = y, Y' = y'\} \\ &= \frac{s!}{x!x'!y!y'!} p_{AB}^x p_{A^cB}^{x'} p_{AB^c}^y p_{A^cB^c}^{y'} \\ &= \frac{s!}{x!x'!y!y'!} p_{A^cB^c}^s \exp\left(x \log \frac{p_{AB}}{p_{A^cB^c}} + x' \log \frac{p_{A^cB}}{p_{A^cB^c}} + y \log \frac{p_{AB^c}}{p_{A^cB^c}}\right). \end{aligned}$$

Lemma 4.4.1 and Theorem 4.4.1 are therefore applicable to any parameter of the form

$$\theta^* = a_0 \log \frac{p_{AB}}{p_{A^cB^c}} + a_1 \log \frac{p_{A^cB}}{p_{A^cB^c}} + a_2 \log \frac{p_{AB^c}}{p_{A^cB^c}}.$$

Putting $a_1 = a_2 = 1$, $a_0 = -1$, $\Delta = e^{\theta^*} = (p_{A^cB} p_{AB^c})/(p_{AB} p_{A^cB^c})$, and denoting the probabilities of A and B in the population by $p_A = p_{AB} + p_{AB^c}$, $p_B = p_{AB} + p_{A^cB}$, one finds

$$\begin{aligned} p_{AB} &= p_A p_B + \frac{1-\Delta}{\Delta} p_{A^cB} p_{AB^c}, \\ p_{A^cB} &= p_{A^c} p_B + \frac{1-\Delta}{\Delta} p_{A^cB} p_{AB^c}, \\ p_{AB^c} &= p_A p_{B^c} + \frac{1-\Delta}{\Delta} p_{A^cB} p_{AB^c}, \\ p_{A^cB^c} &= p_{A^c} p_{B^c} + \frac{1-\Delta}{\Delta} p_{A^cB} p_{AB^c}. \end{aligned}$$

Independence of A and B is therefore equivalent to $\Delta = 1$, and $\Delta < 1$ and $\Delta > 1$ correspond to positive and negative dependence respectively.[8]

The test of the hypothesis of independence, or any of the four hypotheses concerning Δ, is carried out in terms of the conditional distribution of X given $X + X' = m$, $X + Y = t$. Instead of computing this distribution directly, consider first the conditional distribution subject only to the condition $X + X' = m$, and hence $Y + Y' = s - m = n$. This is seen to be

$$\begin{aligned} P\{X &= x, Y = y | X + X' = m\} \\ &= \binom{m}{x}\binom{n}{y}\left(\frac{p_{AB}}{p_B}\right)^x \left(\frac{p_{A^cB}}{p_B}\right)^{m-x} \left(\frac{p_{AB^c}}{p_{B^c}}\right)^y \left(\frac{p_{A^cB^c}}{p_{B^c}}\right)^{n-y}, \end{aligned}$$

which is the distribution of two independent binomial variables, the number of successes in m and n trials with probability $p_1 = p_{AB}/p_B$ and $p_2 = p_{AB^c}/p_{B^c}$. Actually, this is clear without computation, since we are now dealing with samples of fixed size m and n from the subpopulations B and B^c and the probability of A in these subpopulations is p_1 and p_2. If now the additional restriction $X + Y = t$ is imposed, the conditional distribution of X subject to the two conditions $X + X' = m$ and $X + Y = t$ is the same as that of X given $X + Y = t$ in the case of two independent binomials considered in the previous section. It is therefore given by

$$P\{X = x | X + X' = m, X + Y = t\} = C_t(\rho)\binom{m}{x}\binom{n}{t-x}\rho^{t-x}, \qquad x = 0, \ldots, t,$$

that is, by (4.21) expressed in terms of x instead of y. (Here the choice of X as testing variable is quite arbitrary; we could equally well again have chosen Y.) For the parameter ρ one finds

$$\rho = \frac{p_2}{q_2} \Big/ \frac{p_1}{q_1} = \frac{p_{A^cB}\,p_{AB^c}}{p_{AB}\,p_{A^cB^c}} = \Delta.$$

From these considerations it follows that the conditional test given $X + X' = m$, $X + Y = t$, for testing any of the hypotheses concerning Δ is identical with the conditional test given $X + Y = t$ of the same hypothesis concerning $\rho = \Delta$ in the preceding section, in which $X + X' = m$ was given a priori. In particular, the conditional test for testing the hypothesis of independence $\Delta = 1$, Fisher's exact test, is the same as that of testing the equality of two binomial p's and is therefore given in terms of the hypergeometric distribution.

At the beginning of the section it was pointed out that the hypothesis of independence can be tested on the basis of samples obtained in a number of different ways. Either samples of fixed size can be taken from A and A^c or from B and B^c, or the sample can be selected at random from the population at large. Which of these designs is most efficient depends on the cost of sampling from

[8] Δ is equivalent to Yule's measure of association. which is $Q = (1 - \Delta)/(1 + \Delta)$. For a discussion of this and related measures see Goodman and Kruskal (1954, 1959), Edwards (1963), Haberman (1982) and Agresti (2002).

the various categories and from the population at large, and also on the cost of performing the necessary classification of a selected individual with respect to the characteristics in question. Suppose, however, for a moment that these considerations are neglected and that the designs are compared solely in terms of the power that the resulting tests achieve against a common alternative. Then the following results[9] can be shown to hold asymptotically as the total sample size s tends to infinity:

(i) If samples of size m and n $(m+n=s)$ are taken from B and B^c or from A and A^c, the best choice of m and n is $m=n=s/2$.

(ii) It is better to select samples of equal size $s/2$ from B and B^c than from A and A^c provided $|p_B - \frac{1}{2}| > |p_A - \frac{1}{2}|$.

(iii) Selecting the sample at random from the population at large is worse than taking equal samples either from A and A^c or from B and B^c.

These statements, which we shall not prove here, can be established by using the normal approximation for the distribution of the binomial variables X and Y when m and n are fixed, and by noting that under random sampling from the population at large, M/s and N/s tend in probability to p_B and p_{B^c} respectively.

4.7 Alternative Models for 2 × 2 Tables

Conditioning of the multinomial model for the 2×2 table on the row (or column) totals was seen in the last section to lead to the two-binomial model of Section 4.5. Similarly, the multinomial model itself can be obtained as a conditional model in some situations in which not only the marginal totals M, N, T, and T' are random but the total sample size s is also a random variable. Suppose that the occurrence of events (e.g. patients presenting themselves for treatment) is observed over a given period of time, and that the events belonging to each of the categories AB, A^cB, AB^c, A^cB^c are governed by independent Poisson processes, so that by (1.2) the numbers X, X', Y, Y' are independent Poisson variables with expectations λ_{AB}, λ_{A^cB}, λ_{AB^c}, $\lambda_{A^cB^c}$, and hence s is a Poisson variable with expectation $\lambda = \lambda_{AB} + \lambda_{A^cB} + \lambda_{AB^c} + \lambda_{A^cB^c}$.

It may then be of interest to compare the ratio $\lambda_{AB}/\lambda_{A^cB}$ with $\lambda_{AB^c}/\lambda_{A^cB^c}$ and in particular to test the hypothesis $H : \lambda_{AB}/\lambda_{A^cB} \le \lambda_{AB^c}/\lambda_{A^cB^c}$. The joint distribution of X,X',Y,Y' constitutes a four-parameter exponential family, which can be written as

$$\begin{aligned} P(X &= x, X'=x', Y=y, Y'=y') \\ &= \frac{1}{x!x'!y!y'!}\exp\left\{x\log\left(\frac{\lambda_{AB}\lambda_{A^cB^c}}{\lambda_{AB^c}\lambda_{A^cB}}\right) + (x'+x)\log\lambda_{A^cB}\right. \\ &\qquad\qquad \left. +(y+x)\log\lambda_{AB^c} + (y'-x)\log\lambda_{A^cB^c}\right\}. \end{aligned}$$

[9]These results were conjectured by Berkson and proved by Neyman in a course on χ^2.

Thus, UMP unbiased tests exist of the usual one- and two-sided hypotheses concerning the parameter $\theta = \lambda_{AB}\lambda_{A^cB^c}/\lambda_{A^cB}\lambda_{AB^c}$. These are carried out in terms of the conditional distribution of X given

$$X' + X = m, \qquad Y + X = t, \qquad X + X' + Y + Y' = s,$$

where the last condition follows from the fact that given the first two it is equivalent to $Y' - X = s - t - m$. By Problem 2.14, the conditional distribution of X, X', Y given $X + X' + Y + Y' = s$ is the multinomial distribution of Section 4.6 with

$$p_{AB} = \frac{\lambda_{AB}}{\lambda}, \quad p_{A^cB} = \frac{\lambda_{A^cB}}{\lambda}, \quad p_{AB^c} = \frac{\lambda_{AB^c}}{\lambda}, \quad p_{A^cB^c} = \frac{\lambda_{A^cB^c}}{\lambda}.$$

The tests therefore reduce to those derived in Section 4.6.

The three models discussed so far involve different sampling schemes. However, frequently the subjects for study are not obtained by any sampling but are the only ones readily available to the experimenter. To create a probabilistic basis for a test in such situations, suppose that B and B^c are two treatments, either of which can be assigned to each subject, and that A and A^c denote success or failure (e.g. survival, relief of pain, etc.). The hypothesis of no difference in the effectiveness of the two treatments (i.e. independence of A and B) can then be tested by assigning the subjects to the treatments, say m to B and n to B^c, *at random*, i.e. in such a way that all possible $\binom{s}{m}$ assignments are equally likely. It is now this random assignment which takes the place of the sampling process in creating a probability model, thus making it possible to calculate significance.

Under the hypothesis H of no treatment difference, the success or failure of a subject is independent of the treatment to which it is assigned. If the numbers of subjects in categories A and A^c are t and t' respectively ($t + t' = s$), the values of t and t' are therefore fixed, so that we are now dealing with a 2×2 table in which all four margins t, t', m, n are fixed.

Then any one of the four cell counts X, X', Y, Y' determines the other three. Under H, the distribution of Y is the hypergeometric distribution derived as the conditional null distribution of Y given $X + Y = t$ at the end of Section 4.5. The hypothesis is rejected in favor of the alternative that treatment B^c enhances success if Y is sufficiently large. Although this is the natural test under the given circumstances, no optimum property can be claimed for it, since no clear alternative model to H has been formulated.[10]

Consider finally the situation in which the subjects are again given rather than sampled, but B and B^c are attributes (for example, male or female, smoker or nonsmoker) which cannot be assigned to the subjects at will. Then there exists no stochastic basis for answering the question whether observed differences in the rates X/M and Y/N correspond to differences between B and B^c, or whether they are accidental. An approach to the testing of such hypotheses in a nonstochastic setting has been proposed by Freedman and Lane (1982).

---

[10]The one-sided test is of course UMP against the class of alternatives defined by the right side of (4.21), but no reasonable assumptions have been proposed that would lead to this class. For suggestions of a different kind of alternative see Gokhale and Johnson (1978).

The various models for the 2×2 table discussed in Sections 4.6 and 4.7 may be characterized by indicating which elements are random and which fixed:

(i) All margins and s random (Poisson).

(ii) All margins are random, s fixed (multinomial sampling).

(iii) One set of margins random, the other (and then a fortiori s) fixed (binomial sampling).

(iv) All margins fixed. Sampling replaced by random assignment of subjects to treatments.

(v) All aspects fixed; no element of randomness.

In the first three cases there exist UMP unbiased one- and two-sided tests of the hypothesis of independence of A and B. These tests are carried out by conditioning on the values of all elements in (i)–(iii) that are random, so that in the conditional model all margins are fixed. The remaining randomness in the table can be described by any one of the four cell entries; once it is known, the others are determined by the margins. The distribution of such an entry under H has the hypergeometric distribution given at the end of Section 4.5.

The models (i)–(iii) have a common feature. The subjects under observation have been obtained by sampling from a population, and the inference corresponding to acceptance or rejection of H refers to that population. This is not true in cases (iv) and (v).

In (iv) the subjects are given, and a probabilistic basis is created by assigning them at random, m to B and n to $\tilde{B}$. Under the hypothesis H of no treatment difference, the four margins are fixed without any conditioning, and the four cell entries are again determined by any one of them, which under H has the same hypergeometric distribution as before. The present situation differs from the earlier three in that the inference cannot be extended beyond the subjects at hand.[11]

The situation (v) is outside the scope of this book, since it contains no basis for the type of probability calculations considered here. Problems of this kind are however of great importance, since they arise in many observational (as opposed to experimental) studies. For a related discussion, see Finch (1979).

4.8 Some Three-Factor Contingency Tables

When an association between A and B exists in a 2×2 table, it does not follow that one of the factors has a causal influence on the other. Instead, the explanation may, for example, be in the fact that both factors are causally affected by a third factor C. If C has K possible outcomes $C_1, \dots, C_K$, one may then be faced with the apparently paradoxical situation (known as Simpson's paradox) that A and B are independent under each of the conditions C_k $(k = 1, \dots, K)$ but exhibit positive (or negative) association when the tables are aggregated over C that

[11] For a more detailed treatment of the distinction between population models [such as (i)–(iii)] and randomization models [such as (iv)], see Lehmann (1998).

is, when the K separate 2×2 tables are combined into a single one showing the total counts of the four categories. [An interesting example is discussed in Agresti (2002).] In order to determine whether the association of A and B in the aggregated table is indeed "spurious", one would test the hypothesis, (which arises also in other contexts) that A and B are conditionally independent given C_k for all $k = 1, \ldots, K$, against the alternative that there is an association for at least some k.

Let X_k, X_k', Y_k, Y_k' denote the counts in the $4K$ cells of the $2 \times 2 \times K$ table which extends the 2×2 table of Section 4.6 to the present case.

Again, several sampling schemes are possible. Consider first a random sample of size s from the population at large. The joint distribution of the $4K$ cell counts then is multinomial with probabilities p_{ABC_k}, $p_{\tilde{A}BC_k}$, $p_{A\tilde{B}C_k}$, $p_{\tilde{A}\tilde{B}C_k}$ for the outcomes indicated by the subscripts. If Δ_k denotes the AB odds ratio for C_k defined by

$$\Delta_k = \frac{p_{A\tilde{B}C_k} p_{\tilde{A}BC_k}}{p_{ABC_k} p_{\tilde{A}\tilde{B}C_k}} = \frac{p_{A\tilde{B}|C_k} p_{\tilde{A}B|C_k}}{p_{AB|C_k} p_{\tilde{A}\tilde{B}|C_k}},$$

where $p_{AB|C_k} \ldots$ denotes the conditional probability of the indicated event given C_k, then the hypothesis to be tested is $\Delta_k = 1$ for all k.

A second scheme takes samples of size s_k from C_k and classifies the subjects as AB, $\tilde{A}B$, $A\tilde{B}$ or $\tilde{A}\tilde{B}$. This is the case of K independent 2×2 tables, in which one is dealing with K quadrinomial distributions of the kind considered in the preceding sections. Since the kth of these distributions is also that of the same four outcomes in the first model conditionally given C_k, we shall denote the probabilities of these outcomes in the present model again by $p_{AB|C_k}, \ldots$.

To motivate the next sampling scheme, suppose that A and $\tilde{A}$ represent success or failure of a medical treatment, $\tilde{B}$ and B that the treatment is applied or the subject is used as a control, and C_k the kth hospital taking part in this study. If samples of size n_k and m_k are obtained and are assigned to treatment and control respectively, we are dealing with K pairs of binomial distributions. Letting Y_k and X_k denote the number of successes obtained by the treatment subjects and controls in the kth hospital, the joint distribution of these variables by Section 4.5 is

$$\left[\prod \binom{m_k}{x_k} \binom{n_k}{y_k} q_{1k}^{m_k} q_{2k}^{n_k}\right] \exp\left(\sum y_k \log \Delta_k + \sum (x_k + y_k) \log \frac{p_{1k}}{q_{1k}}\right),$$

where p_{1k} and q_{1k}, (p_{2k} and q_{2k}) denote the probabilities of success and failure under B (under $\tilde{B}$).

The above three sampling schemes lead to $2 \times 2 \times K$ tables in which respectively none, one, or two of the margins are fixed. Alternatively, in some situations a model may be appropriate in which the $4K$ variables X_k, X_k', Y_k, Y_k' are independent Poisson with expectations $\lambda_{ABC_k}, \ldots$. In this case, the total sample size s is also random.

For a test of the hypothesis of conditional independence of A and B given C_k for all k (i.e. that $\Delta_1 = \cdots = \Delta_k = 1$), see Problem 12.65. Here we shall consider the problem under the simplifying assumption that the Δ_k have a common value Δ, so that the hypothesis reduces to $H : \Delta = 1$. Applying Theorem 4.4.1 to the third model (K pairs of binomials) and assuming the alternatives to be $\Delta > 1$, we see that a UMP unbiased test exists and rejects H when $\sum Y_k > C(X_1 +$

$Y_1, \ldots, X_K + Y_K$), where C is determined so that the conditional probability of rejection, given that $X_k + Y_k = t_k$, is α for all $k = 1, \ldots, K$. It follows from Section 4.5 that the conditional joint distribution of the Y_k under H is

$$\begin{aligned} P_H[Y_1 &= y_1, \ldots, Y_K = y_K | X_k + Y_k = t_k, k = 1, \ldots, K] \\ &= \prod \frac{\binom{m_k}{t_k - y_k}\binom{n_k}{y_k}}{\binom{m_k + n_k}{t_k}} \end{aligned}$$

The conditional distribution of $\sum Y_k$ can now be obtained by adding the probabilities over all $(y_1, \ldots, y_K)$ whose sum has a given value. Unless the numbers are very small, this is impractical and approximations must be used [see Cox (1966) and Gart (1970)].

The assumption $H' : \Delta_1 = \cdots = \Delta_K = \Delta$ has a simple interpretation when the successes and failures of the binomial trials are obtained by dichotomizing underlying unobservable continuous response variables. In a single such trial, suppose the underlying variable is Z and that success occurs when $Z > 0$ and failure when $Z \leq 0$. If Z is distributed as $F(Z - \zeta)$ with location parameter ζ, we have $p = 1 - F(-\zeta)$ and $q = F(-\zeta)$. Of particular interest is the logistic distribution, for which $F(x) = 1/(1 + e^{-x})$. In this case $p = e^\zeta/(1 + e^\zeta)$, $q = 1/(1+e^\zeta)$, and hence $\log(p/q) = \zeta$. Applying this fact to the success probabilities

$$p_{1k} = 1 - F(-\zeta_{1k}), \qquad p_{2k} = 1 - F(-\zeta_{2k}),$$

we find that

$$\theta_k = \log \Delta_k = \log\left(\frac{p_{2k}}{q_{2k}} \bigg/ \frac{p_{1k}}{q_{1k}}\right) = \zeta_{2k} - \zeta_{1k},$$

so that $\zeta_{2k} = \zeta_{1k} + \theta_k$. In this model, H' thus reduces to the assumption that $\zeta_{2k} = \zeta_{1k} + \theta$, that is, that the treatment shifts the distribution of the underlying response by a constant amount θ.

If it is assumed that F is normal rather than logistic, $F(x) = \Phi(x)$ say, then $\zeta = \Phi^{-1}(p)$, and constancy of $\zeta_{2k} - \zeta_{1k}$ requires the much more cumbersome condition $\Phi^{-1}(p_{2k}) - \Phi^{-1}(p_{1k}) = \text{constant}$. However, the functions $\log(p/q)$ and $\Phi^{-1}(p)$ agree quite well in the range $.1 \leq p \leq .9$ [see Cox (1970, p. 28)], and the assumption of constant Δ_k in the logistic response model is therefore close to the corresponding assumption for an underlying normal response.[12] [The so-called loglinear models, which for contingency tables correspond to the linear models to be considered in Chapter 7 but with a logistic rather than a normal response variable, provide the most widely used approach to contingency tables. See, for example, the books by Cox (1970), Haberman (1974), Bishop, Fienberg, and Holland (1975), Fienberg (1980), Plackett (1981), and Agresti (2002).]

The UMP unbiased test, derived above for the case that the B- and C-margins are fixed, applies equally when any two margins, any one margin, or no margins are fixed, with the understanding that in all cases the test is carried out conditionally, given the values of all random margins.

[12]The problem of discriminating between a logistic and normal response model is discussed by Chambers and Cox (1967).

The test is also used (but no longer UMP unbiased) for testing $H : \Delta_1 = \cdots = \Delta_K = 1$ when the Δ's are not assumed to be equal but when the $\Delta_k - 1$ can be assumed to have the same sign, so that the departure from independence is in the same direction for all the 2×2 tables. A one- or two-sided version is appropriate as the alternatives do or do not specify the direction. For a discussion of this test, the Cochran–Mantel–Haenszel test, and some of its extensions see Agresti (2002, Section 7.4).

Consider now the case $K = 2$, with m_k and n_k fixed, and the problem of testing $H' : \Delta_2 = \Delta_1$ rather than assuming it. The joint distribution of the X's and Y's given earlier can then be written as

$$\left[\prod_{k=1}^{2} \binom{m_k}{x_k}\binom{n_k}{y_k} q_{1k}^{m_k} q_{2k}^{n_k}\right] \times \exp\left(y_2 \log \frac{\Delta_2}{\Delta_1} + (y_1 + y_2)\log \Delta_1 + \sum (x_i + y_i) \log \frac{p_{1i}}{q_{1i}}\right),$$

and H' is rejected in favor of $\Delta_2 > \Delta_1$ if $Y_2 > C$, where C depends on $Y_1 + Y_2$, $X_1 + Y_1$ and $X_2 + Y_2$, and is determined so that the conditional probability of rejection given $Y_1 + Y_2 = w$, $X_1 + Y_1 = t_1$, $X_2 + Y_2 = t_2$ is α. The conditional null distribution of Y_1 and Y_2, given $X_k + Y_k = t_k$ $(k = 1, 2)$, by (4.21) with Δ in place of ρ is

$$C_{t_1}(\Delta)C_{t_2}(\Delta)\binom{m_1}{t_1 - y_1}\binom{n_1}{y_1}\binom{m_2}{t_2 - y_2}\binom{n_2}{y_2}\Delta^{y_1+y_2},$$

and hence the conditional distribution of Y_2, given in addition that $Y_1 + Y_2 = w$, is of the form

$$k(t_1, t_2, w)\binom{m_1}{y + t_1 - w}\binom{n_1}{w - y}\binom{m_2}{t_2 - y}\binom{n_2}{y}.$$

Some approximations to the critical value of this test are discussed by Birch (1964); see also Venable and Bhapkar (1978). [Optimum large-sample tests of some other hypotheses in $2 \times 2 \times 2$ tables are obtained by Cohen, Gatsonis, and Marden (1983).]

4.9 The Sign Test

To test consumer preferences between two products, a sample of n subjects are asked to state their preferences. Each subject is recorded as plus or minus as it favors product B or A. The total number Y of plus signs is then a binomial variable with distribution $b(p, n)$. Consider the problem of testing the hypothesis $p = \frac{1}{2}$ of no difference against the alternatives $p \neq \frac{1}{2}$ (As in previous such problems, we disregard here that in case of rejection it will be necessary to decide which of the two products is preferred.) The appropriate test is the two-sided *sign test*, which rejects when $|Y - \frac{1}{2}n|$ is too large. This is UMP unbiased (Section 4.2).

Sometimes the subjects are also given the possibility of declaring themselves as undecided. If p_-, p_+, and p_0 denote the probabilities of preference for product A, product B, and of no preference respectively, the numbers X, Y, and Z of decisions in favor of these three possibilities are distributed according to the multinomial distribution

$$\frac{n!}{x!y!z!} p_-^x p_+^y p_0^z \quad (x + y + z = n), \tag{4.22}$$

and the hypothesis to be tested is $H : p_+ = p_-$. The distribution (4.22) can also be written as

$$\frac{n!}{x!y!z!} \left(\frac{p_+}{1 - p_0 - p_+} \right)^y \left(\frac{p_0}{1 - p_0 - p_+} \right)^z (1 - p_0 - p_+)^n, \tag{4.23}$$

and is then seen to constitute an exponential family with $U = Y$, $T = Z$, $\theta = \log[p_+/(1 - p_0 - p_+)]$, $\vartheta = \log[p_0/(1 - p_0 - p_+)]$. Rewriting the hypothesis H as $p_+ = 1 - p_0 - p_+$ it is seen to be equivalent to $\theta = 0$. There exists therefore a UMP unbiased test of H, which is obtained by considering z as fixed and determining the best unbiased conditional test of H given $Z = z$. Since the conditional distribution of Y given z is a binomial distribution $b(p, n - z)$ with $p = p_+/(p_+ + p_-)$, the problem reduces to that of testing the hypothesis $p = \frac{1}{2}$ in a binomial distribution with $n - z$ trials, for which the rejection region is $|Y - \frac{1}{2}(n - z)| > C(z)$. The UMP unbiased test is therefore obtained by disregarding the number of cases in which no preference is expressed (the number of *ties*), and applying the sign test to the remaining data.

The power of the test depends strongly on p_0, which governs the distribution of Z. For large p_0, the number $n - z$ of trials in the conditional binomial distribution can be expected to be small, and the test will thus have little power. This may be an advantage in the present case, since a sufficiently high value of p_0, regardless of the value of p_+/p_-, implies that the population as a whole is largely indifferent with respect to the products.

The above conditional sign test applies to any situation in which the observations are the result of n independent trials, each of which is either a success (+), a failure (−), or a tie. As an alternative treatment of ties, it is sometimes proposed to assign each tie at random (with probability $\frac{1}{2}$ each) to either plus or minus. The total number Y' of plus signs after the ties have been broken is then a binomial variable with distribution $b(\pi, n)$, where $\pi = p_+ + \frac{1}{2}p_0$. The hypothesis H becomes $\pi = \frac{1}{2}$, and is rejected when $|Y' - \frac{1}{2}n| > C$, where the probability of rejection is α when $\pi = \frac{1}{2}$. This test can be viewed also as a randomized test based on X, Y, and Z, and it is unbiased for testing H in its original form, since p_+ is $=$ or $\neq p_-$ as π is $=$ or $\neq 1$. Since the test involves randomization other than on the boundaries of the rejection region, it is less powerful than the UMP unbiased test for this situation, so that the random breaking of ties results in a loss of power.

This remark might be thought to throw some light on the question of whether in the determination of consumer preferences it is better to permit the subject to remain undecided or to force an expression of preference. However, here the assumption of a completely random assignment in case of a tie does not apply. Even when the subject is not conscious of a definite preference, there will usually be a slight inclination toward one of the two possibilities, which in a majority of the cases will be brought out by a forced decision. This will be balanced in part by the fact that such forced decisions are more variable than those reached voluntarily. Which of these two factors dominates depends on the strength of the preference.

Frequently, the question of preference arises between a standard product and a possible modification or a new product. If each subject is required to express a definite preference, the hypothesis of interest is usually the one sided hypothesis $p_+ \leq p_-$, where $+$ denotes a preference for the modification. However, if an expression of indifference is permitted the hypothesis to be tested is not $p_+ \leq p_-$ but rather $p_+ \leq p_0 + p_-$, since typically the modification is of interest only if it is actually preferred. As was shown in Example 3.8.1, the one-sided sign test which rejects when the number of plus signs is too large is UMP for this problem.

In some investigations, the subject is asked not only to express a preference but to give a more detailed evaluation, such as a score on some numerical scale. Depending on the situation, the hypothesis can then take on one of two forms. One may be interested in the hypothesis that there is no difference in the consumer's reaction to the two products. Formally, this states that the distribution of the scores $X_1, \ldots, X_n$ expressing the degree of preference of the n subjects for the modified product is symmetric about the origin. This problem, for which a UMP unbiased test does not exist without further assumptions, will be considered in Section 6.10.

Alternatively, the hypothesis of interest may continue to be $H : p_+ = p_-$. Since $p_- = P\{X < 0\}$ and $p_+ = P\{X > 0\}$, this now becomes

$$H : P\{X > 0\} = P\{X < 0\}.$$

Here symmetry of X is no longer assumed even when $P\{X < 0\} = P\{X > 0\}$. If no assumptions are made concerning the distribution of X beyond the fact that the set of its possible values is given, the sign test based on the number of X's that are positive and negative continues to be UMP unbiased.

To see this, note that any distribution of X can be specified by the probabilities

$$p_- = P\{X < 0\}, \qquad p_+ = P\{X > 0\}, \qquad p_0 = P\{X = 0\},$$

and the conditional distributions F_- and F_+ of X given $X < 0$ and $X > 0$ respectively. Consider any fixed distributions F'_-, F'_+, and denote by $\mathcal{F}_0$ the family of all distributions with $F_- = F'_-$, $F_+ = F'_+$ and arbitrary p_-, p_+, p_0. Any test that is unbiased for testing H in the original family of distributions $\mathcal{F}$ in which F_- and F_+ are unknown is also unbiased for testing H in the smaller family $\mathcal{F}_0$. We shall show below that there exists a UMP unbiased test ϕ_0 of H in $\mathcal{F}_0$. It turns out that ϕ_0 is also unbiased for testing H in $\mathcal{F}$ and is independent of F'_-, F'_+. Let ϕ be any other unbiased test of H in $\mathcal{F}$, and consider any fixed alternative, which without loss of generality can be assumed to be in $\mathcal{F}_0$. Since ϕ is unbiased for $\mathcal{F}$, it is unbiased for testing $p_+ = p_-$ in $\mathcal{F}_0$; the power of ϕ_0 against the particular alternative is therefore at least as good as that of ϕ. Hence ϕ_0 is UMP unbiased.

To determine the UMP unbiased test of H in $\mathcal{F}_0$, let the densities of F'_- and F'_+ with respect to some measure μ be f'_- and f'_+. The joint density of the X's at a point $(x_1, \ldots, x_n)$ with

$$x_{i_1}, \ldots, x_{i_r} < 0 = x_{j_1} = \cdots = x_{j_s} < x_{k_i}, \ldots, x_{k_m}$$

is

$$p_-^r p_0^s p_+^m f'_-(x_{i_1}) \ldots f'_-(x_{i_r}) f'_+(x_{k_1}) \ldots f'_+(x_{k_m}).$$

The set of statistics (r, s, m) is sufficient for (p_-, p_0, p_+), and its distribution is given by (4.22) with $x = r$, $y = m$, $z = s$. The sign test is therefore seen to be UMP unbiased as before.

A different application of the sign test arises in the context of a 2×2 table for matched pairs. In Section 4.5, success probabilities for two treatments were compared on the basis of two independent random samples. Unless the population of subjects from which these samples are drawn is fairly homogeneous, a more powerful test can often be obtained by using a sample of matched pairs (for example, twins or the same subject given the treatments at different times). For each pair there are then four possible outcomes: $(0, 0)$, $(0, 1)$, $(1, 0)$, and $(1, 1)$, where 1 and 0 stand for success and failure, and the first and second number in each pair of responses refer to the subject receiving treatment 1 or 2 respectively.

The results of such a study are sometimes displayed in a 2×2 table,

		1st	
		0	1
2nd	0	X	X'
	1	Y	Y'

which despite the formal similarity differs from that considered in Section 4.6. If a sample of s pairs is drawn, the joint distribution of X, Y, X', Y' as before is multinomial, with probabilities p_{00}, p_{01}, p_{10},p_{11}. The success probabilities of the two treatments are $\pi_1 = p_{10} + p_{11}$ for the first and $\pi_2 = p_{01} + p_{11}$ for the second treatment, and the hypothesis to be tested is $H : \pi_1 = \pi_2$ or equivalently $p_{10} = p_{01}$ rather than $p_{10}p_{01} = p_{00}p_{11}$ as it was earlier.

In exponential form, the joint distribution can be written as

$$\frac{s!p_{11}^s}{x!x'!y!y'!} \exp\left(y \log \frac{p_{01}}{p_{10}} + (x' + y) \log \frac{p_{10}}{p_{11}} + x \log \frac{p_{00}}{p_{11}}\right). \tag{4.24}$$

There exists a UMP unbiased test, *McNemar's test*, which rejects H in favor of the alternatives $p_{10} < p_{01}$ when $Y > C(X' + Y, X)$, where the conditional probability of rejection given $X' + Y = d$ and $X = x$ is α for all d and x. Under this condition, the numbers of pairs (0, 0) and (1, 1) are fixed, and the only remaining variables are Y and $X' = d - Y$ which specify the division of the d cases with mixed response between the outcomes (0, 1) and (1, 0). Conditionally, one is dealing with d binomial trials with success probability $p = p_{01}/(p_{01} + p_{10})$, H becomes $p = \frac{1}{2}$, and the UMP unbiased test reduces to the sign test. [The issue of conditional versus unconditional power for this test is discussed by Frisén (1980).]

The situation is completely analogous to that of the sign test in the presence of undecided opinions, with the only difference that there are now two types of ties, (0, 0) and (1, 1), both of which are disregarded in performing the test.

4.10 Problems

Section 4.1

Problem 4.1 *Admissibility.* Any UMP unbiased test ϕ_0, is admissible in the sense that there cannot exist another test ϕ_1 which is at least as powerful as ϕ_0 against all alternatives and more powerful against some.
[If ϕ is unbiased and ϕ' is uniformly at least as powerful as ϕ, then ϕ' is also unbiased.]

Problem 4.2 *p-values.* Consider a family of tests of $H : \theta = \theta_0$ (or $\theta \le \theta_0$), with level-α rejection regions S_α, such that (a) $P_{\theta_0}\{X \in S_\alpha\}$ for all $0 < \alpha < 1$, and (b) $S_\alpha \subset S_{\alpha'}$ for $\alpha < \alpha'$. If the tests S_α are unbiased, the distribution of $\hat{\alpha}$ under any alternative θ satisfies

$$P_\theta\{\hat{\alpha} \le \alpha\} \ge P_{\theta_0}\{\hat{\alpha} \le \alpha\} = \alpha$$

so that it is shifted toward the origin.

Section 4.2

Problem 4.3 Let X have the binomial distribution $b(p, n)$, and consider the hypothesis $H : p = p_0$ at level of significance α. Determine the boundary values of the UMP unbiased test for $n = 10$ with $\alpha = .1$, $p_0 = .2$ and with $\alpha = .05$, $p_0 = .4$, and in each case graph the power functions of both the unbiased and the equal-tails test.

Problem 4.4 Let X have the Poisson distribution $P(\tau)$, and consider the hypothesis $H : \tau = \tau_0$. Then condition (4.6) reduces to

$$\sum_{x=C_1+1}^{C_2-1} \frac{\tau_0^{x-1}}{(x-1)!} e^{-\tau_0} + \sum_{i=1}^{2} (1-\gamma_i) \frac{\tau_0^{C_i-1}}{(C_i-1)!} e^{-\tau_0} = 1 - \alpha,$$

provided $C_1 > 1$.

Problem 4.5 Let T_n/θ have a χ^2-distribution with n degrees of freedom. For testing $H : \theta = 1$ at level of significance $\alpha = .05$, find n so large that the power of the UMP unbiased test is $\ge .9$ against both $\theta \ge 2$ and $\theta \le \frac{1}{2}$. How large does n have to be if the test is not required to be unbiased?

Problem 4.6 Suppose X has density (with respect to some measure μ)

$$p_\theta(x) = C(\theta) \exp[\theta T(x)] h(x) \; ,$$

for some real-valued θ. Assume the distribution of $T(X)$ is continuous under θ (for any θ). Consider the problem of testing $\theta = \theta_0$ versus $\theta \ne \theta_0$. If the null hypothesis is rejected, then a decision is to be made as to whether $\theta > \theta_0$ or $\theta < \theta_0$. We say that a Type 3 (or directional) error is made when it is declared that $\theta > \theta_0$ when in fact $\theta < \theta_0$ (or vice-versa). Consider a level α test that rejects the null hypothesis if $T < C_1$ or $T > C_2$ for constants $C_1 < C_2$. Further suppose that it is declared that $\theta < \theta_0$ if $T < C_1$ and $\theta > \theta_0$ if $T > C_2$.

(i) If the constants are chosen so that the test is UMPU, show that the Type 3 error is controlled in the sense that

$$\sup_{\theta \neq \theta_0} P_\theta\{\text{Type 3 error is made}\} \leq \alpha \ . \tag{4.25}$$

(ii) If the constants are chosen so that the test is equi-tailed in the sense

$$P_{\theta_0}\{T(X) < C_1\} = P_{\theta_0}\{T(X) > C_2\} = \alpha/2 \ ,$$

then show (4.25) holds with α replaced by $\alpha/2$.
(iii) Give an example where the UMPU level α test has the left side of (4.25) strictly $> \alpha/2$. [Confidence intervals for θ after rejection of a two-sided test are discussed in Finner (1994).]

Problem 4.7 Let X and Y be independently distributed according to one-parameter exponential families, so that their joint distribution is given by

$$dP_{\theta_1,\theta_2}(x,y) = C(\theta_1)e^{\theta_1 T(x)}\,d\mu(x)K(\theta_2)e^{\theta_2 U(y)}\,d\nu(y).$$

Suppose that with probability 1 the statistics T and U each take on at least three values and that (a, b) is an interior point of the natural parameter space. Then a UMP unbiased test does not exist for testing $H : \theta_1 = a,\ \theta_2 = b$ against the alternatives $\theta_1 \neq a$ or $\theta_2 \neq b$.[13]
[The most powerful unbiased tests against the alternatives $\theta_1 \neq a,\ \theta_2 \neq b$ have acceptance regions $C_1 < T(x) < C_2$ and $K_1 < U(y) < K_2$ respectively. These tests are also unbiased against the wider class of alternatives $K : \theta_1 \neq a$ or $\theta_2 \neq b$ or both.]

Problem 4.8 Let (X, Y) be distributed according to the exponential family

$$dP_{\theta_1,\theta_2}(x,y) = C(\theta_1,\theta_2)e^{\theta_1 x+\theta_2 y}\,d\mu(x,y) \ .$$

The only unbiased test for testing $H : \theta_1 \leq a, \theta_2 \leq b$ against $K : \theta_1 > a$ or $\theta_2 > b$ or both is $\phi(x,y) \equiv \alpha$.
[Take $a = b = 0$, and let $\beta(\theta_1,\theta_2)$ be the power function of any level-α test. Unbiasedness implies $\beta(0,\theta_2) = \alpha$ for $\theta_2 < 0$ and hence for all θ_2, since $\beta(0,\theta_2)$ is an analytic function of θ_2. For fixed $\theta_2 > 0$, $\beta(\theta_1,\theta_2)$ considered as a function of θ_1 therefore has a minimum at $\theta_1 = 0$, so that $\partial\beta(\theta_1,\theta_2)/\partial\theta_1$ vanishes at $\theta_1 = 0$ for all positive θ_2, and hence for all θ_2. By considering alternatively positive and negative values of θ_2 and using the fact that the partial derivatives of all orders of $\beta(\theta_1,\theta_2)$ with respect to θ_1 are analytic, one finds that for each fixed θ_2 these derivatives all vanish at $\theta_1 = 0$ and hence that the function β must be a constant. Because of the completeness of (X, Y), $\beta(\theta_1,\theta_2) \equiv \alpha$ implies $\phi(x,y) \equiv \alpha$.]

Problem 4.9 For testing the hypothesis $H : \theta = \theta_0$, (θ_0 an interior point of Ω) in the one-parameter exponential family of Section 4.2, let $\mathcal{C}$ be the totality of tests satisfying (4.3) and (4.5) for some $-\infty \leq C_1 \leq C_2 \leq \infty$ and $0 \leq \gamma_1,\ \gamma_2 \leq 1$.

[13] For counterexamples when the conditions of the problem are not satisfied, see Kallenberg et al. (1984).

(i) $\mathcal{C}$ is complete in the sense that given any level-α test ϕ_0 of H there exists $\phi \in \mathcal{C}$ such that ϕ is uniformly at least as powerful as ϕ_0.

(ii) If $\phi_1, \phi_2 \in \mathcal{C}$, then neither of the two tests is uniformly more powerful than the other.

(iii) Let the problem be considered as a two-decision problem, with decisions d_0 and d_1 corresponding to acceptance and rejection of H and with loss function $L(\theta, d_i) = L_i(\theta), i = 0, 1$. Then $\mathcal{C}$ is minimal essentially complete provided $L_1(\theta) < L_0(\theta)$ for all $\theta \neq \theta_0$.

(iv) Extend the result of part (iii) to the hypothesis $H' : \theta_1 \leq \theta \leq \theta_2$. (For more general complete class results for exponential families and beyond, see Brown and Marden (1989).)

[(i): Let the derivative of the power function of ϕ_0 at θ_0 be $\beta'_{\phi_0}(\theta_0) = \rho$. Then there exists $\phi \in \mathcal{C}$ such that $\beta'_\phi(\theta_0) = \rho$ and ϕ is UMP among all tests satisfying this condition.
(ii): See the end of Section 3.7.
(iii): See the proof of Theorem 3.4.2.]

Section 4.3

Problem 4.10 Let $X_1, \ldots, X_n$ be a sample from (i) the normal distribution $N(a\sigma, \sigma^2)$, with a fixed and $0 < \sigma < \infty$; (ii) the uniform distribution $U(\theta - \frac{1}{2}, \theta + \frac{1}{2}), -\infty < \theta < \infty$; (iii) the uniform distribution $U(\theta_1, \theta_2), \infty < \theta_1 < \theta_2 < \infty$. For these three families of distributions the following statistics are sufficient: (i), $T = (\sum X_i, \sum X_i^2)$; (ii) and (iii), $T = (\min(X_1, \ldots, X_n), \max(X_1, \ldots, X_n))$. The family of distributions of T is complete for case (iii), but for (i) and (ii) it is not complete or even boundedly complete.
[(i): The distribution of $\sum X_i / \sqrt{\sum X_i^2}$ does not depend on σ.]

Problem 4.11 Let $X_1, \ldots, X_m$ and $Y_1, \ldots, Y_n$. be samples from $N(\xi, \sigma^2)$ and $N(\xi, \tau^2)$. Then $T = (\sum X_i, \sum Y_j, \sum X_i^2, \sum Y_j^2)$, which in Example 4.3.3 was seen not to be complete, is also not boundedly complete.
[Let $f(t)$ be 1 or -1 as $\bar{y} - \bar{x}$ is positive or not.]

Problem 4.12 *Counterexample.* Let X be a random variable taking on the values -1, 0, 1, 2, ... with probabilities

$$P_\theta\{X = -1\} = \theta; \qquad P_\theta\{X = x\} = (1 - \theta)^2 \theta^x, \quad x = 0, 1, \ldots.$$

Then $\mathcal{P} = \{P_\theta, 0 < \theta < 1\}$ is boundedly complete but not complete. [Girschick et al. (1946)]

Problem 4.13 The completeness of the order statistics in Example 4.3.4 remains true if the family $\mathcal{F}$ is replaced by the family $\mathcal{F}_1$ of all continuous distributions.
[Due to Fraser (1956). To show that for any integrable symmetric function ϕ, $\int \phi(x_1, \ldots, x_n)\, dF(x_1) \ldots$
$dF(x_n) = 0$ for all continuous F implies $\phi = 0$ a.e., replace F by $\alpha_1 F_1 + \cdots + \alpha_n F_n$,

where $0 < \alpha_i < 1, \sum \alpha_i = 1$. By considering the left side of the resulting identity as a polynomial in the α's one sees that $\int \phi(x_1, \dots, x_n)\, dF_1(x_1) \dots dF_n(x_n) = 0$ for all continuous F_i. This last equation remains valid if the F_i are replaced by $I_{a_i}(x)F(x)$, where $I_{a_i}(x) = 1$ if $x \le a_i$ and $= 0$ otherwise. This implies that $\phi = 0$ except on a set which has measure 0 under $F \times \dots \times F$ for all continuous F.]

Problem 4.14 Determine whether T is complete for each of the following situations:

(i) $X_1, \dots, X_n$ are independently distributed according to the uniform distribution over the integers $1, 2, \dots, \theta$ and $T = \max(X_1, \dots, X_n)$.

(ii) X takes on the values 1,2,3,4 with probabilities pq, p^2q, pq^2, $1 - 2pq$ respectively, and $T = X$.

Problem 4.15 Let X, Y be independent binomial $b(p, m)$ and $b(p^2, n)$ respectively. Determine whether (X, Y) is complete when

(i) $m = n = 1$,

(ii) $m = 2$, $n = 1$.

Problem 4.16 Let $X_1, \dots, X_n$ be a sample from the uniform distribution over the integers $1, \dots, \theta$ and let a be a positive integer.

(i) The sufficient statistic $X_{(n)}$ is complete when the parameter space is $\Omega = \{\theta : \theta \le a\}$.

(ii) Show that $X_{(n)}$ is not complete when $\Omega = \{\theta : \theta \ge a\}$, $a \ge 2$, and find a complete sufficient statistic in this case.

Section 4.4

Problem 4.17 Let $X_i (i = 1, 2)$ be independently distributed according to distributions from the exponential families (3.19) with C, Q, T, and h replaced by C_i, Q_i, T_i, and h_i. Then there exists a UMP unbiased test of

(i) $H : Q_2(\theta_2) - Q_1(\theta_1) \le c$ and hence in particular of $Q_2(\theta_2) \le Q_1(\theta_1)$;

(ii) $H : Q_2(\theta_2) + Q_1(\theta_1) \le c$.

Problem 4.18 Let X, Y, Z be independent Poisson variables with means λ, μ, v. Then there exists a UMP unbiased test of $H : \lambda\mu \le v^2$.

Problem 4.19 *Random sample size.* Let N be a random variable with a *power-series* distribution

$$P(N = n) = \frac{a(n)\lambda^n}{C(\lambda)}, \quad n = 0, 1, \dots \quad (\lambda > 0, \text{unknown}).$$

When $N = n$, a sample $X_1, \dots, X_n$ from the exponential family (3.19) is observed. On the basis of $(N, X_1, \dots, X_N)$ there exists a UMP unbiased test of $H : Q(\theta) \le c$.

Problem 4.20 Suppose $P\{I = 1\} = p = 1 - P\{I = 2\}$. Given $I = i$, $X \sim N(\theta, \sigma_i^2)$, where $\sigma_1^2 < \sigma_2^2$ are known. If $p = 1/2$, show that, based on the data (X, I), there does not exist a UMP test of $\theta = 0$ vs $\theta > 0$. However, if p is also unknown, show a UMPU test exists. [See Examples 10.20-21 in Romano and Siegel (1986).]

Problem 4.21 *Measurability of tests of Theorem 4.4.1.* The function ϕ_3 defined by (4.16) and (4.17) is jointly measurable in u and t.
[With $C_1 = v$ and $C_2 = w$, the determining equations for v, w, γ_1, γ_2 are

$$\begin{aligned} F_t(v-) + [1 - F_t(w)] + \gamma_1[F_t(v) - F_t(v-)] \\ +\gamma_2[F_t(w) - F_t(w-)] = \alpha \end{aligned} \tag{4.26}$$

and

$$\begin{aligned} G_t(v-) + [1 - G_t(w)] + \gamma_1[G_t(v) - G_t(v-)] \\ +\gamma_2[G_t(w) - G_t(w-)] = \alpha \end{aligned} \tag{4.27}$$

where

$$F_t(u) = \int_{-\infty}^{u} C_t(\theta_1)e^{\theta_1 y}\, d\nu_t(y), \quad G_t(u) = \int_{-\infty}^{u} C_t(\theta_2)e^{\theta_2 y}\, d\nu_t(y), \tag{4.28}$$

denote the conditional cumulative distribution function of U given t when $\theta = \theta_1$ and $\theta = \theta_2$ respectively.
(1) For each $0 \le y \le \alpha$ let $v(y,t) = F_t^{-1}(y)$ and $w(y,t) = F_t^{-1}(1 - \alpha + y)$, where the inverse function is defined as in the proof of Theorem 4.4.1. Define $\gamma_1(y,t)$ and $\gamma_2(y,t)$ so that for $v = v(y,t)$ and $w = w(y,t)$,

$$\begin{aligned} F_t(v-) + \gamma_1[F_t(v) - F_t(v-)] &= y, \\ 1 - F_t(w) + \gamma_2[F_t(w) - F_t(w-)] &= \alpha - y. \end{aligned}$$

(2) Let $H(y,t)$ denote the left-hand side of (4.27), with $v = v(y,t)$, etc. Then $H(0,t) > \alpha$ and $H(\alpha,t) < \alpha$. This follows by Theorem 3.4.1 from the fact that $v(0,t) = -\infty$ and $w(\alpha,t) = \infty$ (which shows the conditional tests corresponding to $y = 0$ and $y = \alpha$ to be one-sided), and that the left-hand side of (4.27) for any y is the power of this conditional test.
(3) For fixed t, the functions

$$H_1(y,t) = G_t(v-) + \gamma_1[G_t(v) - G_t(v-)]$$

and

$$H_2(y,t) = 1 - G_t(w) + \gamma_2[G_t(w) - G_t(w-)]$$

are continuous functions of y. This is a consequence of the fact, which follows from (4.28), that a.e. $\mathcal{P}^T$ the discontinuities and flat stretches of F_t and G_t coincide.
(4) The function $H(y,t)$ is jointly measurable in y and t. This follows from the continuity of H by an argument similar to the proof of measurability of $F_t(u)$ in the text. Define

$$y(t) = \inf\{y : H(y,t) < \alpha\},$$

and let $v(t) = v[y(t), t]$, etc. Then (4.26) and (4.27) are satisfied for all t. The measurability of $v(t)$, $w(t)$, $\gamma_1(t)$, and $\gamma_2(t)$ defined in this manner will follow from measurability in t of $y(t)$ and $F_t^{-1}[y(t)]$. This is a consequence of the relations, which hold for all real c,

$$\{t : y(t) < c\} = \bigcup_{r<c} \{t : H(r,t) < \alpha\},$$

where r indicates a rational, and

$$\{t : F_t^{-1}[y(t)] \le c\} = \{t : y(t) - F_t(c) \le 0\}.]$$

Problem 4.22 *Continuation.* The function ϕ_4 defined by (4.16), (4.18), and (4.19) is jointly measurable in u and t.
[The proof, which otherwise is essentially like that outlined in the preceding problem, requires the measurability in z and t of the integral

$$g(z,t) = \int_{-\infty}^{z-} u\, dF_t(u).$$

This integral is absolutely convergent for all t, since F_t is a distribution belonging to an exponential family. For any $z < \infty$, $g(z,t) = \lim g_n(z,t)$, where

$$g_n(z,t) = \sum_{j=1}^{\infty} \left(z - \frac{j}{2^n}\right) \left[F_t\left(z - \frac{j-1}{2^n} - 0\right) - F_t\left(z - \frac{j}{2^n} - 0\right)\right],$$

and the measurability of g follows from that of the functions g_n. The inequalities corresponding to those obtained in step (2) of the preceding problem result from the property of the conditional one-sided tests established in Problem 3.45.]

Problem 4.23 The UMP unbiased tests of the hypotheses $H_1, \ldots, H_4$ of Theorem 4.4.1 are unique if attention is restricted to tests depending on U and the T's.

Problem 4.24 The singly truncated normal (STN) distribution, indexed by parameters ν and λ has support the positive real line with density

$$p(x; \nu, \lambda) = C(\nu, \lambda) \exp(-\nu x - \lambda x^2),$$

where $C(\nu, \lambda)$ is a normalizing constant. Based on an i.i.d. sample, show there exists a UMPU test of the null hypothesis that the observations are exponential against the STN alternative, and describe the form of rejection region as explicitly as possible. [See Castillo and Puig (1999).]

Section 4.5

Problem 4.25 *Negative binomial.* Let X, Y be independently distributed according to negative binomial distributions $Nb(p_1, m)$ and $Nb(p_2, n)$ respectively, and let $q_i = 1 - p_i$.

(i) There exists a UMP unbiased test for testing $H : \theta = q_2/q_1 \le \theta_0$ and hence in particular $H' : p_1 \le p_2$.

(ii) Determine the conditional distribution required for testing H' when $m = n = 1$.

Problem 4.26 Let X and Y be independently distributed with Poisson distributions $P(\lambda)$ and $P(\mu)$. Find the power of the UMP unbiased test of $H : \mu \leq \lambda$, against the alternatives $\lambda = .1$, $\mu = .2$; $\lambda = 1$, $\mu = 2$; $\lambda = 10$, $\mu = 20$; $\lambda = .1$, $\mu = .4$; at level of significance $\alpha = .1$.
[Since $T = X + Y$ has the Poisson distribution $P(\lambda + \mu)$, the power is

$$\beta = \sum_{t=0}^{\infty} \beta(t) \frac{(\lambda + \mu)^t}{t!} e^{-(\lambda+\mu)},$$

where $\beta(t)$ is the power of the conditional test given t against the alternative in question.]

Problem 4.27 *Sequential comparison of two binomials.* Consider two sequences of binomial trials with probabilities of success p_1 and p_2 respectively, and let $\rho = (p_2/q_2) \div (p_1/q_1)$.

(i) If $\alpha < \beta$, no test with fixed numbers of trials m and n for testing $H : \rho = \rho_0$ can have power $\geq \beta$ against all alternatives with $\rho = \rho_1$.

(ii) The following is a simple sequential sampling scheme leading to the desired result. Let the trials be performed in pairs of one of each kind, and restrict attention to those pairs in which one of the trials is a success and the other a failure. If experimentation is continued until N such pairs have been observed, the number of pairs in which the successful trial belonged to the first series has the binomial distribution $b(\pi, N)$ with $\pi = p_1 q_2/(p_1 q_2 + P_2 q_1) = 1/(1 + \rho)$. A test of arbitrarily high power against ρ_1 is therefore obtained by taking N large enough.

(iii) If $p_1/p_2 = \lambda$, use inverse binomial sampling to devise a test of $H : \lambda = \lambda_0$ against $K : \lambda > \lambda_0$.

Problem 4.28 *Positive dependence.* Two random variables (X, Y) with c.d.f. $F(x, y)$ are said to be *positively quadrant dependent* if $F(x, y) \geq F(x, \infty)F(\infty, y)$ for all x, y.[14] For the case that (X, Y) takes on the four pairs of values $(0, 0)$, $(0, 1)$, $(1, 0)$, $(1, 1)$ with probabilities p_{00}, p_{01}, p_{10}, p_{11}, (X, Y) are positively quadrant dependent if and only if the odds ratio $\Delta = p_{01}p_{10}/p_{00}p_{11} \leq 1$.

Problem 4.29 *Runs.* Consider a sequence of N dependent trials, and let X_i be 1 or 0 as the i th trial is a success or failure. Suppose that the sequence has the *Markov* property[15]

$$P\{X_i = 1 | x_i, \ldots, x_{i-1}\} = P\{X_i = 1 | x_{i-1}\}$$

and the property of *stationarity* according to which $P\{X_i = 1\}$ and $P\{X_i = 1 | x_{i-1}\}$ are independent of i. The distribution of the X's is then specified by the

[14] For a systematic discussion of this and other concepts of dependence, see Tong (1980, Chapter 5), Kotz, Wang and Hung (1990) and Yanagimoto (1990).

[15] Statistical inference in these and more general Markov chains is discussed, for example, in Bhat and Miller (2002); they provide references at the end of Chapter 5.

probabilities

$$p_1 = P\{X_i = 1|x_{i-1} = 1\} \quad \text{and} \quad p_0 = P\{X_i = 1|x_{i-1} = 0\}$$

and by the initial probabilities

$$\pi_1 = P\{X_1 = 1\} \quad \text{and} \quad \pi_0 = 1 - \pi_1 = P\{X_1 = 0\}$$

(i) Stationarity implies that

$$\pi_1 = \frac{p_0}{p_0 + q_1}, \qquad \pi_0 = \frac{q_1}{p_0 + q_1}.$$

(ii) A set of successive outcomes $x_i, x_{i+1}, \ldots, x_{i+j}$ is said to form a *run* of zeros if $x_i = x_{i+1} = \cdots = x_{i+j} = 0$, and $x_{i-1} = 1$ or $i = 1$, and $x_{i+j+1} = 1$ or $i + j = N$. A run of ones is defined analogously. The probability of any particular sequence of outcomes $(x_1, \ldots, x_N)$ is

$$\frac{1}{p_0 + q_1} p_0^v p_1^{n-v} q_1^u q_0^{m-u},$$

where m and n denote the numbers of zeros and ones, and u and v the numbers of runs of zeros and ones in the sequence.

Problem 4.30 *Continuation.* For testing the hypothesis of independence of the X's, $H : p_0 = p_1$, against the alternatives $K : p_0 < p_1$, consider the *run test*, which rejects H when the total number of runs $R = U + V$ is less than a constant $C(m)$ depending on the number m of zeros in the sequence. When $R = C(m)$, the hypothesis is rejected with probability $\gamma(m)$, where C and γ are determined by

$$P_H\{R < C(m)|m\} + \gamma(m) P_H\{R = C(m)|m\} = \alpha.$$

(i) Against any alternative of K the most powerful similar test (which is at least as powerful as the most powerful unbiased test) coincides with the run test in that it rejects H when $R < C(m)$. Only the supplementary rule for bringing the conditional probability of rejection (given m) up to α depends on the specific alternative under consideration.

(ii) The run test is unbiased against the alternatives K.

(iii) The conditional distribution of R given m, when H is true, is[16]

$$P\{R = 2r\} = \frac{2\binom{m-1}{r-1}\binom{n-1}{r-1}}{\binom{m+n}{m}},$$

$$P\{R = 2r + 1\} = \frac{\binom{m-1}{r-1}\binom{n-1}{r} + \binom{m-1}{r}\binom{n-1}{r-1}}{\binom{m+n}{m}},$$

[(i): Unbiasedness implies that the conditional probability of rejection given m is α for all m. The most powerful conditional level-α test rejects H for those sample

[16] This distribution is tabled by Swed and Eisenhart (1943) and Gibbons and Chakraborti (1992); it can be obtained from the hypergeometric distribution [Guenther (1978)]. For further discussion of the run test, see Lou (1996).

sequences for which $\Delta(u, v) = (p_0/p_1)^v (q_1/q_0)^u$ is too large. Since $p_0 < p_1$ and $q_1 < q_0$ and since $|v - u|$ can only take on the values 0 and 1, it follows that

$$\Delta(1,1) > \Delta(1,2), \quad \Delta(2,1) > \Delta(2,2) > \Delta(2,3), \quad \Delta(3,2) > \cdots.$$

Thus only the relation between $\Delta(i, i+1)$ and $\Delta(i+1, i)$ depends on the specific alternative, and this establishes the desired result.
(ii): That the above conditional test is unbiased for each m is seen by writing its power as

$$\beta(p_0, p_1|m) = (1-\gamma)P\{R < C(m)|m\} + \gamma P\{R \leq C(m)|m\},$$

since by (i) the rejection regions $R < C(m)$ and $R < C(m) + 1$ are both UMP at their respective conditional levels.
(iii): When H is true, the conditional probability given m of any set of m zeros and n ones is $1/\binom{m+n}{m}$. The number of ways of dividing n ones into r groups is $\binom{n-1}{r-1}$, and that of dividing m zeros into $r+1$ groups is $\binom{m-1}{r}$. The conditional probability of getting $r+1$ runs of zeros and r runs of ones is therefore

$$\frac{\binom{m-1}{r}\binom{n-1}{r-1}}{\binom{m+n}{m}}.$$

To complete the proof, note that the total number of runs is $2r+1$ if and only if there are either $r+1$ runs of zeros and r runs of ones or r runs of zeros and $r+1$ runs of ones.]

Problem 4.31 (i) Based on the conditional distribution of $X_2, \ldots, X_n$ given $X_1 = x_1$ in the model of Problem 4.29, there exists a UMP unbiased test of $H : p_0 = p_1$ against $p_0 > p_1$ for every α.

(ii) For the same testing problem, without conditioning on X_1 there exists a UMP unbiased test if the initial probability π_1 is assumed to be completely unknown instead of being given by the value stated in (i) of Problem 4.29.

[The conditional distribution of $X_2, \ldots, X_n$ given x_1 is of the form

$$C(x_1; p_0, p_1, q_0, q_1) p_1^{y_1} p_0^{y_0} q_1^{z_1} q_0^{z_0} (y_1, y_2, z_1, z_2),$$

where y_1 is the number of times a 1 follows a 1, y_0 the number of times a 1 follows a 0, and so on, in the sequence $x_1, X_2, \ldots, X_n$. [See Billingsley (1961, p. 14).]

Problem 4.32 *Rank-sum test.* Let $Y_1, \ldots, Y_N$ be independently distributed according to the binomial distributions $b(p_i, n_i), i = 1, \ldots, N$ where

$$p_i = \frac{1}{1 + e^{-(\alpha + \beta x_i)}}.$$

This is the model frequently assumed in bioassay, where x_i denotes the dose, or some function of the dose such as its logarithm, of a drug given to n_i experimental subjects, and where Y_i is the number among these subjects which respond to the drug at level x_i. Here the x_i are known, and α and β are unknown parameters.

(i) The joint distribution of the Y's constitutes an exponential family, and UMP unbiased tests exist for the four hypotheses of Theorem 4.4.1, concern both α and β.

(ii) Suppose in particular that $x_i = \Delta i$, where Δ is known, and that $n_i = 1$ for all i. Let n be the number of successes in the N trials, and let these successes occur in the s_1st, s_2nd,..., s_nth trial, where $s_1 < s_2 < \cdots < s_n$. Then the UMP unbiased test for testing $H : \beta = 0$ against the alternatives $\beta > 0$ is carried out conditionally, given n, and rejects when the *rank sum* $\sum_{i-1}^{n} s_i$ is too large.

(iii) Let $Y_1, \ldots, Y_M$ and $Z_1, \ldots, Z_N$. be two independent sets of experiments of the type described at the beginning of the problem, corresponding, say, to two different drugs. If Y_i is distributed as $b(p_i, m_i)$ and Z_j as $b(\pi_j, n_j)$, with

$$p_i = \frac{1}{1 + e^{-(\alpha+\beta u_i)}}, \qquad \pi_j = \frac{1}{a + e^{-(\gamma+\beta v_j)}},$$

then UMP unbiased tests exist for the four hypotheses concerning $\gamma - \alpha$ and $\delta - \beta$.

Section 4.8

Problem 4.33 In a $2 \times 2 \times 2$ table with $m_1 = 3$, $n_1 = 4$; $m_2 = 4$, $n_2 = 4$; and $t_1 = 3$, $t_1' = 4$, $t_2 = t_2' = 4$, determine the probabilities that $P(Y_1 + Y_2 \leq K | X_i + Y_i = t_i, i = 1, 2)$ for $k = 0, 1, 2, 3$.

Problem 4.34 In a $2 \times 2 \times K$ table with $\Delta_k = \Delta$, the test derived in the text as UMP unbiased for the case that the B and C margins are fixed has the same property when any two, one, or no margins are fixed.

Problem 4.35 The UMP unbiased test of $H : \Delta = 1$ derived in Section 4.8 for the case that the B- and C-margins are fixed (where the conditioning now extends to all random margins) is also UMP unbiased when

(i) only one of the margins is fixed;

(ii) the entries in the $4K$ cells are independent Poisson variables with means $\lambda_{ABC}, \ldots$, and Δ is replaced by the corresponding cross-ratio of the λ's.

Problem 4.36 Let X_{ijkl} $(i, j, k = 0, 1, \; l = 1, \ldots, L)$ denote the entries in a $2 \times 2 \times 2 \times L$ table with factors A, B, C, and D, and let

$$\Gamma_l = \frac{P_{AB^cCD_l} P_{\tilde{A}BCD_l} P_{A\tilde{B}\tilde{C}D_l} P_{\tilde{A}B\tilde{C}D_l}}{P_{ABCD_l} P_{\tilde{A}\tilde{B}CD_l} P_{AB\tilde{C}D_l} P_{\tilde{A}\tilde{B}\tilde{C}D_l}}.$$

Then

(i) under the assumption $\Gamma_l = \Gamma$ there exists a UMP unbiased test of the hypothesis $\Gamma \leq \Gamma_0$ to for any fixed Γ_0;

(ii) When $l = 2$, there exists a UMP unbiased test of the hypothesis $\Gamma_1 = \Gamma_2$ —in both cases regardless of whether 0, 1, 2 or 3 of the sets of margins are fixed.

Section 4.9

Problem 4.37 In the 2×2 table for matched pairs, show by formal computation that the conditional distribution of Y given $X' + Y = d$ and $X = x$ is binomial with the indicated p.

Problem 4.38 Consider the comparison of two success probabilities in (a) the two-binomial situation of Section 4.5 with $m = n$, and (b) the matched-pairs situation of Section 4.9. Suppose the matching is completely at random, that is, a random sample of $2n$ subjects, obtained from a population of size $N(2n \leq N)$, is divided at random into n pairs, and the two treatments B and B^c are assigned at random within each pair.

(i) The UMP unbiased test for design (a) (Fisher's exact test) is always more powerful than the UMP unbiased test for design (b) (McNemar's test).

(ii) Let X_i (respectively Y_i) be 1 or 0 as the 1st (respectively 2nd) member of the i th pair is a success or failure. Then the correlation coefficient of X_i and Y_i can be positive or negative and tends to zero as $N \to \infty$.

[(ii): Assume that the kth member of the population has probability of success $P_A^{(k)}$ under treatment A and $P_{\tilde{A}}^{(k)}$ under $\tilde{A}$.]

Problem 4.39 In the 2×2 table for matched pairs, in the notation of Section 4.9, the correlation between the responses of the two members of a pair is

$$\rho = \frac{p_{11} - \pi_1 \pi_2}{\sqrt{\pi_1(1 - \pi_1)\pi_2(1 - \pi_2)}}.$$

For any given values of $\pi_1 < \pi_2$, the power of the one-sided McNemar test of $H : \pi_1 = \pi_2$ is an increasing function of ρ.
[The conditional power of the test given $X + Y = d$, $X = x$ is an increasing function $p = p_{0l}/(p_{01} + p_{10})$.]
Note. The correlation ρ increases with the effectiveness of the matching, and McNemar's test under (b) of Problem 4.38 soon becomes more powerful than Fisher's test under (a). For detailed numerical comparisons see Wacholder and Weinberg (1982) and the references given there.

4.11 Notes

The closely related properties of similarity (on the boundary) and unbiasedness are due to Neyman and Pearson (1933, 1936), who applied them to a variety of examples. It was pointed out by Neyman (1937) that similar tests could be obtained through the construction method now called Neyman structure. Theorem 4.3.1 is due to Ghosh (1948) and Hoel (1948). The concepts of completeness and bounded completeness, and the application of the latter to Theorem 4.4.1, were developed by Lehmann and Scheffé (1950).

The sign test, proposed by Arbuthnot (1710) to test that the probability of a male birth is 1/2, may be the first significance test in the literature. The exact test for independence in 2 by 2 table is due to Fisher (1934).

5

Unbiasedness: Applications to Normal Distributions; Confidence Intervals

5.1 Statistics Independent of a Sufficient Statistic

A general expression for the UMP unbiased tests of the hypotheses $H_1 : \theta \leq \theta_0$ and $H_4 : \theta = \theta_0$ in the exponential family

$$dP_{\theta,\vartheta}(x) = C(\theta, \vartheta) \exp\left[\theta U(x) + \sum \vartheta_i T_i(x)\right] d\mu(x) \tag{5.1}$$

was given in Theorem 4.4.1 of the preceding chapter. However, this turns out to be inconvenient in the applications to normal and certain other families of continuous distributions, with which we shall be concerned in the present chapter. In these applications, the tests can be given a more convenient form, in which they no longer appear as conditional tests in terms of U given t, but are expressed unconditionally in terms of a single test statistic. The following are three general methods of achieving this.

(i) In many of the problems to be considered below, the UMP unbiased test ϕ_0, is also UMP invariant, as will be shown in Chapter 6. From Theorem 6.5.3, it is then possible to conclude that ϕ_0 is UMP unbiased. This approach, in which the latter property must be taken on faith during the discussion of the test in the present chapter, is the most economical of the three, and has the additional advantage that it derives the test instead of verifying a guessed solution as is the case with methods (ii) and (iii).

(ii) The conditional descriptions (4.12), (4.14), and (4.16) can be replaced by equivalent unconditional ones, and it is then enough to find an unbiased test which has the indicated structure. This approach is discussed in Pratt (1962).

(iii) Finally, it is often possible to show the equivalence of the test given by Theorem 4.4.1 to a test suspected to be optimal, by means of Theorem 5.1.2

below. This is the course we shall follow here; the alternative derivation (i) will be discussed in Chapter 6.

The reduction by method (iii) depends on the existence of a statistic $V = h(U,T)$, which is independent of T when $\theta = \theta_0$, and which for each fixed t is monotone in U for H_1 and linear in U for H_4. The critical function ϕ_1, for testing H_1 then satisfies

$$\phi(v) = \begin{cases} 1 & \text{when} \quad v > C_0, \\ \gamma_0 & \text{when} \quad v = C_0, \\ 0 & \text{when} \quad v < C_0, \end{cases} \tag{5.2}$$

where C_0 and γ_0 are no longer dependent on t, and are determined by

$$E_{\theta_0}\phi_1(V) = \alpha. \tag{5.3}$$

Similarly the test ϕ_4 of H_4 reduces to

$$\phi(v) = \begin{cases} 1 & \text{when} \quad v < C_1 \text{ or } v > C_2, \\ \gamma_i & \text{when} \quad v = C_i, \quad i = 1, 2, \\ 0 & \text{when} \quad C_1 < v < C_2, \end{cases} \tag{5.4}$$

where the C's and γ's are determined by

$$E_{\theta_0}[\phi_4(V)] = \alpha \tag{5.5}$$

and

$$E_{\theta_0}[V\phi_4(V)] = \alpha E_{\theta_0}(V). \tag{5.6}$$

The corresponding reduction for the hypotheses $H_2 : \theta \leq \theta_1$, or $\theta \geq \theta_2$ and $H_3 : \theta_1 \leq \theta \leq \theta_2$ requires that V be monotone in U for each fixed t, and be independent of T when $\theta = \theta_1$ and $\theta = \theta_2$. The test ϕ_3 is then given by (5.4) with the C's and γ's determined by

$$E_{\theta_1}\phi_3(V) = E_{\theta_2}\phi_3(V) = \alpha. \tag{5.7}$$

The test for H_2 as before has the critical function

$$\phi_2(v; \alpha) = 1 - \phi_3(v; 1 - \alpha).$$

This is summarized in the following theorem.

Theorem 5.1.1 *Suppose that the distribution of X is given by (5.1) and that $V = h(U,T)$ is independent of T when $\theta = \theta_0$. Then ϕ_1 is UMP unbiased for testing H_1 provided the function h is increasing in u for each t, and ϕ_4 is UMP unbiased for H_4 provided*

$$h(u,t) = a(t)u + b(t) \quad \textit{with } a(t) > 0.$$

The tests ϕ_2 and ϕ_3, are UMP unbiased for H_2 and H_3 if V is independent of T when $\theta = \theta_1$ and θ_2, and if h is increasing in u for each t.

PROOF. The test of H_1 defined by (4.12) and (4.13) is equivalent to that given by (5.2), with the constants determined by

$$P_{\theta_0}\{V > C_0(t) \mid t\} + \gamma_0(t)P_{\theta_0}\{V = C_0(t) \mid t\} = \alpha.$$

By assumption, V is independent of T when $\theta = \theta_0$, and C_0 and γ_0 therefore do not depend on t. This completes the proof for H_1, and that for H_2 and H_3 is quite analogous.

The test of H_4 given in Section 4.4 is equivalent to that defined by (5.4) with the constants C_i and γ_i determined by $E_{\theta_0}[\phi_4(V,t) \mid t] = \alpha$ and

$$E_{\theta_0}\left[\phi_4(V,t)\frac{V-b(t)}{a(t)} \,\middle|\, t\right] = \alpha E_{\theta_0}\left[\frac{V-b(t)}{a(t)} \,\middle|\, t\right],$$

which reduces to

$$E_{\theta_0}[V\phi_4(V,t) \mid t] = \alpha E_{\theta_0}[V \mid t].$$

Since V is independent of T for $\theta = \theta_0$, so are the C's and γ's as was to be proved. ■

To prove the required independence of V and T in applications of Theorem 5.1.1 to special cases, the standard methods of distribution theory are available: transformation of variables, characteristic functions, and the geometric method. Alternatively, for a given model $\{P_\vartheta, \vartheta \in \omega\}$, suppose V is any statistic whose distribution does not depend on ϑ; such a statistic is said to be *ancillary*. Then, the following theorem gives sufficient conditions to show V and T are independent.

Theorem 5.1.2 (Basu) *Let the family of possible distributions of X be $\mathcal{P} = \{P_\vartheta, \vartheta \in \omega\}$, let T be sufficient for $\mathcal{P}$, and suppose that the family $\mathcal{P}^T$ of distributions of T is boundedly complete. If V is any ancillary statistic for $\mathcal{P}$, then V is independent of T.*

PROOF. For any critical function ϕ, the expectation $E_\vartheta\phi(V)$ is by assumption independent of ϑ. It therefore follows from Theorem 4.3.2 that $E[\phi(V) \mid t]$ is constant (a.e. $\mathcal{P}^T$) for every critical function ϕ, and hence that V is independent of T. ■

Corollary 5.1.1 *Let $\mathcal{P}$ be the exponential family obtained from (5.1) by letting θ have some fixed value. Then a statistic V is independent of T for all ϑ provided the distribution of V does not depend on ϑ.*

PROOF. It follows from Theorem 4.3.1 that $\mathcal{P}^T$ is complete and hence boundedly complete, and the preceding theorem is therefore applicable. ■

Example 5.1.1 Let $X_1, \ldots, X_n$, be independently, normally distributed with mean ξ and variance σ^2. Suppose first that σ^2 is fixed at σ_0^2. Then the assumptions of Corollary 5.1.1 hold with $T = \bar{X} = \sum X_i/n$ and ϑ proportional to ξ. Let f be any function satisfying

$$f(x_1 + c, \ldots, x_n + c) = f(x_1, \ldots, x_n) \quad \text{for all real } c.$$

If

$$V = f(X_1, \ldots, X_n),$$

then also $V = f(X_1 - \xi, \ldots, X_n - \xi)$. Since the variables $X_i - \xi$ are distributed as $N(0, \sigma_0^2)$, which does not involve ξ, the distribution of V does not depend on ξ. It follows from Corollary 5.1.1 that any such statistic V, and therefore in particular $V = \sum(X_i - \bar{X})^2$, is independent of $\bar{X}$. This is true for all σ.

Suppose, on the other hand, that ξ is fixed at ξ_0. Then Corollary 5.1.1 applies with $T = \sum(X_i - \xi_0)^2$ and $\vartheta = -1/2\sigma^2$. Let f be any function such that

$$f(cx_1, \ldots, cx_n) = f(x_1, \ldots, x_n) \quad \text{for all } c > 0,$$

and let

$$V = f(X_1 - \xi_0, \ldots, X_n - \xi_0).$$

Then V is unchanged if each $X_i - \xi_0$ is replaced by $(X_i - \xi_0)/\sigma$, and since these variables are normally distributed with zero mean and unit variance, the distribution of V does not depend on σ. It follows that all such statistics V, and hence for example

$$\frac{\bar{X} - \xi_0}{\sqrt{\sum(X_i - \bar{X})^2}} \quad \text{and} \quad \frac{\bar{X} - \xi_0}{\sqrt{\sum(X_i - \xi_0)^2}},$$

are independent of $\sum(X_i - \xi_0)^2$. This, however, does not hold for all ξ, but only when $\xi = \xi_0$. ■

Example 5.1.2 Let U_1/σ_1^2 and U_2/σ_2^2 be independently distributed according to χ^2-distributions with f_1 and f_2 degrees of freedom respectively, and suppose that $\sigma_2^2/\sigma_1^2 = a$. The joint density of the U's is then

$$Cu_1^{(f_1/2)-1} u_2^{(f_2/2)-1} \exp\left[-\frac{1}{2\sigma_2^2}(au_1 + u_2)\right]$$

so that Corollary 5.1.1 is applicable with $T = aU_1 + U_2$ and $\vartheta = -1/2\sigma_2^2$. Since the distribution of

$$V = \frac{U_2}{U_1} = a\frac{U_2/\sigma_2^2}{U_1/\sigma_1^2}$$

does not depend on σ_2, V is independent of $aU_1 + U_2$. For the particular case that $\sigma_2 = \sigma_1$, this proves the independence of U_2/U_1 and $U_1 + U_2$. ■

Example 5.1.3 Let $(X_1, \ldots, X_n)$ and $(Y_1, \ldots, Y_n)$ be samples from normal distributions $N(\xi, \sigma^2)$ and $N(\eta, \tau^2)$ respectively. Then $T = (\bar{X}, \sum X_i^2, \bar{Y}, \sum Y_i^2)$ is sufficient for $(\xi, \sigma^2, \eta, \tau^2)$ and the family of distributions of T is complete. Since

$$V = \frac{\sum(X_i - \bar{X})(Y_i - \bar{Y})}{\sqrt{\sum(X_i - \bar{X})^2(Y_i - \bar{Y})^2}}$$

is unchanged when X_i and Y_i are replaced by $(X_i - \xi)/\sigma$ and $(Y_i - \eta)/\tau$, the distribution of V does not depend on any of the parameters, and Theorem 5.1.2 shows V to be independent of T. ■

5.2 Testing the Parameters of a Normal Distribution

The four hypotheses $\sigma \le \sigma_0$, $\sigma \ge \sigma_0$, $\xi \le \xi_0$, $\xi \ge \xi_0$ concerning the variance σ^2 and mean ξ of a normal distribution were discussed in Section 3.9, and it was pointed out there that at the usual significance levels there exists a UMP test only for the first one. We shall now show that the standard (likelihood-ratio) tests are UMP unbiased for the above four hypotheses as well as for some of the corresponding two-sided problems.

For varying ξ and σ, the densities

$$(2\pi\sigma^2)^{-n/2} \exp\left(-\frac{n\xi^2}{2\sigma^2}\right) \exp\left(-\frac{1}{2\sigma^2}\sum x_i^2 + \frac{\xi}{\sigma^2}\sum x_i\right) \tag{5.8}$$

of a sample $X_1, \ldots, X_n$ from $N(\xi, \sigma^2)$ constitute a two-parameter exponential family, which coincides with (5.1) for

$$\theta = -\frac{1}{2\sigma^2}, \quad \vartheta = \frac{n\xi}{\sigma^2}, \quad U(X) = \sum x_i^2, \quad T(x) = \bar{x} = \frac{\sum x_i}{n}.$$

By Theorem 4.4.1, there exists therefore a UMP unbiased test of the hypothesis $\theta \geq \theta_0$, which for $\theta_0 = -1/2\sigma_0^2$ is equivalent to $H : \sigma \geq \sigma_0$. The rejection region of this test can be obtained from (4.12), with the inequalities reversed because the hypothesis is now $\theta \geq \theta_0$. In the present case this becomes

$$\sum x_i^2 \leq C_0(\bar{x})$$

where

$$p_{\sigma_0}\left\{\sum X_i^2 \leq C_0(\bar{x}) \mid \bar{x}\right\} = \alpha.$$

If this is written as

$$\sum x_i^2 - n\bar{x}^2 < C_0'(\bar{x})$$

it follows from the independence of $\sum X_1^2 - n\bar{X}^2 = \sum (X_i - \bar{X})^2$ and $\bar{X}$ (Example 5.1.1) that $C_0'(x)$ does not depend on $\bar{x}$. The test therefore rejects when $\sum (x_i - \bar{x})^2 \leq C_0'$, or equivalently when

$$\frac{\sum (x_i - \bar{x})^2}{\sigma_0^2} \leq C_0, \tag{5.9}$$

with C_0 determined by $P_{\sigma_0}\{\sum (X_i - \bar{X})^2/\sigma_0^2 \leq C_0\} = \alpha$. Since $\sum (X_i - \bar{X})^2/\sigma_0^2$ has a χ^2-distribution with $n-1$ degrees of freedom, the determining condition for C_0 is

$$\int_0^{C_0} \chi_{n-1}^2(y)\, dy = \alpha\ , \tag{5.10}$$

where χ_{n-1}^2 denotes the density of a χ^2 variable with $n-1$ degrees of freedom.

The same result can be obtained through Theorem 5.1.1. A statistic $V = h(U, T)$ of the kind required by the theorem – that is, independent of $\bar{X}$ for $\sigma = \sigma_0$, and all ξ – is

$$V = \sum (X_i - \bar{X})^2 = U - nT^2.$$

This is in fact independent of $\bar{X}$ for all ξ and σ^2. Since $h(u, t)$ is an increasing function of u for each t, it follows that the UMP unbiased test has a rejection region of the form $V \leq C_0'$.

This derivation also shows that the UMP unbiased rejection region for $H: \sigma \le \sigma_1$ or $\sigma \ge \sigma_2$ is

$$C_1 < \sum (x_i - \bar{x})^2 < C_2 \tag{5.11}$$

where the C's are given by

$$\int_{C_1/\sigma_1^2}^{C_2/\sigma_1^2} \chi^2_{n-1}(y)\,dy = \int_{C_1/\sigma_2^2}^{C_2/\sigma_2^2} \chi^2_{n-1}(y)\,dy = \alpha. \tag{5.12}$$

Since $h(u,t)$ is linear in u, it is further seen that the UMP unbiased test of $H: \sigma = \sigma_0$, has the acceptance region

$$C_1' < \frac{\sum (x_i - \bar{x})^2}{\sigma_0^2} < C_2' \tag{5.13}$$

with the constants determined by

$$\int_{C_1'}^{C_2'} \chi^2_{n-1}(y)\,dy = \frac{1}{n-1} \int_{C_2'}^{C_1'} y\chi^2_{n-1}(y)\,dy = 1 - \alpha. \tag{5.14}$$

This is just the test obtained in Example 4.2.2 with $\sum(x_i - \bar{x})^2$ in place of $\sum x_i^2$ and $n-1$ degrees of freedom instead of n, as could have been foreseen. Theorem 5.1.1 shows for this and the other hypotheses considered that the UMP unbiased test depends only on V. Since the distributions of V do not depend on ξ, and constitute an exponential family in σ, the problems are thereby reduced to the corresponding ones for a one-parameter exponential family, which were solved previously.

The power of the above tests can be obtained explicitly in terms of the χ^2-distribution. In the case of the one-sided test (5.9) for example, it is given by

$$\beta(\sigma) = P_\sigma\left\{\frac{\sum(X_i - \bar{X})^2}{\sigma^2} \le \frac{C_0\sigma_0^2}{\sigma^2}\right\} = \int_0^{C_0\sigma_0^2/\sigma^2} \chi^2_{n-1}(y)\,dy.$$

The same method can be applied to the problems of testing the hypotheses $\xi \le \xi_0$ against $\xi > \xi_0$ and $\xi = \xi_0$ against $\xi \ne \xi_0$. As is seen by transforming to the variables $X_i - \xi_0$, there is no loss of generality in assuming that $\xi_0 = 0$. It is convenient here to make the identification of (5.8) with (5.1) through the correspondence

$$\theta = \frac{n\xi}{\sigma^2}, \quad \vartheta = -\frac{1}{2\sigma^2}, \quad U(x) = \bar{x}, \quad T(x) = \sum x_i^2.$$

Theorem 4.4.1 then shows that UMP unbiased tests exist for the hypotheses $\theta \le 0$ and $\theta = 0$, which are equivalent to $\xi \le 0$ and $\xi = 0$. Since

$$V = \frac{\bar{X}}{\sqrt{\sum(X_i - \bar{X})^2}} = \frac{U}{\sqrt{T - nU^2}}$$

is independent of $T = \sum X_i^2$ when $\xi = 0$ (Example 5.1.1), it follows from Theorem 5.1.1 that the UMP unbiased rejection region for $H: \xi \le 0$ is $V \ge C_0'$ or equivalently

$$t(x) \ge C_0, \tag{5.15}$$

where

$$t(x) = \frac{\sqrt{n}\bar{x}}{\sqrt{\frac{1}{n-1}\sum(x_i - \bar{x})^2}}. \tag{5.16}$$

In order to apply the theorem to $H' : \xi = 0$, let $W = \bar{X}/\sqrt{\sum X_i^2}$. This is also independent of $\sum X_i^2$ when $\xi = 0$, and in addition is linear in $U = \bar{X}$. The distribution of W is symmetric about 0 when $\xi = 0$, and conditions (5.4), (5.5), (5.6) with W in place of V are therefore satisfied for the rejection region $|w| \geq C'$ with $P_{\xi=0}\{|W| \geq C'\} = \alpha$. Since

$$t(x) = \frac{\sqrt{(n-1)n}W(x)}{\sqrt{1 - nW^2(x)}},$$

the absolute value of $t(x)$ is an increasing function of $|W(x)|$, and the rejection region is equivalent to

$$|t(x)| \geq C. \tag{5.17}$$

From (5.16) it is seen that $t(X)$ is the ratio of the two independent random $\sqrt{n}\bar{X}/\sigma$ and $\sqrt{\sum(X_i - \bar{X})^2/(n-1)\sigma^2}$. The denominator is distributed as the square root of a χ^2-variable with $n-1$ degrees of freedom, divided by $n-1$; the distribution of the numerator, when $\xi = 0$, is the normal distribution $N(0,1)$. The distribution of such a ratio is *Student's* t-distribution with $n-1$ degrees of freedom, which has probability density (Problem 5.3)

$$t_{n-1}(y) = \frac{1}{\sqrt{\pi(n-1)}} \frac{\Gamma(\frac{1}{2}n)}{\Gamma\left[\frac{1}{2}(n-1)\right]} \frac{1}{\left(1 + \frac{y^2}{n-1}\right)^{\frac{1}{2}n}}. \tag{5.18}$$

The distribution is symmetric about 0, and the constants C_0 and C of the one- and two-sided tests are determined by

$$\int_{C_0}^{\infty} t_{n-1}(y)\,dy = \alpha \quad \text{and} \quad \int_{C}^{\infty} t_{n-1}(y)\,dy = \frac{\alpha}{2}. \tag{5.19}$$

For $\xi \neq 0$, the distribution of $t(X)$ is the so-called *noncentral* t-distribution, which is derived in Problem 5.3. Some properties of the power function of the one- and two-sided t-test are given in Problems 5.1, 5.2, and 5.4. We note here that the distribution of $t(X)$, and therefore the power of the above tests, depends only on the noncentrality parameter $\delta = \sqrt{n}\xi/\sigma$. This is seen from the expression of the probability density given in Problem 5.3, but can also be shown by the following direct argument. Suppose that $\xi'/\sigma' = \xi/\sigma \neq 0$, and denote the common value of ξ'/ξ and σ'/σ by c, which is then also different from zero. If $X_i' = cX_i$ and the X_i are distributed as $N(\xi, \sigma^2)$, the variables X_i' have distribution $N(\xi', \sigma'^2)$. Also $t(X) = t(X')$, and hence $t(X')$ has the same distribution as $t(X)$, as was to be proved. [Tables of the power of the t-test are discussed, for example, in Chapter 31, Section 7 of Johnson, Kotz and Balakrishnan (1995, Vol. 2).]

If ξ_1 denotes any alternative value to $\xi = 0$, the power $\beta(\xi, \sigma) = f(\delta)$ depends on σ. As $\sigma \to \infty$, $\delta \to 0$, and

$$\beta(\xi_1, \sigma) \to f(0) = \beta(0, \sigma) = \alpha,$$

since f is continuous by Theorem 2.7.1. Therefore, regardless of the sample size, the probability of detecting the hypothesis to be false when $\xi \geq \xi_1 > 0$ cannot be made $\geq \beta > \alpha$ for all σ. This is not surprising, since the distributions $N(0, \sigma^2)$ and $N(\xi_1, \sigma^2)$ become practically indistinguishable when σ is sufficiently large. To obtain a procedure with guaranteed power for $\xi \geq \xi_1$, the sample size must be made to depend on σ. This can be achieved by a sequential procedure, with the stopping rule depending on an estimate of σ, but not with a procedure of fixed sample size. (See Problems 5.23 and 5.25.)

The tests of the more general hypotheses $\xi \leq \xi_0$ and $\xi = \xi_0$ are reduced to those above by transforming to the variables $X_i - \xi_0$. The rejection regions for these hypotheses are given as before by (5.15), (5.17), and (5.19), but now with

$$t(x) = \frac{\sqrt{n}(\bar{x} - \xi_0)}{\sqrt{\frac{1}{n-1}\sum(x_i - \bar{x})^2}}.$$

It is seen from the representation of (5.8) as an exponential family with $\theta = n\xi/\sigma^2$ that there exists a UMP unbiased test of the hypothesis $a \leq \xi/\sigma^2 \leq b$, but the method does not apply to the more interesting hypothesis $a \leq \xi \leq b$;[1] nor is it applicable to the corresponding hypothesis for the mean expressed in σ-units: $a \leq \xi/\sigma \leq b$, which will be discussed in Chapter 6. The dual equivalence problem of testing $\xi/\sigma \notin [a, b]$ is treated in Brown, Casella and Hwang (1995), Brown, Hwang, and Munk (1997) and Perlman and Wu (1999).

When testing the mean ξ of a normal distribution, one may from extensive past experience believe σ to be essentially known. If in fact σ is known to be equal to σ_0, it follows from Problem 3.1 that there exists a UMP test ϕ_0 of $H : \xi \leq \xi_0$, against $K : \xi > \xi_0$, which rejects when $(\bar{X} - \xi_0)/\sigma_0$ is sufficiently large, and this test is then uniformly more powerful than the t-test (5.15). On the other hand, if the assumption $\sigma = \sigma_0$ is in error the size of ϕ_0 will differ from α and may greatly exceed it. Whether to take such a risk depends on one's confidence in the assumption and the gain resulting from the use of ϕ_0 when σ is equal to σ_0. A measure of this gain is the *deficiency* d of the t-test with respect to ϕ_0, the number of additional observations required by the t-test to match the power of ϕ_0 when $\sigma = \sigma_0$. Except for very small n, d is essentially independent of sample size and for typical values of α is of the order of 1 to 3 additional observations. [For details see Hodges and Lehmann (1970). Other approaches to such comparisons are reviewed, for example, in Rothenberg (1984).]

5.3 Comparing the Means and Variances of Two Normal Distributions

The problem of comparing the parameters of two normal distributions arises in the comparison of two treatments, products, etc., under conditions similar to those discussed at the beginning of Section 4.5. We consider first the comparison of two variances σ^2 and τ^2, which occurs for example when one is concerned with the variability of analyses made by two different laboratories or by two different methods, and specifically the hypotheses $H : \tau^2/\sigma^2 \leq \Delta_0$ and $H' : \tau^2/\sigma^2 = \Delta_0$.

[1]This problem is discussed in Section 3 of Hodges and Lehmann (1954).

Let $X = (X_1, \dots, X_m)$ and $Y = (Y_1, \dots, Y_n)$ be samples from the normal distributions $N(\xi, \sigma^2)$ and $N(\eta, \tau^2)$ with joint density

$$C(\xi, \eta, \sigma, \tau) \exp\left(-\frac{1}{2\sigma^2}\sum x_i^2 - \frac{1}{2\tau^2}\sum y_j^2 + \frac{m\xi}{\sigma^2}\bar{x} + \frac{n\eta}{\tau^2}\bar{y}\right).$$

This is an exponential family with the four parameters

$$\theta = -\frac{1}{2\tau^2}, \quad \vartheta_1 = -\frac{1}{2\sigma^2}, \quad \vartheta_2 = \frac{n\eta}{\tau^2}, \quad \vartheta_3 = \frac{m\xi}{\sigma^2}$$

and the sufficient statistics

$$U = \sum Y_j^2, \quad T_1 = \sum X_i^2, \quad T_2 = \bar{Y}, \quad T_3 = \bar{X}.$$

It can be expressed equivalently (see Lemma 4.4.1) in terms of the parameters

$$\theta^* = -\frac{1}{2\tau^2} + \frac{1}{2\Delta_0\sigma^2}, \quad \vartheta_i^* = \vartheta_i \quad (i = 1, 2, 3)$$

and the statistics

$$U^* = \sum Y_j^2, \quad T_1^* = \sum X_i^2 + \frac{1}{\Delta_0}\sum Y_j^2, \quad T_2^* = \bar{Y}, \quad T_3^* = \bar{X}.$$

The hypotheses $\theta^* \leq 0$ and $\theta^* = 0$, which are equivalent to H and H' respectively, therefore possess UMP unbiased tests by Theorem 4.4.1.

When $\tau^2 = \Delta_0\sigma^2$, the distribution of the statistic

$$V = \frac{\sum(Y_j - \bar{Y})^2/\Delta_0}{\sum(X_i - \bar{X})^2} = \frac{\sum(Y_j - \bar{Y})^2/\tau^2}{\sum(X_i - \bar{X})^2/\sigma^2}$$

does not depend on σ, ξ, or η, and it follows from Corollary 5.1.1 that V is independent of (T_1^*, T_2^*, T_3^*). The UMP unbiased test of H is therefore given by (5.2) and (5.3), so that the rejection region can be written as

$$\frac{\sum(Y_j - \bar{Y})^2/\Delta_0(n-1)}{\sum(X_i - \bar{X})^2/(m-1)} \geq C_0. \tag{5.20}$$

When $\tau^2 = \Delta_0\sigma^2$, the statistic on the left-hand side of (5.20) is the ratio of the two independent χ^2 variables $\sum(Y_j - \bar{Y})^2/\tau^2$ and $\sum(X_i - \bar{X})^2/\sigma^2$, each divided by the number of its degrees of freedom. The distribution of such a ratio is the *F-distribution* with $n-1$ and $m-1$ degrees of freedom, which has the density

$$F_{n-1,m-1}(y) = \frac{\Gamma\left[\frac{1}{2}(m+n-2)\right]}{\Gamma\left[\frac{1}{2}(m-1)\right]\Gamma\left[\frac{1}{2}(n-1)\right]}\left(\frac{n-1}{m-1}\right)^{\frac{1}{2}(n-1)} \tag{5.21}$$

$$\times \frac{y^{\frac{1}{2}(n-1)-1}}{\left(1 + \frac{n-1}{m-1}y\right)^{\frac{1}{2}(m+n-2)}}.$$

The constant C_0 of (5.20) is then determined by

$$\int_{C_0}^{\infty} F_{n-1,m-1}(y)\,dy = \alpha. \tag{5.22}$$

In order to apply Theorem 5.1.1 to H' let

$$W = \frac{\sum(Y_j - \bar{Y})^2/\Delta_0}{\sum(X_i - \bar{X})^2 + (1/\Delta_0)\sum(Y_j - \bar{Y})^2}.$$

This is also independent of $T^* = (T_1^*, T_2^*, T_3^*)$ when $\tau^2 = \Delta_0\sigma^2$, and is linear in U^*. The UMP unbiased acceptance region of H' is therefore

$$C_1 \le W \le C_2 \tag{5.23}$$

with the constants determined by (5.5) and (5.6) where V is replaced by W. On dividing numerator and denominator of W by σ^2 it is seen that for $\tau^2 = \Delta_0\sigma^2$, the statistic W is a ratio of the form $W_1/(W_1 + W_2)$, where W_1 and W_2 are independent χ^2 variables with $n-1$ and $m-1$ degrees of freedom respectively. Equivalently, $W = Y/(1+Y)$, where $Y = W_1/W_2$ and where $(m-1)Y/(n-1)$ has the distribution $F_{n-1,m-1}$. The distribution of W is the *beta-distribution*[2] with density

$$B_{\frac{1}{2}(n-1),\frac{1}{2}(m-1)}(w) = \frac{\Gamma\left[\frac{1}{2}(m+n-2)\right]}{\Gamma\left[\frac{1}{2}(m-1)\right]\Gamma\left[\frac{1}{2}(n-1)\right]} w^{\frac{1}{2}(n-3)}(1-w)^{\frac{1}{2}(m-3)}, \qquad 0 < w < 1. \tag{5.24}$$

The conditions (5.5) and (5.6), by means of the relations

$$E(W) = \frac{n-1}{m+n-2}$$

and

$$wB_{\frac{1}{2}(n-1),\frac{1}{2}m-1)}(w) = \frac{n-1}{m+n-2} B_{\frac{1}{2}(n+1),\frac{1}{2}(m-1)}(w),$$

become

$$\int_{C_1}^{C_2} B_{\frac{1}{2}(n-1),\frac{1}{2}(m-1)}(w)\,dw = \int_{C_1}^{C_2} B_{\frac{1}{2}(n+1),\frac{1}{2}(m-1)}(w)\,dw = 1-\alpha. \tag{5.25}$$

The definition of V shows that its distribution depends only on the ratio τ^2/σ^2, and so does the distribution of W. The power of the tests (5.20) and (5.23) is therefore also a function only of the variable $\Delta = \tau^2/\sigma^2$; it can be expressed explicitly in terms of the F-distribution, for example in the first case by

$$\begin{aligned}\beta(\Delta) &= P\left\{\frac{\sum(Y_j-\bar{Y})^2/\tau^2(n-1)}{\sum(X_i-\bar{X})^2/\sigma^2(m-1)} \ge \frac{C_0\Delta_0}{\Delta}\right\} \\ &= \int_{C_0\Delta_0/\Delta}^{\infty} F_{n-1,m-1}(y)\,dy.\end{aligned}$$

The hypothesis of equality of the means ξ, η of two normal distributions with unknown variances σ^2 and τ^2, the so-called *Behrens-Fisher problem*, is not accessible by the present method. (See Example 4.3.3; for a discussion of this problem, Section 6.6, Section 11.3.1 and Example 13.5.4.) We shall therefore consider only

---

[2]The relationship $W = Y/(1+Y)$ shows the F- and beta-distributions to be equivalent. Tables of these distributions are discussed in Chapters 24 and 26 of Johnson, Kotz and Balakrishnan (1995. Vol. 2). Critical values of F are tabled by Mardia and Zemroch (1978), who also provide algorithms for the associated computations.

the simpler case in which the two variances are assumed to be equal. The joint density of the X's and Y's is then

$$C(\xi, \eta, \sigma) \exp\left[-\frac{1}{2\sigma^2}\left(\sum x_i^2 + \sum y_j^2\right) + \frac{\xi}{\sigma^2}\sum x_i + \frac{\eta}{\sigma^2}\sum y_j\right], \tag{5.26}$$

which is an exponential family with parameters

$$\theta = \frac{\eta}{\sigma^2}, \quad \vartheta_1 = \frac{\xi}{\sigma^2}, \quad \vartheta_2 = -\frac{1}{2\sigma^2}$$

and the sufficient statistics

$$U = \sum Y_j, \quad T_1 = \sum X_i \quad T_2 = \sum X_i^2 + \sum Y_j^2.$$

For testing the hypotheses

$$H : \eta - \xi \le 0 \quad \text{and} \quad H' : \eta - \xi = 0$$

it is more convenient to represent the densities as an exponential family with the parameters

$$\theta^* = \frac{\eta - \xi}{\left(\frac{1}{m} + \frac{1}{n}\right)\sigma^2}, \quad \vartheta_1^* = \frac{m\xi + n\eta}{(m+n)\sigma^2}, \quad \vartheta_2^* = \vartheta_2$$

and the sufficient statistics

$$U^* = \bar{Y} - \bar{X}, \quad T_1^* = m\bar{X} + n\bar{Y}, \quad T_2^* = \sum X_i^2 + \sum Y_j^2.$$

That this is possible is seen from the identity

$$m\xi\bar{x} + n\eta\bar{y} = \frac{(\bar{y} - \bar{x})(\eta - \xi)}{\frac{1}{m} + \frac{1}{n}} + \frac{(m\bar{x} + n\bar{y})(m\xi + n\eta)}{m+n}.$$

It follows from Theorem 4.4.1 that UMP unbiased tests exist for the hypotheses $\theta^* \le 0$ and $\theta^* = 0$, and hence for H and H'.

When $\eta = \xi$, the distribution of

$$\begin{aligned} V &= \frac{\bar{Y} - \bar{X}}{\sqrt{\sum(X_i - \bar{X})^2 + \sum(Y_j - \bar{Y})^2}} \\ &= \frac{U^*}{\sqrt{T_2^* - \frac{1}{m+n}T_1^{*2} - \frac{mn}{m+n}U^{*2}}} \end{aligned}$$

does not depend on the common mean ξ or on σ, as is seen by replacing X_i with $(X_i - \xi)/\sigma$ and Y_j with $(Y_j - \xi)/\sigma$ in the expression for V, and V is independent of (T_1^*, T_2^*). The rejection region of the UMP unbiased test of H can therefore be written as $V \ge C_0'$ or

$$t(X, Y) \ge C_0, \tag{5.27}$$

where

$$t(X, Y) = \frac{(\bar{Y} - \bar{X})\Big/\sqrt{\frac{1}{m} + \frac{1}{n}}}{\sqrt{\left[\sum(X_i - \bar{X})^2 + \sum(Y_j - \bar{Y})^2\right]/(m+n-2)}}. \tag{5.28}$$

The statistic $t(X, Y)$ is the ratio of the two independent variables

$$\frac{\bar{Y} - \bar{X}}{\sqrt{\left(\frac{1}{m} + \frac{1}{n}\right)\sigma^2}} \quad \text{and} \quad \sqrt{\frac{\sum(X_i - \bar{X})^2 + \sum(Y_j - \bar{Y})^2}{(m + n - 2)\sigma^2}}.$$

The numerator is normally distributed with mean $(\eta - \xi)/\sqrt{m^{-1} + n^{-1}}\sigma$ and unit variance; the square of the denominator as a χ^2 variable with $m + n - 2$ degrees of freedom, divided by $m + n - 2$. Hence $t(X, Y)$ has a noncentral t-distribution with $m + n - 2$ degrees of freedom and noncentrality parameter

$$\delta = \frac{\eta - \xi}{\sqrt{\frac{1}{m} + \frac{1}{n}}\sigma}.$$

When in particular $\eta - \xi = 0$, the distribution of $t(X, Y)$ is Student's t-distribution, and the constant C_0 is determined by

$$\int_{C_0}^{\infty} t_{m+n-2}(y)\, dy = \alpha. \tag{5.29}$$

As before, the assumptions required by Theorem 5.1.1 for H' are not satisfied by V itself but by a function of V,

$$W = \frac{\bar{Y} - \bar{X}}{\sqrt{\sum X_i^2 + \sum Y_j^2 - \frac{\left(\sum X_i + \sum Y_j\right)^2}{m+n}}}$$

which is related to V through

$$V = \frac{W}{\sqrt{1 - \frac{mn}{m+n}W^2}}.$$

Since W is a function of V, it is also independent of (T_1^*, T_2^*) when $\eta = \xi$; in addition it is a linear function of U^* with coefficients dependent only on T^*. The distribution of W being symmetric about 0 when $\eta = \xi$, it follows, as in the derivation of the corresponding rejection region (5.17) for the one-sample problem, that the UMP unbiased test of H' rejects when $|W|$ is too large, or equivalently when

$$|t(X, Y)| > C. \tag{5.30}$$

The constant C is determined by

$$\int_{C}^{\infty} t_{m+n-2}(y)\, dy = \frac{\alpha}{2}.$$

The power of the tests (5.27) and (5.30) depends only on $(\eta - \xi)/\sigma$ and is given in terms of the noncentral t-distribution. Its properties are analogous to those of the one-sample t-test (Problems 5.1, 5.2, and 5.4).

5.4 Confidence Intervals and Families of Tests

Confidence bounds for a parameter θ corresponding to a confidence level $1 - \alpha$ were defined in Section 3.5, for the case that the distribution of the random variable X depends only on θ. When nuisance parameters ϑ are present the defining condition for a lower confidence bound $\underline{\theta}$ becomes

$$P_{\theta,\vartheta}\{\underline{\theta}(X) \leq \theta\} \geq 1 - \alpha \qquad \text{for all } \theta, \vartheta. \tag{5.31}$$

Similarly, confidence intervals for θ at confidence level $1 - \alpha$ are defined as a set of random intervals with end points $\underline{\theta}(X)$, $\bar{\theta}(X)$ such that

$$P_{\theta,\vartheta}\{\underline{\theta}(X) \leq \theta \leq \bar{\theta}(X)\} \geq 1 - \alpha \qquad \text{for all } \theta, \vartheta. \tag{5.32}$$

The infimum over (θ, ϑ) of the left-hand side of (5.31) and (5.32) is the *confidence coefficient* associated with these statements.

As was already indicated in Chapter 3, confidence statements permit a dual interpretation. Directly, they provide bounds for the unknown parameter θ and thereby a solution to the problem of estimating θ. The statement $\underline{\theta} \leq \theta \leq \bar{\theta}$ is not as precise as a point estimate, but it has the advantage that the probability of it being correct can be guaranteed to be at least $1-\alpha$. Similarly, a lower confidence bound can be thought of as an estimate $\underline{\theta}$ which overestimates the true parameter value with probability $\leq \alpha$. In particular for $\alpha = \frac{1}{2}$, if $\underline{\theta}$ satisfies

$$P_{\theta,\vartheta}\{\underline{\theta} \leq \theta\} = P_{\theta,\vartheta}\{\underline{\theta} \geq \theta\} = \frac{1}{2},$$

the estimate is as likely to underestimate as to overestimate and is then said to be *median unbiased.* (See Problem 1.3, for the relation of this property to a more general concept of unbiasedness.) For an exponential family given by (4.10) there exists an estimator of θ which among all median unbiased estimators uniformly minimizes the risk for any loss function $L(\theta, d)$ that is monotone in the sense of the last paragraph of Section 3.5. A full treatment of this result including some probabilistic and measure-theoretic complications, is given by Pfanzagl (1979).

Alternatively, as was shown in Chapter 3, confidence statements can be viewed as equivalent to a family of tests. The following is essentially a review of the discussion of this relationship in Chapter 3, made slightly more specific by restricting attention to the two-sided case. For each θ_0, let $A(\theta_0)$ denote the acceptance region of a level-α test (assumed for the moment to be nonrandomized) of the hypothesis $H(\theta_0) : \theta = \theta_0$. If

$$S(x) = \{\theta : x \in A(\theta)\}$$

then

$$\theta \in S(x) \quad \text{if and only if} \quad x \in A(\theta), \tag{5.33}$$

and hence

$$P_{\theta,\vartheta}\{\theta \in S(X)\} \geq 1 - \alpha \qquad \text{for all } \theta, \vartheta. \tag{5.34}$$

Thus any family of level-α acceptance regions, through the correspondence (5.33), leads to a family of confidence sets at confidence level $1 - \alpha$.

Conversely, given any class of confidence sets $S(x)$ satisfying (5.34), let

$$A(\theta) = \{x : \theta \in S(x)\}. \tag{5.35}$$

Then the sets $A(\theta_0)$ are level-α acceptance regions for testing the hypotheses $H(\theta_0) : \theta = \theta_0$, and the confidence sets $S(x)$ show for each θ_0 whether for the particular x observed the hypothesis $\theta = \theta_0$ is accepted or rejected at level α.

Exactly the same arguments apply if the sets $A(\theta_0)$ are acceptance regions for the hypotheses $\theta \le \theta_0$. As will be seen below, one- and two-sided tests typically, although not always, lead to one-sided confidence bounds and to confidence intervals respectively.

Example 5.4.1 (Normal mean) Confidence intervals for the mean ξ of a normal distribution with unknown variance can be obtained from the acceptance regions $A(\xi_0)$ of the hypothesis $H : \xi = \xi_0$. These are given by

$$\frac{|\sqrt{n}(\bar{x} - \xi_0)|}{\sqrt{\sum(x_i - \bar{x})^2/(n-1)}} \le C,$$

where C is determined from the t-distribution so that the probability of this inequality is $1 - \alpha$ when $\xi = \xi_0$. [See (5.17) and (5.19) of Section 5.2.] The set $S(x)$ is then the set of ξ's satisfying this inequality with $\xi = \xi_0$, that is, the interval

$$\bar{x} - \frac{C}{\sqrt{n}}\sqrt{\frac{1}{n-1}\sum(x_i - \bar{x})^2} \le \xi \le \bar{x} + \frac{C}{\sqrt{n}}\sqrt{\frac{1}{n-1}\sum(x_i - \bar{x})^2}. \tag{5.36}$$

The class of these intervals therefore constitutes confidence intervals for ξ with confidence coefficient $1 - \alpha$.

The length of the intervals (5.36) is proportional to $\sqrt{\sum(x_i - \bar{x})^2}$ and their expected length to σ. For large σ, the intervals will therefore provide little information concerning the unknown ξ. This is a consequence of the fact, which led to similar difficulties for the corresponding testing problem, that two normal distributions $N(\xi_0, \sigma^2)$ and $N(\xi_1, \sigma^2)$ with fixed difference of means become indistinguishable as a tends to infinity. In order to obtain confidence intervals for ξ whose length does not tend to infinity with σ, it is necessary to determine the number of observations sequentially so that it can be adjusted to σ. A sequential procedure leading to confidence intervals of prescribed length is given in Problems 5.23 and 5.24.

However, even such a sequential procedure does not really dispose of the difficulty, but only shifts the lack of control from the length of the interval to the number of observations, As $\sigma \to \infty$, the number of observations required to obtain confidence intervals of bounded length also tends to infinity. Actually, in practice one will frequently have an idea of the order of magnitude of σ. With a sample either of fixed size or obtained sequentially, it is then necessary to establish a balance between the desired confidence $1 - \alpha$, the accuracy given by the length l of the interval, and the number of observations n one is willing to expend. In such an arrangement two of the three quantities $1 - \alpha$, l, and n will be fixed, while the third is a random variable whose distribution depends on σ, so that it will be less well controlled than the others. If $1 - \alpha$ is taken as fixed, the choice between a sequential scheme and one of fixed sample size thus depends essentially on whether it is more important to control l or n.

To obtain lower confidence limits for ξ, consider the acceptance regions

$$\frac{\sqrt{n}(\bar{x} - \xi_0)}{\sqrt{\sum(x_i - \bar{x})^2/(n-1)}} \le C_0$$

for testing $\xi \le \xi_0$ to against $\xi > \xi_0$. The sets $S(x)$ arc then the one-sided intervals

$$\bar{x} - \frac{C_0}{\sqrt{n}}\sqrt{\frac{1}{n-1}\sum(x_i - \bar{x})^2} \le \xi,$$

the left-hand sides of which therefore constitute the desired lower bounds $\underline{\xi}$. If $\alpha = \frac{1}{2}$, the constant C_0 is 0; the resulting confidence bound $\underline{\xi} = \bar{X}$ is a median unbiased estimate of ξ, and among all such estimates it uniformly maximizes

$$P\{-\Delta_1 \le \xi - \underline{\xi} \le \Delta_2\} \qquad \text{for all} \quad \Delta_1, \Delta_2 \ge 0.$$

(For a proof see Section 3.5.) ■

5.5 Unbiased Confidence Sets

Confidence sets can be viewed as a family of tests of the hypotheses $\theta \in H(\theta')$ against alternatives $\theta \in K(\theta')$ for varying θ'. A confidence level of $1 - \alpha$ then simply expresses the fact that all the tests are to be at level α, and the condition therefore becomes

$$P_{\theta,\vartheta}\{\theta' \in S(X)\} \ge 1 - \alpha \qquad \text{for all } \theta \in H(\theta') \text{ and all } \vartheta. \tag{5.37}$$

In the case that $H(\theta')$ is the hypothesis $\theta = \theta'$ and $S(X)$ is the interval $[\underline{\theta}(X), \bar{\theta}(X)]$, this agrees with (5.32). In the one-sided case in which $H(\theta')$ is the hypothesis $\theta \le \theta'$ and $S(X) = \{\theta : \underline{\theta}(X) \le \theta\}$, the condition reduces to $P_{\theta,\vartheta}\{\underline{\theta}(X) \le \theta'\} \ge 1 - \alpha$ for all $\theta' \ge \theta$, and this is seen to be equivalent to (5.31). With this interpretation of confidence sets, the probabilities

$$P_{\theta,\vartheta}\{\theta' \in S(X)\}, \qquad \theta \in K(\theta'), \tag{5.38}$$

are the probabilities of false acceptance of $H(\theta')$ (error of the second kind). The smaller these probabilities are, the more desirable are the tests.

From the point of view of estimation, on the other hand, (5.38) is the probability of covering the wrong value θ'. With a controlled probability of covering the true value, the confidence sets will be more informative the less likely they are to cover false values of the parameter. In this sense the probabilities (5.38) provide a measure of the accuracy of the confidence sets. A justification of (5.38) in terms of loss functions was given for the one-sided case in Section 3.5.

In the presence of nuisance parameters, UMP tests usually do not exist, and this implies the nonexistence of confidence sets that are uniformly most accurate in the sense of minimizing (5.38) for all θ' such that $\theta \in K(\theta')$ and for all ϑ. This suggests restricting attention to confidence sets which in a suitable sense are unbiased. In analogy with the corresponding definition for tests, a family of confidence sets at confidence level $1 - \alpha$ is said to be *unbiased* if

$$P_{\theta,\vartheta}\{\theta' \in S(X)\} \le 1 - \alpha \tag{5.39}$$

$$\text{for all } \theta' \text{ such that } \theta \in K(\theta') \text{ and for all } \vartheta \text{ and } \theta,$$

so that the probability of covering these false values does not exceed the confidence level.

In the two- and one-sided cases mentioned above, the condition (5.39) reduces to

$$P_{\theta,\vartheta}\{\underline{\theta} \le \theta' \le \bar{\theta}\} \le 1-\alpha \qquad \text{for all } \theta' \ne \theta \text{ and all } \vartheta$$

and

$$P_{\theta,\vartheta}\{\underline{\theta} \le \theta'\} \le 1-\alpha \qquad \text{for all } \theta' < \theta \text{ and all } \vartheta.$$

With this definition of unbiasedness, unbiased families of tests lead to unbiased confidence sets and conversely. A family of confidence sets is uniformly most accurate unbiased at confidence level $1-\alpha$ if it minimizes the probabilities

$$P_{\theta,\vartheta}\{\theta' \in S(X)\} \text{ for all } \theta' \text{ such that } \theta \in K(\theta') \text{ and for all } \vartheta \text{ and } \theta,$$

subject to (5.37) and (5.39). The confidence sets obtained on the basis of the UMP unbiased tests of the present and preceding chapter are therefore uniformly most accurate unbiased. This applies in particular to the confidence intervals obtained in the preceding sections. Some further examples are the following.

Example 5.5.1 (Normal variance) If $X_1, \ldots, X_n$ is a sample from $N(\xi, \sigma^2)$, the UMP unbiased test of the hypothesis $\sigma = \sigma_0$ is given by the acceptance region (5.13)

$$C_1' \le \frac{\sum(x_i - \bar{x})^2}{\sigma_0^2} \le C_2',$$

where C_1' and C_2' are determined by (5.14). The most accurate unbiased confidence intervals for σ^2 are therefore

$$\frac{1}{C_2'}\sum(x_i - \bar{x})^2 \le \sigma^2 \le \frac{1}{C_1'}\sum(x_i - \bar{x})^2.$$

[Tables of C_1' and C_2' are provided by Tate and Klett (1959).] Similarly, from (5.9) and (5.10) the most accurate unbiased upper confidence limits for σ^2 are

$$\sigma^2 \le \frac{1}{C_0}\sum(x_i - \bar{x})^2,$$

where

$$\int_{C_0}^{\infty} \chi^2_{n-1}(y)\,dy = 1-\alpha.$$

The corresponding lower confidence limits are uniformly most accurate (without the restriction of unbiasedness) by Section 3.9. ■

Example 5.5.2 (Difference of means) Confidence intervals for the difference $\Delta = \eta - \xi$ of the means of two normal distributions with common variance are obtained from tests of the hypothesis $\eta - \xi = \Delta_0$. If $X_1, \ldots, X_m$ and $Y_1, \ldots, Y_n$ are distributed as $N(\xi, \sigma^2)$ and $N(\eta, \sigma^2)$ respectively, and if $Y_j' = Y_j - \Delta_0$, $\eta' = \eta - \Delta_0$, the hypothesis can be expressed in terms of the variables X_i and Y_j' as $\eta' - \xi = 0$. From (5.28) and (5.30) the UMP unbiased acceptance region is then seen to be

$$\frac{|(\bar{y} - \bar{x} - \Delta_0)| \Big/ \sqrt{\frac{1}{m} + \frac{1}{n}}}{\sqrt{\left[\sum(x_i - \bar{x})^2 + \sum(y_j - \bar{y})^2\right] \Big/ (m+n-2)}} \le C,$$

where C is determined by the equation following (5.30). The most accurate unbiased confidence intervals for $\eta - \xi$ are therefore

$$(\bar{y} - \bar{x}) - CS \leq \eta - \xi \leq (\bar{y} - \bar{x}) + CS \tag{5.40}$$

where

$$S^2 = \left(\frac{1}{m} + \frac{1}{n}\right) \frac{\sum(x_i - \bar{x})^2 + \sum(y_j - \bar{y})^2}{m + n - 2}$$

The one-sided intervals are obtained analogously. ■

Example 5.5.3 (Ratio of variances) If $X_1, \ldots, X_m$ and $Y_1, \ldots, Y_n$ are samples from $N(\xi, \sigma^2)$ and $N(\eta, \tau^2)$, most accurate unbiased confidence intervals for $\Delta = \tau^2/\sigma^2$ are derived from the acceptance region (5.23) as

$$\frac{1 - C_2}{C_2} \frac{\sum(y_j - \bar{y})^2}{\sum(x_i - \bar{x})^2} \leq \frac{\tau^2}{\sigma^2} \leq \frac{1 - C_1}{C_1} \frac{\sum(y_j - \bar{y})^2}{\sum(x_i - \bar{x})^2}, \tag{5.41}$$

where C_1 and C_2 are determined from (5.25).[3] In the particular case that $m = n$, the intervals take on the simpler form

$$\frac{1}{k} \frac{\sum(y_j - \bar{y})^2}{\sum(x_i - \bar{x})^2} \leq \frac{\tau^2}{\sigma^2} \leq k \frac{\sum(y_j - \bar{y})^2}{\sum(x_i - \bar{x})^2}, \tag{5.42}$$

where k is determined from the F-distribution. Most accurate unbiased lower confidence limits for the variance ratio are

$$\underline{\Delta} = \frac{1}{C_0} \frac{\sum(y_j - \bar{y})^2/(n-1)}{\sum(x_i - \bar{x})^2/(m-1)} \leq \frac{\tau^2}{\sigma^2} \tag{5.43}$$

with C_0 given by (5.22). If in (5.22) α is taken to be $\frac{1}{2}$, this lower confidence limit $\underline{\Delta}$ becomes a median unbiased estimate of τ^2/σ^2. Among all such estimates it uniformly minimizes

$$P\left\{-\Delta_1 \leq \frac{\tau^2}{\sigma^2} - \underline{\Delta} \leq \Delta_2\right\} \qquad \text{for all} \quad \Delta_1, \Delta_2 \geq 0.$$

(For a proof see Section 3.5). ■

So far it has been assumed that the tests from which the confidence sets are obtained are nonrandomized. The modifications that are necessary when this assumption is not satisfied were discussed in Chapter 3. The randomized tests can then be interpreted as being nonrandomized in the space of X and an auxiliary variable V which is uniformly distributed on the unit interval. If in particular X is integer-valued as in the binomial or Poisson case, the tests can be represented in terms of the continuous variable $X + V$. In this way, most accurate unbiased confidence intervals can be obtained, for example, for a binomial probability p from the UMP unbiased tests of $H : p = p_0$ (Example 4.2.1). It is not clear a priori that the resulting confidence sets for p will necessarily by intervals. This is, however, a consequence of the following Lemma.

---

[3]A comparison of these limits with those obtained from the equal-tails test is given by Scheffé (1942); some values of C_1 and C_2 are provided by Ramachandran (1958).

Lemma 5.5.1 *Let X be a real-valued random variable with probability density $p_\theta(x)$ which has monotone likelihood ratio in x. Suppose that UMP unbiased tests of the hypotheses $H(\theta_0) : \theta = \theta_0$ exist and are given by the acceptance regions*

$$C_1(\theta_0) \le x \le C_2(\theta_0)$$

and that they are strictly unbiased. Then the functions $C_i(\theta)$ are strictly increasing in θ, and the most accurate unbiased confidence intervals for θ are

$$C_2^{-1}(x) \le \theta \le C_1^{-1}(x).$$

PROOF. Let $\theta_0 < \theta_1$, and let $\beta_0(\theta)$ and $\beta_1(\theta)$ denote the power functions of the above tests ϕ_0 and ϕ_1, for testing $\theta = \theta_0$ and $\theta = \theta_1$. It follows from the strict unbiasedness of the tests that

$$\begin{aligned} E_{\theta_0}\left[\phi_1(X) - \phi_0(X)\right] &= \beta_1(\theta_0) - \alpha > 0 > \alpha - \beta_0(\theta_1) \\ &= E_{\theta_1}\left[\phi_1(X) - \phi_0(X)\right]. \end{aligned}$$

Thus neither of the two intervals $[C_1(\theta_i), C_2(\theta_i)]$ $(i = 0, 1)$ contains the other, and it is seen from Lemma 3.4.2(iii) that $C_i(\theta_0) < C_i(\theta_1)$ for $i = 1, 2$. The functions C_i therefore have inverses, and the inequalities defining the acceptance region for $H(\theta)$ are equivalent to $C_2^{-1}(x) \le \theta \le C_1^{-1}(x)$, as was to be proved. ∎

The situation is indicated in Figure 5.1. From the boundaries $x = C_1(\theta)$ and $x = C_2(\theta)$ of the acceptance regions $A(\theta)$ one obtains for each fixed value of x the confidence set $S(x)$ as the interval of θ's for which $C_1(\theta) \le x \le C_2(\theta)$.

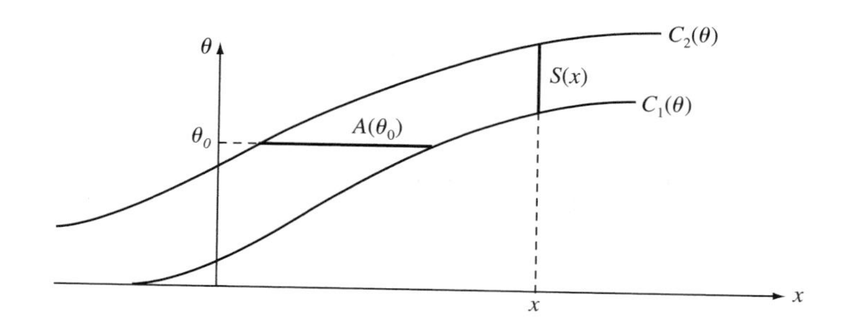

Figure 5.1.

By Section 4.2, the conditions of the lemma are satisfied in particular for a one-parameter exponential family, provided the tests are nonrandomized. In cases such as that of binomial or Poisson distributions, where the family is exponential but X is integer-valued so that randomization is required, the intervals can be obtained by applying the lemma to the variable $X + V$ instead of X, where V is independent of X and uniformly distributed over $(0, 1)$.

Example 5.5.4 In the binomial case, a table of the (randomized) uniformly most accurate unbiased confidence intervals is given by Blyth and Hutchinson (1960). The best choice of nonrandomized intervals and some approximations are discussed (and tables provided) by Blyth and Still (1983) and Blyth (1984). Recent approximations and comparisons are provided by Agresti and Coull (1998) and Brown, Cai and DasGupta (2001, 2002). A large sample approach will be considered in Example 11.2.7. ■

In Lemma 5.5.1, the distribution of X was assumed to depend only on θ. Consider now the exponential family (5.1) in which nuisance parameters are present in addition to θ. The UMP unbiased tests of $\theta = \theta_0$, are then performed as conditional tests given $T = t$, and the confidence intervals for θ will as a consequence also be obtained conditionally. If the conditional distributions are continuous, the acceptance regions will be of the form

$$C_1(\theta; t) \leq u \leq C_2(\theta; t),$$

where for each t the functions C_i are increasing by Lemma 5.5.1. The confidence intervals are then

$$C_2^{-1}(u; t) \leq \theta \leq C_1^{-1}(u; 1).$$

If the conditional distributions are discrete, continuity can be obtained as before through addition of a uniform variable.

Example 5.5.5 (Poisson ratio) Let X and Y be independent Poisson variables with means λ and μ, and let $\rho = \mu/\lambda$. The conditional distribution of Y given $X + Y = t$ is the binomial distribution $b(p, t)$ with

$$p = \frac{\rho}{1 + \rho}.$$

The UMP unbiased test $\phi(y, t)$ of the hypothesis $\rho = \rho_0$ is defined for each t as the UMP unbiased conditional test of the hypothesis $\rho = \rho_0/(1 + \rho_0)$. If

$$\underline{p}(t) \leq p \leq \bar{p}(t)$$

are the associated most accurate unbiased confidence intervals for p given t, it follows that the most accurate unbiased confidence intervals for μ/λ are

$$\frac{\underline{p}(t)}{1 - \underline{p}(t)} \leq \frac{\mu}{\lambda} \leq \frac{\bar{p}(t)}{1 - \bar{p}(t)}.$$

The binomial tests which determine the functions $\underline{p}(t)$ and $\bar{p}(t)$ are discussed in Example 4.2.1. ■

5.6 Regression

The relation between two variables X and Y can be studied by drawing an unrestricted sample and observing the two variables for each subject, obtaining n pairs of measurements $(X_1, Y_1), \ldots, (X_n, Y_n)$ (see Section 5.13 and Problem 5.13). Alternatively, it is frequently possible to control one of the variables such as the age of a subject, the temperature at which an experiment is performed, or the strength of the treatment that is being applied. Observations $Y_1, \ldots, Y_n$ of Y can then be obtained at a number of predetermined levels $x_1, \ldots, x_n$ of x. Suppose that for fixed x the distribution of Y is normal with constant variance

σ^2 and a mean which is a function of x, *the regression of* Y *on* x, and which is assumed to be linear,[4]

$$E[Y|x] = \alpha + \beta x.$$

If we put $v_i = (x_i - \bar{x})/\sqrt{\sum(x_j - \bar{x})^2}$ and $\gamma + \delta v_i = \alpha + \beta x_i$, so that $\sum v_i = 0$, $\sum v_i^2 = 1$, and

$$\alpha = \gamma - \delta \frac{\bar{x}}{\sqrt{\sum(x_j - \bar{x})^2}}, \qquad \beta = \frac{\delta}{\sqrt{\sum(x_j - \bar{x})^2}},$$

the joint density of $Y_1, \ldots, Y_n$ is

$$\frac{1}{(\sqrt{2\pi}\sigma)^n} \exp\left[-\frac{1}{2\sigma^2}\sum(y_i - \gamma - \delta v_i)^2\right].$$

These densities constitute an exponential family (5.1) with

$$U = \sum v_i Y_i, \quad T_1 = \sum Y_i^2, \quad T_2 = \sum Y_i$$

$$\theta = \tfrac{\delta}{\sigma^2}, \qquad \vartheta_1 = -\tfrac{1}{2\sigma^2}, \quad \vartheta_2 = \tfrac{\gamma}{\sigma^2}.$$

This representation implies the existence of UMP unbiased tests of the hypotheses $a\gamma + b\delta = c$ where a, b, and c are given constants, and therefore of most accurate unbiased confidence intervals for the parameter

$$\rho = a\gamma + b\delta.$$

To obtain these confidence intervals explicitly, one requires the UMP unbiased test of $H : \rho = \rho_0$, which is given by the acceptance region

$$\frac{|b\sum v_i Y_i + a\bar{Y} - \rho_0| \Big/ \sqrt{(a^2/n) + b^2}}{\sqrt{\left[\sum(Y_i - \bar{Y})^2 - (\sum v_i Y_i)^2\right] \Big/ (n-2)}} \leq C \tag{5.44}$$

where

$$\int_{-C}^{C} t_{n-2}(y)\, dy = 1 - \alpha \;;$$

see Problem 5.33. The resulting confidence intervals for ρ are centered at $b\sum v_i Y_i + a\bar{Y}$, and their length is

$$L = 2C\sqrt{\left[\frac{a^2}{n} + b^2\right] \frac{\sum(Y_i - \bar{Y})^2 - (\sum v_i Y_i)^2}{n-2}}.$$

It follows from the transformations given in Problem 5.33 that $\left[\sum(Y_i - \bar{Y})^2 - (\sum v_i Y_i)^2\right]/\sigma^2$ has a χ^2-distribution with $n-2$ degrees of freedom and hence that

[4]The literature on regression is enormous and we treat the simplest model. Some texts on the subject include Weisberg (1985), Atkinson and Riani (2000) and Chatterjee, Hadi and Price (2000).

the expected length of the intervals is

$$E(L) = 2C_n\sigma\sqrt{\frac{a^2}{n} + b^2}.$$

In particular applications, a and b typically are functions of the x's. If these are at the disposal of the experimenter and there is therefore some choice with respect to a and b, the expected length of L is minimized by minimizing $(a^2/n) + b^2$. Actually, it is not clear that the expected length is a good criterion for the accuracy of confidence intervals, since short intervals are desirable when they cover the true parameter value but not necessarily otherwise. However, the same result holds for other criteria such as the expected value of $(\bar{\rho} - \rho)^2 + (\rho - \underline{\rho})^2$ or more generally of $f_1(|\bar{\rho} - \rho|) + f_2(|\rho - \underline{\rho}|)$, where f_1 and f_2 are increasing functions of their arguments. (See Problem 5.33.) Furthermore, the same choice of a and b also minimizes the probability of the intervals covering any false value of the parameter. We shall therefore consider $(a^2/n) + b^2$ as an inverse measure of the accuracy of the intervals.

Example 5.6.1 (Slope of regression line) Confidence levels for the slope $\beta = \delta/\sqrt{\sum(x_j - \bar{x})^2}$ are obtained from the above intervals by letting $a = 0$ and $b = 1/\sqrt{\sum(x_j - \bar{x})^2}$. Here the accuracy increases with $\sum(x_j - \bar{x})^2$, and if the x_j must be chosen from an interval $[C_0, C_1]$, it is maximized by putting half of the values at each end point. However, from a practical point of view, this is frequently not a good design, since it permits no check of the linearity of the regression. ■

Example 5.6.2 (Ordinate of regression line) Another parameter of interest is the value $\alpha + \beta x_0$ to be expected from an observation Y at $x = x_0$. Since

$$\alpha + \beta x_0 = \gamma + \frac{\delta(x_0 - \bar{x})}{\sqrt{\sum(x_j - \bar{x})^2}},$$

the constants a and b are $a = 1$, $b = (x_0 - \bar{x})/\sqrt{\sum(x_j - \bar{x})^2}$. The maximum accuracy is obtained by minimizing $|\bar{x} - x_0|$ and, if $\bar{x} = x_0$ cannot be achieved exactly, also maximizing $\sum(x_j - \bar{x})^2$. ■

Example 5.6.3 (Intercept of regression line) Frequently it is of interest to estimate the point x at which $\alpha + \beta x$ has a preassigned value. One may for example wish to find the dosage $x = -\alpha/\beta$ at which $E(Y \mid x) = 0$, or equivalently the value $v = (x - \bar{x})/\sqrt{\sum(x_j - \bar{x})^2}$ at which $\gamma + \delta v = 0$. Most accurate unbiased confidence sets for the solution $-\gamma/\delta$ of this equation can be obtained from the UMP unbiased tests of the hypotheses $-\gamma/\delta = v_0$. The acceptance regions of these tests are given by (5.44) with $a = 1$, $b = v_0$, and $\rho_0 = 0$, and the resulting confidence sets for v are the sets of values v satisfying

$$v^2\left[C^2S^2 - \left(\sum v_iY_i\right)^2\right] - 2v\bar{Y}\left(\sum v_iY_i\right) + \frac{1}{n}(C^2S^2 - n\bar{Y}^2) \geq 0.$$

where $S^2 = [\sum(Y_i - \bar{Y})^2(\sum v_i Y_i)^2]/(n-2)$. If the associated quadratic equation in v has roots $\underline{v}$, $\bar{v}$, the confidence statement becomes

$$\underline{v} \le v \le \bar{v} \qquad \text{when} \qquad \frac{\left|\sum v_i Y_i\right|}{S} > C$$

and

$$v \le \underline{v} \text{ or } v \ge \bar{v} \qquad \text{when} \qquad \frac{\left|\sum v_i Y_i\right|}{S} < C.$$

The somewhat surprising possibility that the confidence sets may be the outside of an interval actually is quite appropriate here. When the line $y = \gamma + \delta v$ is nearly parallel to the v-axis, the intercept with the v-axis will be large in absolute value, but its sign can be changed by a very small change in angle. There is the further possibility that the discriminant of the quadratic polynomial is negative,

$$n\bar{Y}^2 + \left(\sum v_i Y_i\right)^2 < C^2 S^2,$$

in which case the associated quadratic equation has no solutions. This condition implies that the leading coefficient of the quadratic polynomial is positive, so that the confidence set in this case becomes the whole real axis. The fact that the confidence sets are not necessarily finite intervals has led to the suggestion that their use be restricted to the cases in which they do have this form. Such usage will however affect the probability with which the sets cover the true value and hence the validity of the reported confidence coefficient.[5] ■

5.7 Bayesian Confidence Sets

The left side of the confidence statement (5.34) denotes the probability that the random set $S(X)$ will contain the constant point θ. The interpretation of this probability statement, before X is observed, is clear: it refers to the frequency with which this random event will occur. Suppose for example that X is distributed as $N(\theta, 1)$, and consider the confidence interval

$$X - 1.96 < \theta < X + 1.96$$

corresponding to confidence coefficient $\gamma = .95$. Then the random interval $(X - 1.96, X + 1.96)$ will contain θ with probability .95. Suppose now that X is observed to be 2.14. At this point, the earlier statement reduces to the inequality $0.18 < \theta < 4.10$, which no longer involves any random element. Since the only unknown quantity is θ, it is tempting (but not justified) to say that θ lies between 0.18 and 4.10 with probability .95.

To attach a meaningful probability to the event $\theta \in S(x)$ when x is fixed requires that θ be random. Inferences made under the assumption that the parameter θ is itself a random (though unobservable) quantity with a known

[5] A method for obtaining the size of this effect was developed by Neyman, and tables have been computed on its basis by Fix. This work is reported by Bennett (1957).

distribution are called *Bayesian*, and the distribution Λ of θ before any observations are taken its *prior distribution.* After $X = x$ has been observed, inferences concerning θ can be based on its conditional distribution given x, the *posterior distribution.* In particular, any set $S(x)$ with the property

$$P[\theta \in S(x) \mid X = x] \geq \gamma \qquad \text{for all } x$$

is a $100\gamma\%$ Bayesian confidence set or *credible region* for θ. In the rest of this section, the random variable with prior distribution Λ will be denoted by Θ, with θ being the value taken on by Θ in the experiment at hand.

Example 5.7.1 (Normal mean) Suppose that Θ has a normal prior distribution $N(\mu, b^2)$ and that given $\Theta = \theta$, the variables $X_1, \ldots, X_n$. are independent $N(\theta, \sigma^2)$, σ known. Then the posterior distribution of Θ given $x_1, \ldots, x_n$ is normal with mean (Problem 5.34)

$$\eta_x = E[\Theta \mid x] = \frac{n\bar{x}/\sigma^2 + \mu/b^2}{n/\sigma^2 + 1/b^2}$$

and variance

$$\tau_x^2 = Var[\Theta \mid x] = \frac{1}{n/\sigma^2 + 1/b^2}$$

Since $[\Theta - \eta_x]/\tau_x$ then has a standard normal distribution, the interval $I(x)$ with endpoints

$$\frac{n\bar{x}/\sigma^2 + \mu/b^2}{n/\sigma^2 + 1/b^2} \pm \frac{1.96}{\sqrt{n/\sigma^2 + 1/b^2}}$$

satisfies $P[\Theta \in I(x) \mid X = x] = .95$ and is thus a 95% credible region.

For $n = 1$, $\mu = 0$, $\sigma = 1$, the interval reduces to

$$\frac{x}{1 + \frac{1}{b^2}} \pm \frac{1.96}{\sqrt{1 + \frac{1}{b^2}}}$$

which for large b is very close to the confidence interval for θ stated at the beginning of the section. But now the statement that θ lies between these limits with probability .95 is justified, since it is a probability statement concerning the random variable Θ.

The distribution $N(\mu, b^2)$ assigns higher probability to θ-values near μ than to those further away. Suppose instead that no information whatever about θ is available, so that one wishes to model a state of complete ignorance. This could be done by assigning a constant density to all values of θ, that is, by assigning to Θ the density $\pi(\theta) \equiv c$, $-\infty < \theta < \infty$. Unfortunately, the resulting π is not a probability density, since $\int_{-\infty}^{\infty} \pi(\theta)\, d\theta = \infty$. However, if this fact is ignored and the posterior distribution of Θ given x is calculated in the usual way, it turns out (Problem 5.35) that $\pi(\theta \mid x)$ is the density of a genuine probability distribution, namely $N(\mu, \sigma^2/n)$, the limit of the earlier posterior distribution as $b \to \infty$. The *improper* (since it integrates to infinity), *noninformative* prior density $\pi(\theta) \equiv c$ thus leads approximately to the same results as the normal prior $N(\mu, b^2)$ for large b, and can be viewed as an approximation to the latter. ■

Unlike confidence sets, Bayesian credible regions provide exactly the desired kind of probability statement even after the observations are known. They do so, however, at the cost of an additional assumption: that θ is random and has a known prior distribution. Detailed accounts of the Bayesian approach, its application to credible regions, and comparison of the two approaches can be found in Berger (1985a) and Robert (1994). The following examples provide a few illustrations and additional comments.

Example 5.7.2 Let X be binomial $b(p, n)$, and suppose that the prior distribution for p is the beta distribution[6] $B(a, b)$ with density $Cp^{a-1}(1-p)^{b-1}, 0 < p < 1$, $0 < a, b$. Then the posterior distribution of p given $X = x$ is the beta distribution $B(a+x, b+n-x)$ (Problem 5.36). There are of course many sets $S(x)$ whose probability under this distribution is equal to the prescribed coefficient γ. A choice that is frequently recommended is the HPD (highest probability density) region, defined by the requirement that the posterior density of p given x be $\geq k$.

With a beta prior, only the following possibilities can occur: for fixed x,

(a) $\pi(p \mid x)$ is decreasing,

(b) $\pi(p \mid x)$ is increasing,

(c) $\pi(p \mid x)$ is increasing in $(0, p_0)$ and decreasing in $(p_0, 1)$ for some p_0,

(d) $\pi(p \mid x)$ is U-shaped, i.e. decreasing in $(0, p_0)$ and increasing in $(p_0, 1)$ for some p_0.

The HPD region then is of the form

(a) $p < K(-x)$,

(b) $p > K(x)$,

(c) $K_1(x) < p < K_2(x)$,

(d) $p < K_1(x)$ or $p > K_2(x)$,

where the K's are determined by the requirement that the posterior probability of the region, given x, be γ; in cases (c) and (d) this condition must be supplemented by

$$\pi[K_1(x) \mid x] = \pi[K_2(x) \mid x].$$

In general, if $\pi(\theta \mid x)$ denotes the posterior density of θ, the HPD region is defined by

$$\pi(\theta \mid x) \geq k$$

with C determined by the size condition

$$P[\pi(\theta) \mid x) \geq k] = \gamma. \blacksquare$$

[6]This is the so-called conjugate of the binomial distribution; for a more general discussion of conjugate distributions, see Chapter 4 of *TPE2* and Robert (1994), Section 3.2.

Example 5.7.3 (Two-parameter normal mean) Let $X_1, \ldots, X_n$ be independent $N(\xi, \sigma^2)$, and for the sake of simplicity suppose that (ξ, σ) has the joint improper prior density given by

$$\pi(\xi, \sigma)\, d\xi\, d\sigma = d\xi \frac{1}{\sigma}\, d\sigma \qquad \text{for all} \qquad -\infty < \xi < \infty,\ 0 < \sigma,$$

which is frequently used to model absence of information concerning the parameters. Then the joint posterior density of (ξ, σ) given $x = (x_1, \ldots, x_n)$ is of the form

$$\pi(\xi, \sigma \mid x)\, d\xi\, d\sigma = C(x) \frac{1}{\sigma^{n+1}} \exp\left(-\frac{1}{2\sigma^2} \sum_{i=1}^{n} (\xi - x_i)^2 \right) d\xi\, d\sigma.$$

Determination of a credible region for ξ requires the marginal posterior density of given x, which is obtained by integrating the joint posterior density with respect to σ. These densities depend only on the sufficient statistics $\bar{x}$ and $S^2 = \sum(x_i - \bar{x})^2$, and the posterior density of ξ is of the form (Problem 5.37)

$$A(x) \left[\frac{1}{1 + \frac{n(\xi - \bar{x})^2}{S^2}} \right]^{n/2}$$

Here $\bar{x}$ and S enter only as location and scale parameters, and the linear function

$$t = \frac{\sqrt{n}(\xi - \bar{x})}{S/\sqrt{n-1}}$$

of ξ has the t-distribution with $n-1$ degrees of freedom. Since this agrees with the distribution of t for fixed ξ and σ given in Section 5.2, the credible $100(1-\alpha)\%$ region

$$\left| \frac{\sqrt{n}(\xi - \bar{x})}{S/\sqrt{n-1}} \right| \leq C$$

is formally identical with the confidence intervals (5.36). However, they are derived under different assumptions, and their interpretation differs accordingly. The relationship between Bayesian intervals and classical intervals is further explored in Nicolaou (1993) and Severini (1993). ■

Example 5.7.4 (Two-parameter normal: estimating σ) Under the assumptions of the preceding example, credible regions for σ are based on the posterior distribution of σ given x, obtained by integrating the joint posterior density of (ξ, σ) with respect to ξ. Using the fact that $\sum(\xi - x_i)^2 = n(\xi - \bar{x})^2 + \sum(x_i - \bar{x})^2$, it is seen (Problem 5.38) that given x, the conditional (posterior) distribution of $\sum(x_i - \bar{x})^2/\sigma^2$ is χ^2 with $n-1$ degrees of freedom. As in the case of the mean, this agrees with the sampling distribution of the same quantity when a is a (constant) parameter, given in Section 5.2. (The agreement in both cases of two distributions derived under such different assumptions is a consequence of the particular choice of the prior distribution and the fact that it is invariant in the sense of *TPE*2, Section 4.4.) A change of variables now gives the posterior density of σ and shows that $\pi(\sigma \mid x)$ is of the form (c) of Example 5.7.2, so that the HPD region is of the form $K_1(x) < \sigma < K_2(x)$ with $0 < K_1(x) < K_2(x) < \infty$.

Suppose that a credible region is required, not for σ, but for σ^r for some $r > 0$. For consistency, this should then be given by $[K_1(x)]^r < \sigma^r < [K_2(x)]^r$, but this is not the case, since the relative height of the density of a random variable at two points is not invariant under monotone transformations of the variable. In fact, in the present case, the HPD region for σ^r will become one-sided for sufficiently large r although it is two-sided for $r = 1$ (Problem 5.38). ■

Such inconsistencies do not occur if the HPD region is replaced by the equal-tails interval $(C_1(x), C_2(x))$ for which $P[\Theta < C_1(x) \mid X = x] = P[\Theta > C_2(x) \mid X = x] = (1 - \gamma)/2$.[7] More generally inconsistencies under transformations of Θ are avoided when the posterior distribution of Θ is summarized by a number of its percentiles corresponding to the standard confidence points mentioned in Section 3.5. Such a set is a compromise between providing the complete posterior distribution and providing a single interval corresponding to only two percentiles.

Both the confidence and the Bayes approach present difficulties: the first, the problem of postdata interpretation; the second, the choice of a prior distribution and the interpretation of the posterior coverage probabilities if there is no clear basis for this choice. It is therefore not surprising that efforts have been made to find an approach without these drawbacks. The first such attempt, from which most later ones derive, is due to Fisher [1930; for his final account see Fisher (1973)].

To discuss Fisher's concept of fiducial probability, consider once more the example at the beginning of the section, in which X is distributed as $N(\theta, 1)$. Since then $X - \theta$ is distributed as $N(0, 1)$, so is $\theta - X$, and hence

$$P(\theta - X \leq y) = \Phi(y) \qquad \text{for all } y.$$

For fixed $X = x$, this is the formal statement that a random variable θ has distribution $N(x, 1)$. Without assuming θ to be random, Fisher calls $N(x, 1)$ the *fiducial distribution* of θ. Since this distribution is to embody the information about θ provided by the data, it should be unique, and Fisher imposes conditions which he hopes will ensure uniqueness. This leads to some technical difficulties, but more basic is the question of how to interpret fiducial probability. In a series of independent repetitions of the experiment with arbitrarily varying θ_i, the quantities $\theta_1 - X_1, \theta_2 - X_2, \ldots$ will constitute a sequence of independent standard normal variables. From this fact, Fisher attempts to derive the fiducial distribution $N(x, 1)$ of θ as a frequency distribution with respect to an appropriate reference set. However, this argument is difficult to follow and unconvincing. For summaries of the fiducial literature and of later related developments by Dempster, Fraser, and others, see Buehler (1983), Edwards (1983), Seidenfeld (1992), Zabell (1992), Barnard (1995, 1996) and Fraser (1996).

Fisher's effort to define a suitable frame of reference led him to the important concept of *relevant subsets*, which will be discussed in Chapter 10.

To appreciate the differences between the frequentist, Bayesian and Fisherian points of view, see Lehmann (1993), Robert (1994), Berger, Boukai and Wang (1997), Berger (2003) and Bayarri and Berger (2004).

---

[7] They also do not occur when the posterior distribution of Θ is discrete.

5.8 Permutation Tests

For the comparison of a treatment with a control situation in which no treatment is given, it was shown in Section 5.3 that the one-sided t-test is UMP unbiased for testing $H : \eta = \xi$ against $\eta - \xi = \Delta > 0$ when the measurements $X_1, \ldots, X_m$ and $Y_1, \ldots, Y_n$ are samples from normal populations $N(\xi, \sigma^2)$ and $N(\eta, \sigma^2)$. It will be shown in Section 11.3 that the level of this test is (asymptotically) robust against nonnormality – that is, that except for small m or n the level of the test is approximately equal to the nominal level α when the X's and Y's are samples from any distributions with densities $f(x)$ and $f(y - \Delta)$ with finite variance. If such an approximate level is not satisfactory, one may prefer to try to obtain an exact level-α unbiased test (valid for all f) by replacing the original normal model with the nonparametric model for which the joint density of the variables is

$$f(x_1) \ldots f(x_m) f(y_1 - \Delta) \ldots f(y_n - \Delta), \qquad f \in \mathcal{F}, \tag{5.45}$$

where we shall take $\mathcal{F}$ to be the family of all probability densities that are continuous a.e.

If there is much variation in the population being sampled, the sensitivity of the experiment can frequently be increased by dividing the population into more homogeneous subgroups, defined for example by some characteristic such as age or sex. A sample of size $N_i (i = 1, \ldots, c)$ is then taken from the ith subpopulation: m_i to serve as controls, and the other $n_i = N_i - m_i$, to receive the treatment. If the observations in the ith subgroup of such a *stratified sample* are denoted by

$$(X_{i1}, \ldots, X_{im_i}; Y_{i1}, \ldots, Y_{in_i}) = (Z_{i1}, \ldots, Z_{iN_i}),$$

the density of $Z = (Z_{11}, \ldots, Z_{cN_c})$ is

$$p_\Delta(z) = \prod_{i=1}^{c} \left[f_i(x_{i1}) \ldots f_i(x_{im_i}) f_i(y_{i1} - \Delta) \ldots f_i(y_{in_i} - \Delta) \right]. \tag{5.46}$$

Unbiasedness of a test ϕ for testing $\Delta = 0$ against $\Delta > 0$ implies that for all $f_1, \ldots, f_c$,

$$\int \phi(z) p_0(z) \, dz = \alpha \qquad (dz = dz_{11} \ldots dz_{cN_c}). \tag{5.47}$$

Theorem 5.8.1 *If $\mathcal{F}$ is the family of all probability densities f that are continuous a.e., then* (5.47) *holds for all $f_1, \ldots, f_c \in \mathcal{F}$ if and only if*

$$\frac{1}{N_1! \ldots N_c!} \sum_{z' \in S(z)} \phi(z') = \alpha \qquad a.e., \tag{5.48}$$

where $S(z)$ is the set of points obtained from z by permuting for each $i = 1, \ldots, c$ the coordinates $z_{ij} (j = 1, \ldots, N_i)$ within the ith subgroup in all $N_1! \ldots N_c!$ possible ways.

Proof. To prove the result for the case $c = 1$, note that the set of order statistics $T(Z) = (Z_{(1)}, \ldots, Z_{(N)})$ is a complete sufficient statistic for $\mathcal{F}$ (Example 4.3.4). A necessary and sufficient condition for (5.47) is therefore

$$E[\phi(Z) \mid T(z)] = \alpha \qquad \text{a.e.} \tag{5.49}$$

The set $S(z)$ in the present case ($c = 1$) consists of the N points obtained from z through permutation of coordinates, so that $S(z) = \{z' : T(z') = T(z)\}$. It follows from Section 2.4 that the conditional distribution of Z given $T(z)$ assigns probability $1/N!$ to each of the $N!$ points of $S(z)$. Thus (5.49) is equivalent to

$$\frac{1}{N!} \sum_{z' \in S(z)} \phi(z') = \alpha \qquad \text{a.e.,} \tag{5.50}$$

as was to be proved. The proof for general c is completely analogous and is left as an exercise (Problem 5.44.) ∎

The tests satisfying (5.48) are called permutation tests. An extension of this definition is given in Problem 5.54.

5.9 Most Powerful Permutation Tests

For the problem of testing the hypothesis $H : \Delta = 0$ of no treatment effect on the basis of a stratified sample with density (5.46) it was shown in the preceding section that unbiasedness implies (5.48). We shall now determine the test which, subject to (5.48), maximizes the power against a fixed alternative (5.46) or more generally against an alternative with arbitrary fixed density $h(z)$.

The power of a test ϕ against an alternative h is

$$\int \phi(z)h(z)\,dz = \int E[\phi(Z) \mid t]\,dp^T(t).$$

Let $t = T(z) = (z_{(1)}, \ldots, z_{(N)})$, so that $S(z) = S(t)$. As was seen in Example 2.4.1 and Problem 2.6, the conditional expectation of $\phi(Z)$ given $T(Z) = t$ is

$$\psi(t) = \frac{\sum_{z \in S(t)} \phi(z)h(z)}{\sum_{z \in S(t)} h(z)}.$$

To maximize the power of ϕ subject to (5.48) it is therefore necessary to maximize $\psi(t)$ for each t subject to this condition. The problem thus reduces to the determination of a function ϕ which subject to

$$\sum_{z \in S(t)} \phi(z) \frac{1}{N_1! \ldots N_c!} = \alpha,$$

maximizes

$$\sum_{z \in S(t)} \phi(z) \frac{h(z)}{\sum_{z' \in X(t)} h(z')}.$$

By the Neyman–Pearson fundamental lemma, this is achieved by rejecting H for those points z of $S(t)$ for which the ratio

$$\frac{h(z)N_1! \ldots N_c!}{\sum_{z' \in S(t)} h(z')}$$

is too large. Thus the most powerful test is given by the critical function

$$\phi(z) = \begin{cases} 1 & \text{when} \quad h(z) > C[T(z)], \\ \gamma & \text{when} \quad h(z) = C[T(z)], \\ 0 & \text{when} \quad h(z) < C[T(z)]. \end{cases} \tag{5.51}$$

To carry out the test, the $N_1! \dots N_c!$ points of each set $S(z)$ are ordered according to the values of the density h. The hypothesis is rejected for the k largest values and with probability γ for the $(k+1)$st value, where k and γ are defined by

$$k + \gamma = \alpha N_1! \dots N_c!.$$

Consider now in particular the alternatives (5.46). The most powerful permutation test is seen to depend on Δ and the f_i, and is therefore not UMP.

Of special interest is the class of normal alternatives with common variance:

$$f_i = N(\xi_i, \sigma^2).$$

The most powerful test against these alternatives, which turns out to be independent of the ξ_i, σ^2, and Δ, is appropriate when approximate normality is suspected but the assumption is not felt to be reliable. It may then be desirable to control the size of the test at level α regardless of the form of the densities f_i and to have the test unbiased against all alternatives (5.46). However, among the class of tests satisfying these broad restrictions it is natural to make the selection so as to maximize the power against the type of alternative one expects to encounter, that is, against the normal alternatives.

With the above choice of f_i, (5.46) becomes

$$h(z) = \left(\sqrt{2\pi}\sigma\right)^{-N} \times$$

$$\exp\left[-\frac{1}{2\sigma^2}\sum_{i=1}^{c}\left(\sum_{j=1}^{m_i}(z_{ij}-\xi_i)^2 + \sum_{j=m_i+1}^{N_i}(z_{ij}-\xi_i-\Delta)^2\right)\right]. \tag{5.52}$$

Since the factor $\exp[-\sum_i \sum_{j=1}^{N_i}(z_{ij}-\xi_i)^2/2\sigma^2]$ is constant over $S(t)$, the test (5.51) therefore rejects H when $\exp(\Delta \sum_i \sum_{j=m_i+1}^{N_i} z_{ij}) > C[T(z)]$ and hence when

$$\sum_{i=1}^{c}\sum_{j=1}^{n_j} y_{ij} = \sum_{i=1}^{c}\sum_{j=m_i+1}^{N_i} z_{ij} > C[T(z)]. \tag{5.53}$$

Of the $N_1! \dots N_c!$ values that the test statistic takes on over $S(t)$, only

$$\binom{N_1}{n_1} \cdots \binom{N_c}{n_c}$$

are distinct, since the value of the statistic is the same for any two points z' and z'' for which $(z'_{i1}, \dots, z'_{im_i})$ and $(z''_{i1}, \dots, z''_{im_i})$ are permutations of each other for each i. It is therefore enough to compare these distinct values, and to reject H for the k' largest ones and with probability γ' for the $(k'+1)$st, where

$$k' + \gamma' = \alpha \binom{N_1}{n_1} \cdots \binom{N_c}{n_c}.$$

The test (5.53) is most powerful against the normal alternatives under consideration among all tests which are unbiased and of level α for testing $H : \Delta = 0$ in the original family (5.46) with $f_1, \ldots, f_c \in \mathcal{F}$.[8] To complete the proof of this statement it is still necessary to prove the test unbiased against the alternatives (5.46). We shall show more generally that it is unbiased against all alternatives for which $X_{ij} (j = 1, \ldots, m_i)$, $Y_{ik} (k = 1, \ldots, n_i)$ are independently distributed with cumulative distribution functions F_i, G_i respectively such that Y_{ik} is stochastically larger than X_{ij}, that is, such that $G_i(z) \leq F_i(z)$ for all z. This is a consequence of the following lemma.

Lemma 5.9.1 *$X_1, \ldots, X_m$, $Y_1, \ldots, Y_n$ be samples from continuous distributions F, G, and let $\phi(x_1, \ldots, x_m; y_1, \ldots, y_n)$ be a critical function such that* (a) *its expectation is α whenever $G = F$, and* (b) *$y_i \leq y_i'$ for $i = 1, \ldots, n$ implies*

$$\phi(x_1, \ldots, x_m; y_1, \ldots, y_n) \leq \phi(x_1, \ldots, x_m; y_1', \ldots, y_n').$$

Then the expectation $\beta = \beta(F, G)$ of ϕ is $\geq \alpha$ for all pairs of distributions for which Y is stochastically larger than X; it is $\leq \alpha$ if X is stochastically larger than Y.

Proof. By Lemma 3.4.1, there exist functions f, g and independent random variables $V_1, \ldots, V_{m+n}$ such that the distributions of $f(V_i)$ and $g(V_i)$ are F and G respectively and that $f(z) \leq g(z)$ for all z. Then

$$E\phi[f(V_1), \ldots, f(V_m); f(V_{m+1}), \ldots, f(V_{m+n})] = \alpha$$

and

$$E\phi[f(V_1), \ldots, f(V_m); g(V_{m+1}), \ldots, g(V_{m+n})] = \beta.$$

Since for all $(v_1, \ldots, v_{m+n})$,

$$\begin{aligned} &\phi[f(v_1), \ldots, f(v_m);\ f(v_{m+1}), \ldots, f(v_{m+n})] \\ &\quad \leq \phi[f(v_1), \ldots, f(v_m);\ g(v_{m+1}), \ldots, g(v_{m+n})], \end{aligned}$$

the same inequality holds for the expectations of both sides, and hence $\alpha \leq \beta$.

The proof for the case that X is stochastically larger than Y is completely analogous.

The lemma also generalizes to the case of c vectors $(X_{i1}, \ldots, X_{im_i}; Y_{i1}, \ldots, Y_{in_i})$ with distributions (F_i, G_i). If the expectation of a function ϕ is α when $F_i = G_i$ and ϕ is nondecreasing in each y_{ij} when all other variables are held fixed, then it follows as before that the expectation of ϕ is $\geq \alpha$ when the random variables with distribution G_i are stochastically larger than those with distribution F_i.

In applying the lemma to the permutation test (5.53) it is enough to consider the case $c = 1$, the argument in the more general case being completely analogous. Since the rejection probability of the test (5.53) is α whenever $F = G$, it is only necessary to show that the critical function ϕ of the test satisfies (b). Now $\phi = 1$ if $\sum_{i=m+1}^{m+n} z_i$ exceeds sufficiently many of the sums $\sum_{i=m+1}^{m+n} z_{j_i}$, and hence if

[8]For a closely related result. see Odén and Wedel (1975).

sufficiently many of the differences

$$\sum_{i=m+1}^{m+n} z_i - \sum_{i=m+1}^{m+n} z_{j_i}$$

are positive. For a particular permutation $(j_1, \ldots, j_{m+n})$

$$\sum_{i=m+1}^{m+n} z_i - \sum_{i=m+1}^{m+n} z_{j_i} = \sum_{i=1}^{p} z_{s_i} - \sum_{i=1}^{p} z_{r_i},$$

where $r_1 < \cdots < r_p$ denote those of the integers $j_{m+1}, \ldots, j_{m+n}$ that are $\leq m$, and $s_1 < \cdots < s_p$ those of the integers $m+1, \ldots, m+n$ not included in the set $(j_{m+1}, \ldots, j_{m+n})$. If $\sum z_{s_i} - \sum z_{r_i}$ is positive and $y_i \leq y_i'$, that is, $z_i \leq z_i'$ for $i = m+1, \ldots, m+n$, then the difference $\sum z'_{s_i} - \sum z_{r_i}$ is also positive and hence ϕ satisfies (b).

The same argument also shows that the rejection probability of the test is $\leq \alpha$ when the density of the variables is given by (5.46) with $\Delta \leq 0$. The test is therefore equally appropriate if the hypothesis $\Delta = 0$ is replaced by $\Delta \leq 0$.

Except for small values of the sample sizes N_i, the amount of computation required to carry out the permutation test (5.53) is large. Computational methods are discussed by Green (1977), John and Robinson (1983b), Diaconis and Holmes (1994) and Chapter 13 of Good (1994), who has an extensive bibliography.

One can relate the permutation test to the corresponding normal theory t-test as follows. On multiplying both sides of the inequality

$$\sum y_j > C[T(z)]$$

by $(1/m) + (1/n)$ and subtracting $(\sum x_1, + \sum y_j)/m$, the rejection region for $c = 1$ becomes $\bar{y} - \bar{x} > C[T(z)]$ or $W = (\bar{y} - \bar{x})/\sqrt{\sum_{i=1}^{n}(z_i - \bar{z})^2} > C[T(z)]$, since the denominator of W is constant over $S(z)$ and hence depends only on $T(z)$. As was seen at the end of Section 5.3, this is equivalent to

$$\frac{(\bar{y} - \bar{x}) \Big/ \sqrt{\frac{1}{m} + \frac{1}{n}}}{\sqrt{\left[\sum(x_i - \bar{x})^2 + \sum(y_j - \bar{y})^2\right] / (m + n - 2)}} > C[T(z)]. \tag{5.54}$$

The rejection region therefore has the form of a t-test in which the constant cutoff point C_0 of (5.27) has been replaced by a random one. It turns out that when the hypothesis is true, so that the Z's are identically and independently distributed, and m/n is bounded away from zero and infinity as m and n tend to infinity, the difference between the random cutoff point $C[T(Z)]$ and C_0 is small in an appropriate asymptotic sense, and so the permutation test and the t-test given by (5.27) – (5.29) behave similarly in large samples. Such results will be developed in Section 15.2. the *permutation test can be approximated for large samples by the standard t-test.* Exactly analogous results hold for $c > 1$; the appropriate generalization of the two-sample t-test is provided in Problem 7.9. ■

5.10 Randomization As A Basis For Inference

The problem of testing for the effect of a treatment was considered in Section 5.3 under the assumption that the treatment and control measurements $X_1, \ldots, X_m$, and $Y_1, \ldots, Y_n$ constitute samples from normal distributions, and in Sections 5.8 and 5.9 without relying on the assumption of normality. We shall now consider in somewhat more detail the structure of the experiment from which the data are obtained, resuming for the moment the assumption that the distributions involved are normal.

Suppose that the experimental material consists of $m + n$ patients, plants, pieces of material, or the like, drawn at random from the population to which the treatment could be applied. The treatment is given to n of these while the other m serve as controls. The characteristic that is to be influenced by the treatment is then measured in each case, leading to observations $X_1, \ldots, X_m; Y_1, \ldots, Y_n$.

To be specific, suppose that the treatment is carried out by injecting a drug and that $m+n$ ampules are assigned to the $m+n$ patients. The ith measurement can be considered as the sum of two components. One, say U_i, is associated with the ith patient; the other, V_i, with the ith ampule and the circumstances under which it is administered and under which the measurements are taken. The variables U_i and V_i are assumed to be independently distributed, the V's with normal distribution $N(\eta, \sigma^2)$ or $N(\xi, \sigma^2)$ as the ampule contains the drug or is one of those used for control. If in addition the U's are assumed to constitute a random sample from $N(\mu, \sigma_1^2)$, it follows that the X's and Y's are independently normally distributed with common variance $\sigma^2 + \sigma_1^2$ and means

$$E(X) = \mu + \xi, \qquad E(Y) = \mu + \eta.$$

Except for a change of notation their joint distribution is then given by (5.26), and the hypothesis $\eta = \xi$ can be tested by the standard t-test

Unfortunately, under actual experimental conditions, it is frequently not possible to ensure that the patients or other experimental units constitute a random sample from the population of such units. They may be patients in a certain hospital at a given time, or volunteers for an experiment, and may constitute a haphazard rather than a random sample. In this case the U's would have to be considered as unknown constants, since they are not obtained by any definite sampling procedure. This assumption is appropriate also in a different context. Suppose that the experimental units are all the machines in a shop or fields on a farm. If the experiment is performed only to determine the best method for this particular shop or farm, these experimental units are the only relevant ones; that is, a replication of the experiment would consist in comparing the two treatments again for the same machines or fields rather than for a new batch drawn at random from a large population. In this case the units themselves, and therefore the u's, are constant. Under the above assumptions the joint density of the $m + n$ measurements is

$$\frac{1}{(\sqrt{2\pi}\sigma)^{m+n}} \exp\left[-\frac{1}{2\sigma^2}\left(\sum_{i=1}^{m}(x_i - u_i - \xi)^2 + \sum_{j=1}^{n}(y_j - u_{m+j} - \eta)^2\right)\right].$$

Since the u's are completely arbitrary, it is clearly impossible to distinguish between $H : \eta = \xi$ and the alternatives $K : \eta > \xi$. In fact, every distribution of K

also belongs to H and vice versa, and the most powerful level-α test for testing H against any simple alternative specifying ξ, η, σ, and the u's rejects H with probability α regardless of the observations.

Data which could serve as a basis for testing whether or not the treatment has an effect can be obtained through the fundamental device of *randomization.* Suppose that the $N = m + n$ patients are assigned to the N ampules at random, that is, in such a way that each of the $N!$ possible assignments has probability $1/N!$ of being chosen. Then for a given assignment the N measurements are independently normally distributed with variance σ^2 and means $\xi + u_{j_i}$ $(i = 1, \ldots, m)$ and $\eta + u_{j_i}$ $(i = m + 1, \ldots, m + n)$. The overall joint density of the variables

$$(Z_1, \ldots, Z_N) = (X_1, \ldots, X_m; Y_1, \ldots, Y_n)$$

is therefore

$$\frac{1}{N!} \sum_{(j_1, \ldots, j_N)} \frac{1}{(\sqrt{2\pi}\sigma)^N} \tag{5.55}$$
$$\times \exp \left[-\frac{1}{2\sigma^2} \left(\sum_{i=1}^{m} (x_i - u_{j_i} - \xi)^2 + \sum_{i=1}^{n} (y_i - u_{j_{m+i}} - \eta)^2 \right) \right]$$

where the outer summation extends over all $N!$ permutations $(j_1, \ldots, j_N)$ of $(1, \ldots, N)$. Under the hypothesis $\eta = \xi$ this density can be written as

$$\frac{1}{N!} \sum_{(j_1, \ldots, j_N)} \frac{1}{(\sqrt{2\pi}\sigma)^N} \exp \left[-\frac{1}{2\sigma^2} \sum_{i=1}^{N} (z_i - \zeta_{j_i})^2 \right], \tag{5.56}$$

where $\zeta_{j_i} = u_{j_i} + \xi = u_{j_i} + \eta$.

Without randomization a set of y's which is large relative to the x-values could be explained entirely in terms of the unit effects u_i. However, if these are assigned to the y's at random, they will on the average balance those assigned to the x's. As a consequence, a marked superiority of the second sample becomes very unlikely under the hypothesis, and must therefore be attributed to the effectiveness of the treatment.

The method of assigning the treatments to the experimental units completely at random permits the construction of a level-α test of the hypothesis $\eta = \xi$, whose power exceeds α against all alternatives $\eta - \xi > 0$. The actual power of such a test will however depend not only on the alternative value of $\eta - \xi$, which measures the effect of the treatment, but also on the unit effects u_i. In particular, if there is excessive variation among the u's this will swamp the treatment effect (much in the same way as an increase in the variance σ^2 would), and the test will accordingly have little power to detect any given alternative $\eta - \xi$.

In such cases the sensitivity of the experiment can be increased by an approach exactly analogous to the method of stratified sampling discussed in Section 5.8. In the present case this means replacing the process of complete randomization described above by a more restricted randomization procedure. The experimental material is divided into subgroups, which are more homogeneous than the material as a whole, so that within each group the differences among the u's are small. In animal experiments, for example, this can frequently be achieved by a division into litters. Randomization is then applied only within each group. If the ith group contains N_i units, n_i of these are selected at random to receive the treatment, and the remaining $m_i = N_i - n_i$ serve as controls ($\sum N_i = N, \sum m_i = m, \sum n_i = n$).

An example of this approach is the method of *matched pairs*. Here the experimental units are divided into pairs, which are as like each other as possible with respect to all relevant properties, so that within each pair the difference of the u's will be as small as possible. Suppose that the material consists of n such pairs, and denote the associated unit effects (the U's of the previous discussion) by $U_1, U_1'; \ldots; U_n, U_n'$. Let the first and second member of each pair receive the treatment or serve as control respectively, and let the observations for the ith pair be X_i and Y_i. If the matching is completely successful, as may be the case, for example, when the same patient is used twice in the investigation of a sleeping drug, or when identical twins are used, then $U_i' = U_i$ for all i, and the density of the X's and Y's is

$$\frac{1}{(\sqrt{2\pi}\sigma)^2} \exp\left[-\frac{1}{2\sigma^2}\left[\sum(x_i - \xi - u_i)^2 + \sum(y_i - \eta - u_i)^2\right]\right]. \tag{5.57}$$

The UMP unbiased test for testing $H : \eta = \xi$ against $\eta > \xi$ is then given in terms of the differences $W_i = Y_i - X_i$ by the rejection region

$$\sqrt{n}\bar{w} \Big/ \sqrt{\frac{1}{n-1}\sum(w_i - \bar{w})^2} > C. \tag{5.58}$$

(See Problem 5.48.)

However, usually one is not willing to trust the assumption $u_i' = u_i$ even after matching, and it again becomes necessary to randomize. Since as a result of the matching the variability of the u's within each pair is presumably considerably smaller than the overall variation, randomization is carried out only within each pair. For each pair, one of the units is selected with probability $\frac{1}{2}$ to receive the treatment, while the other serves as control. The density of the X's and Y's is then

$$\begin{aligned}\frac{1}{2^n}\frac{1}{(\sqrt{2\pi}\sigma)^{2n}}\prod_{i=1}^{n}\Big\{&\exp\left[-\frac{1}{2\sigma^2}\left[(x_i - \xi - u_i)^2 + (y_i - \eta - u_i')^2\right]\right] \\ &+ \exp\left[-\frac{1}{2\sigma^2}\left[(x_i - \xi - u_i')^2 + (y_i - \eta - u_i)^2\right]\right]\Big\}.\end{aligned} \tag{5.59}$$

Under the hypothesis $\eta = \xi$, and writing

$$z_{i1} = x_i, \quad z_{i2} = y_i, \quad \zeta_{i1} = \xi + u_i, \quad \zeta_{i2} = \eta + u_i' \qquad (i = 1, \ldots, n),$$

this becomes

$$\frac{1}{2^n}\sum \frac{1}{(\sqrt{2\pi}\sigma)^{2n}} \exp\left[-\frac{1}{2\sigma^2}\sum_{i=1}^{n}\sum_{j=1}^{2}(z_{ij} - \zeta_{ij}')^2\right]. \tag{5.60}$$

Here the outer summation extends over the 2^n points $\zeta' = (\zeta_{11}', \ldots, \zeta_{n2}')$ for which $(\zeta_{i1}', \zeta_{i2}')$ is either (ζ_{i1}, ζ_{i2}) or (ζ_{i2}, ζ_{i1})

5.11 Permutation Tests and Randomization

It was shown in the preceding section that randomization provides a basis for testing the hypothesis $\eta = \xi$ of no treatment effect, without any assumptions concerning the experimental units. In the present section, a specific test will be derived for this problem. When the experimental units are treated as constants, the probability density of the observations is given by (5.55) in the case of complete randomization and by (5.59) in the case of matched pairs. More generally, let the experimental material be divided into c subgroups, let the randomization be applied within each subgroup, and let the observations in the ith subgroup be

$$(Z_{i1}, \ldots, Z_{iN_i}) = (X_{i1}, \ldots, X_{im_i}; Y_{i1}, \ldots, Y_{in_i}).$$

For any point $u = (u_{11}, \ldots, u_{cN_c})$, let $S(u)$ denote as before the set of $N_1! \ldots N_c!$ points obtained from u by permuting the coordinates within each subgroup in all $N_1! \ldots N_c!$ possible ways. Then the joint density of the Z's given u is

$$\frac{1}{N_1! \ldots N_c!} \sum_{u' \in S(u)} \frac{1}{(\sqrt{2\pi}\sigma)^N} \tag{5.61}$$

$$\times \exp\left[-\frac{1}{2\sigma^2} \sum_{i=1}^{c} \left(\sum_{j=1}^{m_i} (z_{ij} - \xi - u'_{ij})^2 + \sum_{j=m_i+1}^{N_i} (z_{ij} - \eta - u'_{ij})^2 \right)\right],$$

and under the hypothesis of no treatment effect

$$p_{\sigma,\zeta}(z) = \frac{1}{N_1! \ldots N_c!} \sum_{\zeta' \in S(\zeta)} \frac{1}{(\sqrt{2\pi}\sigma)^N} \exp\left[-\frac{1}{2\sigma^2} \sum_{i=1}^{c} \sum_{j=1}^{N_i} (z_{ij} - \zeta'_{ij})^2\right]. \tag{5.62}$$

It may happen that the coordinates of u or ζ are not distinct. If then some of the points of $S(u)$ or $S(\zeta)$ also coincide, each should be counted with its proper multiplicity. More precisely, if the $N_1! \ldots N_c!$ relevant permutations of $N_1 + \ldots + N_c$ coordinates are denoted by g_k, $k = 1, \ldots, N_1! \ldots N_c!$, then $S(\zeta)$ can be taken to be the ordered set of points $g_k\zeta$, $k = 1, \ldots, N_1! \ldots N_c!$, and (5.62), for example, becomes

$$P_{\sigma,\zeta}(z) = \frac{1}{N_1! \ldots N_c!} \sum_{k=1}^{N_1! \ldots N_c!} \frac{1}{(\sqrt{2\pi}\sigma)^N} \exp\left(-\frac{1}{2\sigma^2} |z - g_k\zeta|^2\right)$$

where $|u|^2$ stands for $\sum_{i=1}^{c} \sum_{j=1}^{N} u_{ij}^2$.

Theorem 5.11.1 *A necessary and sufficient condition for a critical function ϕ to satisfy*

$$\int \phi(z) p_{\sigma,\zeta}(z)\, dz \le \alpha \qquad (dz = dz_{11} \ldots dz_{cN_c}) \tag{5.63}$$

for all $\sigma > 0$ and all vectors ζ is that

$$\frac{1}{N_1! \ldots N_c!} \sum_{z' \in S(z)} \phi(z') \le \alpha \qquad \text{a.e.} \tag{5.64}$$

The proof will be based on the following lemma.

Lemma 5.11.1 *Let A be a set in N-space with positive Lebesgue measure $\mu(A)$. Then for any $\epsilon > 0$ there exist real numbers $\sigma > 0$ and $\xi_1, \ldots, \xi_N$, such that*

$$P\{(X_1, \ldots, X_N) \in A\} \geq 1 - \epsilon,$$

where the X's are independently normally distributed with means $E(X_i) = \xi_i$ and variance $\sigma^2_{X_i} = \sigma^2$.

Proof. Suppose without loss of generality that $\mu(A) < \infty$. Given any $\eta > 0$, there exists a square Q such that

$$\mu(Q \cap A^c) \leq \eta \mu(Q).$$

This follows from the fact that almost every point of A is a density point,[9] or from the more elementary fact that a measurable set can be approximated in measure by unions of disjoint squares. Let a be such that

$$\frac{1}{\sqrt{2\pi}} \int_{-a}^{a} \left(-\frac{t^2}{2}\right) dt = \left(1 - \frac{\epsilon}{2}\right)^{1/N},$$

and let

$$\eta = \frac{\epsilon}{2} \left(\frac{\sqrt{2\pi}}{2a}\right)^N.$$

If $(\xi_1, \ldots, \xi_N)$ is the center of Q, and if $\sigma = b/a = (1/2a)[\mu(Q)]^{1/N}$, where $2b$ is the length of the side of Q, then

$$\begin{aligned}
&\frac{1}{(\sqrt{2\pi}\sigma)^N} \int_{A^c \cap Q^c} \exp\left[-\frac{1}{2\sigma^2} \sum (x_i - \xi_i)^2\right] dx_1 \ldots dx_N \\
&\leq \frac{1}{(\sqrt{2\pi}\sigma)^N} \int_{Q^c} \exp\left[-\frac{1}{2\sigma^2} \sum (x_i - \xi_i)^2\right] dx_1 \ldots dx_N \\
&= 1 - \left[\frac{1}{\sqrt{2\pi}} \int_{-a}^{a} \exp\left(-\frac{t^2}{2}\right) dt\right]^N = \frac{\epsilon}{2}.
\end{aligned}$$

On the other hand,

$$\begin{aligned}
&\frac{1}{(\sqrt{2\pi}\sigma)^N} \int_{A^c \cap Q} \exp\left[-\frac{1}{2\sigma^2} \sum (x_i - \xi_i)^2\right] dx_1 \ldots dx_N \\
&\leq \frac{1}{(\sqrt{2\pi}\sigma)^N} \mu(A^c \cap Q) < \frac{\epsilon}{2},
\end{aligned}$$

and by adding the two inequalities one obtains the desired result. ■

Proof.[Proof of the theorem] Let ϕ be any critical function, and let

$$\psi(z) = \frac{1}{N_1! \ldots N_c!} \sum_{z' \in S(z)} \phi(z').$$

If (5.64) does not hold, there exists $\eta > 0$ such that $\phi(z) > \alpha + \eta$ on a set A of positive measure. By the Lemma there exists $\sigma > 0$ and $\zeta = (\zeta_{11}, \ldots, \zeta_{cN_c})$

[9] See, for example, Billingsley (1995), p.417.

such that $P\{Z \in A\} > 1 - \eta$ when $Z_{11}, \dots, Z_{cN_c}$ are independently normally distributed with common variance σ^2 and means $E(Z_{ij}) = \zeta_{ij}$. It follows that

$$\begin{aligned} \int \phi(z) p_{\sigma,\zeta}(z)\, dz &= \int \psi(z) p_{\sigma,\zeta}(z)\, dz \\ &\geq \int_A \psi(z) \frac{1}{(\sqrt{2\pi}\sigma)^N} \exp\left[-\frac{1}{2\sigma^2} \sum\sum (z_{ij} - \zeta_{ij})^2\right] dz \\ &> (\alpha + \eta)(1 - \eta), \end{aligned} \tag{5.65}$$

which is $> \alpha$, since $\alpha + \eta < 1$. This proves that (5.63) implies (5.64). The converse follows from the first equality in (5.65). ■

Corollary 5.11.1 *Let H be the class of densities*

$$\{p_{\sigma,\zeta}(z) : \sigma > 0, -\infty < \zeta_{ij} < \infty\}.$$

A complete family of tests for H at level of significance α is the class of tests $\mathcal{C}$ satisfying

$$\frac{1}{N_1! \dots N_c!} \sum_{z' \in S(z)} \phi(z') = \alpha \qquad \text{a.e.} \tag{5.66}$$

Proof. The corollary states that for any given level-α test ϕ_0 there exists an element ϕ of $\mathcal{C}$ which is uniformly at least as powerful as ϕ_0. By the preceding theorem the average value of ϕ_0 over each set $S(z)$ is $\leq \alpha$. On the sets for which this inequality is strict, one can increase ϕ_0 to obtain a critical function ϕ satisfying (5.66), and such that $\phi_0(z) \leq \phi(z)$ for all z. Since against all alternatives the power of ϕ is at least that of ϕ_0, this establishes the result. An explicit construction of ϕ, which shows that it can be chosen to be measurable, is given in Problem 5.51.

This corollary shows that the normal randomization model (5.61) leads exactly to the class of tests that was previously found to be relevant when the U's constituted a sample but the assumption of normality was not imposed. It therefore follows from Section 5.9 that the most powerful level-α test for testing (5.62) against a simple alternative (5.61) is given by (5.51) with $h(z)$ equal to the probability density (5.61). If $\eta - \xi = \Delta$, the rejection region of this test reduces to

$$\sum_{u' \in S(u)} \exp\left[\frac{1}{\sigma^2} \sum_{i=1}^{c} \left(\sum_{j=1}^{N_i} z_{ij} u'_{ij} + \Delta \sum_{j=m_i+1}^{N_i} (z_{ij} - u'_{ij}) \right)\right] > C[T(z)], \tag{5.67}$$

since both $\sum\sum z_{ij}$ and $\sum\sum z_{ij}^2$ are constant on $S(z)$ and therefore functions only of $T(z)$. It is seen that this test depends on Δ and the unit effects u_{ij}, so that a UMP test does not exist.

Among the alternatives (5.61) a subclass occupies a central position and is of particular interest. This is the class of alternatives specified by the assumption that the unit effects u_i constitute a sample from a normal distribution. Although this assumption cannot be expected to hold exactly – in fact, it was just as a safeguard against the possibility of its breakdown that randomization was introduced – it is in many cases reasonable to suppose that it holds at least approximately. The resulting subclass of alternatives is given by the probability densities

$$\frac{1}{(\sqrt{2\pi}\sigma)^N} \tag{5.68}$$

$$\times \exp\left[-\frac{1}{2\sigma^2}\sum_{i=1}^{c}\left(\sum_{j=1}^{m_i}(z_{ij}-u_i-\xi)^2+\sum_{j=m_i+1}^{N_i}(z_{ij}-u_i-\eta)^2\right)\right].$$

These alternatives are suggestive also from a slightly different point of view. The procedure of assigning the experimental units to the treatments at random within each subgroup was seen to be appropriate when the variation of the u's is small within these groups and is employed when this is believed to be the case. This suggests, at least as an approximation, the assumption of constant $u_{ij} = u_i$, which is the limiting case of a normal distribution as the variance tends to zero, and for which the density is also given by (5.68).

Since the alternatives (5.68) are the same as the alternatives (5.52) of Section 5.9 with $u_i - \xi = \xi_i$, $u_i - \eta = \xi_i - \Delta$, *the permutation test* (5.53) *is seen to be most powerful for testing the hypothesis* $\eta = \xi$ *in the normal randomization model* (5.61) *against the alternatives* (5.68) *with* $\eta - \xi > 0$. The test retains this property in the still more general setting in which neither normality nor the sample property of the U's is assumed to hold. Let the joint density of the variables be

$$\sum_{u' \in S(u)} \prod_{i=1}^{c}\left[\prod_{j=1}^{m_i} f_i(z_{ij}-u'_{ij}-\xi)\prod_{j=m_i+1}^{N_i} f_i(z_{ij}-u'_{ij}-\eta)\right), \tag{5.69}$$

with f_i continuous a.e. but otherwise unspecified.[10] Under the hypothesis $H : \eta = \xi$, this density is symmetric in the variables $(z_{i1}, \ldots, z_{iN_i})$ of the ith subgroup for each i, so that any permutation test (5.48) has rejection probability α for all distributions of H. By Corollary 5.11.1, these permutation tests therefore constitute a complete class, and the result follows. ■

5.12 Randomization Model and Confidence Intervals

In the preceding section, the unit responses u_i were unknown constants (parameters) which were observed with error, the latter represented by the random terms V_i. A limiting case assumes that the variation of the V's is so small compared with that of the u's that these error variables can be taken to be constant, i.e. that $V_i = v$. The constant v can then be absorbed into the u's, and can therefore be assumed to be zero. This leads to the following two-sample *randomization model*:

N subjects would give "true" responses $u_1, \ldots, u_N$ if used as controls. The subjects are assigned at random, n to treatment and m to control. If the responses

[10] Actually, all that is needed is that $f_1, \ldots, f_c \in \mathcal{F}$, where $\mathcal{F}$ is any family containing all normal distributions.

are denoted by $X_1, \ldots, X_m$ and $Y_1, \ldots, Y_n$ as before, then under the hypothesis H of no treatment effect, the X's and Y's are a random permutation of the u's. Under this model, in which the random assignment of the subjects to treatment and control constitutes the only random element, the probability of the rejection region (5.54) is the same as under the more elaborate models of the preceding sections.

The corresponding limiting model under the alternatives assumes that the treatment has the effect of adding a constant amount Δ to the unit response, so that the X's and Y's are given by $(u_{i_1}, \ldots ; u_{i_m}; u_{i_{m+1}} + \Delta, \ldots, u_{i_{m+n}} + \Delta)$ for some permutation $(i_1, \ldots, i_N)$ of $(1, \ldots, N)$.

These models generalize in the obvious way to stratified samples. In particular, for paired comparisons it is assumed under H that the unit effects (u_i, u_i') are constants, of which one is assigned at random to treatment and the other to control. Thus the pair (X_i, Y_i) is equal to (u_i, u_i') or (u_i', u_i) with probability $\frac{1}{2}$ each, and the assignments in the n pairs are independent; the sample space consists of 2^n points each of which has probability $(\frac{1}{2})^n$. Under the alternative, it is assumed as before that Δ is added to each treated subject, so that $P(X_i = u_i,\ Y_i = u_i' + \Delta) = P(X_i = u_i',\ Y_i = u_i + \Delta) = \frac{1}{2}$. The distribution generated for the observations by such a randomization model is exactly the conditional distribution given $T(z)$ of the preceding sections. In the two-sample case, for example, this common distribution is specified by the fact that all permutations of $(X_1, \ldots, X_m;\ Y_1 - \Delta, \ldots, Y_n - \Delta)$ are equally likely. As a consequence, the power of the test (5.54) in the randomization model is also the conditional power in the two-sample model (5.45). As was pointed out in Section 4.4, the conditional power $\beta(\Delta \mid T(z))$ can be interpreted as an unbiased estimate of the unconditional power $\beta_F(\Delta)$ in the two-sample model. The advantage of $\beta(\Delta \mid T(z))$ is that it depends only on Δ, not on the unknown F. Approximations to $\beta(\Delta \mid T(z))$ are discussed by J. Robinson (1973), G. Robinson (1982), John and Robinson (1983a), and Gabriel and Hsu (1983).

The tests (5.53), which apply to all three models – the sampling model (5.46), the randomization model, and the intermediate model (5.69) – can be inverted in the usual way to produce confidence sets for Δ. We shall now determine these sets explicitly for the paired comparisons and the two-sample case. The derivations will be carried out in the randomization model. However, they apply equally in the other two models, since the tests, and therefore the associated confidence sets, are identical for the three models.

Consider first the case of paired observations (x_i, y_i), $i = 1, \ldots, n$. The one-sided test rejects $H : \Delta = 0$ in favor of $\Delta > 0$ when $\sum_{i=1}^n y_i$ is among the K largest of the 2^n sums obtained by replacing y_i by x_i for all, some, or none of the values $i = 1, \ldots, n$. (It is assumed here for the sake of simplicity that $\alpha = K/2^n$, so that the test requires no randomization to achieve the exact level α.) Let $d_i = y_i - x_i = 2y_i - t_i$, where $t_i = x_i + y_i$ is fixed. Then the test is equivalent to rejecting when $\sum d_i$ is one of the K largest of the 2^n values $\sum \pm d_i$, since an interchange of y_i with x_i is equivalent to replacing d_i by $-d_i$. Consider now testing $H : \Delta = \Delta_0$ against $\Delta > \Delta_0$. The test then accepts when $\sum(d_i - \Delta_0)$ is one of the $l = 2^n - K$ smallest of the 2^n sums $\sum \pm(d_i - \Delta_0)$, since it is now $y_i - \Delta_0$ that is being interchanged with x_i. We shall next invert this statement, replacing Δ_0 by Δ, and see that it is equivalent to a lower confidence bound for Δ.

In the inequality

$$\sum (d_i - \Delta) < \sum [\pm(d_i - \Delta)], \tag{5.70}$$

suppose that on the right side the minus sign attaches to the $(d_i - \Delta)$ with $i = i_1, \ldots, i_r$ and the plus sign to the remaining terms. Then (5.70) is equivalent to

$$d_{i_1} + \cdots + d_{i_r} - r\Delta < 0, \quad \text{or} \quad \frac{d_{i_1} + \cdots + d_{i_r}}{r} < \Delta.$$

Thus, $\sum(d_i - \Delta)$ is among the l smallest of the $\sum \pm(d_i - \Delta)$ if and only if at least $2^n - l$ of the $M = 2^n - 1$ averages $(d_{i_1} + \cdots + d_{i_r})/r$ are $< \Delta$, i.e. if and only if $\delta_{(K)} < \Delta$, where $\delta_{(1)} < \cdots < \delta_{(M)}$ is the ordered set of averages $(d_{i_1} + \cdots + d_{i_r})/r$, $r = 1, \ldots, M$. This establishes $\delta_{(K)}$ as a lower confidence bound for Δ at confidence level $\gamma = K/2^n$. [Among all confidence sets that are unbiased in the model (5.46) with $m_i = n_i = 1$ and $c = n$, these bounds minimize the probability of falling below any value $\Delta' < \Delta$ for the normal model (5.52).]

By putting successively $K = 1, 2, \ldots, 2^n$, it is seen that the $M + 1$ intervals

$$(-\infty, \delta_{(1)}), (\delta_{(1)}, \delta_{(2)}), \ldots, (\delta_{(M-1)}, \delta_{(M)}), (\delta_M, \infty) \tag{5.71}$$

each have probability $1/(M+1) = 1/2^n$ of containing the unknown Δ. The two-sided confidence intervals $(\delta_{(K)}, \delta_{(2^n-K)})$ with $\gamma = (2^{n-1} - K)/2^{n-1}$ correspond to the two-sided version of the test (5.53) with error probability $(1-\gamma)/2$ in each tail. A suitable subset of the points $\delta_{(1)}, \ldots, \delta_{(M)}$ constitutes a set of confidence points in the sense of Section 3.5.

The inversion procedure for the two-group case is quite analogous. Let $(x_1, \ldots, x_m, y_1, \ldots, y_n)$ denote the m control and n treatment observations, and suppose without loss of generality that $m \le n$. Then the hypothesis $\Delta = \Delta_0$ is accepted against $\Delta > \Delta_0$ if $\sum_{j=1}^n (y_j - \Delta_0)$ is among the l smallest of the $\binom{m+n}{n}$ sums obtained by replacing a subset of the $(y_j - \Delta_0)$'s with x's. The inequality

$$\sum (y_j - \Delta_0) < (x_{i_1} + \cdots + x_{i_r}) + [y_{j_1} + \cdots + y_{j_{n-r}} - (n-r)\Delta],$$

with $(i_1, \ldots, i_r, j_1, \ldots, j_{n-r})$ a permutation of $(1, \ldots, n)$, is equivalent to $y_{i_1} + \cdots + y_{i_r} - r\Delta_0 < x_{i_1} + \cdots + x_{i_r}$, or

$$\bar{y}_{i_1,\ldots,i_r} - \bar{x}_{i_1,\ldots,i_r} < \Delta_0. \tag{5.72}$$

Note that the number of such averages with $r \ge 1$ (i.e. omitting the empty set of subscripts) is equal to

$$\sum_{K=1}^{m} \binom{m}{K}\binom{n}{K} = \binom{m+n}{n} - 1 = M$$

(Problem 5.57). Thus, $H : \Delta = \Delta_0$ is accepted against $\Delta > \Delta_0$ at level $\alpha = 1 - l/(M+1)$ if and only if at least K of the M differences (5.72) are less than Δ_0, and hence if and only if $\delta_{(K)} < \Delta_0$, where $\delta_{(1)} < \cdots < \delta_{(M)}$ denote the ordered set of differences (5.72). This establishes $\delta_{(K)}$ as a lower confidence bound for Δ with confidence coefficient $\gamma = 1 - \alpha$.

As in the paired comparisons case, it is seen that the intervals (5.71) each have probability $1/(M+1)$ of containing Δ. Thus, two-sided confidence intervals and standard confidence points can be derived as before. For the generalization to stratified samples, see Problem 5.58.

Algorithms for computing the order statistics $\delta_{(1)}, \ldots, \delta_{(M)}$ in the paired-comparison and two-sample cases are discussed by Tritchler (1984); also see Garthwaite (1996). If M is too large for the computations to be practicable, reduced analyses based on either a fixed or random subset of the set of all $M+1$ permutations are discussed, for example, by Gabriel and Hall (1983) and Vadiveloo (1983). [See also Problem 5.60(i).] Different such methods are compared by Forsythe and Hartigan (1970). For some generalizations, and relations to other subsampling plans, see Efron (1982, Chapter 9).

5.13 Testing for Independence in a Bivariate Normal Distribution

So far, the methods of the present chapter have been illustrated mainly by the two-sample problem. As a further example, we shall now apply two of the formulations that have been discussed, the normal model of Section 5.3 and the nonparametric one of Section 5.8, to the hypothesis of independence in a bivariate distribution.

The probability density of a sample $(X_1, Y_1), \ldots, (X_n, Y_n)$ from a bivariate normal distribution is

$$\frac{1}{(2\pi\sigma\tau\sqrt{1-\rho^2})^n} \exp\left[-\frac{1}{2(1-\rho^2)}\left(\frac{1}{\sigma^2}\sum(x_i-\xi)^2 \right.\right. \tag{5.73}$$
$$\left.\left. -\frac{2\rho}{\sigma\tau}\sum(x_i-\xi)(y_i-\eta)+\frac{1}{\tau^2}\sum(y_i-\eta)^2\right)\right].$$

Here (ξ, σ^2) and (η, τ^2) are the mean and variance of X and Y respectively, and ρ is the correlation coefficient between X and Y. The hypotheses $\rho \le \rho_0$ and $\rho = \rho_0$ for arbitrary ρ_0 cannot be treated by the methods of the present chapter, and will be taken up in Chapter 6. For the present, we shall consider only the hypothesis $\rho = 0$ that X and Y are independent, and the corresponding one-sided hypothesis $\rho \le 0$.

The family of densities (5.73) is of the exponential form (1) with

$$U = \sum X_i Y_i, \quad T_1 = \sum X_i^2, \quad T_2 = \sum Y_i^2, \quad T_3 = \sum X_i, \quad T_4 = \sum Y_i$$

and

$$\theta = \tfrac{\rho}{\sigma\tau(1-\rho^2)}, \qquad \vartheta_1 = \tfrac{-1}{2\sigma^2(1-\rho^2)}, \qquad \vartheta_2 = \tfrac{-1}{2\tau^2(1-\rho^2)},$$
$$\vartheta_3 = \tfrac{1}{1-\rho^2}\left(\tfrac{\xi}{\sigma^2} - \tfrac{\eta\rho}{\sigma\tau}\right), \qquad \vartheta_4 = \tfrac{1}{1-\rho^2}\left(\tfrac{\eta}{\tau^2} - \tfrac{\xi\rho}{\sigma\tau}\right),$$

The hypothesis $H : \rho \le 0$ is equivalent to $\theta < 0$. Since the sample correlation coefficient

$$R = \frac{\sum(X_i - \bar{X})(Y_i - \bar{Y})}{\sqrt{\sum(X_i - \bar{X})^2 \sum(Y_i - \bar{Y})^2}}$$

is unchanged when the X_i and Y_i are replaced by $(X_i - \xi)/\sigma$ and $(Y_i - \eta)/\tau$, the distribution of R does not depend on ξ, η, σ, or τ, but only on ρ. For $\theta = 0$ it therefore does not depend on $\vartheta_1, \ldots, \vartheta_4$, and hence by Theorem 5.1.2, R is independent of $(T_1, \ldots, T_4)$ when $\theta = 0$. It follows from Theorem 5.1.1 that the UMP unbiased test of H rejects when

$$R \geq C_0, \tag{5.74}$$

or equivalently when

$$\frac{R}{\sqrt{(1 - R^2)/(n-2)}} > K_0. \tag{5.75}$$

The statistic R is linear in U, and its distribution for $\rho = 0$ is symmetric about 0. The UMP unbiased test of the hypothesis $\rho = 0$ against the alternative $\rho \neq 0$ therefore rejects when

$$\frac{|R|}{\sqrt{(1 - R^2)/(n-2)}} > K_1. \tag{5.76}$$

Since $\sqrt{n-2}R/\sqrt{1-R^2}$ has the t-distribution with $n-2$ degrees of freedom when $\rho = 0$ (Problem 5.64), the constants K_0 and K_1 in the above tests are given by

$$\int_{K_0}^{\infty} t_{n-2}(y)\,dy = \alpha \quad \text{and} \quad \int_{K_1}^{\infty} t_{n-2}(y)\,dy = \frac{\alpha}{2} \tag{5.77}$$

Since the distribution of R depends only on the correlation coefficient ρ, the same is true of the power of these tests.

Some large sample properties of the above test will be examined in Problem (11.64). In particular, if (X_i, Y_i) is not bivariate normal, the level of the above test is approximately α in large samples under the hypothesis H_1 that X_i and Y_i are independent, but not necessarily under the hypothesis H_2 that the correlation between X_i and Y_i is 0. For the nonparametric model H_1, one can obtain an exact level-α unbiased test of independence in analogy to the permutation test of Section 5.8. For any bivariate distribution of (X, Y), let Y_x denote a random variable whose distribution is the conditional distribution of Y given x. We shall say that there is *positive regression dependence* between X and Y if for any $x < x'$ the variable $Y_{x'}$ is stochastically larger than Y_x. Generally speaking, larger values of Y will then correspond to larger values of X; this is the intuitive meaning of positive dependence. An example is furnished by any normal bivariate distribution with $\rho > 0$. (See Problem 5.68.) Regression dependence is a stronger requirement than positive quadrant dependence, which was defined in Problem 4.28. However, both reflect the intuitive meaning that large (small) values of Y will tend to correspond to large (small) values of X.

As alternatives to H_1 consider positive regression dependence in a general bivariate distribution possessing a density. To see that unbiasedness implies similarity, let F_1, F_2 be any two univariate distributions with densities f_1, f_2 and consider the one-parameter family of distribution functions

$$F_1(x)F_2(y)\{1 + \Delta[1 - F_1(x)][1 - F_2(y)]\}, \qquad 0 \leq \Delta \leq 1. \tag{5.78}$$

This is positively regression dependent (Problem 5.69), and by letting $\Delta \to 0$ one sees that unbiasedness of ϕ against these distributions implies that the rejection probability is α when X and Y are independent, and hence that

$$\int \phi(x_1, \ldots, x_n; y_1, \ldots, y_n) f_1(x_1) \cdots f_1(x_n) f_2(y_1) \cdots f_2(y_n)\,dx\,dy = \alpha$$

for all probability densities f_1 and f_2. By Theorem 5.8.1 this in turn implies

$$\frac{1}{(n!)^2} \sum \phi(x_{i_1}, \ldots, x_{i_n}; y_{j_1}, \ldots, y_{j_n}) = \alpha.$$

Here the summation extends over the $(n!)^2$ points of the set $S(x, y)$, which is obtained from a fixed point (x, y) with $x = (x_1, \ldots, x_n), y = (y_1, \ldots, y_n)$ by permuting the x-coordinates and the y-coordinates, each among themselves in all possible ways.

Among all tests satisfying this condition, the most powerful one against the normal alternatives (5.73) with $\rho > 0$ rejects for the k' largest values of (5.73) in each set $S(x, y)$, where $k'/(n!)^2 = \alpha$. Since $\sum x_i^2$, $\sum y_i^2$, $\sum x_i$, $\sum y_i$, are all constant on $S(x, y)$, the test equivalently rejects for the k' largest values of $\sum x_i y_i$ in each $S(x, y)$.

Of the $(n!)^2$ values that the statistic $\sum X_i Y_i$ takes on over $S(x, y)$, only $n!$ are distinct, since the statistic remains unchanged if the X's and Y's are subjected to the same permutation. A simpler form of the test is therefore obtained, for example by rejecting H_1 for the k largest values of $\sum x_{(i)} y_{j_i}$, of each set $S(x, y)$, where $x_{(i)} < \cdots < x_{(n)}$ and $k/n! = \alpha$. The test can be shown to be unbiased against all alternatives with positive regression dependence. (See Problem 6.62.)

In order to obtain a comparison of the permutation test with the standard normal test based on the sample correlation coefficient R, let $T(X, Y)$ denote the set of ordered X's and Y's

$$T(X, Y) = (X_{(1)}, \ldots, X_{(n)}; Y_{(1)}, \ldots, Y_{(n)}).$$

The rejection region of the permutation test can then be written as

$$\sum X_i Y_i > C[T(X, Y)].$$

or equivalently as $R > K[T(X, Y)]$. It again turns out that the difference between $K[T(X, Y)]$ and the cutoff point C_0 of the corresponding normal test (5.74) tends to zero in an appropriate sense. Such results are developed in Section 15.2; also see Problem 15.13. For large n, the standard normal test (5.74) therefore serves as an approximation for the permutation test.

5.14 Problems

Section 5.2

Problem 5.1 Let $X_1, \ldots, X_n$ be a sample from $N(\xi, \sigma^2)$. The power of Student's t-test is an increasing function of ξ/σ in the one-sided case $H : \xi \leq 0, K : \xi > 0$, and of $|\xi|/\sigma$ in the two-sided case $H : \xi = 0, K : \xi \neq 0$.

[If

$$S = \sqrt{\frac{1}{n-1} \sum (X_i - \bar{X})^2},$$

the power in the two-sided case is given by

$$1 - P\left\{-\frac{CS}{\sigma} - \frac{\sqrt{n}\xi}{\sigma} \leq \frac{\sqrt{n}(\bar{X} - \xi)}{\sigma} \leq \frac{CS}{\sigma} - \frac{\sqrt{n}\xi}{\sigma}\right\}$$

and the result follows from the fact that it holds conditionally for each fixed value of S/σ.]

Problem 5.2 In the situation of the previous problem there exists no test for testing $H : \xi = 0$ at level α, which for all σ has power $\geq \beta > \alpha$ against the alternatives (ξ, σ) with $\xi = \xi_1 > 0$.

[Let $\beta(\xi_1, \sigma)$ be the power of any level α test of H, and let $\beta(\sigma)$ denote the power of the most powerful test for testing $\xi = 0$ against $\xi = \xi_1$ when σ is known. Then $\inf_\sigma \beta(\xi_1, \sigma) \leq \inf_\sigma \beta(\sigma) = \alpha$.]

Problem 5.3 (i) Let Z and V be independently distributed as $N(\delta, 1)$ and χ^2 with f degrees of freedom respectively. Then the ratio $Z \div \sqrt{V/f}$ has the noncentral t-distribution with f degrees of freedom and noncentrality parameter δ, the probability density of which is [11]

$$\begin{aligned} p_\delta(t) &= \frac{1}{2^{\frac{1}{2}(f-1)}\Gamma(\frac{1}{2}f)\sqrt{\pi f}} \int_0^\infty y^{\frac{1}{2}(f-1)} \\ &\quad \times \exp\left(-\frac{1}{2}y\right) \exp\left[-\frac{1}{2}\left(t\sqrt{\frac{y}{f}} - \delta\right)^2 dy\right] dy \end{aligned} \tag{5.79}$$

or equivalently

$$\begin{aligned} p_\delta(t) &= \frac{1}{2^{\frac{1}{2}(f-1)}\Gamma(\frac{1}{2}f)\sqrt{\pi f}} \exp\left(-\frac{1}{2}\frac{f\delta^2}{f+t^2}\right) \\ &\quad \times \left(\frac{f}{f+t^2}\right)^{\frac{1}{2}(f+1)} \int_0^\infty v^f \exp\left[-\frac{1}{2}\left(v - \frac{\delta t}{\sqrt{f+t^2}}\right)^2\right] dv. \end{aligned}$$

Another form is obtained by making the substitution $w = t\sqrt{y}/\sqrt{f}$ in (5.79).

(ii) If $X_1, \ldots, X_n$ are independently distributed as $N(\xi, \sigma^2)$, then $\sqrt{n}\bar{X} \div \sqrt{\sum(X_1 - \bar{X})^2/(n-1)}$ has the noncentral t-distribution with $n-1$ degrees of freedom and noncentrality parameter $\delta = \sqrt{n}\xi/\sigma$. In the case $\delta = 0$, show that t-distribution with $n-1$ degrees of freedom is given by (5.18).

[(i): The first expression is obtained from the joint density of Z and V by transforming to $t = z \div \sqrt{v/f}$ and v.]

Problem 5.4 Let $X_1, \ldots, X_n$ be a sample from $N(\xi, \sigma^2)$. Denote the power of the one-sided t-test of $H : \xi \leq 0$ against the alternative ξ/σ by $\beta(\xi/\sigma)$, and by $\beta^*(\xi/\sigma)$ the power of the test appropriate when σ is known. Determine $\beta(\xi/\sigma)$ for $n = 5, 10, 15$, $\alpha = .05$, $\xi/\sigma = .07, 0.8, 0.9, 1.0, 1.1, 1.2$, and in each case compare it with $\beta^*(\xi/\sigma)$. Do the same for the two-sided case.

Problem 5.5 Let $Z_1, \ldots, Z_n$ be independently normally distributed with common variance σ^2 and means $E(Z_i) = \zeta_i (i = 1, \ldots, s)$, $E(Z_i) = 0$ $(i = s+1, \ldots, n)$.

[11] A systematic account of this distribution can be found in in Owen (1985) and Johnson, Kotz and Balakrishnan (1995).

There exist UMP unbiased tests for testing $\zeta_1 \leq \zeta_1^0$ and $\zeta_1 = \zeta_1^0$ given by the rejection regions

$$\frac{Z_1 - \zeta_1^0}{\sqrt{\sum_{i=s+1}^{n} Z_i^2/(n-s)}} > C_0 \quad \text{and} \quad \frac{|Z_1 - \zeta_1^0|}{\sqrt{\sum_{i=s+1}^{n} Z_i^2/(n-s)}} > C.$$

When $\zeta_1 = \zeta_1^0$, the test statistic has the t-distribution with $n - s$ degrees of freedom.

Problem 5.6 Let $X_1, \ldots, X_n$ be independently normally distributed with common variance σ^2 and means $\zeta_1, \ldots, \zeta_n$, and let $Z_i = \sum_{j=1}^n a_{ij} X_j$, be an orthogonal transformation (that is, $\sum_{i=1}^n a_{ij} a_{ik} = 1$ or 0 as $j = k$ or $j \neq k$). The Z's are normally distributed with common variance σ^2 and means $\zeta_i = \sum a_{ij} \xi_j$.

[The density of the Z's is obtained from that of the X's by substituting $x_i = \sum b_{ij} z_j$, where (b_{ij}) is the inverse of the matrix (a_{ij}), and multiplying by the Jacobian, which is 1.]

Problem 5.7 If $X_1, \ldots, X_n$ is a sample from $N(\xi, \sigma^2)$, the UMP unbiased tests of $\xi \leq 0$ and $\xi = 0$ can be obtained from Problems 5.5 and 5.6 by making an orthogonal transformation to variables $Z_1, \ldots, Z_n$ such that $Z_1 = \sqrt{n}\bar{X}$.

[Then

$$\sum_{i=2}^n Z_i^2 = \sum_{i=1}^n Z_i^2 - Z_1^2 = \sum_{i=1}^n X_i^2 - n\bar{X}^2 = \sum_{i=1}^n (X_i - \bar{X})^2.]$$

Problem 5.8 Let $X_1, X_2, \ldots$ be a sequence of independent variables distributed as $N(\xi, \sigma^2)$, and let $Y_n = [nX_{n+1} - (X_1 + \cdots + X_n)]/\sqrt{n(n+1)}$. Then the variables $Y_1, Y_2, \ldots$ are independently distributed as $N(0, \sigma^2)$.

Problem 5.9 Let N have the binomial distribution based on 10 trials with success probability p. Given $N = n$, let $X_1, \cdots, X_n$ be i.i.d. normal with mean θ and variance one. The data consists of $(N, X_1, \cdots, X_N)$.
(i). If p has a known value p_0, show there does not exist a UMP test of $\theta = 0$ versus $\theta > 0$. [In fact, a UMPU test does not exist either.]
(ii). If p is unknown (taking values in (0,1)), find a UMPU test of $\theta = 0$ versus $\theta > 0$.

Problem 5.10 As in Example 3.9.2, suppose X is multivariate normal with unknown mean $\xi = (\xi_1, \ldots, \xi_k)^T$ and known positive definite covariance matrix Σ. Assume $a = (a_1, \ldots, a_k)^T$ is a fixed vector. The problem is to test

$$H : \sum_{i=1}^k a_i \xi_i = \delta \quad \text{vs.} \quad K : \sum_{i=1}^k a_k \xi_i \neq \delta .$$

Find a UMPU level α test. *Hint:* First consider $\Sigma = I_k$, the identity matrix.

Problem 5.11 Let $X_i = \xi + U_i$, and suppose that the joint density f of the U's is *spherically symmetric*, that is, a function of $\sum U_i^2$ only,

$$f(u_1, \ldots, u_n) = q(\sum u_i^2) .$$

Show that the null distribution of the one-sample t-statistic is independent of q and hence is the same as in the normal case, namely Student's t with $n-1$ degrees of freedom. *Hint:* Write t_n as

$$\frac{n^{1/2}\bar{X}_n/\sqrt{\sum X_j^2}}{\sqrt{\sum(X_i-\bar{X}_n)^2/(n-1)\sum X_j^2}},$$

and use the fact that when $\xi=0$, the density of $X_1,\dots,X_n$ is constant over the spheres $\sum x_j^2=c$ and hence the conditional distribution of the variables $X_i/\sqrt{\sum X_j^2}$ given $\sum X_j^2=c$ is uniform over the conditioning sphere and hence independent of q. *Note.* This model represents one departure from the normal-theory assumption, which does not affect the level of the test. The effect of a much weaker symmetry condition more likely to arise in practice is investigated by Efron (1969).

Section 5.3

Problem 5.12 Let $X_1,\dots,X_n$ and $Y_1,\dots,Y_n$ be independent samples from $N(\xi,\sigma^2)$ and $N(\eta,\tau^2)$ respectively. Determine the sample size necessary to obtain power $\ge\beta$ against the alternatives $\tau/\sigma>\Delta$ when $\alpha=.05,\beta=.9,\Delta=1.5,2,3$, and the hypothesis being tested is $H:\tau/\sigma\le 1$.

Problem 5.13 If $m=n$, the acceptance region (5.23) can be written as

$$\max\left(\frac{S_Y^2}{\Delta_0 S_X^2},\frac{\Delta_0 S_X^2}{S_Y^2}\right)\le\frac{1-C}{C},$$

where $S_X^2=\sum(X_i-\bar{X})^2$, $S_Y^2=\sum(Y_i-\bar{Y})^2$ and where C is determined by

$$\int_0^C B_{n-1,n-1}(w)\,dw=\frac{\alpha}{2}.$$

Problem 5.14 Let $X_1,\dots,X_m$ and $Y_1,\dots,Y_n$ be samples from $N(\xi,\sigma^2)$ and $N(\eta,\sigma^2)$. The UMP unbiased test for testing $\eta-\xi=0$ can be obtained through Problems 5.5 and 5.6 by making an orthogonal transformation from $(X_1,\dots X_m,Y_1,\dots Y_n)$ to $(Z_1,\dots,Z_{m+n})$ such that $Z_1=(\bar{Y}-\bar{X})/\sqrt{1/m+(1/n)}$, $Z_2=(\sum X_i+\sum Y_i)/\sqrt{m+n}$.

Problem 5.15 *Exponential densities.* Let $X_1,\dots,X_n$, be a sample from a distribution with exponential density $a^{-1}e^{-(x-b)/a}$ for $x\ge b$.

(i) For testing $a=1$ there exists a UMP unbiased test given by the acceptance region

$$C_1\le 2\sum[x_i-\min(x_1,\dots,x_n)]\le C_2,$$

where the test statistic has a χ^2-distribution with $2n-2$ degrees of freedom when $\alpha=1$, and C_1,C_2 are determined by

$$\int_{C_1}^{C_2}\chi^2_{2n-2}(y)\,dy=\int_{C_1}^{C_2}\chi^2_{2n}(y)\,dy=1-\alpha.$$

(ii) For testing $b = 0$ there exists a UMP unbiased test given by the acceptance region

$$0 \leq \frac{n \min(x_1, \dots, x_n)}{\sum [x_i - \min(x_i, \dots, x_n)]} \leq C.$$

When $b = 0$, the test statistic has probability density

$$p(u) = \frac{n-1}{(1+u)^n}, \qquad u \geq 0.$$

[These distributions for varying b do not constitute an exponential family, and Theorem 4.4.1 is therefore not directly applicable. For (i), one can restrict attention to the ordered variables $X_{(1)} < \cdots < X_{(n)}$, since these are sufficient for a and b, and transform to new variables $Z_1 = nX_{(1)}, Z_i = (n-i+1)[X_{(i)} - X_{(i-1)}]$ for $i = 2, \dots, n$, as in Problem 2.15. When $a = 1$, Z_1 is a complete sufficient statistic for b, and the test is therefore obtained by considering the conditional problem given z_1. Since $\sum_{i=2}^n Z_i$, is independent of Z_1, the conditional UMP unbiased test has the acceptance region $C_1 \leq \sum_{i=2}^n Z_i \leq C_2$ for each z_1, and the result follows.

For (ii), when $b = 0$, $\sum_{i=1}^n Z_i$, is a complete sufficient statistic for a, and the test is therefore obtained by considering the conditional problem given $\sum_{i=1}^n z_i$. The remainder of the argument uses the fact that $Z_1 / \sum_{i=1}^n Z_i$ is independent of $\sum_{i=1}^n Z_i$, when $b = 0$, and otherwise is similar to that used to prove Theorem 5.1.1.]

Problem 5.16 Let $X_1, \dots, X_n$ be a sample from the Pareto distribution $P(c, \tau)$, both parameters unknown. Obtain UMP unbiased tests for the parameters c and τ. [Problems 5.15 and 3.8.]

Problem 5.17 Extend the results of the preceding problem to the case, considered in Problem 3.29, that observation is continued only until $X_{(1)}, \dots, X_{(r)}$ have been observed.

Problem 5.18 *Gamma two-sample problem.* Let $X_1, \dots X_m$; $Y_1, \dots, Y_n$ be independent samples from gamma distributions $\Gamma(g_1, b_1), \Gamma(g_2, b_2)$ respectively.

(i) If g_1, g_2 are known, there exists a UMP unbiased test of $H : b_2 = b_1$ against one- and two-sided alternatives, which can be based on a beta distribution. [Some applications and generalizations are discussed in Lentner and Buehler (1963).]

(ii) If g_1, g_2 are unknown, show that a UMP unbiased test of H continues to exist, and describe its general form.

(iii) If $b_2 = b_1 = b$ (unknown), there exists a UMP unbiased test of $g_2 = g_1$ against one- and two-sided alternatives; describe its general form.

[(i): If $Y_i (i = 1, 2)$ are independent $\Gamma(g_i, b)$, then $Y_1 + Y_2$ is $\Gamma(g_1 + g_2, b)$ and $Y_1/(Y_1 + Y_2)$ has a beta distribution.]

Problem 5.19 *Inverse Gaussian distribution.*[12] Let $X_1, \dots, X_n$ be a sample from the inverse Gaussian distribution $I(\mu, \tau)$, both parameters unknown.

(i) There exists a UMP unbiased test of $\mu \le \mu_0$ against $\mu > \mu_0$, which rejects when $\bar{X} > C[\sum(X_i + 1/X_i)]$, and a corresponding UMP unbiased test of $\mu = \mu_0$ against $\mu_0 \neq \mu_0$.
[The conditional distribution needed to carry out this test is given by Chhikara and Folks (1976).]

(ii) There exist UMP unbiased tests of $H : \tau = \tau_0$ against both one- and two-sided hypotheses based on the statistic $V = \sum(1/X_i - 1/\bar{X})$.

(iii) When $\tau = \tau_0$, the distribution of $\tau_0 V$ is χ^2_{n-1}.

[Tweedie (1957).]

Problem 5.20 Let $X_1, \dots, X_m$ and $Y_1, \dots, Y_n$ be independent samples from $I(\mu, \sigma)$ and $I(\nu, \tau)$ respectively.

(i) There exist UMP unbiased tests of τ_2/τ_1 against one- and two-sided alternatives.

(ii) If $\tau = \sigma$, there exist UMP unbiased tests of ν/μ against one- and two-sided alternatives.

[Chhikara (1975).]

Problem 5.21 Suppose X and Y are independent, normally distributed with variance 1, and means ξ and η, respectively. Consider testing the simple null hypothesis $\xi = \eta = 0$ against the composite alternative hypothesis $\xi > 0,\ \eta > 0$. Show that a UMPU test does not exist.

Section 5.4

Problem 5.22 On the basis of a sample $X = (X_1, \dots, X_n)$ of fixed size from $N(\xi, \sigma^2)$ there do not exist confidence intervals for ξ with positive confidence coefficient and of bounded length.[13]
[Consider any family of confidence intervals $\delta(X) \pm L/2$ of constant length L. Let $\xi_1, \dots \xi_{2n}$ be such that $|\xi_i - \xi_j| > L$ whenever $i \neq j$. Then the sets $S_i\{x : |\delta(x) - \xi_i| \le L/2\}$ $(i = 1, \dots, 2N)$ are mutually exclusive. Also, there exists $\sigma_0 > 0$ such that

$$|P_{\xi_i,\sigma}\{X \in S_i\} - P_{\xi_1,\sigma}\{X \in S_i\}| \le \frac{1}{2N} \qquad \text{for} \quad \sigma > \sigma_0,$$

[12] For additional information concerning inference in inverse Gaussian distributions, see Folks and Chhikara (1978) and Johnson, Kotz and Balakrishnan (1994, volume 1).

[13] A similar conclusion holds in the problem of constructing a confidence interval for the ratio of normal means (Fieller's problem), as discussed in Koschat (1987). For problems where it is impossible to construct confidence intervals with finite expected length, see Gleser and Hwang (1987).

as is seen by transforming to new variables $Y_j = (X_j - \xi_1)/\sigma$ and applying Lemmas 5.5.1 and 5.11.1 of the Appendix. Since $\min_i P_{\xi_1,\sigma}\{X \in S_i\} \le 1/(2N)$, it follows for $\sigma > \sigma_0$ that $\min_i P_{\xi_1,\sigma}\{X \in S_i\} \le 1/N$, and hence that

$$\inf_{\xi,\sigma} P_{\xi,\sigma}\left\{|\delta(X) - \xi| \le \frac{L}{2}\right\} \le \frac{1}{N}$$

The confidence coefficient associated with the intervals $\delta(X) \pm L/2$ is therefore zero, and the same must be true a fortiori of any set of confidence intervals of length $\le L$.]

Problem 5.23 *Stein's two-stage procedure.*

(i) If mS^2/σ^2 has a $\chi^2 =$ distribution with m degrees of freedom, and if the conditional distribution of Y given $S = s$ is $N(0, \sigma^2/S^2)$, then Y has Student's t-distribution with m degrees of freedom.

(ii) Let $X_1, X_2, \ldots$ be independently distributed as $N(\xi, \sigma^2)$. Let $\bar{X}_0 = \sum_{i=1}^{n_0} X_i/n_0$, $S^2 = \sum_{i=1}^{n_0}(X_i - \bar{X}_0)^2/(n_0 - 1)$, and let $a_1 = \cdots = a_{n_0} = a$, $a_{n_0+1} = \cdots = a_n = b$ and $n \ge n_0$ be measurable functions of S. Then

$$Y = \frac{\sum_{i=1}^{n} a_i(X_i - \xi)}{\sqrt{S^2 \sum_{i=1}^{n} a_i^2}}$$

has Student's distribution with $n_0 - 1$ degrees of freedom.

(iii) Consider a two-stage sampling scheme $\prod_1$, in which S^2 is computed from an initial sample of size n_0, and then $n - n_0$ additional observations are taken. The size of the second sample is such that

$$n = \max\left\{n_0 + 1, \left[\frac{S^2}{c}\right] + 1\right\}$$

where c is any given constant and where $[y]$ denotes the largest integer $\ge y$. There then exist numbers $a_1, \ldots, a_n$ such that $a_1 = \cdots = a_{n_0}, a_{n_0+1} = \cdots a_n$, $\sum_{i=1}^n a_i = 1$, $\sum_{i=1}^n a_i^2 = c/S^2$. It follows from (ii) that $\sum_{i=1}^n a_i(X_i - \xi)/\sqrt{c}$ has Student's t-distribution with $n_0 - 1$ degrees of freedom.

(iv) The following sampling scheme $\prod_2$, which does not require that the second sample contain at least one observation, is slightly more efficient than $\prod_1$, for the applications to be made in Problems 5.24 and 5.25. Let n_0, S^2, and c be defined as before; let

$$n = \max\left\{n_0, \left[\frac{S^2}{c}\right] + 1\right\}$$

$a_i = 1/n$ $(i = 1, \ldots, n)$, and $\bar{X} = \sum_{i=1}^n a_i X_i$. Then $\sqrt{n}(\bar{X} - \xi)/S$ has again the t-distribution with $n_0 - 1$ degrees of freedom.

[(ii): Given $S = s$, the quantities a, b, and n are constants, $\sum_{i=1}^n a_i(X_i - \xi) = n_0 a(\bar{X}_0 - \xi)$ is distributed as $N(0, n_0 a^2 \sigma^2)$, and the numerator of Y is therefore normally distributed with zero mean and variance $\sigma^2 \sum_{i=1}^n a_i^2$. The result now follows from (i).]

Problem 5.24 Confidence intervals of fixed length for a normal mean.

(i) In the two-stage procedure $\prod_1$, defined in part (iii) of the preceding problem, let the number c be determined for any given $L > 0$ and $0 < \gamma < 1$ by

$$\int_{-L/2\sqrt{c}}^{L/2\sqrt{c}} t_{n_0-1}(y)\,dy = \gamma,$$

where t_{n_0-1} denotes the density of the t-distribution with $n_0 - 1$ degrees of freedom. Then the intervals $\sum_{i=1}^n a_i X_i \pm L/2$ are confidence intervals for ξ of length L and with confidence coefficient γ.

(ii) Let c be defined as in (i), and let the sampling procedure be $\prod_2$ as defined in part (iv) of Problem 5.23. The intervals $\bar{X} \pm L/2$ are then confidence intervals of length L for ξ with confidence coefficient $\geq \gamma$, while the expected number of observations required is slightly lower than under $\prod_1$.

[(i): The probability that the intervals cover ξ equals

$$P_{\xi,\sigma}\left\{-\frac{L}{2\sqrt{c}} \leq \frac{\sum_{i=1}^n a_i(X_i - \xi)}{\sqrt{c}} \leq \frac{L}{2\sqrt{c}}\right\} = \gamma$$

(ii): The probability that the intervals cover ξ equals

$$P_{\xi,\sigma}\left\{\frac{\sqrt{n}|\bar{X} - \xi|}{S} \leq \frac{\sqrt{n}L}{2S}\right\} \geq \left\{\frac{\sqrt{n}|\bar{X} - \xi|}{S} \leq \frac{L}{2\sqrt{c}}\right\} = \gamma.]$$

Problem 5.25 *Two-stage t-tests with power independent of σ.*

(i) For the procedure $\prod_1$ with any given c, let C be defined by

$$\int_C^\infty t_{n_0-1}(y)\,dy = \alpha.$$

Then the rejection region $(\sum_{i=1}^n a_i X_i - \xi_0)/\sqrt{c} > C$ defines a level-α test of $H : \xi \leq \xi_0$ with strictly increasing power function $\beta_c(\xi)$ depending only on ξ.

(ii) Given any alternative ξ_1 and any $\alpha < \beta < 1$, the number c can be chosen so that $\beta_c(\xi_1) = \beta$.

(iii) The test with rejection region $\sqrt{n}(\bar{X} - \xi_0)/S > C$ based on $\prod_2$ and the same c as in (i) is a level-α test of H which is uniformly more powerful than the test given in (i).

(iv) Extend parts (i)–(iii) to the problem of testing $\xi = \xi_0$ against $\xi \neq \xi_0$.

[(i) and (ii): The power of the test is

$$\beta_c(\xi) = \int_{C-(\xi-\xi_0)/\sqrt{c}} t_{n_0-1}(y)\,dy.$$

(iii): This follows from the inequality $\sqrt{n}|\xi - \xi_0|/S \geq |\xi - \xi_0|/\sqrt{c}$.]

Problem 5.26 Let $S(x)$ be a family of confidence sets for a real-valued parameter θ, and let $\mu[S(x)]$ denote its Lebesgue measure. Then for every fixed distribution Q of X (and hence in particular for $Q = P_{\theta_0}$ where θ_0 is the true value of θ)

$$E_Q\{\mu[S(X)]\} = \int_{\theta \neq \theta_0} Q\{\theta \in S(X)\}\, d\theta$$

provided the necessary measurability conditions hold.
[The identity is known as the Ghosh-Pratt identity; see Ghosh (1961) and Pratt (1961a). To prove it, write the expectation on the left side as a double integral, apply Fubini's theorem, and note that the integral on the right side is unchanged if the point $\theta = \theta_0$ is added to the region of integration.]

Problem 5.27 Use the preceding problem to show that uniformly most accurate confidence sets also uniformly minimize the expected Lebesgue measure (length in the case of intervals) of the confidence sets.[14]

Section 5.5

Problem 5.28 Let $X_1, \ldots, X_n$ be distributed as in Problem 5.15. Then the most accurate unbiased confidence intervals for the scale parameter a are

$$\frac{2}{C_2} \sum [x_i - \min(x_1, \ldots, x_n)] \le a \le \frac{2}{C_1} \sum [x_i - \min(x_1, \ldots, x_n)].$$

Problem 5.29 Most accurate unbiased confidence intervals exist in the following situations:

(i) If X, Y are independent with binomial distributions $b(p_1, m)$ and $b(p_2, m)$, for the parameter $p_1 q_2 / p_2 q_1$.

(ii) In a 2×2 table, for the parameter Δ of Section 4.6.

Problem 5.30 *Shape parameter of a gamma distribution.* Let $X_1, \ldots, X_n$ be a sample from the gamma distribution $\Gamma(g, b)$ defined in Problem 3.34.

(i) There exist UMP unbiased tests of $H : g \le g_0$ against $g > g_0$ and of $H' : g = g_0$ against $g \neq g_0$, and their rejection regions are based on $W = \prod (X_i / \bar{X})$.

(ii) There exist uniformly most accurate confidence intervals for g based on W.

[Shorack (1972).]
Notes.

(1) The null distribution of W is discussed in Bain and Engelhardt (1975), Glaser (1976), and Engelhardt and Bain (1978).

(2) For $g = 1$, $\Gamma(g, b)$ reduces to an exponential distribution, and (i) becomes the UMP unbiased test for testing that a distribution is exponential against the alternative that it is gamma with $g > 1$ or with $g \neq 1$.

---

[14] For the corresponding result concerning one-sided confidence bounds, see Madansky (1962).

(3) An alternative treatment of this and some of the following problems is given by Bar-Lev and Reiser (1982).

Problem 5.31 *Scale parameter of a gamma distribution.* Under the assumptions of the preceding problem, there exists

(i) A UMP unbiased test of $H : b \le b_0$ against $b > b_0$ which rejects when $\sum X_i > C(\prod, X_i)$.

(ii) Most accurate unbiased confidence intervals for b.

[The conditional distribution of $\sum X_i$ given $\prod X_i$, which is required for carrying out this test, is discussed by Engelhardt and Bain (1977).]

Problem 5.32 In Example 5.5.1, consider a confidence interval for σ^2 of the form $I = [d_n^{-1} S_n^2, c_n^{-1} S_n^2]$, where $S_n^2 = \sum_i (X_i - \bar{X})^2$ and $c_n < d_n$ are constants. Subject to the level constraint, choose c_n and d_n to minimize the length of I. Argue that the solution has shorter length that the uniformly most accurate one; however, it is biased and so does not uniformly improve the probability of covering false values. [The solution, given in Tate and Klett (1959), satisfies $\chi^2_{n+3}(c_n) = \chi^2_{n+3}(d_n)$ and $\int_{c_n}^{d_n} \chi^2_{n-1}(y)dy = 1 - \alpha$, where $\chi^2_n(y)$ denotes the Chi-squared density with n degrees of freedom. Improvements of this interval which incorporate $\bar{X}$ into their construction are discussed in Cohen (1972) and Shorrock (1990); also see Goutis and Casella (1991).]

Section 5.6

Problem 5.33 (i) Under the assumptions made at the beginning of Section 5.6, the UMP unbiased test of $H : \rho = \rho_0$ is given by (5.44).

(ii) Let $(\underline{\rho}, \bar{\rho})$ be the associated most accurate unbiased confidence intervals for $\rho = a\gamma + b\delta$, where $\underline{\rho} = \underline{\rho}(a, b)$, $\bar{\rho} = \bar{\rho}(a, b)$. Then if f_1 and f_2 are increasing functions, the expected value of $f_1(|\bar{\rho} - \rho|) + f_2(|\rho - \underline{\rho}|)$ is an increasing function of $a^2/n + b^2$.

[(i): Make any orthogonal transformation from $y_1, \ldots, y_n$ to new variables $z_1, \ldots, z_n$, such that $z_1 = \sum_i [bv_i + (a/n)]y_i / \sqrt{(a^2/n) + b^2}$, $z_2 = \sum_i (av_i - b)y_i / \sqrt{a^2 + nb^2}$, and apply Problems 5.5 and 5.6.

(ii): If $a_1^2/n + b_1^2 < a_2^2/n + b_2^2$, the random variable $|\bar{\rho}(a_2, b_2) - \rho|$ is stochastically larger than $|\bar{\rho}(a_1, b_1) - \rho|$, and analogously for $\underline{\rho}$.]

Section 5.7

Problem 5.34 Verify the posterior distribution of Θ given x in Example 5.7.1.

Problem 5.35 If $X_1, \ldots, X_n$, are independent $N(\theta, 1)$ and θ has the improper prior $\pi(\theta) \equiv 1$, determine the posterior distribution of θ given the X's.

Problem 5.36 Verify the posterior distribution of p given x in Example 5.7.2.

Problem 5.37 In Example 5.7.3, verify the marginal posterior distribution of ξ given x.

Problem 5.38 In Example 5.7.4, show that

(i) the posterior density $\pi(\sigma \mid x)$ is of type (c) of Example 5.7.2;

(ii) for sufficiently large r, the posterior density of σ^r given x is no longer of type (c).

Problem 5.39 If X is normal $N(\theta, 1)$ and θ has a Cauchy density $b/\{\pi[b^2+(\theta-\mu)^2]\}$, determine the possible shapes of the HPD regions for varying μ and b.

Problem 5.40 Let $\theta = (\theta_1, \ldots, \theta_s)$ with θ_i real-valued, X have density $p_\theta(x)$, and Θ a prior density $\pi(\theta)$. Then the $100\gamma\%$ HPD region is the $100\gamma\%$ credible region R that has minimum volume.
[Apply the Neyman–Pearson fundamental lemma to the problem of minimizing the volume of R.]

Problem 5.41 Let $X_1, \ldots, X_m$ and $Y_1, \ldots, Y_n$ be independently distributed as $N(\xi, \sigma^2)$ and $N(\eta, \sigma^2)$ respectively, and let (ξ, η, σ) have the joint improper prior density given by

$$\pi(\xi, \eta, \sigma)\, d\xi\, d\eta\, d\sigma = d\xi\; d\eta \cdot \frac{1}{\sigma}\, d\sigma \qquad \text{for all} \quad -\infty < \xi, \eta < \infty, \quad 0 < \sigma.$$

Under these assumptions, extend the results of Examples 5.7.3 and 5.7.4 to inferences concerning (i) $\eta - \xi$ and (ii) σ.

Problem 5.42 Let $X_1, \ldots, X_m$ and $Y_1, \ldots, Y_n$ be independently distributed as $N(\xi, \sigma^2)$ and $N(\eta, \tau^2)$, respectively and let $(\xi, \eta, \sigma, \tau)$ have the joint improper prior density $\pi(\xi, \eta, \sigma, \tau)\, d\xi\, d\eta\, d\sigma\, d\tau = d\xi\, d\eta(1/\sigma)\, d\sigma(1/\tau)\, d\tau$. Extend the result of Example 5.7.4 to inferences concerning τ^2/σ^2.
Note. The posterior distribution of $\eta - \xi$ in this case is the so-called Behrens–Fisher distribution. The credible regions for $\eta - \xi$ obtained from this distribution do not correspond to confidence intervals with fixed coverage probability, and the associated tests of $H : \eta = \xi$ thus do not have fixed size (which instead depends on τ/σ). From numerical evidence [see Robinson (1976) for a summary of his and earlier results] it appears that the confidence intervals are conservative, that is, the actual coverage probability always exceeds the nominal one.

Problem 5.43 Let $T_1, \ldots, T_{s-1}$ have the multinomial distribution (2.34), and suppose that $(p_1, \ldots, p_{s-1})$ has the Dirichlet prior density $D(a_1, \ldots, a_s)$ with density proportional to $p_1^{a_1-1} \ldots p_s^{a_s-1}$, where $p_s = 1-(p_1+\cdots+p_{s-1})$. Determine the posterior distribution of $(p_1, \ldots, p_{s-1})$ given the T's.

Section 5.8

Problem 5.44 Prove Theorem 5.8.1 for arbitrary values of c.

Section 5.9

Problem 5.45 If $c = 1$, $m = n = 4$, $\alpha = .1$ and the ordered coordinates $z_{(1)}, \dots, z_{(N)}$ of a point z are 1.97, 2.19, 2.61, 2.79, 2.88, 3.02, 3.28, 3.41, determine the points of $S(z)$ belonging to the rejection region (5.53).

Problem 5.46 *Confidence intervals for a shift.* [Maritz (1979)]

(i) Let $X_1, \dots, X_m; Y_1, \dots, Y_n$ be independently distributed according to continuous distributions $F(x)$ and $G(y) = F(y - \Delta)$ respectively. Without any further assumptions concerning F, confidence intervals for Δ can be obtained from permutation tests of the hypotheses $H(\Delta_0) : \Delta = \Delta_0$. Specifically, consider the point $(z_1, \dots, z_{m+n}) = (x_1, \dots, x_m, y_1 - \Delta, \dots, y_n - \Delta)$ and the $\binom{m+n}{m}$ permutations $i_1 < \cdots < i_m; i_{m+1} < \cdots < i_{m+n}$ of the integers $1, \dots, m + n$. Suppose that the hypothesis $H(\Delta)$ is accepted for the k of these permutations which lead to the smallest values of

$$\left| \sum_{j=m+1}^{m+n} z_{i_j}/n - \sum_{j=1}^{m} z_{i_j}/m \right|$$

where

$$k = (1 - \alpha)\binom{m+n}{m}.$$

Then the totality of values Δ for which $H(\Delta)$ is accepted constitute an interval, and these intervals are confidence intervals for Δ at confidence level $1 - \alpha$.

(ii) Let $Z_1, \dots, Z_N$ be independently distributed, symmetric about θ, with distribution $F(z - \theta)$, where $F(z)$ is continuous and symmetric about 0. Without any further assumptions about F, confidence intervals for θ can be obtained by considering the 2^N points $Z'_1, \dots, Z'_N$ where $Z'_i = \pm(Z_i - \theta_0)$, and accepting $H(\theta_0) : \theta = \theta_0$ for the k of these points which lead to the smallest values of $|\sum Z'_i|$, where $k = (1 - \alpha)2^N$.

[(i): A point is in the acceptance region for $H(\Delta)$ if

$$\left| \frac{\sum(y_j - \Delta)}{n} - \frac{\sum x_i}{m} \right| = |\bar{y} - \bar{x} - \Delta|$$

is exceeded by at least $\binom{m+n}{n} - k$ of the quantities $|\bar{y}' - \bar{x}' - \gamma\Delta|$, where $(x'_1, \dots, x'_m, y'_1, \dots, y'_n)$ is a permutation of $(x_1, \dots, x_m, y_1, \dots, y_n)$, the quantity γ is determined by this permutation, and $|\gamma| \le 1$. The desired result now follows from the following facts (for an alternative proof, see Section 14): (a) The set of Δ's for which $(\bar{y} - \bar{x} - \Delta)^2 \le (\bar{y}' - \bar{x}' - \gamma\Delta)^2$ is, with probability one, an interval containing $\bar{y} - \bar{x}$. (b) The set of Δ's for which $(\bar{y} - \bar{x} - \Delta)^2$ is exceeded by a particular set of at least $\binom{m+n}{m} - k$ of the quantities $(\bar{y}' - \bar{x}' - \gamma\Delta)^2$ is the intersection of the corresponding intervals (a) and hence is an interval containing $\bar{y} - \bar{x}$. (c) The set of Δ's of interest is the union of the intervals (b) and, since they have a nonempty intersection, also an interval.]

Section 5.10

Problem 5.47 In the matched-pairs experiment for testing the effect of a treatment, suppose that only the differences $Z_i = Y_i - X_i$ are observable. The Z's are assumed to be a sample from an unknown continuous distribution, which under the hypothesis of no treatment effect is symmetric with respect to the origin. Under the alternatives it is symmetric with respect to a point $\zeta > 0$. Determine the test which among all unbiased tests maximizes the power against the alternatives that the Z's are a sample from $N(\zeta, \sigma^2)$ with $\zeta > 0$.

[Under the hypothesis, the set of statistics $(\sum_{i=1}^n Z_i^2, \ldots, \sum_{i=1}^n Z_i^{2n})$ is sufficient; that it is complete is shown as the corresponding result in Theorem 5.8.1. The remainder of the argument follows the lines of Section 11.]

Problem 5.48 (i) If $X_1, \ldots, X_n;\ Y_1, \ldots, Y_n$ are independent normal variables with common variance σ^2 and means $E(X_i) = \xi_i$, $E(Y_i) = \xi_i + \Delta$, the UMP unbiased test of $\Delta = 0$ against $\Delta > 0$ is given by (5.58).

(ii) Determine the most accurate unbiased confidence intervals for Δ.

[(i): The structure of the problem becomes clear if one makes the orthogonal transformation $X_i' = (Y_i - X_i)/\sqrt{2}, Y_i' = (X_i + Y_i)/\sqrt{2}$.]

Problem 5.49 *Comparison of two designs.* Under the assumptions made at the beginning of Section 12, one has the following comparison of the methods of complete randomization and matched pairs. The unit effects and experimental effects U_i and V_i are independently normally distributed with variances σ_1^2, σ^2 and means $E(U_i) = \mu$ and $E(V_i) = \xi$ or η as V_i corresponds to a control or treatment. With complete randomization, the observations are $X_i = U_i + V_i$ $(i = 1, \ldots, n)$ for the controls and $Y_i = U_{n+i} + V_{n+i}$ $(i = 1, \ldots, n)$ for the treated cases, with $E(X_i) = \mu + \xi$, $E(Y_i) = \mu + \eta$. For the matched pairs, if the matching is assumed to be perfect, the X's are as before, but $Y_i = U_i + V_{m+i}$. UMP unbiased tests are given by (5.27) for complete randomization and by (5.58) for matched pairs. The distribution of the test statistic under an alternative $\Delta = \eta - \xi$ is the noncentral t-distribution with noncentrality parameter $\sqrt{n}\Delta/\sqrt{2(\sigma^2 + \sigma_1^2)}$ and $2n - 2$ degrees of freedom in the first case, and with noncentrality parameter $\sqrt{n}\Delta/\sqrt{2}\sigma$ and $n - 1$ degrees of freedom in the second. Thus the method of matched pairs has the disadvantage of a smaller number of degrees of freedom and the advantage of a larger noncentrality parameter. For $\alpha = .05$ and $\Delta = 4$, compare the power of the two methods as a function of n when σ_1, $\sigma = 2$ and when $\sigma_1 = 2$, $\sigma = 1$.

Problem 5.50 *Continuation.* An alternative comparison of the two designs is obtained by considering the expected length of the most accurate unbiased confidence intervals for $\Delta = \eta - \xi$ in each case. Carry this out for varying n and confidence coefficient $1 - \alpha = .95$ when $\sigma_1 = 1$, $\sigma = 2$ and when $\sigma_1 = 2$, $\sigma = 1$.

Section 5.11

Problem 5.51 Suppose that a critical function ϕ_0 satisfies (5.64) but not (5.66), and let $\alpha < \frac{1}{2}$. Then the following construction provides a measurable critical function ϕ satisfying (5.66) and such that $\phi_0(z) \leq \phi(z)$ for all z Inductively, sequences of functions $\phi_1, \phi_2, \dots$ and $\psi_0, \psi_1, \dots$ are defined through the relations

$$\psi_m(z) = \sum_{z' \in S(z)} \frac{\phi_m(z')}{N_1! \dots N_c!}, \qquad m = 0, 1, \dots,$$

and

$$\phi_m(z) = \begin{cases} \phi_{m-1}(z) + [\alpha - \psi_{m-1}(z)] \\ \qquad\qquad \text{if both } \phi_{m-1}(z) \text{ and } \psi_{m-1}(z) \text{ are } < \alpha, \\ \phi_{m-1}(z) \qquad \text{otherwise.} \end{cases}$$

The function $\phi(z) = \lim \phi_m(z)$ then satisfies the required conditions.

[The functions ϕ_m are nondecreasing and between 0 and 1. It is further seen by induction that $0 \leq \alpha - \psi_m(z) \leq (1 - \gamma)^m [\alpha - \psi_0(z)]$, where $\gamma = 1/(N_1! \dots N_c!)$.]

Problem 5.52 Consider the problem of testing $H : \eta = \xi$ in the family of densities (5.61) when it is given that $\sigma > c > 0$ and that the point $(\zeta_{11}, \dots, \zeta_{cN_c}$ of (5.62) lies in a bounded region R containing a rectangle, where c and R are known. Then Theorem 5.11.1 is no longer applicable. However, unbiasedness of a test ϕ of H implies (5.66), and therefore reduces the problem to the class of permutation tests.

[Unbiasedness implies $\int (\phi(z) p_{\sigma,\zeta}(z)\, dz = \alpha$ and hence

$$\alpha = \int \psi(z) p_{\sigma,\zeta}(z)\, dz = \int \psi(z) \frac{1}{(\sqrt{2\pi}\sigma)^N} \exp\left[-\frac{1}{2\sigma^2} \sum\sum (z_{ij} - \zeta_{ij})^2\right] dz$$

for all $\sigma > c$ and ζ in R. The result follows from completeness of this last family.]

Problem 5.53 To generalize Theorem 5.11.1 to other designs, let $Z = (Z_1, \dots, Z_N)$ and let $G = \{g_1, \dots, g_r\}$ be a group of permutations of N coordinates or more generally a group of orthogonal transformations of N-space If

$$P_{\sigma,\zeta}(z) = \frac{1}{r} \sum_{k=1}^{r} \frac{1}{(\sqrt{2\pi}\sigma)^N} \exp\left(-\frac{1}{2\sigma^2} |z - g_k \zeta|^2\right), \tag{5.80}$$

where $|z|^2 = \sum z_i^2$, then $\int \phi(z) p_{\sigma,\zeta}(z)\, dz \leq \alpha$ for all $\sigma > 0$ and all ζ implies

$$\frac{1}{r} \sum_{z' \in S(z)} \phi(z') \leq \alpha \qquad \text{a.e.,} \tag{5.81}$$

where $S(z)$ is the set of points in N-space obtained from z by applying to it all the transformations g_k, $k = 1, \dots, r$.

Problem 5.54 *Generalization of Corollary 5.11.1.* Let H be the class of densities (5.80) with $\sigma > 0$ and $-\infty < \zeta_i < \infty$ $(i = 1, \dots, N)$. A complete family of tests of H at level of significance α is the class of permutation tests satisfying

$$\frac{1}{r} \sum_{z' \in S(z)} \phi(z') = \alpha \qquad \text{a.e.} \tag{5.82}$$

Section 5.12

Problem 5.55 If $c = 1$, $m = n = 3$, and if the ordered x's and y's are respectively 1.97, 2.19, 2.61 and 3.02, 3.28, 3.41, determine the points $\delta_{(1)}, \ldots, \delta_{(19)}$ defined as the ordered values of (5.72).

Problem 5.56 If $c = 4$, $m_i = n_i = 1$, and the pairs (x_i, y_i) are (1.56,2.01), (1.87,2.22), (2.17,2.73), and (2.31,2.60), determine the points $\delta_{(1)}, \ldots, \delta_{(15)}$ which define the intervals (5.71).

Problem 5.57 If m, n are positive integers with $m \leq n$, then

$$\sum_{K=1}^{m} \binom{m}{K} \binom{n}{K} = \binom{m+n}{m} - 1$$

Problem 5.58 (i) Generalize the randomization models of Section 14 for paired comparisons ($n_1 = \cdots = n_c = 2$) and the case of two groups ($c = 1$) to an arbitrary number c of groups of sizes $n_1, \ldots, n_c$.

(ii) Generalize the confidence intervals (5.71) and (5.72) to the randomization model of part (i).

Problem 5.59 Let $Z_1, \ldots, Z_n$ be i.i.d. according to a continuous distribution symmetric about θ, and let $T_{(1)} < \cdots < T_{(M)}$ be the ordered set of $M = 2^n - 1$ subsamples; $(Z_{i_1} + \cdots + Z_{i_r})/r$, $r \geq 1$. If $T_{(0)} = -\infty$, $T_{(M+1)} = \infty$, then

$$P_\theta[T_{(i)} < \theta < T_{(i+1)}] = \frac{1}{M+1} \qquad \text{for all} \quad i = 0, 1, \ldots, M.$$

[Hartigan (1969).]

Problem 5.60 (i) Given n pairs $(x_1, y_1), \ldots, (x_n, y_n)$, let G be the group of 2^n permutations of the 2^n variables which interchange x_i and y_i in all, some, or none of the n pairs. Let G_0 be any subgroup of G, and let e be the number of elements in G_0. Any element $g \in G_0$ (except the identity) is characterized by the numbers $i_1, \ldots, i_r$ ($r \geq 1$) of the pairs in which x_i and y_i have been switched. Let $d_i = y_i - x_i$, and let $\delta_{(1)} < \cdots < \delta_{(e-1)}$, denote the ordered values $(d_{i_1} + \cdots + d_{i_r})/r$ corresponding to G_0. Then (5.71) continues to hold with $e - 1$ in place of M.

(ii) State the generalization of Problem 5.59 to the situation of part (i). [Hartigan (1969).]

Problem 5.61 The preceding problem establishes a $1:1$ correspondence between $e - 1$ permutations T of G_0 which are not the identity and $e - 1$ nonempty subsets $\{i_1, \ldots, i_r\}$ of the set $\{1, \ldots, n\}$. If the permutations T and T' correspond respectively to the subsets $R = \{i_1, \ldots, i_r\}$ and $R' = \{j_1, \ldots, j_s\}$, then the group product $T'T$ corresponds to the subset $(R \cap \tilde{S}) \cup (\tilde{R} \cap S) = (R \cup S) - (R \cap S)$. [Hartigan (1969).]

Problem 5.62 Determine for each of the following classes of subsets of $\{1, \ldots, n\}$ whether (together with the empty subset) it forms a group under the group operation of the preceding problem: All subsets $\{i_1, \ldots, i_r\}$ with

(i) $r = 2$;

(ii) $r =$ even;

(iii) r divisible by 3.

(iv) Give two other examples of subgroups G_0 of G.
Note. A class of such subgroups is discussed by Forsythe and Hartigan (1970).

Problem 5.63 Generalize Problems 5.60(i) and 5.61 to the case of two groups of sizes m and n ($c = 1$).

Section 5.13

Problem 5.64 (i) If the joint distribution of X and Y is the bivariate normal distribution (5.69), then the conditional distribution of Y given x is the normal distribution with variance $\tau^2(1-\rho^2)$ and mean $\eta + (\rho\tau/\sigma)(x - \xi)$.

(ii) Let $(X_1, Y_1), \ldots, (X_n, Y_n)$ be a sample from a bivariate normal distribution, let R be the sample correlation coefficient, and suppose that $\rho = 0$. Then the conditional distribution of $\sqrt{n-2}R/\sqrt{1-R^2}$ given $x_1, \ldots, x_n$, is Student's t-distribution with $n-2$ degrees of freedom provided $\sum(x_i - \bar{x})^2 > 0$. This is therefore also the unconditional distribution of this statistic.

(iii) The probability density of R itself is then

$$p(r) = \frac{1}{\sqrt{n}} \frac{\Gamma[\frac{1}{2}(n-1)]}{\Gamma[\frac{1}{2}(n-2)]}(1-r^2)^{\frac{1}{2}n-2}. \tag{5.83}$$

[(ii): If $v_i = (x_1 - \bar{x})/\sqrt{\sum(x_j - \bar{x})^2}$ so that $\sum v_i = 0$, $\sum v_1^2 = 1$, the statistic can be written as

$$\frac{\sum v_i Y_i}{\sqrt{\left[\sum Y_i^2 - n\bar{Y}^2 - (\sum v_i Y_i)^2\right]/(n-2)}}.$$

Since its distribution depends only on ρ one can assume $\eta = 0$, $\tau = 1$. The desired result follows from Problem 5.6 by making an orthogonal transformation from $(Y_1, , \ldots, Y_n)$ to $(Z_1, \ldots, Z_n)$ such that $Z_1 = \sqrt{n}\bar{Y}$, $Z_2 = \sum v_i Y_i$.]

Problem 5.65 (i) Let $(X_1, Y_1), \ldots, (X_n, Y_n)$ be a sample from the bivariate normal distribution (5.69), and let $S_1^2 = \sum(X_i - \bar{X})^2$, $S_2^2 = \sum(Y_i - \bar{Y})^2$, $S_{12} = \sum(X_i - \bar{X})(Y_i - \bar{Y})$. There exists a UMP unbiased test for testing the hypothesis $\tau/\sigma = \Delta$. Its acceptance region is

$$\frac{|\Delta^2 S_1^2 - S_2^2|}{\sqrt{(\Delta^2 S_1^2 + S_2^2)^2 - 4\Delta^2 S_{12}^2}} \leq C,$$

and the probability density of the test statistic is given by (5.83) when the hypothesis is true.

(ii) Under the assumption $\tau = \sigma$, there exists a UMP unbiased test for testing $\eta = \xi$, with acceptance region $|\bar{Y} - \bar{X}|/\sqrt{S_1^2 + S_2^2 - 2S_{12}} \leq C$. On multiplication by a suitable constant the test statistic has Student's t-distribution with $n-1$ degrees of freedom when $\eta = \xi$.

[Due to Morgan (1939) and Hsu (1940). (i): The transformation $U = \Delta X + Y$, $V = X - (1/\Delta)Y$ reduces the problem to that of testing that the correlation coefficient in a bivariate normal distribution is zero.
(ii): Transform to new variables $V_i = Y_i - X_i$, $U_i = Y_i + X_i$.]

Problem 5.66 (i) Let $(X_1, Y_1), \ldots, (X_n, Y_n)$ be a sample from the bivariate normal distribution (5.73), and let $S_1^2 = \sum(X_i - \bar{X})^2$, $S_{12} = \sum(X_i - \bar{X})(Y_i - \bar{Y})$, $S_2^2 = \sum(Y_i - \bar{Y})^2$.
Then (S_1^2, S_{12}, S_2^2) are independently distributed of $(\bar{X}, \bar{Y})$, and their joint distribution is the same as that of $(\sum_{i=1}^{n-1} X_i'^2, \sum_{i=1}^{n-1} X_i'Y_i', \sum_{i=1}^{n-1} Y_i'^2)$, where $(X_i', Y_i'), i = 1, \ldots, n-1$, are a sample from the distribution (5.73) with $\xi = \eta = 0$.

(ii) Let $X_1, \ldots, X_m$ and $Y_1, \ldots, Y_m$ be two samples from $N(0,1)$. Then the joint density of $S_1^2 = \sum X_i^2$, $S_{12} = \sum X_iY_i$, $S_2^2 = \sum Y_i^2$ is

$$\frac{1}{4\pi\Gamma(m-1)}(s_1^2 s_2^2 - s_{12}^2)^{\frac{1}{2}(m-3)} \exp\left[-\frac{1}{2}(s_1^2 + s_2^2)\right]$$

for $s_{12}^2 \leq s_1^2 s_2^2$, and zero elsewhere.

(iii) The joint density of the statistics (S_1^2, S_{12}, S_2^2) of part (i) is

$$\frac{(s_1^2 s_2^2 - s_{12}^2)^{\frac{1}{2}(n-4)}}{4\pi\Gamma(n-2)\left(\sigma\tau\sqrt{1-\rho^2}\right)^{n-1}} \exp\left[-\frac{1}{2(1-\rho^2)}\left(\frac{s_1^2}{\sigma^2} - \frac{2\rho s_{12}}{\sigma\tau} + \frac{s_2^2}{\tau^2}\right)\right] \tag{5.84}$$

for $s_{12}^2 \leq s_1^2 s_2^2$ and zero elsewhere.

[(i): Make an orthogonal transformation from $X_1, \ldots, X_n$ to $X_1', \ldots, X_n'$ such that $X_n' = \sqrt{n}\bar{X}$, and apply the same orthogonal transformation also to $Y_1, \ldots, Y_n$. Then

$$\begin{aligned} Y_n' &= \sqrt{n}\bar{Y}, \qquad \sum_{i=1}^{n-1} X_i'Y_i' = \sum_{i=1}^{n}(X_i - \bar{X})(Y_i - \bar{Y}), \\ \sum_{i=1}^{n-1} X_i'^2 &= \sum_{i=1}^{n}(X_i - \bar{X})^2, \qquad \sum_{i=1}^{n-1} Y_i'^2 = \sum_{i=1}^{n}(Y_i - \bar{Y})^2. \end{aligned}$$

The pairs of variables $(X_1', Y_1'), \ldots, (X_n', Y_n')$ are independent, each with a bivariate normal distribution with the same variances and correlation as those of (X, Y) and with means $E(X_i') - E(Y_i') = 0$ for $i = 1, \ldots, n-1$.
(ii): Consider first the joint distribution of $S_{12} = \sum x_iY_i$ and $S_2^2 = \sum Y_i^2$ given $x_1 \ldots, x_m$. Letting $Z_1 = S_{12}/\sqrt{\sum x_i^2}$ and making an orthogonal transformation from $Y_1, \ldots, Y_m$ to $Z_1, \ldots, Z_m$ so that $S_2^2 = \sum_{i=1}^m Z_i^2$, the variables Z_1 and $\sum_{i=2}^m Z_i^2 = S_2^2 - Z_1^2$ are independently distributed as $N(0,1)$ and χ_{m-1}^2 respectively. From this the joint conditional density of $S_{12} = s_1 Z_1$ and S_2^2 is obtained by a simple transformation of variables. Since the conditional distribution depends on the x's only through s_1^2, the joint density of S_1^2, S_{12}, S_2^2 is found by multiplying

the above conditional density by the marginal one of S_1^2, which is χ^2_m. The proof is completed through use of the identity

$$\Gamma\left[\tfrac{1}{2}(m-1)\right]\Gamma\left(\tfrac{1}{2}m\right) = \frac{\sqrt{\pi}\Gamma(m-1)}{2^{m-2}}.$$

(iii): If $(X', Y') = (X_1', Y_1'; \ldots; X_m', Y_m')$ is a sample from a bivariate normal distribution with $\xi = \eta = 0$, then $T = (\sum X_i'^2, \sum X_i'Y_i', \sum Y_i'^2)$ is sufficient for $\theta(\sigma, \rho, \tau)$, and the density of T is obtained from that given in part (ii) for $\theta_0 = (1, 0, 1)$ through the identity [Problem 3.39 (i)]

$$p_\theta^T(t) = p_{\theta_0}^T(t)\frac{p_\theta^{X',Y'}(x',y')}{p_{\theta_0}^{X',Y'}(x',y')}.$$

The result now follows from part (i) with $m = n - 1$.]

Problem 5.67 If $(X_1, Y_1), \ldots, (X_n, Y_n)$ is a sample from a bivariate normal distribution, the probability density of the sample correlation coefficient R is[15]

$$\begin{aligned} p_\rho(r) &= \frac{2^{n-3}}{\pi(n-3)!}(1-\rho^2)^{\frac{1}{2}(n-1)}(1-r^2)^{\frac{1}{2}(n-4)} \\ &\quad \times \sum_{k=0}^{\infty} \Gamma^2\left[\tfrac{1}{2}(n+k-1)\right]\frac{(2\rho r)^k}{k!} \end{aligned} \tag{5.85}$$

or alternatively

$$\begin{aligned} p_\rho(r) &= \frac{n-2}{\pi}(1-\rho^2)^{\frac{1}{2}(n-1)}(1-r^2)^{\frac{1}{2}(n-4)} \\ &\quad \times \int_0^1 \frac{t^{n-2}}{(1-\rho r t)^{n-1}}\frac{1}{\sqrt{1-t^2}}\,dt. \end{aligned} \tag{5.86}$$

Another form is obtained by making the transformation $t = (1-v)/(1-\rho r v)$ in the integral on the right-hand side of (5.86). The integral then becomes

$$\frac{1}{(1-\rho r)^{\frac{1}{2}(2n-3)}}\int_0^1 \frac{(1-v)^{n-2}}{\sqrt{2v}}\left[1 - \tfrac{1}{2}v(1+\rho r)\right]^{-1/2} dv. \tag{5.87}$$

Expanding the last factor in powers of v, the density becomes

$$\begin{aligned} &\frac{n-2}{\sqrt{2\pi}}\frac{\Gamma(n-1)}{\Gamma(n-\frac{1}{2})}(1-\rho^2)^{\frac{1}{2}(n-1)}(1-r^2)^{\frac{1}{2}(n-4)}(1-\rho r)^{-n+\frac{3}{2}} \\ &\quad \times F\left(\tfrac{1}{2}; \tfrac{1}{2}; n-\tfrac{1}{2}; \frac{1+\rho r}{2}\right), \end{aligned} \tag{5.88}$$

where

$$F(a,b,c,x) = \sum_{j=0}^{\infty} \frac{\Gamma(a+j)}{\Gamma(a)}\frac{\Gamma(b+j)}{\Gamma(b)}\frac{\Gamma(c)}{\Gamma(c+j)}\frac{x^j}{j!} \tag{5.89}$$

is a hypergeometric function.

[15] The distribution of R is reviewed by Johnson and Kotz (1970, Vol. 2, Section 32) and Patel and Read (1982).

[To obtain the first expression make a transformation from (S_1^2, S_2^2, S_{12}) with density (5.84) to (S_1^2, S_2^2, R) and expand the factor $\exp\{\rho s_{12}/(1-\rho^2)\sigma\tau\} = \exp\{\rho r s_1 s_2/(1-\rho^2)\sigma\tau\}$ into a power series. The resulting series can be integrated term by term with respect to s_1^2 and s_2^2. The equivalence with the second expression is seen by expanding the factor $(1-\rho r t)^{-(n-1)}$ under the integral in (5.86) and integrating term by term.]

Problem 5.68 If X and Y have a bivariate normal distribution with correlation coefficient $\rho > 0$, they are positively regression-dependent.
[The conditional distribution of Y given x is normal with mean $\eta + \rho\tau\sigma^{-1}(x-\xi)$ and variance $\tau^2(l-\rho^2)$. Through addition to such a variable of the positive quantity $\rho\tau\sigma^{-1}(x'-x)$ it is transformed into one with the conditional distribution of Y given $x' > x$.]

Problem 5.69 (i) The functions (5.78) are bivariate cumulative distributions functions.

(ii) A pair of random variables with distribution (5.78) is positively regression-dependent. [The distributions (5.78) were introduced by Morgenstem (1956).]

Problem 5.70 If X, Y are positively regression dependent, they are positively quadrant dependent.
[Positive regression dependence implies that

$$P[Y \le y \mid X \le x] \ge P[Y \le y \mid X \le x'] \quad \text{for all} \quad x < x' \text{ and } y, \tag{5.90}$$

and (5.90) implies positive quadrant dependence.]

5.15 Notes

The optimal properties of the one- and two-sample normal-theory tests were obtained by Neyman and Pearson (1933) as some of the principal applications of their general theory. Theorem 5.1.2 is due to Basu (1955), and its uses are reviewed in Boos and Hughes-Oliver (1998). For converse aspects of this theorem see Basu (1958), Koehn and Thomas (1975), Bahadur (1979), Lehmann (1980) and Basu (1982). An interesting application is discussed in Boos and Hughes-Oliver (1998). In some exponential family regression models where UMPU tests do not exist, classes of admissible, unbiased tests are obtained in Cohen, Kemperman and Sackrowitz (1994).

The roots of the randomization model of Section 5.10 can be traced to Neyman (1923); see Speed (1990) and Fienberg and Tanur (1996). Permutation tests, as alternatives to the standard tests having fixed critical levels, were initiated by Fisher (1935a) and further developed, among others, by Pitman (1937, 1938a), Lehmann and Stein (1949), Hoeffding (1952), and Box and Andersen (1955). Some aspects of these tests are reviewed in Bell and Sen (1984) and Good (1994). Applications to various experimental designs are given in Welch (1990). Optimality of permutation tests in a multivariate nonparametric two-sample setting are studied in Runger and Eaton (1992). Explicit confidence intervals based on subsampling were given by Hartigan (1969). The theory of unbiased confidence sets and its relation to that of unbiased tests is due to Neyman (1937a).

6
Invariance

6.1 Symmetry and Invariance

Many statistical problems exhibit symmetries, which provide natural restrictions to impose on the statistical procedures that are to be employed. Suppose, for example, that $X_1, \ldots, X_n$ are independently distributed with probability densities $p_{\theta_1}(x_1), \ldots, p_{\theta_n}(x_n)$. For testing the hypothesis $H : \theta_1 = \cdots = \theta_n$ against the alternative that the θ's are not all equal, the test should be symmetric in $x_1, \ldots, x_n$, since otherwise the acceptance or rejection of the hypothesis would depend on the (presumably quite irrelevant) numbering of these variables.

As another example consider a circular target with center O, on which are marked the impacts of a number of shots. Suppose that the points of impact are independent observations on a bivariate normal distribution centered on O. In testing this distribution for circular symmetry with respect to O, it seems reasonable to require that the test itself exhibit such symmetry. For if it lacks this feature, a two-dimensional (for example, Cartesian) coordinate system is required to describe the test, and acceptance or rejection will depend on the choice of this system, which under the assumptions made is quite arbitrary and has no bearing on the problem.

The mathematical expression of symmetry is invariance under a suitable group of transformations. In the first of the two examples above the group is that of all permutations of the variables $x_1, \ldots, x_n$ since a function of n variables is symmetric if and only if it remains invariant under all permutations of these variables. In the second example, circular symmetry with respect to the center O is equivalent to invariance under all rotations about O.

In general, let X be distributed according to a probability distribution $P_\theta, \theta \in \Omega$, and let g be a transformation of the sample space $\mathcal{X}$. All such transformations considered in connection with invariance will be assumed to be 1 : 1 transformations of $\mathcal{X}$ onto itself. Denote by gX the random variable that takes on the value gx when $X = x$, and suppose that when the distribution of X is $P_\theta, \theta \in \Omega$, the distribution of gX is $P_{\theta'}$ with θ' also in Ω. The element θ' of Ω which is associated with θ in this manner will be denoted by $\bar{g}\theta$, so that

$$P_\theta\{gX \in A\} = P_{\bar{g}\theta}\{X \in A\}. \tag{6.1}$$

Here the subscript θ on the left member indicates the distribution of X, not that of gX. Equation (6.1) can also be written as $P_\theta(g^{-1}A) = P_{\bar{g}\theta}(A)$ and hence as

$$P_{\bar{g}\theta}(gA) = P_\theta(A). \tag{6.2}$$

The parameter set Ω remains invariant under g (or is preserved by g) if $\bar{g}\theta \in \Omega$ for all $\theta \in \Omega$, and if in addition for any $\theta' \in \Omega$ there exists $\theta \in \Omega$ such that $\bar{g}\theta = \theta'$. These two conditions can be expressed by the equation

$$\bar{g}\Omega = \Omega. \tag{6.3}$$

The transformation $\bar{g}$ of Ω onto itself defined in this way is 1 : 1 provided the distributions P_θ corresponding to different values of θ are distinct. To see this let $\bar{g}\theta_1 = \bar{g}\theta_2$. Then $P_{\bar{g}\theta_1}(gA) = P_{\bar{g}\theta_2}(gA)$ and therefore $P_{\theta_1}(A) = P_{\theta_2}(A)$ for all A, so that $\theta_1 = \theta_2$.

Lemma 6.1.1 *Let g, g' be two transformations preserving Ω. Then the transformations $g'g$ and g^{-1} defined by*

$$(g'g)x = g'(gx) \quad \textit{and} \quad g(g^{-1}x) = x \quad \textit{for all} \quad x \in \mathcal{X}$$

also preserve Ω and satisfy

$$\overline{g'g} = \overline{g'} \cdot \bar{g} \qquad \textit{and} \qquad (\overline{g^{-1}}) = (\bar{g})^{-1}. \tag{6.4}$$

Proof. If the distribution of X is P_θ then that of gX is $P_{\bar{g}\theta}$ and that of $g'gX = g'(gX)$ is therefore $P_{\bar{g}'\bar{g}\theta}$. This establishes the first equation of (6.4); the proof of the second one is analogous. ■

We shall say that *the problem of testing $H : \theta \in \Omega_H$* against $K : \theta \in \Omega_K$ *remains invariant* under a transformation g if $\bar{g}$ preserves both Ω_H and Ω_K, so that the equation

$$\bar{g}\Omega_H = \Omega_H \tag{6.5}$$

holds in addition to (6.3). Let $\mathcal{C}$ be a class of transformations satisfying these two conditions, and let G be the smallest class of transformations containing $\mathcal{C}$ such that $g, g' \in G$ implies that $g'g$ and g^{-1} belong to G. Then G is a group of transformations, all of which by Lemma 6.1.1 preserve both Ω and Ω_H. Any class $\mathcal{C}$ of transformations leaving the problem invariant can therefore be extended to a group G. It follows further from Lemma 6.1.1 that the class of induced transformations $\bar{g}$ form a group $\bar{G}$. The two equations (6.4) express the fact that $\bar{G}$ is a homomorphism of G.

In the presence of symmetries in both sample and parameter space represented by the groups G and $\bar{G}$, it is natural to restrict attention to tests ϕ which are also symmetric, that is, which satisfy

$$\phi(gx) = \phi(x) \qquad \text{for all} \quad x \in X \quad \text{and } g \in G. \tag{6.6}$$

A test ϕ satisfying (6.6) is said to be *invariant under* G. The restriction to invariant tests is a particular case of the principle of invariance formulated in Section 1.5. As was indicated there and in the examples above, a transformation g can be interpreted as a change of coordinates. From this point of view, a test is invariant if it is independent of the particular coordinate system in which the data are expressed.[1]

A transformation g, in order to leave a problem invariant, must in particular preserve the class $\mathcal{A}$ of measurable sets over which the distributions P_θ are defined. This means that any set $A \in \mathcal{A}$ is transformed into a set of $\mathcal{A}$ and is the image of such a set, so that gA and $g^{-1}A$ both belong to $\mathcal{A}$. Any transformation satisfying this condition is said to be *bimeasurable*. Since a group with each element g also contains g^{-1} its elements are automatically bimeasurable if all of them are measurable. If g' and g are bimeasurable, so are $g'g$ and g^{-1}. The transformations of the group G above generated by a class $\mathcal{C}$ are therefore all bimeasurable provided this is the case for the transformations of $\mathcal{C}$.

6.2 Maximal Invariants

If a problem is invariant under a group of transformations, the *principle of invariance* restricts attention to invariant tests. In order to obtain the best of these, it is convenient first to characterize the totality of invariant tests.

Let two points x_1, x_2 be considered equivalent under G,

$$x_1 \sim x_2 (\bmod G),$$

if there exists a transformation $g \in G$ for which $x_2 = gx_1$. This is a true equivalence relation, since G is a group and the sets of equivalent points, the *orbits* of G, therefore constitute a partition of the sample space. (Cf. Appendix, Section A.1.) A point x traces out an orbit as all transformations g of G are applied to it; this means that the orbit containing x consists of the totality of points gx with $g \in G$. It follows from the definition of invariance that a function is invariant if and only if it is constant on each orbit.

A function M is said to be *maximal invariant* if it is invariant and if

$$M(x_1) = M(x_2) \quad \text{implies} \quad x_2 = gx_1 \quad \text{for some } g \in G, \tag{6.7}$$

that is, if it is constant on the orbits but for each orbit takes on a different value. All maximal invariants are equivalent in the sense that their sets of constancy coincide.

Theorem 6.2.1 *Let $M(x)$ be a maximal invariant with respect to G. Then, a necessary and sufficient condition for ϕ to be invariant is that it depends on x only through $M(x)$; that is, that there exists a function h for which $\phi(x) = h[M(x)]$ for all x.*

---

[1]The relationship between this concept of invariance under reparametrization and that considered in differential geometry is discussed in Barndorff-Nielson, Cox and Reid (1986).

PROOF. If $\phi(x) = h[M(x)]$ for all x, then $\phi(gx) = h[M(gx)] = h[M(x)] = \phi(x)$ so that ϕ is invariant. On the other hand, if ϕ is invariant and if $M(x_1) = M(x_2)$, then $x_2 = gx_1$ for some g and therefore $\phi(x_2) = \phi(x_1)$. ■

Example 6.2.1 (i) Let $x = (x_1, \ldots, x_n)$, and let G be the group of translations

$$gx = (x_1 + c, \ldots, x_n + c), \qquad -\infty < c < \infty.$$

Then the set of differences $y = (x_1 - x_n, \ldots, x_{n-1} - x_n)$ is invariant under G. To see that it is maximal invariant suppose that $x_i - x_n = x_i' - x_n'$ for $i = 1, \ldots, n-1$. Putting $x_n' - x_n = c$, one has $x_i' = x_i + c$ for all i, as was to be shown. The function y is of course only one representation of the maximal invariant. Others are for example $(x_1 - x_2, x_2 - x_3, \ldots, x_{n-1} - x_n)$ or the redundant $(x_1 - \bar{x}, \ldots, x_n - \bar{x})$. In the particular case that $n = 1$, there are no invariants. The whole space is a single orbit, so that for any two points there exists a transformation of G taking one into the other. In such a case the transformation group G is said to be *transitive*. The only invariant functions are then the constant functions $\phi(x) \equiv c$.

(ii) if G is the group of transformations

$$gx = (cx_1, \ldots, cx_n), \qquad c \neq 0,$$

a special role is played by any zero coordinates. However, in statistical applications the set of points for which none of the coordinates is zero typically has probability 1; attention can then be restricted to this part of the sample space, and the set of ratios $x_1/x_n, \ldots, x_{n-1}/x_n$ is a maximal invariant. Without this restriction, two points x, x' are equivalent with respect to the maximal invariant partition if among their coordinates there are the same number of zeros (if any), if these occur at the same places, and if for any two nonzero coordinates x_i, x_j the ratios x_j/x_i and x_j'/x_i' are equal.

(iii) Let $x = (x_1, \ldots, x_n)$, and let G be the group of all orthogonal transformations $x' = \Gamma_x$ of n-space. Then $\sum x_i^2$ is maximal invariant, that is, two points x and x^* can be transformed into each other by an orthogonal transformation if and only if they have the same distance from the origin. The proof of this is immediate if one restricts attention to the plane containing the points x, x^* and the origin. ■

Example 6.2.2 (i) Let $x = (x_1, \ldots, x_n)$, and let G be the set of $n!$ permutations of the coordinates of x. Then the set of ordered coordinates (*order statistics*) $x_{(1)} \leq \cdots \leq x_{(n)}$ is maximal invariant. A permutation of the x_i obviously does not change the set of values of the coordinates and therefore not the $x_{(i)}$. On the other hand, two points with the same set of ordered coordinates can be obtained from each other through a permutation of coordinates.

(ii) Let G be the totality of transformations $x_i' = f(x_i), i = 1, \ldots, n$, such that f is continuous and strictly increasing, and suppose that attention can be restricted to the points that have n distinct coordinates. If the x_i are considered as n points on the real line, any such transformation preserves their order. Conversely, if $x_1, \ldots, x_n$ and $x_1', \ldots, x_n'$ are two sets of points in the same order, say $x_{i_1} < \cdots < x_{i_n}$ and $x_{i_1}' < \cdots < x_{i_n}'$, there exists a transformation f satisfying the required conditions and such that $x_i' = f(x_i)$ for all i. It can be defined for example as $f(x) = x + (x_{i_1}' - x_{i_1})$ for $x \leq x_{i_1}$, $f(x) = x + (x_{i_n}' - x_{i_n})$ for $x \geq x_{i_n}$, and to be linear between x_{i_k} and $x_{i_{k+1}}$ for $k = 1, \ldots, n-1$. A formal expression for the maximal invariant in this case is the set of *ranks* $(r_1, \ldots, r_n)$ of $(x_1, \ldots, x_n)$. Here the rank r_i of x_i is defined through

$$X_i = X_{(r_i)}$$

so that r_i is the number of x's $\leq x_i$. In particular, $r_i = 1$ if x_i is the smallest $x, r_i = 2$ if it is the second smallest, and so on. ■

Example 6.2.3 Let x be an $n \times s$ matrix $(s \leq n)$ of rank s, and let G be the group of linear transformations $gx = xB$, where B is any nonsingular $s \times s$ matrix. Then a maximal invariant under G is the matrix $t(x) = x(x^T x)^{-1} x^T$, where x^T denotes the transpose of x. Here $(x^T x)^{-1}$ is meaningful because the $s \times s$ matrix $x^T x$ is nonsingular; see Problem 6.3. That $t(x)$ is invariant is clear, since

$$t(gx) = xB(B^T x^T xB)^{-1} B^T x^T = x(x^T x)^{-1} x^T = t(x).$$

To see that $t(x)$ is maximal invariant, suppose that

$$x_1(x_1^T x_1)^{-1} x_1^T = x_2(x_2^T x_2)^{-1} x_2.$$

Since $(x_i^T x_i)^{-1}$ is positive definite, there exist nonsingular matrices C_i such that $(x_i^T x_i)^{-1} = C_i C_i^T$ and hence

$$(x_1 C_1)(x_1 C_1)^T = (x_2 C_2)(x_2 C_2)^T.$$

This implies the existence of an orthogonal matrix Q such that $x_2 C_2 = x_1 C_1 Q$ and thus $x_2 = x_1 B$ with $B = C_1 Q C_2^{-1}$, as was to be shown.

In the special case $s = n$, we have $t(x) = I$, so that there are no nontrivial invariants. This corresponds to the fact that in this case G is transitive, since any two nonsingular $n \times n$ matrices x_1 and x_2 satisfy $x_2 = x_1 B$ with $B = x_1^{-1} x_2$. This result can be made more intuitive through a geometric interpretation. Consider the s-dimensional subspace S of R^n spanned by the s columns of x. Then $P = x(x^T x)^{-1} x^T$ has the property that for any y in R^n, the vector Py is the projection of y onto S. (This will be proved in Section 7.2.) The invariance of P expresses the fact that the projection of y onto S is independent of the choice of vectors spanning S. To see that it is maximal invariant, suppose that the projection of every y onto the spaces S_1 and S_2 spanned by two different sets of s vectors is the same. Then $S_1 = S_2$, so that the two sets of vectors span the same space. There then exists a nonsingular transformation taking one of these sets into the other. ■

A somewhat more systematic way of determining maximal invariants is obtained by selecting, by means of a specified rule, a unique point $M(x)$ on each orbit. Then clearly $M(X)$ is maximal invariant. To illustrate this method, consider once more two of the earlier examples.

Example 6.2.1(i) (continued). The orbit containing the point $(a_1, \ldots, a_n)$ under the group of translations is the set $(a_1 + c, \ldots, a_n + c), -\infty < c < \infty\}$, which is a line in E_n.

(a) As representative point $M(x)$ on this line, take its intersection with the hyperplane $x_n = 0$. Since then $a_n + c = 0$, this point corresponds to the value $c = -a_n$ and thus has coordinates $(a_1 - a_n, \ldots, a_{n-1} - a_n, 0)$. This leads to the maximal invariant $(x_1 - x_n, \ldots, x_{n-1} - x_n)$.

(b) An alternative point on the line is its intersection with the hyperplane $\sum x_i = 0$. Then $c = -\bar{a}$, and $M(a) = (a_1 - \bar{a}, \dots, a_n - \bar{a})$.

(c) The point need not be specified by an intersection property. It can for instance be taken as the point on the line that is closest to the origin. Since the value of c minimizing $\sum(a_i + c)^2$ is $c = -\bar{a}$, this leads to the same point as (b). ■

Example 6.2.1(iii) (continued). The orbit containing the point $(a_1, \dots, a_n)$ under the group of orthogonal transformations is the hypersphere containing $(a_1, \dots, a_n)$ and with center at the origin. As representative point on this sphere, take its north pole, i.e. the point with $a_1 = \cdots = a_{n-1} = 0$. The coordinates of this point are $(0, \dots, 0, \sqrt{\sum a_i^2})$ and hence lead to the maximal invariant $\sum x_i^2$. (Note that in this example, the determination of the orbit is essentially equivalent to the determination of the maximal invariant.) ■

Frequently, it is convenient to obtain a maximal invariant in a number of steps, each corresponding to a subgroup of G. To illustrate the process and a difficulty that may arise in its application, let $x = (x_1, \dots, x_n)$, suppose that the coordinates are distinct, and consider the group of transformations

$$gx = (ax_1 + b, \dots, ax_n + b), \qquad a \neq 0, \qquad -\infty < b < \infty.$$

Applying first the subgroup of translations $x_i' = xi + b$, a maximal invariant is $y = (y_1, \dots, y_{n-1})$ with $y_i = x_i - x_n$. Another subgroup consists of the scale changes $x_i'' = ax_i$. This induces a corresponding change of scale in the y's: $y_i'' = ay_i$, and a maximal invariant with respect to this group acting on the y-space is $z = (z_1, \dots, z_{n-2})$ with $z_i = y_i/y_{n-1}$. Expressing this in terms of the x's, we get $z_i = (x_i - x_n)/(x_{n-1} - x_n)$, which is maximal invariant with respect to G.

Suppose now the process is carried out in the reverse order. Application first of the subgroup $x_i'' = ax_i$ yields as maximal invariant $u = (u_1, \dots, u_{n-1})$ with $u_i = x_i/x_n$. However, the translations $x_i' = x_i + b$ do not induce transformations in u-space, since $(x_i + b)/(x_n + b)$ is not a function of x_i/x_n.

Quite generally, let a transformation group G be *generated* by two subgroups D and E in the sense that it is the smallest group containing D and E. Then G consists of the totality of products $e_m d_m \dots e_1 d_1$ for $m = 1, 2, \dots,$ with $d_i \in D$, $e_i \in E$ $(i = 1, \dots, m)$.[2] The following theorem shows that whenever the process of determining a maximal invariant in steps can be carried out at all, it leads to a maximal invariant with respect to G.

Theorem 6.2.2 *Let G be a group of transformations, and let D and E be two subgroups generating G. Suppose that $y = s(x)$ is maximal invariant with respect to D, and that for any $e \in E$*

$$s(x_i) = s(x_2) \qquad \textit{implies} \qquad s(ex_1) = s(ex_2). \tag{6.8}$$

If $z = t(y)$ is maximal invariant under the group E^* of transformations e^* defined by

$$e^* y = s(ex) \qquad \textit{when} \qquad y = s(x),$$

[2]See Section A.1 of the Appendix.

then $z = t[s(x)]$ is maximal invariant with respect to G.

PROOF. To show that $t[s(x)]$ is invariant, let $x' = gx, g = e_m d_m \cdots e_1 d_1$. Then

$$t[s(x')] = t[s(e_m d_m \cdots e_1 d_1 x)] = t[e_m^* s(d_m \cdots e_1 d_1 x)] = t[s(e_{m-1} d_{m-1} \cdots e_1 d_1 x)],$$

and the last expression can be reduced by induction to $t[s(x)]$. To see that $t[s(x)]$ is in fact maximal invariant, suppose that $t[s(x')] = t[s(x)]$. Setting $y' = s(x')$, $y = s(x)$, one has $t(y') = t(y)$, and since $t(y)$ is maximal invariant with respect to E^*, there exists e^* such that $y' = e^* y$. Then $s(x') = e^* s(x) = s(ex)$, and by the maximal invariance of $s(x)$ with respect to D there exists $d \in D$ such that $x' = dex$. Since de is an element of G this completes the proof. ■

Techniques for obtaining the distribution of maximal invariants are discussed by Andersson (1982), Eaton (1983, 1989), Farrell (1985), Wijsman (1990) and Anderson (2003).

6.3 Most Powerful Invariant Tests

In the presence of symmetries, one may wish to restrict attention to invariant tests, and it then becomes of interest to determine the most powerful invariant test. The following is a simple example.

Example 6.3.1 Let $X_1, \ldots, X_n$ be i.i.d. on $(0, 1)$ and consider testing the hypothesis H_0 that the the common distribution of the X's is uniform on $(0, 1)$ against the two alternatives H_1:

$$p_1(x_1, \ldots, x_n) = f(x_1) \cdots f(x_n)$$

and

$$p_2(x_1, \ldots, x_n) = f(1 - x_1) \cdots f(1 - x_n) \ ,$$

where f is a fixed (known) density.
(i) This problem remains invariant under the 2 element group G consisting of the transformations

$$g: \ x_i' = 1 - x_i \ , \qquad i = 1, \ldots, n$$

and the identity transformation $x_i' = x_i$ for $i = 1, \ldots, n$.
(ii) The induced transformation $\bar{g}$ is the space of alternatives takes p_1 into p_2 and p_2 into p_1.
(iii) A test $\phi(x_1, \ldots, x_n)$ remains invariant under G if and only if

$$\phi(x_1, \ldots, x_n) = \phi(1 - x_1, \ldots, 1 - x_n) \ .$$

(iv) There exists a UMP invariant test (i.e. an invariant test which is simultaneously most powerful against both p_1 and p_2), and it rejects H_0 when the average

$$\bar{p}(x_1, \ldots, x_n) = \frac{1}{2} [p_1(x_1, \ldots, x_n) + p_2(x_1, \ldots, x_n)]$$

is sufficiently large.

We leave the proof of (i)-(iii) to Problem 6.5. To prove (iv), note that any invariant test satisfies

$$E_{p_1}[\phi(X_1,\ldots,X_n)] = E_{p_2}[\phi(X_1,\ldots,X_n)] = E_{\bar{p}}[\phi(X_1,\ldots,X_n)] \ .$$

Therefore, maximizing the power against p_1 or p_2 is equivalent to maximizing the power under $\bar{p}$, and the result follows from the Neyman-Pearson Lemma. ■

This example is a special case of the following result.

Theorem 6.3.1 *Suppose the problem of testing Ω_0 against Ω_1 remains invariant under a finite group $G = \{g_1,\ldots,g_N\}$ and that $\bar{G}$ is transitive over Ω_0 and over Ω_1. Then there exists a UMP invariant test of Ω_0 against Ω_1, and it rejects Ω_0 when*

$$\frac{\sum_{i=1}^N p_{\bar{g}_i\theta_1}(x)/N}{\sum_{i=1}^N p_{\bar{g}_i\theta_0}(x)/N} \tag{6.9}$$

is sufficiently large, where θ_0 and θ_1 are any elements of Ω_0 and Ω_1, respectively.

The proof is exactly analogous to that of the preceding example; see Problem 6.6.

The results of the previous section provide an alternative approach to the determination of most powerful invariant tests. By Theorem 6.2.1, the class of all invariant functions can be obtained as the totality of functions of a maximal invariant $M(x)$. Therefore, in particular the class of all invariant tests is the totality of tests depending only on the maximal invariant statistic M. The latter statement, while correct for all the usual situations, actually requires certain qualifications regarding the class of measurable sets in M-space. These conditions will be discussed at the end of the section; they are satisfied in the examples below.

Example 6.3.2 Let $X = (X_1,\ldots,X_n)$, and suppose that the density of X is $f_i(x_1-\theta,\ldots,x_n-\theta)$ under H_i ($i = 0,1$), where θ ranges from $-\infty$ to ∞. The problem of testing H_0 against H_1 is invariant under the group G of transformations

$$gx = (x_1+c,\ldots,x_n+c), \qquad -\infty < c < \infty.$$

which in the parameter space induces the transformations

$$\bar{g}\theta = \theta + c.$$

By Example 6.2.1, a maximal invariant under G is $Y = (X_1 - X_n,\ldots,X_{n-1}-X_n)$. The distribution of Y is independent of θ and under H_i has the density

$$\int_{-\infty}^{\infty} f_i(y_1+z,\ldots,y_{n-1}+z,z)\,dz.$$

When referred to Y, the problem of testing H_0 against H_1 therefore becomes one of testing a simple hypothesis against a simple alternative. The most powerful test is then independent of θ, and therefore UMP among all invariant tests. Its rejection region by the Neyman–Pearson lemma is

$$\frac{\int_{\infty}^{\infty} f_1(y_1+z,\ldots,y_{n-1}+z,z)\,dz}{\int_{\infty}^{\infty} f_0(y_1+z,\ldots,y_{n-1}+z,z)\,dz} = \frac{\int_{-\infty}^{\infty} f_1(x_1+u,\ldots,x_n+u)\,du}{\int_{-\infty}^{\infty} f_0(x_1+u,\ldots,x_n+u)\,du} > C. \tag{6.10}$$

A general theory of *separate families of hypotheses* (in which the family K of alternatives does not adjoin the hypothesis H but, as above, is separated from it) was initiated by Cox (1961, 1962). A bibliography of the subject is given in Pereira (1977); see also Loh (1985), Pace and Salvan (1990) and Rukhin (1993). ■

Example 6.3.2 illustrates the fact, also utilized in Theorem 6.3.1, that if the group $\bar{G}$ is transitive over both Ω_0 and Ω_1, then the problem reduces to one of testing a simple hypothesis against a simple alternative, and a UMP invariant test is then obtained by the Neyman-Pearson Lemma. Note also the close similarity between Theorem 6.3.1 and Example 6.3.2 shown by a comparison of (6.9) and the right side of (6.10), where the summation in (6.9) is replaced by integration with respect to Lebesgue measure.

Before applying invariance, it is frequently convenient first to reduce the data to a sufficient statistic T. If there exists a test $\phi_0(T)$ that is UMP among all invariant tests depending only on T, one would like to be able to conclude that $\phi_0(T)$ is also UMP among all invariant tests based on the original X. Unfortunately, this does not follow, since it is not clear that for any invariant test based on X there exists an equivalent test based on T, which is also invariant. Sufficient conditions for $\phi_0(T)$ to have this property are provided by Hall, Wijsman, and Ghosh (1965) and Hooper (1982a), and a simple version of such a result (applicable to Examples 6.3.3 and 6.3.4 below) will be given by Theorem 6.5.3 in Section 6.5. For a review and clarification of this and later work on invariance and sufficiency see Berk, Nogales, and Oyola (1996), Nogales and Oyola (1996) and Nogales, Oyola and Pérez (2000).

Example 6.3.3 If $X_1, \dots, X_n$ is a sample from $N(\xi, \sigma^2)$, the hypothesis $H : \sigma \geq \sigma_0$ remains invariant under the transformations $X_i' = X_i + c$, $-\infty < c < \infty$. In terms of the sufficient statistics $Y = \bar{X}, S^2 = \Sigma(X_i - \bar{X})^2$ these transformations become $Y' = Y + c, (S^2)' = S^2$, and a maximal invariant is S^2. The class of invariant tests is therefore the class of tests depending on S^2. It follows from Theorem 3.4.1 that there exists a UMP invariant test, with rejection region $\Sigma(X_i - \bar{X})^2 \leq C$. This coincides with the UMP unbiased test (6.11). ■

Example 6.3.4 If $X_1, \dots, X_m$ and $Y_1, \dots, Y_n$ are samples from $N(\xi, \sigma^2)$ and $N(\eta, \tau^2)$, a set of sufficient statistics is $T_1 = \bar{X}$, $T_2 = \bar{Y}$, $T_3 = \sqrt{\Sigma(X_i - \bar{X})^2}$, and $T_4 = \sqrt{\Sigma(Y_j - \bar{Y})^2}$. The problem of testing $H : \tau^2/\sigma^2 \leq \Delta_0$ remains invariant under the transformations $T_1' = T_1 + c_1$, $T_2' = T_2 + c_2$, $T_3' = T_3$, $T_4' = T_4$, $-\infty < c_1, c_2 < \infty$, and also under a common change of scale of all four variables. A maximal invariant with respect to the first group is (T_3, T_4). In the space of this maximal invariant, the group of scale changes induces the transformations $T_3'' = cT_3$, $T_4'' = cT_4$, $0 < c$, which has as maximal invariant the ratio T_4/T_3. The statistic $Z = [T_4^2/(n-1)] \div [T_3^2/(m-1)]$ on division by $\Delta = \tau^2/\sigma^2$ has an F-distribution with density given by (5.21), so that the density of Z is

$$\frac{c(\Delta) z^{\frac{1}{2}(n-3)}}{\left(\Delta + \dfrac{n-1}{m-1} z\right)^{\frac{1}{2}(m+n-2)}}, \qquad z > 0.$$

For varying Δ, these densities constitute a family with monotone likelihood ratio, so that among all tests of H based on Z, and therefore among all invariant tests, there exists a UMP one given by the rejection region $Z > C$. This coincides with the UMP unbiased test (5.20). ■

Example 6.3.5 In the method of *paired comparisons* for testing whether a treatment has a beneficial effect, the experimental material consists of n pairs of subjects. From each pair, a subject is selected at random for treatment while the other serves as control. Let X_i be 1 or 0 as for the ith pair the experiment turns out in favor of the treated subject or the control, and let $p_i = P\{X_i = 1\}$. The hypothesis of no effect, $H : p_i = \frac{1}{2}$ for $i = 1, \ldots, n$, is to be tested against the alternatives that $p_i > \frac{1}{2}$ for all i.

The problem remains invariant under all permutations of the n variables $X_1, \ldots, X_n$, and a maximal invariant under this group is the total number of successes $X = X_1 + \cdots + X_n$. The distribution of X is

$$P\{X = k\} = q_1 \cdots q_n \sum \frac{p_{i_1}}{q_{i_1}} \cdots \frac{p_{i_k}}{q_{i_k}},$$

where $q_i = 1 - p_i$ and where the summation extends over all $\binom{n}{k}$ choices of subscripts $i_1 < \cdots < i_k$. The most powerful invariant test against an alternative $(p'_1, \ldots, p'_n)$ rejects H when

$$f(k) = \frac{1}{\binom{n}{k}} \sum \frac{p'_{i_1}}{q'_{i_1}} \cdots \frac{p'_{i_k}}{q'_{i_k}} > C.$$

To see that f is an increasing function of k, note that $a_i = p'_i/q'_i > 1$, and that

$$\sum_j \sum a_j a_{i_1} \cdots a_{i_k} = (k+1) \sum a_{i_1} \cdots a_{i_{k+1}}$$

and

$$\sum_j \sum a_{i_1} \cdots a_{i_k} = (n-k) \sum a_{i_1} \cdots a_{i_{k_1}}.$$

Here, in both equations, the second summation on the left-hand side extends over all subscripts $i_1 < \cdots < i_k$ of which none is equal to j, and the summation on the right-hand side extends over all subscripts $i_1 < \cdots < i_{k+1}$ and $i_1 < \cdots < i_k$ respectively without restriction. Then

$$\begin{aligned} f(k+1) = \frac{1}{\binom{n}{k+1}} \sum a_{i_1} \cdots a_{i_{k+1}} &= \frac{1}{(n-k)\binom{n}{k}} \sum_j \sum a_j a_{i_1} \cdots a_{i_k} \\ &> \frac{1}{\binom{n}{k}} \sum a_{i_1} \cdots a_{i_k} = f(k), \end{aligned}$$

as was to be shown. Regardless of the alternative chosen, the test therefore rejects when $k > C$, and hence is UMP invariant. If the ith comparison is considered plus or minus as X_i is 1 or 0, this is seen to be another example of the sign test. (Cf. Example 3.8.1 and Section 4.9.) ■

Sufficient statistics provide a simplification of a problem by reducing the sample space; this process involves no change in the parameter space. Invariance, on the other hand, by reducing the data to a maximal invariant statistic M, whose distribution may depend only on a function of the parameter, typically also shrinks the parameter space. The details are given in the following theorem.

Theorem 6.3.2 *If $M(x)$ is invariant under G, and if $v(\theta)$ maximal invariant under the induced group $\bar{G}$, then the distribution of $M(X)$ depends only on $v(\theta)$.*

PROOF. Let $v(\theta_1) = v(\theta_2)$. Then $\theta_2 = \bar{g}\theta_1$, and hence

$$\begin{aligned} P_{\theta_2}\{M(X) \in B\} &= P_{\bar{g}\theta_1}\{M(X) \in B\} = P_{\theta_1}\{M(gX) \in B\} \\ &= P_{\theta_1}\{M(X) \in B\}. \end{aligned}$$

This result can be paraphrased by saying that the principle of invariance identifies all parameter points that are equivalent with respect to $\bar{G}$. ■

In application, for instance in Examples 6.3.3 and 6.3.4, the maximal invariants $M(x)$ and $\delta = v(\theta)$ under G and $\bar{G}$ are frequently real-valued, and the family of probability densities $p_\delta(m)$ of M has monotone likelihood ratio. For testing the hypothesis $H : \delta \leq \delta_0$ there exists then a UMP test among those depending only on M, and hence a UMP invariant test. Its rejection region is $M \geq C$, where

$$\int_C^\infty P_{\delta_0}(m)\, dm = \alpha. \tag{6.11}$$

Consider this problem now as a two-decision problem with decisions d_0 and d_1 of accepting or rejecting H, and a loss function $L(\theta, d_i) = L_i(\theta)$. Suppose that $L_i(\theta)$ depends only on the parameter δ, $L_i(\theta) = L_i'(\delta)$ say, and satisfies

$$L_1'(\delta) - L_0'(\delta) \gtrless 0 \qquad \text{as} \quad \delta \lesseqgtr \delta_0. \tag{6.12}$$

It then follows from Theorem 3.4.2 that the family of rejection regions $M \geq C(\alpha)$, as α varies from 0 to 1, forms a complete family of decision procedures among those depending only on M, and hence a complete family of invariant procedures. As before, the choice of a particular significance level α can be considered as a convenient way of specifying a test from this family.

At the beginning of the section it was stated that the class of invariant tests coincides with the class of tests based on a maximal invariant statistic $M = M(X)$. However, a statistic is not completely specified by a function, but requires also specification of a class $\mathcal{B}$ of measurable sets. If in the present case $\mathcal{B}$ is the class of all sets B for which $M^{-1}(B) \in \mathcal{A}$, the desired statement is correct. For let $\phi(x) = \psi[M(x)]$ and ϕ by $\mathcal{A}$-measurable, and let C be a Borel set on the line. Then $\phi^{-1}(C) = M^{-1}[\psi^{-1}(C)] \in \mathcal{A}$ and hence $\psi^{-1}(C) \in \mathcal{B}$, so that ψ is $\mathcal{B}$-measurable and $\phi(x) = \psi[M(x)]$ is a test based on the statistic M.

In most applications, $M(x)$ is a measurable function taking on values in a Euclidean space and it is convenient to take $\mathcal{B}$ as the class of Borel sets. If $\phi(x) = \psi[M(x)]$ is then an arbitrary measurable function depending only on $M(x)$, it is not clear that $\psi(m)$ is necessarily $\mathcal{B}$-measurable. This measurability can be concluded if $\mathcal{X}$ is also Euclidean with $\mathcal{A}$ the class of Borel sets, and if the range of M is a Borel set. We shall prove it here only under the additional assumption (which in applications is usually obvious, and which will not be verified explicitly in each case) that there exists a vector-valued Borel-measurable function $Y(x)$ such that $[M(x), Y(x)]$ maps $\mathcal{X}$ onto a Borel subset of the product space $\mathcal{M} \times \mathcal{Y}$, that this mapping is $1:1$, and that the inverse mapping is also Borel-measurable. Given any measurable function ϕ of x, there exists then a measurable function ϕ' of (m, y) such that $\phi(x) \equiv \phi'[M(x), Y(x)]$. If ϕ depends only on $M(x)$, then ϕ' depends only on m, so that $\phi'(m, y) = \psi(m)$ say, and ψ is a measurable function of m.[3] In Example 6.2.1(i) for instance, where $x = (x_1, \ldots x_n)$ and $M(x) = (x_1 - x_n, \ldots, x_{n-1} - x_n)$, the function $Y(x)$ can be taken as $Y(x) = x_n$.

6.4 Sample Inspection by Variables

A sample is drawn from a lot of some manufactured product in order to decide whether the lot is of acceptable quality. In the simplest case, each sample item is classified directly as satisfactory or defective (*inspection by attributes*), and the decision is based on the total number of defectives. More generally, the quality of an item is characterized by a variable Y (*inspection by variables*), and an item is considered satisfactory if Y exceeds a given constant u. The probability of a defective is then

$$p = P\{Y \leq u\}$$

and the problem becomes that of testing the hypothesis $H : p \geq p_0$.

As was seen in Example 3.8.1, no use can be made of the actual value of Y unless something is known concerning the distribution of Y. In the absence of such information, the decision will be based, as before, simply on the number of defectives in the sample. We shall consider the problem now under the assumption that the measurements $Y_1, \ldots, Y_n$ constitute a sample from $N(\eta, \sigma^2)$. Then

$$p = \int_{-\infty}^{u} \frac{1}{\sqrt{2\pi}\sigma} \exp\left[-\frac{1}{2\sigma^2}(y-\eta)^2\right] dy = \Phi\left(\frac{u-\eta}{\sigma}\right),$$

where

$$\Phi(y) = \int_{-\infty}^{y} \frac{1}{\sqrt{2\pi}} \exp\left(-\tfrac{1}{2}t^2\right) dt$$

denotes the cumulative distribution function of a standard normal distribution, and the hypothesis H becomes $(u - \eta)/\sigma \geq \Phi^{-1}(p_0)$. In terms of the variables $X_1 = Y_i - u$, which have mean $\xi = \eta - u$ and variance σ^2, this reduces to

$$H : \frac{\xi}{\sigma} \leq \theta_0$$

with $\theta_0 = -\Phi^{-1}(p_0)$. This hypothesis, which was considered in Section 5.2, for $\theta_0 = 0$, occurs also in other contexts. It is appropriate when one is interested in the mean ξ of a normal distribution, expressed in σ units rather than on a fixed scale.

For testing H, attention can be restricted to the pair of variables $\bar{X}$ and $S = \sqrt{\sum(X_i - \bar{X})^2}$, since they form a set of sufficient statistics for (ξ, σ), which satisfy the conditions of Theorem 6.5.3 of the next section. These variables are independent, the distribution of $\bar{X}$ being $N(\xi, \sigma^2/n)$ and that of S/σ being χ_{n-1}. Multiplication of $\bar{X}$ and S by a common constant $c > 0$ transforms the parameters into $\xi' = c\xi, \sigma' = c\sigma$, so that ξ/σ and hence the problem of testing H remain

[3]The last statement follows, for example, from Theorem 18.1 of Billingsley (1995).

invariant. A maximal invariant under these transformations is $\bar{x}/s$ or

$$t = \frac{\sqrt{n}\bar{x}}{s/\sqrt{n-1}},$$

the distribution of which depends only on the maximal invariant in the parameter space $\theta = \xi/\sigma$ (cf. Section 5.2). Thus, the invariant tests are those depending only on t, and it remains to find the most powerful test of $H : \theta \le \theta_0$ within this class.

The probability density of t is (Problem 5.3)

$$p_\delta(t) = C \int_0^\infty \exp\left[-\frac{1}{2}\left(t\sqrt{\frac{w}{n-1}} - \delta\right)^2\right] w^{\frac{1}{2}(n-2)} \exp\left(-\tfrac{1}{2}w\right) dw,$$

where $\delta = \sqrt{n}\theta$ is the noncentrality parameter, and this will now be shown to constitute a family with monotone likelihood ratio. To see that the ratio

$$r(t) = \frac{\int_0^\infty \exp\left[-\frac{1}{2}\left(t\sqrt{\frac{w}{n-1}} - \delta_1\right)^2\right] w^{\frac{1}{2}(n-2)} \exp(-\frac{1}{2}w)\, dw}{\int_0^\infty \exp\left[-\frac{1}{2}\left(t\sqrt{\frac{w}{n-1}} - \delta_0\right)^2\right] w^{\frac{1}{2}(n-2)} \exp(-\frac{1}{2}w)\, dw}$$

is an increasing function of t for $\delta_0 < \delta_1$, suppose first that $t < 0$ and let $v = -t\sqrt{w/(n-1)}$. The ratio then becomes proportional to

$$\frac{\int_0^\infty f(v) \exp\left[-(\delta_1-\delta_0)v - \frac{(n-1)v^2}{2t^2}\right] dv}{\int_0^\infty f(v) \exp\left[-\frac{(n-1)v^2}{2t^2}\right] dv}$$
$$= \int \exp[-(\delta_1 - \delta_0)v] g_{t^2}(v)\, dv$$

where

$$f(v) = \exp(-\delta_0 v) v^{n-1} \exp(-v^2/2)$$

and

$$g_{t^2}(v) = \frac{f(v) \exp\left[-\frac{(n-1)v^2}{2t^2}\right]}{\int_0^\infty f(z) \exp\left[-\frac{(n-1)z^2}{2t^2}\right] dz}.$$

Since the family of probability densities $g_{t^2}(v)$ is a family with monotone likelihood ratio, the integral of $\exp[-(\delta_1 - \delta_0)v]$ with respect to this density is a decreasing function of t^2 (Problem 3.39), and hence an increasing function of t for $t < 0$. Similarly one finds that $r(t)$ is an increasing function of t for $t > 0$ by making the transformation $v = t\sqrt{w/(n-1)}$. By continuity it is then an increasing function of t for all t.

There exists therefore a UMP invariant test of $H : \xi/\sigma \le \theta_0$, which rejects when $t > C$, where C is determined by (6.11). In terms of the original variables Y_i the rejection region of the UMP invariant test of $H : p \ge p_0$ becomes

$$\frac{\sqrt{n}(\bar{y} - u)}{\sqrt{\sum(y_i - \bar{y})^2/(n-1)}} > C. \tag{6.13}$$

If the problem is considered as a two-decision problem with losses $L_0(p)$ and $L_1(p)$ for accepting or rejecting $p \ge p_0$, which depend only on p and satisfy the condition corresponding to (6.12), the class of tests (6.13) constitutes a complete family of invariant procedures as C varies from $-\infty$ to ∞.

Consider next the comparison of two products on the basis of samples $X_1, \ldots, X_m; Y_1, \ldots, Y_n$ from $N(\xi, \sigma^2)$ and $N(\eta, \sigma^2)$. If

$$p = \Phi\left(\frac{u-\xi}{\sigma}\right), \qquad \pi = \Phi\left(\frac{u-\eta}{\sigma}\right),$$

one wishes to test the hypothesis $p \le \pi$, which is equivalent to

$$H : \eta \le \xi.$$

The statistics $\bar{X}, \bar{Y}$, and $S = \sqrt{\sum(X_i - \bar{X})^2 + \sum(Y_j - \bar{Y})^2}$ are a set of sufficient statistics for ξ, η, σ. The problem remains invariant under the addition of an arbitrary common constant to $\bar{X}$ and $\bar{Y}$, which leaves $\bar{Y} - \bar{X}$ and S as maximal invariants. It is also invariant under multiplication of $\bar{X}, \bar{Y}$, and S, and hence of $\bar{Y} - \bar{X}$ and S, by a common positive constant, which reduces the data to the maximal invariant $(\bar{Y} - \bar{X})/S$. Since

$$t = \frac{(\bar{y} - \bar{x})/\sqrt{\frac{1}{m} + \frac{1}{n}}}{s/\sqrt{m+n-2}}$$

has a noncentral t-distribution with noncentrality parameter $\delta = \sqrt{mn}(\eta - \xi)/\sqrt{m+n}\sigma$, the UMP invariant test of $H : \eta - \xi \le 0$ rejects when $t > C$. This coincides with the UMP unbiased test (5.27). Analogously, the corresponding two-sided test (5.30), with rejection region $|t| \ge C$, is UMP invariant for testing the hypothesis $p = \pi$ against the alternatives $p \ne \pi$ (Problem 6.18).

6.5 Almost Invariance

Let G be a group of transformations leaving a family $\mathcal{P} = \{P_\theta, \theta \in \otimes\}$ of distributions of X invariant. A test ϕ is said to be *equivalent to an invariant test* if there exists an invariant test ϕ such that $\phi(x) = \psi(x)$ for all x except possibly on a $\mathcal{P}$-null set N; ϕ is said to be *almost invariant with respect to G if*

$$\phi(gx) = \phi(x) \qquad \text{for all} \quad x \in \mathcal{X} - N_g, \quad g \in G \tag{6.14}$$

where the exceptional null set N_g is permitted to depend on g. This concept is required for investigating the relationship of invariance to unbiasedness and to certain other desirable properties. In this connection it is important to know whether a UMP invariant test is also UMP among almost invariant tests. This turns out to be the case under assumptions which are made precise in Theorem 6.5.1 below and which are satisfied in all the usual applications.

If ϕ is equivalent to an invariant test, then $\phi(gx) = \phi(x)$ for all $x \notin N \cup g^{-1}N$. Since $P_\theta(g^{-1}N) = P_{\bar{g}\theta}(N) = 0$, it follows that ϕ is then almost invariant. The following theorem gives conditions under which conversely any almost invariant test is equivalent to an invariant one.

Theorem 6.5.1 *Let G be a group of transformations of $\mathcal{X}$, and let $\mathcal{A}$ and $\mathcal{B}$ be σ-fields of subsets of $\mathcal{X}$ and G such that for any set $A \in \mathcal{A}$ the set of pairs (x, g)*

for which $gx \in A$ is measurable $\mathcal{A} \times \mathcal{B}$. Suppose further that there exists a σ-finite measure ν over G such that $\nu(B) = 0$ implies $\nu(Bg) = 0$ for all $g \in G$. Then any measurable function that is almost invariant under G (where "almost" refers to some σ-finite measure μ) is equivalent to an invariant function.

PROOF. Because of the measurability assumptions, the function $\phi(gx)$ considered as a function of the two variables x and g is measurable $\mathcal{A} \times \mathcal{B}$. It follows that $\phi(gx) - \phi(x)$ is measurable $\mathcal{A} \times \mathcal{B}$, and so therefore is the set S of points (x, g) with $\phi(gx) \neq \phi(x)$. If ϕ is almost invariant, any section of S with fixed g is a μ-null set. By Fubini's theorem (Theorem 2.2.4), there exists therefore a μ-null set N such that for all $x \in \mathcal{X} - N$

$$\phi(gx) = \phi(x) \quad \text{a.e. } \nu.$$

Without loss of generality suppose that $\nu(G) = 1$, and let A be the set of points x for which

$$\int \phi(g'x)\, d\nu(g') = \phi(gx) \quad \text{a.e. } \nu.$$

If

$$f(x, g) = \left| \int \phi(g'x)\, d\nu(g') - \phi(gx) \right|$$

then A is the set of points x for which

$$\int f(x, g)\, d\nu(g) = 0.$$

Since this integral is a measurable function of x, it follows that A is measurable. Let

$$\psi(x) = \begin{cases} \int \phi(gx) d\nu(g) & \text{if} \quad x \in A, \\ 0 & \text{if} \quad x \notin A. \end{cases}$$

Then ψ is measurable and $\psi(x) = \phi(x)$ for $x \notin N$, since $\phi(gx) = \phi(x)$ a.e. ν implies that $\int \phi(g'x)\, d\nu(g') = \phi(x)$ and that $x \in A$. To show that ψ is invariant it is enough to prove that the set A is invariant. For any point $x \in A$, the function $\phi(gx)$ is constant except on a null subset N_x of G. Then $\phi(ghx)$ has the same constant value for all $g \notin N_x h^{-1}$, which by assumption is again a ν-null set; and hence $hx \in A$, which completes the proof. ■

Additional results concerning the relation of invariance and almost invariance are given by Berk and Bickel (1968) and Berk (1970). In particular, the basic idea of the following example is due to Berk (1970).

Example 6.5.1 (Counterexample) Let $Z, Y_1, \ldots, Y_n$ be independently distributed as $N(\theta, 1)$, and consider the 1 : 1 transformations $y_i' = y_i (i = 1, \ldots, n)$ and

$z' = z$ except for a finite number of points $a_1, \ldots, a_k$ for which
$a_i' = a_{j_i}$, for some permutation $(j_1, \ldots, j_k)$ of $(1, \ldots, k)$.

If the group G is generated by taking for $(a_1, \ldots, a_k)$, $k = 1, 2, \ldots$, all finite sets and for $(j_1, \ldots, j_k)$ all permutations of $(1, \ldots, k)$, then $(z, y_1, \ldots, y_n)$ is almost invariant It is however not equivalent to an invariant function, since $(y_1, \ldots, y_n)$ is maximal invariant. ■

Corollary 6.5.1 *Suppose that the problem of testing $H : \theta \in \omega$ against $K : \theta \in \Omega - \omega$ remains invariant under G and that the assumptions of Theorem 6.5.1 hold. Then if ϕ_0 is UMP invariant, it is also UMP within the class of almost invariant tests.*

PROOF. If ϕ is almost invariant, it is equivalent to an invariant test ψ by Theorem 6.5.1. The tests ϕ and ψ have the same power function, and hence ϕ_0 is uniformly at least as powerful as ϕ. ■

In applications, $\mathcal{P}$ is usually a dominated family, and μ any σ-finite measure equivalent to $\mathcal{P}$ (which exists by Theorem A.4.2 of the Appendix). If ϕ is almost invariant with respect to $\mathcal{P}$, it is then almost invariant with respect to μ and hence equivalent to an invariant test. Typically, the sample space $\mathcal{X}$ is an n-dimensional Euclidean space, $\mathcal{A}$ is the class of Borel sets, and the elements of G are transformations of the form $y = f(x, \tau)$, where τ ranges over a set of positive measure in an m-dimensional space and f is a Borel-measurable vector-valued function of $m+n$ variables. If $\mathcal{B}$ is taken as the class of Hotel sets in m-space the measurability conditions of the theorem are satisfied.

The requirement that for all $g \in G$ and $B \in \mathcal{B}$

$$\nu(B) = 0 \quad \text{implies} \quad \nu(Bg) = 0 \tag{6.15}$$

is satisfied in particular when

$$\nu(Bg) = \nu(B) \qquad \text{for all} \quad g \in G, \quad B \in \mathcal{B}. \tag{6.16}$$

The existence of such a *right invariant measure* is guaranteed for a large class of groups by the theory of Haar measure. (See, for example, Eaton (1989).) Alternatively, it is usually not difficult to check the condition (6.15) directly.

Example 6.5.2 Let G be the group of all nonsingular linear transformations of n-space. Relative to a fixed coordinate system the elements of G can be represented by nonsingular $n \times n$ matrices $A = (a_{ij}), A' = (a'_{ij}), \ldots$ with the matrix product serving as the group product of two such elements. The σ-field $\mathcal{B}$ can be taken to be the class of Borel sets in the space of the n^2 elements of the matrices, and the measure ν can be taken as Lebesgue measure over $\mathcal{B}$. Consider now a set S of matrices with $\nu(S) = 0$, and the set S^* of matrices $A'A$ with $A' \in S$ and A fixed. If $a = \max |a_{ij}|$, $C' = A'A$, and $C'' = A''A$, the inequalities $|a''_{ij} - a'_{ij}| \leq \epsilon$ for all i, j imply $|c''_{ij} - c'_{ij}| \leq na\epsilon$. Since a set has ν-measure zero if and only if it can be covered by a union of rectangles whose total measure does not exceed any given $\epsilon > 0$, it follows that $\nu(S^*) = 0$, as was to be proved. ■

In the preceding chapters, tests were compared purely in terms of their power functions (possibly weighted according to the seriousness of the losses involved). Since the restriction to invariant tests is a departure from this point of view, it is of interest to consider the implications of applying invariance to the power functions rather than to the tests themselves. Any test that is invariant or almost invariant under a group G has a power function which is invariant under the group $\bar{G}$ induced by G in the parameter space.

To see that the converse is in general not true, let X_1, X_2, X_3 be independently, normally distributed with mean ξ and variance σ^2, and consider the hypothesis

$\sigma \geq \sigma_0$. The test with rejection region

$$|X_2 - X_1| > k \qquad \text{when} \quad \bar{X} < 0,$$
$$|X_3 - X_2| > k \qquad \text{when} \quad \bar{X} \geq 0$$

is not invariant under the group G of transformations $X_i' = X_i + c$, but its power function is invariant under the associated group $\bar{G}$.

The two properties, almost invariance of a test ϕ and invariance of its power function, become equivalent if before the application of invariance considerations the problem is reduced to a sufficient statistic whose distributions constitute a boundedly complete family.

Lemma 6.5.1 *Let the family $\mathcal{P}^T = \{P_\theta^T, \theta \in \Omega\}$ of distributions of T be boundedly complete, and let the problem of testing $H : \theta \in \Omega_H$ remain invariant under a group G of transformations of T. Then a necessary and sufficient condition for the power function of a test $\psi(t)$ to be invariant under the induced group $\bar{G}$ over Ω is that $\psi(t)$ is almost invariant under G.*

PROOF. For all $\theta \in \Omega$ we have $E_{\bar{g}\theta}\psi(T) = E_\theta\psi(gT)$. If ψ is almost invariant, $E_\theta\psi(T) = E_\theta\psi(gT)$ and hence $E_{\bar{g}\theta}\psi(T) = E_\theta\psi(T)$, so that the power function of ψ is invariant. Conversely, if $E_\theta\psi(T) = E_{\bar{g}\theta}\psi(T)$, then $E_\theta\psi(T) = E_\theta\psi(gT)$, and by the bounded completeness of $\mathcal{P}^T$, we have $\psi(gt) = \psi(t)$ a.e. $\mathcal{P}^T$. ■

As a consequence, it is seen that UMP almost invariant tests also possess the following optimum property.

Theorem 6.5.2 *Under the assumptions of Lemma 6.5.1, let $v(\theta)$ be maximal invariant with respect to $\bar{G}$, and suppose that among the tests of H based on the sufficient statistic T there exists a UMP almost invariant one, say $\psi_0(t)$. Then $\psi_0(t)$ is UMP in the class of all tests based on the original observations X, whose power function depends only on $v(\theta)$.*

PROOF. Let $\phi(x)$ be any such test, and let $\psi(t) = E[\phi(X)|t]$. The power function of $\psi(t)$, being identical with that of $\phi(x)$, depends then only on $v(\theta)$, and hence is invariant under $\bar{G}$. It follows from Lemma 6.5.1 that $\psi(t)$ is almost invariant under G, and $\psi_0(t)$ is uniformly at least as powerful as $\psi(t)$ and therefore as $\phi(x)$. ■

Example 6.5.3 For the hypothesis $\tau^2 \leq \sigma^2$ concerning the variances of two normal distributions, the statistics $(\bar{X}, \bar{Y}, S_x^2, S_Y^2)$ constitute a complete set of sufficient statistics. It was shown in Example 6.3.4 that there exists a UMP invariant test with respect to a suitable group G, which has rejection region $S_Y^2/S_X^2 > C_0$. Since in the present case almost invariance of a test with respect to G implies that it is equivalent to an invariant one (Problem 6.21), Theorem 6.5.2 is applicable with $v(\theta) = \Delta = \tau^2/\sigma^2$, and the test is therefore UMP among all tests whose power function depends only on Δ. ■

Theorem 6.5.1 makes it possible to establish a simple condition under which reduction to sufficiency before the application of invariance is legitimate.

Theorem 6.5.3 *Let X be distributed according to P_θ, $\theta \in \Omega$, and let T be sufficient for θ. Suppose G leaves invariant the problem of testing $H : \theta \in \Omega_H$, and that T satisfies*

$$T(x_1) = T(x_2) \quad \textit{implies} \quad T(gx_1) = T(gx_2) \quad \textit{for all} \quad g \in G,$$

so that G induces a group $\tilde{G}$ of transformations of T-space through

$$\tilde{g}T(x) = T(gx).$$

(i) *If $\varphi(x)$ is any invariant test of H, there exists an almost invariant test ψ based on T, which has the same power function as φ.*

(ii) *If in addition the assumptions of Theorem 6.5.1 are satisfied, the test ψ of (i) can be taken to be invariant.*

(iii) *If there exists a test $\psi_0(T)$ which is UMP among all $\tilde{G}$-invariant tests based on T, then under the assumptions of (ii), ψ_0, is also UMP among all G-invariant tests based on X.*

This theorem justifies the derivation of the UMP invariant tests of Examples 6.3.3 and 6.3.4.

PROOF. (i): Let $\psi(t) = E[\varphi(X)|t]$. Then ψ has the same power function as φ. To complete the proof, it suffices to show that $\psi(t)$ is almost invariant, i.e. that

$$\psi(\tilde{g}t) = \psi(t) \qquad (\text{a.e. } \mathcal{P}^T).$$

It follows from (1) that

$$E_\theta[\varphi(gX)|\tilde{g}t] = E_{\bar{g}\theta}[\varphi(X)|t] \qquad (\text{a.e. } P_\theta).$$

Since T is sufficient, both sides of this equation are independent of θ. Furthermore $\varphi(gx) = \varphi(x)$ for all x and g, and this completes the proof. ■

Part (ii) follows immediately from (i) and Theorem 6.5.1, and part (iii) from (ii).

6.6 Unbiasedness and Invariance

The principles of unbiasedness and invariance complement each other in that each is successful in cases where the other is not. For example, there exist UMP unbiased tests for the comparison of two binomial or Poisson distributions, problems to which invariance considerations are not applicable. UMP unbiased tests also exist for testing the hypothesis $\sigma = \sigma_0$ against $\sigma \neq \sigma_0$ in a normal distribution, while invariance does not reduce this problem sufficiently far. Conversely, there exist UMP invariant tests of hypotheses specifying the values of more than one parameter (to be considered in Chapter 7) but for which the class of unbiased tests has no UMP member. There are also hypotheses, for example the one-sided hypothesis $\xi/\sigma \leq \theta_0$ in a univariate normal distribution or $\rho \leq \rho_0$ in a bivariate one (Problem 6.19) with $\theta_0, \rho_0 \neq 0$, where a UMP invariant test exists but the existence of a UMP unbiased test does not follow by the methods of Chapter 5 and is an open question.

On the other hand, to some problems both principles have been applied successfully. These include Student's hypotheses $\xi \leq \xi_0$ and $\xi = \xi_0$ concerning the mean of a normal distribution, and the corresponding two sample problems $\eta - \xi \le \Delta_0$ and $\eta - \xi = \Delta_0$ when the variances of the two samples are assumed equal. Other examples are the one-sided hypotheses $\sigma^2 \ge \sigma_0^2$ and $\tau^2/\sigma^2 \ge \Delta_0$ concerning the variances of one or two normal distributions. The hypothesis of independence $\rho = 0$ in a bivariate normal distribution is still another case in point (Problem 6.19). In all these examples the two optimum procedures coincide. We shall now show that this is not accidental but is the case whenever the UMP invariant test is UMP also among all almost invariant tests and the UMP unbiased test is unique. In this sense, the principles of unbiasedness and of almost invariance are consistent.

Theorem 6.6.1 *Suppose that for a given testing problem there exists a UMP unbiased test ϕ^* which is unique (up to sets of measure zero), and that there also exists a UMP almost invariant test with respect to some group G. Then the latter is also unique (up to sets of measure zero), and the two tests coincide a.e.*

PROOF. If $U(\alpha)$ is the class of unbiased level-α tests, and if $g \in G$, then $\phi \in U(\alpha)$ if and only if $\phi g \in U(\alpha)$.[4] Denoting the power function of the test ϕ by $\beta_\phi(\theta)$, we thus have

$$\begin{aligned}\beta_{\phi^* g}(\theta) &= \beta_{\phi^*}(\bar{g}\theta) = \sup_{\phi \in U(\alpha)} \beta_\phi(\bar{g}\theta) = \sup_{\phi \in U(\alpha)} \beta_{\phi g}(\theta) \\ &= \sup_{\phi g \in U(\alpha)} \beta_{\phi g}(\theta) = \beta_{\phi^*}(\theta).\end{aligned}$$

It follows that ϕ^* and $\phi^* g$ have the same power function, and, because of the uniqueness assumption, that ϕ^* is almost invariant. Therefore, if ϕ' is UMP almost invariant, we have $\beta_{\phi'}(\theta) \ge \beta_{\phi^*}(\theta)$ for all θ. On the other hand, ϕ' is unbiased, as is seen by comparing it with the invariant test $\phi(x) \equiv \alpha$, and hence $\beta_{\phi'}(\theta) \le \beta_{\phi^*}(\theta)$ for all θ. Since ϕ' and ϕ^* therefore have the same power function, they are equal a.e. because of the uniqueness of ϕ^*, as was to be proved. ■

This theorem provides an alternative derivation for some of the tests of Chapter 5. In Theorem 4.4.1, the existence of UMP unbiased tests was established for one- and two-sided hypotheses concerning the parameter θ of the exponential family (4.10). For this family, the statistics (U, T) are sufficient and complete, and in terms of these statistics the UMP unbiased test is therefore unique. Convenient explicit expressions for some of these tests, which were derived in Chapter 5, can instead be obtained by noting that when a UMP almost invariant test exists, the same test by Theorem 6.6.1 must also be UMP unbiased. This proves for example that the tests of Examples 6.3.3 and 6.3.4 are UMP unbiased.

The principles of unbiasedness and invariance can be used to supplement each other in cases where neither principle alone leads to a solution but where they do so when applied in conjunction. As an example consider a sample $X_1, \ldots, X_n$ from $N(\xi, \sigma^2)$ and the problem of testing $H : \xi/\sigma = \theta_0 \ne 0$ against the two-sided alternatives that $\xi/\sigma \ne \theta_0$. Here sufficiency and invariance reduce the problem to the consideration of $t = \sqrt{n}\bar{x}/\sqrt{\sum(x_i - \bar{x})^2/(n-1)}$. The distribution of this statistic is the noncentral t-distribution with noncentrality parameter $\delta = \sqrt{n}\xi/\sigma$ and $n - 1$ degrees of freedom. For varying δ, the family of these distributions can

[4] ϕg denotes the critical function which assigns to x the value $\phi(gx)$.

be shown to be STP_∞. [Karlin (1968, pp. 118–119; see Problem 3.50] and hence in particular STP_3. It follows by Problem 6.42 that among all tests of H based on t, there exists a UMP unbiased one with acceptance region $C_1 \leq t \leq C_2$, where C_1, C_2 are determined by the conditions

$$P_{\delta_0}\{C_1 \leq t \leq C_2\} = 1 - \alpha \quad \text{and} \quad \left.\frac{\partial P_\delta\{C_1 \leq t \leq C_2\}}{\partial \delta}\right|_{\delta=\delta_0} = 0.$$

In terms of the original observations, this test then has the property of being UMP among all tests that are unbiased and invariant. Whether it is also UMP unbiased without the restriction to invariant tests is an open problem.

An analogous example occurs in the testing of the hypotheses $H : \rho = \rho_0$ and $H' : \rho_1 \leq \rho \leq \rho_2$ against two-sided alternatives on the basis of a sample from a bivariate normal distribution with correlation coefficient ρ. (The testing of $\rho \leq \rho_0$ against $\rho > \rho_0$ is treated in Problem 6.19.) The distribution of the sample correlation coefficient has not only monotone likelihood ratio as shown in Problem 6.19, but is in fact STP_∞. [Karlin (1968, Section 3.4)]. Hence there exist tests of both H and H' which are UMP among all tests that are both invariant and unbiased.

Another case in which the combination of invariance and unbiasedness appears to offer a promising approach is the *Behrens–Fisher problem*. Let $X_1, \ldots, X_m$ and $Y_1, \ldots, Y_n$ be samples from normal distributions $N(\xi, \sigma^2)$ and $N(\eta, \tau^2)$ respectively. The problem is that of testing $H : \eta \leq \xi$ (or $\eta = \xi$) without assuming equality of the variances σ^2 and τ^2. A set of sufficient statistics for $(\xi, \eta, \sigma, \tau)$ is then $(\bar{X}, \bar{Y}, S_X^2, S_Y^2)$, where $S_X^2 = \sum(X_i - \bar{X})^2/(m-1)$ and $S_Y^2 = \sum(Y_j - \bar{Y})^2/(n-1)$. Adding the same constant to $\bar{X}$ and $\bar{Y}$ reduces the problem to $\bar{Y} - \bar{X}$, S_X^2, S_Y^2, and multiplication of all variables by a common positive constant to $(\bar{Y} - \bar{X})/\sqrt{S_X^2 + S_Y^2}$ and S_Y^2/S_X^2. One would expect any reasonable invariant rejection region to be of the form

$$\frac{\bar{Y} - \bar{X}}{\sqrt{S_X^2 + S_Y^2}} \geq g\left(\frac{S_Y^2}{S_X^2}\right) \tag{6.17}$$

for some suitable function g. If this test is also to be unbiased, the probability of (6.17) must equal α when $\eta = \xi$ for all values of τ/σ. It has been shown by Linnik and others that only pathological functions g with this property can exist. [This work is reviewed by Pfanzagl (1974).] However, approximate solutions are available which provide tests that are satisfactory for all practical purposes. These are the Welch approximate t-solution described in Section 11.3, and the Welch–Aspin test. Both are discussed, and evaluated, in Scheffé (1970) and Wang (1971); see also Chernoff (1949), Wallace (1958), Davenport and Webster (1975) and Robinson (1982). The Behrens-Fisher problem will be revisited in Examples 13.5.4 and 15.6.3 and Section 15.2.

The property of a test ϕ_1 being UMP invariant is relative to a particular group G_1, and does not exclude the possibility that there might exist another test ϕ_2 which is UMP invariant with respect to a different group G_2. Simple instances can be obtained from Examples 6.5.1 and 6.6.11.

Example 6.6.8 (continued) If G_1 is the group G of Example 6.5.1, a UMP invariant test of $H : \theta \leq \theta_0$ against $\theta > \theta_0$ rejects when $Y_1 + \cdots + Y_n > C$.

Let G_2 be the group obtained by interchanging the role of Z and Y_1. Then a UMP invariant test with respect to G_2 rejects when $Z + Y_2 + \cdots + Y_n > C$. Analogous UMP invariant tests are obtained by interchanging the role of Z and any one of the other Y's and further examples by applying the transformations of G in Example 6.5.1 to more than one variable. In particular, if it is applied independently to all $n+1$ variables, only the constants remain invariant, and the test $\phi \equiv \alpha$ is UMP invariant. ■

Example 6.6.11 For another example (due to Charles Stein), let (X_{11}, X_{12}) and (X_{21}, X_{22}) be independent and have bivariate normal distributions with zero means and covariance matrices

$$\begin{pmatrix} \sigma_1^2 & \rho\sigma_1\sigma_2 \\ \rho\sigma_1\sigma_2 & \sigma_2^2 \end{pmatrix} \quad \text{and} \quad \begin{pmatrix} \Delta\sigma_1^2 & \Delta\rho\sigma_1\sigma_2 \\ \Delta\rho\sigma_1\sigma_2 & \Delta\sigma_2^2 \end{pmatrix}.$$

Suppose that these matrices are nonsingular, or equivalently that $|\rho| \neq 1$, but that all σ_1, σ_2, ρ, and Δ are otherwise unknown. The problem of testing $\Delta = 1$ against $\Delta > 1$ remains invariant under the group G_1 of all nonsingular transformations

$$\begin{aligned} X'_{i1} &= bX_{i1} \\ X'_{i2} &= a_1X_{i1} + a_2X_{i2} \end{aligned}, \qquad (a_2, b > 0).$$

Since the probability is 0 that $X_{11}X_{22} = X_{12}X_{21}$, the 2×2 matrix (X_{ij}) is nonsingular with probability 1, and the sample space can therefore be restricted to be the set of all nonsingular such matrices. A maximal invariant under the subgroup corresponding to $b = 1$ is the pair (X_{11}, X_{21}). The argument of Example 6.3.4 then shows that there exists a UMP invariant test under G_1 which rejects when $X_{21}^2 X_{11}^2 > C$.

By interchanging 1 and 2 in the second subscript of the X's one sees that under the corresponding group G_2 the UMP invariant test rejects when $X_{22}^2 X_{12}^2 > C$.

A third group leaving the problem invariant is the smallest group containing both G_1 and G_2, namely the group G of all common nonsingular transformations

$$\begin{aligned} X'_{i1} &= a_{11}X_{i1} + a_{12}X_{i2} \\ X'_{i2} &= a_{21}X_{i1} + a_{22}X_{i2} \end{aligned}, \qquad (i = 1, 2).$$

Given any two nonsingular sample points $Z = (X_{ij})$ and $Z' = (X'_{ij})$, there exists a nonsingular linear transformation A such that $Z' = AZ$. There are therefore no invariants under G, and the only invariant size-α test is $\phi \equiv \alpha$. It follows vacuously that this is UMP invariant under G. ■

6.7 Admissibility

Any UMP unbiased test has the important property of admissibility (Problem 4.1), in the sense that there cannot exist another test which is uniformly at least as powerful and against some alternatives actually more powerful than the given one. The corresponding property does not necessarily hold for UMP invariant tests, as is shown by the following example.

Example 6.7.11 (continued) Under the assumptions of Example 6.6.11 it was seen that the UMP invariant test under G is the test $\varphi \equiv \alpha$ which has power

$\beta(\Delta) \equiv \alpha$. On the other hand, X_{11} and X_{21} are independently distributed as $N(0, \sigma_1^2)$ and $N(0, \Delta\sigma_1^2)$. On the basis of these observations there exists a UMP test for testing $\Delta = 1$ against $\Delta > 1$ with rejection region $X_{21}^2/X_{11}^2 > C$ (Problem 3.62). The power function of this test is strictly increasing in Δ and hence $> \alpha$ for all $\Delta > 1$. ■

Admissibility of optimum invariant tests therefore cannot be taken for granted but must be established separately for each case.

We shall distinguish two slightly different concepts of admissibility. A test φ_0 will be called *α-admissible* for testing $H : \theta \in \Omega_H$ against a class of alternatives $\theta \in \Omega'$ if for any other level-α test φ

$$E_\theta \varphi(X) \geq E_\theta \varphi_0(X) \qquad \text{for all} \quad \theta \in \Omega' \tag{6.18}$$

implies $E_\theta \varphi(X) = E_\theta \varphi_0(X)$ for all $\theta \in \Omega'$. This definition takes no account of the relationship of $E_\theta \varphi(X)$ and $E_\theta \varphi_0(X)$ for $\theta \in \Omega_H$ beyond the requirement that both tests are of level α. For some unexpected, and possibly undesirable consequences of α-admissibility, see Perlman and Wu (1999). A concept closer to the decision-theoretic notion of admissibility discussed in Section 1.8, defines φ_0 to be *d-admissible* for testing H against Ω' if (6.18) and

$$E_\theta \varphi(X) \leq E_\theta \varphi_0(X) \qquad \text{for all} \quad \theta \in \Omega_H \tag{6.19}$$

jointly imply $E_\theta \varphi(X) = E_\theta \varphi_0(X)$ for all $\theta \in \Omega_H \cup \Omega'$ (see Problem 6.32).

Any level-α test φ_0 that is α-admissible is also d-admissible provided no other test φ exists with $E_\theta \varphi(X) = E_\theta \varphi_0(X)$ for all $\theta \in \Omega'$ but $E_\theta \varphi(X) \neq E_\theta \varphi_0(X)$ for some $\theta \in \Omega_H$. That the converse does not hold is shown by the following example.

Example 6.7.12 Let X be normally distributed with mean ξ and known variance σ^2. For testing $H : \xi \leq -1$ or ≥ 1 against $\Omega' : \xi = 0$, there exists a level-α test φ_0, which rejects when $C_1 \leq X \leq C_2$ and accepts otherwise, such that (Problem 6.33)

$$E_\xi \varphi_0(X) \leq E_{\xi=-1} \varphi_0(X) = \alpha \qquad \text{for} \quad \xi \leq -1$$

and

$$E_\xi \varphi_0(X) \leq E_{\xi=+1} \varphi_0(X) = \alpha' < \alpha \qquad \text{for} \quad \xi \geq +1.$$

A slight modification of the proof of Theorem 3.7.1 shows that φ_0 is the unique test maximizing the power at $\xi = 0$ subject to

$$E_\xi \varphi(X) \leq \alpha \quad \text{for} \quad \xi \leq -1 \quad \text{and} \quad E_\xi \varphi(X) \leq \alpha' \quad \text{for} \quad \xi \geq 1,$$

and hence that φ_0 is d-admissible.

On the other hand, the test φ with rejection region $|X| \leq C$, where $E_{\xi=-1} \varphi(X) = E_{\xi=1} \varphi(X) = \alpha$, is the unique test maximizing the power at $\xi = 0$ subject to $E_\xi \varphi(X) \leq \alpha$ for $\xi \leq -1$ or ≥ 1, and hence is more powerful against Ω' than φ_0, so that φ_0 is not α-admissible. ■

A test that is admissible under either definition against Ω' is also admissible against any Ω'' containing Ω' and hence in particular against the class of all alternatives $\Omega_K = \Omega - \Omega_H$. The terms α- and d-admissible without qualification will be reserved for admissibility against Ω_K. Unless a UMP test exists, any α-admissible test will be admissible against some $\Omega' \subset \Omega_K$ and inadmissible against others. Both the strength of an admissibility result and the method of proof will depend on the set Ω'.

Consider in particular the admissibility of a UMP unbiased test mentioned at the beginning of the section. This does not rule out the existence of a test with greater power for all alternatives of practical importance and smaller power only for alternatives so close to H that the value of the power there is immaterial. In the present section, we shall discuss two methods for proving admissibility against various classes of alternatives.

Theorem 6.7.1 *Let X be distributed according to an exponential family with density*

$$p_\theta(x) = C(\theta) \exp\left(\sum_{j=1}^{s} \theta_j T_j(x)\right)$$

with respect to a σ-finite measure μ over a Euclidean sample space $(\mathcal{X}, \mathcal{A})$, and let Ω be the natural parameter space of this family. Let Ω_H and Ω' be disjoint nonempty subsets of Ω, and suppose that φ_0 is a test of $H : \theta \in \Omega_H$ based on $T = (T_1, \ldots, T_s)$ with acceptance region A_0 which is a closed convex subset of R^s possessing the following property: If $A_0 \cap \{\sum a_i t_i > c\}$ is empty for some c, there exists a point $\theta^* \in \Omega$ and a sequence $\lambda_n \to \infty$ such that $\theta^* + \lambda_n a \in \Omega'$ [where λ_n is a scalar and $a = (a_1, \ldots, a_s)$]. Then if A is any other acceptance region for H satisfying

$$P_\theta(X \in A) \le P_\theta(X \in A_0) \qquad \textit{for all} \quad \theta \in \Omega',$$

A is contained in A_0, except for a subset of measure 0, i.e. $\mu(A \cap \tilde{A}_0) = 0$.

Proof. Suppose to the contrary that $\mu(A \cap \tilde{A}_0) > 0$. Then it follows from the closure and convexity of A_0, that there exist $a \in R^s$ and a real number c such that

$$A_0 \cap \left\{t : \sum a_i t_i > c\right\} \text{ is empty} \tag{6.20}$$

and

$$A \cap \left\{t : \sum a_i t_i > c\right\} \text{ has positive } \mu\text{-measure,} \tag{6.21}$$

that is, the set A protrudes in some direction from the convex set A_0. We shall show that this fact and the exponential nature of the densities imply that

$$P_\theta(A) > P_\theta(A_0) \qquad \text{for some} \quad \theta \in \Omega', \tag{6.22}$$

which provides the required contradiction. Let φ_0 and φ denote the indicators of $\tilde{A}_0$ and $\tilde{A}$ respectively, so that (6.22) is equivalent to

$$\int [\varphi_0(t) - \varphi(t)] \, dP_\theta(t) > 0 \qquad \text{for some} \quad \theta \in \Omega'.$$

If $\theta = \theta^* + \lambda_n a \in \Omega'$, the left side becomes

$$\frac{C(\theta^* + \lambda_n a)}{C(\theta^*)} e^{c\lambda_n} \int [\varphi_0(t) - \varphi(t)] e^{\lambda_n(\sum a_i t_i - c)} \, dP_{\theta^*}(t).$$

Let this integral be $I_n^+ + I_n^-$, where I_n^+ and I_n^- denote the contributions over the regions of integration $\{t : \sum a_i t_i > c\}$ and $\{t : \sum a_i t_i \le c\}$ respectively. Since I_n^- is bounded, it is enough to show that $I_n^+ \to \infty$ as $n \to \infty$. By (6.20), $\varphi_0(t) = 1$ and hence $\varphi_0(t) - \varphi(t) \ge 0$ when $\sum a_i t_i > c$, and by (6.21)

$$\mu\left\{\varphi_0(t) - \varphi(t) > 0 \quad \text{and} \quad \sum a_i t_i > c\right\} > 0.$$

This shows that $I_n^+ \to \infty$ as $\lambda_n \to \infty$ and therefore completes the proof. ■

Corollary 6.7.1 *Under the assumptions of Theorem 6.7.1, the test with acceptance region A_0 is d-admissible. If its size is α and there exists a finite point θ_0 in the closure $\bar{\Omega}_H$ of Ω_H for which $E_{\theta_0}\varphi_0(X) = \alpha$, then φ_0 is also α-admissible.*

PROOF.

(i) Suppose φ satisfies (6.18). Then by Theorem 6.7.1, $\varphi_0(x) \le \varphi(x)$ (a.e. μ). If $\varphi_0(x) < \varphi(x)$ on a set of positive measure, then $E_\theta\varphi_0(X) < E_\theta\varphi(X)$ for all θ and hence (6.19) cannot hold.

(ii) By the argument of part (i), (6.18) implies $\alpha = E_{\theta_0}\varphi_0(X) < E_{\theta_0}\varphi(X)$, and hence by the continuity of $E_\theta\varphi(X)$ there exists a point $\theta \in \Omega_H$ for which $\alpha < E_\theta\varphi(X)$. Thus φ is not a level-α test. ■

Theorem 6.7.1 and the corollary easily extend to the case where the competitors φ of φ_0 are permitted to be randomized but the assumption that φ_0 is nonrandomized is essential. Thus, the main applications of these results are to the case that μ is absolutely continuous with respect to Lebesgue measure. The boundary of A_0 will then typically have measure zero, so that the closure requirement for A_0 can be dropped.

Example 6.7.13 (Normal mean) If $X_1, \ldots, X_n$ is a sample from the normal distribution $N(\xi, \sigma^2)$, the family of distributions is exponential with $T_1 = \bar{X}$, $T_2 = \sum X_i^2$, $\theta_1 = n\xi/\sigma^2$, $\theta_2 = -1/2\sigma^2$. Consider first the one-sided problem $H : \theta_1 \le 0$, $K : \theta_1 > 0$ with $\alpha < \frac{1}{2}$. Then the acceptance region of the t-test is $A : T_1/\sqrt{T_2} \le C$ $(C > 0)$, which is convex [Problem 6.34(i)]. The alternatives $\theta \in \Omega' \subset K$ will satisfy the conditions of Theorem 6.7.1 if for any half plane $a_1t_1 + a_2t_2 > c$ that does not intersect the set $t_1 \le C\sqrt{t_2}$ there exists a ray $(\theta_1^* + \lambda a_1, \theta_2^* + \lambda a_2)$ in the direction of the vector (a_1, a_2) for which $(\theta_1^* + \lambda a_1, \theta_2^* + \lambda a_2) \in \Omega'$ for all sufficiently large λ. In the present case, this condition must hold for all $a_1 > 0 > a_2$. Examples of sets Ω' satisfying this requirement (and against which the t-test is therefore admissible) are

$$\Omega_1' : \theta_1 > k_1 \text{ or } \frac{\xi}{\sigma^2} > k_1'$$

and

$$\Omega_2' : \frac{\theta_1}{\sqrt{-\theta_2}} > k_2 \text{ or } \frac{\xi}{\sigma} > k_2'.$$

On the other hand, the condition is not satisfied for $\Omega' : \xi > k$ (Problem 6.34).

Analogously, the acceptance region $A : T_1^2 \le CT_2$ of the two-sided t-test for testing $H : \theta_1 = 0$ against $\theta_1 \ne 0$ is convex, and the test is admissible against $\Omega_1' : |\xi/\sigma^2| > k_1$ and $\Omega_2' : |\xi/\sigma| > k_2$. ■

In decision theory, a quite general method for proving admissibility consists in exhibiting a procedure as a unique Bayes solution. In the present case, this is justified by the following result, which is closely related to Theorem 3.8.1.

Theorem 6.7.2 *Suppose the set* $\{x : f_\theta(x) > 0\}$ *is independent of* θ, *and let a* σ*-field be defined over the parameter space* Ω, *containing both* Ω_H *and* Ω_K *and such that the densities* $f_\theta(x)$ *(with respect to* μ*) of* X *are jointly measurable in* θ *and* x. *Let* Λ_0 *and* Λ_1 *be probability distributions over this* σ*-field with* $\Lambda_0(\Omega_H) = \Lambda_1(\Omega_K) = 1$, *and let*

$$h_i(x) = \int f_\theta(x)\, d\Lambda_i(\theta).$$

Suppose φ_0 *is a nonrandomized test of* H *against* K *defined by*

$$\varphi_0(x) = \begin{cases} 1 & \\ 0 & \end{cases} \quad \text{if} \quad \frac{h_1(x)}{h_0(x)} \gtrless k,$$

and that $\mu\{x : h_1(x)/h_0(x) = k\} = 0$.

(i) *Then* φ_0 *is d-admissible for testing* H *against* K.

(ii) *Let* $\sup_{\Omega_H} E_\theta\varphi_0(X) = \alpha$ *and* $\omega = \{\theta : E_\theta\varphi_0(X) = \alpha\}$. *If* $\omega \subset \Omega_H$ *and* $\Lambda_0(\omega) = 1$, *then* φ_0 *is also* α*-admissible.*

(iii) *If* Λ_1 *assigns probability 1 to* $\Omega' \subset \Omega_K$, *the conclusions of (i) and (ii) apply with* Ω' *in place of* Ω_K.

PROOF. (i): Suppose φ is any other test, satisfying (6.18) and (6.19) with $\Omega' = \Omega_K$. Then also

$$\int E_\theta\varphi(X)\, d\Lambda_0(\theta) \le \int E_\theta\varphi_0(X)\, d\Lambda_0(\theta)$$

and

$$\int E_\theta\varphi(X)\, d\Lambda_1(\theta) \ge \int E_\theta\varphi_0(X)\, d\Lambda_1(\theta).$$

By the argument of Theorem 3.8.1, these inequalities are equivalent to

$$\int \varphi(x)h_0(x)\, d\mu(x) \le \int \varphi_0(x)h_0(x)\, d\mu(x)$$

and

$$\int \varphi(x)h_1(x)\, d\mu(x) \ge \int \varphi_0(x)h_1(x)\, d\mu(x),$$

and the $h_i(x)$ $(i = 0, 1)$ are probability densities with respect to μ. This contradicts the uniqueness of the most powerful test of h_0 against h_1 at level $\int \varphi(x)h_0(x)\, d\mu(x)$.

(ii): By assumption, $\int E_\theta\varphi_0(x)\, d\Lambda_0(\theta) = \alpha$, so that φ_0 is a level-α test of h_0. If φ is any other level-α test of H satisfying (6.18) with $\Omega' = \Omega_K$, it is also a level-α test of h_0 and the argument of part (i) can be applied as before.

(iii): This follows immediately from the proofs of (i) and (ii). ■

Example 6.7.13 (continued) In the two-sided normal problem of Example 6.7.13 with $H : \xi = 0$, $K : \xi \neq 0$ consider the class $\Omega'_{a,b}$ of alternatives (ξ, σ)

satisfying

$$\sigma^2 = \frac{1}{a+\eta^2}, \quad \xi = \frac{b\eta}{a+\eta^2}, \qquad -\infty < \eta < \infty \tag{6.23}$$

for some fixed a, $b > 0$, and the subset ω, of Ω_H of points $(0, \sigma^2)$ with $\sigma^2 < 1/a$. Let Λ_0, Λ_1 be distributions over ω and $\Omega'_{a,b}$ defined by the densities [Problem 6.35(i)]

$$\lambda_0(\eta) = \frac{C_0}{(a+\eta^2)^{n/2}}$$

and

$$\lambda_1(\eta) = \frac{C_1 e^{(n/2)b^2\eta^2/(a+\eta^2)}}{(a+\eta^2)^{n/2}}.$$

Straightforward calculation then shows [Problem 6.35(ii)] that the densities h_0 and h_1 of Theorem 6.7.2 become

$$h_0(x) = \frac{C_0 e^{-(a/2)\sum x_i^2}}{\sqrt{\sum x_i^2}}$$

and

$$h_1(x) = \frac{C_1 \exp\left(-\frac{a}{2}\sum x_i^2 + \frac{b^2(\sum x_i)^2}{2\sum x_i^2}\right)}{\sqrt{\sum x_i^2}},$$

so that the Bayes test φ_0 of Theorem 6.7.2 rejects when $\bar{x}^2/\sum x_i^2 > k$ and hence reduces to the two-sided t-test.

The condition of part (ii) of the theorem is clearly satisfied so that the t-test is both d- and α-admissible against $\Omega'_{a,b}$.

When dealing with invariant tests, it is of particular interest to consider admissibility against invariant classes of alternatives. In the case of the two-sided test φ_0, this means sets Ω' depending only on $|\xi/\sigma|$. It was seen in Example 6.7.13 that φ_0 is admissible against $\Omega' : |\xi/\sigma| \geq B$ for any B, that is, against distant alternatives, and it follows from the test being UMP unbiased or from Example 6.7.13 (continued) that φ_0, is admissible against $\Omega' : |\xi/\sigma| \leq A$ for any $A > 0$, that is, against alternatives close to H. This leaves open the question whether φ_0 is admissible against sets $\Omega' : 0 < A < |\xi/\sigma| < B < \infty$, which include neither nearby nor distant alternatives. It was in fact shown by Lehmann and Stein (1953) that φ_0 is admissible for testing H against $|\xi|/\sigma = \delta$ for any $\delta > 0$ and hence that it is admissible against any invariant Ω'. It was also shown there that the one-sided t-test of $H : \xi = 0$ is admissible against $\xi/\sigma = \delta'$ for any $\delta' > 0$. These results will not be proved here. The proof is based on assigning to log σ the uniform density on $(-N, N)$ and letting $N \to \infty$, thereby approximating the "improper" prior distribution which assigns to log a the uniform distribution on $(-\infty, \infty)$, that is, Lebesgue measure.

That the one-sided t-test φ_1 of $H : \xi < 0$ is not admissible against all Ω' is shown by Brown and Sackrowitz (1984), who exhibit a test φ satisfying

$$E_{\xi,\sigma}\varphi(X) < E_{\xi,\sigma}\varphi_1(X) \quad \text{for all} \quad \xi < 0,\ 0 < \sigma < \infty$$

and

$$E_{\xi,\sigma}\varphi(X) > E_{\xi,\sigma}\varphi_1(X) \quad \text{for all} \quad 0 < \xi_1 < \xi < \xi_2 < \infty,\ 0 < \sigma < \infty. \blacksquare$$

Example 6.7.14 (Normal variance) For testing the variance σ^2 of a normal distribution on the basis of a sample $X_1, \ldots, X_n$ from $N(\xi, \sigma^2)$, the Bayes approach of Theorem 6.7.2 easily proves α-admissibility of the standard test against any location invariant set of alternatives Ω', that is, any set Ω' depending only on σ^2. Consider first the one-sided hypothesis $H : \sigma \leq \sigma_0$ and the alternatives $\Omega' : \sigma = \sigma_1$ for any $\sigma_1 > \sigma_0$. Admissibility of the UMP invariant (and unbiased) rejection region $\sum(X_i - \bar{X})^2 > C$ follows immediately from Section 3.9, where it was shown that this test is Bayes for a pair of prior distributions (Λ_0, Λ_1): namely, Λ_1 assigning probability 1 to any point (ξ_1, σ_1), and Λ_0 putting $\sigma = \sigma_0$ and assigning to ξ the normal distribution $N(\xi_1, (\sigma_1^2 - \sigma_0^2)/n)$. Admissibility of $\sum(X_i - \bar{X})^2 \leq C$ when the hypothesis is $H : \sigma \geq \sigma_0$ and $\Omega' = \{(\xi, \sigma) : \sigma = \sigma_1\}$, $\sigma_1 < \sigma_0$, is seen by interchanging Λ_0 and Λ_1, σ_0 and σ_1.

A similar approach proves α-admissibility of any size-α rejection region

$$\sum(X_i - \bar{X})^2 \leq C_1 \text{ or } \geq C_2 \tag{6.24}$$

for testing $H : \sigma = \sigma_0$ against $\Omega' : \{\sigma = \sigma_1\} \cup \{\sigma = \sigma_2\}$ $(\sigma_1 < \sigma_0 < \sigma_2)$. On Ω_H, where the only variable is ξ, the distribution Λ_0 for ξ can be taken as the normal distribution with an arbitrary mean ξ_1 and variance $(\sigma_2^2 - \sigma_0^2)/n$. On Ω', let the conditional distribution of ξ given $\sigma = \sigma_2$ assign probability 1 to the value ξ_1, and let the conditional distribution of ξ given $\sigma = \sigma_1$ be $N(\xi_1, (\sigma_2^2 - \sigma_1^2)/n)$. Finally, let Λ_1 assign probabilities p and $1-p$ to $\sigma = \sigma_1$ and $\sigma = \sigma_2$, respectively. Then the rejection region satisfies (6.24), and any constants C_1 and C_2 for which the test has size a can be attained by proper choice of p [Problem 6.36(i)]. ■

The results of Examples 6.7.13 and 6.7.14 can be used as the basis for proving admissibility results in many other situations involving normal distributions. The main new difficulty tends to be the presence of additional (nuisance) means. These can often be eliminated by use of the following lemma.

Lemma 6.7.1 *For any given σ^2 and $M^2 > \sigma^2$ there exists a distribution Λ_σ such that*

$$I(z) = \int \frac{1}{\sqrt{2\pi}\sigma} e^{-(1/2\sigma^2)(z-\zeta)^2} d\Lambda_\sigma(\zeta)$$

is the normal density with mean zero and variance M^2.

PROOF. Let $\theta = \zeta/\sigma$, and let θ be normally distributed with zero mean and variance τ^2. Then it is seen [Problem 6.36(ii)] that

$$I(z) = \frac{1}{\sqrt{2\pi}\sigma\sqrt{1+\tau^2}} \exp\left[-\frac{1}{2\sigma^2(1+\tau^2)} z^2\right].$$

The result now follows by letting $\tau^2 = (M^2/\sigma^2) - 1$, so that $\sigma^2(1+\tau^2) = M^2$. ■

Example 6.7.15 Let $X_1, \ldots, X_m$; $Y_1, \ldots, Y_n$ be samples from $N(\xi, \sigma^2)$ and $N(\eta, \tau^2)$ respectively, and consider the problem of testing $H : \tau/\sigma = 1$ against $\tau/\sigma = \Delta > 1$.

(i) Suppose first that $\xi = \eta = 0$. If Λ_0 and Λ_1 assign probability 1 to the points $(\sigma_0, \tau_0 = \sigma_0)$ and $(\sigma_1, \tau_1 = \Delta\sigma_1)$ respectively, the ratio h_1/h_0 of Theorem

6.7.2 is proportional to

$$\exp\left\{-\frac{1}{2}\left[\left(\frac{1}{\Delta^2\sigma_1^2}-\frac{1}{\sigma_0^2}\right)\sum y_j^2-\left(\frac{1}{\sigma_0^2}-\frac{1}{\sigma_1^2}\right)\sum x_i^2\right]\right\},$$

and for suitable choice of critical value and $\sigma_1 < \sigma_0$, the rejection region of the Bayes test reduces to

$$\frac{\sum y_j^2}{\sum x_i^2} > \frac{\Delta^2\sigma_1^2 - \sigma_0^2}{\sigma_0^2 - \sigma_1^2}.$$

The values σ_0^2 and σ_1^2 can then be chosen to e this test any preassigned size α.

(ii) If ξ and η are unknown, then $\bar{X}$, $\bar{Y}$, $S_X^2 = \sum(X_i - \bar{X})^2$, $S_Y^2 = \sum(Y_j - \bar{Y})^2$ are sufficient statistics, and S_X^2 and S_Y^2 can be represented as $S_X^2 = \sum_{i=1}^{m-1} U_i^2$, $S_Y^2 = \sum_{j=1}^{n-1} V_j^2$, with the U_i, V_j independent normal with means 0 and variances σ^2 and τ^2 respectively.

To σ and τ assign the distributions Λ_0 and Λ_1 of part (i) and conditionally, given σ and τ, let ξ and η be independently distributed according to $\Lambda_{0\sigma}$, $\Lambda_{0\tau}$, over Ω_H and $\Lambda_{1\sigma}$, $\Lambda_{1\tau}$ over Ω_K, with these four conditional distributions determined from Lemma 6.7.1 in such a way that

$$\int \frac{\sqrt{m}}{\sqrt{2\pi}\sigma_0} e^{-(m/2\sigma_0^2)(\bar{x}-\xi)^2}\, d\Lambda_{0\sigma_0}(\xi) = \int \frac{\sqrt{m}}{\sqrt{2\pi}\sigma_1} e^{-(m/2\sigma_1^2)(\bar{x}-\xi)^2}\, d\Lambda_{0\sigma_1}(\xi),$$

and analogously for η. This is possible by choosing the constant M^2 of Lemma 6.7.1 greater than both σ_0^2 and σ_1^2. With this choice of priors, the contribution from $\bar{x}$ and $\bar{y}$ to the ratio h_1/h_0 of Theorem 6.7.2 disappears, so that h_1/h_0 reduces to the expression for this ratio in part (i), with $\sum x_i^2$ and $\sum y_j^2$ replaced by $\sum(x_i - \bar{x})^2$ and $\sum(y_j - \bar{y})^2$ respectively. ■

This approach applies quite generally in normal problems with nuisance means, provided the prior distribution of the variances σ^2, τ^2, ... assigns probability 1 to a bounded set, so that M^2 can be chosen to exceed all possible values of these variances.

Admissibility questions have been considered not only for tests but also for confidence sets. These will not be treated here (but see Example 8.5.4); convenient entries to the literature are Cohen and Strawderman (1973) and Joshi (1982). For additional results, see Hooper (1982b) and Arnold (1984).

6.8 Rank Tests

One of the basic problems of statistics is the two-sample problem of testing the equality of two distributions. A typical example is the comparison of a treatment with a control, where the hypothesis of no treatment effect is tested against the alternatives of a beneficial effect. This was considered in Chapter 5 under the assumption of normality, and the appropriate test was seen to be based on Student's t. It was also shown that when approximate normality is suspected but the assumption cannot be trusted, one is led to replacing the t-test by its permutation analogue, which in turn can be approximated by the original t-test.

We shall consider the same problem below without, at least for the moment, making any assumptions concerning even the approximate form of the underlying distributions, assuming only that they are continuous. The observations then consist of samples $X_1, \ldots, X_m$ and $Y_1, \ldots, Y_n$ from two distributions with continuous cumulative distribution functions F and G, and the problem becomes that of testing the hypothesis

$$H_1 : G = F.$$

If the treatment effect is assumed to be additive, the alternatives are $G(y) = F(y - \Delta)$. We shall here consider the more general possibility that the size of the effect may depend on the value of y (so that Δ becomes a nonnegative function of y) and therefore test H_1 against the one-sided alternatives that the Y's are stochastically larger than the X's,

$$K_1 : G(z) \leq F(z) \quad \text{for all } z, \quad \text{and} \quad G \neq F.$$

An alternative experiment that can be performed to test the effect of a treatment consists of the comparison of N pairs of subjects, which have been matched so as to eliminate as far as possible any differences not due to the treatment. One member of each pair is chosen at random to receive the treatment while the other serves as control. If the normality assumption of Section 5.10 is dropped and the pairs of subjects can be considered to constitute a sample, the observations $(X_1, Y_1), \ldots, (X_N, Y_N)$ are a sample from a continuous bivariate distribution F. The hypothesis of no effect is then equivalent to the assumption that F is symmetric with respect to the line $y = x$:

$$H_2 : F(x, y) = F(y, x).$$

Another basic problem, which occurs in many different contexts, concerns the dependence or independence of two variables. In particular, if $(X_1, Y_1), \ldots, (X_N, Y_N)$ is a sample from a bivariate distribution F, one will be interested in the hypothesis

$$H_3 : F(x, y) = G_1(x) G_2(y)$$

that X and Y are independent, which was considered for normal distributions in Section 5.13. The alternatives of interest may, for example, be that X and Y are positively dependent. An alternative formulation results when x, instead of being random, can be selected for the experiment. If the chosen values are $x_1 < \cdots < x_N$ and F_i denotes the distribution of Y given x_i, the Y's are independently distributed with continuous cumulative distribution functions $F_1, \ldots, F_N$. The hypothesis of independence of Y from x becomes

$$H_4 : F_1 = \cdots = F_N,$$

while under the alternatives of positive regression dependence the variables Y_i are stochastically increasing with i.

In these and other similar problems, invariance reduces the data so completely that the actual values of the observations are discarded and only certain order relations between different groups of variables are retained. It is nevertheless possible on this basis to test the various hypotheses in question, and the resulting tests frequently are nearly as powerful as the standard normal tests. We shall now carry out this reduction for the four problems above.

The two-sample problem of testing H_1 against K_1 remains invariant under the group G of all transformations

$$x'_i = \rho(x_i), \quad y'_j = \rho(y_j) \qquad (i = 1, \ldots, m, \quad j = 1, \ldots, n)$$

such that ρ is continuous and strictly increasing. This follows from the fact that these transformations preserve both the continuity of a distribution and the property of two variables being either identically distributed or one being stochastically larger than the other. As was seen (with a different notation) in Example 6.2.3, a maximal invariant under G is the set of ranks

$$(R'; S') = (R'_1, \ldots, R'_m; S'_1, \ldots, S'_n)$$

of $X_1, \ldots, X_m; Y_1, \ldots, Y_n$ in the combined sample. Since the distribution of $(R'_1, \ldots, R'_m; S'_1, \ldots, S'_n)$ is symmetric in the first m and in the last n variables for all distributions F and G, a set of sufficient statistics for (R', S') is the set of the X-ranks and that of the Y-ranks without regard to the subscripts of the X's and Y's This can be represented by the ordered X-ranks and Y-ranks

$$R_1 < \cdots < R_m \quad \text{and} \quad S_1 < \cdots < S_n,$$

and therefore by one of these sets alone since each of them determines the other. Any invariant test is thus a *rank test*, that is, it depends only on the ranks of the observations, for example on $(S_1, \ldots, S_n)$.

That almost invariant tests are equivalent to invariant ones in the present context was shown first by Bell (1964). A streamlined and generalized version of his approach is given by Berk and Bickel (1968) and Berk (1970), who also show that the conclusion of Theorem 6.5.3 remains valid in this case.

To obtain a similar reduction for H_2, it is convenient first to make the transformation $Z_i = Y_i - X_i$, $W_i = X_i + Y_i$. The pairs of variables (Z_i, W_i) are then again a sample from a continuous bivariate distribution. Under the hypothesis this distribution is symmetric with respect to the w-axis, while under the alternatives the distribution is shifted in the direction of the positive z-axis The problem is unchanged if all the w's are subjected to the same transformation $w'_i = \lambda(w_i)$, where λ is $1:1$ and has at most a finite number of discontinuities, and $(Z_1, \ldots, Z_N)$ constitutes a maximal invariant under this group. [Cf. Problem 6.2(ii).]

The Z's are a sample from a continuous univariate distribution D, for which the hypothesis of symmetry with respect to the origin,

$$H'_2 : D(z) + D(-z) = 1 \quad \text{for all } z,$$

is to be tested against the alternatives that the distribution is shifted toward positive z-values This problem is invariant under the group G of all transformations

$$z'_i = \rho(z_i) \qquad (i = 1, \ldots, N)$$

such that ρ is continuous, odd, and strictly increasing. If $z_{i_1}, \ldots, z_{i_m} < 0 < z_{j_1}, \ldots, z_{j_n}$, where $i_1 < \cdots < i_m$ and $j_1 < \cdots < j_n$, let $s'_1, \ldots, s'_n$ denote the ranks of $z_{j_i}, \ldots, z_{j_n}$, among the absolute values $|z_1|, \ldots, |z_N|$, and $r'_1, \ldots, r'_m$ the ranks of $|z_{i_1}|, \ldots, |z_{i_m}|$ among $|z_1|, \ldots, |z_N|$. The transformations ρ preserve the sign of each observation, and hence in particular also the numbers m and n. Since ρ is a continuous, strictly increasing function of $|z|$, it leaves the order of the absolute values invariant and therefore the ranks r'_i and s'_j. To see that the latter are maximal invariant, let $(z_1, \dots, z_N)$ and $(z'_1, \dots, z'_N)$ be two sets of points with $m' = m$, $n' = n$, and the same r'_i and s'_j. There exists a continuous, strictly increasing function on the positive real axis such that $|z'_i| = \rho(|z_i|)$ and $\rho(0) = 0$. If ρ is defined for negative z by $\rho(-z) = -\rho(z)$, it belongs to G and $z'_i = \rho(z_i)$ for all i, as was to be proved. As in the preceding problem, sufficiency permits the further reduction to the ordered ranks $r_1 < \cdots < r_m$ and $s_1 < \cdots < s_n$. This retains the information for the rank of each absolute value whether it belongs to a positive or negative observation, but not with which positive or negative observation it is associated.

The situation is very similar for the hypotheses H_3 and H_4. The problem of testing for independence in a bivariate distribution against the alternatives of positive dependence is unchanged if the X_i and Y_i are subjected to transformations $X'_i = \rho(X_i), Y'_i = \lambda(Y_i)$ such that ρ and λ are continuous and strictly increasing. This leaves as maximal invariant the ranks $(R'_1, \dots, R'_N)$ of $(X_1, \dots, X_N)$ among the X's and the ranks $(S'_1, \dots, S'_N)$ of $(Y_1, \dots, Y_N)$ among the Y's. The distribution of $(R'_1, S'_1), \dots, (R'_N, S'_N)$ is symmetric in these N pairs for all distributions of (X, Y). It follows that a sufficient statistic is $(S_1, \dots, S_N)$ where $(1, S_1), \dots, (N, S_N)$ is a permutation of $(R'_1, S'_1), \dots, (R'_N, S'_N)$ and where therefore S_i is the rank of the variable Y associated with the ith smallest X.

The hypothesis H_4 that $Y_1, \dots, Y_n$ constitutes a sample is to be tested against the alternatives K_4 that the Y_i are stochastically increasing with i. This problem is invariant under the group of transformations $y'_i = \rho(y_i)$ where ρ is continuous and strictly increasing. A maximal invariant under this group is the set of ranks $S_1, \dots, S_N$ of $Y_1, \dots, Y_N$.

Some invariant tests of the hypotheses H_1 and H_2 will be considered in the next two sections. Corresponding results concerning H_3 and H_4 are given in Problems 6.60–6.62.

6.9 The Two-Sample Problem

The problem of testing the two-sample hypothesis $H : G = F$ against the one-sided alternatives K that the Y's are stochastically larger than the X's is reduced by the principle of invariance to the consideration of tests based on the ranks $S_1 < \cdots < S_n$ of the Y's. The specification of the S_i is equivalent to specifying for each of the $N = m + n$ positions within the combined sample (the smallest, the next smallest, etc.) whether it is occupied by an x or a y. Since for any set of observations n of the N positions are occupied by y's and since the $\binom{N}{n}$ possible assignments of n positions to the y's are all equally likely when $G = F$, the joint distribution of the S_i under H is

$$P\{S_1 = s_1, \dots, S_n = s_n\} = 1 \Big/ \binom{N}{n} \tag{6.25}$$

for each set $1 \le s_1 < s_2 < \cdots < s_n \le N$. Any rank test of H of size

$$\alpha = k \Big/ \binom{N}{n}$$

therefore has a rejection region consisting of exactly k points $(s_1, \ldots, s_n)$.

For testing H against K there exists no UMP rank test, and hence no UMP invariant test. This follows for example from a consideration of two of the standard tests for this problem, since each is most powerful among all rank tests against some alternative. The two tests in question have rejection regions of the form

$$h(s_1) + \cdots + h(s_n) > C. \tag{6.26}$$

One, the Wilcoxon *two-sample test*, is obtained from (6.26) by letting $h(s) = s$, so that it rejects H when the sum of the y-ranks is too large. We shall show below that for sufficiently small Δ, this is most powerful against the alternatives that F is the logistic distribution $F(x) = 1/(1+e^{-x})$, and that $G(y) = F(y-\Delta)$. The other test, the *normal-scores test*, has the rejection region (6.26) with $h(s) = E(W_{(s)})$, where $W_{(1)} < \cdots < W_{(N)}$, is an ordered sample of size N from a standard normal distribution.[5] This is most powerful against the alternatives that F and G are normal distributions with common variance and means ξ and $\eta = \xi + \Delta$, when Δ is sufficiently small.

To prove that these tests have the stated properties it is necessary to know the distribution of $(S_1, \ldots, S_n)$ under the alternatives. If F and G have densities f and g such that f is positive whenever g is, the joint distribution of the S_i is given by

$$P\{S_1 = s_1, \ldots, S_n = s_n\} = E\left[\frac{g(V_{(s_1)})}{f(V_{(s_1)})} \cdots \frac{g(V_{(s_n)})}{f(V_{(s_n)})}\right] \Big/ \binom{N}{n}, \tag{6.27}$$

where $V_{(1)} < \cdots < V_{(N)}$ is an ordered sample of size N from the distribution F. (See Problem 6.42.) Consider in particular the translation (or shift) alternatives

$$g(y) = f(y - \Delta),$$

and the problem of maximizing the power for small values of Δ. Suppose that f is differentiable and that the probability (6.27), which is now a function of Δ, can be differentiated with respect to Δ under the expectation sign. The derivative of (6.27) at $\Delta = 0$ is then

$$\frac{\partial}{\partial \Delta} P_\Delta\{S_1 = s_1, \ldots, S_n = S_n\}\Big|_{\Delta=0} = -\sum_{i=1}^{n} E\left[\frac{f'(V_{(s_i)})}{f(V_{(s_i)})}\right] \Big/ \binom{N}{n}.$$

Since under the hypothesis the probability of any ranking is given by (6.25), it follows from the Neyman–Pearson lemma in the extended form of Theorem 3.6.1, that the derivative of the power function at $\Delta = 0$ is maximized by the rejection region

$$-\sum_{i=1}^{n} E\left[\frac{f'(V_{(s_i)})}{f(V_{(s_i)})}\right] > C. \tag{6.28}$$

The same test maximizes the power itself for sufficiently small Δ. To see this let s denote a general rank point $(s_1, \ldots, s_n)$, and denote by $s^{(j)}$ the rank point

[5] Tables of the expected order statistics from a normal distribution are given in *Biometrika Tables for Statisticians*, Vol. 2, Cambridge U. P., 1972, Table 9. For additional references, see David (1981, Appendix, Section 3.2).

giving the jth largest value to the left-hand side of (6.28). If

$$\alpha = k \Big/ \binom{N}{n},$$

the power of the test is then

$$\beta(\Delta) = \sum_{j=1}^{k} P_\Delta(s^{(j)}) = \sum_{j=1}^{k} \left[\frac{1}{\binom{N}{n}} + \Delta \frac{\partial}{\partial \Delta} P_\Delta(s^{(j)}) \bigg|_{\Delta=0} + \cdots \right].$$

Since there is only a finite number of points s, there exists for each j a number $\Delta_j > 0$ such that the point $s^{(j)}$ also gives the jth largest value to $P_\Delta(s)$ for all $\Delta < \Delta_j$. If Δ is less than the smallest of the numbers

$$\Delta_j, \qquad j = 1, \ldots, \binom{N}{n},$$

the test also maximizes $\beta(\Delta)$.

If $f(x)$ is the normal density $N(\xi, \sigma^2)$, then

$$-\frac{f'(x)}{f(x)} = -\frac{d}{dx} \log f(x) = \frac{x - \xi}{\sigma^2},$$

and the left-hand side of (6.28) becomes

$$\sum E \frac{V_{(s_i)} - \xi}{\sigma^2} = \frac{1}{\sigma} \sum E(W_{(s_i)})$$

where $W_{(1)} < \cdots < W_{(N)}$ is an ordered sample from $N(0, 1)$. The test that maximizes the power against these alternatives (for sufficiently small Δ) is therefore the normal-scores test.

In the case of the logistic distribution,

$$F(x) = \frac{1}{1 + e^{-x}}, \qquad f(x) = \frac{e^{-x}}{(1 + e^{-x})^2},$$

and hence

$$-\frac{f'(x)}{f(x)} = 2F(x) - 1.$$

The locally most powerful rank test therefore rejects when $\sum E[F(V_{(x_i)})] > C$. If V has the distribution F, then $U = F(V)$ is uniformly distributed over $(0, 1)$ (Problem 3.22). The rejection region can therefore be written as $\sum E(U_{(s_i)}) > C$, where $U_{(1)} < \cdots < U_{(N)}$ is an ordered sample of size N from the uniform distribution $U(0, 1)$. Since $E(U_{(s_i)}) = s_i/(N + 1)$, the test is seen to be the Wilcoxon test.

Both the normal-scores test and the Wilcoxon test are unbiased against the one-sided alternatives K. In fact, let ϕ be the critical function of any test determined by (6.26) with h nondecreasing. Then ϕ is nondecreasing in the y's and the probability of rejection is α for all $F = G$. By Lemma 5.9.1 the test is therefore unbiased against all alternatives of K.

It follows from the unbiasedness properties of these tests that the most powerful invariant tests in the two cases considered are also most powerful against their respective alternatives among all tests that are invariant and unbiased. The nonexistence of a UMP test is thus not relieved by restricting the tests to be unbiased as well as invariant. Nor does the application of the unbiasedness principle alone lead to a solution, as was seen in the discussion of permutation tests in Section 5.9. With the failure of these two principles, both singly and in conjunction, the problem is left not only without a solution but even without a formulation. A possible formulation (stringency) will be discussed in Chapter 8. However, the determination of a most stringent test for the two-sample hypothesis is an open problem.

For testing $H : G = F$ against the two-sided alternatives that the Y's are either stochastically smaller or larger than the X's two-sided versions of the rank tests of this section can be used. In particular, suppose that h is increasing and that $h(s)+h(N+1-s)$ is independent of s, as is the case for the Wilcoxon and normal-scores statistics. Then under H, the statistic $\Sigma h(s_j)$ is symmetrically distributed about $n\Sigma_{i=1}^N h(i)/N = \mu$, and (6.26) suggests the rejection region

$$\left|\sum h(s_j) - \mu\right| = \frac{1}{N}\left|m\sum_{j=1}^{n} h(s_j) - n\sum_{i=1}^{m} h(r_i)\right| > C.$$

The theory here is still less satisfactory than in the one-sided case. These tests need not even be unbiased [Sugiura (1965)], and it is not known whether they are admissible within the class of all rank tests. On the other hand, the relative asymptotic efficiencies are the same as in the one-sided case.

The two-sample hypothesis $G = F$ can also be tested against the general alternatives $G \neq F$. This problem arises in deciding whether two products, two sets of data, or the like can be pooled when nothing is known about the underlying distributions. Since the alternatives are now unrestricted, the problem remains invariant under all transformations $x_i' = f(x_i)$, $y_j' = f(y_j)$, $i = 1, \dots, m$, $j = 1, \dots, n$, such that f has only a finite number of discontinuities. There are no invariants under this group, so that the only invariant test is $\phi(x, y) \equiv \alpha$. This is however not admissible, since there do exist tests of H that are strictly unbiased against all alternatives $G \neq F$ (Problem 6.54). One of the tests most commonly employed for this problem is the *Smirnov test.* Let the *empirical distribution functions* of the two samples be defined by

$$S_{x_1,\dots,x_m}(z) = \frac{a}{m}, \qquad S_{y_1,\dots,y_n}(z) = \frac{b}{n},$$

where a and b are the numbers of x's and y's less or equal to z respectively. Then H is rejected according to this test when

$$\sup_z |S_{x_1,\dots,x_m}(z) - S_{y_1,\dots,y_n}(z)| > C.$$

Accounts of the theory of this and related tests are given, for example, in Durbin (1973), Serfling (1980), Gibbons and Chakraborti (1992) and Hájek, Sidák, and Sen (1999).

Two-sample rank tests are distribution-free for testing $H : G = F$ but not for the nonparametric: Behrens-Fisher situation of testing $H : \eta = \xi$ when the X's and Y's are samples from $F((x - \xi)/\sigma)$ and $F((y - \eta)/\tau)$ with σ, τ unknown. A detailed study of the effect of the difference in scales on the levels of the Wilcoxon and normal-scores tests is provided by Pratt (1964).

6.10 The Hypothesis of Symmetry

When the method of paired comparisons is used to test the hypothesis of no treatment effect, the problem was seen in Section 6.8 to reduce through invariance to that of testing the hypothesis

$$H_2' : D(z) + D(-z) = 1 \text{ for all } z,$$

which states that the distribution D of the differences $Z_i = Y_i - X_i$ $(i = 1, \ldots, N)$ is symmetric with respect to the origin. The distribution D can be specified by the triple (ρ, F, G) where

$$\rho = P\{Z \le 0\}, \qquad F(z) = P\{|Z| \le z \mid Z > 0\},$$
$$G(z) = P\{Z \le z \mid Z > 0\},$$

and the hypothesis of symmetry with respect to the origin then becomes

$$H : p = \tfrac{1}{2}, G = F.$$

Invariance and sufficiency were shown to reduce the data to the ranks $S_1 < \cdots < S_n$ of the positive Z's among the absolute values $|Z_1|, \ldots, |Z_N|$. The probability of $S_1 = s_1, \ldots, S_n = s_n$ is the probability of this event given that there are n positive observations multiplied by the probability that the number of positive observations is n. Hence

$$P\{S_1 = s_1, \ldots, S_n = s_n\}$$
$$= \binom{N}{n} (1-\rho)^n \rho^{N-n} P_{F,G}\{S_1 = s_1, \ldots, S_n = s_n \mid n\}$$

where the second factor is given by (6.27). Under H, this becomes

$$P\{S_1 = s_1, \ldots, S_n = s_n\} = \frac{1}{2^N}$$

for each of the

$$\sum_{n=0}^{N} \binom{N}{n} = 2^N$$

n-tuples $(s_1, \ldots, s_n)$ satisfying $1 \le s_1 < \cdots < s_n \le N$. Any rank test of size $\alpha = k/2^N$ therefore has a rejection region containing exactly k such points $(s_1, \ldots, s_n)$.

The alternatives K of a beneficial treatment effect are characterized by the fact that the variable Z being sampled is stochastically larger than some random variable which is symmetrically distributed about 0. It is again suggestive to use rejection regions of the form $h(s_1) + \cdots + h(s_n) > C$, where however n is no longer a constant as it was in the two-sample problem, but depends on the observations. Two particular cases are the *Wilcoxon one-sample test*, which is obtained by putting $h(s) = s$, and the analogue of the normal-scores test with $h(s) = E(W_{(s)})$ where $W_{(1)} < \cdots < W_{(N)}$ are the ordered values of $|V_1|, \ldots, |V_N|$, the V's being a sample from $N(0, 1)$. The W's are therefore an ordered sample of size N from a distribution with density $\sqrt{2/\pi} e^{-w^2/2}$ for $w \ge 0$.

As in the two-sample problem, it can be shown that each of these tests is most powerful (among all invariant tests) against certain alternatives, and that they are both unbiased against the class K. Their asymptotic efficiencies relative to the t-test for testing that the mean of Z is zero have the same values $3/\pi$ and 1 as the corresponding two-sample tests, when the distribution of Z is normal.

In certain applications, for example when the various comparisons are made under different experimental conditions or by different methods, it may be unrealistic to assume that the variables $Z_1, \ldots, Z_N$ have a common distribution. Suppose instead that the Z_i are still independently distributed but with arbitrary continuous distributions D_i. The hypothesis to be tested is that each of these distributions is symmetric with respect to the origin.

This problem remains invariant under all transformations $z'_i = f_i(z_i)$ $i = 1, \ldots, N$, such that each f_i is continuous, odd, and strictly increasing. A maximal invariant is then the number n of positive observations, and it follows from Example 6.5.1 that there exists a UMP invariant test, the *sign test*, which rejects when n is too large. This test reflects the fact that the magnitude of the observations or of their absolute values can be explained entirely in terms of the spread of the distributions D_i, so that only the signs of the Z's are relevant.

Frequently, it seems reasonable to assume that the Z's are identically distributed, but the assumption cannot be trusted. One would then prefer to use the information provided by the ranks s_i but require a test which controls the probability of false rejection even when the assumption fails. As is shown by the following lemma, this requirement is in fact satisfied for every (symmetric) rank test. Actually, the lemma will not require even the independence of the Z's; it will show that any symmetric rank test continues to correspond to the stated level of significance provided only the treatment is assigned at random within each pair.

Lemma 6.10.1 *Let $\phi(z_1, \ldots, z_N)$ be symmetric in its N variables and such that*

$$E_D \phi(Z_1, \ldots, Z_N) = \alpha \tag{6.29}$$

when the Z's are a sample from any continuous distribution D which is symmetric with respect to the origin. Then

$$E\phi(Z_1, \ldots, Z_N) = \alpha \tag{6.30}$$

if the joint distribution of the Z's is unchanged under the 2^N transformations $Z'_1 = \pm Z_1, \ldots, Z'_N = \pm Z_N$.

PROOF. The condition (6.29) implies

$$\sum_{(j_1,\ldots,j_N)} \sum \frac{\phi(\pm z_{j_1}, \ldots, \pm z_{j_N})}{2^N \cdot N!} = \alpha \quad \text{a.e.,} \tag{6.31}$$

where the outer summation extends over all $N!$ permutations $(j_1, \ldots, j_N)$ and the inner one over all 2^N possible choices of the signs $+$ and $-$. This is proved exactly as was Theorem 5.8.1. If in addition ϕ is symmetric, (6.31) implies

$$\sum \frac{\phi(\pm z_1, \ldots, \pm z_N)}{2^N} = \alpha. \tag{6.32}$$

Suppose that the distribution of the Z's is invariant under the 2^N transformations in question. Then the conditional probability of any sign combination of

$Z_1, \ldots, Z_N$ given $|Z_1|, \ldots, |Z_N|$ is $1/2^N$. Hence (6.32) is equivalent to

$$E[\phi(Z_1, \ldots, Z_N) \mid |Z_1|, \ldots, |Z_N|] = \alpha \quad \text{a.e.}, \tag{6.33}$$

and this implies (6.30) which was to be proved. ∎

The tests discussed above can be used to test symmetry about any known value θ_0 by applying them to the variables $Z_i - \theta_0$. The more difficult problem of testing for symmetry about an unknown point θ will not be considered here. Tests of this hypothesis are discussed, among others, by Antille, Kersting, and Zucchini (1982), Bhattacharya, Gastwirth, and Wright (1982), Boos (1982), and Koziol (1983).

As will be seen in Section 11.3.1, the one-sample t-test is not robust against dependence. Unfortunately, this is also true-although to a somewhat lesser extent—of the sign and one-sample Wilcoxon tests [Gastwirth and Rubin (1971)].

6.11 Equivariant Confidence Sets

Confidence sets for a parameter θ in the presence of nuisance parameters ϑ were discussed in Chapter 5 (Sections 5.4 and 5.5) under the assumption that θ is real-valued. The correspondence between acceptance regions $A(\theta_0)$ of the hypotheses $H(\theta_0) : \theta = \theta_0$ and confidence sets $S(x)$ for θ given by (5.33) and (5.34) is, however, independent of this assumption; it is valid regardless of whether θ is real-valued, vector-valued, or possibly a label for a completely unknown distribution function (in the latter case, confidence intervals become confidence bands for the distribution function). This correspondence, which can be summarized by the relationship

$$\theta \in S(x) \quad \text{if and only if} \quad x \in A(\theta), \tag{6.34}$$

was the basis for deriving uniformly most accurate and uniformly most accurate unbiased confidence sets. In the present section, it will be used to obtain uniformly most accurate equivariant confidence sets.

We begin by defining equivariance for confidence sets. Let G be a group of transformations of the variable X preserving the family of distributions $\{P_{\theta,\vartheta}, (\theta, \vartheta) \in \Omega\}$ and let $\bar{G}$ be the induced group of transformations of Ω. If $\bar{g}(\theta, \vartheta) = (\theta', \vartheta')$, we shall suppose that θ' depends only on $\bar{g}$ and θ and not on ϑ, so that $\bar{g}$ induces a transformation in the space of θ. In order to keep the notation from becoming unnecessarily complex, it will then be convenient to write also $\theta' = \bar{g}\theta$. For each transformation $g \in G$, denote by g^* the transformation acting on sets S in θ-space and defined by

$$g^* S = \{\bar{g}\theta : \theta \in S\}, \tag{6.35}$$

so that $g^* S$ is the set obtained by applying the transformation $\bar{g}$ to each point θ of S. The invariance argument of Section 1.5, then suggests restricting consideration to confidence sets satisfying

$$g^* S(x) = S(gx) \qquad \text{for all} \quad x \in \mathcal{X}, \quad g \in G. \tag{6.36}$$

We shall say that such confidence sets are *equivariant* under G. This terminology is preferable to the older term invariance which creates the impression that the confidence sets remain unchanged under the transformation $X' = gX$. If the transformation g is interpreted as a change of coordinates, (6.36) means that the confidence statement does not depend on the coordinate system used to express the data. The statement that the transformed parameter $\bar{g}\theta$ lies in $S(gx)$ is equivalent to stating that $\theta \in g^{*-1}S(gx)$, which is equivalent to the original statement $\theta \in S(x)$ provided (6.36) holds.

Example 6.11.1 Let X, Y be independently normally distributed with means ξ, η and unit variance, and let G be the group of all rigid motions of the plane, which is generated by all translations and orthogonal transformations. Here $\bar{g} = g$ for all $g \in G$. An example of an equivariant class of confidence sets is given by

$$S(x, y) = \left\{(\xi, \eta) : (x - \xi)^2 + (y - \eta)^2 \le C\right\},$$

the class of circles with radius $\sqrt{C}$ and center (x, y). The set $g^*S(x, y)$ is the set of all points $g(\xi, \eta)$ with $(\xi, \eta) \in S(x, y)$ and hence is obtained by subjecting $S(x, y)$ to the rigid motion g. The result is the circle with radius $\sqrt{C}$ and center $g(x, y)$, and (6.36) is therefore satisfied. ■

In accordance with the definitions given in Chapters 3 and 5, a class of confidence sets for θ will be said to be *uniformly most accurate equivariant* at confidence level $1 - \alpha$ if among all equivariant classes of sets $S(x)$ at that level it minimizes the probability

$$P_{\theta,\vartheta}\{\theta' \in S(X)\} \qquad \text{for all} \quad \theta' \neq \theta.$$

In order to derive confidence sets with this property from families of UMP invariant tests, we shall now investigate the relationship between equivariance of confidence sets and invariance of the associated tests.

Suppose that for each θ_0 there exists a group of transformations G_{θ_0} which leaves invariant the problem of testing $H(\theta_0) : \theta = \theta_0$, and denote by G the group of transformations generated by the totality of groups G_θ.

Lemma 6.11.1 (i) *Let $S(x)$ be any class of confidence sets that is equivariant under G, and let $A(\theta) = \{x : \theta \in S(x)\}$; then the acceptance region $A(\theta)$ is invariant under G_θ for each θ.*

(ii) *If in addition, for each θ_0 the acceptance region $A(\theta_0)$ is UMP invariant for testing $H(\theta_0)$ at level α, the class of confidence sets $S(x)$ is uniformly most accurate among all equivariant confidence sets at confidence level* $1 - \alpha$.

Proof. (i): Consider any fixed θ, and let $g \in G_\theta$. Then

$$\begin{aligned} gA(\theta) &= \{gx : \theta \in S(x)\} = \{x : \theta \in S(g^{-1}x)\} = \{x : \theta \in g^{*-1}S(x)\} \\ &= \{x : \bar{g}\theta \in S(x)\} = \{x : \theta \in S(x)\} = A(\theta). \end{aligned}$$

Here the third equality holds because $S(x)$ is equivariant, and the fifth one because $g \in G_\theta$ and therefore $\bar{g}\theta = \theta$.

(ii): If $S'(x)$ is any other equivariant class of confidence sets at the prescribed level, the associated acceptance regions $A'(\theta)$ by (i) define invariant tests of the hypotheses $H(\theta)$. It follows that these tests are uniformly at most as powerful as those with acceptance regions $A(\theta)$ and hence that

$$P_{\theta,\vartheta}\{\theta' \in S(X)\} \le P_{\theta,\vartheta}\{\theta' \in S'(X)\} \qquad \text{for all} \quad \theta' \neq \theta,$$

as was to be proved. ■

It is an immediate consequence of the lemma that if UMP invariant acceptance regions $A(\theta)$ have been found for each hypothesis $H(\theta)$ (invariant with respect to G_θ), and if the confidence sets $S(x) = \{\theta : x \in A(\theta)\}$ are equivariant under G, then they are uniformly most accurate equivariant.

Example 6.11.2 Under the assumptions of Example 6.11.1, the problem of testing $\xi = \xi_0$, $\eta = \eta_0$ is invariant under the group G_{ξ_0,η_0} of orthogonal transformations about the point (ξ_0, η_0):

$$\begin{aligned} X' - \xi_0 &= a_{11}(X - \xi_0) + a_{12}(Y - \eta_0), \\ Y' - \eta_0 &= a_{21}(X - \xi_0) + a_{22}(Y - \eta_0), \end{aligned}$$

where the matrix (a_{ij}) is orthogonal. There exists under this group a UMP invariant test, which has the acceptance region (Problem 7.8)

$$(X - \xi_0)^2 + (Y - \eta_0)^2 \le C.$$

Let G_0 be the smallest group containing the groups $G_{\xi,\eta}$, for all ξ, η. Since this is a subgroup of the group G of Example 6.11.1 (the two groups actually coincide, but this is immaterial for the argument), the confidence sets $(X - \xi)^2 + (Y - \eta)^2 \le C$ are equivariant under G_0 and hence uniformly most accurate equivariant. ■

Example 6.11.3 Let $X_1, \dots, X_n$ be independently normally distributed with mean ξ and variance σ^2. Confidence intervals for ξ are based on the hypotheses $H(\xi_0) : \xi = \xi_0$, which are invariant under the groups G_{ξ_0} of transformations $X_i' = a(X_i - \xi_0) + \xi_0$ $(a \ne 0)$. The UMP invariant test of $H(\xi_0)$ has acceptance region

$$\frac{\sqrt{(n-1)n}|\bar{X} - \xi_0|}{\sqrt{\sum(X_i - \bar{X})^2}} \le C,$$

and the associated confidence intervals are

$$\bar{X} - \frac{C}{\sqrt{n(n-1)}}\sqrt{\sum(X_i - \bar{X})^2} \le \xi \le \bar{X} + \frac{C}{\sqrt{n(n-1)}}\sqrt{\sum(X_i - \bar{X})^2}. \quad (6.37)$$

The group G in the present case consists of all transformations $g : X_i' = aX_i + b$ $(a \ne 0)$, which on ξ induces the transformation $\bar{g} : \xi' = a\xi + b$. Application of the associated transformation g^* to the interval (6.37) takes it into the set of points $a\xi + b$ for which ξ satisfies (6.37), that is, into the interval with end points

$$a\bar{X} + b - \frac{|a|C}{\sqrt{n(n-1)}}\sqrt{\sum(X_i - \bar{X})^2}, \qquad a\bar{X} + b + \frac{|a|C}{\sqrt{n(n-1)}}\sqrt{\sum(X_i - \bar{X})^2}$$

Since this coincides with the interval obtained by replacing X_i in (6.37) with $aX_i + b$, the confidence intervals (6.37) are equivariant under G_0 and hence uniformly most accurate equivariant. ■

Example 6.11.4 In the two-sample problem of Section 6.9, assume the shift model in which the X's and Y's have densities $f(x)$ and $g(y) = f(y - \Delta)$ respectively, and consider the problem of obtaining confidence intervals for the shift parameter Δ which are distribution-free in the sense that the coverage probability is independent of the true f. The hypothesis $H(\Delta_0) : \Delta = \Delta_0$ can be tested, for example, by means of the Wilcoxon test applied to the observations X_i, $Y_j - \Delta_0$, and confidence sets for Δ can then be obtained by the usual inversion process. The resulting confidence intervals are of the form $D_{(k)} < \Delta < D_{(mn+1-k)}$ where $D_{(1)} < \cdots < D_{(mn)}$ are the mn ordered differences $Y_j - X_i$. [For details see Problem 6.52 and for fuller accounts nonparametric books such as Randles and Wolfe (1979), Gibbons and Chakraborti (1992) and Lehmann (1998).] By their construction, these intervals have coverage probability $1 - \alpha$, which is independent of f. However, the invariance considerations of Sections 6.8 and 6.9 do not apply. The hypothesis $H(\Delta_0)$ is invariant under the transformations $X_i' = \rho(X_i)$, $Y_j' = \rho(Y_j - \Delta_0) + \Delta_0$ with ρ continuous and strictly increasing, but the shift model, and hence the problem under consideration, is not invariant under these transformations. ■

6.12 Average Smallest Equivariant Confidence Sets

In the examples considered so far, the invariance and equivariance properties of the confidence sets corresponded to invariant properties of the associated tests. In the following examples this is no longer the case.

Example 6.12.1 Let $X_1, \ldots, X_n$, be a sample from $N(\xi, \sigma^2)$, and consider the problem of estimating σ^2.

The model is invariant under translations $X_i' = X_i + a$, and sufficiency and invariance reduce the data to $S^2 = \sum(X_i - \bar{X})^2$. The problem of estimating σ^2 by confidence sets also remains invariant under scale changes $X_i' = bX_i$, $S' = bS$, $\sigma' = b\sigma$ $(0 < b)$, although these do not leave the corresponding problem of testing the hypothesis $\sigma = \sigma_0$ invariant. (Instead, they leave invariant the *family* of these testing problems, in the sense that they transform one such hypothesis into another.) The totality of equivariant confidence sets based on S is given by

$$\frac{\sigma^2}{S^2} \in A, \tag{6.38}$$

where A is any fixed set on the line satisfying

$$P_{\sigma=1}\left(\frac{1}{S^2} \in A\right) = 1 - \alpha. \tag{6.39}$$

That any set $\sigma^2 \in S^2 \cdot A$ is equivariant is obvious. Conversely, suppose that $\sigma^2 \in C(S^2)$ is an equivariant family of confidence sets for σ^2. Then $C(S^2)$ must satisfy $b^2 C(S^2) = C(b^2 S^2)$ and hence

$$\sigma^2 \in C(S^2) \quad \text{if and only if} \quad \frac{\sigma^2}{S^2} \in \frac{1}{S^2} C(S^2) = C(1),$$

which establishes (6.38) with $A = C(1)$.

Among the confidence sets (6.38) with A satisfying (6.39) there does not exist one that uniformly minimizes the probability of covering false values (Problem 6.73). Consider instead the problem of determining the confidence sets that are physically smallest in the sense of having minimum Lebesgue measure. This requires minimizing $\int_A dv$ subject to (6.39). It follows from the Neyman-Pearson lemma that the minimizing A^* is

$$A^* = \{v : p(v) > C\}, \tag{6.40}$$

where $p(v)$ is the density of $V = 1/S^2$ when $\sigma = 1$, and where C is determined by (6.39). Since $p(v)$ is unimodal (Problem 6.74), these smallest confidence sets are intervals, $aS^2 < \sigma^2 < bS^2$. Values of a and b are tabled by Tate and Klett (1959), who also table the corresponding (different) values a', b' for the uniformly most accurate unbiased confidence intervals $a'S^2 < \sigma^2 < b'S^2$ (given in Example 5.5.1).

Instead of minimizing the Lebesgue measure $\int_A dv$ of the confidence sets A, one may prefer to minimize the scale-invariant measure

$$\int_A \frac{1}{v}\, dv. \tag{6.41}$$

To an interval (a, b), (6.41) assigns, in place of its length $b - a$, its logarithmic length $\log b - \log a = \log(b/a)$. The optimum solution A^{**} with respect to this new measure is again obtained by applying the Neyman Pearson lemma, and is given by

$$A^{**} = \{v : vp(v) > C\}, \tag{6.42}$$

which coincides with the uniformly most accurate unbiased confidence sets [Problem 6.75(i)].

One advantage of minimizing (6.41) instead of Lebesgue measure is that it then does not matter whether one estimates σ or σ^2 (or σ^r for some other power of r), since under (6.41), if (a, b) is the best interval for σ, then (a^r, b^r) is the best interval for σ^r [Problem 6.75(ii)]. ■

Example 6.12.2 Let X_i $(i = 1, \ldots, r)$ be independently normally distributed as $N(\xi, 1)$. A slight generalization of Example 6.11.2 shows that uniformly most accurate equivariant confidence sets for $(\xi_1, \ldots, \xi_r)$ exist with respect to the group G of all rigid transformations and are given by

$$\sum (X_i - \xi_i)^2 \le C. \tag{6.43}$$

Suppose that the context of the problem does not possess the symmetry which would justify invoking invariance with respect to G, but does allow the weaker assumption of invariance under the group G_0 of translations $X_i' = X_i + a_i$. The totality of equivariant confidence sets with respect to G_0 is given by

$$(X_1 - \xi_1, \ldots, X_r - \xi_r) \in A, \tag{6.44}$$

where A is any fixed set in r-space satisfying

$$P_{\xi_1 = \cdots = \xi_r = 0}((X_1, \ldots, X_r) \in A) = 1 - \alpha. \tag{6.45}$$

Since uniformly most accurate equivariant confidence sets do not exist (Problem 6.73), let us consider instead the problem of determining the confidence sets of smallest Lebesgue measure. (This measure is invariant under G_0.) This is given by (6.40) with $v = (v_1, \ldots, v_r)$ and $p(v)$ the density of $(X_1, \ldots, X_r)$ when $\xi_1 = \cdots = \xi_r = 0$, and hence coincides with (6.43).

Quite surprisingly, the confidence sets (6.43) are inadmissible if and only if $r \ge 3$. A further discussion of this fact and references are deferred to Example 8.5.4. ■

Example 6.12.3 In the preceding example, suppose that the X_i are distributed as $N(\xi_i, \sigma^2)$ with σ^2 unknown, and that a variable S^2 is available for estimating σ^2. Of S^2 assume that it is independent of the X's and that S^2/σ^2 has a χ^2 -distribution with f degrees of freedom.

The estimation of $(\xi_1, \ldots, \xi_r)$ by confidence sets on the basis of X's and S^2 remains invariant under the group G_0 of transformations

$$X_i' = bX_i + a_i, \qquad S' = bS, \qquad \xi_i' = b\xi_i + a_i, \quad \sigma' = b\sigma,$$

and the most general equivariant confidence set is of the form

$$\left(\frac{X_1 - \xi_1}{S}, \ldots, \frac{X_r - \xi_r}{S}\right) \in A, \tag{6.46}$$

where A is any fixed set in r-space satisfying

$$P_{\xi_1 = \cdots = \xi_r = 0}\left[\left(\frac{X_1}{S}, \ldots, \frac{X_r}{S}\right) \in A\right] = 1 - \alpha. \tag{6.47}$$

The confidence sets (6.46) can be written as

$$(\xi_1, \ldots, \xi_r) \in (X_1, \ldots, X_r) - SA, \tag{6.48}$$

where $-SA$ is the set obtained by multiplying each point of A by the scalar $-S$.

To see (6.48), suppose that $C(X_1, \ldots, X_r; S)$ is an equivariant confidence set for $(\xi_1, \ldots, \xi_r)$. Then the r-dimensional set C must satisfy

$$C(bX_1 + a_1, \ldots, bX_r + a_r; bS) = b[C(X_1, \ldots, X_r; S)] + (a_1, \ldots, a_r)$$

for all $a_1, \ldots, a_r$ and all $b > 0$. It follows that $(\xi_1, \ldots, \xi_r) \in C$ if and only if

$$\begin{aligned}\left(\frac{X_1 - \xi_1}{S}, \ldots, \frac{X_r - \xi_r}{S}\right) \in \frac{(X_1, \ldots, X_r) - C(X_1, \ldots, X_r; S)}{S} &= C(0, \ldots, 0; 1) \\ &= A.\end{aligned}$$

The equivariant confidence sets of smallest volume are obtained by choosing for A the set A^* given by (6.40) with $v = (v_1, \ldots, v_r)$ and $p(v)$ the joint density of $(X_1/S, \ldots, X_r/S)$ when $\xi_1 = \cdots = \xi_r = 0$. This density is a decreasing function of $\sum v_i^2$ (Problem 6.76), and the smallest equivariant confidence sets are therefore given by

$$\sum (X_i - \xi_i)^2 \leq CS^2. \tag{6.49}$$

[Under the larger group G generated by all rigid transformations of $(X_1, \ldots, X_r)$ together with the scale changes $X_i' = bX_i$, $S' = bS$, the same sets have the stronger property of being uniformly most accurate equivariant; see Problem 6.77.] ■

Examples 6.12.1–6.12.3 have the common feature that the equivariant confidence sets $S(X)$ for $\theta = (\theta_1, \ldots, \theta_r)$ are characterized by an r-valued *pivotal quantity*, that is, a function $h(X, \theta) = (h_1(X, \theta), \ldots, h_r(X, \theta))$ of the observations X and parameters θ being estimated that has a fixed distribution, and such that the most general equivariant confidence sets are of the form

$$h(X, \theta) \in A \tag{6.50}$$

for some fixed set A.[6] When the functions h_i are linear in θ, the confidence sets $C(X)$ obtained by solving (6.50) for θ are linear transforms of A (with random coefficients), so that the volume or invariant measure of $C(X)$ is minimized by minimizing

$$\int_A \rho(v_1, \ldots, v_r)\, dv_1 \ldots dv_r \tag{6.51}$$

for the appropriate ρ. The problem thus reduces to that of minimizing (6.51) subject to

$$P_{\theta_0}\{h(X, \theta_0) \in A\} = \int_A p(v_1, \ldots, v_r)\, dv_1 \ldots dv_r = 1 - \alpha, \tag{6.52}$$

where $p(v_1, \ldots, v_r)$ is the density of the pivotal quantity $h(X, \theta)$. The minimizing A is given by

$$A^* = \left\{ v : \frac{p(v_1, \ldots, v_r)}{\rho(v_1, \ldots, v_r)} > C \right\}, \tag{6.53}$$

with C determined by (6.52).

The following is one more illustration of this approach.

Example 6.12.4 Let $X_1, \ldots, X_m$ and $Y_1, \ldots, Y_n$ be samples from $N(\xi, \sigma^2)$ and $N(\eta, \tau^2)$ respectively, and consider the problem of estimating $\Delta = \tau^2/\sigma^2$. Sufficiency and invariance under translations $X_i' = X_i + a_1$, $Y_j' = Y_j + a_2$ reduce the data to $S_X^2 = \sum(X_i, -\bar{X})^2$ and $S_Y^2 = \sum(Y_j - \bar{Y})^2$. The problem of estimating Δ also remains invariant under the scale changes

$$X_i' = b_1 X_i, \quad Y_j' = b_2 Y_j, \qquad 0 < b_1, b_2 < \infty,$$

which induce the transformations

$$S_X' = b_1 S_X, \qquad S_Y' = b_2 S_Y, \qquad \sigma' = b_1 \sigma, \qquad \tau' = b_2 \tau. \tag{6.54}$$

The totality of equivariant confidence sets for Δ is given by $\Delta/V \in A$, where $V = S_Y^2/S_X^2$ and A is any fixed set on the line satisfying

$$P_{\Delta=1}\left(\frac{1}{V} \in A\right) = 1 - \alpha. \tag{6.55}$$

To see this, suppose that $C(S_X, S_Y)$ are any equivariant confidence sets for Δ. Then C must satisfy

$$C(b_1 S_X, b_2 S_Y) = \frac{b_2^2}{b_1^2} C(S_X, S_Y), \tag{6.56}$$

and hence $\Delta \in C(S_X, S_Y)$ if and only if the pivotal quantity V/Δ satisfies

$$\frac{\Delta}{V} = \frac{S_X^2 \Delta}{S_Y^2} \in \frac{S_X^2}{S_Y^2} C(S_X, S_Y) = C(1, 1) = A.$$

As in Example 6.12.1, one may now wish to choose A so as to minimize either its Lebesgue measure $\int_A dv$ or the invariant measure $\int_A (1/v)\, dv$. The resulting

[6] More general results concerning the relationship of equivariant confidence sets and pivotal quantities are given in Problems 6.69–6.72.

confidence sets are of the form

$$p(v) > C \quad \text{and} \quad vp(v) > C \tag{6.57}$$

respectively. In both cases, they are intervals $V/b < \Delta < V/a$ [Problem 6.78(i)]. The values of a and b minimizing Lebesgue measure are tabled by Levy and Narula (1974); those for the invariant measure coincide with the uniformly most accurate unbiased intervals [Problem 6.78(ii)]. ■

6.13 Confidence Bands for a Distribution Function

Suppose that $X = (X_1, \ldots, X_n)$ is a sample from an unknown continuous cumulative distribution function F, and that lower and upper bounds L_X and M_X are to be determined such that with preassigned probability $1 - \alpha$ the inequalities

$$L_X(u) \le F(u) \le M_X(u) \qquad \text{for all } u$$

hold for all continuous cumulative distribution functions F. This problem is invariant under the group G of transformations

$$X_i' = g(X_i), \qquad i = 1, \ldots, n,$$

where g is any continuous strictly increasing function. The induced transformation in the parameter space is $\bar{g}F = F(g^{-1})$.

If $S(x)$ is the set of continuous cumulative distribution functions

$$S(x) = \{F : L_x(u) \le F(u) \le M_x(u) \text{ for all } u\},$$

then

$$\begin{aligned} g^*S(x) &= \{\bar{g}F : L_x(u) \le F(u) \le M_x(u) \text{ for all } u\} \\ &= \{F : L_x[g^{-1}(u)] \le F(u) \le M_x[g^{-1}(u)] \text{ for all } u\}. \end{aligned}$$

For an equivariant procedure, this must coincide with the set

$$S(gx) = \left\{F : L_{g(x_1),\ldots,g(x_n)}(u) \le F(u) \le M_{g(x_1),\ldots,g(x_n)}(u) \text{ for all } u\right\}.$$

The condition of equivariance is therefore

$$\begin{aligned} L_{g(x_1),\ldots,g(x_n)}[g(u)] &= L_x(u), \\ M_{g(x_1),\ldots,g(x_n)}[g(u)] &= M_x(u) \qquad \text{for all } x \text{ and } u. \end{aligned}$$

To characterize the totality of equivariant procedures, consider the *empirical distribution function* (EDF) T_x given by

$$T_x(u) = \frac{i}{n} \quad \text{for} \quad x_{(i)} \le u < x_{(i+1)}, \quad i = 0, \ldots, n,$$

where $x_{(1)} < \cdots < x_{(n)}$ is the ordered sample and where $x_{(0)} = -\infty$, $x_{(n+1)} = \infty$. Then a necessary and sufficient condition for L and M to satisfy the above equivariance condition is the existence of numbers $a_0, \ldots, a_n$; $a_0', \ldots, a_n'$ such that

$$L_x(u) = a_i, \quad M_x(u) = a_i' \qquad \text{for} \quad x_{(i)} < u < x_{(i+1)}.$$

That this condition is sufficient is immediate. To see that it is also necessary, let u, u' be any two points satisfying $x_{(i)} < u < u' < x_{(i+1)}$. Given any $y_1, \dots, y_n$ and v with $y_{(i)} < v < y_{(i+1)}$, there exist g, $g' \in G$ such that

$$g(y_{(i)}) = g'(y_{(i)}) = x_{(i)}, \quad g(v) = u, \quad g'(v) = u'.$$

If L_x, M_x are equivariant, it then follows that $L_x(u') = L_y(v)$ and $L_x(u) = L_y(v)$, and hence that $L_x(u') = L_x(u)$ and similarly $M_x(u') = M_x(u)$, as was to be proved. This characterization shows L_x and M_x to be step functions whose discontinuity points are restricted to those of T_x.

Since any two continuous strictly increasing cumulative distribution functions can be transformed into one another through a transformation $\bar{g}$, it follows that all these distributions have the same probability of being covered by an equivariant confidence band. (See Problem 6.84.) Suppose now that F is continuous but no longer strictly increasing. If I is any interval of constancy of F, there are no observations in I, so that I is also an interval of constancy of the sample cumulative distribution function. It follows that the probability of the confidence band covering F is not affected by the presence of I and hence is the same for all continuous cumulative distribution functions F.

For any numbers a_i, a'_i let Δ_i, Δ'_i be determined by

$$a_i = \frac{i}{n} - \Delta_i, \qquad a'_i = \frac{i}{n} - \Delta'_i$$

Then it was seen above that any numbers $\Delta_0, \dots, \Delta_n; \Delta'_0, \dots, \Delta'_n$ define a confidence band for F, which is equivariant and hence has constant probability of covering the true F. From these confidence bands a test can be obtained of the hypothesis of *goodness of fit* $F = F_0$ that the unknown F equals a hypothetical distribution F_0. The hypothesis is accepted if F_0 ties entirely within the band, that is, if

$$-\Delta_i < F_0(u) - T_x(u) < \Delta'_i$$
$$\text{for all} \quad x_{(i)} < u < x_{(i+1)} \quad \text{and all} \quad i = 1, \dots, n.$$

Within this class of tests there exists no UMP member, and the most common choice of the Δ's is $\Delta_i = \Delta'_i = \Delta$ for all i. The acceptance region of the resulting *Kolmogorov-Smirnov test* can be written as

$$\sup_{-\infty < u < \infty} |F_0(u) - T_x(u)| < \Delta. \tag{6.58}$$

Tables of the null distribution of the Kolmogorov-Smirnov statistic are given by Birnbaum (1952). For large n, approximate critical values can be obtained from the limit distribution K of $\sqrt{n} \sup |F_0(u) - T_x(u)|$, due to Kolmogorov and tabled by Smirnov (1948). Derivations of K can be found, for example, in Feller (1948), Billingsley (1968), and Hájek, Sidák and Sen (1999). The large sample properties of this test will be studied in Example 11.2.12 and Section 14.2. The more general problem of testing goodness-of-fit will be presented in Chapter 14.

6.14 Problems

Section 6.1

Problem 6.1 Let G be a group of measurable transformations of $(\mathcal{X}, \mathcal{A})$ leaving $\mathcal{P} = \{P_\theta, \theta \in \Omega\}$ invariant, and let $T(x)$ be a measurable transformation to $(\mathcal{T}, \mathcal{B})$. Suppose that $T(x_1) = T(x_2)$ implies $T(gx_1) = T(gx_2)$ for all $g \in G$, so that G induces a group G^* on $\mathcal{T}$ through $g^*T(x) = T(gx)$, and suppose further that the induced transformations g^* are measurable $\mathcal{B}$. Then G^* leaves the family $\mathcal{P}^T = \{P_\theta^T, \theta \in \Omega\}$ of distributions of T invariant.

Section 6.2

Problem 6.2 (i) Let $\mathcal{X}$ be the totality of points $x = (x_1, \ldots, x_n)$ for which all coordinates are different from zero, and let G be the group of transformations $x_i' = cx_i, c > 0$. Then a maximal invariant under G is $(\text{sgn } x_n, x_1/x_n, \ldots, x_{n-1}/x_n)$ where sgn x is 1 or -1 as x is positive or negative.

(ii) Let $\mathcal{X}$ be the space of points $x = (x_1, \ldots, x_n)$ for which all coordinates are distinct, and let G be the group of all transformations $x_i' = f(x_i), i = 1, \ldots, n$, such that f is a $1:1$ transformation of the real line onto itself with at most a finite number of discontinuities. Then G is transitive over $\mathcal{X}$.

[(ii): Let $x = (x_1, \ldots, x_n)$ and $x' = (x_1', \ldots, x_n')$ be any two points of $\mathcal{X}$. Let $I_1, \ldots, I_n$ be a set of mutually exclusive open intervals which (together with their end points) cover the real line and such that $x_j \in I_j$. Let $I_1', \ldots, I_n'$ be a corresponding set of intervals for $x_1', \ldots, x_n'$. Then there exists a transformation f which maps each I_j continuously onto I_j', maps x_j into x_j', and maps the set of $n-1$ end points of $I_1, \ldots, I_n$ onto the set of end points of $I_1', \ldots, I_n'$.]

Problem 6.3 Suppose M is any $m \times p$ matrix. Show that $M^T M$ is positive semidefinite. Also, show the rank of $M^T M$ equals the rank of M, so that in particular $M^T M$ is nonsingular if and only if $m \geq p$ and M is of rank p.

Problem 6.4 (i) A sufficient condition for (6.8) to hold is that D is a normal subgroup of G.

(ii) If G is the group of transformations $x' = ax + b, a \neq 0, -\infty < b < \infty$, then the subgroup of translations $x' = x + b$ is normal but the subgroup $x' = ax$ is not.

[The defining property of a normal subgroup is that given $d \in D, g \in G$, there exists $d' \in D$ such that $gd = d'g$. The equality $s(x_1) = s(x_2)$ implies $x_2 = dx_1$ for some $d \in D$, and hence $ex_2 = edx_1 = d'ex_1$. The result (i) now follows, since s is invariant under D.]

Section 6.3

Problem 6.5 Prove statements (i)-(iii) of Example 6.3.1.

Problem 6.6 Prove Theorem 6.3.1
(i) by analogy with Example 6.3.1, and
(ii) by the method of Example 6.3.2. [*Hint:* A maximal invariant under G is the set $\{g_1 x, \ldots, g_N x\}$.

Problem 6.7 Consider the situation of Example 6.3.1 with $n = 1$, and suppose that f is strictly increasing on $(0, 1)$.
(i) The likelihood ratio test rejects if $X < \alpha/2$ or $X > 1 - \alpha/2$.
(ii) The MP invariant test agrees with the likelihood ratio test when f is convex.
(iii) When f is concave, the MP invariant test rejects when

$$\frac{1}{2} - \frac{\alpha}{2} < X < \frac{1}{2} + \frac{\alpha}{2},$$

and the likelihood ratio test is the least powerful invariant test against both alternatives and has power $< \alpha$.

Problem 6.8 Let X, Y have the joint probability density $f(x, y)$. Then the integral $h(z) = \int_{-\infty}^{\infty} f(y - z, y) dy$ is finite for almost all z, and is the probability density of $Z = Y - X$.
[Since $P\{Z \le b\} = \int_{-\infty}^{b} h(z) dz$, it is finite and hence h is finite almost everywhere.]

Problem 6.9 (i) Let $X = (X_1, \ldots, X_n)$ have probability density $(1/\theta^n) f[(x_1 - \xi)/\theta, \ldots, (x_n - \xi)/\theta]$, where $-\infty < \xi < \infty, 0 < \theta$ are unknown, and where f is even. The problem of testing $f = f_0$ against $f = f_1$ remains invariant under the transformations $x'i = a x_i + b$ $(i = 1, \ldots, n)$, $a \ne 0$, $-\infty < b < \infty$ and the most powerful invariant test is given by the rejection region

$$\int_{-\infty}^{\infty} \int_{0}^{\infty} v^{n-2} f_1(v x_1 + u, \ldots, v x_n + u)\, dv\, du$$
$$> C \int_{-\infty}^{\infty} \int_{0}^{\infty} v^{n-2} f_0(v x_1 + u, \ldots, v x_n + u)\, dv\, du.$$

(ii) Let $X = (X_1, \ldots, X_n)$ have probability density $f(x_1 - \sum_{j=1}^{k} w_{1j}\beta_j, \ldots, x_n - \sum_{j=1}^{k} w_{nj}\beta_j)$ where $k < n$, the w's are given constants, the matrix (w_{ij}) is of rank k, the β's are unknown, and we wish to test $f = f_0$ against $f = f_1$. The problem remains invariant under the transformations $x_i' = x_i + \Sigma_{j=1}^{k} w_{ij}\gamma_j$, $-\infty < \gamma_1, \ldots, \gamma_k < \infty$, and the most powerful invariant test is given by the rejection region

$$\frac{\int \cdots \int f_1(x_1 - \sum w_{1j}\beta_j, \ldots, x_n - \sum w_{nj}\beta_j) d\beta_1, \ldots, d\beta_k}{\int \cdots \int f_0(x_1 - \sum w_{1j}\beta_j, \ldots, x_n - \sum w_{nj}\beta_j) d\beta_1, \ldots, d\beta_k} > C.$$

[A maximal invariant is given by $y =$

$$\left(x_1 - \sum_{r=n-k+1}^{n} a_{1_r} x_r,\ x_2 - \sum_{r=n-k+1}^{n} a_{2_r} x_r, \ldots, x_{n-k} - \sum_{r=n-k+1}^{n} a_{n-k,r} x_r \right)$$

for suitably chosen constants a_{ir}.]

Problem 6.10 Let $X_1, \dots, X_m; Y_1, \dots, Y_n$ be samples from exponential distributions with densities for $\sigma^{-1}e^{-(x-\xi)/\sigma}$, for $x \geq \xi$, and $\tau^{-1}e^{-(y-n)/\tau}$ for $y \geq \eta$.

(i) For testing $\tau/\sigma \leq \Delta$ against $\tau/\sigma > \Delta$, there exists a UMP invariant test with respect to the group $G: X_i' = aX_i + b, Y_j' = aY_j + c, a > 0, -\infty < b, c < \infty$, and its rejection region is

$$\frac{\sum[y_j - \min(y_1, \dots, y_n)]}{\sum[x_i - \min(x_1, \dots, x_m)]} > C.$$

(ii) This test is also UMP unbiased.

(iii) Extend these results to the case that only the r smallest X's and the s smallest Y's are observed.

[(ii): See Problem 5.15.]

Problem 6.11 If $X_1, \dots, X_n$ and $Y_1, \dots, Y_n$ are samples from $N(\xi, \sigma^2)$ and $N(\eta, \tau^2)$ respectively, the problem of testing $\tau^2 = \sigma^2$ against the two-sided alternatives $\tau^2 \neq \sigma^2$ remains invariant under the group G generated by the transformations $X_i' = aX_i + b$, $Y_i' = aY_i + c$, $(a \neq 0)$, and $X_i' = Y_i$, $Y_i' = X_i$. There exists a UMP invariant test under G with rejection region

$$W = \max\left\{\frac{\sum(Y_i - \bar{Y})^2}{\sum(X_i = \bar{X})}, \frac{\sum(X_i = \bar{X})^2}{\sum(Y_i - \bar{Y})^2}\right\} \geq k.$$

[The ratio of the probability densities of W for $\tau^2/\sigma^2 = \Delta$ and $\tau^2/\sigma^2 = 1$ is proportional to $[(1 + w)/(\Delta + w)]^{n-1} + [(1 + w)/(1 + \Delta w)]^{n-1}$ for $w \geq 1$. The derivative of this expression is ≥ 0 for all Δ.]

Problem 6.12 Let $X_1, \dots, X_n$ be a sample from a distribution with density

$$\frac{1}{\tau^n} f\left(\frac{x_1}{\tau}\right) \dots f\left(\frac{x_n}{\tau}\right),$$

where $f(x)$ is either zero for $x < 0$ or symmetric about zero. The most powerful scale-invariant test for testing $H: f = f_0$ against $K: f = f_1$ rejects when

$$\frac{\int_0^\infty v^{n-1} f_1(vx_1) \dots f_1(vx_n)\, dv}{\int_0^\infty v^{n-1} f_0(vx_1) \dots f_0(vx_n)\, dv} > C.$$

Problem 6.13 *Normal vs. double exponential.* For $f_0(x) = e^{-x^2/2}/\sqrt{2\pi}$, $f_1(x) = e^{-|x|}/2$, the test of the preceding problem reduces to rejecting when $\sqrt{\sum x_i^2}/\sum |x_i| < C$.

(Hogg, 1972.)

Note. The corresponding test when both location and scale are unknown is obtained in Uthoff (1973). Testing normality against Cauchy alternatives is discussed by Franck (1981).

Problem 6.14 *Uniform vs. triangular.*

(i) For $f_0(x) = 1$ $(0 < x < 1)$, $f_1(x) = 2x$ $(0 < x < 1)$, the test of Problem 6.12 reduces to rejecting when $T = x_{(n)}/\bar{x} < C$.

(ii) Under f_0, the statistic $2n \log T$ is distributed as χ^2_{2n}.

(Quesenberry and Starbuck, 1976.)

Problem 6.15 Show that the test of Problem 6.9(i) reduces to

(i) $[x_{(n)} - x_{(1)}]/S < c$ for normal vs. uniform;

(ii) $[\bar{x} - x_{(1)}]/S < c$ for normal vs. exponential;

(iii) $[\bar{x} - x_{(1)}]/[x_{(n)} - x_{(1)}] < c$ for uniform vs. exponential.

(Uthoff, 1970.)

Note. When testing for normality, one is typically not interested in distinguishing the normal from some other given shape but would like to know more generally whether the data are or are not consonant with a normal distribution. This is a special case of the problem of testing for goodness of fit, which is briefly discussed at the end of Section 6.13 and forms the topic of Chapter 14; also, see the many references in the notes to Chapter 14.

Problem 6.16 Let $X_1, \ldots, X_n$ be independent and normally distributed. Suppose X_i has mean μ_i and variance σ^2 (which is the same for all i). Consider testing the null hypothesis that $\mu_i = 0$ for all i. Using invariance considerations, find a UMP invariant test with respect to a suitable group of transformations in each of the following cases:
(i). σ^2 is known and equal to one.
(ii). σ^2 is unknown.

Section 6.4

Problem 6.17 (i) When testing $H : p \leq p_0$ against $K : p > p_0$ by means of the test corresponding to (6.13), determine the sample size required to obtain power β against $p = p_1$, $\alpha = .05$, $\beta = .9$ for the cases $p_0 = .1$, $p_1 = .15, .20, .25$; $p_0 = .05$, $p_1 = .10, .15, .20, .25$; $p_0 = .01$, $p_1 = .02, .05, .10, .15, .20$.

(ii) Compare this with the sample size required if the inspection is by attributes and the test is based on the total number of defectives.

Problem 6.18 *Two-sided t-test.*

(i) Let $X_1, \ldots, X_n$ be a sample from $N(\xi, \sigma^2)$. For testing $\xi = 0$ against $\xi \neq 0$, there exists a UMP invariant test with respect to the group $X_i' = cX_i$, $c \neq 0$, given by the two-sided t-test (5.17).

(ii) Let $X_1, \ldots, X_m$, and $Y_1, \ldots, Y_n$ be samples from $N(\xi, \sigma^2)$ and $N(\eta, \sigma^2)$ respectively. For testing $\eta = \xi$ against $\eta \neq \xi$ there exists a UMP invariant test with respect to the group $X_i' = aX_i + b, Y_j' = aY_j + b, a \neq 0$, given by the two-sided t-test (5.30).

(ii) Let Ω and ω be invariant under $\bar{G}$, and countable. Then the likelihood ratio $\sup_\Omega p_\theta(x)/\sup_\omega p_\theta(x)$ is almost invariant under G.

(iii) Suppose that $p_\theta(x)$ is continuous in θ for all x, that Ω is a separable pseudometric space, and that Ω and ω are invariant. Then the likelihood ratio is almost invariant under G.

Problem 6.28 *Inadmissible likelihood-ratio test.* In many applications in which a UMP invariant test exists, it coincides with the likelihood-ratio test. That this is, however, not always the case is seen from the following example. Let $P_1, \ldots, P_n$ be n equidistant points on the circle $x^2 + y^2 = 4$, and $Q_1, \ldots, Q_n$ on the circle $x^2 + y^2 = 1$. Denote the origin in the (x, y) plane by O, let $0 < \alpha \le \frac{1}{2}$ be fixed, and let (X, Y) be distributed over the $2n + 1$ points $P_1, \ldots, P_n, Q_1, \ldots, Q_n$, O with probabilities given by the following table:

	P_i	Q_i	O
H	α/n	$(1-2\alpha)/n$	α
K	p_i/n	0	$(n-1)/n$

where $\sum p_i = 1$. The problem remains invariant under rotations of the plane by the angles $2k\pi/n$ $(k = 0, 1, \ldots, n-1)$. The rejection region of the likelihood-ratio test consists of the points $P_1, \ldots, P_n$, and its power is $1/n$. On the other hand, the UMP invariant test rejects when $X = Y = 0$, and has power $(n-1)/n$.

Problem 6.29 Let G be a group of transformations of $\mathcal{X}$, and let $\mathcal{A}$ be a σ-field of subsets of $\mathcal{X}$, and μ a measure over $(\mathcal{X}, \mathcal{A})$. Then a set $A \in \mathcal{A}$ is said to be almost invariant if its indicator function is almost invariant.

(i) The totality of almost invariant sets forms a σ-field $\mathcal{A}_0$, and a critical function is almost invariant if and only if it is $\mathcal{A}_0$-measurable.

(ii) Let $\mathcal{P} = \{P_\theta, \theta \in \Omega\}$ be a dominated family of probability distributions over $(\mathcal{X}, \mathcal{A})$, and suppose that $\bar{g}\theta = \theta$ for all $\bar{g} \in \bar{G}, \theta \in \Omega$. Then the σ-field $\mathcal{A}_0$ of almost invariant sets is sufficient for $\mathcal{P}$.

[Let $\lambda = \sum c_i P_{\theta_i}$, be equivalent to $\mathcal{P}$. Then

$$\frac{dP_\theta}{d\lambda}(gx) = \frac{dP_{g^{-1}\theta}}{\sum c_i\, dP_{g^{-1}\theta_i}}(x) = \frac{dP_\theta}{d\lambda}(x) \qquad \text{(a.e. } \lambda),$$

so that $dP_\theta/d\lambda$ is almost invariant and hence $\mathcal{A}_0$-measurable.]

Problem 6.30 The UMP invariant test of Problem 6.13 is also UMP similar.

[Consider the problem of testing $\alpha = 0$ vs. $\alpha > 0$ in the two-parameter exponential family with density

$$C(\alpha, \tau) \exp\left(-\frac{\alpha}{2\tau^2}\sum x_i^2 - \frac{1-\alpha}{\tau}\sum |x_i|\right), \qquad 0 \le \alpha < 1.]$$

Note. For the analogous result for the tests of Problem 6.14, 6.15, see Quesenberry and Starbuck (1976).

Problem 6.31 The following UMP unbiased tests of Chapter 5 are also UMP invariant under change in scale:

(i) The test of $g \le g_0$ in a gamma distribution (Problem 5.30).

(ii) The test of $b_1 \le b_2$ in Problem 5.18(i).

Section 6.7

Problem 6.32 The definition of d-admissibility of a test coincides with the admissibility definition given in Section 1.8 when applied to a two-decision procedure with loss 0 or 1 as the decision taken is correct or false.

Problem 6.33 (i) The following example shows that α-admissibility does not always imply d-admissibility. Let X be distributed as $U(0, \theta)$, and consider the tests φ_1 and φ_2 which reject when respectively $X < 1$ and $X < \frac{3}{2}$ for testing $H : \theta = 2$ against $K : \theta = 1$. Then for $\alpha = \frac{3}{4}$, φ_1 and φ_2 are both α-admissible but φ_2 is not d-admissible.

(ii) Verify the existence of the test φ_0 of Example 6.7.12.

Problem 6.34 (i) The acceptance region $T_1/\sqrt{T_2} \le C$ of Example 6.7.13 is a convex set in the (T_1, T_2) plane.

(ii) In Example 6.7.13, the conditions of Theorem 6.7.1 are not satisfied for the sets $A : T_1/\sqrt{T_2} \le C$ and $\Omega' : \xi > k$.

Problem 6.35 (i) In Example 6.7.13 (continued) show that there exist C_0, C_1 such that $\lambda_0(\eta)$ and $\lambda_1(\eta)$ are probability densities (with respect to Lebesgue measure).

(ii) Verify the densities h_0 and h_1.

Problem 6.36 Verify

(i) the admissibility of the rejection region (6.24);

(ii) the expression for $I(z)$ given in the proof of Lemma 6.7.1.

Problem 6.37 Let $X_1, \ldots, X_m; Y_1, \ldots, Y_n$ be independent $N(\xi, \sigma^2)$ and $N(\eta, \sigma^2)$ respectively. The one-sided t-test of $H : \delta = \xi/\sigma \le 0$ is admissible against the alternatives (i) $0 < \delta < \delta_1$ for any $\delta_1 > 0$; (ii) $\delta > \delta_2$ for any $\delta_2 > 0$.

Problem 6.38 For the model of the preceding problem, generalize Example 6.7.13 (continued) to show that the two-sided t-test is a Bayes solution for an appropriate prior distribution.

Problem 6.39 Suppose $X = (X_1, \ldots, X_k)^T$ is multivariate normal with unknown mean vector $(\theta_1, \ldots, \theta_k)^T$ and known nonsingular covariance matrix Σ. Consider testing the null hypothesis $\theta_i = 0$ for all i against $\theta_i \ne 0$ for some i. Let C be any closed convex subset of k-dimensional Euclidean space, and let ϕ be the test that accepts the null hypothesis if X falls in C. Show that ϕ is admissible. *Hint*: First assume Σ is the identity and use Theorem 6.7.1. [An alternative proof is provided by Strasser (1985, Theorem 30.4).]

Section 6.9

Problem 6.40 *Wilcoxon two-sample test.* Let $U_{ij} = 1$ or 0 as $X_i < Y_j$ or $X_i > Y_j$, and let $U = \sum\sum U_{ij}$ be the number of pairs X_i, Y_j with $X_i < Y_j$.

(i) Then $U = \sum S_i - \frac{1}{2}n(n+1)$, where $S_1 < \cdots < S_n$ are the ranks of the Y's so that the test with rejection region $U > C$ is equivalent to the Wilcoxon test.

(ii) Any given arrangement of x's and y's can be transformed into the arrangement $x \dots xy \dots y$ through a number of interchanges of neighboring elements. The smallest number of steps in which this can be done for the observed arrangement is $mn - U$.

Problem 6.41 *Expectation and variance of Wilcoxon statistic.* If the X's and Y's are samples from continuous distributions F and G respectively, the expectation and variance of the Wilcoxon statistic U defined in the preceding problem are given by

$$E\left(\frac{U}{mn}\right) = P\{X < Y\} = \int F\,dG \tag{6.59}$$

and

$$\begin{aligned} mnVar\left(\frac{U}{mn}\right) &= \int F\,dG + (n-1)\int (1-G)^2\,dF \\ &\quad +(m-1)\int F^2\,dG - (m+n-1)\left(\int F\,dG\right)^2 . \end{aligned} \tag{6.60}$$

Under the hypothesis $G = F$, these reduce to

$$E\left(\frac{U}{mn}\right) = \frac{1}{2}, \qquad Var\left(\frac{U}{mn}\right) = \frac{m+n+1}{12mn}. \tag{6.61}$$

Problem 6.42 (i) Let $Z_1, \dots, Z_N$ be independently distributed with densities $f_1, \dots, f_N$, and let the rank of Z_i be denoted by T_i. If f is any probability density which is positive whenever at least one of the f_i is positive, then

$$P\{T_1 = t_1, \dots, T_N = t_n\} = \frac{1}{N!}E\left[\frac{f_1\left(V_{(t_1)}\right)}{f\left(V_{(t_1)}\right)} \cdots \frac{f_N\left(V_{(t_N)}\right)}{f\left(V_{(t_N)}\right)}\right]. \tag{6.62}$$

where $V_{(1)} < \cdots < V_{(N)}$ is an ordered sample from a distribution with density f.

(ii) If $N = m + n$, $f_1 = \cdots = f_m = f$, $f_{m+1} = \cdots = f_{m+n} = g$, and $S_1 < \cdots < S_n$ denote the ordered ranks of $Z_{m+1}, \dots, Z_{m+n}$ among all the Z's, the probability distribution of $S_1, \dots, S_n$ is given by (6.27).

[(i): The probability in question is $\int \dots \int f_1(z_1) \dots f_N(z_N)\,dz_1 \cdots dz_N$ integrated over the set in which z_i is the t_ith smallest of the z's for $i = 1, \dots, N$. Under the transformation $w_{t_i} = z_i$ the integral becomes $\int \dots \int f_1(w_{t_1}) \dots f_N(w_{t_N})\,dw_1 \cdots dw_N$ integrated over the set $w_1 < \cdots < w_N$. The desired result now follows from the fact that the probability density of the order statistics $V_{(1)} < \cdots < V_{(N)}$ is $N!f(w_1) \cdots f(w_N)$ for $w_1 < \dots < w_N$.]

Problem 6.43 (i) For any continuous cumulative distribution function F, define $F^{-1}(0) = -\infty$, $F^{-1}(y) = \inf\{x : F(x) = y\}$ for $0 < y < 1$, $F^{-1}(1) = \infty$ if $F(x) < 1$ for all finite x, and otherwise $\inf\{x : F(x) = 1\}$. Then $F[F^{-1}(y)] = y$ for all $0 \le y \le 1$, but $F^{-1}[F(y)]$ may be $< y$.

(ii) Let Z have a cumulative distribution function $G(z) = h[F(z)]$, where F and h are continuous cumulative distribution functions, the latter defined over (0,1). If $Y = F(Z)$, then $P\{Y < y\} = h(y)$ for all $0 \le y \le 1$.

(iii) If Z has the continuous cumulative distribution function F, then $F(Z)$ is uniformly distributed over (0, 1).

[(ii): $P\{F(Z) < y\} = P\{Z < F^{-1}(y)\} = F[F^{-1}(y)] = y$.]

Problem 6.44 Let Z_i have a continuous cumulative distribution function F_i $(i = 1, \ldots, N)$, and let G be the group of all transformations $Z_i' = f(Z_i)$ such that f is continuous and strictly increasing.

(i) The transformation induced by f in the space of distributions is $F_i' = F_i(f^{-1})$.

(ii) Two N-tuples of distributions $(F_1, \ldots, F_N)$ and $(F_1', \ldots, F_N')$ belong to the same orbit with respect to $\bar{G}$ if and only if there exist continuous distribution functions $h_1, \ldots, h_N$ defined on (0,1) and strictly increasing continuous distribution functions F and F' such that $F_i = h_i(F)$ and $F_i' = h_i(F')$.

[(i): $P\{f(Z_i) \le y\} = P\{Z_i \le f^{-1}(y)\} = F_i[f^{-1}(y)]$.
(ii): If $F_i = h_i(F)$ and the F_i' are on the same orbit, so that $F_i' = F_i(f^{-1})$, then $F_i' = h_i(F')$ with $F' = F(f^{-1})$. Conversely, if $F_i = h_i(F)$, $F_i' = h_i(F')$, then $F_i' = F_i(f^{-1})$ with $f = F'^{-1}(F)$.]

Problem 6.45 Under the assumptions of the preceding problem, if $F_i = h_i(F)$, the distribution of the ranks $T_1, \ldots, T_N$ of $Z_1, \ldots, Z_N$ depends only on the h_i, not on F. If the h_i are differentiable, the distribution of the T_i is given by

$$P\{T_1 = t_1, \ldots, T_N = t_n\} = \frac{E\left[h_1'\left(U_{(t_1)}\right) \ldots h_N'\left(U_{(t_N)}\right)\right]}{N!}, \tag{6.63}$$

where $U_{(1)} < \cdots < U_{(N)}$ is an ordered sample of size N from the uniform distribution $U(0,1)$. [The left-hand side of (6.63) is the probability that of the quantities $F(Z_1), \ldots, F(Z_N)$, the ith one is the t_ith smallest for $i = 1, \ldots, N$. This is given by $\int \ldots \int h_1'(y_1) \ldots h_N'(y_N)\, dy$ integrated over the region in which y_i is the t_ith smallest of the y's for $i = 1, \ldots, N$. The proof is completed as in Problem 6.42.]

Problem 6.46 *Distribution of order statistics.*

(i) If $Z_1, \ldots, Z_N$ is a sample from a cumulative distribution function F with density f, the joint density of $Y_i = Z_{(s_i)}$, $i = 1, \ldots, n$, is

$$\frac{N! f(y_1) \ldots f(y_n)}{(s_1 - 1)!(s_2 - s_1 - 1)! \ldots (N - s_n)!} \tag{6.64}$$

$$\times [F(y_1)]^{s_1 - 1} [F(y_2) - F(y_1)]^{s_2 - s_1 - 1} \ldots [1 - F(y_n)]^{N - s_n}$$

for $y_1 < \cdots < y_n$.

(ii) For the particular case that the Z's are a sample from the uniform distribution on (0,1), this reduces to

$$\frac{N!}{(s_1-1)!(s_2-s_1-1)!\dots(N-s_n)!} \tag{6.65}$$

$$y_1^{s_1-1}(y_2-y_1)^{s_2-s_1-1}\dots(1-y_n)^{N-s_n}.$$

For $n=1$, (6.65) is the density of the beta-distribution $B_{s,N-s+1}$, which therefore is the distribution of the single order statistic $Z_{(s)}$ from $U(0,1)$.

(iii) Let the distribution of $Y_1,\dots,Y_n$ be given by (6.65), and let V_i be defined by $Y_i = V_i V_{i+1}\dots V_n$ for $i=1,\dots,n$. Then the joint distribution of the V_i is

$$\frac{N!}{(s_1-1)!\dots(N-s_n)!}\prod_{i=1}^{n} v_i^{s_i-1}(1-v_i)^{s_{i+1}-s_i-1} \qquad (s_{n+1}=N+1),$$

so that the V_i are independently distributed according to the beta-distribution $B_{s_i,s_{i+1}-s_i}$.

[(i): If $Y_1 = Z_{(s_1)},\dots,Y_n = Z_{(s_n)}$ and $Y_{n+1},\dots,Y_N$ are the remaining Z's in the original order of their subscripts, the joint density of $Y_1,\dots,Y_n$ is $N(N-1)\dots(N-n+1)\int\dots\int f(y_{n+1})\dots f(y_N)\,dy_{n+1}\dots dy_N$ integrated over the region in which s_1-1 of the y's are $< y_1$, s_2-s_1-1 between y_1 and $y_2,\dots,$ and $N-s_n > y_n$. Consider any set where a particular s_1-1 of the y's is $< y_1$, a particular s_2-s_1-1 of them is between y_1 and y_2, and so on, There are $N!/(s_1-1)!\dots(N-s_n)!$ of these regions, and the integral has the same value over each of them, namely $[F(y_1)]^{s_1-1}[F(y_2)-F(y_1)]^{s_2-s_1-1}\dots[1-F(y_n)]^{N-s_n}$.]

Problem 6.47 (i) If $X_1,\dots,X_m$ and $Y_1,\dots,Y_n$ are samples with continuous cumulative distribution functions F and $G=h(F)$ respectively, and if h is differentiable, the distribution of the ranks $S_1<\dots<S_n$ of the Y's is given by

$$P\{S_1=s_1,\dots,S_n=s_n\} = \frac{E\left[h'\left(U_{(s_1)}\right)\dots h'\left(U_{(s_n)}\right)\right]}{\binom{m+n}{m}} \tag{6.66}$$

where $U_{(1)}<\dots<U_{(m+n)}$ is an ordered sample from the uniform distribution $U(0,1)$.

(ii) If in particular $G=F^k$, where k is a positive integer, (6.66) reduces to

$$\begin{aligned} P\{S_1 &= s_1,\dots,S_n=s_n\} \\ &= \frac{k^n}{\binom{m+n}{m}}\prod_{j=1}^{n}\frac{\Gamma(s_j+jk-j)}{\Gamma(s_j)}\cdot\frac{\Gamma(s_{j+1})}{\Gamma(s_{j+1}+jk-j)}. \end{aligned} \tag{6.67}$$

Problem 6.48 For sufficiently small $\theta>0$, the Wilcoxon test at level

$$\alpha = k\Big/\binom{N}{n}, \qquad k \text{ a positive integer,}$$

maximizes the power (among rank tests) against the alternatives (F,G) with $G=(1-\theta)F+\theta F^2$.

Problem 6.49 An alternative proof of the optimum property of the Wilcoxon test for detecting a shift in the logistic distribution is obtained from the preceding problem by equating $F(x-\theta)$ with $(1-\theta)F(x)+\theta F^2(x)$, neglecting powers of θ higher than the first. This leads to the differential equation $F - \theta F' = (1-\theta)F + \theta F^2$, the solution of which is the logistic distribution.

Problem 6.50 Let $\mathcal{F}_0$ be a family of probability measures over $(\mathcal{X}, \mathcal{A})$, and let $\mathcal{C}$ be a class of transformations of the space $\mathcal{X}$. Define a class $\mathcal{F}_1$ of distributions by $F_1 \in \mathcal{F}_1$ if there exists $F_0 \in \mathcal{F}_0$ and $f \in \mathcal{C}$ such that the distribution of $f(X)$ is F_1 when that of X is F_0. If ϕ is any test satisfying (a) $E_{F_0}\phi(X) = \alpha$ for all $F_0 \in \mathcal{F}_0$, and (b) $\phi(x) \le \phi[f(x)]$ for all x and all $f \in \mathcal{C}$, then ϕ is unbiased for testing $\mathcal{F}_0$ against $\mathcal{F}_1$

Problem 6.51 Let $X_1, \ldots, X_m;\ Y_1, \ldots, Y_n$ be samples from a common continuous distribution F. Then the Wilcoxon statistic U defined in Problem 6.40 is distributed symmetrically about $\frac{1}{2}mn$ even when $m \ne n$.

Problem 6.52 (i) If $X_1, \ldots, X_m$ and $Y_1, \ldots, Y_n$ are samples from $F(x)$ and $G(y) = F(y - \Delta)$ respectively (F continuous), and $D_{(1)} < \cdots < D_{(mn)}$ denote the ordered differences $Y_j - X_i$, then

$$P\left[D_{(k)} < \Delta < D_{(mn+1-k)}\right] = P_0[k \le U \le mn - k],$$

where U is the statistic defined in Problem 6.40 and the probability on the right side is calculated for $\Delta = 0$.

(ii) Determine the above confidence interval for Δ when $m = n = 6$, the confidence coefficient is $\frac{20}{21}$, and the observations are x : .113, .212, .249, .522, .709, .788, and y : .221, .433, .724, .913, .917, 1.58.

(iii) For the data of (ii) determine the confidence intervals based on Student's t for the case that F is normal.

Hint: $D_{(i)} \le \Delta < D_{(i+1)}$ if and only if $U_\Delta = mn - i$, where U_Δ is the statistic U of Problem 6.40 calculated for the observations

$$X_1, \ldots, X_m; Y_1 - \Delta, \ldots, Y_n - \Delta.$$

[An alternative measure of the amount by which G exceeds F (without assuming a location model) is $p = P\{X < Y\}$. The literature on confidence intervals for p is reviewed in Mee (1990).]

Problem 6.53 (i) Let X, X' and Y, Y' be independent samples of size 2 from continuous distributions F and G respectively. Then

$$\begin{aligned} p &= P\{\max(X, X') < \min(Y, Y')\} + P\{\max(Y, Y') < \min(X, X')\} \\ &= \tfrac{1}{3} + 2\Delta, \end{aligned}$$

where $\Delta = \int (F - G)^2\, d[(F + G)/2]$.

(ii) $\Delta = 0$ if and only if $F = G$.

[(i): $p = \int (1 - F)^2\, dG^2 + \int (1 - G)^2\, dF^2$ which after some computation reduces to the stated form.

(ii): $\Delta = 0$ implies $F(x) = G(x)$ except on a set N which has measure zero both under F and G. Suppose that $G(x_1) - F(x_1) = \eta > 0$. Then there exists x_0 such that $G(x_0) = F(x_0) + \frac{1}{2}\eta$ and $F(x) < G(x)$ for $x_0 \le x \le x_1$. Since $G(x_1) - G(x_0) > 0$, it follows that $\Delta > 0$.]

Problem 6.54 *Continuation.*

(i) There exists at every significance level α a test of $H : G = F$ which has power $> \alpha$ against all continuous alternatives (F, G) with $F \neq G$.

(ii) There does not exist a nonrandomized unbiased rank test of H against all $G \neq F$ at level

$$\alpha = 1 \Big/ \binom{m+n}{n}.$$

[(i): let $X_i, X_i'; Y_i, Y_i'$ $(i = 1, \ldots, n)$ be independently distributed, the X's with distribution F, the Y's with distribution G, and let $V_i = 1$ if $\max(X_i, X_1') < \min(Y_i, Y_i')$ or $\max(Y_i, Y_i') < \min(X_i, X_i')$, and $V_i = 0$ otherwise. Then $\sum V_i$ has a binomial distribution with the probability p defined in Problem 6.53, and the problem reduces to that of testing $p = \frac{1}{3}$ against $p > \frac{1}{3}$.
(ii): Consider the particular alternatives for which $P\{X < Y\}$ is either 1 or 0.]

Problem 6.55 (i) Let $X_1, \ldots, X_m; Y_1, \ldots, Y_n$ be i.i.d. according to a continuous distribution F, let the ranks of the Y's be $S_1 < \cdots < S_n$, and let $T = h(S_1) + \cdots + h(S_n)$. Then if either $m = n$ or $h(s) + h(N + 1 - s)$ is independent of s, the distribution of T is symmetric about $n \sum_{i=1}^N h(i)/N$.

(ii) Show that the two-sample Wilcoxon and normal-scores statistics are symmetrically distributed under H, and determine their centers of symmetry.

[(i): Let $S_i' = N + 1 - S_i$, and use the fact that $T' = \sum h(S_j')$ has the same distribution under H as T.]

Section 6.10

Problem 6.56 (i) Let m and n be the numbers of negative and positive observations among $Z_1, \ldots, Z_N$, and let $S_1 < \cdots < S_n$ denote the ranks of the positive Z's among $|Z_1|, \ldots |Z_N|$. Consider the $N + \frac{1}{2}N(N-1)$ distinct sums $Z_i + Z_j$ with $i = j$ as well as $i \neq j$. The Wilcoxon signed rank statistic $\sum S_j$, is equal to the number of these sums that are positive.

(ii) If the common distribution of the Z's is D, then

$$E\left(\sum S_j\right) = \tfrac{1}{2}N(N+1) - ND(0) - \tfrac{1}{2}N(N-1)\int D(-z)\,dD(z).$$

[(i) Let K be the required number of positive sums. Since $Z_i + Z_j$ is positive if and only if the Z corresponding to the larger of $|Z_i|$ and $|Z_j|$ is positive, $K = \sum_{i=1}^N \sum_{j=1}^N U_{ij}$ where $U_{ij} = 1$ if $Z_j > 0$ and $|Z_i| \le Z_j$ and $U_{ij} = 0$ otherwise.]

Problem 6.57 Let $Z_1, \ldots, Z_N$ be a sample from a distribution with density $f(z-\theta)$, where $f(z)$ is positive for all z and f is symmetric about 0, and let m, n, and the S_j be defined as in the preceding problem.

(i) The distribution of n and the S_j is given by

$$P\{\text{the number of positive } Z\text{'s is } n \text{ and } S_1 = s_1, \ldots, S_n = s_n\} \tag{6.68}$$
$$= \frac{1}{2^N} E\left[\frac{f\left(V_{(r_1)}+\theta\right) \ldots f\left(V_{(r_m)}+\theta\right) f\left(V_{(s_1)}-\theta\right) \ldots f\left(V_{(s_n)}-\theta\right)}{f\left(V_{(1)}\right) \ldots f\left(V_{(N)}\right)}\right],$$

where $V_{(1)} < \cdots < V_{(N)}$, is an ordered sample from a distribution with density $2f(v)$ for $v > 0$, and 0 otherwise.

(ii) The rank test of the hypothesis of symmetry with respect to the origin, which maximizes the derivative of the power function at $\theta = 0$ and hence maximizes the power for sufficiently small $\theta > 0$, rejects, under suitable regularity conditions, when

$$-E\left[\sum_{j=1}^{n} \frac{f'(V_{(s_j)})}{f(V_{(s_j)})}\right] > C.$$

(iii) In the particular case that $f(z)$ is a normal density with zero mean, the rejection region of (ii) reduces to $\sum E(V(s_j) > C$, where $V_{(1)} < \cdots < V_{(N)}$ is an ordered sample from a χ-distribution with 1 degree of freedom.

(iv) Determine a density f such that the one-sample Wilcoxon test is most powerful against the alternatives $f(z-\theta)$ for sufficiently small positive θ.

[(i): Apply Problem 6.42(i) to find an expression for $P\{S_1 = s_1, \ldots, S_n = s_n$ given that the number of positive Z's is $n\}$.]

Problem 6.58 An alternative expression for (6.68) is obtained if the distribution of Z is characterized by (ρ, F, G). If then $G = h(F)$ and h is differentiable, the distribution of n and the S_j is given by

$$\rho^m (1-\rho)^n E\left[h'(U_{(s_1)}) \cdots h'(U_{(s_n)})\right], \tag{6.69}$$

where $U_{(1)}, < \cdots < U_{(N)}$ is an ordered sample from $U(0,1)$.

Problem 6.59 *Unbiased tests of symmetry.* Let $Z_1, \ldots, Z_N$, be a sample, and let ϕ be any rank test of the hypothesis of symmetry with respect to the origin such that $z_i \le z_i'$ for all i implies $\phi(z_1, \ldots, z_N) \le \phi(z_1', \ldots, z'N)$. Then ϕ is unbiased against the one-sided alternatives that the Z's are stochastically larger than some random variable that has a symmetric distribution with respect to the origin.

Problem 6.60 *The hypothesis of randomness.*[7] Let $Z_1, \ldots, Z_N$ be independently distributed with distributions $F_1, \ldots, F_N$, and let T_i denote the rank of Z_i among the Z's For testing the *hypothesis of randomness* $F_1 = \cdots = F_N$ against

[7]Some tests of randomness are treated in Diaconis (1988).

the alternatives K of an *upward trend*, namely that Z_i is stochastically increasing with i, consider the rejection regions

$$\sum it_i > C \tag{6.70}$$

and

$$\sum iE(V_{(t_i)}) > C, \tag{6.71}$$

where $V_{(1)} < \cdots < V_{(N)}$ is an ordered sample from a standard normal distribution and where t_i is the value taken on by T_i.

(i) The second of these tests is most powerful among rank tests against the normal alternatives $F = N(\gamma + i\delta, \sigma^2)$ for sufficiently small δ.

(ii) Determine alternatives against which the first test is a most powerful rank test.

(iii) Both tests are unbiased against the alternatives of an upward trend; so is any rank test ϕ satisfying $\phi(z_1, \ldots, z_N) \le \phi(z'_1, \ldots, z'_N)$ for any two points for which $i < j$, $z_i < z_j$ implies $z'_i < z'_j$ for all i and j.

[(iii): Apply Problem 6.50 with $\mathcal{C}$ the class of transformations $z'_1 = z_1, z'_i = f_i(z_i)$ for $i > 1$, where $z < f_2(z) < \cdots < f_N(z)$ and each f_i is nondecreasing. If $\mathcal{F}_0$ is the class of N-tuples $(F_1, \ldots, F_N)$ with $F_1 = \cdots = F_N$, then $\mathcal{F}_1$ coincides with the class K of alternatives.]

Problem 6.61 In the preceding problem let $U_{ij} = 1$ if $(j - i)(Z_j - Z_i) > 0$, and $= 0$ otherwise.

(i) The test statistic $\sum iT_i$, can be expressed in terms of the U's through the relation

$$\sum_{i=1}^{N} iT_i = \sum_{i<j} (j - i)U_{ij} + \frac{N(N+1)(N+2)}{6},$$

(ii) The smallest number of steps [in the sense of Problem 6.40(ii)] by which $(Z_1, \ldots, Z_N)$ can be transformed into the ordered sample $(Z_{(1)}, \ldots, Z_{(N)})$ is $[N(N-1)/2] - U$, where $U = \sum_{i<j} U_{ij}$. This suggests $U > C$ as another rejection region for the preceding problem.

[(i): Let $V_{ij} = 1$ or 0 as $Z_i \le Z_i$ or $Z_i > Z_j$. Then $T_j = \sum_{i=1}^N V_{ij}$, and $V_{ij} = U_{ij}$ or $1 - U_{ij}$ as $i < j$ or $i \ge j$. Expressing $\sum_{j=1}^N jT_j = \sum_{j=1}^N j \sum_{i=1}^N V_{ij}$ in terms of the U's and using the fact that $U_{ij} = U_{ji}$, the result follows by a simple calculation.]

Problem 6.62 *The hypothesis of independence.* Let $(X_1, Y_1), \ldots, (X_N, Y_N)$ be a sample from a bivariate distribution, and $(X_{(1)}, Z_1), \ldots, (X_{(N)}, Z_N)$ be the same sample arranged according to increasing values of the X's so that the Z's are a permutation of the Y's. Let R_i be the rank of X_i among the X's, S_i the rank of Y_i among the Y's, and T_i the rank of Z_i among the Z's, and consider the hypothesis of independence of X and Y against the alternatives of positive regression dependence.

(i) Conditionally, given $(X_{(1)}, \dots, X_{(N)})$, this problem is equivalent to testing the hypothesis of randomness of the Z's against the alternatives of an upward trend.

(ii) The test (6.70) is equivalent to rejecting when the *rank correlation coefficient*

$$\frac{\sum(R_i - \bar{R})(S_i - \bar{S})}{\sqrt{\sum(R_i - \bar{R}^2)\sum(S_i - \bar{S})^2}} = \frac{12}{N^3 - N}\sum\left(R_i - \frac{N+1}{2}\right)\left(S_i - \frac{N+1}{2}\right)$$

is too large.

(iii) An alternative expression for the rank correlation coefficient[8] is

$$1 - \frac{6}{N^3 - N}\sum(S_i - R_i)^2 = 1 - \frac{6}{N^3 - N}\sum(T_i - i)^2.$$

(iv) The test $U > C$ of Problem 6.61(ii) is equivalent to rejecting when Kendall's t-statistic $\sum_{i<j} V_{ij}/N(N-1)$ is too large where V_{ij} is $+1$ or -1 as $(Y_j - Y_i)(X_j - X_i)$ is positive or negative.

(v) The tests (ii) and (iv) are unbiased against the alternatives of positive regression dependence.

Section 6.11

Problem 6.63 In Example 6.11.1, a family of sets $S(x, y)$ is a class of equivariant confidence sets if and only if there exists a set $\mathcal{R}$ of real numbers such that

$$S(x, y) = \bigcup_{r \in \mathcal{R}} \{(\xi, \eta) : (x - \xi)^2 + (y - \eta)^2 = r^2\}.$$

Problem 6.64 Let $X_1, \dots, X_n$; $Y_1, \dots, Y_n$ be samples from $N(\xi, \sigma^2)$ and $N(\eta, \tau^2)$ respectively. Then the confidence intervals (5.42) for τ^2/σ^2, which can be written as

$$\frac{\sum(Y_j - \bar{Y})^2}{k\sum(X_i - \bar{X})^2} \le \frac{\tau^2}{\sigma^2} \le \frac{k\sum(Y_j - \bar{Y})^2}{\sum(X_i - \bar{X})^2},$$

are uniformly most accurate equivariant with respect to the smallest group G containing the transformations $X_i' = aX + b$, $Y_i' = aY + c$ for all $a \ne 0$, b, c and the transformation $X_i' = dY_i$, $Y_i' = X_i/d$ for all $d \ne 0$.
[Cf. Problem 6.11.]

Problem 6.65 (i) *One-sided equivariant confidence limits.* Let θ be real-valued, and suppose that, for each θ_0, the problem of testing $\theta \le \theta_0$ against $\theta > \theta_0$ (in the presence of nuisance parameters ϑ) remains invariant under a group G_{θ_0} and that $A(\theta_0)$ is a UMP invariant acceptance region for this hypothesis at level α. Let the associated confidence sets $S(x) = \{\theta : x \in A(\theta)\}$

---

[8] For further material on these and other tests of independence, see Kendall (1970), Aiyar, Guillier, and Albers (1979), Kallenberg and Ledwina (1999).

be one-sided intervals $S(x) = \{\theta : \underline{\theta}(x) \le \theta\}$, and suppose they are equivariant under all G_θ and hence under the group G generated by these. Then the lower confidence limits $\underline{\theta}(X)$ are uniformly most accurate equivariant at confidence level $1-\alpha$ in the sense of minimizing $P_{\theta,\vartheta}\{\underline{\theta}(X) \le \theta'\}$ for all $\theta' < \theta$.

(ii) Let $X_1, \ldots, X_n$ be independently distributed as $N(\xi, \sigma^2)$. The upper confidence limits $\sigma^2 \le \sum(X_i - \bar{X})^2/C_0$ of Example 5.5.1 are uniformly most accurate equivariant under the group $X_i' = X_i + c$, $-\infty < c < \infty$. They are also equivariant (and hence uniformly most accurate equivariant) under the larger group $X_i' = aX_i + c$, $-\infty < a,\ c < \infty$.

Problem 6.66 *Counterexample.* The following example shows that the equivariance of $S(x)$ assumed in the paragraph following Lemma 6.11.1 does not follow from the other assumptions of this lemma. In Example 6.5.1, let $n = 1$, let $G^{(1)}$ be the group G of Example 6.5.1, and let $G^{(2)}$ be the corresponding group when the roles of Z and $Y = Y_1$ are reversed. For testing $H(\theta_0) : \theta = \theta_0$ against $\theta \ne \theta_0$ let G_{θ_0} be equal to $G^{(1)}$ augmented by the transformation $Y' = \theta_0 - (Y_1 - \theta_0)$ when $\theta \le 0$, and let G_{θ_0} be equal to $G^{(2)}$ augmented by the transformation $Z' = \theta_0 - (Z - \theta_0)$ when $\theta > 0$. Then there exists a UMP invariant test of $H(\theta_0)$ under G_{θ_0} for each θ_0, but the associated confidence sets $S(x)$ are not equivariant under $G = \{G_\theta, -\infty < \theta < \infty\}$.

Problem 6.67 (i) Let $X_1, \ldots, X_n$ be independently distributed as $N(\xi, \sigma^2)$, and let $\theta = \xi/\sigma$. The lower confidence bounds $\underline{\theta}$ for θ, which at confidence level $1-\alpha$ are uniformly most accurate invariant under the transformations $X_i' = aX_i$, are

$$\underline{\theta} = C^{-1}\left(\frac{\sqrt{n}\bar{X}}{\sqrt{\sum(X_i - \bar{X})^2/(n-1)}}\right)$$

where the function $C(\theta)$ is determined from a table of noncentral t so that

$$P_\theta\left\{\frac{\sqrt{n}\bar{X}}{\sqrt{\sum(X_i - \bar{X})^2/(n-1)}} \le C(\theta)\right\} = 1 - \alpha.$$

(ii) Determine $\underline{\theta}$ when the x's are 7.6, 21.2, 15.1, 32.0, 19.7, 25.3, 29.1, 18.4 and the confidence level is $1 - \alpha = .95$.

Problem 6.68 (i) Let $(X_1, Y_1), \ldots, (X_n, Y_n)$ be a sample from a bivariate normal distribution, and let

$$\underline{\rho} = C^{-1}\left(\frac{\sum(X_i - \bar{X})(Y_i - \bar{Y})}{\sqrt{\sum(X_i - \bar{X})^2 \sum(Y_i - \bar{Y})^2}}\right),$$

where $C(\rho)$ is determined such that

$$P_\theta\left\{\frac{\sum(X_i - \bar{X})(Y_i - \bar{Y})}{\sqrt{\sum(X_i - \bar{X})^2 \sum(Y_i - \bar{Y})^2}} \le C(\rho)\right\} = 1 - \alpha.$$

Then $\underline{\rho}$ is a lower confidence limit for the population correlation coefficient ρ at confidence level $1 - \alpha$; it is uniformly most accurate invariant with respect to the group of transformations $X_i' = aX_i + b$, $Y_i' = cY_i + d$, with $ac > 0$, $-\infty < b$, $d < \infty$.

(ii) Determine $\underline{\rho}$ at level $1 - \alpha = .95$ when the observations are (12.9,.56), (9.8,.92), (13.1,.42), (12.5,1.01), (8.7,.63), (10.7,.58), (9.3,.72), (11.4,.64).

Note. The following problems explore the relationship between pivotal quantities and equivariant confidence sets. For more details see Arnold (1984).

Let X be distributed according $P_{\theta,\vartheta}$, and consider confidence sets for θ that are equivariant under a group G^*, as in Section 6.11. If w is the set of possible θ-values, define a group $\tilde{G}$ on $\mathcal{X} \times w$ by $\tilde{g}(\theta, x) = (gx, \bar{g}\theta)$.

Problem 6.69 Let $V(X, \theta)$ be any pivotal quantity [i.e. have a fixed probability distribution independent of (θ, ϑ)], and let B be any set in the range space of V with probability $P(V \in B) = 1 - \alpha$. Then the sets $S(x)$ defined by

$$\theta \in S(x) \quad \text{if and only if} \quad V(\theta, x) \in B \tag{6.72}$$

are confidence sets for θ with confidence coefficient $1 - \alpha$.

Problem 6.70 (i) If $\tilde{G}$ is transitive over $\mathcal{X} \times w$ and $V(X, \theta)$ is maximal invariant under $\tilde{G}$, then $V(X, \theta)$ is pivotal.

(ii) By (i), any quantity $W(X, \theta)$ which is invariant under $\tilde{G}$ is pivotal; give an example showing that the converse need not be true.

Problem 6.71 Under the assumptions of the preceding problem, the confidence set $S(x)$ is equivariant under G^*.

Problem 6.72 Under the assumptions of Problem 6.70, suppose that a family of confidence sets $S(x)$ is equivariant under G^*. Then there exists a set B in the range space of the pivotal V such that (6.72) holds. In this sense, all equivariant confidence sets can be obtained from pivotals.

[Let A be the subset of $\mathcal{X} \times w$ given by $A = \{(x, \theta) : \theta \in S(x)\}$. Show that $\tilde{g}A = A$, so that any orbit of $\tilde{G}$ is either in A or in the complement of A. Let the maximal invariant $V(x, \theta)$ be represented as in Section 6.2 by a uniquely defined point on each orbit, and let B be the set of these points whose orbits are in A. Then $V(x, \theta) \in B$ if and only if $(x, \theta) \in A$.] *Note.* Problem 6.72 provides a simple check of the equivariance of confidence sets. In Example 6.12.2, for instance, the confidence sets (6.43) are based on the pivotal vector $(X_1 - \xi_1, \ldots, X_r - \xi_r)$, and hence are equivariant.

Section 6.12

Problem 6.73 In Examples 6.12.1 and 6.12.2 there do not exist equivariant sets that uniformly minimize the probability of covering false values.

Problem 6.74 In Example 6.12.1, the density $p(v)$ of $V = 1/S^2$ is unimodal.

Problem 6.75 Show that in Example 6.12.1,

(i) the confidence sets $\sigma^2/S^2 \in A^{**}$ with A^{**} given by (6.42) coincide with the uniformly most accurate unbiased confidence sets for σ^2;

(ii) if (a, b) is best with respect to (6.41) for σ, then (a^r, b^r) is best for σ^r $(r > 0)$.

Problem 6.76 Let $X_1, \ldots, X_r$ be i.i.d. $N(0,1)$, and let S^2 be independent of the X's and distributed as χ^2_ν. Then the distribution of $(X_1/S\sqrt{\nu}, \ldots, X_r/S\sqrt{\nu})$ is a central multivariate t-distribution, and its density is

$$p(v_1, \ldots, v_r) = \frac{\Gamma(\frac{1}{2}(\nu + r))}{(\pi\nu)^{r/2}\Gamma(\nu/2)}\left(1 + \frac{1}{\nu}\sum v_i^2\right)^{-\frac{1}{2}(\nu+r)} .$$

Problem 6.77 The confidence sets (6.49) are uniformly most accurate equivariant under the group G defined at the end of Example 6.12.3.

Problem 6.78 In Example 6.12.4, show that

(i) both sets (6.57) are intervals;

(ii) the sets given by $vp(v) > C$ coincide with the intervals (5.41).

Problem 6.79 Let $X_1, \ldots, X_m$; $Y_1, \ldots, Y_n$ be independently normally distributed as $N(\xi, \sigma^2)$ and $N(\eta, \sigma^2)$ respectively. Determine the equivariant confidence sets for $\eta - \xi$ that have smallest Lebesgue measure when

(i) σ is known;

(ii) σ is unknown.

Problem 6.80 Generalize the confidence sets of Example 6.11.3 to the case that the X_i are $N(\xi_i, d_i\sigma^2)$ where the d's are known constants.

Problem 6.81 Solve the problem corresponding to Example 6.12.1 when

(i) $X_1, \ldots, X_n$ is a sample from the exponential density $E(\xi, \sigma)$, and the parameter being estimated is σ;

(ii) $X_1, \ldots, X_n$ is a sample from the uniform density $U(\xi, \xi + \tau)$, and the parameter being estimated is τ.

Problem 6.82 Let $X_1, \ldots, X_n$ be a sample from the exponential distribution $E(\xi, \sigma)$. With respect to the transformations $X_i' = bX_i + a$ determine the smallest equivariant confidence sets

(i) for σ, both when size is defined by Lebesgue measure and by the equivariant measure (6.41);

(ii) for ξ.

Problem 6.83 Let X_{ij} $(j = 1, \ldots, n_i;\ i = 1, \ldots, s)$ be samples from the exponential distribution $E(\xi_i, \sigma)$. Determine the smallest equivariant confidence sets for $(\xi_1, \ldots, \xi_r)$ with respect to the group $X_{ij}' = bX_{ij} + a_i$.

Section 6.13

Problem 6.84 If the confidence sets $S(x)$ are equivariant under the group G, then the probability $P_\theta\{\theta \in S(X)\}$ of their covering the true value is invariant under the induced group $\bar{G}$.

Problem 6.85 Consider the problem of obtaining a (two-sided) confidence band for an unknown continuous cumulative distribution function F.

(i) Show that this problem is invariant both under strictly increasing and strictly decreasing continuous transformations $X_i' = f(X_i)$, $i = 1, \ldots, n$, and determine a maximal invariant with respect to this group.

(ii) Show that the problem is not invariant under the transformation

$$X_i' = \begin{cases} X_i & \text{if } |X_i| \geq 1, \\ X_i - 1 & \text{if } 0 < X_i < 1, \\ X_i + 1 & \text{if } -1 < X_i < 0. \end{cases}$$

[(ii): For this transformation g, the set $g^*S(x)$ is no longer a band.]

6.15 Notes

Invariance considerations were introduced for particular classes of problems by Hotelling (1936) and Pitman (1939b). The general theory of invariant and almost invariant tests, together with its principal parametric applications, was developed by Hunt and Stein (1946) in an unpublished paper. In their paper, invariance was not proposed as a desirable property in itself but as a tool for deriving most stringent tests (cf. Chapter 8). Apart from this difference in point of view, the present account is based on the ideas of Hunt and Stein, about which E. L. Lehmann learned through conversations with Charles Stein during the years 1947–1950.

Of the admissibility results of Section 6.7, Theorem 6.7.1 is due to Birnbaum (1955) and Stein (1956a); Example 6.7.13 (continued) and Lemma 6.7.1, to Kiefer and Schwartz (1965).

The problem of minimizing the volume or diameter of confidence sets is treated in DasGupta (1991).

Deuchler (1914) appears to contain the first proposal of the two-sample procedure known as the Wilcoxon test, which was later discovered independently by many different authors. A history of this test is given by Kruskal (1957). Hoeffding (1951) derives a basic rank distribution of which (6.20) is a special case, and from it obtains locally optimum tests of the type (6.21).

7
Linear Hypotheses

7.1 A Canonical Form

Many testing problems concern the means of normal distributions and are special cases of the following *general univariate linear hypothesis*. Let $X_1, \ldots, X_n$ be independently normally distributed with means $\xi_1, \ldots, \xi_n$ and common variance σ^2. The vector of means[1] $\underset{\sim}{\xi}$ is known to lie in a given s-dimensional linear subspace $\prod_\Omega$ $(s < n)$, and the hypothesis H to be tested is that $\underset{\sim}{\xi}$ lies in a given $(s - r)$-dimensional subspace $\prod_\omega$ of $\prod_\Omega$ $(r \leq s)$.

Example 7.1.1 In the two-sample problem of testing equality of two normal means (considered with a different notation in Section 5.3), it is given that $\xi_i = \xi$ for $i = 1, \ldots, n_1$ and $\xi_i = \eta$ for $i = n_1 + 1, \ldots, n_1 + n_2$, and the hypothesis to be tested is $\eta = \xi$. The space $\prod_\Omega$ is then the space of vectors

$$(\xi, \ldots, \xi, \eta, \ldots, \eta) = \xi(1, \ldots, 1, 0, \ldots, 0) + \eta(0, \ldots, 0, 1, \ldots, 1)$$

spanned by $(1, \ldots, 1, 0, \ldots, 0)$ and $(0, \ldots, 0, 1, \ldots, 1)$, so that $s = 2$. Similarly, $\prod_\omega$ is the set of all vectors $(\xi, \ldots, \xi) = \xi(1, \ldots, 1)$ and hence $r = 1$.

Another hypothesis that can be tested in this situation is $\eta = \xi = 0$. The space $\prod_\omega$ is then the origin, $s - r = 0$ and hence $r = 2$. The more general hypothesis $\xi = \xi_0, \eta = \eta_0$ is not a linear hypothesis, since $\prod_\omega$ does not contain the origin. However, it reduces to the previous case through the transformation $X_i' = X_i - \xi_0$ $(i = 1, \ldots, n_1)$, $X_i' = X_i - \eta_0$ $(i = n_1 + 1, \ldots, n_1 + n_2)$.

[1]Throughout this chapter, a fixed coordinate system is assumed given in n-space. A vector with components $\xi_1, \ldots, \xi_n$ is denoted by $\underset{\sim}{\xi}$, and an $n \times 1$ column matrix with elements $\xi_1, \ldots, \xi_n$ by ξ.

Example 7.1.2 The regression problem of Section 5.6 is essentially a linear hypothesis. Changing the notation to make it conform with that of the present section, let $\xi_i = \alpha + \beta t_i$, where α, β are unknown, and the t_i known and not all equal. Since $\prod_\Omega$ is the space of all vectors $\alpha(1, \ldots, 1) + \beta(t_1, \ldots, t_n)$, it has dimension $s = 2$. The hypothesis to be tested may be $\alpha = \beta = 0$ $(r = 2)$ or it may only specify that one of the parameters is zero $(r = 1)$. The more general hypotheses $\alpha = \alpha_0$, $\beta = \beta_0$ can be reduced to the previous case by letting $X'_i = X_i - \alpha_0, -\beta_0 t_i$, since then $E(X'_i) = \alpha' + \beta' t_i$ with $\alpha' = \alpha - \alpha_0, \beta' = \beta - \beta_0$.

Higher polynomial regression and regression in several variables also fall under the linear-hypothesis scheme. Thus if $\xi_i = \alpha + \beta t_i + \gamma t_i^2$ or more generally $\xi_i = \alpha + \beta t_i + \gamma u_i$, where the t_i and u_i are known, it can be tested whether one or more of the regression coefficients α, β, γ are zero, and by transforming to the variables $X'_i = X_i - \alpha_0 - \beta_0 t_i - \gamma_0 u_i$ also whether these coefficients have specified values other than zero. ■

In the general case, the hypothesis can be given a simple form by making an orthogonal transformation to variables $Y_1, \ldots, Y_n$

$$Y = CX, \qquad C = (c_{ij}) \quad i, j = 1, \ldots, n, \tag{7.1}$$

such that the first s row vectors $\underset{\sim}{c}_1, \ldots, \underset{\sim}{c}_s$ of the matrix C span $\prod_\Omega$, with $\underset{\sim}{c}_{r+1}, \ldots, \underset{\sim}{c}_s$, spanning $\prod_\omega$. Then $Y_{s+1} = \cdots = Y_n = 0$ if and only if $\underline{X}$ is in $\prod_\Omega$, and $Y_1 = \cdots = Y_r = Y_{s+1} = \cdots = Y_n = 0$ if and only if $\underline{X}$ is in $\prod_\omega$. Let $\eta_i = E(Y_i)$, so that $\eta = C\xi$. Then since $\underset{\sim}{\xi}$ lies in $\prod_\Omega$ a priori and in $\prod_\omega$ under H, it follows that $\eta_i = 0$ for $i = s + 1, \ldots, n$ in both cases, and $\eta_i = 0$ for $i = 1, \ldots, r$ when H is true. Finally, since the transformation is orthogonal, the variables $Y_1, \ldots, Y_n$ are again independent and normally distributed with common variance σ^2, and the problem reduces to the following canonical form.

The variables $Y_1, \ldots, Y_n$ are independently, normally distributed with common variance σ^2 and means $E(Y_i) = \eta_i$ for $i = 1, \ldots, s$ and $E(Y_i) = 0$ for $i = s + 1, \ldots, n$, so that their joint density is

$$\frac{1}{(\sqrt{2\pi}\sigma)^n} \exp\left[-\frac{1}{2\sigma^2}\left(\sum_{i=1}^{s}(y_i - \eta_i)^2 + \sum_{i=s+1}^{n} y_i^2\right)\right]. \tag{7.2}$$

The η's and σ^2 are unknown, and the hypothesis to be tested is

$$H : \eta_1 = \cdots = \eta_r = 0 \qquad (r \le s < n). \tag{7.3}$$

Example 7.1.3 To illustrate the determination of the transformation (7.1), consider once more the regression model $\xi_i = \alpha + \beta t_i$, of Example 7.1.2. It was seen there that $\prod_\Omega$ is spanned by $(1, \ldots, 1)$ and $(t_1, \ldots, t_n)$. If the hypothesis being tested is $\beta = 0$, $\prod_\omega$ is the one-dimensional space spanned by the first of these vectors. The row vector $\underset{\sim}{c}_2$ is in $\prod_\omega$ and of length 1, and hence $\underset{\sim}{c}_2 = (1/\sqrt{n}, \ldots, 1/\sqrt{n})$. Since $\underset{\sim}{c}_1$ is in $\prod_\Omega$, of length 1, and orthogonal to $\underset{\sim}{c}_2$, its coordinates are of the form $a + bt_i$, $i = 1, \ldots, n$, where a and b are determined by the conditions $\sum(a + bt_i) = 0$ and $\sum(a + bt_i)^2 = 1$. The solutions of these equations are $a = -b\bar{t}$, $b = 1/\sqrt{\sum(t_j - \bar{t})^2}$, and therefore $a + bt_i = (t_i - \bar{t})/\sqrt{\sum(t_j - \bar{t})^2}$, and

$$Y_1 = \frac{\sum X_i(t_i - \bar{t})}{\sqrt{\sum(t_j - \bar{t})^2}} = \frac{\sum (X_i - \bar{X})(t_i - \bar{t})}{\sqrt{\sum(t_j - \bar{t})^2}}.$$

The remaining row vectors of C can be taken to be any set of orthogonal unit vectors that are orthogonal to $\prod_\Omega$; it turns out not to be necessary to determine them explicitly.

If the hypothesis to be tested is $\alpha = 0$, $\prod_\omega$ is spanned by $(t_1, \dots, t_n)$, so that the ith coordinate of $\underset{\sim}{c}_2$ is $t_i/\sqrt{\sum t_j^2}$. The coordinates of $\underset{\sim}{c}_1$ are again of the form $a + bt_i$ with a and b now determined by the equations $\sum(a + bt_i)t_i = 0$ and $\sum(a + bt_i)^2 = 1$. The solutions are $b = -an\bar{t}/\sum t_j^2$, $a = \sqrt{\sum t_j^2/n\sum(t_j - \bar{t})^2}$, and therefore

$$Y_1 = \sqrt{\frac{n\sum t_j^2}{\sum(t_j - \bar{t})^2}} \left(\bar{X} - \frac{\bar{t}}{\sum t_j^2} \sum t_i X_i \right).$$

In the case of the hypothesis $\alpha = \beta = 0$, $\prod_\omega$ is the origin, and $\underset{\sim}{c}_1, \underset{\sim}{c}_2$ can be taken as any two orthogonal unit vectors in $\prod_\Omega$. One possible choice is that appropriate to the hypothesis $\beta = 0$, in which case Y_1 is the linear function given there and $Y_2 = \sqrt{x}\bar{X}$. ■

The general linear-hypothesis problem in terms of the Y's remains invariant under the group G_1 of transformations $Y_i' = Y_i + c_i$ for $i = r + 1, \dots, s$; $Y_i' = Y_i$ for $i = 1, \dots, r$; $s + 1, \dots, n$. This leaves $Y_1, \dots, Y_r$ and $Y_{s+1}, \dots, Y_n$ as maximal invariants. Another group of transformations leaving the problem invariant is the group G_2 of all orthogonal transformations of $Y_1, \dots, Y_r$. The middle set of variables having been eliminated, it follows from Example 6.2.1(iii) that a maximal invariant under G_2 is $U = \sum_{i=1}^r Y_i^2, Y_{s+1}, \dots, Y_n$. This can be reduced to U and $V = \sum_{i=s+1}^n Y_i^2$ by sufficiency. Finally, the problem also remains invariant under the group G_3 of scale changes $Y_i' = cY_i, c \neq 0$, for $i = 1, \dots, n$. In the space of U and V this induces the transformation $U^* = c^2U, V^* = c^2V$, under which $W = U/V$ is maximal invariant. Thus the principle of invariance reduces the data to the single statistic [2]

$$W = \frac{\sum\limits_{i=1}^r Y_i^2}{\sum\limits_{i=s+1}^n Y_i^2}. \tag{7.4}$$

Each of the three transformation groups G_i $(i = 1, 2, 3)$ which lead to the above reduction induces a corresponding group $\bar{G}_i$ in the parameter space. The group $\bar{G}_1$ consists of the translations $\eta_i' = \eta_i + c_i$ $(i = r+1, \dots, s)$, $\eta_i' = \eta_i$ $(i = 1, \dots, r)$, $\sigma' = \sigma$, which leaves $(\eta_1, \dots, \eta_r, \sigma)$ as maximal invariants. Since any orthogonal transformation of $Y_1, \dots, Y_r$ induces the same transformation on $\eta_1, \dots, \eta_r$ and leaves σ^2 unchanged, a maximal invariant under $\bar{G}_2$ is $\left(\sum_{i=1}^r \eta_i^2, \sigma^2\right)$. Finally the elements of $\bar{G}_3$ are the transformations $\eta_i' = c\eta_i$, $\sigma' = |c|\sigma$, and hence a maximal invariant with respect to the totality of these transformations is

$$\psi^2 = \frac{\sum\limits_{i=1}^r \eta_i^2}{\sigma^2}. \tag{7.5}$$

[2] A corresponding reduction without assuming normality is discussed by Jagers (1980).

It follows from Theorem 6.3.2 that the distribution of W depends only on ψ^2, so that the principle of invariance reduces the problem to that of testing the simple hypothesis $H: \psi = 0$. More precisely, the probability density of W is (cf. Problems 7.2 and 7.3)

$$p_\psi(w) = e^{-\frac{1}{2}\psi^2} \sum_{k=0}^{\infty} c_k \frac{(\frac{1}{2}\psi^2)^k}{k!} \frac{w^{\frac{1}{2}r-1+k}}{(1+w)^{\frac{1}{2}(r+n-s)+k}}, \tag{7.6}$$

where

$$c_k = \frac{\Gamma\left[\frac{1}{2}(r+n-s)+k\right]}{\Gamma\left(\frac{1}{2}r+k\right)\Gamma[\frac{1}{2}(n-s)]}.$$

For any ψ_1 the ratio $p_{\psi_1}(w)/p_o(w)$ is an increasing function of w, and it follows from the Neyman-Pearson fundamental lemma that the most powerful invariant test for testing $\psi = 0$ against $\psi = \psi_1$ rejects when W is too large, or equivalently when

$$W^* = \frac{\sum_{i=1}^{r} Y_i^2/r}{\sum_{i=s+1}^{n} Y_i^2/(n-s)} > C. \tag{7.7}$$

The cutoff point C is determined so that the probability of rejection is α when $\psi = 0$. Since in this case W^* is the ratio of two independent χ^2 variables, each divided by the number of its degrees of freedom, the distribution of W^* is the F-distribution with r and $n-s$ degrees of freedom, and hence C is determined by

$$\int_C^\infty F_{r,n-s}(y)dy = \alpha. \tag{7.8}$$

The test is independent of ψ_1, and hence is UMP among all invariant tests. By Theorem 6.5.2, it is also UMP among all tests whose power function depends only on ψ^2.

The rejection region (7.7) can also be expressed in the form

$$\frac{\sum_{i=1}^{r} Y_i^2}{\sum_{i=1}^{r} Y_i^2 + \sum_{i=s+1}^{n} Y_i^2} > C'. \tag{7.9}$$

When $\psi = 0$, the left-hand side is distributed according to the beta-distribution with r and $n-s$ degrees of freedom [defined through (5.24)], so that C' is determined by

$$\int_{C'}^1 B_{\frac{1}{2}r,\frac{1}{2}(n-s)}(y)\,dy = \alpha. \tag{7.10}$$

For an alternative value of ψ, the left-hand side of (7.9) is distributed according to the *noncentral beta-distribution* with noncentrality parameter ψ, the density of which is (Problem 7.3)

$$g_\psi(y) = e^{-\frac{1}{2}\psi^2} \sum_{k=0}^{\infty} \frac{\left(\frac{1}{2}\psi^2\right)^k}{k!} B_{\frac{1}{2}r+k,\frac{1}{2}(n-s)}(y). \tag{7.11}$$

The power of the test against an alternative ψ is therefore [3]

$$\beta(\psi) = \int_{C'}^{1} g_\psi(y)\,dy.$$

In the particular case $r = 1$ the rejection region (7.7) reduces to

$$\frac{|Y_1|}{\sqrt{\sum\limits_{i=s+1}^{n} Y_i^2/(n-s)}} > C_0. \tag{7.12}$$

This is a two-sided t-test which by the theory of Chapter 5 (see for example Problem 5.5) is UMP unbiased. On the other hand, no UMP unbiased test exists for $r > 1$.

The F-test (7.7) shares the admissibility properties of the two-sided t-test discussed in Section 6.7. In particular, the test is admissible against distant alternatives $\psi^2 \geq \psi_1^2$ (Problem 7.6) and against nearby alternatives $\psi^2 \leq \psi_2^2$ (Problem 7.7). It was shown by Lehmann and Stein (1953) that the test is in fact admissible against the alternatives $\psi^2 \leq \psi_1^2$ for any ψ_1 and hence against all invariant alternatives.

7.2 Linear Hypotheses and Least Squares

In applications to specific problems it is usually not convenient to carry out the reduction to canonical form explicitly. The test statistic W can be expressed in terms of the original variables by noting that $\sum_{i=s+1}^{n} Y_i^2$ is the minimum value of

$$\sum_{i=1}^{s}(Y_i - \eta_i)^2 + \sum_{i=s+1}^{n} Y_i^2 = \sum_{i=1}^{n}[Y_i - E(Y_i)]^2$$

under unrestricted variation of the η's. Also, since the transformation $Y = CX$ is orthogonal and orthogonal transformations leave distances unchanged,

$$\sum_{i=1}^{n}[Y_i - E(Y_i)]^2 = \sum_{i=1}^{n}(X_i - \xi_i)^2.$$

Furthermore, there is a $1:1$ correspondence between the totality of s-tuples $(\eta_1, \ldots, \eta_s)$ and the totality of vectors $\underline{\xi}$ in $\prod_\Omega$. Hence

$$\sum_{i=s+1}^{n} Y_i^2 = \sum_{i=1}^{n}(X_i - \hat{\xi}_i)^2, \tag{7.13}$$

where the $\hat{\xi}$'s are the least-squares estimates of the ξ's under Ω, that is, the values that minimize $\sum_{i=1}^{n}(X_i - \xi_i)^2$ subject to $\underline{\xi}$ in $\prod_\Omega$.

---

[3]Tables of the power of the F-test are provided by Tiku (1967, 1972) [reprinted in Graybill (1976)] and Cohen (1977); charts are given in Pearson and Hartley (1972). Various approximations are discussed by Johnson, Kotz and Balakrishnan (1995).

In the same way it is seen that

$$\sum_{i=1}^{r} Y_i^2 + \sum_{i=s+1}^{n} Y_i^2 = \sum_{i=1}^{n} (X_i - \hat{\hat{\xi}}_i)^2$$

where the $\hat{\hat{\xi}}$'s are the values that minimize $\sum(X_i - \xi_i)^2$ subject to $\underline{\xi}$ in $\prod_\omega$. The test (7.7) therefore becomes

$$W^* = \frac{\left[\sum_{i=1}^{n}(X_i - \hat{\hat{\xi}}_i)^2 - \sum_{i=1}^{n}(X_i - \hat{\xi}_i)^2\right]\Big/ r}{\sum_{i=1}^{n}(X_i - \hat{\xi}_i)^2/(n-s)} > C, \tag{7.14}$$

where C is determined by (7.8). Geometrically the vectors $\underline{\hat{\xi}}$ and $\underline{\hat{\hat{\xi}}}$ are the projections of $\underline{X}$ on $\prod_\Omega$ and $\prod_\omega$, so that the triangle formed by $\underline{X}$, $\underline{\hat{\xi}}$, and $\underline{\hat{\hat{\xi}}}$ has a right angle at $\underline{\hat{\xi}}$ (see Figure 7.1).

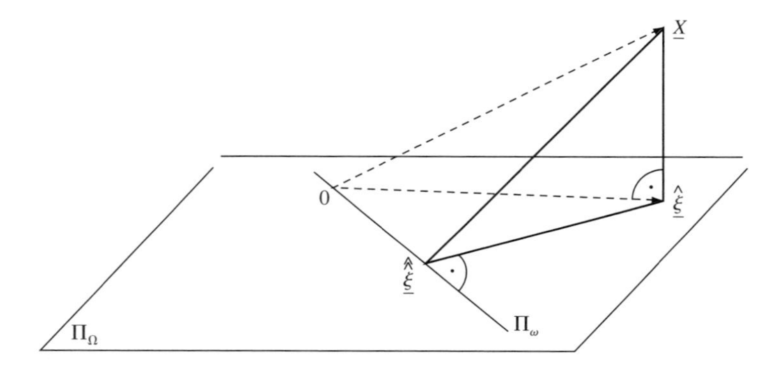

Figure 7.1.

Thus the denominator and numerator of W^*, except for the factors $1/(n-s)$ and $1/r$, are the squares of the distances between $\underline{X}$ and $\underline{\hat{\xi}}$ and between $\underline{\hat{\xi}}$ and $\underline{\hat{\hat{\xi}}}$ respectively. An alternative expression for W^* is therefore

$$W^* = \frac{\sum_{i=1}^{n}(\hat{\xi}_i - \hat{\hat{\xi}}_i)^2\Big/ r}{\sum_{i=1}^{n}(X_i - \hat{\xi}_i)^2/(n-s)}. \tag{7.15}$$

It is desirable to express also the *noncentrality parameter* $\psi^2 = \sum_{i=1}^r \eta_i^2/\sigma^2$ in terms of the ξ's. Now $X = C^{-1}Y$, $\xi = C^{-1}\eta$, and

$$\sum_{i=1}^{r} Y_i^2 = \sum_{i=1}^{n}(X_i - \hat{\hat{\xi}}_i)^2 - \sum_{i=1}^{n}(X_i - \hat{\xi}_i)^2. \tag{7.16}$$

If the right-hand side of (7.16) is denoted by $f(X)$, it follows that $\sum_{i=1}^r \eta_i^2 = f(\xi)$.

A slight generalization of a linear hypothesis is the inhomogeneous hypothesis which specifies for the vector of means $\underline{\xi}$ a subhyperplane $\prod'_\omega$ of $\prod_\Omega$ not passing through the origin. Let $\prod_\omega$ denote the subspace of $\prod_\Omega$ which passes through the origin and is parallel to $\prod'_\omega$. If $\underline{\xi}^0$ is any point of $\prod'_\omega$, the set $\prod'_\omega$ consists of the totality of points $\underline{\xi} = \underline{\xi}^* + \underline{\xi}^0$ as $\underline{\xi}^*$ ranges over $\prod_\omega$. Applying the transformation (7.1) with respect to $\prod_\omega$, the vector of means $\underline{\eta}$ for $\underline{\xi} \in \prod'_\omega$ is then given by $\eta = C\xi = C\xi^* + C\xi^0$ in the canonical form (7.2), and the totality of these vectors is therefore characterized by the the equations $\eta_1 = \eta_1^0, \ldots, \eta_r = \eta_r^0$, $\eta_{s+1} = \cdots = \eta_n = 0$, where η_i^0 is the ith coordinate of $C\xi^0$. In the canonical form, the inhomogeneous hypothesis $\underline{\xi} \in \prod'_\omega$ therefore becomes $\eta_i = \eta_i^0$ $(i = 1, \ldots, r)$. This reduces to the homogeneous case on replacing Y_i with $Y_i - \eta_i^0$, and it follows from (7.7) that the UMP invariant test has the rejection region

$$\frac{\sum_{i=1}^{r} (Y_i - \eta_i^o)^2 / r}{\sum_{i=s+1}^{n} Y_i^2/(n-s)} > C , \tag{7.17}$$

and that the noncentrality parameter is $\psi^2 = \sum_{i=1}^r (\eta_i - \eta_i^0)^2/\sigma^2$.

In applications it is usually most convenient to apply the transformation $X_i - \xi_i^0$ directly to (7.14) or (7.15). It follows from (7.17) that such a transformation always leaves the denominator unchanged. This can also be seen geometrically, since the transformation is a translation of n-space parallel to $\prod_\Omega$ and therefore leaves the distance $\sum(X_i - \hat{\xi}_i)^2$ from $\underline{X}$ to $\prod_\Omega$ unchanged. The noncentrality parameter can be computed as before by replacing X with ξ in the transformed numerator (7.16).

Some examples of linear hypotheses, all with $r = 1$, were already discussed in Chapter 5. The following treats two of these from the present point of view.

Example 7.2.1 Let $X_1, \ldots, X_n$ be independently, normally distributed with common mean μ and variance σ^2, and consider the hypothesis $H : \mu = 0$. Here $\prod_\Omega$ is the line $\xi_i = \cdots = \xi_n$, $\prod_\omega$ is the origin, and $s = r = 1$. Let $\bar{X} = n^{-1} \sum_i X_i$. From the identity

$$\sum (X_i - \mu)^2 = \sum (X_i - \bar{X})^2 + n(\bar{X} - \mu)^2 ,$$

it is seen that $\hat{\xi}_i = \bar{X}$, while $\hat{\hat{\xi}}_i = 0$. The test statistic and ψ^2 are therefore given by

$$W = \frac{n\bar{X}^2}{\sum(X_i - \bar{X})^2} \quad \text{and} \quad \psi^2 = \frac{n\mu^2}{\sigma^2}.$$

Under the hypothesis, the distribution of $(n-1)W$ is that of the square of a variable having Student's t-distribution with $n - 1$ degrees of freedom. ■

Example 7.2.2 In the two-sample problem considered in Example 7.1.1 with $n = n_1 + n_2$, the sum of squares

$$\sum_{i=1}^{n_1} (X_i - \xi)^2 + \sum_{i=n_1+1}^{n} (X_i - \eta)^2$$

is minimized by

$$\hat{\xi} = X_{\cdot}^{(1)} = \sum_{i=1}^{n_1} \frac{X_i}{n_1}, \qquad \hat{\eta} = X_{\cdot}^{(2)} = \sum_{i=n_1+1}^{n} \frac{X_i}{n_2},$$

while, under the hypothesis $\eta - \xi = 0$,

$$\hat{\hat{\xi}} = \hat{\hat{\eta}} = \bar{X} = \frac{n_1 X_{\cdot}^{(1)} + n_2 X_{\cdot}^{(2)}}{n}.$$

The numerator of the test statistic (7.15) is therefore

$$n_1(X_{\cdot}^{(1)} - \bar{X})^2 + n_2(X_{\cdot}^{(2)} - \bar{X})^2 = \frac{n_1 n_2}{n_1 + n_2}\left[X_{\cdot}^{(2)} - X_{\cdot}^{(1)}\right]^2.$$

The more general hypothesis $\eta - \xi = \theta_0$ reduces to the previous case on replacing X_i with $X_i - \theta_0$ for $i = n_1 + 1, \ldots, n$, and is therefore rejected when

$$\frac{\left(X_{\cdot}^{(2)} - X_{\cdot}^{(1)} - \theta_0\right)^2 \Big/ \left(\frac{1}{n_1} + \frac{1}{n_2}\right)}{\left[\sum\limits_{i=1}^{n_1}\left(X_i - X_{\cdot}^{(1)}\right)^2 + \sum\limits_{i=n_1+1}^{n}\left(X_i - X_{\cdot}^{(2)}\right)^2\right] \Big/ (n_1 + n_2 - 2)} > C.$$

The noncentrality parameter is $\psi^2 = (\eta - \xi - \theta_0)^2/(1/n_1 + 1/n_2)\sigma^2$. Under the hypothesis, the square root of the test statistic has the t-distribution with $n_1 + n_2 - 2$ degrees of freedom. ■

Explicit formulae for the $\hat{\xi}_i$ and $\hat{\hat{\xi}}_i$ can be obtained by introducing a coordinate system into the parameter space. Suppose that, in such a system, $\prod_\Omega$ is defined by the equations

$$\xi_i = \sum_{j=1}^{s} a_{ij}\beta_j, \qquad i = 1, \ldots, n,$$

or, in matrix notation,

$$\underset{n\times 1}{\xi} = \underset{n\times s}{A} \; \underset{s\times 1}{B}, \tag{7.18}$$

where A is known and of rank s, and $\beta_1, \ldots, \beta_s$ are unknown parameters. If $\hat{\beta}_1, \ldots, \hat{\beta}_s$ are the least-squares estimators minimizing $\sum_i (X_i - \sum_j a_{ij}\beta_j)^2$, it is seen by differentiation that the $\hat{\beta}_j$ are the solutions of the equations

$$A^T A \beta = A^T X$$

and hence are given by

$$\hat{\beta} = (A^T A)^{-1} A^T X.$$

(That $A^T A$ is nonsingular follows by Problem 6.3.) Thus, we obtain

$$\hat{\xi} = A(A^T A)^{-1} A^T X.$$

Since $\hat{\xi} = \hat{\xi}(X)$ is the projection of X into the space $\prod_\Omega$ spanned by the s columns of A, the formula $\hat{\xi} = A(A^T A)^{-1} A^T X$ shows that $P = A(A^T A)^{-1} A^T$ has the property claimed for it in Example 6.2.3, that for any X in R^n, PX is the projection of X into $\prod_\Omega$.

Schrader (1982), Ronchetti (1982) and Hettmansperger, McKean and Sheather (2000). [For a simple alternative of this kind to Student's t-test, see Prescott (1975).]

Sometimes, it is of interest to test the hypothesis $H : \mu_1 = \cdots = \mu_s$ considered at the beginning of the section, against only the ordered alternatives $\mu_1 \leq \cdots \leq \mu_s$ rather than against the general alternatives of any inequalities among the μ's. Then the F-test (7.19) is no longer reasonable; more powerful alternative tests for this and other problems involving ordered alternatives are discussed by Robertson, Wright and Dykstra (1988). The problem of testing H against one-sided alternatives such as $K : \xi_i \geq 0$ for all i, with at least one inequality strict, is treated by Perlman (1969) and in Barlow et al. (1972), which gives a survey of the literature; also see Tang (1994), Liu and Berger (1995) and Perlman and Wu (1999). Minimal complete classes and admissibility for this and related problems are discussed by Marden (1982a) and Cohen and Sackrowitz (1992).

7.4 Two-Way Layout: One Observation per Cell

The hypothesis of equality of several means arises when a number of different treatments, procedures, varieties, or manifestations of some other factors are to be compared. Frequently one is interested in studying the effects of more than one factor, or the effects of one factor as certain other conditions of the experiment vary, which then play the role of additional factors. In the present section we shall consider the case that the number of factors affecting the outcomes of the experiment is two.

Suppose that one observation is obtained at each of a number of levels of these factors, and denote by X_{ij} ($i = 1, \ldots, a$; $j = 1, \ldots, b$) the value observed when the first factor is at the ith and the second at the jth level. It is assumed that the X_{ij} are independently normally distributed with constant variance σ^2, and for the moment also that the two factors act independently (they are then said to be *additive*), so that ξ_{ij} is of the form $\alpha'_i + \beta'_j$. Putting $\mu = \alpha'_{\cdot} + \beta'_{\cdot}$ and $\alpha_i = \alpha'_i - \alpha'_{\cdot}$, $\beta_j = \beta'_j - \beta'_{\cdot}$, this can be written as

$$\xi_{ij} = \mu + \alpha_i + \beta_j, \qquad \sum \alpha_i = \sum \beta_j = 0, \tag{7.20}$$

where the α's and β's (the *main effects* of A and B) and μ are uniquely determined by (7.20) as[6]

$$\alpha_i = \xi_{i\cdot} - \xi_{\cdot\cdot}, \qquad \beta_j = \xi_{\cdot j} - \xi_{\cdot\cdot}, \qquad \mu = \xi_{\cdot\cdot}. \tag{7.21}$$

Consider the hypothesis

$$H : \alpha_1 = \cdots = \alpha_a = 0 \tag{7.22}$$

that the first factor has no effect on the outcome being observed. This arises in two quite different contexts. The factor of interest, corresponding say to a number of treatments, may be β, while α corresponds to a classification according to,

[6]The replacing of a subscript by a dot indicates that the variable has been averaged with respect to that subscript.

for example, the site on which the observations are obtained (farm, laboratory, city, etc.). The hypothesis then represents the possibility that this subsidiary classification has no effect on the experiment so that it need not be controlled. Alternatively, α may be the (or a) factor of primary interest. In this case, the formulation of the problem as one of hypothesis testing would usually be an oversimplification, since in case of rejection of H, one would require estimates of the α's or at least a grouping according to high and low values.

The hypothesis H is a linear hypothesis with $r = a-1$, $s = 1+(a-1)+(b-1) = a+b-1$, and $n-s = (a-1)(b-1)$. The least-squares estimates of the parameters under Ω can be obtained from the identity

$$\begin{aligned}\sum\sum (X_{ij} - \xi_{ij})^2 &= \sum\sum (X_{ij} - \mu - \alpha_i - \beta_j)^2 \\ &= \sum\sum [(X_{ij} - X_{i\cdot} - X_{\cdot j} + X_{\cdot\cdot}) + (X_{i\cdot} - X_{\cdot\cdot} - \alpha_i) \\ &\qquad + (X_{\cdot j} - X_{\cdot\cdot} - \beta_j) + (X_{\cdot\cdot} - \mu)]^2 \\ &= \sum\sum (X_{ij} - X_{i\cdot} - X_{\cdot j} + X_{\cdot\cdot})^2 \\ &\qquad + b\sum (X_{i\cdot} - X_{\cdot\cdot} - \alpha_i)^2 \\ &\qquad + a\sum (X_{\cdot j} - X_{\cdot\cdot} - \beta_j)^2 + ab\,(X_{\cdot\cdot} - \mu)^2,\end{aligned}$$

which is valid because in the expansion of the third sum of squares the cross-product terms vanish. It follows that

$$\hat{\alpha}_i = X_{i\cdot} - X_{\cdot\cdot}, \qquad \hat{\beta}_j = X_{\cdot j} - X_{\cdot\cdot}, \qquad \hat{\mu} = X_{\cdot\cdot}, \tag{7.23}$$

and that

$$\sum\sum \left(X_{ij} - \hat{\xi}_{ij}\right)^2 = \sum\sum (X_{ij} - X_{i\cdot} - X_{\cdot j} + X_{\cdot\cdot})^2 .$$

Under the hypothesis H we still have $\hat{\hat{\beta}}_j = X_{\cdot j} - X_{\cdot\cdot}$ and $\hat{\hat{\mu}} = X_{\cdot\cdot}$, and hence $\hat{\xi}_{ij} - \hat{\hat{\xi}}_{ij} = X_{i\cdot} - X_{\cdot\cdot}$. The best invariant test therefore rejects when

$$W^* = \frac{b\sum (X_{i\cdot} - X_{\cdot\cdot})^2 /(a-1)}{\sum\sum (X_{ij} - X_{i\cdot} - X_{\cdot j} + X_{\cdot\cdot})^2 /(a-1)(b-1)} > C. \tag{7.24}$$

The noncentrality parameter, on which the power of the test depends, is given by

$$\psi^2 = \frac{b\sum(\xi_{i\cdot} - \xi_{\cdot\cdot})^2}{\sigma^2} = \frac{b\sum \alpha_i^2}{\sigma^2}. \tag{7.25}$$

This problem provides another example of an analysis of variance. The total variation can be broken into three components,

$$\begin{aligned}\sum\sum(X_{ij} - X_{\cdot\cdot})^2 &= b\sum(X_{i\cdot} - X_{\cdot\cdot})^2 + a\sum(X_{\cdot j} - X_{\cdot\cdot})^2 \\ &\qquad + \sum\sum(X_{ij} - X_{i\cdot} - X_{\cdot j} + X_{\cdot\cdot})^2.\end{aligned}$$

Of these, the first contains the variation due to the α's, the second that due to the β's. The last component, in the canonical form of Section 7.1, is equal to $\sum_{i=s+1}^n Y_i^2$. It is therefore the sum of squares of those variables whose means are zero even under Ω. Since this residual part of the variation, which on division by $n-s$ is an estimate of σ^2, cannot be attributed to any effects such as the α's or

β's, it is frequently labeled "error," as an indication that it is due solely to the randomness of the observations, not to any differences of the means. Actually, the breakdown is not quite as sharp as is suggested by the above description. Any component such as that attributed to the α's always also contains some "error," as is seen for example from its expectation, which is

$$E\sum(X_{i\cdot} - X_{\cdot\cdot})^2 = (a-1)\sigma^2 + b\sum \alpha_i^2.$$

Instead of testing whether a certain factor has any effect, one may wish to estimate the size of the effect at the various levels of the factor. Other parameters that are sometimes interesting to estimate are the average outcomes (for example yields) $\xi_{1\cdot}, \ldots, \xi_{a\cdot}$ when the factor is at the various levels. If $\theta_i = \mu + \alpha_i = \xi_{i\cdot}$, confidence sets for $(\theta_1, \ldots, \theta_a)$ are obtained by considering the hypotheses $H(\theta^0)$: $\theta_i = \theta_i^0 (i = 1, \ldots, a)$. For testing $\theta_1 = \cdots = \theta_a = 0$, the least-squares estimates of the ξ_{ij} are $\hat{\xi}_{ij} = X_{i\cdot} + X_{\cdot j} - X_{\cdot\cdot}$ and $\hat{\hat{\xi}}_{ij} = X_{\cdot j} - X_{\cdot\cdot}$. The denominator sum of squares is therefore $\sum\sum(X_{ij} - X_{i\cdot} - X_{\cdot j} + X_{\cdot\cdot})^2$ as before, while the numerator sum of squares is

$$\sum\sum\left(\hat{\xi}_{ij} - \hat{\hat{\xi}}_{ij}\right)^2 = b\sum X_{i\cdot}^2.$$

The general hypothesis reduces to this special case on replacing X_{ij} with the variable $X_{ij} - \theta_i^0$. Since $s = a + b - 1$ and $r = a$, the hypothesis $H(\theta^0)$ is rejected when

$$\frac{b\sum(X_{i\cdot} - \theta_i^0)^2/a}{\sum\sum(X_{ij} - X_{i\cdot} - X_{\cdot j} + X_{\cdot\cdot})^2/(a-1)(b-1)} > C.$$

The associated confidence sets for $(\theta_1, \ldots, \theta_a)$ are the spheres

$$\sum(\theta_i - X_{i\cdot})^2 \le \frac{aC\sum\sum(X_{ij} - X_{i\cdot} - X_{\cdot j} + X_{\cdot\cdot})^2}{(a-1)(b-1)b}.$$

When considering confidence sets for the effects $\alpha_1, \ldots, \alpha_a$, one must take account of the fact that the α's are not independent. Since they add up to zero, it would be enough to restrict attention to $\alpha_1, \ldots, \alpha_{a-1}$. However, an easier and more symmetric solution is found by retaining all the α's. The rejection region of $H : \alpha_i = \alpha_i^0$ for $i = 1, \ldots, a$ (with $\sum \alpha_i^0 = 0$) is obtained from (7.24) by letting $X'_{ij} = X_{ij} - \alpha_i^0$, and hence is given by

$$b\sum(X_{i\cdot} - X_{\cdot\cdot} - \alpha_i^0)^2 > \frac{C\sum\sum(X_{ij} - X_{i\cdot} - X_{\cdot j} + X_{\cdot\cdot})^2}{(b-1)}.$$

The associated confidence set consists of the totality of points $(\alpha_1, \ldots, \alpha_a)$ satisfying $\sum \alpha_i = 0$ and

$$\sum[\alpha_i - (X_{i\cdot} - X_{\cdot\cdot})]^2 \le \frac{C\sum\sum(X_{ij} - X_{i\cdot} - X_{\cdot j} + X_{\cdot\cdot})^2}{b(b-1)}.$$

In the space of $(\alpha_1, \ldots, \alpha_a)$, this inequality defines a sphere whose center $(X_{1\cdot} - X_{\cdot\cdot}, \ldots, X_{a\cdot} - X_{\cdot\cdot})$ lies on the hyperplane $\sum \alpha_i = 0$. The confidence sets for the α's therefore consist of the interior and surface of the great hyperspheres obtained by cutting the a-dimensional spheres with the hyperplane $\sum \alpha_i = 0$.

In both this and the previous case, the usual method shows the class of confidence sets to be invariant under the appropriate group of linear transformations, and the sets are therefore uniformly most accurate invariant.

A rank test of (7.22) analogous to the Kruskal–Wallis test for the one-way layout is Friedman's test, obtained by ranking the s observations $X_{1j}, \dots, X_{sj}$ separately from 1 to s at each level j of the second factor. If these ranks are denoted by $R_{1j}, \dots, R_{sj}$, Friedman's test rejects for large values of $\sum(R_{i\cdot} - R_{\cdot\cdot})^2$. Unless s is large, this test suffers from the fact that comparisons are restricted to observations at the same level of factor 2. The test can be improved by "aligning" the observations from different levels, for example, by subtracting from each observation at the jth level its mean $X_{\cdot j}$ for that level, and then ranking the aligned observations from 1 to ab. For a discussion of these tests and their efficiency see Lehmann (1998, Chapter 6), and for an extension to tests of (7.22) in the model (7.20) when there are several observations per cell, Mack and Skillings (1980). Further discussion is provided by Hettmansperger (1984) and Gibbons and Chakraborti (1992).

That in the experiment described at the beginning of the section there is only one observation per cell, and that as a consequence hypotheses about the α's and β's cannot be tested without some restrictions on the means ξ_{ij}, does not of course justify the assumption of additivity. Rather, it is the other way around: the experiment should not be performed with just one observation per cell unless the factors can safely be assumed to be additive. Faced with such an experiment without prior assurance that the assumption holds, one should test the hypothesis of additivity. A number of tests for this purpose are discussed, for example, in Hegemann and Johnson (1976) and Marasinghe and Johnson (1981).

7.5 Two-Way Layout: m Observations Per Cell

In the preceding section it was assumed that the effects of the two factors α and β are independent and hence additive. The factors may, however, interact in the sense that the effect of one depends on the level of the other. Thus the effectiveness of a teacher depends for example on the quality or the age of the students, and the benefit derived by a crop from various amounts of irrigation depends on the type of soil as well as on the variety being planted. If the additivity assumption is dropped, the means ξ_{ij} of X_{ij} are no longer given by (7.20) under Ω but are completely arbitrary. More than ab observations, one for each combination of levels, are then required, since otherwise $s = n$. We shall here consider only the simple case in which the number of observations is the same at each combination of levels.

Let X_{ijk} $(i = 1, \dots, a; j = 1, \dots, b; k = 1, \dots, m)$ be independent normal with common variance σ^2 and mean $E(X_{ijk}) = \xi_{ij}$. In analogy with the previous notation we write

$$\begin{aligned}\xi_{ij} &= \xi_{\cdot\cdot} + (\xi_{i\cdot} - \xi_{\cdot\cdot}) + (\xi_{\cdot j} - \xi_{\cdot\cdot}) + (\xi_{ij} - \xi_{i\cdot} - \xi_{\cdot j} + \xi_{\cdot\cdot}) \\ &= \mu + \alpha_i + \beta_j + \gamma_{ij}\end{aligned}$$

with $\sum_i \alpha_i = \sum_j \beta_j = \sum_i \gamma_{ij} = \sum_j \gamma_{ij} = 0$. Then α_i is the average effect of factor 1 at level i, averaged over the b levels of factor 2, and a similar interpretation holds for the β's. The γ's are called *interactions*, since γ_{ij} measures the extent to which the joint effect $\xi_{ij} - \xi_{\cdot\cdot}$ of factors 1 and 2 at levels i and j exceeds the sum $(\xi_{i\cdot} - \xi_{\cdot\cdot}) + (\xi_{\cdot j} - \xi_{\cdot\cdot})$ of the individual effects. Consider again the hypothesis that the α's are zero. Then $r = a - 1$, $s = ab$, and $n - s = (m-1)ab$. From the decomposition

$$\sum\sum\sum(X_{ijk} - \xi_{ij})^2 = \sum\sum\sum(X_{ijk} - X_{ij\cdot})^2 + m\sum\sum(X_{ij\cdot} - \xi_{ij})^2$$

and

$$\begin{aligned}\sum\sum(X_{ij\cdot} - \xi_{ij})^2 &= \sum\sum(X_{ij\cdot} - X_{i\cdot\cdot} - X_{\cdot j\cdot} + X_{\cdots} - \gamma_{ij})^2 \\ &\quad + b\sum(X_{i\cdot\cdot} - X_{\cdots} - \alpha_i)^2 + a\sum(X_{\cdot j\cdot} - X_{\cdots} - \beta_j)^2 \\ &\quad + ab(X_{\cdots} - \mu)^2\end{aligned}$$

it follows that

$$\begin{gathered}\hat{\mu} = \hat{\hat{\mu}} = \hat{\xi}_{\cdot\cdot} = X_{\cdots}, \qquad \hat{\alpha}_i = \hat{\xi}_{i\cdot} - \hat{\xi}_{\cdot\cdot} = X_{i\cdot\cdot} - X_{\cdots}, \\ \hat{\beta}_j = \hat{\hat{\beta}}_j = \hat{\xi}_{\cdot j} - \hat{\xi}_{\cdot\cdot} = X_{\cdot j\cdot} - X_{\cdots}, \\ \hat{\gamma}_{ij} = \hat{\hat{\gamma}}_{ij} = X_{ij\cdot} - X_{i\cdot\cdot} - X_{\cdot j\cdot} + X_{\cdots},\end{gathered}$$

and hence that

$$\sum\sum\sum(X_{ijk} - \hat{\xi}_{ij})^2 = \sum\sum\sum(X_{ijk} - X_{ij\cdot})^2,$$

$$\sum\sum\sum(\hat{\xi}_{ij} - \hat{\hat{\xi}}_{ij})^2 = mb\sum(X_{i\cdot\cdot} - X_{\cdots})^2.$$

The most powerful invariant test therefore rejects when

$$W^* = \frac{mb\sum(X_{i\cdot\cdot} - X_{\cdots})^2/(a-1)}{\sum\sum\sum(X_{ijk} - X_{ij\cdot})^2/(m-1)ab} > C, \tag{7.26}$$

and the noncentrality parameter in the distribution of W^* is

$$\frac{mb\sum(\xi_{i\cdot} - \xi_{\cdot\cdot})^2}{\sigma^2} = \frac{mb\sum\alpha_i^2}{\sigma^2}. \tag{7.27}$$

Another hypothesis of interest is the hypothesis H' that the two factors are additive,[7]

$$H' : \gamma_{ij} = 0 \quad \text{for all } i, j.$$

The least-squares estimates of the parameters are easily derived as before, and the UMP invariant test is seen to have the rejection region (Problem 7.13)

$$W^* = \frac{m\sum\sum(X_{ij\cdot} - X_{i\cdot\cdot} - X_{\cdot j\cdot} + X_{\cdots})^2/(a-1)(b-1)}{\sum\sum\sum(X_{ijk} - X_{ij\cdot})^2/(m-1)ab} > C. \tag{7.28}$$

[7]A test of H' against certain restricted alternatives has been proposed for the case of one observation per cell by Tukey (1949a); see Hegemann and Johnson (1976) for further discussion.

Under H', the statistic W^* has the F-distribution with $(a-1)(b-1)$ and $(m-1)ab$ degrees of freedom; the noncentrality parameter for any alternative set of γ's is

$$\psi^2 = \frac{m \sum\sum \gamma_{ij}^2}{\sigma^2}. \tag{7.29}$$

The decomposition of the total variation into its various components, in the present case, is given by

$$\begin{aligned}\sum\sum\sum(X_{ijk} - X_{\cdot\cdot\cdot})^2 &= mb\sum(X_{i\cdot\cdot} - X_{\cdot\cdot\cdot})^2 + ma\sum(X_{\cdot j\cdot} - X_{\cdot\cdot\cdot})^2 \\ &\quad + m\sum\sum(X_{ij\cdot} - X_{i\cdot\cdot} - X_{\cdot j\cdot} + X_{\cdot\cdot\cdot})^2 \\ &\quad + \sum\sum\sum(X_{ijk} - X_{ij\cdot})^2.\end{aligned}$$

Here the first three terms contain the variation due to the α's, β's and γ's respectively, and the last component corresponds to error. The tests for the hypotheses that the α's, β's, or γ's are zero, the first and third of which have the rejection regions (7.26) and (7.28), are then obtained by comparing the α, β, or γ sum of squares with that for error.

An analogous decomposition is possible when the γ's are assumed a priori to be equal to zero. In that case, the third component which previously was associated with γ represents an additional contribution to error, and the breakdown becomes

$$\begin{aligned}\sum\sum\sum(X_{ijk} - X_{\cdot\cdot\cdot})^2 &= mb\sum(X_{i\cdot\cdot} - X_{\cdot\cdot\cdot})^2 + ma\sum(X_{\cdot j\cdot} - X_{\cdot\cdot\cdot})^2 \\ &\quad + \sum\sum\sum(X_{ijk} - X_{i\cdot\cdot} - X_{\cdot j\cdot} + X_{\cdot\cdot\cdot})^2,\end{aligned}$$

with the last term corresponding to error. The hypothesis $H : \alpha_1 = \cdots = \alpha_a = 0$ is then rejected when

$$\frac{mb\sum(X_{i\cdot\cdot} - X_{\cdot\cdot\cdot})^2/(a-1)}{\sum\sum\sum(X_{ijk} - X_{i\cdot\cdot} - X_{\cdot j\cdot} + X_{\cdot\cdot\cdot})^2/(abm - a - b + 1)} > C.$$

Suppose now that the assumption of no interaction, under which this test was derived, is not justified. The denominator sum of squares then has a noncentral χ^2-distribution instead of a central one; and is therefore stochastically larger than was assumed (Problem 7.15). It follows that the actual rejection probability is less than it would be for $\sum\sum\gamma_{ij}^2 = 0$. This shows that the probability of an error of the first kind will not exceed the nominal level of significance, regardless of the values of the γ's. However, the power also decreases with increasing $\sum\sum\gamma_{ij}^2/\sigma^2$ and tends to zero as this ratio tends to infinity.

The analysis of variance and the associated tests derived in this section for two factors extend in a straightforward manner to a larger number of factors (see for example Problem 7.16). On the other hand, if the number of observations is not the same for each combination of levels (each *cell*), explicit formulae for the least-squares estimators may no longer be available, but there is no difficulty in computing these estimators and the associated UMP invariant tests numerically. However, in applications it is then not always clear how to define main effects, interactions, and other parameters of interest, and hence what hypothesis to test. These issues are discussed, for example, in Hocking and Speed (1975) and Speed, Hocking, and Hackney (1979). See also *TPE*2, Chapter 3, Example 4.9, Arnold

(1981, Section 7.4), Searle (1987), McCulloch and Searle (2001) and Hocking (2003).

Of great importance are arrangements in which only certain combinations of levels occur, since they permit reducing the size of the experiment. Thus for example three independent factors, at m levels each, can be analyzed with only m^2 observations, instead of the m^3 required if 1 observation were taken at each combination of levels, by adopting a Latin-square design (Problem 7.17).

The class of problems considered here contains as a special case the two-sample problem treated in Chapter 5, which concerns a single factor with only two levels. The questions discussed in that connection regarding possible inhomogeneities of the experimental material and the randomization required to offset it are of equal importance in the present, more complex situations. If inhomogeneous material is subdivided into more homogeneous groups, this classification can be treated as constituting one or more additional factors. The choice of these groups is an important aspect in the determination of a suitable experimental design.[8] A very simple example of this is discussed in Problems 5.49 and 5.50.

Multiple comparison procedures for two-way (and higher) layouts are discussed by Spjøtvoll (1974); additional references can be obtained from Miller (1977b, 1986) and Westfall and Young (1993). The more general problem of multiple testing will be treated in Chapter 9.

7.6 Regression

Hypotheses specifying one or both of the regression coefficients α, β when $X_1, \dots, X_n$ are independently normally distributed with common variance σ^2 and means

$$\xi_i = \alpha + \beta t_i \tag{7.30}$$

are essentially linear hypotheses, as was pointed out in Example 7.1.2. The hypotheses $H_1 : \alpha = \alpha_0$ and $H_2 : \beta = \beta_0$ were treated in Section 5.6, where they were shown to possess UMP unbiased tests. We shall now consider H_1 and H_2, as well as the hypothesis $H_3 : \alpha = \alpha_0$, $\beta = \beta_0$, from the present point of view. By the general theory of Section 7.1, the resulting tests will be UMP invariant under suitable groups of linear transformations. For the first two cases, in which $r = 1$, this also provides, by the argument of Section 6.6, an alternative proof of their being UMP unbiased.

The space $\prod_\Omega$ is the same for all three hypotheses. It is spanned by the vectors $(1, \dots, 1)$ and $(t_1, \dots, t_n)$ and therefore has dimension $s = 2$ unless the t_i are all

[8] For a discussion of various designs and the conditions under which they are appropriate see, for example, Box, Hunter, and Hunter (1978), Montgomery (2001) and Wu and Hamada (2000). Optimum properties of certain designs, proved by Wald, Ehrenfeld, Kiefer, and others, are discussed by Kiefer (1958), Silvey (1980), Atkinson and Donev (1992) and Pukelsheim (1993). The role of randomization, treated for the two-sample problem in Section 5.10, is studied by Kempthorne (1955), Wilk and Kempthorne (1955), Scheffé (1959), and others; see, for example, Lorenzen (1984) and Giesbrecht and Gumpertz (2004).

equal, which we shall assume not to be the case. The least-squares estimates α and β under Ω are obtained by minimizing $\sum(X_i-\alpha-\beta t_i)^2$. For any fixed value of β, this is achieved by the value $\alpha = \bar{X} - \beta\bar{t}$, for which the sum of squares reduces to $\sum[(X_i - \bar{X}) - \beta(t_i - \bar{t})]^2$. By minimizing this with respect to β one finds

$$\hat{\beta} = \frac{\sum(X_i - \bar{X})(t_i - \bar{t})}{\sum(t_j - \bar{t})^2}, \qquad \hat{\alpha} = \bar{X} - \hat{\beta}\bar{t}; \tag{7.31}$$

and

$$\sum(X_i - \hat{\alpha} - \hat{\beta}t_i)^2 = \sum(X_i - \bar{X})^2 - \hat{\beta}^2 \sum(t_i - \bar{t})^2$$

is the denominator sum of squares for all three hypotheses. The numerator of the test statistic (7.7) for testing the two hypotheses $\alpha = 0$ and to $\beta = 0$ is Y_1^2, and for testing $\alpha = \beta = 0$ is $Y_1^2 + Y_2^2$.

For the hypothesis $\alpha = 0$, the statistic Y_1 was shown in Example 7.1.3 to be equal to

$$\left(\bar{X} - \bar{t}\frac{\sum t_i X_i}{\sum t_j^2}\right)\sqrt{n\frac{\sum t_j^2}{\sum(t_j - \bar{t})^2}} = \hat{\alpha}\sqrt{n\frac{\sum(t_j - \bar{t})^2}{\sum t_j^2}}.$$

Since then

$$E(Y_1) = \alpha\sqrt{n\frac{\sum(t_j - \bar{t})^2}{\sum t_j^2}},$$

the hypothesis $\alpha = \alpha_0$ is equivalent to the hypothesis

$$E(Y_1) = \eta_1^0 = \alpha_0\sqrt{n\sum(t_j - \bar{t})^2/\sum t_j^2}\,,$$

for which the rejection region (7.17) is

$$(n-s)(Y_1 - \eta_1^0)^2/\sum_{i=s+1}^{n} Y_i^2 > C_0$$

and hence

$$\frac{|\hat{\alpha} - \alpha_0|\sqrt{n\sum(t_j - \bar{t})^2/\sum t_j^2}}{\sqrt{\sum(X_i - \hat{\alpha} - \hat{\beta}t_i)^2/(n-2)}} > C_0. \tag{7.32}$$

For the hypothesis $\beta = 0$, Y_1 was shown to be equal to

$$\frac{\sum(X_i - \bar{X})(t_i - \bar{t})}{\sqrt{\sum(t_j - \bar{t})^2}} = \hat{\beta}\sqrt{\sum(t_j - \bar{t})^2}.$$

Since then $E(Y_1) = \beta\sqrt{\sum(t_j - \bar{t})^2}$, the hypothesis $\beta = \beta_0$ is equivalent to $E(Y_1) = \eta_1^0 = \beta_0\sqrt{\sum(t_j - \bar{t})^2}$ and the rejection region is

$$\frac{|\hat{\beta} - \beta_0|\sqrt{\sum(t_j - \bar{t})^2}}{\sqrt{\sum(X_i - \hat{\alpha} - \hat{\beta}t_i)^2/(n-2)}} > C_0. \tag{7.33}$$

For testing $\alpha = \beta = 0$, it was shown in Example 7.1.3 that

$$Y_1 = \hat{\beta}\sqrt{\sum(t_j - \bar{t})^2}, \qquad Y_2 = \sqrt{n}\bar{X} = \sqrt{n}(\hat{\alpha} + \hat{\beta}\bar{t});$$

the numerator of (7.7) is therefore

$$\frac{Y_1^2 + Y_2^2}{2} = \frac{n(\hat{\alpha} + \hat{\beta}\bar{t})^2 + \hat{\beta}^2 \sum (t_j - \bar{t})^2}{2}.$$

The more general hypothesis $\alpha = \alpha_0$, $\beta = \beta_0$ is equivalent to $E(Y_1) = \eta_1^0$, $E(Y_2) = \eta_2^0$, where $\eta_1^0 = \beta_0\sqrt{\sum(t_j - \bar{t})^2}$, $\eta_2^0 = \sqrt{n}(\alpha_0 + \beta_0\bar{t})$; and the rejection region (7.17) can therefore be written as

$$\frac{\left[n(\hat{\alpha} - \alpha_0)^2 + 2n\bar{t}(\hat{\alpha} - \alpha_0)(\hat{\beta} - \beta_0) + \sum t_i^2(\hat{\beta} - \beta_0)^2\right]/2}{\sum(X_i - \hat{\alpha} - \hat{\beta}t_i)^2/(n-2)} > C. \tag{7.34}$$

The associated confidence sets for (α, β) are obtained by reversing this inequality and replacing α_0 and β_0 by α and β. The resulting sets are ellipses centered at $(\hat{\alpha}, \hat{\beta})$.

The simple regression model (7.30) can be generalized in many directions; the means ξ_i may for example be polynomials in t_1 of higher than the first degree (see Problem 7.20), or more complex functions such as trigonometric polynomials; or they may be functions of several variables, t_i, u_i, v_i. Some further extensions will now be illustrated by a number of examples.

Example 7.6.1 A variety of problems arise when there is more than one regression-line. Suppose that the variables X_{ij} are independently normally distributed with common variance and means

$$\xi_{ij} = \alpha_i + \beta_i t_{ij} \qquad (j = 1, \ldots, n_i; \quad i = 1, \ldots, b). \tag{7.35}$$

The hypothesis that these regression lines have equal slopes

$$H : \beta_1 = \cdots = \beta_b$$

may occur for example when the equality of a number of growth rates is to be tested. The parameter space $\prod_\Omega$ has dimension $s = 2b$ provided none of the sums $\sum_j (t_{ij} - t_{i\cdot})^2$ is zero; the number of constraints imposed by the hypothesis is $r = b - 1$. The minimum value of $\sum\sum(X_{ij} - \xi_{ij})^2$ under Ω is obtained by minimizing $\sum_j (X_{ij} - \alpha_i - \beta_i t_{ij})^2$ for each i, so that by (7.31),

$$\hat{\beta}_i = \frac{\sum_j (X_{ij} - X_{i\cdot})(t_{ij} - t_{i\cdot})}{\sum_j (t_{ij} - t_{i\cdot})^2}, \qquad \hat{\alpha}_i = X_{i\cdot} - \hat{\beta}_i t_{i\cdot}\,.$$

Under H, one must minimize $\sum\sum(X_{ij} - \alpha_i - \beta t_{ij})^2$, which for any fixed β leads to $\alpha_i = X_{i\cdot} - \beta t_{i\cdot}$ and reduces the sum of squares to $\sum\sum[(X_{ij} - X_{i\cdot}) - \beta(t_{ij} - t_{i\cdot})]^2$. Minimizing this with respect to β, one finds

$$\hat{\hat{\beta}} = \frac{\sum\sum(X_{ij} - X_{i\cdot})(t_{ij} - t_{i\cdot})}{\sum\sum(t_{ij} - t_{i\cdot})^2}, \qquad \hat{\hat{\alpha}}_i = X_{i\cdot} - \hat{\hat{\beta}}_{i\cdot}.$$

Since

$$X_{ij} - \hat{\xi}_{ij} = X_{ij} - \hat{\alpha}_i - \hat{\beta}_i t_{ij} = (X_{ij} - X_{i\cdot}) - \hat{\beta}_i(t_{ij} - t_{i\cdot})$$

and

$$\hat{\xi}_{ij} - \hat{\hat{\xi}}_{ij} = (\hat{\alpha}_i - \hat{\hat{\alpha}}_i) + t_{ij}(\hat{\beta}_i - \hat{\hat{\beta}}) = (\hat{\beta}_i - \hat{\hat{\beta}})(t_{ij} - t_{i\cdot}),$$

the rejection region (7.15) is

$$\frac{\sum_i(\hat{\beta}_i - \hat{\hat{\beta}})^2 \sum_j (t_{ij} - t_{i\cdot})^2/(b-1)}{\sum\sum\left[(X_{ij} - X_{i\cdot}) - \hat{\beta}_i(t_{ij} - t_{i\cdot})\right]^2/(n-2b)} > C, \tag{7.36}$$

where the left-hand side under H has the F-distribution with $b-1$ and $n-2b$ degrees of freedom.

Since

$$E(\hat{\beta}_i) = \beta_i \quad \text{and} \quad E(\hat{\hat{\beta}}) = \frac{\sum_i \beta_i \sum_j (t_{ij} - t_{i\cdot})^2}{\sum\sum(t_{ij} - t_{i\cdot})^2},$$

the noncentrality parameter of the distribution for an alternative set of β's is $\psi^2 = \sum_i(\beta_i - \tilde{\beta})^2 \sum_j (t_{ij} - t_{i\cdot})^2/\sigma^2$, where $\tilde{\beta} = E(\hat{\hat{\beta}})$. In the particular case that the n_i and the t_{ij} are independent of i, $\tilde{\beta}$ reduces to $\bar{\beta} = \sum \beta_j/b$. ■

Example 7.6.2 The regression model (7.35) arises in the comparison of a number of treatments when the experimental units are treated as fixed and the unit effects u_{ij} (defined in Section 5.9) are proportional to known constants t_{ij}. Here t_{ij} might for example be a measure of the fertility of the i, jth piece of land or the weight of the i, jth experimental animal prior to the experiment. It is then frequently possible to assume that the proportionality factor β_i does not depend on the treatment, in which case (7.35) reduces to

$$\xi_{ij} = \alpha_i + \beta t_{ij} \tag{7.37}$$

and the hypothesis of no treatment effect becomes

$$H : \alpha_1 = \cdots = \alpha_b.$$

The space $\prod_\Omega$ coincides with $\prod_\omega$ of the previous example, so that $s = b+1$ and

$$\hat{\beta} = \frac{\sum\sum(X_{ij} - X_{i\cdot})(t_{ij} - t_{i\cdot})}{\sum\sum(t_{ij} - t_{i\cdot})^2}, \qquad \hat{\alpha}_i = X_{i\cdot} - \hat{\beta} t_{i\cdot}.$$

Minimization of $\sum\sum(X_{ij} - \alpha - \beta t_{ij})^2$ gives

$$\hat{\hat{\beta}} = \frac{\sum\sum(X_{ij} - X_{\cdot\cdot})(t_{ij} - t_{\cdot\cdot})}{\sum\sum(t_{ij} - t_{\cdot\cdot})^2}, \qquad \hat{\hat{\alpha}} = X_{\cdot\cdot} - \hat{\hat{\beta}} t_{\cdot\cdot},$$

where $X_{\cdot\cdot} = \sum\sum X_{ij}/n$, $t_{\cdot\cdot} = \sum\sum t_{ij}/n$, $n = \sum n_i$. The sum of squares in the numerator of W^* in (7.15) is thus

$$\sum\sum\left(\hat{\xi}_{ij} - \hat{\hat{\xi}}_{ij}\right)^2 = \sum\sum\left[(X_{i\cdot} - X_{\cdot\cdot}) + \hat{\beta}(t_{ij} - t_{i\cdot}) - \hat{\hat{\beta}}(t_{ij} - t_{\cdot\cdot})\right]^2.$$

The hypothesis H is therefore rejected when

$$\frac{\sum\sum\left[(X_{i\cdot} - X_{\cdot\cdot}) + \hat{\beta}(t_{ij} - t_{i\cdot}) - \hat{\hat{\beta}}(t_{ij} - t_{\cdot\cdot})\right]^2/(b-1)}{\sum\sum\left[(X_{ij} - X_{i\cdot}) - \hat{\beta}(t_{ij} - t_{i\cdot})\right]^2/(n-b-1)} > C, \tag{7.38}$$

where under H the left-hand side has the F-distribution with $b-1$ and $n-b-1$ degrees of freedom.

The hypothesis H can be tested without first ascertaining the values of the t_{ij}; it is then the hypothesis of no effect in a one-way classification considered in Section 7.3, and the test is given by (7.19). Actually, since the unit effects u_{ij} are assumed to be constants, which are now completely unknown, the treatments are assigned to the units either completely at random or at random within subgroups. The appropriate test is then a randomization test for which (7.19) is an approximation. ■

Example 7.6.2 illustrates the important class of situations in which an analysis of variance (in the present case concerning a one-way classification) is combined with a regression problem (in the present case linear regression on the single "concomitant variable" t). Both parts of the problem may of course be considerably more complex than was assumed here. Quite generally, in such combined problems one can test (or estimate) the treatment effects as was done above, and a similar analysis can be given for the regression coefficients. The breakdown of the variation into its various treatment and regression components is the so-called *analysis of covariance.*

7.7 Random-Effects Model: One-way Classification

In the factorial experiments discussed in Sections 7.3, 7.4, and 7.5, the factor levels were considered fixed, and the associated effects (the μ's in Section 7.3, the α's, β's and γ's in Sections 7.4 and 7.5) to be unknown constants. However, in many applications, these levels and their effects instead are (unobservable) random variables. If all the effects are constant or all random, one speaks of *fixed-effects model* (*model I*) or *random-effects model* (*model II*) respectively, and the term *mixed model* refers to situations in which both types occur.[9] Of course, only the model I case constitutes a linear hypothesis according to the definition given at the beginning of the chapter. In the present section we shall treat as model II the case of a single factor (one-way classification), which was analyzed under the model I assumption in Section 7.3.

As an illustration of this problem, consider a material such as steel, which is manufactured or processed in batches. Suppose that a sample of size n is taken from each of s batches and that the resulting measurements X_{ij} ($j = 1, \ldots, n$; $i = 1, \ldots, s$) are independently normally distributed with variance σ^2 and mean ξ_i. If the factor corresponding to i were constant, with the same effect α_i in each replication of the experiment, we would have

$$\xi_i = \mu + \alpha_i \qquad \left(\sum \alpha_i = 0\right)$$

and

$$X_{ij} = \mu + \alpha_i + U_{ij} \ ,$$

where the U_{ij} are independently distributed as $N(0, \sigma^2)$. The hypothesis of no effect is $\xi_1 = \cdots = \xi_s$, or equivalently $\alpha_1 = \cdots = \alpha_s = 0$. However, the effect is

[9]For a recent exposition of random effects models, see Sahai and Ojeda (2004).

associated with the batches, of which a new set will be involved in each replication of the experiment; the effect therefore does not remain constant. Instead, we shall suppose that the batch effects constitute a sample from a normal distribution, and to indicate their random nature we shall write A_i for α_i, so that

$$X_{ij} = \mu + A_i + U_{ij}. \tag{7.39}$$

The assumption of additivity (lack of interaction) of batch and unit effect, in the present model, implies that the A's and U's are independent. If the expectation of A_i is absorbed into μ, it follows that the A's and U's are independently normally distributed with zero means and variances σ_A^2 and σ^2 respectively. The X's of course are no longer independent.

The hypothesis of no batch effect, that the A's are zero and hence constant, takes the form

$$H : \sigma_A^2 = 0$$

This is not realistic in the present situation, but is the limiting case of the hypothesis

$$H(\Delta_0) : \frac{\sigma_A^2}{\sigma^2} \leq \Delta_0$$

that the batch effect is small relative to the variation of the material within a batch. These two hypotheses correspond respectively to the model I hypotheses $\sum \alpha_i^2 = 0$ and $\sum \alpha_i^2/\sigma^2 \leq \Delta_0$.

To obtain a test of $H(\Delta_0)$ it is convenient to begin with the same transformation of variables that reduced the corresponding model I problem to canonical form. Each set $(X_{i1}, \ldots, X_{in})$ is subjected to an orthogonal transformation $Y_{ij} = \sum_{k=1}^n c_{jk} X_{ik}$ such that $Y_{i1} = \sqrt{n} X_{i\cdot}$. Since $c_{1k} = 1/\sqrt{n}$ for $k = 1, \ldots, n$ (see Example 7.1.3), it follows from the assumption of orthogonality that $\sum_{k=1}^n c_{jk} = 0$ for $j = 2, \ldots, n$ and hence that $Y_{ij} = \sum_{k=1}^n c_{jk} U_{ik}$ for $j > 1$. The Y_{ij} with $j > 1$ are therefore independently normally distributed with zero mean and variance σ^2. They are also independent of $U_{i\cdot}$ since $(\sqrt{n} U_{i\cdot} - Y_{i2} \ldots Y_{in})' = C(U_{i1} U_{i2} \ldots U_{in})'$ (a prime indicates the transpose of a matrix). On the other hand, the variables $Y_{i1} = \sqrt{n} X_{i\cdot} = \sqrt{n}(\mu + A_i + U_{i\cdot})$ are also independently normally distributed but with mean $\sqrt{n}\mu$ and variance $\sigma^2 + n\sigma_A^2$. If an additional orthogonal transformation is made from $(Y_{11}, \ldots, Y_{s1})$ to $(Z_{11}, \ldots, Z_{s1})$ such that $Z_{11} = \sqrt{s} Y_{\cdot 1}$, the Z's are independently normally distributed with common variance $\sigma^2 + n\sigma_A^2$ and means $E(Z_{11}) = \sqrt{sn}\mu$ and $E(Z_{i1}) = 0$ for $i > 1$. Putting $Z_{ij} = Y_{ij}$ for $j > 1$ for the sake of conformity, the joint density of the Z's is then

$$(2\pi)^{-ns/2} \sigma^{-(n-1)s} \left(\sigma^2 + n\sigma_A^2\right)^{-s/2} \tag{7.40}$$
$$\times \exp\left[-\frac{1}{2\left(\sigma^2 + n\sigma_A^2\right)} \left(\left(z_{11} - \sqrt{sn}\mu\right)^2 + \sum_{i=2}^s z_{i1}^2\right) - \frac{1}{2\sigma^2} \sum_{i=1}^s \sum_{j=2}^n z_{ij}^2\right].$$

The problem of testing $H(\Delta_0)$ is invariant under addition of an arbitrary constant to Z_{11}, which leaves the remaining Z's as a maximal set of invariants. These constitute samples of size $s(n-1)$ and $s-1$ from two normal distributions with means zero and variances σ^2 and $\tau^2 = \sigma^2 + n\sigma_A^2$.

The hypothesis $H(\Delta_0)$ is equivalent to $\tau^2/\sigma^2 \le 1 + \Delta_0 n$, and the problem reduces to that of comparing two normal variances, which was considered in Example 6.3.4 without the restriction to zero means. The UMP invariant test, under multiplication of all Z_{ij} by a common positive constant, has the rejection region

$$W^* = \frac{1}{1+\Delta_0 n} \cdot \frac{S_A^2/(s-1)}{S^2/(n-1)s} > C, \tag{7.41}$$

where

$$S_A^2 = \sum_{i=2}^{s} Z_{i1}^2 \quad \text{and} \quad S^2 = \sum_{i=1}^{s}\sum_{j=2}^{n} Z_{ij}^2 = \sum_{i=1}^{s}\sum_{j=2}^{n} Y_{ij}^2.$$

The constant C is determined by

$$\int_C^\infty F_{s-1,(n-1)s}(y)\,dy = \alpha.$$

Since

$$\sum_{j=1}^{n} Y_{ij}^2 - Y_{i1}^2 = \sum_{j=1}^{n} U_{ij}^2 - nU_{i\cdot}^2$$

and

$$\sum_{i=1}^{s} Z_{i1}^2 - Z_{11}^2 = \sum_{i=1}^{s} Y_{i1}^2 - Y_{\cdot 1}^2,$$

the numerator and denominator sums of squares of W^*, expressed in terms of the X's, become

$$S_A^2 = n\sum_{i=1}^{s}(X_{i\cdot} - X_{\cdot\cdot})^2 \quad \text{and} \quad S^2 = \sum_{i=1}^{s}\sum_{j=1}^{n}(X_{ij} - X_{i\cdot})^2.$$

In the particular case $\Delta_0 = 0$, the test (7.41) is equivalent to the corresponding model I test (7.19), but they are of course solutions of different problems, and also have different power functions. Instead of being distributed according to a noncentral χ^2-distribution as in model I, the numerator sum of squares of W^* is proportional to a central χ^2-variable even when the hypothesis is false, and the power of the test (7.41) against an alternative value of Δ is obtained from the F-distribution through

$$\beta(\Delta) = P_\Delta\{W^* > C\} = \int_{\frac{1+\Delta_0 n}{1+\Delta n}C}^\infty F_{s-1,(n-1)s}(y)\,dy.$$

The family of tests (7.41) for varying Δ_0 is equivalent to the confidence statements

$$\underline{\Delta} = \frac{1}{n}\left[\frac{S_A^2/(s-1)}{CS^2/(n-1)s} - 1\right] \le \Delta. \tag{7.42}$$

The corresponding upper confidence bounds for Δ are obtained from the tests of the hypotheses $\Delta \ge \Delta_0$. These have the acceptance regions $W^* \ge C'$, where W^* is given by (7.41) and C' is determined by

$$\int_{C'}^\infty F_{s-1,(n-1)s} = 1 - \alpha\ .$$

The resulting confidence bounds are

$$\Delta \leq \frac{1}{n}\left[\frac{S_A^2/(s-1)}{C'S^2/(n-1)s} - 1\right] = \bar{\Delta}. \tag{7.43}$$

Both the confidence sets (7.42) and (7.43) are equivariant with respect to the group of transformations generated by those considered for the testing problems, and hence are uniformly most accurate equivariant.

When $\underline{\Delta}$ is negative, the confidence set $(\underline{\Delta}, \infty)$ contains all possible values of the parameter Δ. For small Δ, this will happen with high probability ($1-\alpha$ for $\Delta = 0$), as must be the case, since $\underline{\Delta}$ is then required to be a safe lower bound for a quantity which is equal to or near zero. Even more awkward is the possibility that $\bar{\Delta}$ is negative, so that the confidence set $(-\infty, \bar{\Delta})$ is empty. An interpretation is suggested by the fact that this occurs if and only if the hypothesis $\Delta \geq \Delta_0$ is rejected for all positive values of Δ_0. This may be taken as an indication that the assumed model is not appropriate, [10] although it must be realized that for small Δ the probability of the event $\bar{\Delta} < 0$ is near α even when the assumptions are satisfied, so that this outcome will occasionally be observed.

The tests of $\Delta \leq \Delta_0$ and $\Delta \geq \Delta_0$ are not only UMP invariant but also UMP unbiased, and UMP unbiased tests also exist for testing $\Delta = \Delta_0$ against the two-sided alternatives $\Delta \neq \Delta_0$. This follows from the fact that the joint density of the Z's constitutes an exponential family. The confidence sets associated with these three families of tests are then uniformly most accurate unbiased (Problem 7.21). That optimum unbiased procedures exist in the model II case but not in the corresponding model I problem is explained by the different structure of the two hypotheses. The model II hypothesis $\sigma_A^2 = 0$ imposes one constraint, since it concerns the single parameter σ_A^2. On the other hand, the corresponding model I hypothesis $\sum_{i=1}^s \alpha_i^2 = 0$ specifies the values of the s parameters $\alpha_1, \ldots, \alpha_s$, and since $s-1$ of these are independent, imposes $s-1$ constraints.

A UMP invariant test of $\Delta \leq \Delta_0$ does not exist if the sample sizes n_i are unequal. An invariant test with a weaker optimum property for this case is obtained by Spjøtvoll (1967).

Since Δ is a ratio of variances, it is not surprising that the test statistic W^* is quite sensitive to the assumption of normality; such robustness issues are discussed in Section 11.3.1). More robust alternatives are discussed, for example, by Arvesen and Layard (1975). Westfall (1989) compares invariant variance ratio tests in mixed models.

Optimality of standard F tests in balanced ANOVA models with mixed effects is derived in Mathew and Sinha (1988a) and optimal tests in some unbalanced designs are derived in Mathew and Sinha (1988b).

7.8 Nested Classifications

The theory of the preceding section does not carry over even to so simple a situation as the general one-way classification with unequal numbers in the different

[10] For a discussion of possibly more appropriate alternative models, see Smith and Murray (1984).

classes (Problem 7.24). However, the unbiasedness approach does extend to the important case of a *nested* (hierarchical) classification with equal numbers in each class. This extension is sufficiently well indicated by carrying it through for the case of two factors; it follows for the general case by induction with respect to the number of factors.

Returning to the illustration of a batch process, suppose that a single batch of raw material suffices for several batches of the finished product. Let the experimental material consist of ab batches, b coming from each of a batches of raw material, and let a sample of size n be taken from each. Then (7.39) becomes

$$X_{ijk} = \mu + A_i + B_{ij} + U_{ijk} \qquad (i = 1, \ldots, a; \quad j = 1, \ldots, b; \quad k = 1, \ldots, n) \tag{7.44}$$

where A_i denotes the effect of the ith batch of raw material, B_{ij} that of the jth batch of finished product obtained from this material, and U_{ijk} the effect of the kth unit taken from this batch. All these variables are assumed to be independently normally distributed with zero means and with variances σ_A^2, σ_B^2, and σ^2 respectively. The main part of the induction argument consists of proving the existence of an orthogonal transformation to variables Z_{ijk}, the joint density of which, except for a constant, is

$$\exp\left[-\frac{1}{2\left(\sigma^2 + n\sigma_B^2 + bn\sigma_A^2\right)}\left(\left(z_{111} - \sqrt{abn}\mu\right)^2 + \sum_{i=2}^{a} z_{i11}^2\right) - \frac{1}{2\left(\sigma^2 + n\sigma_B^2\right)}\sum_{i=1}^{a}\sum_{j=2}^{b} z_{ij1}^2 - \frac{1}{2\sigma^2}\sum_{i=1}^{a}\sum_{j=1}^{b}\sum_{k=2}^{n} z_{ijk}^2\right]. \tag{7.45}$$

As a first step, there exists for each fixed i, j an orthogonal transformation from $(X_{ij1}, \ldots, X_{ijn})$ to $(Y_{ij1}, \ldots, Y_{ijn})$ such that

$$Y_{ij1} = \sqrt{n}X_{ij\cdot} = \sqrt{n}\mu + \sqrt{n}(A_i + B_{ij} + U_{ij\cdot}).$$

As in the case of a single classification, the variables Y_{ijk} with $k > 1$ depend only on the U's, are independently normally distributed with zero mean and variance σ^2, and are independent of the $U_{ij\cdot}$. On the other hand, the variables Y_{ij1} have exactly the structure of the Y_{ij} in the one-way classification,

$$Y_{ij1} = \mu' + A_i' + U_{ij}',$$

where $\mu' = \sqrt{n}\mu$, $A_i' = \sqrt{n}A_i$, $U_{ij}' = \sqrt{n}(B_{ij} + U_{ij\cdot})$, and where the variances of A_i' and U_{ij}' are ${\sigma'_A}^2 = n\sigma_A^2$ and $\sigma'^2 = \sigma^2 + n\sigma_B^2$ respectively. These variables can therefore be transformed to variables Z_{ij1} whose density is given by (7.40) with Z_{ij1} in place of Z_{ij}. Putting $Z_{ijk} = Y_{ijk}$ for $k > 1$, the joint density of all Z_{ijk} is then given by (7.45).

Two hypotheses of interest can be tested on the basis of (7.45)—$H_1 : \sigma_A^2/(\sigma^2 + n\sigma_B^2) \le \Delta_0$ and $H_2 : \sigma_B^2/\sigma^2 \le \Delta_0$. Both state that one or the other of the classifications has little effect on the outcome. Let

$$S_A^2 = \sum_{i=2}^{a} Z_{i11}^2, \qquad S_B^2 = \sum_{i=1}^{a}\sum_{j=2}^{b} Z_{ij1}^2, \qquad S^2 = \sum_{i=1}^{a}\sum_{j=1}^{b}\sum_{k=2}^{n} Z_{ijk}^2.$$

To obtain a test of H_1, one is tempted to eliminate S^2 through invariance under multiplication of Z_{ijk} for $k > 1$ by an arbitrary constant. However, these transformations do not leave (7.45) invariant, since they do not always preserve the fact that σ^2 is the smallest of the three variances σ^2, $\sigma^2 + n\sigma_B^2$, and $\sigma^2 + n\sigma_B^2 + bn\sigma_A^2$. We shall instead consider the problem from the point of view of unbiasedness. For any unbiased test of H_1, the probability of rejection is α whenever $\sigma_A^2/(\sigma^2 + n\sigma_B^2) = \Delta_0$, and hence in particular when the three variances are σ^2, τ_0^2, and $(1 + bn\Delta_0)\tau_0^2$ for any fixed τ_0^2 and all $\sigma^2 < \tau_0^2$. It follows by the techniques of Chapter 4 that the conditional probability of rejection given $S^2 = s^2$ must be equal to α for almost all values of s^2. With S^2 fixed, the joint distribution of the remaining variables is of the same type as (7.45) after the elimination of Z_{111}, and a UMP unbiased conditional test given $S^2 = s^2$ has the rejection region

$$W_1^* = \frac{1}{1 + bn\Delta_0} \cdot \frac{S_A^2 \big/ (a-1)}{S_B^2 \big/ (b-1)a} \geq C_1. \tag{7.46}$$

Since S_A^2 and S_B^2 are independent of S^2, the constant C_1 is determined by the fact that when $\sigma_A^2/(\sigma^2 + n\sigma_B^2) = \Delta_0$, the statistic W_1^* is distributed as $F_{a-1,(b-1)a}$ and hence in particular does not depend on s. The test (7.46) is clearly unbiased and hence UMP unbiased.

An alternative proof of this optimality property can be obtained using Theorem 6.6.1. The existence of a UMP unbiased test follows from the exponential family structure of the density (7.45), and the test is the same whether τ^2 is equal to $\sigma^2 + n\sigma_B^2$ and hence $\geq \sigma^2$, or whether it is unrestricted. However, in the latter case, the test (7.46) is UMP invariant and therefore is UMP unbiased even when $\tau^2 \geq \sigma^2$.

The argument with respect to H_2 is completely analogous and shows the UMP unbiased test to have the rejection region

$$W_2^* = \frac{1}{1 + n\Delta_0} \cdot \frac{S_B^2 \big/ (b-1)a}{S^2 \big/ (n-1)ab} \geq C_2, \tag{7.47}$$

where C_2 is determined by the fact that for $\sigma_B^2/\sigma^2 = \Delta_0$, the statistic W_2^* is distributed as $F_{(b-1)a,(n-1)ab}$.

It remains to express the statistics S_A^2, S_B^2, and S^2 in terms of the X's. From the corresponding expressions in the one-way classification, it follows that

$$\begin{aligned}
S_A^2 &= \sum_{i=1}^{a} Z_{i11}^2 - Z_{111}^2 = b\sum (Y_{i\cdot 1} - Y_{\cdot\cdot 1})^2, \\
S_B^2 &= \sum_{i=1}^{a} \left[\sum_{j=1}^{b} Z_{ij1}^2 - Z_{i11}^2 \right] = \sum\sum (Y_{ij1} - Y_{i\cdot 1})^2,
\end{aligned}$$

and

$$\begin{aligned}
S^2 &= \sum_{i=1}^{a}\sum_{j=1}^{b} \left[\sum_{k=1}^{n} Y_{ijk}^2 - Y_{ij1}^2 \right] = \sum_i \sum_j \left[\sum_{k=1}^{n} U_{ijk}^2 - nU_{ij\cdot}^2 \right] \\
&= \sum_i \sum_j \sum_k (U_{ijk} - U_{ij\cdot})^2.
\end{aligned}$$

Hence

$$S_A^2 = bn\sum(X_{i\cdot\cdot} - X_{\cdot\cdot\cdot})^2, \qquad S_B^2 = n\sum\sum(X_{ij\cdot} - X_{i\cdot\cdot})^2, \qquad (7.48)$$
$$S^2 = \sum\sum\sum(X_{ijk} - X_{ij\cdot})^2.$$

It is seen from the expression of the statistics in terms of the Z's that their expectations are $E[S_A^2/(a-1)] = \sigma^2 + n\sigma_B^2 + bn\sigma_A^2$, $E[S_B^2/(b-1)a] = \sigma^2 + n\sigma_B^2$, and $E[S^2/(n-1)ab] = \sigma^2$. The decomposition

$$\sum\sum\sum(X_{ijk} - X_{\cdot\cdot\cdot})^2 = S_A^2 + S_B^2 + S^2$$

therefore forms a basis for the analysis of the variance of X_{ijk},

$$Var(X_{ijk}) = \sigma_A^2 + \sigma_B^2 + \sigma^2$$

by providing estimates of the *components of variance* σ_A^2, σ_B^2, and σ^2, and tests of certain ratios of these components.

Nested two-way classifications also occur as mixed models. Suppose for example that a firm produces the material of the previous illustrations in different plants. If α_i denotes the effect of the ith plant (which is fixed, since the plants do not change in the replication of the experiment), B_{ij} the batch effect, and U_{ijk} the unit effect, the observations have the structure

$$X_{ijk} = \mu + \alpha_i + B_{ij} + U_{ijk}. \qquad (7.49)$$

Instead of reducing the X's to the fully canonical form in terms of the Z's as before, it is convenient to carry out only the reduction to the Y's (such that $Y_{ij1} = \sqrt{n}X_{ij\cdot}$) and the first of the two transformations which take the Y's into the Z's. If the resulting variables are denoted by W_{ijk}, they satisfy $W_{i11} = \sqrt{b}Y_{i\cdot 1}$, $W_{ijk} = Y_{ijk}$ for $k > 1$ and

$$\sum_{i=1}^{a}(W_{i11} - W_{\cdot 11})^2 = S_A^2, \qquad \sum_{i=1}^{a}\sum_{j=2}^{b} W_{ij1}^2 = S_B^2, \qquad \sum_{i=1}^{a}\sum_{j=1}^{b}\sum_{k=2}^{n} W_{ijk}^2 = S^2\ ,$$

where S_A^2, S_B^2, and S^2 are given by (7.48). The joint density of the W's is, except for a constant,

$$\exp\left[-\frac{1}{2(\sigma^2 + n\sigma_B^2)}\left(\sum_{i=1}^{a}(w_{i11} - \mu - \alpha_i)^2 + \sum_{i=1}^{a}\sum_{j=2}^{b} w_{ij1}^2\right) \qquad (7.50)\right.$$
$$\left. - \frac{1}{2\sigma^2}\sum_{i=1}^{a}\sum_{j=1}^{b}\sum_{k=2}^{n} w_{ijk}^2\right].$$

This shows clearly the different nature of the problem of testing that the plant effect is small,

$$H : \alpha_1 = \cdots = \alpha_a = 0 \quad \text{or} \quad H' : \frac{\sum \alpha_i^2}{\sigma^2 + n\sigma_B^2} \le \Delta_0\ ,$$

and testing the corresponding hypothesis for the batch effect: $\sigma_B^2/\sigma^2 \le \Delta_0$. The first of these is essentially a model I problem (linear hypothesis). As before, unbiasedness implies that the conditional rejection probability given $S^2 = s^2$ is equal to α a.e. With S^2 fixed, the problem of testing H is a linear hypothesis, and the rejection region of the UMP invariant conditional test given $S^2 = s^2$ has the rejection region (7.46) with $\Delta_0 = 0$. The constant C_1 is again independent of S^2, and the test is UMP among all tests that are both unbiased and invariant. A test with the same property also exists for testing H'. Its rejection region is

$$\frac{S_A^2 \Big/ (a-1)}{S_B^2 \Big/ (b-1)a} \geq C',$$

where C' is determined from the noncentral F-distribution instead of, as before, the (central) F-distribution.

On the other hand, the hypothesis $\sigma_B^2/\sigma^2 \leq \Delta_0$ is essentially model II. It is invariant under addition of an arbitrary constant to each of the variables W_{i11}, which leaves $\sum_{i=1}^a \sum_{j=2}^b W_{ij1}^2$ and $\sum_{i=1}^a \sum_{j=1}^b \sum_{k=2}^n W_{ijk}^2$ as maximal invariants, and hence reduces the structure to pure model II with one classification. The test is then given by (7.47) as before. It is both UMP invariant and UMP unbiased.

Very general mixed models (containing general type II models as special cases) are discussed, for example, by Harville (1978), J. Miller (1977a), and Brown (1984), but see the note following Problem 7.36.

The different one- and two-factor models are discussed from a Bayesian point of view, for example, in Box and Tiao (1973) and Broemeling (1985). In distinction to the approach presented here, the Bayesian treatment also includes inferences concerning the values of the individual random components such as the batch means ξ_i of Section 7.7.

7.9 Multivariate Extensions

The univariate linear models studied so far in this chapter arise in the study of the effects of various experimental conditions (factors) on a single characteristic such as yield, weight, length of life, or blood pressure. This characteristic is assumed to be normally distributed with a mean that depends on the various factors under investigation, and a variance that is independent of these factors. We shall now consider the multivariate analogue of this model, which is appropriate when one is concerned with the effect of one or more factors simultaneously on several characteristics, for example the effect of a change in the diet of dairy cows on both fat content and quantity of milk.

A random vector $(X_1, \ldots, X_p)$ has a multivariate normal density if its density is of the form

$$\frac{\sqrt{|A|}}{(2\pi)^{\frac{1}{2}p}} \exp\left[-\tfrac{1}{2} \sum \sum a_{ij}(x_i - \xi_i)(x_j - \xi_j)\right], \tag{7.51}$$

where the matrix $A = (a_{ij})$ is positive definite, and $|A|$ denotes its determinant. The means and covariance matrix of the X's are given by

$$E(X_i) = \xi_i, \qquad E(X_i - \xi_i)(X_j - \xi_j) = \sigma_{ij}, \quad (\sigma_{ij}) = A^{-1}. \tag{7.52}$$

Such a model was previously introduced in Section 3.9.2.

Consider now n i.i.d. multivariate normal vectors $X_k = (X_{k,1}, \ldots, X_{k,p})$, $k = 1, \ldots, n$, with means $E(X_{k,i}) = \xi_i$ and covariance matrix A^{-1}. A natural extension of the one-sample problem of testing the mean ξ of a normal distribution with unknown variance is that of testing the hypothesis

$$\xi_1 = \xi_{1,0}, \ \ldots \ , \xi_p = \xi_{p,0} \ ;$$

without loss of generality, assume $\xi_{k,0} = 0$ for all k. The joint density of $X_1, \ldots, X_n$ is

$$\frac{|A|^{n/2}}{(2\pi)^{np/2}} \exp\left[-\frac{1}{2}\sum_{k=1}^{n}\sum_{i=1}^{p}\sum_{j=1}^{p} a_{i,j}(x_{k,i} - \xi_i)(x_{k,j} - \xi_j)\right] .$$

Writing the exponent as

$$\sum_{i=1}^{p}\sum_{j=1}^{p} a_{i,j} \sum_{k=1}^{n}(x_{k,i} - \xi_i)(x_{k,j} - \xi_j) ,$$

it is seen that the vector of sample means $(\bar{X}_1, \ldots, \bar{X}_p)$ together with

$$S_{i,j} = \sum_{k=1}^{n}(X_{k,i} - \bar{X}_i)(X_{k,j} - \bar{X}_j) , \quad i, j = 1, \ldots p \tag{7.53}$$

are sufficient for the unknown mean vector ξ and unknown covariance matrix $\Sigma = A^{-1}$ (assumed positive definite). For the remainder of this section, assume $n > p$, so that the matrix S with (i, j) component $S_{i,j}$ is nonsingular with probability one (Problem 7.38).

We shall now consider the group of transformations

$$X'_k = CX_k \quad (C \text{ nonsingular}) .$$

This leaves the problem invariant, since it preserves the normality of the variables and their means. It simply replaces the unknown covariance matrix by another one. In the space of sufficient statistics, this group induces the transformations

$$\bar{X}^* = C\bar{X} \quad \text{and} \quad S^* = CSC^T , \ \text{where} \ S = (S_{i,j}) . \tag{7.54}$$

Under this group, the statistic

$$W = \bar{X}^T S^{-1} \bar{X} \tag{7.55}$$

is maximal invariant (Problem 7.39).

The distribution of W depends only on the maximal invariant in the parameter space; this is found to be

$$\psi^2 = \sum_{i=1}^{p}\sum_{j=1}^{p} a_{ij}\xi_i\xi_j, \tag{7.56}$$

and the probability density of W is given by (Problem 7.40)

$$p_\psi(w) = e^{-\frac{1}{2}\psi^2} \sum_{k=0}^{\infty} \frac{(\frac{1}{2}\psi^2)^k}{k!} c_k \frac{w^{\frac{1}{2}p-1+k}}{(1+w)^{\frac{1}{2}n+k}}. \tag{7.57}$$

This is the same as the density of the test statistic in the univariate case, given as (7.6), with r and s there replaced by p. For any $\psi_0 < \psi_1$ the ratio $p_{\psi_1}(w)/p_{\psi_0}(w)$ is an increasing function of w, and it follows from the Neyman–Pearson Lemma that the most powerful invariant test for testing $H : \xi_1 = \cdots = \xi_p = 0$ rejects when W is too large, or equivalently when

$$\frac{n-p}{p} W > C. \tag{7.58}$$

The quantity $(n-1)W$, which for $p = 1$ reduces to the square of Student's t, is Hotelling's T^2-statistic. The constant C is determined from the fact that for $\psi = 0$ the statistic $(n-p)W/p$ has the F-distribution with p and $n-p$ degrees of freedom. As in the univariate case, there also exists a UMP invariant test of the more general hypothesis $H' : \psi^2 \le \psi_0^2$, with rejection region $W > C'$.

The T^2-test was shown by Stein (1956) to be admissible against the class of alternatives $\psi^2 \ge c$ for any $c > 0$ by the method of Theorem 6.7.1. Against the class of alternatives $\psi^2 \le c$ admissibility was proved by Kiefer and Schwartz (1965) [see Problem 7.44 and Schwartz (1967, 1969)].

Most accurate equivariant confidence sets for the unknown mean vector $(\xi_1, \ldots, \xi_p)$ are obtained from the UMP invariant test of $H : \xi_i = \xi_{i0}$ $(i = 1, \ldots, p)$, which has acceptance region

$$n \sum \sum (\bar{X}_i - \xi_{i0})(n-1)S^{i,j}(\bar{X}_j - \xi_{j0}) \le C ,$$

where $S^{i,j}$ are the elements of S^{-1}. The associated confidence sets are therefore ellipsoids

$$n \sum \sum (\xi_i - \bar{X}_i)(n-1)S^{ij}(\xi_j - \bar{X}_j) \le C \tag{7.59}$$

centered at $(\bar{X}_1, \ldots, \bar{X}_p)$. These confidence sets are equivariant under the group of transformations considered in this section (Problem 7.41), and by Lemma 6.10.1 are therefore uniformly most accurate among all equivariant confidence sets at the specified level.

The result extends to the two-sample problem with equal covariances (Problem 7.43), but the situation becomes more complicated for multivariate generalizations of univariate linear hypotheses with $r > 1$. Then, the maximal invariant is no longer univariate and a UMP invariant test no longer exists. For a discussion of this case, see Anderson (2003), Section 8.10.

7.10 Problems

Section 7.1

Problem 7.1 *Expected sums of squares.* The expected values of the numerator and denominator of the statistic W^* defined by (7.7) are

$$E\left(\sum_{i=1}^{r} \frac{Y_i^2}{r}\right) = \sigma^2 + \frac{1}{r}\sum_{i=1}^{r} \eta_i^2 \quad \text{and} \quad E\left[\sum_{i=s+1}^{n} \frac{Y_i^2}{n-s}\right] = \sigma^2.$$

Problem 7.2 *Noncentral χ^2-distribution.*[11]

(i) If X is distributed as $N(\psi, 1)$, the probability density of $V = X^2$ is $P_\psi^V(v) = \sum_{k-0}^{\infty} P_k(\psi) f_{2k+1}(v)$, where $P_k(\psi) = (\psi^2/2)^k e^{-(1/2)\psi^2}/k!$ and where f_{2k+1} is the probability density of a χ^2-variable with $2k+1$ degrees of freedom.

[11] The literature on noncentral χ^2, including tables, is reviewed in Tiku (1985a), Chou, Arthur, Rosenstein, and Owen (1994), and Johnson, Kotz and Balakrishnan (1995).

(ii) Let $Y_1, \ldots, Y_r$ be independently normally distributed with unit variance and means $\eta_1, \ldots, \eta_r$. Then $U = \sum Y_i^2$ is distributed according to the noncentral χ^2-distribution with r degrees of freedom and noncentrality parameter $\psi^2 = \sum_{i=1}^r \eta_i^2$, which has probability density

$$p_\psi^U(u) = \sum_{k=0}^{\infty} P_k(\psi) f_{r+2k}(u). \tag{7.60}$$

Here $P_k(\psi)$ and $f_{r+2k}(u)$ have the same meaning as in (i), so that the distribution is a mixture of χ^2-distributions with Poisson weights.

[(i): This is seen from

$$p_\psi^V(v) = \frac{e^{-\frac{1}{2}(\psi^2+v)}(e^{\psi\sqrt{v}} + e^{-\psi\sqrt{v}})}{2\sqrt{2\pi v}}$$

by expanding the expression in parentheses into a power series, and using the fact that $\Gamma(2k) = 2^{2k-1}\Gamma(k)\Gamma(k+\frac{1}{2})/\sqrt{\pi}$.
(ii): Consider an orthogonal transformation to $Z_1, \ldots, Z_r$ such that $Z_1 = \sum \eta_i Y_i/\psi$. Then the Z's are independent normal with unit variance and means $E(Z_1) = \psi$ and $E(Z_i) = 0$ for $i > 1$.]

Problem 7.3 *Noncentral F- and beta-distribution.*[12] Let $Y_1, \ldots, Y_r; Y_{s+1}, \ldots, Y_n$ be independently normally distributed with common variance σ^2 and means $E(Y_i) = \eta_i$ $(i = 1, \ldots, r)$; $E(Y_i) = 0$ $(i = s+1, \ldots, n)$.

(i) The probability density of $W = \sum_{i=1}^r Y_i^2 / \sum_{i=s+1}^n Y_i^2$ is given by (7.6). The distribution of the constant multiple $(n-s)W/r$ of W is the *noncentral F-distribution.*

(ii) The distribution of the statistic $B = \sum_{i=1}^r Y_i^2/(\sum_{i=1}^r Y_i^2 + \sum_{i=s+1}^n Y_i^2)$ is the *noncentral beta-distribution*, which has probability density

$$\sum_{k=0}^{\infty} P_k(\psi) g_{\frac{1}{2}r+k, \frac{1}{2}(n-s)}(b), \tag{7.61}$$

where

$$g_{p,q}(b) = \frac{\Gamma(p+q)}{\Gamma(p)\Gamma(q)} b^{p-1}(1-b)^{q-1}, \qquad 0 \le b \le 1 \tag{7.62}$$

is the probability density of the (central) beta-distribution.

Problem 7.4 (i) The noncentral χ^2 and F distributions have strictly monotone likelihood ratio.

(ii) Under the assumptions of Section 7.1, the hypothesis $H' : \psi^2 \le \psi_0^2$ ($\psi_0 > 0$ given) remains invariant under the transformations $G_i (i = 1, 2, 3)$ that were used to reduce $H : \psi = 0$, and there exists a UMP invariant test with rejection region $W > C'$. The constant C' is determined by $P_{\psi_0}\{W > C'\} = \alpha$, with the density of W given by (7.6).

[12] For literature on noncentral F, see Tiku (1985b) and Johnson, Kotz and Balakrishnan (1995).

[(i): Let $f(z) = \sum_{k=0}^{\infty} b_k z^k / \sum_{k=0}^{\infty} a_k z^k$ where the constants a_k, b_k are > 0 and $\sum a_k z^k$ and $\sum b_k z^k$ converge for all $z > 0$, and suppose that $b_k/a_k < b_{k+1}/a_{k+1}$ for all k. Then

$$f'(z) = \frac{\sum\sum_{k<n}(n-k)(a_k b_n - a_n b_k) z^{k+n-1}}{\left(\sum_{k=0}^{\infty} a_k z^k\right)^2}$$

is positive, since $(n-k)(a_k b_n - a_n b_k) > 0$ for $k < n$, and hence f is increasing.]
Note. The noncentral χ^2 and F-distributions are in fact STP_∞ [see for example Marshall and Olkin (1979) and Brown, Johnstone and MacGibbon (1981)], and there thus exists a test of $H : \psi = \psi_0$ against $\psi = \psi_0$ which is UMP among all tests that are both invariant and unbiased.

Problem 7.5 *Best average power.*

(i) Consider the general linear hypothesis H in the canonical form given by (7.2) and (7.3) of Section 7.1, and for any $\eta_{r+1}, \ldots, \eta_s$, σ, and ρ let $S = S(\eta_{r+1}, \ldots, \eta_s, \sigma : \rho)$ denote the sphere $\{(\eta_1, \ldots, \eta_r) : \sum_{i=1}^r \eta_i^2/\sigma^2 = \rho^2\}$. If $\beta_\phi(\eta_1, \ldots, \eta_r, \sigma)$ denotes the power of a test ϕ of H, then the test (7.9) maximizes the average power

$$\frac{\int_S \beta_\phi(\eta_1, \ldots, \eta_r, \sigma)\, dA}{\int_S dA}$$

for every $\eta_{r+1}, \ldots, \eta_s, \sigma$, and ρ among all unbiased (or similar) tests. Here dA denotes the differential of area on the surface of the sphere.

(ii) The result (i) provides an alternative proof of the fact that the test (7.9) is UMP among all tests whose power function depends only on $\sum_{i=1}^r \eta_i^2/\sigma^2$.

[(i): if $U = \sum_{i=1}^r Y_i^2, V = \sum_{i=s+1}^n Y_i^2$, unbiasedness (or similarity) implies that the conditional probability of rejection given $Y_{r+1}, \ldots, Y_s$, and $U + V$ equals α a.e. Hence for any given $\eta_{r+1}, \ldots, \eta_s$, σ, and ρ, the average power is maximized by rejecting when the ratio of the average density to the density under H is larger than a suitable constant $C(y_{r+1}, \ldots, y_s, u + v)$, and hence when

$$g(y_1, \ldots, y_r; \eta_1, \ldots, \eta_r) = \int_S \exp\left(\sum_{i=1}^r \frac{\eta_i y_i}{\sigma^2}\right) dA > C(y_{r+1}, \ldots, y_s, u + v).$$

As will be indicated below, the function g depends on $y_1, \ldots, y_r$ only through u and is an increasing function of u. Since under the hypothesis $U/(U + V)$ is independent of $Y_{r+1}, \ldots, Y_s$ and $U + V$, it follows that the test is given by (7.9). The exponent in the integral defining g can be written as $\sum_{i=1}^r \eta_i y_i/\sigma^2 = (\rho\sqrt{u}\cos\beta)/\sigma$, where β is the angle $(0 \le \beta \le \pi)$ between $(\eta_1, \ldots, \eta_r)$ and $(y_1, \ldots, y_r)$. Because of the symmetry of the sphere, this is unchanged if β is replaced by the angle γ between $(\eta_1, \ldots, \eta_r)$ and an arbitrary fixed vector. This shows that g depends on the y's only through u: for fixed $\eta_1, \ldots, \eta_r, \sigma$ denote it by $h(u)$. Let S' be the subset of S in which $0 \le \gamma \le \pi/2$. Then

$$h(u) = \int_{S'} \left[\exp\left(\frac{\rho\sqrt{u}\cos\gamma}{\sigma}\right) + \exp\left(\frac{-\rho\sqrt{u}\cos\gamma}{\sigma}\right)\right] dA,$$

which proves the desired result.]

Problem 7.6 Use Theorem 6.7.1 to show that the F-test (7.7) is α-admissible against $\Omega' : \psi \geq \psi_1$ for any $\psi_1 > 0$.

Problem 7.7 Given any $\psi_2 > 0$, apply Theorem 6.7.2 and Lemma 6.7.1 to obtain the F-test (7.7) as a Bayes test against a set Ω' of alternatives contained in the set $0 < \psi \leq \psi_2$.

Section 7.2

Problem 7.8 Under the assumptions of Section 7.1 suppose that the means ξ_i are given by

$$\xi_i = \sum_{j=1}^{s} a_{ij}\beta_j,$$

where the constants a_{ij} are known and the matrix $A = (a_{ij})$ has full rank, and where the β_j are unknown parameters. Let $\theta = \sum_{j=1}^{s} e_j\beta_j$ be a given linear combination of the β_j.

(i) If $\hat{\beta}_j$ denotes the values of the β_j minimizing $\sum(X_i - \xi_i)^2$ and if $\hat{\theta} = \sum_{j=1}^{s} e_j\hat{\beta}_j = \sum_{j=1}^{n} d_i X_i$, the rejection region of the hypothesis $H : \theta = \theta_0$ is

$$\frac{|\hat{\theta} - \theta_0| / \sqrt{\sum d_i^2}}{\sqrt{\sum \left(X_i - \hat{\xi}_i\right)^2 / (n-s)}} > C_0 \, , \tag{7.63}$$

where the left-hand side under H has the distribution of the absolute value of Student's t with $n - s$ degrees of freedom.

(ii) The associated confidence intervals for θ are

$$\hat{\theta} - k\sqrt{\frac{\sum \left(X_i - \hat{\xi}_i\right)^2}{n-s}} \leq \theta \leq \hat{\theta} + k\sqrt{\frac{\sum \left(X_i - \hat{\xi}_i\right)^2}{n-s}} \tag{7.64}$$

with $k = C_0\sqrt{\sum d_i^2}$. These intervals are uniformly most accurate equivariant under a suitable group of transformations.

[(i): Consider first the hypothesis $\theta = 0$, and suppose without loss of generality that $\theta = \beta_1$; the general case can be reduced to this by making a linear transformation in the space of the β's. If $\underline{a}_1, \ldots, \underline{a}_s$ denote the column vectors of the matrix A which by assumption span Π_Ω, then $\underline{\xi} = \beta_1\underline{a}_1 + \cdots + \beta_s\underline{a}_s$, and since $\hat{\underline{\xi}}$ is in Π_Ω also $\hat{\underline{\xi}} = \hat{\beta}_1\underline{a}_1 + \cdots + \hat{\beta}_s\underline{a}_s$. The space Π_ω defined by the hypothesis $\beta_1 = 0$ is spanned by the vectors $\underline{a}_2, \ldots, \underline{a}_s$ and also by the row vectors $\underline{c}_2, \ldots, \underline{c}_s$ of the matrix C of (7.1), while $\underline{c}_1$ is orthogonal to Π_ω. By (7.1), the vector $\underline{X}$ is given by $\underline{X} = \sum_{i=1}^{n} Y_i\underline{c}_i$, and its projection $\hat{\underline{\xi}}$ on Π_Ω therefore satisfies $\hat{\underline{\xi}} = \sum_{i=1}^{s} Y_i\underline{c}_i$. Equating the two expressions for $\hat{\underline{\xi}}$ and taking the inner product of both sides of this equation with $\underline{c}_i$ gives $Y_1 = \hat{\beta}_1 \sum_{i=1}^{n} a_{i1}c_{i1}$, since the $\underline{c}$'s are an orthogonal set of unit vectors. This shows that Y_1 is proportional to $\hat{\beta}_1$ and, since the variance of Y_1 is the same as that of the X's, that $|Y_1| = |\hat{\beta}_1| / \sqrt{\sum d_i^2}$. The result for testing

$\beta_1 = 0$ now follows from (7.12) and (7.13). The test for $\beta_1 = \beta_1^0$ is obtained by making the transformation $X_i^* = X_i - a_i\beta_1^0$.
(ii): The invariance properties of the intervals (7.64) can again be discussed without loss of generality by letting θ be the parameter β_1. In the canonical form of Section 7.1, one then has $E(Y_1) = \eta_1 = \lambda\beta_1$ with $|\lambda| = 1/\sqrt{\sum d_1^2}$ while $\eta_2, \ldots, \eta_s$ do not involve β_1. The hypothesis $\beta_1 = \beta_1^0$ is therefore equivalent to $\eta_1 = \eta_1^0$, with $\eta_1^0 = \lambda\beta_1^0$. This is invariant (a) under addition of arbitrary constants to $Y_2 \ldots, Y_s$; (b) under the transformations $Y_1^* = -(Y_1 - \eta_1^0) + \eta_1^0$; (c) under the scale changes $Y_i^* = cY_i$ $(i = 2, \ldots, n), Y_1^* - \eta_1^{0*} = c(Y_1 - \eta_1^0)$. The confidence intervals for $\theta = \beta_1$ are then uniformly most accurate equivariant under the group obtained from (a), (b), and (c) by varying η_1^0.]

Problem 7.9 Let X_{ij} $(j = 1, \ldots, m_i)$ and Y_{ik} $(k = 1, \ldots, n_i)$ be independently normally distributed with common variance σ^2 and means $E(X_{ij}) = \xi_i$ and $E(Y_{ij}) = \xi_i + \Delta$. Then the UMP invariant test of $H : \Delta = 0$ is given by (7.63) with $\theta = \Delta$, $\theta_0 = 0$ and

$$\hat{\theta} = \frac{\sum_i \frac{m_i n_i}{N_i}(Y_{i\cdot} - X_{i\cdot})}{\sum_i \frac{m_i n_i}{N_i}}, \qquad \hat{\xi}_i = \frac{\sum_{j=1}^{m_i} X_{ij} + \sum_{k=1}^{n_i}(Y_{ik} - \hat{\theta})}{N_i},$$

where $N_i = m_i + n_i$.

Problem 7.10 Let $X_1, \ldots, X_n$ be independently normally distributed with known variance σ_0^2 and means $E(X_i) = \xi_i$, and consider any linear hypothesis with $s \le n$ (instead of $s < n$ which is required when the variance is unknown). This remains invariant under a subgroup of that employed when the variance was unknown, and the UMP invariant test has rejection region

$$\sum \left(X_i - \hat{\hat{\xi}}_i\right)^2 - \left(X_i - \hat{\xi}_i\right)^2 = \left(\hat{\xi}_i - \hat{\hat{\xi}}_i\right)^2 > C\sigma_0^2 \tag{7.65}$$

with C determined by

$$\int_C^\infty \chi_r^2(y)\,dy = \alpha. \tag{7.66}$$

Section 7.3

Problem 7.11 If the variables X_{ij} $(j = 1, \ldots, n_i; i = 1, \ldots, s)$ are independently distributed as $N(\mu_i, \sigma^2)$, then

$$E\left[\sum n_i (X_{i\cdot} - X_{\cdot\cdot})^2\right] = (s-1)\sigma^2 + \sum n_i (\mu_i - \mu_\cdot)^2,$$
$$E\left[\sum\sum (X_{ij} - X_{i\cdot})^2\right] = (n-s)\sigma^2.$$

Problem 7.12 Let $Z_1, \ldots, Z_s$ be independently distributed as $N(\zeta_i, a_i^2), i = 1, \ldots, s$, where the a_i are known constants.

(i) With respect to a suitable group of linear transformations there exists a UMP invariant test of $H : \zeta_1 = \cdots = \zeta_s$ given by the rejection region

$$\sum \frac{1}{a_i^2}\left(Z_i - \frac{\sum Z_j/a_j^2}{\sum 1/a_j^2}\right)^2 = \sum \left(\frac{Z_i}{a_i}\right)^2 - \frac{(\sum Z_j/a_j^2)^2}{\sum (1/a_j^2)} > C \qquad (7.67)$$

(ii) The power of this test is the integral from C to ∞ of the noncentral χ^2-density with $s-1$ degrees of freedom and noncentrality parameter λ^2 obtained by substituting ζ_i for Z_i in the left-hand side of (7.67).

Section 7.5

Problem 7.13 The linear-hypothesis test of the hypothesis of no interaction in a two-way layout with m observations per cell is given by (7.28).

Problem 7.14 In the two-way layout of Section 7.5 with $a = b = 2$, denote the first three terms in the partition of $\sum\sum\sum (X_{ijk} - X_{ij\cdot})^2$ by S_A^2, S_B^2, and S_{AB}^2, corresponding to the A, B, and AB effects (i.e. the α's, β's, and γ's), and denote by H_A, H_B, and H_{AB} the hypotheses of these effects being zero. Define a new two-level factor B' which is at level 1 when A and B are both at level 1 or both at level 2, and which is at level 2 when A and B are at different levels. Then

$$H_{B'} = H_{AB}, \qquad S_{B'} = S_{AB}, \qquad H_{AB'} = H_B, \qquad S_{AB'} = S_B,$$

so that the B-effect has become an interaction, and the AB-interaction the effect of the factor B'. [Shaffer (1977b).]

Problem 7.15 Let X_λ denote a random variable distributed as noncentral χ^2 with f degrees of freedom and noncentrality parameter λ^2. Then $X_{\lambda'}$ is stochastically larger than X_λ if $\lambda < \lambda'$.
[It is enough to show that if Y is distributed as $N(0,1)$, then $(Y + \lambda')^2$ is stochastically larger than $(Y + \lambda)^2$. The equivalent fact that for any $z > 0$,

$$P\{|Y + \lambda'| \le z\} \le P\{|Y + \lambda| \le z\},$$

is an immediate consequence of the shape of the normal density function. An alternative proof is obtained by combining Problem 7.4 with Lemma 3.4.2.]

Problem 7.16 Let X_{ijk} $(i = 1,\ldots,a; j = 1,\ldots,b; k = 1,\ldots,m)$ be independently normally distributed with common variance σ^2 and mean

$$E(X_{ijk}) = \mu + \alpha_i + \beta_j + \gamma_k \qquad \left(\sum \alpha_i = \sum \beta_j = \sum \gamma_k = 0\right).$$

Determine the linear hypothesis test for testing $H : \alpha_i = \ldots \alpha_a = 0$.

Problem 7.17 In the three-factor situation of the preceding problem, suppose that $a = b = m$. The hypothesis H can then be tested on the basis of m^2 observations as follows. At each pair of levels (i,j) of the first two factors one observation is taken, to which we refer as being in the ith row and the jth column. If the levels of the third factor are chosen in such a way that each of them occurs once and only once in each row and column, the experimental design is a *Latin square*. The m^2 observations are denoted by $X_{ij(k)}$, where the third subscript indicates the level of the third factor when the first two are at levels i and j. It is assumed that $E(X_{ij(k)}) = \xi_{ij(k)} = \mu + \alpha_i + \beta_j + \gamma_k$, with $\sum \alpha_i = \sum \beta_j = \sum \gamma_k = 0$.

(i) The parameters are determined from the ξ's through the equations

$$\xi_{i\cdot(\cdot)} = \mu + \alpha_i, \qquad \xi_{\cdot j(\cdot)} = \mu + \beta_j, \qquad \xi_{\cdot\cdot(k)} = \mu + \gamma_k, \qquad \xi_{\cdot\cdot(\cdot)} = \mu.$$

(Summation over j with i held fixed automatically causes summation also over k.)

(ii) The least-squares estimates of the parameters may be obtained from the identity

$$\begin{aligned} \sum_i \sum_j & \left[x_{ij(k)} - \xi_{ij(k)}\right]^2 \\ = {} & m \sum \left[x_{i\cdot(\cdot)} - x_{\cdot\cdot(\cdot)} - \alpha_i\right]^2 + m \sum \left[x_{\cdot j(\cdot)} - x_{\cdot\cdot(\cdot)} - \beta_j\right]^2 \\ & + m \sum \left[x_{\cdot\cdot(k)} - x_{\cdot\cdot(\cdot)} - \gamma_k\right]^2 + m^2 \left[x_{\cdot\cdot(\cdot)} - \mu\right]^2 \\ & + \sum_i \sum_k \left[x_{ij(k)} - x_{i\cdot(\cdot)} - x_{\cdot j(\cdot)} - x_{\cdot\cdot(k)} + 2x_{\cdot\cdot(\cdot)}\right]^2 . \end{aligned}$$

(iii) For testing the hypothesis $H : \alpha_1 = \cdots = \alpha_m = 0$, the test statistic W^* of (7.15) is

$$\frac{m \sum \left[X_{i\cdot(\cdot)} - X_{\cdot\cdot(\cdot)}\right]^2}{\sum \sum \left[X_{ij(k)} - X_{i\cdot(\cdot)} - X_{\cdot j(\cdot)} - X_{\cdot\cdot(k)} + 2X_{\cdot\cdot(\cdot)}\right]^2 / (m-2)}.$$

The degrees of freedom are $m-1$ for the numerator and $(m-1)(m-2)$ for the denominator, and the noncentrality parameter is $\psi^2 = m \sum \alpha_i^2 / \sigma^2$.

Section 7.6

Problem 7.18 In a regression situation, suppose that the observed values X_j and Y_j of the independent and dependent variable differ from certain true values X_j' and Y_j' by errors U_j, V_j which are independently normally distributed with zero means and variances σ_U^2 and σ_V^2. The true values are assumed to satisfy a linear relation: $Y_j' = \alpha + \beta X_j'$. However, the variables which are being controlled, and which are therefore constants, are the X_j rather than the X_j'. Writing x_j for X_j, we have $x_j = X_j' + U_j$, $Y_j = Y_j' + V_j$, and hence $Y_j = \alpha + \beta x_j + W_j$, where $W_j = V_j - \beta U_j$. The results of Section 7.6 can now be applied to test that β or $\alpha + \beta x_0$ has a specified value.

Problem 7.19 Let $X_1, \ldots, X_m$; $Y_1, \ldots, Y_n$ be independently normally distributed with common variance σ^2 and means $E(X_i) = \alpha + \beta(u_i - \bar{u})$, $E(Y_j) = \gamma + \delta(v_j - \bar{v})$, where the u's and v's are known numbers. Determine the UMP invariant tests of the linear hypotheses $H : \beta = \delta$ and $H : \alpha = \gamma$, $\beta = \delta$.

Problem 7.20 Let $X_1, \ldots, X_n$ be independently normally distributed with common variance σ^2 and means $\xi_i = \alpha + \beta t_i + \gamma t_i^2$, where the t_i are known. If the coefficient vectors $(t_1^k, \ldots, t_n^k)$, $k = 0, 1, 2$, are linearly independent, the parameter space Π_Ω has dimension $s = 3$, and the least-squares estimates $\hat{\alpha}, \hat{\beta}, \hat{\gamma}$ are the unique solutions of the system of equations

$$\alpha \sum t_i^k + \beta \sum t_i^{k+1} + \gamma \sum t_i^{k+2} = \sum t_i^k X_i \qquad (k = 0, 1, 2).$$

The solutions are linear functions of the X's, and if $\hat{\gamma} = \sum c_i X_i$, the hypothesis $\gamma = 0$ is rejected when

$$\frac{|\hat{\gamma}| / \sqrt{\sum c_i^2}}{\sqrt{\sum \left(X_i - \hat{\alpha} - \hat{\beta} t_i - \hat{\gamma} t_i^2\right)^2 / (n-3)}} > C_0.$$

Section 7.7

Problem 7.21 (i) The test (7.41) of $H : \Delta \leq \Delta_0$ is UMP unbiased.

(ii) Determine the UMP unbiased test of $H : \Delta = \Delta_0$ and the associated uniformly most accurate unbiased confidence sets for Δ.

Problem 7.22 In the model (7.39), the correlation coefficient ρ between two observations X_{ij}, X_{ik} belonging to the same class, the so-called *intraclass correlation coefficient*, is given by $\rho = \sigma_A^2 / (\sigma_A^2 + \sigma^2)$.

Section 7.8

Problem 7.23 The tests (7.46) and (7.47) are UMP unbiased.

Problem 7.24 If X_{ij} is given by (7.39) but the number n_i of observations per batch is not constant, obtain a canonical form corresponding to (7.40) by letting $Y_{i1} = \sqrt{n_i} X_{i\cdot}$. Note that the set of sufficient statistics has more components than when n_i is constant.

Problem 7.25 The general nested classification with a constant number of observations per cell, under model II, has the structure

$$X_{ijk\cdots} = \mu + A_i + B_{ij} + C_{ijk} + \cdots + U_{ijk\cdots},$$
$$i = 1, \ldots, a; j = 1, \ldots, b; k = 1, \ldots, c; \ldots .$$

(i) This can be reduced to a canonical form generalizing (7.45).

(ii) There exist UMP unbiased tests of the hypotheses

$$H_A : \frac{\sigma_A^2}{cd \ldots \sigma_B^2 + d \ldots \sigma_C^2 + \cdots + \sigma^2} \leq \Delta_0,$$

$$H_B : \frac{\sigma_B^2}{d \ldots \sigma_C^2 + \cdots + \sigma^2} \leq \Delta_0.$$

Problem 7.26 Consider the model II analogue of the two-way layout of Section 7.5, according to which

$$X_{ijk} = \mu + A_i + B_j + C_{ij} + E_{ijk} \qquad (7.68)$$
$$(i = 1, \ldots, a; \quad j = 1, \ldots, b; \quad k = 1, \ldots, n),$$

where the A_i, B_j, C_{ij}, and E_{ijk} are independently normally distributed with mean zero and with variances σ_A^2, σ_B^2, σ_C^2 and σ^2 respectively. Determine tests which are UMP among all tests that are invariant (under a suitable group) and unbiased of the hypotheses that the following ratios do not exceed a given constant (which may be zero):

(i) σ_C^2/σ^2;

(ii) $\sigma_A^2/(n\sigma_C^2 + \sigma^2)$;

(iii) $\sigma_B^2/(n\sigma_C^2 + \sigma^2)$.

Note that the test of (i) requires $n > 1$, but those of (ii) and (iii) do not. [Let $S_A^2 = nb\sum(X_{i\cdot\cdot} - X_{\cdot\cdot\cdot})^2$, $S_B^2 = na\sum(X_{\cdot j\cdot} - X_{\cdot\cdot\cdot})^2$, $S_C^2 = n\sum\sum(X_{ij\cdot} - X_{i\cdot\cdot} - X_{\cdot j\cdot} + X_{\cdot\cdot\cdot})^2$, $S^2 = \sum\sum\sum(X_{ijk} - X_{ij\cdot})^2$, and make a transformation to new variables Z_{ijk} (independent, normal, and with mean zero except when $i = j = k = 1$) such that

$$S_A^2 = \sum_{i=2}^{a} Z_{i11}^2, \qquad S_B^2 = \sum_{j=2}^{b} Z_{1j1}^2, \qquad S_C^2 = \sum_{i=2}^{a}\sum_{j=2}^{b} Z_{ij1}^2,$$

$$S^2 = \sum_{i=1}^{a}\sum_{j=1}^{b}\sum_{k=2}^{n} Z_{ijk}^2.]$$

Problem 7.27 Consider the mixed model obtained from (7.68) by replacing the random variables A_i by unknown constants α_i satisfying $\sum \alpha_i = 0$. With (ii) replaced by (ii′) $\sum \alpha_i^2/(n\sigma_C^2 + \sigma^2)$, there again exist tests which are UMP among an tests that are invariant and unbiased, and in cases (i) and (iii) these coincide with the corresponding tests of Problem 7.26.

Problem 7.28 Consider the following generalization of the univariate linear model of Section 7.1. The variables X_i $(i = 1, \ldots, n)$ are given by $X_i = \xi_i + U_i$, where $(U_1, \ldots, U_n)$ have a joint density which is *spherical*, that is, a function of $\sum_{i=1}^n u_i^2$, say

$$f(U_1, \ldots, U_n) = q\left(\sum U_i^2\right).$$

The parameter spaces Π_Ω and Π_ω and the hypothesis H are as in Section 7.1.

(i) The orthogonal transformation (7.1) reduces $(X_1, \ldots, X_n)$ to canonical variables $(Y_1, \ldots, Y_n)$ with $Y_i = \eta_i + V_i$, where $\eta_i = 0$ for $i = s+1, \ldots, n$, H reduces to (7.3), and the V's have joint density $q(v_1, \ldots, v_n)$.

(ii) In the canonical form of (i), the problem is invariant under the groups G_1, G_2, and G_3 of Section 7.1, and the statistic W^* given by (7.7) is maximal invariant.

Problem 7.29 Under the assumptions of the preceding problem, the null distribution of W^* is independent of q and hence the same as in the normal case, namely, F with r and $n-s$ degrees of freedom. [See Problem 5.11]. *Note.* The analogous multivariate problem is treated by Kariya (1981); also see Kariya (1985) and Kariya and Sinha (1985). For a review of work on spherically and elliptically symmetric distributions, see Chmielewski (1981).

Problem 7.30 Consider the additive random-effects model

$$X_{ijk} = \mu + A_i + B_j + U_{ijk} \quad (i = 1, \ldots, a; \quad j = 1, \ldots, b; \quad k = 1, \ldots, n),$$

where the A's, B's, and U's are independent normal with zero means and variances σ_A^2, σ_B^2, and σ^2 respectively. Determine

(i) the joint density of the X's,

(ii) the UMP unbiased test of $H : \sigma_B^2/\sigma^2 \leq \delta$.

Problem 7.31 For the mixed model

$$X_{ij} = \mu + \alpha_i + B_j + U_{ij} \qquad (i = 1, \ldots, a; \quad j = 1, \ldots, n),$$

where the B's and U's are as in Problem 7.30 and the α's are constants adding to zero, determine (with respect to a suitable group leaving the problem invariant)

(i) a UMP invariant test of $H : \alpha_1 = \cdots = \alpha_a$;

(ii) a UMP invariant test of $H : \xi_1 = \cdots = \xi_a = 0 \quad (\xi_i = \mu + \alpha_i)$;

(iii) a test of $H : \sigma_B^2/\sigma^2 \leq \delta$ which is both UMP invariant and UMP unbiased.

Problem 7.32 Let $(X_{1j}, \ldots, X_{pj})$, $j = 1, \ldots, n$, be a sample from a p-variate normal distribution with mean $(\xi_1, \ldots, \xi_p)$ and covariance matrix $\Sigma = (\sigma_{ij})$, where $\sigma_{ij}^2 = \sigma^2$ when $j = i$, and $\sigma_{ij}^2 = \rho\sigma^2$ when $j \neq i$. Show that the covariance matrix is positive definite if and only if $\rho > -1/(p-1)$.
[For fixed σ and $\rho < 0$, the quadratic form $(1/\sigma^2)\sum\sum \sigma_{ij} y_i y_j = \sum y_i^2 + \rho \sum\sum y_i y_j$ takes on its minimum value over $\sum y_i^2 = 1$ when all the y's are equal.]

Problem 7.33 Under the assumptions of the preceding problem, determine the UMP invariant test (with respect to a suitable G) of $H : \xi_i = \ldots = \xi_p$.
[Show that this model agrees with that of Problem 7.31 if $\rho = \sigma_b^2/(\sigma_b^2 + \sigma^2)$, except that instead of being positive, ρ now only needs to satisfy $\rho > -1/(p-1)$.]

Problem 7.34 Permitting interactions in the model of Problem 7.30 leads to the model

$$X_{ijk} = \mu + A_i + B_j + C_{ij} + U_{ijk} \quad (i = 1, \ldots, a; j = 1, \ldots, b, k = 1, \ldots, n).$$

where the A's, B's, C's, and U's are independent normal with mean zero and variances σ_A^2, σ_B^2, σ_C^2 and σ^2.

(i) Give an example of a situation in which such a model might be appropriate.

(ii) Reduce the model to a convenient canonical form along the lines of Section 7.4.

(iii) Determine UMP unbiased tests of (a) $H_1 : \sigma_B^2 = 0$; (b) $H_2 : \sigma_C^2 = 0$.

Problem 7.35 Formal analogy with the model of Problem 7.34 suggests the mixed model

$$X_{ijk} = \mu + \alpha_i + B_j + C_{ij} + U_{ijk}$$

with the B's, C's, and U's as in Problem 7.34. Reduce this model to a canonical form involving $X_{...}$ and the sums of squares

$$\frac{\sum(X_{i..}-X_{...}-\alpha_i)^2}{n\sigma_C^2+\sigma^2}, \quad \frac{\sum\sum(X_{.j.}-X_{...})^2}{an\sigma_B^2+n\sigma_C^2+\sigma^2},$$

$$\frac{\sum\sum(X_{ij.}-X_{i..}-X_{.j.}+X_{...})^2}{n\sigma_C^2+\sigma^2}, \quad \frac{\sum\sum\sum(X_{ijk}-X_{i..}-X_{.j.}+X_{...})^2}{\sigma^2}.$$

Problem 7.36 Among all tests that are both unbiased and invariant under suitable groups under the assumptions of Problem 7.35, there exist UMP tests of

(i) $H_1 : \alpha_1 = \cdots = \alpha_a = 0$;

(ii) $H_2 : \sigma_B^2/(n\sigma_C^2 + \sigma^2) \leq C$;

(iii) $H_3 : \sigma_C^2/\sigma^2 \leq C$.

Note. The independence assumptions of Problems 7.35 and 7.36 often are not realistic. For alternative models, derived from more basic assumptions, see Scheffé (1956, 1959). Relations between the two types of models are discussed in Hocking (1973), Cohen and Miller (1976), and Stuart and Ord (1991).

Problem 7.37 Let $(X_{1j1}, \ldots, X_{1jn}; X_{2j1}, \ldots, X_{2jn}; \ldots; X_{aj1}, \ldots, X_{ajn})$, $j = 1, \ldots, b$, be a sample from an an-variate normal distribution. Let $E(X_{ijk}) = \xi_i$, and denote by $\sum_{ii'}$ the matrix of covariances of $(X_{ij1}, \ldots, X_{ijn})$ with $(X_{i'j1}, \ldots, X_{i'jn})$. Suppose that for all i, the diagonal elements of $\sum_{ii}$ are $= \tau^2$ and the off-diagonal elements are $= \rho_1\tau^2$, and that for $i \neq i'$ all n^2 elements of $\sum_{ii'}$ are $= \rho_2\tau^2$.

(i) Find necessary and sufficient conditions on ρ_1 and ρ_2 for the overall $abn \times abn$ covariance matrix to be positive definite.

(ii) Show that this model agrees with that of Problem 7.35 for suitable values of ρ_1 and ρ_2.

Section 7.9

Problem 7.38 If $n \leq p$, the matrix S with (i, j) component $S_{i,j}$ defined in (7.53) is singular. If $n > p$, it is nonsingular with probability 1. If $n \leq p$, the test $\phi \equiv \alpha$ is the only test that is invariant under the group of nonsingular linear transformations.

Problem 7.39 Show that the statistic W given in (7.55) is maximal invariant. [*Hint:* If $(\bar{X}, S)$ and $(\bar{Y}, T)$ are such that

$$\bar{X}^T S^{-1} \bar{X} = \bar{Y}^T T^{-1} \bar{Y} ,$$

then a transformation C that transforms one to the other is given by $C = Y(X^T S^{-1} X)^{-1} X^T S^{-1}$.]

Problem 7.40 Verify that the density of W is given by (7.55).

Problem 7.41 The confidence ellipsoids (7.59) for $(\xi_1, \ldots, \xi_p)$ are equivariant under the group of Section 7.9.

Problem 7.42 For testing a multivariate mean vector ξ is zero in the case where Σ is known, derive a UMPI test.

Problem 7.43 Extend the one-sample problem to the two-sample problem for testing whether two multivariate normal distributions with common unknown covariance matrix have the same mean vectors.

Problem 7.44 *Bayes character and admissibility of Hotelling's T^2.*

(i) Let $(X_{\alpha 1}, \ldots, X_{\alpha p})$, $\alpha = 1, \ldots, n$, be a sample from a p-variate normal distribution with unknown mean $\xi = (\xi_1, \ldots, \xi_p)$ and covariance matrix $\Sigma = A^{-1}$, and with $p \le n-1$. Then the one-sample T^2-test of $H : \xi = 0$ against $K : \xi \ne 0$ is a Bayes test with respect to prior distributions Λ_0 and Λ_1 which generalize those of Example 6.7.13 (continued).

(ii) The test of part (i) is admissible for testing H against the alternatives $\psi^2 \le c$ for any $c > 0$.

[If ω is the subset of points $(0, \Sigma)$ of Ω_H satisfying $\Sigma^{-1} = A + \eta'\eta$ for some fixed positive definite $p \times p$ matrix A and arbitrary $\eta = (\eta_1, \ldots, \eta_p)$, and $\Omega'_{A,b}$ is the subset of points (ξ, Σ) of Ω_K satisfying $\Sigma^{-1} = A + \eta'\eta$, $\xi' = b\Sigma\eta'$ for the same A and some fixed $b > 0$, let Λ_0 and Λ_1 have densities defined over ω and $\Omega_{A,b}$, respectively by

$$\lambda_0(\eta) = C_0 |A + \eta'\eta|^{-n/2}$$

and

$$\lambda_1(\eta) = C_1 |A + \eta'\eta|^{-n/2} \exp\left\{ \frac{nb^2}{2} \left[\eta(A + \eta'\eta)^{-1}\eta' \right] \right\}.$$

(Kiefer and Schwartz, 1965).]

Problem 7.45 Suppose $(X_1, \ldots, X_p)$ have the multivariate normal density (7.51), so that $E(X_i) = \xi_i$ and A^{-1} is the known positive definite covariance matrix. The vector of means $\xi = (\xi_1, \ldots, \xi_p)$ is known to lie in a given s-dimensional linear space Π_Ω with $s \le p$; the hypothesis to be tested is that ξ lies in a given $(s-r)$-dimensional linear subspace Π_ω of $\Pi_\Omega (r \le s)$.
(i) Determine the UMPI test under a suitable group of transformations as explicitly as possible. Find an expression for the power function.
(ii) Specialize to the case of a simple null hypothesis.

7.11 Notes

The general linear model in the parametric form (7.18) was formulated at the beginning of the 19th century by Legendre and Gauss, who were concerned with estimating the unknown parameters. [For an account of its history, see Seal (1967).] The canonical form (7.2) of the model is due to Kolodziejczyk (1935). The analysis of variance, including the concept of interaction, was developed by Fisher in the 1920s and 1930s, and a systematic account is provided by Scheffé

(1959) in a book that includes a careful treatment of alternative models and of robustness questions.

Different approaches to analysis of variance than that given here are considered in Speed (1987) and the discussion following this paper, and in Diaconis (1988, Section 8C). Rank tests are discussed in Marden and Muyot (1995). Admissibility results for testing homogeneity of variances in a normal balanced one-way layout are given in Cohen and Marden (1989). Linear models have been generalized in many directions. Loglinear models provide extensions to important discrete data. [Both are reviewed in Christensen (2000).] These two classes of models are subsumed in generalized linear models discussed for example in McCullagh and Nelder (1983), Dobson (1990) and Agresti (2002), and they in turn are a subset of additive linear models which are discussed in Hastie and Tibshirani (1990, 1997). Modern treatments of regression analysis can be found, for example, in Weisberg (1985), Atkinson and Riani (2000) and Ruppert, Wand and Carroll (2003). UMPI tests can be constructed for tests of lack of fit in some regression models; see Christensen (1989) and Miller, Neill and Sherfey (1998).

Hsu (1941) shows that the test (7.7) is UMP among all tests whose power function depends only on the noncentrality parameter. Hsu (1945) obtains a result on best average power for the T^2-test analogous to that of Chapter 7, Problem 7.5.

Tests of multivariate linear hypotheses and the associated confidence sets have their origin in the work of Hotelling (1931). More details on these procedures and discussion of other multivariate techniques can be found in the comprehensive books by Anderson (2003) and Seber (1984). A more geometric approach stressing invariance is provided by Eaton (1983).

For some recent work on using rank tests in multivariate problems, see Choi and Marden (1997), Hettmansperger, Möttönen and Oja (1997), and Akritas, Arnold and Brunner (1997).

8

The Minimax Principle

8.1 Tests with Guaranteed Power

The criteria discussed so far, unbiasedness and invariance, suffer from the disadvantage of being applicable, or leading to optimum solutions, only in rather restricted classes of problems. We shall therefore turn now to an alternative approach, which potentially is of much wider applicability. Unfortunately, its application to specific problems is in general not easy, unless there exists a UMP invariant test.

One of the important considerations in planning an experiment is the number of observations required to insure that the resulting statistical procedure will have the desired precision or sensitivity. For problems of hypothesis testing this means that the probabilities of the two kinds of errors should not exceed certain preassigned bounds, say α and $1-\beta$, so that the tests must satisfy the conditions

$$\begin{aligned} E_\theta\varphi(X) &\leq \alpha \quad \text{for } \theta \in \Omega_H, \\ E_\theta\varphi(X) &\geq \beta \quad \text{for } \theta \in \Omega_K. \end{aligned} \tag{8.1}$$

If the power function $E_\theta\varphi(X)$ is continuous and if $\alpha < \beta$, (8.1) cannot hold when the sets Ω_H and Ω_K are contiguous. This mathematical difficulty corresponds in part to the fact that the division of the parameter values θ into the classes Ω_H and Ω_K for which the two different decisions are appropriate is frequently not sharp. Between the values for which one or the other of the decisions is clearly correct there may lie others for which the relative advantages and disadvantages of acceptance and rejection are approximately in balance. Accordingly we shall assume that Ω is partitioned into three sets

$$\Omega = \Omega_H + \Omega_I + \Omega_K,$$

of which Ω_I designates the *indifference zone*, and Ω_K the class of parameter values differing so widely from those postulated by the hypothesis that false acceptance of H is a serious error, which should occur with probability at most $1 - \beta$.

To see how the sample size is determined in this situation, suppose that $X_1, X_2, \ldots$ constitute the sequence of available random variables, and for a moment let n be fixed and let $X = (X_1, \ldots, X_n)$. In the usual applications (for a more precise statement, see Problem 8.1), there exists a test φ_n which maximizes

$$\inf_{\Omega_k} E_\theta \varphi(X) \tag{8.2}$$

among all level-α tests based on X. Let $\beta_n = \inf_{\Omega_K} E_\theta \varphi_n(X)$, and suppose that for sufficiently large n there exists a test satisfying (8.1). [Conditions under which this is the case are given by Berger (1951a) and Kraft (1955).] The desired sample size, which is the smallest value of n for which $\beta_n \geq \beta$, is then obtained by trial and error. This requires the ability of determining for each fixed n the test that maximizes (8.2) subject to

$$E_\theta \varphi(X) \leq \alpha \qquad \text{for} \quad \theta \in \Omega_H. \tag{8.3}$$

A method for determining a test with this *maximin* property (of maximizing the minimum power over Ω_K) is obtained by generalizing Theorem 3.8.1. It will be convenient in this discussion to make a change of notation, and to denote by ω and ω' the subsets of Ω previously denoted by Ω_H and Ω_K. Let $\mathcal{P} = \{P_\theta, \theta \in \omega \cup \omega'\}$ be a family of probability distributions over a sample space $(\mathcal{X}, \mathcal{A})$ with densities $p_\theta = dP_\theta/d\mu$ with respect to a σ-finite measure μ, and suppose that the densities $p_\theta(x)$ considered as functions of the two variables (x, θ) are measurable $(\mathcal{A} \times \mathcal{B})$ and $(\mathcal{A} \times \mathcal{B}')$, where $\mathcal{B}$ and $\mathcal{B}'$ are given σ-fields over ω and ω'. Under these assumptions, the following theorem gives conditions under which a solution of a suitable Bayes problem provides a test with the required properties.

Theorem 8.1.1 *For any distributions Λ and Λ' over $\mathcal{B}$ and $\mathcal{B}'$, let $\varphi_{\Lambda,\Lambda'}$ be the most powerful test for testing*

$$h(x) = \int_\omega p_\theta(x)\, d\Lambda(\theta)$$

at level α against

$$h'(x) = \int_{\omega'} p_\theta(x)\, d\Lambda'(\theta)$$

and let $\beta_{\Lambda,\Lambda'}$, be its power against the alternative h'. If there exist Λ and Λ' such that

$$\begin{aligned} \sup_\omega E_\theta \varphi_{\Lambda,\Lambda'}(X) &\leq \alpha, \\ \inf_{\omega'} E_\theta \varphi_{\Lambda,\Lambda'}(X) &= \beta_{\Lambda,\Lambda'}, \end{aligned} \tag{8.4}$$

then:

(i) *$\varphi_{\Lambda,\Lambda'}$ maximizes $\inf_{\omega'} E_\theta \varphi(X)$ among all level-α tests of the hypothesis $H : \theta \in \omega$ and is the unique test with this property if it is the unique most powerful level-α test for testing h against h'.*

(ii) *The pair of distributions* Λ, Λ' *is least favorable in the sense that for any other pair* ν, ν' *we have*

$$\beta_{\Lambda,\Lambda'} \le \beta_{\nu,\nu'}.$$

PROOF. (i): If φ^* is any other level-α test of H, it is also of level α for testing the simple hypothesis that the density of X is h, and the power of φ^* against h' therefore cannot exceed $\beta_{\Lambda,\Lambda'}$. It follows that

$$\inf_{\omega'} E_\theta \varphi^*(X) \le \int_{\omega'} E_\theta \varphi^*(X)\, d\Lambda'(\theta) \le \beta_{\Lambda,\Lambda'} = \inf_{\omega'} E_\theta \varphi_{\Lambda\Lambda'}(X),$$

and the second inequality is strict if $\varphi_{\Lambda\Lambda'}$ is unique.

(ii): Let ν, ν' be any other distributions over $(\omega, \mathcal{B})$ and $(\omega', \mathcal{B}')$, and let

$$g(x) = \int_\omega p_\theta(x) d\nu(\theta), \qquad g'(x) = \int_{\omega'} p_\theta(x)\, d\nu'(\theta).$$

Since both $\varphi_{\Lambda,\Lambda'}$ and $\varphi_{\nu,\nu'}$ are level-α tests of the hypothesis that $g(x)$ is the density of X, it follows that

$$\beta_{\nu,\nu'} \ge \int \varphi_{\Lambda,\Lambda'}(x) g'(x)\, d\mu(x) \ge \inf_{\omega'} E_\theta \varphi_{\Lambda,\Lambda'}(X) = \beta_{\Lambda,\Lambda'}. \; \blacksquare$$

Corollary 8.1.1 *Let* Λ, Λ' *be two probability distributions and* C *a constant such that*

$$\varphi_{\Lambda,\Lambda'}(x) = \begin{cases} 1 & \text{if} \quad \int_{\omega'} p_\theta(x)\, d\Lambda'(\theta) > C \int_\omega p_\theta(x)\, d\Lambda(\theta) \\ \gamma & \text{if} \quad \int_{\omega'} p_\theta(x)\, d\Lambda'(\theta) = C \int_\omega p_\theta(x)\, d\Lambda(\theta) \\ 0 & \text{if} \quad \int_{\omega'} p_\theta(x)\, d\Lambda'(\theta) < C \int_\omega p_\theta(x)\, d\Lambda(\theta) \end{cases} \tag{8.5}$$

is a size-α test for testing that the density of X *is* $\int_\omega p_\theta(x)\, d\Lambda(\theta)$ *and such that*

$$\Lambda(\omega_0) = \Lambda'(\omega_0') = 1, \tag{8.6}$$

where

$$\begin{aligned} \omega_0 &= \Big\{\theta : \theta \in \omega \text{ and } E_\theta \varphi_{\Lambda,\Lambda'}(X) = \sup_{\theta' \in \omega} E_{\theta'} \varphi_{\Lambda,\Lambda'}(X)\Big\} \\ \omega_0' &= \Big\{\theta : \theta \in \omega' \text{ and } E_\theta \varphi_{\Lambda,\Lambda'}(X) = \inf_{\theta' \in \omega'} E_{\theta'} \varphi_{\Lambda,\Lambda'}(X)\Big\}. \end{aligned}$$

Then the conclusions of Theorem 8.1.1 *hold.*

PROOF. If h, h', and $\beta_{\Lambda,\Lambda'}$ are defined as in Theorem 8.1.1, the assumptions imply that $\varphi_{\Lambda,\Lambda'}$ is a most powerful level-α test for testing h against h', that

$$\sup_\omega E_\theta \varphi_{\Lambda,\Lambda'}(X) = \int_\omega E_\theta \varphi_{\Lambda,\Lambda'}(X)\, d\Lambda(\theta) = \alpha,$$

and that

$$\inf_{\omega'} E_\theta \varphi_{\Lambda,\Lambda'}(X) = \int_{\omega'} E_\theta \varphi_{\Lambda,\Lambda'}(X)\, d\Lambda'(\theta) = \beta_{\Lambda,\Lambda'}.$$

The condition (8.4) is thus satisfied and Theorem 8.1.1 applies. $\blacksquare$

Suppose that the sets Ω_H, Ω_I, and Ω_K are defined in terms of a nonnegative function d, which is a measure of the distance of θ from H, by

$$\begin{aligned}\Omega_H &= \{\theta : d(\theta) = 0\}, \qquad \Omega_I = \{\theta : 0 < d(\theta) < \Delta\},\\ \Omega_K &= \{0 : d(\theta) \geq \Delta\}.\end{aligned}$$

Suppose also that the power function of any test is continuous in θ. In the limit as $\Delta = 0$, there is no indifference zone. Then Ω_K becomes the set $\{\theta : d(\theta) > 0\}$, and the infimum of $\beta(\theta)$ over Ω_K is $\leq \alpha$ for any level-α test. This infimum is therefore maximized by any test satisfying $\beta(\theta) \geq \alpha$ for all $\theta \in \Omega_K$, that is, by any unbiased test, so that unbiasedness is seen to be a limiting form of the maximin criterion. A more useful limiting form, since it will typically lead to a unique test, is given by the following definition. A test φ_0 is said to *maximize the minimum power locally*[1] if, given any other test φ, there exists Δ_0 such that

$$\inf_{\omega_\Delta} \beta_{\varphi_0}(\theta) \geq \inf_{\omega_\Delta} \beta_\varphi(\theta) \qquad \text{for all} \quad 0 < \Delta < \Delta_0, \tag{8.7}$$

where ω_Δ is the set of θ's for which $d(\theta) \geq \Delta$.

8.2 Examples

In Chapter 3 it was shown for a family of probability densities depending on a real parameter θ that a UMP test exists for testing $H : \theta \leq \theta_0$ against $\theta > \theta_0$ provided for all $\theta < \theta'$ the ratio $p_{\theta'}(x)/p_\theta(x)$ is a monotone function of some real-valued statistic. This assumption, although satisfied for a one-parameter exponential family, is quite restrictive, and a UMP test of H will in fact exist only rarely. A more general approach is furnished by the formulation of the preceding section. If the indifference zone is the set of θ's with $\theta_0 < \theta < \theta_1$, the problem becomes that of maximizing the minimum power over the class of alternatives $\omega' : \theta \geq \theta_1$. Under appropriate assumptions, one would expect the least favorable distributions Λ and Λ' of Theorem 8.1.1 to assign probability 1 to the points θ_0 and θ_1, and hence the maximin test to be given by the rejection region $p_{\theta_1}(x)/p_{\theta_0}(x) > C$. The following lemma gives sufficient conditions for this to be the case.

Lemma 8.2.1 *Let $X_1, \ldots, X_n$ be identically and independently distributed with probability density $f_\theta(x)$, where θ and x are real-valued, and suppose that for any $\theta < \theta'$ the ratio $f_{\theta'}(x)/f_\theta(x)$ is a nondecreasing function of x. Then the level-α test φ of H which maximizes the minimum power over ω' is given by*

$$\varphi(x_1, \ldots, x_1) = \begin{cases} 1 & \text{if} \quad r(x_1, \ldots, x_n) > C, \\ \gamma & \text{if} \quad r(x_1, \ldots, x_n) = C, \\ 0 & \text{if} \quad r(x_1, \ldots, x_n) < C, \end{cases} \tag{8.8}$$

where $r(x_1, \ldots, x_n) = f_{\theta_1}(x_1) \ldots f_{\theta_1}(x_n)/f_{\theta_0}(x_1) \ldots f_{\theta_0}(x_n)$ and where C and γ are determined by

$$E_{\theta_0}\varphi(X_1, \ldots, X_n) = \alpha. \tag{8.9}$$

---

[1] A different definition of local minimaxity is given by Giri and Kiefer (1964).

PROOF. The function $\varphi(x_1, \ldots, x_n)$ is nondecreasing in each of its arguments, so that by Lemma 3.4.2,

$$E_\theta \varphi(X_1, \ldots, X_n) \leq E_{\theta'} \varphi(X_1, \ldots, X_n)$$

when $\theta < \theta'$. Hence the power function of φ is monotone and φ is a level-α test. Since $\varphi = \varphi_{\Lambda,\Lambda'}$, where Λ and Λ' are the distributions assigning probability 1 to the points θ_0 and θ_1, the condition (8.4) is satisfied, which proves the desired result as well as the fact that the pair of distributions (Λ, Λ') is least favorable. ■

Example 8.2.1 Let θ be a location parameter, so that $f_\theta(x) = g(x - \theta)$, and suppose for simplicity that $g(x) > 0$ for all x. We will show that a necessary and sufficient condition for $f_\theta(x)$ to have monotone likelihood ratio in x is that $-\log g$ is convex. The condition of monotone likelihood ratio in x,

$$\frac{g(x - \theta')}{g(x - \theta)} \leq \frac{g(x' - \theta')}{g(x' - \theta)} \qquad \text{for all} \quad x < x', \quad \theta < \theta',$$

is equivalent to

$$\log g(x' - \theta) + \log g(x - \theta') \leq \log g(x - \theta) + \log g(x' - \theta').$$

Since $x - \theta = t(x - \theta') + (1 - t)(x' - \theta)$ and $x' - \theta' = (1 - t)(x - \theta') + t(x' - \theta)$, where $t = (x' - x)/(x' - x + \theta' - \theta)$, a sufficient condition for this to hold is that the function $-\log g$ is convex. To see that this condition is also necessary, let $a < b$ be any real numbers, and let $x - \theta' = a$, $x' - \theta = b$, and $x' - \theta' = x - \theta$. Then $x - \theta = \frac{1}{2}(x' - \theta + x - \theta') = \frac{1}{2}(a + b)$, and the condition of monotone likelihood ratio implies

$$\tfrac{1}{2}[\log g(a) + \log g(b)] \leq \log g\left[\tfrac{1}{2}(a + b)\right].$$

Since $\log g$ is measurable, this in turn implies that $-\log g$ is convex.[2]

A density g for which $-\log g$ is convex is called *strongly unimodal.* Basic properties of such densities were obtained by Ibragimov (1956). Strong unimodality is a special case of total positivity. A density of the form $g(x - \theta)$ which is totally positive of order r is said to be a Polya frequency function of order r. It follows from Example 8.2.1 that $g(x - \theta)$ is a Polya frequency function of order 2 if and only if it is strongly unimodal. [For further results concerning Polya frequency functions and strongly unimodal densities, see Karlin (1968), Marshall and Olkin (1979), Huang and Ghosh (1982), and Loh (1984a, b).]

Two distributions which satisfy the above condition [besides the normal distribution, for which the resulting densities $p_\theta(x_1, \ldots, x_n)$ form an exponential family] are the *double exponential distribution* with

$$g(x) = \tfrac{1}{2} e^{-|x|}$$

and the *logistic distribution*, whose cumulative distribution function is

$$G(x) = \frac{1}{1 + e^{-x}},$$

so that the density is $g(x) = e^{-x}/(1 + e^{-x})^2$. ■

---

[2]See Sierpinski (1920).

Example 8.2.2 To consider the corresponding problem for a scale parameter, let $f_\theta(x) = \theta^{-1}h(x/\theta)$ where h is an even function. Without loss of generality one may then restrict x to be nonnegative, since the absolute values $|X_1|, \ldots, |X_n|$ form a set of sufficient statistics for θ. If $Y_i = \log X_i$ and $\eta = \log \theta$, the density of Y_i is

$$h(e^{y-\eta})e^{y-\eta}.$$

By Example 8.2.1, if $h(x) > 0$ for all $x \geq 0$, a necessary and sufficient condition for $f_{\theta'}(x)/f_\theta(x)$ to be a nondecreasing function of x for all $\theta < \theta'$ is that $-\log[e^y h(e^y)]$ or equivalently $-\log h(e^y)$ is a convex function of y. An example in which this holds—in addition to the normal and double-exponential distributions, where the resulting densities form an exponential family—is the *Cauchy distribution* with

$$h(x) = \frac{1}{\pi}\frac{1}{1+x^2}.$$

Since the convexity of $-\log h(y)$ implies that of $-\log h(e^y)$, it follows that if h is an even function and $h(x-\theta)$ has monotone likelihood ratio, so does $h(x/\theta)$. When h is the normal or double-exponential distribution, this property of $h(x/\theta)$ also follows from Example 8.2.1. That monotone likelihood ratio for the scale-parameter family does not conversely imply the same property for the associated location parameter family is illustrated by the Cauchy distribution. The condition is therefore more restrictive for a location than for a scale parameter. ■

The chief difficulty in the application of Theorem 8.1.1 to specific problems is the necessity of knowing, or at least being able to guess correctly, a pair of least favorable distributions (Λ, Λ'). Guidance for obtaining these distributions is sometimes provided by invariance considerations. If there exists a group G of transformations of X such that the induced group $\bar{G}$ leaves both ω and ω' invariant, the problem is symmetric in the various θ's that can be transformed into each other under $\bar{G}$. It then seems plausible that unless Λ and Λ' exhibit the same symmetries, they will make the statistician's task easier, and hence will not be least favorable.

Example 8.2.3 In the problem of paired comparisons considered in Example 6.3.5, the observations X_i $(i = 1, \ldots, n)$ are independent variables taking on the values 1 and 0 with probabilities p_i and $q_i = 1 - p_i$. The hypothesis H to be tested specifies the set $\omega : \max p_i \leq \frac{1}{2}$. Only alternatives with $p_i \geq \frac{1}{2}$ for all i are considered, and as ω' we take the subset of those alternatives for which $\max p_i \geq \frac{1}{2} + \delta$. One would expect Λ to assign probability 1 to the point $p_1 = \cdots p_n = \frac{1}{2}$, and Λ' to assign positive probability only to the n points $(p_1, \ldots, p_n)$ which have $n-1$ coordinates equal to $\frac{1}{2}$ and the remaining coordinate equal to $\frac{1}{2} + \delta$. Because of the symmetry with regard to the n variables, it seems plausible that Λ' should assign equal probability $1/n$ to each of these n points. With these choices, the test $\varphi_{\Lambda,\Lambda'}$ rejects when

$$\sum_{i=1}^{n} \left(\frac{\frac{1}{2}+\delta}{\frac{1}{2}}\right)^{x_i} > C.$$

This is equivalent to $\sum_{i=1}^{n} x_i > C$, which had previously been seen to be UMP invariant for this problem. Since the critical function $\varphi_{\Lambda,\Lambda'}(x_1, \ldots, x_n)$ is nondecreasing in each of its arguments, it follows from Lemma 3.4.2 that $p_i \le p'_i$ for $i = 1, \ldots, n$ implies

$$E_{p_1,\ldots,p_n}\varphi_{\Lambda,\Lambda'}(X_1,\ldots,X_n) \le E_{p'_1,\ldots,p'_n}\varphi_{\Lambda,\Lambda'}(X_1,\ldots,X_n)$$

and hence the conditions of Theorem 8.1.1 are satisfied. ■

Example 8.2.4 Let $X = (X_1, \ldots, X_n)$ be a sample from $N(\xi, \sigma^2)$, and consider the problem of testing $H : \sigma = \sigma_0$ against the set of alternatives $\omega' : \sigma \le \sigma_1$ or $\sigma \ge \sigma_2$ $(\sigma_1 < \sigma_0 < \sigma_2)$. This problem remains invariant under the transformations $X'_i = X_i + c$, which in the parameter space induce the group $\bar{G}$ of transformations $\xi' = \xi + c$, $\sigma' = \sigma$. One would therefore expect the least favorable distribution Λ over the line $\omega : -\infty < \xi < \infty$, $\sigma = \sigma_0$, to be invariant under $\bar{G}$. Such invariance implies that Λ assigns to any interval a measure proportional to the length of the interval. Hence Λ cannot be a probability measure and Theorem 8.1.1 is not directly applicable. The difficulty can be avoided by approximating Λ by a sequence of probability distributions, in the present case for example by the sequence of normal distributions $N(0, k)$, $k = 1, 2, \ldots$.

In the particular problem under consideration, it happens that there also exist least favorable distributions Λ and Λ', which are true probability distributions and therefore not invariant. These distributions can be obtained by an examination of the corresponding one-sided problem in Section 3.9, as follows. On ω, where the only variable is ξ, the distribution Λ of ξ is taken as the normal distribution with an arbitrary mean ξ_1 and with variance $(\sigma_2^2 - \sigma_0^2)/n$. Under Λ' all probability should be concentrated on the two lines $\sigma = \sigma_1$ and $\sigma = \sigma_2$ in the (ξ, σ) plane, and we put $\Lambda' = p\Lambda'_1 + q\Lambda'_2$, where Λ'_1 is the normal distribution with mean ξ_1 and variance $(\sigma_2^2 - \sigma_1^2)/n$, while Λ'_2 assigns probability 1 to the point (ξ_1, σ_2). A computation analogous to that carried out in Section 3.9 then shows the acceptance region to be given by

$$\frac{\dfrac{p}{\sigma_1^{n-1}\sigma_2}\exp\left[\dfrac{-1}{2\sigma_1^2}\sum(x_i - \bar{x})^2 - \dfrac{n}{2\sigma_2^2}(\bar{x} - \xi_1)^2\right] + \dfrac{q}{\sigma_2^n}\exp\left[\dfrac{-1}{2\sigma_2^2}\left\{\sum(x_i - \bar{x})^2 + n(\bar{x} - \xi_1)^2\right\}\right]}{\dfrac{1}{\sigma_0^{n-1}\sigma_2}\exp\left[\dfrac{-1}{2\sigma_0^2}\sum(x_i - \bar{x})^2 - \dfrac{n}{2\sigma_2^2}(\bar{x} - \xi_1)^2\right]} < C,$$

which is equivalent to

$$C_1 \le \sum(x_i - \bar{x})^2 \le C_2.$$

The probability of this inequality is independent of ξ, and hence C_1 and C_2 can be determined so that the probability of acceptance is $1 - \alpha$ when $\sigma = \sigma_0$, and is equal for the two values $\sigma = \sigma_1$ and $\sigma = \sigma_2$.

It follows from Section 3.7 that there exist p and C which lead to these values of C_1 and C_2 and that the above test satisfies the conditions of Corollary 8.1.1 with $\omega_0 = \omega$, and with ω'_0 consisting of the two lines $\sigma = \sigma_1$ and $\sigma = \sigma_2$. ■

8.3 Comparing Two Approximate Hypotheses

As in Section 3.2, let $P_0 \neq P_1$ be two distributions possessing densities p_0 and p_1 with respect to a measure μ. Since distributions even at best are known only approximately, let us assume that the true distributions are approximately P_0 or P_1 in the sense that they lie in one of the families

$$\mathcal{P}_i = \{Q : Q = (1-\epsilon_i)P_i + \epsilon_i G_i\}, \qquad i = 0, 1, \tag{8.10}$$

with ϵ_0, ϵ_1 given and the G_i arbitrary unknown distributions. We wish to find the level-α test of the hypothesis H that the true distribution lies in $\mathcal{P}_0$, which maximizes the minimum power over $\mathcal{P}_1$. This is the problem considered in Section 8.1 with θ indicating the true distribution, $\Omega_H = \mathcal{P}_0$, and $\Omega_K = \mathcal{P}_1$.

The following theorem shows the existence of a pair of least favorable distributions Λ and Λ' satisfying the conditions of Theorem 8.1.1, each assigning probability 1 to a single distribution, Λ to $Q_0 \in \mathcal{P}_0$ and Λ' to $Q_1 \in \mathcal{P}_1$, and exhibits the Q_i explicitly.

Theorem 8.3.1 *Let*

$$\begin{aligned} q_0(x) &= \begin{cases} (1-\epsilon_0)p_0(x) & \text{if } \frac{p_1(x)}{p_0(x)} < b, \\ \frac{(1-\epsilon_0)p_1(x)}{b} & \text{if } \frac{p_1(x)}{p_0(x)} \geq b, \end{cases} \\ q_1(x) &= \begin{cases} (1-\epsilon_1)p_1(x) & \text{if } \frac{p_1(x)}{p_0(x)} > a, \\ a(1-\epsilon_1)p_0(x) & \text{if } \frac{p_1(x)}{p_0(x)} \leq a. \end{cases} \end{aligned} \tag{8.11}$$

(i) *For all $0 < \epsilon_i < 1$, there exist unique constants a and b such that q_0 and q_1 are probability densities with respect to μ; the resulting q_i are members of $\mathcal{P}_i$ $(i = 0, 1)$.*

(ii) *There exist δ_0, δ_1 such that for all $\epsilon_i \leq \delta_i$ the constants a and b satisfy $a < b$ and that the resulting q_0 and q_1 are distinct.*

(iii) *If $\epsilon_i \leq \delta_i$ for $i = 0, 1$, the families $\mathcal{P}_0$ and $\mathcal{P}_1$ are nonoverlapping and the pair (q_0, q_1) is least favorable, so that the maximin test of $\mathcal{P}_0$ against $\mathcal{P}_1$ rejects when $q_1(x)/q_0(x)$ is sufficiently large.*

Note. Suppose $a < b$, and let

$$r(x) = \frac{p_1(x)}{p_0(x)}, \quad r^*(x) = \frac{q_1(x)}{q_0(x)}, \quad \text{and} \quad k = \frac{1-\epsilon_1}{1-\epsilon_0}.$$

Then

$$r^*(x) = \begin{cases} ka & \text{when } r(x) \leq a, \\ kr(x) & \text{when } a < r(x) < b, \\ kb & \text{when } b \leq r(x). \end{cases} \tag{8.12}$$

The maximin test thus replaces the original probability ratio with a censored version.

Proof. The proof will be given under the simplifying assumption that $p_0(x)$ and $p_1(x)$ are positive for all x in the sample space.

(i): For q_1 to be a probability density, a must satisfy the equation

$$P_1[r(X) > a] + aP_0[r(X) \le a] = \frac{1}{1-\epsilon_1}. \tag{8.13}$$

If (8.13) holds, it is easily checked that $q_1 \in \mathcal{P}_1$ (Problem 8.12). To prove existence and uniqueness of a solution a of (8.13), let

$$\gamma(c) = P_1[r(X) > c] + cP_0[r(X) \le c].$$

Then

$$\gamma(0) = 1 \quad \text{and} \quad \gamma(c) \to \infty \quad \text{as } c \to \infty. \tag{8.14}$$

Furthermore (Problem 8.14)

$$\begin{aligned} \gamma(c+\Delta) - \gamma(c) &= \Delta \int_{r(x)\le c} p_0(x)\, d\mu(x) \\ &\quad + \int_{c<r(x)\le c+\Delta} [c+\Delta - r(x)] p_0(x)\, d\mu(x). \end{aligned} \tag{8.15}$$

It follows from (8.15) that $0 \le \gamma(c+\Delta) - \gamma(c) \le \Delta$, so that $-\gamma$ is continuous and nondecreasing. Together with (8.14) this establishes the existence of a solution. To prove uniqueness, note that

$$\gamma(c+\Delta) - \gamma(c) \ge \Delta \int_{r(x)<c} p_0(x)\, d\mu(x) \tag{8.16}$$

and that $\gamma(c) = 1$ for all c for which

$$P_i[r(x) \le c] = 0 \qquad (i = 0, 1). \tag{8.17}$$

If c_0 is the supremum of the values for which (8.17) holds, (8.16) shows that γ is strictly increasing for $c > c_0$ and this proves uniqueness. The proof for b is exactly analogous (Problem 8.13).

(ii): As $\epsilon_1 \to 0$, the solution a of (8.13) tends to c_0. Analogously, as $\epsilon_1 \to 0$, $b \to \infty$ (Problem 8.13).

(iii): This will follow from the following facts:

(a) When X is distributed according to a distribution in $\mathcal{P}_0$, the statistic $r^*(X)$ is stochastically largest when the distribution of X is Q_0.

(b) When X is distributed according to a distribution in $\mathcal{P}_1$, $r^*(X)$ is stochastically smallest for Q_1.

(c) $r^*(X)$ is stochastically larger when the distribution of X is Q_1 than when it is Q_0.

These statements are summarized in the inequalities

$$Q_0'[r^*(X) < t] \ge Q_0[r^*(X) < t] \ge Q_1[r^*(X) < t] \ge Q_1'[r^*(X) < t] \tag{8.18}$$

for all t and all $Q_i' \in \mathcal{P}_i$.

From (8.12), it is seen that (8.18) is obvious when $t \le ka$ or $t > kb$. Suppose therefore that $ak < t \le bk$, and denote the event $r^*(X) < t$ by E. Then $Q_0'(E) \ge (1-\epsilon_0)P_0(E)$ by (8.10). But $r^*(x) < t < kb$ implies $r(X) < b$ and hence $Q_0(E) = (1-\epsilon)P_0(E)$. Thus $Q_0'(E) \ge Q_0(E)$, and analogously $Q_1'(E) \le Q_1(E)$. Finally, the middle inequality of (8.18) follows from Corollary 3.2.1.

If the ϵ's are sufficiently small so that $Q_0 \neq Q_1$, it follows from (a)–(c) that $\mathcal{P}_0$ and $\mathcal{P}_1$ are nonoverlapping.

That (Q_0, Q_1) is least favorable and the associated test φ is maximin now follows from Theorem 8.1.1, since the most powerful test φ for testing Q_0 against Q_1 is a nondecreasing function of $q_1(X)/q_0(X)$. This shows that $E\varphi(X)$ takes on its sup over $\mathcal{P}_0$ at Q_0 and its inf over $\mathcal{P}_1$ at Q_1, and this completes the proof. ■

Generalizations of this theorem are given by Huber and Strassen (1973, 1974). See also Rieder (1977) and Bednarski (1984). An optimum permutation test, with generalizations to the case of unknown location and scale parameters, is discussed by Lambert (1985).

When the data consist of n identically, independently distributed random variables $X_1, \ldots, X_n$, the neighborhoods (8.10) may not be appropriate, since they do not preserve the assumption of independence. If P_i has density

$$p_i(x_1, \ldots, x_n) = f_i(x_1) \ldots f_i(x_n) \qquad (i = 0, 1), \tag{8.19}$$

a more appropriate model approximating (8.19) may then assign to $X = (X_1, \ldots, X_n)$ the family $\mathcal{P}_i^*$ of distributions according to which the X_j are independently distributed, each with distribution

$$(1 - \epsilon_i)F_i(x_j) + \epsilon_i G_i(x_j), \tag{8.20}$$

where F_i has density f_i and where as before the G_i are arbitrary.

Corollary 8.3.1 *Suppose q_0 and q_1 defined by (8.11) with $x = x_j$ satisfy (8.18) and hence are a least favorable pair for testing $\mathcal{P}_0$ against $\mathcal{P}_1$ on the basis of the single observation X_j. Then the pair of distributions with densities $q_i(x_1) \ldots q_i(x_n)$ $(i = 0, 1)$ is least favorable for testing $\mathcal{P}_0^*$ against $\mathcal{P}_1^*$, so that the maximin test is given by*

$$\varphi(x_1, \ldots, x_n) = \begin{cases} 1 & \\ \gamma & \text{if } \prod_{j=1}^{n} \left[\dfrac{q_1(x_j)}{q_0(x_j)} \right] \gtreqless c. \\ 0 & \end{cases} \tag{8.21}$$

PROOF. By assumption, the random variables $Y_j = q_1(X_j)/q_0(X_j)$ are stochastically increasing as one moves successively from $Q_0' \in \mathcal{P}_0$ to Q_0 to Q_1 to $Q_1' \in \mathcal{P}_1$. The same is then true of any function $\psi(Y_1, \ldots, Y_n)$ which is nondecreasing in each of its arguments by Lemma 3.4.1, and hence of φ defined by (8.21). The proof now follows from Theorem 8.3.1. ■

Instead of the problem of testing P_0 against P_1, consider now the situation of Lemma 8.2.1 where $H : \theta \leq \theta_0$ is to be tested against $\theta \geq \theta_1$ $(\theta_0 < \theta_1)$ on the basis of n independent observations X_j, each distributed according to a distribution $F_\theta(x_j)$ whose density $f_\theta(x_j)$ is assumed to have monotone likelihood ratio in x_j.

A robust version of this problem is obtained by replacing F_θ with

$$(1 - \epsilon)F_\theta(x_j) + \epsilon G(x_j), \qquad j = 1, \ldots, n, \tag{8.22}$$

where ϵ is given and for each θ the distribution G is arbitrary. Let $\mathcal{P}_0^{**}$ and $\mathcal{P}_1^{**}$ be the classes of distributions (8.22) with $\theta \leq \theta_0$ and $\theta \geq \theta_1$ respectively; and let $\mathcal{P}_0^*$ and $\mathcal{P}_1^*$ be defined as in Corollary 8.3.1 with f_{θ_i} in place of f_i. Then the maximin test (8.21) of $\mathcal{P}_0^*$ against $\mathcal{P}_1^*$ retains this property for testing $\mathcal{P}_0^{**}$ against $\mathcal{P}_1^{**}$.

This is proved in the same way as Corollary 8.3.1, using the additional fact that if $F_{\theta'}$ is stochastically larger than F_θ, then $(1-\epsilon)F_{\theta'} + \epsilon G$ is stochastically larger than $(1-\epsilon)F_\theta + \epsilon G$.

8.4 Maximin Tests and Invariance

When the problem of testing Ω_H against Ω_K remains invariant under a certain group of transformations, it seems reasonable to expect the existence of an invariant pair of least favorable distributions (or at least of sequences of distributions which in some sense are least favorable and invariant in the limit), and hence also of a maximin test which is invariant. This suggests the possibility of bypassing the somewhat cumbersome approach of the preceding sections. If it could be proved that for an invariant problem there always exists an invariant test that maximizes the minimum power over Ω_K, attention could be restricted to invariant tests; in particular, a UMP invariant test would then automatically have the desired maximin property (although it would not necessarily be admissible). These speculations turn out to be correct for an important class of problems, although unfortunately not in general. To find out under what conditions they hold, it is convenient first to separate out the statistical aspects of the problem from the group-theoretic ones by means of the following lemma.

Lemma 8.4.1 *Let $\mathcal{P} = \{P_\theta, \theta \in \Omega\}$ be a dominated family of distributions on $(\mathcal{X}, \mathcal{A})$, and let G be a group of transformations of $(\mathcal{X}, \mathcal{A})$, such that the induced group $\bar{G}$ leaves the two subsets Ω_H and Ω_K of Ω invariant. Suppose that for any critical function φ there exists an (almost) invariant critical function ψ satisfying*

$$\inf_{\bar{G}} E_{\bar{g}\theta}\varphi(X) \le E_\theta\psi(X) \le \sup_{\bar{G}} E_{\bar{g}\theta}\varphi(X) \tag{8.23}$$

for all $\theta \in \Omega$. Then if there exists a level-α test φ_0 maximizing $\inf_{\Omega_k} E_\theta\varphi(X)$, there also exists an (almost) invariant test with this property.

Proof. Let $\inf_{\Omega_K} E_\theta\varphi_0(X) = \beta$, and let ψ_0 be an (almost) invariant test such that (8.23) holds with $\varphi = \varphi_0$, $\psi = \psi_0$. Then

$$E_\theta\psi_0(X) \le \sup_{\bar{G}} E_{\bar{g}\theta}\varphi_0(X) \le \alpha \qquad \text{for all} \quad \theta \in \Omega_H$$

and

$$E_\theta\psi_0(X) \ge \inf_{\bar{G}} E_{\bar{g}\theta}\varphi_0(X) \ge \beta \qquad \text{for all} \quad \theta \in \Omega_K,$$

as was to be proved. ∎

To determine conditions under which there exists an invariant or almost invariant test ψ satisfying (8.23), consider first the simplest case that G is a finite group, $G = \{g_1, \dots, g_N\}$ say. If ψ is then defined by

$$\psi(x) = \frac{1}{N}\sum_{i=1}^{N} \varphi(g_i x), \tag{8.24}$$

it is clear that ψ is again a critical function, and that it is invariant under G. It also satisfies (8.23), since $E_\theta \varphi(gX) = E_{\bar{g}\theta}\varphi(X)$ so that $E_\theta \psi(X)$ is the average of a number of terms of which the first and last member of (8.23) are the minimum and maximum respectively.

An illustration of the finite case is furnished by Example 8.2.3. Here the problem remains invariant under the $n!$ permutations of the variables $(X_1, \ldots, X_n)$. Lemma 8.4.1 is applicable and shows that there exists an invariant test maximizing $\inf_{\Omega_K} E_\theta \varphi(X)$. Thus in particular the UMP invariant test obtained in Example 6.3.5 has this maximin property and therefore constitutes a solution of the problem.

It also follows that, under the setting of Theorem 6.3.1, the UMPI test given by (6.9) is maximin.

The definition (8.24) suggests the possibility of obtaining $\psi(x)$ also in other cases by averaging the values of $\varphi(gx)$ with respect to a suitable probability distribution over the group G. To see what conditions would be required of this distribution, let $\mathcal{B}$ be a σ-field of subsets of G and ν a probability distribution over $(G, \mathcal{B})$. Disregarding measurability problems for the moment, let ψ be defined by

$$\psi(x) = \int \varphi(gx)\, d\nu(g). \tag{8.25}$$

Then $0 \le \psi \le 1$, and (8.23) is seen to hold by applying Fubini's theorem (Theorem 2.2.4) to the integral of ψ with respect to the distribution P_θ. For any $g_0 \in G$,

$$\psi(g_0 x) = \int \varphi(g g_0 x)\, d\nu(g) = \int \varphi(hx)\, d\nu^*(h)\ ,$$

where $h = gg_0$ and where ν^* is the measure defined by

$$\nu^*(B) = \nu(Bg_0^{-1}) \qquad \text{for all} \quad B \in \mathcal{B},$$

into which ν is transformed by the transformation $h = gg_0$. Thus ψ will have the desired invariance property, $\psi(g_0 x) = \psi(x)$ for all $g_0 \in G$, if ν is *right invariant*, that is, if it satisfies

$$\nu(Bg) = \nu(B) \qquad \text{for all} \quad B \in \mathcal{B}, \quad g \in G. \tag{8.26}$$

Such a condition was previously used in (6.16).

The measurability assumptions required for the above argument are: (i) For any $A \in \mathcal{A}$, the set of pairs (x, g) with $gx \in A$ is measurable $(\mathcal{A} \times \mathcal{B})$. This insures that the function ψ defined by (8.25) is again measurable. (ii) For any $B \in \mathcal{B}$, $g \in G$, the set Bg belongs to $\mathcal{B}$.

Example 8.4.1 If G is a finite group with elements $g_1, \ldots, g_N$, let $\mathcal{B}$ be the class of all subsets of G and ν the probability measure assigning probability $1/N$ to each of the N elements. The condition (8.26) is then satisfied, and the definition (8.25) of ψ in this case reduces to (8.24). ■

Example 8.4.2 Consider the group G of orthogonal $n \times n$ matrices Γ, with the group product $\Gamma_1\Gamma_2$ defined as the corresponding matrix product. Each matrix can be interpreted as the point in n^2-dimensional Euclidean space whose coordinates are the n^2 elements of the matrix. The group then defines a subset of this space; the Borel subsets of G will be taken as the σ-field $\mathcal{B}$. To prove the existence of a right invariant probability measure over $(G, \mathcal{B})$, we shall define a random orthogonal matrix whose probability distribution satisfies (8.26) and is therefore the required measure. With any nonsingular matrix $x = (x_{ij})$, associate the orthogonal matrix $y = f(x)$ obtained by applying the following Gram–Schmidt orthogonalization process to the n row vectors $x_i = (x_{i1}, \ldots, x_{in})$ of $x : y_1$ is the unit vector in the direction of x_1; y_2 the unit vector in the plane spanned by x_1 and x_2 which is orthogonal to y_1 and forms an acute angle with x_2; and so on. Let $y = (y_{ij})$ be the matrix whose ith row is y_i.

Suppose now that the variables X_{ij} $(i, j = 1, \ldots, n)$ are independently distributed as $N(0, 1)$, let X denote the random matrix (X_{ij}), and let $Y = f(X)$. To show that the distribution of the random orthogonal matrix Y satisfies (8.26), consider any fixed orthogonal matrix Γ and any fixed set $B \in \mathcal{B}$. Then $P\{Y \in B\Gamma\} = P\{Y\Gamma' \in B\}$ and from the definition of f it is seen that $Y\Gamma' = f(X\Gamma')$. Since the n^2 elements of the matrix $X\Gamma'$ have the same joint distribution as those of the matrix X, the matrices $f(X\Gamma')$ and $f(X)$ also have the same distribution, as was to be proved. ■

Examples 8.4.1 and 8.4.2 are sufficient for the applications to be made here. General conditions for the existence of an invariant probability measure, of which these examples are simple special cases, are given in the theory of Haar measure. [This is treated, for example, in the books by Halmos (1974), Loomis (1953), and Nachbin (1965). For a discussion in a statistical setting, see Eaton (1983, 1989), Farrell (1985a), and Wijsman (1990), and for a more elementary treatment Berger (1985a).]

8.5 The Hunt–Stein Theorem

Invariant measures exist (and are essentially unique) for a large class of groups, but unfortunately they are frequently not finite and hence cannot be taken to be probability measures. The situation is similar and related to that of the nonexistence of a least favorable pair of distributions in Theorem 8.1.1. There it is usually possible to overcome the difficulty by considering instead a sequence of distributions which has the desired property in the limit. Analogously we shall now generalize the construction of ψ as an average with respect to a right-invariant probability distribution, by considering a sequence of distributions over G which are approximately right-invariant for n sufficiently large.

Let $\mathcal{P} = \{P_\theta, \theta \in \Omega\}$ be a family of distributions over a Euclidean space $(\mathcal{X}, \mathcal{A})$ dominated by a σ-finite measure μ, and let G be a group of transformations of $(\mathcal{X}, \mathcal{A})$ such that the induced group $\bar{G}$ leaves Ω invariant.

Theorem 8.5.1 (Hunt–Stein.) *Let $\mathcal{B}$ be a σ-field of subsets of G such that for any $A \in \mathcal{A}$ the set of pairs (x, g) with $gx \in A$ is in $\mathcal{A} \times \mathcal{B}$ and for any $B \in \mathcal{B}$ and $g \in G$ the set Bg is in $\mathcal{B}$. Suppose that there exists a sequence of probability distributions ν_n over $(G, \mathcal{B})$ which is asymptotically right-invariant in the sense that for any $g \in G$, $B \in \mathcal{B}$,*

$$\lim_{n\to\infty} |\nu_n(Bg) - \nu_n(B)| = 0. \tag{8.27}$$

Then given any critical function φ, there exists a critical function ψ which is almost invariant and satisfies (8.23).

PROOF. Let

$$\psi_n(x) = \int \varphi(gx)\,d\nu_n(g),$$

which as before is measurable and between 0 and 1. By the weak compactness theorem (Theorem A.5.1 of the Appendix) there exists a subsequence $\{\psi_{n_i}\}$ and a measurable function ψ between 0 and 1 satisfying

$$\lim_{i\to\infty} \int \psi_{n_i} p\,d\mu = \int \psi p\,d\mu$$

for all μ-integrable functions p, so that in particular

$$\lim_{i\to\infty} E_\theta \psi_{n_i}(X) = E_\theta \psi(X)$$

for all $\theta \in \Omega$. By Fubini's theorem,

$$E_\theta \psi_{n_i}(X) = \int [E_\theta \varphi(gX)]\,d\nu_{n_i}(g) = \int E_{\bar{g}\theta}\varphi(X)\,d\nu_{n_i}(g)\ ,$$

so that

$$\inf_{\bar{G}} E_{\bar{g}\theta}\varphi(X) \le E_\theta \psi_{n_i}(X) \le \sup_{\bar{G}} E_{\bar{g}\theta}\varphi(X),$$

and ψ satisfies (8.23).

In order to prove that ψ is almost invariant we shall show below that for all x and g,

$$\psi_{n_i}(gx) - \psi_{n_i}(x) \to 0. \tag{8.28}$$

Let $I_A(x)$ denote the indicator function of a set $A \in \mathcal{A}$. Using the fact that $I_{gA}(gx) = I_A(x)$, we see that (8.28) implies

$$\begin{aligned}
\int_A \psi(x)\,dP_\theta(x) &= \lim_{i\to\infty} \int \psi_{n_i}(x) I_A(x)\,dP_\theta(x) \\
&= \lim_{i\to\infty} \int \psi_{n_i}(gX) I_{gA}(gx)\,dP_\theta(x) \\
&= \int \psi(x) I_{gA}(x)\,dP_{\bar{g}\theta}(x) = \int_A \psi(gx)\,dP_\theta(x)\ ,
\end{aligned}$$

and hence $\psi(gx) = \psi(x)$ (a.e. $\mathcal{P}$), as was to be proved.

To prove (8.28), consider any fixed x and any integer m, and let G be partitioned into the mutually exclusive sets

$$B_k = \left\{h \in G : a_k < \varphi(hx) \le a_k + \frac{1}{m}\right\}, \qquad k = 0, \ldots, m,$$

where $a_k = (k-1)/m$. In particular, B_0 is the set $\{h \in G : \varphi(hx) = 0\}$. It is seen from the definition of the sets B_k that

$$\begin{aligned}
\sum_{k=0}^m a_k \nu_{n_i}(B_k) \le \sum_{k=0}^m \int_{B_k} \varphi(hx)\,d\nu_{n_i}(h) &\le \sum_{k=0}^m \left(a_k + \frac{1}{m}\right)\nu_{n_i}(B_k) \\
&\le \sum_{k=0}^m a_k \nu_{n_i}(B_k) + \frac{1}{m}\ ,
\end{aligned}$$

and analogously that

$$\left| \sum_{k=0}^{m} \int_{B_k g^{-1}} \varphi(hgx)\, d\nu_{n_i}(h) - \sum_{k=0}^{m} a_k \nu_{n_i}(B_k g^{-1}) \right| \le \frac{1}{m},$$

from which it follows that

$$\psi_{n_i}(gx) - \psi_{n_i}(x) :\le \sum |a_k| \cdot |\nu_{n_i}(B_k g^{-1}) - \nu_{n_i}(B_k)| + \frac{2}{m}.$$

By (8.27) the first term of the right-hand side tends to zero as i tends to infinity, and this completes the proof. ■

When there exist a right-invariant measure ν over G and a sequence of subsets G_n of G with $G_n \subseteq G_{n+1}$, $\cup G_n = G$, and $\nu(G_n) = c_n < \infty$, it is suggestive to take for the probability measures ν_n of Theorem 8.5.1 the measures ν/c_n truncated on G_n. This leads to the desired result in the example below. On the other hand, there are cases in which there exists such a sequence of subsets of G_n but no invariant test satisfying (8.23) and hence no sequence ν_n satisfying (8.27).

Example 8.5.1 Let $x = (x_1, \ldots, x_n)$, $\mathcal{A}$ be the class of Borel sets in n-space, and G the group of translations $(x_1 + g, \ldots, x_n + g)$, $-\infty < g < \infty$. The elements of G can be represented by the real numbers, and the group product gg' is then the sum $g + g'$. If $\mathcal{B}$ is the class of Borel sets on the real line, the measurability assumptions of Theorem 8.5.1 are satisfied. Let ν be Lebesgue measure, which is clearly invariant under G, and define ν_n to be the uniform distribution on the interval $I(-n, n) = \{g : -n \le g \le n\}$. Then for all $B \in \mathcal{B}$, $g \in G$,

$$|\nu_n(B) - \nu_n(Bg)| = \frac{1}{2n} |\nu[B \cap I(-n, n)] - \nu[B \cap I(-n-g, n-g)]| \le \frac{|g|}{2n},$$

so that (8.27) is satisfied.

This argument also covers the group of scale transformations $(ax_1, \ldots, ax_n)$, $0 < a < \infty$, which can be transformed into the translation group by taking logarithms. ■

When applying the Hunt–Stein theorem to obtain invariant minimax tests, it is frequently convenient to carry out the calculation in steps, as was done in Theorem 6.6.1. Suppose that the problem remains invariant under two groups D and E, and denote by $y = s(x)$ a maximal invariant with respect to D and by E^* the group defined in Theorem 6.2.2, which E induces in y-space. If D and E^* satisfy the conditions of the Hunt–Stein theorem, it follows first that there exists a maximin test depending only on $y = s(x)$, and then that there exists a maximin test depending only on a maximal invariant $z = t(y)$ under E^*.

Example 8.5.2 Consider a univariate linear hypothesis in the canonical form in which $Y_1, \ldots, Y_n$ are independently distributed as $N(\eta_i, \sigma^2)$, where it is given that $\eta_{s+1} = \cdots = \eta_n = 0$, and where the hypothesis to be tested is $\eta_1 = \cdots = \eta_r = 0$. It was shown in Section 7.1 that this problem remains invariant under certain groups of transformations and that with respect to these groups there exists a UMP invariant test. The groups involved are the group of orthogonal transformations, translation groups of the kind considered in Example 8.5.1, and a group of scale changes. Since each of these satisfies the assumptions of the Hunt–Stein theorem, and since they leave invariant the problem of maximizing the minimum power over the set of alternatives

$$\sum_{i=1}^{r} \frac{\eta_i^2}{\sigma^2} \geq \psi_1^2 \qquad (\psi_1 > 0), \tag{8.29}$$

it follows that the UMP invariant test of Chapter 7 is also the solution of this maximin problem. It is also seen slightly more generally that the test which is UMP invariant under the same groups for testing

$$\sum_{i=1}^{r} \frac{\eta_i^2}{\sigma^2} \leq \psi_0^2$$

(Problem 7.4) maximizes the minimum power over the alternatives (8.29) for $\psi_0 < \psi_1$. ■

Example 8.5.3 (Stein) Let G be the group of all nonsingular linear transformations of p-space. That for $p > 1$ this does not satisfy the conditions of Theorem 8.5.1 is shown by the following problem, which is invariant under G but for which the UMP invariant test does not maximize the minimum power. Generalizing Example 6.2.1, let $X = (X_1, \ldots, X_p)$, $Y = (Y_1, \ldots, Y_p)$ be independently distributed according to p-variate normal distributions with zero means and nonsingular covariance matrices $E(X_iX_j) = \sigma_{ij}$ and $E(Y_iY_j) = \Delta\sigma_{ij}$, and let $H : \Delta \leq \Delta_0$ be tested against $\Delta \geq \Delta_1$ $(\Delta_0 < \Delta_1)$, the σ_{ij} being unknown.

This problem remains invariant if the two vectors are subjected to any common nonsingular transformation, and since with probability 1 this group is transitive over the sample space, the UMP invariant test is trivially $\varphi(x, y) \equiv \alpha$. The maximin power against the alternatives $\Delta \geq \Delta_1$ that can be achieved by invariant tests is therefore α. On the other hand, the test with rejection region $Y_1^2/X_1^2 > C$ has a strictly increasing power function $\beta(\Delta)$, whose minimum over the set of alternatives $\Delta \geq \Delta_1$ is $\beta(\Delta_1) > \beta(\Delta_0) = \alpha$. ■

It is a remarkable feature of Theorem 8.5.1 that its assumptions concern only the group G and not the distributions P_θ.[3] When these assumptions hold for a certain G it follows from (8.23) as in the proof of Lemma 8.4.1 that for any testing problem which remains invariant under G and possesses a UMP invariant test, this test maximizes the minimum power over any invariant class of alternatives. Suppose conversely that a UMP invariant test under G has been shown in a particular problem not to maximize the minimum power, as was the case for the group of linear transformations in Example 8.5.3. Then the assumptions of Theorem 8.5.1 cannot be satisfied. However, this does not rule out the possibility that for another problem remaining invariant under G, the UMP invariant test may maximize the minimum power. Whether or not it does is no longer a property of the group alone but will in general depend also on the particular distributions.

[3]These assumptions are essentially equivalent to the condition that the group G is *amenable*. Amenability and its relationship to the Hunt–Stein theorem are discussed by Bondar and Milnes (1982) and (with a different terminology) by Stone and von Randow (1968).

Consider in particular the problem of testing $H : \xi_1 = \cdots = \xi_p = 0$ on the basis of a sample $(X_{\alpha 1}, \ldots, X_{\alpha p})$, $\alpha = 1, \ldots, n$, from a p-variate normal distribution with mean $E(X_{\alpha i}) = \xi_i$ and common covariance matrix $(\sigma_{ij}) = (a_{ij})^{-1}$. This problem remains invariant under a number of groups, including that of all nonsingular linear transformations of p-space, and a UMP invariant test exists. An invariant class of alternatives under these groups is

$$\sum \sum \frac{a_{ij}\xi_i\xi_j}{\sigma^2} \geq \psi_1^2. \tag{8.30}$$

Here, Theorem 8.5.1 is not applicable, and the question of whether the T^2-test of $H : \psi = 0$ maximizes the minimum power over the alternatives

$$\sum \sum a_{ij}\xi_i\xi_j = \psi_1^2 \tag{8.31}$$

[and hence a fortiori over the alternatives (8.30)] presents formidable difficulties. The minimax property was proved for the case $p = 2$, $n = 3$ by Giri, Kiefer, and Stein (1963), for the case $p = 2$, $n = 4$ by Linnik, Pliss, and Salaevskii (1968), and for $p = 2$ and all $n \geq 3$ by Salaevskii (1971). The proof is effected by first reducing the problem through invariance under the group G_1 of Example 6.6.11, to which Theorem 8.5.1 is applicable, and then applying Theorem 8.1.1 to the reduced problem. It is a consequence of this approach that it also establishes the admissibility of T^2 as a test of H against the alternatives (8.31). In view of the inadmissibility results for point estimation when $p \geq 3$ (see *TPE2*, Sections 5.4-5.5, it seems unlikely that T^2 is admissible for $p \geq 3$, and hence that the same method can be used to prove the minimax property in this situation.

The problem becomes much easier when the minimax property is considered against local or distant alternatives rather than against (8.31). Precise definitions and proofs of the fact that T^2 possesses these properties for all p and n are provided by Giri and Kiefer (1964) and in the references given in Section 7.9.

The theory of this and the preceding section can be extended to confidence sets if the accuracy of a confidence set at level $1 - \alpha$ is assessed by its volume or some other appropriate measure of its size. Suppose that the distribution of X depends on the parameters θ to be estimated and on nuisance parameters ϑ, and that μ is a σ-finite measure over the parameter set $\omega = \{\theta : (\theta, \vartheta) \in \Omega\}$, with ω assumed to be independent of ϑ. Then the confidence sets $S(X)$ for θ are minimax with respect to μ at level $1 - \alpha$ if they minimize

$$\sup E_{\theta,\vartheta}\mu[S(X)]$$

among all confidence sets at the given level.

The problem of minimizing $E\mu[S(X)]$ is related to that of minimizing the probability of covering false values (the criterion for accuracy used so far) by the relation (Problem 8.34)

$$E_{\theta_0,\vartheta}\mu[S(X)] = \int_{\theta \neq \theta_0} P_{\theta_0,\vartheta}[\theta \in S(X)]\, d\mu(\theta), \tag{8.32}$$

which holds provided μ assigns measure zero to the set $\{\theta = \theta_0\}$. (For the special case that θ is real-valued and μ Lebesgue measure, see Problem 5.26.)

Suppose now that the problem of estimating θ is invariant under a group G in the sense of Section 6.11 and that it satisfies the invariance condition

$$\mu[S(gx)] = \mu[S(x)]. \tag{8.33}$$

If uniformly most accurate equivariant confidence sets exist, they minimize (8.32) among all equivariant confidence sets at the given level, and one may hope that under the assumptions of the Hunt–Stein theorem, they will also be minimax with respect to μ among the class of all (not necessarily equivariant) confidence sets at the given level. Such a result does hold and can be used to show for example that the most accurate equivariant confidence sets of Examples 6.11.2 and 6.11.3 minimize their maximum expected Lebesgue measure. A more general class of examples is provided by the confidence intervals derived from the UMP invariant tests of univariate linear hypotheses such as the confidence spheres for $\theta_i = \mu + \alpha_i$ or for α_i given in Section 7.4.

Minimax confidence sets $S(x)$ are not necessarily admissible; that is, there may exist sets $S'(x)$ having the same confidence level but such that

$$E_{\theta,\vartheta}\mu[S'(X)] \leq E_{\theta,\vartheta}\mu[S(X)] \qquad \text{for all} \quad \theta, \vartheta$$

with strict inequality holding for at least some (θ, ϑ).

Example 8.5.4 Let X_i $(i = 1, \ldots, s)$ be independently normally distributed with mean $E(X_i) = \theta_i$ and variance 1, and let G be the group generated by translations $X_i + c_i$ $(i = 1, \ldots, s)$ and orthogonal transformations of $(X_1, \ldots, X_s)$. (G is the Euclidean group of rigid motions in s-space.) In Example 6.12.2, it was argued that the confidence sets

$$C_0 = \{(\theta_1, \ldots, \theta_s) : \sum (\theta_i - X_i)^2 \leq c\} \tag{8.34}$$

are uniformly most accurate equivariant. The volume $\mu[S(X)]$ of any confidence set $S(X)$ remains invariant under the transformations $g \in G$, and it follows from the results of Problems 8.26 and 8.4 and Examples 8.5.1 and 8.5.2 that the confidence sets (8.34) minimize the maximum expected volume.

However, very surprisingly, they are not admissible unless $s = 1$ or 2. In the case $s \geq 3$, Stein (1962) suggested the region (8.34) can be improved by recentered regions of the form

$$C_1 = \{(\theta_1, \ldots, \theta_s) : (\theta_i - \hat{b}X_i)^2 \leq c\} , \tag{8.35}$$

where $\hat{b} = \max(0, 1 - (s-2)/\sum_i X_i^2)$. In fact, Brown (1966) proved that, for $s \geq 3$,

$$P_\theta\{\theta \in C_1\} > P_\theta\{\theta \in C_0\}$$

for all θ. This result, which will not be proved here, is closely related to the inadmissibility of $X_1, \ldots, X_s$ as a point estimator of $(\theta_1, \ldots, \theta_s)$ for a wide variety of loss functions. The work on point estimation, which is discussed in *TPE*2, Sections 5.4-5.6, for squared error loss, provides easier access to these ideas than the present setting. Further entries into the literature on admissibility are Stein (1981), Hwang and Casella (1982), and Tseng and Brown (1997); additional references are provided in *TPE*2, p.423.

The inadmissibility of the confidence sets (8.34) is particularly surprising in that the associated UMP invariant tests of the hypotheses $H : \theta_i = \theta_{i_0}$ $(i = 1, \ldots, s)$ are admissible (Problems 8.24, 8.25). ■

8.6 Most Stringent Tests

One of the practical difficulties in the consideration of tests that maximize the minimum power over a class Ω_K of alternatives is the determination of an appropriate Ω_K. If no information is available on which to base the choice of this set, and if a natural definition is not imposed by invariance arguments, a frequently reasonable definition can be given in terms of the power that can be achieved against the various alternatives. The *envelope power function* β_α^* was defined in Problem 6.25 by

$$\beta_\alpha^*(\theta) = \sup \beta_\varphi(\theta),$$

where β_φ denotes the power of a test φ and where the supremum is taken over all level-α tests of H. Thus $\beta_\alpha^*(\theta)$ is the maximum power that can be attained at level α against the alternative θ. (That it can be attained follows under mild restrictions from Theorem A.5.1 of the Appendix.) If

$$S_\Delta^* = \{\theta : \beta_\alpha^*(\theta) = \Delta\},$$

then of two alternatives $\theta_1 \in S_{\Delta_1}^*$, $\theta_2 \in S_{\Delta_2}^*$, θ_1 can be considered closer to H, equidistant, or further away than θ_2 as Δ_1 is $<$, $=$, or $> \Delta_2$.

The idea of measuring the distance of an alternative from H in terms of the available information has been encountered before. If for example $X_1, \ldots, X_n$ is a sample from $N(\xi, \sigma^2)$, the problem of testing $H : \xi \le 0$ was discussed (Section 5.2) both when the alternatives ξ are measured in absolute units and when they are measured in σ-units. The latter possibility corresponds to the present proposal, since it follows from invariance considerations (Problem 6.25) that $\beta_\alpha^*(\xi, \sigma)$ is constant on the lines $\xi/\sigma = \text{constant}$.

Fixing a value of Δ and taking as Ω_K the class of alternatives θ for which $\beta_\alpha^*(\theta) \ge \Delta$, one can determine the test that maximizes the minimum power over Ω_K. Another possibility, which eliminates the need of selecting a value of Δ, is to consider for any test φ the difference $\beta_\alpha^*(\theta) - \beta_\varphi(\theta)$. This difference measures the amount by which the actual power $\beta_\varphi(\theta)$ falls short of the maximum power attainable. A test that minimizes

$$\sup_{\Omega-\omega} [\beta_\alpha^*(\theta) - \beta_\varphi(\theta)] \tag{8.36}$$

is said to be *most stringent*. Thus a test is most stringent if it minimizes its maximum shortcoming.

Let φ_Δ be a test that maximizes the minimum power over S_Δ^*, and hence minimizes the maximum difference between $\beta_\alpha^*(\theta)$ and $\beta_\varphi(\theta)$ over S_Δ^*. If φ_Δ happens to be independent of Δ, it is most stringent. This remark makes it possible to apply the results of the preceding sections to the determination of most stringent tests. Suppose that the problem of testing $H : \theta \in \omega$ against the alternatives $\theta \in \Omega - \omega$ remains invariant under a group G, that there exists a UMP almost invariant test φ_0 with respect to G, and that the assumptions of Theorem 8.5.1 hold. Since $\beta_\alpha^*(\theta)$ and hence the set S_Δ^* is invariant under $\bar{G}$ (Problem 6.25), it follows that φ_0 maximizes the minimum power over S_Δ^* for each Δ, and φ_0 is therefore most stringent.

As an example of this method consider the problem of testing $H : p_1, \ldots, p_n \le \frac{1}{2}$ against the alternative $K : p_i > \frac{1}{2}$ for all i, where p_i is the probability of success in the ith trial of a sequence of n independent trials. If X_i is 1 or 0 as the ith trial is a success or failure, then the problem remains invariant under permutations of the X's, and the UMP invariant test rejects (Example 6.3.5) when $\sum X_i > C$. It now follows from the remarks above that this test is also most stringent.

Another illustration is furnished by the general univariate linear hypothesis. Here it follows from the discussion in Example 8.5.2 that the standard test for testing $H : \eta_1 = \cdots = \eta_r = 0$ or $H' : \sum_{i=1}^r \eta_i^2/\sigma^2 \le \psi_0^2$ is most stringent.

When the invariance approach is not applicable, the explicit determination of most stringent tests typically is difficult. The following is a class of problems for which they are easily obtained by a direct approach. Let the distributions of X constitute a one-parameter exponential family, the density of which is given by (3.19), and consider the hypothesis $H : \theta = \theta_0$. Then according as $\theta > \theta_0$ or $\theta < \theta_0$, the envelope power $\beta_\alpha^*(\theta)$ is the power of the UMP one-sided test for testing H against $\theta > \theta_0$ or $\theta < \theta_0$. Suppose that there exists a two-sided test φ_0 given by (4.3), such that

$$\sup_{\theta<\theta_0} [\beta_\alpha^*(\theta) - \beta_{\varphi_0}(\theta)] = \sup_{\theta>\theta_0} [\beta_\alpha^*(\theta) - \beta_{\varphi_0}(\theta)], \tag{8.37}$$

and that the supremum is attained on both sides, say at points $\theta_1 < \theta_0 < \theta_2$. If $\beta_{\varphi_0}(\theta_i) = \beta_i$, $i = 1, 2$, an application of the fundamental lemma [Theorem 3.6.1(iii)] to the three points θ_1, θ_2, θ_0 shows that among all tests φ with $\beta_\varphi(\theta_1) \ge \beta_1$ and $\beta_\varphi(\theta_2) \ge \beta_2$, only φ_0 satisfies $\beta_\varphi(\theta_0) \le \alpha$. For any other level-$\alpha$ test, therefore, either $\beta_\varphi(\theta_1) < \beta_1$ or $\beta_\varphi(\theta_2) < \beta_2$, and it follows that φ_0 is the unique most stringent test. The existence of a test satisfying (8.37) can be proved by a continuity consideration [with respect to variation of the constants C_i and γ_i which define the boundary of the test (4.3)] from the fact that for the UMP one-sided test against the alternatives $\theta > \theta_0$ the right-hand side of (8.37) is zero and the left-hand side positive, while the situation is reversed for the other one-sided test.

8.7 Problems

Section 8.1

Problem 8.1 *Existence of maximin tests.*[4] Let $(\mathcal{X}, \mathcal{A})$ be a Euclidean sample space, and let the distributions P_θ, $\theta \in \Omega$, be dominated by a σ-finite measure over $(\mathcal{X}, \mathcal{A})$. For any mutually exclusive subsets Ω_H, Ω_K of Ω there exists a level-α test maximizing (8.2).
[Let $\beta = \sup[\inf_{\Omega_k} E_\theta\varphi(X)]$, where the supremum is taken over all level-α tests of $H : \theta \in \Omega_H$. Let φ_n be a sequence of level-α tests such that $\inf_{\Omega_K} E_\theta\varphi_n(X)$ tends to β. If φ_{n_i} is a subsequence and φ a test (guaranteed by Theorem 8.5.1 of the Appendix) such that $E_\theta\varphi_{n_i}(X)$ tends to $E_\theta\varphi(X)$ for all $\theta \in \Omega$, then φ is a level-α test and $\inf_{\Omega_k} E_\theta\varphi(X) = \beta$.]

---

[4] The existence of maximin tests is established in considerable generality in Cvitanic and Karatzas (2001).

Problem 8.2 *Locally most powerful tests.* [5] Let d be a measure of the distance of an alternative θ from a given hypothesis H. A level-α test φ_0 is said to be *locally most powerful* (LMP) if, given any other level-α test φ, there exists Δ such that

$$\beta_{\varphi_0}(\theta) \geq \beta_{\varphi}(\theta) \quad \text{for all } \theta \text{ with } 0 < d(\theta) < \Delta. \tag{8.38}$$

Suppose that θ is real-valued and that the power function of every test is continuously differentiable at θ_0.

(i) If there exists a unique level-α test φ_0 of $H : \theta = \theta_0$, maximizing $\beta'_\varphi(\theta_0)$, then φ_0 is the unique LMP level-α test of H against $\theta > \theta_0$ for $d(\theta) = \theta - \theta_0$.

(ii) To see that (i) is not correct without the uniqueness assumption, let X take on the values 0 and 1 with probabilities $P_\theta(0) = \frac{1}{2} - \theta^3$, $P_\theta(1) = \frac{1}{2} + \theta^3$, $-\frac{1}{2} < \theta^3 < \frac{1}{2}$, and consider testing $H : \theta = 0$ against $K : \theta > 0$. Then every test φ of size α maximizes $\beta'_\varphi(0)$, but not every such test is LMP. [Kallenberg et al. (1984).]

(iii) The following[6] is another counterexample to (i) without uniqueness, in which in fact no LMP test exists. Let X take on the values 0, 1, 2 with probabilities

$$\begin{aligned} P_\theta(x) &= \alpha + \epsilon\left[\theta + \theta^2 \sin\left(\frac{x}{\theta}\right)\right] \qquad \text{for} \quad x = 1, 2, \\ P_\theta(0) &= 1 - p_\theta(1) - p_\theta(2), \end{aligned}$$

where $-1 \leq \theta \leq 1$ and ϵ is a sufficiently small number. Then a test φ at level α maximizes $\beta'(0)$ provided

$$\varphi(1) + \varphi(2) = 1\ ,$$

but no LMP test exists.

(iv) A unique LMP test maximizes the minimum power locally provided its power function is bounded away from α for every set of alternatives which is bounded away from H.

(v) Let $X_1, \ldots, X_n$ be a sample from a Cauchy distribution with unknown location parameter θ, so that the joint density of the X's is $\pi^{-n}\prod_{i=1}^n[1 + (x_i - \theta)^2]^{-1}$. The LMP test for testing $\theta = 0$ against $\theta > 0$ at level $\alpha < \frac{1}{2}$ is not unbiased and hence does not maximize the minimum power locally.

[(iii): The unique most powerful test against θ is

$$\left\{ \begin{array}{c} \varphi(1) \\ \varphi(2) \end{array} \right. = 1 \quad \text{if } \sin\left(\frac{1}{\theta}\right) \gtrless \sin\left(\frac{2}{\theta}\right),$$

and each of these inequalities holds at values of θ arbitrarily close to 0. (v): There exists M so large that any point with $x_i \geq M$ for all $i = 1, \ldots, n$ lies in the acceptance region of the LMP test. Hence the power of the test tends to zero as θ tends to infinity.]

[5] Locally optimal tests for multiparameter hypotheses are given in Gupta and Vermeire (1986).

[6] Due to John Pratt.

Problem 8.3 A level-α test φ_0 is locally unbiased (loc. unb.) if there exists $\Delta_0 > 0$ such that $\beta_{\varphi_0}(\theta) \geq \alpha$ for all θ with $0 < d(\theta) < \Delta_0$; it is LMP loc. unb. if it is loc. unb. and if, given any other loc. unb. level-α test φ, there exists Δ such that (8.38) holds. Suppose that θ is real-valued and that $d(\theta) = |\theta - \theta_0|$, and that the power function of every test is twice continuously differentiable at $\theta = \theta_0$.

(i) If there exists a unique test φ_0 of $H : \theta = \theta_0$ against $K : \theta \neq \theta_0$ which among all loc. unb. tests maximizes $\beta''(\theta_0)$, then φ_0 is the unique LMP loc. unb. level-α test of H against K.

(ii) The test of part (i) maximizes the minimum power locally provided its power function is bounded away from α for every set of alternatives that is bounded away from H.

[(ii): A necessary condition for a test to be locally minimax is that it is loc. unb.]

Problem 8.4 *Locally uniformly most powerful tests.* If the sample space is finite and independent of θ, the test φ_0 of Problem 8.2(i) is not only LMP but also locally uniformly most powerful (LUMP) in the sense that there exists a value $\Delta > 0$ such that φ_0 maximizes $\beta_\varphi(\theta)$ for all θ with $0 < \theta - \theta_0 < \Delta$.
[See the argument following (6.21) of Section 6.9.]

Problem 8.5 The following two examples show that the assumption of a finite sample space is needed in Problem 8.4.

(i) Let $X_1, \ldots, X_n$ be i.i.d. according to a normal distribution $N(\sigma, \sigma^2)$ and test $H : \sigma = \sigma_0$ against $K : \sigma > \sigma_0$.

(ii) Let X and Y be independent Poisson variables with $E(X) = \lambda$ and $E(Y) = \lambda + 1$, and test $H : \lambda = \lambda_0$ against $K : \lambda > \lambda_0$. In each case, determine the LMP test and show that it is not LUMP.

[Compare the LMP test with the most powerful test against a simple alternative.]

Section 8.2

Problem 8.6 Let the distribution of X depend on the parameters $(\theta, \vartheta) = (\theta_1, \ldots, \theta_r, \vartheta_1, \ldots, \vartheta_s)$. A test of $H : \theta = \theta^0$ is *locally strictly unbiased* if for each φ, (a) $\beta_\varphi(\theta^0, \varphi) = \alpha$, (b) there exists a θ-neighborhood of θ^0 in which $\beta_\varphi(\theta, \vartheta) > \alpha$ for $\theta \neq \theta^0$.

(i) Suppose that the first and second derivatives

$$\beta_\varphi^i(\vartheta) = \frac{\partial}{\partial\theta_i}\beta_\varphi(\theta, \vartheta)\bigg|_{\theta^0} \quad \text{and} \quad \beta_\varphi^{ij}(\vartheta) = \frac{\partial^2}{\partial\theta_i\partial\theta_j}\beta_\varphi(\theta, \vartheta)\bigg|_{\theta^0}$$

exist for all critical functions φ and all ϑ. Then a necessary and sufficient condition for φ to be locally strictly unbiased is that $\beta_\varphi' = 0$ for all i and ϑ, and that the matrix $(\beta_\varphi^{ij}(\vartheta))$ is positive definite for all ϑ.

(ii) A test of H is said to be of *type E* (*type D* is $s = 0$ so that there are no nuisance parameters) if it is locally strictly unbiased and among all tests with this property maximizes the determinant $|(\beta_\varphi^{ij})|$.[7] (This determinant under the stated conditions turns out to be equal to the Gaussian curvature of the power surface at θ^0.) Then the test φ_0 given by (7.7) for testing the general linear univariate hypothesis (7.3) is of type E.

[(ii): With $\theta = (\eta_1, \ldots, \eta_r)$ and $\vartheta = (\eta_{r+1}, \ldots, n_s, \sigma)$, the test φ_0, by Problem 7.5, has the property of maximizing the surface integral

$$\int_S [\beta_\varphi(\eta, \sigma^2) - \alpha]\, dA$$

among all similar (and hence all locally unbiased) tests where $S = \{(\eta_1, \ldots, \eta_r) : \sum_{i=1}^r \eta_i^2 = \rho^2\sigma^2\}$. Letting ρ tend to zero and utilizing the conditions

$$\beta_\varphi^i(\vartheta) = 0, \qquad \int_S \eta_i \eta_j\, dA = 0 \quad \text{for } i \neq j, \qquad \int_S \eta_i^2\, dA = k(\rho\sigma),$$

one finds that φ_0 maximizes $\sum_{i=1}^r \beta_\varphi^{ii}(\eta, \sigma^2)$ among all locally unbiased tests. Since for any positive definite matrix, $|(\beta_\varphi^{ij})| \leq \prod \beta_\varphi^{ii}$, it follows that for any locally strictly unbiased test φ,

$$|(\beta_\varphi^{ij})| \leq \prod \beta_\varphi^{ii} \leq \left[\frac{\Sigma \beta_\varphi^{ii}}{r}\right]^r \leq \left[\frac{\Sigma \beta_{\varphi_0}^{ii}}{r}\right]^r = [\beta_{\varphi_0}^{11}]^r = |(\beta_{\varphi_0}^{ij})|.]$$

Problem 8.7 Let $Z_1, \ldots, Z_n$ be identically independently distributed according to a continuous distribution D, of which it is assumed only that it is symmetric about some (unknown) point. For testing the hypothesis $H : D(0) = \frac{1}{2}$, the sign test maximizes the minimum power against the alternatives $K : D(0) \leq q (q < \frac{1}{2})$. [A pair of least favorable distributions assign probability 1 respectively to the distributions $F \in H$, $G \in K$ with densities

$$f(x) = \frac{1 - 2q}{2(1-q)} \left(\frac{q}{1-q}\right)^{[|x|]}, \qquad g(x) = (1 - 2q) \left(\frac{q}{1-q}\right)^{|[x]|}$$

where for all x (positive, negative, or zero) $[x]$ denotes the largest integer $\leq x$.]

Problem 8.8 Let $f_\theta(x) = \theta g(x) + (1 - \theta) h(x)$ with $0 \leq \theta \leq 1$. Then $f_\theta(x)$ satisfies the assumptions of Lemma 8.2.1 provided $g(x)/h(x)$ is a nondecreasing function of x.

Problem 8.9 Let $x = (x_1, \ldots, x_n)$, and let $g_\theta(x, \xi)$ be a family of probability densities depending on $\theta = (\theta_1, \ldots, \theta_r)$ and the real parameter ξ, and jointly measurable in x and ξ. For each θ, let $h_\theta(\xi)$ be a probability density with respect to a σ-finite measure ν such that $p_\theta(x) = \int g_\theta(x, \xi) h_\theta(\xi)\, d\nu(\xi)$ exists. We shall say that a function f of two arguments $u = (u_1, \ldots, u_r)$, $v = (v_1, \ldots, v_s)$ is nondecreasing in (u, v) if $f(u', v)/f(u, v) \leq f(u', v')/f(u, v')$ for all (u, v) satisfying $u_i \leq u_i'$, $v_j \leq v_j'$ $(i = 1, \ldots, r; j = 1, \ldots, s)$. Then $p_\theta(x)$ is nondecreasing in (x, θ) provided the product $g_\theta(x, \xi) h_\theta(\xi)$ is (a) nondecreasing in (x, θ) for each fixed ξ;

[7] An interesting example of a type-D test is provided by Cohen and Sackrowitz (1975), who show that the χ^2-test of Chapter 14.3 has this property. Type D and E tests were introduced by Isaacson (1951).

(b) nondecreasing in (θ, ξ) for each fixed x; (c) nondecreasing in (x, ξ) for each fixed θ.
[Interpreting $g_\theta(x, \xi)$ as the conditional density of x given ξ, and $h_\theta(\xi)$ as the a priori density of ξ, let $\rho(\xi)$ denote the a posteriori density of ξ given x, and let $\rho'(\xi)$ be defined analogously with θ' in place of θ. That $p_\theta(x)$ is nondecreasing in its two arguments is equivalent to

$$\int \frac{g_\theta(x', \xi)}{g_\theta(x, \xi)} \rho(\xi) \, d\nu(\xi) \leq \int \frac{g_{\theta'}(x', \xi)}{g_{\theta'}(x, \xi)} \rho'(\xi) \, d\nu(\xi).$$

By (a) it is enough to prove that

$$D = \int \frac{g_\theta(x', \xi)}{g_\theta(x, \xi)} [\rho'(\xi) - \rho(\xi)] \, d\nu(\xi) \geq 0.$$

Let $S_- = \{\xi : \rho'(\xi)/\rho(\xi) < 1\}$ and $S_+ = \{\xi : \rho(\xi)/\rho(\xi) \geq 1\}$. By (b) the set S_- lies entirely to the left of S_+. It follows from (c) that there exists $a \leq b$ such that

$$D = a \int_{S_-} [\rho'(\xi) - \rho(\xi)] \, d\nu(\xi) + b \int_{S_+} [\rho'(\xi) - \rho(\xi)] \, d\nu(\xi),$$

and hence that $D = (b - a) \int_{S_+} [\rho'(\xi) - \rho(\xi)] \, d\nu(\xi) \geq 0$.]

Problem 8.10 (i) Let X have binomial distribution $b(p, n)$, and consider testing $H : p = p_0$ at level α against the alternatives $\Omega_K : p/q \leq \frac{1}{2} p_0/q_0$ or $\geq 2p_0/q_0$. For $\alpha = .05$ determine the smallest sample size for which there exists a test with power $\geq .8$ against Ω_K if $p_0 = .1, .2, .3, .4, .5$.

(ii) Let $X_1, \ldots, X_n$ be independently distributed as $N(\xi, \sigma^2)$. For testing $\sigma = 1$ at level $\alpha = .05$, determine the smallest sample size for which there exists a test with power $\geq .9$ against the alternatives $\sigma^2 \leq \frac{1}{2}$ and $\sigma^2 \geq 2$.

[See Problem 4.5.]

Problem 8.11 *Double-exponential distribution.* Let $X_1, \ldots, X_n$ be a sample from the double-exponential distribution with density $\frac{1}{2} e^{-|x-\theta|}$. The LMP test for testing $\theta \leq 0$ against $\theta > 0$ is the sign test, provided the level is of the form

$$\alpha = \frac{1}{2^n} \sum_{k=0}^{m} \binom{n}{k},$$

so that the level-α sign test is nonrandomized.
[Let R_k $(k = 0, \ldots, n)$ be the subset of the sample space in which k of the X's are positive and $n - k$ are negative. Let $0 \leq k < l < n$, and let S_k, S_l be subsets of R_k, R_l such that $P_0(S_k) = P_0(S_l) \neq 0$. Then it follows from a consideration of $P_\theta(S_k)$ and $P_0(S_l)$ for small θ that there exists Δ such that $P_\theta(S_k) < P_\theta(S_l)$ for $0 < \theta < \Delta$. Suppose now that the rejection region of a nonrandomized test of $\theta = 0$ against $\theta > 0$ does not consist of the upper tail of a sign test. Then it can be converted into a sign test of the same size by a finite number of steps, each of which consists in replacing an S_k by an S_l with $k < l$, and each of which therefore increases the power for θ sufficiently small.]

Section 8.3

Problem 8.12 If (8.13) holds, show that q_1 defined by (8.11) belongs to $\mathcal{P}_1$.

Problem 8.13 Show that there exists a unique constant b for which q_0 defined by (8.11) is a probability density with respect to μ, that the resulting q_0 belongs to $\mathcal{P}_0$, and that $b \to \infty$ as $\epsilon_0 \to 0$.

Problem 8.14 Prove the formula (8.15).

Problem 8.15 Show that if $\mathcal{P}_0 \neq \mathcal{P}_1$ and ϵ_0, ϵ_1 are sufficiently small, then $Q_0 \neq Q_1$.

Problem 8.16 Evaluate the test (8.21) explicitly for the case that P_i is the normal distribution with mean ξ_i and known variance σ^2, and when $\epsilon_0 = \epsilon_1$.

Problem 8.17 Determine whether (8.21) remains the maximin test if in the model (8.20) G_i is replaced by G_{ij}.

Problem 8.18 Write out a formal proof of the maximin property outlined in the last paragraph of Section 8.3.

Section 8.4

Problem 8.19 Let $X_1, \ldots, X_n$ be independently normally distributed with means $E(X_i) = \mu_i$ and variance 1. The test of $H : \mu_1 = \cdots = \mu_n = 0$ that maximizes the minimum power over $\omega' : \sum \mu_i \geq d$ rejects when $\sum X_i \geq C$.
[If the least favorable distribution assigns probability 1 to a single point, invariance under permutations suggests that this point will be $\mu_1 = \cdots = \mu_n = d/n$].

Problem 8.20 [8] (i) In the preceding problem determine the maximin test if ω' is replaced by $\sum a_i \mu_i \geq d$, where the a's are given positive constants.

(ii) Solve part (i) with $Var(X_i) = 1$ replaced by $Var(X_i) = \sigma_i^2$ (known).
[(i): Determine the point $(\mu_1^*, \ldots, \mu_n^*)$ in ω' for which the MP test of H against $K : (\mu_1^*, \ldots, \mu_n^*)$ has the smallest power, and show that the MP test of H against K is a maximin solution.]

Problem 8.21 Let $X_1, \ldots, X_n$ be independent normal variables with variance 1 and means $\xi_1, \ldots, \xi_n$, and consider the problem of testing $H : \xi_1 = \cdots = \xi_n = 0$ against the alternatives $K = \{K_1, \ldots, K_n\}$, where $K_i : \xi_j = 0$ for $j \neq i$, $\xi_i = \xi$ (known and positive). Show that the problem remains invariant under permutation of the X's and that there exists a UMP invariant test ϕ_0 which rejects when $\sum e^{-\xi x_j} > C$, by the following two methods.

(i) The order statistics $X_{(1)} < \cdots < X_{(n)}$ constitute a maximal invariant.

[8]Due to Fritz Scholz.

(ii) Let f_0 and f_i denote the densities under H and K_i respectively. Then the level-α test ϕ_0 of H vs. $K' : f = (1/n) \sum f_i$ is UMP invariant for testing H vs. K.

[(ii): If ϕ_0 is not UMP invariant for H vs. K, there exists an invariant test ϕ_1 whose (constant) power against K exceeds that of ϕ_0. Then ϕ_1 is also more powerful against K'.]

Problem 8.22 The UMP invariant test ϕ_0 of Problem 8.21

(i) maximizes the minimum power over K;

(ii) is admissible.

(iii) For testing the hypothesis H of Problem 8.21 against the alternatives $K' = \{K_1, \dots, K_n, K'_1, \dots, K'_n\}$, where under $K'_i : \xi_j = 0$ for all $j \neq i$, $\xi_i = -\xi$, determine the UMP test under a suitable group G', and show that it is both maximin and invariant.

[ii): Suppose ϕ' is uniformly at least as powerful as ϕ_0, and more powerful for at least one K_i, and let

$$\phi^*(x_1, \dots, x_n) = \frac{\sum \phi'(x_{i_1}, \dots, x_{i_n})}{n!},$$

where the summation extends over all permutations. Then ϕ^* is invariant, and its power is independent of i and exceeds that of ϕ_0.]

Problem 8.23 For testing $H : f_0$ against $K : \{f_1, \dots, f_s\}$, suppose there exists a finite group $G = \{g_1, \dots, g_N\}$ which leaves H and K invariant and which is transitive in the sense that given f_j, $f_{j'} (1 \leq j, j')$ there exists $g \in G$ such that $\bar{g} f_j = f_{j'}$. In generalization of Problems 8.21, 8.22, determine a UMP invariant test, and show that it is both maximin against K and admissible.

Problem 8.24 To generalize the results of the preceding problem to the testing of $H : f$ vs. $K : \{f_\theta, \theta \in \omega\}$, assume:

(i) There exists a group G that leaves H and K invariant.

(ii) $\bar{G}$ is transitive over ω.

(iii) There exists a probability distribution Q over G which is right-invariant in the sense of Section 8.4.

Determine a UMP invariant test, and show that it is both maximin against K and admissible.

Problem 8.25 Let $X_1, \dots, X_n$ be independent normal with means $\theta_1, \dots, \theta_n$ and variance 1.

(i) Apply the results of the preceding problem to the testing of $H : \theta_1 = \cdots = \theta_n = 0$ against $K : \sum \theta_i^2 = r^2$, for any fixed $r > 0$.

(ii) Show that the results of (i) remain valid if H and K are replaced by $H' : \sum \theta_i^2 \leq r_0^2$, $K' : \sum \theta_i^2 \geq r_1^2$ $(r_0 < r_1)$.

Problem 8.26 Suppose in Problem 8.25(i) the variance σ^2 is unknown and that the data consist of $X_1, \ldots, X_n$ together with an independent random variable S^2 for which S^2/σ^2 has a χ^2-distribution. If K is replaced by $\sum \theta_i^2/\sigma^2 = r^2$, then

(i) the confidence sets $\sum(\theta_i - X_i)^2/S^2 \leq C$ are uniformly most accurate equivariant under the group generated by the n-dimensional generalization of the group G_0 of Example 6.11.2, and the scale changes $X_i' = cX_i$, $S'^2 = c^2S^2$.

(ii) The confidence sets of (i) are minimax with respect to the measure μ given by

$$\mu[C(X, S^2)] = \frac{1}{\sigma^2}[\text{ volume of } C(X, S^2)].$$

[Use polar coordinates with $\theta^2 = \sum \theta_i^2$.]

Section 8.5

Problem 8.27 Let $X = (X_1, \ldots, X_p)$ and $Y = (Y_1, \ldots, Y_p)$ be independently distributed according to p-variate normal distributions with zero means and covariance matrices $E(X_iX_j) = \sigma_{ij}$ and $E(Y_iY_j) = \Delta\sigma_{ij}$.

(i) The problem of testing $H : \Delta \leq \Delta_0$ remains invariant under the group G of transformations $X^* = XA$, $Y^* = YA$, where $A = (a_{ij})$ is any nonsingular $p \times p$ matrix with $a_{ij} = 0$ for $i > j$, and there exists a UMP invariant test under G with rejection region $Y_1^2/X_1^2 > C$.

(ii) The test with rejection region $Y_1^2/X_1^2 > C$ maximizes the minimum power for testing $\Delta \leq \Delta_0$ against $\Delta \geq \Delta_1$ $(\Delta_0 < \Delta_1)$.
[(ii): That the Hunt–Stein theorem is applicable to G can be proved in steps by considering the group G_q of transformations $X_q' = \alpha_1X_1 + \cdots + \alpha_qX_q$, $X_i' = X_i$ for $i = 1, \ldots, q-1, q+1, \ldots, p$, successively for $q = 1, \ldots, p-1$. Here $\alpha_q \neq 0$, since the matrix A is nonsingular if and only if $a_{ii} \neq 0$ for all i. The group product $(\gamma_1, \ldots, \gamma_q)$ of two such transformations $(\alpha_1, \ldots, \alpha_q)$ and $(\beta_1, \ldots, \beta_q)$ is given by $\gamma_1 = \alpha_q + \beta_1$, $\gamma_2 = a_2\beta_q + \beta_2, \ldots,$ $\gamma_{q-1} = \alpha_{q-1}\beta_q + \beta_{q-1}$, $\gamma_q = \alpha_q, \beta_q$, which shows G_q to be isomorphic to a group of scale changes (multiplication of all components by β_q) and translations [addition of $(\beta_1, \ldots, \beta_{q-1}, 0)$]. The result now follows from the Hunt–Stein theorem and Example 8.5.1, since the assumptions of the Hunt–Stein theorem, except for the easily verifiable measurability conditions, concern only the abstract structure $(G, \mathcal{B})$, and not the specific realization of the elements of G as transformations of some space.]

Problem 8.28 Suppose that the problem of testing $\theta \in \Omega_H$ against $\theta \in \Omega_K$ remains invariant under G, that there exists a UMP almost invariant test φ_0 with respect to G, and that the assumptions of Theorem 8.5.1 hold. Then φ_0 maximizes $\inf_{\Omega_K}[w(\theta)E_\theta\varphi(X) + u(\theta)]$ for any weight functions $w(\theta) \geq 0$, $u(\theta)$ that are invariant under $\bar{G}$.

Problem 8.29 Suppose X has the multivariate normal distribution in $\mathbf{R}^k$ with unknown mean vector h and known positive definite covariance matrix C^{-1}.

Consider testing $h = 0$ versus $|C^{1/2}h| \geq b$ for some $b > 0$, where $|\cdot|$ denotes the Euclidean norm.
(i) Show the test that rejects when $|C^{1/2}X|^2 > c_{k,1-\alpha}$ is maximin, where $c_{k,1-\alpha}$ denotes the $1-\alpha$ quantile of the Chi-squared distribution with k degrees of freedom.
(ii) Show that the maximin power of the above test is given $P\{\chi_k^2(b^2) > c_{k,1-\alpha}\}$, where $\chi_k^2(b^2)$ denotes a random variable that has the noncentral Chi-squared distribution with k degrees of freedom and noncentrality parameter b^2.

Problem 8.30 Suppose $X_1, \ldots, X_k$ are independent, with $X_i \sim N(\theta_i, 1)$. Consider testing the null hypothesis $\theta_1 = \cdots = \theta_k = 0$ against $\max|\theta_i| \geq \delta$, for some $\delta > 0$. Find a maximin level α test as explicitly as possible. Compare this test with the maximin test if the alternative parameter space were $\sum_i \theta_i^2 \geq \delta^2$. Argue they are quite similar for small δ. Specifically, consider the power of each test against $(\delta, 0, \ldots, 0)$ and show that it is equal to $\alpha + C_\alpha \delta^2 + o(\delta^2)$ as $\delta \to 0$, and the constant C_α is the same for both tests.

Section 8.6

Problem 8.31 *Existence of most stringent tests.* Under the assumptions of Problem 8.1 there exists a most stringent test for testing $\theta \in \Omega_H$ against $\theta \in \Omega - \Omega_H$.

Problem 8.32 Let $\{\Omega_\Delta\}$ be a class of mutually exclusive sets of alternatives such that the envelope power function is constant over each Ω_Δ and that $\cup\Omega_\Delta = \Omega - \Omega_H$, and let φ_Δ maximize the minimum power over Ω_Δ. If $\varphi_\Delta = \varphi$ is independent of Δ, then φ is most stringent for testing $\theta \in \Omega_H$.

Problem 8.33 Let $(Z_1, \ldots, Z_N) = (X_1, \ldots, X_m, Y_1, \ldots, Y_n)$ be distributed according to the joint density (5.55), and consider the problem of testing $H : \eta = \xi$ against the alternatives that the X's and Y's are independently normally distributed with common variance σ^2 and means $\eta \neq \xi$. Then the permutation test with rejection region $|\bar{Y} - \bar{X}| > C[T(Z)]$, the two-sided version of the test (5.54), is most stringent.
[Apply Problem 8.32 with each of the sets Ω_Δ consisting of two points (ξ_1, η_1, σ), (ξ_2, η_2, σ) such that

$$\xi_1 = \zeta - \frac{n}{m+n}\delta, \qquad \eta_1 = \zeta + \frac{m}{m+n}\delta;$$
$$\xi_2 = \zeta + \frac{n}{m+n}\delta, \qquad \eta_2 = \zeta - \frac{m}{m+n}\delta$$

for some ζ and δ.]

Problem 8.34 Show that the UMP invariant test of Problem 8.21 is most stringent.

8.8 Notes

The concepts and results of Section 8.1 are essentially contained in the minimax theory developed by Wald for general decision problems. An exposition of this theory and some of its applications is given in Wald's book (1950). For more recent assessments of the important role of the minimax approach, see Brown (1994, 2000). The ideas of Section 8.3, and in particular Theorem 8.3.1, are due to Huber (1965) and form the core of his theory of robust tests [Huber (1981, Chapter 10)]. The material of sections 8.4 and 8.5, including Lemma 8.4.1, Theorem 8.5.1, and Example 8.5.2, constitutes the main part of an unpublished paper of Hunt and Stein (1946).

9

Multiple Testing and Simultaneous Inference

9.1 Introduction and the FWER

When testing more than one parameter, say

$$H: \theta_1 = \cdots = \theta_s = 0 \tag{9.1}$$

against the alternatives that one or more of the θ's are positive, it is typically not enough simply to accept or reject H. In case of acceptance, nothing more is required: the finding is that none of the parameter values are significant. However, when H is rejected, one will in most cases want to know just which of the parameters θ are significant. And when H is tested against the two-sided alternatives that one or more of the θ's are different from 0, one would in case of rejection usually want to know the signs of the significant θ's.[1]

Example 9.1.1 (Normal one-sample problem) Suppose that $X_1, \ldots, X_n$ is a sample from $N(\xi, \sigma^2)$ and consider the hypothesis H: $\xi \le \xi_0, \sigma \le \sigma_0$. In case of rejection one would want to know whether it is the mean or the variance that is rejected, or perhaps both. ■

Example 9.1.2 (Comparing several treatments with a control) When testing several treatments against a control, the overall null hypothesis states that none of the treatments is an improvement over, or differs from, the control. In case of rejection one will wish to know just which of the treatments show a significant difference. ■

---

[1]We shall here disregard this latter issue, but see Comment 2 at the end of Section 9.3.

Example 9.1.3 (Testing equality of several treatments) Instead of comparing several treatments with a control, one may wish to compare a number of possible alternative situations with each other. If the quality of the ith of s alternatives is measured by a parameter θ_i, the hypothesis is

$$H\colon \theta_1 = \cdots = \theta_s \ . \ \blacksquare \tag{9.2}$$

Since most multiple testing problems, like those in Examples 9.1.2 and 9.1.3, are concerned with *multiple comparisons*, the whole subject of multiple testing is frequently, and somewhat inaccurately, called multiple comparisons.

When comparing several medical, agricultural, or industrial treatments, the numbers of treatments is typically fairly small, say, in the single digits. Larger numbers occur in some educational studies, where for example it may be desired to compare performance in the 50 of the U.S. states. A fairly recent application of multiple comparison theory occurs in microarrays where thousands or even tens of thousands of genes are tested simultaneously. Each microarray corresponds to one unit (plant, animal or person) and in these experiments the sample size (the number of such units) is typically of a much smaller order of magnitude (in the tens) than the number of comparisons being tested.

Let us now consider the general problem of simultaneously testing a finite numbers of hypotheses H_i $(i = 1, \ldots, s)$. We shall assume that tests for the individual hypotheses are available and the problem is how to combine them into a simultaneous test procedure.

The easiest approach is to disregard the multiplicity and simply test each hypothesis at level α. However, with such a procedure the probability of one or more false rejections rapidly increases with s. When the number of true hypotheses is large, we shall be nearly certain to reject some of them. To get a numerical idea of this phenomenon, the following Table shows (to 2 decimals) the probability of one or more false rejections when all of the hypotheses $H_1, \ldots, H_s$ are true, when the test statistics used for testing $H_1, \ldots, H_s$ are independent, and when the level at which each of the s hypotheses is tested is $\alpha = .05$.

s	1	2	5	10	50
P(at least one false rejection)	.05	.10	.23	.40	.92

In this sense the claim that the procedure controls the probability of false rejections at level .05 is clearly very misleading.

We shall therefore in the present chapter replace the usual condition for testing a single hypothesis, that the probability of a false rejection not exceed α, by the requirement, when testing several hypotheses, that the probability of one or more false rejections, not exceed a given level. This probability is called the family-wise error rate (FWER). Here the term "family" refers to the collection of hypotheses $H_1, \ldots, H_s$ that is being considered for joint testing. In a laboratory testing blood samples, this might be all the tests performed in a day, or those performed in a day by a given tester. Alternatively, the tests given in the morning and afternoon might be considered as separate families, and so on. Which tests are to be treated jointly as a family depends on the situation.

Once the family has been defined, we shall require that

$$FWER \leq \alpha \tag{9.3}$$

for all possible constellations of true and false hypotheses. This is sometimes called *strong* error control to distinguish it from the much weaker (and typically not very meaningful) condition of *weak* control which requires (9.3) to hold only when all the hypotheses of the family are true.

Methods that control the FWER are often described by the p-values of the individual tests, which were introduced in Section 3.2. We now present two simple methods that control the FWER which can be stated easily in terms of p-values. Each hypothesis H_i can be viewed as a subset, ω_i, of Ω. Assume that $\hat{p}_i$ is a p-value for testing H_i; specifically, we assume

$$P\{\hat{p}_i \leq u\} \leq u \tag{9.4}$$

for any $u \in (0,1)$ and any $P \in \omega_i$. Note that it is not required that the distribution of $\hat{p}_i$ be uniform on $(0,1)$ whenever H_i is true. (For example, if H_i corresponds to testing $\theta_i \leq 0$ but the true θ_i is < 0, exact uniformity is too strong. Also, even if the null hypothesis is simple, the p-value may have a discrete distribution.)

Theorem 9.1.1 *(Bonferroni Procedure) If, for $i = 1, \ldots, s$, hypothesis H_i is rejected when $\hat{p}_i \leq \alpha/s$, then the FWER for the simultaneous testing of $H_1, \ldots, H_s$ satisfies (9.3).*

PROOF. Suppose hypotheses H_i with $i \in I$ are true and the remainder false, with $|I|$ denoting the cardinality of I. From the Bonferroni inequality it follows that

$$FWER = P\{\text{reject any } H_i \text{ with } i \in I\} \leq \sum_{i \in I} P\{\text{reject } \ H_i\}$$

$$= \sum_{i \in I} P\{\hat{p}_i \leq \frac{\alpha}{s}\} \leq \sum_{i \in I} \frac{\alpha}{s} \leq |I|\alpha/s \leq \alpha \ . \ \blacksquare$$

While such Bonferroni based procedures satisfactorily control the FWER, their ability to detect cases in which H_i is false will typically be very low since H_i is tested at level α/s which - particularly if s is large - is orders smaller than the conventional α levels.

For this reason procedures are prized for which the levels of the individual tests are increased over α/s without an increase in the FWER. It turns out that such a procedure due to Holm (1979) is available under the present minimal assumptions.

The Holm procedure can conveniently be stated in terms of the p-values $\hat{p}_1, \ldots, \hat{p}_s$ of the s individual tests. Let the ordered p-values be denoted by $\hat{p}_{(1)} \leq \ldots \leq \hat{p}_{(s)}$, and the associated hypotheses by $H_{(1)}, \ldots, H_{(s)}$. Then the Holm procedure is defined stepwise as follows:

Step 1. If $\hat{p}_{(1)} \geq \alpha/s$, accept $H_1, \ldots, H_s$ and stop. If $\hat{p}_{(1)} < \alpha/s$ reject $H_{(1)}$ and test the remaining $s-1$ hypotheses at level $\alpha/(s-1)$.

Step 2. If $\hat{p}_{(1)} < \alpha/s$ but $\hat{p}_{(2)} \geq \alpha/(s-1)$, accept $H_{(2)}, \ldots, H_{(s)}$ and stop. If $\hat{p}_{(1)} < \alpha/s$ and $\hat{p}_{(2)} < \alpha/(s-1)$, reject $H_{(2)}$ in addition to $H_{(1)}$ and test the remaining $s-2$ hypotheses at level $\alpha/(s-2)$.

And so on.

Theorem 9.1.2 *The Holm procedure satisfies (9.3).*

Proof. Suppose H_i with $i \in I$ is the set of true hypotheses, so $P \in \omega_i$ if and only if $i \in I$. Let j be the smallest (random) index satisfying

$$\hat{p}_{(j)} = \min_{i \in I} \hat{p}_i \ .$$

Note that $j \le s - |I| + 1$. Now, the Holm procedure commits a false rejection if

$$\hat{p}_{(1)} \le \alpha/s, \hat{p}_{(2)} \le \alpha/(s-1), \ldots, \hat{p}_{(j)} \le \alpha/(s-j+1) \ ,$$

which certainly implies that

$$\min_{i \in I} \hat{p}_i = \hat{p}_{(j)} \le \alpha/(s-j+1) \le \alpha/|I| \ .$$

Therefore, by the Bonferroni inequality, the probability of a false rejection is bounded above by

$$P\{\min_{i \in I} \hat{p}_i \le \alpha/|I|\} \le \sum_{i \in I} P\{\hat{p}_i \le \alpha/|I|\} \le \alpha \ . \blacksquare$$

The Bonferroni method is an example of a *single-step* procedure, meaning any hypothesis is rejected if its corresponding p-value is less than a common cutoff value (which in the Bonferroni case is α/s). The Holm procedure is a special case of a class of *stepdown* procedures, which we now briefly describe. Roughly speaking, stepdown procedures begin by determining whether the test that looks most significant should be rejected. If each individual test is summarized by a p-value, this can be described as follows. Let

$$\alpha_1 \le \alpha_2 \le \cdots \le \alpha_s \tag{9.5}$$

be constants. If $\hat{p}_{(1)} \ge \alpha_1$, accept all hypotheses. Otherwise, for $r = 1, \ldots, s$, reject hypotheses $H_{(1)}, \ldots, H_{(r)}$ if

$$\hat{p}_{(1)} < \alpha_1, \ldots, \hat{p}_{(r)} < \alpha_r \ . \tag{9.6}$$

That is, a stepdown procedure starts with the most significant p-value and continues rejecting hypotheses as long as their corresponding p-values are small. The Holm procedure uses $\alpha_i = \alpha/(s-i+1)$. (Alternatively, if the rejection region of each test corresponds to large value of a test statistic, a stepdown procedure begins by determining whether or not the hypothesis corresponding to the largest test statistic should be rejected; see Procedure 9.1.1 below.)

On the other hand, *stepup* procedures begin by looking at the least significant p-value (or the smallest value of a test statistic when the individual tests reject for large values). For a given set of constants (9.5), reject all hypotheses if $\hat{p}_{(s)} < \alpha_s$. Otherwise, for $r = s, \ldots, 1$, reject hypotheses $H_{(1)}, \ldots, H_{(r)}$ if

$$\hat{p}_{(s)} \ge \alpha_s, \ldots, \hat{p}_{(r+1)} \ge \alpha_{r+1} \ \text{ but } \ \hat{p}_{(r)} < \alpha_r \ . \tag{9.7}$$

Safeguards against false rejections are of course not the only concern of multiple testing procedures. Corresponding to the power of a single test one must also consider the ability of a multiple test procedure to detect departures from the hypotheses when they do occur. For certain parametric models, optimality results for some stepwise procedures will be developed in the next section. For now, we show that it is possible to improve upon the Holm method by incorporating the dependence structure of the individual tests.

To see how, suppose that a test of the individual hypothesis H_j is based on a test statistic $T_{n,j}$, with large values indicating evidence against H_j. (The use of the subscript n in the test statistics will be for asymptotic purposes later on.)

If P is the true probability distribution generating the data, let $I = I(P) \subset \{1, \ldots, s\}$ denote the indices of the set of true hypotheses; that is, $i \in I$ if and only $P \in \omega_i$. For $K \subset \{1, \ldots, s\}$, let H_K denote the intersection hypothesis that all H_i with $i \in K$ are true; that is, H_K is equivalent to $P \in \bigcap_{i \in K} \omega_i$. In order to improve upon the Holm method, the basic idea is to use critical values that more accurately approximate the distribution of $\max_{j \in K} T_{n,j}$ when testing H_K, at least when K is in fact true. Let

$$T_{n,r_1} \geq T_{n,r_2} \geq \cdots \geq T_{n,r_s} \tag{9.8}$$

denote the observed ordered test statistics, and let $H_{(1)}, H_{(2)}, \ldots, H_{(s)}$ be the corresponding hypotheses. A stepdown procedure begins with the most significant test statistic. First, test the joint null hypothesis $H_{\{1,\ldots,s\}}$ that all hypotheses are true. This hypothesis is rejected if T_{n,r_1} is large. If it is not large, accept all hypotheses; otherwise, reject the hypothesis corresponding to the largest test statistic. Once a hypothesis is rejected, remove it and test the remaining hypotheses by rejecting for large values of the maximum of the remaining test statistics, and so on. To be specific, consider the following generic procedure, based on critical values $\hat{c}_{n,K}(1-\alpha)$, where $\hat{c}_{n,K}(1-\alpha)$ is designed for testing the intersection hypothesis H_K at nominal level α. Although we are not specifying the constants at this point, we note that they could be nonrandom or data-dependent.

Procedure 9.1.1 (Generic Stepdown Method)

1. Let $K_1 = \{1, \ldots, s\}$. If $T_{n,r_1} \leq \hat{c}_{n,K_1}(1-\alpha)$, then accept all hypotheses and stop; otherwise, reject $H_{(1)}$ and continue.

2. Let K_2 be the indices of the hypotheses not previously rejected. If $T_{n,r_2} \leq \hat{c}_{n,K_2}(1-\alpha)$, then accept all remaining hypotheses and stop; otherwise, reject $H_{(2)}$ and continue.

   ⋮

j. Let K_j be the indices of the hypotheses not previously rejected. If $T_{n,r_j} \leq \hat{c}_{n,K_j}(1-\alpha)$, then accept all remaining hypotheses and stop; otherwise, reject $H_{(j)}$ and continue.

   ⋮

s. If $T_{n,s} \leq \hat{c}_{n,K_s}(1-\alpha)$, then accept $H_{(s)}$; otherwise, reject $H_{(s)}$.

The problem now is how to construct the $\hat{c}_{n,K}(1-\alpha)$ so that the FWER is controlled. The following result reduces the multiple testing problem of controlling the FWER to that of constructing single tests that control the probability of a Type 1 error.

Theorem 9.1.3 *Let P denote the true distribution generating the data. Consider Procedure 9.1.1 based on critical values $\hat{c}_{n,K}(1-\alpha)$ which satisfy the monotonicity requirement: for any $K \supset I(P)$,*

$$\hat{c}_{n,K}(1-\alpha) \geq \hat{c}_{n,I(P)}(1-\alpha) . \tag{9.9}$$

(i) Then,

$$FWER_P \leq P\{\max(T_{n,j} : j \in I(P)) > \hat{c}_{n,I(P)}(1-\alpha)\} . \tag{9.10}$$

(ii) Also suppose that if $\hat{c}_{n,K}(1-\alpha)$ is used to test the intersection hypothesis H_K, then it is level α when $K = I(P)$; that is,

$$P\{\max(T_{n,j} : j \in I(P)) > \hat{c}_{n,I(P)}(1-\alpha)\} \leq \alpha . \tag{9.11}$$

Then $FWER_P \leq \alpha$.

PROOF. Consider the event that a true hypothesis is rejected, so that for some $i \in I(P)$, hypothesis H_i is rejected. Let $\hat{j}$ be the smallest index j in the method where this occurs, so that

$$\max\{T_{n,j} : j \in I(P)\} > \hat{c}_{n,K_{\hat{j}}}(1-\alpha) . \tag{9.12}$$

Since $K_{\hat{j}} \supset I(P)$, assumption (9.9) implies

$$\hat{c}_{n,K_{\hat{j}}}(1-\alpha) \geq \hat{c}_{n,I(P)}(1-\alpha) \tag{9.13}$$

and so (i) follows. Part (ii) follows immediately from (i). ■

Example 9.1.4 (Multivariate Normal Mean) Suppose $(X_1, \ldots, X_s)$ is multivariate normal with unknown mean $\mu = (\mu_1, \ldots, \mu_s)$ and known covariance matrix Σ having (i,j) component $\sigma_{i,j}$. Consider testing $H_j : \mu_j \leq 0$ versus $\mu_j > 0$. Let $T_{n,j} = X_j/\sqrt{\sigma_{j,j}}$, since the test that rejects for large $X_j/\sqrt{\sigma_{j,j}}$ is UMP for testing H_j. To apply Theorem 9.1.3, let $\hat{c}_{n,K}(1-\alpha)$ be the $1-\alpha$ quantile of the distribution of $\max(X_j : j \in K)$ when $\mu = 0$. Since

$$\max(X_j : j \in I) \leq \max(X_j : j \in K)$$

whenever $I \subset K$, the monotonicity requirement (9.9) is satisfied. Moreover, the resulting test procedure rejects at least as many hypotheses as the Holm procedure (Problem 9.5) In the special case when $\sigma_{i,i} = \sigma^2$ is independent of i and $\sigma_{i,j}$ as the product structure $\sigma_{i,j} = \lambda_i \lambda_j$, then Appendix 3 (p.374) of Hochberg and Tamhane (1987) reduces the problem of determining the distribution of the maximum of a multivariate normal vector to a univariate integral. In general, one can resort to simulation to approximate the critical values; see Example 11.2.13. ■

Example 9.1.5 (One-way Layout) Suppose for $i = 1, \ldots, s$ and $j = 1, \ldots, n_i$, $X_{i,j} = \mu_i + \epsilon_{i,j}$, where the $\epsilon_{i,j}$ are i.i.d. $N(0, \sigma^2)$; the vector $\mu = (\mu_1, \ldots, \mu_s)$ and σ^2 are unknown. Consider testing $H_i : \mu_i = 0$ against $\mu_i \neq 0$. Let $t_{n,i} = n_i^{1/2} \bar{X}_{i \cdot}/S$, where

$$\bar{X}_{i \cdot} = n_i^{-1} \sum_{j=1}^{n_i} X_{i,j}, \qquad S^2 = \sum_{i=1}^{s} \sum_{j=1}^{n_i} (X_{i,j} - \bar{X}_{i \cdot})^2/\nu ,$$

and $\nu = \sum_i (n_i - 1)$. Under H_i, $t_{n,i}$ has a t-distribution with ν degrees of freedom. Let $T_{n,i} = |t_{n,i}|$, and let $\hat{c}_{n,K}(1-\alpha)$ denote the $1-\alpha$ quantile of the distribution of $\max(T_{n,i} : i \in K)$ when $\mu = 0$ and $\sigma = 1$. Since

$$\max(T_{n,i} : i \in I) \leq \max(T_{n,i} : i \in K) ,$$

the monotonicity requirement (9.9) follows. Note that the joint distribution of $(t_{n,1}, \ldots, t_{n,s})$ follows an s-variate multivariate t-distribution with ν degrees of freedom; see Hochberg and Tamhane (1987, p.374-5). ■

When the number of tests is in the tens or hundreds of thousands, control of the FWER at conventional levels becomes so stringent that individual departures from the hypothesis have little chance of being detected, and it is unreasonable to control the probability of even one false rejection. A radical weakening of the FWER was proposed by Benjamini and Hochberg (1995), who suggested the following. For a given multiple testing decision rule, let N be the total number of rejections and let F be the number of false rejections, i.e., the number of rejections among the N rejections corresponding to true null hypotheses. Define Q to be F/N (and defined to be 0 if $N = 0$). Thus Q is the proportion of rejected hypotheses that are rejected erroneously. When none of the hypotheses are rejected, both numerator and denominator of that proportion are 0, and Q is then defined to be 0. The *false discovery rate* (FDR) is

$$FDR = E(Q). \tag{9.14}$$

When all hypotheses are true, $FDR = FWER$. In general, $FDR \leq FWER$ (Problem 9.9), and typically this inequality is strict, so that the FDR is more liberal (in the sense of permitting more rejections) than the FWER. The FDR is a fairly recent idea, and its properties and behavior are the subject of very active research. We shall here only mention some recent papers on this topic: Finner and Roters (2001), Benjamini and Yekutielli (2001) and Sarkar (2002).

9.2 Maximin Procedures

In the present section we shall obtain optimal procedures for a class of problems of the kind illustrated in Examples 9.1.1 and 9.1.2.

Consider the general problem of testing simultaneously s hypotheses H_i: $\theta_i \leq 0$ against the alternatives $\theta_i > 0$, $(i = 1, \ldots, s)$ and suppose that we would reject the individual hypotheses H_i if a test statistic T_i were sufficiently large. The joint c.d.f. of $(T_1, \ldots, T_s)$ will be denoted by F_θ, $\theta = (\theta_1, \ldots, \theta_s)$, and we shall assume that the marginal distribution of T_i depends only on θ_i. The parameter and sample space will be assumed to be finite or infinite open rectangles $\underline{\theta}_i < \theta_i < \overline{\theta}_i$ and $\underline{t}_i < t_i < \overline{t}_i$ respectively. For ease of notation we shall suppose that

$$\underline{\theta}_i = \underline{t}_i = -\infty \quad \text{and} \quad \overline{\theta}_i = \overline{t}_i = \infty \quad \text{for all } i .$$

We shall assume further that, for any B,

$$P_{\theta_i}\{T_i \leq B\} \to 1 \text{ as } \theta_i \to -\infty \quad \text{and} \quad P_{\theta_i}\{T_i \geq B\} \to 1 \text{ as } \theta_i \to +\infty .$$

A crucial assumption will be that the distributions F_θ are stochastically increasing in the following sense, which generalizes the univariate definition in

Section 3.4 to s dimensions. A set ω in $\mathbb{R}^s$ is said to be monotone increasing if

$$t = (t_1, \ldots, t_s) \in \omega \text{ and } t_i \leq t_i' \text{ for all } i \text{ implies } t' \in \omega \ ,$$

and the distributions F_θ will be called stochastically increasing if $\theta_i \leq \theta_i'$ for all i implies

$$\int_\omega dF_\theta \leq \int_\omega dF_{\theta'} \tag{9.15}$$

for every monotone increasing set ω.

The condition will be assumed not only for the distributions of $(T_1, \ldots, T_s)$ but also for $(\pm T_1, \ldots, \pm T_s)$. Thus, for example, for $(-T_1, \ldots, -T_s)$ it means that for any decreasing region the inequality (9.15) will be reversed. A class of models for which (9.15) holds is given in Problem 9.10.

For the sake of simplicity, we shall suppose that when $\theta_1 = \ldots = \theta_s$, the variables $(T_1, \ldots, T_s)$ are *exchangeable*, i.e., that the joint distribution is invariant under permutations of the components. In addition, we assume that the joint distribution of $(T_1, \ldots, T_s)$ has a density with respect to Lebesgue measure.[2] In order for the critical constants to be uniquely defined, we further assume that the joint density is positive on its (assumed rectangular) region of support, but this can be weakened.

Under these assumptions we shall restrict attention to decision rules satisfying the following monotonicity condition. A decision procedure E for the simultaneous testing of $H_1, \ldots, H_s$ based on $T = (T_1, \ldots, T_s)$ states for each possible observation vector t the subset I_t of $\{1, \ldots, s\}$ of values i for which the hypothesis H_i is rejected. A decision rule E is said to be monotone increasing if $t_i \leq t_i'$ for $i \in I_t$ and $t_i' < t_i$ for $i \notin I_t$ implies that $I_t = I_{t'}$.

The ordered T-values will be denoted by $T_{(1)} \leq T_{(2)} \leq \cdots \leq T_{(s)}$ and the corresponding hypotheses by $H_{(1)}, \ldots, H_{(s)}$. Consider the following monotone decision procedure D, which can be viewed as an application of Procedure 9.1.1.

The Stepdown Procedure D:
Step 1. If $T_{(s)} < C_1$, accept $H_1, \ldots, H_s$. If $T_{(s)} \geq C_1$ but $T_{(s-1)} < C_2$, reject $H_{(s)}$ and accept $H_{(1)}, \ldots, H_{(s-1)}$.
Step 2. If $T_{(s)} \geq C_1$, and $T_{(s-1)} \geq C_2$, but $T_{(s-2)} < C_3$ reject $H_{(s)}$ and $H_{(s-1)}$ and accept $H_{(1)}, \ldots, H_{(s-2)}$.
And so on. The C's are determined by

$$P_{\underbrace{0, \ldots, 0}_{j}}\{\max(T_1, \ldots, T_j) \geq C_{s-j+1}\} = \alpha \ , \tag{9.16}$$

and therefore the C's are nonincreasing.

Lemma 9.2.1 *Under the above assumptions, the procedure D with critical constants given by (9.16) controls the FWER in the strong sense.*

[2] This assumption is used only so that the critical constants of the optimal procedures lead to control at exact level α.

PROOF. Apply Theorem 9.1.3 with $\hat{c}_{n,K}(1-\alpha) = C_{s-|K|+1}$, where $|K|$ is the cardinality of K. Then, by the monotonicity of the Cs, condition (9.9) holds. We must verify (9.11) for every P_θ. Suppose θ is such that exactly p hypotheses are true. By exchangeability, we can assume $H_1, \ldots, H_p$ are true and $H_{p+1}, \ldots, H_s$ are false. A false rejection occurs if and only if at least one of $H_1, \ldots, H_p$ is rejected. Since D is monotone, the probability of this event is largest when

$$\theta_1 = \cdots = \theta_p = 0 \quad \text{and} \quad \theta_{p+1} \to \infty, \cdots, \theta_s \to \infty \ ,$$

and, by (9.16), the sup of this probability is equal to

$$P_{\underbrace{0,\ldots,0}_{p}}\{T_i \geq C_{s-p+1} \quad \text{for some } i = 1, \ldots, p\} = \alpha \ . \ \blacksquare$$

The procedure D defined above is an example of a stepdown procedure in that it starts with the most significant (or, in this case, the largest) test statistic and continues rejecting hypotheses as long as their corresponding test statistics are large. In contrast, stepup procedures begin with the least significant test statistic. Consider the following monotone stepup procedure U.

The Stepup Procedure U:
Step 1. If $T_{(1)} > C_1^*$ reject $H_1, \ldots, H_s$. If $T_{(1)} \leq C_1^*$ but $T_{(2)} > C_2^*$, accept $H_{(1)}$ and reject $H_{(2)}, \ldots, H_{(s)}$.
Step 2. If $T_{(1)} \leq C_1^*$, and $T_{(2)} \leq C_2^*$ but $T_{(3)} > C_3^*$, accept $H_{(1)}$ and $H_{(2)}$ and reject $H_{(3)}, \ldots, H_{(s)}$.
And so on. The C^*'s are determined by

$$P_{\underbrace{0,\ldots,0}_{j}}\{L_j\} = 1 - \alpha \ , \tag{9.17}$$

where

$$L_j = \{T_{\pi(1)} \leq C_1^*, \ldots, T_{\pi(j)} \leq C_j^* \quad \text{for some permutation of } \{1, \ldots, j\}\} \ .$$

The following lemma proves control of the FWER and is left as an exercise (Problem 9.11).

Lemma 9.2.2 *Under the above assumptions, the stepup procedure U with critical constants given by (9.17) controls the FWER in the strong sense.*

Subject to controlling the FWER we want to maximize what corresponds to the power of a single test, i.e., the probability of rejecting hypotheses that are in fact false. Let

$$\beta_i(\theta) = P_\theta\{\text{reject at least } i \text{ hypotheses}\}$$

and, for any $\epsilon > 0$, let $A_i(\epsilon)$ denote the set in the parameter space for which at least i of the θ's are $> \epsilon$. Then we shall be interested in maximizing

$$\inf_{\theta \in A_i(\epsilon)} \beta_i(\theta) \quad \text{for } i = 1, 2, \ldots, s. \tag{9.18}$$

This is in the same spirit as the maximin criterion of Chapter 8. However, it is the false hypotheses we should like to reject, and so we also consider maximizing

$$\inf_{\theta \in A_i(\epsilon)} P_\theta\{\text{reject at least } i \text{ false hypotheses}\} \ . \tag{9.19}$$

We note the following obvious fact.

Lemma 9.2.3 *Under (9.15), for any monotone increasing procedure E, the functions $\beta_i(\theta_1, \ldots, \theta_s)$ are nondecreasing in each of the variables $\theta_1, \ldots, \theta_s$.*

For the sake of simplicity we shall now consider the maximin problem first for the case $s = 2$. Corresponding to any decision rule E, let $e_{0,0}$ denote the part of the sample space where both hypotheses are accepted, $e_{0,1}$ where H_1 is accepted and H_2 is rejected, $e_{1,0}$ where H_1 is rejected and H_2 is accepted, and $e_{1,1}$ where both H_1 and H_2 are rejected. The following is an optimality result for the stepdown procedure D. It will be convenient in the following theorem to restate the procedure D in the case $s = 2$.

Theorem 9.2.1 *Assume the conditions described at the beginning of this section.*
(i) A monotone increasing decision procedure with FWER $\le \alpha$ will maximize (9.18) for $i = 1$ if and only if it rejects at least one hypothesis when

$$\max(T_1, T_2) \ge C_1 \ , \tag{9.20}$$

in which case H_i is rejected if $T_i > C_1$; in the contrary case, both hypotheses are accepted. The constant C_1 is determined by

$$P_{0,0}\{\max(T_1, T_2) \ge C_1\} = \alpha \tag{9.21}$$

The minimum value of $\beta_1(\theta)$ over $A_1(\epsilon)$ is $P_\epsilon\{T_i \ge C_1\}$.
(ii) A monotone increasing decision rule with FWER $\le \alpha$ and satisfying (9.20) will maximize (9.18) for $i = 2$ if and only if it takes the following decisions:
$d_{0,0}$: accept H_1 and H_2 when $\max(T_1, T_2) < C_1$
$d_{1,0}$: reject H_1 and accept H_2 when $T_1 \ge C_1$ and $T_2 < C_2$
$d_{0,1}$: accept H_1 and reject H_2 when $T_1 < C_2$ and $T_2 \ge C_1$
$d_{1,1}$: reject both H_1 and H_2 when both T_1 and T_2 are $\ge C_2$ (and when 9.20 holds).
Here C_2 is determined by

$$P_0\{T_i \ge C_2\} = \alpha, \tag{9.22}$$

and hence $C_2 < C_1$.

The minimum probability over $A_2(\epsilon)$ of rejecting both hypotheses is

$$P_{\epsilon,\epsilon}\{\text{at least one } T_i \text{ is } \ge C_1 \text{ and both are } \ge C_2\} \ .$$

(iii) The result (i) holds if the criterion (9.18) is replaced by (9.19) with $i = 1$, and $P_\epsilon\{T_i \ge C_1\}$ is also the maximum value of criterion (9.19).

PROOF. To prove (i), note that the claimed optimal solution has minimum power when $\theta = (\epsilon, -\infty)$ and D has $P_\epsilon\{T_1 \ge C_1\}$ for the claimed optimal value of $\beta_1(\theta)$. Now, suppose that E is any other monotone decision rule with FWER $\le \alpha$. Assume there exists $(t_1, t_2) \notin d_{0,0}$, i.e., rejecting at least one hypothesis, but $(t_1, t_2) \in e_{0,0}$. Then, there exists at least one component of (t_1, t_2) that is $\ge C_1$, say $t_1 \ge C_1$. It follows that

$$P_{\epsilon,-\infty}\{e_{0,0}\} \ge P_{\epsilon,-\infty}\{T_1 < t_1, \ T_2 < t_2\} = P_\epsilon\{T_1 < t_1\} > P_\epsilon\{T_1 < C_1\}$$

and hence

$$P_{\epsilon,-\infty}\{e_{0,0}^c\} < P_{\epsilon,-\infty}\{T_1 \ge C_1\} = P_\epsilon\{T_1 \ge C_1\} \ .$$

Thus, E has a smaller value of criterion (9.18) than does the claimed optimal D. Therefore, $e_{0,0}$ cannot have points outside of $d_{0,0}$, i.e., $e_{0,0}$ must be a proper subset of $d_{0,0}$. But then, since both procedures are monotone, $e_{0,0}^c$ is bigger than $d_{0,0}^c$ on a set of positive Lebesgue measure and so

$$P_{0,0}\{e_{0,0}^c\} > P_{0,0}\{d_{0,0}^c\} = \alpha \ .$$

It follows that for the maximin procedure, the region $d_{0,0}^c$ must be given by (9.20).

To prove (ii), the goal now is to show that, among all monotone nondecreasing procedures which control the FWER and satisfy (9.20), D maximizes

$$\inf_{A_2(\epsilon)} \beta_2(\theta) = \inf_{A_2(\epsilon)} P_\theta\{d_{1,1}\} \ .$$

To prove this, consider any other monotone procedure E which controls the FWER and satisfying $e_{0,0} = d_{0,0}$, and suppose that $e_{1,1}$ contains a point $(t_1,\ t_2)$ with $t_i < C_2$ for some i, say $t_1 < C_2$. Then, since E is monotone, it contains the quadrant $\{T_1 \geq t_1,\ T_2 \geq t_2\}$, and hence

$$P_{0,\infty}\{e_{1,1}\} \geq P_{0,\infty}\{T_1 \geq t_1,\ T_2 \geq t_2\} = P_0\{T_1 \geq t_1\} > P_0\{T_1 \geq C_2\} = \alpha \ ,$$

which contradicts strong control. It follows that $e_{1,1}$ is a proper subset of $d_{1,1}$, and

$$P_\theta\{e_{1,1}\} < P_\theta\{d_{1,1}\} \qquad \text{for all } \theta \ .$$

Since the inf over $A_2(\epsilon)$ of both sides is attained at (ϵ, ϵ),

$$\inf_{A_2(\epsilon)} P_\theta\{e_{1,1}\} < \inf_{A_2(\epsilon)} P_\theta\{d_{1,1}\} \ ,$$

as was to be proved.

To prove (iii), observe that, for any θ,

$$P_\theta\{\text{rejecting at least one false } H_i\} \leq P_\theta\{\text{rejecting at least one } H_i\} \ ,$$

and so

$$\inf_{\theta \in A_1(\epsilon)} P_\theta\{\text{rejecting at least one false } H_i\} \leq \inf_{\theta \in A_1(\epsilon)} P_\theta\{\text{rejecting at least one } H_i\} \ .$$

But, the right side is $P_\epsilon\{T_1 > C_1\}$, and so it suffices to show that D satisfies

$$\inf_{\theta \in A_1(\epsilon)} P_\theta\{D \text{ rejects at least one false } H_i\} = P_\epsilon\{T_1 > C_1\} \ .$$

But, this last result is easily checked.

Finally, once $d_{0,0}$ and $d_{1,1}$ are determined, so are $d_{0,1}$ and $d_{1,0}$ by monotonicity, and this completes the proof. ■

Theorem 9.2.1 provides the maximin test which first maximizes $\inf \beta_1(\theta)$ and then $\inf \beta_2(\theta)$. In the next result, the order in which these aspects are maximized is reversed, which results in the stepup procedure U being optimal.

Theorem 9.2.2 *Assume the conditions described at the beginning of this section. (i) A monotone decision rule with FWER $\leq \alpha$ will maximize (9.18) for $i = 2$ if and only if it rejects both hypotheses, i.e., takes decision $u_{1,1}$, when*

$$\min(T_1, T_2) \geq C_1^* \tag{9.23}$$

and accepts H_i if $T_i < C_1^*$, where $C_1^* = C_2$ is determined by (9.22). Its minimum power $\beta_2(\theta)$ over $A_2(\epsilon)$ is

$$P_\epsilon\{\min(T_1, T_2) \geq C_1^*\} . \tag{9.24}$$

(ii) The monotone procedure with FWER $\leq \alpha$ and satisfying (9.23) maximizes (9.18) for $i = 1$ if and only it takes the following decisions:

$$u_{0,1} = \{T_1 < C_1^*,\ T_2 \geq C_2^*\}$$

$$u_{1,0} = \{T_1 \geq C_2^*,\ T_2 < C_1^*\}$$

$$u_{0,0} = \{T_1 < C_1^*,\ T_2 < C_2^*\} \bigcap u_{1,1}^c ,$$

where C_2^* is determined by

$$P_{0,0}\{u_{0,0}^c\} = \alpha . \tag{9.25}$$

Its minimum power $\beta_1(\theta)$ over $A_1(\epsilon)$ is

$$P_\epsilon\{T_i \geq C_2^*\} . \tag{9.26}$$

(iii) The result (ii) holds if criterion (9.18) with $i = 1$ is replaced by (9.19) with $i = 1$.

Note that

$$C_1^* = C_2 < C_1 < C_2^* . \tag{9.27}$$

Also, the best minimum power $\beta_1(\theta)$ over $A_1(\epsilon)$ for the procedure of Theorem 9.2.1 exceeds that for Theorem 9.2.2, while the situation is reversed for the best minimum power of $\beta_2(\theta)$ over $A_2(\epsilon)$. This is, of course, as it must be since the first of these two procedures maximized the minimum value of $\beta_1(\theta)$ over $A_1(\epsilon)$ while the second maximized the minimum value of $\beta_2(\theta)$ over $A_2(\epsilon)$.

PROOF. (i) Suppose that E is any other monotone procedure with FWER $\leq \alpha$. Assume there exists $(t_1,\ t_2) \in e_{1,1}$ such that $t_i < C_1^*$ for some i, say $t_1 < C_1^*$. Then,

$$P_{0,\infty}\{e_{1,1}\} \geq P_{0,\infty}\{T_1 \geq t_1,\ T_2 \geq t_2\} = P_0\{T_1 \geq t_1\} > P_0\{T_1 \geq C_1^*\} = \alpha ,$$

which would violate the FWER condition. Therefore, $e_{1,1} \subset u_{1,1}$. But then

$$\inf_{A_2(\epsilon)} \beta_2(\theta)$$

is smaller for E than for U, as was to be proved.

(ii) Note that the claimed solution $\inf_{A_1(\epsilon)} \beta(\theta)$ is given by

$$\inf_{\theta \in A_1(\epsilon)} P_\theta\{u_{0,0}^c\} = P_{\epsilon,-\infty}\{u_{0,0}^c\} = P_\epsilon\{T_1 \geq C_1^*\} .$$

We now seek to determine $u_{0,0}$, as in Theorem 9.2.1, but with the added constraint that $u_{0,0} \subset u_{1,1}^c$.

To prove optimality for the claimed solution, suppose that E is another monotone procedure controlling FWER at α, and satisfying $e_{1,1} = u_{1,1}$ with $u_{1,1}$ given by (9.23). Assume $(t_1,\ t_2) \in e_{0,0}$ but $\notin u_{0,0}$, so that $T_i > C_2^*$ for some i, say $i = 1$. Then,

$$P_{\epsilon,-\infty}\{e_{0,0}\} \geq P_{\epsilon,-\infty}\{T_1 \leq t_1,\ T_2 \leq t_2\} = P_\epsilon\{T_1 \leq t_1\} > P_\epsilon\{T_1 > C_2^*\} .$$

Hence,

$$P_{\epsilon,-\infty}\{e_{0,0}^c\} < P_\epsilon\{T_1 > C_2^*\} \, ,$$

so that E cannot be optimal. It follows that $e_{0,0} \subset u_{0,0}$. But if $e_{0,0}$ is a proper subset of $u_{0,0}$, the set $e_{0,0}^c$ in which E rejects at least one hypothesis contains $u_{0,0}^c$ and so

$$P_{0,0}\{e_{0,0}^c\} > P_{0,0}\{u_{0,0}^c\} = \alpha \, ,$$

and E does not control the FWER at α.

Finally, the proof of (iii) is analogous to the proof of (iii) in Theorem 9.2.1. ■

Theorems 9.2.1 and 9.2.2 have natural extensions to the case of s hypotheses where the aim is to maximize the s quantities (9.18). As in the case $s = 2$, these maximizations lead to different procedures, and one must choose their order of importance. The two most natural choices are the following:

(a) Begin by maximizing $\inf \beta_1(\theta)$, which will lead to an optimal choice for $d_{0,0,\ldots,0}$, the decision to accept all hypotheses. With $d_{0,\ldots,0}$ fixed, the partition of $d_{0,\ldots,0}^c$ into the subsets in which the remaining decisions should be taken is begun by maximizing the minimum of $\beta_2(\theta)$ over the part of the parameter space in which at least 2 hypotheses are false, and so on.

(b) Alternatively, we may start at the other end by maximizing $\inf \beta_s(\theta)$, and from there proceed downward.

We shall here only state the result for case (a). For its proof and the statement and proof for case (b), see Lehmann, Romano, and Shaffer (2003).

Theorem 9.2.3 *Under the assumptions made at the beginning of this section, among all monotone procedures E with FWER $\leq \alpha$, the stepdown procedure D with critical constants given by (9.16), has the following properties:*
(i) it maximizes $\inf \beta_1(\theta)$ *over* $A_1(\epsilon)$
(ii) it maximizes $\inf \beta_2(\theta)$ *over* $A_2(\epsilon)$ *subject to the additional condition $e_{s,2} \subset d_{s,1}$, where $e_{s,i}$ and $d_{s,i}$ denote the events that the procedures E and D reject at least i of the hypotheses $H_1, \ldots, H_s$.*
(iii) Quite generally, it maximizes both (9.18) and (9.19) among all monotone procedures E with FWER $\leq \alpha$ and satisfying $e_{s,i} \subset d_{s,i-1}$.

We shall now provide a canonical form for certain stepdown procedures, and particularly for the maximin procedure D of Theorem 9.2.3, that provides additional insights.

Let $\hat{p}_1, \ldots, \hat{p}_s$ be the p-values of the statistics $T_1, \ldots, T_s$, and denote the ordered p-values by $\hat{p}_{(1)} \leq \cdots \leq \hat{p}_{(s)}$. If F denotes the common marginal distribution of T_i under $\theta_i = 0$, we have that

$$\hat{p}_i = 1 - F(T_i) \tag{9.28}$$

and hence that

$$\hat{p}_{(1)} = 1 - F(T_{(s)}) \, . \tag{9.29}$$

In terms of the $\hat{p}$'s, the steps of the stepdown procedure

$$T_{(s)} \geq C_1, \; T_{(s-1)} \geq C_2, \ldots \tag{9.30}$$

are equivalent respectively to

$$\hat{p}_{(1)} \leq \alpha_1, \; \hat{p}_{(2)} \leq \alpha_2, \ldots \tag{9.31}$$

for suitable α's. In particular, $T_{(s)} \geq C_1$ is equivalent to $\hat{p}_{(1)} \leq \alpha_1$. Thus, by (9.29), $T_{(s)} < C_1$ is equivalent to $F(T_{(s)}) < 1 - \alpha_1$, so that

$$C_1 = F^{-1}(1 - \alpha_1) \; .$$

On the other hand, if G_s denotes the distribution of $T_{(s)}$ when all the θ_i are 0, it follows from (9.16) that $C_1 = G_s^{-1}(1 - \alpha)$ and hence that

$$1 - \alpha_1 = F[G_s^{-1}(1 - \alpha)] \; , \tag{9.32}$$

which gives α_1 as a function of α.

It is of interest to determine the ranges of the step levels $\alpha_1, \ldots, \alpha_s$. Since $G_s(t) \leq F(t)$ for all t, it follows from (9.32) that $1 - \alpha_1 \geq 1 - \alpha$ for all F, or

$$\alpha_1 \leq \alpha \quad \text{for all } F \; , \tag{9.33}$$

with equality when $F = G$, i.e., when $T_1 = \cdots T_s$. To find a lower bound for α_1, put $u = G^{-1}(1 - \alpha)$ in (9.32) so that

$$1 - \alpha_1 = F(u) \quad \text{with} \;\; 1 - \alpha = G_s(u) \tag{9.34}$$

and note that for all u

$$1 - G_s(u) = P\{\text{at least one } T_i \geq u\} \leq \sum P\{T_i \geq u\} = s[1 - F(u)] \; .$$

Thus,

$$F(u) \leq 1 - \frac{1}{s}[1 - G(u)] = 1 - \frac{\alpha}{s}$$

and hence

$$\alpha_1 \geq \frac{\alpha}{s} \; . \tag{9.35}$$

We shall now show that the lower bound (9.35) is sharp by giving an example of a joint distribution of $(T_1, \ldots, T_s)$ for which it is attained.

Example 9.2.1 (A Least Favorable Distribution) Let U be uniformly distributed on $(0, 1)$ and suppose that when $H_1, \ldots, H_s$ are all true,

$$Y_1 = U, \;\; Y_2 = U + \frac{1}{s}(mod\ 1), \ldots, Y_s = U + \frac{s-1}{s}(mod\ 1) \; .$$

Since $(Y_1, \ldots, Y_s)$ does not satisfy our assumption of exchangeability, replace it by the exchangeable set of variables $(X_1, \ldots, X_s) = (Y_{\pi(1)}, \ldots, Y_{\pi(s)})$, where $(\pi(1), \ldots, \pi(s))$ is a random permutation of $(1, \ldots, s)$ (and independent of U). Let $T_i = 1 - X_i$ and suppose that H_i is rejected when T_i is large. To show that

$$F[G_s^{-1}(1 - \alpha)] = 1 - \frac{\alpha}{s} \; , \tag{9.36}$$

note that the T's are uniformly distributed on $(0, 1)$ so that (9.36) becomes

$$G_s(1 - \frac{\alpha}{s}) = 1 - \alpha \; .$$

Now

$$1 - G_s(1 - \frac{\alpha}{s}) = P\{\text{at least one } T_i \geq 1 - \frac{\alpha}{s}\} = P\{\text{at least one } X_i \leq \frac{\alpha}{s}\} \; .$$

But the events $\{X_i \le \alpha/s\}$ are mutually exclusive, and therefore

$$P\{\text{at least one } X_i \le \frac{\alpha}{s}\} = \sum_{i=1}^{s} P\{X_i \le \frac{\alpha}{s}\} = s \cdot \frac{\alpha}{s} = \alpha ,$$

which implies (9.36). ■

We shall now briefly sketch the corresponding development for α_2, defined by the fact that $\hat{p}_{(2)} \le \alpha_2$ is equivalent to $T_{(s-1)} \ge C_2$, where C_2 is determined by (9.16) so that

$$G_{s-1}(C_2) = 1 - \alpha .$$

Note that G_{s-1} is *not* the distribution of $T_{(s-1)}$, i.e., of the 2nd largest of s T's, but of the largest of $T_1, \ldots, T_{s-1}$ (i.e., the largest of $s-1$ T's). In exact analogy with the derivation of (9.32) it now follows that

$$1 - \alpha_2 = F[G_{s-1}^{-1}(1-\alpha)] . \tag{9.37}$$

The maximum value of α_2, as in the case of α_1, is equal to α and is attained when $T_1 = \cdots = T_{s-1}$.

The argument giving the lower bound shows that $\alpha_2 \ge \alpha/(s-1)$. To show that this value is attained, we must find an example for which

$$G_{s-1}(1 - \frac{\alpha}{s-1}) = 1 - \alpha .$$

Example 9.2.1 will serve this purpose since in that case

$$1 - G_{s-1}(1 - \frac{\alpha}{s-1}) = P\{\text{at least one of } T_1, \ldots, T_{s-1} \ge 1 - \frac{\alpha}{s-1}\}$$

$$= \sum_{i=1}^{s-1} P\{X_i \le \frac{\alpha}{s-1}\} = (s-1) \cdot \frac{\alpha}{s-1} = \alpha$$

for any α satisfying $\alpha/(s-1) < 1/s$, i.e., $\alpha < (s-1)/s$.

Continuing in this way we arrive at the following result.

Theorem 9.2.4 *(i) The step levels α_i defined by the procedure D with critical constants given by (9.16) and the equivalence of (9.30) and (9.31) are given by*

$$1 - \alpha_i = F[G_{s-i+1}(1-\alpha)] , \tag{9.38}$$

where G_j is the distribution of $\max(T_1, \ldots, T_j)$.

(ii) The range of α_i is

$$\frac{\alpha}{s-i+1} \le \alpha_i \le \alpha . \tag{9.39}$$

Furthermore, the upper bound α is attained when $T_1 = \cdots = T_s$, i.e., when there really is no multiplicity. The lower bound $\alpha/(s-i+1)$ is attained when the distribution of $T_1, \ldots, T_{s-i+1}$ is that of Example 9.2.1.

Not all points in the s-dimensional rectangle (9.39) are possible for $(\alpha_1, \ldots, \alpha_s)$. In particular, since for all t

$$G_i(t) \ge G_j(t) \quad \text{when } i < j ,$$

it follows that

$$\alpha_1 \le \alpha_2 \le \cdots \le \alpha_s \; . \tag{9.40}$$

The values of α_i given by (9.38) can be determined when the joint distribution of $(T_1, \ldots, T_s)$ (and hence the distributions G_s) is known. Consider, however, the situation in which the common marginal distribution F of the statistics T_i needed to carry out the tests of the individual hypotheses H_i at a given level is known, but the joint distribution of the T's is unknown. Then, we are unable to determine the step levels (9.38).

It follows, however, from (9.39) that the procedure (9.31) with

$$\alpha_i = \alpha/(s - i + 1) \quad \text{for } i = 1, \ldots, s \tag{9.41}$$

will control the FWER for all joint distributions of $(T_1, \ldots, T_s)$, since these levels are conservative in all cases. This is just the Holm procedure of Theorem 9.1.2.

Also, none of the levels α_i can be larger than $\alpha/(s-i+1)$ without violating the FWER condition for some distribution. To see this, note that if levels α_i are used in Example 9.2.1, it follows from the discussion of this example that when i of the hypotheses are true, the probability of at least one false rejection is $(s-i+1)\alpha_i$. Thus, if α_i exceeds $\alpha/(s-i+1)$, the FWER condition will be violated.

Of course, if the class of joint distributions of the T's is restricted, the range of α_i may be smaller than (9.39). For example, suppose that the T's are independent. Then, putting $u = G_s^{-1}(1-\alpha)$ as before, we see from (9.34) that

$$1 - \alpha_1 = F(u) \quad \text{and} \quad 1 - \alpha = F^s(u)$$

so that

$$\alpha_1 = 1 - (1-\alpha)^{1/s} \; ,$$

and more generally that

$$\alpha_i = 1 - (1-\alpha)^{1/(s-i+1)} \; .$$

In this case, the range reduces to a single point.

More interesting is the case of positive quadrant dependence when

$$G_s(u) \ge F^s(u)$$

and hence

$$1 - \alpha \ge (1-\alpha_1)^{1/s}$$

and

$$1 - (1-\alpha)^s \le \alpha_1 \le \alpha \; . \tag{9.42}$$

The bounds are sharp since the upper bound is attained when $T_1 = \cdots = T_s$ and the lower bound is attained in the case of independence.

9.3 The Hypothesis of Homogeneity

The previous section dealt with situations in which each of the parameters varies independently, so that any subset of the hypotheses $H_1, \ldots, H_s$ can be true with the remaining ones being false. This condition is not satisfied, for example, when the set of hypotheses is

$$H_{i,j} : \theta_i = \theta_j , \quad i < j \tag{9.43}$$

for all $\binom{s}{2}$ pairs $i < j$. Then, for instance, the set $\{H_{1,2}, H_{2,3}\}$ can not be the set of all true hypotheses since the truth of $H_{1,2}$ and $H_{2,3}$ implies the truth of $H_{1,3}$. It follows from this transitivity that the set of true hypotheses constitutes a partition of the μ's, say

$$\mu_{i_1} = \cdots = \mu_{i_r} ; \quad \mu_{i_{r+1}} = \cdots = \mu_{i_{r+k}} ; \cdots . \tag{9.44}$$

All pairs within a set of the partition are equal, and two μs in different sets are unequal. We shall therefore use the statement $\mu_{i_1} = \cdots = \mu_{i_r}$ as shorthand for the statement that all hypotheses $H_{k,l}$ with (k, l) any pair of subscripts from the set $\{i_1, \ldots, i_r\}$ are true.

Unfortunately, the results of the tests of the hypotheses (9.43) do not share this simple structure since it is possible to accept $H_{1,2} : \mu_1 = \mu_2$ and $H_{2,3} : \mu_2 = \mu_3$ while rejecting $H_{1,3} : \mu_1 = \mu_3$. We shall return to this point at the end of the section.

We shall now consider the simultaneous testing of the $\binom{s}{2}$ hypotheses (9.43) by means of a Holm type stepdown procedure, as in the preceding section. We assume that statistics $T_{i,j}$ are available for testing the individual hypotheses $H_{i,j}$. In the case of normal variables with sample means $\bar{X}_i$ and common variance σ^2, these would be the statistics $T_{i,j} = |\bar{X}_i - \bar{X}_j|/\hat{\sigma}$. The procedure begins with the largest of the T's corresponding to the pair (i, j) with the largest difference $|\bar{X}_i - \bar{X}_j|$. This would be tested at level $\alpha/\binom{s}{2}$, since $\binom{s}{2}$ is the total number of hypotheses being tested. If this hypothesis is accepted, all the hypotheses (9.43) are accepted and the procedure is terminated. In the contrary case, we next test the second largest of the T's at level $\alpha/(\binom{s}{2} - 1)$, and so on. By Theorem 9.1.2, this procedure controls the FWER, regardless of the joint distribution of the $T_{i,j}$.

However, the fact that the parameters $\theta_{i,j} = \mu_i - \mu_j$ do not vary independently but are subject to certain logical restrictions enables us to do better. To illustrate the situation, suppose that $s = 6$. Let

$$\bar{X}_{(1)} \leq \cdots \leq \bar{X}_{(s)}$$

denote the ordered values of the sample means, and let $\mu_{(i)}$ be the mean corresponding to $\bar{X}_{(i)}$. At the first stage, we test $\mu_{(1)} = \mu_{(6)}$. If $(\bar{X}_{(6)} - \bar{X}_{(1)})/\hat{\sigma} < C$, we accept all the hypotheses $H_{i,j}$ and terminate the procedure. If $(\bar{X}_{(6)} - \bar{X}_{(1)})/\hat{\sigma} \geq C$, we reject the hypothesis $\mu_{(1)} = \mu_{(6)}$ and test the largest of the differences $\bar{X}_{(6)} - \bar{X}_{(2)}$ and $\bar{X}_{(5)} - \bar{X}_{(1)}$.

Let us now express the rule in terms of the p-values. By (9.28),

$$\hat{p}_{i,j} = 1 - F(T_{i,j}) , \tag{9.45}$$

where F is the distribution of $|\bar{X}_i - \bar{X}_j|/\hat{\sigma}$, and the rejection region $|\bar{X}_{(6)} - \bar{X}_{(1)}|/\hat{\sigma} \geq C$ becomes $\min_{i,j} \hat{p}_{i,j} \leq \alpha/\binom{s}{2}$. If the next largest difference is $(\bar{X}_{(5)} - \bar{X}_{(1)})/\hat{\sigma}$, say, we would at the next step compare $1 - F[(\bar{X}_{(5)} - \bar{X}_{(1)})/\hat{\sigma}]$ with $\alpha/(\binom{s}{2} - 1)$, and so on.

However, using the relations between the differences $|\bar{X}_j - \bar{X}_i|$, we can in the present situation do considerably better than that.

To see this, consider the case where one hypothesis is false, say $\mu_1 \neq \mu_4$. Then, μ_2 cannot be equal to both μ_1 and μ_4; thus, one of the hypotheses $\mu_1 = \mu_2$ or $\mu_2 = \mu_4$ must be false, and similarly for μ_3, μ_5 and μ_6. Therefore, at step 2 when one hypothesis is false, at least 5 must be false, and the number of possible true hypotheses is not $\binom{6}{2} - 1 = 14$ but instead is $\binom{6}{2} - 5 = 10$.

An argument similar to that of Theorem 9.1.2 shows that at the second step of the Holm procedure, we can increase $\alpha/14$ to $\alpha/10$ without violating the FWER. Indeed, suppose that at least one hypothesis is false, and so at most 10 are true. Let I be the set (i, j) of true hypothesis $H_{i,j}$, and let

$$\hat{p}_{\min} = \min\{\hat{p}_{i,j} : \ (i, j) \in I\} \ .$$

Then, if a false rejection occurs, it occurs at step 1 or step 2, but in either case, it must be that $\hat{p}_{\min} \leq \alpha/10$. But, by Bonferroni,

$$P\{\hat{p}_{\min} \leq \frac{\alpha}{10}\} \leq \sum_{(i,j)\in I} P\{\hat{p}_{i,j} \leq \frac{\alpha}{10}\} \leq |I| \cdot \frac{\alpha}{10} \leq \alpha \ .$$

Similar improvements are possible at the succeeding steps.

As pointed out at the beginning of the section, each set of true hypotheses (9.44) corresponds to a partition of the integers $\{1, \ldots, s\}$ and determines the corresponding number of possible true hypotheses

$$\binom{r}{2} + \binom{k}{2} + \cdots \ .$$

The following table, adapted from Shaffer (1986), where this improvement was first proposed, shows for $s = 3$ to 10 the maximum possible number of true hypotheses.

Table 9.1.
Possible Number of True Hypotheses

s	Total # of Hypotheses $H_{i,j}$	Possible Number of True Hypotheses
3	3	0, 1, 3
4	6	0-3, 6
5	10	0-4, 6, 10
6	15	0-4, 6, 7, 10, 15
7	21	0-7, 9, 10, 11, 15, 21
8	28	0-13, 15, 16, 21, 28
9	36	0-13, 15, 16, 18, 21, 22, 28, 36
10	45	0-18, 20, 21, 22, 24, 28, 29, 36, 45

Here, for example, the entries 0-4, 6, 10 for $s = 5$ correspond to the numbers of possible true pairs $\mu_i = \mu_j$ for the given partitions. Thus, the case $\mu_1 = \mu_2 = \mu_3 = \mu_4 = \mu_5$ corresponds to the partition $(\mu_1, \ldots, \mu_5)$ and allows $\binom{5}{2} = 10$ true pairs $\mu_i = \mu_j$. The case $\mu_1 \neq \mu_2 = \mu_3 = \mu_4 = \mu_5$ corresponds to the partition $\{\mu_1\}$, $\{\mu_2, \mu_3, \mu_4, \mu_5\}$ and allows $\binom{4}{2} = 6$ true pairs $\mu_i = \mu_j$. The case $\mu_1 = \mu_2 \neq \mu_3 = \mu_4 = \mu_5$ corresponds to the partition $\{\mu_1, \mu_2\}$, $\{\mu_3, \mu_4, \mu_5\}$ and allows $\binom{2}{2} + \binom{3}{2} = 4$ true pairs, and so on.

The reductions are substantial. At the second step, for example, $\binom{s}{2} - 1 = \frac{(s-2)(s+1)}{2}$ is decreased to $\binom{s-1}{2} = \frac{(s-2)(s-1)}{2}$; the difference is $s - 2$ and hence tends to ∞ as $s \to \infty$.

Shaffer gave a simple algorithm for finding the maximum number of true hypotheses given that i hypotheses have been declared false. Use of the procedure based on these numbers has been called S_1 (Donoghue (2004)). A more powerful procedure, called S_2, uses the maximum number of true hypotheses given the particular hypotheses that have been declared false. A difficulty with the S_2 procedure, particularly when s gets large, is to determine the maximum numbers of true hypotheses that are possible at any given step. An algorithm to deal with this problem has been developed by Donoghue (2004).

Like the Holm procedure itself, this modification only utilizes the marginal distributions of the statistics $T_{i,j} = |\bar{X}_i - \bar{X}_j|/\hat{\sigma}$, which are proportional to t-statistics. However, under the assumption of normality, the joint distribution of these statistics is also known, and so the levels (9.38) could be used - with $s-i+1$ replaced by the number of true hypotheses possible at this stage - to achieve a further improvement. Note, however, that this can be difficult because the set of possible true hypotheses is not unique, so a number of joint distributions would have to be determined. An alternative approach that incorporates logical constraints and dependence among the test statistics is described in Westfall (1997).

Multiple comparison procedures, many of them going back to the 1950's, employ not only tests based on ranges, but also the corresponding procedures based on F-tests. Most of them are special cases of a general class of stagewise step-down procedures which we shall now consider for testing homogeneity of s normal populations with common variance based on samples of equal size $n_i = n$.

For this purpose, we require a slight shift of point of view. The hypothesis $H : \mu_{i_1} = \cdots = \mu_{i_r}$ was previously considered as shorthand for the hypothesis that all pairs within this set are equal, and the problem as that of testing these $\binom{r}{2}$ separate hypotheses. Now we shall also admit the more traditional interpretation of H as a hypothesis in its own right for which a global test such as an F-test might be appropriate. It should be emphasized that, logically, the two interpretations are of course equivalent; they differ only in the way they are analyzed.

The first step in the class of procedures to be considered is to test the hypothesis

$$H_s : \mu_1 = \cdots = \mu_s \tag{9.46}$$

either with a range test or an F-test at a critical value C_s corresponding to some level α_s. In case of acceptance, the means are judged to exhibit no significant differences, the set $\{\mu_1, \ldots, \mu_s\}$ is declared homogeneous, and the procedure terminates. If H_1 is rejected, a search for the source of the differences is initiated by proceeding to the second stage, which consists in testing the s hypotheses

$$H_{s-1,i} : \mu_1 = \cdots = \mu_{i-1} = \mu_{i+1} = \cdots = \mu_s$$

each by means of a range or an F test at a common critical value corresponding to a common level α_{s-1}. For any hypothesis that is accepted, the associated set of means (and all of its subsets) are judged not to have shown any significant differences and are not tested further. For any rejected hypothesis, the $s - 1$ subsets of size $s - 2$ are tested (except those that are subsets of an $(s - 1)$-set whose homogeneity has been accepted). At stage i, the $k = s - i + 1$ differences would be tested for all subsets that are not included in an $(s - i + 2)$-set whose homogeneity has been accepted. Moreover, assume that all tests at stage i are performed at the same level, and denote this level by α_k corresponding to a critical value C_k. The procedure is continued in this way until no hypotheses are left to be tested.

To see the relation of this stagewise procedure to the fully sequential approach described at the beginning of the section which is based on the ordered differences $|\bar{X}_i - \bar{X}_j|/\hat{\sigma}$, let us compare the two procedures when all the tests of the stagewise procedure are based on standardized ranges. In both cases the first step is based on $|\bar{X}_{(s)} - \bar{X}_{(1)}|/\hat{\sigma}$ and rejects the homogeneity of $\{\mu_1, \ldots, \mu_s\}$ if this statistic is $\geq$ some constant C_s. The stagewise procedure next compares the two subranges

$$|\bar{X}_{(s)} - \bar{X}_{(2)}|/\hat{\sigma} \quad \text{and} \quad |\bar{X}_{(s-1)} - \bar{X}_{(1)}|/\hat{\sigma}$$

with a common critical value C_{s-1}. Note, however, that if the larger of the two is $< C_{s-1}$, this will a fortiori be true of the smaller one. This second step could thus equally well be described as comparing the second largest of the ranges $|\bar{X}_i - \bar{X}_j|/\hat{\sigma}$ with C_{s-1}, and in case of acceptance terminating the procedure. In case of rejection, we would next compare the smaller of the two $(s-1)$-ranges with C_{s-1}. Continuing in this way, C_i would be used to test all eligible i-ranges.

The fully sequential procedure described at the beginning of the section also would terminate at the second step if the larger of the two $(s - 1)$ ranges is too small. But if it is large enough for rejection, the next step would differ in two ways: (i) the critical level would be lowered further; (ii) the next test statistic would be the 3rd-largest of the differences $|\bar{X}_i - \bar{X}_j|/\hat{\sigma}$, which may but need not coincide with the smaller of the $(s - 1)$-ranges. Thus, the two procedures differ slightly, although they are very much in the same spirit.

To complete the description of a stagewise procedure, once the test statistics have been chosen, it is necessary to specify the critical values $C_2, \ldots, C_s$ for the successive stages or equivalently the levels $\alpha_2, \ldots, \alpha_s$ at which the tests are performed. Note that there is no α_1 of C_1 since at the sth stage only singlets are left, and hence there are no longer any hypotheses to be tested.

Before discussing the best choice of α's let us consider some specific methods that have been proposed in the literature. Additional properties and uses of some of these will be mentioned at the end of the section.

(i) *Tukey's T-method.* This procedure employs the Studentized range test at each stage with a common critical value $C_k = C$ for all k. The method has an unusual feature which makes it particularly simple to apply. In general, in order to determine whether a particular subset S_0 of means should be called nonhomogeneous, it is necessary to proceed stagewise since the homogeneity of S_0 itself is not tested unless homogeneity has been rejected for all sets containing S_0. However, with Tukey's T-method it is only necessary to test S_0 itself. If the Studentized range of S_0 exceeds C, so will that of any set containing S_0, and S_0 is declared nonhomogeneous. In the contrary case, homogeneity of S_0 is accepted. The two facts which jointly eliminate the need for a stagewise procedure in this case are (a) that the range, and hence the Studentized range, of S_0 cannot exceed that of any set S containing S_0, and (b) the constancy of the critical value. The next method applies this idea to a procedure based on F-tests.

(ii) *Gabriel's simultaneous test procedure.* F-statistics do not have property (a) above. However, this property is possessed by the statistics νF, where ν is the number of numerator degrees of freedom (Problem 9.17). Hence a procedure based on F-statistics with critical values $C_k = C/(k-1)$ satisfies both (a) and (b), since $k-1$ is the number of numerator degrees of freedom when k means are tested, that is, at the $s-k+1$st stage. This procedure, which in this form was proposed by Gabriel (1964), permits the testing of many additional hypotheses and when these are included becomes Scheffé's S-method, which will be discussed in Sections 9.4 and 9.5.

(iii) *Fisher's least-significant-difference method.* This procedure employs an F-test at the first stage, and Studentized range tests with a common critical value $C_2 = \cdots = C_s$ at all succeeding stages. The constants C_s and C_2 are related by the fact that the first stage F-test and the pairwise t-test of the last stage have the same level.

The usual descriptions of (i) and (iii) consider only the first and last stages of these procedures, and omit the conclusions which can be drawn from the intermediate stages.

Several classes of procedures have been defined by prescribing the significance levels α_k, which can then be applied to the chosen test statistics at each stage. Examples are:

(iv) *The Newman–Keuls levels:*

$$\alpha_k = \alpha.$$

(v) *The Duncan levels:*

$$\alpha_k = 1 - \gamma^{k-1}.$$

(vi) *The Tukey levels:*

$$\alpha_k = \begin{cases} 1 - \gamma^{k/2}, & 2 < k < s-1 \\ 1 - \gamma^{s/2}, & k = s-1, s. \end{cases}$$

In both (v) and (vi), $\gamma = 1 - \alpha_2$.

Most of the above methods and some others are reviewed in the books by Hochberg and Tamhane (1987) and Hsu (1996).

Let us now consider the choice of the levels α_k more systematically. For this purpose, denote the probability of at least one false rejection, that is, of rejecting homogeneity of at least one set of μ's which in fact is homogeneous, by $\alpha(\mu_1, \ldots, \mu_s)$. As before we impose the restriction that the FWER should not exceed α, so that

$$\alpha(\mu_1, \ldots, \mu_s) \le \alpha \quad \text{for all } (\mu_1, \ldots, \mu_s)\ . \tag{9.47}$$

In order to study the best choice of $\alpha_2, \ldots \alpha_s$ subject to (9.47), let us begin by assuming σ^2 to be known, say $\sigma^2 = 1$. Then the F-tests are replaced by χ^2-tests and the Studentized range tests by range tests; the latter reject when the range of the subgroup being tested is too large.

To evaluate the maximum of the left side of (9.47), suppose that the μ's fall into r distinct subgroups of sizes $v_1, \ldots, v_r$ ($\sum v_i = s$), say

$$\mu_{i_1} = \cdots = \mu_{i_{v_1}}; \quad \mu_{i_{v_1+1}} = \cdots = \mu_{i_{v_1+v_2}}; \ldots, \tag{9.48}$$

where $(i_1, \ldots, i_s)$ is a permutation of $(1, \ldots, s)$. Then, both χ^2 and range statistics for testing the r hypotheses

$$H_1' : \mu_{i_1} = \cdots = \mu_{i_{v_1}}; \qquad H_2' : \mu_{i_{v_1+1}} = \cdots = \mu_{i_{v_1+v_2}}; \cdots \tag{9.49}$$

are independent. The following result then gives conditions on the individual levels α_i so that the FWER is controlled.

Lemma 9.3.1 *If the test statistics for testing the r hypotheses (9.49) are independent, then the sup of $\alpha(\mu_1, \ldots, \mu_s)$ over all $(\mu_1, \ldots, \mu_s)$ satisfying (9.48) is given by*

$$\sup \alpha(\mu_1, \ldots, \mu_s) = 1 - \prod_{i=1}^{r} (1 - \alpha_{v_i}) \ , \tag{9.50}$$

where $\alpha_1 = 0$.

Proof. Since false rejection can occur only when at least one of the hypotheses (9.49) is rejected,

$$\begin{aligned} \alpha(\mu_1, \ldots, \mu_s) &\le P \text{ (rejecting at least one } H_i') \\ &= 1 - P \text{ (accepting all the } H_i') \\ &= 1 - \prod_{i=1}^{r} (1 - \alpha_{v_i}) \ , \end{aligned}$$

the last equality following from the assumption of independence.

To see that the upper bound (9.50) is sharp, let the distances between the different groups of means (9.48) all tend to infinity. Then the probability of accepting homogeneity of any set containing $\{\mu_{i_1}, \ldots, \mu_{i_{v_1}}\}$ as a proper subset, and therefore not reaching the stage at which H_1' is tested, tends to zero. The same is true for $H_2', \ldots, H_r'$, and hence $\alpha(\mu_1, \ldots, \mu_s)$ tends to the right side of (9.50). ■

It is interesting to note that $\sup \alpha(\mu_1, \ldots, \mu_s)$ depends only on $\alpha_2, \ldots, \alpha_s$ and not on whether χ^2- or range statistics are used at the various stages. In fact, Lemma 9.3.1 remains true for many other statistics (Problem 9.18).

It follows from Lemma 9.3.1 that a procedure with levels $(\alpha_2, \ldots, \alpha_s)$ satisfies (9.47) if and only if

$$\prod_{i=1}^{r} (1 - \alpha_{v_i}) \ge 1 - \alpha \qquad \text{for all} \quad (v_1, \ldots, v_r) \ \text{ with } \ \sum v_i = s. \tag{9.51}$$

To see how to choose $\alpha_2, \ldots, \alpha_s$, subject to (9.47) or (9.51), let us say that $(\alpha_2, \ldots, \alpha_s)$ is *inadmissible* if there exists another set of levels $(\alpha_2', \ldots, \alpha_s')$ satisfying (9.51) and such that

$$\alpha_i \le \alpha_i' \qquad \text{for all } i, \text{with strict inequality for some } i. \tag{9.52}$$

These inequalities imply that the procedure with the levels α_i' has uniformly better chance of detecting existing inhomogeneities than the procedure with levels α_i. The definition is thus in the spirit of α-admissibility discussed in Chapter 6, Section 6.7.

Lemma 9.3.2 *Under the assumptions of Lemma 9.3.1, necessary conditions for* $(\alpha_2, \ldots, \alpha_s)$ *to be admissible are*
(i) $\alpha_2 \leq \cdots \leq \alpha_s$ *and*
(ii) $\alpha_s = \alpha_{s-1} = \alpha$.

PROOF. (i) Suppose to the contrary that there exists k such that $\alpha_{k+1} < \alpha_k$, and consider the procedure in which $\alpha'_i = \alpha_i$ for $i \neq k+1$ and $\alpha'_{k+1} = \alpha_k$. It suffices to show that $\prod(1 - \alpha'_{v_i}) \geq 1 - \alpha$ for all $(v_1, \ldots, v_r)$. If none of the v's is equal to $k+1$, then $\alpha'_{v_i} = \alpha_{v_i}$ for all i, and the result follows. Otherwise replace each v that is equal to $k+1$ by two v's—one equal to k and one equal to 1—and denote the resulting set of v's by $w_1, \ldots, w_{r'}$. Then

$$\prod_{i=1}^{r}(1 - \alpha'_{v_i}) = \prod_{i=1}^{r'}(1 - \alpha_{w_i}) \geq 1 - \alpha \, .$$

(ii) The left side of (9.51) involves α_s if and only if $r = 1$, $v_1 = s$. Thus the only restriction on α_s is $\alpha_s \leq \alpha$, and the only admissible choice is $\alpha_s = \alpha$. The argument for α_{s-1} is analogous (Problem 9.19). ■

Part (ii) of this lemma shows that Tukey's T-method and Gabriel's simultaneous test procedure are inadmissible since in both $\alpha_{s-1} < \alpha_s$. The same argument shows Duncan's set of levels to be inadmissible. [These choices can however be justified from other points of view; see for example Spjøtvoll (1974) and the comments at the end of the section.] It also follows from the lemma that for $s = 3$ there is a unique best choice of levels, namely

$$\alpha_2 = \alpha_3 = \alpha \, . \tag{9.53}$$

Having fixed $\alpha_s = \alpha_{s-1} = \alpha$, how should we choose the remaining α's? In order to have a reasonable chance of detecting existing inhomogeneities for all patterns, we should like to have none of the α's too small. In view of part (i) of Lemma 9.3.2, this aim is perhaps best achieved by maximizing α_2, the level at the last stage when individual pairs are being tested.

Lemma 9.3.3 *Under the assumptions of Lemma 9.3.1, the maximum value of* α_2 *subject to (9.47) is*

$$\alpha_2 = 1 - (1 - \alpha)^{[s/2]^{-1}} \tag{9.54}$$

where $[A]$ *denotes the largest integer* $\leq A$.

PROOF. Instead of fixing α and maximizing α_2, it is more convenient to fix α_2, say at α^*, and then to minimize α. The lemma will be proved by showing that the resulting minimum value of α is

$$\alpha = 1 - (1 - \alpha^*)^{[s/2]} \, . \tag{9.55}$$

Suppose first that s is even. Since α_2 is fixed at α^*, it follows from Lemma 9.3.1 that the right side of (9.50) can be made arbitrarily close to α given by (9.55). This is seen by letting $\nu_1 = \cdots = \nu_{s/2} = 2$. When s is odd, the same argument applies if we put an additional ν equal to 1. ■

Lemmas 9.3.2 and 9.3.3 show that any procedure with $\alpha_s = \alpha_2$, and hence Fisher's least-significant-difference procedure and the Newman–Keuls choice of levels, is admissible for $s = 3$ but inadmissible for $s \geq 4$. The second of these statements is seen from the fact that (9.47) implies $\alpha_2 \leq 1 - (1-\alpha)^{[s/2]^{-1}} < \alpha$ when $s \geq 4$. The choice $\alpha_2 = \alpha_s$ thus violates Lemma 9.3.2(ii).

Once α_2 has been fixed at the value given by (9.54), it turns out that subject to (9.47) there exists a unique optimal choice of the remaining α's when s is odd, and a narrow range of choices when s is even.

Theorem 9.3.1 *When s is odd, then $\alpha_3, \ldots, \alpha_s$ are maximized, subject to (9.47) and (9.54), by*

$$\alpha_i^* = 1 - (1-\alpha_2)^{[i/2]} , \tag{9.56}$$

and these values can be attained simultaneously.

Proof. If we put $\gamma_i = 1 - \alpha_i$ and $\gamma = 1 - \alpha_2$, then by (9.49) and (9.56) any procedure satisfying the conditions of the theorem must satisfy

$$\prod \gamma_{v_i} \geq \gamma^{[s/2]} = \gamma^{(s-1)/2}$$

Let i be odd, and consider any configuration in which $v_1 = i$ and all the remaining v's are equal to 2. Then

$$\gamma_i \gamma^{(s-i)/2} \geq \gamma^{(s-1)/2},$$

and hence

$$\gamma_i \geq \gamma_i^* = 1 - \alpha_i^* . \tag{9.57}$$

An analogous argument proves (9.56) for even i.

Consider now the procedure defined by (9.56). This clearly satisfies (9.54), and it only remains to check that it also satisfies (9.47) or equivalently (9.51), and hence that

$$\prod \gamma^{[v_i/2]} \geq \gamma^{(s-1)/2}$$

or that

$$\sum_{i=1}^{r} \left[\frac{v_i}{2}\right] \leq \frac{s-1}{2}$$

Now $\sum[v_i/2] = (s-b)/2$, where b is the number of odd v's (including ones). Since s is odd, $b \geq 1$, and this completes the proof. ■

Note that the levels (9.56) are close to the Tukey levels (vi), which are admissible but do not satisfy (9.54).

When s is even, a uniformly best choice is not available. In this case, the Tukey levels (vi) satisfy (9.54), are admissible, and constitute a reasonable choice. [See Lehmann and Shaffer (1979).]

So far we have assumed $\sigma^2 = 1$ in order to get independence of the r test statistics used for testing the hypotheses H_i', $i = 1, \ldots, r$. If σ^2 is unknown, the χ^2 and range statistics are replaced by F and studentized range statistics. These are no longer independent but are positively quadrant dependent in the sense that

$$P\{T_1' \leq t_1, \ldots, T_r' \leq t_r\} \geq \prod_{i=1}^{r} P\{T_i' \leq t_i\} , \tag{9.58}$$

where $T_1', \ldots, T_r'$ are the test statistics used for testing $H_1', \ldots, H_r'$. This follows from the following lemmas.

Lemma 9.3.4 *Let $F_1(S), \ldots, F_r(S)$ be nondecreasing functions of a random variable S. Then,*

$$E\{\prod_{i=1}^{r} F_i(S)\} \geq \prod_{i=1}^{s} E\{F_i(S)\} , \tag{9.59}$$

provided the expectations exist.

PROOF. By induction, it suffices to consider $r = 2$. To show $Cov[F_1(S), F_2(S)] \geq 0$, assume without loss of generality that $E[F_2(S)] = 0$. Let x be such that $F_2(x) = 0$; if no such x exists, let x be any point satisfying $F_2(y) \geq 0$ if $y > x$ and $F_2(y) \leq 0$ if $y < x$. Now,

$$Cov[F_1(S), F_2(S)] = E\{[F_1(S) - F_1(x)] \cdot F_2(S)\} .$$

If $S \geq x$, $F_1(S) - F_1(x) \geq 0$ and $F_2(S) \geq 0$, and so the quantity inside the expectation is ≥ 0. Similarly, if $S < x$, $F_1(S) - F_1(x) \leq 0$ and $F_2(S) \leq 0$ and so the quantity inside the expectation is ≥ 0. ■

Lemma 9.3.5 *Assume $Y_1, \ldots, Y_r, S$ are independent, where S is a nonnegative random variable. Then, $T_i' = Y_i/S$ satisfy (9.58).*

PROOF. Let G_i denote the distribution of Y_i. Fix $t_1, \ldots, t_r$. By conditioning on S,

$$P\{T_1' \leq t_1, \ldots, T_r' \leq t_r\} = E[\prod_i G_i(t_i S)] .$$

Apply Lemma 9.3.4 with $F_i(s) = G_i(t_i s)$ to get the last quantity is an upper bound for

$$\prod_i E[G_i(t_i S)] = \prod_i P\{T_i' \leq t_i\} . ■$$

For this situation, we have the following result.

Theorem 9.3.2 *If the test statistics for testing the r hypotheses (9.49) are positively quadrant dependent in the sense of (9.58), then*

$$\sup \alpha(\mu_1, \ldots, \mu_s) \leq 1 - \prod_{i=1}^{r}(1 - \alpha_{v_i}) , \tag{9.60}$$

where, as before, $\alpha_1 = 0$.

PROOF. That the right side of (9.60) is an upper bound for $\alpha(\mu_1, \ldots, \mu_s)$ follows from the proof of Lemma 9.3.1 and the assumption of positive quadrant dependence. ■

Note, however, that we can no longer assert that the upper bound is sharp. For the F and Studentized range tests, the sharp upper bound will depend on the total sample size n.

Theorem 9.3.2 guarantees that the procedures using the α-levels derived under the assumption of independence, continue to control the FWER even in the case of positive dependence. The proof of Lemma 9.3.2 shows that $\alpha_s = \alpha_{s-1} = \alpha$ continues to be necessary for admissibility even in the positively dependent case. However, the maximization results for $\alpha_2, \ldots, \alpha_s$ can then no longer be asserted. They nevertheless have the great advantage that they define procedures that do not require detailed knowledge of the joint distribution of the various test statistics.

Even in the simplified version with known variance the multiple testing problem considered in the present section is clearly much more difficult than the testing of a single hypothesis; the procedures presented above still ignore many important aspects of the problem.

1. *Choice of test statistic.* The most obvious feature that has not been dealt with is the choice of test statistics. Unfortunately it does not appear that the invariance considerations which were so helpful in the case of a single hypothesis play a similar role here.

2. *Order relation of significant means.* Whenever two means μ_i and μ_j are judged to differ, we should like to state not only that $\mu_i \neq \mu_j$, but that if $\bar{X}_i < \bar{X}_j$ then also $\mu_i < \mu_j$. Such additional statements introduce the possibility of additional errors (stating $\mu_i < \mu_j$ when in fact $\mu_i > \mu_j$), and it is not obvious that when these are included, the probability of at least one error is still bounded by α. [For recent work on directional errors, see Finner (1999) and Shaffer (1990, 2002).]

3. *Nominal versus true levels.* The levels $\alpha_2, \ldots, \alpha_s$, sometimes called *nominal levels*, are the levels at which the hypotheses $\mu_i = \mu_j$, $\mu_i = \mu_j = \mu_k, \ldots$ are tested. They are however not the true probabilities of falsely rejecting the homogeneity of these sets, but only the upper bounds of these probabilities with respect to variation of the remaining μ's. The true probabilities tend to be much smaller (particularly when s is large), since they take into account that homogeneity of a set S_0 is rejected only if it is also rejected for all sets S containing S_0.

4. *Interpretability.* As pointed out at the beginning of the section, the totality of acceptance and rejection statements resulting from a multiple comparison procedure typically does not lead to a simple partition of means. This is illustrated by the possibility that the hypothesis of homogeneity is rejected for a set S but for none of its subsets. As another example, consider the case $s = 3$, where it may happen that the hypotheses $\mu_i = \mu_j$ and $\mu_j = \mu_k$ are accepted but $\mu_i = \mu_k$ is rejected. The number of such "inconsistencies" and the corresponding difficulty of interpreting the results may be formidable. Measures of the complexity of the totality of statements as a third criterion (besides level and power) are discussed by Shaffer (1981). The inconsistencies and resulting difficulties of interpretation suggest the consideration of an alternative formulation of the problem which avoids this difficulty. Instead of testing the $\binom{s}{2}$ hypotheses $H_{i,j} : \mu_i = \mu_j$, estimate the (unknown) partition of the μ's defined by (9.48). Possible approaches to such procedures are discussed for example in Hochberg and Tamhane (1987, Chapter 10, Section 6) and by Dayton (2003).

5. Procedures (i) and (ii) can be inverted to provide simultaneous confidence intervals for all differences $\mu_j - \mu_i$. The T-method (discussed in Problems 9.29–9.32) was designed to give simultaneous intervals for all differences $\mu_j - \mu_i$; it can be extended to cover also all *contrasts* in the μ's, that is, all linear functions $\sum c_i \mu_i$ with $\sum c_i = 0$, but against more complex contrasts the intervals tend to be longer than those of Scheffés S-method, which was intended for the simultaneous consideration of all contrasts. [For a comparison of the two methods, see for example Scheffé (1959, Section 3.7) and Arnold (1981, Chapter 12).] It is a disadvantage of the remaining (truly stagewise or sequential) procedures of this section that the problem of corresponding confidence sets is considerably more complicated. For a discussion of such confidence methods, see Holm (1999) and the references cited there.

6. To control the rate of false rejections, we have restricted attention to procedures controlling the FWER, the probability of at least one error. Instead, one might wish to control the false discovery rate as defined at the end of Section 9.1; see Benjamini and Hochberg (1995). Alternatively, an optimality theory based on the number of false rejections is given in Spjøtvoll (1972). Another possibility is the control the k-FWER, the probability of making k or more false rejections, as well as the probability that the false discovery proportion exceeds some threshold; see Korn et al. (2004), Romano and Shaikh (2004) and Lehmann and Romano (2005).

7. The optimal choice of the α_k discussed in this section can be further improved, at the cost of considerable additional complication, by permitting the α's to depend on the outcomes of the other tests. This possibility is discussed, for example, in Marcus, Peritz, and Gabriel (1976); see also Holm (1979) and Shaffer (1984).

The procedures discussed in this section were concerned with testing the equality of means. In more complex situations, further problems arise. Consider, for example, the two-way layout of 7.5 with

$$\mu_{i,j} = \mu + \alpha_i + \beta_j + \gamma_{i,j} \qquad (\sum \alpha_i = \sum \beta_j = \sum_i \gamma_{i,j} = \sum_j \gamma_{i,j} = 0) \ .$$

If we are interested in multiple testing of the α's, β's, and γ's, the first question that arises is whether we want to treat these three cases (α's, β's, γ's) as a single family, as two families (the main effects forming one family, the interactions the other), or as three families in which each of the three sets is handled separately.

The most appropriate designation of what constitutes a family depends very much on context. Consider, for example, the National Assessment of Educational Progress which makes it possible to compare the progress made by any two states. For a federal report, the set of all $\binom{50}{2}$ possible hypotheses would constitute an appropriate family. However, a particular state would be interested primarily in the comparison of its performance with those of the other 49 states, thus leading to a family of size 49. A comparison which is not significant in the federal report might then turn out to be significant in the state report. Some of the issues concerning the most suitable definition of family are discussed in Tukey (1991) and in the books by Hochberg and Tamhane (1987), and Westfall and Young (1993).

We shall in the next two sections consider simultaneous inferences for various families of linear functions of means in normal linear models. However, since we are assuming fully articulated parametric models, we shall consider the slightly more demanding problem of obtaining simultaneous confidence intervals rather than restricting attention to hypothesis testing.

As the simplest example, suppose that $X_1, \ldots, X_s$ are normal variables with means $\mu_1, \ldots, \mu_s$ and unit variance. We can then apply to the hypotheses $H_i : \mu_i = \mu_{i,0}$ the approach of Section 9.1 and test these hypotheses by means of a stepdown procedure. The resulting acceptance regions can then be converted in the usual way into confidence sets. It is shown in Holm (1999) that these sets are rather complicated and not rectangular, so that they do not consist of intervals for the individual μ_i's. (They can, of course, be enclosed in a larger rectangle, but the intervals obtained by such a process tend to be unnecessarily large.)

9.4 Scheffé's S-Method: A Special Case

If $X_1, \ldots, X_r$ are independent normal with common variance σ^2 and expectations $E(X_i) = \alpha + \beta t_i$, confidence sets for (α, β) were obtained in Section 7.6. A related problem is that of determining confidence bands for the whole regression line $\xi = \alpha + \beta t$, that is, functions $L'(t; X)$, $M'(t; X)$ such that

$$P\{L'(t; X) \le \alpha + \beta t \le M'(t; X) \text{ for all } t\} = \gamma. \tag{9.61}$$

The problem of obtaining simultaneous confidence intervals for a continuum of parametric functions arises also in other contexts. In the present section, a general problem of this kind will be considered for linear models. Confidence bands for an unknown distribution function were treated in Section 6.13.

Suppose first that $X_1, \ldots, X_r$ are independent normal with variance $\sigma^2 = 1$ and with means $E(X_i) = \xi_i$, and that simultaneous confidence intervals are required for all linear functions $\sum u_i \xi_i$. No generality is lost by dividing $\sum u_i \xi_i$ and its lower and upper bound by $\sqrt{\sum u_i^2}$, so that attention can be restricted to confidence sets

$$S(x) = \{\xi : \; L(u; x) \le \sum u_i \xi_i \le M(u; x) \quad \text{for all} \;\; u \in U\}\;, \tag{9.62}$$

where x, u denote both the vectors with coordinates x_i, u_i and the $r \times 1$ column matrices with these elements, and where U is the set of all u with $\sum u_i^2 = 1$. The sets $S(x)$ are to satisfy

$$P_\xi\{\xi \in S(X)\} = \gamma \;\text{ for all }\; \xi = (\xi_1, \ldots, \xi_r). \tag{9.63}$$

Since $u = (u_1, \ldots, u_r) \in U$ if and only if $-u = (-u_1, \ldots, -u_r) \in U$, the simultaneous inequalities (9.62) imply $L(-u; x) \le -\sum u_i \xi_i \le M(-u; x)$, and hence

$$-M(-u; x) \le \sum u_i \xi_i \le -L(-u; x)$$

and

$$\max(L(u; x), -M(-u; x)) \le \sum u_i \xi_i \le \min(M(u; x), -L(-u; x)).$$

Nothing is therefore lost by assuming that L and M satisfy

$$L(u;x) = -M(-u;x). \tag{9.64}$$

The problem of determining suitable confidence bounds $L(u;x)$ and $M(u;x)$ is invariant under the group G_1 of orthogonal transformations

$$G_1 : gx = Qx, \bar{g}\xi = Q\xi \qquad (Q \text{ an orthogonal } r \times r \text{ matrix}).$$

Writing $\sum u_i\xi_i = u'\xi$, we have

$$\begin{aligned} g^*S(x) &= \{Q\xi : L(u;x) \le u'\xi \le M(u;x) \text{ for all } u \in U\} \\ &= \{\xi : L(u;x) \le u'(Q^{-1}\xi) \le M(u;x) \text{ for all } u \in U\} \\ &= \{\xi : L(Q^{-1}u;x) \le u'\xi \le M(Q^{-1}u;x) \text{ for all } u \in U\}, \end{aligned}$$

where the last equality uses the fact that U is invariant under orthogonal transformations of u.

Since

$$S(gx) = \{\xi : L(u;Qx) \le u'\xi \le M(u;Qx) \text{ for all } u \in U\},$$

the confidence sets $S(x)$ are equivariant under G_1 if and only if

$$L(u;Qx) = L(Q^{-1}u;x), \qquad M(u;Qx) = M(Q^{-1}u;x),$$

or equivalently if

$$L(Qu;Qx) = L(u;x), \qquad M(Qu;Qx) = M(u;x) \quad \text{for all} \quad x,\ Q \text{ and } u \in U, \tag{9.65}$$

that is, if L and M are invariant under common orthogonal transformations of u and x.

A function L of u and x is invariant under these transformations if and only if it depends on u and x only through $u'x$, $x'x$, and $u'u$ [Problem 9.23(i)] and hence (since $u'u = 1$) if there exists h such that

$$L(u;x) = h(u'x, x'x). \tag{9.66}$$

A second group of transformations leaving the problem invariant is the group of translations

$$G_2 : gx = x + a, \bar{g}\xi = \xi + a$$

where $x + a = (x_1 + a_1, \ldots, x_r + a_r)$. An argument paralleling that leading to (9.65) shows that $L(u;x)$ is equivariant under G_2 if and only if [Problem 9.23(ii)]

$$L(u;x+a) = L(u;x) + \sum a_i u_i \quad \text{for all} \quad x, a, \text{ and } u. \tag{9.67}$$

The function h of (9.66) must therefore satisfy

$$h[u'(x+a), (x+a)'(x+a)] = h(u'x, x'x) + a'u \quad \text{for all} \quad a, x \text{ and } u \in U,$$

and hence, putting $x = 0$,

$$h(u'a, a'a) = a'u + h(0,0).$$

A necessary condition (which clearly is also sufficient) for $S(x)$ to be equivariant under both G_1 and G_2 is therefore the existence of constants c and d such that

$$S(x) = \left\{\xi : \sum u_i x_i - c \leq \sum u_i \xi_i \leq \sum u_i x_i + d \text{ for all } u \in U\right\}$$

From (9.64) it follows that $c = d$, so that the only equivariant families $S(x)$ are given by

$$S(x) = \left\{\xi : \left|\sum u_i(x_i - \xi_i)\right| \leq c \text{ for all } u \in U\right\} \tag{9.68}$$

The constant c is determined by (9.63), which now reduces to

$$P_0 \left\{\left|\sum u_i X_i\right| \leq c \text{ for all } u \in U\right\} = \gamma. \tag{9.69}$$

By the Schwarz inequality $(\sum u_i X_i)^2 \leq \sum X_i^2$, since $\sum u_i^2 = 1$, and hence

$$\left|\sum u_i X_i\right| \leq c \text{ for all } u \in U \text{ if and only if } \sum X_i^2 \leq c^2. \tag{9.70}$$

The constant c in (9.68) is therefore given by

$$P(\chi_r^2 \leq c^2) = \gamma. \tag{9.71}$$

In (9.68), it is of course possible to drop the restriction $u \in U$ by writing (9.68) in the equivalent form

$$S(x) = \left\{\xi : \left|\sum u_i(x_i - \xi_i)\right| \leq c\sqrt{\sum u_i^2} \text{ for all } u\right\}. \tag{9.72}$$

So far attention has been restricted to the confidence bands (9.62). However, confidence sets do not have to be intervals, and it may be of interest to consider more general simultaneous confidence sets

$$S(x) : \sum u_i \xi_i \in A(u, x) \text{ for all } u \in U. \tag{9.73}$$

For these sets, the equivariance conditions (9.65) and (9.67) become respectively (Problem 9.24)

$$A(Qu, Qx) = A(u, x) \text{ for all } x, Q \text{ and } u \in U \tag{9.74}$$

and

$$A(u, x + a) = A(u, x) + u'a \text{ for all } u, x, \text{ and } a. \tag{9.75}$$

The first of these is equivalent to the condition that the set $A(u, x)$ depends on $u \in U$ and x only through $u'x$ and $x'x$. On the other hand putting $x = 0$ in (9.75) gives

$$A(u, a) = A(u, 0) + u'a.$$

It follows from (9.74) that $A(u, 0)$ is a fixed set A_1 independent of u, so that

$$A(u, x) = A_1 + u'x. \tag{9.76}$$

The most general equivariant sets (under G_1 and G_2) are therefore of the form

$$\sum u_i(x_i - \xi_i) \in A \text{ for all } u \in U, \tag{9.77}$$

where $A = -A_i$.

We shall now suppose that $r > 1$ and then show that among all A which define confidence sets (9.77) with confidence coefficient $\geq \gamma$, the sets (9.68) are smallest[3] in the very strong sense that if $A_0 = [-c_0, c_0]$ denotes the set (9.68) with confidence coefficient γ, then A_0 is a subset of A.

To see this, note that if $Y_i = X_i - \xi_i$, the sets A are those satisfying

$$P\left(\sum u_i Y_i \in A \text{ for all } u \in U\right) \geq \gamma. \tag{9.78}$$

Now the set of values taken on by $\sum u_i y_i$ for a fixed $y = (y_1, \ldots, y_r)$ as u ranges over U is the interval (Problem 9.24)

$$I(y) = \left[-\sqrt{\sum y_i^2}, +\sqrt{\sum y_i^2}\right].$$

Let c^* be the largest value of c for which the interval $[-c, c]$ is contained in A. Then the probability (9.78) is equal to

$$P\{I(Y) \subset A\} = P\{I(Y) \subset [-c^*, c^*]\}.$$

Since $P\{I(Y) \subset A\} \geq \gamma$, it follows that $c^* \geq c_0$, and this completes the proof.

It is of interest to compare the simultaneous confidence intervals (9.68) for all $\sum u_i \xi_i$, $u \in U$, with the joint confidence spheres for $(\xi_1, \ldots, \xi_r)$ given by (6.43). These two sets of confidence statements are equivalent in the following sense.

Theorem 9.4.1 *The parameter vector $(\xi_1, \ldots, \xi_r)$ satisfies $\sum (X_i - \xi_i)^2 \leq c^2$ if and only if it satisfies (9.68).*

PROOF. The result follows immediately from (9.70) with X_i replaced by $X_i - \xi_i$. ■

Another comparison of interest is that of the simultaneous confidence intervals (9.72) for all u with the corresponding interval

$$S'(x) = \left\{\xi : \left|\sum u_i(x_i - \xi_i)\right| \leq c'\sqrt{\sum u_i^2}\right\} \tag{9.79}$$

for a single given u. Since $\sum u_i(X_i - \xi_i)/\sqrt{\sum u_i^2}$ has a standard normal distribution, the constant c' is determined by $P(\chi_1^2 \leq c'^2) = \gamma$ instead of by (9.71). If $r > 1$, the constant $c^2 = c_r^2$ is clearly larger than $c'^2 = c_1^2$. The lengthening of the confidence intervals by the factor c_r/c_1 in going from (9.79) to (9.72) is the price one must pay for asserting confidence γ for all $\sum u_i \xi_i$ instead of a single one.

In (9.79), it is assumed that the vector u defines the linear combination of interest and is given before any observations are available. However, it often happens that an interesting linear combination $\sum \hat{u}_i \xi_i$ to be estimated is suggested by the data. The intervals

$$\left|\sum \hat{u}_i(x_i - \xi_i)\right| \leq c\sqrt{\sum \hat{u}_i^2} \tag{9.80}$$

with c given by (9.71) then provide confidence limits for $\sum \hat{u}_i \xi_i$ at confidence level γ, since they are included in the set of intervals (9.72). [The notation $\hat{u}_i$

[3]A more general definition of smallness is due to Wijsman (1979). It has been pointed out by Professor Wijsman that his concept is equivalent to that of tautness defined by Wynn and Bloomfield (1971).

in (9.80) indicates that the u's were suggested by the data rather than fixed in advance.]

Example 9.4.1 (Two groups) Suppose the data exhibit a natural split into a lower and upper group, say $\xi_{i_1}, \ldots, \xi_{i_k}$, and $\xi_{j_1}, \ldots, \xi_{j_{r-k}}$, with averages $\bar{\xi}_-$ and $\bar{\xi}_+$, and that confidence limits are required for $\bar{\xi}_+ - \bar{\xi}_-$. Letting $\bar{X}_- = (X_{i_1} + \cdots + X_{i_k})/k$ and $\bar{X}_+ = (X_{j_1} + \cdots + X_{j_{r-k}})/(r-k)$ denote the associated averages of the X's we see that

$$\bar{X}_+ - \bar{X}_- - c\sqrt{\frac{1}{k} + \frac{1}{r-k}} \le \bar{\xi}_+ - \bar{\xi}_- \le \bar{X}_+ - \bar{X}_- + c\sqrt{\frac{1}{k} + \frac{1}{r-k}} \tag{9.81}$$

with c given by (9.71) provide the desired limits. Similarly

$$\bar{X}_- - \frac{c}{\sqrt{k}} \le \bar{\xi}_- \le \bar{X}_- + \frac{c}{\sqrt{k}}, \qquad \bar{X}_+ - \frac{c}{\sqrt{r-k}} \le \bar{\xi}_+ \le \bar{X}_+ + \frac{c}{\sqrt{r-k}} \tag{9.82}$$

provide simultaneous confidence intervals for the two group means separately, with c again given by (9.71). For a discussion of related examples and issues see Peritz (1965). ■

Instead of estimating a data-based function $\sum \hat{u}_i \xi_i$, one may be interested in testing it. At level $\alpha = 1 - \gamma$, the hypothesis $\sum \hat{u}_i \xi_i = 0$ is rejected when the confidence intervals (9.80) do not cover the origin, i.e., when

$$\left|\sum \hat{u}_i x_i\right| \ge c\sqrt{\sum \hat{u}_i^2}.$$

Equivariance with respect to the group G_1 of orthogonal transformations assumed at the beginning of this section is appropriate only when all linear combinations $\sum u_i \xi_i$ with $u \in U$ are of equal importance. Suppose instead that interest focuses on the individual means, so that simultaneous confidence intervals are required for $\xi_1, \ldots, \xi_r$. This problem remains invariant under the translation group G_2. However, it is no longer invariant under G_1, but only under the much smaller subgroup G_0 generated by the $n!$ permutations and the 2^n changes of sign of the X's. The only simultaneous intervals that are equivariant under G_0 and G_2 are given by [Problem 9.25(i)]

$$S(x) = \{\xi : x_i - \Delta \le \xi_i \le x_i + \Delta \text{ for all } i\} \tag{9.83}$$

where Δ is determined by

$$P[S(X)] = P(\max |Y_i| \le \Delta) = \gamma \tag{9.84}$$

with $Y_1, \ldots, Y_r$ being independent $N(0, 1)$.

These *maximum-modulus* intervals for the ξ's can be extended to all linear combinations $\sum u_i \xi_i$ of the ξ's by noting that the right side of (9.83) is equal to the set [Problem 9.25(ii)]

$$\left\{\xi : \left|\sum u_i(X_i - \xi_i)\right| \le \Delta \sum |u_i| \text{ for all } u\right\}, \tag{9.85}$$

which therefore also has probability γ, but which is not equivariant under G_1. A comparison of the intervals (9.85) with the Scheffé intervals (9.72) shows [Problem 9.25(iii)] that the intervals (9.85) are shorter when $\sum u_j \xi_j = \xi_i$ (i.e. when $u_j = 1$ for $j = i$, and $u_j = 0$ otherwise), but that they are longer for example when $u_1 = \cdots = u_r$.

9.5 Scheffé's S-Method for General Linear Models

The results obtained in the preceding section for the simultaneous estimation of all linear functions $\sum u_i \xi_i$ when the common variance of the variables X_i is known easily extend to the general linear model of Section 7.1. In the canonical form (7.2), the observations are n independent normal random variables with common unknown variance σ^2 and with means $E(Y_i) = \eta_i$ for $i = 1, \ldots, s$ and $E(Y_i) = 0$ for $i = s+1, \ldots, n$. Simultaneous confidence intervals are required for all linear functions $\sum_{i=1}^r u_i n_i$ with $u \in U$, where U is the set of all $u = (u_1, \ldots, u_r)$ with $\sum_{i=1}^r u_i^2 = 1$. Invariance under the translation group $Y_i' = Y_i + a_i$, $i = r+1, \ldots, s$, leaves $Y_1, \ldots, Y_r$; $Y_{s+1}, \ldots, Y_n$ as maximal invariants, and sufficiency justifies restricting attention to $Y = (Y_1, \ldots, Y_r)$ and $S^2 = \sum_{j=s+1}^n Y_j^2$. The confidence intervals corresponding to (9.62) are therefore of the form

$$L(u; y, S) \leq \sum_{i=1}^{r} u_i \eta_i \leq M(u; y, S) \qquad \text{for all} \quad u \in U, \tag{9.86}$$

and in analogy to (9.64) may be assumed to satisfy

$$L(u; y, S) = -M(-u; y, S). \tag{9.87}$$

By the argument leading to (9.66), it is seen in the present case that equivariance of $L(u; y, S)$ under G_1 requires that

$$L(u; y, S) = h(u'y, y'y, S),$$

and equivariance under G_2 requires that L be of the form

$$L(u; y, S) = \sum_{i=1}^{r} u_i y_i - c(S).$$

Since σ^2 is unknown, the problem is now also invariant under the group of scale changes

$$G_3 : y_i' = b y_i \ (i = 1, \ldots, r), S' = bS \ (b > 0).$$

Equivariance of the confidence intervals under G_3 leads to the condition [Problem 9.26(i)]

$$L(u; by, bS) = bL(u; y, S) \qquad \text{for all} \quad b > 0,$$

and hence to

$$b \sum u_i y_i - c(bS) = b\left[\sum u_i y_i - c(S)\right],$$

or $c(bS) = bc(S)$. Putting $S = 1$ shows that $c(S)$ is proportional to S. Thus

$$L(u; y, S) = \sum u_i y_i - cS, \qquad M(u; y, S) = \sum u_i y_i + dS,$$

and by (9.87), $c = d$, so that the equivariant simultaneous intervals are given by

$$\sum u_i y_i - cS \leq \sum u_i \eta_i \leq \sum u_i y_i + cS \ \text{ for all } \ u \in U. \tag{9.88}$$

Since (9.88) is equivalent to

$$\frac{\sum (y_i - \eta_i)^2}{S^2} \leq c^2,$$

the constant c is determined from the F-distribution by

$$P_0\left\{\frac{\sum Y_i^2/r}{S^2/(n-s)} \le \frac{n-s}{r}c^2\right\} = P_0\left\{F_{r,n-s} \le \frac{n-s}{r}c^2\right\} = \gamma. \tag{9.89}$$

As in (9.72), the restriction $u \in U$ can be dropped; this only requires replacing c in (9.88) and (9.89) by $c\sqrt{\sum u_i^2} = c\sqrt{\operatorname{Var}\sum u_i Y_i/\sigma^2}$.

As in the case of known variance, instead of restricting attention to the confidence bands (9.88), one may wish to permit more general simultaneous confidence sets

$$\sum u_i \eta_i \in A(u; y, S). \tag{9.90}$$

The most general equivariant confidence sets are then of the form [Problem 9.26(ii)]

$$\frac{\sum u_i(y_i - \eta_i)}{S} \in A \quad \text{for all} \quad u \in U, \tag{9.91}$$

and for a given confidence coefficient, the set A is minimized by $A_0 = [-c, c]$, so that (9.91) reduces to (9.88).

For applications, it is convenient to express the intervals (9.88) in terms of the original variables X_i and ξ_i. Suppose as in Section 7.1 that $X_1, \ldots, X_n$ are independently distributed as $N(\xi_i, \sigma^2)$, where $\underset{\sim}{\xi} = (\xi_1, \ldots, \xi_n)$ is assumed to lie in a given s-dimensional linear subspace $\prod_\Omega$ $(s < n)$. Let V be an r-dimensional subspace of $\prod_\Omega$ $(r \le s)$, let $\hat{\xi}_i$ be the least squares estimates of the ξ's under $\prod_\Omega$, and let $S^2 = \sum (X_i - \hat{\xi}_i)^2$. Then the inequalities

$$\sum v_i\hat{\xi}_i - cS\sqrt{\frac{\operatorname{Var}\left(\sum v_i\hat{\xi}_i\right)}{\sigma^2}} \le \sum v_i \xi_i \le \sum v_i\hat{\xi}_i + cS\sqrt{\frac{\operatorname{Var}\left(\sum v_i\hat{\xi}_i\right)}{\sigma^2}}$$
$$\text{for all} \quad v \in V, \tag{9.92}$$

with c given by (9.89), provide simultaneous confidence intervals for $\sum v_i\xi_i$ for all $v \in V$ with confidence coefficient γ.

This result is an immediate consequence of (9.88) and (9.89) together with the following three facts, which will be proved below:

(i) If $\sum_{i=1}^s u_i\eta_i = \sum_{j=1}^n v_j\xi_j$, then $\sum_{i=1}^s u_i Y_i = \sum_{j=1}^n v_j\hat{\xi}_j$;

(ii) $\sum_{i=s+1}^n Y_i^2 = \sum_{j=1}^n (X_j - \hat{\xi}_j)^2$,

To state (iii), note that the η's are obtained as linear functions of the ξ's through the relationship

$$(\eta_1, \ldots, \eta_r, \eta_{r+1}, \ldots, \eta_s, 0, \ldots, 0)' = C(\xi_1, \ldots, \xi_n)' \tag{9.93}$$

where C is defined by (7.1) and the prime indicates a transpose. This is seen by taking the expectation of both sides of (7.1). For each vector $u = (u_1, \ldots, u_r)$, (9.93) expresses $\sum u_i\eta_i$ as a linear function $\sum v_j^{(u)}\xi_j$ of the ξ's.

(iii) As u ranges over r-space, $v^{(u)} = (v_1^{(u)}, \ldots, v_n^{(u)})$ ranges over V.

PROOF OF (i) Recall from Section 7.2 that

$$\sum_{j=1}^{n}(X_j - \xi_j)^2 = \sum_{i=1}^{s}(Y_i - \eta_i)^2 + \sum_{j=s+1}^{n} Y_j^2.$$

Since the right side is minimized by $\eta_i = Y_i$ and the left side by $\xi_j = \hat{\xi}_j$, this shows that

$$(Y_1 \cdots Y_s 0 \cdots 0)' = C(\hat{\xi}_1 \cdots \hat{\xi}_j)',$$

and the result now follows from comparison with (9.93).

PROOF OF (ii) This is just equation (7.13).

PROOF OF (iii) Since $\eta_i = \sum_{j=1}^{n} c_{ij}\xi_j$, we have $\sum u_i\eta_i = \sum v_j^{(u)}\xi_j$ with $v_j^{(u)} = \sum_{i=1}^{r} u_i c_{ij}$. Thus, the vectors $v^{(u)} = (v_1^{(u)}, \ldots, v_n^{(u)})$ are linear combinations, with weights $u_1, \ldots, u_r$, of the first r row vectors of C. Since the space spanned by these row vectors is V, the result follows.

The set of linear functions $\sum v_i\xi_i$, $v \in V$, for which the interval (9.92) does not cover the origin—that is, for which v satisfies

$$\left|\sum v_i\hat{\xi}_i\right| > cS\sqrt{\frac{\operatorname{Var}\left(\sum v_i\hat{\xi}_i\right)}{\sigma^2}} \tag{9.94}$$

—is declared significantly different from 0 by the intervals (9.92). Thus (9.94) is a rejection region at level $\alpha = 1 - \gamma$ of the hypothesis $H : \sum v_i\xi_i = 0$ for all $v \in V$ in the sense that H is rejected if and only if at least one $v \in V$ satisfies (9.94). If $\prod_\omega$ denotes the $(s-r)$-dimensional space of vectors $v \in \prod_\Omega$ which are orthogonal to V, then H states that $\xi \in \prod_\omega$, and the rejection region (9.94) is in fact equivalent to the F-test of $H : \xi \in \prod_\omega$ of Section 7.1. In canonical form, this was seen in the sentence following (9.88).

To implement the intervals (9.92) in specific situations in which the corresponding intervals for a single given function $\sum v_i\xi_i$ are known, it is only necessary to designate the space V and to obtain its dimension r, the constant c then being determined by (9.89).

Example 9.5.1 (All contrasts) Let X_{ij} $(j = 1, \ldots, n_i; i = 1, \ldots, s)$ be independently distributed as $N(\xi_i, \sigma^2)$, and is suppose V is the space of all vectors $v = (v_1, \ldots, v_n)$ satisfying

$$\sum v_i = 0. \tag{9.95}$$

Any function $\sum v_i\xi_i$ with $v \in V$ is called a *contrast* among the ξ_i. The set of contrasts includes in particular the differences $\bar{\xi}_+ - \bar{\xi}_-$ discussed in Example 9.4.1. The space $\prod_\Omega$ is the set of all vectors $(\xi_1, \ldots, \xi_1; \xi_2, \ldots, \xi_2; \xi_s, \ldots, \xi_s)$ and has dimension s, while V is the subspace of vectors $\prod_\Omega$ that are orthogonal to $(1, \ldots, 1)$ and hence has dimension $r = s - 1$. It was seen in Section 7.3 that

$\hat{\xi}_i = X_{i\cdot}$, and if the vectors of V are denoted by

$$\left(\frac{w_1}{n_1}, \ldots, \frac{w_1}{n_1}; \frac{w_2}{n_2}, \ldots, \frac{w_2}{n_2}; \frac{w_s}{n_s}, \ldots, \frac{w_s}{n_s}\right),$$

the simultaneous confidence intervals (9.92) become (Problem 9.28)

$$\sum w_i X_{i\cdot} - cS\sqrt{\frac{\sum w_i^2}{n_i}} \le \sum w_i \xi_i \le \sum w_i X_{i\cdot} + cS\sqrt{\frac{\sum w_i^2}{n_i}} \tag{9.96}$$

$$\text{for all } (w_1, \ldots, w_s) \text{ satisfying } \sum w_i = 0,$$

with $S^2 = \sum\sum (X_{ij} - X_{i\cdot})^2$.

In the present case the space $\prod_\omega$ is the set of vectors with all coordinates equal, so that the associated hypothesis is $H : \xi_1 = \cdots = \xi_s$. The rejection region (9.94) is thus equivalent to that given by (7.19).

Instead of testing the overall homogeneity hypothesis H, we may be interested in testing one or more subhypotheses suggested by the data. In the situation corresponding to that of Example 9.4.1 (but with replications), for instance, interest may focus on the hypotheses $H_1 : \xi_{i_1} = \cdots = \xi_{i_k}$ and $H_2 : \xi_{j_1} = \cdots = \xi_{j_{s-k}}$. A level α simultaneous test of H_1 and H_2 is given by the rejection region

$$\frac{\sum^{(1)} n_i (X_{i\cdot} - X_{\cdot\cdot}^{(1)})^2/(k-1)}{S^2/(n-s)} > C, \qquad \frac{\sum^{(2)} n_i (X_{i\cdot} - X_{\cdot\cdot}^{(2)})^2/(s-k-1)}{S^2/(n-s)} > C,$$

where $\sum^{(1)}$, $\sum^{(2)}$, $X_{\cdot\cdot}^{(1)}$, $X_{\cdot\cdot}^{(2)}$ indicate that the summation or averaging extends over the sets $(i_1, \ldots, i_k)$ and $(j_1, \ldots, j_{s-k})$ respectively, $S^2 = \sum\sum (X_{ij} - X_{i\cdot})^2$, $\alpha = 1 - \gamma$, and the constant C is given by (9.89) with $r = s$ and is therefore the same as in (7.19), rather than being determined by the $F_{k-1,n-s}$ and $F_{s-k-1,n-s}$ distributions. The reason for this larger critical value is, of course, the fact the H_1 and H_2 were suggested by the data. The present procedure is an example of Gabriel's simultaneous test procedure mentioned in Section 9.3. ■

Example 9.5.2 (Two-way layout) As a second example, consider first the additive model in the two-way classification of Section 7.4 or 7.5, and then the more general interaction model of Section 7.5.

Suppose X_{ij} are independent $N(\xi_{ij}, \sigma^2)$ $(i = 1, \ldots, a;\ j = 1, \ldots, b)$, with ξ_{ij} given by (7.20), and let V be the space of all linear functions $\sum w_i \alpha_i = \sum w_i(\xi_{i\cdot} - \xi_{\cdot\cdot})$. As was seen in Section 7.4, $s = a + b - 1$. To determine r, note that V can also be represented as $\sum_{i=1} w_i \xi_{i\cdot}$ with $\sum w_i = 0$ [Problem 9.27(i)], which shows that $r = a - 1$. The least-squares estimators $\hat{\xi}_i$ were found in Section 7.4 to be $\hat{\xi}_{ij} = X_{i\cdot} + X_{\cdot j} - X_{\cdot\cdot}$, so that $\hat{\xi}_{i\cdot} = X_{i\cdot}$ and $S^2 = \sum\sum (X_{ij} - X_{i\cdot} - X_{\cdot j} + X_{\cdot\cdot})^2$. The simultaneous confidence intervals (9.92) therefore can be written as

$$\sum w_i X_{i\cdot} - cS\sqrt{\frac{\sum w_i^2}{b}} \le \sum w_i \xi_{i\cdot} \le \sum w_i X_{i\cdot} + cS\sqrt{\frac{\sum w_i^2}{b}}$$

$$\text{for all } w \text{ with } \sum_{i=1}^{a} w_i = 0.$$

If there are m observations in each cell, and the model is additive as before, the only changes required are to replace $X_{i\cdot}$ by $X_{i\cdot\cdot}$, S^2 by $\sum\sum\sum (X_{ijk} - X_{i\cdot\cdot} - X_{\cdot j\cdot} + X_{\cdot\cdot\cdot})^2$, and the expression under the square root by $\sum w_i^2/bm$.

Let us now drop the assumption of additivity and consider the general linear model $\xi_{ijk} = \mu + \alpha_i + \beta_j + \gamma_{ij}$, with μ and the α's, β's, and γ's defined as in Section 7.5. The dimension s of $\prod_\Omega$ is then ab, and the least squares estimators of the parameters were seen in Section 7.5 to be

$$\hat{\mu} = X_{\cdots}, \qquad \hat{\alpha}_i = X_{i\cdot\cdot} - X_{\cdots}, \qquad \hat{\beta}_j = X_{\cdot j\cdot} - X_{\cdots},$$
$$\hat{\gamma}_{ij} = X_{ij\cdot} - X_{i\cdot\cdot} - X_{\cdot j\cdot} + X_{\cdots}$$

The simultaneous intervals for all $\sum w_i\alpha_i$, or for all $\sum w_i\xi_{i\cdot\cdot}$ with $\sum w_i = 0$, are therefore unchanged except for the replacement of $S^2 = \sum(X_{ijk} - X_{i\cdot\cdot} - X_{\cdot j\cdot} + X_{\cdots})^2$ by $S^2 = \sum(X_{ijk} - X_{ij\cdot})^2$ and of $n - s = n - a - b + 1$ by $n - s = n - ab = (m-1)ab$ in (9.89).

Analogously, one can obtain simultaneous confidence intervals for the totality of linear functions $\sum w_{ij}\gamma_{ij}$, or equivalently the set of functions $\sum w_{ij}\xi_{ij\cdot}$ for the totality of w's satisfying $\sum_i w_{ij} = \sum_j w_{ij} = 0$ [Problem 9.27(ii), (iii)]. ■

Example 9.5.3 (Regression line) As a last example consider the problem of obtaining confidence bands for a regression line, mentioned at the beginning of the section. The problem was treated for a single value t_0 in Section 5.6 (with a different notation) and in Section 7.6. The simultaneous confidence intervals in the present case become

$$\begin{aligned} \hat{\alpha} + \hat{\beta}t - cS\left[\frac{1}{n} + \frac{(t-\bar{t})^2}{\sum(t_i - \bar{t})^2}\right]^{1/2} &\leq \alpha + \beta t \\ &\leq \hat{\alpha} + \hat{\beta}t + cS\left[\frac{1}{n} + \frac{(t-\bar{t})^2}{\sum(t_i - \bar{t})^2}\right]^{1/2}, \end{aligned} \tag{9.97}$$

where $\hat{\alpha}$ and $\hat{\beta}$ are given by (7.23),

$$S^2 = \sum(X_i - \hat{\alpha} - \hat{\beta}t_i)^2 = \sum(X_i - \bar{X})^2 - \hat{\beta}^2\sum(t_i - \bar{t})^2$$

and c is determined by (9.89) with $r = s = 2$. This is the Working–Hotelling confidence band for a regression line. ■

At the beginning of the section, the Scheffé intervals were derived as the only confidence bands that are equivariant under the indicated groups. If the requirement of equivariance (particular under orthogonal transformations) is dropped, other bounds exist which are narrower for certain sets of vectors u at the cost of being wider for others [Problems 9.26(iii) and 9.32]. A general method that gives special emphasis to a given subset is described by Richmond (1982). Some optimality results not requiring equivariance but instead permitting bands which are narrower for some values of t at the expense of being wider for others are provided, among others, by Bohrer (1973), Cima and Hochberg (1976), Richmond (1982), Naiman (1984a,b), and Piegorsch (1985a, b). If bounds are required only for a subset, it may be possible that intervals exist at the prescribed confidence level, which are uniformly narrower than the Scheffé intervals. This is the case for example for the intervals (9.97) when t is restricted to a given finite interval. For a discussion of this and related problems, and references to the literature, see for example Wynn and Bloomfield (1971) and Wynn (1984).

9.6 Problems

Section 9.1

Problem 9.1 Show the Bonferroni procedure, while generally conservative, can have FWER $= \alpha$ by exhibiting a joint distribution for $(\hat{p}_1, \ldots, \hat{p}_s)$ and satisfying (9.4) such that $P\{\min_i \hat{p}_i \leq \alpha/s\} = \alpha$.

Problem 9.2 (i) Under the assumptions of Theorem 9.1.1, suppose also that the p-values are mutually independent. Then, the procedure which rejects any H_i for which $\hat{p}_i < c(\alpha, s) = 1 - (1-\alpha)^{1/s}$ controls the FWER.
(i) Compare α/s with $c(\alpha, s)$ and show

$$\lim_{s \to \infty} \frac{c(\alpha, s)}{(\alpha/s)} = \frac{-\log(1-\alpha)}{\alpha} .$$

For $\alpha = .05$, this limiting value to 3 decimals is 1.026, so the increase in cutoff value is not substantial.

Problem 9.3 Show that, under the assumptions of Theorem 9.1.2, it is not possible to increase any of the critical values $\alpha_i = \alpha/(s-i+1)$ in the Holm procedure (9.6) without violating the FWER.

Problem 9.4 Under the assumptions of Theorem 9.1.2 and independence of the p-values, the critical values $\alpha/(s-i+1)$ can be increased to $1-(1-\alpha)^{1/(s-i+1)}$. For any i, calculate the limiting value of the ratio of these critical values, as $s \to \infty$.

Problem 9.5 In Example 9.1.4, verify that the stepdown procedure based on the maximum of $X_j/\sqrt{\sigma_{j,j}}$ improves upon the Holm procedure. By Theorem 9.1.3, the procedure has FWER $\leq \alpha$. Compare the two procedures in the case $\sigma_{i,i} = 1$, $\sigma_{i,j} = \rho$ if $i \neq j$; consider $\rho = 0$ and $\rho \to \pm 1$.

Problem 9.6 Suppose H_i is specifies the unknown probability P belongs to a subset of the parameter space ω_i, for $i = 1, \ldots, s$. For any $K \subset \{1, \ldots, k\}$, let H_K be the intersection hypothesis $P \in \bigcap_{j \in K} \omega_j$. Suppose ϕ_K is level α for testing H_K. Consider the multiple testing procedures that rejects H_i if ϕ_K rejects H_K whenever $i \in K$. Show, the FWER $\leq \alpha$. [This method of constructing tests that control the FWER is called the closure method of Marcus, Peritz and Gabriel (1976).]

Problem 9.7 As in Procedure 9.1.1, suppose that a test of the individual hypothesis H_j is based on a test statistic $T_{n,j}$, with large values indicating evidence against the H_j. Assume $\bigcap_{j=1}^{s} \omega_j$ is not empty. For any subset K of $\{1, \ldots, s\}$, let $c_{n,K}(\alpha, P)$ denote an α-quantile of the distribution of $\max_{j \in K} T_{n,j}$ under P.

Concretely,

$$c_{n,K}(\alpha, P) = \inf\{x : P\{\max_{j \in K} T_{n,j} \le x\} \ge \alpha\} . \tag{9.98}$$

For testing the intersection hypothesis H_K, it is only required to approximate a critical value for $P \in \bigcap_{j \in K} \omega_j$. Because there may be many such P, we define

$$c_{n,K}(1-\alpha) = \sup\{c_{n,K}(1-\alpha, P) : P \in \bigcap_{j \in K} \omega_j\} . \tag{9.99}$$

(i) In Procedure 9.1.1, show that the choice $\hat{c}_{n,K}(1-\alpha) = c_{n,K}(1-\alpha)$ controls the FWER, as long as (9.9) holds.
(ii) Further assume that for every subset $K \subset \{1, \ldots, k\}$, there exists a distribution P_K which satisfies

$$c_{n,K}(1-\alpha, P) \le c_{n,K}(1-\alpha, P_K) \tag{9.100}$$

for all P such that $I(P) \supset K$. Such a P_K may be referred to being least favorable among distributions P such that $P \in \bigcap_{j \in K} \omega_j$. (For example, if H_j corresponds to a parameter $\theta_j \le 0$, then intuition suggests a least favorable configuration should correspond to $\theta_j = 0$.) In addition, assume the subset pivotality condition of Westfall and Young (1993); that is, assume there exists a P_0 with $I(P_0) = \{1, \ldots, s\}$ such that the joint distribution of $\{T_{n,j} : j \in I(P_K)\}$ under P_K is the same as the distribution of $\{T_{n,j} : j \in I(P_K)\}$ under P_0. This condition says the (joint) distribution of the test statistics used for testing the hypotheses H_j, $j \in I(P_K)$ is unaffected by the truth or falsehood of the remaining hypotheses (and therefore we assume all hypotheses are true by calculating the distribution of the maximum under P_0). Show we can use $\hat{c}_{n,K}(1-\alpha, P_0)$ for $\hat{c}_{n,K}(1-\alpha)$.
(iii) Further assume the distribution of $(T_{n,1}, \ldots, T_{n,s})$ under P_0 is invariant under permutations (or exchangeable). Then, the critical values $\hat{c}_{n,K}(1-\alpha)$ can be chosen to depend only on $|K|$.

Problem 9.8 Rather than finding multiple tests that control the FWER, consider the k-FWER, the probability of rejecting k or more false hypotheses. For a given k, if there are s hypotheses, consider the procedure that rejects any hypothesis whose p-value is $\le k\alpha/s$. Show that the resulting procedure controls the k-FWER. [Additional stepdown procedures that control the number of false rejections, as well as the probability that the proportion of false rejections exceeds a given bound, are obtained in Lehmann and Romano (2005).]

Problem 9.9 In general, show that $FDR \le FWER$, and equality holds when all hypotheses are true. Therefore, control of the FWER at level α implies control of the FDR.

Section 9.2

Problem 9.10 . Suppose $(X_1, \ldots, X_k)^T$ has a multivariate c.d.f. $F(\cdot)$. For $\theta \in \mathbb{R}^k$, let $F_\theta(x) = F(x - \theta)$ define a multivariate location family. Show that (9.15) is satisfied for this family. (In particular, it holds if F is any multivariate normal distribution.)

Problem 9.11 Prove Lemma 9.2.2.

Problem 9.12 We have suppressed the dependence of the critical constants $C_1, \ldots, C_s$ in the definition of the stepdown procedure D, and now more accurately call them $C_{s,1}, \ldots, C_{s,s}$. Argue that, for fixed s, $C_{s,j}$ is nonincreasing in j and only depends on $s - j$.

Problem 9.13 Under the assumptions of Theorem 9.2.1, suppose there exists another monotone rule E that strongly controls the FWER, and such that

$$P_\theta\{d_{0,0}^c\} \leq P_\theta\{e_{0,0}^c\} \quad \text{for all } \theta \in \omega_{0,0}^c \;, \tag{9.101}$$

with strict inequality for some $\theta \in \omega_{0,0}^c$. Argue that the $\leq$ in (9.101) is an equality, and hence $e_{0,0} \triangle d_{0,0}$ has Lebesgue measure 0, where $A \triangle B$ denotes the symmetric difference between sets A and B. A similar result for the region $d_{1,1}$ can be made as well.

Problem 9.14 In general, the optimality results of Section 9.2 require the procedures to be monotone. To see why this is required, consider 9.2.2 (i). Show the procedure E to be inadmissible. *Hint: One can always add large negative values of T_1 and T_2 to the region $u_{1,1}$ without violating the FWER.*

Problem 9.15 Prove part (i) of Theorem 9.2.3.

Problem 9.16 In general, show $C_s = C_1^*$. In the case $s = 2$, show (9.27).

Section 9.3

Problem 9.17 Show that

$$\sum_{i=1}^{r+1} \left(Y_i - \frac{Y_1 + \cdots + Y_{r+1}}{r+1} \right)^2 - \sum_{i=1}^{r} \left(Y_i - \frac{Y_1 + \cdots + Y_r}{r} \right)^2 \geq 0.$$

Problem 9.18 (i) For the validity of Lemma 9.3.1 it is only required that the probability of rejecting homogeneity of any set containing $\{\mu_{i_1}, \ldots, \mu_{i_{v_1}}\}$ as a proper subset tends to 1 as the distance between the different groups (9.48) all $\to \infty$, with the analogous condition holding for $H_2', \ldots, H_r'$.

(ii) The condition of part (i) is satisfied for example if homogeneity of a set S is rejected for large values of $\sum |X_{i\cdot} - X_{\cdot\cdot}|$, where the sum extends over the subscripts i for which $\mu_i \in S$.

Problem 9.19 In Lemma 9.3.2, show that $\alpha_{s-1} = \alpha$ is necessary for admissibility.

Problem 9.20 Prove Lemma 9.3.3 when s is odd.

Problem 9.21 Show that the Tukey levels (vi) satisfy (9.54) when s is even but not when s is odd.

Problem 9.22 The Tukey T-method leads to the simultaneous confidence intervals

$$|(X_{j\cdot} - X_{i\cdot}) - (\mu_j - \mu_i)| \leq \frac{CS}{\sqrt{sn(n-1)}} \qquad \text{for all } i, j. \tag{9.102}$$

[The probability of (9.102) is independent of the μ's and hence equal to $1 - \alpha_s$.]

Section 9.4

Problem 9.23 (i) A function L satisfies the first equation of (9.65) for all u, x, and orthogonal transformations Q if and only if it depends on u and x only through $u'x$, $x'x$, and $u'u$.

(ii) A function L is equivariant under G_2 if and only if it satisfies (9.67).

Problem 9.24 (i) For the confidence sets (9.73), equivariance under G_1 and G_2 reduces to (9.74) and (9.75) respectively.

(ii) For fixed $(y_1, \ldots, y_r)$, the statements $\sum u_i y_i \in A$ hold for all $(u_1, \ldots, u_r)$ with $\sum u_i^2 = 1$ if and only if A contains the interval $I(y) = [-\sqrt{\sum Y_i^2}, +\sqrt{\sum Y_i^2}]$.

(iii) Show that the statement following (9.77) ceases to hold when $r = 1$.

Problem 9.25 Let X_i $(i = 1, \ldots, r)$ be independent $N(\xi_i, 1)$.

(i) The only simultaneous confidence intervals equivariant under G_0 are those given by (9.83).

(ii) The inequalities (9.83) and (9.85) are equivalent.

(iii) Compared with the Scheffé intervals (9.72), the intervals (9.85) for $\sum u_j \xi_j$ are shorter when $\sum u_j \xi_j = \xi_i$ and longer when $u_1 = \cdots = u_r$.

[(ii): For a fixed $u = (u_1, \ldots, u_r)$, $\sum u_i y_i$ is maximized subject to $|y_i| \leq \Delta$ for all i, by $y_i = \Delta$ when $u_i > 0$ and $y_i = -\Delta$ when $u_i < 0$.]

Section 9.5

Problem 9.26 (i) The confidence intervals $L(u; y, S) = \sum u_i y_i - c(S)$ are equivariant under G_3 if and only if $L(u; by, bS) = bL(u; y, S)$ for all $b > 0$.

(ii) The most general confidence sets (9.90) which are equivariant under G_1, G_2, and G_3 are of the form (9.91).

Problem 9.27 (i) In Example 9.5.2, the set of linear functions $\sum w_i \alpha_i = \sum w_i (\xi_{i\cdot} - \xi_{\cdot\cdot})$ for all w can also be represented as the set of functions $\sum w_i \xi_{i\cdot}$ for all w satisfying $\sum w_i = 0$.

(ii) The set of linear functions $\sum\sum w_{ij}\gamma_{ij} = \sum\sum w_{ij}(\xi_{ij\cdot} - \xi_{i\cdot\cdot} - \xi_{\cdot j\cdot} + \xi_{\cdot\cdot\cdot})$ for all w is equivalent to the set $\sum\sum w_{ij}\xi_{ij\cdot}$ for all w satisfying $\sum_i w_{ij} = \sum_j w_{ij} = 0$.

(iii) Determine the simultaneous confidence intervals (9.92) for the set of linear functions of part (ii).

Problem 9.28 (i) In Example 9.5.1, the simultaneous confidence intervals (9.92) reduce to (9.96).

(ii) What change is needed in the confidence intervals of Example 9.5.1 if the v's are not required to satisfy (9.95), i.e., if simultaneous confidence intervals are desired for all linear functions $\sum v_i \xi_i$ instead of all contrasts? Make a table showing the effect of this change for $s = 2, 3, 4, 5; n_i = n = 3, 5, 10$.

Problem 9.29 *Tukey's T-Method.* Let X_i $(i = 1, \ldots, r)$ be independent $N(\xi_i, 1)$, and consider simultaneous confidence intervals

$$L[(i,j);x] \le \xi_j - \xi_i \le M[(i,j);x] \quad \text{for all } i \neq j. \tag{9.103}$$

The problem of determining such confidence intervals remains invariant under the group G_0' of all permutations of the X's and under the group G_2 of translations $gx = x + a$.

(i) In analogy with (9.64), attention can be restricted to confidence bounds satisfying

$$L[(i,j);x] = -M[(j,i);x]. \tag{9.104}$$

(ii) The only simultaneous confidence intervals satisfying (9.104) and equivariant under G_0' and G_2 are those of the form

$$S(x) = \{\xi : x_j - x_i - \Delta < \xi_j - \xi_i < x_j - x_i + \Delta \text{ for all } i \neq j\}. \tag{9.105}$$

(iii) The constant Δ for which (9.105) has probability γ is determined by

$$P_0\{\max |X_j - X_i| < \Delta\} = P_0\{X_{(n)} - X_{(1)} < \Delta\} = \gamma, \tag{9.106}$$

where the probability P_0 is calculated under the assumption that $\xi_1 = \cdots = \xi_r$.

Problem 9.30 In the preceding problem consider arbitrary contrasts $\sum c_i \xi_i$ with $\sum c_i = 0$. The event

$$|(X_j - X_i) - (\xi_j - \xi_i)| \le \Delta \qquad \text{for all } i \neq j \tag{9.107}$$

is equivalent to the event

$$\left|\sum c_i X_i - \sum c_i \xi_i\right| \le \frac{\Delta}{2} \sum |c_i| \qquad \text{for all } c \text{ with} \sum c_i = 0, \tag{9.108}$$

which therefore also has probability γ. This shows how to extend the Tukey intervals for all pairs to all contrasts.
[That (9.108) implies (9.107) is obvious. To see that (9.107) implies (9.108), let $y_i = x_i - \xi_i$ and maximize $|\sum c_i y_i|$ subject to $|y_j - y_i| \le \Delta$ for all i and j. Let P and N denote the sets $\{i : c_i > 0\}$ and $\{i : c_i < 0\}$, so that

$$\sum c_i y_i = \sum_{i \in P} c_i y_i - \sum_{i \in N} |c_i|\, y_i.$$

Then for fixed c, the sum $\sum c_i y_i$ is maximized by maximizing the y_i's for $i \in P$ and minimizing those for $i \in N$. Since $|y_j - y_i| \le \Delta$, it is seen that $\sum c_i y_i$ is maximized by $y_i = \Delta/2$ for $i \in P$, $y_i = -\Delta/2$ for $i \in N$. The minimization of $\sum c_i y_i$ is handled analogously.]

Problem 9.31 (i) Let X_{ij} ($j = 1, \dots n; i = 1, \dots, s$) be independent $N(\xi_i, \sigma^2)$, σ^2 unknown. Then the problem of obtaining simultaneous confidence intervals for all differences $\xi_j - \xi_i$ is invariant under G_0', G_2, and the scale changes G_3.

(ii) The only equivariant confidence bounds based on the sufficient statistics $X_{i\cdot}$ and $S^2 = \sum\sum(X_{ij} - X_{i\cdot})^2$ and satisfying the condition corresponding to (9.104) are those given by

$$\begin{aligned} S(x) &= \Big\{ x : x_{j\cdot} - x_{i\cdot} - \frac{\Delta'}{\sqrt{n-s}} S \leq \xi_j - \xi_i \\ &\leq x_{j\cdot} - x_{i\cdot} + \frac{\Delta'}{\sqrt{n-s}} S \qquad \text{for all } i \neq j \Big\} \end{aligned} \tag{9.109}$$

with Δ' determined by the null distribution of the *Studentized range*

$$P_0 \left\{ \frac{\max |X_{j\cdot} - X_{i\cdot}|}{S/\sqrt{n-s}} < \Delta' \right\} = \gamma. \tag{9.110}$$

(iii) Extend the results of Problem 9.30 to the present situation.

Problem 9.32 Construct an example [i.e., choose values $n_1 = \cdots = n_s = n$ and α particular contrast $(c_1, \dots, c_s)$] for which the Tukey confidence intervals (9.108) are shorter than the Scheffé intervals (9.96), and an example in which the situation is reversed.

Problem 9.33 *Dunnett's method.* Let X_{0j} ($j = 1, \dots, m$) and X_{ik} ($i = 1, \dots, s; k = 1, \dots, n$) represent measurements on a standard and s competing new treatments, and suppose the X's are independently distributed as $N(\xi_0, \sigma^2)$ and $N(\xi_i, \sigma^2)$ respectively. Generalize Problems 9.29 and 9.31 to the problem of obtaining simultaneous confidence intervals for the s differences $\xi_i - \xi_0$ ($i = 1, \dots, s$).

Problem 9.34 In generalization of Problem 9.30, show how to extend the Dunnett intervals of Problem 9.33 to the set of all contrasts.
[Use the fact that the event $|y_i - y_0| \leq \Delta$ for $i = 1, \dots, s$ is equivalent to the event $|\sum_{i=0}^s c_i y_i| \leq \Delta \sum_{i=1}^s |c_i|$ for all $(c_0, \dots, c_s)$ satisfying $\sum_{i=0}^s c_i = 0$.]
Note. As is pointed out in Problems 9.26(iii) and 9.32, the intervals resulting from the extension of the Tukey (and Dunnett) methods to all contrasts are shorter than the Scheffé intervals for the differences for which these methods were designed and for contrasts close to them, and longer for some other contrasts. For details and generalizations, see for example Miller (1981), Richmond (1982), and Shaffer (1977a).

Problem 9.35 In the regression model of Problem 7.8, generalize the confidence bands of Example 9.5.3 to the regression surfaces

(i) $h_1(e_1, \dots, e_s) = \sum_{j=1}^s e_j \beta_j$;

(ii) $h_2(e_2, \dots, e_s) = \beta_1 + \sum_{j=2}^s e_j \beta_j$.

9.7 Notes

Many of the basic ideas for making multiple inferences were pioneered by Tukey (1953); see Tukey (1991), Braun (1994), and Shaffer (1995). See Duncan (1955) for an exposition of the ideas of one of the early workers in the area of multiple comparisons.

Comprehensive accounts on the theory and methodology of multiple testing can be found in Hochberg and Tamhane (1987), Westfall and Young (1993), and Hsu (1996) and Dudoit, Shaffer and Boldrick (2003). Some recent work on stepwise procedures includes Troendle (1995), Finner and Roters (1998, 2002), and Romano and Wolf (2004). Confidence sets based on multiple tests are studied in Haytner and Hsu (1994), Miwa and Hayter (1999) and Holm (1999).

The first simultaneous confidence intervals (for a regression line) were obtained by Working and Hotelling (1929). Scheffé's approach was generalized in Roy and Bose (1953). The optimal property of the Scheffé intervals presented in Section 9.4 is a special case of results of Wijsman (1979, 1980). A review of the literature on the relationship of tests and confidence sets for a parameter vector with the associated simultaneous confidence intervals for functions of its components can be found in Kanoh and Kusunoki (1984). Some alternative methods to construct confidence bands in regression contexts are given in Faraway and Sun (1995) and Spurrier (1999).

10
Conditional Inference

10.1 Mixtures of Experiments

The present chapter has a somewhat different character from the preceding ones. It is concerned with problems regarding the proper choice and interpretation of tests and confidence procedures, problems which—despite a large literature—have not found a definitive solution. The discussion will thus be more tentative than in earlier chapters, and will focus on conceptual aspects more than on technical ones.

Consider the situation in which either the experiment $\mathcal{E}$ of observing a random quantity X with density p_θ (with respect to μ) or the experiment $\mathcal{F}$ of observing an X with density q_θ (with respect to ν) is performed with probability p and $q = 1 - p$ respectively. On the basis of X, and knowledge of which of the two experiments was performed, it is desired to test $H_0 : \theta = \theta_0$ against $H_1 : \theta = \theta_1$. For the sake of convenience it will be assumed that the two experiments have the same sample space and the same σ-field of measurable sets. The sample space of the overall experiment consists of the union of the sets

$$\mathcal{X}_0 = \{(I, x) : I = 0,\ x \in \mathcal{X}\} \quad \text{and} \quad \mathcal{X}_1 = \{(I, x) : I = 1,\ x \in \mathcal{X}\}$$

where I is 0 or 1 as $\mathcal{E}$ or $\mathcal{F}$ is performed.

A level-α test of H_0 is defined by its critical function

$$\phi_i(x) = \phi(i, x)$$

and must satisfy

$$pE_0\Big[\phi_0(X) \mid \mathcal{E}\Big] + qE_0\Big[\phi_1(X) \mid \mathcal{F}\Big] = p\int \phi_0 p_{\theta_0}\, d\mu + q\int \phi_1 q_{\theta_0}\, d\nu \le \alpha. \quad (10.1)$$

Suppose that p is unknown, so that H_0 is composite. Then a level-α test of H_0 satisfies (10.1) for all $0 < p < 1$, and must therefore satisfy

$$\alpha_0 = \int \phi_0 p_{\theta_0}\, d\mu \le \alpha \quad \text{and} \quad \alpha_1 = \int \phi_1 q_{\theta_0}\, d\nu \le \alpha. \quad (10.2)$$

As a result, a UMP test against H_1 exists and is given by

$$\phi_0(x) = \begin{cases} 1 \\ \gamma_0 & \text{if } \dfrac{p_{\theta_1}(x)}{p_{\theta_0}(x)} \gtreqless c_0, \\ 0 \end{cases} \qquad \phi_1(x) = \begin{cases} 1 \\ \gamma_1 & \text{if } \dfrac{q_{\theta_1}(x)}{q_{\theta_0}(x)} \gtreqless c_1, \\ 0 \end{cases} \quad (10.3)$$

where the c_i and γ_i are determined by

$$E_{\theta_0}\Big[\phi_0(X) \mid \mathcal{E}\Big] = E_{\theta_0}\Big[\phi_1(X) \mid \mathcal{F}\Big] = \alpha. \quad (10.4)$$

The power of this test against H_1 is

$$\beta(p) = p\beta_0 + q\beta_1 \quad (10.5)$$

with

$$\beta_0 = E_{\theta_1}\Big[\phi_0(X) \mid \mathcal{E}\Big], \qquad \beta_1 = E_{\theta_1}\Big[\phi_1(X) \mid \mathcal{F}\Big]. \quad (10.6)$$

The situation is analogous to that of Section 4.4 and, as was discussed there, it may be more appropriate to consider the conditional power β_i when $I = i$, since this is the power pertaining to the experiment that has been performed. As in the earlier case, the conditional power β_I can also be interpreted as an estimate of the unknown $\beta(p)$, which is unbiased, since

$$E(\beta_I) = p\beta_0 + q\beta_1 = \beta(p).$$

So far, the probability p of performing experiment $\mathcal{E}$ has been assumed to be unknown. Suppose instead that the value of p is known, say $p = \frac{1}{2}$. The hypothesis H can be tested at level α by means of (10.3) as before, but the power of the test is now known to be $\frac{1}{2}(\beta_0 + \beta_1)$. Suppose that $\beta_0 = .3$, $\beta_1 = .9$, so that at the start of the experiment the power is $\frac{1}{2}(.3 + .9) = .6$. Now a fair coin is tossed to decide whether to perform $\mathcal{E}$ (in case of heads) or $\mathcal{F}$ (in case of tails). If the coin shows heads, should the power be reassessed and scaled down to .3?

Let us postpone the answer and first consider another change resulting from the knowledge of p. A level-α test of H now no longer needs to satisfy (10.2) but only the weaker condition

$$\frac{1}{2}\left[\int \phi_0 p_{\theta_0}\,d\mu + \int \phi_1 q_{\theta_0}\,d\nu\right] \le \alpha. \tag{10.7}$$

The most powerful test against K is then again given by (10.3), but now with $c_0 = c_1 = c$ and $\gamma_0 = \gamma_1 = \gamma$ determined by (Problem 10.3)

$$\tfrac{1}{2}(\alpha_0 + \alpha_1) = \alpha, \tag{10.8}$$

where

$$\alpha_0 = E_{\theta_0}\Big[\phi_0(X) \mid \mathcal{E}\Big], \qquad \alpha_1 = E_{\theta_0}\Big[\phi_1(X) \mid \mathcal{F}\Big]. \tag{10.9}$$

As an illustration of the change, suppose that experiment $\mathcal{F}$ is reasonably informative, say that the power β_1 given by (10.6), is .8, but that $\mathcal{E}$ has little ability to distinguish between p_{θ_0} and p_{θ_1}. Then it will typically not pay to put much of the rejection probability into α_0; if β_0 [given by (10.6)] is sufficiently small, the best choice of α_0 and α_1 satisfying (10.8) is approximately $\alpha_0 \approx 0$, $\alpha_1 \approx 2\alpha$. The situation will be reversed if $\mathcal{F}$ is so informative that $\mathcal{F}$ can attain power close to 1 with an α_1 much smaller than $\alpha/2$.

When p is known, there are therefore two issues. Should the procedure be chosen which is best on the average over both experiments, or should the best conditional procedure be preferred; and, for a given test or confidence procedure, should probabilities such as level, power, and confidence coefficient be calculated conditionally, given the experiment that has been selected, or unconditionally? The underlying question is of course the same: Is a conditional or unconditional point of view more appropriate?

The answer cannot be found within the model but depends on the context. If the overall experiment will be performed many times, for example in an industrial or agricultural setting, the average performance may be the principal feature of interest, and an unconditional approach suitable. However, if repetitions refer to different clients, or are potential rather than actual, interest will focus on the particular event at hand, and conditioning seems more appropriate. Unfortunately, as will be seen in later sections, it is then often not clear how the conditioning events should be chosen.

The difference between the conditional and the unconditional approach tends to be most striking, and a choice between them therefore most pressing, when the two experiments $\mathcal{E}$ and $\mathcal{F}$ differ sharply in the amount of information they contain, if for example the difference $|\beta_1 - \beta_0|$ in (10.6) is large. To illustrate an extreme situation in which this is not the case, suppose that $\mathcal{E}$ and $\mathcal{F}$ consist in observing X with distribution $N(\theta, 1)$ and $N(-\theta, 1)$ respectively, that one of them is selected with known probabilities p and q respectively, and that it is desired to test $H : \theta = 0$ against $K : \theta > 0$. Here $\mathcal{E}$ and $\mathcal{F}$ contain exactly the same amount of information about θ. The unconditional most powerful level-α test of H against $\theta_1 > 0$ is seen to reject (Problem 10.5) when $X > c$ if $\mathcal{E}$ is performed, and when $X < -c$ if $\mathcal{F}$ is performed, where $P_0(X > c) = \alpha$. The test is UMP against $\theta > 0$, and happens to coincide with the UMP conditional test.

The issues raised here extend in an obvious way to mixtures of more than two experiments. As an illustration of a mixture over a continuum, consider a regression situation. Suppose that $X_1, \dots, X_n$ are independent, and that the conditional density of X_i given t_i is

$$\frac{1}{\sigma} f\left(\frac{x_i - \alpha - \beta t_i}{\sigma}\right).$$

The t_i themselves are obtained with error. They may for example be independently normally distributed with mean c_i and known variance τ^2, where the c_i are the intended values of the t_i. Then it will again often be the case that the most appropriate inference concerning α, β, and σ is conditional on the observed values of the t's (which represent the experiment actually being performed). Whether this is the case will, as before, depend on the context.

The argument for conditioning also applies when the probabilities of performing the various experiments are unknown, say depend on a parameter ϑ, provided ϑ is unrelated to θ, so that which experiment is chosen provides no information concerning θ. A more precise statement of this generalization is given at the end of the next section.

10.2 Ancillary Statistics

Mixture models can be described in the following general terms. Let $\{\mathcal{E}_z, z \in \mathcal{Z}\}$ denote a collection of experiments of which one is selected according to a known probability distribution over $\mathcal{Z}$. For any given z, the experiment $\mathcal{E}_z$ consists in observing a random quantity X, which has a distribution $P_\theta(\cdot \mid z)$. Although this structure seems rather special, it is common to many statistical models.

Consider a general statistical model in which the observations X are distributed according to P_θ, $\theta \in \Omega$, and suppose there exists an *ancillary statistic*, that is, a statistic Z whose distribution F does not depend on θ. Then one can think of X as being obtained by a two-stage experiment: Observe first a random quantity Z with distribution F; given $Z = z$, observe a quantity X with distribution $P_\theta(\cdot \mid z)$. The resulting X is distributed according to the original distribution P_θ. Under these circumstances, the argument of the preceding section suggests that it will frequently be appropriate to take the conditional point of view.[1] (Unless Z is discrete, these definitions involve technical difficulties concerning sets of measure zero and the existence of conditional distributions, which we shall disregard.)

An important class of models in which ancillary statistics exist is obtained by invariance considerations. Suppose the model $\mathcal{P} = \{P_\theta, \theta \in \Omega\}$ remains invariant under the transformations

$$X \to gX, \quad \theta \to \bar{g}\theta; \qquad g \in G, \quad \bar{g} \in \bar{G},$$

and that $\bar{G}$ is transitive over Ω.[2]

Theorem 10.2.1 *If $\mathcal{P}$ remains invariant under G and if $\bar{G}$ is transitive over Ω, then a maximal invariant T (and hence any invariant) is ancillary.*

---

[1] A distinction between experimental mixtures and the present situation, relying on aspects outside the model, is discussed by Basu (1964) and Kalbfleisch (1975).

[2] The family $\mathcal{P}$ is then a group family; see *TPE2*, Section 1.3.

PROOF. It follows from Theorem 6.3.2 that the distribution of a maximal invariant under G is invariant under $\bar{G}$. Since $\bar{G}$ is transitive, only constants are invariant under $\bar{G}$. The probability $P_\theta(T \in B)$ is therefore constant, independent of θ, for all B, as was to be proved. ■

As an example, suppose that $X = (X_1, \ldots, X_n)$ is distributed according to a location family with joint density $f(x_1 - \theta, \ldots, x_n - \theta)$. The most powerful test of $H : \theta = \theta_0$ against $K : \theta = \theta_1 > \theta_0$ rejects when

$$\frac{f(x_1 - \theta_1, \ldots, x_n - \theta_1)}{f(x_1 - \theta_0, \ldots, x_n - \theta_0)} \geq c. \tag{10.10}$$

Here the set of differences $Y_i = X_i - X_n$ $(i = 1, \ldots, n-1)$ is ancillary. This is obvious by inspection and follows from Theorem 10.2.1 in conjunction with Example 6.2.1(i). It may therefore be more appropriate to consider the testing problem conditionally given $Y_1 = y_1, \ldots, Y_{n-1} = y_{n-1}$. To determine the most powerful conditional test, transform to $Y_1, \ldots, Y_n$, where $Y_n = X_n$. The conditional density of Y_n given $y_1, \ldots, y_{n-1}$ is

$$p_\theta(y_n \mid y_1, \ldots, y_{n-1}) = \frac{f(y_1 + y_n - \theta, \ldots, y_{n-1} + y_n - \theta, y_n - \theta)}{\int f(y_1 + u, \ldots, y_{n-1} + u, u)\, du}. \tag{10.11}$$

and the most powerful conditional test rejects when

$$\frac{p_{\theta_1}(y_n \mid y_1, \ldots, y_{n-1})}{p_{\theta_0}(y_n \mid y_1, \ldots, y_{n-1})} > c(y_1, \ldots, y_{n-1}). \tag{10.12}$$

In terms of the original variables this becomes

$$\frac{f(x_1 - \theta_1, \ldots, x_n - \theta_1)}{f(x_1 - \theta_0, \ldots, x_n - \theta_0)} > c(x_1 - x_n, \ldots, x_{n-1} - x_n). \tag{10.13}$$

The constant $c(x_1 - x_n, \ldots, x_{n-1} - x_n)$ is determined by the fact that the conditional probability of (10.13), given the differences of the x's, is equal to α when $\theta = \theta_0$.

For describing the conditional test (10.12) and calculating the critical value $c(y_1, \ldots, y_{n-1})$, it is useful to note that the statistic $Y_n = X_n$ could be replaced by any other Y_n satisfying the equivariance condition[3]

$$Y_n(x_1 + a, \ldots, x_n + a) = Y_n(x_1, \ldots, x_n) + a \quad \text{for all } a. \tag{10.14}$$

This condition is satisfied for example by the mean of the X's, the median, or any of the order statistics. As will be shown in the following Lemma 10.2.1, any two statistics Y_n and Y_n' satisfying (10.14) differ only by a function of the differences $Y_i = X_i - X_n$ $(i = 1, \ldots, n-1)$. Thus conditionally, given the values $y_1, \ldots, y_{n-1}$, Y_n and Y_n' differ only by a constant, and their conditional distributions (and the critical values $c(y_1, \ldots, y_{n-1})$) differ by the same constant. One can therefore choose Y_n, subject to (10.14), to make the conditional calculations as convenient as possible.

Lemma 10.2.1 *If Y_n and Y_n' both satisfy* (10.14), *then their difference $\Delta = Y_n' - Y_n$ depends on $(x_1, \ldots, x_n)$ only through the differences $(x_1 - x_n, \ldots, x_{n-1} - x_n)$.*

[3]For a more detailed discussion of equivariance, see *TPE*2, Chapter 3.

PROOF. Since Y_n and Y_n' satisfy (10.14),

$$\Delta(x_1 + a, \dots, x_n + a) = \Delta(x_1, \dots, x_n) \qquad \text{for all } a.$$

Putting $a = -x_n$, one finds

$$\Delta(x_1, \dots, x_n) = \Delta(x_1 - x_n, \dots, x_{n-1} - x_n, 0),$$

which is a function of the differences. ■

The existence of ancillary statistics is not confined to models that remain invariant under a transitive group $\bar{G}$. The mixture and regression examples of Section 10.1 provide illustrations of ancillaries without the benefit of invariance. Further examples are given in Problems 10.8–10.13.

If conditioning on an ancillary statistic is considered appropriate because it makes the inference more relevant to the situation at hand, it is desirable to carry the process as far as possible and hence to condition on a *maximal* ancillary. An ancillary Z is said to be maximal if there does not exist an ancillary U such that $Z = f(U)$ without Z and U being equivalent. [For a more detailed treatment, which takes account of the possibility of modifying statistics on sets of measure zero without changing their probabilistic properties, see Basu (1959).]

Conditioning, like sufficiency and invariance, leads to a reduction of the data. In the conditional model, the ancillary is no longer part of the random data but has become a constant. As a result, conditioning often leads to a great simplification of the inference. Choosing a maximal ancillary for conditioning thus has the additional advantage of providing the greatest reduction of the data.

Unfortunately, maximal ancillaries are not always unique, and one must then decide which maximal ancillary to choose for conditioning. [This problem is discussed by Cox (1971) and Becker and Gordon (1983).] If attention is restricted to ancillary statistics that are invariant under a given group G, the maximal ancillary of course coincides with the maximal invariant.

Another issue concerns the order in which to apply reduction by sufficiency and ancillarity.

Example 10.2.1 Let (X_i, Y_i), $i = 1, \dots, n$, be independently distributed according to a bivariate normal distribution with $E(X_i) = E(Y_i) = 0$, $\operatorname{Var}(X_i) = \operatorname{Var}(Y_i) = 1$, and unknown correlation coefficient ρ. Then $X_1, \dots, X_n$ are independently distributed as $N(0,1)$ and are therefore ancillary. The conditional density of the Y's given $X_1 = x_i, \dots, X_n = x_n$ is

$$C \exp\left(-\frac{1}{2(1-\rho^2)} \sum (y_i - \rho x_i)^2\right),$$

with the sufficient statistics $(\sum Y_i^2, \sum x_i Y_i)$.

Alternatively, one could begin by noticing that $(Y_1, \dots, Y_n)$ is ancillary. The conditional distribution of the X's given $Y_1 = y_1, \dots, Y_n = y_n$ then admits the sufficient statistics $(\sum X_i^2, \sum X_i y_i)$. A unique maximal ancillary V does not exist in this case, since both the X's and Y's would have to be functions of V. Thus V would have to be equivalent to the full sample $(X_1, Y_1), \dots, (X_n, Y_n)$, which is not ancillary.

Suppose instead that the data are first reduced to the sufficient statistics $T = (\sum X_i^2 + \sum Y_i^2, \sum X_i Y_i)$. Based on T, no nonconstant ancillaries appear to exist.[4] This example and others like it suggest that it is desirable to reduce the data as far as possible through sufficiency, before attempting further reduction by means of ancillary statistics. ■

Note that contrary to this suggestion, in the location example at the beginning of the section, the problem was not first reduced to the sufficient statistics $X_{(1)} < \cdots < X_{(n)}$. The omission can be justified in hindsight by the fact that the optimal conditional tests are the same whether or not the observations are first reduced to the order statistics.

In the structure described at the beginning of the section, the variable Z that labels the experiment was assumed to have a known distribution. The argument for conditioning on the observed value of Z does not depend on this assumption. It applies also when the distribution of Z depends on an unknown parameter ϑ, which is independent of θ and hence by itself contains no information about θ, that is, when the distribution of Z depends only on ϑ, the conditional distribution of X given $Z = z$ depends only on θ, and the parameter space Ω for (θ, ϑ) is a Cartesian product $\Omega = \Omega_1 \times \Omega_2$, with

$$(\theta, \vartheta) \in \Omega \quad \Leftrightarrow \quad \theta \in \Omega_1 \quad \text{and} \quad \vartheta \in \Omega_2 \,. \tag{10.15}$$

(the parameters θ and ϑ are then said to be *variation-independent*, or unrelated.)

Statistics Z satisfying this more general definition are called *partial ancillary* or *S-ancillary*. (The term ancillary without modification will be reserved here for a statistic that has a known distribution.) Note that if $X = (T, Z)$ and Z is a partial ancillary, then T is a partial sufficient statistic in the sense of Problem 3.60. For a more detailed discussion of this and related concepts of partial ancillarity, see for example Basu (1978) and Barndorff–Nielsen (1978).

Example 10.2.2 Let X and Y be independent with Poisson distributions $P(\lambda)$ and $P(\mu)$, and let the parameter of interest be $\theta = \mu/\lambda$. It was seen in Section 10.4 that the conditional distribution of Y given $Z = X + Y = z$ is binomial $b(p, z)$ with $p = \mu/(\lambda + \mu) = \theta/(\theta + 1)$ and therefore depends only on θ, while the distribution of Z is Poisson with mean $\vartheta = \lambda + \mu$. Since the parameter space $0 < \lambda$, $\mu < \infty$ is equivalent to the Cartesian product of $0 < \theta < \infty$, $0 < \vartheta < \infty$, it follows that Z is S-ancillary for θ.

The UMP unbiased level-α test of $H : \mu \leq \lambda$ against $\mu > \lambda$ is UMP also among all tests whose conditional level given z is α for all z. (The class of conditional tests coincides exactly with the class of all tests that are similar on the boundary $\mu = \lambda$.) ■

When Z is S-ancillary for θ in the presence of a nuisance parameter ϑ, the unconditional power $\beta(\theta, \vartheta)$ of a test φ of $H : \theta = \theta_0$ may depend on ϑ as well as on θ. The conditional power $\beta(\vartheta \mid z) = E_\theta[\varphi(X) \mid z]$ can then be viewed as an unbiased estimator of the (unknown) $\beta(\theta, \vartheta)$, as was discussed at the end of Section 4.4. On the other hand, if no nuisance parameters ϑ are present and Z

[4]So far, nonexistence has not been proved. It seems likely that a proof can be obtained by the methods of Unni (1978).

is ancillary for θ, the unconditional power $\beta(\theta) = E_\theta \varphi(X)$ and the conditional power $\beta(\theta \mid z)$ provide two alternative evaluations of the power of φ against θ, which refer to different sampling frameworks, and of which the latter of course becomes available only after the data have been obtained.

Surprisingly, the S-ancillarity of $X + Y$ in Example 10.2.2 does not extend to the corresponding binomial problem.

Example 10.2.3 Let X and Y have independent binomial distributions $b(p_1, m)$ and $b(p_2, n)$ respectively. Then it was seen in Section 4.5 that the conditional distribution of Y given $Z = X + Y = z$ depends only on the crossproduct ratio $\Delta = p_2 q_1 / p_1 q_2$ $(q_i = 1 - p_i)$. However, Z is not S-ancillary for Δ. To see this, note that S-ancillarity of Z implies the existence of a parameter ϑ unrelated to Δ and such that the distribution of Z depends only on ϑ. As Δ changes, the family of distributions $\{P_\vartheta, \vartheta \in \Omega_2\}$ of Z would remain unchanged. This is not the case, since Z is binomial when $\Delta = 1$ and not otherwise (Problem 10.15). Thus Z is not S-ancillary.

In this example, all unbiased tests of $H : \Delta = \Delta_0$ have a conditional level given z that is independent of z, but conditioning on z cannot be justified by S-ancillarity. ■

Closely related to this example is the situation of the multinomial 2×2 table discussed from the point of view of unbiasedness in Section 4.6.

Example 10.2.4 In the notation of Section 4.6, let the four cell entries of a 2×2 table be X, X', Y, Y' with row totals $X + X' = M$, $Y + Y' = N$, and column totals $X + Y = T$, $X' + Y' = T'$, and with total sample size $M + N = T + T' = s$. Here it is easy to check that (M, N) is S-ancillary for $\theta = (\theta_1, \theta_2) = (p_{AB}/p_B, p_{A\tilde{B}}/p_{\tilde{B}})$ with $\vartheta = p_B$. Since the cross-product ratio Δ can be expressed as a function of (θ_1, θ_2), it may be appropriate to condition a test of $H : \Delta = \Delta_0$ on (M, N). Exactly analogously one finds that (T, T') is S-ancillary for $\theta' = (\theta_1', \theta_2') = (p_{AB}/p_A, p_{\tilde{A}B}/p_{\tilde{A}})$, and since Δ is also a function of (θ_1', θ_2'), it may be equally appropriate to condition a test of H on (T, T'). One might hope that the set of all four marginals $(M, N, T, T') = Z$ would be S-ancillary for Δ. However, it is seen from the preceding example that this is not the case.

Here, all unbiased tests have a constant conditional level given z. However, S-ancillarity permits conditioning on only one set of margins (without giving any guidance as to which of the two to choose), not on both. ■

Despite such difficulties, the principle of carrying out tests and confidence estimation conditionally on ancillaries or S-ancillaries frequently provides an attractive alternative to the corresponding unconditional procedures, primarily because it is more appropriate for the situation at hand. However, insistence on such conditioning leads to another difficulty, which is illustrated by the following example.

Example 10.2.5 Consider N populations $\prod_i$, and suppose that an observation X_i from $\prod_i$ has a normal distribution $N(\xi_i, 1)$. The hypothesis to be tested is $H : \xi_1 = \cdots = \xi_N$. Unfortunately, N is so large that it is not practicable to take an observation from each of the populations; the total sample size is restricted to be $n < N$. A sample $\prod_{J_1}, \ldots, \prod_{J_n}$ of n of the N populations is therefore selected at random, with probability $1/\binom{N}{n}$ for each set of n, and an observation X_{j_i} is obtained from each of the populations $\prod_{j_i}$, in the sample.

Here the variables $J_1, \ldots, J_n$ are ancillary, and the requirement of conditioning on ancillaries would restrict any inference to the n populations from which observations are taken. Systematic adherence to this requirement would therefore make it impossible to test the original hypothesis H.[5] Of course, rejection of the partial hypothesis $H_{j_1,\ldots,j_n} : \xi_{j_1} = \cdots = \xi_{j_n}$ would imply rejection of the original H. However, acceptance of $H_{j_1,\ldots,j_n}$ would permit no inference concerning H.

The requirement to condition in this case runs counter to the belief that a sample may permit inferences concerning the whole set of populations, which underlies much of statistical practice.

With an unconditional approach such an inference is provided by the test with rejection region

$$\sum \left[X_{j_i} - \left(\frac{1}{n} \sum_{k=1}^{n} X_{j_k} \right) \right]^2 \geq c,$$

where c is the upper α-percentage point of χ^2 with $n-1$ degrees of freedom. Not only does this test actually have unconditional level α, but its conditional level given $J_1 = j_1, \ldots, J_n = j_n$ also equals α for all $(j_1, \ldots, j_n)$. There is in fact no difference in the present case between the conditional and the unconditional test: they will accept or reject for the same sample points. However, as has been pointed out, there is a crucial difference between the conditional and unconditional interpretations of the results.

If $\beta_{j_1,\ldots,j_n}(\xi_{j_1}, \ldots, \xi_{j_n})$ denotes the conditional power of this test given $J_1 = j_1, \ldots, J_n = j_n$, its unconditional power is

$$\frac{\sum \beta_{j_1,\ldots,j_n}(\xi_{j_1}, \ldots, \xi_{j_n})}{\binom{N}{n}}$$

summed over all $\binom{N}{n}$ n-tuples $j_1 < \ldots < j_n$. As in the case with any test, the conditional power given an ancillary (in the present case $J_1, \ldots, J_n$) can be viewed as an unbiased estimate of the unconditional power. ■

10.3 Optimal Conditional Tests

Although conditional tests are often sensible and are beginning to be employed in practice [see for example Lawless (1972, 1973, 1978) and Kappenman (1975)], not much theory has been developed for the resulting conditional models. Since the conditional model tends to be simpler than the original unconditional one, the conditional point of view will frequently bring about a simplification of the theory. This possibility will be illustrated in the present section on some simple examples.

---

[5]For other implications of this requirement, called the weak conditionality principle, see Birnbaum (1962) and Berger and Wolpert (1988).

Example 10.3.1 Specializing the example discussed at the beginning of Section 10.1, suppose that a random variable is distributed according to $N(\theta, \sigma_1^2)$ or $N(\theta, \sigma_0^2)$ as $I = 1$ or 0, and that $P(I = 1) = P(I = 0) = \frac{1}{2}$. Then the most powerful test of $H : \theta = \theta_0$ against $\theta = \theta_1 (> \theta_0)$ based on (I, X) rejects when

$$\frac{x - \frac{1}{2}(\theta_0 + \theta_1)}{2\sigma_i^2} \geq k.$$

A UMP test against the alternatives $\theta > \theta_0$ therefore does not exist. On the other hand, if H is tested conditionally given $I = i$, a UMP conditional test exists and rejects when $X > c_i$ where $P(X > c_i \mid I = i) = \alpha$ for $i = 0,\ 1$. ■

The nonexistence of UMP unconditional tests found in this example is typical for mixtures with known probabilities of two or more families with monotone likelihood ratio, despite the existence of UMP conditional tests in these cases.

Example 10.3.2 Let $X_1, \ldots, X_n$ be a sample from a normal distribution $N(\xi, a^2\xi^2)$, $\xi > 0$, with known coefficient of variation $a > 0$, and consider the problem of testing $H : \xi = \xi_0$ against $K : \xi > \xi_0$. Here $T = (T_1, T_2)$ with $T_1 = \bar{X}$, $T_2 = \sqrt{(1/n)\sum X_i^2}$ is sufficient, and $Z = T_1/T_2$ is ancillary. If we let $V = \sqrt{n}T_2/a$, the conditional density of V given $Z = z$ is equal to (Problem 10.18)

$$p_\xi(v \mid z) = \frac{k}{\xi^n} v^{n-1} \exp\left\{ -\frac{1}{2}\left[\frac{v}{\xi} - \frac{z\sqrt{n}}{a}\right]^2 \right\}. \tag{10.16}$$

The density has monotone likelihood ratio, so that the rejection region $V > C(z)$ constitutes a UMP conditional test.

Unconditionally, $Y = \bar{X}$ and $S^2 = \sum(X_i - \bar{X})^2$ are independent with joint density

$$cs^{(n-3)/2} \exp\left(-\frac{n}{2a^2\xi^2}(y - \xi)^2 - \frac{1}{2a^2\xi^2}s^2 \right), \tag{10.17}$$

and a UMP test does not exist. [For further discussion of this example, see Hinkley (1977).] ■

An important class of examples is obtained from situations in which the model remains invariant under a group of transformations that is transitive over the parameter space, that is, when the given class of distributions constitutes a group family. The maximal invariant V then provides a natural ancillary on which to condition, and an optimal conditional test may exist even when such a test does not exist unconditionally. Perhaps the simplest class of examples of this kind are provided by location families under the conditions of the following lemma.

Lemma 10.3.1 *Let $X_1, \ldots, X_n$ be independently distributed according to $f(x_i - \theta)$, with f strongly unimodal. Then the family of conditional densities of $Y_n = X_n$ given $Y_i = X_i - X_n$ $(i = 1, \ldots, n-1)$ has monotone likelihood ratio.*

PROOF. The conditional density (10.11) is proportional to

$$f(y_n + y_1 - \theta) \cdots f(y_n + y_{n-1} - \theta) f(y_n - \theta) \tag{10.18}$$

By taking logarithms and using the fact that each factor is strongly unimodal, it is seen that the product is also strongly unimodal, and the result follows from Example 8.2.1. ■

Lemma 10.3.1 shows that for strongly unimodal f there exists a UMP conditional test of $H : \theta \leq \theta_0$ against $K : \theta > \theta_0$ which rejects when

$$X_n > c(X_1 - X_n, \ldots, X_{n-1} - X_n). \tag{10.19}$$

Conditioning has reduced the model to a location family with sample size one. The double-exponential and logistic distributions are both strongly unimodal (Section 9.2), and thus provide examples of UMP conditional tests. In neither case does there exist a UMP unconditional test unless $n = 1$.

As a last class of examples, we shall consider a situation with a nuisance parameter. Let $X_1, \ldots, X_m$ and $Y_1, \ldots, Y_n$ be independent samples from location families with densities $f(x_1 - \xi, \ldots, x_m - \xi)$ and $g(y_1 - \eta, \ldots, y_n - \eta)$ respectively, and consider the problem of testing $H : \eta \leq \xi$ against $K : \eta > \xi$. Here the differences $U_i = X_i - X_m$ and $V_j = Y_j - Y_n$ are ancillary. The conditional density of $X = X_m$ and $Y = Y_n$ given the u's and v's is seen from (10.18) to be of the form

$$f_u^*(x - \xi) g_v^*(y - \eta), \tag{10.20}$$

where the subscripts u and v indicate that f^* and g^* depend on the u's and v's respectively. The problem of testing H in the conditional model remains invariant under the transformations: $x' = x + c$, $y' = y + c$, for which $Y - X$ is maximal invariant. A UMP invariant conditional test will then exist provided the distribution of $Z = Y - X$, which depends only on $\Delta = \eta - \xi$, has monotone likelihood ratio. The following lemma shows that a sufficient condition for this to be the case is that f_u^* and g_v^* have monotone likelihood ratio in x and y respectively.

Lemma 10.3.2 *Let X, Y be independently distributed with densities $f^*(x - \xi)$, $g^*(y - \eta)$ respectively. If f^* and g^* have monotone likelihood with respect to ξ and η, then the family of densities of $Z = Y - X$ has monotone likelihood ratio with respect to $\Delta = \eta - \xi$.*

PROOF. The density of Z is

$$h_\Delta(z) = \int g^*(y-\Delta) f^*(y-z)\, dy. \tag{10.21}$$

To see that $h_\Delta(z)$ has monotone likelihood ratio, one must show that for any $\Delta < \Delta'$, $h_{\Delta'}(z)/h_\Delta(z)$ is an increasing function of z. For this purpose, write

$$\frac{h_{\Delta'}(z)}{h_\Delta(z)} = \int \frac{g^*(y-\Delta')}{g^*(y-\Delta)} \cdot \frac{g^*(y-\Delta)f^*(y-z)}{\int g^*(u-\Delta)f(u-z)\,du}\,dy.$$

The second factor is a probability density for Y,

$$p_z(y) = C_z g^*(y-\Delta) f^*(y-z), \tag{10.22}$$

which has monotone likelihood ratio in the parameter z by the assumption made about f^*. The ratio

$$\frac{h_{\Delta'}(z)}{h_\Delta(z)} = \int \frac{g^*(y-\Delta')}{g^*(y-\Delta)} p_z(y)\,dy \tag{10.23}$$

is the expectation of $g^*(Y-\Delta')/g^*(Y-\Delta)$ under the distribution $p_z(y)$. By the assumption about g^*, $g^*(y-\Delta')/g^*(y-\Delta)$ is an increasing function of y, and it follows from Lemma 3.4.2 that its expectation is an increasing function of z. ■

It follows from (10.18) that $f_u^*(x-\xi)$ and $g_v^*(y-\eta)$ have monotone likelihood ratio provided this condition holds for $f(x-\xi)$ and $g(y-\eta)$, i.e. provided f and g are strongly unimodal. Under this assumption, the conditional distribution $h_\Delta(z)$ then has monotone likelihood ratio by Lemma 10.3.2, and a UMP conditional test exists and rejects for large values of Z. (This result also follows from Problem 8.9.)

The difference between conditional tests of the kind considered in this section and the corresponding (e.g., locally most powerful) unconditional tests typically disappears as the sample size(s) tend(s) to infinity. Some results in this direction are given by Liang (1984); see also Barndorff–Nielsen (1983).

The following multivariate example provides one more illustration of a UMP conditional test when unconditionally no UMP test exists. The results will only be sketched. The details of this and related problems can be found in the original literature reviewed by Marden and Perlman (1980) and Marden (1983).

Example 10.3.3 Suppose you observe $m+1$ independent normal vectors of dimension $p = p_1 + p_2$,

$$Y = (Y_1 \;\; Y_2) \quad \text{and} \quad Z_1, \dots, Z_m,$$

with common covariance matrix Σ and expectations

$$E(Y_1) = \eta_1, \qquad E(Y_2) = E(Z_1) = \cdots = E(Z_m) = 0.$$

(The normal multivariate two-sample problem with covariates can be reduced to this canonical form.) The hypothesis being tested is $H : \eta_1 = 0$. Without the restriction $E(Y_2) = 0$, the model would remain invariant under the group G of transformations: $Y^* = YB$, $Z^* = ZB$, where B is any nonsingular $p \times p$ matrix. However, the stated problem remains invariant only under the subgroup G' in which B is of the form [Problem 10.22(i)]

$$B = \begin{pmatrix} B_{11} & 0 \\ B_{21} & B_{22} \end{pmatrix} \begin{matrix} p_1 \\ p_2 \end{matrix} .$$
$$\qquad\qquad p_1 \quad p_2$$

If

$$Z'Z = S = \begin{pmatrix} S_{11} & S_{12} \\ S_{21} & S_{22} \end{pmatrix} \quad \text{and} \quad \Sigma = \begin{pmatrix} \Sigma_{11} & \Sigma_{12} \\ \Sigma_{21} & \Sigma_{22} \end{pmatrix},$$

the maximal invariants under G' are the two statistics $D = Y_2 S_{22}^{-1} Y_2'$ and

$$N = \frac{(Y_1 - S_{12}S_{22}^{-1}Y_2)(S_{11} - S_{12}S_{22}^{-1}S_{21})^{-1}(Y_1 - S_{12}S_{22}^{-1}Y_2)'}{1+D},$$

and the joint distribution of (N, D) depends only on the maximal invariant under G',

$$\Delta = \eta_1(\Sigma_{11} - \Sigma_{12}\Sigma_{22}^{-1}\Sigma_{21})^{-1}\eta_1'.$$

The statistic D is ancillary [Problem 10.22(ii)], and the conditional distribution of N given $D = d$ is that of the ratio of two independent χ^2-variables: the numerator noncentral χ^2 with p degrees of freedom and noncentrality parameter $\Delta/(1+d)$, and the denominator central χ^2 with $m+1-p$ degrees of freedom. It follows from Section 7.1 that the conditional density has monotone likelihood ratio. A conditionally UMP invariant test therefore exists, and rejects H when $(m+1-p)N/p > C$, where C is the critical value of the F-distribution with p and $m+1-p$ degrees of freedom. On the other hand, a UMP invariant (unconditional) test does not exist; comparisons of the optimal conditional test with various competitors are provided by Marden and Perlman (1980). ■

10.4 Relevant Subsets

The conditioning variables considered so far have been ancillary statistics, i.e. random variables whose distribution is fixed, independent of the parameters governing the distribution of X, or at least of the parameter of interest. We shall now examine briefly some implications of conditioning without this constraint. Throughout most of the section we shall be concerned with the simple case in which the conditioning variable is the indicator of some subset C of the sample space, so that there are only two conditioning events $I = 1$ (i.e. $X \in C$) and $I = 0$ (i.e. $X \in C^c$, the complement of C). The mixture problem at the beginning of Section 10.1, with $\mathcal{X}_1 = C$ and $\mathcal{X}_0 = C^c$, is of this type.

Suppose X is distributed with density p_θ, and R is a level-α rejection region for testing the simple hypothesis $H : \theta = \theta_0$ against some class of alternatives. For any subset C of the sample space, consider the conditional rejection probabilities

$$\alpha_C = P_{\theta_0}(X \in R \mid C) \quad \text{and} \quad \alpha_{C^c} = P_{\theta_0}(X \in R \mid C^c), \tag{10.24}$$

and suppose that $\alpha_C > \alpha$ and $\alpha_{C^c} < \alpha$. Then we are in the difficulty described in Section 10.1. Before X was observed, the probability of falsely rejecting H was stated to be α. Now that X is known to have fallen into C (or C^c), should the original statement be adjusted and the higher value α_C (or lower value α_{C^c}) be quoted? An extreme case of this possibility occurs when C is a subset of R or R^c, since then $P(X \in R \mid X \in C) = 1$ or 0.

It is clearly always possible to choose C so that the conditional level α_C exceeds the stated α. It is not so clear whether the corresponding possibility always exists for the levels of a family of confidence sets for θ, since the inequality must now hold for all θ.

Definition 10.4.1 A subset C of the sample space is said to be a *negatively biased relevant subset* for a family of confidence sets $S(X)$ with unconditional confidence level $\gamma = 1 - \alpha$ if for some $\epsilon > 0$

$$\gamma_C(\theta) = P_\theta[\theta \in S(X) \mid X \in C] \leq \gamma - \epsilon \qquad \text{for all } \theta, \tag{10.25}$$

and a *positively biased relevant subset* if

$$P_0[\theta \in S(X) \mid X \in C] \geq \gamma + \epsilon \qquad \text{for all } \theta. \tag{10.26}$$

The set C is *semirelevant, negatively or positively biased*, if respectively

$$P_\theta[\theta \in S(X) \mid X \in C] \leq \gamma \qquad \text{for all } \theta \tag{10.27}$$

or

$$P_\theta[\theta \in S(X) \mid X \in C] \geq \gamma \qquad \text{for all } \theta, \tag{10.28}$$

with strict inequality holding for at least some θ.

Obvious examples of relevant subsets are provided by the subsets $\mathcal{X}_0$ and $\mathcal{X}_1$ of the two-experiment example of Section 10.1.

Relevant subsets do not always exist. The following four examples illustrate the various possibilities.

Example 10.4.1 Let X be distributed as $N(\theta, 1)$, and consider the standard confidence intervals for θ:

$$S(X) = \{\theta : X - c < \theta < X + c\},$$

where $\Phi(c) - \Phi(-c) = \gamma$. In this case, there exists not even a semirelevant subset.

To see this, suppose first that a positively biased semirelevant subset C exists, so that

$$A(\theta) = P_\theta[X - c < \theta < X + c \text{ and } X \in C] - \gamma P_\theta[X \in C] \geq 0$$

for all θ, with strict inequality for some θ_0. Consider a prior normal density $\lambda(\theta)$ for θ with mean 0 and variance τ^2, and let

$$\beta(x) = P[x - c < \Theta < x + c \mid x],$$

where Θ has density $\lambda(\theta)$. The posterior distribution of Θ given x is then normal with mean $\tau^2 x/(1+\tau^2)$ and variance $\tau^2/(1+\tau^2)$ [Problem 10.24(i)], and it follows that

$$\beta(x) = \Phi\left[\frac{x}{\tau\sqrt{1+\tau^2}} + \frac{c\sqrt{1+\tau^2}}{\tau}\right] - \Phi\left[\frac{x}{\tau\sqrt{1+\tau^2}} - \frac{c\sqrt{1+\tau^2}}{\tau}\right]$$

$$\leq \quad \Phi\left[\frac{c\sqrt{1+\tau^2}}{\tau}\right] - \Phi\left[\frac{-c\sqrt{1+\tau^2}}{\tau}\right] \leq \gamma + \frac{c}{\sqrt{2\pi\tau^2}}.$$

Next let $h(\theta) = \sqrt{2\pi}\tau\lambda(\theta) = e^{-\theta^2/2\tau^2}$ and

$$D = \int h(\theta)A(\theta)\,d\theta \leq \sqrt{2\pi}\tau \int \lambda(\theta)\{P_\theta[X - c < \theta < X + c \text{ and } X \in C] \\ -E_\theta[\beta(X)I_C(X)]\}\,d\theta + \frac{c}{\tau}.$$

The integral on the right side is the difference of two integrals each of which equals $P[X - c < \Theta < X + c$ and $X \in C]$, and is therefore 0, so that $D \leq c/\tau$.

Consider now a sequence of normal priors $\lambda_m(\theta)$ with variances $\tau_m^2 \to \infty$, and the corresponding sequences $h_m(\theta)$ and D_m. Then $0 \leq D_m \leq c/\tau_m$ and hence $D_m \to 0$. On the other hand, D_m is of the form $D_m = \int_{-\infty}^{\infty} A(\theta)h_m(\theta)\,d\theta$, where $A(\theta)$ is continuous, nonnegative, and > 0 for some θ_0. There exists $\delta > 0$ such that $A(\theta) \leq \frac{1}{2}A(\theta_0)$ for $|\theta - \theta_0| < \delta$ and hence

$$D_m \geq \int_{\theta_0-\delta}^{\theta_0+\delta} \frac{1}{2}A(\theta_0)h_m(\theta)\,d\theta \to \delta A(\theta_0) > 0 \qquad \text{as} \quad m \to \infty.$$

This provides the desired contradiction. ■

That also no negatively semirelevant subsets exist is a consequence of the following result.

Theorem 10.4.2 *Let $S(x)$ be a family of confidence sets for θ such that $P_\theta[\theta \in S(X)] = \gamma$ for all θ, and suppose that $0 < P_\theta(C) < 1$ for all θ.*

(i) If C is semirelevant, then its complement C^c is semirelevant with opposite bias.

(ii) If there exists a constant a such that

$$1 > P_\theta(C) > a > 0 \qquad \text{for all } \theta$$

and C is relevant, then C^c is relevant with opposite bias.

PROOF. The result is an immediate consequence of the identity

$$P_\theta(C)[\gamma_C(\theta) - \gamma] = [1 - P_\theta(C)][\gamma - \gamma_{C^c}(\theta)]. \;■$$

The next example illustrates the situation in which a semirelevant subset exists but no relevant one.

Example 10.4.2 Let X be $N(\theta, 1)$, and consider the uniformly most accurate lower confidence bounds $\underline{\theta} = X - c$ for θ, where $\Phi(c) = \gamma$. Here $S(X)$ is the interval $[X - c, \infty)$ and it seems plausible that the conditional probability of $\theta \in S(X)$ will be lowered for a set C of the form $X \geq k$. In fact

$$P_\theta(X - c \leq \theta \mid X \geq k) = \begin{cases} \frac{\Phi(c)-\Phi(k-\theta)}{1-\Phi(k-\theta)} & \text{when} \quad \theta > k - c, \\ 0 & \text{when} \quad \theta < k - c. \end{cases} \tag{10.29}$$

The probability (10.29) is always $< \gamma$, and tends to γ as $\theta \to \infty$. The set $X \geq k$ is therefore semirelevant negatively biased for the confidence sets $S(X)$.

We shall now show that no relevant subset C with $P_\theta(C) > 0$ exists in this case. It is enough to prove the result for negatively biased sets; the proof for positive bias is exactly analogous. Let A be the set of x-values $-\infty < x < c + \theta$, and suppose that C is negatively biased and relevant, so that

$$P_\theta[X \in A \mid C] \leq \gamma - \epsilon \qquad \text{for all } \theta.$$

If

$$a(\theta) = P_\theta(X \in C), \qquad b(\theta) = P_\theta(X \in A \cap C),$$

then

$$b(\theta) \leq (y - \epsilon)\, a(\theta) \qquad \text{for all } \theta. \tag{10.30}$$

The result is proved by comparing the integrated coverage probabilities

$$A(R) = \int_{-R}^{R} a(\theta)\, d\theta, \qquad B(R) = \int_{-R}^{R} b(\theta)\, d\theta$$

with the Lebesgue measure of the intersection $C \cap (-R, R)$,

$$\mu(R) = \int_{-R}^{R} I_C(x)\, dx,$$

where $I_C(x)$ is the indicator of C, and showing that

$$\frac{A(R)}{\mu(R)} \to 1, \quad \frac{B(R)}{\mu(R)} \to \gamma \quad \text{as} \quad R \to \infty. \tag{10.31}$$

This contradicts the fact that by (10.30),

$$B(R) \leq (\gamma - \epsilon) A(R) \qquad \text{for all } R,$$

and so proves the desired result.

To prove (10.31), suppose first that $\mu(\infty) < \infty$. Then if ϕ is the standard normal density

$$A(\infty) = \int_{-\infty}^{\infty} d\theta \int_C \phi(x - \theta)\, dx = \int_C dx = \mu(\infty),$$

and analogously $B(\infty) = \gamma\mu(\infty)$, which establishes (10.31).

When $\mu(\infty) = \infty$, (10.31) will be proved by showing that

$$A(R) = \mu(R) + K_1(R), \qquad B(R) = \gamma\mu(R) + K_2(R), \tag{10.32}$$

where $K_1(R)$ and $K_2(R)$ are bounded. To see (10.32), note that

$$\begin{aligned} \mu(R) = \int_{-R}^{R} I_C(x)\, dx &= \int_{-R}^{R} I_C(x) \left[\int_{-\infty}^{\infty} \phi(x - \theta)\, d\theta\right] dx \\ &= \int_{-\infty}^{\infty} \left[\int_{-R}^{R} I_C(x)\phi(x - \theta)\, dx\right] d\theta, \end{aligned}$$

while

$$A(R) = \int_{-R}^{R} \left[\int_{-\infty}^{\infty} I_C(x)\phi(x - \theta)\, dx\right] d\theta. \tag{10.33}$$

A comparison of each of these double integrals with that over the region $-R < x < R$, $-R < \theta < R$, shows that the difference $A(R) - \mu(R)$ is made up of four integrals, each of which can be seen to be bounded by using the fact that $\int |t|\phi(t)\, dt < \infty$ [Problem 10.24(ii)]. This completes the proof. ■

Example 10.4.3 Let $X_1, \ldots, X_n$ be independently normally distributed as $N(\xi, \sigma^2)$, and consider the uniformly most accurate equivariant (and unbiased) confidence intervals for ξ given by (5.36).

It was shown by Buehler and Feddersen (1963) and Brown (1967) that in this case there exist positively biased relevant subsets of the form

$$C : \frac{|\bar{X}|}{S} \le k. \tag{10.34}$$

In particular, for confidence level $\gamma = .5$ and $n = 2$, Brown shows that with $C : |\bar{X}|/|X_2 - X_1| \le \frac{1}{2}(1+\sqrt{2})$, the conditional level is $> \frac{2}{3}$ for all values of ξ and σ. Goutis and Casella (1992) provide detailed values for general n.

It follows from Theorem 10.4.2 that C^c is negatively biased semirelevant, and Buehler (1959) shows that any set $C^* : S \le k$ has the same property. These results are intuitively plausible, since the length of the confidence intervals is proportional to S, and one would expect short intervals to cover the true value less often than long ones.

Theorem 10.4.2 does not show that C^c is negatively biased relevant, since the probability of the set (10.34) tends to zero as $\xi/\sigma \to \infty$. It was in fact proved by Robinson (1976) that no negatively biased relevant subset exists in this case.

The calculations for C^c throw some light on the common practice of stating confidence intervals for ξ only when a preliminary test of $H : \xi = 0$ rejects the hypothesis. For a discussion of this practice see Olshen (1973), and Meeks and D'Agostino (1983). ■

The only type of example still missing is that of a negatively biased relevant subset. It was pointed out by Fisher (1956a,b) that the Welch–Aspin solution of the Behrens–Fisher problem (discussed in Sections 6.6 and 11.3) provides an illustration of this possibility. The following are much simpler examples of both negatively and positively biased relevant subsets.

Example 10.4.4 An extreme form of both positively and negatively biased subsets was encountered in Section 7.7, where lower and upper confidence bounds $\underline{\Delta} < \Delta$ and $\Delta < \bar{\Delta}$ were obtained in (7.42) and (7.43) for the ratio $\Delta = \sigma_A^2/\sigma^2$ in a model II one-way classification. Since

$$P(\underline{\Delta} \le \Delta \mid \underline{\Delta} < 0) = 1 \quad \text{and} \quad P(\Delta \le \bar{\Delta} \mid \bar{\Delta} < 0) = 0,$$

the sets $C_1 : \underline{\Delta} < 0$ and $C_2 : \bar{\Delta} < 0$ are relevant subsets with positive and negative bias respectively. ■

The existence of conditioning sets C for which the conditional coverage probability of level-γ confidence sets is 0 or 1, such as in Example 10.4.4 or Problems 10.27, 10.28 are an embarrassment to confidence theory, but fortunately they are rare. The significance of more general relevant subsets is less clear,[6] particularly when a number of such subsets are available. Especially awkward in this connection is the possibility [discussed by Buehler (1959)] of the existence of two relevant subsets C and C' with nonempty intersection and opposite bias.

[6]For a discussion of this issue, see Buehler (1959), Robinson (1976, 1979a), and Bondar (1977).

If a conditional confidence level is to be cited for some relevant subset C, it seems appropriate to take account also of the possibility that X may fall into C^c and to state in advance the three confidence coefficients γ, γ_C, and γ_{C^c}. The (unknown) probabilities $P_\theta(C)$ and $P_\theta(C^c)$ should also be considered. These points have been stressed by Kiefer, who has also suggested the extension to a partition of the sample space into more than two sets. For an account of these ideas see Kiefer (1977a,b), Brownie and Kiefer (1977), and Brown (1978).

Kiefer's theory does not consider the choice of conditioning set or statistic. The same question arose in Section 10.2 with respect to conditioning on ancillaries. The problem is similar to that of the choice of model. The answer depends on the context and purpose of the analysis, and must be determined from case to case.

10.5 Problems

Section 10.1

Problem 10.1 Let the experiments of $\mathcal{E}$ and $\mathcal{F}$ consist in observing $X : N(\xi, \sigma_0^2)$ and $X : N(\xi, \sigma_1^2)$ respectively $(\sigma_0 < \sigma_1)$, and let one of the two experiments be performed, with $P(\mathcal{E}) = P(\mathcal{F}) = \frac{1}{2}$. For testing $H : \xi = 0$ against $\xi = \xi_1$, determine values σ_0, σ_1, ξ_1, and α such that

$$\text{(i)} \quad \alpha_0 < \alpha_1; \qquad \text{(ii)} \quad \alpha_0 > \alpha_1,$$

where the α_i are defined by (10.9).

Problem 10.2 Under the assumptions of Problem 10.1, determine the most accurate invariant (under the transformation $X' = -X$) confidence sets $S(X)$ with

$$P(\xi \in S(X) \mid \mathcal{E}) + P(\xi \in S(X) \mid \mathcal{F}) = 2\gamma.$$

Find examples in which the conditional confidence coefficients γ_0 given $\mathcal{E}$ and γ_1 given $\mathcal{F}$ satisfy

$$\text{(i)} \quad \gamma_0 < \gamma_1; \qquad \text{(ii)} \quad \gamma_0 > \gamma_1.$$

Problem 10.3 The test given by (10.3), (10.8), and (10.9) is most powerful under the stated assumptions.

Problem 10.4 Let $X_1, \ldots, X_n$ be independently distributed, each with probability p or q as $N(\xi, \sigma_0^2)$ or $N(\xi, \sigma_1^2)$.

(i) If p is unknown, determine the UMP unbiased test of $H : \xi = 0$ against $K : \xi > 0$.

(ii) Determine the most powerful test of H against the alternative ξ_1 when it is known that $p = \frac{1}{2}$, and show that a UMP unbiased test does not exist in this case.

(iii) Let α_k $(k = 0, \ldots, n)$ be the conditional level of the unconditional most powerful test of part (ii) given that k of the X's came from $N(\xi, \sigma_0^2)$ and $n - k$ from $N(\xi, \sigma_1^2)$. Investigate the possible values $\alpha_0, \alpha_1, \ldots, \alpha_n$.

Problem 10.5 With known probabilities p and q perform either $\mathcal{E}$ or $\mathcal{F}$, with X distributed as $N(\theta, 1)$ under $\mathcal{E}$ or $N(-\theta, 1)$ under $\mathcal{F}$. For testing $H : \theta = 0$ against $\theta > 0$ there exist a UMP unconditional and a UMP conditional level-α test. These coincide and do not depend on the value of p.

Problem 10.6 In the preceding problem, suppose that the densities of X under $\mathcal{E}$ and $\mathcal{F}$ are $\theta e^{-\theta x}$ and $(1/\theta)e^{-x/\theta}$ respectively. Compare the UMP conditional and unconditional tests of $H : \theta = 1$ against $K : \theta > 1$.

Section 10.2

Problem 10.7 Let X, Y be independently normally distributed as $N(\theta, 1)$, and let $V = Y - X$ and

$$W = \begin{cases} Y - X & \text{if} \quad X + Y > 0, \\ X - Y & \text{if} \quad X + Y \leq 0. \end{cases}$$

(i) Both V and W are ancillary, but neither is a function of the other.

(ii) (V, W) is not ancillary. [Basu (1959).]

Problem 10.8 An experiment with n observations $X_1, \ldots, X_n$ is planned, with each X_i distributed as $N(\theta, 1)$. However, some of the observations do not materialize (for example, some of the subjects die, move away, or turn out to be unsuitable). Let $I_j = 1$ or 0 as X_j is observed or not, and suppose the I_j are independent of the X's and of each other and that $P(I_j = 1) = p$ for all j.

(i) If p is known, the effective sample size $M = \sum I_j$ is ancillary.

(ii) If p is unknown, there exists a UMP unbiased level-α test of $H : \theta \leq 0$ vs. $K : \theta > 0$. Its conditional level (given $M = m$) is $\alpha_m = \alpha$ for all $m = 0, \ldots, n$.

Problem 10.9 Consider n tosses with a biased die, for which the probabilities of $1, \ldots, 6$ points are given by

1	2	3	4	5	6
$\frac{1-\theta}{12}$	$\frac{2-\theta}{12}$	$\frac{3-\theta}{12}$	$\frac{1+\theta}{12}$	$\frac{2+\theta}{12}$	$\frac{3+\theta}{12}$

and let X_i be the number of tosses showing i points.

(i) Show that the triple $Z_1 = X_1 + X_5$, $Z_2 = X_2 + X_4$, $Z_3 = X_3 + X_6$ is a maximal ancillary; determine its distribution and the distribution of $X_1, \ldots, X_6$ given $Z_1 = z_1$, $Z_2 = z_2$, $Z_3 = z_3$.

(ii) Exhibit five other maximal ancillaries. [Basu (1964).]

Problem 10.10 In the preceding problem, suppose the probabilities are given by

1	2	3	4	5	6
$\frac{1-\theta}{6}$	$\frac{1-2\theta}{6}$	$\frac{1-3\theta}{6}$	$\frac{1+\theta}{6}$	$\frac{1+2\theta}{6}$	$\frac{1+3\theta}{6}$

Exhibit two different maximal ancillaries.

Problem 10.11 Let X be uniformly distributed on $(\theta, \theta+1)$, $0 < \theta < \infty$, let $[X]$ denote the largest integer $\leq X$, and let $V = X - [X]$.

(i) The statistic $V(X)$ is uniformly distributed on $(0, 1)$ and is therefore ancillary.

(ii) The marginal distribution of $[X]$ is given by

$$[X] = \begin{cases} [\theta] & \text{with probability } 1 - V(\theta), \\ [\theta] + 1 & \text{with probability } V(\theta). \end{cases}$$

(iii) Conditionally, given that $V = v$, $[X]$ assigns probability 1 to the value $[\theta]$ if $V(\theta) \leq v$ and to the value $[\theta] + 1$ if $V(\theta) > v$. [Basu (1964).]

Problem 10.12 Let X, Y have joint density

$$p(x, y) = 2f(x)f(y)F(\theta xy),$$

where f is a known probability density symmetric about 0, and F its cumulative distribution function. Then

(i) $p(x, y)$ is a probability density.

(ii) X and Y each have marginal density f and are therefore ancillary, but (X, Y) is not.

(iii) $X \cdot Y$ is a sufficient statistic for θ. [Dawid (1977).]

Problem 10.13 A sample of size n is drawn with replacement from a population consisting of N distinct unknown values $\{a_1, \ldots, a_N\}$. The number of distinct values in the sample is ancillary.

Problem 10.14 Assuming the distribution (4.22) of Section 4.9, show that Z is S-ancillary for $p = p_+/(p_+ + p_-)$.

Problem 10.15 In the situation of Example 10.2.3, $X + Y$ is binomial if and only if $\Delta = 1$.

Problem 10.16 In the situation of Example 10.2.2, the statistic Z remains S-ancillary when the parameter space is $\Omega = \{(\lambda, \mu) : \mu \leq \lambda\}$.

Problem 10.17 Suppose $X = (U, Z)$, the density of X factors into

$$p_{\theta,\vartheta}(x) = c(\theta, \vartheta)g_\theta(u; z)h_\vartheta(z)k(u, z),$$

and the parameters θ, ϑ are unrelated. To see that these assumptions are not enough to insure that Z is S-ancillary for θ, consider the joint density

$$C(\theta, \vartheta)e^{-\frac{1}{2}(u-\theta)^2-\frac{1}{2}(z-\vartheta)^2}I(u, z),$$

where $I(u, z)$ is the indicator of the set $\{(u, z) : u \leq z\}$. [Basu (1978).]

Section 10.3

Problem 10.18 Verify the density (10.16) of Example 10.3.2.

Problem 10.19 Let the real-valued function f be defined on an open interval.

(i) If f is logconvex, it is convex.

(ii) If f is strongly unimodal, it is unimodal.

Problem 10.20 Let $X_1, \ldots, X_m$ and $Y_1, \ldots, Y_n$ be positive, independent random variables distributed with densities $f(x/\sigma)$ and $g(y/\tau)$ respectively. If f and g have monotone likelihood ratios in (x, σ) and (y, τ) respectively, there exists a UMP conditional test of $H : \tau/\sigma \leq \Delta_0$ against $\tau/\sigma > \Delta_0$ given the ancillary statistics $U_i = X_i/X_m$ and $V_j = Y_j/Y_n$ $(i = 1, \ldots, m-1;\ j = 1, \ldots, n-1)$.

Problem 10.21 Let $V_1, \ldots, V_n$ be independently distributed as $N(0, 1)$, and given $V_1 = v_1, \ldots, V_n = v_n$, let X_i $(i = 1, \ldots, n)$ be independently distributed as $N(\theta v_i, 1)$.

(i) There does not exist a UMP test of $H : \theta = 0$ against $K : \theta > 0$.

(ii) There does exist a UMP conditional test of H against K given the ancillary $(V_1, \ldots, V_n)$. [Buehler (1982).]

Problem 10.22 In Example 10.3.3,

(i) the problem remains invariant under G' but not under G;

(ii) the statistic D is ancillary.

Section 10.4

Problem 10.23 In Example 10.4.1, check directly that the set $C = \{x : x \leq -k \text{ or } x \geq k\}$ is not a negatively biased semirelevant subset for the confidence intervals $(X - c, X + c)$.

Problem 10.24 (i) Verify the posterior distribution of Θ given x claimed in Example 10.4.1.

(ii) Complete the proof of (10.32).

Problem 10.25 Let X be a random variable with cumulative distribution function F. If $E|X| < \infty$, then $\int_{-\infty}^{0} F(x)\,dx$ and $\int_0^{\infty}[1 - F(x)]\,dx$ are both finite. [Apply integration by parts to the two integrals.]

Problem 10.26 Let X have probability density $f(x - \theta)$, and suppose that $E|X| < \infty$. For the confidence intervals $X - c < \theta$ there exist semirelevant but no relevant subsets. [Buehler (1959).]

Problem 10.27 Let $X_1, \ldots, X_n$ be independently distributed according to the uniform distribution $U(\theta, \theta + 1)$.

(i) Uniformly most accurate lower confidence bounds $\underline{\theta}$ for θ at confidence level $1 - \alpha$ exist and are given by

$$\underline{\theta} = \max(X_{(1)} - k, X_{(n)} - 1),$$

where $X_{(1)} = \min(X_1, \ldots, X_n)$, $X_{(n)} = \max(X_1, \ldots, X_n)$, and $(1 - k)^n = \alpha$.

(ii) The set $C : x_{(n)} - x_{(1)} \geq 1 - k$ is a relevant subset with $P_\theta(\underline{\theta} \leq \theta \mid C) = 1$ for all θ.

(iii) Determine the uniformly most accurate conditional lower confidence bounds $\underline{\theta}(v)$ given the ancillary statistic $V = X_{(n)} - X_{(1)} = v$, and compare them with $\underline{\theta}$. [The conditional distribution of $Y = X_{(1)}$ given $V = v$ is $U(\theta, \theta + 1 - v)$.]

[Pratt (1961a), Barnard (1976).]

Problem 10.28 (i) Under the assumptions of the preceding problem, the uniformly most accurate unbiased (or invariant) confidence intervals for θ at confidence level $1 - \alpha$ are

$$\underline{\theta} = \max(X_{(1)} + d, X_{(n)}) - 1 < \theta < \min(X_{(1)}, X_{(n)} - d) = \bar{\theta},$$

where d is the solution of the equation

$$\begin{aligned} 2d^n &= \alpha && \text{if} \quad \alpha < 1/2^{n-1}, \\ 2d^n - (2d - 1)^n &= \alpha && \text{if} \quad \alpha > 1/2^{n-1}. \end{aligned}$$

(ii) The sets $C_1 : X_{(n)} - X_{(1)} > d$ and $C_2 : X_{(n)} - X_{(1)} < 2d - 1$ are relevant subsets with coverage probability

$$P_\theta[\underline{\theta} < \theta < \bar{\theta} \mid C_1] = 1 \quad \text{and} \quad P_\theta[\underline{\theta} < \theta < \bar{\theta} \mid C_2] = 0.$$

(iii) Determine the uniformly most accurate unbiased (or invariant) conditional confidence intervals $\underline{\theta}(v) < \theta < \bar{\theta}(v)$ given $V = v$ at confidence level $1 - \alpha$, and compare $\underline{\theta}(v)$, $\bar{\theta}(v)$, and $\bar{\theta}(v) - \underline{\theta}(v)$ with the corresponding unconditional quantities.

[Welch (1939), Pratt (1961a), Kiefer (1977a).]

Problem 10.29 Suppose X_1 and X_2 are i.i.d. with

$$P\{X_i = \theta - 1\} = P\{X_i = \theta + 1\} = \frac{1}{2}\,.$$

Let C be the confidence set consisting of the single point $(X_1 + X_2)/2$ if $X_1 \neq X_2$ and $X_1 - 1$ if $X_1 = X_2$. Show that, for all θ,

$$P_\theta\{\theta \in C\} = .75\ ,$$

but

$$P_\theta\{\theta \in C | X_1 = X_2\} = .5$$

and

$$P_\theta\{\theta \in C | X_1 \neq X_2\} = 1 .$$

[Berger and Wolpert (1988)]

Problem 10.30 Instead of conditioning the confidence sets $\theta \in S(X)$ on a set C, consider a randomized procedure which assigns to each point x a probability $\psi(x)$ and makes the confidence statement $\theta \in S(x)$ with probability $\psi(x)$ when x is observed.[7]

(i) The randomized procedure can be represented by a nonrandomized conditioning set for the observations (X, U), where U is uniformly distributed on $(0, 1)$ and independent of X, by letting $C = \{(x, u) : u < \psi(x)\}$.

(ii) Extend the definition of relevant and semirelevant subsets to randomized conditioning (without the use of U).

(iii) Let $\theta \in S(X)$ be equivalent to the statement $X \in A(\theta)$. Show that ψ is positively biased semirelevant if and only if the random variables $\psi(X)$ and $I_{A(\theta)}(X)$ are positively correlated, where I_A denotes the indicator of the set A.

Problem 10.31 The nonexistence of (i) semirelevant subsets in Example 10.4.1 and (ii) relevant subsets in Example 10.4.2 extends to randomized conditioning procedures.

10.6 Notes

Conditioning on ancillary statistics was introduced by Fisher (1934, 1935, 1936).[8] The idea was emphasized in Fisher (1956b) and by Cox (1958), who motivated it in terms of mixtures of experiments providing different amounts of information. The consequences of adopting a general principle of conditioning in mixture situations were explored by Birnbaum (1962) and Durbin (1970). Following Fisher's suggestion (1934), Pitman (1938b) developed a theory of conditional tests and confidence intervals for location and scale parameters. For recent paradox concerning conditioning on an ancillary statistic, see Brown (1990) and Wang (1999).

The possibility of relevant subsets was pointed out by Fisher (1956a,b) (who called them *recognizable*. Its implications (in terms of betting procedures) were developed by Buehler (1959), who in particular introduced the distinction between relevant and semirelevant, positively and negatively biased subsets, and proved

[7]Randomized and nonrandomized conditioning is interpreted in terms of betting strategies by Buehler (1959) and Pierce (1973).

[8]Fisher's contributions to this topic are discussed in Savage (1976, pp. 467–469).

the nonexistence of relevant subsets in location models. The role of relevant subsets in statistical inference, and their relationship to Bayes and admissibility properties, was discussed by Pierce (1973), Robinson (1976, 1979a,b), Bondar (1977), and Casella (1988), among others.

Fisher (1956a, b) introduced the idea of relevant subsets in the context of the Behrens–Fisher problem. As a criticism of the Welch–Aspin solution, he established the existence of negatively biased relevant subsets for that procedure. It was later shown by Robinson (1976) that no such subsets exist for Fisher's preferred solution, the so-called Behrens–Fisher intervals. This fact may be related to the conjecture [supported by substantial numerical evidence in Robinson (1976) but so far unproved] that the unconditional coverage probability of the Behrens–Fisher intervals always exceeds the nominal level. For a review of these issues, see Wallace (1980) and Robinson (1982).

Maata and Casella (1987) examine the conditional properties of some confidence intervals for the variance in the one-sample normal problem. The conditional properties of some confidence sets for the multivariate normal mean, including confidence sets centered at James-Stein or shrinkage estimators, see Casella (1987) and George and Casella (1994). The conditional properties of the standard confidence sets in a normal linear model are studied in Hwang and Brown (1991).

In testing a simple hypothesis against a simple alternative, Berger, Brown and Wolpert (1994) present a conditional frequentist methodology that agrees with a Bayesian approach.

Part II

Large-Sample Theory

11

Basic Large Sample Theory

11.1 Introduction

Chapters 3-7 were concerned with the derivation of UMP, UMP unbiased, and UMP invariant tests. Unfortunately, the existence of such tests turned out to be restricted essentially to one-parameter families with monotone likelihood ratio, exponential families, and group families, respectively. Tests maximizing the minimum or average power over suitable classes of alternatives exist fairly generally, but are difficult to determine explicitly, and their derivation in Chapter 8 was confined primarily to situations in which invariance considerations apply.

Despite their limitations, these approaches have proved their value by application to large classes of important situations. On the other hand, they are unlikely to be applicable to complex new problems. What is needed for such cases is a simpler, less detailed, more generally applicable formulation. The development and implementation of such an approach will be the subject of the remaining chapters. It replaces optimality by asymptotic optimality obtained by embedding the actual situation in a sequence of situations of increasing sample size, and applying optimality to the limit situation. These limits tend to be of the simple type for which optimality has been established in earlier chapters.

A feature of asymptotic optimality is that it refers not to a single test but to a sequence of tests, although this distinction will often be suppressed. An important consequence is that asymptotically optimal procedures - unlike most optimal procedures in the small sample approach - are not unique since many different sequences have the same limit. In fact, quite different methods of construction may lead to procedures which are asymptotically optimal.

The following are some specific examples to keep in mind where finite sample considerations fail to provide optimal procedures, but for which a large sample approach will seen to be more successful.

Example 11.1.1 (One parameter families) Suppose $X_1, \ldots, X_n$ are i.i.d. according to some family of distributions P_θ indexed by a real-valued parameter θ. Then, it was mentioned after Corollary 3.4.1 that UMP tests for testing $\theta = \theta_0$ against $\theta > \theta_0$ exist for all sample sizes (under weak regularity conditions) only when the distributions P_θ constitute an exponential family. For example, location models typically do not have monotone likelihood ratio, and so UMP tests rarely exist in this situation, though the normal location model is a happy exception. On the other hand, we shall see that under weak assumptions, there generally exist tests for one-parameter families which are asymptotically UMP in a suitable sense; see Section 13.3. For example, we shall derive an asymptotically optimal one-sided test in the Cauchy location model, among others. ■

Example 11.1.2 (Behrens-Fisher Problem) Consider testing the equality of means for two independent samples, from normal distributions with possibly different (unknown) variances. As previously mentioned, finite sample optimality considerations such as unbiasedness or invariance do not lead to an optimal test, even though the setting is a multiparameter exponential family. An optimal test sequence will be derived in Example 13.5.4. ■

Example 11.1.3 (The Chi-squared Test) Consider n multinomial trials with $k+1$ possible outcomes, labelled 1 to $k+1$. Suppose p_j denotes the probability of a result in the jth category. Let Y_j denote the number of trials resulting in category j, so that $(Y_1, \ldots, Y_{k+1})$ has the multinomial distribution with joint density obtained in Example 2.7.2. Suppose the null hypothesis is that $p = \pi = (\pi_1, \ldots, \pi_{k+1})$. The alternative hypothesis is unrestricted and includes all $p \neq \pi$ (with $\sum_{j=1}^{k+1} p_j = 1$). The class of alternatives is too large for a UMP test to exist, nor do unbiasedness or invariance considerations rescue the problem. The usual Chi-squared test, which is based on the test statistic Q_n given by

$$Q_n = \sum_{j=1}^{k+1} \frac{(Y_j - n\pi_j)^2}{n\pi_j} , \tag{11.1}$$

will be seen to posses an asymptotic maximin property; see Section 14.3. ■

Example 11.1.4 (Nonparametric Mean) Suppose $X_1, \ldots, X_n$ are i.i.d. from a distribution F with finite mean μ and finite variance. The problem is to test $\mu = 0$. Except when F is assumed to belong to a number of simple parametric families, optimal tests for the mean rarely exist. Moreover, if we assume only a second moment, it is impossible to construct reasonable tests that are of a given size (Theorem 11.4.6). But, by making a weak restriction on the family, we will see that it is possible to construct tests that are approximately level α and that in addition possess an asymptotic maximin property; see Section 11.4. ■

In the remaining chapters, we shall consider hypothesis testing and estimation by confidence sets from a large sample or asymptotic point of view. In this approach, exact results are replaced by approximate ones that have the advantage of both greater simplicity and generality. But, the large sample approach is not just restricted to situations where no finite sample optimality approach works. As the following example shows, limit theorems often provide an easy way to approximate the critical value and power of a test (whether it has any optimality properties or not).

Example 11.1.5 (Simple vs. Simple) Suppose that $X_1, \ldots, X_n$ are i.i.d. with common distribution P. The problem is to test the simple null hypothesis $P = P_0$ versus the simple alternative $P = P_1$. Let p_i denote the density of P_i with respect to a measure μ. By the Neyman-Pearson Lemma, the optimal test rejects for large values of $\sum_{i=1}^{n} \log[p_1(X_i)/p_0(X_i)]$. The exact null distribution of this test statistic may be difficult to obtain since, in general, an n-fold integration is required. On the other hand, since the statistic takes the simple form of a sum of i.i.d. variables, large sample approximations to the critical value and power are easily obtained from the Central Limit Theorem (Theorem 11.2.4).■

Another application of the large sample approach (discussed in Section 11.3) is the study of the robustness of tests when the assumptions under which they are derived do not hold. Here, asymptotic considerations have been found to be indispensable. The problem is just too complicated for the more detailed small sample methods to provide an adequate picture. In general, two distinct types of robustness considerations arise, which may be termed robustness of validity and robustness of efficiency; this distinction has been pointed out by Tukey and McLaughin (1963), Box and Tiao (1964), and Mosteller and Tukey (1977). For robustness of validity, the issue is whether a level α test retains its level and power if the parameter space is enlarged to include a wider class of distributions. For example, in testing whether the mean of a normal population is zero, we may wish to consider the validity of a test without assuming normality. However, even when a test possesses a robustness of validity, are its optimality properties preserved when the parameter space is enlarged? This question is one of robustness of efficiency (or inference robustness). In the context of the one-sample normal location model, for example, one would study the behavior of procedures (such as a one-sample t-test) when the underlying distribution has thicker tails than the normal, or perhaps when the observations are not assumed independent. Large sample theory offers valuable insights into these issues, as will be seen in Section 11.3.

When finite and large sample optimal procedures do not exist for a given problem, it becomes important to determine procedures which have at least reasonable performance characteristics. Large sample considerations often lead to suitable definitions and methods of construction. An example of this nature that will be treated later is the problem of testing whether an i.i.d. sample is uniformly distributed or, more generally, of goodness of fit.

As the starting point of a large sample theory of inference, we now define asymptotic analogs of the concepts of size, level of significance, confidence coefficient and confidence level. Suppose that data $X^{(n)}$ comes from a model indexed by a parameter $\theta \in \Omega$. Typically, $X^{(n)}$ refers to an i.i.d. sample of n observations, and an asymptotic approach assumes that $n \to \infty$. Of course, two-sample problems can be considered in this setup, as well as more complex data structures. Nothing is assumed about the family Ω, so that the problem may be parametric or nonparametric. First, consider testing a null hypothesis H that $\theta \in \Omega_H$ versus the alternative hypothesis K that $\theta \in \Omega_K$, where Ω_H and Ω_K are two mutually exclusive subsets of Ω. We will be studying sequences of tests $\phi_n(X^{(n)})$.

Definition 11.1.1 For a given level α, a sequence of tests $\{\phi_n\}$ is *pointwise asymptotically level* α if, for any $\theta \in \Omega_H$,

$$\limsup_{n\to\infty} E_\theta[\phi_n(X^{(n)})] \le \alpha \ . \tag{11.2}$$

Condition (11.2) guarantees that for any $\theta \in \Omega_H$ and any $\epsilon > 0$, the level of the test will be less than or equal to $\alpha + \epsilon$ when n is sufficiently large. However, the condition does not guarantee the existence of an n_0 (independent of θ) such that

$$E_\theta[\phi_n(X^{(n)})] \le \alpha + \epsilon$$

for all $\theta \in \Omega_H$ and all $n \ge n_0$. We can therefore not guarantee the behavior of the size

$$\sup_{\theta\in\Omega_H} E_\theta[\phi_n(X^{(n)})]$$

of the test, no matter how large n is.

Example 11.1.6 (Uniform versus Pointwise Convergence) To illustrate the above point, consider the function

$$f(n,\theta) = \alpha + (1-\alpha)\exp(-n/\theta) \ ,$$

defined for positive integers n and $\theta > 0$. Then, for any $\theta > 0$, $f(n,\theta) \to \alpha$ as $n \to \infty$; that is, $f(n,\theta)$ converges to α pointwise in θ. However, this convergence is not uniform in θ because

$$\sup_{\theta>0} f(n,\theta) = \alpha + (1-\alpha)\sup_{\theta>0}\exp(-n/\theta) = 1 \ .$$

To cast this example in the context of hypothesis testing, assume $X_1, \ldots, X_n$ are i.i.d. with the exponential distribution function

$$F_\theta(t) = P_\theta\{X_i \le t\} = 1 - \exp(-t/\theta) \ .$$

Define

$$\phi_n(X_1,\ldots,X_n) = \alpha + (1-\alpha)I\{\min(X_1,\ldots,X_n) > 1\} \ .$$

Here and throughout, the notation $I\{E\}$ denotes an *indicator* random variable that is 1 if the event E occurs and is 0 otherwise. Then, $E_\theta[\phi_n(X_1,\ldots,X_n)] = f(n,\theta)$. Hence, if Ω_H is the positive real line, the test sequence ϕ_n satisfies (11.2), but its size is 1 for every n. ■

In order to guarantee the behavior of the limiting size of a test sequence, we require the following stronger condition.

Definition 11.1.2 The sequence $\{\phi_n\}$ is *uniformly asymptotically level* α if

$$\limsup_{n\to\infty} \sup_{\theta\in\Omega_H} E_\theta[\phi_n(X^{(n)})] \le \alpha \ . \tag{11.3}$$

If instead of (11.3), the sequence $\{\phi_n\}$ satisfies

$$\lim_{n\to\infty} \sup_{\theta\in\Omega_H} E_\theta[\phi_n(X^{(n)})] = \alpha \ , \tag{11.4}$$

then this value of α is called the limiting size of $\{\phi_n\}$.

Of course, we also will study the behavior of tests under the alternative hypothesis. The following is a weak condition that we expect reasonable tests to satisfy.

Definition 11.1.3 The sequence $\{\phi_n\}$ is *pointwise consistent in power* if, for any θ in Ω_K,

$$E_\theta[\phi_n(X^{(n)})] \to 1 \tag{11.5}$$

as $n \to \infty$.

Example 11.1.7 (One-parameter families, Example 11.1.1, continued) Let $T_n = T_n(X_1, \ldots, X_n)$ be a sequence of statistics, with distributions depending on a real-valued parameter θ. For testing $H: \ \theta = \theta_0$ against $K: \ \theta > \theta_0$, consider the tests ϕ_n that reject H when $T_n \geq C_n$. In many applications, it will turn out that, when $\theta = \theta_0$, $n^{1/2}(T_n - \theta_0)$ has a limiting normal distribution with mean 0 and variance $\tau^2(\theta_0)$ in the sense that, for any real number t,

$$P_{\theta_0}\{n^{1/2}(T_n - \theta_0) \leq t\} \to \Phi(t/\tau(\theta_0)) \ , \tag{11.6}$$

where $\Phi(\cdot)$ is the standard normal c.d.f. Let z_α satisfy $\Phi(z_\alpha) = \alpha$. Then, the test with

$$C_n = \theta_0 + \frac{\tau(\theta_0)}{n^{1/2}} z_{1-\alpha}$$

has limiting size α, since

$$P_{\theta_0}\{T_n \geq \theta_0 + \frac{\tau(\theta_0)}{n^{1/2}} z_{1-\alpha}\} \to \alpha \ .$$

Consider next the power of ϕ_n under the assumption that not only (11.6) holds, but that it remains valid when θ_0 is replaced by any $\theta > \theta_0$. Then, the power of ϕ_n against θ is

$$\beta_n(\theta) = P_\theta\{n^{1/2}(T_n - \theta) \geq z_{1-\alpha}\tau(\theta_0) - n^{1/2}(\theta - \theta_0)\}$$

and hence $\beta_n(\theta) \to 1$ for any $\theta > \theta_0$, so that the test sequence is pointwise consistent in power. ■

Similar definitions apply to the construction of confidence sets. Let $g = g(\theta)$ be the parameter function of interest, for some mapping g from Ω to some space Ω_g. Let $S_n = S_n(X^{(n)}) \in \Omega_g$ denote a sequence of confidence sets for $g(\theta)$.

Definition 11.1.4 A sequence of confidence sets S_n is *pointwise asymptotically level* $1 - \alpha$ if, for any $\theta \in \Omega$,

$$\liminf_{n\to\infty} P_\theta\{g(\theta) \in S_n(X^{(n)})\} \geq 1 - \alpha \ . \tag{11.7}$$

The sequence $\{S_n\}$ is uniformly asymptotically level $1 - \alpha$ if

$$\liminf_{n \to \infty} \inf_{\theta \in \Omega} P_\theta\{g(\theta) \in S_n(X^{(n)})\} \geq 1 - \alpha \ . \tag{11.8}$$

If the lim inf in the left hand side of (11.8) can be replaced by a lim, then the left hand side is called the limiting confidence coefficient for $\{S_n\}$.

Most of the asymptotic theory we shall consider is local in a sense that we now briefly describe. In the hypothesis testing context, any reasonable test sequence ϕ_n is pointwise consistent in power. However, any actual situation has finite sample size n and its power against any fixed alternative is typically less than one. In order to obtain a meaningful assessment of power, one therefore considers sequences of alternatives θ_n tending to Ω_H at a suitable rate, so that the limiting power of ϕ_n against θ_n is less than one. (See Example 11.2.5 for a simple example of such a local approach.)

An alternative to the local approach is to consider the rate at which the power tends to one against a fixed alternative. Although there exists a large literature on this approach based on large-deviation theory, the resulting approximations tend to be less accurate and we shall not treat this topic here.

It is also important to mention that asymptotic results may provide poor approximations to the actual finite sample setting. Furthermore, convergence to a limit as $n \to \infty$ certainly does not guarantee that the approximation will improve with increasing n; an example is provided by Hodges (1957). Any asymptotic result should therefore be accompanied by an investigation of its reliability for finite sample sizes. Such checks can be carried out by simulations studies or higher order asymptotic analysis.

The concepts and definitions presented in this introduction will be explored more fully in the remaining chapters. First, we need techniques to be able to approximate significance levels, power functions, and confidence coefficients. To this end, the next section is devoted to useful results from the theory of weak convergence and other convergence concepts.

11.2 Basic Convergence Concepts

11.2.1 Weak Convergence and Central Limit Theorems

In this section, the basic notation, definitions and results from the theory of weak convergence are introduced. The main theorems will be presented without proof, but we will provide illustrations of their use. For a more complete background, the reader is referred to Pollard (1984), Dudley (1989) or Billingsley (1995).

Let X denote a $k \times 1$ random vector (which is just a vector-valued random variable), so that the ith component X_i of X is a real-valued random variable. Then, $X^T = (X_1, \ldots, X_k)$. The (multivariate) cumulative distribution function (c.d.f.) of X is defined to be:

$$F_X(x_1, \ldots, x_k) = P\{X_1 \leq x_1, \ldots, X_k \leq x_k\} \ .$$

Here, the probability P refers to the probability on whatever space X is defined. A point $x^T = (x_1, \ldots, x_k)$ at which the c.d.f. $F_X(\cdot)$ is continuous is called a *continuity point* of F_X. Alternatively, x is a continuity point of F_X if the boundary of the set of $(y_1, \ldots, y_k)$ such that $y_i \le x_i$ for all i has probability 0 under the distribution of X.[1] As an example, the multivariate normal distribution was first studied in Section 3.9.2.

Definition 11.2.1 A sequence of random vectors $\{X_n\}$ with c.d.f.s $\{F_{X_n}(\cdot)\}$ is said to *converge in distribution* (or *in law*) to a random vector X with c.d.f. $F_X(\cdot)$ if

$$F_{X_n}(x_1, \ldots, x_k) \to F_X(x_1, \ldots, x_k)$$

at all continuity points $(x_1, \ldots, x_k)$ of $F_X(\cdot)$. This convergence will also be denoted $X_n \stackrel{d}{\to} X$. Because it really only has to do with the laws of the random variables (and not with the random variables themselves), we may also equivalently say F_{X_n} converges weakly to F_X, written $F_{X_n} \stackrel{d}{\to} F_X$.[2]

The limiting random vector X plays an auxiliary role, since any random variable with the same distribution would serve the same purpose. Therefore, the notation will sometimes be abused so that we also say X_n converges in distribution to the c.d.f. F, written $X_n \stackrel{d}{\to} F$.

There are many equivalent characterizations of weak convergence, some of which are recorded in the next theorem.

Theorem 11.2.1 (Portmanteau Theorem) *Suppose X_n and X are random vectors in $\mathbb{R}^k$. The following are equivalent:*

(i) $X_n \stackrel{d}{\to} X$.

(ii) $Ef(X_n) \to Ef(X)$ *for all bounded, continuous real-valued functions* f.

(iii) For any open set O in $\mathbb{R}^k$, $\liminf P(X_n \in O) \ge P(X \in O)$.

(iv) For any closed set G in $\mathbb{R}^k$, $\limsup P(X_n \in G) \le P(X \in G)$.

(v) For any set E in $\mathbb{R}^k$ for which ∂E, the boundary of E, satisfies $P(X \in \partial E) = 0$, $P(X_n \in E) \to P(X \in E)$.

(vi) $\liminf Ef(X_n) \ge Ef(X)$ *for any nonnegative continuous* f.

---

[1]In general, the *boundary* of a set E in $\mathbb{R}^k$, denoted ∂E is defined as follows. The closure of E, denoted $\bar{E}$, is the set of $x \in \mathbb{R}^k$ for which there exists a sequence $x_n \in E$ with $x_n \to x$. The set E is *closed* if $E = \bar{E}$. The *interior* of E, denoted E°, is the set of x such that, for some $\epsilon > 0$, the *Euclidean ball* with center x and radius ϵ, defined by $\{y \in \mathbb{R}^k : |y - x| < \epsilon\}$, is contained in E. Here $|\cdot|$ denotes the usual Euclidean norm. The set E is *open* if $E = E^\circ$. If E^c denotes the complement of a set E, then evidently, E° is the complement of the closure of E^c, and so E is open if and only if E^c is closed. The boundary ∂E of a set E is then defined to be $\bar{E} - E^\circ = \bar{E} \cap (E^\circ)^c$.

[2]The term *weak convergence* (also sometimes called weak star convergence) distinguishes this type of convergence from stronger convergence concepts to be discussed later. However, the term is used because it is a special case of convergence in the weak star topology for elements in a Banach space (such as the space of signed measures on $\mathbb{R}^k$), though we will make no direct use of any such topological notions.

Another equivalent characterization of weak convergence is based on the notion of the characteristic function of a random vector.

Definition 11.2.2 The *characteristic function* of a random vector X (taking values in $\mathbb{R}^k$) is the function $\zeta_X(\cdot)$ from $\mathbb{R}^k$ to the complex plane given by

$$\zeta_X(t) = E(e^{i\langle t,X\rangle}).$$

In the definition, $\langle t,X\rangle$ refers to the usual inner product, so that $\langle t,X\rangle = \sum_{j=1}^k t_j X_j$. Two important properties of characteristic functions are the following. First, the distribution of X is uniquely determined by its characteristic function. Second, the characteristic function of a sum of independent real-valued random variables is the product of the individual characteristic functions (Problem 11.7).

Example 11.2.1 (Multivariate Normal Distribution) Suppose a random vector $X^T = (X_1, \ldots, X_k)$ is $N(\mu, \Sigma)$, the multivariate normal distribution with mean vector $\mu^T = (\mu_1, \ldots, \mu_k)$ and covariance matrix Σ. In the case $k = 1$, if X is normally distributed with mean μ and variance σ^2, its characteristic function is:

$$E(e^{itX}) = \int_{-\infty}^{\infty} e^{itx} \frac{1}{\sqrt{2\pi}\sigma} e^{[-(x-\mu)^2/2\sigma^2]} dx = \exp(it\mu - \frac{1}{2}\sigma^2 t^2) \ , \tag{11.9}$$

which can be verified by a simple integration (Problem 11.8). To obtain the characteristic function for $k > 1$, note that

$$\zeta_X(t) = E(e^{i\langle t,X\rangle})$$

is the characteristic function

$$\zeta_{\langle t,X\rangle}(\lambda) = E(e^{\lambda i\langle t,X\rangle})$$

of $\langle t,X\rangle$ evaluated at $\lambda = 1$. Now if X is multivariate normal $N(\mu, \Sigma)$, then $\langle t,X\rangle$ is univariate normal with mean $\langle t,\mu\rangle$ and variance $\langle \Sigma t, t\rangle = t^T \Sigma t$. Therefore, by the case $k = 1$, we find that

$$E(e^{i\langle t,X\rangle}) = \exp(i\langle t,\mu\rangle - \frac{1}{2}\langle \Sigma t, t\rangle) \ . \ \blacksquare \tag{11.10}$$

Theorem 11.2.2 (Continuity Theorem) *$X_n \xrightarrow{d} X$ in $\mathbb{R}^k$ if and only if*

$$\zeta_{X_n}(t) \to \zeta_X(t)$$

for all t in $\mathbb{R}^k$.

Note that it is not enough to assume $\zeta_{X_n}(t) \to \zeta(t)$ for some limit function $\zeta(\cdot)$ in order to conclude $X_n \xrightarrow{d} X$; one must know that $\zeta(\cdot)$ is the characteristic function of some random variable (or that $\zeta(\cdot)$ is continuous at 0) (Problem 11.9).

Weak convergence of random vectors on $\mathbb{R}^k$ can be reduced to studying weak convergence on the real line by means of the following result, the proof of which follows immediately from Theorem 11.2.2 (Problem 11.10).

Theorem 11.2.3 (Cramér-Wold Device) *A sequence of random vectors X_n on $\mathbb{R}^k$ satisfies $X_n \xrightarrow{d} X$ iff $\langle t, X_n\rangle \xrightarrow{d} \langle t, X\rangle$ for every $t \in \mathbb{R}^k$.*

The following result is crucial for this and the following chapters.

Theorem 11.2.4 (Multivariate Central Limit Theorem) *Let* $X_n^T = (X_{n,1}, \ldots, X_{n,k})$ *be a sequence of i.i.d. random vectors with mean vector* $\mu^T = (\mu_1, \ldots, \mu_k)$ *and covariance matrix* Σ. *Let* $\overline{X}_{n,j} = \frac{1}{n}\sum_{i=1}^n X_{i,j}$. *Then*

$$(n^{1/2}(\overline{X}_{n,1} - \mu_1), \ldots, n^{1/2}(\overline{X}_{n,k} - \mu_k))^T \xrightarrow{d} N(0, \Sigma) .$$

To cover situations in which the distribution varies with sample size, we will deal with a *triangular array* of variables $\{X_{n,i} : \ 1 \le i \le r_n, \ n = 1, 2, \ldots\}$, where it is assumed $r_n \to \infty$ as $n \to \infty$. Typically, $r_n = n$, and so the term triangular array is an appropriate description, but note that the term triangular array is used even if $r_n \neq n$. The following limit theorem provides sufficient conditions for asymptotic normality for a normalized sum of real-valued variables making up a triangular array. (See Billingsley (1995), p. 369.)

Theorem 11.2.5 (Lindeberg Central Limit Theorem) *Suppose, for each* n, $X_{n,1}, \ldots, X_{n,r_n}$ *are independent real-valued random variables. Assume* $E(X_{n,i}) = 0$ *and* $\sigma_{n,i}^2 = E(X_{n,i}^2) < \infty$. *Let* $s_n^2 = \sum_{i=1}^{r_n} \sigma_{n,i}^2$. *Suppose, for each* $\epsilon > 0$,

$$\sum_{i=1}^{r_n} \frac{1}{s_n^2} E[X_{n,i}^2 I\{|X_{n,i}| > \epsilon s_n\}] \to 0 \qquad \text{as } n \to \infty. \tag{11.11}$$

Then, $\sum_{i=1}^{r_n} X_{n,i}/s_n \xrightarrow{d} N(0,1)$.

For most applications, *Lindeberg's Condition* (11.11) can be verified by *Lyapounov's Condition*, which says that, for some $\delta > 0$, $|X_{n,i}|^{2+\delta}$ are integrable and

$$\lim_{n\to\infty} \sum_{i=1}^{r_n} \frac{1}{s_n^{2+\delta}} E[|X_{n,i}|^{2+\delta}] = 0 . \tag{11.12}$$

Indeed, (11.12) implies (11.11) (Problem 11.11), and the result may be stated as follows.

Corollary 11.2.1 (Lyapounov Central Limit Theorem). *Suppose, for each* n, $X_{n,1}, \ldots, X_{n,r_n}$ *are independent. Assume* $E(X_{n,i}) = 0$ *and* $\sigma_{n,i}^2 = E(X_{n,i}^2) < \infty$. *Let* $s_n^2 = \sum_{i=1}^{r_n} \sigma_{n,i}^2$. *Suppose, for some* $\delta > 0$, *(11.12) holds. Then,* $\sum_{i=1}^{r_n} X_{n,i}/s_n \xrightarrow{d} N(0,1)$.

There also exists a partial converse to Lindeberg's Central Limit Theorem, due to Feller and Lévy. (See Billingsley (1995), p. 574.)

Theorem 11.2.6 *Suppose, for each* n, $X_{n,1}, \ldots, X_{n,r_n}$ *are independent, mean* 0, $\sigma_{n,i}^2 = E(X_{n,i}^2) < \infty$ *and* $s_n^2 = \sum_{i=1}^{r_n} \sigma_{n,i}^2$. *Also, assume the array is uniformly asymptotically negligible; that is,*

$$\max_{1 \le i \le r_n} P\{|X_{n,i}/s_n| \ge \epsilon\} \to 0 \tag{11.13}$$

for any $\epsilon > 0$. *If* $\sum_{i=1}^{r_n} X_{n,i}/s_n \xrightarrow{d} N(0,1)$, *then the Lindeberg Condition (11.11) is satisfied.*

Corollary 11.2.2 *Suppose, for each n, $X_{n,1}, \ldots, X_{n,n}$ are i.i.d. with mean 0 and variance σ_n^2. Let $s_n^2 = n\sigma_n^2$. Assume $\sum_{i=1}^n X_{n,i}/s_n \xrightarrow{d} N(0,1)$. Then, the Lindeberg Condition (11.11) is satisfied.*

Corollary 11.2.2 follows from Theorem 11.2.6 because the assumption that the nth row of the triangular array is i.i.d. implies the array is uniformly asymptotically negligible, so that the condition (11.13) holds. Indeed,

$$P\{|X_{n,i}|/s_n \geq \epsilon\} \leq \frac{E(|X_{n,i}|^2)}{s_n^2 \epsilon^2} = \frac{1}{n\epsilon^2} \to 0 .$$

The following Berry-Esseen Theorem gives information on the error in the normal approximation provided by the Central Limit Theorem.

Theorem 11.2.7 *Suppose $X_1, \ldots, X_n$ are i.i.d. real-valued random variables with c.d.f. F. Let $\mu(F)$ denote the mean of F and let $\sigma^2(F)$ denote the variance of F, assumed finite and nonzero. Let $S_n = \sum_{i=1}^n X_i$. Then, there exists a universal constant C (not depending on F, n, or x) such that*

$$\left| P\left\{ \frac{S_n - n\mu(F)}{n^{1/2}\sigma(F)} \leq x \right\} - \Phi(x) \right| \leq \frac{C}{n^{1/2}} \frac{E_F[|X_1 - \mu(F)|^3]}{\sigma(F)^3} , \tag{11.14}$$

where $\Phi(\cdot)$ denotes the standard normal c.d.f.

The Berry-Esseen Theorem holds if $C = 0.7975$. The smallest value of C for which the result holds is unknown, but it is known that it fails for $C < 0.4097$ (van Beek (1972)).

If F is a fixed distribution with finite third moment and nonzero variance, the right side of (11.14) tends to zero and hence the left side of (11.14) tends to zero uniformly in x. Furthermore, if $\mathbf{F}$ is the family of distributions F with

$$\frac{E_F[|X - \mu(F)|^3]}{\sigma^3(F)} < B , \tag{11.15}$$

for some fixed $B < \infty$, then this convergence is also uniform in F as F varies in $\mathbf{F}$. Thus, if S_n is the sum of n i.i.d. variables with distribution F_n in $\mathbf{F}$, then

$$\sup_x \left| P\left\{ \frac{S_n - n\mu(F_n)}{n^{1/2}\sigma(F_n)} \leq x \right\} - \Phi(x) \right| \to 0 . \tag{11.16}$$

Example 11.2.2 Suppose $X_1, \ldots, X_n$ are i.i.d. Bernoulli trials with probability of success p. Then, $S_n = \sum_i X_i$ is binomial based on n trials and success probability p, and the usual Central Limit Theorem asserts that the probability that $(S_n - np)/[np(1-p)]^{1/2}$ is less or equal to x converges to $\Phi(x)$, if p is not zero or one. It follows from the Berry-Esseen theorem that this convergence is uniform in both x and p as long as $p \in [\epsilon, 1-\epsilon]$ for some $\epsilon > 0$. To see why, we show that condition (11.15) is satisfied. Observe that

$$E[|X_1 - p|^3] = p(1-p)[(1-p)^2 + p^2] \leq p(1-p) .$$

Thus,

$$E[|X_1 - p|^3]/[p(1-p)]^{3/2} \leq [\epsilon(1-\epsilon)]^{-1/2} ,$$

so that (11.15) holds with $B^2 = \epsilon(1-\epsilon)$. Thus, (11.16) holds, so that if S_n is binomial based on n trials and success probability $p_n \to p \in (0,1)$, then

$$P\{\frac{S_n - np_n}{[np_n(1-p_n)]^{1/2}} \leq x_n\} \to \Phi(x) \tag{11.17}$$

whenever $x_n \to x$. ■

Example 11.2.3 (The Sample Median) As an application of the Berry-Esseen theorem and the previous example, the following result establishes the asymptotic normality of the sample median. Given a sample $X_1, \ldots, X_n$ with order statistics $X_{(1)} \leq \cdots \leq X_{(n)}$, the median $\tilde{X}_n$ is defined to be the middle order statistic $X_{(k)}$ if $n = 2k-1$ is odd and the average of $X_{(k)}$ and $X_{(k+1)}$ if $n = 2k$ is even.

Theorem 11.2.8 *Suppose $X_1, \ldots, X_n$ are i.i.d. real-valued random variables with c.d.f. F. Assume $F(\theta) = 1/2$, and that F is differentiable at θ with $F' = f$ and $f(\theta) > 0$. Let $\tilde{X}_n$ denote the sample median. Then*

$$n^{1/2}(\tilde{X}_n - \theta) \stackrel{d}{\to} \mathrm{N}(0, \frac{1}{4f^2(\theta)}) .$$

PROOF. Assume first that n tends to ∞ through odd values and, without loss of generality, that $\theta = 0$. Fix any real number a and let S_n be the number of X_i that exceed $a/n^{1/2}$. Then the event $\{\tilde{X}_n \leq a/n^{1/2}\}$ is equivalent to the event $\{S_n \leq (n-1)/2\}$. But, S_n is binomial with parameters n and success probability $p_n = 1 - F(a/n^{1/2})$. Thus,

$$P\{n^{1/2}\tilde{X}_n \leq a\} = P\{S_n \leq \frac{n-1}{2}\} = P\{\frac{S_n - np_n}{[np_n(1-p_n)]^{1/2}} \leq x_n\} ,$$

where

$$x_n = \frac{\frac{1}{2}(n-1) - np_n}{[np_n(1-p_n)]^{1/2}} = \frac{n^{1/2}(\frac{1}{2} - p_n) - 1/(2n^{1/2})}{[p_n(1-p_n)]^{1/2}} .$$

As $n \to \infty$, $p_n \to 1/2$ and

$$n^{1/2}(\frac{1}{2} - p_n) = a \cdot \frac{F(a/n^{1/2}) - F(0)}{a/n^{1/2}} \to af(0) ,$$

which implies $x_n \to 2af(0)$. Therefore, by (11.17),

$$P\{n^{1/2}\tilde{X}_n \leq a\} \to \Phi[2f(0)a] ,$$

which completes the proof for odd n. For the case of even n, see Problem 11.15. ■

Another result concerning uniformity in weak convergence is the following theorem of Polyá.

Theorem 11.2.9 (Polyá's Theorem) *Suppose $X_n \stackrel{d}{\to} X$ and X has a continuous c.d.f F_X. Let F_{X_n} denote the c.d.f. of X_n. Then, $F_{X_n}(x)$ converges to $F_X(x)$, uniformly in x.*

It is interesting and important to know that weak convergence of F_n to F can be expressed in terms of $\rho(F_n, F)$, where ρ is a metric on the space of distributions.

(Some basic properties of metrics are reviewed in the appendix, Section A.2.) To be specific, on the real line, define the Lévy distance between distributions F and G as follows.

Definition 11.2.3 Let F and G be distribution functions on the real line. The *Lévy distance* between F and G, denoted $\rho_L(F,G)$ is defined by

$$\rho_L(F,G) = \inf\{\epsilon > 0: \ F(x-\epsilon) - \epsilon \leq G(x) \leq F(x+\epsilon) + \epsilon \quad \text{for all } x\} .$$

The definition implies that $\rho_L(F,G) = \rho_L(G,F)$ and that ρ_L is a metric on the space of distribution functions (Problem 11.20). Moreover, if F_n and F are distribution functions, then weak convergence of F_n to F is equivalent to $\rho_L(F_n, F) \to 0$ (Problem 11.22). In this sense, ρ_L metrizes weak convergence.

We shall next consider the implication of weak convergence for the convergence of quantiles. Ideally, the $(1-\alpha)$ quantile $x_{1-\alpha}$ of a distribution F is defined by

$$F(x_{1-\alpha}) = 1 - \alpha . \tag{11.18}$$

For the solutions of (11.18), it is necessary to distinguish three cases. First, if F is continuous and strictly increasing, the equation (11.18) has a unique solution. Second, if F is not strictly increasing, it may happen that $F(x) = 1-\alpha$ on an interval $[a,b)$ or $[a,b]$, so that any x in such an interval could serve as a $1-\alpha$ quantile. Then, we shall define the $1-\alpha$ quantile as the left hand endpoint of the interval. Third, if F has discontinuities, then (11.18) may have no solutions. This happens if $F(x) > 1-\alpha$ and $\sup\{F(y): \ y < x\} \leq 1-\alpha$, but in this case we would call x the $1-\alpha$ quantile of F. A general definition encompassing all these possibilities is given by

$$x_{1-\alpha} = \inf\{x: \ F(x) \geq 1-\alpha\} . \tag{11.19}$$

This is also sometimes written as $x_{1-\alpha} = F^{-1}(1-\alpha)$ although F may not have a proper inverse function.

Weak convergence of F_n to F is not enough to guarantee that $F_n^{-1}(1-\alpha)$ converges to $F^{-1}(1-\alpha)$, but the following result shows this is true if F is continuous and strictly increasing at $F^{-1}(1-\alpha)$. Part (ii) of the lemma will be used later (and depends on the notion of convergence in probability introduced below).

Lemma 11.2.1 *(i) Let $\{F_n\}$ be a sequence of distribution functions on the real line converging weakly to a distribution function F. Assume F is continuous and strictly increasing at $y = F^{-1}(1-\alpha)$. Then,*

$$F_n^{-1}(1-\alpha) \to F^{-1}(1-\alpha) .$$

(ii). More generally, suppose $\{\hat{F}_n\}$ is a sequence of random distribution functions satisfying $\hat{F}_n(x) \xrightarrow{P} F(x)$ at all x which are continuity points of a fixed distribution function F. Assume F is continuous and strictly increasing at $F^{-1}(1-\alpha)$. Then,

$$\hat{F}_n^{-1}(1-\alpha) \xrightarrow{P} F^{-1}(1-\alpha) .$$

PROOF. To prove (i), fix $\delta > 0$. Let $y-\epsilon$ and $y+\epsilon$ be continuity points of F for some $0 < \epsilon \leq \delta$. Then,

$$F_n(y-\epsilon) \to F(y-\epsilon) < 1-\alpha$$

and

$$F_n(y + \epsilon) \to F(y + \epsilon) > 1 - \alpha.$$

Hence, for all sufficiently large n,

$$y - \epsilon \leq F_n^{-1}(1 - \alpha) \leq y + \epsilon ,$$

and so, $|F_n^{-1}(1 - \alpha) - y| \leq \delta$ for all sufficiently large n. Since δ was arbitrary, the result (i) is proved. The proof of (ii) is similar. ■

11.2.2 Convergence in Probability and Applications

As pointed out earlier, convergence in law of X_n to X asserts only that the distribution of X_n tends to that of X, but says nothing about X_n itself becoming close to X. The following stronger form of convergence provides that X_n and X themselves are close for large n.

Definition 11.2.4 A sequence of random vectors $\{X_n\}$ *converges in probability* to X, written $X_n \xrightarrow{P} X$, if, for every $\epsilon > 0$,

$$P\{|X_n - X| > \epsilon\} \to 0 \qquad \text{as } n \to \infty.$$

Convergence in probability implies convergence in distribution (Problem 11.30); the converse is false in general. However, if X_n converges in distribution to a distribution assigning probability one to a constant vector c, then X_n converges in probability to c, and conversely. Note that, unlike weak convergence, X_n and X must be defined on the same probability space in order for Definition 11.2.4 to make sense.

Convergence in probability of a sequence of random vectors X_n is equivalent to convergence in probability of their components. That is, if $X_n = (X_{n,1}, \ldots, X_{n,k})^T$ and $X = (X_1, \ldots, X_k)^T$, then $X_n \xrightarrow{P} X$ iff for each $i = 1, \ldots, k$, $X_{n,i} \xrightarrow{P} X_i$. Moreover, $X_n \xrightarrow{P} 0$ if and only if $|X_n| \xrightarrow{P} 0$ (Problem 11.31).

A sequence of real-valued random variables X_n converges in probability to infinity, written $X_n \xrightarrow{P} \infty$ if, for any real number B,

$$P\{X_n < B\} \to 0$$

as $n \to \infty$.

The next result and the later Theorem 11.2.16 deal with the convergence of the average of i.i.d. random variables toward their expectation, and are known as the weak and strong laws of large numbers. The terminology reflects the fact that the strong law asserts a stronger conclusion than the weak law.

Theorem 11.2.10 (Weak Law of Large Numbers) *Let X_i be i.i.d. real-valued random variables with mean μ. Then,*

$$\bar{X}_n \equiv \frac{1}{n} \sum_{i=1}^{n} X_i \xrightarrow{P} \mu .$$

Note that it is possible for $\bar{X}_n$ to converge in probability to a constant even if the mean does not exist (Problem 11.28). Also, if the X_i are nonnegative and the mean is not finite, then $\bar{X}_n \xrightarrow{P} \infty$ (Problem 11.32).

Suppose $X_1, \ldots, X_n$ are i.i.d. according to a model $\{P_\theta,\ \theta \in \Omega\}$. A sequence of estimators $T_n = T_n(X_1, \ldots, X_n)$ is said to be a weakly consistent (or just consistent) estimator sequence of $g(\theta)$ if, for each $\theta \in \Omega$,

$$T_n \xrightarrow{P} g(\theta) \ .$$

Thus, the consistency of an estimator sequence merely asserts convergence in probability for each value of the parameter. For example, the Weak Law of Large Numbers asserts that the sample mean is a consistent estimator of the population mean whenever the population mean exists.

Example 11.2.4 Suppose $X_1, \ldots, X_n$ are i.i.d. according to either P_0 or P_1. If p_i denotes the density of P_i with respect to a dominating measure, then by the Neyman-Pearson Lemma, an optimal test rejects for large values of

$$T_n \equiv \frac{1}{n} \sum_{i=1}^{n} \log[p_1(X_i)/p_0(X_i)] \ .$$

By the Weak Law of Large Numbers, under P_0,

$$T_n \xrightarrow{P} -K(P_0, P_1) \ , \tag{11.20}$$

where $K(P_0, P_1)$ is the so-called Kullback-Leibler Information, defined as

$$K(P_0, P_1) = -E_{P_0}[\log(p_1(X_1)/p_0(X_1))] \ . \tag{11.21}$$

The convergence (11.20) assumes $K(P_0, P_1)$ is well-defined in the sense that the expectation in (11.21) exists. But, by Jensen's inequality (since the negative log is convex),

$$K(P_0, P_1) \geq -\log[E_{P_0}(p_1(X_1)/p_0(X_1))] \geq 0 \ .$$

If P_0 and P_1 are distinct, then, the first inequality is strict, so that $K(P_0, P_1) \geq 0$ with equality iff $P_0 = P_1$. Note, however, that $K(P_0, P_1)$ may be ∞, but even in this case, the convergence (11.20) holds; see Problem 11.33. Similarly, under the alternative hypothesis P_1,

$$T_n \xrightarrow{P} E_{P_1}[\log(p_1(X_1)/p_0(X_1)] = K(P_1, P_0) \geq 0 \ .$$

Note that $K(P_0, P_1)$ need not equal $K(P_1, P_0)$.

In summary, T_n converges in probability, under P_0, to a negative constant (possibly $-\infty$), while, under P_1, T_n converges in probability to a positive constant (assuming P_0 and P_1 are distinct). Therefore, for testing P_0 versus P_1, the test that rejects when $T_n > 0$ is *asymptotically perfect* in the sense that both error probabilities tend to zero; that is, $P_0\{T_n > 0\} \to 0$ and $P_1\{T_n \leq 0\} \to 0$. It also follows that, for fixed $\alpha \in (0,1)$, if ϕ_n is a most powerful level α test sequence for testing P_0 versus P_1 based on n i.i.d. observations, then the power of ϕ_n against P_1 tends to one. Thus, if P_0 and P_1 are fixed with $n \to \infty$, the problem is degenerate from an asymptotic point of view. ■

For convergence in probability to a constant, it is not necessary for the X_n to be defined on the same probability space. Suppose P_n is a probability on a probability space $(\Omega_n, \mathcal{F}_n)$, and let X_n be a random vector from Ω_n to $\mathbb{R}^k$. Then, if c is a fixed constant vector in $\mathbb{R}^k$, we say that X_n converges to c in P_n-probability if, for every $\epsilon > 0$,

$$P_n\{|X_n - c| > \epsilon\} \to 0 \qquad \text{as } n \to \infty .$$

Alternatively, we may say X_n converges to c in probability if it is understood that the law of X_n is determined by P_n.

For a sequence of numbers x_n and y_n, the notation $x_n = o(y_n)$ means $x_n/y_n \to 0$ as $n \to \infty$. For random variables X_n and Y_n, the notation $X_n = o_P(Y_n)$ means $X_n/Y_n \stackrel{P}{\to} 0$. Similarly, $X_n = o_{P_n}(Y_n)$ means $X_n/Y_n \to 0$ in P_n-probability.

The following theorem is very useful for proving limit theorems.

Theorem 11.2.11 (Slutsky's Theorem) *Suppose $\{X_n\}$ is a sequence of real-valued random variables such that $X_n \stackrel{d}{\to} X$. Further, suppose $\{A_n\}$ and $\{B_n\}$ satisfy $A_n \stackrel{P}{\to} a$, and $B_n \stackrel{P}{\to} b$, where a and b are constants. Then, $A_n X_n + B_n \stackrel{d}{\to} aX + b$.*

The conclusion in Slutsky's Theorem may be strengthened to convergence in probability if it is assumed that $X_n \stackrel{P}{\to} X$. The following corollary to Slutsky's Theorem is also fundamental.

Corollary 11.2.3 *Suppose $\{X_n\}$ is a sequence of real-valued random variables such that X_n tends to X in distribution, where X has a continuous cumulative distribution function F. If $C_n \to c$ in probability, where c is a constant, then*

$$P\{X_n \le C_n\} \to F(c) .$$

Corollary 11.2.3 is useful even when C_n are nonrandom constants tending to c. Also, the corollary holds even if $c = \infty$ or $c = -\infty$ (Problem 11.36), with the interpretation $F(\infty) = 1$ and $F(-\infty) = 0$.

Note that Slutsky's theorem holds more generally if the convergence in probability assumptions are replaced by convergence in P_n-probability.

Example 11.2.5 (Local Power Calculation) Suppose S_n is binomial based on n trials and success probability p. Consider testing $p = 1/2$ versus $p > 1/2$. The uniformly most powerful test rejects for large values of S_n. By Example 11.2.2,

$$Z_n \equiv (S_n - \frac{n}{2})/(n/4)^{1/2} \stackrel{d}{\to} N(0,1) ,$$

and so the test that rejects the null hypothesis when this quantity exceeds the normal critical value $z_{1-\alpha}$ is asymptotically level α. Let $\beta_n(p)$ denote the power of this test against a fixed alternative $p > 1/2$. Then, $(S_n - np)/[np(1-p)]^{1/2}$ is asymptotically standard normal if p is the true value. Hence,

$$\beta_n(p) = P_p\{Z_n > z_{1-\alpha}\} = P_p\{\frac{S_n - np}{[np(1-p)]^{1/2}} > d_n(p)\} ,$$

where

$$d_n(p) = \frac{z_{1-\alpha}}{[4p(1-p)]^{1/2}} + n^{1/2} \frac{\frac{1}{2} - p}{[p(1-p)]^{1/2}} \to -\infty$$

if $p > 1/2$. Thus, $\beta_n(p) \to 1$ as $n \to \infty$ for any $p > 1/2$, and so the test sequence is pointwise consistent.

This result does not distinguish between alternative values of p. Better discrimination is obtained by considering alternatives for which the power tends to a value less than 1. This is achieved by replacing a fixed alternative p by a sequence p_n tending to $1/2$, so that the task of distinguishing between $1/2$ and p_n becomes more difficult as information accumulates with increasing n. It turns out that the power will tend to a limit less than one but greater than α if $p_n = 1/2 + hn^{-1/2}$ if $h > 0$. To see this, note that, by Example 11.2.2, under p_n, $(S_n - np_n)/[np_n(1-p_n)]^{1/2}$ is asymptotically standard normal. Then,

$$\beta_n(p_n) = P_{p_n}\{Z_n > z_{1-\alpha}\} = P_{p_n}\{\frac{S_n - np_n}{[np_n(1-p_n)]^{1/2}} > d_n(p_n)\} .$$

But, $d_n(p_n) \to z_{1-\alpha} - 2h$. Hence, if Z denotes a standard normal variable,

$$\beta_n(p_n) \to P\{Z > z_{1-\alpha} - 2h\} = 1 - \Phi(z_{1-\alpha} - 2h) .$$

Also, note that $\beta_n(p_n) \to 1$ if $n^{1/2}(p_n - 1/2) \to \infty$ and $\beta_n(p_n) \to \alpha$ if $n^{1/2}(p_n - 1/2) \to 0$ (Problem 11.37). ■

The following is another useful result concerning convergence in probability.

Theorem 11.2.12 *Suppose X_n and X are random vectors in $\mathbb{R}^k$ with $X_n \overset{P}{\to} X$. Let g be a continuous function from $\mathbb{R}^k$ to $\mathbb{R}^s$. Then, $g(X_n) \overset{P}{\to} g(X)$.*

Example 11.2.6 (Sample Standard Deviation) Let $X_1, \ldots, X_n$ be i.i.d. real-valued random variables with common mean μ and finite variance σ^2. The usual unbiased sample variance estimator is given by

$$S_n^2 = \frac{1}{n-1} \sum_{i=1}^{n} (X_i - \bar{X}_n)^2 , \tag{11.22}$$

where $\bar{X}_n = n^{-1} \sum_{i=1}^{n} X_i$ is the sample mean. By the weak law of large numbers, $\bar{X}_n \to \mu$ in probability and $n^{-1} \sum_{i=1}^{n} X_i^2 \to E(X_1^2) = \mu^2 + \sigma^2$ in probability. Hence,

$$\frac{n-1}{n} S_n^2 = n^{-1} \sum_{i=1}^{n} X_i^2 - \bar{X}_n^2 \to \sigma^2$$

in probability, by Slutsky's Theorem. Thus, $S_n^2 \to \sigma^2$ in probability, which implies $S_n \to \sigma$ in probability, by Theorem 11.2.12. ■

Example 11.2.7 (Confidence Intervals for A Binomial p) Suppose S_n is binomial based on n trials and unknown success probability p. Let $\hat{p}_n = S_n/n$. By Example 11.2.2, for any $p \in (0,1)$, $n^{1/2}(\hat{p}_n - p)$ converges in distribution to $N(0, p(1-p))$. This implies $\hat{p}_n \overset{P}{\to} p$ and so

$$[\hat{p}_n(1-\hat{p}_n)]^{1/2} \overset{P}{\to} [p(1-p)]^{1/2}$$

as well. Therefore, by Slutsky's Theorem, for any $p \in (0,1)$,

$$\frac{n^{1/2}(\hat{p}_n - p)}{[\hat{p}_n(1-\hat{p}_n)]^{1/2}} \xrightarrow{d} N(0,1) .$$

This implies that the confidence interval

$$\hat{p}_n \pm z_{1-\frac{\alpha}{2}} \left[\frac{\hat{p}_n(1-\hat{p}_n)}{n} \right]^{1/2} \tag{11.23}$$

is pointwise consistent in level, for any fixed p in $(0,1)$, where z_β is the β quantile of $N(0,1)$. Note, however, that this confidence interval is not uniformly consistent in level; in fact, for any n, the coverage probability can be arbitrarily close to 0 (Problem 11.38).

Unfortunately, an accumulating literature has shown that the coverage of the interval in (11.23) is quite unreliable even for large values of n or $np(1-p)$, and varies quite erratically as the sample size increases. To cite just one example, the probability of the interval (11.23) covering the true p when $p = .2$ and $1-\alpha = .95$ is .946 when $n = 30$, and it is .928 when $n = 98$. This example is taken from Table 1 of Brown, Cai and DasGupta (2001), who survey the literature and recommend more reliable alternatives. Because of the great practical importance of the problem, we summarize some of their principal recommendations.

For small n, the authors recommend two procedures. The first, which goes back to Wilson (1927), is based on the quadratic inequality

$$|\hat{p}_n - p| \le z_{1-\frac{\alpha}{2}} \left[\frac{p(1-p)}{n} \right]^{1/2} , \tag{11.24}$$

which has probability under p tending to $1-\alpha$. So, if we were testing the simple null hypothesis that p is true, we can invert the test with acceptance region (11.24). Solving for p in (11.24), one obtains the Wilson interval (Problem 11.39)

$$\tilde{p}_n \pm z_{1-\frac{\alpha}{2}} \frac{n^{1/2}}{\tilde{n}} \left[\hat{p}_n \hat{q}_n + \frac{z^2_{1-\frac{\alpha}{2}}}{4n} \right]^{1/2} , \tag{11.25}$$

where $\tilde{p}_n = \tilde{S}_n/\tilde{n}$, $\tilde{S}_n = S_n + \frac{1}{2} z^2_{1-\frac{\alpha}{2}}$, $\tilde{n} = n + z^2_{1-\frac{\alpha}{2}}$, and $\hat{q}_n = 1 - \hat{p}_n$. As an alternative, the authors recommend an equal-tailed Bayes interval based on the Beta prior with $a = b = 1/2$; see Example 5.7.2.

Theoretical and additional numerical support are provided in Brown, Cai and DasGupta (2002). Other approximations are reviewed in Johnson, Kotz and Kemp (1992). ■

Theorem 11.2.13 (Continuous Mapping Theorem) *Suppose $X_n \xrightarrow{d} X$. Let g be a (measurable) map from $\mathbb{R}^k$ to $\mathbb{R}^s$. Let C be the set of points in $\mathbb{R}^k$ for which g is continuous. If $P(X \in C) = 1$, then $g(X_n) \xrightarrow{d} g(X)$.*

Example 11.2.8 Suppose X_n is a sequence of real-valued random variables such that $X_n \xrightarrow{d} N(0, \sigma^2)$. By the Continuous Mapping Theorem, it follows that

$$\frac{X_n^2}{\sigma^2} \xrightarrow{d} \chi^2_1 ,$$

where χ_k^2 denotes the Chi-squared distribution with k degrees of freedom. More generally, suppose X_n is a sequence of $k \times 1$ vector-valued random variables such that

$$X_n \xrightarrow{d} N(0, \Sigma) ,$$

where Σ is assumed positive definite. Then, there exists a unique positive definite symmetric matrix C such that $C \cdot C = \Sigma$ and we write $C = \Sigma^{1/2}$. (For the construction of the square root of a positive definite symmetric matrix, see Lehmann (1999), p.306.) By the Continuous Mapping Theorem, it follows that

$$|C^{-1} X_n|^2 \xrightarrow{d} \chi_k^2 . \blacksquare$$

The following method is often used to prove limit theorems, especially asymptotic normality.

Theorem 11.2.14 (Delta Method) *Suppose $X_1, X_2, \ldots$ and X are random vectors in $\mathbb{R}^k$. Assume $\tau_n(X_n - \mu) \xrightarrow{d} X$ where μ is a constant vector and $\{\tau_n\}$ is a sequence of constants $\tau_n \to \infty$.*
(i) Suppose g is a function from $\mathbb{R}^k$ to $\mathbb{R}$ which is differentiable at μ with gradient (vector of first partial derivatives) of dimension $1 \times k$ at μ equal to $\dot{g}(\mu)$.[3] *Then,*

$$\tau_n[g(X_n) - g(\mu)] \xrightarrow{d} \dot{g}(\mu) X . \tag{11.26}$$

In particular, if X is multivariate normal in $\mathbb{R}^k$ with mean vector 0 and covariance matrix Σ, then

$$\tau_n[g(X_n) - g(\mu)] \xrightarrow{d} N(0, \dot{g}(\mu)\Sigma\dot{g}(\mu)^T) . \tag{11.27}$$

(ii) More generally, suppose $g = (g_1, \ldots, g_q)^T$ is a mapping from $\mathbb{R}^k$ to $\mathbb{R}^q$, where g_i is a function from $\mathbb{R}^k$ to $\mathbb{R}$ which is differentiable at μ. Let D be the $q \times k$ matrix with (i, j) entry equal to $\partial g_i(y_1, \ldots, y_k)/\partial y_j$ evaluated at μ. Then,

$$\tau_n[g(X_n) - g(\mu)] = \tau_n[g_1(X_n) - g_1(\mu), \ldots, g_q(X_n) - g_q(\mu)]^T \xrightarrow{d} DX .$$

In particular, if X is multivariate normal in $\mathbb{R}^k$ with mean vector 0 and covariance matrix Σ, then

$$\tau_n[g(X_n) - g(\mu)] \xrightarrow{d} N(0, D\Sigma D^T) .$$

PROOF. We prove (i) with (ii) left as an exercise (Problem 11.44). Note that $X_n - \mu = o_P(1)$. Differentiability of g at μ implies

$$g(x) = g(\mu) + \dot{g}(\mu)(x - \mu) + R(x - \mu) ,$$

where $R(y) = o(|y|)$ as $|y| \to 0$. Now,

$$\tau_n[g(X_n) - g(\mu)] - \dot{g}(\mu)\tau_n(X_n - \mu) = \tau_n R(X_n - \mu) .$$

By Slutsky's Theorem, it suffices to show $\tau_n R(X_n - \mu) = o_P(1)$. But,

$$\tau_n R(X_n - \mu) = \tau_n |X_n - \mu| \cdot h(X_n - \mu) ,$$

---

[3] When $k = 1$, we may also use the notation $g'(\mu)$ for the ordinary first derivative of g with respect to μ, as well as $g''(\mu)$ for the second derivative.

where $h(y) = R(y)/|y|$ and $h(0)$ is defined to be 0, so that h is continuous at 0. The weak convergence hypothesis and the Continuous Mapping Theorem imply $\tau_n|X_n - \mu|$ has a limiting distribution. So, by Slutsky's Theorem, it is enough to show $h(X_n - \mu) = o_P(1)$. But, this follows by the Continuous Mapping Theorem as well. ■

Note that (11.26) and (11.27) remain true if $\dot{g}(\mu) = 0$ with the interpretation that the limit distribution places all its mass at zero, in which case we can conclude

$$\tau_n[g(X_n) - g(\mu)] \stackrel{P}{\to} 0 .$$

Example 11.2.9 (Binomial Variance) Suppose S_n is binomal based on n trials and success probability p. Let $\hat{p}_n = S_n/n$. By the Central Limit Theorem,

$$n^{1/2}(\hat{p}_n - p) \stackrel{d}{\to} N(0, p(1-p)) .$$

Consider estimating $g(p) = p(1-p)$. By the Delta Method,

$$n^{1/2}[g(\hat{p}_n) - g(p)] \stackrel{d}{\to} N(0, (1-2p)^2 p(1-p)) .$$

If $p = 1/2$, then $\dot{g}(1/2) = 0$, so that

$$n^{1/2}[g(\hat{p}_n) - g(p)] \stackrel{P}{\to} 0 .$$

In order to obtain a nondegenerate limit distribution in this case, note that

$$n[g(\hat{p}_n) - \frac{1}{4}] = -[n^{1/2}(\hat{p}_n - \frac{1}{2})]^2 .$$

Therefore, by the Continuous Mapping Theorem,

$$n[g(\hat{p}_n) - \frac{1}{4}] \stackrel{d}{\to} -X^2 ,$$

where X is $N(0, 1/4)$, or

$$n[g(\hat{p}_n) - \frac{1}{4}] \stackrel{d}{\to} -\frac{1}{4}\chi_1^2 ,$$

where χ_1^2 is a random variable distributed as Chi-squared with one degree of freedom. ■

In the case $\dot{g}(\mu) = 0$, it is not surprising that the limit distribution is a multiple of a Chi-squared variable with one degree of freedom. Indeed, suppose $k = 1$ and g is twice differentiable at μ with second derivative $g''(\mu)$, so that

$$g(x) = g(\mu) + \frac{1}{2}g''(\mu)(x-\mu)^2 + R(x-\mu) ,$$

where $R(x - \mu) = o[(x-\mu)^2]$ as $x \to \mu$. Arguing as in the proof of Theorem 11.2.14 yields

$$\tau_n^2[g(X_n) - g(\mu)] - \tau_n^2 \frac{g''(\mu)}{2}(X_n - \mu)^2 = \tau_n^2 R(X_n - \mu) = o_P(1) \qquad (11.28)$$

(Problem 11.46). By the Continuous Mapping Theorem,

$$\tau_n(X_n - \mu) \stackrel{d}{\to} X$$

implies

$$\tau_n^2 \frac{g''(\mu)}{2}(X_n - \mu)^2 \xrightarrow{d} \frac{g''(\mu)}{2} X^2 .$$

By Slutsky's Theorem, $\tau_n^2[g(X_n) - g(\mu)]$ has this same limiting distribution. Of course, if X is $N(\mu, \sigma^2)$, then this limiting distribution is $\frac{g''(\mu)\sigma^2}{2}\chi_1^2$.

Example 11.2.10 (Sample Correlation) Let (U_i, V_i) be i.i.d. bivariate random vectors in the plane, with both U_i and V_i assumed to have finite nonzero variances. Let $\sigma_U^2 = Var(U_i)$, $\sigma_V^2 = Var(V_i)$, $\mu_U = E(U_i)$, $\mu_V = E(V_i)$ and let $\rho = Cov(U_i, V_i)/(\sigma_U \sigma_V)$ be the population correlation coefficient. The usual sample correlation coefficient is given by

$$\hat{\rho}_n = \frac{\sum_{i=1}^n (U_i - \bar{U}_n)(V_i - \bar{V}_n)/n}{S_U S_V} , \tag{11.29}$$

where $\bar{U}_n = \sum U_i/n$, $\bar{V}_n = \sum V_i/n$, $S_U^2 = \sum(U_i - \bar{U}_n)^2/n$ and $S_V^2 = \sum(V_i - \bar{V}_n)^2/n$. Then, $n^{1/2}(\hat{\rho}_n - \rho)$ is asymptotically normal. The important observation is that $\hat{\rho}_n$ is a smooth function of the vector of means $\bar{X}_n$, where X_i is the vector $X_i = (U_i, V_i, U_i^2, V_i^2, U_i V_i)^T$. In fact, $\hat{\rho}_n = g(\bar{X}_n)$, where

$$g((y_1, y_2, y_3, y_4, y_5)^T) = \frac{y_5 - y_1 y_2}{(y_3 - y_1^2)^{1/2}(y_4 - y_2^2)^{1/2}} .$$

Note that g is smooth and $\dot{g}$ is readily computed. Let $\mu = E(X_i)$ denote the mean vector. Further assume that U_i and V_i have finite fourth moments. Then, by the multivariate CLT,

$$n^{1/2}(\bar{X}_n - \mu) \xrightarrow{d} N(0, \Sigma) ,$$

where Σ is the covariance matrix of X_1. For example, the $(1, 5)$ component of Σ is $Cov(U_1, U_1 V_1)$. Hence, by the delta method,

$$n^{1/2}[g(\bar{X}_n) - g(\mu)] = n^{1/2}(\hat{\rho}_n - \rho) \xrightarrow{d} N(0, \dot{g}(\mu)\Sigma\dot{g}(\mu)^T) . \tag{11.30}$$

As an example, suppose that (U_i, V_i) is bivariate normal; in this case, (11.30) reduces to (Problem 11.47)

$$n^{1/2}(\hat{\rho}_n - \rho) \xrightarrow{d} N(0, (1 - \rho^2)^2) . \tag{11.31}$$

This implies $(1 - \hat{\rho}_n^2) \xrightarrow{P} 1 - \rho^2$. Then, by Slutsky's theorem,

$$n^{1/2}(\hat{\rho}_n - \rho)/(1 - \hat{\rho}_n^2) \xrightarrow{d} N(0, 1) ,$$

and so the confidence interval

$$\hat{\rho}_n \pm n^{-1/2} z_{1-\frac{\alpha}{2}}(1 - \hat{\rho}_n^2)$$

is a pointwise asymptotically level $1-\alpha$ confidence interval for ρ. The error in this asymptotic approximation derives from both the normal approximation to the distribution of $\hat{\rho}_n$ and the fact that one is approximating the limiting variance. To counter the second of these effects, the following variance stabilization technique can be used. By the delta method, if h is differentiable, then

$$n^{1/2}[h(\hat{\rho}_n) - h(\rho)] \xrightarrow{d} N(0, [h'(\rho)]^2(1 - \rho^2)^2) .$$

The idea is to choose h so that the limiting variance does not depend on ρ and is a constant; such a transformation is then called a *variance stabilizing transformation.* The solution is known as Fisher's z-transformation and is given by

$$h(\rho) = \frac{1}{2}\log(\frac{1+\rho}{1-\rho}) = \text{arctanh}(\rho) .$$

Then,

$$h(\hat{\rho}_n) \pm n^{-1/2} z_{1-\frac{\alpha}{2}}$$

is a pointwise asymptotically level $1-\alpha$ confidence interval for $h(\rho)$. The inverse function of h is the hyperbolic tangent function

$$\tanh(y) = h^{-1}(y) = \frac{e^y - e^{-y}}{e^y + e^{-y}} ,$$

so that

$$[\tanh(\text{arctanh}(\hat{\rho}_n) - n^{-1/2} z_{1-\frac{\alpha}{2}}), \tanh(\text{arctanh}(\hat{\rho}_n) + n^{-1/2} z_{1-\frac{\alpha}{2}})] \qquad (11.32)$$

is also a pointwise asymptotically level $1-\alpha$ confidence interval for ρ.[4] ■

Sometimes, $\{X_n\}$ may not have a limiting distribution, but the weaker property of *tightness* may hold, which only requires that no probability escapes to $\pm\infty$.

Definition 11.2.5 A sequence of random vectors $\{X_n\}$ is *tight* (or *uniformly tight*) if $\forall \epsilon > 0$, there exists a constant B such that

$$\inf_n P\{|X_n| \le B\} \ge 1 - \epsilon .$$

A bounded sequence of numbers $\{x_n\}$ is sometimes written $x_n = O(1)$; more generally $x_n = O(y_n)$ if $x_n/y_n = O(1)$. If $\{X_n\}$ is tight, we sometimes also say X_n is bounded in probability, and write $|X_n| = O_P(1)$. If X_n is tight and $Y_n \xrightarrow{P} 0$ (sometimes written $Y_n = o_P(1)$), then $|X_n Y_n| \xrightarrow{P} 0$ (Problem 11.55). The notation $|X_n| = O_P(|Y_n|)$ means $|X_n|/|Y_n|$ is tight.

Tightness of a sequence of random vectors in $\mathbb{R}^k$ is equivalent to each of the component variables being tight $\mathbb{R}$ (Problem 11.40). Note that tightness, like convergence in distribution, really refers to the sequence of laws of X_n, denoted $\mathcal{L}(X_n)$. Thus, we shall interchangeably refer to tightness of a sequence of random variables or the sequence of their distributions.

In a statistical context, suppose $X_1, \ldots, X_n$ are i.i.d. according to a model $\{P_\theta,\ \theta \in \Omega\}$. Recall that an estimator sequence T_n is a (weakly) consistent estimator of $g(\theta)$ if, for every $\theta \in \Omega$,

$$T_n - g(\theta) \to 0$$

---

[4]For discussion of this transformation, see Mudholkar (1983), Stuart and Ord, Vol. 1 (1987) and Efron and Tibshirani (1993), p.54. Numerical evidence supports replacing n by $n-3$ in (11.32).

in probability when P_θ is true. An estimator sequence T_n is said to be τ_n-consistent for $g(\theta)$ if, for every $\theta \in \Omega$,

$$\tau_n[T_n - g(\theta)]$$

is tight when P_θ is true. For example, if the underlying population has a finite variance, it follows from the Central Limit Theorem that the sample mean is a $n^{1/2}$-consistent estimator of the population mean.

Whenever X_n converges in distribution to a limit distribution, then $\{X_n\}$ is tight, and the following partial converse is true. Just as any bounded sequence of real numbers has a subsequence which converges, so does any sequence of random variables X_n that is $O_P(1)$. This important result is stated next.

Theorem 11.2.15 (Prohorov's Theorem) *Suppose $\{X_n\}$ is tight on $\mathbb{R}^k$. Then, there exists a subsequence n_j and a random vector X such that $X_{n_j} \stackrel{d}{\to} X$.*

11.2.3 Almost Sure Convergence

On occasion, we shall utilize a form of convergence of X_n to X stronger than convergence in probability.

Definition 11.2.6 Suppose X_n and X are random vectors in $\mathbb{R}^k$, defined on a common probability space $(\mathcal{X}, \mathcal{F})$. Then, X_n is said to *converge almost surely* (a.s.) to X if $X_n(\omega) \to X(\omega)$ on a set of points ω which has probability one; that is, if

$$P\{\omega \in \mathcal{X} : \lim_{n\to\infty} |X_n(\omega) - X(\omega)| = 0\} = 1 .$$

This is denoted by $X_n \to X$ a.s..

Equivalently, we say that X_n converges to X with probability one, since there is a set of outcomes ω having probability one such that $X_n(\omega) \to X(\omega)$. If X_n converges almost surely to X, then X_n converges in probability to X, but the converse is false (but see Problem 11.61). Indeed, convergence in probability does not even guarantee $X_n(\omega) \to X(\omega)$ for any outcome ω. The following provides a classic counterexample.

Example 11.2.11 (Convergence in probability, but not a.s.) Suppose U is uniformly distributed on $[0, 1)$, so that $\mathcal{X}$ is [0,1), F is the class of Borel sets, $U = U(\omega) = \omega$, and P is the uniform probability measure. For $m = 1, 2, \ldots$ and $j = 1, \ldots, m$, let $Y_{m,j}$ be one if $U \in [(j-1)/m, j/m)$ and zero otherwise. For any m, exactly one of the $Y_{m,j}$ is one and the rest are zero; also, $P\{Y_{m,j} = 1\} = 1/m \to 0$ as $m \to \infty$. String together all the variables so that $X_1 = Y_1$, $X_2 = Y_{2,1}$, $X_3 = Y_{2,2}$, $X_4 = Y_{3,1}$, $X_5 = Y_{3,2}$, etc. Then, $X_n \to 0$ in probability. But X_n does not converge to 0 for any outcome U since X_n oscillates infinitely often between 0 and 1. ■

Theorem 11.2.16 (Strong Law of Large Numbers) *Let X_i be i.i.d. real-valued random variables with mean μ. Then*

$$\bar{X}_n \equiv \tfrac{1}{n} \sum_{i=1}^{n} X_i \to \mu \quad \text{a.s.}$$

Conversely, if $\overline{X}_n \to \mu$, a.s. with $|\mu| < \infty$, then $E|X_1| < \infty$.

In a statistical context, suppose $X_1, \ldots, X_n$ are i.i.d. according to a model $\{P_\theta,\ \theta \in \Omega\}$. Suppose, under each θ, $T_n = T_n(X_1, \ldots, X_n)$ converges almost surely to $g(\theta)$. Then, T_n is said to be strongly consistent estimator of $g(\theta)$.

One of the most fundamental examples of almost sure convergence is provided by the Glivenko-Cantelli theorem. To state the result, first define the Kolmogorov-Smirnov distance between c.d.f.s F and G as

$$d_K(F, G) = \sup_t |F(t) - G(t)| \ . \tag{11.33}$$

Theorem 11.2.17 (Glivenko-Cantelli Theorem) *Suppose $X_1, \ldots, X_n$ are i.i.d. real-valued random variables with c.d.f. F. Let $\hat{F}_n$ be the empirical c.d.f. defined by*

$$\hat{F}_n(t) = \frac{1}{n} \sum_{i=1}^{n} I\{X_i \le t\} \ . \tag{11.34}$$

Then,

$$d_K(\hat{F}_n, F) \to 0 \quad a.s.$$

To prove the Glivenko-Cantelli Theorem, note that, for every fixed t, $\hat{F}_n(t) \to F(t)$ almost surely, by the Strong Law of Large Numbers. That this convergence is uniform in t follows from the fact that F is monotone (Problem 11.53).

Example 11.2.12 (Kolmogorov-Smirnov Test) The Glivenko-Cantelli Theorem 11.2.17 forms the basis for the Kolmogorov-Smirnov goodness of fit test, previously introduced in Section 6.13. Specifically, consider the problem of testing the simple null hypothesis that $F = F_0$ versus $F \neq F_0$. The Glivenko-Cantelli Theorem implies that, under F,

$$d_K(\hat{F}_n, F_0) \to d_K(F, F_0) \quad a.s.$$

(and hence in probability as well), where the right side is zero if and only if $F = F_0$. Thus, the statistic $d_K(\hat{F}_n, F_0)$ tends to be small under the null hypothesis and large under the alternative. In order for this statistic to have a nondegenerate limit distribution under F_0, we normalize by multiplication of $n^{1/2}$ and the Kolmogorov-Smirnov goodness of fit test statistic is given by

$$T_n \equiv \sup_{t \in \mathbb{R}} n^{1/2} |\hat{F}_n(t) - F_0(t)| = n^{1/2} d_K(\hat{F}_n, F_0) \ . \tag{11.35}$$

The Kolmogorov-Smirnov test rejects the null hypothesis if $T_n > s_{n,1-\alpha}$, where $s_{n,1-\alpha}$ is the $1 - \alpha$ quantile of the null distribution of T_n when F_0 is the uniform $U(0, 1)$ distribution. Recall from Section 6.13 that the finite sampling distribution of T_n under F_0 is the same for all continuous F_0 (also see Problem 11.57), but its exact form is difficult to express. Some approaches to obtaining this distribution are discussed in Durbin (1973) and Section 4.3 of Gibbons and Chakraborti (1992). Values for $s_{n,1-\alpha}$ have been tabled in Birnbaum (1952). For exact power calculations in both the continuous and discrete case, see Niederhausen (1981) and Gleser (1985).

By the duality of tests and confidence regions, the Kolmogorov-Smirnov test can be inverted to yield uniform confidence bands for F, given by

$$R_{n,1-\alpha} = \{F : \ n^{1/2} \sup_t |\hat{F}_n(t) - F(t)| \le s_{n,1-\alpha}\} \ . \tag{11.36}$$

By construction, $P_F\{F \in R_{n,1-\alpha}\} = 1 - \alpha$ if F is continuous; furthermore, the confidence band is conservative if F is not continuous (Problem 11.58).

The limiting behavior of T_n will be discussed in Section 14.2. In fact, when $F = F_0$, T_n has a continuous strictly increasing limiting distribution with $1-\alpha$ quantile $s_{1-\alpha}$ (and so $s_{n,1-\alpha} \to s_{1-\alpha}$). It follows that the width of the band (11.36) is $O(n^{-1/2})$. Alternatives to the Kolmogorov-Smirnov bands that are more narrow in the tails and wider in the middle are discussed in Owen (1995). ■

The following useful inequality, which holds for finite sample sizes, actually implies the Glivenko-Cantelli Theorem (Problem 11.59).

Theorem 11.2.18 (Dvoretzky, Kiefer, Wolfowitz Inequality) *Suppose $X_1, \ldots, X_n$ are i.i.d. real-valued random variables with c.d.f. F. Let $\hat{F}_n$ be the empirical c.d.f. (11.34). Then, for any $d > 0$ and any positive integer n,*

$$P\{d_K(\hat{F}_n, F) > d\} \le C \exp(-2nd^2) \ , \tag{11.37}$$

where C is a universal constant.

Massart (1990) shows that we can take $C = 2$, which greatly improves the original value obtained by Dvoretzky, Kiefer, and Wolfowitz (1956).

Example 11.2.13 (Monte Carlo Simulation) Suppose $X_1, \ldots, X_n$ are i.i.d. observations with common distribution P. Assume P is known. The problem is to determine the distribution or quantile of some real-valued statistic $T_n(X_1, \ldots, X_n)$ for a fixed finite sample size n. Denote this distribution by $J_n(t)$, so that

$$J_n(t) = P\{T_n(X_1, \ldots, X_n) \le t\} \ .$$

This distribution may not have a tractable form or may not be explicitly computable, but the following simulation scheme allows the distribution $J(t)$ to be estimated to any desired level of accuracy. For $j = 1, \ldots, B$, let $X_{j,1}, \ldots, X_{j,n}$ be a sample of size n from P; then, one simply evaluates $T_n(X_{j,1}, \ldots, X_{j,n})$, and the empirical distribution of these B values serves as an approximation to the true sampling distribution $J_n(t)$. Specifically, $J_n(t)$ is approximated by

$$\hat{J}_{n,B}(t) = B^{-1} \sum_{j=1}^{B} I\{T_n(X_{j,1}, \ldots, X_{j,n}) \le t\} \ .$$

For large B, $\hat{J}_{n,B}(t)$ will be a good approximation to the true sampling distribution $J_n(t, P)$. One (though perhaps crude) way of quantifying the closeness of this approximation is the following. By the Dvoretsky, Kiefer, Wolfowitz inequality (11.37) (with B now taking over the role of n), there exists a universal constant C so that

$$P\{d_K(\hat{J}_{n,B}, J_n) > d\} \le C \exp(-2Bd^2).$$

Hence, if we desire the probability of the supremum distance between $\hat{J}_{n,B}(\cdot)$ and $J_n(\cdot, P)$ to be greater than d with probability less than ϵ, all we need to do is ensure that B is large enough so that $C \exp(-2Bd^2) \le \epsilon$. Since B, the number of simulations, is determined by the statistician (assuming enough computing power), the desired accuracy can be obtained. Further results on the choice of B are given in Jockel (1986).

Here, we are tacitly assuming that one can easily accomplish the sampling of observations from P. Of course, when P corresponds to a cumulative distribution function F on the real line, one can usually just obtain observations from F by $F^{-1}(U)$, where U is a random variable having the uniform distribution on $(0, 1)$. This construction assumes an ability to calculate an inverse function $F^{-1}(\cdot)$. A sample $X_{j,1}, \ldots, X_{j,n}$ of n i.i.d. F variables can then be obtained from n i.i.d. Uniform (0,1) observations $U_{j,1}, \ldots, U_{j,n}$ by the prescription $X_{j,n} = F^{-1}(U_{j,n})$. If F^{-1} is not tractable, other methods for generating observations with prescribed distributions are available in statistical software packages, such as S-plus, Excel, or Maple.

Note, however, that we have ignored any error from the use of a pseudo-random number generator, which presumably would be needed to generate the Uniform (0,1) variables. The above idea forms the basis of many approximation schemes; for some general references on Monte Carlo simulation, see Devroye (1986) and Ripley (1987). ■

Almost sure convergence is the strongest type of convergence we have introduced and it has many consequences. For example, suppose $X_n \to X$ almost surely and $|X_n| \le 1$ with probability one. Then, $|X| \le 1$ with probability one, and so $E(|X|) \le 1$; by the Lebesgue dominated convergence Theorem (Theorem 2.2.2), it follows that $E(X_n) \to E(X)$. If the assumption that $X_n \to X$ almost surely is replaced by the weaker condition that X_n converges in distribution to X, then the argument to show $E(X_n) \to E(X)$ breaks down. However, we shall now show that the result continues to hold since the conclusion pertains only to distributional properties of X_n and X. The argument is based on the following theorem.

Theorem 11.2.19 (Almost Sure Representation Theorem) *Suppose $X_n \stackrel{d}{\to} X$ in $\mathbb{R}^k$. Then, there exist random vectors $\tilde{X}_n$ and $\tilde{X}$ defined on some common probability space such that $\tilde{X}_n$ has the same distribution as X_n and $\tilde{X}_n \to \tilde{X}$ a.s. (and so $\tilde{X}$ has the same distribution as X).*

Example 11.2.14 (Convergence of Moments) Suppose X_n and X are real-valued random variables and $X_n \stackrel{d}{\to} X$. If the X_n are uniformly bounded, then $E(X_n) \to E(X)$. To see why, construct $\tilde{X}_n$ and $\tilde{X}$ by the Almost Sure Representation Theorem and then apply the Dominated Convergence Theorem (Theorem 2.2.2) to the $\tilde{X}_n$ to conclude

$$E(X_n) = E(\tilde{X}_n) \to E(\tilde{X}) = E(X) . \tag{11.38}$$

If the X_n are not uniformly bounded, but $X_n \geq 0$, then by Fatou's Lemma (Theorem 2.2.1), we may conclude

$$E(X) = E(\tilde{X}) \leq \liminf_n E(\tilde{X}_n) = \liminf_n E(X_n) .$$

As a final result, suppose $X_n \xrightarrow{d} X$ and $|X|$ has distribution F which is continuous at t. Then, by the Continuous Mapping Theorem,

$$|X_n| I\{|X_n| \leq t\} \xrightarrow{d} |X| I\{|X| \leq t\} .$$

By (11.38), we may conclude

$$E[|X_n| I\{|X_n| \leq t\}] \to E[|X| I\{|X| \leq t\}] . \tag{11.39}$$

If, in addition, $E|X_n| \to E|X|$, then

$$E[|X_n| I\{|X_n| > t\}] \to E[|X| I\{|X| > t\}] . \blacksquare \tag{11.40}$$

11.3 Robustness of Some Classical Tests

Optimality theory postulates a statistical model and then attempts to determine a best procedure for that model. Since model assumptions tend to be unreliable, it is necessary to go a step further and ask how sensitive the procedure and its optimality are to the assumptions. In the normal models of Chapters 4-7, three assumptions are made: independence, identity of distribution, and normality. In the two-sample t-test, there is the additional assumption of equality of variance. We shall consider the effects of nonnormality and inequality of variance in the first subsection, and that of dependence in the next subsection.

The natural first question to ask about the robustness of a test concerns the behavior of the significance level. If an assumption is violated, is the significance level still approximately valid? Such questions are typically answered by combining two methods of attack: The actual significance level under some alternative distribution is either calculated exactly or, more usually, estimated by simulation. In addition, asymptotic results are obtained which provide approximations to the true significance level for a wide variety of models. We here restrict ourselves to a brief sketch of the latter approach.

11.3.1 Effect of Distribution

Consider the one-sample problem where $X_1, \ldots, X_n$ are independently distributed as $N(\xi, \sigma^2)$. Tests of $H : \xi = \xi_0$ are based on the test statistic

$$t_n = t_n(X_1, \ldots, X_n) = \frac{\sqrt{n}(\bar{X}_n - \xi_0)}{S_n} = \frac{\sqrt{n}(\bar{X}_n - \xi_0)}{\sigma} \Big/ \frac{S_n}{\sigma}, \tag{11.41}$$

where $S_n^2 = \sum(X_i - \bar{X}_n)^2/(n-1)$; see Section 5.2. When $\xi = \xi_0$ and the X's are normal, t_n has the t-distribution with $n-1$ degrees of freedom. Suppose, however, that the normality assumption fails and the X's instead are distributed according to some other distribution F with mean ξ_0 and finite variance. Then by the Central Limit Theorem, $\sqrt{n}(\bar{X}_n - \xi_0)/\sigma$ has the limit distribution $N(0,1)$;

furthermore S_n/σ tends to 1 in probability by Example 11.2.6. Therefore, by Slutsky's theorem, t_n has the limit distribution $N(0,1)$ regardless of F. This shows in particular that the t-distribution with $n-1$ degrees of freedom tends to $N(0,1)$ as $n \to \infty$.

To be specific, consider the one-sided t-test which rejects when $t_n \ge t_{n-1,1-\alpha}$, where $t_{n-1,1-\alpha}$ is the $1-\alpha$ quantile of the t-distribution with $n-1$ degrees of freedom. It follows from Corollary 11.2.3 and the asymptotic normality of the t-distribution that (see Problem 11.42 (ii))

$$t_{n-1,1-\alpha} \to z_{1-\alpha} = \Phi^{-1}(1-\alpha) \ .$$

In fact, the difference $t_{n-1,1-\alpha} - z_{1-\alpha}$ is $O(n^{-1})$, as will be seen in Section 11.4.1.

Let $\alpha_n(F)$ be the true probability of the rejection region $t_n \ge t_{n-1,1-\alpha}$ when the distribution of the X's is F. Then $\alpha_n(F) = P_F\{t_n \ge t_{n-1,1-\alpha}\}$ has the same limit as $P_\Phi\{t_n \ge z_{1-\alpha}\}$, which is α. Thus, the t-test is pointwise asymptotically level α, assuming the underlying distribution has a finite nonzero variance. However, the t-test is not uniformly asymptotically level α. This issue will be studied more closely in Section 11.4. For sufficiently large n, the actual rejection probability $\alpha_n(F)$ will be close to the nominal level α; how close depends on F and n. For entries to the literature dealing with this dependence, see Cressie (1980), Tan (1982), Benjamini (1983), and Edelman (1990). Other robust approaches for testing the mean are discussed in Sutton (1993) and Chen (1995). The use of resampling will be deferred to Chapter 15.

To study the corresponding test of variance, suppose first that the mean ξ is 0. When F is normal, the UMP test of $H : \sigma = \sigma_0$ against $\sigma > \sigma_0$ rejects when $\sum X_i^2/\sigma_0^2$ is too large, where the null distribution of $\sum X_i^2/\sigma_0^2$ is χ_n^2. By the Central Limit theorem, $\sqrt{n}(\sum X_i^2 - n\sigma_0^2)/n$ tends in law to $N(0, 2\sigma_0^4)$ as $n \to \infty$, since $\mathrm{Var}(X_i^2) = 2\sigma_0^4$. If the rejection region is written as

$$\frac{\sum X_i^2 - n\sigma_0^2}{\sqrt{2n}\sigma_0^2} \ge C_n \ ,$$

it follows that $C_n \to z_{1-\alpha}$.

Suppose now instead that the X's are distributed according to a distribution F with $E(X_i) = 0$, $E(X_i^2) = Var(X_i) = \sigma^2$, and $Var(X_i^2) = \gamma^2$. Then $\sum(X_i^2 - n\sigma_0^2)/\sqrt{n}$ tends in law to $N(0,\gamma^2)$ when $\sigma = \sigma_0$, and the rejection probability $\alpha_n(F)$ of the test tends to

$$\lim P\left\{\frac{\sum X_i^2 - n\sigma_0^2}{\sqrt{2n}\sigma_0^2} \ge z_{1-\alpha}\right\} = 1 - \Phi\left(\frac{z_{1-\alpha}\sqrt{2}\sigma_0^2}{\gamma}\right) .$$

Depending on γ, which can take on any positive value, the sequence $\alpha_n(F)$ can thus tend to any limit $< \frac{1}{2}$. Even asymptotically and under rather small departures from normality (if they lead to big changes in γ), the size of the χ^2-test is thus completely uncontrolled.

For sufficiently large n, the difficulty can be overcome by Studentization[5], where one divides the test statistic by a consistent estimate of the asymptotic standard deviation. Letting $Y_i = X_i^2$ and $E(Y_i) = \eta = \sigma^2$, the test statistic then reduces to $\sqrt{n}(\bar{Y} - \eta_0)$. To obtain an asymptotically valid test, it is only

[5]Studentization is defined in a more general context at the end of Section 7.3.

necessary to divide by a suitable estimator of $\sqrt{VarY_i}$ such as $\sqrt{\sum(Y_i - \bar{Y})^2/n}$. (However, since $Y_i^2 = X_i^4$, small changes in the tail of X_i may have large effects on Y_i^2, and n may have to be rather large for the asymptotic result to give a good approximation.)

When ξ is unknown, the normal theory test for σ^2 is based on $\sum(X_i - \bar{X}_n)^2$, and the sequence

$$\frac{1}{\sqrt{n}}\left[\sum(X_i - \bar{X}_n)^2 - n\sigma_0^2\right] = \frac{1}{\sqrt{n}}\left(\sum X_i^2 - n\sigma_0^2\right) - \frac{1}{\sqrt{n}}n\bar{X}^2$$

again has the limit distribution $N(0, \gamma^2)$. To see this, note that the distribution of $\sum(X_i - \bar{X}_n)^2$ is independent of ξ and put $\xi = 0$. Since $\sqrt{n}\bar{X}$ has a (normal) limit distribution, $n\bar{X}^2$ is bounded in probability and so $n\bar{X}^2/\sqrt{n}$ tends to zero in probability. The result now follows from that for $\xi = 0$ and Slutsky's theorem.

The above results carry over to the corresponding two-sample problems that were considered in Section 5.3. Consider the two-sample t-statistic given by (5.28). An extension of the one-sample argument shows that as $m,\ n \to \infty$, $(\bar{Y}_n - \bar{X}_m)/\sigma\sqrt{1/m + 1/n}$ tends in law to $N(0, 1)$ while $[\sum(X_i - \bar{X}_m)^2 + \sum(Y_j - \bar{Y}_n)^2]/(m+n-2)\sigma^2$ tends in probability to 1 for samples $X_1, \ldots, X_m;\ Y_1, \ldots, Y_n$ from any common distribution F with finite variance. Thus, the rejection probability $\alpha_{m,n}(F)$ tends to α for any such F. As will be seen in Section 11.3.3, the same robustness property for the UMP invariant test of equality of s means also holds.

On the other hand, the F-test for variances, just like the one-sample χ^2-test, is extremely sensitive to the assumption of normality. To see this, express the rejection region in terms of $\log S_Y^2 - \log S_X^2$, where $S_X^2 = \sum(X_i - \bar{X}_m)^2/(m-1)$ and $S_Y^2 = \sum(Y_j - \bar{Y}_n)^2/(n-1)$, and suppose that as m and $n \to \infty$, $m/(m+n)$ remains fixed at ρ. By the result for the one-sample problem and the delta method with $g(u) = \log u$ (Theorem 11.2.14), it is seen that $\sqrt{m}[\log S_X^2 - \log \sigma^2]$ and $\sqrt{n}[\log S_Y^2 - \log \sigma^2]$ both tend in law to $N(0, \gamma^2/\sigma^4)$ when the X's and Y's are distributed as F, and hence that $\sqrt{m+n}[\log S_Y^2 - \log S_X^2]$ tends in law to the normal distribution with mean 0 and variance

$$\frac{\gamma^2}{\sigma^4}\left(\frac{1}{\rho} + \frac{1}{1-\rho}\right) = \frac{\gamma^2}{\rho(1-\rho)\sigma^4}\ .$$

In the particular case that F is normal, $\gamma^2 = 2\sigma^4$ and the variance of the limit distribution is $2/\rho(1-\rho)$. For other distributions γ^2/σ^4 can take on any positive value and, as in the one-sample case, $\alpha_n(F)$ can tend to any limit less than $\frac{1}{2}$. [For an entry into the extensive literature on more robust alternatives, see for example Conover, Johnson, and Johnson (1981), Tiku and Balakrishnan (1984), Boos and Brownie (1989), Baker (1995), Hall and Padmanabhan (1997), and Section 2.10 of Hettmansperger and McKean (1998).]

Having found that the rejection probability of the one- and two-sample t-tests is relatively insensitive to nonnormality (at least for large samples), let us turn to the corresponding question concerning the power of these tests. By similar asymptotic calculations, it can be shown that the same conclusion holds: Power values of the t-tests obtained under normality are asymptotically valid also for all other distributions with finite variance. This is a useful result if it has been decided to employ a t-test and one wishes to know what power it will have against a given alternative ξ/σ or $(\eta - \xi)/\sigma$, or what sample sizes are required to obtain a given power.

Recall that there exists a modification of the t-test, the *permutation* version of the t-test discussed in Section 5.9, whose size is independent of F not only asymptotically but exactly. Moreover, we will see in Section 15.2 that its asymptotic power is equal to that of the t-test. It may seem that the permutation t-test has all the properties one could hope for. However, this overlooks the basic question of whether the t-test itself, which is optimal under normality, will retain a high standing with respect to its competitors under other distributions. The t-tests are in fact not robust in this sense. Some tests which are preferable when a broad spectrum of distributions F is considered possible were discussed in Section 6.9. A permutation test with this property has been proposed by Lambert (1985).

As a last problem, consider the level of the two-sample t-test when the variances $\text{Var}(X_i) = \sigma^2$ and $\text{Var}(Y_j) = \tau^2$ may differ (as in the Behrens-Fisher problem), and the assumption of normality may fail as well. As before, one finds that $(\bar{Y}_m - \bar{X}_n)/\sqrt{\sigma^2/m + \tau^2/n}$ tends in law to $N(0,1)$ as $m, n \to \infty$, while $S_X^2 = \sum(X_i - \bar{X}_m)^2/(m-1)$ and $S_Y^2 = \sum(Y_i - \bar{Y}_n)^2/(n-1)$ respectively tend to σ^2 and τ^2 in probability. If m and n tend to ∞ through a sequence with fixed proportion $m/(m+n) = \rho$, the squared denominator of the t-statistic,

$$D^2 = \frac{m-1}{m+n-2} S_X^2 + \frac{n-1}{m+n-2} S_Y^2 ,$$

tends in probability to $\rho\sigma^2 + (1-\rho)\tau^2$, and the limit of

$$t = \frac{1}{\sqrt{\frac{1}{m} + \frac{1}{n}}} \left(\frac{\bar{Y}_n - \bar{X}_m}{\sqrt{\frac{\sigma^2}{m} + \frac{\tau^2}{n}}} \cdot \frac{\sqrt{\frac{\sigma^2}{m} + \frac{\tau^2}{n}}}{D} \right)$$

is normal with mean zero and variance

$$\frac{(1-\rho)\sigma^2 + \rho\tau^2}{\rho\sigma^2 + (1-\rho)\tau^2} . \tag{11.42}$$

When $m = n$, so that $\rho = \frac{1}{2}$, the t-test thus has approximately the right level even if σ and τ are far apart. The accuracy of this approximation for different values of $m = n$ and τ/σ is discussed by Ramsey (1980) and Posten, Yeh, and Owen (1982). However, when $\rho \neq \frac{1}{2}$, the actual size of the test can differ greatly from the nominal level α even for large m and n. An approximate test of the hypothesis $H: \eta = \xi$ when σ, τ are not assumed equal, which asymptotically is free of this difficulty, can be obtained through Studentization, i.e., by replacing D^2 with $(1/m)S_X^2 + (1/n)S_Y^2$ and referring the resulting statistic to the standard normal distribution. This approximation is very crude, and not reliable unless m and n are fairly large. A refinement, the *Welch approximate t-test*, refers the resulting statistic not to the standard normal but to the t-distribution with a random number of degrees of freedom f given by

$$\frac{1}{f} = \left(\frac{R}{1+R}\right)^2 \frac{1}{m-1} + \frac{1}{(1+R)^2} \cdot \frac{1}{n-1} ,$$

where $R = (nS_X^2)/(mS_Y^2)$.[6] When the X's and Y's are normal, the actual level of this test has been shown to be quite close to the nominal level for sample sizes as small as $m = 4$, $n = 8$ and $m = n = 6$ [see Wang (1971)]. A further refinement will be mentioned in Section 15.6. A simple but crude approach that controls the level is to use as degrees of freedom the smaller of $n - 1$ and $m - 1$, as remarked by Scheffé (1970).

The robustness of the level of Welch's test against nonnormality is studied by Yuen (1974), who shows that for heavy-tailed distributions the actual level tends to be considerably smaller than the nominal level (which leads to an undesirable loss of power), and who proposes an alternative. Some additional results are discussed in Scheffé (1970) and in Tiku and Singh (1981). The robustness of some quite different competitors of the t-test is investigated in Pratt (1964).

For testing the equality of s normal means with $s > 2$, the classical test based on the F-statistic (7.19) is not robust, even if all the observations are normally distributed, regardless of the sample sizes (Scheffé (1959), Problem 11.86); again, the problem is due to the assumption of a common variance. More appropriate test for this generalized Behrens-Fisher problem have been proposed by Welch (1951), James (1951), and Brown and Forsythe (1974a), and are further discussed by Clinch and Kesselman (1982), Hettmansperger and McKean (1998) and Chapter 10 of Pesarin (2001). The corresponding robustness problem for more general linear hypotheses is treated by James (1954) and Johansen (1980); see also Rothenberg (1984).

11.3.2 Effect of Dependence

The one-sample t-test arises when a sequence of measurements $X_1, \ldots, X_n$, is taken of a quantity ξ, and the X's are assumed to be independently distributed as $N(\xi, \sigma^2)$. The effect of nonnormality on the level of the test was discussed in the preceding subsection. Independence may seem like a more innocuous assumption. However, it has been found that observations occurring close in time or space are often positively correlated [Student (1927), Hotelling (1961), Cochran (1968)]. The present section will therefore be concerned with the effect of this type of dependence.

Lemma 11.3.1 *Let $X_1, \ldots, X_n$ be jointly normally distributed with common marginal distribution $N(0, \sigma^2)$ and with correlation coefficients $\rho_{i,j} = \operatorname{corr}(X_i, X_j)$. Assume that*

$$\frac{1}{n} \sum_{i \neq j} \sum \rho_{i,j} \to \gamma \tag{11.43}$$

and

$$\frac{1}{n^2} \sum_{i \neq j} \sum \rho_{i,j}^2 \to 0 \tag{11.44}$$

as $n \to \infty$. Then,

[6] For a variant see Fenstad (1983).

(i) the distribution of the t-statistic t_n defined in equation (11.41) (with $\xi_0 = 0$) tends to the normal distribution $N(0, 1+\gamma)$;
(ii) if $\gamma \neq 0$, the level of the t-test is not robust even asymptotically as $n \to \infty$. Specifically, if $\gamma > 0$, the asymptotic level of the t-test carried out at nominal level α is

$$1 - \Phi\left(\frac{z_{1-\alpha}}{\sqrt{1+\gamma}}\right) > 1 - \Phi(z_{1-\alpha}) = \alpha .$$

PROOF. (i): Since the X_i are jointly normal, the numerator $\sqrt{n}\bar{X}_n$ of t_n is also normal, with mean zero and variance

$$Var\left(\sqrt{n}\bar{X}\right) = \sigma^2\left[1 + \frac{1}{n}\sum\sum_{i \neq j} \rho_{i,j}\right] \to \sigma^2(1+\gamma) , \qquad (11.45)$$

and hence tends in law to $N(0, \sigma^2(1+\gamma))$. The denominator of t_n is the square root of

$$S_n^2 = \frac{1}{n-1}\sum X_i^2 - \frac{n}{n-1}\bar{X}_n^2 .$$

By (11.45), $Var(\bar{X}_n) \to 0$ and so $\bar{X}_n \xrightarrow{P} 0$. A calculation similar to (11.45) shows that $Var(n^{-1}\sum_{i=1}^n X_i^2) \to 0$ (Problem 11.65). Thus, $n^{-1}\sum_{i=1}^n X_i^2 \xrightarrow{P} \sigma^2$ and so $S_n \xrightarrow{P} \sigma$. By Slutsky's theorem, the distribution of t_n therefore tends to $N(0, 1+\gamma)$.

The implications (ii) are obvious. ■

Under the assumptions of Lemma 11.3.1, the joint distribution of the X's is determined by σ^2 and the correlation coefficients $\rho_{i,j}$, with the asymptotic level of the t-test depending only on γ. The following examples illustrating different correlation structures show that even under rather weak dependence of the observations, the assumptions of Lemma 11.3.1 are satisfied with $\gamma \neq 0$, and hence that the level of the t-test is quite sensitive to the assumption of independence.

MODEL A. (CLUSTER SAMPLING). Suppose the observations occur in s groups (or clusters) of size m, and that any two observations within a group have a common correlation coefficient ρ, while those in different groups are independent. (This may be the case, for instance, when the observations within a group are those taken on the same day or by the same observer, or involve some other common factor.) Then (Problem 11.67),

$$Var(\bar{X}) = \frac{\sigma^2}{ms}[1 + (m-1)\rho] ,$$

which tends to zero as $s \to \infty$. The conditions of the lemma hold with $\gamma = (m-1)\rho$, and the level of the t-test is not asymptotically robust as $s \to \infty$. In particular, the test overstates the significance of the results when $\rho > 0$.

To provide a specific structure leading to this model, denote the observations in the ith group by $X_{i,j}$ $(j = 1, \ldots, m)$, and suppose that $X_{i,j} = A_i + U_{i,j}$, where A_i is a factor common to the observations in the ith group. If the A's and U's (none of which are observable) are all independent with normal distributions $N(\xi, \sigma_A^2)$ and $N(0, \sigma_0^2)$ respectively, then the joint distribution of the X's is that prescribed by Model A with $\sigma^2 = \sigma_A^2 + \sigma_0^2$ and $\rho = \sigma_A^2/\sigma^2$.

MODEL B. (MOVING-AVERAGE PROCESS). When the dependence of nearby observations is not due to grouping as in Model A, it is often reasonable to assume that $\rho_{i,j}$ depends only on $|j-i|$ and is nonincreasing in $|j-i|$. Let $\rho_{i,i+k}$ then be denoted by ρ_k, and suppose that the correlation between X_i and X_{i+k} is negligible for $k > m$ (m an integer $< n$), so that one can put $\rho_k = 0$ for $k > m$. Then the conditions for Lemma 11.3.1 are satisfied with

$$\gamma = 2\sum_{k=1}^{m} \rho_k \ .$$

In particular, if $\rho_1, \ldots, \rho_m$ are all positive, the t-test is again too liberal.

A specific structure leading to Model B is given by the moving-average process

$$X_i = \xi + \sum_{j=0}^{m} \beta_j U_{i+j} \ ,$$

where the U's are independent $N(0, \sigma_0^2)$. The variance σ^2 of the X's is then $\sigma^2 = \sigma_0^2 \sum_{j=0}^{m} \beta_j^2$ and

$$\rho_k = \begin{cases} \dfrac{\sum_{i=0}^{m-k} \beta_i \beta_{i+k}}{\sum_{j=0}^{m} \beta_j^2} & \text{for} \quad k \le m, \\ 0 & \text{for} \quad k > m. \end{cases}$$

MODEL C. (FIRST-ORDER AUTOREGRESSIVE PROCESS). A simple model for dependence in which the $|\rho_k|$ are decreasing in k but $\neq 0$ for all k is the *first-order autoregressive process* defined by

$$X_{i+1} = \xi + \beta(X_i - \xi) + U_{i+1}, \qquad |\beta| < 1, \quad i = 1, \ldots, n \ ,$$

with the U_i independent $N(0, \sigma_0^2)$. If X_1 is $N(\xi, \tau^2)$, the marginal distribution of X_i for $i > 1$ is normal with mean ξ and variance $\sigma_i^2 = \beta^2\sigma_{i-1}^2 + \sigma_0^2$. The variance of X_i will thus be independent of i provided $\tau^2 = \sigma_0^2/(1-\beta^2)$. For the sake of simplicity we shall assume this to be the case, and take ξ to be zero. From

$$X_{i+k} = \beta^k X_i + \beta^{k-1} U_{i+1} + \beta^{k-2} U_{i+2} + \cdots + \beta U_{i+k-1} + U_{i+k}$$

it then follows that $\rho_k = \beta^k$, so that the correlation between X_i and X_j decreases exponentially with increasing $|j-i|$. The assumptions of Lemma 11.3.1 are again satisfied, and $\gamma = 2\beta/(1-\beta)$. Thus, in this case too, the level of the t-test is not asymptotically robust. [Some values of the actual asymptotic level when the nominal level is .05 or .01 are given by Gastwirth and Rubin (1971).]

It is seen that in general the effect of dependence on the level of the t-test is more serious than that of nonnormality. In order to robustify the test against general dependence through studentization (as was done in the two-sample case with unequal variances), it is necessary to consistently estimate γ, which implicitly depends on estimation of all the $\rho_{i,j}$. Unfortunately, the number of parameters $\rho_{i,j}$ exceeds the number of observations. However, robustification is possible against some types of dependence. For example, it may be reasonable to assume a model such as A–C so that it is only required to estimate a reduced number of correlations.[7] Some specific procedures of this type are discussed by Albers (1978), [and for an associated sign test by Falk and Kohne (1984)]. Such robust procedures will in fact often also be insensitive to the assumption of normality, as can be shown by appealing to an appropriate Central Limit Theorem for dependent variables [see e.g. Billingsley (1995, Section 27)]. The validity of these procedures is of course limited to the particular model assumed, including the value of a parameter such as m in Models A and B. In fact, robustification is achievable for fairly general classes of models with dependence by using an appropriate bootstrap method; see Problem 15.33 and Lahiri (2003). Alternatively, one can use subsampling, as in Romano and Thombs (1996); see Section 15.7.

The results of the present section easily extend to the case of the two-sample t-test, when each of the two series of observations shows dependence of the kind considered here.

11.3.3 Robustness in Linear Models

In this section, we consider the large sample robustness properties of some of the linear model tests discussed in Chapter 7. As in Section 11.3.1, we focus on the effect of distribution.

A large class of these testing situations is covered by the following general model, which was discussed in Problem 7.8. Let $X_1, \dots, X_n$ be independent with $E(X_i) = \xi_i$ and $Var(X_i) = \sigma^2 < \infty$, where we assume the vector ξ to lie in an s-dimensional subspace Π_Ω of $\mathbb{R}^n$, defined by the following parametric set of equations

$$\xi_i = \sum_{j=1}^{s} a_{i,j}\beta_j \ , \qquad i = 1, \dots, n. \tag{11.46}$$

Here the $a_{i,j}$ are known coefficients and the β_j are unknown parameters. In matrix form, the $n \times 1$ vector ξ with ith component ξ_i satisfies $\xi = A\beta$, where A is an $n \times s$ matrix having (i, j) entry $a_{i,j}$ and β is an $s \times 1$ vector with jth component β_j. It is assumed A is known and of rank s. In the asymptotics below, the $a_{i,j}$ may depend on n, but s remains fixed. Throughout, the notation will suppress this dependence on n.

The least squares estimators $\hat{\xi}_1, \dots, \hat{\xi}_n$ of $\xi_1, \dots, \xi_n$ are defined as the values of ξ_i minimizing

$$\sum_{i=1}^{n} (X_i - \xi_i)^2$$

subject to $\xi \in \Pi_\Omega$, where Π_Ω is the space spanned by the s columns of A. Correspondingly, the least squares estimators $\hat{\beta}_1, \dots, \hat{\beta}_s$ of $\beta_1, \dots, \beta_s$ are the values of β_j minimizing

$$\sum_{i=1}^{n} (X_i - \sum_{j=1}^{s} a_{i,j}\beta_j)^2 \ .$$

[7] Models of a sequence of dependent observations with various covariance structures are discussed in books on time series such as Brockwell and Davis (1991), Hamilton (1994) or Fuller (1996).

By taking partial derivatives of of this last expression with respect to the β_j, it is seen that that $\hat{\beta}_j$ are solutions of the equations

$$A^T A\beta = A^T X$$

and so

$$\hat{\beta} = (A^T A)^{-1} A^T X \ .$$

(The fact that $A^T A$ is nonsingular follows from Problem 6.3.) Thus,

$$\hat{\xi} = PX \ ,$$

where

$$P = A(A^T A)^{-1} A^T \ . \tag{11.47}$$

In fact, $\hat{\xi}$ is the projection of X into the space Π_Ω. (These estimators formed the basis of optimal invariant tests studied in Chapter 7.) Some basic properties of P and $\hat{\xi}$ are recorded in the following lemma.

Lemma 11.3.2 *(i) The matrix P defined by (11.47) is symmetric ($P = P^T$) and idempotent ($P^2 = P$).*
(ii) $X - \hat{\xi}$ is orthogonal to $\hat{\xi}$; that is,

$$\hat{\xi}^T(X - \hat{\xi}) = 0 \ .$$

PROOF. The proof of (i) follows by matrix algebra (Problem 11.71). To prove (ii), note that

$$\hat{\xi}^T(X - \hat{\xi}) = (PX)^T(X - PX) = X^T P^T (X - PX)$$
$$= X^T P^T X - X^T P^T PX = 0 \ ,$$

since by (i) $P^T P = P^T$. ■

Note that $\hat{\beta}_j$ is a linear combination of the X_i. Thus, if the X_i are normally distributed, so are the $\hat{\beta}_j$. Without the assumption of normality, the asymptotic normality of $\hat{\beta}_j$ can be established by the following lemma, which can be obtained as a consequence of the Lindeberg Central Limit Theorem (Problem 11.72).

Lemma 11.3.3 *Let $Y_1, Y_2, \ldots$ be independently identically distributed with mean zero and finite variance σ^2. (i) Let $c_1, c_2, \ldots$ be a sequence of constants. Then a sufficient condition for $\sum_{i=1}^n c_i Y_i / \sqrt{\sum c_i^2}$ to tend in law to $N(0, \sigma^2)$ is that*

$$\frac{\max_{i=1,\ldots,n} c_i^2}{\sum_{j=1}^n c_j^2} \to 0 \quad \text{as } n \to \infty \ . \tag{11.48}$$

(ii) More generally, suppose $C_{n,1}, \ldots, C_{n,n}$ is a sequence of random variables, independent of $Y_1, \ldots, Y_n$. Then, a sufficient condition for $\sum_{i=1}^n C_{n,i} Y_i / \sqrt{\sum C_{n,i}^2}$ to tend in law to $N(0, \sigma^2)$ is

$$\frac{\max_{i=1,\ldots,n} C_{n,i}^2}{\sum_{j=1}^n C_{n,j}^2} \xrightarrow{P} 0 \quad \text{as } n \to \infty \ .$$

The condition (11.48) prevents the c's from increasing so fast that the last term essentially dominates the sum, in which case there is no reason to expect asymptotic normality.

Example 11.3.1 Suppose $U_1, U_2, \ldots$ are i.i.d. with mean 0 and finite nonzero variance σ^2. Consider the simple regression model

$$X_i = \alpha + \beta t_i + U_i \ ,$$

where the t_i are known and not all equal. The least squares estimator $\hat{\beta}$ of β satisfies

$$\hat{\beta} - \beta = \frac{\sum (X_i - \alpha - \beta t_i)(t_i - \bar{t})}{\sum (t_i - \bar{t})^2} \ .$$

By Lemma 11.3.3,

$$\frac{(\hat{\beta} - \beta)\sqrt{\sum (t_i - \bar{t})^2}}{\sigma} \xrightarrow{d} N(0,1)$$

provided

$$\frac{\max(t_i - \bar{t})^2}{\sum (t_j - \bar{t})^2} \to 0 \ . \tag{11.49}$$

Condition (11.49) holds in the case of equal spacing $t_i = a + i\Delta$, but not when the t's grow exponentially, for example, when $t_i = 2^i$ (Problem 11.73). ■

Consider the hypothesis

$$H : \theta = \sum_{j=1}^{s} b_j \beta_j = 0 \ , \tag{11.50}$$

where the b's are known constants with $\sum b_j^2 = 1$. Assume without loss of generality that $A^T A = I$, the identity matrix, so that the columns of A are mutually orthogonal and of length one. The least squares estimator of θ is given by

$$\hat{\theta} = \sum_{j=1}^{s} b_j \hat{\beta}_j = \sum_{i=1}^{n} d_i X_i \ , \tag{11.51}$$

where by (11.46)

$$d_i = \sum_{j=1}^{s} a_{i,j} b_j \tag{11.52}$$

(Problem 11.74). By the orthogonality of A, $\sum d_i^2 = \sum b_j^2 = 1$, so that under H,

$$E(\hat{\theta}) = \sum_{j=1}^{s} E(b_j \hat{\beta}_j) = \sum_{j=1}^{s} b_j \beta_j = 0$$

and

$$Var(\hat{\theta}) = Var(\sum_{i=1}^{n} d_i X_i) = \sigma^2 \sum_{i=1}^{n} d_i^2 = \sigma^2 \ .$$

Consider the uniformly most powerful invariant test that rejects H when the t-statistic

$$\frac{|\hat{\theta}|}{\sqrt{\sum(X_i - \hat{\xi}_i)^2/(n-s)}} \geq C\ . \tag{11.53}$$

Now, the denominator of (11.53) tends in probability to σ. To see why, with s fixed, it suffices to show

$$\frac{1}{n}\sum(X_i - \hat{\xi}_i)^2 \xrightarrow{P} \sigma^2\ .$$

But, the left side is

$$\frac{\sum(X_i - \xi_i)^2}{n} + \frac{2\sum(X_i - \xi_i)(\xi_i - \hat{\xi}_i)}{n} + \frac{\sum(\xi_i - \hat{\xi}_i)^2}{n}\ .$$

The first term tends in probability to σ^2, by the Weak Law of Large Numbers. By the Cauchy-Schwarz Inequality, half the middle term is bounded by the square root of the product of the first and third terms. Therefore, it suffices to show the third term tends to 0 in probability. Since this term is nonnegative, it suffices to show its expectation tends to 0, by Markov's Inequality (Problem 11.26). But its expectation is the trace of the covariance matrix of $\hat{\xi}$ divided by n. Letting I_n denote the $n \times n$ identity matrix, the covariance matrix of $\hat{\xi} = PX$ is

$$\sigma^2 P I_n P^T = \sigma^2 P P^T = \sigma^2 P\ .$$

But, the trace of P is

$$tr(P) = tr(A(A^TA)^{-1}A^T) = tr(A^TA(A^TA)^{-1}) = tr(I_s) = s\ ,$$

since $tr(BC) = tr(CB)$ for any $n \times s$ matrix B and $s \times n$ matrix C. Hence, the denominator of (11.53) converges in probability to σ. By Lemma 11.3.3, the numerator of (11.53) converges in distribution to $N(0, \sigma^2)$ provided

$$\max d_i^2 \to 0 \quad as\ \ n \to \infty\ . \tag{11.54}$$

Under this condition, the level of the t-test is therefore robust against nonnormality.

So far, $b = (b_1, \ldots, b_s)^T$ has been fixed. To determine when the level of (11.53) is robust for all b with $\sum b_j^2 = 1$, it is only necessary to find the maximum value of d_i^2 as b varies. By the Schwarz inequality

$$d_i^2 = \left(\sum_j a_{i,j} b_j\right)^2 \leq \sum_{j=1}^{s} a_{i,j}^2\ ,$$

with equality holding when $b_j = a_{i,j}/\sqrt{\sum_k a_{i,k}^2}$. ,The desired maximum of d_i^2 is therefore $\sum_j a_{i,j}^2$, and

$$\max_i \sum_{j=1}^{s} a_{i,j}^2 \to 0 \quad as\ \ n \to \infty \tag{11.55}$$

is a sufficient condition for the asymptotic normality of every $\hat{\theta}$ of the form (11.51).

The condition (11.55) depends on the particular parametrization (11.46) chosen for Π_Ω. Note however that

$$\sum_{j=1}^{s} a_{i,j}^2 = \Pi_{i,i} \ , \tag{11.56}$$

where $\Pi_{i,j}$ is the (i,j) element of the projection matrix P.

This shows that the value of $\Pi_{i,i}$ is coordinate free, i.e. it is unchanged by an arbitrary change of coordinates $\beta^* = B^{-1}\beta$, where B is a nonsingular matrix, since

$$\xi = A\beta = AB\beta^* = A^*\beta^*$$

with $A^* = AB$, and

$$P^* = AB(B^T A^T AB)^{-1} B^T A^T = ABB^{-1}(A^T A)^{-1}(B^T)^{-1}BA = P \ .$$

Hence, (11.55) is equivalent to the coordinate-free Huber condition

$$\max_i \Pi_{i,i} \to 0 \quad as \ \ n \to \infty \ . \tag{11.57}$$

For evaluating $\Pi_{i,i}$, it is helpful to note that

$$\hat{\xi}_i = \sum_{j=1}^{n} \Pi_{i,j} X_j \quad (i = 1, \ldots, n),$$

so that $\Pi_{i,i}$ is simply the coefficient of X_i in $\hat{\xi}_i$, which must be calculated in any case to carry out the test.

If $\Pi_{i,i} \le M_n$ for all $i = 1, \ldots, n$, then also $\Pi_{i,j} \le M_n$ for all i and j. This follows from the fact that there exists a nonsingular E with $P = EE^T$, on applying the Cauchy-Schwarz inequality to the (i,j) element of EE^T. Condition (11.57) is therefore equivalent to

$$\max_{i,j} \Pi_{i,j} \to 0 \quad as \ \ n \to \infty \ . \tag{11.58}$$

Example 11.3.2 (Example 11.3.1, continued) In Example 11.3.1, the coefficient of X_i in $\hat{\xi}_i = \hat{\alpha} + \hat{\beta} t_i$ is

$$\Pi_{i,i} = \frac{1}{n} + \frac{(t_i - \bar{t})^2}{\sum (t_j - \bar{t})^2}$$

and the Huber condition reduces to the condition (11.49) found earlier. ■

Example 11.3.3 (Two-way Layout) Consider the two-way layout with m observations per cell and the additive model

$$\xi_{i,j,k} = E(X_{i,j,k}) = \mu + \alpha_i + \beta_j$$

with

$$\sum_i \alpha_i = \sum_j \beta_j = 0 \ ,$$

$i = 1, \ldots, a$; $j = 1, \ldots b$; $k = 1, \ldots m$. It is easily seen (Problem 11.75) that, for fixed a and b, the Huber condition is satisfied as $m \to \infty$. ■

Let us next generalize the hypothesis (11.50) to hypotheses which impose several linear constraints such as (11.50). Without loss of generality, choose the parametrization in (11.46) in such a way that the s columns of A are orthogonal and of length one and make the transformation

$$Y = CX$$

(used in (7.1), where C is orthogonal and the first s rows of C are equal to those of A^T, say

$$C = \begin{pmatrix} A^T \\ D \end{pmatrix} \tag{11.59}$$

for some $(n-s) \times n$ matrix D. If $\eta_i = E(Y_i)$, we then have that

$$\eta = \begin{pmatrix} A^T \\ D \end{pmatrix} A\beta = (\beta_1, \ldots, \beta_s, 0, \ldots, 0)^T . \tag{11.60}$$

By the orthogonality of C, the Y_i are independent with Y_i distributed as $N(\eta_i, \sigma^2)$, where $\eta_i = \beta_i$ for $i = 1, \ldots, s$ and $\eta_i = 0$ for $i = s+1, \ldots, n$. We want to test

$$H: \sum_{j=1}^{s} \alpha_{i,j}\eta_j = 0 ; \qquad i = 1, \ldots, r$$

where we shall assume that the r vectors $(\alpha_{i,1}, \ldots, \alpha_{i,s})^T$ are orthogonal and of length one. Then the variables

$$Z_i = \begin{cases} \sum_{j=1}^{n} \alpha_{i,j} Y_j & i = 1, \ldots, r \\ Y_i & i = s+1, \ldots, n \end{cases} \tag{11.61}$$

are independent $N(\zeta_i, \sigma^2)$ with

$$\zeta_i = \begin{cases} \sum_{j=1}^{s} \alpha_{i,j}\eta_j & i = 1, \ldots, r \\ \eta_i & i = r+1, \ldots, s \\ 0 & i = s+1, \ldots, n \end{cases} \tag{11.62}$$

The standard UMPI test of $H: \zeta_1 = \cdots = \zeta_r = 0$ rejects when

$$\frac{\sum_{i=1}^{r} Z_i^2 / r}{\sum_{j=s+1}^{n} Z_j^2/(n-s)} > k , \tag{11.63}$$

where k is determined so that the probability of (11.63) is α when the Zs are normal and H holds.

We shall now suppose that the model (11.46) is embedded in a sequence of such models defined by matrices $A_{i,j}^{(n)}$, with s fixed and $n \to \infty$. Suppose that the Xs are not normal but given by

$$X_i = U_i + \xi_i ,$$

where the Us are i.i.d. according to a distribution F with mean 0 and variance $\sigma^2 < \infty$. We then have the following robustness result.

Theorem 11.3.1 *Let $\alpha_n(F)$ denote the rejection probability of the test (11.63) when the Us have distribution F and the null hypothesis constraints are satisfied.*

Then, $\alpha_n(F) \to \alpha$ *provided*

$$\max_i \sum_{j=1}^{s} (a_{i,j}^{(n)})^2 \to 0 \tag{11.64}$$

or equivalently

$$\max \Pi_{i,i}^{(n)} \to 0 ,$$

where $\Pi_{i,i}^{(n)}$ *is the ith diagonal element of* $P = A(A^T A)^{-1} A^T$.

PROOF. We must show that the limiting distribution of (11.63) is the same as when F is normal. First, we shall show that the denominator of (11.63) satisfies

$$\frac{1}{n-s} \sum_{j=s+1}^{n} Z_j^2 \xrightarrow{P} \sigma^2 . \tag{11.65}$$

Note that $X = C^T Y$ and $Y = QZ$ where C^T and Q are both orthogonal. Therefore,

$$\frac{1}{n-s} \sum_{j=s+1}^{n} Z_j^2 = \frac{n}{n-s} \left[\frac{1}{n} \sum_{i=1}^{n} Z_i^2 \right] - \frac{1}{n-s} \sum_{i=1}^{s} Z_i^2$$

$$= \frac{n}{n-s} \cdot \frac{1}{n} \sum_{i=1}^{n} X_i^2 - \frac{1}{n-s} \sum_{i=1}^{s} Z_i^2 .$$

To see that this tends to σ^2 in probability, we first show that

$$\frac{1}{n} \sum_{i=1}^{n} X_i^2 \xrightarrow{P} \sigma^2 .$$

But,

$$\frac{\sum_{i=1}^{n} X_i^2}{n} = \frac{\sum_{i=1}^{n} (X_i - \xi_i)^2}{n} - \frac{2\sum_{i=1}^{n} \xi_i X_i}{n} + \frac{\sum_{i=1}^{n} \xi_i^2}{n} .$$

The first term on the right tends to σ^2 in probability, by the Weak Law of Large Numbers. By the orthogonality of C, the last term is equal to $\sum_{i=1}^{s} \beta_i^2 / n$, which tends to 0 since s is fixed. It is easily checked that the middle term has a mean and variance which tend to 0. Hence, $\sum X_i^2/n$ tends in probability to σ^2. Next, we show that

$$\frac{\sum_{i=1}^{s} Z_i^2}{n} \xrightarrow{P} 0 .$$

It suffices to show

$$\frac{\sum_{i=1}^{s} E(Z_i^2)}{n} = \frac{\sum_{i=1}^{s} Var(Z_i)}{n} + \frac{\sum_{i=1}^{s} [E(Z_i)]^2}{n} \to 0 .$$

Since s is fixed and $Var(Z_i) = \sigma^2$, we only need to show

$$\frac{\sum_{i=1}^{s} [E(Z_i)]^2}{n} \to 0 .$$

For $i \le r$,

$$E(Z_i) = \sum_{j=1}^{s} \alpha_{i,j} \eta_j = \sum_{j=1}^{s} \alpha_{i,j} \beta_j$$

and

$$[E(Z_i)]^2 \le \sum_{j=1}^{s} \alpha_{i,j}^2 \sum_{j=1}^{s} \beta_j^2 = \sum_{j=1}^{s} \beta_j^2 \ .$$

For $r+1 \le i \le s$, $E(Z_i) = \beta_i$, in which case the same bound holds. Therefore,

$$\frac{\sum_{i=1}^{s} [E(Z_i)]^2}{n} \le \frac{s \sum_{j=1}^{s} \beta_j^2}{n} \to 0 \ ,$$

and the result (11.65) follows.

Next, we consider the numerator of (11.63). We show the joint asymptotic normality of $(Z_1, \ldots, Z_r)$. By the Cramér-Wold device, it suffices to show that, for any constants $\gamma_1, \ldots, \gamma_r$ with $\sum_i \gamma_i^2 = 1$,

$$\sum_{i=1}^{r} \gamma_i Z_i \xrightarrow{d} N(0, \sigma^2) \ .$$

Indeed, since the columns of A are orthogonal, $\hat{\beta}_i = Y_i$ for $1 \le i \le s$ and so Z_i is a linear combination of $\hat{\beta}_1, \ldots, \hat{\beta}_s$. But then so is $\sum_i \gamma_i Z_i$ and asymptotic normality follows from the argument for $\hat{\theta}$ of the form (11.51). ■

Example 11.3.4 (Test of Homogeneity) Let $X_{i,j}$ ($j = 1, \ldots n_i$; $i = 1, \ldots, s$) be independently distributed as $N(\mu_i, \sigma^2)$. The problem is to test the null hypothesis

$$H: \ \mu_1 = \cdots = \mu_s \ .$$

In this case, the test (11.63) is UMP invariant and reduces to

$$W^* = \frac{\sum n_i (X_{i\cdot} - X_{\cdot\cdot})^2 / (s-1)}{\sum \sum (X_{i,j} - X_{i\cdot})^2 / (n-s)} \ , \tag{11.66}$$

where

$$X_{i\cdot} = \sum_j X_{i,j} / n_i \ , \qquad X_{\cdot\cdot} = \sum_i \sum_j X_{i,j} / n$$

and $n = \sum_i n_i$. If instead of $X_{i,j}$ being $N(\mu_i, \sigma^2)$, assume that $X_{i,j}$ has a distribution $F(x - \mu_i)$, where F is an arbitrary distribution with finite variance. Then, the theorem implies that, if $\min_i n_i \to \infty$, then the rejection probability tends to α. In fact, the distributions may even vary within each sample, but it is important that the different samples have a common variance or the result fails; see Problems 11.85 and 11.86. ■

11.4 Nonparametric Mean

11.4.1 Edgeworth Expansions

Suppose $X_1, \ldots, X_n$ are i.i.d. with c.d.f. F. Let $\mu(F)$ denote the mean of F, and consider the problem of testing $\mu(F) = 0$. As in Section 11.3.1, let $\alpha_n(F)$ denote the actual rejection probability of the one-sided t-test under F. It was seen that the t-test is pointwise consistent in level in the sense that $\alpha_n(F) \to \alpha$ whenever F has a finite nonzero variance $\sigma^2(F)$. We shall now examine the rate at which the difference $\alpha_n(F) - \alpha$ tends to 0.

In order to study this problem, we will consider expansions of the distribution function of the sample mean, as well as its studentized version. Such expansions are known as *Edgeworth expansions*. Let $\Phi(\cdot)$ denote the standard normal c.d.f. and $\varphi(\cdot)$ the standard normal density. Also let

$$\gamma = \gamma(F) = \frac{E_F[(X_i - \mu(F))^3]}{\sigma^3(F)}$$

and

$$\kappa = \kappa(F) = \frac{E_F[X_i - \mu(F))^4]}{\sigma^4(F)} - 3 .$$

The values γ and κ are known as the *skewness* and *kurtosis* of F, respectively.

Theorem 11.4.1 *Assume $E_F(|X_i|^{k+2}) < \infty$. Let ψ_F denote the characteristic function of F, and assume*

$$\limsup_{|s| \to \infty} |\psi_F(s)| < 1 . \tag{11.67}$$

Then,

$$P_F\{\frac{n^{1/2}[\bar{X}_n - \mu(F)]}{\sigma(F)} \le x\} = \Phi(x) + \sum_{j=1}^{k} n^{-j/2} \varphi(x) p_j(x, F) + r_n(x, F) , \tag{11.68}$$

where $r_n(x, F) = o(n^{-k/2})$ and $p_j(x, F)$ is a polynomial in x of degree $3j - 1$ which depends on F through its first $j + 2$ moments. In particular,

$$p_1(x, F) = -\frac{1}{6}\gamma(x^2 - 1) , \tag{11.69}$$

and

$$p_2(x, F) = -x\left[\frac{1}{24}\kappa(x^2 - 3) + \frac{1}{72}\gamma^2(x^4 - 10x^2 + 15)\right] . \tag{11.70}$$

Moreover, the expansion holds uniformly in x in the sense that, for fixed F,

$$n^{-k/2} \sup_x |r_n(x, F)| \to 0 \quad \text{as } n \to \infty.$$

The assumption (11.67) is known as *Cramér's condition* and can be viewed as a smoothness assumption on F; it holds, for example, if F is absolutely continuous (or more generally is nonsingular) but fails if F is a lattice distribution, i.e. X_1 can only take on values of the form $a + jb$ for some fixed a and b as j varies through the integers. A proof of Theorem 11.4.1 can be found in Feller (1971, Section XVI.4) or Bhattacharya and Rao (1976), who also provide formulae for the $p_j(x, F)$ when $j > 2$. The proofs hinge on expansions of characteristic function.

Note that the term of order $n^{-1/2}$ is zero if and only if the underlying skewness $\gamma(F)$ is zero. This shows that the dominant error in using a standard normal approximation to the distribution of the standardized sample mean is due to skewness of the underlying distribution. Expansions such as these hold for many classes of statistics and provide more information than a weak convergence result, such as that provided by the Central Limit Theorem. As an example, the following result provides an Edgeworth expansion for the studentized sample mean. Let $S_n^2 = \sum_i (X_i - \bar{X}_n)^2/(n-1)$.

Theorem 11.4.2 *Assume $E_F(|X_i|^{k+2}) < \infty$ and that F is absolutely continuous.*[8] *Then, uniformly in t,*

$$P_F\{\frac{n^{1/2}[\bar{X}_n - \mu(F)]}{S_n} \le t\} = \Phi(t) + \sum_{j=1}^{k} n^{-j/2}\varphi(t)q_j(t, F) + \bar{r}_n(t, F) \ , \quad (11.71)$$

where $n^{-k/2} \sup_t |\bar{r}_n(t, F)| \to 0$ and $q_j(t, F)$ is a polynomial which depends on F through its first $j+2$ moments. In particular,

$$q_1(t, F) = \frac{1}{6}\gamma(2t^2 + 1) \ , \quad (11.72)$$

and

$$q_2(t, F) = t\left[\frac{1}{12}\kappa(t^2 - 3) - \frac{1}{18}\gamma^2(t^4 + 2t^2 - 3) - \frac{1}{4}(t^2 + 1)\right] \ . \quad (11.73)$$

Example 11.4.1 (Expansion for the t-distribution) Suppose F is normal $N(\mu, \sigma^2)$. Let $t_n = n^{1/2}(\bar{X}_n - \mu)/S_n$. Then, $\gamma(F) = \kappa(F) = 0$. By Theorem 11.4.2,

$$P_F\{t_n \le t\} = \Phi(t) - \frac{1}{4n}(t + t^3)\varphi(t) + o(n^{-1}) \ . \quad (11.74)$$

This result implies a corresponding expansion for the quantiles of the t-distribution, known as a *Cornish-Fisher expansion.* Specifically, let $t = t_{n-1,1-\alpha}$ be the $1-\alpha$ quantile of the t-distribution with $n-1$ degrees of freedom. We would like to determine $c = c_{1-\alpha}$ such that

$$t_{n-1,1-\alpha} = z_{1-\alpha} + \frac{c_{1-\alpha}}{n} + o(n^{-1}) \ .$$

When $t = t_{n-1,1-\alpha}$, the left side of (11.74) is $1-\alpha$ and the right side is by a Taylor expansion,

$$\Phi(z) + \frac{c}{n}\varphi(z) - \frac{1}{4n}(z + z^3)\varphi(z) + o(n^{-1}) \ ,$$

where $z = z_{1-\alpha}$. Since $\Phi(z) = 1 - \alpha$, we must have

$$\frac{c}{n}\varphi(z) - \frac{1}{4n}(z + z^3)\varphi(z) = o(n^{-1})$$

---

[8] Alternatively, one can assume $E_F(|X_i|^{2j+2}) < \infty$ and the distribution of (X_i, X_i^2) satisfies the multivariate analogue of Cramér's condition; see Hall (1992), Chapter 2.

so that

$$c = c_{1-\alpha} = \frac{1}{4} z_{1-\alpha}(1 + z_{1-\alpha}^2) .$$

Therefore,

$$n(t_{n-1,1-\alpha} - z_{1-\alpha}) \to \frac{1}{4} z_{1-\alpha}(1 + z_{1-\alpha}^2) . \blacksquare \tag{11.75}$$

In Section 11.3.1, we showed that the t-test has error in rejection probability tending to 0 as long as the underlying distribution has a finite nonzero variance. We will now make use of Edgeworth expansions in order to determine the orders of error in rejection probability for tests of the mean. All tests considered are based on the t-statistic t_n. In order to study this problem, we consider three factors: the one-sided case which rejects for large t_n versus the two-sided case which rejects for large $|t_n|$; the use of a normal critical value versus a t critical value; and the dependence on F, especially whether $\gamma(F)$ is 0 or not. For $j = 1, 2$, let $\alpha_{n,j}^z(F)$ denote the error in rejection probability under F of the j-sided test using the normal quantile, and let $\alpha_{n,j}^t(F)$ denote the analogous quantity using the appropriate t-quantile. For example,

$$\alpha_{n,2}^t(F) = P_F\{|t_n| \geq t_{n-1,1-\frac{\alpha}{2}}\} .$$

We assume $E_F(X_i^4) < \infty$ and that F is absolutely continuous so that we can apply the Edgeworth expansions in Theorems 11.4.1 and 11.4.2 with $k = 2$.

The One-sided Case. First, consider the test using the normal quantile. By (11.71),

$$\alpha_{n,1}^z(F) - \alpha = n^{-1/2}\varphi(z_{1-\alpha})q_1(z_{1-\alpha}, F) + n^{-1}\varphi(z_{1-\alpha})q_2(z_{1-\alpha}, F) + o(n^{-1}) .$$

It follows that

$$\alpha_{n,1}^z(F) - \alpha = O(n^{-1/2}) .$$

However, if $\gamma(F) = 0$, then $q_1(z_{1-\alpha}, F) = 0$ and so

$$\alpha_{n,1}^z(F) - \alpha = O(n^{-1})$$

in this case. Using the t-quantiles instead of the normal quantiles yields

$$\alpha_{n,1}^t(F) - \alpha = \Phi(t_{n-1,\alpha}) - \alpha + n^{-1/2}\varphi(t_{n-1,1-\alpha})q_1(t_{n-1,1-\alpha}, F) + O(n^{-1}) .$$

Then, applying (11.75), $t_{n-1,1-\alpha} - z_{1-\alpha} = O(n^{-1})$, so that a Taylor's expansion yields

$$\alpha_{n,1}^t(F) - \alpha = n^{-1/2}\varphi(z_{1-\alpha})q_1(z_{1-\alpha}, F) + O(n^{-1}) .$$

Therefore,

$$\alpha_{n,1}^t(F) - \alpha = O(n^{-1/2}) ,$$

but the error in rejection probability is $O(n^{-1})$ if $\gamma(F) = 0$.

The Two-sided Case. Let $z = z_{1-\frac{\alpha}{2}}$. Then, using the fact that $\varphi(z) = \varphi(-z)$,

$$\alpha_{n,2}^z(F) = P_F\{|t_n| \geq z\} = 1 - [P_F\{t_n \leq z\} - P_F\{t_n \leq -z\}]$$

$$= \alpha + n^{-1/2}\varphi(z)[q_1(z,F) - q_1(-z,F)] + O(n^{-1}) .$$

But, $q_1(\cdot, F)$ is an even function, which implies

$$\alpha^z_{n,2}(F) - \alpha = O(n^{-1}) ,$$

even if $\gamma(F)$ is not zero. Similarly, it can be shown that (Problem 11.90)

$$\alpha^t_{n,2}(F) - \alpha = O(n^{-1}) . \tag{11.76}$$

11.4.2 The t-test

It was seen in Section 11.3.1 that the classical t-test of the mean is asymptotically pointwise consistent in level for the class $\mathbf{F}$ of all distributions with finite nonzero variance. In Section 11.4.1, the orders of error in rejection probability were obtained for a given F. However, these results are not reassuring unless the convergence is uniform in F. If it is not, then for any n, no matter how large, there will exist F in $\mathbf{F}$ for which the rejection probability under F, $\alpha_n(F)$, is not even close to α. We shall show below that the convergence is not uniform and that the situation is even worse than what this negative result suggests. Namely, we shall show that for any n, there exist distributions F for which $\alpha_n(F)$ is arbitrarily close to 1; that is, the size of the t-test is 1.

Suppose $X_1, \ldots, X_n$ are i.i.d. real-valued random variables with unknown c.d.f. $F \in \mathbf{F}$, where $\mathbf{F}$ is a large nonparametric class of distributions. Let $\mu(F)$ denote the mean of F and $\sigma^2(F)$ the variance of F. The goal is to test the null hypothesis $\mu(F) = 0$ versus $\mu(F) > 0$, or perhaps the two-sided alternative $\mu(F) \neq 0$.

Theorem 11.4.3 *For every n, the size of the t-test is 1 for the family $\mathbf{F}_0$ of all distributions with finite variance.*

PROOF. Let c be an arbitrary positive constant less than one and let $p_n = 1 - c^{1/n}$ so that $(1-p_n)^n = c$. Let $F = F_{n,c}$ be the distribution that places mass $1 - p_n$ at p_n and mass p_n at $p_n - 1$, so that $\mu(F) = 0$. With probability c, we have all observations equal to p_n. For such a sample, the numerator $n^{1/2}\bar{X}_n$ of the t-statistic is $n^{1/2}p_n > 0$ while the denominator is 0. Thus, the t-statistic blows up and the hypothesis will be rejected. The probability of rejection is therefore $\geq c$, and by taking c arbitrarily close to 1 the theorem is proved. (Note that one can modify the distributions $F_{n,c}$ used in the proof to be continuous rather than discrete.) ■

It follows that the t-test is not even uniformly asymptotically level α for the family $\mathbf{F}_0$.

Instead of $\mathbf{F}_0$, one may wish to consider the behavior of the t-test against other nonparametric families. If $\mathbf{F}_2$ is the family of all symmetric distributions with finite variance, it turns out that the t-test is still not uniformly level α, and this is true even if the symmetric distributions have their support on $(-1, 1)$ or any other fixed compact set; see Romano (2004). In fact, the size of the t-test under symmetry is one for moderate values of α; see Basu and DasGupta (1995). However, it can be shown that the size of the t-test is bounded away from 1 for small values of α, by a result of Edelman (1990). Basu and DasGupta (1995) also show that if $\mathbf{F}_3$ is the family of all symmetric unimodal distributions (with no moment restrictions), then the largest rejection probability under F of the t-test occurs when F is uniform on $[-1, 1]$, at least in the case of very small α.

On the other hand, we will now show that the t-test is uniformly consistent over certain large subfamilies of distributions with two finite moments. For this purpose, consider a family of distributions $\tilde{\mathbf{F}}$ on the real line satisfying

$$\lim_{\lambda \to \infty} \sup_{F \in \tilde{\mathbf{F}}} E_F \left[\frac{|X - \mu(F)|^2}{\sigma^2(F)} I \left\{ \frac{|X - \mu(F)|}{\sigma(F)} > \lambda \right\} \right] = 0 . \tag{11.77}$$

For example, for any $\epsilon > 0$ and $b > 0$, let $\mathbf{F}_b^{2+\epsilon}$ be the set of distributions satisfying

$$E_F \left[\frac{|X - \mu(F)|^{2+\epsilon}}{\sigma^{2+\epsilon}(F)} \right] \leq b .$$

Then, $\tilde{\mathbf{F}} = \mathbf{F}_b^{2+\epsilon}$ satisfies (11.77). To see why, take expectations of both sides of the inequality

$$\lambda^\epsilon Y^2 I\{|Y| > \lambda\} \leq |Y|^{2+\epsilon} .$$

Lemma 11.4.1 *Suppose $X_{n,1}, \ldots, X_{n,n}$ are i.i.d. F_n with $F_n \in \tilde{\mathbf{F}}$, where $\tilde{\mathbf{F}}$ satisfies (11.77). Let $\bar{X}_n = \sum_{i=1}^n X_{n,i}/n$. Then, under F_n,*

$$\frac{n^{1/2}[\bar{X}_n - \mu(F_n)]}{\sigma(F_n)} \xrightarrow{d} N(0, 1) .$$

PROOF. Let $Y_{n,i} = [X_{n,i} - \mu(F_n)]/\sigma(F_n)$. We verify the Lindeberg Condition (11.11), which in the case of n i.i.d. variables reduces to showing

$$\limsup_n E[Y_{n,i}^2 I\{|Y_{n,i}| > \epsilon n^{1/2}\}] = 0$$

for every $\epsilon > 0$. But, for every $\lambda > 0$,

$$\limsup_n E[Y_{n,i}^2 I\{|Y_{n,i}| > \epsilon n^{1/2}\}] \leq \limsup_n E[Y_{n,i}^2 I\{|Y_{n,i}| > \lambda\}] .$$

Let $\lambda \to \infty$ and the right side tends to zero. ■

Lemma 11.4.2 *Let $Y_{n,1}, \ldots, Y_{n,n}$ be i.i.d. with c.d.f. G_n and finite mean $\mu(G_n)$ satisfying*

$$\lim_{\beta \to \infty} \limsup_{n \to \infty} E_{G_n} [|Y_{n,i} - \mu(G_n)| I\{|Y_{n,i} - \mu(G_n)| \geq \beta\}] = 0 . \tag{11.78}$$

Let $\bar{Y}_n = \sum_{i=1}^n Y_{n,i}/n$. Then, under G_n, $\bar{Y}_n - \mu(G_n) \to 0$ in probability.

PROOF. Without loss of generality, assume $\mu(G_n) = 0$. Define

$$Z_{n,i} = Y_{n,i} I\{|Y_{n,i}| \leq n\} .$$

Let $m_n = E(Z_{n,i})$ and $\bar{Z}_n = \sum_{i=1}^n Z_{n,i}/n$. Then, the event $\{|\bar{Y}_n - m_n| > \epsilon\}$ implies either $\{|\bar{Z}_n - m_n| > \epsilon\}$ occurs or $\{\bar{Y}_n \neq \bar{Z}_n\}$ occurs. Hence, for any $\epsilon > 0$,

$$P\{|\bar{Y}_n - m_n| > \epsilon\} \leq P\{|\bar{Z}_n - m_n| > \epsilon\} + P\{\bar{Y}_n \neq \bar{Z}_n\} . \tag{11.79}$$

The last term is bounded above by

$$P\{\bigcup_{i=1}^n \{Y_{n,i} \neq Z_{n,i}\}\} \leq \sum_{i=1}^n P\{Y_{n,i} \neq Z_{n,i}\} = nP\{|Y_{n,i}| > n\} .$$

The first term on the right side of (11.79) can be bounded by Chebyshev's inequality, so that

$$P\{|\bar{Y}_n - m_n| > \epsilon\} \le (n\epsilon^2)^{-1} E(Z_{n,1}^2) + nP\{|Y_{n,1}| > n\} . \tag{11.80}$$

For $t > 0$, let

$$\tau_n(t) = t[1 - G_n(t) + G_n(-t)]$$

and

$$\kappa_n(t) = \frac{1}{t}\int_{-t}^{t} x^2 dG_n(t) = -\tau_n(t) + \frac{2}{t}\int_0^t \tau_n(x)dx ; \tag{11.81}$$

the last equality follows by integration by parts (Problem 11.96) and corrects (7.7), p.235 of Feller (1971). Hence,

$$P\{|\bar{Y}_n - m_n| > \epsilon\} \le \epsilon^{-2}\kappa_n(n) + \tau_n(n) . \tag{11.82}$$

But, for any $t > 0$,

$$\tau_n(t) \le E[|Y_{n,1}|I\{|Y_{n,1}| \ge t\}] ,$$

so $\tau_n(n) \to 0$ by (11.78). Fix any $\delta > 0$ and let β_0 be such that

$$\limsup_n E\left[|Y_{n,1}|I\{|Y_{n,1}| > \beta_0\}\right] < \frac{\delta}{4} .$$

Then, there is an n_0 such that, for all $n \ge n_0$,

$$E\left[|Y_{n,1}|I\{|Y_{n,1}| > \beta_0\}\right] < \frac{\delta}{2} ,$$

and so

$$E|Y_{n,1}| \le \beta_0 + \frac{\delta}{2}$$

for all $n \ge n_0$ as well. Then, if $n \ge n_0 > \beta_0$,

$$\frac{1}{n}\int_0^n \tau_n(x)dx \le \frac{1}{n}\int_0^n E\left[|Y_{n,1}|I\{|Y_{n,1}| \ge x\}\right]dx$$

$$\le \frac{1}{n}\int_0^{\beta_0} E|Y_{n,1}|dx + \frac{1}{n}\int_{\beta_0}^n \frac{\delta}{2}dx \le \frac{\beta_0(\beta_0 + \frac{\delta}{2})}{n} + \frac{\delta}{2} ,$$

which is less than δ for all sufficiently large n. Thus, $\kappa_n(n) \to 0$ as $n \to \infty$ and so (11.82) tends to 0 as well. Therefore, $\bar{Y}_n - m_n \to 0$ in probability. Finally, $m_n \to 0$; to see why, observe

$$0 = E(Y_{n,i}) = m_n + E\left[Y_{n,1}I\{|Y_{n,1}| > n\}\right] ,$$

so that

$$|m_n| \le E\left[|Y_{n,1}|I\{|Y_{n,1}| > n\}\right] \to 0 ,$$

by assumption (11.78). ■

Lemma 11.4.3 *Let $\tilde{\mathbf{F}}$ be a family of distributions satisfying (11.77). Suppose $X_{n,1}, \ldots, X_{n,n}$ are i.i.d. $F_n \in \tilde{\mathbf{F}}$ and $\mu(F_n) = 0$. Then, under F_n,*

$$\frac{\frac{1}{n}\sum_{i=1}^n X_{n,i}^2}{\sigma^2(F_n)} \to 1 \quad \textit{in probability.}$$

PROOF. Apply Lemma 11.4.2 to $Y_{n,i} = [X_{n,i}^2/\sigma^2(F_n)] - 1$. To see that Lemma 11.4.2 applies, note that if $\beta > 1$, then the event $\{|Y_{n,i}| > \beta\}$ implies $X_{n,i}^2/\sigma^2(F_n) > \beta + 1$ (since $X_{n,i}^2/\sigma^2(F_n) > 0$) and also $|Y_{n,i}| < X_{n,i}^2/\sigma^2(F_n)$. Hence, for $\beta > 1$,

$$E\left[|Y_{n,i}|I\{|Y_{n,i}| \geq \beta\}\right] \leq E\left[\frac{X_{n,i}^2}{\sigma^2(F_n)} I\{\frac{|X_{n,i}|}{\sigma(F_n)} > \sqrt{\beta+1}\}\right] .$$

The sup over n then tends to 0 as $\beta \to \infty$ by the assumption $F_n \in \tilde{\mathbf{F}}$. ■

We are now in a position to study the behavior of the t-test uniformly across a fairly large class of distributions.

Theorem 11.4.4 *Let $F_n \in \tilde{\mathbf{F}}$, where $\tilde{\mathbf{F}}$ satisfies (11.77). Assume*

$$n^{1/2}\mu(F_n)/\sigma(F_n) \to \delta \quad \text{as } n \to \infty$$

(where $|\delta|$ is allowed to be ∞). Let $X_1, \ldots, X_n$ be i.i.d. with c.d.f F_n, and consider the t-statistic

$$t_n = n^{1/2}\bar{X}_n/S_n ,$$

where $\bar{X}_n$ is the sample mean and S_n^2 is the sample variance. If $|\delta| < \infty$, then under F_n,

$$t_n \xrightarrow{d} N(\delta, 1) .$$

If $\delta \to \infty$ (respectively, $-\infty$), then $t_n \to \infty$ (respectively, $-\infty$) in probability under F_n.

PROOF. Write

$$t_n = \frac{n^{1/2}[\bar{X}_n - \mu(F_n)]}{S_n} + \frac{n^{1/2}\mu(F_n)/\sigma(F_n)}{S_n/\sigma(F_n)} .$$

The proof will follow if we show $S_n/\sigma(F_n) \to 1$ in probability under F_n and if

$$\frac{n^{1/2}[\bar{X}_n - \mu(F_n)]}{\sigma(F_n)} \xrightarrow{d} N(0,1) . \tag{11.83}$$

But the latter follows by Lemma 11.4.1. To show $S_n^2/\sigma^2(F_n) \to 1$ in probability, use Lemma 11.4.3 (Problem 11.93). ■

Theorem 11.4.4 now allows us to deduce that the t-test is uniformly consistent in level, and it also yields a limiting power calculation.

Theorem 11.4.5 *Let $\tilde{\mathbf{F}}$ satisfy (11.77) and let $\tilde{\mathbf{F}}_0$ be the set of F in $\tilde{\mathbf{F}}$ with $\mu(F) = 0$. For testing $\mu(F) = 0$ versus $\mu(F) > 0$, the t-test that rejects when $t_n > z_{1-\alpha}$ (or $t_{n-1,1-\alpha}$) is uniformly asymptotically level α over $\tilde{\mathbf{F}}_0$; that is,*

$$|\sup_{F \in \tilde{\mathbf{F}}_0} P_F\{t_n > z_{1-\alpha}\} - \alpha| \to 0 \tag{11.84}$$

as $n \to \infty$. Also, the limiting power against $F_n \in \tilde{\mathbf{F}}$ with $n^{1/2}\mu(F_n)/\sigma(F_n) \to \delta$ is given by

$$\lim_n P_{F_n}\{t_n > z_{1-\alpha}\} = 1 - \Phi(z_{1-\alpha} - \delta) . \tag{11.85}$$

Furthermore,

$$\inf_{\{F \in \tilde{\mathbf{F}}:\ n^{1/2}\mu(F)/\sigma(F) \ge \delta\}} P_F\{t_n > z_{1-\alpha}\} \to 1 - \Phi(z_{1-\alpha} - \delta) \ . \tag{11.86}$$

PROOF. To prove (11.84), if the result failed, one could extract a subsequence $\{F_n\}$ with $F_n \in \tilde{\mathbf{F}}_0$ such that

$$P_{F_n}\{t_n > z_{1-\alpha}\} \to \beta \neq \alpha \ .$$

But this contradicts Theorem 11.4.4 since t_n is asymptotically standard normal under F_n. The proof of (11.85) follows from Theorem 11.4.4 as well. To prove (11.86), again argue by contradiction and assume there exists a subsequence $\{F_n\}$ with $n^{1/2}\mu(F_n)/\sigma(F_n) \ge \delta$ such that

$$P_{F_n}\{t_n > z_{1-\alpha}\} \to \gamma < 1 - \Phi(z_{1-\alpha} - \delta) \ .$$

The result follows from (11.85) if $n^{1/2}\mu(F_n)/\sigma(F_n)$ has a limit; otherwise, pass to any convergent subsequence and apply the same argument. ■

Note that (11.86) does not hold if $\tilde{\mathbf{F}}$ is replaced by all distributions with finite second moments or finite fourth moments, or even the more restricted family of distributions supported on a compact set. In fact, there exists a sequence of distributions $\{F_n\}$ supported on a fixed compact set and satisfying $n^{1/2}\mu(F_n)/\sigma(F_n) \ge \delta$ such that the limiting power of the t-test against this sequence of alternatives is α; see Problem 11.97 for a construction. Nevertheless, the t-test behaves well for typical distributions, as demonstrated in Theorem 11.4.5. However, it is important to realize the t-test does not behave uniformly well across distributions with large skewness, as the limiting normal theory fails.

11.4.3 A Result of Bahadur and Savage

The negative results for the t-test under the families of all distributions with finite variance, or even the family of symmetric distributions with infinitely many moments are perhaps unexpected in view of the fact that the t-test is pointwise consistent in level for any distribution with finite (nonzero) variance, but they should not really be surprising. After all, the t-test was designed for the family of normal distributions and not for nonparametric families. This raises the question whether there do exist more satisfactory tests of the mean for nonparametric families.

For the family of distributions with finite variance and for some related families, this question was answered by Bahadur and Savage (1956). The desired results follows from the following basic lemma.

Lemma 11.4.4 *Let* $\mathbf{F}$ *be a family of distributions on* $\mathbb{R}$ *satisfying:*

(i) For every $F \in \mathbf{F}$, $\mu(F)$ *exists and is finite.*

(ii) For every real m, *there is an* $F \in \mathbf{F}$ *with* $\mu(F) = m$.

(iii) The family $\mathbf{F}$ *is convex in the sense that, if* $F_i \in \mathbf{F}$ *and* $\gamma \in [0,1]$, *then* $\gamma F_1 + (1-\gamma)F_2 \in \mathbf{F}$.

Let $X_1, \ldots, X_n$ be i.i.d. $F \in \mathbf{F}$ and let $\phi_n = \phi_n(X_1, \ldots, X_n)$ be any test function. Let $\mathbf{G}_m$ denote the set of distributions $F \in \mathbf{F}$ with $\mu(F) = m$. Then,

$$\inf_{F \in \mathbf{G}_m} E_F(\phi_n) \quad \text{and} \quad \sup_{F \in \mathbf{G}_m} E_F(\phi_n)$$

are independent of m.

PROOF. To show the result for the sup, fix m_0 and let $F_j \in \mathbf{G}_{m_0}$ be such that

$$\lim_j E_{F_j}(\phi_n) = \sup_{F \in \mathbf{G}_{m_0}} E_F(\phi_n) \equiv s .$$

Fix m_1. The goal is to show

$$\sup_{F \in \mathbf{G}_{m_1}} E_F(\phi_n) = s .$$

Let H_j be a distribution in $\mathbf{F}$ with mean h_j satisfying

$$m_1 = (1 - \frac{1}{j})m_0 + \frac{1}{j}h_j$$

and define

$$G_j = (1 - \frac{1}{j})F_j + \frac{1}{j}H_j .$$

Thus, $G_j \in \mathbf{G}_{m_1}$. An observation from G_j can be obtained through a two-stage procedure. First, a coin is flipped with probability of heads $1/j$. If the outcome is a head, then the observation has the distribution H_j; otherwise, the observation is from F_j. So, with probability $[1 - (1/j)]^n$, a sample of size n from G_j is just a sample from F_j. Then,

$$\sup_{G \in \mathbf{G}_{m_1}} E_G(\phi_n) \geq E_{G_j}(\phi_n) \geq (1 - \frac{1}{j})^n E_{F_j}(\phi_n) \to s$$

as $j \to \infty$. Thus,

$$\sup_{G \in \mathbf{G}_{m_1}} E_G(\phi_n) \geq \sup_{G \in \mathbf{G}_{m_0}} E_G(\phi_n) .$$

Interchanging the roles of m_0 and m_1 and applying the same argument makes the last inequality an equality. The result for the inf can be obtained by applying the argument to $1 - \phi_n$. ■

Theorem 11.4.6 *Let $\mathbf{F}$ satisfy (i)-(iii) of Lemma 11.4.4.*
(i) Any test of $H: \ \mu(F) = 0$ which has size α for the family $\mathbf{F}$ has power $\leq \alpha$ for any alternative F in $\mathbf{F}$.
(ii) Any test of $H: \ \mu(F) = 0$ which has power β against some alternative F in $\mathbf{F}$ has size $\geq \beta$.

Among the families satisfying (i)-(iii) of Lemma 11.4.4 is the family $\mathbf{F}_0$ of distributions with finite second moment and that with infinitely many moments. Part (ii) of the above theorem provides an alternative proof of Theorem 11.4.3 since the power of the t-test against the normal alternatives $N(\mu, 1)$ tends to 1 as $\mu \to \infty$. Theorem 11.4.6 now shows that the failure of the t-test for the family of all distributions with finite variance is not the fault of the t-test; in this setting, there exists no reasonable test of the mean. The reason is that slight changes in the tails of the distribution can result in enormous changes in the mean.

11.4.4 Alternative Tests

Another family satisfying conditions (i)-(iii) of Theorem 11.4.6 is the family of all distributions with compact support. However, the family of all distributions on a fixed compact set is excluded because it does not satisfy Condition (ii). In fact, the following construction due to Anderson (1967), shows that reasonable tests of the mean do exist if we assume the family of distributions is supported on a specified compact set. Specifically, let $\mathbf{G}$ be the family of distributions supported on $[-1, 1]$, and let $\mathbf{G}_0$ be the set of distributions on $[-1, 1]$ having mean 0. We will exhibit a test that has size α for any fixed sample size n and all $F \in \mathbf{G}_0$, and is pointwise consistent in power. First, recall the Kolmogorov-Smirnov confidence band $R_{n,1-\alpha}$ given by (11.36). This leads to a conservative confidence interval $I_{n,1-\alpha}$ for $\mu(F)$ as follows. Include the value μ in $I_{n,1-\alpha}$ if and only if there exists some G in $R_{n,1-\alpha}$ with $\mu(G) = \mu$. Then,

$$\{F \in R_{n,1-\alpha}\} \subset \{\mu(F) \in I_{n,1-\alpha}\}$$

and so

$$P_F\{\mu(F) \in I_{n,1-\alpha}\} \geq P_F\{F \in R_{n,1-\alpha}\} \geq 1 - \alpha ,$$

where the last inequality follows by construction of the Kolmogorov-Smirnov confidence bands. Finally, for testing $\mu(F) = 0$ versus $\mu(F) \neq 0$, let ϕ_n be the test that accepts the null hypothesis if and only if the value 0 falls in $I_{n,1-\alpha}$. By construction,

$$\sup_{F \in \mathbf{G}_0} E_F(\phi_n) \leq \alpha .$$

We claim that

$$I_{n,1-\alpha} \subset \bar{X}_n \pm 2n^{-1/2} s_{n,1-\alpha} , \tag{11.87}$$

where $s_{n,1-\alpha}$ is the $1 - \alpha$ quantile of the null distribution of the Kolmogorov-Smirnov test statistic. The result (11.87) follows from the following lemma.

Lemma 11.4.5 *Suppose F and G are distributions on $[-1, 1]$ with*

$$\sup_t |F(t) - G(t)| \leq \epsilon .$$

Then, $|\mu(F) - \mu(G)| \leq 2\epsilon$.

For a proof, see Problem 11.94. The result (11.87) now follows by applying the lemma to F and the empirical cdf $\hat{F}_n$.

Let F be a distribution with mean $\mu(F) \neq 0$. Suppose without loss of generality that $\mu(F) > 0$. Also, let $L_{n,1-\alpha}$ be the lower endpoint of the interval $I_{n,1-\alpha}$. Then,

$$E_F(\phi_n) \geq P_F\{L_{n,1-\alpha} > 0\} \geq P_F\{\bar{X}_n > 2n^{-1/2} s_{n,1-\alpha}\} \to 1 , \tag{11.88}$$

by Slutsky's theorem, since $\bar{X}_n \to \mu(F) > 0$ and $n^{-1/2} s_{n,1-\alpha} \to 0$. Thus, the test is pointwise consistent in power against any distribution in $\mathbf{G}$ having nonzero mean. In fact, if $\{F_n\}$ is such that $|n^{1/2}\mu(F_n)| \to \infty$, then the limiting power against such a sequence is one (Problem 11.95). ■

While Anderson's method controls the level and is pointwise consistent in power, it is not efficient; an efficient test construction which is of exact level α can be based on the confidence interval construction of Romano and Wolf (2000).

Let us next consider the family of symmetric distributions. Here the mean coincides with the center of symmetry, and reasonable level α tests for this center exist. They can, for example, be based on the signed ranks. The one-sample Wilcoxon test is an example. A large family of randomization tests that control the level is discussed in 15.2.

Finally, we mention a quite different approach to the problem considered in this section concerning the validity of the t-test in a nonparametric setting. Originally, the t-test was derived for testing the mean, μ, on the basis of a sample $X_1, \ldots, X_n$ from $N(\mu, \sigma^2)$. But, μ is not only the mean of the normal distribution but it is also, for example, its median. Instead of embedding the normal family in the family of all distributions with finite mean (and perhaps finite variance), we could obtain a different viewpoint by embedding it in the family of all continuous distributions F, and then test the hypothesis that the median of F is 0. A suitable test is then the sign test.

11.5 Problems

Section 11.1

Problem 11.1 For each $\theta \in \Omega$, let $f_n(\theta)$ be a real-valued sequence. We say $f_n(\theta)$ converges uniformly (in θ) to $f(\theta)$ if

$$\sup_{\theta \in \Omega} |f_n(\theta) - f(\theta)| \to 0$$

as $n \to \infty$. If Ω if a finite set, show that the pointwise convergence $f_n(\theta) \to f(\theta)$ for each fixed θ implies uniform convergence. However, show the converse can fail even if Ω is countable.

Section 11.2

Problem 11.2 For a univariate c.d.f. F, show that the set of points of discontinuity is countable.

Problem 11.3 Let X be $N(0,1)$ and $Y = X$. Determine the set of continuity points of the bivariate distribution of (X, Y).

Problem 11.4 Show that $x = (x_1, \ldots, x_k)^T$ is a continuity point of the distribution F_X of X if the boundary of the set of $(y_1, \ldots, y_k)$ such that $y_i \leq x_i$ for all i has probability 0 under the distribution of X. Show by example that it is not sufficient for x to have probability 0 under F_X in order for x to be a continuity point.

Problem 11.5 Prove the equivalence of (i) and (vi) in the Portmanteau Theorem (Theorem 11.2.1).

Problem 11.6 Suppose $X_n \xrightarrow{d} X$. Show that $Ef(X_n)$ need not converge to $Ef(X)$ if f is unbounded and continuous, or if f is bounded but discontinuous.

Problem 11.7 Show that the characteristic function of a sum of independent real-valued random variables is the product of the individual characteristic functions. (The converse is false; counterexamples are given in Romano and Siegel (1986), Examples 4.29-4.30.)

Problem 11.8 Verify (11.9).

Problem 11.9 Let X_n have characteristic function ζ_n. Find a counterexample to show that it is not enough to assume $\zeta_n(t)$ converges (pointwise in t) to a function $\zeta(t)$ in order to conclude that X_n converges in distribution.

Problem 11.10 Show that Theorem 11.2.3 follows from Theorem 11.2.2.

Problem 11.11 Show that Lyapounov's Central Limit Theorem (Corollary 11.2.1) follows from the Lindeberg Central Limit Theorem (Theorem 11.2.5).

Problem 11.12 Suppose X_k is a noncentral chi-squared variable with k degrees of freedom and noncentrality parameter δ_k^2. Show that $(X_k - k)/(2k)^{1/2} \xrightarrow{d} N(\mu, 1)$ if $\delta_k^2/(2k)^{1/2} \to \mu$ as $k \to \infty$.

Problem 11.13 Suppose $X_{n,1}, \ldots, X_{n,n}$ are i.i.d. Bernoulli trials with success probability p_n. If $p_n \to p \in (0, 1)$, show that

$$n^{1/2}[\bar{X}_n - p_n] \xrightarrow{d} N(0, p(1-p)) \; .$$

Is the result true even if p is 0 or 1?

Problem 11.14 Let $X_1, \ldots, X_n$ be i.i.d. with density p_0 or p_1, and consider testing the null hypothesis H that p_0 is true. The MP level-α test rejects when $\Pi_{i=1}^n r(X_i) \geq C_n$, where $r(X_i) = p_i(X_i)/p_0(X_i)$, or equivalently when

$$\frac{1}{\sqrt{n}} \left\{ \sum \log r(X_i) - E_0[\log r(X_i)] \right\} \geq k_n. \tag{11.89}$$

(i) Show that, under H, the left side of (11.89) converges in distribution to $N(0, \sigma^2)$ with $\sigma^2 = \mathrm{Var}_0[\log r(X_i)]$, provided $\sigma < \infty$.

(ii) From (i) it follows that $k_n \to \sigma z_{1-\alpha}$, where z_α is the α quantile of $N(0, 1)$.

(iii) The power of the test (11.89) against p_1 tends to 1 as $n \to \infty$. *Hint*: Use Problem 3.39(iv).

Problem 11.15 Complete the proof of Theorem 11.2.8 by considering n even.

Problem 11.16 Generalize Theorem 11.2.8 to the case of the pth sample quantile.

Problem 11.17 Let $X_1, \ldots, X_n$ be i.i.d. normal with mean θ and variance 1. Let $\bar{X}_n$ be the usual sample mean and let $\tilde{X}_n$ be the sample median. Let p_n be the probability that $\bar{X}_n$ is closer to θ than $\tilde{X}_n$ is. Determine $\lim_{n\to\infty} p_n$.

Problem 11.18 Suppose $X_1, \ldots, X_n$ are i.i.d. real-valued random variables with c.d.f. F. Assume $\exists \theta_1 < \theta_2$ such that $F(\theta_1) = 1/4$, $F(\theta_2) = 3/4$, and F is differentiable, with density f taking positive values at θ_1 and θ_2. Show that the sample inter-quartile range (defined as the difference between the .75 quantile and .25 quantile) is a $\sqrt{n}$- consistent estimator of the population inter-quartile range $(\theta_2 - \theta_1)$.

Problem 11.19 Prove Polyá's Theorem 11.2.9. *Hint:* First consider the case of distributions on the real line.

Problem 11.20 Show that $\rho_L(F, G)$ defined in Definition 11.2.3 is a metric; that is, show $\rho_L(F, G) = \rho_L(G, F)$, $\rho_L(F, G) = 0$ if and only if $F = G$, and

$$\rho_L(F, G) \le \rho_L(F, H) + \rho_L(H, G) .$$

Problem 11.21 For cumulative distribution functions F and G on the real line, define the Kolmogorov-Smirnov distance between F and G to be

$$d_K(F, G) = \sup_x |F(x) - G(x)| .$$

Show that $d_K(F, G)$ defines a metric on the space of distribution functions; that is, show $d_K(F, G) = d_K(G, F)$, $d_K(F, G) = 0$ implies $F = G$ and

$$d_K(F, G) \le d_K(F, H) + d_K(H, G) .$$

Also, show that $\rho_L(F, G) \le d_K(F, G)$, where ρ_L is the Lévy metric. Construct a sequence F_n such that $\rho_L(F_n, F) \to 0$ but $d_K(F_n, F)$ does not converge to zero.

Problem 11.22 Let F_n and F be c.d.f.s on $\mathbb{R}$. Show that weak convergence of F_n to F is equivalent to $\rho_L(F_n, F) \to 0$, where ρ_L is the Lévy metric.

Problem 11.23 Suppose F and G are two probability distributions on $\mathbb{R}^k$. Let L be the set of (measurable) functions f from $\mathbb{R}^k$ to $\mathbb{R}$ satisfying $|f(x) - f(y)| \le |x - y|$, where $|\cdot|$ is the usual Euclidean norm. Define the Bounded-Lipschitz Metric as

$$\lambda(F, G) = \sup\{|E_F f(X) - E_G f(X)| : f \in \mathcal{L}\} .$$

Show that $F_n \xrightarrow{d} F$ is equivalent to $\lambda(F_n, F) \to 0$. Thus, weak convergence on $\mathbb{R}^k$ is metrizable. [See examples 21-22 in Pollard (1984).]

Problem 11.24 Construct a sequence of distribution functions $\{F_n\}$ on the real line such that F_n converges in distribution to F, but the convergence $F_n^{-1}(1 -$

$\alpha) \to F^{-1}(1-\alpha)$ fails, even if F is assumed continuous. On the other hand, if F is assumed continuous (but not necessarily strictly increasing), show that

$$F_n(F_n^{-1}(1-\alpha)) \to F(F^{-1}(1-\alpha)) = 1-\alpha \ .$$

[Note the left side need not be $1-\alpha$ since F_n is not assumed continuous.]

Problem 11.25 Prove part (ii) of Lemma 11.2.1.

Problem 11.26 (Markov's Inequality) Let X be a real-valued random variable with $X \geq 0$. Show that, for any $t > 0$,

$$P\{X \geq t\} \leq \frac{E[XI\{X \geq t\}]}{t} \leq \frac{E(X)}{t} \ ;$$

here $I(X \geq t)$ is the indicator variable that is 1 if $X \geq t$ and is 0 otherwise.

Problem 11.27 (Chebyshev's Inequality). (i) Show that, for any real-valued random variable X and any constants $a > 0$ and c,

$$E(X-c)^2 \geq a^2 P\{|X-c| \geq a\} \ .$$

(ii). Hence, if X_n is any sequence of random variables and c is a constant such that $E(X_n - c)^2 \to 0$, then $X_n \to c$ in probability. Give a counterexample to show the converse is false.

Problem 11.28 Give an example of an i.i.d. sequence of real-valued random variables such that the sample mean converges in probability to a finite constant, yet the mean of the sequence does not exist.

Problem 11.29 If $X_n \xrightarrow{P} 0$ and

$$\sup_n E[|X_n|^{1+\delta}] < \infty \quad \text{for some } \delta > 0 \ , \tag{11.90}$$

then show $E[|X_n|] \to 0$. More generally, if the X_n are *uniformly integrable* in the sense $\sup_n E[|X_n| I\{|X_n| > t\}] \to 0$ as $t \to \infty$, then $E[|X_n|] \to 0$. [A converse is given in Dudley (1989), p.279.]

Problem 11.30 Suppose X_n and X are real-valued random variables (defined on a common probability space). Prove that, if X_n converges to X in probability, then X_n converges in distribution to X. Show by counterexample that the converse is false. However, show that if X is a constant with probability one, then X_n converging to X in distribution implies X_n converges to X in probability.

Problem 11.31 Suppose X_n is a sequence of random vectors.
(i). Show $X_n \xrightarrow{P} 0$ if and only if $|X_n| \xrightarrow{P} 0$ (where the first zero refers to the zero vector and the second to the real number zero).
(ii). Show that convergence in probability of X_n to X is equivalent to convergence in probability of their components to the respective components of X.

Problem 11.32 Suppose $X_1, \ldots, X_n$ are i.i.d. real-valued random variables. Write $X_i = X_i^+ - X_i^-$, where $X_i^+ = \max(X_i, 0)$. Suppose X_i^- has a finite mean, but X_i^+ does not. Let $\bar{X}_n$ be the sample mean. Show $\bar{X}_n \stackrel{P}{\to} \infty$. *Hint:* For $B > 0$, let $Y_i = X_i$ if $X_i \le B$ and $Y_i = B$ otherwise; apply the Weak Law to $\bar{Y}_n$.

Problem 11.33 (i) Let $K(P_0, P_1)$ be the Kullback-Leibler Information, defined in (11.21). Show that $K(P_0, P_1) \ge 0$ with equality iff $P_0 = P_1$.
(ii) Show the convergence (11.20) holds even when $K(P_0, P_1) = \infty$. *Hint:* Use Problem 11.32.

Problem 11.34 As in Example 11.2.4, consider the problem of testing $P = P_0$ versus $P = P_1$ based on n i.i.d. observations. The problem is an alternative way to show that a most powerful level α $(0 < \alpha < 1)$ test sequence has limiting power one. If P_0 and P_1 are distinct, there exists E such that $P_0(E) \neq P_1(E)$. Let $\hat{p}_n$ denote the proportion of observations in E and construct a level α test sequence based on $\hat{p}_n$ which has power tending to one.

Problem 11.35 If X_n is a sequence of real-valued random variables, prove that $X_n \to 0$ in P_n-probability if and only if $E_{P_n}[\min(|X_n|, 1)] \to 0$.

Problem 11.36 (i) Prove Corollary 11.2.3.
(ii) Suppose $X_n \stackrel{d}{\to} X$ and $C_n \stackrel{P}{\to} \infty$. Show $P\{X_n \le C_n\} \to 1$.

Problem 11.37 In Example 11.2.5, show that $\beta_n(p_n) \to 1$ if $n^{1/2}(p_n - 1/2) \to \infty$ and $\beta_n(p_n) \to \alpha$ if $n^{1/2}(p_n - 1/2) \to 0$.

Problem 11.38 In Example 11.2.7, let I_n be the interval (11.23). Show that, for any n,

$$\inf_p P_p\{p \in \hat{I}_n\} = 0 \ .$$

Hint: Consider p positive but small enough so that the chance that a sample of size n results in 0 successes is nearly 1.

Problem 11.39 Show how the interval (11.25) is obtained from (11.24).

Problem 11.40 Show that tightness of a sequence of random vectors in $\mathbb{R}^k$ is equivalent to each of the component variables being tight $\mathbb{R}$.

Problem 11.41 Suppose P_n is a sequence of probabilities and X_n is a sequence of real-valued random variables; the distribution of X_n under P_n is denoted $\mathcal{L}(X_n|P_n)$. Prove that $\mathcal{L}(X_n|P_n)$ is tight if and only if $X_n/a_n \to 0$ in P_n-probability for every sequence $a_n \uparrow \infty$.

Problem 11.42 Suppose $X_n \stackrel{d}{\to} N(\mu, \sigma^2)$. (i). Show that, for any sequence of numbers c_n, $P(X_n = c_n) \to 0$. (ii). If c_n is any sequence such that $P(X_n > c_n) \to \alpha$, then $c_n \to \mu + \sigma z_{1-\alpha}$, where $z_{1-\alpha}$ is the $1-\alpha$-quantile of $N(0,1)$.

Problem 11.43 Let $X_1, \cdots, X_n$ be i.i.d. normal with mean θ and variance 1. Suppose $\hat{\theta}_n$ is a location equivariant sequence of estimators such that, for every fixed θ, $n^{1/2}(\hat{\theta}_n - \theta)$ converges in distribution to the standard normal distribution (if θ is true). Let $\bar{X}_n$ be the usual sample mean. Show that, if θ is fixed at the true value, then $n^{1/2}(\hat{\theta}_n - \bar{X}_n)$ tends to 0 in probability under θ.

Problem 11.44 Prove part (ii) of Theorem 11.2.14.

Problem 11.45 Suppose R is a real-valued function on $\mathbb{R}^k$ with $R(y) = o(|y|^p)$ as $|y| \to 0$, for some $p > 0$. If Y_n is a sequence of random vectors satisfying $|Y_n| = o_P(1)$, then show $R(Y_n) = o_P(|Y_n|^p)$. *Hint:* Let $g(y) = R(y)/|y|^p$ with $g(0) = 0$ so that g is continuous at 0; apply the Continuous Mapping Theorem.

Problem 11.46 Use Problem 11.45 to prove (11.28).

Problem 11.47 Assume (U_i, V_i) is bivariate normal with correlation ρ. Let $\hat{\rho}_n$ denote the sample correlation given by (11.29). Verify the limit result (11.31).

Problem 11.48 (i) If $X_1, \ldots, X_n$ is a sample from a Poisson distribution with mean $E(X_i) = \lambda$, then $\sqrt{n}(\sqrt{\bar{X}} - \sqrt{\lambda})$ tends in law to $N(0, \frac{1}{4})$ as $n \to \infty$.
(ii) If X has the binomial distribution $b(p, n)$, then $\sqrt{n}[\arcsin\sqrt{X/n} - \arcsin\sqrt{p}]$ tends in law to $N(0, \frac{1}{4})$ as $n \to \infty$.
Note. Certain refinements of variance stabilizing transformations are discussed by Anscombe (1948), Freeman and Tukey (1950), and Hotelling (1953). Transformations of data to achieve approximately a normal linear model are considered by Box and Cox (1964); for later developments stemming from this work see Bickel and Doksum (1981), Box and Cox (1982), and Hinkley and Runger (1984).

Problem 11.49 Suppose $X_{i,j}$ are independently distributed as $N(\mu_i, \sigma_i^2)$; $i = 1, \ldots, s$; $j = 1, \ldots, n_i$. Let $S_{n,i}^2 = \sum_j (X_{i,j} - \bar{X}_i)^2$, where $\bar{X}_i = n_i^{-1} \sum_j X_{i,j}$. Let $Z_{n,i} = \log[S_{n,i}^2/(n_i - 1)]$. Show that, as $n_i \to \infty$,

$$\sqrt{n_i - 1}[Z_{n,i} - \log(\sigma_i^2)] \xrightarrow{d} N(0, 2) .$$

Thus, for large n_i, the problem of testing equality of all the σ_i can be approximately viewed as testing equality of means of normally distributed variables with known (possibly different) variances. Use Problem 7.12 to suggest a test.

Problem 11.50 Let $X_1, \cdots, X_n$ be i.i.d. Poisson with mean λ. Consider estimating $g(\lambda) = e^{-\lambda}$ by the estimator $T_n = e^{-\bar{X}_n}$. Find an approximation to the bias of T_n; specifically, find a function $b(\lambda)$ satisfying

$$E_\lambda(T_n) = g(\lambda) + n^{-1}b(\lambda) + O(n^{-2})$$

as $n \to \infty$. Such an expression suggests a new estimator $T_n - n^{-1}b(\lambda)$, which has bias $O(n^{-2})$. But, $b(\lambda)$ is unknown. Show that the estimator $T_n - n^{-1}b(\bar{X}_n)$ has bias $O(n^{-2})$.

Problem 11.51 Let $X_1, \ldots, X_n$ be a random sample from the Poisson distribution with unknown mean λ. The uniformly minimum variance unbiased estimator (UMVUE) of $exp(-\lambda)$ is known to be $[(n-1)/n]^{T_n}$, where $T_n = \sum_{i=1}^n X_i$. Find the asymptotic distribution of the UMVUE (appropriately normalized). *Hint:* It may be easier to first find the asymptotic distribution of $exp(-T_n/n)$.

Problem 11.52 Let $X_{i,j}$, $1 \le i \le I$, $1 \le j \le n$ be independent with $X_{i,j}$ Poisson with mean λ_i. The problem is to test the null hypothesis that the λ_i are all the same versus they are not all the same. Consider the test that rejects the null hypothesis iff

$$T \equiv \frac{n \sum_{i=1}^I (\bar{X}_i - \bar{X})^2}{\bar{X}}$$

is large, where $\bar{X}_i = \sum_j X_{i,j}/n$ and $\bar{X} = \sum_i \bar{X}_i / I$.
(i) How large should the critical values be so that, if the null hypothesis is correct, the probability of rejecting the null hypothesis tends (as $n \to \infty$ with I fixed) to the nominal level α.
(ii) Show that the test is pointwise consistent in power against any $(\lambda_1, \ldots, \lambda_I)$, as long as the λ_i are not all equal.

Problem 11.53 Prove the Glivenko-Cantelli Theorem. *Hint:* Use the Strong Law of Large Numbers and the monotonicity of F.

Problem 11.54 Let $X_1, \ldots, X_n$ be i.i.d. P on S. Suppose S is countable and let $\mathcal{E}$ be the collection of *all* subsets of S. Let $\hat{P}_n$ be the *empirical measure*; that is, for any subset E of $\mathcal{E}$, $\hat{P}_n(E)$ is the proportion of observations X_i that fall in E. Prove, with probability one,

$$\sup_{E \in \mathcal{E}} |\hat{P}_n(E) - P(E)| \to 0 \ .$$

Problem 11.55 Suppose X_n is a tight sequence and $Y_n \xrightarrow{P} 0$. Show that $X_n Y_n \xrightarrow{P} 0$. If it is assumed $Y_n \to 0$ almost surely, can you conclude $X_n Y_n \to 0$ almost surely?

Problem 11.56 For a c.d.f. F, define the quantile transformation Q by

$$Q(u) = \inf\{t : \ F(t) \ge u\} \ .$$

(i) Show the event $\{F(t) \ge u\}$ is the same as $\{Q(u) \le t\}$.
(ii) If U is uniformly distributed on $(0, 1)$, show the distribution of $Q(U)$ is F.

Problem 11.57 Let $U_1, \ldots, U_n$ be i.i.d. with c.d.f. $G(u) = u$ and let $\hat{G}_n$ denote the empirical c.d.f. of $U_1, \ldots, U_n$. Define

$$B_n(u) = n^{1/2}[\hat{G}_n(u) - u] \ .$$

(Note that $B_n(\cdot)$ is a random function, called the *uniform empirical process*).
(i) Show that the distribution of the Kolmogorov-Smirnov test statistic $n^{1/2} d_K(\hat{G}_n, G)$ under G is that of $\sup_u |B_n(u)|$.

(ii) Suppose $X_1, \ldots, X_n$ are i.i.d. F (not necessarily continuous), and let $\hat{F}_n$ denote the empirical c.d.f. of $X_1, \ldots, X_n$. Show that the distribution of the Kolmogorov-Smirnov test statistic $n^{1/2} d_K(\hat{F}_n, F)$ under F is that of $\sup_t |B_n(F(t))|$, where B_n is defined in (i). Deduce that this distribution does not depend on F when F is continuous.

Problem 11.58 Consider the uniform confidence band $R_{n,1-\alpha}$ for F given by (11.36). Let $\mathbf{F}$ be the set of all distributions on $\mathbb{R}$. Show,

$$\inf_{F \in \mathbf{F}} P_F\{F \in R_{n,1-\alpha}\} \geq 1 - \alpha \ .$$

Problem 11.59 Show how Theorem 11.2.18 implies Theorem 11.2.17. *Hint:* Use the Borel-Cantelli Lemma; see Billingsley (1995, Theorem 4.3).

Problem 11.60 (i) If $X_1, \ldots, X_n$ are i.i.d. with c.d.f. F and empirical distribution $\hat{F}_n$, use Theorem 11.2.18 to show that $n^{1/2} \sup |\hat{F}_n(t) - F(t)|$ is a tight sequence.

(ii) Let F_n be any sequence of distributions, and let $\hat{F}_n$ be the empirical distribution based on a sample of size n from F_n. Show that $n^{1/2} \sup |\hat{F}_n(t) - F_n(t)|$ is a tight sequence.

Problem 11.61 Show that $X_n \to X$ in probability is equivalent to the statement that, for any subsequence X_{n_j}, there exists a further subsequence $X_{n_{j_k}}$ such that $X_{n_{j_k}} \to X$ with probability one.

Section 11.3

Problem 11.62 (i) Let $X_1, \ldots, X_n$ be a sample from $N(\xi, \sigma^2)$. For testing $\xi = 0$ against $\xi > 0$, show that the power of the one-sided one-sample t-test against a sequence of alternatives $N(\xi_n, \sigma^2)$ for which $n^{1/2}\xi_n/\sigma \to \delta$ tends to $1 - \Phi(z_{1-\alpha} - \delta)$.

(ii) The result of (i) remains valid if $X_1, \ldots, X_n$ are a sample from any distribution with mean ξ and finite variance σ^2.

Problem 11.63 Generalize the previous problem to the two-sample t-test.

Problem 11.64 Let (Y_i, Z_i) be i.i.d. bivariate random vectors in the plane, with both Y_i and Z_i assumed to have finite nonzero variances. Let $\mu_Y = E(Y_1)$ and $\mu_Z = E(Z_1)$, let ρ denote the correlation between Y_1 and Z_1, and let $\hat{\rho}_n$ denote the sample correlation, as defined in (11.29).

(i). Under the assumption $\rho = 0$, show directly (without appealing to Example 11.2.10) that $n^{1/2}\hat{\rho}_n$ is asymptotically normal with mean 0 and variance

$$\tau^2 = Var[(Y_1 - \mu_Y)(Z_1 - \mu_Z)]/Var(Y_1)Var(Z_1).$$

(ii). For testing that Y_1 and Z_1 are independent, consider the test that rejects when $n^{1/2}|\hat{\rho}_n| > z_{1-\frac{\alpha}{2}}$. Show that the asymptotic rejection probability is α, without assuming normality, but under the sole assumption that Y_1 and Z_1 have arbitrary distributions with finite nonzero variances.

(iii). However, for testing $\rho = 0$, the above test is not asymptotically robust. Show that there exist bivariate distributions for (Y_1, Z_1) for which $\rho = 0$ but the limiting variance τ^2 can take on any given positive value.
(iv). For testing $\rho = 0$ against $\rho > 0$, define a denominator D_n and a critical value c_n such that the rejection region $n^{1/2}\hat{\rho}_n/D_n \geq c_n$ has probability tending to α, under any bivariate distribution with $\rho = 0$ and finite, nonzero marginal variances.

Problem 11.65 Under the assumptions of Lemma 11.3.1, compute $Cov(X_i^2, X_j^2)$ in terms of $\rho_{i,j}$ and σ^2. Show that $Var(n^{-1}\sum_{i=1}^n X_i^2) \to 0$ and hence $n^{-1}\sum_{i=1}^n X_i^2 \xrightarrow{P} \sigma^2$.

Problem 11.66 (i) Given ρ, find the smallest and largest value of (11.42) as σ^2/τ^2 varies from 0 to ∞.
(ii) For nominal level $\alpha = .05$ and $\rho = .1, .2, .3, .4$, determine the smallest and the largest asymptotic level of the t-test as σ^2/τ^2 varies from 0 to ∞.

Problem 11.67 Verify the formula for $Var(\bar{X})$ in Model A.

Problem 11.68 In Model A, suppose that the number of observations in group i is n_i. if $n_i \leq M$ and $s \to \infty$, show that the assumptions of Lemma 11.3.1 are satisfied and determine γ.

Problem 11.69 Show that the conditions of Lemma 11.3.1 are satisfied and γ has the stated value: (i) in Model B; (ii) in Model C.

Problem 11.70 Determine the maximum asymptotic level of the one-sided t-test when $\alpha = .05$ and $m = 2, 4, 6$: (i) in Model A; (ii) in Model B.

Problem 11.71 Prove (i) of Lemma 11.3.2.

Problem 11.72 Prove Lemma 11.3.3. *Hint:* For part (ii), use Problem 11.61.

Problem 11.73 Verify the claims made in Example 11.3.1.

Problem 11.74 Verify (11.52).

Problem 11.75 In Example 11.3.3, verify the Huber Condition holds.

Problem 11.76 Let X_{ijk} $(k = 1, \ldots, n_{ij};\ i = 1, \ldots, a;\ j = 1, \ldots, b)$ be independently normally distributed with mean $E(X_{ijk}) = \xi_{ij}$ and variance σ^2. Then the test of any linear hypothesis concerning the ξ_{ij} has a robust level provided $n_{ij} \to \infty$ for all i and j.

Problem 11.77 In the two-way layout of the preceding problem give examples of submodels $\Pi_\Omega^{(1)}$ and $\Pi_\Omega^{(2)}$ of dimensions s_1 and s_2, both less than ab, such that in one case the condition (11.57) continues to require $n_{ij} \to \infty$ for all i and j but becomes a weaker requirement in the other case.

Problem 11.78 Suppose (11.57) holds for some particular sequence $\Pi_\Omega^{(n)}$ with fixed s. Then it holds for any sequence $\Pi_\Omega'^{(n)} \subset \Pi_\Omega^{(n)}$ of dimension $s' < s$. *Hint:* If Π_Ω is spanned by the s columns of A, let Π'_Ω be spanned by the first s' columns of A.

Problem 11.79 Show that (11.48) holds whenever c_n tends to a finite nonzero limit, but the condition need not hold if $c_n \to 0$.

Problem 11.80 Let $\{c_n\}$ and $\{c'_n\}$ be two increasing sequences of constants such that $c'_n/c_n \to 1$ as $n \to \infty$. Then $\{c_n\}$ satisfies (11.48) if and only if $\{c'_n\}$ does.

Problem 11.81 Let $c_n = u_0 + u_1 n + \cdots + u_k n^k$, $u_i \geq 0$ for all i. Then c_n satisfies (11.48). What if $c_n = 2^n$? *Hint:* Apply Problem 11.80 with $c'_n = n^k$.

Problem 11.82 If $\xi_i = \alpha + \beta t_i + \gamma u_i$, express the condition (11.57) in terms of the t's and u's.

Problem 11.83 If $\Pi_{i,i}$ are defined as in (11.56), show that $\sum_{i=1}^n \Pi_{i,i}^2 = s$. *Hint:* Since the $\Pi_{i,i}$ are independent of A, take A to be orthogonal.

Problem 11.84 The size of each of the following tests is robust against nonnormality:

(i) the test (7.24) as $b \to \infty$,

(ii) the test (7.26) as $mb \to \infty$,

(iii) the test (7.28) as $m \to \infty$.

Problem 11.85 For $i = 1, \ldots, s$ and $j = 1, \ldots, n_i$, let $X_{i,j}$ be independent, with $X_{i,j}$ having distribution F_i, where F_i is an arbitrary distribution with mean μ_i and finite common variance σ^2. Consider testing $\mu_1 = \cdots = \mu_s$ based on the test statistic (11.66), which is UMPI under normality. Show the test remains robust with respect to the rejection probability under H_0 even if the F_i differ and are not normal.

Problem 11.86 In the preceding problem, investigate the rejection probability when the F_i have different variances. Assume $\min n_i \to \infty$ and $n_i/n \to \rho_i$.

Problem 11.87 Show that the test derived in Problem 11.49 is not robust against nonnormality.

Problem 11.88 Let $X_1, \ldots, X_n$ be a sample from $N(\xi, \sigma^2)$, and consider the UMP invariant level-α test of $H : \xi/\sigma \leq \theta_0$ (Section 6.4). Let $\alpha_n(F)$ be the actual significance level of this test when $X_1, \ldots, X_n$ is a sample from a distribution F with $E(X_i) = \xi$, $Var(X_i) = \sigma^2 < \infty$. Then the relation $\alpha_n(F) \to \alpha$ will not in general hold unless $\theta_0 = 0$. *Hint:* First find the limiting joint distribution of $\sqrt{n}(\bar{X} - \xi)$ and $\sqrt{n}(S^2 - \sigma^2)$.

Section 11.4

Problem 11.89 When sampling from a normal distribution, one can derive an Edgeworth expansion for the t-statistic as follows. Suppose $X_1, \ldots, X_n$ are i.i.d. $N(\mu, \sigma^2)$ and let $t_n = n^{1/2}(\bar{X}_n - \mu)/S_n$, where S_n^2 is the usual unbiased estimate of σ^2. Let Φ be the standard normal c.d.f. and let $\Phi' = \varphi$. Show

$$P\{t_n \le t\} = \Phi(t) - \frac{1}{4n}(t + t^3)\varphi(t) + O(n^{-2}) \qquad (11.91)$$

as follows. It suffices to let $\mu = 0$ and $\sigma = 1$. By conditioning on S_n, we can write

$$P\{t_n \le t\} = E\{\Phi[t(1 + S_n^2 - 1)^{1/2}]\} .$$

By Taylor expansion inside the expectation, along with moments of S_n^2, one can deduce (11.91).

Problem 11.90 Assuming F is absolutely continuous with 4 moments, verify (11.76).

Problem 11.91 Let ϕ_n be the classical t-test for testing the mean is zero versus the mean is positive, based on n i.i.d. observations from F. Consider the power of this test against the distribution $N(\mu, 1)$. Show the power tends to one as $\mu \to \infty$.

Problem 11.92 Suppose $\mathbf{F}$ satisfies the conditions of Theorem 11.4.6. Assume there exists ϕ_n such that

$$\sup_{F \in \mathbf{F}:\ \mu(F) = 0} E_F(\phi_n) \to \alpha .$$

Show that

$$\limsup_n E_F(\phi_n) \le \alpha$$

for every $F \in \mathbf{F}$.

Problem 11.93 In the proof of Theorem 11.4.4, prove $S_n/\sigma(F_n) \to 1$ in probability.

Problem 11.94 Prove Lemma 11.4.5.

Problem 11.95 Consider the problem of testing $\mu(F) = 0$ versus $\mu(F) \ne 0$, for $F \in \mathbf{F_0}$, the class of distributions supported on $[0, 1]$. Let ϕ_n be Anderson's test.
(i) If

$$|n^{1/2}\mu(F_n)| \ge \delta > 2s_{n,1-\alpha} ,$$

then show that

$$E_{F_n}(\phi_n) \ge 1 - \frac{1}{2(2s_{n,1-\alpha} - \delta)^2} ,$$

where $s_{n,1-\alpha}$ is the $1 - \alpha$ quantile of the null distribution of the Kolmogorov-Smirnov statistic. *Hint:* Use (11.88) and Chebyshev's inequality.
(ii) Deduce that the minimum power of ϕ_n over $\{F :\ n^{1/2}\mu(F)| \ge \delta\}$ is at least $1 - [2(2s_{n,1-\alpha} - \delta)^{-2}]$ if $\delta > 2s_{n,1-\alpha}$.

(iii) Use (ii) to show that, if $F_n \in \mathbf{F_0}$ is any sequence of distributions satisfying $n^{1/2}|\mu(F_n)| \to \infty$, then $E_{F_n}(\phi_n) \to 1$.

Problem 11.96 Prove the second equality in (11.81). In the proof of Lemma 11.4.2, show that $\kappa_n(n) \to 0$.

Problem 11.97 Let $Y_{n,1}, \ldots, Y_{n,n}$ be i.i.d. bernoulli variables with success probability p_n, where $np_n = \lambda$ and $\lambda^{1/2} = \delta$. Let $U_{n,1}, \ldots, U_{n,n}$ be i.i.d. uniform variables on $(-\tau_n, \tau_n)$, where $\tau_n^2 = 3p_n^2$. Then, let $X_{n,i} = Y_{n,i} + U_i$, so that F_n is the distribution of $X_{n,i}$. (Note that $n^{1/2}\mu(F_n)/\sigma(F_n) = \delta$.)
(i) If t_n is the t-statistic, show that, under F_n, $t_n \xrightarrow{d} V^{1/2}$, where V is Poisson with mean δ^2, and so if $z_{1-\alpha}$ is not an integer,

$$P_{F_n}\{t_n > t_{n-1,1-\alpha}\} \to P\{V^{1/2} > z_{1-\alpha}\} \ .$$

(ii) Show, for $\alpha < 1/2$, the limiting power of the t-test against F_n satisfies

$$P\{V^{1/2} > z_{1-\alpha}\} \le 1 - P\{V = 0\} = \exp(-\delta^2) \ .$$

This is strictly smaller than $1 - \Phi(z_{1-\alpha} - \delta)$ if and only if

$$\Phi(z_{1-\alpha} - \delta) < \exp(-\delta^2) \ .$$

Certainly, for small δ, this inequality holds, since the left hand side tends to $1-\alpha$ as $\delta \to 0$ while the right hand side tends to 1.

11.6 Notes

The convergence concepts in Section 11.2 are classical and can be found in most graduate probability texts such as Billingsley (1995) or Dudley (1989). The Central Limit Theory for Bernoulli trials dates back to de Moivre (1733) and for more general distributions to Laplace (1812). Their treatment was probabilistic and did not involve problems in inference. Normal experiments were first treated in Gauss (1809). Further history is provided in Stigler (1986) and Hald (1990, 1998).

Concern about the robustness of classical normal theory tests began to be voiced in the 1920s (Neyman and Pearson (1928), Shewhart and Winters (1928), Sophister (1928), and Pearson (1929)) and has been an important topic ever since. Particularly influential were Box (1953), where the term *robustness* was introduced; also see Scheffé (1959, Chapter 10), Tukey (1960) and Hotelling (1961). The robustness of regression tests studied in Section 11.3.3 is based on Huber (1973).

As remarked in Example 11.3.4, the F-test for testing equality of means is not robust if the underlying variances differ, even if the sample sizes are equal and $s > 2$; see Scheffé (1959). More appropriate tests for this generalized Behrens–Fisher problem have been proposed by Welch (1951), James (1951), and Brown and Forsythe (1974b), and are further discussed by Clinch and Kesselman (1982). The corresponding robustness problem for more general linear hypotheses is treated by James (1954) and Johansen (1980); see also Rothenberg (1984).

The linear model F-test—as was seen to be the case for the t-test—is highly nonrobust against dependence of the observations. Tests of the hypothesis that the covariance matrix is proportional to the identity against various specified forms of dependence are considered in King and Hillier (1985). For recent work on robust testing in linear models, see Müller (1998) and the references cited there.

The usual test for equality of variances is Bartlett's test, which is discussed in Cyr and Monoukian (1982) and Glaser (1982). Bartlett's test is highly sensitive to the assumption of normality, and therefore is rarely appropriate. More robust tests for this latter hypothesis are reviewed in Conover, Johnson, and Johnson (1981). For testing homogeneity of covariance matrices, see Beran and Srivastava (1985) and Zhang and Boos (1992).

Robustness properties of the t-test are studied in Efron (1969), Lehmann and Loh (1990), Basu and DasGupta (1995), Basu (1999) and Romano (2004). The nonexistence results of Bahadur and Savage (1956), and also Hoeffding (1956), have been generalized to other problems; see Donoho (1988) and Romano (2004) and the references there.

The idea of expanding the distribution of the sample mean in order to study the error in normal approximation can be traced to Chebyshev (1890) and Edgeworth (1905). But it was not until Cramér (1928, 1937) provided some rigorous results. The fundamental theory of Edgeworth expansions is developed in Bhattacharya and Rao (1976); also see Bickel (1974), Bhattacharya and Ghosh (1978), Hall (1992) and Hall and Jing (1995).

12

Quadratic Mean Differentiable Families

12.1 Introduction

As mentioned at the beginning of Chapter 11, the finite sample theory of optimality for hypothesis testing applied only to rather special parametric families, primarily exponential families and group families. On the other hand, asymptotic optimality will apply more generally to parametric families satisfying smoothness conditions. In particular, we shall assume a certain type of differentiability condition, called *quadratic mean differentiability.* Such families will be considered in Section 12.2. In Section 12.3, the notion of *contiguity* will be developed, primarily as a technique for calculating the limiting distribution or power of a test statistic under an alternative sequence, especially when the limiting distribution under the null hypothesis is easy to obtain. In Section 12.4, these techniques will then be applied to classes of tests based on the likelihood function, namely the Wald, Rao, and likelihood ratio tests. The asymptotic optimality of these tests will be established in Chapter 13.

12.2 Quadratic Mean Differentiability (q.m.d.)

Consider a parametric model $\{P_\theta, \theta \in \Omega\}$, where, throughout this section, Ω is assumed to be an open subset of $\mathbb{R}^k$. The probability measures P_θ are defined on some measurable space $(\mathcal{X}, \mathcal{C})$. Assume each P_θ is absolutely continuous with respect to a σ-finite measure μ, and set $p_\theta(x) = dP_\theta(x)/d\mu(x)$. In this section, smooth parametric models will be considered. To motivate the smoothness condition given in Definition 12.2.1 below, consider the case of n i.i.d. random variables $X_1, \ldots, X_n$ and the problem of testing a simple null hypothesis $\theta = \theta_0$ against a simple alternative θ_1 (possibly depending on n). The most powerful test rejects when the loglikelihood ratio statistic

$$\log[L_n(\theta_1)/L_n(\theta_0)]$$

is sufficiently large, where

$$L_n(\theta) = \prod_{i=1}^{n} p_\theta(X_i) \tag{12.1}$$

denotes the likelihood function. We would like to obtain certain expansions of the loglikelihood ratio, and the smoothness condition we impose will ensure the existence of such an expansion.

Example 12.2.1 (Normal Location Model) Suppose P_θ is $N(\theta, \sigma^2)$, where σ^2 is known. It is easily checked that

$$\log[L_n(\theta_1)/L_n(\theta_0)] = \frac{n}{\sigma^2}[(\theta_1 - \theta_0)\bar{X}_n - \frac{1}{2}(\theta_1^2 - \theta_0^2)] , \tag{12.2}$$

where $\bar{X}_n = \sum_{i=1}^{n} X_i/n$. By the Weak Law of Large Numbers, under θ_0,

$$(\theta_1 - \theta_0)\bar{X}_n - \frac{1}{2}(\theta_1^2 - \theta_0^2) \xrightarrow{P} (\theta_1 - \theta_0)\theta_0 - \frac{1}{2}(\theta_1^2 - \theta_0^2) = -\frac{1}{2}(\theta_1 - \theta_0)^2 ,$$

and so $\log[L_n(\theta_1)/L_n(\theta_0)] \xrightarrow{P} -\infty$. Therefore, $\log[L_n(\theta_1)/L_n(\theta_0)]$ is asymptotically unbounded in probability under θ_0. As in Example 11.2.5, a more useful result is obtained if θ_1 in (12.2) is replaced by $\theta_0 + hn^{-1/2}$. We then find

$$\log[L_n(\theta_0 + hn^{-1/2})/L_n(\theta_0)] = \frac{hn^{1/2}(\bar{X}_n - \theta_0)}{\sigma^2} - \frac{h^2}{2\sigma^2} = hZ_n - \frac{h^2}{2\sigma^2} , \tag{12.3}$$

where $Z_n = n^{1/2}(\bar{X}_n - \theta_0)/\sigma^2$ is $N(0, 1/\sigma^2)$. Notice that the expansion (12.3) is a linear function of Z_n and a simple quadratic function of h, with the coefficient of h^2 nonrandom. Furthermore, $\log[L_n(\theta_0 + hn^{-1/2})/L_n(\theta_0)]$ is distributed as $N(-h^2/2\sigma^2, h^2/\sigma^2)$ under θ_0 for every n. (The relationship that the mean is the negative of half the variance will play a key role in the next section.) ■

The following more general family permits an asymptotic version of (12.3).

Example 12.2.2 (One-parameter Exponential Family) Let $X_1, \ldots, X_n$ be i.i.d. having density

$$p_\theta(x) = \exp[\theta T(x) - A(\theta)]$$

with respect to a σ-finite measure μ. Assume θ_0 lies in the interior of the natural parameter space. Then,

$$\log[L_n(\theta_0 + hn^{-1/2})/L_n(\theta_0)] = hn^{-1/2}\sum_{i=1}^{n} T(X_i) - n[A(\theta_0 + hn^{-1/2}) - A(\theta_0)] .$$

Recall (Problem 2.16) that $E_{\theta_0}[T(X_i)] = A'(\theta_0)$ and $Var_{\theta_0}[T(X_i)] = A''(\theta_0)$. By a Taylor expansion,

$$n[A(\theta_0 + hn^{-1/2}) - A(\theta_0)] = hn^{1/2}A'(\theta_0) + \frac{1}{2}h^2 A''(\theta_0) + o(1)$$

as $n \to \infty$, so that

$$\log[L_n(\theta_0 + hn^{-1/2})/L_n(\theta_0)] = hZ_n - \frac{1}{2}h^2 A''(\theta_0) + o(1) , \tag{12.4}$$

where, under θ_0,

$$Z_n = n^{-1/2} \sum_{i=1}^{n} \{T(X_i) - E_{\theta_0}[T(X_i)]\} \xrightarrow{d} N(0, A''(\theta_0)) .$$

Thus, the loglikelihood ratio (12.4) behaves asymptotically like the loglikelihood ratio (12.3) from a normal location model. As we will see, such approximations allow one to deduce asymptotic optimality properties for the exponential model (or any model whose likelihood ratios satisfy an appropriate generalization of (12.4)) from optimality properties of the simple normal location model. ■

We would like to obtain an approximate result like (12.4) for more general families. Classical smoothness conditions usually assume that, for fixed x, the function $p_\theta(x)$ is differentiable in θ at θ_0; that is, for some function $\dot{p}_\theta(x)$,

$$p_{\theta_0+h}(x) - p_{\theta_0}(x) - \langle \dot{p}_{\theta_0}(x), h \rangle = o(|h|)$$

as $|h| \to 0$. In addition, higher order differentiability is typically assumed with further assumptions on the remainder terms. In order to avoid such strong assumptions, it turns out to be useful to work with square roots of densities. For fixed x, differentiability of $p_\theta^{1/2}(x)$ at $\theta = \theta_0$ requires the existence of a function $\eta(x, \theta_0)$ such that

$$R(x, \theta_0, h) \equiv p_{\theta_0+h}^{1/2}(x) - p_{\theta_0}^{1/2}(x) - \langle \eta(x, \theta_0), h \rangle = o(|h|) .$$

To obtain a weaker, more generally applicable condition, we will not require $R^2(x, \theta_0, h) = o(|h|^2)$ for every x, but we will impose the condition that $R^2(X, \theta_0, h)$ averaged with respect to μ is $o(|h|^2)$. Let $L^2(\mu)$ denote the space of functions g such that $\int g^2(x)\, d\mu(x) < \infty$. The convenience of working with square roots of densities is due in large part to the fact that $p_\theta^{1/2}(\cdot) \in L^2(\mu)$, a fact first exploited by Le Cam; see Pollard (1997) for an explanation. The desired smoothness condition is now given by the following definition.

Definition 12.2.1 The family $\{P_\theta, \theta \in \Omega\}$ is *quadratic mean differentiable* (abbreviated q.m.d.) at θ_0 if there exists a vector of real-valued functions $\eta(\cdot, \theta_0) = (\eta_1(\cdot, \theta_0), \ldots, \eta_k(\cdot, \theta_0))^T$ such that

$$\int_{\mathcal{X}} \left[\sqrt{p_{\theta_0+h}(x)} - \sqrt{p_{\theta_0}(x)} - < \eta(x, \theta_0), h > \right]^2 d\mu(x) = o(|h|^2) \tag{12.5}$$

as $|h| \to 0$.[1]

The vector-valued function $\eta(\cdot, \theta_0)$ will be called the quadratic mean derivative of P_θ at θ_0. Clearly, $\eta(x, \theta_0)$ is not unique since it can be changed on a set of x values having μ-measure zero. If q.m.d. holds at all θ_0, then we say the family is q.m.d.

---

[1] The definition of q.m.d. is a special case of Fréchet differentiability of the map $\theta \to p_\theta^{1/2}(\cdot)$ from Ω to $L^2(\mu)$.

The following are useful facts about q.m.d. families.

Lemma 12.2.1 *Assume* $\{P_\theta, \theta \in \Omega\}$ *is q.m.d. at* θ_0. *Let* $h \in \mathbb{R}^k$.

(i) Under P_{θ_0}, $\langle \frac{\eta(X,\theta_0)}{p_{\theta_0}^{1/2}(X)}, h\rangle$ *is a random variable with mean 0; i.e., satisfying*

$$\int p_{\theta_0}^{1/2}(x)\langle \eta(x,\theta_0), h\rangle d\mu(x) = 0 .$$

(ii) The components of $\eta(\cdot, \theta_0)$ *are in* $L^2(\mu)$*; that is, for* $i = 1, \ldots, k$,

$$\int \eta_i^2(x,\theta_0)\, d\mu(x) < \infty .$$

PROOF. In the definition of q.m.d., replace h by $hn^{-1/2}$ to deduce that

$$\int \left\{ n^{1/2} \left[p_{\theta_0+hn^{-1/2}}^{1/2}(x) - p_{\theta_0}^{1/2}(x) \right] - \langle \eta(x,\theta_0), h\rangle \right\}^2 d\mu(x) \to 0$$

as $n \to \infty$. But, if $\int (g_n - g)^2\, d\mu \to 0$ and $\int g_n^2 d\mu < \infty$, then $\int g^2 d\mu < \infty$ (Problem 12.3). Hence, for any $h \in \mathbb{R}^k$, $\langle \eta(x,\theta_0), h\rangle \in L^2(\mu)$. Taking h equal to the vector of zeros except for a 1 in the ith component yields (*ii*). Also, if $\int (g_n - g)^2 d\mu \to 0$ and $\int p^2 d\mu < \infty$ then $\int p g_n d\mu \to \int pg\, d\mu$ (Problem 12.4). Taking $p = p_{\theta_0}^{1/2}$ and $g_n = n^{1/2} \left[p_{\theta_0+hn^{-1/2}}^{1/2}(x) - p_{\theta_0}^{1/2}(x) \right]$ yields

$$\begin{aligned}
&\int p_{\theta_0}^{1/2}(x)\langle \eta(x,\theta_0), h\rangle d\mu(x) \\
&= \lim_{n\to\infty} n^{1/2} \int p_{\theta_0}^{1/2}(x) [p_{\theta_0+hn^{-1/2}}^{1/2}(x) - p_{\theta_0}^{1/2}(x)]\, d\mu(x) \\
&= \lim_{n\to\infty} n^{1/2} \left[\int p_{\theta_0}^{1/2}(x)\, p_{\theta_0+hn^{-1/2}}^{1/2}(x)\, d\mu(x) - 1 \right] \\
&= -\tfrac{1}{2} \lim n^{-1/2}\, n \int [p_{\theta_0}^{1/2}(x) - p_{\theta_0+hn^{-1/2}}^{1/2}(x)]^2 d\mu(x) .
\end{aligned}$$

But,

$$\begin{aligned}
n \int &\left[p_{\theta_0}^{1/2}(x) - p_{\theta_0+hn^{-1/2}}^{1/2}(x) \right]^2 d\mu(x) \\
&\to \int |\langle \eta(x,\theta_0), h\rangle|^2 d\mu(x) < \infty , \qquad (12.6)
\end{aligned}$$

and (*i*) follows. ■

Note that Lemma 12.2.1 (*i*) asserts that the finite-dimensional set of vectors $\{\langle \eta(\cdot,\theta_0), h\rangle,\ h \in \mathbb{R}^k\}$ in $L^2(\mu)$ is orthogonal to $p_{\theta_0}^{1/2}(\cdot)$.

It turns out that, when q.m.d. holds, the integrals of products of the components of $\eta(\cdot,\theta)$ play a vital role in the theory of asymptotic efficiency. Such values (multiplied by 4 for convenience) are gathered into a matrix, which we call the *Fisher Information matrix*. The use of the term *information* is justified by Problem 12.5.

Definition 12.2.2 For a q.m.d. family with derivative $\eta(\cdot, \theta)$, define the *Fisher Information matrix* to be the matrix $I(\theta)$ with (i, j) entry

$$I_{i,j}(\theta) = 4 \int \eta_i(x, \theta)\eta_j(x, \theta)\, d\mu(x) \ .$$

The existence of $I(\theta)$ follows from Lemma 12.2.1 (ii) and the Cauchy-Schwarz inequality. Furthermore, $I(\theta)$ does not depend on the choice of dominating measure μ (Problem 12.8).

Lemma 12.2.2 *For any* $h \in \mathbb{R}^k$,

$$\int |\langle h, \eta(x, \theta_0)\rangle|^2\, d\mu(x) = \tfrac{1}{4}\langle h, I(\theta_0)h\rangle \ .$$

PROOF. Of course

$$\langle h, \eta(x, \theta_0)\rangle = \Sigma h_i \eta_i(x, \theta_0) \ .$$

Square it and integrate. ■

Next, we would like to determine simple sufficient conditions for q.m.d. to hold. Assuming that the pointwise derivative of $p_\theta(x)$ with respect to θ exists, one would expect that the quadratic mean derivative $\eta(\cdot, \theta_0)$ is given by

$$\eta_i(\cdot, \theta) = \frac{\partial}{\partial \theta_i} p_\theta^{1/2}(x) = \frac{1}{2} \frac{\frac{\partial}{\partial \theta_i} p_\theta(x)}{p_\theta^{1/2}(x)} \ . \tag{12.7}$$

In fact, Hájek (1972) gave sufficient conditions where this is the case, and the following result for the case $k = 1$ is based on his argument.

Theorem 12.2.1 *Suppose* Ω *is an open subset of* $\mathbb{R}$ *and fix* $\theta_0 \in \Omega$. *Assume* $p_\theta^{1/2}(x)$ *is an absolutely continuous function of* θ *in some neighborhood of* θ_0, *for* μ*-almost all* x.[2] *Also, assume for* μ*-almost all* x, *the derivative* $p'_\theta(x)$ *of* $p_\theta(x)$ *with respect to* θ *exists at* $\theta = \theta_0$. *Define*

$$\eta(x, \theta) = \frac{p'_\theta(x)}{2p_\theta^{1/2}(x)} \tag{12.8}$$

if $p_\theta(x) > 0$ *and* $p'_\theta(x)$ *exists and define* $\eta(x, \theta) = 0$ *otherwise. Also, assume the Fisher Information* $I(\theta)$ *is finite and continuous in* θ *at* θ_0. *Then,* $\{P_\theta\}$ *is q.m.d. at* θ_0 *with quadratic mean derivative* $\eta(\cdot, \theta_0)$.

PROOF. If $p_\theta(x) > 0$ and $p'_\theta(x)$ exists, then from standard calculus it follows that

$$\frac{d}{d\theta} p_\theta^{1/2}(x) = \eta(x, \theta) \ .$$

[2] A real-valued function g defined on an interval $[a, b]$ is absolutely continuous if $g(\theta) = g(a) + \int_a^\theta h(x)dx$ for some integrable function h and all $\theta \in [a, b]$; Problem 2 on p.182 of Dudley (1989) clarifies the relationship between this notion of absolute continuity of a function and the general notion of a measure being absolute continuous with respect to another measure, as defined in Section 2.2.

We are now in a position to obtain an asymptotic expansion of the loglikelihood ratio whose asymptotic form corresponds to that of the normal location model in Example 12.2.1. First, define the *score function* (or *score vector*) $\tilde{\eta}(x, \theta)$ by

$$\tilde{\eta}(x, \theta) = \frac{2\eta(x, \theta)}{p_\theta^{1/2}(x)} \tag{12.10}$$

if $p_\theta(x) > 0$ and $\tilde{\eta}(x, \theta) = 0$ otherwise. Under the conditions of Theorem 12.2.2, $\tilde{\eta}(x, \theta)$ can often be computed as the gradient vector of $\log p_\theta(x)$. Also, define the normalized score vector Z_n by

$$Z_n = Z_{n,\theta_0} = n^{-1/2} \sum_{i=1}^{n} \tilde{\eta}(X_i, \theta_0) \ . \tag{12.11}$$

The following theorem, due to Le Cam, is the main result of this section.

Theorem 12.2.3 *Suppose $\{P_\theta, \theta \in \Omega\}$ is q.m.d. at θ_0 with derivative $\eta(\cdot, \theta_0)$ and Ω is an open subset of $\mathbb{R}^k$. Suppose $I(\theta_0)$ is nonsingular. Fix θ_0 and consider the likelihood ratio $L_{n,h}$ defined by*

$$L_{n,h} = \frac{L_n(\theta_0 + hn^{-1/2})}{L_n(\theta_0)} = \prod_{i=1}^{n} \frac{p_{\theta_0 + hn^{-1/2}}(X_i)}{p_{\theta_0}(X_i)} \ , \tag{12.12}$$

where the likelihood function $L_n(\cdot)$ is defined in (12.1).

(i) Then, as $n \to \infty$,

$$\log(L_{n,h}) - \left[\langle h, Z_n \rangle - \frac{1}{2} \langle h, I(\theta_0) h \rangle \right] = o_{P_{\theta_0}^n}(1). \tag{12.13}$$

(ii) Under $P_{\theta_0}^n$, $Z_n \xrightarrow{d} N(0, I(\theta_0))$ and so

$$\log(L_{n,h}) \xrightarrow{d} N\left(-\tfrac{1}{2}\langle h, I(\theta_0) h\rangle, \langle h, I(\theta_0) h \rangle\right) . \tag{12.14}$$

PROOF. Consider the triangular array $Y_{n,1}, \ldots, Y_{n,n}$, where

$$Y_{n,i} = \frac{p_{\theta_0 + hn^{-1/2}}^{1/2}(X_i)}{p_{\theta_0}^{1/2}(X_i)} - 1.$$

Note that $E_{\theta_0}(Y_{n,i}^2) \le 2 < \infty$ and

$$\log(L_{n,h}) = 2 \sum_{i=1}^{n} \log(1 + Y_{n,i}) \ . \tag{12.15}$$

But,

$$\log(1 + y) = y - \tfrac{1}{2} y^2 + y^2 r(y) \ ,$$

where $r(y) \to 0$ as $y \to 0$, so that

$$\log(L_{n,h}) = 2 \sum_{i=1}^{n} Y_{n,i} - \sum_{i=1}^{n} Y_{n,i}^2 + 2 \sum_{i=1}^{n} Y_{n,i}^2 \, r(Y_{n,i}) \ .$$

The idea of expanding the likelihood ratio in terms of variables involving square roots of densities is known as Le Cam's square root trick; see Le Cam (1969).

The proof of (i) will follow from the following four convergence results:

$$\sum_{i=1}^{n} E_{\theta_0}(Y_{n,i}) \to -\tfrac{1}{8}\langle h, I(\theta_0)h\rangle \tag{12.16}$$

$$\sum_{i=1}^{n}[Y_{n,i} - E_{\theta_0}(Y_{n,i})] - \frac{1}{2}\langle h, Z_n\rangle \stackrel{P^n_{\theta_0}}{\to} 0 \tag{12.17}$$

$$\sum_{i=1}^{n} Y^2_{n,i} \stackrel{P^n_{\theta_0}}{\to} \frac{1}{4}\langle h, I(\theta_0)h\rangle \tag{12.18}$$

$$\sum Y^2_{n,i}\, r(Y_{n,i}) \stackrel{P^n_{\theta_0}}{\to} 0\ . \tag{12.19}$$

Once these four convergences have been established, part (ii) of the theorem follows by the Central Limit Theorem and the facts that

$$E_{\theta_0}[\langle\tilde{\eta}(X_1,\theta_0), h\rangle] = 0 \qquad \text{by Lemma 12.2.1 } (i)$$

and

$$\text{Var}_{\theta_0}[\langle\tilde{\eta}(X_1,\theta_0), h\rangle] = \langle h, I(\theta_0)h\rangle \qquad \text{by Lemma 12.2.2.}$$

(a) To show (12.16),

$$\begin{aligned}\sum_{i=1}^{n} E_{\theta_0}(Y_{n,i}) &= n\int\left[\frac{p^{1/2}_{\theta_0+hn^{-1/2}}(x)}{p^{1/2}_{\theta_0}(x)} - 1\right] p_{\theta_0}(x)\,d\mu(x)\\ &= -\tfrac{n}{2}\int\left[p^{1/2}_{\theta_0+hn^{-1/2}}(x) - p^{1/2}_{\theta_0}(x)\right]^2 d\mu(x)\\ &\to -\tfrac{1}{2}\int|\langle\eta(x,\theta_0), h\rangle|^2 d\mu(x)\end{aligned}$$

by (12.6). This last expression is equal to $-\frac{1}{8}\langle h, I(\theta_0)h\rangle$ by Lemma 12.2.2, and (12.16) follows.

(b) To show (12.17), write

$$Y_{n,i} = \frac{1}{2}n^{-1/2}\langle h, \tilde{\eta}(X_i,\theta_0)\rangle + n^{-1/2}\frac{R_n(X_i)}{p^{1/2}_{\theta_0}(X_i)}\ , \tag{12.20}$$

where $\int R^2_n(x)\,d\mu(x) \to 0$ (by q.m.d.). Hence,

$$\sum_{i=1}^{n}[Y_{n,i} - E_{\theta_0}(Y_{n,i})] = \frac{1}{2}\langle h, Z_n\rangle + hn^{-1/2}\sum_{i=1}^{n}\left[\frac{R_n(X_i)}{p^{1/2}_{\theta_0}(X_i)} - E_{\theta_0}\left(\frac{R_n(X_i)}{p^{1/2}_{\theta_0}(X_i)}\right)\right].$$

The last term, under $P^n_{\theta_0}$, has mean 0 and variance bounded by

$$h^2 E_{\theta_0}\left[\frac{R^2_n(X_i)}{p_{\theta_0}(X_i)}\right] = h^2\int R^2_n(x)\,d\mu(x) \to 0\ .$$

So, (12.17) follows.

(c) To prove (12.18), by the Weak Law of Large Numbers, under θ_0,

$$\tfrac{1}{n}\sum_{i=1}^{n}[\langle h, \tilde{\eta}(X_i,\theta_0)\rangle]^2 \stackrel{P}{\to} E_{\theta_0}\{[\langle h, \tilde{\eta}(X_1,\theta_0)\rangle]^2\} = \langle h, I(\theta_0)h\rangle\ . \tag{12.21}$$

Now using equation (12.20), we get

$$\sum_{i=1}^{n} Y_{n,i}^2 = \tfrac{1}{4n} \sum_{i=1}^{n} [\langle h, \tilde{\eta}(X_i, \theta_0) \rangle]^2 + \tfrac{1}{n} \sum_{i=1}^{n} \frac{R_n^2(X_i)}{p_{\theta_0}(X_i)}$$

$$+ \tfrac{1}{n} \sum_{i=1}^{n} [\langle h, \tilde{\eta}(X_i, \theta_0) \rangle] \sum_{j=1}^{n} \frac{R_n(X_j)}{p_{\theta_0}^{1/2}(X_j)} \ . \tag{12.22}$$

By (12.21), the first term converges in probability under θ_0 to $\frac{1}{4}\langle h, I(\theta_0) h \rangle$. The second term is nonnegative and has expectation under θ_0 equal to

$$\int R_n^2(x) \mu(dx) \to 0 \ ;$$

hence, the second term goes to 0 in probability under $P_{\theta_0}^n$ by Markov's inequality. The last term goes to 0 in probability under $P_{\theta_0}^n$ by the Cauchy-Schwarz inequality and the convergences of the first two terms. Thus, (12.18) follows. By taking expectations in (12.22), a similar argument shows

$$n E_{\theta_0}(Y_{n,i}^2) = \frac{1}{4} \langle h, I(\theta_0) h \rangle + o(1) \tag{12.23}$$

as $n \to \infty$, which also implies $E_{\theta_0}(Y_{n,i}) \to 0$.
(d) Finally, to prove (12.19), note that

$$\left| \sum_{i=1}^{n} Y_{n,i}^2 \, r(Y_{n,i}) \right| \le \max_{1 \le i \le n} |r(Y_{n,i})| \sum_{i=1}^{n} Y_{n,i}^2 .$$

So, it suffices to show $\max_i |r(Y_{n,i})| \to 0$ in probability under θ_0, which follows if we can show

$$\max_{1 \le i \le n} |Y_{n,i}| \overset{P_{\theta_0}^n}{\to} 0 \ . \tag{12.24}$$

But, $\sum_{i=1}^{n} [Y_{n,i} - E_{\theta_0}(Y_{n,i})]$ is asymptotically normal by (12.17) and the Central Limit Theorem. Hence, Corollary 11.2.2 is applicable with $s_n^2 = O(1)$, which yields the Lindeberg Condition

$$n E_{\theta_0} [|Y_{n,i} - E_{\theta_0}(Y_{n,i})|^2 I\{|Y_{n,i} - E_{\theta_0}(Y_{n,i})| \ge \epsilon\}] \to 0 \tag{12.25}$$

for any $\epsilon > 0$. But then,

$$P_{\theta_0} \{ \max_{1 \le i \le n} |Y_{n,i} - E_{\theta_0}(Y_{n,i})| > \epsilon \} \le n P_{\theta_0} \{ |Y_{n,i} - E_{\theta_0}(Y_{n,i})|^2 > \epsilon^2 \} \ ,$$

which can be bounded by the expression on the left side of (12.25) divided by ϵ^2, and so $\max_{1 \le i \le n} |Y_{n,i} - E_{\theta_0}(Y_{n,i})| \to 0$ in probability under θ_0. The result (12.24) follows, since $E_{\theta_0}(Y_{n,i}) \to 0$. ■

Remark 12.2.1 Since the theorem concerns the local behavior of the likelihood ratio near θ_0, it is not entirely necessary to assume Ω is open. However, it is important to assume θ_0 is an interior point; see Problem 12.14.

Remark 12.2.2 The theorem holds if h is replaced by h_n on the left side of each part of the theorem where $h_n \to h$. Under further assumptions, it is plausible that the left side of (12.13) tends to 0 in probability uniformly in h as long as h varies in a compact set; that is, for any $c > 0$, the supremum over h such that $|h| \le c$ of the absolute value of the left side of (12.13) tends to 0 in probability under θ_0; see Problem 13.12.

12.3 Contiguity

Contiguity is an asymptotic form of a probability measure Q being absolutely continuous with respect to another probability measure P. In order to motivate the concept, suppose P and Q are two probability measures on some measurable space $(\mathcal{X}, \mathcal{F})$. Assume that Q is absolutely continuous with respect to P. This means that $E \in \mathcal{F}$ and $P(E) = 0$ implies $Q(E) = 0$.

Suppose $T = T(X)$ is a random vector from $\mathcal{X}$ to $\mathbb{R}^k$, such as an estimator, test statistic, or test function. How can one compute the distribution of T under Q if you know how to compute probabilities or expectations under P? Specifically, suppose it is required to compute $E_Q[f(T)]$, where f is some measurable function from $\mathbb{R}^k$ to $\mathbb{R}$. Let p and q denote the densities of P and Q with respect to a common measure μ. Then, assuming Q is absolutely continuous with respect to P,

$$E_Q[f(T(X))] = \int_{\mathcal{X}} f(T(x))dQ(x) \tag{12.26}$$

$$= \int_{\mathcal{X}} f(T(x))\frac{q(x)}{p(x)}p(x)d\mu(x) = E_P[f(T(X))L(X)] \ , \tag{12.27}$$

where $L(X)$ is the usual likelihood ratio statistic:

$$L(X) = \frac{q(X)}{p(X)} \ . \tag{12.28}$$

Hence, the distribution of $T(X)$ under Q can be computed if the joint distribution of $(T(X), L(X))$ under P is known. Let $F^{T,L}$ denote the joint distribution of $(T(X), L(X))$ under P. Then, by taking f to be the indicator function $f(T(X)) = I_B[T(X)]$ defined to be equal to one if $T(X)$ falls in B and equal to zero otherwise, we obtain:

$$Q\{T(X) \in B\} = \int_{\mathcal{X}} I(T(x) \in B)L(x)p(x)\mu(dx) \tag{12.29}$$

$$= E_P[I(T(X) \in B)L(X)] = \int_{B \times \mathbb{R}} r dF^{T,L}(t,r) \ . \tag{12.30}$$

Thus, under absolute continuity of Q with respect to P, the problem of finding the distribution of $T(X)$ under Q can in principle be obtained from the joint distribution of $T(X)$ and $L(X)$ under P.

More generally, if $f = f(t,r)$ is a function from $\mathbb{R}^k \times \mathbb{R}$ to $\mathbb{R}$,

$$E_Q[f(T(X), L(X))] = \int_{\mathbb{R}^k \times \mathbb{R}} f(t,r) r dF^{T,L}(t,r) \tag{12.31}$$

(Problem 12.18).

Contiguity is an asymptotic version of absolute continuity that permits an analogous asymptotic statement. Consider sequences of pairs of probabilities $\{P_n, Q_n\}$, where P_n and Q_n are probabilities on some measurable space $(\mathcal{X}_n, \mathcal{F}_n)$. Let T_n be some random vector from $\mathcal{X}_n$ to $\mathbb{R}^k$. Suppose the asymptotic distribution of T_n under P_n is easily obtained, but the behavior of T_n under Q_n is also required. For example, if T_n represents a test function for testing P_n versus Q_n, the power of T_n is the expectation of T_n under Q_n. Contiguity provides a means of performing the required calculation. An example may help fix ideas.

Example 12.3.1 (The Wilcoxon Signed Rank Statistic) Let $X_1, \ldots, X_n$ be i.i.d. real-valued random variables with common density $f(\cdot)$. Assume that $f(\cdot)$ is symmetric about θ. The problem is to test the null hypothesis that $\theta = 0$ against the alternative hypothesis that $\theta > 0$. Consider the Wilcoxon signed rank statistic defined by:

$$W_n = W_n(X_1, \ldots, X_n) = n^{-3/2} \sum_{i=1}^{n} R_{i,n}^{+} \text{sign}(X_i) , \tag{12.32}$$

where $\text{sign}(X_i)$ is 1 if $X_i \geq 0$ and is -1 otherwise, and $R_{i,n}^{+}$ is the rank of $|X_i|$ among $|X_1|, \ldots, |X_n|$. Under the null hypothesis, the behavior of W_n is fairly easy to obtain. If $\theta = 0$, the variables $\text{sign}(X_i)$ are i.i.d., each 1 or -1 with probability 1/2, and are independent of the variables $R_{i,n}^{+}$. Hence, $E_{\theta=0}(W_n) = 0$. Define $\tilde{I}_k$ to be 1 if the kth largest $|X_i|$ corresponds to a positive observation and -1 otherwise. Then, we have

$$Var_{\theta=0}(W_n) = n^{-3} Var(\sum_{k=1}^{n} k\tilde{I}_k) \tag{12.33}$$

$$= n^{-3} \sum_{k=1}^{n} k^2 = n^{-3} \frac{n(n+1)(2n+1)}{6} \to \frac{1}{3} \tag{12.34}$$

as $n \to \infty$. Not surprisingly, $W_n \xrightarrow{d} N(0, \frac{1}{3})$. To see why, note that (Problem 12.19)

$$W_n - n^{-1/2} \sum_{i=1}^{n} U_i \text{sign}(X_i) = o_P(1) , \tag{12.35}$$

where $U_i = G(|X_i|)$ and G is the c.d.f. of $|X_i|$. But, under the null hypothesis, U_i and $\text{sign}(X_i)$ are independent. Moreover, the random variables $U_i \text{sign}(X_i)$ are i.i.d., and so the Central Limit Theorem is applicable. Thus, W_n is asymptotically normal with mean 0 and variance 1/3, and this is true whenever the underlying distribution has a symmetric density about 0. Indeed, the exact distribution of W_n is the same for all distributions symmetric about 0. Hence, the test that rejects the null hypothesis if W_n exceeds $3^{-1/2} z_{1-\alpha}$ has limiting level $1 - \alpha$. Of course, for finite n, critical values for W_n can be obtained exactly. Suppose now that we want to approximate the power of this test. The above argument does not generalize to even close alternatives since it heavily uses the fact that the variables are symmetric about zero. Contiguity provides a fairly simple means of attacking this problem, and we will reconsider this example later. ■

We now return to the general setup.

Definition 12.3.1 Let P_n and Q_n be probability distributions on $(\mathcal{X}_n, \mathcal{F}_n)$. The sequence $\{Q_n\}$ is *contiguous* to the sequence $\{P_n\}$ if $P_n(E_n) \to 0$ implies $Q_n(E_n) \to 0$ for every sequence $\{E_n\}$ with $E_n \in \mathcal{F}_n$.

The following equivalent definition is sometimes useful. The sequence $\{Q_n\}$ is contiguous to $\{P_n\}$ if for every sequence of real-valued random variables T_n such that $T_n \to 0$ in P_n-probability we also have $T_n \to 0$ in Q_n-probability.

If $\{Q_n\}$ is contiguous to $\{P_n\}$ and $\{P_n\}$ is contiguous to $\{Q_n\}$, then we say the sequences $\{P_n\}$ and $\{Q_n\}$ are *mutually contiguous*, or just contiguous.

Example 12.3.2 Suppose P_n is the standard normal distribution $N(0, 1)$ and Q_n is $N(\xi_n, 1)$. Unless ξ_n is bounded, P_n and Q_n cannot be contiguous. Indeed, suppose $\xi_n \to \infty$ and consider $E_n = \{x : \; |x - \xi_n| < 1\}$. Then, $Q_n(E_n) \approx 0.68$ for all n, but $P_n(E_n) \to 0$. Note that, regardless of the values of ξ_n, P_n and Q_n are mutually absolutely continuous for every n. ∎

Example 12.3.3 Suppose P_n is the joint distribution of n i.i.d. observations $X_1, \ldots, X_n$ from $N(0, 1)$ and Q_n is the joint distribution of n i.i.d. observations from $N(\xi_n, 1)$. Unless $\xi_n \to 0$, P_n and Q_n cannot be contiguous. For example, suppose $\xi_n > \epsilon > 0$ for all large n. Let $\bar{X}_n = n^{-1} \sum_{i=1}^n X_i$ and consider $E_n = \{\bar{X}_n > \epsilon/2\}$. By the law of large numbers, $P_n(E_n) \to 0$ but $Q_n(E_n) \to 1$. As will be seen shortly, in order for P_n and Q_n to be contiguous, it will be necessary and sufficient for $\xi_n \to 0$ in such a way so that $n^{1/2}\xi_n$ remains bounded. ∎

We now would like a useful means of determining whether or not Q_n is contiguous to P_n. Suppose P_n and Q_n have densities p_n and q_n with respect to μ_n. For $x \in \mathcal{X}_n$, define the *likelihood ratio* of Q_n with respect to P_n by

$$L_n(x) = \begin{cases} \frac{q_n(x)}{p_n(x)} & \text{if } p_n(x) > 0 \\ \infty & \text{if } p_n(x) = 0 < q_n(x) \\ 1 & \text{if } p_n(x) = q_n(x) = 0. \end{cases} \tag{12.36}$$

Under P_n or Q_n, the event $\{p_n = q_n = 0\}$ has probability 0, so it really doesn't matter how L_n is defined in this case (as long as it is measurable). Note that L_n is regarded as an extended random variable, which means it is allowed to take on the value ∞, at least under Q_n. Of course, under P_n, L_n is finite with probability one.

Observe that

$$E_{P_n}(L_n) = \int_{\mathcal{X}_n} L_n(x) p_n(x) \mu_n(dx) = \int_{\{x: \; p_n(x) > 0\}} q_n(x) \mu_n(dx)$$

$$= Q_n\{x : \; p_n(x) > 0\} = 1 - Q_n\{x : \; p_n(x) = 0\} \le 1 \; , \tag{12.37}$$

with equality if and only if Q_n is absolutely continuous with respect to P_n.

Example 12.3.4 (Contiguous but not absolutely continuous sequence) Suppose P_n is uniformly distributed on $[0, 1]$ and Q_n is uniformly distributed on $[0, \theta_n]$, where $\theta_n > 1$. Then, Q_n is not absolutely continuous with respect to P_n.

Note that the likelihood ratio L_n is equal to $1/\theta_n$ with probability one under P_n, and so

$$E_{P_n}(L_n) = \frac{1}{\theta_n} < 1 \ .$$

It will follow from Theorem 12.3.1 that Q_n is contiguous to P_n if $\theta_n \to 1$.

The notation $\mathcal{L}(T|P)$ refers to the distribution of a random variable (or possibly an extended random variable) $T = T(X)$ when X is governed by P. Let $G_n = \mathcal{L}(L_n|P_n)$, the distribution of the likelihood ratio under P_n. Note that G_n is a tight sequence, because by Markov's inequality,

$$P_n\{L_n > c\} \le \frac{E_{P_n}(L_n)}{c} \le \frac{1}{c} \ , \tag{12.38}$$

where the last inequality follows from (12.37).

The statement that $E_{P_n}(L_n) = 1$ implies that Q_n is absolutely continuous with respect to P_n, by (12.37). The following result, known as Le Cam's First Lemma, may be regarded as an asymptotic version of this statement.

Theorem 12.3.1 *Given P_n and Q_n, consider the likelihood ratio L_n defined in (12.36). Let G_n denote the distribution of L_n under P_n. Suppose G_n converges weakly to a distribution G. If G has mean 1, then Q_n is contiguous to P_n.*

PROOF. Suppose $P_n(E_n) = \alpha_n \to 0$. Let ϕ_n be a most powerful level α_n test of P_n versus Q_n. By the Neyman-Pearson Lemma, the test is of the form

$$\phi_n = \begin{cases} 1 & \text{if } L_n > k_n \\ 0 & \text{if } L_n < k_n, \end{cases} \tag{12.39}$$

for some k_n chosen so the test is level α_n. Since ϕ_n is at least as powerful as the test that has rejection region E_n,

$$Q_n\{E_n\} \le \int \phi_n dQ_n \ ,$$

so it suffices to show the right side tends to zero. Now, for any $y < \infty$,

$$\int \phi_n dQ_n = \int_{L_n \le y} \phi_n dQ_n + \int_{L_n > y} \phi_n dQ_n$$

$$\le y \int \phi_n dP_n + \int_{L_n > y} dQ_n \le y \int \phi_n dP_n + 1 - \int_{L_n \le y} dQ_n$$

$$= y\alpha_n + 1 - \int_{L_n \le y} L_n dP_n = y\alpha_n + 1 - \int_0^y x dG_n(x) \ .$$

Fix any $\epsilon > 0$ and take y to be a continuity point of G with

$$\int_0^y x dG(x) > 1 - \frac{\epsilon}{2} \ ,$$

which is possible since G has mean 1. But G_n converges weakly to G implies

$$\int_0^y x dG_n(x) \to \int_0^y x dG(x) \ , \tag{12.40}$$

by an argument like that in Example 11.2.14 (Problem 12.27). Thus, for sufficiently large n,

$$1 - \int_0^y x dG_n(x) < \frac{\epsilon}{2}$$

and $y\alpha_n < \epsilon/2$. It follows that, for sufficiently large n,

$$\int \phi_n dQ_n < \epsilon \ ,$$

as was to be proved. ■

The following result summarizes some equivalent characterizations of contiguity. The notation $\mathcal{L}(T|P)$ refers to the distribution (or law) of a random variable T under P.

Theorem 12.3.2 *The following are equivalent characterizations of $\{Q_n\}$ being contiguous to $\{P_n\}$.*

(i) For every sequence of real-valued random variables T_n such that $T_n \to 0$ in P_n-probability, it also follows that $T_n \to 0$ in Q_n-probability.

(ii) For every sequence T_n such that $\mathcal{L}(T_n|P_n)$ is tight, it also follows that $\mathcal{L}(T_n|Q_n)$ is tight.

(iii) If G is any limit point [3] of $\mathcal{L}(L_n|P_n)$, then G has mean 1.

PROOF. First, we show that (ii) implies (i). Suppose $T_n \to 0$ in P_n-probability; that is, $P_n\{|T_n| > \delta\} \to 0$ for every $\delta > 0$. Then, there exists $\epsilon_n \downarrow 0$ such that $P_n\{|T_n| > \epsilon_n\} \to 0$. So, $|T_n|/\epsilon_n$ is tight under $\{P_n\}$. By hypothesis, $|T_n|/\epsilon_n$ is also tight under $\{Q_n\}$. Assume the conclusion that $T_n \to 0$ in Q_n-probability fails; then, one could find $\epsilon > 0$ such that $Q_n\{|T_n| > \epsilon\} > \epsilon$ for infinitely many n. Then, of course, $Q_n\{|T_n| > \sqrt{\epsilon_n}\} > \epsilon$ for infinitely many n. Since $1/\sqrt{\epsilon_n} \uparrow \infty$, it follows that $|T_n|/\epsilon_n$ cannot be tight under $\{Q_n\}$, which is a contradiction.

Conversely, to show that (i) implies (ii), assume that $\mathcal{L}(T_n|P_n)$ is tight. Then, given $\epsilon > 0$, there exists k such that $P_n\{|T_n| > k\} < \epsilon/2$ for all n. If $\mathcal{L}(T_n|Q_n)$ is not tight, then for every j, $Q_n\{|T_n| > j\} > \epsilon$ for some n. That is, there exists a subsequence n_j such that $Q_{n_j}\{|T_{n_j}| > j\} > \epsilon$ for every j. As soon as $j > k$,

$$P_{n_j}\{|T_{n_j}| > j\} \le P_{n_j}\{|T_{n_j}| > k\} < \frac{\epsilon}{2} \ ,$$

a contradiction.

To show (iii) implies (i), first recall (12.38), which implies G_n is tight. Assuming $P_n\{A_n\} \to 0$, we must show $Q_n\{A_n\} \to 0$. Assume that this is not the case. Then, there exists a subsequence n_j and $\epsilon > 0$ such that $Q_{n_j}\{A_{n_j}\} \ge \epsilon$ for all n_j. But, there exists a further subsequence n_{j_k} such that $G_{n_{j_k}}$ converges to some G. Assuming (iii), G has mean 1. By Theorem 12.3.1, $P_{n_{j_k}}$ and $Q_{n_{j_k}}$ are contiguous. Since $Q_{n_{j_k}}\{A_{n_{j_k}}\} \to 0$, this is a contradiction.

[3] G is a limit point of a sequence G_n of distributions if G_{n_j} converges in distribution to G for some subsequence n_j.

Conversely, suppose (i) and that G_n converges weakly to G (or apply the following argument to any convergent subsequence). By Example 11.2.14, it follows that

$$\int x dG(x) \le \liminf_n E_{P_n}(L_n) \le 1 ,$$

so it suffices to show $\int x dG(x) \ge 1$. Let t be a continuity point of G. Then, also by Example 11.2.14 (specifically (11.39)),

$$\int x dG(x) \ge \int_{\{x \le t\}} x dG(x) = \lim_n E_{P_n}(L_n 1\{L_n \le t\}) = \lim_n Q_n\{L_n \le t\} .$$

So, it suffices to show that, given any $\epsilon > 0$, there exists a t such that $Q_n\{L_n > t\} < \epsilon$ for all large n. If this fails, then for every j, there exists n_j such that $Q_{n_j}\{L_{n_j} > j\} > \epsilon$. But, by (12.38),

$$P_{n_j}\{L_{n_j} > j\} \le \frac{1}{j} \to 0$$

as $j \to \infty$, which would contradict (i). ■

As will be seen in many important examples, loglikelihood ratios are typically asymptotically normally distributed, and the following corollary is useful.

Corollary 12.3.1 *Consider a sequence $\{P_n, Q_n\}$ with likelihood ratio L_n defined in (12.36). Assume*

$$\mathcal{L}(L_n|P_n) \stackrel{d}{\to} \mathcal{L}(e^Z) , \tag{12.41}$$

where Z is distributed as $N(\mu, \sigma^2)$. Then, Q_n and P_n are mutually contiguous if and only if $\mu = -\sigma^2/2$.

PROOF. To show Q_n is contiguous to P_n, apply part (iii) of Theorem 12.3.2 by showing $E(e^Z) = 1$. But, recalling the characteristic function of Z from equation (11.10), it follows that

$$E(e^Z) = \exp(\mu + \frac{1}{2}\sigma^2) ,$$

which equals 1 if and only if $\mu = -\sigma^2/2$. That P_n is contiguous to Q_n follows by Problem 12.23. ■

We may write (12.41) equivalently as

$$\mathcal{L}(\log(L_n)|P_n) \stackrel{d}{\to} \mathcal{L}(Z) .$$

However, since $P_n\{L_n = 0\}$ may be positive, we may have $\log(L_n) = -\infty$ with positive probability, in which case $\log(L_n)$ is regarded as an extended real-valued random variable taking values in $\mathbb{R} \bigcup \{\pm\infty\}$. If X_n is an extended real-valued random variable and X is a real-valued random variable with c.d.f. F, we say (as in Definition 11.2.1) X_n converges in distribution to X if

$$P_n\{X_n \in (-\infty, t]\} \to F(t)$$

whenever t is a continuity point of F. It follows that if X_n converges in distribution to a random variable that is finite (with probability one), then the probability that X_n is finite must tend to 1.

Example 12.3.5 (Example 12.3.2, continued). Again, suppose that $P_n = N(0,1)$ and $Q_n = N(\xi_n, 1)$. In this case,

$$L_n = L_n(X) = \exp(\xi_n X - \frac{1}{2}\xi_n^2) \ .$$

Thus,

$$\mathcal{L}(\log(L_n)|P_n) = N\left(-\frac{\xi_n^2}{2}, \xi_n^2\right) \ .$$

Such a sequence of distributions will converge weakly along a subsequence n_j if and only if $\xi_{n_j} \to \xi$ (for some $|\xi| < \infty$), in which case, the limiting distribution is $N(\frac{-\xi^2}{2}, \xi^2)$ and the relationship between the mean and the variance ($\mu = -\sigma^2/2$) is satisfied. Hence, Q_n is contiguous to P_n if and only if ξ_n is bounded. ■

Example 12.3.6 (Example 12.3.3, continued). Suppose $X_1, \ldots, X_n$ are i.i.d. with common distribution $N(\xi, 1)$. Let P_n represent the joint distribution when $\xi = 0$ and let Q_n represent the joint distribution when $\xi = \xi_n$. Then,

$$\log(L_n(X_1, \ldots, X_n)) = \xi_n \sum_{i=1}^{n} X_i - \frac{n\xi_n^2}{2} \ , \tag{12.42}$$

and so

$$\mathcal{L}(\log(L_n)|P_n) = N\left(-\frac{n\xi_n^2}{2}, n\xi_n^2\right) \ .$$

By an argument similar to that of the previous example, Q_n is contiguous to P_n if and only if $n\xi_n^2$ remains bounded, i.e. $\xi_n = O(n^{-1/2})$. Note that, even if $\xi_n \to 0$, but at a rate slower than $n^{-1/2}$, Q_n is not contiguous to P_n. This is related to the assertion that the problem of testing P_n versus Q_n is degenerate unless $\xi_n \asymp n^{-1/2}$, in the sense that the most powerful level α test ϕ_n has asymptotic power satisfying $E_{\xi_n}(\phi_n) \to 1$ if $n^{1/2}|\xi_n| \to \infty$ and $E_{\xi_n}(\phi_n) \to \alpha$ if $n^{1/2}\xi_n \to 0$.[4] Indeed, suppose without loss of generality that $\xi_n > 0$. Then, the most powerful level α test rejects when $n^{1/2}\bar{X}_n > z_{1-\alpha}$, where $\bar{X}_n = \sum_{i=1} X_i/n$ and $z_{1-\alpha}$ denotes the $1-\alpha$ quantile of the standard normal distribution. The power of ϕ_n against ξ_n is then

$$P_{\xi_n}\{n^{1/2}\bar{X}_n > z_{1-\alpha}\} = P_{\xi_n}\{n^{1/2}(\bar{X}_n - \xi_n) > z_{1-\alpha} - n^{1/2}\xi_n\}$$

$$= P\{Z > z_{1-\alpha} - n^{1/2}\xi_n\},$$

where Z is a standard normal variable. Clearly, the last expression tends to 1 if and only if $n^{1/2}\xi_n \to \infty$; furthermore, it tends to α if and only if $n^{1/2}\xi_n \to 0$. The limiting power is bounded away from α and 1 if and only if $\xi_n \asymp n^{-1/2}$. ■

Example 12.3.7 (Q.m.d. families) Let $\{P_\theta, \ \theta \in \Omega\}$ with Ω an open subset of $\mathbb{R}^k$ be q.m.d., with corresponding densities $p_\theta(\cdot)$. By Theorem 12.2.3, under

[4]Two real-valued sequences $\{a_n\}$ and $\{b_n\}$ are said to be of the same order, written $a_n \asymp b_n$ if $|a_n/b_n|$ is bounded away from 0 and ∞.

θ_0,

$$\log(\frac{dP^n_{\theta_0+hn^{-1/2}}}{dP^n_{\theta_0}}) = n^{-1/2}\sum_{i=1}^{n}\langle h, \tilde{\eta}(X_i,\theta_0)\rangle - \frac{1}{2}\langle h, I(\theta_0)h\rangle + o_{P^n_{\theta_0}}(1), \quad (12.43)$$

where $\tilde{\eta}(x,\theta) = 2\eta(x,\theta)/p_\theta^{1/2}(x)$, $\eta(\cdot,\theta)$ is the quadratic mean derivative at θ, and $I(\theta)$ is the Information matrix at θ. Hence, by Corollary 12.3.1, $P^n_{\theta_0+hn^{-1/2}}$ and $P^n_{\theta_0}$ are mutually contiguous. ■

Suppose Q_n is contiguous to P_n. As before, let L_n be the likelihood ratio defined by (12.28). Let T_n be an arbitrary sequence of real-valued statistics. The following theorem allows us to determine the asymptotic behavior of (T_n, L_n) under Q_n from the behavior of (T_n, L_n) under P_n.

Theorem 12.3.3 *Suppose Q_n is contiguous to P_n. Let T_n be a sequence of real-valued random variables. Suppose, under P_n, (T_n, L_n) converges in distribution to a limit law $F(\cdot,\cdot)$; that is, for any bounded continuous function f on $(-\infty,\infty)\times[0,\infty)$,*

$$E_{P_n}[f(T_n, L_n)] \to \int\int f(t,r)dF(t,r)\ . \quad (12.44)$$

Then, the limiting distribution of (T_n, L_n) under Q_n has density $rdF(t,r)$; that is,

$$E_{Q_n}[f(T_n, L_n)] \to \int\int f(t,r)rdF(t,r) \quad (12.45)$$

for any bounded continuous f. Equivalently, if under P_n $(T_n, \log(L_n))$ converges weakly to a limit law $\bar{F}(\cdot,\cdot)$, then

$$E_{Q_n}[f(T_n, \log(L_n))] \to \int\int f(t,r)e^r d\bar{F}(t,r) \quad (12.46)$$

for any bounded continuous f.

Note that equation (12.45) is simply an asymptotic version of (12.31).

Remark 12.3.1 The result is also true if T_n is vector-valued, and the proof is the same.

PROOF. Let $F_n = \mathcal{L}((T_n, L_n)|P_n)$ and $G_n = \mathcal{L}((T_n, L_n)|Q_n)$. Since L_n converges in distribution under P_n, contiguity and Theorem 12.3.2 (iii) imply that

$$\int rdF(t,r) = 1\ .$$

Thus, $rdF(t,r)$ defines a probability distribution on $(-\infty,\infty)\times[0,\infty)$.

Let f be a nonnegative, continuous function on $(-\infty,\infty)\times[0,\infty]$. By the Portmanteau Theorem (Theorem 11.2.1 (vi)), it suffices to show that

$$\liminf_n \int f(t,r)dG_n(t,r) \ge \int f(t,r)rdF(t,r)\ .$$

Note that

$$\int f(t,r)dG_n(t,r) = E_{Q_n}[f(T_n, L_n)] = \int f(T_n, L_n)dQ_n$$

$$\geq \int_{\{p_n>0\}} f(T_n, L_n)dQ_n = \int f(T_n, L_n)L_n dP_n = \int f(t,r)rdF_n(t,r) \ .$$

So, it suffices to show

$$\liminf_n \int f(t,r)rdF_n(t,r) \geq \int rf(t,r)dF(t,r) \ .$$

But, $rf(t,r)$ is a nonnegative, continuous function, and so the result follows again by the Portmanteau Theorem. ■

The following special case is often referred to as Le Cam's Third Lemma.

Corollary 12.3.2 *Assume that, under P_n, $(T_n, \log(L_n)) \xrightarrow{d} (T,Z)$, where (T,Z) is bivariate normal with $E(T) = \mu_1$, $Var(T) = \sigma_1^2$, $E(Z) = \mu_2$, $Var(Z) = \sigma_2^2$ and $Cov(T,Z) = \sigma_{1,2}$. Assume $\mu_2 = -\sigma_2^2/2$, so that Q_n is contiguous to P_n. Then, under Q_n, T_n is asymptotically normal:*

$$\mathcal{L}(T_n|Q_n) \xrightarrow{d} N(\mu_1 + \sigma_{1,2}, \sigma_1^2) \ .$$

PROOF. Let $\bar{F}(\cdot,\cdot)$ denote the bivariate normal distribution of (T,Z). By Theorem 12.3.3, the limiting distribution of $\mathcal{L}((T_n, \log(L_n))|Q_n)$ has density $e^r d\bar{F}(x,r)$; let $\tilde{T}$ denote a random variable having this distribution. The characteristic function of $\tilde{T}$ is given by:

$$E(e^{i\lambda\tilde{T}}) = \int e^{i\lambda x} e^r d\bar{F}(x,r) = E(e^{i\lambda T + Z}) \ , \tag{12.47}$$

which is the characteristic function of (T,Z) evaluated at $t = (t_1,t_2)^T = (\lambda, -i)^T$. By Example 11.2.1, this is given by

$$\exp(i\langle\mu,t\rangle - \frac{1}{2}\langle\Sigma t, t\rangle) = \exp(i\mu_1\lambda + \mu_2 - \frac{1}{2}\langle\Sigma(\lambda,-i)^T, (\lambda,-i)^T\rangle)$$

$$= \exp(i\mu_1\lambda + \mu_2 - \frac{1}{2}\lambda^2\sigma_1^2 + \lambda i\sigma_{1,2} + \frac{\sigma_2^2}{2}) = \exp[i(\mu_1 + \sigma_{1,2})\lambda - \frac{1}{2}\lambda^2\sigma_1^2] \ ,$$

the last equality following from the fact that $\mu_2 = -\sigma_2^2/2$ (by contiguity). But, this last expression is indeed the characteristic function of the normal distribution with mean $\mu_1 + \sigma_{1,2}$ and variance σ_1^2. ■

Example 12.3.8 (Asymptotically Linear Statistic) Let $\{P_\theta,\ \theta \in \Omega\}$ with Ω an open subset of $\mathbb{R}^k$ be q.m.d., with corresponding densities $p_\theta(\cdot)$. Recall Example 12.3.7, which shows that $P^n_{\theta_0+hn^{-1/2}}$ and $P^n_{\theta_0}$ are mutually contiguous. The expansion (12.43) shows a lot more. For example, suppose an estimator (sequence) $\hat{\theta}_n$ is asymptotically linear in the following sense: under θ_0,

$$n^{1/2}(\hat{\theta}_n - \theta_0) = n^{-1/2}\sum_{i=1}^n \psi_{\theta_0}(X_i) + o_{P^n_{\theta_0}}(1) \ , \tag{12.48}$$

where $E_{\theta_0}[\psi_{\theta_0}(X_1)] = 0$ and $\tau^2 \equiv Var_{\theta_0}[\psi_{\theta_0}(X_1)] < \infty$. Thus, under θ_0,

$$n^{1/2}(\hat{\theta}_n - \theta_0) \stackrel{d}{\to} N(0, \tau^2) .$$

Then, the joint behavior of $\hat{\theta}_n$ with the likelihood ratio satisfies

$$(n^{1/2}(\hat{\theta}_n - \theta_0), \frac{dP^n_{\theta_0+hn^{-1/2}}}{dP^n_{\theta_0}}) \tag{12.49}$$

$$= [n^{-1/2}\sum_{i=1}^n (\psi_{\theta_0}(X_i), \langle h, \tilde{\eta}(X_i,\theta_0)\rangle)] + (0, -\frac{1}{2}\langle h, I(\theta_0)h\rangle) + o_{P^n_{\theta_0}}(1) .$$

By the bivariate Central Limit Theorem, this converges under θ_0 to a bivariate normal distribution with covariance

$$\sigma_{1,2} \equiv Cov_{\theta_0}(\psi_{\theta_0}(X_1), \langle h, \tilde{\eta}(X_i,\theta_0)\rangle) . \tag{12.50}$$

Hence, under $P^n_{\theta_0+hn^{-1/2}}$, $n^{1/2}(\hat{\theta}_n - \theta_0)$ converges in distribution to $N(\sigma_{1,2}, \tau^2)$, by Corollary 12.3.2. It follows that, under $P^n_{\theta_0+hn^{-1/2}}$,

$$n^{1/2}(\hat{\theta}_n - (\theta_0 + hn^{-1/2})) \stackrel{d}{\to} N(\sigma_{1,2} - h, \tau^2) . \blacksquare$$

Example 12.3.9 (t-statistic) Consider a location model $f(x-\theta)$ for which $f(x)$ has mean 0 and variance σ^2, and which satisfies the assumptions of Corollary 12.2.1, which imply this family is q.m.d. For testing $\theta = \theta_0 = 0$, consider the behavior of the usual t-statistic

$$t_n = \frac{n^{1/2}\bar{X}_n}{S_n} = \frac{n^{1/2}\bar{X}_n}{\sigma} + o_{P_{\theta_0}}(1) .$$

Then, (12.48) holds with $\psi_{\theta_0}(X_i) = X_i/\sigma$. We seek the behavior of t_n under $\theta_n = h/n^{1/2}$. Although this can be obtained by direct means, let us obtain the results by contiguity. Note that (12.43) holds with

$$\tilde{\eta}(X_i, \theta_0) = -\frac{f'(x)}{f(x)} .$$

Thus, $\sigma_{1,2}$ in (12.50) reduces to

$$\sigma_{1,2} = -\frac{h}{\sigma} Cov_{\theta_0=0}\left(X_i, \frac{f'(X_i)}{f(X_i)}\right) = -\frac{h}{\sigma}\int_{-\infty}^{\infty} x f'(x)dx = \frac{h}{\sigma} .$$

Hence, under $\theta_n = h/n^{1/2}$,

$$t_n \stackrel{d}{\to} N(\frac{h}{\sigma}, 1) . \blacksquare$$

Example 12.3.10 (Sign Test) As in the previous example, consider a location model $f(x-\theta)$, where f is a density with respect to Lebesgue measure. Assume the conditions in Corollary 12.2.1, so that the family is q.m.d. Further suppose that $f(x)$ is continuous at $x = 0$ and $P_{\theta=0}\{X_i > 0\} = 1/2$. For testing $\theta = \theta_0 = 0$, consider the (normalized) sign statistic

$$S_n = n^{-1/2}\sum_{i=1}^n [I\{X_i > 0\} - \frac{1}{2}] ,$$

where $I\{X_i > 0\}$ is one if $X_i > 0$ and is 0 otherwise. Then, (12.48) holds with $\psi_0(X_i) = I\{X_i > 0\} - \frac{1}{2}$ and so

$$S_n \xrightarrow{d} N(0, \frac{1}{4}) .$$

Under $\theta_n = h/n^{1/2}$, $S_n \xrightarrow{d} N(\sigma_{1,2}, 1/4)$, where $\sigma_{1,2}$ is given by (12.50) and equals

$$\sigma_{1,2} = -hCov_0\left[I\{X_i > 0\}, \frac{f'(X_i)}{f(X_i)}\right] = -h\int_0^\infty f'(x)dx = hf(0) .$$

Hence, under $\theta_n = h/n^{1/2}$,

$$S_n \xrightarrow{d} N(hf(0), \frac{1}{4}) . \blacksquare$$

Example 12.3.11 (Example 12.3.1, continued). Recall the Wilcoxon signed rank statistic W_n given by (12.32). For illustration, suppose the underlying density $f(\cdot)$ of the observations is normal with mean θ and variance 1. Under the null hypothesis $\theta = 0$, W_n is asymptotically normal $N(0, \frac{1}{3})$. The problem now is to compute the asymptotic power against the sequence of alternatives $\theta_n = h/n^{1/2}$ for some $h > 0$. Under the null hypothesis, by (12.35) and (12.42),

$$(W_n, \log(L_n)) = (n^{-1/2}\sum_{i=1}^n U_i \text{sign}(X_i), hn^{-1/2}\sum_{i=1}^n X_i - \frac{h^2}{2}) + o_{P_0^n}(1) , \quad (12.51)$$

where $U_i = G(|X_i|)$ and G is the c.d.f. of $|X_i|$. This last expression is asymptotically bivariate normal with covariance under $\theta = 0$ equal to

$$\sigma_{1,2} = hCov_0[G(|X_1|)\text{sign}(X_1), X_1] = hE_0[G(|X_1|)|X_1|] , \quad (12.52)$$

and thus $\sigma_{1,2}$ is equal to $h/\sqrt{\pi}$ (Problem 12.28). Hence, under $\theta_n = h/n^{1/2}$, W_n is asymptotically normal with mean $h/\sqrt{\pi}$ and variance 1/3. Thus, the asymptotic power of the test that rejects when $W_n > 3^{-1/2}z_{1-\alpha}$ is

$$\lim_{n\to\infty} P_{\theta_n}\{W_n - \frac{h}{\sqrt{\pi}} > 3^{-1/2}z_{1-\alpha} - \frac{h}{\sqrt{\pi}}\} = 1 - \Phi(z_{1-\alpha} - (3/\pi)^{1/2}h) ,$$

where $\Phi(\cdot)$ is the standard normal c.d.f.

More generally, assume the underlying model is a location model $f(x - \theta)$, where $f(x)$ is assumed symmetric about zero. Assume $f'(x)$ exists for Lebesgue almost all x and

$$0 < I \equiv \int \frac{[f'(x)]^2}{f(x)}dx < \infty .$$

Then, by Corollary 12.2.1, this model is q.m.d. and (12.43) holds with

$$\tilde{\eta}(x, 0) = -\frac{f'(x)}{f(x)} .$$

Under the null hypothesis $\theta = 0$, $W_n \xrightarrow{d} N(0, 1/3)$, as in the normal case. Under the sequence of alternatives $\theta_n = h/n^{1/2}$,

$$W_n \xrightarrow{d} N(\sigma_{1,2}, \frac{1}{3}) ,$$

where $\sigma_{1,2}$ is given by (12.50). In this case,

$$\sigma_{1,2} = Cov_{\theta=0}[U\text{sign}(X), -h\frac{f'(X)}{f(X)}] ,$$

where $U = G(|X|)$ and G is the c.d.f. of $|X|$ when X has density $f(\cdot)$. So, $G(x) = 2F(x) - 1$, where F is the c.d.f. of X. By an integration by parts (see Problem 12.29),

$$\sigma_{1,2} = -hE_{\theta=0}[G(|X|)\text{sign}(X)\frac{f'(X)}{f(X)}] = 2h\int_{-\infty}^{\infty} f^2(x)dx \ . \tag{12.53}$$

Thus, under $\theta_n = h/n^{1/2}$,

$$W_n \xrightarrow{d} N(2h\int_{-\infty}^{\infty} f^2(x)dx, \frac{1}{3}) \ . \ \blacksquare$$

Example 12.3.12 (Neyman-Pearson Statistic) Assume $\{P_\theta,\ \theta \in \Omega\}$ is q.m.d. at θ_0, where Ω is an open subset of $\mathbb{R}^k$ and $I(\theta_0)$ is nonsingular, so that the assumptions behind Theorem 12.2.3 are in force. Let $p_\theta(\cdot)$ be the corresponding density of P_θ. Consider the likelihood ratio statistic based on n i.i.d. observations $X_1, \ldots, X_n$ given by

$$L_{n,h} = \frac{dP^n_{\theta_0+hn^{-1/2}}}{dP^n_{\theta_0}} = \prod_{i=1}^{n} \frac{p_{\theta_0+hn^{-1/2}}(X_i)}{p_{\theta_0}(X_i)} \ . \tag{12.54}$$

By Theorem 12.2.3, under P_{θ_0},

$$\log(L_{n,h}) \xrightarrow{d} N(-\frac{\sigma_h^2}{2}, \sigma_h^2) \ , \tag{12.55}$$

where $\sigma_h^2 = \langle h, I(\theta_0)h\rangle$. Apply Corollary 12.3.2 with $T_n \equiv \log(L_{n,h})$, so that $T = Z$ and $\sigma_{1,2} = \sigma_h^2$. Then, under $P^n_{\theta_0+hn^{-1/2}}$, $\log(L_{n,h})$ is asymptotically $N(\frac{\sigma_h^2}{2}, \sigma_h^2)$. Hence, the test that rejects when $\log(L_{n,h})$ exceeds $-\frac{1}{2}\sigma_h^2 + z_{1-\alpha}\sigma_h$ is asymptotically level α for testing $\theta = \theta_0$ versus $\theta = \theta_0 + hn^{-1/2}$, where $z_{1-\alpha}$ denotes the $1-\alpha$ quantile of $N(0,1)$. Then, the limiting power of this test sequence for testing $\theta = \theta_0$ versus $\theta = \theta_0 + hn^{-1/2}$ is $1 - \Phi(z_{1-\alpha} - \sigma_h)$ (Problem 12.30). $\blacksquare$

12.4 Likelihood Methods in Parametric Models

The goal of this section is to study some classical large sample methods based on the likelihood function. The classical likelihood ratio test, as well as the tests of Wald and Rao will be introduced, but optimality of these tests will be deferred until the next chapter. Throughout this section, we will assume that $X_1, \ldots, X_n$ are i.i.d. with common distribution P_θ, where $\theta \in \Omega$ and Ω is an open subset of $\mathbb{R}^k$. We will also assume each P_θ is absolutely continuous with respect to a common σ-finite measure μ, so that p_θ denotes the density of P_θ with respect to μ. The *likelihood function* is defined by

$$L_n(\theta) = \prod_{i=1}^{n} p_\theta(X_i) \ . \tag{12.56}$$

It is thus the (joint) probability density of the observations at fixed values of $X_1, \ldots, X_n$, viewed as a function of θ. Note that, for the sake of simplicity, the dependence of $L_n(\theta)$ on $X_1, \ldots, X_n$ has been suppressed. (In the case that $X_1, \ldots, X_n$ are not i.i.d., $L_n(\theta)$ is modified so that the joint density of the X_i's is used rather than the product of the marginal densities.)

12.4.1 Efficient Likelihood Estimation

In preparation for the construction of reasonable large sample tests and confidence regions, we begin by studying some efficient point estimators of θ which will serve as a basis for such tests. If the likelihood $L_n(\theta)$ has a unique maximum $\hat{\theta}_n$, then $\hat{\theta}_n$ is called the *maximum likelihood estimator* (MLE) of θ. If, in addition, $L_n(\theta)$ is differentiable in θ, $\hat{\theta}_n$ will be a solution of the *likelihood equations*

$$\frac{\partial}{\partial \theta_j} \log L_n(\theta) = 0 \qquad j = 1, \ldots, k \ .$$

Example 12.4.1 (Normal Family) Suppose $X_1, \ldots, X_n$ is an i.i.d. sample from $N(\mu, \sigma^2)$, with both parameters unknown, so $\theta = (\mu, \sigma^2)^T$. In this case, the log likelihood function is

$$\log L_n(\mu, \sigma^2) = -\frac{n}{2} \log(2\pi) - n \log(\sigma) - \frac{1}{2\sigma^2} \sum_{i=1}^{n} (X_i - \mu)^2 \ ,$$

and the likelihood equations reduce to

$$\frac{1}{\sigma^2} \sum_{i=1}^{n} (X_i - \mu) = 0$$

and

$$-\frac{n}{2\sigma^2} + \frac{1}{2\sigma^4} \sum_{i=1}^{n} (X_i - \mu)^2 = 0 \ .$$

These equations have a unique solution, given by the maximum likelihood estimator $(\hat{\mu}_n, \hat{\sigma}_n^2)$, where $\hat{\mu}_n = \bar{X}_n$ is the usual sample mean and $\hat{\sigma}_n^2$ is the biased version of the sample variance given by

$$\hat{\sigma}_n^2 = n^{-1} \sum (X_i - \bar{X}_n)^2$$

(Problem 12.35). By the weak law of large numbers, $\bar{X}_n \to \mu$ in probability; by Example 11.2.6, $\hat{\sigma}_n^2 \to \sigma^2$ in probability as well. A direct argument easily establishes the joint limiting distribution of the MLE. First note that

$$n^{1/2}[\hat{\sigma}_n^2 - n^{-1} \sum_{i=1}^{n} (X_i - \mu)^2] = n^{1/2} (\bar{X}_n - \mu)^2 \stackrel{P}{\to} 0$$

since $n^{1/2}(\bar{X}_n - \mu)$ is $N(0, \sigma^2)$ and $\bar{X}_n - \mu \stackrel{P}{\to} 0$. Hence, by Slutsky's Theorem, $n^{1/2}((\bar{X}_n, \hat{\sigma}_n^2)^T - (\mu, \sigma^2)^T)$ has the same limiting distribution as

$$n^{1/2}[(\bar{X}_n, n^{-1} \sum_{i=1}^{n} (X_i - \mu)^2)^T - (\mu, \sigma^2)^T] \ ,$$

which by the multivariate CLT tends in distribution to $N(0, \Sigma)$, where Σ is the 2×2 diagonal matrix with (i,j) entry $\sigma_{i,j}$ given by $\sigma_{1,1} = \sigma^2$ and $\sigma_{2,2} = Var[(X_1 - \mu)^2] = 2\sigma^4$. In fact, $\Sigma = I^{-1}(\theta)$ in this case. ■

Example 12.4.2 (MLE for a one-parameter exponential family) Suppose $X_1, \ldots, X_n$ is an i.i.d. sample from a one-parameter exponential family with common density with respect to a σ-finite measure μ given by

$$p_\theta(x) = \exp[\theta T(x) - A(\theta)] .$$

Here, θ is assumed to be an interior point of the natural parameter space. From Problem 2.16, recall that $E_\theta[T(X_i)] = A'(\theta)$ and $Var_\theta[T(X_i)] = A''(\theta)$. To show the maximum likelihood estimator is well-defined and to find an expression for it, we examine the derivative of the log of $L_n(\theta)$, which is equal to

$$\frac{\partial \log L_n(\theta)}{\partial \theta} = \sum_{i=1}^n [T(X_i) - A'(\theta)] .$$

The likelihood equation sets this equal to zero, which reduces to the equation $\bar{T}_n = A'(\theta)$, where $\bar{T}_n = n^{-1}\sum_{i=1}^n T(X_i)$. Hence, the MLE is found by equating the sample mean of the $T(X_i)$ values to its expected value. Assuming the equation $\bar{T}_n = A'(\theta)$ can be solved for θ, it must be the maximum likelihood estimator. Indeed, the second derivative of the log likelihood is $-nA''(\theta) < 0$, which also shows there can at be at most one solution to the likelihood equation. Furthermore, by the law of large numbers, $\bar{T}_n \xrightarrow{P} A'(\theta)$, which combined with the fact that $A''(\theta) > 0$ yields that, with probability tending to one, there exists exactly one solution to the likelihood equation. Thus, $\hat{\theta}_n$ is well-defined with probability tending to one. To determine its limiting distribution, first note that

$$n^{1/2}[\bar{T}_n - A'(\theta)] \xrightarrow{d} N(0, A''(\theta)) ,$$

by the Central Limit Theorem. Since A' is strictly increasing, we can define the inverse function B of A' so that $B(A'(\theta)) = \theta$. Then, $\hat{\theta}_n = B(A'(\hat{\theta}_n)) = B(\bar{T}_n)$. By the delta method,

$$n^{1/2}(\hat{\theta}_n - \theta) \xrightarrow{d} N(0, \tau^2) ,$$

where

$$\tau^2 = A''(\theta)[B'(A'(\theta))]^2 .$$

But using the chain rule to differentiate both sides of the identity $B(A'(\theta)) = \theta$ yields $B'(A'(\theta))A''(\theta) = 1$, so that

$$n^{1/2}(\hat{\theta}_n - \theta) \xrightarrow{d} N\left(0, \frac{1}{A''(\theta)}\right) .$$

In fact, the asymptotic variance $[A''(\theta)]^{-1}$ is $I^{-1}(\theta)$, where $I(\theta)$ is the Fisher Information. ■

Problem 12.37 generalizes the previous example to multiparameter exponential families.

The general theory of asymptotic normality of the MLE is much more difficult and we shall here only give a heuristic treatment. For precise conditions and rigorous proofs, see Lehmann and Casella (1998), Chapter 6 and Ibragimov and Has'minskii (1981), Section 3.3. Let $X_1, \ldots, X_n$ be i.i.d. according to a family $\{P_\theta\}$ which is q.m.d. at θ_0 with nonsingular Fisher Information matrix $I(\theta_0)$ and quadratic mean derivative $\eta(\cdot, \theta_0)$. Define

$$L_{n,h} = \frac{L_n(\theta_0 + hn^{-1/2})}{L_n(\theta_0)} . \tag{12.57}$$

By Theorem 12.2.3,

$$\log(L_{n,h}) = \langle h, Z_n \rangle - \frac{1}{2}\langle h, I(\theta_0)h \rangle + o_{P_{\theta_0}^n}(1) , \tag{12.58}$$

where Z_n is the normalized score vector

$$Z_n = Z_n(\theta_0) = 2n^{-1/2} \sum_{i=1}^{n} [\eta(X_i, \theta_0)/p_{\theta_0}^{1/2}(X_i)] \tag{12.59}$$

and satisfies, under θ_0,

$$Z_n \xrightarrow{d} N(0, I(\theta_0)) .$$

Note that $Z_n = Z_n(\theta_0)$ depends on θ_0, but we will usually omit this dependence in the notation.

If the MLE $\hat{\theta}_n$ is well-defined, then $\hat{\theta}_n = \theta_0 + \hat{h}_n n^{-1/2}$, where $\hat{h}_n$ is the value of h maximizing $L_{n,h}$. The result (12.58) suggests that, if θ_0 is the true value, $\hat{h}_n$ is approximately equal to $\tilde{h}_n$ which maximizes

$$\log(\tilde{L}_{n,h}) \equiv \langle h, Z_n \rangle - \frac{1}{2}\langle h, I(\theta_0)h \rangle . \tag{12.60}$$

Since $\log(\tilde{L}_{n,h})$ is a simple (quadratic) function of h, it is easily checked (Problem 12.44) that

$$\tilde{h}_n = I^{-1}(\theta_0) Z_n . \tag{12.61}$$

It then follows that

$$n^{1/2}(\hat{\theta}_n - \theta_0) = \hat{h}_n \approx \tilde{h}_n = I^{-1}(\theta_0) Z_n \xrightarrow{d} N(0, I^{-1}(\theta_0)) .$$

The symbol $\approx$ is used to indicate an approximation based on heuristic considerations. Unfortunately, the above approximation is not rigorous without further conditions. In fact, without further conditions, the maximum likelihood estimator may not even be consistent. Indeed, an example of Le Cam (presented in Example 4.1 of Chapter 6 in Lehmann and Casella (1998)) shows that the maximum likelihood estimator $\hat{\theta}_n$ may exist and be unique but does not converge to the true value θ in probability (i.e., it is inconsistent). Moreover, the example shows this can happen even in very smooth families in which good estimators do exist. Rigorous conditions for the MLE to be consistent were given by Wald (1949), and have since then been weakened (for a survey, see Perlman (1972)). Cramér (1946) derived good asymptotic behavior of the maximum likelihood estimator under just certain smoothness conditions, often known as *Cramér type conditions*. Furthermore, he gave conditions under which there exists a consistent sequence of roots $\hat{\theta}_n$ of the likelihood equations (not necessarily the MLE) satisfying

$$n^{1/2}(\hat{\theta}_n - \theta_0) = I^{-1}(\theta_0) Z_n + o_{P_{\theta_0}^n}(1) , \tag{12.62}$$

from which asymptotic normality follows. Cramér's conditions required that the underlying family of densities were three times differentiable with respect to θ, as well as further technical assumptions on differentiability inside the integral signs; see Chapter 6 of Lehmann and Casella (1998). Estimators satisfying (12.62) are called *efficient*. In the case where $\hat{\theta}_n$ is a solution to the likelihood equations, it is called an *efficient likelihood estimator* (ELE) sequence.

Determination of an efficient sequence of roots of the likelihood equations tends to be difficult when the equations have multiple roots. Asymptotically equivalent estimators can be constructed by starting with any estimator $\tilde{\theta}_n$ that is $n^{1/2}$-consistent, i.e. for which $n^{1/2}(\tilde{\theta}_n - \theta)$ is bounded in probability. The resulting estimator can be taken to be the root closest to $\tilde{\theta}_n$, or an approximation to it based on a Newton-Raphson linearization method; for more details, see Section 6.4 of Lehmann and Casella (1998), Gan and Jiang (1999) and Small, Wang and Yang (2000). A similar, but distinct, approach based on discretization of an initial estimator, leads to Le Cam's (1956, 1969) *one-step maximum likelihood estimator*, which satisfies (12.62) under fairly weak conditions.

If $\hat{\theta}_n$ is any estimator sequence (not necessarily the MLE or an ELE) which satisfies (12.62), it follows that, under θ_0,

$$n^{1/2}(\hat{\theta}_n - \theta_0) \xrightarrow{d} N(0, I^{-1}(\theta_0)) .$$

For the remainder of this section, we will assume such an estimator sequence $\hat{\theta}_n$ is available, by means of verification of Cramér type assumptions presented in Lehmann and Casella (1998), or by direct verification as in the case of exponential families of Example 12.4.2 and Problem 12.37. For testing applications, it is also important to study the behavior of the estimator under contiguous alternatives. The following theorem assumes the expansion (12.62) (which is only assumed to hold under θ_0) in order to derive the limiting behavior of $\hat{\theta}_n$ under contiguous sequences θ_n.

Theorem 12.4.1 *Assume $X_1, \ldots, X_n$ are i.i.d. according to a q.m.d. model $\{P_\theta,\ \theta \in \Omega\}$ with nonsingular Information matrix $I(\theta)$, $\theta \in \Omega$, an open subset of $\mathbb{R}^k$. Suppose an estimator $\hat{\theta}_n$ has the expansion (12.62) when $\theta = \theta_0$. Let $\theta_n = \theta_0 + h_n n^{-1/2}$, where $h_n \to h \in \mathbb{R}^k$. Then, under $P^n_{\theta_n}$,*

$$n^{1/2}(\hat{\theta}_n - \theta_n) \xrightarrow{d} N(0, I^{-1}(\theta_0)) ; \tag{12.63}$$

equivalently, under $P^n_{\theta_n}$,

$$n^{1/2}(\hat{\theta}_n - \theta_0) \xrightarrow{d} N(h, I^{-1}(\theta_0)) . \tag{12.64}$$

Furthermore, if $g(\theta)$ is a differentiable map from Ω to $\mathbb{R}$ with nonzero gradient $\dot{g}(\theta)$ of dimension $1 \times k$, then under $P^n_{\theta_n}$,

$$n^{1/2}(g(\hat{\theta}_n) - g(\theta_n)) \xrightarrow{d} N(0, \sigma^2_{\theta_0}) , \tag{12.65}$$

where

$$\sigma^2_{\theta_0} = \dot{g}(\theta_0) I^{-1}(\theta_0) \dot{g}(\theta_0)^T . \tag{12.66}$$

Proof. We prove the result in the case $h_n = h$, the more general case deferred to Problem 13.13. We will first show (12.64). By the Cramér-Wold device, it is enough to show that, for any $t \in \mathbb{R}^k$, under $P^n_{\theta_n}$,

$$\langle n^{1/2}(\hat{\theta}_n - \theta_0), t\rangle \xrightarrow{d} N(\langle h, t\rangle, \langle t, I^{-1}(\theta_0)t\rangle) .$$

By the assumption (12.62), we only need to show that, under $P^n_{\theta_n}$,

$$\langle I^{-1}(\theta_0)Z_n, t\rangle \xrightarrow{d} N(\langle h, t\rangle, \langle t, I^{-1}(\theta_0)t\rangle) .$$

By Example 12.3.7, $P^n_{\theta_n}$ is contiguous to $P^n_{\theta_0}$, so we can apply Corollary 12.3.2 with $T_n = \langle I^{-1}(\theta_0)Z_n, t\rangle$. Then,

$$(T_n, \log(L_{n,h})) = (\langle I^{-1}(\theta_0)Z_n, t\rangle, \langle h, Z_n\rangle - \frac{1}{2}\langle h, I(\theta_0)h\rangle) + o_{P^n_{\theta_0}}(1) .$$

But, under θ_0, Z_n converges in law to Z, where Z is distributed as $N(0, I(\theta_0))$. By Slutsky's Theorem and the Continuous Mapping Theorem (or the bivariate Central Limit Theorem), under θ_0,

$$(T_n, \log(L_{n,h})) \xrightarrow{d} (\langle I^{-1}(\theta_0)Z, t\rangle, \langle h, Z\rangle - \frac{1}{2}\langle h, I(\theta_0)h\rangle) .$$

This limiting distribution is bivariate normal with covariance

$$\sigma_{1,2} = Cov(\langle I^{-1}(\theta_0)Z, t\rangle, \langle h, Z\rangle) = E[(h^T Z)(I^{-1}(\theta_0)Z)^T t]$$

$$= h^T E(Z_1 Z_1^T) I^{-1}(\theta_0)t = h^T I(\theta_0) I^{-1}(\theta_0)t = \langle h, t\rangle .$$

The result (12.64) follows from Corollary 12.3.2. The assertion (12.65) follows from (12.63) and the delta method. ■

Under the conditions of the previous theorem, the estimator sequence $g(\hat{\theta}_n)$ possesses a weak robustness property in the sense that its limiting distribution is unchanged by small perturbations of the parameter values. In the literature, such estimator sequences are sometimes called *regular.*

Corollary 12.4.1 *Assume $X_1, \ldots, X_n$ are i.i.d. according to a q.m.d. model $\{P_\theta,\ \theta \in \Omega\}$ with normalized score vector Z_n given by (12.59) and nonsingular Information matrix $I(\theta_0)$. Let $\theta_n = \theta_0 + h_n n^{-1/2}$, where $h_n \to h \in \mathbb{R}^k$. Then, under $P^n_{\theta_n}$,*

$$Z_n \xrightarrow{d} N(I(\theta_0)h, I(\theta_0)) . \tag{12.67}$$

The proof is left as an exercise (Problem 12.38).

12.4.2 Wald Tests and Confidence Regions

Wald proposed tests and confidence regions based on the asymptotic distribution of the maximum likelihood estimator. In this section, we introduce these methods and study their large sample behavior; some optimality properties will be discussed in Sections 13.3 and 13.4. We assume $\hat{\theta}_n$ is any estimator satisfying (12.62). Let $g(\theta)$ be a mapping from Ω to the real line, assumed differentiable with nonzero gradient vector $\dot{g}(\theta)$ of dimension $1 \times k$. Suppose the problem is to test the null hypothesis $g(\theta) = 0$ versus the alternative $g(\theta) > 0$. Let θ_0 denote the true value of θ. Under the assumptions of Theorem 12.4.1, under θ_0,

$$n^{1/2}[g(\hat{\theta}_n) - g(\theta_0)] \xrightarrow{d} N(0, \sigma^2_{\theta_0}) ,$$

where

$$\sigma_{\theta_0}^2 = \dot{g}(\theta_0) I^{-1}(\theta_0) \dot{g}(\theta_0)^T .$$

Assuming that $\dot{g}(\cdot)$ and $I(\cdot)$ are continuous, the asymptotic variance can be consistently estimated by

$$\hat{\sigma}_n^2 \equiv \dot{g}(\hat{\theta}_n) I^{-1}(\hat{\theta}_n) \dot{g}(\hat{\theta}_n)^T .$$

Hence, the test that rejects when

$$n^{1/2} g(\hat{\theta}_n) > \hat{\sigma}_n z_{1-\alpha}$$

is pointwise asymptotically level α.

We can also calculate the limiting power against a sequence of alternatives $\theta_n = \theta_0 + hn^{-1/2}$. Assume $g(\theta_0) = 0$. Then,

$$P_{\theta_n}\{n^{1/2} g(\hat{\theta}_n) > \hat{\sigma}_n z_{1-\alpha}\} = P_{\theta_n}\{n^{1/2}[g(\hat{\theta}_n) - g(\theta_n)] > \hat{\sigma}_n z_{1-\alpha} - n^{1/2} g(\theta_n)\} .$$

By Theorem 12.4.1, $n^{1/2}[g(\hat{\theta}_n) - g(\theta_n)]$ is asymptotically $N(0, \sigma_{\theta_0}^2)$, under θ_n. Also, $\hat{\sigma}_n \to \sigma_{\theta_0}$ in probability under θ_n (since this convergence holds under θ_0 and therefore under θ_n by contiguity). Finally, $n^{1/2} g(\theta_n) \to \dot{g}(\theta_0) h$. Hence, the limiting power is

$$\lim_{n\to\infty} P_{\theta_n}\{n^{1/2} g(\hat{\theta}_n) > \hat{\sigma}_n z_{1-\alpha}\} = 1 - \Phi(z_{1-\alpha} - \sigma_{\theta_0}^{-1} \dot{g}(\theta_0) h) . \quad (12.68)$$

Similarly, a pointwise asymptotically level $1-\alpha$ level confidence interval for $g(\theta)$ is given by

$$g(\hat{\theta}_n) \pm z_{1-\frac{\alpha}{2}} n^{-1/2} \hat{\sigma}_n .$$

Example 12.4.3 (Normal Coefficient of Variation) Let $X_1, \ldots, X_n$ be i.i.d. $N(\mu, \sigma^2)$ with both parameters unknown, as in Example 12.4.1. Consider inferences for $g((\mu, \sigma^2)^T) = \mu/\sigma$, the coefficient of variation. Recall that a uniformly most accurate invariant one-sided confidence bound exists for μ/σ; however, it is quite complicated to compute since it involves the noncentral t-distribition and no explicit formula is available. However, a normal approximation leads to an interval that is asymptotically valid. Note that

$$\dot{g}((\mu, \sigma^2)^T) = (\frac{1}{\sigma}, -\frac{\mu}{2\sigma^3}) .$$

By Example 12.4.1, $n^{1/2}[(\bar{X}_n, S_n^2)^T - (\mu, \sigma^2)^T]$ is asymptotically bivariate normal with asymptotic covariance matrix Σ, where Σ is the diagonal matrix with $(1,1)$ entry σ^2 and $(2,2)$ entry $2\sigma^4$. Then, the delta method implies that

$$n^{1/2}(\frac{\bar{X}_n}{S_n} - \frac{\mu}{\sigma}) \xrightarrow{d} N(0, 1 + \frac{\mu^2}{2\sigma^2}) .$$

Thus, the interval

$$\frac{\bar{X}_n}{S_n} \pm n^{-1/2}(1 + \frac{\bar{X}_n^2}{2S_n^2}) z_{1-\frac{\alpha}{2}}$$

is asymptotically pointwise level $1-\alpha$. ■

Consider now the general problem of constructing a confidence region for θ, under the assumptions of Theorem 12.4.1. The convergence

$$n^{1/2}(\hat{\theta}_n - \theta) \stackrel{d}{\to} N(0, I^{-1}(\theta)) \tag{12.69}$$

implies that

$$I^{1/2}(\theta) n^{1/2}(\hat{\theta}_n - \theta) \stackrel{d}{\to} N(0, I_k) ,$$

the multivariate normal distribution in $\mathbb{R}^k$ with mean 0 and identity covariance matrix I_k. Hence, by the Continuous Mapping Theorem 11.2.13 and Example 11.2.8,

$$n(\hat{\theta}_n - \theta)^T I(\theta)(\hat{\theta}_n - \theta) \stackrel{d}{\to} \chi_k^2 ,$$

the Chi-squared distribution with k degrees of freedom. Thus, a pointwise asymptotic level $1-\alpha$ confidence region for θ is

$$\{\theta : n(\hat{\theta}_n - \theta)^T I(\theta)(\hat{\theta}_n - \theta) \le c_{k,1-\alpha}\} , \tag{12.70}$$

where $c_{k,1-\alpha}$ is the $1-\alpha$ quantile of χ_k^2. In (12.70), $I(\theta)$ is often replaced by a consistent estimator, such as $I(\hat{\theta}_n)$ (assuming $I(\cdot)$ is continuous), and the resulting confidence region is known as Wald's confidence ellipsoid.

By the duality between confidence regions and tests, this leads to an asymptotic level α test of $\theta = \theta_0$ versus $\theta \neq \theta_0$, known as Wald tests. Specifically, for testing $\theta = \theta_0$ versus $\theta \neq \theta_0$, Wald's test rejects if

$$n(\hat{\theta}_n - \theta_0) I(\hat{\theta}_n)(\hat{\theta}_n - \theta_0) > c_{k,1-\alpha} . \tag{12.71}$$

Alternatively, $I(\hat{\theta}_n)$ may be replaced by $I(\theta_0)$ or any consistent estimator of $I(\theta_0)$. Under $\theta_n = \theta_0 + hn^{-1/2}$, the limiting distribution of the Wald statistic given by the left side of (12.71) is $\chi_k^2(|I^{1/2}(\theta_0)h|^2)$, the noncentral Chi-squared distribution with k degrees of freedom and noncentrality parameter $|I^{1/2}(\theta_0)h|^2$ (Problem 12.45).

More generally, consider inference for $g(\theta)$, where $g = (g_1, \ldots, g_q)^T$ is a mapping from $\mathbb{R}^k$ to $\mathbb{R}^q$. Assume g_i is differentiable and let $D = D(\theta)$ denote the $q \times k$ matrix with (i, j) entry $\partial g_i(y_1, \ldots, y_k)/\partial y_j$ evaluated at θ. Then, the Delta Method and (12.69) imply that

$$n^{1/2}[g(\hat{\theta}_n) - g(\theta)] \stackrel{d}{\to} N(0, V(\theta)) , \tag{12.72}$$

where $V(\theta) = D(\theta) I^{-1}(\theta) D^T(\theta)$. Assume $V(\theta)$ is positive definite and continuous in θ. By the Continuous Mapping Theorem,

$$n[g(\hat{\theta}_n) - g(\theta)]^T V^{-1}(\theta)[g(\hat{\theta}_n) - g(\theta)] \stackrel{d}{\to} \chi_q^2 .$$

Hence, a pointwise asymptotically level $1-\alpha$ confidence region for $g(\theta)$ is

$$\{\theta : \; n[g(\hat{\theta}_n) - g(\theta)]^T V^{-1}(\hat{\theta}_n)[g(\hat{\theta}_n) - g(\theta)] \le \chi_q^2(1-\alpha)\} .$$

Next, suppose it is desired to test $g(\theta) = 0$. The Wald test rejects when

$$W_n = n g(\hat{\theta}_n) V^{-1}(\hat{\theta}_n) g^T(\hat{\theta}_n)$$

exceeds $\chi_q^2(1-\alpha)$, and it is pointwise asymptotically level α.

12.4.3 Rao Score Tests

Instead of the Wald tests, it is possible to construct tests based directly on Z_n in (12.59), which have the advantage of not requiring computation of a maximum likelihood estimator. Assume q.m.d. holds at θ_0, with derivative $\eta(\cdot, \theta_0)$ and, as usual, set

$$\tilde{\eta}(x, \theta_0) = 2\eta(x, \theta_0)/p_{\theta_0}^{1/2}(x) .$$

Under the assumptions of Theorem 12.2.2, the quadratic mean derivative $\eta(\cdot, \theta_0)$ is given by (12.9) and $n^{1/2} Z_n$ can then be computed by

$$n^{1/2} Z_n = \sum_{i=1}^{n} \tilde{\eta}(X_i, \theta_0) =$$

$$\sum_{i=1}^{n} \frac{\dot{p}_{\theta_0}(X_i)}{p_{\theta_0}(X_i)} = \left(\frac{\partial}{\partial \theta_1} \log L_n(\theta), \ldots, \frac{\partial}{\partial \theta_k} \log L_n(\theta)\right)\Bigg|_{\theta=\theta_0} . \tag{12.73}$$

As mentioned earlier, the statistic Z_n is known as the normalized *score* vector. Its use stems from the fact that inference can be based on Z_n, which involves differentiating the log likelihood at a single point θ_0, avoiding the problem of maximizing the likelihood. Even if the ordinary differentiability conditions assumed in Theorem 12.2.2 fail, inference can be based on Z_n, as we will now see.

Suppose for the moment that θ is real-valued and consider testing $\theta = \theta_0$ versus $\theta > \theta_0$. For a given test $\phi = \phi(X_1, \ldots, X_n)$, let

$$\beta_\phi(\theta) = E_\theta[\phi(X_1, \ldots, X_n)]$$

denote its power function. By Problem 12.17, assuming q.m.d., $\beta_\phi(\theta)$ is differentiable at θ_0 with

$$\beta'_\phi(\theta_0) = \int \cdots \int \phi(x_1, \ldots, x_n) \sum_{i=1}^{n} \tilde{\eta}(x_i, \theta_0) \prod_{i=1}^{n} p_{\theta_0}(x_i) \mu(dx_1) \cdots \mu(dx_n) .$$

Consider the problem of finding the level α test ϕ that maximizes $\beta'_\phi(\theta_0)$. By the general form of the Neyman-Pearson Lemma, the optimal test rejects for large values of $\sum_i \tilde{\eta}(X_i, \theta_0)$, or equivalently, large values of Z_n. By Problem 8.2, if this is the unique test maximizing the slope of the power function at θ_0, then it is also locally most powerful. Thus, tests based on Z_n are appealing from this point of view.

We turn now to the asymptotic behavior of tests based on Z_n. Assume the assumptions of quadratic mean differentiability hold for general k, so that under θ_0,

$$Z_n \xrightarrow{d} N(0, I(\theta_0)) .$$

By Corollary 12.4.1, under $\theta_n = \theta_0 + hn^{-1/2}$,

$$Z_n \xrightarrow{d} N(I(\theta_0)h, I(\theta_0)) .$$

It follows that, under $\theta_n = \theta_0 + hn^{-1/2}$,

$$I^{-1/2}(\theta_0) Z_n \xrightarrow{d} N(I^{1/2}(\theta_0)h, I_k) . \tag{12.74}$$

Now, suppose $k = 1$ and the problem is to test $\theta = \theta_0$ versus $\theta > \theta_0$. Rao's score test rejects when the one-sided *score statistic* $I^{-1/2}(\theta_0)Z_n$ exceeds $z_{1-\alpha}$ and is asymptotically level α. In this case, the Wald test that rejects when $I^{1/2}(\theta_0)n^{1/2}(\hat{\theta}_n - \theta_0)$ exceeds $z_{1-\alpha}$ and the score test are asymptotically equivalent, in the sense that the probability that the two tests yield the same decision tends to one, both under the null hypothesis $\theta = \theta_0$ and under a sequence of alternatives $\theta_0 + hn^{-1/2}$. The equivalence follows from contiguity, the expansion (12.62), and the fact that $I(\hat{\theta}_n) \to I(\theta_0)$ in probability under θ_0 and under $\theta_0 + hn^{-1/2}$. Note that the two tests may differ greatly for alternatives far from θ_0; see Example 13.3.3.

Example 12.4.4 (Bivariate Normal Correlation) Assume $X_i = (U_i, V_i)$ are i.i.d. according to the bivariate normal distribution with means zero and variances one, so that the only unknown parameter is ρ, the correlation. In this case,

$$\log L_n(\rho) = -n\log(2\pi) - \frac{n}{2}\log(1-\rho^2) - \sum_{i=1}^{n}[\frac{1}{2(1-\rho^2)}(U_i^2 - 2\rho U_i V_i + V_i^2)]$$

and so

$$\frac{\partial}{\partial\rho}\log L_n(\rho) = \frac{n\rho}{1-\rho^2} + \frac{1}{1-\rho^2}\sum_{i=1}^{n} U_i V_i - \frac{\rho}{(1-\rho^2)^2}\sum_{i=1}^{n}(U_i^2 - 2\rho U_i V_i + V_i^2) .$$

In the special case $\theta_0 = \rho_0 = 0$,

$$Z_n = n^{-1/2}\sum_{i=1}^{n} U_i V_i \xrightarrow{d} N(0,1) .$$

For other values of ρ_0, the statistic is more complicated; however, we have bypassed maximizing the likelihood, which may have multiple roots in this example. ■

For general k, consider testing a simple null hypothesis $\theta = \theta_0$ versus a multi-sided alternative $\theta \neq \theta_0$. Then, assuming the expansion (12.62), we can replace $n^{1/2}(\hat{\theta}_n - \theta_0)$ in the Wald statistic (12.70) by $I^{-1}(\theta_0)Z_n$. In this case, the *score test* rejects the null hypothesis when the multi-sided *score statistic* $Z_n^T I^{-1}(\theta_0)Z_n$ exceeds $c_{k,1-\alpha}$, and is asymptotically level α. Again, the Wald test and Rao's score test are asymptotically equivalent in the sense described above.

Next, we consider a composite null hypothesis. Interest focuses on $\theta_1, \ldots, \theta_r$, the first r components of θ with the remaining $k - r$ components viewed as nuisance parameters. Let $\theta_{1,0}, \ldots, \theta_{r,0}$ be fixed and consider testing the null hypothesis $\theta_i = \theta_{i,0}$ for $i = 1, \ldots, r$. The Wald test is based on the limit

$$n^{1/2}(\hat{\theta}_{n,1} - \theta_1, \ldots, \hat{\theta}_{n,r} - \theta_r) \xrightarrow{d} N\left(0, \Sigma^{(r)}(\theta)\right) ,$$

where $\Sigma(\theta) = I^{-1}(\theta)$ and $\Sigma^{(r)}(\theta)$ is the $r \times r$ matrix formed by the intersection of the first r rows and columns of $\Sigma(\theta)$. Similarly, define $I^{(r)}(\theta)$ as the $r \times r$ matrix formed by the intersection of the first r rows and columns of $I(\theta)$. Partition $I(\theta)$ as

$$I(\theta) = \begin{pmatrix} I^{(r)}(\theta) & I_{12}(\theta) \\ I_{21}(\theta) & I_{22}(\theta) \end{pmatrix} . \tag{12.75}$$

Note that (Problem 12.49)

$$[\Sigma^{(r)}(\theta)]^{-1} = [I^{(r)}(\theta)] \; - I_{12}(\theta)I_{22}^{-1}(\theta)I_{21}(\theta) \; . \tag{12.76}$$

The score test is based on $Z_n^{(r)}(\theta)$, the r-vector obtained as the first r components of $Z_n(\theta)$, where $Z_n(\theta)$ is defined in (12.59). Under q.m.d. at θ,

$$Z_n^{(r)}(\theta) \xrightarrow{d} N\left(0, I^{(r)}(\theta)\right) \; ,$$

and so,

$$S_n(\theta) = [Z_n^{(r)}(\theta)]^T [I^{(r)}(\theta)]^{-1} [Z_n^{(r)}(\theta)] \xrightarrow{d} \chi_r^2 \; .$$

However, when the null hypothesis is not completely specified, the Rao score test statistic is $S_n(\hat{\theta}_{n,0})$, where

$$\hat{\theta}_{n,0} = (\theta_{1,0}, \ldots, \theta_{r,0}, \hat{\theta}_{r+1,0}, \ldots, \hat{\theta}_{k,0})$$

is an efficient likelihood estimator of θ under the restricted parameter space satisfying the constraints of the null hypothesis. In fact, as argued by Hall and Mathiason (1990), any $n^{1/2}$-consistent estimator can be used in the score statistic. One-sided score tests are studied in Silvapulle and Silvapulle (1995).

12.4.4 Likelihood Ratio Tests

In addition to that Wald and Rao scores tests of Sections 12.4.2 and 12.4.3, let us now consider a third test of $\theta \in \Omega_0$ versus $\theta \notin \Omega_0$, based on the *likelihood ratio statistic* $2\log(R_n)$, where

$$R_n = \frac{\sup_{\theta \in \Omega} L_n(\theta)}{\sup_{\theta \in \Omega_0} L_n(\theta)} \; . \tag{12.77}$$

The *likelihood ratio test* rejects for large values of $2\log(R_n)$. If $\hat{\theta}_n$ and $\hat{\theta}_{n,0}$ are MLEs for θ as θ varies in Ω and Ω_0 respectively, then

$$R_n = L_n(\hat{\theta}_n)/L_n(\hat{\theta}_{n,0}) \; . \tag{12.78}$$

Example 12.4.5 (Multivariate Normal Mean) Suppose $X = (X_1, \ldots, X_k)^T$ is multivariate normal with unknown mean vector θ and known positive definite covariance matrix Σ. The likelihood function is given by

$$\frac{|\Sigma|^{-1/2}}{(2\pi)^{k/2}} \exp\left[-\frac{1}{2}(X-\theta)^T\Sigma^{-1}(X-\theta)\right] \; .$$

Assume $\theta \in \mathbb{R}^k$ and that the null hypothesis asserts $\theta_i = 0$ for $i = 1, \ldots, k$. Then,

$$2\log(R_1) = -\inf_\theta (X-\theta)^T\Sigma^{-1}(X-\theta) + X^T\Sigma^{-1}X = X^T\Sigma^{-1}X = |\Sigma^{-1/2}X|^2 \; .$$

Under the null hypothesis, $\Sigma^{-1/2}X$ is exactly standard multivariate normal, and so the null distribution of $2\log(R_1)$ is exactly χ_k^2 in this case.

Now, consider testing the composite hypothesis $\theta_i = 0$ for $i = 1, \ldots, p$, with the remaining parameters $\theta_{p+1}, \ldots, \theta_k$ regarded as nuisance parameters. More generally, suppose

$$\Omega_0 = \{\theta = (\theta_1, \ldots, \theta_k) : \; A(\theta - a) = 0\} \; , \tag{12.79}$$

where A is a $p \times k$ matrix of rank p and a is some fixed $k \times 1$ vector. Then,

$$2\log(R_1) = -\inf_{\theta \in \mathbb{R}^k} (X-\theta)^T \Sigma^{-1}(X-\theta) + \inf_{\theta \in \Omega_0} (X-\theta)^T \Sigma^{-1}(X-\theta)$$

$$= \inf_{\theta \in \Omega_0} (X-\theta)^T \Sigma^{-1}(X-\theta) . \tag{12.80}$$

The null distribution of (12.80) is χ^2_p (Problem 12.50). ∎

Let us now consider the large sample behavior of the likelihood ratio test in greater generality. First, suppose $\Omega_0 = \{\theta_0\}$ is simple. Then,

$$\log(R_n) = \sup_h [\log(L_{n,h})] ,$$

where $L_{n,h}$ is defined in (12.57). If the family is q.m.d. at θ_0, then

$$\log(R_n) = \sup_h [\langle h, Z_n \rangle - \frac{1}{2}\langle h, I(\theta_0)h\rangle + o_{P^n_{\theta_0}}(1)] .$$

It is then plausible that $\log(R_n)$ should behave like

$$\log \tilde{R}_n \equiv \sup_h [\log(\tilde{L}_{n,h})] ,$$

where $\tilde{L}_{n,h}$ is defined by (12.60). But $\tilde{L}_{n,h}$ is maximized at $\tilde{h}_n = I^{-1}(\theta_0)Z_n$ and so

$$\log(R_n) \approx \log(\tilde{R}_n) = \log(\tilde{L}_{n,\tilde{h}_n}) = \frac{1}{2} Z_n I^{-1}(\theta_0) Z_n .$$

Since, $2\log(\tilde{R}_n) \xrightarrow{d} \chi^2_k$, the heuristics suggest that $2\log(R_n) \xrightarrow{d} \chi^2_k$ as well. In fact, $2\log(\tilde{R}_n)$ is Rao's score test statistic, and so these heuristics also suggest that Rao's score test, the likelihood ratio test, and Wald's test, are all asymptotically equivalent in the sense described earlier in comparing the Wald test and the score test. Note, however, that the tests are not always asymptotically equivalent; some striking differences will be presented in Section 13.3.

These heuristics can be made rigorous under stronger assumptions, such as Cramér type differentiability conditions used in proving asymptotic normality of the MLE or an ELE; see Theorem 7.7.2 in Lehmann (1999). Alternatively, once the general heuristics point toward the limiting behavior, the approximations may be made rigorous by direct calculation in a particular situation. A general theorem based on the existence of efficient likelihood estimators will be presented following the next example.

Example 12.4.6 (Multinomial Goodness of Fit) Consider a sequence of n independent trials, each resulting in one of $k+1$ outcomes $1, \ldots, k+1$. Outcome j occurs with probability p_j on any given trial. Let Y_j be the number of trials resulting in outcome j. Consider testing the simple null hypothesis $p_j = \pi_j$ for $j = 1, \ldots, k+1$. The parameter space Ω is

$$\Omega = \{(p_1, \ldots, p_k) \in \mathbb{R}^k : p_i \geq 0, \sum_{j=1}^k p_j \leq 1\} \tag{12.81}$$

since p_{k+1} is determined as $1-\sum_{j=1}^k p_j$. In this case, the likelihood can be written as

$$L_n(p_1, \ldots, p_k) = \frac{n!}{Y_1! \cdots Y_{k+1}!} p_1^{Y_1} \cdots p_{k+1}^{Y_{k+1}} .$$

By solving the likelihood equations, it is easily checked that the unique MLE is given by $\hat{p}_j = Y_j/n$ (Problem 12.55 (i)). Hence, the likelihood ratio statistic is

$$R_n = \frac{L_n(Y_1/n, \ldots, Y_k/n)}{L_n(\pi_1, \ldots, \pi_k)} ,$$

and so (Problem 12.55 (ii))

$$\log(R_n) = n \sum_{j=1}^{k+1} \hat{p}_j \log(\frac{\hat{p}_j}{\pi_j}) . \tag{12.82}$$

The previous heuristics suggest that $2\log(R_n)$ converges in distribution to χ_k^2, which will be proved in Theorem 12.4.2 below. Note that the Taylor expansion

$$f(x) = x\log(x/x_0) = (x - x_0) + \frac{1}{2x_0}(x - x_0)^2 + o[(x - x_0)^2]$$

as $x \to x_0$ implies $2\log(R_n) \approx Q_n$, where Q_n is Pearson's Chi-squared statistic given by

$$Q_n = \sum_{j=1}^{k+1} \frac{(Y_j - n\pi_j)^2}{n\pi_j} . \tag{12.83}$$

Indeed $2\log(R_n) - Q_n \xrightarrow{P} 0$, under the null hypothesis (Problem 12.57) and so they have the same limiting distribution. Moreover, it can be checked (Problem 12.56) that Rao's Score test statistic is exactly Q_n. The Chi-squared test will be treated more fully in Section 14.3. ■

Next, we present a fairly general result on the asymptotic distribution of the likelihood ratio statistic. Actually, we consider a generalization of the likelihood ratio statistic. Rather than having to compute the maximum likelihood estimators $\hat{\theta}_n$ and $\hat{\theta}_{n,0}$ in (12.78), we assume these estimators satisfy (12.62) under the models with parameter spaces Ω and Ω_0, respectively.

Theorem 12.4.2 *Assume $X_1, \ldots, X_n$ are i.i.d. according to q.m.d. family $\{P_\theta,\ \theta \in \Omega\}$, where Ω is an open subset of $\mathbb{R}^k$ and $I(\theta)$ is positive definite. Further assume, for θ in a neighborhood of θ_0 and a (measurable) function $M(x)$ satisfying $E_{\theta_0}[M(X_i)] < \infty$,*

$$|\log p_\theta(x) - \log p_{\theta_0}(x) - (\theta - \theta_0)\tilde{\eta}_{\theta_0}(x)| \le M(x)|\theta - \theta_0|^2 . \tag{12.84}$$

(i) Consider testing the simple null hypothesis $\theta = \theta_0$. Suppose $\hat{\theta}_n$ is an efficient estimator for θ assuming $\theta \in \Omega$ in the sense that it satisfies (12.62) when $\theta = \theta_0$. Then, the likelihood ratio $R_n = L_n(\hat{\theta}_n)/L_n(\theta_0)$ satisfies, under θ_0,

$$2\log(R_n) \xrightarrow{d} \chi_k^2 .$$

(ii) Consider testing the composite null hypothesis $\theta \in \Omega_0$, where

$$\Omega_0 = \{\theta = (\theta_1, \ldots, \theta_k) :\ A(\theta - a) = 0\} ,$$

and A is a $p \times k$ matrix of rank p and a is a fixed $k \times 1$ vector. Let $\hat{\theta}_{n,0}$ denote an efficient estimator of θ assuming $\theta \in \Omega_0$; that is, assume the expansion (12.62) holds based on the model $\{P_\theta,\ \theta \in \Omega_0\}$ and any $\theta \in \Omega_0$. Then, the likelihood ratio $R_n = L_n(\hat{\theta}_n)/L_n(\hat{\theta}_{n,0})$ satisfies, under any $\theta_0 \in \Omega_0$,

$$2\log(R_n) \xrightarrow{d} \chi_p^2 \ .$$

(iii) More generally, suppose Ω_0 is represented as

$$\Omega_0 = \{\theta :\ g = (g_1(\theta), \ldots, g_p(\theta))^T = 0\} \ ,$$

where $g_i(\theta)$ is a continuously differentiable function from $\mathbb{R}^k$ to $\mathbb{R}$. Let $D = D(\theta)$ be the $p \times k$ matrix with (i,j) entry $\partial g_i(\theta)/\partial \theta_j$, assumed to have rank p. Then, $2\log(R_n) \xrightarrow{d} \chi_p^2$.

Proof. First, consider (i). Let $\hat{h}_n = n^{1/2}(\hat{\theta}_n - \theta_0)$ so that $2\log(R_n) = 2\log(L_{n,\hat{h}_n})$. Fix any $c > 0$ and define

$$\epsilon_{n,c} = \sup_{|h| \le c} |\log(L_{n,h}) - [\langle h, Z_n \rangle - \frac{1}{2}\langle h, I(\theta_0)h \rangle]| \ ;$$

by Problem 13.12, $\epsilon_{n,c} \to 0$ in probability under θ_0. By the triangle inequality,

$$2\log(L_{n,\hat{h}_n}) \le 2[\langle \hat{h}_n, Z_n \rangle - \frac{1}{2}\langle \hat{h}_n, I(\theta_0)\hat{h}_n \rangle + \epsilon_{n,c}]$$

if $|\hat{h}_n| \le c$. But, using (12.62),

$$2[\langle \hat{h}_n, Z_n \rangle - \frac{1}{2}\langle \hat{h}_n, I(\theta_0)\hat{h}_n \rangle] = Z_n^T I^{-1}(\theta_0) Z_n + o_{P_{\theta_0}}(1) \ ;$$

so,

$$2\log(L_{n,\hat{h}_n}) \le Z_n^T I^{-1}(\theta_0) Z_n + \tilde{\epsilon}_{n,c}$$

if $|\hat{h}_n| \le c$, where $\tilde{\epsilon}_{n,c} \to 0$ in probability under θ_0 for any $c > 0$. Therefore,

$$P\{2\log(L_{n,\hat{h}_n}) \ge x\} \le P\{Z_n^T I^{-1}(\theta_0) Z_n + \tilde{\epsilon}_{n,c} \ge x,\ |\hat{h}_n| \le c\} + P\{|\hat{h}_n| > c\}$$

$$\le P\{Z_n^T I^{-1}(\theta_0) Z_n + \tilde{\epsilon}_{n,c} \ge x\} + P\{|\hat{h}_n| > c\} \ . \qquad (12.85)$$

But, under θ_0 , $Z_n^T I^{-1}(\theta_0) Z_n$ is asymptotically χ_k^2 and $\hat{h}_n \xrightarrow{d} Z$ where Z is $N(0, I^{-1}(\theta_0))$, so (12.85) tends to

$$P\{\chi_k^2 \ge x\} + P\{|Z| > c\} \ .$$

Let $c \to \infty$ to conclude

$$\limsup_n P\{2\log(L_{n,\hat{h}_n}) \ge x\} \le P\{\chi_k^2 \ge x\} \ .$$

A similar argument yields

$$\liminf_n P\{2\log(L_{n,\hat{h}_n}) \ge x\} \ge P\{\chi_k^2 \ge x\} \ , \qquad (12.86)$$

and (i) is proved.

The proof of (ii) is based on a similar argument, combined with the results of Example 12.4.5 for testing a composite null hypothesis about a multivariate normal mean vector. The proof of (iii) is left as an exercise (Problem 12.60). ■

In the special case where the null hypothesis is specified by $\theta_i = \theta_{i,0}$ for $i = 1, \ldots, p$, with θ_i regarded as a nuisance parameter for $\theta_i > p$, the degrees of freedom can be remembered as the dimension of Ω minus the dimension of Ω_0.

Example 12.4.7 (One-sample Normal Mean) Suppose $X_1, \ldots, X_n$ are i.i.d. $N(\mu, \sigma^2)$ with both parameters unknown. Consider testing $\mu = 0$ versus $\mu \neq 0$. Then (Problem 12.46),

$$2\log(R_n) = n\log(1 + \frac{t_n^2}{n-1}) , \tag{12.87}$$

where $t_n^2 = n\bar{X}_n^2/S_n^2$ is the one-sample t-statistic. By Problem 11.89, one can deduce the following Edgeworth expansion for $2\log(R_n)$ (Problem 12.47):

$$P\{2\log(R_n) \leq r\} = 1 - 2[\Phi(-z) + \frac{3}{4n} z\phi(z)] + O(n^{-2}) , \tag{12.88}$$

where $z = \sqrt{r}$, Φ is the standard normal c.d.f. and $\Phi' = \phi$. This implies that the test that rejects when $2\log(R_n) > z_{1-\frac{\alpha}{2}}$ has rejection probability equal to $\alpha + O(n^{-1})$. But, a simple correction, known as a *Bartlett correction*, can improve the χ_1^2 approximation. Indeed, (12.88) and a Taylor expansion implies

$$P\{2\log(R_n)(1 + \frac{b}{n}) > z_{1-\frac{\alpha}{2}}\} = \alpha + O(n^{-2}) , \tag{12.89}$$

if we take $b = 3/2$. Thus, the error in rejection probability of the Bartlett-corrected test is $O(n^{-2})$. Of course, in this example, the exact two-sided t-test is available. ■

It is worth knowing that, quite generally, a simple multiplicative correction to the likelihood ratio statistic greatly improves the quality of the approximation. Specifically, for an appropriate choice of b, comparing $2\log(R_n)(1+\frac{b}{n})$ to the usual limiting χ_p^2 reduces the error in rejection probability from $O(n^{-1})$ to $O(n^{-2})$. In practice, b can be derived by analytical means or estimated. The idea for such a Bartlett correction originated in Bartlett (1937). For appropriate regularity conditions that imply a Bartlett correction works, see Barndorff-Nielsen and Hall (1988), Bickel and Ghosh (1990), Jensen (1993) and DiCiccio and Stern (1994).

12.5 Problems

Section 12.2

Problem 12.1 Generalize Example 12.2.1 to the case where X is multivariate normal with mean vector θ and nonsingular covariance matrix Σ.

Problem 12.2 Generalize Example 12.2.2 to the case of a multiparameter exponential family. Compare with the result of Problem 12.1.

Problem 12.3 Suppose g_n is a sequence of functions in $L^2(\mu)$; that is, $\int g_n^2 d\mu < \infty$. Assume, for some function g, $\int (g_n - g)^2 d\mu \to 0$. Prove that $\int g^2 d\mu < \infty$.

Problem 12.4 Suppose g_n is a sequence of functions in $L^2(\mu)$ and, for some function g, $\int (g_n - g)^2 d\mu \to 0$. If $\int h^2 d\mu < \infty$, show that $\int h g_n d\mu \to \int h g d\mu$.

Problem 12.5 Suppose X and Y are independent, with X distributed as P_θ and Y as $\bar{P}_\theta$, as θ varies in a common index set Ω. Assume the families $\{P_\theta\}$ and $\{\bar{P}_\theta\}$ are q.m.d. with Fisher Information matrices $I_X(\theta)$ and $I_Y(\theta)$, respectively. Show that the model based on the joint data (X, Y) is q.m.d. and its Fisher Information matrix is given by $I_X(\theta) + I_Y(\theta)$.

Problem 12.6 Fix a probability P. Let $u(x)$ satisfy

$$\int u(x) dP(x) = 0 \ .$$

(i) Assume $\sup_x |u(x)| < \infty$, so that

$$p_\theta(x) = [1 + \theta u(x)]$$

defines a family of densities (with respect to P) for all small $|\theta|$. Show this family is q.m.d. at $\theta = 0$. Calculate the quadratic mean derivative, score function, and $I(0)$.
(ii) Alternatively, if u is unbounded, define $p_\theta(x) = C(\theta) \exp(\theta u(x))$, assuming $\int \exp(\theta u(x)) dx$ exists for all small $|\theta|$. For this family, argue the family is q.m.d. at $\theta = 0$, and calculate the score function and $I(0)$.
(iii) Suppose $\int u^2(x) dP(x) < \infty$. Define

$$p_\theta(x) = C(\theta) 2[1 + \exp(-2\theta u(x))]^{-1} \ .$$

Show this family is q.m.d. at $\theta = 0$, and calculate the score function and $I(0)$. [The constructions in this problem are important for nonparametric applications, used later in Chapters 13 and 14. The last construction is given in van der Vaart (1998).]

Problem 12.7 Fix a probability P on S and functions $u_i(x)$ such that $\int u_i(x) dP(x) = 0$ and $\int u_i^2(x) dP(x) < \infty$, for $i = 1, 2$. Adapt Problem 12.6 to construct a family of distributions P_θ with $\theta \in \mathbb{R}^2$, defined for all small $|\theta|$, such that $P_{0,0} = P$, the family is q.m.d. at $\theta = (0, 0)$ with score vector at $\theta = (0, 0)$ given by $(u_1(x), u_2(x))$. If S is the real line, construct the P_θ that works even if P_θ is required to be smooth if P and the u_i are smooth (i.e. having differentiable densities) or subject to moment constraints (i.e. having finite pth moments).

Problem 12.8 Show that the definition of $I(\theta)$ in Definition 12.2.2 does not depend on the choice of dominating measure μ.

Problem 12.9 In Examples 12.2.3 and 12.2.4, find the quadratic mean derivative and $I(\theta)$.

Problem 12.10 In Example 12.2.5, show that $\int \{[f'(x)]^2 / f(x)\} dx$ is finite iff $\beta > 1/2$.

Problem 12.11 Prove Theorem 12.2.2 using an argument similar to the proof of Theorem 12.2.1.

Problem 12.12 Suppose $\{P_\theta\}$ is q.m.d. at θ_0 with derivative $\eta(\cdot, \theta_0)$. Show that, on $\{x : p_{\theta_0}(x) = 0\}$, we must have $\eta(x, \theta_0) = 0$, except possibly on a μ-null set. *Hint:* On $\{p_{\theta_0}(x) = 0\}$, write

$$0 \le n^{1/2} p^{1/2}_{\theta_0 + hn^{-1/2}}(x) = \langle h, \eta(x, \theta_0)\rangle + r_{n,h}(x) ,$$

where $\int r^2_{n,h}(x)\mu(dx) \to 0$. This implies, with h fixed, that $r_{n,h}(x) \to 0$ except for x in μ-null set, at least along some subsequence.

Problem 12.13 Suppose $\{P_\theta\}$ is q.m.d. at θ_0. Show

$$P_{\theta_0 + h}\{x : p_{\theta_0}(x) = 0\} = o(|h|^2)$$

as $|h| \to 0$. Hence, if $X_1, \ldots, X_n$ are i.i.d. with likelihood ratio $L_{n,h}$ defined by (12.12), show that

$$P^n_{\theta_0 + hn^{-1/2}}\{L_{n,h} = \infty\} \to 0 .$$

Problem 12.14 To see what might happen when the parameter space is not open, let

$$f_0(x) = xI\{0 \le x \le 1\} + (2 - x)I\{1 < x \le 2\} .$$

Consider the family of densities indexed by $\theta \in [0, 1)$ defined by

$$p_\theta(x) = (1 - \theta^2) f_0(x) + \theta^2 f_0(x - 2) .$$

Show that the condition (12.5) holds when $\theta_0 = 0$, if it is only required that h tends to 0 through positive values. Investigate the behavior of the likelihood ratio (12.12) for such a family. (For a more general treatment, consult Pollard (1997).)

Problem 12.15 Suppose $X_1, \ldots, X_n$ are i.i.d. and uniformly distributed on $(0, \theta)$. Let $p_\theta(x) = \theta^{-1} I\{0 < x < \theta\}$. and $L_n(\theta) = \prod_i p_\theta(X_i)$. Fix p and θ_0. Determine the limiting behavior of $L_n(\theta_0 + hn^{-p})/L_n(\theta_0)$ under θ_0. For what p is the limiting distribution nondegenerate?

Problem 12.16 Suppose $\{P_\theta, \theta \in \Omega\}$ is a model with Ω an open subset of $\mathbb{R}^k$, and having densities $p_\theta(x)$ with respect to μ. Define the model to be L_1-differentiable at θ_0 if there exists a vector of real-valued functions $\zeta(\cdot, \theta_0)$ such that

$$\int |p_{\theta_0 + h}(x) - p_{\theta_0}(x) - \langle \zeta(x, \theta_0), h\rangle| d\mu(x) = o(|h|) \tag{12.90}$$

as $|h| \to 0$. Show that, if the family is q.m.d. at θ_0 with q.m. derivative $\eta(\cdot, \theta_0)$, then it is L_1-differentiable with

$$\zeta(x, \theta_0) = 2\eta(x, \theta_0) p^{1/2}_{\theta_0}(x) ,$$

but the converse is false.

Problem 12.17 Assume $\{P_\theta, \theta \in \Omega\}$ is L_1-differentiable, so that (12.90) holds. For simplicity, assume $k = 1$ (but the problem generalizes). Let $\phi(\cdot)$ be uniformly bounded and set $\beta(\theta) = E_\theta[\phi(X)]$. Show, $\beta'(\theta)$ exists at θ_0 and

$$\beta'(\theta_0) = \int \phi(x)\zeta(x, \theta_0)\mu(dx) . \tag{12.91}$$

Hence, if $\{P_\theta\}$ is q.m.d. at θ_0 with derivative $\eta(\cdot, \theta_0)$, then,

$$\beta'(\theta_0) = \int \phi(x)\tilde{\eta}(x, \theta_0)p_{\theta_0}(x)\mu(dx) \ , \tag{12.92}$$

where $\tilde{\eta}(x, \theta_0) = 2\eta(x, \theta_0)/p_{\theta_0}^{1/2}(x)$. More generally, if $X_1, \ldots, X_n$ are i.i.d. P_θ and $\phi(X_1, \ldots, X_n)$ is uniformly bounded, then $\beta(\theta) = E_\theta[\phi(X_1, \ldots, X_n)]$ is differentiable at θ_0 with

$$\beta'(\theta_0) = \int \cdots \int \phi(x_1, \ldots, x_n) \sum_{i=1}^n \tilde{\eta}(x_i, \theta_0) \prod_{i=1}^n p_{\theta_0}(x_i)\mu(dx_1) \cdots \mu(dx_n) \ . \tag{12.93}$$

Section 12.3

Problem 12.18 Prove (12.31).

Problem 12.19 Show the convergence (12.35).

Problem 12.20 Fix two probabilities P and Q and let $P_n = P$ and $Q_n = Q$. Show that $\{P_n\}$ and $\{Q_n\}$ are contiguous iff P and Q are absolutely continuous.

Problem 12.21 Fix two probabilities P and Q and let $P_n = P^n$ and $Q_n = Q^n$. Show that $\{P_n\}$ and $\{Q_n\}$ are contiguous iff $P = Q$.

Problem 12.22 Suppose Q_n is contiguous to P_n and let L_n be the likelihood ratio defined by (12.36). Show that $E_{P_n}(L_n) \to 1$. Is the converse true?

Problem 12.23 Consider a sequence $\{P_n, Q_n\}$ with likelihood ratio L_n defined in (12.36). Assume

$$\mathcal{L}(L_n|P_n) \xrightarrow{d} W \ ,$$

where $P\{W = 0\} = 0$. Also, under (12.41), deduce that P_n is contiguous to Q_n and hence P_n and Q_n are mutually contiguous if and only if $\mu = -\sigma^2/2$.

Problem 12.24 Suppose, under P_n, $X_n = Y_n + o_{P_n}(1)$; that is, $X_n - Y_n \to 0$ in P_n-probability. Suppose Q_n is contiguous to P_n. Show that $X_n = Y_n + o_{Q_n}(1)$.

Problem 12.25 Suppose X_n has distribution P_n or Q_n and $T_n = T_n(X_n)$ is sufficient. Let P_n^T and Q_n^T denote the distribution of T_n under P_n and Q_n, respectively. Prove or disprove: Q_n is contiguous to P_n if and only if Q_n^T is contiguous to P_n^T.

Problem 12.26 Suppose Q is absolutely continuous with respect to P. If $P\{E_n\} \to 0$, then $Q\{E_n\} \to 0$.

Problem 12.27 Prove the convergence (12.40).

Problem 12.28 Show that $\sigma_{1,2}$ in (12.52) reduces to $h/\sqrt{\pi}$.

Problem 12.29 Verify (12.53) and evaluate it in the case where $f(x) = \exp(-|x|)/2$ is the double exponential density.

Problem 12.30 Suppose $X_1, \ldots, X_n$ are i.i.d. according to a model which is q.m.d. at θ_0. For testing $\theta = \theta_0$ versus $\theta = \theta_0 + hn^{-1/2}$, consider the test ψ_n that rejects H if $\log(L_{n,h})$ exceeds $z_{1-\alpha}\sigma_h - \frac{1}{2}\sigma_h^2$, where $L_{n,h}$ is defined by (12.54) and $\sigma_h^2 = \langle h, I(\theta_0)h \rangle$. Find the limiting value of $E_{\theta_0 + hn^{-1/2}}(\psi_n)$.

Problem 12.31 Suppose P_θ is the uniform distribution on $(0, \theta)$. Fix h and determine whether or not P_1^n and $P_{1+h/n}^n$ are mutually contiguous. Consider both $h > 0$ and $h < 0$.

Problem 12.32 Assume $X_1, \ldots, X_n$ are i.i.d. according to a family $\{P_\theta\}$ which is q.m.d. at θ_0. Suppose, for some statistic $T_n = T_n(X_1, \ldots, X_n)$ and some function $\mu(\theta)$ assumed differentiable at θ_0, $n^{1/2}(T_n - \mu(\theta_n)) \stackrel{d}{\to} N(0, \sigma^2)$ under θ_n whenever $\theta_n = \theta_0 + hn^{-1/2}$. Show the same result holds, first whenever h is replaced by $h_n \to h$, and then whenever $n^{1/2}(\theta_n - \theta_0) = O(1)$.

Problem 12.33 Generalize Corollary 12.3.2 in the following way. Suppose $T_n = (T_{n,1}, \ldots, T_{n,k}) \in \mathbb{R}^k$. Assume that, under P_n,

$$(T_{n,1}, \ldots, T_{n,k}, \log(L_n)) \stackrel{d}{\to} (T_1, \ldots, T_k, Z) ,$$

where $(T_1, \ldots, T_k, Z)$ is multivariate normal with $Cov(T_i, Z) = c_i$. Then, under Q_n,

$$(T_{n,1}, \ldots, T_{n,k}) \stackrel{d}{\to} (T_1 + c_1, \ldots, T_k + c_k) .$$

Problem 12.34 Suppose $X_1, \ldots, X_n$ are i.i.d. according to a model $\{P_\theta : \theta \in \Omega\}$, where Ω is an open subset of $\mathbf{R}^k$. Assume that the model is q.m.d. Show that there cannot exist an estimator sequence T_n satisfying

$$\lim_{n \to \infty} \sup_{|\theta - \theta_0| \le n^{-1/2}} P_\theta^n(n^{1/2}|T_n - \theta| > \epsilon) = 0 \tag{12.94}$$

for every $\epsilon > 0$ and any θ_0. (Here P_θ^n means the joint probability distribution of $(X_1, \ldots, X_n)$ under θ). Suppose the above condition (12.94) only holds for some $\epsilon > 0$. Does the same conclusion hold?

Section 12.4

Problem 12.35 In Example 12.4.1, show that the likelihood equations have a unique solution which corresponds to a global maximum of the likelihood function.

Problem 12.36 Suppose $X_1, \ldots, X_n$ are i.i.d. P_θ according to the lognormal model of Example 12.2.7. Write down the likelihood function and show that it is unbounded.

Problem 12.37 Generalize Example 12.4.2 to multiparameter exponential families.

Problem 12.38 Prove Corollary 12.4.1. *Hint:* Simply define $\hat{\theta}_n = \theta_0 + n^{-1/2} I^{-1}(\theta_0) Z_n$ and apply Theorem 12.4.1.

Problem 12.39 Let (X_i, Y_i), $i = 1 \ldots n$ be i.i.d. such that X_i and Y_i are independent and normally distributed, X_i has variance σ^2, Y_i has variance τ^2 and both have common mean μ.
(i) If σ and τ are known, determine an efficient likelihood estimator (ELE) $\hat{\mu}$ of μ and find the limit distribution of $n^{1/2}(\hat{\mu} - \mu)$.
(ii) If σ and τ are unknown, provide an estimator $\bar{\mu}$ for which $n^{1/2}(\bar{\mu} - \mu)$ has the same limit distribution as $n^{1/2}(\hat{\mu} - \mu)$.
(iii) What can you infer from your results (i) and (ii) regarding the Information matrix $I(\theta)$, $\theta = (\mu, \sigma, \tau)$?

Problem 12.40 Let $X_1, \ldots, X_n$ be a sample from a Cauchy location model with density $f(x - \theta)$, where

$$f(z) = \frac{1}{\pi(1 + z^2)}.$$

Compare the limiting distribution of the sample median with that of an efficient likelihood estimator.

Problem 12.41 Let $X_1, \ldots, X_n$ be i.i.d. $N(\theta, \theta^2)$. Compare the asymptotic distribution of $\bar{X}_n^2$ with that of an efficient likelihood estimator sequence.

Problem 12.42 Let $X_1, \cdots, X_n$ be i.i.d. with density

$$f(x, \theta) = [1 + \theta \cos(x)]/2\pi,$$

where the parameter θ satisfies $|\theta| < 1$ and x ranges between 0 and 2π. (The observations X_i may be interpreted as directional data. The case $\theta = 0$ corresponds to the uniform distribution on the circle.) Construct an efficient likelihood estimator of θ, as explicitly as possible.

Problem 12.43 Suppose $X_1, \ldots, X_n$ are i.i.d., uniformly distributed on $[0, \theta]$. Find the maximum likelihood estimator $\hat{\theta}_n$ of θ. Determine a sequence τ_n such that $\tau_n(\hat{\theta}_n - \theta)$ has a limiting distribution, and determine the limit law.

Problem 12.44 Verify that $\tilde{h}_n$ in (12.61) maximizes $\tilde{L}_{n,h}$.

Problem 12.45 For a q.m.d. model with $\hat{\theta}_n$ satisfying (12.62), find the limiting behavior of the Wald statistic given in the left side of (12.71) under $\theta_n = \theta_0 + hn^{-1/2}$.

Problem 12.46 Suppose $X_1, \ldots, X_n$ are i.i.d. $N(\mu, \sigma^2)$ with both parameters unknown. Consider testing $\mu = 0$ versus $\mu \neq 0$. Find the likelihood ratio test statistic, and determine its limiting distribution under the null hypothesis. Calculate the limiting power of the test against the sequence of alternatives $(\mu, \sigma^2) = (h_1 n^{-1/2}, \sigma^2 + h_2 n^{-1/2})$.

Problem 12.47 In Example 12.4.7, verify (12.88) and (12.89).

Problem 12.48 Suppose $X_1, \ldots, X_n$ are i.i.d. P_θ, with $\theta \in \Omega$, an open subset of $\mathbb{R}^k$. Assume the family is q.m.d. at θ_0 and consider testing the simple null hypothesis $\theta = \theta_0$. Suppose $\hat{\theta}_n$ is an estimator sequence satisfying (12.62), and consider the Wald test statistic $n(\hat{\theta}_n - \theta_0)^T I(\theta_0)(\hat{\theta}_n - \theta_0)$. Find its limiting distribution against the sequence of alternatives $\theta_0 + hn^{-1/2}$, as well as an expression for its limiting power against such a sequence of alternatives.

Problem 12.49 Prove (12.76). Then, show that

$$[\Sigma^{(r)}(\theta)]^{-1} \leq [I^{(r)}(\theta)] \ .$$

What is the statistical interpretation of this inequality?

Problem 12.50 In Example 12.4.5, consider the case of a composite null hypothesis with Ω_0 given by (12.79). Show that the null distribution of the likelihood ratio statistic given by (12.80) is χ_p^2. *Hint:* First consider the case $a = 0$ so that Ω_0 is a linear subspace of dimension $k - p$. Let $Z = \Sigma^{-1/2} X$, so that

$$2\log(R_n) = \inf_{\theta \in \Omega_0} |Z - \Sigma^{-1/2}\theta|^2 \ .$$

As θ varies in Ω_0, $\Sigma^{-1/2}\theta$ varies in a subspace L of dimension $k - p$. If P is the projection matrix onto L and I is the identity matrix, then $2\log(R_n) = |(I - P)Z|^2$.

Problem 12.51 In Example 12.4.5, determine the distribution of the likelihood ratio statistic against an alternative, both for the simple and composite null hypotheses.

Problem 12.52 Suppose $X_1, \ldots, X_n$ are i.i.d. $N(\mu, \sigma^2)$ with both parameters unknown. Consider testing the simple null hypothesis $(\mu, \sigma^2) = (0, 1)$. Find and compare the Wald test, Rao's Score test, and the likelihood ratio test.

Problem 12.53 Suppose $X_1, \ldots, X_n$ are i.i.d. with the gamma $\Gamma(g, b)$ density

$$f(x) = \frac{1}{\Gamma(g) b^g} x^{g-1} e^{-x/b} \quad x > 0 \ ,$$

with both parameters unknown (and positive). Consider testing the null hypothesis that $g = 1$, i.e., under the null hypothesis the underlying density is exponential. Determine the likelihood ratio test statistic and find its limiting distribution.

Problem 12.54 Suppose $(X_1, Y_1), \ldots, (X_n, Y_n)$ are i.i.d., with X_i also independent of Y_i. Further suppose X_i is normal with mean μ_1 and variance 1, and Y_i is normal with mean μ_2 and variance 1. It is known that $\mu_i \geq 0$ for $i = 1, 2$. The problem is to test the null hypothesis that at most one μ_i is positive versus the alternative that both μ_1 and μ_2 are positive.
(i) Determine the likelihood ratio statistic for this problem.
(ii) In order to carry out the test, how would you choose the critical value (sequence) so that the size of the test is α?

Problem 12.55 (i) In Example 12.4.6, check that the MLE is given by $\hat{p}_j = Y_j/n$. (ii) Show (12.82).

Problem 12.56 In Example 12.4.6, show that Rao's Score test is exactly Pearson's Chi-squared test.

Problem 12.57 In Example 12.4.6, show that $2\log(R_n) - Q_n \xrightarrow{P} 0$ under the null hypothesis.

Problem 12.58 Prove (12.86).

Problem 12.59 Provide the details of the proof to part (ii) of Theorem 12.4.2.

Problem 12.60 Prove (iii) of Theorem 12.4.2. *Hint:* If θ_0 satisfies the null hypothesis $g(\theta_0) = 0$, then testing Ω_0 behaves asymptotically like testing the null hypothesis $D(\theta_0)(\theta - \theta_0) = 0$, which is a hypothesis of the form considered in part (ii) of the theorem.

Problem 12.61 The problem is to test independence in a contingency table. Specifically, suppose $X_1, \ldots, X_n$ are i.i.d., where each X_i is cross-classified, so that $X_i = (r, s)$ with probability $p_{r,s}$, $r = 1, \ldots, R$, $s = 1, \ldots, S$. Under the full model, the $p_{r,s}$ vary freely, except they are nonnegative and sum to 1. Let $p_{r\cdot} = \sum_s p_{r,s}$ and $p_{\cdot s} = \sum_r p_{r,s}$. The null hypothesis asserts $p_{r,s} = p_{r\cdot} p_{\cdot s}$ for all r and s. Determine the likelihood ratio test and its limiting null distribution.

Problem 12.62 Consider the following model which therefore generalizes model (iii) of Section 4.7. A sample of n_i subjects is obtained from class $A_i (i = 1, \ldots, a)$, the samples from different classes being independent. If $Y_{i,j}$ is the number of subjects from the ith sample belonging to $B_j (j = 1, \ldots, b)$, the joint distribution of $(Y_{i,1}, \ldots, Y_{i,b})$ is multinomial, say,

$$M(n_i; p_{1|i}, \ldots, p_{b|i}) \; .$$

Determine the likelihood ratio statistic for testing the hypothesis of *homogeneity* that the vector $(p_{1|i}, \ldots, p_{b|i})$ is independent of i, and specify its asymptotic distribution.

Problem 12.63 The *hypothesis of symmetry* in a square two-way contingency table arises when one of the responses $A_1, \ldots, A_a$ is observed for each of n subjects on two occasions (e.g. before and after some intervention). If $Y_{i,j}$ is the number of subjects whose responses on the two occasions are (A_i, A_j), the joint distribution of the $Y_{i,j}$ is multinomial, with the probability of a subject response of (A_i, A_j) denoted by $p_{i,j}$. The hypothesis H of *symmetry* states that $p_{i,j} = p_{j,i}$ for all i and j; that is, that the intervention has not changed the probabilities. Determine the likelihood ratio statistic for testing H, and specify its asymptotic distribution. [Bowker (1948).]

Problem 12.64 In the situation of Problem 12.63, consider the *hypothesis of marginal homogeneity* $H' : p_{i+} = p_{+i}$ for all i, where $p_{i+} = \sum_{j=1}^{a} p_{iij}$, $p_{+i} = \sum_{j=1}^{a} p_{jii}$.

(i) The maximum-likelihood estimates of the p_{iij} under H' are given by $\hat{\hat{p}}_{ij} = Y_{ij}/(1+\lambda_i-\lambda_j)$, where the λ's are the solutions of the equations $\sum_j Y_{ij}/(1+\lambda_i-\lambda_j) = \sum_j Y_{ij}/(1+\lambda_j-\lambda_i)$. (These equations have no explicit solutions.)

(ii) Determine the number of degrees of freedom for the limiting χ^2-distribution of the likelihood ratio criterion.

Problem 12.65 Consider the third of the three sampling schemes for a $2\times 2\times K$ table discussed in Section 4.8, and the two hypotheses

$$H_1 : \Delta_1 = \cdots = \Delta_K = 1 \quad \text{and} \quad H_2 : \Delta_1 = \cdots = \Delta_K.$$

(i) Obtain the likelihood-ratio test statistic for testing H_1.
(ii) Obtain equations that determine the maximum likelihood estimates of the parameters under H_2. (These equations cannot be solved explicitly.)
(iii) Determine the number of degrees of freedom of the limiting χ^2-distribution of the likelihood ratio test for testing (a) H_1, (b) H_2.
[For a discussion of these and related hypotheses, see for example Shaffer (1973), Plackett (1981), or Bishop, Fienberg, and Holland (1975), and the recent study by Liang and Self (1985).]

Problem 12.66 Suppose $X_1, \ldots, X_n$ are i.i.d. $N(\theta, 1)$. Consider Hodges' superefficient estimator of θ (unpublished, but cited in Le Cam (1953)), defined as follows Let $\hat{\theta}_n$ be 0 if $|\bar{X}_n| \le n^{-1/4}$; otherwise, let $\hat{\theta}_n = \bar{X}_n$. For any fixed θ, determine the limiting distribution of $n^{1/2}(\hat{\theta}_n - \theta)$. Next, determine the limiting distribution of $n^{1/2}(\hat{\theta}_n - \theta_n)$ under $\theta_n = hn^{-1/2}$.

Problem 12.67 Let $(X_{j,1}, X_{j,2})$, $j = 1, \ldots, n$ be independent pairs of independent exponentially distributed random variables with $E(X_{j,1}) = \theta\lambda_j$ and $E(X_{j,2}) = \lambda_j$. Here, θ and the λ_j are all unknown. The problem is to test $\theta = 1$ against $\theta > 1$. Compare the Rao, Wald, and likelihood ratio tests for this problem. Without appealing to any general results, find the limiting distribution of your statistics, as well as the limiting power against suitable local alternatives. (Note: the number of parameters is increasing with n so you can't directly appeal to our previous large sample results.)

12.6 Notes

According to Le Cam and Yang (2000), the notion of quadratic mean differentiability was initiated in conversations between Hájek and Le Cam in 1962. Hájek (1962) appears to be the first publication making use of this notion. The importance of q.m.d. was prominent in the fundamental works of Le Cam (1969, 1970) and Hajek (1972), and has been used extensively ever since.

The notion of (mutual) contiguity is due to Le Cam (1960). Its usefulness was soon recognized by Hájek (1962), who first considered the one-sided version. Three of Le Cam's fundamental lemmas concerning contiguity became known as Le Cam's three lemmas, largely due to their prominence in Hájek and Sidák (1967). Further results can be found in Roussas (1972), Le Cam (1986), Chapter 6, Hájek, Sidák, and Sen (1999), and Le Cam and Yang (2000), Chapter 3.

The methods studied in Section 12.4 are based on the notion of *likelihood*, whose general importance was recognized in Fisher (1922, 1925). Rigorous approaches were developed by Wald (1939, 1943) and Cramér (1946). Cramér defined the asymptotic efficiency of an asymptotically normal estimator to be the ratio of its asymptotic variance to the Fisher Information; that such a definition is flawed even for asymptotically normal estimators was made clear by Hodges superefficient estimator (Problem 12.66). Le Cam (1956) introduced the *one-step maximum likelihood estimator*, which is based on a discretization trick coupled with a Newton-Raphson approximation. Such estimators satisfy (12.62) under weak assumptions and enjoy other optimality properties; for example, see Section 7.3 of Millar (1983). The notion of a *regular* estimator sequence introduced at the end of Section 12.4.1 plays an important role in the theory of efficient estimation and the Hajék-Inagaki Convolution Theorem; see Hajék (1970), Le Cam (1979), Beran (1999), Millar (1985), and van der Vaart (1988).

The asymptotic behavior of the likelihood ratio statistic was studied in Wilks (1938) and Chernoff (1954). Pearson's Chi-squared statistic was introduced in Pearson (1900) and the Rao score tests by Rao (1947). In fact, the Rao score test was actually introduced in the univariate case by Wald (1941b). The asymptotic equivalence of many of the classical tests is explored in Hall and Mathiason (1990). Methods based on integrated likelihoods are reviewed in Berger, Liseo and Wolpert (1999). Caveats about the finite sample behavior of Rao and Wald tests are given in Le Cam (1990); also see Fears, Benichou and Gail (1996) and Pawitan (2000). The behavior of likelihood ratio tests under nonstandard conditions is studied in Vu and Zhou (1997). Extensions of likelihood methods to semiparametric and nonparametric models are developed in Murphy and van der Vaart (1997), Owen (1988, 2001) and Fan, Zhang and Zhang (2001). Robust version of the Wald, likelihood, and score tests are given in Heritier and Ronchetti (1994).

13

Large Sample Optimality

13.1 Testing Sequences, Metrics, and Inequalities

In this chapter, some asymptotic optimality theory of hypothesis testing is developed. We consider testing one sequence of distributions against another (the asymptotic version of testing a simple hypothesis against a simple alternative). It turns out that this problem degenerates if the two sequences are too close together or too far apart. The non-degenerate situation can be characterized in terms of a suitable distance or metric between the distributions of the two sequences. Two such metrics, the total variation and the Hellinger metric, will be introduced below.

We begin by considering some of the basic metrics for probability distributions that are useful in statistics. Fundamental inequalities relating these metrics are developed, from which some large sample implications can be derived. We now recall the definition of a metric space; also see Section A.2 in the appendix.

Definition 13.1.1 A set $\mathcal{P}$ is a *metric space* if there exists a real-valued function d defined on $\mathcal{P} \times \mathcal{P}$ such that, for all points p, q, and r in $\mathcal{P}$, $d(p, q) \geq 0$, $d(p, q) = d(q, p)$ and $d(p, q) \leq d(p, r) + d(r, q)$. A function d satisfying these conditions is called a *metric.*

In the present context, $\mathcal{P}$ will be a collection of probabilities on a (measurable) space $\mathcal{X}$ (endowed with a σ-field). We have already encountered two metrics on the collection of probability distributions on $\mathbb{R}$. One is the Lévy distance $\rho_L(F, G)$, defined in Definition 11.2.3. The other, used in Example 11.2.12, is the Kolmogorov-Smirnov distance between distribution functions F and G on the real line, defined as

$$d_K(F,G) = \sup_t |F(t) - G(t)| \ . \tag{13.1}$$

It is easy to see that d_K is indeed a metric (Problem 11.21). In the context of hypothesis testing, two additional distances arise naturally, the total variation distance and the Hellinger distance.

Before considering the asymptotic problem, consider the problem of testing a simple hypothesis P_0 against a simple alternative P_1. Here, P_i is a probability measure on $(\mathcal{X}, \mathcal{F})$ and p_i will denote the density of P_i with respect to a dominating measure μ.

In contrast to previous chapters where the hypothesis and alternative were treated asymmetrically, consider the problem of finding the test $\phi = \phi(X)$ that minimizes the sum of the error probabilities. For a test ϕ, denote the sum of the probability of rejecting P_0 when P_0 is true and the probability of rejecting P_1 when P_1 is true by

$$S_{P_0,P_1}(\phi) = \int_{\mathcal{X}} \phi(x) dP_0(x) + \int_{\mathcal{X}} (1 - \phi(x)) dP_1(x) \ . \tag{13.2}$$

and let

$$S(P_0, P_1) = \inf_\phi \ [S_{P_0,P_1}(\phi)] \ . \tag{13.3}$$

The following theorem gives the test ϕ^* that minimizes $S_{P_0,P_1}(\phi)$ over all possible tests ϕ, as well as a simple expression for $S(P_0, P_1)$. Just as in the Neyman-Pearson setup where the level α is fixed, the optimal test ϕ^* is based on comparing p_0 with p_1 according to the likelihood ratio $p_1(x)/p_0(x)$, so that the only difference is the choice of critical value.

Theorem 13.1.1 *$S_{P_0,P_1}(\phi)$ is minimized by taking $\phi = \phi^*$ a.e. μ, where ϕ^* is any test satisfying $\phi^*(x) = 1$ if $p_1(x) > p_0(x)$ and $\phi^*(x) = 0$ if $p_1(x) < p_0(x)$. Furthermore,*

$$S(P_0, P_1) = S_{P_0,P_1}(\phi^*) = 1 - \frac{1}{2} \int_{\mathcal{X}} |p_1(x) - p_0(x)| \mu(dx) \ . \tag{13.4}$$

PROOF. For any test ϕ,

$$S_{P_0,P_1}(\phi) = \int_{\mathcal{X}} \phi(x)(p_0(x) - p_1(x)) \mu(dx) + 1 \ . \tag{13.5}$$

Let $D_- = \{x : \ p_0(x) - p_1(x) < 0\}$. On D_-, the integrand is minimized by taking $\phi^*(x) = 1$ (since the only constraint on ϕ^* is that it take values in $[0,1]$). Similarly, on $D_+ \equiv \{x : \ p_0(x) - p_1(x) > 0\}$, the integrand is minimized by taking $\phi^*(x) = 0$. On the set $\{x : \ p_0(x) = p_1(x)\}$, it does not matter how $\phi^*(x)$ is defined. Thus, for any minimizing ϕ^*,

$$S_{P_0,P_1}(\phi^*) = \int_{D_-} [p_0(x) - p_1(x)] \mu(dx) + 1 \ . \tag{13.6}$$

Reversing the roles of P_0 and P_1 yields

$$S_{P_1,P_0}(\phi^*) = \int_{D_+} [p_1(x) - p_0(x)] \mu(dx) + 1 \ . \tag{13.7}$$

By symmetry, both expressions are the same, so summing the last two equations and then dividing by two yields

$$S_{P_0,P_1}(\phi^*) = 1 + \frac{1}{2}[\int_{D_-} [p_0(x) - p_1(x)]\mu(x) + \int_{D_+} [p_1(x) - p_0(x)]\mu(dx)] \quad (13.8)$$

$$= 1 - \frac{1}{2}\int_{\mathcal{X}} |p_1(x) - p_0(x)|\mu(dx) \ . \ \blacksquare \quad (13.9)$$

The integral appearing in the last expression leads us to the so-called *total variation* distance between P_0 and P_1.

Definition 13.1.2 The *total variation distance* between P_0 and P_1, denoted $\|P_1 - P_0\|_1$, is given by

$$\|P_1 - P_0\|_1 = \int |p_1 - p_0| d\mu \ , \quad (13.10)$$

where p_i is the density of P_i with respect to any measure μ dominating both P_0 and P_1.

It is easy to see that this distance defines a metric (Problem 13.1) and that this distance is independent of the choice of dominating measure μ. For alternative characterizations of the total variation distance, see Problem 13.2. Equation (13.9) can be restated as

$$S_{P_0,P_1}(\phi^*) = 1 - \frac{1}{2}\|P_1 - P_0\|_1 \ . \quad (13.11)$$

If $X_1, \ldots, X_n$ are i.i.d. P, let P^n denote their joint distribution. We will next consider a sequence of tests ϕ_n for testing P_n^n against Q_n^n. The minimum sum of error probabilities is then $S(P_n^n, Q_n^n)$. The test (sequence) that minimizes the sum of error probabilities is connected with the more usual test in which probability of false rejection of P_n^n is fixed at α by the following lemma. The proof is left as an exercise (Problem 13.5).

Lemma 13.1.1 *(i) If there exists a sequence of tests ϕ_n for which the sum of error probabilities tends to 0, then given any fixed α $(0 < \alpha < 1)$ and n sufficiently large, the level of ϕ_n will be less than α, and its power will tend to 1 as $n \to \infty$. (ii). If for every sequence $\{\phi_n\}$, the sum of the error probabilities tends to 1, then for any sequence whose rejection probability under P_n^n tends to α, the limiting power is α, and hence is no better than that of a test that rejects P_n^n with probability α independent of the data.*

We would like to determine conditions for which the limiting sum of error probabilities is zero or one, as well as for the more important intermediate situation. In order to determine the limiting behavior of $S(P_n^n, Q_n^n)$, we need to study the behavior of $\|P_n^n - Q_n^n\|_1$. Unfortunately, this quantity is often difficult to compute, but it is related to another distance which is easier to manage. This is the following Hellinger distance.

Definition 13.1.3 Let P_0 and P_1 be probabilities on $(\mathcal{X}, \mathcal{F})$. The *Hellinger distance* $H(P_0, P_1)$ between P_0 and P_1 is given by

$$H^2(P_0, P_1) = \frac{1}{2} \int_{\mathcal{X}} [\sqrt{p_1(x)} - \sqrt{p_0(x)}]^2 d\mu(x) , \tag{13.12}$$

where p_i is the density of P_i with respect to any measure μ dominating P_0 and P_1.

The value of $H(P_0, P_1)$ is independent of the choice of μ (Problem 13.1) and one can, for example, always use $\mu = P_0 + P_1$. It is also easy to see that this distance defines a metric.[1] By squaring the integrand and using the fact that the densities p_i must integrate to one, it follows that

$$H^2(P_0, P_1) = 1 - \rho(P_0, P_1) , \tag{13.13}$$

where $\rho(P_0, P_1)$ is known as the *affinity* between P_0 and P_1 and is given by

$$\rho(P_0, P_1) = \int_{\mathcal{X}} \sqrt{p_0(x) p_1(x)} d\mu(x) . \tag{13.14}$$

Note that, by Cauchy-Schwarz, $0 \le \rho(P_0, P_1) \le 1$ and $\rho(P_0, P_1) = 1$ if and only if $P_0 = P_1$. Furthermore, $\rho(P_0, P_1) = 0$ if and only if P_0 and P_1 are *mutually singular*, i.e., there exists a (measurable) set E with $P_0(E) = 1$ and $P_1(E) = 0$. It follows, for example, that $H(P_0, P_1) = 0$ if and only if $P_0 = P_1$.

From equation (13.14), it immediately follows that

$$\rho(P_0^n, P_1^n) = \rho^n(P_0, P_1) \tag{13.15}$$

and hence

$$H^2(P_0^n, P_1^n) = 1 - \rho^n(P_0, P_1) = 1 - [1 - H^2(P_0, P_1)]^n . \tag{13.16}$$

Therefore, the behavior of $H^2(P_0^n, P_1^n)$ with increasing n can be obtained from n and $H(P_0, P_1)$ in a simple way.

Next, we will relate $H(P_0, P_1)$ to $\|P_0 - P_1\|_1$, which was already seen to have a clear statistical interpretation.

Theorem 13.1.2 *The following relationships hold between Hellinger distance and total variation distance:*

$$H^2(P_0, P_1) \le \frac{1}{2} \|P_0 - P_1\|_1$$

$$\le H(P_0, P_1)[2 - H^2(P_0, P_1)]^{1/2} = [1 - \rho^2(P_0, P_1)]^{1/2} . \tag{13.17}$$

PROOF. To prove the first inequality, note that

$$H^2(P_0, P_1) = \frac{1}{2} \int [\sqrt{p_1} - \sqrt{p_0}]^2 d\mu \le \frac{1}{2} \int |\sqrt{p_1} - \sqrt{p_0}| \cdot |\sqrt{p_1} + \sqrt{p_0}| d\mu$$

---

[1] Some authors prefer to leave out the constant 1/2 in their definition. Using Definition 13.1.3, the square of the Hellinger distance between P_0 and P_1 is just one-half the square of the $L_2(\mu)$-distance between $\sqrt{p_0}$ and $\sqrt{p_1}$. Using the Hellinger distance makes it unnecessary to choose a particular μ, and the Hellinger distance is even defined for all pairs of probabilities on a space where no single dominating measure exists.

$$= \frac{1}{2} \int |p_1 - p_0| d\mu = \frac{1}{2} \|P_0 - P_1\|_1 .$$

To prove the second inequality, apply the Cauchy-Schwarz inequality to get

$$\frac{1}{2} \|P_0 - P_1\|_1 = \frac{1}{2} \int |\sqrt{p_1} - \sqrt{p_0}| \cdot |\sqrt{p_1} + \sqrt{p_0}| d\mu$$

$$\leq [\frac{1}{2} \int (\sqrt{p_1} - \sqrt{p_0})^2 d\mu]^{1/2} [\frac{1}{2} \int (\sqrt{p_1} + \sqrt{p_0})^2 d\mu]^{1/2}$$

$$= H(P_0, P_1) [\frac{1}{2} \int (\sqrt{p_1} + \sqrt{p_0})^2 d\mu]^{1/2}$$

$$= H(P_0, P_1)[1 + \rho(P_0, P_1)]^{1/2} = H(P_0, P_1)[2 - H^2(P_0, P_1)]^{1/2} ,$$

with the last equality following from the definition $H^2(P_0, P_1) = 1 - \rho(P_0, P_1)$; the last equality in the statement of the theorem follows immediately from this definition as well. ■

Consider now the problem of deciding between P_0^n and P_1^n based on n i.i.d. observations from P_0 or P_1. Theorems 13.1.1 and 13.1.2 immediately yield the following result.

Corollary 13.1.1 *Fix any P_0 and P_1 with $P_0 \neq P_1$. Then, $S(P_0^n, P_1^n)$ tends to 0 exponentially fast; more specifically,*

$$S(P_0^n, P_1^n) \leq \rho^n(P_0, P_1) \to 0 \qquad as \quad n \to \infty . \tag{13.18}$$

PROOF. By Theorem 13.1.2 and equation (13.16),

$$\frac{1}{2} \|P_0^n - P_1^n\|_1 \geq H^2(P_0^n, P_1^n) = 1 - \rho^n(P_0, P_1) . \tag{13.19}$$

Hence, by Theorem 13.1.1 and (13.19),

$$S(P_0^n, P_1^n) = 1 - \frac{1}{2} \|P_0^n - P_1^n\|_1 \leq \rho^n(P_0, P_1) \to 0 \tag{13.20}$$

as $n \to \infty$, since $\rho(P_0, P_1) < 1$ as $P_0 \neq P_1$. ■

Thus, we can conclude there always exists a perfectly discriminating sequence of tests for testing P_0 against P_1 based on n i.i.d. observations in the sense that the sum of the error probabilities tends to 0.

Since, for any fixed n, the probabilities of error in testing P_0^n against P_1^n are not zero (unless P_0 and P_1 are singular), such asymptotic convergence is of limited value. To obtain a more discriminating result, we will consider the problem of testing $P_{\theta_0}^n$ against $P_{\theta_n}^n$ based on n i.i.d. observations, where P_{θ_n} is a sequence of probability distributions getting *closer* to P_{θ_0}. Closeness here will conveniently be expressed by the Hellinger metric. We would like to consider P_{θ_n} close enough to P_{θ_0} as $n \to \infty$ so that the testing problem becomes difficult for the statistician in the sense that there does not exist a test sequence whose error probabilities both tend to zero. On the other hand, we would also not want P_{θ_n} and P_{θ_0} to be so close that no sequence of tests will have any reasonable amount of power. The following theorem characterizes this situation and shows that the intermediate situation occurs if and only if $nH^2(P_{\theta_0}, P_{\theta_n}) \asymp 1$.

Theorem 13.1.3 *Suppose*

$$c_1 = \liminf nH^2(P_{\theta_0}, P_{\theta_n}) \le \limsup nH^2(P_{\theta_0}, P_{\theta_n}) = c_2 \ . \tag{13.21}$$

Then,

$$1 - [1 - \exp(-2c_2)]^{1/2} \le \liminf S(P^n_{\theta_0}, P^n_{\theta_n}) \tag{13.22}$$

$$\le \limsup S(P^n_{\theta_0}, P^n_{\theta_n}) \le \exp(-c_1) \ .$$

Proof. To prove $\limsup S(P^n_{\theta_0}, P^n_{\theta_n}) \le \exp(-c_1)$, assume first that

$$nH^2(P_{\theta_0}, P_{\theta_n}) \to c \ge c_1 \ .$$

By Corollary 13.1.1,

$$S(P^n_{\theta_0}, P^n_{\theta_n}) \le \rho^n(P_{\theta_0}, P_{\theta_n}) = [1 - H^2(P_{\theta_0}, P_{\theta_n})]^n \to \exp(-c) \le \exp(-c_1) \ .$$

By applying this argument to subsequences θ_{n_j} such that $n_j H^2(P_{\theta_0}, P_{\theta_{n_j}})$ converges, the last inequality in (13.22) follows. Similarly, suppose

$$nH^2(P_{\theta_0}, P_{\theta_n}) \to c \le c_2 \ .$$

The first inequality follows if we show that

$$1 - [1 - \exp(-2c)]^{1/2} \le \liminf_n S(P^n_{\theta_0}, P^n_{\theta_n}) \ .$$

By Theorem 13.1.1 and then Theorem 13.1.2,

$$S(P^n_{\theta_0}, P^n_{\theta_n}) = 1 - \frac{1}{2}\|P^n_{\theta_n} - P^n_{\theta_0}\|_1 \ge 1 - [1 - \rho^2(P^n_{\theta_0}, P^n_{\theta_n})]^{1/2} \ .$$

By (13.15), this becomes

$$1-[1-\rho^{2n}(P_{\theta_0}, P_{\theta_n})]^{1/2} = 1-\{1-[1-H^2(P_{\theta_0}, P_{\theta_n})]^{2n}\}^{1/2} \to 1-[1-\exp(-2c)]^{1/2} \ ,$$

and the result follows. ■

Thus, from an asymptotic point of view, it is reasonable to consider alternatives θ_n to θ_0 such that $nH^2(P_{\theta_0}, P_{\theta_n})$ is bounded away from 0 and ∞. Otherwise, the problem is asymptotically degenerate in the sense that, either there exists a test sequence ϕ_n for testing θ_0 versus θ_n such that the probability of a type 1 error tends to zero and the power at θ_n tends to one, or no sequence of level α tests will have asymptotic power greater than α. We next consider what the condition on $nH^2(P_{\theta_0}, P_{\theta_n})$ becomes in some classical examples.

Example 13.1.1 (Quadratic Mean Differentiable Families) Assume that $\{P_\theta, \theta \in \Omega\}$ is q.m.d. with derivative $\eta(\cdot, \theta_0)$ at θ_0 and positive definite $I(\theta_0)$. Suppose $n^{1/2}(\theta_n - \theta_0) \to h$. By equation (12.6) and Lemma 12.2.2,

$$2nH^2(P_{\theta_0}, P_{\theta_n}) = n \int [\sqrt{p_{\theta_n}} - \sqrt{p_{\theta_0}}]^2 d\mu$$

$$\to \int |\langle \eta(x, \theta_0), h\rangle|^2 d\mu(x) = \frac{1}{4}\langle h, I(\theta_0) h\rangle < \infty \ . \tag{13.23}$$

Thus, the nondegenerate situation occurs when $|\theta_n - \theta_0| = O(n^{-1/2})$. Note that the limiting value (13.23) is never 0 unless $h = 0$ (Problem 13.8). ■

Example 13.1.2 (Uniform Family; Example 12.2.8, continued) Let P_θ be the uniform distribution on $(0, \theta)$. Then, $nH^2(P_{\theta_0}, P_{\theta_n})$ tends to a finite, positive limit if and only if $n(\theta_n - \theta_0) \to h < \infty$ (Problem 13.4). Hence, alternatives θ_n such that $\theta_n - \theta_0 \asymp n^{-1}$ cannot be perfectly discriminated, yet tests can be constructed that have reasonable power against these alternatives. ■

To clarify the difference between the previous two examples, note that in Example 13.1.1 we have

$$H^2(P_{\theta_0}, P_{\theta_n}) \asymp (\theta_n - \theta_0)^2$$

while in Example 13.1.2 we have

$$H^2(P_{\theta_0}, P_{\theta_n}) \asymp |\theta_n - \theta_0| \ .$$

Example 13.1.3 (Example 12.2.5, continued) Consider densities

$$p_\theta(x) = C(\beta) \exp\{-|x - \theta|^\beta\}$$

and set $\theta_0 = 0$. In this example, the following can be shown (see Le Cam and Yang (1990), Lemma 5 in Section 7.3). If $\beta > 1/2$, the family is q.m.d. and so $H^2(P_0, P_\delta)/\delta^2$ tends to a finite limit as $\delta \to 0$; thus, the right rate to keep the problem nondegenerate is $\delta \asymp n^{-1/2}$. If $\beta = 1/2$, $H^2(P_0, P_\delta)/[\delta^2 |\log(\delta)|]$ tends to a finite limit as $\delta \to 0$, and so the corresponding nondegenerate rate is $\delta \asymp (n \log n)^{-1/2}$. If $0 < \beta < 1/2$, $H^2(P_0, P_\delta)/\delta^{1+2\beta}$ tends to a finite limit, in which case the corresponding nondegenerate rate is $\delta \asymp n^{-1/(1+2\beta)}$. ■

Even though the above asymptotic development studies the limiting behavior of tests based on the criterion of minimum sum of error probabilities, it is also relevant to the usual Neyman-Pearson formulation when we we consider tests whose level is α for some fixed $\alpha > 0$. For, if $nH^2(P_{\theta_0}, P_{\theta_n}) \to \infty$, then $S(P^n_{\theta_0}, P^n_{\theta_n}) \to 0$, by Theorem 13.1.3. Thus, by Lemma 13.1.1, given $\epsilon > 0$, for large enough n there exists a test sequence ϕ_n whose level is less than ϵ and whose power against θ_n is at least $1 - \epsilon$. So clearly, there exist level α test sequences whose power against θ_n tend to one.

On the other hand, if $nH^2(P_{\theta_0}, P_{\theta_n}) \to 0$, then no sequence of level α tests has limiting power against θ_n greater than α (Problem 13.6).

As before, the interesting nondegenerate asymptotic situation occurs when $nH^2(P_{\theta_0}, P_{\theta_n}) \to c$ for some finite positive c. In this case, there exists a level α test sequence whose limiting power against θ_n exceeds α. Typically, the value of the limiting power is strictly less than one, but in some cases it may equal one (which does not contradict Theorem 13.1.3 because the sum of the errors is tending to $\alpha > 0$); see Problem 13.9.

The following theorem clarifies the relationship between P_n and Q_n being contiguous and the Hellinger metric between P_n and Q_n.

Theorem 13.1.4 *(i) If $nH^2(P_n, Q_n) \to 0$, then $\|Q^n_n - P^n_n\|_1 \to 0$ and $\{P^n_n\}$ and $\{Q^n_n\}$ are contiguous.*
(ii) If $nH^2(P_n, Q_n) \to \infty$, then $S(P^n_n, Q^n_n) \to 0$ and $\{P^n_n\}$ and $\{Q^n_n\}$ are not contiguous.

Proof. To prove (i), note that Theorem 13.1.3 holds if P_{θ_0} is allowed to vary with n, with no change in the argument or the conclusion. Thus, by (13.21) with $c_2 = 0$, $nH^2(P_n, Q_n) \to 0$ implies $S(P_n^n, Q_n^n) \to 1$. Therefore, by Problem 13.10, $\|P_n^n - Q_n^n\|_1 \to 0$. To prove (ii), assume $nH^2(P_n, Q_n) \to \infty$. By Theorem 13.1.3, $S(P_n^n, Q_n^n) \to 0$. Hence, there exists a test sequence ϕ_n^* such that $E_{P_n^n}(\phi_n^*) \to 0$ and $E_{Q_n^n}(\phi_n^*) \to 1$. Let L_n denote the likelihood ratio of Q_n^n with respect to P_n^n. But, Theorem 13.1.1 shows that ϕ_n^* can be taken to be the indicator of the set $A_n \equiv L_n > 1$. Then, $P_n^n(A_n) \to 0$ but $Q_n^n(A_n) \to 1$. ■

Example 13.1.4 (Example 13.1.1, continued) Assume $\{P_\theta, \theta \in \Omega\}$ is q.m.d. at θ_0, and $h_n \to h$. Then, by a calculation similar to that in Example 13.1.1, $nH^2(P_{\theta_0+hn^{-1/2}}, P_{\theta_0+h_n n^{-1/2}}) \to 0$ (Problem 13.11). Therefore, by Theorem 13.1.4(i), $P^n_{\theta_0+h_n n^{-1/2}}$ is contiguous to $P^n_{\theta_0}$. This result forms the basis for generalizing results such as Theorem 12.2.3, Theorem 12.4.1 and Corollary 12.4.1, which have been shown to be true when $h_n = h$, to the more general case when $h_n \to h$; see Problems 13.12 and 13.13. ■

In the intermediate situation $nH^2(P_n, Q_n) \asymp 1$, P_n^n and Q_n^n may or may not be contiguous. Example 13.1.1 provides an example where contiguity holds. However reconsider Example 13.1.2, where P_n is uniform on $[0, 1]$ and Q_n is uniform on $[0, 1 + hn^{-1}]$, where $h > 0$. Then, $nH^2(P_n, Q_n) \asymp 1$, but Q_n^n is not contiguous with respect to P_n^n. To see why, let A_n be the event that the maximum of n i.i.d. observations exceeds 1. Then, $P_n^n(A_n) = 0$, while $Q_n^n(A_n) \to 1 - e^{-h}$. For a sharp result on the relationship between contiguity and Hellinger distance, see Oosterhoff and van Zwet (1979).

13.2 Asymptotic Relative Efficiency

Consider the problem of testing $H : \theta \in \Omega_0$ against $\theta \notin \Omega_0$ when $X_1, \ldots, X_n$ are i.i.d. according to a model $\{P_\theta,\ \theta \in \Omega\}$. Our main goal is to derive tests that are asymptotically optimal. However, other considerations (such as robustness) may suggest using non-optimal tests. It is then important to know how much is lost by the use of such sub-optimal tests. In this section, we shall therefore compare the performance of two test procedures ϕ_n and $\tilde{\phi}_n$. In this context, performance is measured in terms of power. Roughly speaking, the relative efficiency of $\tilde{\phi}_n$ with respect to ϕ_n is defined to be $n/\tilde{n}$, where n and $\tilde{n}$ are the sample sizes required for ϕ_n and $\tilde{\phi}_{\tilde{n}}$ to have the same power at the same level against the same alternative. For instance, a ratio of 2 would indicate that $\tilde{\phi}_n$ is twice as efficient as ϕ_n because twice as many observations are required for ϕ_n to have the same power at a given alternative as $\tilde{\phi}_n$. Such a comparison can be based on the following result.

Theorem 13.2.1 *Suppose $X_1, \ldots, X_n$ are i.i.d. according to a q.m.d. family indexed by a real parameter θ, and consider testing $\theta = \theta_0$ versus $\theta > \theta_0$. Assume the sequence $\phi = \{\phi_n\}$ is based on test statistics T_n satisfying the following: there exists a function $\mu(\cdot)$ and a number $\sigma^2 > 0$ such that, under any sequence θ_n*

satisfying $n^{1/2}(\theta_n - \theta_0) = O(1)$,

$$n^{1/2}[T_n - \mu(\theta_n)] \xrightarrow{d} N(0, \sigma^2) \quad ; \tag{13.24}$$

moreover, $\mu(\cdot)$ *is assumed to have a right-hand derivative* $\mu'(\theta_0) > 0$ *at* θ_0. *Suppose* ϕ_n *rejects when* $n^{1/2}[T_n - \mu(\theta_0)] > \hat{c}_n$, *where*

$$\hat{c}_n \to z_{1-\alpha}\sigma \tag{13.25}$$

in probability under θ_0. *Then, the following is true.*
(i) $E_{\theta_0}(\phi_n) \to \alpha$ *as* $n \to \infty$.
(ii) The limiting power of ϕ_n *against* θ_n *satisfying* $n^{1/2}(\theta_n - \theta_0) \to h$ *is*

$$\lim_n E_{\theta_n}(\phi_n) = 1 - \Phi\left[z_{1-\alpha} - h\frac{\mu'(\theta_0)}{\sigma}\right] \ . \tag{13.26}$$

(iii) Fix $0 < \alpha < \beta < 1$. *Let* θ_k *be any sequence satisfying* $\theta_k > \theta_0$ *and* $\theta_k \to \theta_0$ *as* $k \to \infty$ *and let* n_k *be any sequence for which* $E_{\theta_k}(\phi_{n_k}) \geq \beta$. *Then,*[2]

$$n_k \sim \frac{(z_{1-\alpha} - z_{1-\beta})^2\sigma^2}{[(\theta_k - \theta_0)\mu'(\theta_0)]^2} \ . \tag{13.27}$$

Proof. Part (i) follows by Slutsky's Theorem. To prove (ii), let θ_n satisfy $n^{1/2}(\theta_n - \theta_0) \to h$. By contiguity (Example 13.1.4, it follows that $\hat{c}_n \to z_{1-\alpha}\sigma$ in probability under θ_n. Also,

$$n^{1/2}[\mu(\theta_n) - \mu(\theta_0)] \to h\mu'(\theta_0) \ .$$

Letting Z denote a standard normal variable, by Slutsky's Theorem,

$$E_{\theta_n}(\phi_n) = P_{\theta_n}\{n^{1/2}[T_n - \mu(\theta_n)] > \hat{c}_n - n^{1/2}[\mu(\theta_n) - \mu(\theta_0)]\}$$

$$\to P\{\sigma Z > z_{1-\alpha}\sigma - h\mu'(\theta_0)\} \ ,$$

implying (ii).

To prove (iii), choose $h = h_\beta$ so that the right side of (13.26) is β, and hence

$$h_\beta = (z_{1-\alpha} - z_{1-\beta}) \cdot \frac{\sigma}{\mu'(\theta_0)} \ .$$

By (ii), if θ_n satisfies $n^{1/2}(\theta_n - \theta_0) \to h_\beta$, then the limiting power of ϕ_n against θ_n is β. It follows that the limiting power of ϕ_n against θ_n is β if and only if θ_n satisfies

$$n \sim \frac{(z_{1-\alpha} - z_{1-\beta})^2\sigma^2}{[(\theta_n - \theta_0)\mu'(\theta_0)]^2} \ .$$

For an arbitrary sequence $\theta_k \to \theta_0$, let m_k satisfy $m_k^{1/2}(\theta_k - \theta_0) \to h_\beta$. Then, since $m_k^{1/2}(\theta_k - \theta_0) = O(1)$, the asymptotic normality assumption for T_{m_k} holds, and the above argument shows the limiting power of ϕ_{m_k} against θ_k is β iff

$$m_k \sim \frac{(z_{1-\alpha} - z_{1-\beta})^2\sigma^2}{[(\theta_k - \theta_0)\mu'(\theta_0)]^2} \ .$$

---

[2]The notation $a_k \sim b_k$ means $a_k/b_k \to 1$.

To show $n_k \sim m_k$, we first show that $\limsup(n_k/m_k) \geq 1$. But, the q.m.d. assumption precludes n_k being bounded (Problem 13.17), while the above argument shows the limiting power against n_k would be bounded above by β if $n_k \to \infty$. So, it suffices to show $\liminf(n_k/m_k) \leq 1$. Fix $\epsilon > 0$ and let s_k satisfy

$$s_k^{1/2}(\theta_k - \theta_0) \to (z_{1-\alpha} - z_{1-\beta}) \cdot \frac{\sigma}{\mu'(\theta_0)} + \epsilon \ .$$

Note that $s_k/m_k < 1 + C\epsilon$ for some C. Then, the limiting power of ϕ_{s_k} against θ_k is, by the above argument, strictly greater than β. Hence, for large enough n, $n_k \leq s_k$, and so

$$\liminf \frac{n_k}{m_k} \leq \liminf \frac{s_k}{m_k} \leq 1 + C\epsilon \ .$$

Since ϵ was arbitrary, the result follows. ■

Inspection of (13.26) shows that, the larger the value $\mu'(\theta_0)/\sigma$, the smaller is the sample size required to achieve a given power β. A test sequence generated by T_n will therefore be more efficient the larger its value of $[\mu'(\theta_0)/\sigma]$. This value is called the *efficacy* of the test sequence. Under some regularity conditions, Rao (1963) proved that

$$[\mu'(\theta_0)/\sigma(\theta_0)]^2 \leq I(\theta_0) \ ,$$

where $I(\theta_0)$ is the usual Fisher Information. Such a result will follow from the results in Section 13.3 under the assumption of quadratic mean differentiability.

Example 13.2.1 (Wald and Rao Tests) Under the assumptions of Theorem 13.2.1, suppose $\hat{\theta}_n$ satisfies (12.62),and consider the Wald test that rejects for large values of $\hat{\theta}_n - \theta_0$. By Theorem 12.4.1, the assumptions of Theorem 13.2.1 hold with $\mu(\theta) = \theta$ and $\sigma^2 = I^{-1}(\theta_0)$. (The theorem establishes asymptotic normality under sequences θ_n of the form $\theta_0 + hn^{-1/2}$, but it holds more generally for sequences θ_n satisfying $n^{1/2}(\theta_n - \theta_0) = O(1)$, by Problem 12.32.) Hence, the squared efficacy of the Wald test is $I(\theta_0)$. The same is true for Rao's score test (Problem 13.18). ■

Corollary 13.2.1 *Assume the conditions of Theorem 13.2.1 hold for $\phi = \{\phi_n\}$ and consider a competing test sequence $\tilde{\phi} = \{\tilde{\phi}_n\}$ based on a test statistic $\tilde{T}_n$ satisfying (13.24) with μ and σ replaced by $\tilde{\mu}$ and $\tilde{\sigma}$. Fix $0 < \alpha < \beta < 1$ and for $\theta > \theta_0$, let $N(\theta)$ and $\tilde{N}(\theta)$ be the smallest sample sizes necessary for ϕ and $\tilde{\phi}$ to have power at least β against θ. Then,*

$$\lim_{\theta \downarrow \theta_0} \frac{N(\theta)}{\tilde{N}(\theta)} = \left[\frac{\tilde{\mu}'(\theta_0)/\tilde{\sigma}}{\mu'(\theta_0)/\sigma} \right]^2 \ , \tag{13.28}$$

and the right hand side is called the (Pitman) Asymptotic Relative Efficiency (ARE) of $\tilde{\phi}$ with respect to ϕ.

PROOF. Apply (iii) of Theorem 13.2.1. ■

Notice that the ARE is independent of α and β. Also, the tests are only required to be asymptotically level α, and the critical values may be random. Thus, we can, for example, compare tests based on an exact critical value, such as one obtained from the exact sampling distribution of T_n under θ_0, with tests based on asymptotic normality, possibly combined with an estimate of the asymptotic variance. Another possibility is to use a critical value obtained from a permutation distribution, such as the tests studied in Section 5.12. Nevertheless, under the assumptions stated, the resulting efficacy of a test is unchanged whether a test is based on an exact critical value or an approximate one. This implies the ARE is one when comparing two tests based on the same test statistic but with different critical values, as long as (13.25) is satisfied.

The ARE provides a single number for comparing two tests, independent of α and β. However, for finite samples, the relative efficiency depends on both α and β. Thus, the asymptotic measure may not give a very good picture of the actual finite-sample situation.

The following lemma facilitates the computation of the efficacy of a test sequence.

Lemma 13.2.1 *Assume $X_1, \ldots, X_n$ are i.i.d. according to a family which is q.m.d. at θ_0 and that the unknown parameter θ varies in an open subset of $\mathbb{R}$. Suppose, under $\theta_n = \theta_0 + h/n^{1/2}$, we have*

$$n^{1/2} T_n \stackrel{d}{\to} N(hm, \sigma^2) .$$

Then, the assumptions in Theorem 13.2.1 hold for T_n and the efficacy of T_n is m/σ.

PROOF. Let $\mu(\theta) = m(\theta - \theta_0)$. The assumptions imply

$$n^{1/2}(T_n - \mu(\theta_n)) \stackrel{d}{\to} N(0, \sigma^2)$$

under θ_n whenever θ_n is of the form $\theta_n = \theta_0 + hn^{-1/2}$. By Problem 12.32, the same result holds whenever $n^{1/2}(\theta_n - \theta_0) = O(1)$, so that the asymptotic normality assumption holds for T_n with $\mu'(\theta) = m$. Thus, the efficacy of T_n is m/σ. ■

Example 13.2.2 (One-sample Tests of Location) Suppose $X_1, \ldots, X_n$ are i.i.d. according to a location model with density $f(x - \theta)$, where f is assumed to be symmetric about 0. Assume $f'(x)$ exists for almost all x, and the Fisher Information is positive and finite, so that the family is q.m.d. We would like to compare competing tests for testing $\theta = 0$ versus $\theta > 0$. Consider the three tests that reject for large values of t_n, S_n, and W_n, the classical t-statistic t_n, the sign test statistic S_n, and the Wilcoxon signed rank statistic W_n studied in Examples 12.3.9, 12.3.10, and 12.3.11, respectively. Regardless of whether or not f is known, all three tests can be used to yield tests that are pointwise consistent in level as long as f is symmetric and has finite variance. Let σ_f^2 denote the variance of f. Under $\theta_n = h/n^{1/2}$, we have

$$n^{1/2} t_n \stackrel{d}{\to} N(\frac{h}{\sigma_f}, 1) ,$$

$$n^{1/2} S_n \stackrel{d}{\to} N(hf(0), \frac{1}{4}) ,$$

and

$$n^{1/2} W_n \stackrel{d}{\to} N(2h \int_{-\infty}^{\infty} f^2(x) dx, \frac{1}{3}) .$$

Thus, the efficacies of t, S, and W are $1/\sigma$, $2f(0)$, and $(12)^{1/2} \int f^2$, respectively. Therefore (with an obvious change of notation that shows the dependence on f),

$$e_{S,t}(f) = [2f(0)\sigma_f]^2$$

and

$$e_{W,t}(f) = 12\sigma_f^2[\int f^2]^2 . \tag{13.29}$$

In particular, when f is the normal density φ, $e_{S,t}(\varphi) = 2/\pi \approx 0.637$ and $e_{W,t}(\varphi) = 3/\pi \approx 0.955$. Thus, under normality, the sign test requires a sample size that is about 57 percent greater than the t-test to achieve the same power.

On the other hand, the efficiency loss for the Wilcoxon test is less than 5 percent. When f is not normal, the efficiency of both the sign test and the Wilcoxon test with respect to the t-test can be arbitrary large. To see this, modify φ by moving small masses out in the tails of the distribution so that σ_f becomes quite large but $f(0)$ and $\int f^2$ remain about the same. Moreover, the Wilcoxon test can never be much less efficient than the t-test, regardless of f; in fact (Problem 13.21),

$$e_{W,t}(f) \geq 0.864 \quad \text{for all } f . \tag{13.30}$$

Interestingly, when f is the double exponential density, the sign test is the most efficient of the three. In fact, it will later be seen in Section 13.3 that the sign test is asymptotically uniformly most powerful for testing the location parameter in a double exponential location model. ■

Example 13.2.3 (Two-Sample Tests of Shift) Suppose $X_1, \ldots, X_m$ are i.i.d. with c.d.f. F and, independently, $Y_1, \ldots, Y_n$ are i.i.d. with c.d.f. G. Assume

$$G(x) = F(x - \theta) \tag{13.31}$$

for some θ. If F is unknown, such a nonparametric two-sample shift model was studied in Section 5.8, where the class of permutation tests was introduced. Consider the problem of testing $\theta = 0$ versus $\theta > 0$. We would like to compare the normal scores test and the Wilcoxon test W introduced in Section 6.9, as well as the two-sample t-test and the permutation t-test. It turns out that, even when F and G are normal with a common variance, the normal scores test and the Wilcoxon test are nearly as powerful as the t-test. To obtain a numerical comparison, suppose $m = n$. Then, the notion of relative efficiency applies with no changes (by viewing the observations as pairs (X_i, Y_i)), and so the (Pitman) asymptotic relative efficiencies can be computed for test statistics satisfying the assumptions of Theorem 13.2.1.

In the particular case of the Wilcoxon test, $e_{W,t} = 3/\pi$ when F and G are normal with equal variance. Some numerical evidence supports the fact that the relative efficiency is nearly independent of α and β in this context; see Lehmann (1998), p.79. As in the one-sample case, the (Pitman) asymptotic relative efficiency is always $\geq .864$, but may exceed 1 and can be infinite. The situation is even more favorable for the normal-scores test. Its asymptotic relative efficiency, relative to the t-test, is always ≥ 1 under the model (13.31); moreover, it is 1 only when F is normal. Thus, while the t-test is performance robust in the sense that its level and power is asymptotically independent of F as discussed in Section 11.3, the present results show that the efficiency and optimality properties of the t-test are quite nonrobust. The same comments apply to the permutation t-test (whose asymptotic properties will be discussed in Section 15.2.

The above results do not depend on the assumption of equal sample sizes; they are also valid if $m/n \asymp 1$. At least in the case that F is normal, the asymptotic results given by the (Pitman) efficiencies agree well with those found for small samples. The results also extend to testing the equality of s means, and the asymptotic relative efficiency of the Kruskal-Wallis test to the normal theory F-test is the same as the Wilcoxon to the t-test in the case $s = 2$. For a more detailed discussion of these and related efficiency results, see for example, Lehmann (1998), Randles and Wolfe (1979), Blair and Higgins (1980), and Groeneboom (1980).

The most ambitious goal in the nonparametric two-sample shift model would be to find a test which does not depend on F, yet would have asymptotic efficiency at least 1 with respect to any other test, for all F (or at least all F in a nonparametric family). Such *adaptive* tests (which achieve simultaneous optimality by adapting themselves to the unknown F) do in fact exist if F is sufficiently smooth. Their possibility was first suggested by Stein (1956b), and has been carried out for point estimation problems by Beran (1974), Stone (1975) and Bickel (1982). ■

We now briefly mention some other notions of asymptotic relative efficiency. Consider two test sequences $\phi = \{\phi_n\}$ and $\tilde{\phi} = \{\tilde{\phi}_n\}$, each indexed by the sample size n. For simplicity, suppose ϕ is determined by a test statistic $T = \{T_n\}$ which rejects for large values. Then, ϕ_n is really a family of tests indexed by n and α, where the value α determines the size of the test. Define $N(\alpha, \beta, \theta)$ to be the sample size necessary for the test ϕ_n to have power $\geq \beta$ against the fixed alternative θ, subject to the constraint that the size of ϕ_n is α. Thus, N is the smallest sample size n such that, for some critical value $c = c(n, \alpha)$, we have

$$\sup_{\theta_0 \in \Omega_0} P_{\theta_0}\{T_n > c\} \leq \alpha \tag{13.32}$$

and

$$P_\theta\{T_n > c\} \geq \beta \ .$$

Similarly, define $\tilde{N}(\alpha, \beta, \theta)$ corresponding to a test $\tilde{\phi}_n$ based on a test statistic $\tilde{T}_n$. Then, the *relative efficiency* of $\tilde{\phi}$ with respect to ϕ is defined to be

$$e_{\tilde{T},T}(\alpha, \beta, \theta) = N(\alpha, \beta, \theta)/\tilde{N}(\alpha, \beta, \theta) \ .$$

While this measure has a useful statistical interpretation, its value depends on three arguments α, β and θ; moreover, it is typically quite difficult to compute $N(\alpha, \beta, \theta)$ for a given test ϕ. However, it is often possible to calculate the limiting values of $e_{\tilde{T},T}(\alpha, \beta, \theta)$ as $\alpha \to 0$, $\beta \to 1$, or $\theta \to \theta_0 \in \Omega_0$, with the remaining two arguments kept fixed. The case $\alpha \to 0$ is known as the Bahadur efficiency, the case $\beta \to 1$ as the Hodges-Lehmann efficiency, and the case $\theta \to \theta_0$ coincides with the (Pitman) ARE already introduced. These various types of efficiency are reviewed in Serfling (1980, Chapter 10) and Nikitin (1995, Chapter 1). While each of these notions of asymptotic relative efficiency have some merit, we argue that the Pitman ARE has the most practical significance. In practice, α, though small, is regarded as fixed, and so comparisons based on the Bahadur efficiency with $\alpha \to 0$ may be questionable. On the other hand, with α fixed, comparing procedures with power tending to 1 seems inappropriate since then the probability of an error of the second kind now becomes smaller than the probability of an error if the first kind. Typically, for values of the parameter at a fixed distance from Ω_0, any reasonable test will have power tending to one. It then becomes more important to choose a test that is better equipped to deal with the more difficult situation when θ is near Ω_0, and the Pitman asymptotic relative efficiency provides a useful measure in this situation. Numerical evidence for the superiority of Pitman over Bahadur efficiency is provided in Groeneboom and Oosterhoff (1981).

13.3 AUMP Tests in Univariate Models

Suppose $X_1, \ldots, X_n$ are i.i.d. P_θ, with θ real-valued, and consider testing the hypothesis $\theta = \theta_0$ against $\theta > \theta_0$. As was discussed in Section 3.4, even in this one-parameter model, UMP tests rarely exist. In the present section we shall show that under weak smoothness assumptions, asymptotically optimal tests do exist.

As we saw in Section 13.1, when the q.m.d. assumption holds, informative power calculations for large samples are obtained not against fixed alternatives (for which the power tends to 1) but against sequences of alternatives of the form

$$\theta_{n,h} = \theta_0 + hn^{-1/2} \qquad h > 0 \; , \tag{13.33}$$

for which the power tends to a value strictly between α and 1. Asymptotic optimality is most naturally studied in terms of these alternatives.

Let $\{\alpha_n\}$ be a sequence of levels tending to α. By the Neyman-Pearson Lemma, the most powerful test $\phi_{n,h}$ for testing $\theta = \theta_0$ against $\theta_{n,h}$ at level α_n rejects when

$$L_{n,h} = \prod_{i=1}^n [p_{\theta_0 + hn^{-1/2}}(X_i)/p_{\theta_0}(X_i)]$$

is sufficiently large; more specifically, it is given by

$$\phi_{n,h} = \begin{cases} 1 & \text{if } \log(L_{n,h}) > c_{n,h} \\ \gamma_{n,h} & \text{if } \log(L_{n,h}) = c_{n,h} \\ 0 & \text{if } \log(L_{n,h}) < c_{n,h}, \end{cases} \tag{13.34}$$

where the constants $c_{n,h}$ and $\gamma_{n,h}$ are determined so that $E_{\theta_0}(\phi_{n,h}) = \alpha_n$.

The limits of the critical values $c_{n,h}$ and the power of the tests (13.34) against the alternatives (13.33) are given in the following lemma, under the assumption of quadratic mean differentiability.

Lemma 13.3.1 *Assume $\{P_\theta, \; \theta \in \Omega\}$ is q.m.d. at θ_0 with Ω an open subset of $\mathbb{R}$. Consider testing $\theta = \theta_0$ against $\theta_{n,h} = \theta_0 + hn^{-1/2}$ at level $\alpha_n \to \alpha \in (0,1)$.*
(i) As $n \to \infty$, the critical values $c_{n,h}$ of the most powerful test sequence $\phi_{n,h}$ defined in (13.34) satisfy

$$c_{n,h} \to \frac{-h^2 I(\theta_0)}{2} + hI^{1/2}(\theta_0) z_{1-\alpha} \; , \tag{13.35}$$

where $I(\theta_0)$ is the Fisher Information at θ_0 and $z_{1-\alpha} = \Phi^{-1}(1-\alpha)$ is the $1-\alpha$ quantile of $N(0,1)$. Moreover,

$$P_{\theta_0}\{\log(L_{n,h}) > c_{n,h}\} \to \alpha \tag{13.36}$$

and

$$P_{\theta_0}\{\log(L_{n,h}) = c_{n,h}\} \to 0 \ . \tag{13.37}$$

(ii) The power of $\phi_{n,h}$ satisfies

$$E_{\theta_0+hn^{-1/2}}(\phi_{n,h}) \to 1 - \Phi[z_{1-\alpha} - hI^{1/2}(\theta_0)] \ . \tag{13.38}$$

(iii) More generally, consider testing $\theta = \theta_0$ against θ_{n,h_n} where $h_n \to h$, with $|h| < \infty$. Then, the power of ϕ_{n,h_n} against θ_{n,h_n} converges to the right side of (13.38), i.e., it has the same limiting power as $\phi_{n,h}$.

Proof. By Theorem 12.2.3, under θ_0, $\log(L_{n,h})$ converges weakly to $N(-\sigma_h^2/2, \sigma_h^2)$, where $\sigma_h^2 = h^2 I(\theta_0)$. Then, (13.37) follows by Problem 11.42(i). Hence,

$$\alpha_n = E_{\theta_0}(\phi_{n,h}) = P_{\theta_0}\{\log(L_{n,h}) > c_{n,h}\} + o(1) \ ,$$

and so (13.36) follows. By Problem 11.42(ii), it follows that $c_{n,h}$ tends to the $1-\alpha$ quantile of $N(-\sigma_h^2/2, \sigma_h^2)$, and so (13.35) follows.

To prove (ii), under $\theta_{n,h}$, $\log(L_{n,h})$ converges in distribution to a variable Y_h distributed as $N(\sigma_h^2/2, \sigma_h^2)$, as shown in Example 12.3.12 by a contiguity argument. Hence, under $\theta_0 + hn^{-1/2}$, the probability that $\log(L_{n,h}) = c_{n,h}$ tends to 0, again by Problem 11.42(i). Letting Z denote a standard normal variable,

$$E_{\theta_{n,h}}(\phi_{n,h}) = P_{\theta_{n,h}}\{\log(L_{n,h}) > c_{n,h}\} + o(1)$$

$$\to P\{Y_h > \frac{-\sigma_h^2}{2} + \sigma_h z_{1-\alpha}\} = P\{Z > -\sigma_h + z_{1-\alpha}\} = 1 - \Phi(z_{1-\alpha} - hI^{1/2}(\theta_0)),$$

and (ii) follows.

The proof of (iii) is left to Problem 13.27. ■

Next, we consider the notion of an asymptotically most powerful test sequence for testing a simple hypothesis $\theta = \theta_0$ against a simple alternative sequence θ_n.

Definition 13.3.1 For testing $\theta = \theta_0$ against $\theta = \theta_n$, $\{\phi_n\}$ is *asymptotically most powerful* (AMP) at (asymptotic) level α if $\limsup_n E_{\theta_0}(\phi_n) \le \alpha$ and if for any other sequence of test functions $\{\psi_n\}$ satisfying $\limsup_n E_{\theta_0}(\psi_n) \le \alpha$,

$$\limsup_n E_{\theta_n}(\psi_n) - E_{\theta_n}(\phi_n) \le 0 \ . \tag{13.39}$$

For q.m.d. families, Lemma 13.3.1 implies the following result (Problem 13.28).

Theorem 13.3.1 *Assume $\{P_\theta,\ \theta \in \Omega\}$ is q.m.d. at θ_0 with Ω an open subset of $\mathbb{R}$ and Fisher Information $I(\theta_0)$. Given $X_1, \ldots, X_n$ i.i.d. P_θ, consider testing $\theta = \theta_0$ against $\theta_n = \theta_0 + h_n n^{-1/2}$, where $h_n \to h > 0$. Then, $\phi_n = \phi_n(X_1, \ldots, X_n)$ is AMP level α if and only if $E_{\theta_0}(\phi_n) \to \alpha$ and*

$$\limsup_n E_{\theta_0+hn^{-1/2}}(\phi_n) = [1 - \Phi(z_{1-\alpha} - hI^{1/2}(\theta_0))]. \tag{13.40}$$

Of course, for testing a simple null hypothesis against a simple alternative, one always has available the optimal finite sample Neyman Pearson test sequence $\phi_{n,h}$ given by (13.34). However, the tests $\phi_{n,h}$ will typically depend on h and therefore will not be uniformly best against all alternatives. However, at this point, there is a profound difference between the finite sample and the asymptotic theory. Most powerful tests typically are unique while this is not true for asymptotically most powerful tests, since they can be changed on sets whose probability tends to zero without changing the asymptotic power. This difference opens up the possibility that among the set of AMP tests there may be one that is AMP simultaneously for all values of h. This possibility will be explored in the remainder of this section.

For this purpose, recall the expansion of $\log(L_{n,h})$. By Theorem 12.2.3,

$$\log(L_{n,h}) - [hZ_n - \frac{1}{2}h^2 I(\theta_0)] = o_{P^n_{\theta_0}}(1) , \tag{13.41}$$

where $\tilde{\eta}(x,\theta) = 2\eta(x,\theta)/p_\theta^{1/2}(x)$, $\eta(\cdot,\theta)$ is the quadratic mean derivative at θ, and Z_n is the score statistic given by

$$Z_n \equiv n^{-1/2} \sum_{i=1}^{n} \tilde{\eta}(X_i, \theta_0), \tag{13.42}$$

By Problem 12.24, the left hand side of (13.41) tends in probability to 0 not only under the null hypothesis but also under the alternative sequence $P^n_{\theta_0+hn^{-1/2}}$ as well. Hence, the test that rejects for large values of $\log(L_{n,h})$ should behave approximately like the test that rejects for large values of $hZ_n - \frac{1}{2}h^2 I(\theta_0)$. But, this latter test is equivalent to rejecting for large values of Z_n, regardless of the value of h.

Consider therefore the Rao's score test $\tilde{\phi}_n$ given by

$$\tilde{\phi}_n = \begin{cases} 1 & \text{if } Z_n \geq I^{1/2}(\theta_0) z_{1-\alpha} \\ 0 & \text{otherwise.} \end{cases} \tag{13.43}$$

As discussed in Section 12.4.3, $\tilde{\phi}_n$ maximizes the derivative of the power function at θ_0, and we will soon see that the limiting power of $\tilde{\phi}_n$ against alternatives of the form $\theta_0 + hn^{-1/2}$ is the optimal value given by the right side of (13.38).

We now derive the asymptotic properties of $\tilde{\phi}_n$. Although we could argue by comparing $\tilde{\phi}_n$ with $\phi_{n,h}$, we proceed instead with a direct calculation. First observe that, under θ_0, $E_{\theta_0}(\tilde{\phi}_n) \to \alpha$. To see why, note that, under θ_0, $Z_n \stackrel{d}{\to} N(0, I(\theta_0))$, by Theorem 12.2.3. The asymptotic consistency in level follows by Slutsky's Theorem.

Next, we calculate the limiting power of $\tilde{\phi}_n$ against an alternative sequence θ_{n,h_n} with $h_n \to h < \infty$. By Corollary 12.4.1, under the alternative sequence $\theta_0 + h_n n^{-1/2}$,

$$Z_n \stackrel{d}{\to} N(hI(\theta_0), I(\theta_0)) . \tag{13.44}$$

Therefore,

$$E_{\theta_0+h_n n^{-1/2}}(\tilde{\phi}_n) = P_{\theta_0+h_n n^{-1/2}}\{Z_n \geq I^{-1/2}(\theta_0) z_{1-\alpha}\}$$

$$= P_{\theta_0+h_n n^{-1/2}}\{\frac{Z_n - hI(\theta_0)}{I^{1/2}(\theta_0)} \geq \frac{I^{1/2}(\theta_0) z_{1-\alpha} - hI(\theta_0)}{I^{1/2}(\theta_0)}\}$$

$$\to P\{Z > z_{1-\alpha} - hI^{1/2}(\theta_0)\} = 1 - \Phi[z_{1-\alpha} - hI^{1/2}(\theta_0)]. \tag{13.45}$$

Thus, $\tilde{\phi}_n$ has the same limiting power against θ_{n,h_n} as ϕ_{n,h_n}. Moreover, the convergence to the limiting power is uniform over h in $[0, c]$ for any $c < \infty$; that is,

$$\sup_{0 \le h \le c} \left| E_{\theta_0 + hn^{-1/2}}(\tilde{\phi}_n) - \{1 - \Phi[z_{1-\alpha} - hI^{1/2}(\theta_0)]\} \right| \to 0 \tag{13.46}$$

as $n \to \infty$. For if not, there would exist a sequence $h_n \in [0, c]$ for which

$$E_{\theta_0 + h_n n^{-1/2}}(\tilde{\phi}_n) - \{1 - \Phi[z_{1-\alpha} - hI^{1/2}(\theta_0)]\} \tag{13.47}$$

does not converge to 0. Then, there exists a subsequence h_{n_j} for which (13.47) converges along this subsequence to $\delta \neq 0$. Take a further subsequence $h_{n_{j_k}}$ which converges to a limit, say h. But by (13.45), along every subsequence $h_{n_{j_k}}$ which converges to h, we have

$$E_{\theta_0 + h_{n_{j_k}} n_{j_k}^{-1/2}}(\tilde{\phi}_{n_{j_k}}) \to 1 - \Phi[z_{1-\alpha} - hI^{1/2}(\theta_0)] ,$$

which renders a contradiction. In summary, we have proved the following.

Lemma 13.3.2 *Under the assumption of Lemma 13.3.1, let $\tilde{\phi}_n$ be the test (13.43). Then, $\tilde{\phi}_n$ is asymptotically level α and its limiting power against $\theta_0 + hn^{-1/2}$ converges to the optimal limiting power uniformly in $h \in [0, c]$ for any $c > 0$; specifically, (13.46) holds.*

Lemma 13.3.2 asserts an optimality property for $\tilde{\phi}_n$. This notion of optimality is appropriate for q.m.d. families since the optimal limiting power against sequences of the form $\theta_0 + hn^{-1/2}$ is nondegenerate, i.e., strictly between α and 1. Even for q.m.d. families, the conclusion of Lemma 13.3.2 does not imply uniform optimality against all alternative sequences with h unrestricted to all of $\mathbb{R}$. We would now like to define a general notion of asymptotically uniformly most powerful of a test sequence ϕ_n satisfying $\limsup E_{\theta_0}(\phi_n) \le \alpha$. A natural definition might be to require that, for any other test sequence ψ_n satisfying $\limsup E_{\theta_0}(\psi_n) \le \alpha$, we have

$$\limsup_n [E_\theta(\psi_n) - E_\theta(\phi_n)] \le 0$$

for all θ. This definition does not work because most tests are consistent, i.e., for any fixed θ, both $E_\theta(\phi_n)$ and $E_\theta(\psi_n)$ tend to one, and hence the difference will tend to zero. To avoid this difficulty, we will require ϕ_n to behave well uniformly across θ, which implies that ϕ_n must behave well against local alternatives θ_n converging to θ_0 at an appropriate rate. Of course, under the q.m.d. assumption, it was seen in Section 13.1 and in Lemma 13.3.1 that the nondegenerate rate corresponds to $\theta_n - \theta_0 \asymp n^{-1/2}$.

Following Wald (1941a, 1943) and Roussas (1972), we therefore define an asymptotically uniformly most powerful (AUMP) test sequence.

Definition 13.3.2 For testing $\theta = \theta_0$ against $\theta > \theta_0$, a sequence of tests $\{\phi_n\}$ is called *asymptotically uniformly most powerful* (AUMP) at (asymptotic) level

α if $\limsup_n E_{\theta_0}(\phi_n) \le \alpha$ and if for any other sequence of test functions $\{\psi_n\}$ satisfying $\limsup_n E_{\theta_0}(\psi_n) \le \alpha$,

$$\limsup_n \ \sup\{E_\theta(\psi_n) - E_\theta(\phi_n) : \ \theta > \theta_0\} \le 0 \ . \tag{13.48}$$

Equivalently, ϕ_n is AUMP level α if $\limsup_n E_{\theta_0}(\phi_n) \le \alpha$ and ϕ_n is AMP against *any* sequence of alternatives $\{\theta_n\}$ with $\theta_n > 0$ (Problem 13.29). Note that this definition is not restricted to q.m.d. families; it also easily generalizes further to problems with nuisance parameters; see (13.71). Also, note that the definition differs slightly from those of Wald and Roussas in that we allow tests that are not exactly level α for finite n, as long as the lim sup of the size is bounded above by α. Of course, we will typically consider tests meeting the stronger requirement $E_{\theta_0}(\phi_n) \to \alpha$, but we prefer not to rule out a priori tests that do not satisfy this convergence.

A slightly weaker notion than Definition 13.3.2 is the following.

Definition 13.3.3 For testing $\theta = \theta_0$ against $\theta > \theta_0$, a sequence of tests $\{\phi_n\}$ is called *locally asymptotically uniformly most powerful* (LAUMP) at level α if $\limsup_n E_{\theta_0}(\phi_n) \le \alpha$ and for any other sequence of test functions $\{\psi_n\}$ satisfying $\limsup_n E_{\theta_0}(\psi_n) \le \alpha$,

$$\limsup_n \ \sup\{E_\theta(\psi_n) - E_\theta(\phi_n) : \ 0 < n^{1/2}(\theta - \theta_0) \le c\} \le 0 \tag{13.49}$$

for any $c > 0$.

In (13.48), the sup over $\{\theta : \theta > \theta_0\}$ can be reparametrized as the sup over $\{h : \theta_0 + hn^{-1/2} > 0\}$. Hence, condition (13.48) can be rewritten as

$$\limsup_n \ \sup\{E_{\theta_0 + hn^{-1/2}}(\psi_n) - E_{\theta_0 + hn^{-1/2}}(\phi_n) : \ h > 0\} \le 0$$

and (13.49) can be rewritten as this same expression with the sup over $h > 0$ replaced by the sup over $\{0 < h \le c\}$. In view of Lemma 13.3.1, under q.m.d., we can express the conditions for a test sequence ϕ_n to be AUMP or LAUMP in terms of the limiting values of its power against local alternatives.

Theorem 13.3.2 *Consider testing $\theta = \theta_0$ against $\theta > \theta_0$ in a q.m.d. family with nonzero Fisher Information $I(\theta_0)$. If $\phi_n = \phi_n(X_1, \ldots, X_n)$ is any sequence of tests based on n i.i.d. observations such that $E_{\theta_0}(\phi_n) \to \alpha$, then*

$$\limsup_n E_{\theta_0 + hn^{-1/2}}(\phi_n) \le [1 - \Phi(z_{1-\alpha} - hI^{1/2}(\theta_0))]. \tag{13.50}$$

Moreover, ϕ_n is AUMP at level α if and only if

$$\sup_{h>0} |E_{\theta_0 + hn^{-1/2}}(\phi_n) - [1 - \Phi(z_{1-\alpha} - hI^{1/2}(\theta_0))]| \to 0 \tag{13.51}$$

and ϕ_n is LAUMP if and only if, for every $c > 0$,

$$\sup_{c \ge h > 0} |E_{\theta_0 + hn^{-1/2}}(\phi_n) - [1 - \Phi(z_{1-\alpha} - hI^{1/2}(\theta_0))]| \to 0 \ . \tag{13.52}$$

Lemma 13.3.2 asserts that $\tilde{\phi}_n$ defined by (13.43) is not only AMP, but LAUMP. We now obtain necessary and sufficient conditions for a test to be LAUMP, as well as a sufficient condition for a test to be AUMP. The results are summarized as follows.

Theorem 13.3.3 *Consider testing $\theta = \theta_0$ against $\theta > \theta_0$ in a q.m.d. family with nonzero Fisher Information $I(\theta_0)$. Let $\tilde{\phi}_n$ be the test defined by (13.43).*
(i). Then, $\tilde{\phi}_n$ satisfies (13.52) and so is LAUMP at level α.
(ii). Any test sequence ϕ_n satisfying, under θ_0,

$$\phi_n - \tilde{\phi}_n \xrightarrow{P} 0 \tag{13.53}$$

is also LAUMP at level α.
(iii). For ϕ_n to be LAUMP at level α, the condition (13.53) is also necessary.
(iv). If, in addition, $Z_n \to \infty$ in $P^n_{\theta_n}$-probability whenever $n^{1/2}(\theta_n - \theta_0) \to \infty$, then $\tilde{\phi}_n$ is also AUMP at level α.

Proof. The proof of (i) follows from Lemma 13.3.2 and Theorem 13.3.2. To prove (ii), the condition (13.53) ensures the limiting size requirement. By contiguity, under θ_{n,h_n}, $\phi_n - \tilde{\phi}_n \to 0$ in probability whenever $h_n \le c$. It follows that

$$E_{\theta_0 + h_n n^{-1/2}}(\phi_n) - E_{\theta_0 + h_n n^{-1/2}}(\tilde{\phi}_n) \to 0$$

whenever $h_n \le c$, which implies

$$\sup_{0 \le h \le c} \left| E_{\theta_0 + hn^{-1/2}}(\phi_n) - E_{\theta_0 + hn^{-1/2}}(\tilde{\phi}_n) \right| \to 0 ,$$

and (ii) follows.

To prove (iii), fix $h > 0$ and consider the sequence of alternatives $\theta_{n,h}$. Let $\bar{\phi}_n$ be the indicator of the event

$$L_{n,h} > k \equiv \exp(\frac{-\sigma_h^2}{2} + \sigma_h z_{1-\alpha}) ,$$

where $\sigma_h^2 = h^2 I(\theta_0)$. Then, $\bar{\phi}_n$ is LAUMP level α by (ii) (from the asymptotic normality of $\log(L_n)$). Suppose ϕ_n^* is also LAUMP level α. By Problem 13.30, $E_{\theta_0}(\phi_n^*) \to \alpha$. Then, letting p_θ^n denote the joint density under θ and letting μ_n denote a measure dominating $p_{\theta_0}^n$ and $p_{\theta_{n,h}}^n$,

$$\int (\bar{\phi}_n - \phi_n^*)(p_{\theta_{n,h}}^n - k p_{\theta_0}^n) d\mu_n \to 0 .$$

But, the integrand in the above equation is always nonnegative. Hence, the integral over the set where $\{p_{\theta_0}^n > 0\}$ also tends to 0, so that

$$\int (\bar{\phi}_n - \phi_n^*)(L_{n,h} - k) p_{\theta_0}^n d\mu_n \to 0 .$$

Since the integrand is nonnegative, it follows (by Markov's inequality) that for every $\eta > 0$, under θ_0,

$$P_{\theta_0}\{|\bar{\phi}_n - \phi_n^*| \cdot |L_{n,h} - k| > \eta\} \to 0 . \tag{13.54}$$

We want to conclude that, for any $\epsilon > 0$,

$$P_{\theta_0}\{|\bar{\phi}_n - \phi_n^*| > \epsilon\} \to 0 .$$

But, for any $\delta > 0$,

$$P_{\theta_0}\{|\bar{\phi}_n - \phi_n^*| > \epsilon\} = P_{\theta_0}\{|\bar{\phi}_n - \phi_n^*| > \epsilon,\ |L_{n,h} - k| > \delta\}$$

$$+P_{\theta_0}\{|\bar{\phi}_n - \phi_n^*| > \epsilon,\ |L_{n,h} - k| \le \delta\}\ . \tag{13.55}$$

As $n \to \infty$, the last term tends to a limit $c(\delta)$; moreover, $c(\delta) \to 0$ as $\delta \to 0$ since $L_{n,h}$ has a continuous limiting distribution under θ_0. Thus, the last term in (13.55) can be made arbitrarily small if δ is chosen small enough, whereas the first term is bounded above by

$$P_{\theta_0}\{|\bar{\phi}_n - \phi_n^*| \cdot |L_{n,h} - k| > \epsilon\delta\} \to 0$$

by (13.54) with $\eta = \epsilon\delta$, and the result follows.

To prove (iv), if the result were false, there would exist a sequence θ_n such that $n^{1/2}(\theta_n - \theta_0) \to \infty$ and $E_{\theta_n}(\tilde{\phi}_n)$ does not converge to one. But,

$$E_{\theta_n}(\tilde{\phi}_n) = P_{\theta_n}\{Z_n > I^{1/2}(\theta_0)z_{1-\alpha}\} \to 1$$

by the added assumption. ■

Example 13.3.1 (Location Models) Suppose P_θ has density with respect to Lebesgue measure on the real line given by $f(x - \theta)$, for some fixed f. Assume the conditions of Corollary 12.2.1 to ensure the family is q.m.d., so that f' exists almost everywhere (with respect to Lebesgue measure),

$$I = I(\theta) = \int_{-\infty}^{\infty} \frac{[f'(x)]^2}{f(x)} dx$$

is finite and positive, and the quadratic mean derivative is

$$\eta(x, \theta) = -\frac{1}{2} \frac{f'(x-\theta)}{f^{1/2}(x-\theta)}\ .$$

Then, the score statistic reduces to

$$Z_n = -n^{-1/2} \sum_{i=1}^{n} \frac{f'(X_i - \theta_0)}{f(X_i - \theta_0)}\ .$$

The test (13.43) is LAUMP level α. It is also AUMP level α if f is strongly unimodal (Problem 13.36); in this case, Example 1 of Section 8.2 shows that the test is also UMP if $n = 1$. ■

Example 13.3.2 (Double Exponential Location Family) As a special case of the previous example, let $f(x) = \frac{1}{2}\exp(-|x|)$. Then, $I(\theta) = 1$. Without loss of generality, consider $\theta_0 = 0$. Then,

$$Z_n = n^{-1/2} \sum_{i=1}^{n} \text{sign}(X_i)\ ,$$

where we take $\text{sign}(x) = 1$ if $x \ge 0$ and $\text{sign}(x) = -1$ otherwise. The resulting test which rejects when $Z_n > z_{1-\alpha}$ is LAUMP at level α. Moreover, this test is AUMP at level α as well. Although this follows from the previous example (since f is strongly unimodal), we give a direct proof. Note that

$$Var_\theta(Z_n) = Var_\theta[\text{sign}(X_1)] \le E_\theta\{[\text{sign}(X_i)]^2\} = 1\ .$$

Hence, to show $Z_n \to \infty$ in $P_{\theta_n}^n$-probability if $n^{1/2}\theta_n \to \infty$, it is enough to show that $E_{\theta_n}(Z_n) \to \infty$ (by Chebyshev's inequality and the previous bound for

$Var_\theta(Z_n)$; see Problem 13.31). Letting F denote the c.d.f. with density f, we have

$$E_{\theta_n}(Z_n) = 2n^{1/2}[F(\theta_n) - F(0)] = n^{1/2}[1 - \exp(-\theta_n)] \to \infty ,$$

and the result follows.

In the double exponential location model, a MLE is a sample median $\hat{\theta}_n$; the test that rejects the null hypothesis if $n^{1/2}\hat{\theta}_n > z_{1-\alpha}$ is also AUMP and is asymptotically equivalent to the test based on Z_n in the sense that the probability that both tests lead to the same conclusion tends to 1, both under the null hypothesis and against a sequence of contiguous alternatives (Problem 13.32). ■

The following example shows that, without strong unimodality, a LAUMP test need not be AUMP in the location model of Example 13.3.1.

Example 13.3.3 (Cauchy Location Model) Here, $f(x) = [\pi(1+x^2)]^{-1}$ and $f'(x) = -2x\pi^{-1}(1+x^2)^{-2}$. Let $\theta_0 = 0$. Then,

$$Z_n = 2n^{-1/2}\sum_{i=1}^{n} \frac{X_i}{1+X_i^2} .$$

By Theorem 13.3.3, since $I(\theta) = 1/2$, the Rao score test that rejects when Z_n exceeds $z_{1-\alpha}/\sqrt{2}$ is LAUMP at level α. However, this test is *not* AUMP at level α. To see why, first note that, for any large $B > 0$, $P_\theta\{X_i > B\} \to 1$ as $\theta \to \infty$, and so, with n fixed,

$$P_\theta\{\min(X_1, \ldots, X_n) > B\} \to 1$$

as $\theta \to \infty$. Since, $x/(1+x^2)$ is decreasing in x on the set $\{x \geq 1\}$, this implies that, for any $z > 0$,

$$P_\theta\{Z_n > z\} \to 0 \quad as \quad \theta \to \infty \tag{13.56}$$

and thus, for any $c > 0$,

$$\lim_{n\to\infty} \inf_{n^{1/2}\theta \geq c} P_\theta\{Z_n > z\} = 0 . \tag{13.57}$$

But, even the worst case power cannot be below α for an AUMP test.

Thus, the score test based on Z_n cannot be AUMP. Next, compare the test based on Z_n with the test that rejects for large values of $\tilde{X}_n$, the sample median. By Theorem 11.2.8, under P_θ^n,

$$n^{1/2}(\tilde{X}_n - \theta) \xrightarrow{d} N\left(0, \frac{\pi^2}{4}\right) .$$

Furthermore, since $\tilde{X}_n$ is location equivariant, the distribution of $n^{1/2}(\tilde{X}_n - \theta)$ under θ does not depend on θ. Consider the asymptotically level α test that rejects when $n^{1/2}\tilde{X}_n > \frac{\pi}{2}z_{1-\alpha}$. We have

$$\inf_{n^{1/2}\theta \geq c} P_\theta\{n^{1/2}\tilde{X}_n > \frac{\pi}{2}z_{1-\alpha}\} = \inf_{n^{1/2}\theta \geq c} P_\theta\{n^{1/2}(\tilde{X}_n - \theta) > \frac{\pi}{2}z_{1-\alpha} - n^{1/2}\theta\}$$

$$= \inf_{n^{1/2}\theta \geq c} P_0\{n^{1/2}\tilde{X}_n > \frac{\pi}{2}z_{1-\alpha} - n^{1/2}\theta\} = P_0\{n^{1/2}\tilde{X}_n > \frac{\pi}{2}z_{1-\alpha} - c\} ,$$

which, as $n \to \infty$, tends to

$$1 - \Phi\left(z_{1-\alpha} - \frac{2c}{\pi}\right) > \alpha > 0 .$$

Note, however, the test based on $\tilde{X}_n$ is neither LAUMP nor AUMP, though its power tends to one uniformly over $\{\theta : \theta > \delta\}$ for any $\delta > 0$.

However, AUMP tests do exist in the present situation. One such test is the Wald test based on an efficient likelihood estimator. Actually, all that is required is a location equivariant estimator $\hat{\theta}_n$ which satisfies

$$n^{1/2}(\hat{\theta}_n - \theta) \xrightarrow{d} N(0, I^{-1}(\theta)) , \tag{13.58}$$

where in this case $I^{-1}(\theta) = 2$. Indeed, the above argument with $\hat{\theta}_n$ replacing $\tilde{X}_n$ applies with the asymptotic variance of $\tilde{X}_n$ of $\pi^2/4$ replaced by 2.

As mentioned in Section 12.4.1, a difficulty in constructing an efficient likelihood estimator is due to the fact that the likelihood equation may have multiple roots. In order to deal with this situation, let $\ell_n(\theta) = \log(L_n(\theta))$. Define

$$\hat{\theta}_n = \tilde{X}_n + \frac{\ell'_n(\tilde{X}_n)}{nI(\tilde{X}_n)} . \tag{13.59}$$

The construction is based on the fact that the nearest root to a consistent estimator is efficient (under regularity conditions which hold for this model). Instead of determining the closest root exactly, which involves solving $\ell'_n(\theta) = 0$, a linear approximation to $\ell'_n(\theta)$ (expanded about $\tilde{X}_n$) is used; see Section 6.4 of Lehmann and Casella (1998). By Corollary 4.4 in Section 6.4 of Lehmann and Casella (1998), $\hat{\theta}_n$ satisfies (13.58). The test that rejects when $n^{1/2}\hat{\theta}_n > 2^{1/2}z_{1-\alpha}$ therefore is AUMP (Problem 13.33). ■

Example 13.3.4 (Wald Tests) As Example 13.3.3 shows, a AUMP test can be based on an efficient estimator, resulting in the Wald tests introduced in Subsection 12.4.2. Actually, this holds more generally. Assume the conditions of Theorem 13.3.3. Suppose $\hat{\theta}_n$ satisfies (12.62). For testing $\theta = \theta_0$ versus $\theta > \theta_0$, the test ϕ_n that rejects when $n^{1/2}(\hat{\theta}_n - \theta_0) > z_{1-\alpha}I^{-1/2}(\theta_0)$ is LAUMP level α. Indeed, the expansion (12.62) implies that $\phi_n - \tilde{\phi}_n \to 0$ in probability under θ_0, so that ϕ_n is LAUMP by (ii) of Theorem 13.3.3. To show ϕ_n is AUMP as well, it is enough to show $n^{1/2}(\hat{\theta}_n - \theta_0) \to \infty$ under θ_n whenever $n^{1/2}(\theta_n - \theta_0) \to \infty$; the argument is similar to (iv) of Theorem 13.3.3. This last condition holds in any location model if $\hat{\theta}_n$ is location equivariant (Problem 13.34). ■

Example 13.3.5 (Correlation Coefficient) Let $X_i = (U_i, V_i)$ be i.i.d. bivariate normal with zero means, unit variances, and unknown correlation ρ. For testing $\rho = 0$ versus $\rho > 0$, we saw in Example 12.4.4 that Rao's score test rejects for large values of

$$Z_n = n^{-1/2} \sum_{i=1}^{n} U_i V_i .$$

By Theorem 13.3.3, this test is LAUMP. To show it is also AUMP, we must show $Z_n \to \infty$ in probability under ρ_n whenever $n^{1/2}\rho_n \to \infty$. Now,

$$E_{\rho_n}(Z_n) = n^{1/2}\rho_n \to \infty$$

and

$$Var_{\rho_n}(Z_n) = Var_{\rho_n}(U_1 V_1) \le E_{\rho_n}(U_1^2 V_1^2) = E_{\rho_n}[V_1^2 E_{\rho_n}(U_1^2 | V_1)] \ .$$

But, the conditional distribution of U_1 given V_1 is $N(\rho_n V_1, 1 - \rho_n^2)$ and so

$$E_{\rho_n}(U_1^2 | V_1) = \rho_n^2 V_1^2 + (1 - \rho_n^2) \le V_1^2 + 1 \ .$$

Hence,

$$Var_{\rho_n}(Z_n) \le E_{\rho_n}(V_1^4 + V_1^2) \le 4 \ .$$

The result now follows by Chebyshev's inequality; see Problem 13.31. ■

It is important to recognize that no asymptotic method, efficient or not, can perform well in all situations. Some anomalies with the Wald test are discussed in Vaeth (1985), Mantel (1987), Le Cam (1990), Benichou, Fears and Gail (1996) and Pawitan (2000). We also remark that, for two-sided hypotheses, AUMP tests, or even LAUMP tests, typically do not exist (Problem 13.39), but an asymptotic approach based on asymptotic unbiasedness is fruitful (Problem 13.55).

When $\theta = (\theta_1, \ldots, \theta_k)$, it is natural to next consider one-sided tests of θ_1 in the presence of nuisance parameters $\theta_2, \ldots, \theta_k$. One approach to finding an upper bound for the limiting power of a test sequence is to fix the nuisance parameters and apply the results of this section. The resulting bounds need not be attainable by any method. A more general approach that leads to bounds which are attainable is discussed in Section 13.5.

13.4 Asymptotically Normal Experiments

In the previous section, a fairly direct approach was taken to compute the best limiting power of a sequence of tests. Since the problem there was reduced to testing a simple hypothesis versus a simple alternative, an optimal test could be derived via the Neyman-Pearson Lemma for finite sample sizes, which resulted in a calculation of the optimal limiting power. Implicit in the calculation was the fact that the likelihood ratios behave approximately like those in a normal location model. More explicitly, given n i.i.d. observations from a q.m.d. family $\{P_\theta\}$, when testing $\theta = \theta_0$ versus $\theta = \theta_0 + hn^{-1/2}$, the optimal test rejects for large values of the likelihood ratio $L_{n,h}$. By Theorem 12.2.3, $L_{n,h}$ satisfies

$$\log(L_{n,h}) - [hZ_n - \frac{1}{2} h^2 I(\theta_0)] = o_{P_{\theta_0}^n}(1) \ , \tag{13.60}$$

where Z_n is the score vector

$$Z_n = 2n^{-1/2} \sum_{i=1}^n \eta(X_i, \theta_0) / p_{\theta_0}^{1/2}(X_i)$$

and $\eta(\cdot, \theta_0)$ is the quadratic mean derivative at θ_0. By contiguity, the left side of this expression tends to 0 in probability under $P_{\theta_0 + hn^{-1/2}}^n$ as well. The asymptotic power calculations flow from these results.

An alternative (and more general) approach is based upon a deeper connection between the expansion (13.60) and the exact likelihood ratios for a particular normal location model. Specifically, consider the normal location model where you observe an observation X from the normal location family $\{Q_h,\ h \in \mathbb{R}\}$, where Q_h is the normal distribution with unknown mean h and known variance $I^{-1}(\theta_0)$. Let L_h denote the likelihood ratio $dQ_h/dQ_0(X)$. Then,

$$\log(L_h) = hZ - \frac{1}{2}h^2 I(\theta_0) , \tag{13.61}$$

where $Z = I(\theta_0)X$. Hence, the loglikelihood $\log(L_{n,h})$ given by (13.60) behaves similarly to $\log(L_h)$; the former is approximately quadratic in h, it is linear in Z_n, the coefficient of h^2 is nonrandom, and Z_n is asymptotically normal $N(0, I(\theta_0))$. These approximations are exact for the normal experiment with Z_n replaced by Z. In a certain sense, the experiments $\{P^n_{\theta_0+hn^{-1/2}},\ h \in \mathbb{R}\}$ and $\{Q_h, h \in \mathbb{R}\}$ are close to each other. Le Cam (1964) formalized the notion of experiments being close, and he showed some profound consequences.[3] For our purposes, we would like to show that, corresponding to any test ϕ_n based on $X_1, \ldots, X_n$ from $\{P^n_{\theta+hn^{-1/2}}\}$, there exists a test ϕ for the normal location problem such that the power functions are approximately the same, as functions of the local parameter h. Then, since an optimality result is available for the normal location model (like a UMP test in the one-sided testing problem), this will directly lead to an upper bound for what is achievable asymptotically in terms of power for the testing problem based on n observations from $\{P_\theta\}$.

Consider the approximating normal experiment consisting of observing one observation X from $N(h, I^{-1}(\theta_0))$, for which θ_0 is viewed as fixed. If $Z = I(\theta_0)X$, then Z is an observation from $\tilde{Q}_h$, where $\tilde{Q}_h = N(hI(\theta_0), I(\theta_0))$. Clearly, the Information contained in X is the same as that of Z. Thus, we could equally well view the two experiments $\{N(h, I^{-1}(\theta_0)), h \in \mathbb{R}\}$ or $\{N(I(\theta_0)h, I(\theta_0))\}$ as limiting approximations to the experiment $\{P^n_{\theta_0+hn^{-1/2}}, h \in \mathbb{R}\}$. The former representation consisting of observing X from $N(h, I^{-1}(\theta_0))$ seems more natural since the unknown parameter h refers to the mean of X. On the other hand, the experiment of observing Z from $N(I(\theta_0)h, I(\theta_0))$ directly matches Z_n in (13.60). The point is that either experiment applies since they are equivalent.

This approach works, not only for one-parameter problems with no nuisance parameters, but also for more general testing problems where the hypothesis concerns a real-valued parameter in the presence of nuisance parameters, and multiparameter problems. For this purpose, we first give the definition of an asymptotically normal sequence of experiments. Consider a sequence of statistical models $\{Q_{n,h}, h \in \mathbb{R}^k\}$. (This can easily be generalized to the case where h is only defined for a subset Ω_n of $\mathbb{R}^k$ which can vary with n.) Thus, for a given n, there is available data on the (measure) space $(\mathcal{X}_n, \mathcal{F}_n)$ where the probability distributions $Q_{n,h}$ live.

[3] The term *experiment* rather than *model* was used by Le Cam, but the terms are essentially synonymous. While a model postulates a family of probability distributions from which data can be observed, an experiment additionally specifies the exact amount of data (or sample size) that is observed. Thus, if $\{P_\theta,\ \theta \in \mathbb{R}\}$ is the family of normal distributions $N(\theta, 1)$ which serves as a model for some data, the experiment $\{P_\theta,\ \theta \in \mathbb{R}\}$ implicitly means one observation is observed from $N(\theta, 1)$; if an experiment consists of n observations from $N(\theta, 1)$, then this is denoted $\{P^n_\theta, \theta \in \mathbb{R}\}$.

Definition 13.4.1 For a sequence of experiments $\{Q_{n,h}, h \in \mathbb{R}^k\}$, let $L_{n,h}$ denote the likelihood ratio of $Q_{n,h}$ with respect to $Q_{n,0}$, defined by (12.36). Suppose there exists a sequence of random k-vectors Z_n mapping $\mathcal{X}_n$ to $\mathbb{R}^k$ and a $k \times k$ positive definite symmetric matrix C such that

$$\log(L_{n,h}) = \langle h, Z_n \rangle - \frac{1}{2}\langle h, Ch \rangle + o_{Q_{n,0}}(1) \tag{13.62}$$

and $Z_n \stackrel{d}{\to} N(0, C)$ under $Q_{n,0}$. Then, the sequence $\{Q_{n,h}, h \in \mathbb{R}^k\}$ is called *asymptotically normal.*

If $\{Q_h\}$ denotes $N(Ch, C)$, the k-variate normal distribution with mean vector Ch and covariance matrix C, then we also say that $\{Q_{n,h}, h \in \mathbb{R}^k\}$ converges to the limiting experiment $\{Q_h\}$. The terminology may be confusing, since $Q_{n,h}$ is not asymptotically normal (and, in fact, $Q_{n,h}$ typically has a distribution on a space that varies with n); it is the log likelihood ratios from the experiment that are asymptotically normal. In particular, note that if $L(h)$ denotes the likelihood of an observation Z from Q_h, then

$$\log(L(h)/L(0)) = \langle h, Z \rangle - \frac{1}{2}\langle h, Ch \rangle \ ;$$

that is, the right side of (13.62) without the error term is exact for the (multivariate) normal location model.

Example 13.4.1 (Quadratic Mean Differentiable Families) Suppose the family $\{P_\theta,\ \theta \in \Omega\}$ is q.m.d. at θ_0. Let $Q_{n,h} = P^n_{\theta_0 + hn^{-1/2}}$ and $C = I(\theta_0)$. By Theorem 12.2.3, $\{Q_{n,h}\}$ is asymptotically normal with covariance C and Z_n the score vector as defined in (12.59). Because we are now parametrizing by the local parameter h, we sometimes speak of $\{P^n_{\theta_0 + hn^{-1/2}}\}$ as being *locally* asymptotically normal at θ_0, and the terms asymptotically normal and locally asymptotically normal are used interchangeably. ■

The random vector (sequence) Z_n defined by (13.62) is called the score vector. Note, however, that any $\bar{Z}_n$ for which $Z_n - \bar{Z}_n \to 0$ in probability under $Q_{n,0}$ also satisfies (13.62).

Example 13.4.2 (Two-Sample Problems) Suppose that $X_1, \ldots, X_m$ are i.i.d. according to P_θ, $\theta \in \Omega$, where Ω is an open subset of $\mathbb{R}^k$. Independently of the X's, suppose $Y_1, \ldots, Y_n$ are i.i.d. according to $\bar{P}_\theta$, $\theta \in \Omega$. Suppose both families are q.m.d. at θ_0. Thus, $\{P^m_{\theta_0 + hm^{-1/2}}, h \in \mathbb{R}^k\}$ and $\{\bar{P}^n_{\theta_0 + hn^{-1/2}}, h \in \mathbb{R}^k\}$ are each asymptotically normal with corresponding Z_m and $\bar{Z}_n$ satisfying as $m, n \to \infty$, $Z_m \stackrel{d}{\to} N(0, I(\theta_0))$ and $\bar{Z}_n \stackrel{d}{\to} N(0, \bar{I}(\theta_0))$ under θ_0. Let $L_{m,h}$ be the likelihood ratio $dP^m_{\theta_0 + hm^{-1/2}}/dP^m_{\theta_0}$ based on $X_1, \ldots, X_m$, and let $\bar{L}_{n,h}$ be the corresponding likelihood ratio based on $Y_1, \ldots, Y_n$. Then, for the combined experiment (and noting $hn^{-1/2} = hm^{-1/2}(m/n)^{1/2}$),

$$\log\left(\frac{d(P^m_{\theta_0 + hn^{-1/2}} \times \bar{P}^n_{\theta_0 + hn^{-1/2}})}{d(P^m_{\theta_0} \times \bar{P}^n_{\theta_0})}\right) = \log(L_{m,h(m/n)^{1/2}}) + \log(\bar{L}_{n,h}) \tag{13.63}$$

$$= \langle h(m/n)^{1/2}, Z_m \rangle - \frac{1}{2}\frac{m}{n}\langle h, I(\theta_0)h \rangle + \langle h, \bar{Z}_n \rangle - \frac{1}{2}\langle h, \bar{I}(\theta_0)h \rangle + o_{P^m_{\theta_0} \times \bar{P}^n_{\theta_0}}(1)$$

$$= \langle h, (m/n)^{1/2} Z_m + \bar{Z}_n \rangle - \frac{1}{2} \langle h, (\frac{m}{n} I(\theta_0) + \bar{I}(\theta_0)) h \rangle + o_{P^m_{\theta_0} \times \bar{P}^n_{\theta_0}}(1) \ .$$

If we assume that $m/n \to \lambda < \infty$, this last expression equals

$$\langle h, \lambda^{1/2} Z_m + \bar{Z}_n \rangle - \frac{1}{2} \langle h, \lambda I(\theta_0) + \bar{I}(\theta_0) \rangle + o_{P^m_{\theta_0} \times \bar{P}^n_{\theta_0}}(1) \ .$$

Thus, the experiment sequence $\{P^m_{\theta_0 + hm^{-1/2}} \times \bar{P}^n_{\theta_0 + hn^{-1/2}}\}$ is asymptotically normal with covariance $C = C(\theta_0) = \lambda I(\theta_0) + \bar{I}(\theta_0)$. ■

Some properties of an asymptotically normal experiment sequence are the following. First, $Q_{n,h}$ is contiguous to $Q_{n,0}$, since under $Q_{n,0}$, $\log(L_{n,h}) \xrightarrow{d} N(-\frac{\sigma^2}{2}, \sigma^2)$, where $\sigma^2 = \langle h, Ch \rangle$, so that Corollary 12.3.1 applies. In fact, the expansion (13.62) implies that Q_{n,h_1} and Q_{n,h_2} are mutually contiguous for any h_1 and h_2 (Problem 13.41). It also follows by Corollary 12.3.2 that, under $Q_{n,h}$, $Z_n \xrightarrow{d} N(Ch, C)$ (Problem 13.42).

We are now in a position to relate a testing problem for an asymptotically normal $\{Q_{n,h}\}$ to one for the normal experiment $\{N(Ch, C)\}$.

Theorem 13.4.1 *Suppose $\{Q_{n,h}, h \in \mathbb{R}^k\}$ is an asymptotically normal sequence of models with covariance matrix C. Let ϕ_n be a test, i.e., a function defined on $\mathcal{X}_n$, the space where the probabilities $Q_{n,h}$ live, with values in $[0,1]$. Let $\beta_n(h)$ denote the power of ϕ_n against $Q_{n,h}$. Then, for every subsequence $\{n_j\}$, there exists a further subsequence $\{n_{j_m}\}$ and a test ϕ in the limiting experiment $\{N(Ch, C)\}$ (or equivalently, the experiment $\{N(h, C^{-1})\}$) such that, for every h,*

$$\beta_{n_{j_m}}(h) \to \beta(h) \ ,$$

where $\beta(h)$ is the power of ϕ.

Proof. Let Z_n be the vector appearing in the definition (13.4.1), so that under $Q_{n,0}$, $Z_n \xrightarrow{d} N(0, C)$. Since $\phi_n \in [0,1]$, $\{\phi_n\}$ is tight. Hence, under $Q_{n,0}$, (ϕ_n, Z_n) is tight. By Prohorov's Theorem 11.2.15, given any subsequence $\{n_j\}$, there exists a further subsequence $\{n_{j_m}\}$ such that

$$(\phi_{n_{j_m}}, Z_{n_{j_m}}) \xrightarrow{d} (\bar{\phi}, \bar{Z})$$

under $Q_{n_{j_m},0}$, where $\bar{Z}$ denotes a random variable with distribution $N(0, C)$ (independent of h) and $\bar{\phi} \in [0,1]$. Let $L_{n,h}$ denote the likelihood ratio of $Q_{n,h}$ with respect to $Q_{n,0}$. Then, by (13.62), under $Q_{n_{j_m},0}$,

$$(\phi_{n_{j_m}}, L_{n_{j_m},h}) \xrightarrow{d} \mathcal{L}(\bar{\phi}, \exp(\langle h, \bar{Z} \rangle - \frac{1}{2} \langle h, Ch \rangle)) \ .$$

If $F(\cdot, \cdot)$ denotes this limit law, then under $Q_{n_{j_m},h}$, we have by Theorem 12.3.3, $(\phi_{n_{j_m}}, L_{n_{j_m},h})$ converges to a limit law with density $r dF(t, r)$. But since $\phi_n \in [0,1]$, weak convergence implies convergence of moments, so that

$$\int \phi_{n_{j_m}} dQ_{n_{j_m},h} \to \int\int tr dF(t,r) = E[\bar{\phi} \exp(\langle h, \bar{Z} \rangle - \frac{1}{2} \langle h, Ch \rangle)]. \quad (13.64)$$

Define $\phi(\bar{Z}) = E(\bar{\phi} | \bar{Z})$, i.e., the conditional expectation under the (fixed) joint distribution of $(\bar{\phi}, \bar{Z})$. Then, the right side of (13.64) is equal to

$$E[\phi(\bar{Z}) \exp(\langle h, \bar{Z} \rangle - \frac{1}{2} \langle h, Ch \rangle)]$$

$$= \int \phi(\bar{z}) \exp(\langle h, \bar{z} \rangle - \frac{1}{2}\langle h, Ch \rangle) dN(0, C)(\bar{z}) .$$

But, $\exp(\langle h, \bar{z} \rangle - \frac{1}{2}\langle h, Ch \rangle) dN(0, C)(\bar{z})$ is actually the density of $N(Ch, C)$ (Problem 13.43). Hence, if the experiment consists of observing $Z \sim N(Ch, C)$, then the last expression is

$$E_h[\phi(Z)] = \int \phi(z) dN(Ch, C)(z) . \blacksquare$$

Theorem 13.4.1 suggests the following strategy for obtaining asymptotically optimal tests in a variety of situations. First, an optimal test, say a UMP test, is derived (or quoted from an earlier chapter) and its power computed from an appropriate normal experiment. Second, the actual experiment sequence is shown (or known) to converge to the normal limiting experiment; as a result, the power of the normal model serves as an upper bound for the asymptotic power of the actual sequence. Finally, a test sequence is constructed whose asymptotic power attains the upper bound and which is therefore asymptotically optimal. A similar strategy will apply to constructing tests that are asymptotically maximin. The remainder of this section will illustrate this approach.

13.5 Applications to Parametric Models

13.5.1 One-sided Hypotheses

We now apply Theorem 13.4.1 to the following situation. Suppose $X_1, \ldots, X_n$ are i.i.d. P_θ, where θ varies in Ω, an open subset of $\mathbb{R}^k$. Assume the family is q.m.d. with positive definite Information matrix $I(\theta)$.

First suppose $\theta = (\theta_1, \ldots, \theta_k)$ and consider testing $\theta_1 \le 0$ against $\theta_1 > 0$ in the presence of nuisance parameters $\theta_2, \ldots, \theta_k$. Fix $\theta_0 = (\theta_{0,1}, \ldots, \theta_{0,k})$ with $\theta_{0,1} = 0$. We now derive an upper bound for the limiting power of a test ϕ_n satisfying, for $h_1 \le 0$,

$$\limsup_{n \to \infty} E_{\theta_0 + hn^{-1/2}}(\phi_n) \le \alpha . \tag{13.65}$$

By Theorem 13.4.1, we can approximate the power of ϕ_n by the power of $\phi = \phi(X)$, where $X \sim N(h, I^{-1}(\theta_0))$. But then (13.65) implies

$$E_h \phi(X) \le \alpha \quad \text{if } h_1 \le 0 ,$$

i.e., $\phi(X)$ is a level α test for testing $h_1 \le 0$ against $h_1 > 0$ in the limit experiment. But, by Example 3.9.2, a UMP level α test exists for this problem and has power $1 - \Phi(z_{1-\alpha} - h_1[I^{-1}(\theta_0)_{1,1}]^{-1/2})$. By Theorem 13.4.1 with $h_1 > 0$,

$$\limsup_{n} E_{\theta_0 + hn^{-1/2}}(\phi_n) \le 1 - \Phi(z_{1-\alpha} - h_1[I^{-1}(\theta_0)_{1,1}]^{-1/2}) .$$

More generally, let g be a function from Ω to $\mathbb{R}$, and assume g is differentiable with gradient vector $\dot{g}(\theta)$ of dimension $1 \times k$. The problem is to test $g(\theta) \le 0$ against $g(\theta) > 0$. Suppose ϕ_n is a test based on $X_1, \ldots, X_n$ whose limiting size is $\bar{\alpha} \le \alpha$ (see Definition 11.1.2). Fix θ_0 such that $g(\theta_0) = 0$. We will derive an upper bound for the limiting power of ϕ_n under $\theta_0 + hn^{-1/2}$ and then obtain tests for which this limiting power is attained. First, note that

$$g(\theta_0 + hn^{-1/2}) = n^{-1/2}\langle \dot{g}(\theta_0)^T, h\rangle + o(n^{-1/2}) .$$

If h is such that $\langle \dot{g}(\theta_0)^T, h\rangle < 0$, then $g(\theta_0 + hn^{-1/2}) < 0$ for all sufficiently large n. The assumption on the limiting level of ϕ_n implies that, for such an h,

$$\limsup_{n\to\infty} E_{\theta_0+hn^{-1/2}}(\phi_n) \le \alpha . \tag{13.66}$$

Now, according to Theorem 13.4.1, we can approximate the power of a test sequence ϕ_n by the power of a test $\phi = \phi(X)$ for the experiment based on X from the model $N(h, I^{-1}(\theta_0))$. Let $\beta_\phi(h)$ denote the power of $\phi(X)$ when $X \sim N(h, I^{-1}(\theta_0))$. Then, (13.66) implies that $\beta_\phi(h) \le \alpha$ if $\langle \dot{g}(\theta_0)^T, h\rangle < 0$. Since $\beta_\phi(\cdot)$ is continuous, it follows that $\beta_\phi(h) \le \alpha$ if $\langle \dot{g}(\theta_0)^T, h\rangle \le 0$. Now, fix an alternative h_1 with $\langle \dot{g}(\theta_0)^T, h_1\rangle > 0$. Theorem 13.4.1 implies that

$$\limsup_{n\to\infty} E_{\theta_0+h_1n^{-1/2}}(\phi_n) \le \sup_{\phi\in A_\alpha} \beta_\phi(h_1) , \tag{13.67}$$

where $A_\alpha = \{\phi : \ \beta_\phi(h) \le \alpha \text{ whenever } \langle \dot{g}(\theta_0)^T, h\rangle \le 0\}$. But then, the right side of (13.67) is maximized when ϕ is the most powerful level α test for testing $\langle \dot{g}(\theta_0)^T, h\rangle \le 0$ against $h = h_1$. In fact, for the problem of testing $\langle \dot{g}(\theta_0)^T, h\rangle \le 0$ versus $\langle \dot{g}(\theta_0)^T, h\rangle > 0$, there exists a uniformly most powerful test based on X which rejects for large values of $\langle \dot{g}(\theta_0)^T, X\rangle$; see Section 3.9.2. But,

$$\langle \dot{g}(\theta_0)^T, X\rangle \sim N(\langle \dot{g}(\theta_0)^T, h\rangle, \sigma^2_{\theta_0}) ,$$

where

$$\sigma^2_{\theta_0} = \dot{g}(\theta_0) I^{-1}(\theta_0) \dot{g}(\theta_0)^T .$$

Hence, for testing $\langle \dot{g}(\theta_0)^T, h\rangle \le 0$ at level α, the UMP test rejects when $\langle \dot{g}(\theta_0)^T, X\rangle > z_{1-\alpha}\sigma_{\theta_0}$. The power of this test against h is then

$$1 - \Phi(z_{1-\alpha} - \sigma^{-1}_{\theta_0}\langle \dot{g}(\theta_0)^T, h\rangle) .$$

Therefore, Theorem 13.4.1 implies that, for any h such that $\langle \dot{g}(\theta_0)^T, h\rangle$,

$$\limsup_{n} E_{\theta_0+hn^{-1/2}}(\phi_n) \le 1 - \Phi(z_{1-\alpha} - \sigma^{-1}_{\theta_0}\langle \dot{g}(\theta_0)^T, h\rangle) . \tag{13.68}$$

The above development is summarized in (i) of the following theorem. Part (ii) asserts that an optimal test sequence may be constructed if an efficient estimator sequence is available.

Theorem 13.5.1 *Suppose $X_1, \ldots, X_n$ are i.i.d. according to P_θ, $\theta \in \Omega$, where Ω is assumed to be an open subset of $\mathbb{R}^k$. Let Ω_0 denote the set of θ with $g(\theta) \le 0$, for some function g from $\mathbb{R}^k$ to $\mathbb{R}$ which is assumed differentiable with gradient $\dot{g}(\theta)$. Consider testing the null hypothesis $\theta \in \Omega_0$ versus $g(\theta) > 0$. Assume the family $\{P_\theta, \theta \in \Omega\}$ is q.m.d. at every θ for which $g(\theta) = 0$ with nonsingular Fisher Information matrix $I(\theta)$.*
(i) Let $\phi_n = \phi_n(X_1, \ldots, X_n)$ be a uniformly asymptotically level α sequence of tests, so that

$$\limsup_{n\to\infty} \sup_{\Omega_0} E_\theta(\phi_n) \le \alpha , \tag{13.69}$$

and suppose that $g(\theta_0) = 0$. *Then, for any* h *such that* $\langle \dot{g}(\theta_0)^T, h \rangle > 0$, *(13.68) holds.*

(ii) Let $\hat{\theta}_n$ *be any estimator satisfying (12.62) (such as an efficient likelihood estimator). Suppose* $I(\theta)$ *is continuous in* θ *and* $\dot{g}(\theta)$ *is continuous at* θ_0. *Then, the test sequence* ϕ_n *that rejects when* $n^{1/2} g(\hat{\theta}_n) \geq z_{1-\alpha} \hat{\sigma}_n$, *where*

$$\hat{\sigma}_n^2 = \dot{g}(\hat{\theta}_n) I^{-1}(\hat{\theta}_n) \dot{g}(\hat{\theta}_n)^T ,$$

is pointwise asymptotically level α. *Moreover, the inequality (13.68) becomes an equality, and the limsup on the left side of (13.68) may be replaced by a lim.*

PROOF. The proof of (i) follows from the discussion preceding the theorem (applying that argument to subsequences for which limits exist). The proof of (ii) follows from Theorem 12.4.1 and the discussion in Subsection 12.4.2. ■

In fact, the properties claimed in (ii) above hold more generally for any test sequence that rejects if $T_n > t_n$, if T_n satisfies

$$T_n = \dot{g}(\theta_0) I^{-1}(\theta_0) Z_{n,\theta_0} + o_{P_{\theta_0}^n}(1)$$

for every $\theta_0 \in \Omega_0$, where Z_{n,θ_0} is the score vector defined in (12.59), and if

$$t_n \xrightarrow{P} z_{1-\alpha} \sigma_{\theta_0}$$

under θ_0, where σ_{θ_0} is given by (12.66).

Example 13.5.1 (One-sample Normal Model) Let $X_1, \ldots, X_n$ be i.i.d. normal with mean μ and variance σ^2 so that $\theta = (\mu, \sigma^2)$. Consider testing $\mu \leq 0$ versus $\mu > 0$. Of course, the usual t-test is UMPU and UMPI. Theorem 13.5.1 applies immediately to the test ϕ_n that rejects when $n^{1/2} \bar{X}_n / S_n$ exceeds $z_{1-\alpha}$. Therefore, for any σ,

$$\lim_n E_{h_1 n^{-1/2}, \sigma + h_2 n^{-1/2}}(\phi_n) = 1 - \Phi(z_{1-\alpha} - h_1 \sigma^{-1}) , \tag{13.70}$$

and so ϕ_n is LAUMP. Equation (13.70) also holds for the t-test, i.e., when the normal critical value $z_{1-\alpha}$ is replaced by the corresponding critical value obtained from the t-distribution with $n-1$ degrees of freedom, which gives an asymptotic optimality property for the t-test that does not depend on the restriction to unbiased or invariant tests. In fact, we now show ϕ_n is AUMP. Specifically, in the case where there is a nuisance parameter σ, it is natural to define a test ϕ_n to be AUMP level α if ϕ_n is uniformly asymptotically level α and for any other uniformly asymptotically level α test ψ, we have

$$\limsup_n \sup\{E_{\mu,\sigma}(\psi_n) - E_{\mu,\sigma}(\phi_n) : \ \mu > 0, \sigma > 0\} \leq 0 . \tag{13.71}$$

(Obviously, we would modify this definition if the nuisance parameter σ varied in a parameter space different from the positive reals.) To see that ϕ_n possesses this property, argue as follows. If it did not, there would exist $\mu_n > 0$ and $\sigma_n > 0$ such that

$$\limsup_n \{E_{\mu_n,\sigma_n}(\psi_n) - E_{\mu_n,\sigma_n}(\phi_n)\} > 0 .$$

With σ_n now fixed, let $\tilde{\psi}_n$ the UMP test for testing $\mu \le 0$ versus $\mu > 0$ if $\sigma = \sigma_n$ is known. Since $\tilde{\psi}_n$ has greater power than ψ_n, it follows that

$$\limsup_n \{E_{\mu_n,\sigma_n}(\tilde{\psi}_n) - E_{\mu_n,\sigma_n}(\phi_n)\} > 0 .$$

But,

$$E_{\mu_n,\sigma_n}(\tilde{\psi}_n) = 1 - \Phi(z_{1-\alpha} - n^{1/2}\mu_n/\sigma_n) .$$

Since the power of the t-test and the power of the test ϕ_n depend on (μ, σ) only through μ/σ,

$$E_{\mu_n,\sigma_n}(\phi_n) = E_{\frac{\mu_n}{\sigma_n},1}(\phi_n) .$$

So, it suffices to show, uniformly in μ and $\sigma = 1$, that

$$\sup_{\mu>0} |1 - \Phi(z_{1-\alpha} - n^{1/2}\mu) - E_{\mu,1}(\phi_n)| \to 0 ,$$

or, for any sequence μ_n with $\mu_n > 0$,

$$E_{\mu_n,1}(\phi_n) - [1 - \Phi(z_{1-\alpha} - n^{1/2}\mu_n)] \to 0 . \tag{13.72}$$

But,

$$E_{\mu_n,1}(\phi_n) = P_{\mu_n,1}\{n^{1/2}(\bar{X}_n - \mu_n)/S_n > z_{1-\alpha} - n^{1/2}\mu_n/S_n\} .$$

Under $\mu = \mu_n$ and $\sigma = 1$, the left hand side $n^{1/2}(\bar{X}_n - \mu_n)/S_n$ has the t-distribution with $n-1$ degrees of freedom, and so tends in distribution to Z which has the standard normal distribution. Also, $S_n \to 1$ in probability. By Slutsky's theorem, if $n^{1/2}\mu_n \to \delta$, then

$$E_{\mu_n,1}(\phi_n) \to P\{Z > z_{1-\alpha} - \delta\}$$

and (13.72) holds. If $n^{1/2}\mu_n \to \infty$, then $n^{1/2}(\bar{X}_n - \mu_n)/S_n$ is still asymptotically standard normal, while $z_{1-\alpha} - n^{1/2}\mu_n/S_n \to -\infty$ in probability; then, $E_{\mu_n,1}(\phi_n) \to 1$ and (13.72) holds. To complete the argument, one must pass to subsequences such that $n^{1/2}\mu_n$ converges (possibly to ∞) and apply the previous argument along such subsequences. The conclusion is that ϕ_n is AUMP. ■

Consider the following special case of Theorem 13.5.1. Suppose $\theta = (\theta_1, \ldots, \theta_k)$ and interest focuses on inference for θ_1 in the presence of the nuisance parameters $\theta_2, \ldots, \theta_k$. Specifically, consider testing $\theta_1 = \theta_{1,0}$ versus $\theta_1 > \theta_{1,0}$. As usual, let $I(\theta)$ denote the Fisher Information matrix with (i,j) entry denoted $I_{i,j}(\theta)$; it is assumed $I(\theta)$ is invertible with inverse $I^{-1}(\theta)$ having (i,j) entry $[I^{-1}(\theta)]_{i,j}$. It is interesting to compare the power of the asymptotically optimal tests when the nuisance parameters are unknown with the situation in which they are known. If $\theta_2, \ldots, \theta_k$ are fixed and known, then the best limiting power against the sequence of alternatives $\theta_{1,0} + h_1 n^{-1/2}$ of an asymptotically level α test was obtained in Theorem 13.3.2, and is equal to

$$1 - \Phi(z_{1-\alpha} - h_1 I_{1,1}^{1/2}(\theta_{1,0}, \theta_2, \ldots, \theta_k)) .$$

If the nuisance parameters are unknown, the best limiting power was obtained in Theorem 13.5.1; simply apply the theorem with $g(\theta) = \theta_1$, $\dot{g}(\theta) = (1, 0, \ldots, 0)$ and $h = (h_1, 0, \ldots, 0)$. The resulting limiting power value is equal to

$$1 - \Phi(z_{1-\alpha} - h_1\{I^{-1}(\theta_{1,0}, \theta_2, \ldots, \theta_k)_{1,1}\}^{-1/2}) .$$

Comparing these situations, we see that

$$\frac{1}{I_{1,1}(\theta)} \le [I^{-1}(\theta)]_{1,1} ,$$

since the power of the test when $(\theta_2, \ldots, \theta_k)$ are known exceeds that when $(\theta_2, \ldots, \theta_k)$ are unknown. Equality holds if $I_{1,j}(\theta) = 0$ for all $j \neq 1$. Since the same argument applies to any of the components of θ, there is no loss in power when testing any component in the presence of the remaining parameters if and only if $I(\theta)$ is a diagonal matrix.

Example 13.5.2 (Location Scale Models) Suppose $X_1, \ldots, X_n$ are i.i.d. with density $\sigma^{-1} f((x-\mu)/\sigma)$, where f is absolutely continuous. Both the location parameter μ and the scale parameter σ are unknown. If $\theta = (\mu, \sigma)$, then the components of the Information matrix are given by (Problem 13.44)

$$I_{1,1} = \sigma^{-2} \int \left[\frac{f'(x)}{f(x)} \right]^2 f(x) dx ,$$

$$I_{2,2} = \sigma^{-2} \int \left[\frac{x f'(x)}{f(x)} + 1 \right]^2 f(x) dx$$

and

$$I_{1,2} = \sigma^{-2} \int x \left[\frac{f'(x)}{f(x)} \right]^2 f(x) dx .$$

It follows that the off-diagonal element $I_{1,2}$ is equal to 0 if f is symmetric.

We specialize further and let $f(x) = C(\beta) \exp(-|x|^\beta)$ for some fixed β. Recall from Example 12.2.5 that, if $\beta > 1/2$, then f generates a location model which is q.m.d.; the location scale model with σ unknown is also q.m.d. (Problem 13.45). For $\beta > 1$, the MLE $\hat{\mu}_n$ for μ is the unique minimizer of $\sum_i |X_i - \mu|^\beta$; for $\beta = 1$, any value between the middle order statistics is an MLE. Moreover, the unique MLE $\hat{\sigma}_n$ for σ is given by

$$\hat{\sigma}_n = \beta^{1/\beta} \left[\frac{\sum_i |X_i - \hat{\mu}_n|^\beta}{n} \right]^{1/\beta} . \tag{13.73}$$

For testing $\mu \le 0$ against $\mu > 0$, the Wald test which rejects for large values of $\hat{\mu}_n / \hat{\sigma}_n$ is LAUMP; If $1/2 < \beta < 1$, Rao's score test is more convenient to apply and is LAUMP (Problem 13.46). ■

Example 13.5.3 (Bivariate Normal Correlation) As in Example 13.3.5, let $X_i = (U_i, V_i)$ be i.i.d. bivariate normal with unknown correlation ρ. However, here we assume the means and variances of U_i and V_i are unknown as well. The MLE $\hat{\rho}_n$ is given by the sample correlation (11.29). A LAUMP test rejects when $n^{1/2} \hat{\rho}_n > z_{1-\alpha}$. Note that, in this case, the Information is not diagonal and the optimal limiting power is strictly smaller than the case where only ρ is unknown (Problem 13.47). ■

Theorem 13.5.1 can be generalized to two-sample problems, since the proof essentially only depends on Theorem 13.4.1 and the assumption that the experiment is asymptotically normal. By Example 13.4.2, asymptotic normality holds for two-sample models if each of the one-sample models is quadratic mean differentiable. Specifically, suppose $X_1, \ldots, X_m$ are i.i.d. P_θ, $\theta \in \Omega$, Ω an open subset of $\mathbb{R}^k$. Also, suppose $Y_1, \ldots, Y_n$ are i.i.d. $\bar{P}_\theta$, $\theta \in \Omega$. Let $I(\theta)$ denote the Information an X_i contains about θ; similarly, let $\bar{I}(\theta)$ be the Information a Y_j contains about θ. Assume these Information matrices are nonsingular and continuous. Fix any θ_0 and assume both models are q.m.d. at θ_0 with corresponding score statistics Z_m and $\bar{Z}_n$ (in the notation of Example 13.4.2). Then, the combined experiment is asymptotically normal with score statistic

$$Z_{m,n} = (m/n)^{1/2} Z_m + \bar{Z}_n \ .$$

If we also assume $m/n \to \lambda < \infty$, then the joint experiment is asymptotically normal with covariance

$$C(\theta_0) = \lambda I(\theta_0) + \bar{I}(\theta_0) \ .$$

Consider testing $g(\theta) = 0$ versus $g(\theta) > 0$, for some continuously differentiable g with gradient $\dot{g}(\theta)$. A generalization of (13.68) yields for any uniformly asymptotically level α test sequence that (Problem 13.48)

$$\limsup_n E_{\theta_0 + hn^{-1/2}}(\phi_n) \le 1 - \Phi(z_{1-\alpha} - \sigma_{\theta_0}^{-1} \langle \dot{g}(\theta_0)^T, h \rangle) \ , \tag{13.74}$$

where

$$\sigma_{\theta_0}^2 = \dot{g}(\theta_0) C^{-1}(\theta_0) \dot{g}(\theta_0)^T \ .$$

To find such a test, assume there exists an estimator sequence $\hat{\theta}_n$ satisfying

$$n^{1/2}(\hat{\theta}_n - \theta_0) = C^{-1}(\theta_0) Z_{m,n} + o_{P_{\theta_0}^m \times \bar{P}_{\theta_0}^n}(1) \ . \tag{13.75}$$

Then, the test that rejects when $n^{1/2} g(\hat{\theta}_n) > z_{1-\alpha} \hat{\sigma}_n$, where

$$\hat{\sigma}_n^2 = \dot{g}(\hat{\theta}_n) C^{-1}(\hat{\theta}_n) \dot{g}(\hat{\theta}_n)^T$$

is pointwise asymptotically level α and the inequality (13.74) is an equality (Problem 13.49).

Example 13.5.4 (Behrens-Fisher Problem) As a special case of the above, assume P_θ is $N(\xi, \sigma^2)$ and $\bar{P}_\theta$ is $N(\eta, \tau^2)$ so that $\theta = (\xi, \eta, \sigma^2, \tau^2)$, and all four parameters vary freely. Consider testing $\eta - \xi = 0$ versus $\eta - \xi > 0$, so that $g(\theta) = \eta - \xi$ and $\dot{g}(\theta) = (-1, 1, 0, 0)$. For this problem, neither invariance nor unbiasedness considerations reduce the problem sufficiently to obtain any kind of optimal test. However, a large sample optimality result is easily obtained. Fix $\theta_0 = (\xi_0, \xi_0, \sigma^2, \tau^2)$. Assume $m/n \to \lambda < \infty$. Then, it is easy to check that the covariance matrix C in definition 13.4.1 is the diagonal matrix with diagonal elements λ/σ^2, $1/\tau^2$, $2\lambda/\sigma^2$, and $2/\tau^2$. Hence,

$$\sigma_{\theta_0}^2 = \dot{g}(\theta_0) C^{-1}(\theta_0) \dot{g}(\theta_0)^T = \frac{\sigma^2}{\lambda} + \tau^2 \ . \tag{13.76}$$

Thus, the bound in (13.74) with $h = (h_1, h_2, 0, 0)$ reduces to

$$1 - \Phi\left[z_{1-\alpha} - (\frac{\sigma^2}{\lambda} + \tau^2)^{-1/2}(h_2 - h_1) \right] \ .$$

It is easy to construct a test sequence that achieves this bound. Consider the test that rejects the null hypothesis when

$$n^{1/2}(\bar{Y}_n - \bar{X}_m) > z_{1-\alpha} \left[S_Y^2 + (\frac{n}{m}) S_X^2 \right]^{1/2} ,$$

where $\bar{Y}_n = n^{-1} \sum_j Y_j$, $S_Y^2 = (n-1)^{-1} \sum_j (Y_j - \bar{Y}_n)^2$, and similarly for $\bar{X}_m$ and S_X^2. This test is pointwise consistent in level; the order of error in the rejection probability will be revisited in Example 15.6.3. The limiting power of this test against the sequence of parameter values $(\xi_0 + h_1 n^{-1/2}, \xi_0 + h_2 n^{-1/2}, \sigma^2, \tau^2)$ is given by

$$P\left\{ \frac{n^{1/2}[(\bar{Y}_n - h_2 n^{1/2}) - (\bar{X}_m - h_1 n^{-1/2})]}{(S_Y^2 + \frac{n}{m} S_X^2)^{1/2}} > z_{1-\alpha} - \frac{h_2 - h_1}{(S_Y^2 + \frac{n}{m} S_X^2)^{1/2}} \right\} .$$

But, $S_Y^2 \to \tau^2$ in probability, $S_X^2 \to \sigma^2$ in probability, and the left hand side is asymptotically standard normal. The result follows by Slutsky's theorem. ■

13.5.2 Equivalence Hypotheses

In this section, we will apply Theorem 13.4.1 to the following situation. Suppose $X_1, \ldots, X_n$ are i.i.d. P_θ where $\theta \in \Omega$ and Ω is an open subset of $\mathbb{R}^k$. Interest focuses on $g(\theta)$, where g is a function from Ω to $\mathbb{R}$. Assume g is differentiable with gradient vector $\dot{g}(\theta)$ of dimension $1 \times k$. We wish to test the null hypothesis $|g(\theta)| \geq \Delta$ against the alternative $|g(\theta)| < \Delta$. (We are tacitly assuming there exists values of θ satisfying $g(\theta) \geq \Delta$ and $g(\theta) \leq \Delta$.) This problem was studied in Theorem 3.7.1, where a UMP test was derived for a one-parameter exponential family. A UMP equivalence test for a linear combination of means of a multivariate normal distribution was obtained in Example 3.9.3.

We will formulate the asymptotic problem in two distinct ways. First, we will consider the case when the null hypothesis parameter space is the complement of a fixed interval $(-\Delta, \Delta)$. Then, we will also consider the case when this interval shrinks with n.

(i). *Fixed* Δ. Suppose $\Delta > 0$ is fixed and the problem is to test $|g(\theta)| \geq \Delta$ versus $|g(\theta)| < \Delta$. For any fixed alternative value θ with $|g(\theta)| < \Delta$, the power of any reasonable test against θ will tend to one. Therefore, just as we did for one-sided hypotheses, we compare power functions against local alternatives. Consider any fixed θ_0 satisfying $|g(\theta_0)| = \Delta$. For sake of argument, consider the case $g(\theta_0) = -\Delta$. We wish to derive an (obtainable) upper bound for the limiting power of a test sequence ϕ_n under $\theta_0 + hn^{-1/2}$. But a crude way to bound the power is based on the simple fact that any level α test for testing $|g(\theta)| \geq \Delta$ versus $|g(\theta)| < \Delta$ is also level α for testing $g(\theta) \leq -\Delta$ versus $g(\theta) > -\Delta$. Since upper bounds for the asymptotic power were obtained in Theorem 13.5.1, an immediate result follows. In this asymptotic setup, the statistical problem is somewhat degenerate as it becomes one of testing a one-sided hypothesis. For example, suppose $X_1, \ldots, X_n$ are i.i.d. $N(\theta, 1)$ Then for large n, one can distinguish $\theta \leq -\Delta$ and $\theta > -\Delta$ with error probabilities that are uniformly small and tend to zero exponentially fast with n. In essence, the statistical issue arises only if the true θ is near the boundary of $[-\Delta, \Delta]$, in which case determining significance essentially becomes one of testing a one-sided hypothesis.

Theorem 13.5.2 *Suppose $X_1, \ldots, X_n$ are i.i.d. according to P_θ, $\theta \in \Omega$, where Ω is assumed to be an open subset of $\mathbb{R}^k$. Consider testing the null hypothesis*

$$\theta \in \Omega_0 = \{\theta : \ |g(\theta)| \geq \Delta\}$$

versus $|g(\theta)| < \Delta$, where the function g from $\mathbb{R}^k$ to $\mathbb{R}$ is assumed differentiable with gradient $\dot{g}(\theta)$. Assume the family $\{P_\theta, \theta \in \Omega\}$ is q.m.d. at every θ with $|g(\theta)| = \Delta$ and assume the Fisher Information matrix $I(\theta)$ is nonsingular for such θ. Let $\phi_n = \phi_n(X_1, \ldots, X_n)$ be a uniformly asymptotically level α sequence of tests; that is,

$$\limsup_{n \to \infty} \sup_{\Omega_0} E_\theta(\phi_n) \leq \alpha \ . \tag{13.77}$$

(i) Assume θ_0 satisfies $g(\theta_0) = -\Delta$. Then, for any h such that $\langle \dot{g}(\theta_0)^T, h \rangle > 0$,

$$\limsup_n E_{\theta_0 + hn^{-1/2}}(\phi_n) \leq 1 - \Phi(z_{1-\alpha} - \sigma_{\theta_0}^{-1} \langle \dot{g}(\theta_0)^T, h \rangle) \ , \tag{13.78}$$

where

$$\sigma_{\theta_0}^2 = \dot{g}(\theta_0) I^{-1}(\theta_0) \dot{g}(\theta_0)^T \ . \tag{13.79}$$

(ii) Assume θ_0 satisfies $g(\theta_0) = \Delta$. Then, for any h such that $\langle \dot{g}(\theta_0)^T, h \rangle < 0$,

$$\limsup_n E_{\theta_0 + hn^{-1/2}}(\phi_n) \leq 1 - \Phi(z_{1-\alpha} - \sigma_{\theta_0}^{-1} |\langle \dot{g}(\theta_0)^T, h \rangle|) \ , \tag{13.80}$$

(iii) Let $\hat{\theta}_n$ be any estimator satisfying (12.62). Suppose $I(\theta)$ is continuous in θ and $\dot{g}(\theta)$ is continuous at θ_0. Then, the test sequence ϕ_n that rejects when $|g(\hat{\theta}_n)| < \Delta - n^{-1/2} \hat{\sigma}_n z_{1-\alpha}$, where

$$\hat{\sigma}_n^2 = \dot{g}(\hat{\theta}_n) I^{-1}(\hat{\theta}_n) \dot{g}(\hat{\theta}_n)^T \tag{13.81}$$

is pointwise asymptotically level α and is locally asymptotically UMP in the sense that the inequality (13.78) is an equality. In fact, the same properties hold for any test sequence that rejects if $|T_n| < \Delta - n^{-1/2} \hat{\sigma} z_{1-\alpha}$, if T_n satisfies

$$T_n = \dot{g}(\theta_0) I^{-1}(\theta_0) Z_{n,\theta_0} + o_{P_{\theta_0}^n}(1) \tag{13.82}$$

for every $\theta_0 \in \Omega_0$, where Z_{n,θ_0} is the score vector defined in (12.59).

PROOF. As remarked above, (13.78) follows because ϕ_n is also a uniformly asymptotically level α test for testing $g(\theta) \leq -\Delta$ versus $g(\theta) > -\Delta$. For this one-sided testing problem, the optimal bound was obtained in Theorem 13.5.1. The same argument applies to (13.80). To prove (iii), let $\theta_n = \theta_0 + hn^{-1/2}$. Then, assumption (12.62) and contiguity arguments imply that, under θ_n,

$$n^{1/2}(\hat{\theta}_n - \theta_n) \xrightarrow{d} N(0, I^{-1}(\theta_0)) \ .$$

Thus, under θ_n, $\hat{\sigma}_n$ tends in probability to σ_{θ_0}. Moreover, the Delta method implies, under θ_n,

$$n^{1/2}(g(\hat{\theta}_n) - g(\theta_n)) \xrightarrow{d} N(0, \sigma_{\theta_0}^2) \ .$$

Now, if $g(\theta_0) = -\Delta$ and $\langle \dot{g}(\theta_0)^T, h \rangle > 0$, then

$$g(\theta_n) = -\Delta + n^{-1/2} \langle \dot{g}(\theta_0)^T, h \rangle + o(n^{-1/2}) \ .$$

So, under θ_n,

$$n^{1/2}[g(\hat{\theta}_n) + \Delta] \xrightarrow{d} N\left(\langle \dot{g}(\theta_0)^T, h\rangle, \sigma^2_{\theta_0}\right) . \tag{13.83}$$

Therefore,

$$E_{\theta_n}(\phi_n) = P_{\theta_n}\{|g(\hat{\theta}_n)| < \Delta - n^{-1/2}\hat{\sigma}_n z_{1-\alpha}\} ,$$

which tends to the right side of (13.78) by (13.83) and Slutsky's Theorem. The same proof works for any estimator T_n of the form (13.82). ■

Example 13.5.5 (Normal One-Sample Problem) Suppose $X_1, \ldots, X_n$ are i.i.d. $N(\mu, \sigma^2)$, with both parameters unknown. Consider testing $|\mu| \geq \Delta$ versus $|\mu| < \Delta$. The standard t-test for testing the one-sided hypothesis $\mu \leq \Delta$ against $\mu > -\Delta$ rejects if

$$n^{1/2}(\bar{X}_n + \Delta)/S_n > t_{n-1,1-\alpha} ,$$

where S_n^2 is the (unbiased) sample variance and $t_{n-1,1-\alpha}$ is the $1-\alpha$ quantile of the t-distribution with $n-1$ degrees of freedom. Similarly, the standard t-test of the hypothesis $\mu \geq \Delta$ rejects if

$$n^{1/2}(\bar{X}_n - \Delta)/S_n < -t_{n-1,1-\alpha} .$$

The intersection of these rejection regions is therefore a level α test of the null hypothesis $|\mu| \geq \Delta$. Such a construction that intersects the rejection regions of two one-sided tests (TOST) was proposed in Westlake (1981) and Schuirmann (1981), and can be viewed as a special case of Berger's (1982) intersection-union tests. The resulting test is denoted ϕ_n^{TOST} that rejects when $|\bar{X}_n| < \Delta - n^{-1/2}S_n t_{n-1,1-\alpha}$. (In fact, we see here that our general asymptotic construction in (iii) of the above theorem merely replaces the t_{n-1} quantiles by the standard normal quantiles; that is, the intersection two rejection regions, each of *asymptotic* size α yields a rejection region whose asymptotic size is bounded above by α.) In general, by combining two one-sided tests, the resulting TOST can be quite conservative in that its size can be quite less than α. However, in this example, the size of ϕ_n^{TOST} is actually α, as can be seen by calculating the rejection probability under (μ, σ) with $\mu = \Delta$ and $\sigma \to 0$ (Problem 13.53). The asymptotic power of ϕ_n^{TOST} against a sequence with mean $-\Delta + hn^{-1/2}$ $(h > 0)$ and variance fixed at σ^2 is obtained by the previous theorem or calculated directly as

$$P_{\Delta+hn^{-1/2},\sigma}\{|\bar{X}_n| < \Delta - n^{-1/2}S_n t_{n-1,1-\alpha}\} = \Phi(z_{1-\alpha} - \frac{h}{\sigma}) ,$$

which is the optimal bound when (13.78) is specialized to this situation. A similar calculation applies to sequences of the form $\Delta - hn^{-1/2}$. Thus, the TOST is asymptotically optimal in this setup. It should be remarked that the TOST has been criticized because it is biased (in finite samples) and tests have been proposed that have greater power; some proposals are discussed in Brown, Casella, and Hwang (1995), Berger and Hsu (1996), and Perlman and Wu (1999). Such tests cannot have greater asymptotic power against local alternatives, at least under the setup of Theorem 13.5.2. On the other hand, the TOST will be seen to be inefficient under the asymptotic formulation treated below. ■

(ii) Shrinking Δ. We now consider a second asymptotic formulation of the problem, in which the null hypothesis $|g(\theta)| \geq \delta n^{-1/2}$ is tested against the alternative hypothesis $|g(\theta)| < \delta n^{-1/2}$. Notice that now the parameter spaces (or hypotheses) are changing with n. Of course, a given hypothesis testing situation deals with a particular n, and there is flexibility in how the problem is embedded into a sequence of similar problems to get a useful approximation. In particular, if equivalence corresponds to $|g(\theta)| < \Delta$, we can always make the identification $\delta = \Delta n^{1/2}$. From an asymptotic point of view, it makes sense to allow the null hypothesis parameter space to change with n, since otherwise the problem becomes degenerate in the sense that the values of Δ and $-\Delta$ for $g(\theta)$ can be perfectly distinguished asymptotically. In testing for bioequivalence, for example, Δ is chosen so small that a value of $|g(\theta)| \leq \Delta$ is deemed to be essentially zero. In a particular situation such as Example 13.5.5 with σ not too small, if a value for μ of Δ cannot be perfectly tested against a value for μ of 0, then Δ and $-\Delta$ cannot be perfectly tested as well, and the asymptotic setup should reflect this.

The main result of this subsection is the following theorem.

Theorem 13.5.3 *Suppose $X_1, \ldots, X_n$ are i.i.d. according to P_θ, $\theta \in \Omega$, where Ω is assumed to be an open subset of $\mathbb{R}^k$. Consider testing the null hypothesis*

$$\theta \in \Omega_{0,n} = \{\theta : \ |g(\theta)| \geq \delta n^{-1/2}\}$$

versus $|g(\theta)| < \delta n^{-1/2}$, where the function g from $\mathbb{R}^k$ to $\mathbb{R}$ is assumed differentiable with gradient $\dot{g}(\theta)$. Assume for every θ with $g(\theta) = 0$ that the family $\{P_\theta, \theta \in \Omega\}$ is q.m.d. at θ and $I(\theta)$ is nonsingular.
(i) Let $\phi_n = \phi_n(X_1, \ldots, X_n)$ be a uniformly asymptotically level α sequence of tests, so that

$$\limsup_{n\to\infty} \sup_{\Omega_{0,n}} E_\theta(\phi_n) \leq \alpha .$$

Assume θ_0 satisfies $g(\theta_0) = 0$. Then, for any h such that $|\langle \dot{g}(\theta_0)^T, h\rangle| = \delta' < \delta$,

$$\limsup_{n\to\infty} E_{\theta_0 + hn^{-1/2}}(\phi_n) \leq \Phi\left(\frac{C-\delta'}{\sigma_{\theta_0}}\right) - \Phi\left(\frac{-C-\delta'}{\sigma_{\theta_0}}\right) , \tag{13.84}$$

where $\sigma^2_{\theta_0}$ is given by

$$\sigma^2_{\theta_0} = \dot{g}(\theta_0) I^{-1}(\theta_0) \dot{g}(\theta_0)^T \tag{13.85}$$

and $C = C(\alpha, \delta, \sigma_{\theta_0})$ satisfies

$$\Phi\left(\frac{C-\delta}{\sigma_{\theta_0}}\right) - \Phi\left(\frac{-C-\delta}{\sigma_{\theta_0}}\right) = \alpha \tag{13.86}$$

(ii) Let $\hat{\theta}_n$ be any estimator satisfying (12.62). Suppose $I(\theta)$ is continuous in θ and $\dot{g}(\theta)$ is continuous at θ_0. Then, the test sequence ϕ_n that rejects when $n^{1/2}|g(\hat{\theta}_n)| \leq C(\alpha, \delta, \hat{\sigma}_n)$, where

$$\hat{\sigma}^2_n = \dot{g}(\hat{\theta}_n) I^{-1}(\hat{\theta}_n) \dot{g}(\hat{\theta}_n)^T ,$$

is pointwise asymptotically level α and is locally asymptotically UMP in the sense that the inequality (13.84) is an equality. In fact, the same properties hold for any *test sequence that rejects if* $|T_n| < C(\alpha, \delta, \hat{\sigma}_n)$, *if* T_n *satisfies*

$$T_n = \dot{g}(\theta_0) I^{-1}(\theta_0) Z_{n,\theta_0} + o_{P^n_{\theta_0}}(1)$$

for every $\theta_0 \in \Omega_0$, *where* Z_{n,θ_0} *is the score vector defined in (12.59).*

PROOF. Fix θ_0 satisfying $g(\theta_0) = 0$. We will derive an upper bound for the limiting power of a test sequence ϕ_n under $\theta_0 + hn^{-1/2}$. Note that

$$g(\theta_0 + hn^{-1/2}) = n^{-1/2} \langle \dot{g}(\theta_0)^T, h \rangle + o(n^{-1/2}) .$$

So, if h is such that $|\langle \dot{g}(\theta_0)^T, h \rangle| > \delta$, then $|g(\theta_0 + hn^{-1/2})| > \delta n^{-1/2}$ for all sufficiently large n. Hence, if ϕ_n has limiting size α, then for such an h,

$$\limsup_{n \to \infty} E_{\theta_0 + hn^{-1/2}}(\phi_n) \leq \alpha \quad . \tag{13.87}$$

By Theorem 13.4.1, we can approximate the power of a test sequence ϕ_n by the power of a test $\phi = \phi(X)$ for the (limit) experiment based on X from the model $N(h, I^{-1}(\theta_0))$. Let $\beta_\phi(h)$ denote the power function of $\phi(X)$ when $X \sim N(h, I^{-1}(\theta_0))$. Then, (13.87) implies $\beta_\phi(h) \leq \alpha$ if $|\langle \dot{g}(\theta_0)^T, h \rangle| > \delta$. By continuity of $\beta_\phi(h)$, $\beta_\phi(h) \leq \alpha$ for any h with $|\langle \dot{g}(\theta_0)^T, h \rangle| \geq \delta$. The test ϕ that maximizes $\beta_\phi(h)$ for this limiting normal problem was given in Example 3.9.3 with $\Sigma = I^{-1}(\theta_0)$, $\xi = h$, and $a^T = \dot{g}(\theta_0)$. Thus, if ϕ is level α for testing $|\langle \dot{g}(\theta_0)^T, h \rangle| \geq \delta$ and h satisfies $|\langle \dot{g}(\theta_0)^T, h \rangle| = \delta' < \delta$, then

$$\beta_\phi(h) \leq \Phi\left(\frac{C - \delta'}{\sigma_{\theta_0}}\right) - \Phi\left(\frac{-C - \delta'}{\sigma_{\theta_0}}\right).$$

and $C = C(\alpha, \delta, \sigma_{\theta_0})$ satisfies (13.86).

To prove (ii), consider the test that rejects when $n^{1/2}|g(\hat{\theta}_n)| \leq C(\alpha, \delta, \hat{\sigma}_n)$. Fix h such that $|\langle \dot{g}(\theta_0)^T, h \rangle| < \delta$ and let $\theta_n = \theta_0 + hn^{-1/2}$. Then, as in the proof of Theorem 13.5.2 (iii), under θ_n,

$$n^{1/2}[g(\hat{\theta}_n) - g(\theta_n)] \xrightarrow{d} N(0, \sigma^2_{\theta_0}) .$$

But,

$$n^{1/2} g(\theta_n) = \langle h, \dot{g}(\theta_0)^T \rangle + o(1) .$$

Therefore, under θ_n,

$$n^{1/2} g(\hat{\theta}_n) \xrightarrow{d} N\left(\langle h, \dot{g}(\theta_0)^T \rangle, \sigma^2_{\theta_0}\right) .$$

Also, under θ_n, $\hat{\sigma}_n$ tends in probability to σ_{θ_0}, and so $C(\alpha, \delta, \hat{\sigma}_n)$ tends in probability to $C(\alpha, \delta, \sigma_{\theta_0})$. Hence, letting Z denote a standard normal variable,

$$P_{\theta_n}\{n^{1/2}|g(\hat{\theta}_n)| \leq C(\alpha, \delta, \hat{\sigma}_n)\} \to P\{|\sigma_{\theta_0} Z + \langle h, \dot{g}(\theta_0)^T \rangle| \leq C(\alpha, \delta, \sigma_{\theta_0})\} ,$$

which agrees with the right hand side of (13.84). ∎

Example 13.5.6 (Normal Problem, Example 13.5.5, continued) Suppose $X_1, \ldots, X_n$ are i.i.d. $N(\mu, \sigma^2)$ with both parameters unknown, so that $\theta = (\mu, \sigma)$. Let $g(\theta) = \mu$ and consider testing $|\mu| \geq \delta n^{-1/2}$ versus $|\mu| < \delta n^{-1/2}$. By the previous theorem, for any test sequence ϕ_n with limiting size bounded by α and any h with $|h| < \delta$,

$$E_{hn^{-1/2},\sigma}(\phi_n) \leq \Phi\left(\frac{C - h}{\sigma}\right) - \Phi\left(\frac{-C - h}{\sigma}\right) , \tag{13.88}$$

where $C = C(\alpha, \delta, \sigma)$ satisfies (13.86). A test whose limiting power achieves this bound is given by the test ϕ_n^* that rejects when

$$n^{1/2}|\bar{X}_n| \le C(\alpha, \delta, S_n) ,$$

where S_n^2 is the (unbiased) sample variance (or any consistent estimator of σ^2). On the other hand, the test ϕ_n^{TOST} given in example 13.5.5 is *no longer* asymptotically efficient. This test (with $\Delta = \delta n^{-1/2}$) rejects when

$$n^{1/2}|\bar{X}_n| < \delta - S_n t_{n-1,1-\alpha}$$

and has power against $(\mu, \sigma) = (hn^{-1/2}, \sigma)$ given by

$$P_{hn^{-1/2},\sigma}\left\{\frac{-\delta + S_n t_{n-1,1-\alpha} - h}{\sigma} < Z_n < \frac{\delta - S_n t_{n-1,1-\alpha} - h}{\sigma}\right\} , \tag{13.89}$$

where

$$Z_n = n^{1/2}(\bar{X}_n - hn^{-1/2})/\sigma \sim N(0,1) .$$

Also, $S_n \to \sigma$ in probability and $t_{n-1,1-\alpha} \to z_{1-\alpha}$. By Slutsky's Theorem, (13.89) converges to

$$P\left\{\frac{-\delta}{\sigma} + z_{1-\alpha} - \frac{h}{\sigma} < Z < \frac{\delta}{\sigma} - z_{1-\alpha} - \frac{h}{\sigma}\right\} , \tag{13.90}$$

where $Z \sim N(0,1)$. Observe that this last expression is positive only if $\sigma z_{1-\alpha} < \delta$; otherwise, the limiting power is zero! On the other hand, the limiting optimal power of ϕ_n^* is always positive (and greater than α when $|h| < \delta$). Even when the limiting power of ϕ_n^{TOST} is positive, it is always strictly less than that of ϕ_n^*.

Note that the limiting expression (13.90) for the power of ϕ_n^{TOST} corresponds exactly to using a TOST test in the limiting experiment $N(h, \sigma^2)$ for testing $|h| \ge \delta$ versus $|h| < \delta$ with σ known based on one observation X. In the limit experiment, the TOST procedure corresponds to the test that rejects if $|X| < \delta - \sigma z_{1-\alpha}$ (which can be viewed as a TOST construction because its rejection region is the intersection of the rejection regions of the two one-sided tests of $h < \delta$ and $h > -\delta$). But, for this limit experiment, the optimal UMP procedure of Section 3.7 rejects when $|X| < C(\alpha, \delta, \sigma)$. In general,

$$C(\alpha, \delta, \sigma) > \delta - \sigma z_{1-\alpha}$$

(Problem 13.54), which shows that the test ϕ_n^* of Theorem 13.5.3 is always more powerful than the asymptotic TOST construction of Theorem 13.5.2. ■

13.5.3 Multi-sided Hypotheses

We now consider the problem of testing $\theta = \theta_0$ versus $\theta \ne \theta_0$ as θ varies in an open subset of $\mathbb{R}^k$. Theorem 13.4.1 relates this problem to testing $h = 0$ versus $h \ne 0$ based on an observation X from the normal model $N(h, I^{-1}(\theta_0))$, where, as usual, $I(\theta_0)$ is the Fisher Information. For this normal model, no UMP test exists, and Theorem 13.4.1 does not lead to an asymptotically UMP test sequence for the original problem. However, we will obtain an optimality result based on the maximin approach. Indeed, for this limiting normal model, an optimal maximin test exists, which allows one to construct an asymptotically maximin test sequence.

In order to have a nondegenerate asymptotically maximin procedure, it is necessary to consider alternatives at some distance from the null hypothesis, just as in the finite sample maximin theory. When testing based on n i.i.d. observations, this distance must shrink with n, in order to avoid a degenerate asymptotic theory, since there will typically exist test sequences whose asymptotic power tends to one uniformly over alternatives whose distance from θ_0 is fixed. It is convenient to consider this fixed distance as given by $|I^{1/2}(\theta_0)(\theta - \theta_0)|$, where $|\cdot|$ denotes the usual Euclidean norm of a vector in $\mathbb{R}^k$. For q.m.d. models, it will be seen that it is necessary to let this distance shrink at rate $n^{-1/2}$ in order to obtain a limiting minimum power greater than α and less than 1.

In the following theorem, $c_{k,1-\alpha}$ denotes the upper $1-\alpha$ quantile of the Chi-squared distribution with k degrees of freedom.

Theorem 13.5.4 *Assume $X_1, \ldots, X_n$ are i.i.d. P_θ, where θ varies in an open subset Ω of $\mathbb{R}^k$. Assume this family is q.m.d. at θ_0 with positive definite Information matrix $I(\theta_0)$. The problem is to test the null hypothesis $\theta = \theta_0$ against $\theta \neq \theta_0$. Let $\phi_n = \phi_n(X_1, \ldots, X_n)$ be any sequence of tests such that $E_{\theta_0}(\phi_n) \to \alpha$. Then, for any $b > 0$,*

$$\limsup_{n\to\infty} \inf\{E_{\theta_0+hn^{-1/2}}(\phi_n) : |I^{1/2}(\theta_0)h| \geq b\} \leq P\{\chi_k^2(b^2) \geq c_{k,1-\alpha}\} \ , \quad (13.91)$$

where $\chi_k^2(b^2)$ denotes a random variable that has the noncentral Chi-squared distribution with k degrees of freedom and noncentrality parameter b^2.

PROOF. Denote by $\beta_n(h)$ the rescaled power function of ϕ_n, i.e.,

$$\beta_n(h) \equiv E_{\theta_0+hn^{-1/2}}(\phi_n) \ .$$

By assumption, $\beta_n(0) \to \alpha$. Denote by $R = R(\alpha, b)$ the right hand side of (13.91). Now, argue by contradiction; that is, assume for the test sequence ϕ_n and some subsequence $\{n_j\}$,

$$\lim_{n_j\to\infty} \inf\{\beta_{n_j}(h) : \ |I^{1/2}(\theta_0)h| \geq b\} > R \ .$$

Then, by Theorem 13.4.1, there exists a further subsequence n_{j_m} such that

$$\beta_{n_{j_m}}(h) \to \beta(h)$$

for every h, where $\beta(h)$ corresponds to a level α test of $h = 0$ versus $h \neq 0$ in the (limiting) experiment consisting of observing an X which is $N(h, I^{-1}(\theta_0))$. Thus, $\beta(h) > R$ for every h such that $|I^{1/2}(\theta_0)h| \geq b$, which implies

$$\inf\{\beta(h) : \ |I^{1/2}(\theta_0)h| \geq b\} > R \ .$$

This is a contradiction, since R is the maximin power for testing $h = 0$ versus $|I^{1/2}(\theta_0)h| \geq b$ based on X (Problem 8.29). ■

We first illustrate the theorem in the case $k = 1$.

Example 13.5.7 (Simple vs Two-sided Alternative) Suppose $X_1, \ldots, X_n$ are i.i.d. P_θ, $\theta \in \mathbb{R}$. Consider testing $\theta = \theta_0$ versus $\theta \neq \theta_0$. Assume the family is q.m.d. at θ_0. Let ϕ_n be any test sequence satisfying $E_{\theta_0}(\phi_n) \to \alpha$. By Theorem

13.5.4 with $d = I^{-1/2}(\theta_0)b$, an upper bound for the limiting maximin power over the complement of shrinking neighborhoods is given by

$$\limsup_n \inf\{E_{\theta_0+hn^{-1/2}}(\phi_n) : \ |h| \geq d\} \leq P\{\chi_1^2(I(\theta_0)d^2) \geq c_{1,1-\alpha}\} .$$

In the one-sided case, an AUMP level α test (13.43) rejects for large values of the score statistic Z_n given by (13.42). Consider the two-sided version $\phi_{n,2}$ of this test which rejects when $I^{-1}(\theta_0)Z_n^2 > c_{1,1-\alpha}$. Since $I^{-1}(\theta_0)Z_n^2$ is asymptotically Chi-squared with one degree of freedom, this test is consistent in level. Moreover, its power function satisfies, for any $0 < d < D < \infty$,

$$\inf[P_{\theta_0+hn^{-1/2}}\{I^{-1}(\theta_0)Z_n^2 > c_{1,1-\alpha}\} : \ d \leq h \leq D]$$

$$\to P\{\chi_1^2(I(\theta_0)d^2) \geq c_{1,1-\alpha}\} . \tag{13.92}$$

To see why, the convergence (13.44) implies that, under $\theta_n = \theta_0 + h_n n^{-1/2}$,

$$I^{-1}(\theta_0)Z_n^2 \stackrel{d}{\to} \chi_1^2(I(\theta_0)h^2) .$$

If (13.92) failed, there would exist h_n satisfying $h_n \to h \in [d, D]$ such that the limiting power of $\phi_{n,2}$ against θ_n tends to

$$P\{\chi_1^2(I(\theta_0)h^2) > c_{1,1-\alpha}\} < P\{\chi_1^2(I(\theta_0)d^2) > c_{1,1-\alpha}\} .$$

But, this last inequality is a contradiction since $h \geq d$ and the family of $\chi_1(\psi^2)$ with ψ^2 varying has monotone likelihood ratio (see Problem 7.4). It is typically possible to prove the stronger result with D in (13.92) replaced by ∞. This technical issue is the same as encountered in the one-sided case in Section 13.3 when determining whether or not Rao's score test is not only LAUMP but AUMP; see Theorem 13.3.3 (iv). For an alternative asymptotic optimality approach in the two-sided case, see Problem 13.55. ■

By a similar argument, we can prove the following optimality result for Rao's test in the general k multi-sided testing problem. Analogous results hold for both the Wald and likelihood ratio tests (Problem 13.57).

Theorem 13.5.5 *Assume the conditions of Theorem 13.5.4. For testing $\theta = \theta_0$ versus $\theta \neq \theta_0$, consider the test ϕ_n^* that rejects when $Z_n^T I^{-1}(\theta_0)Z_n > c_{k,1-\alpha}$. Then, $E_{\theta_0}(\phi_n^*) \to \alpha$ and for any b and B satisfying $0 < b < B < \infty$,*

$$\inf\{E_{\theta_0+hn^{-1/2}}(\phi_n^*) : \ b \leq |I^{1/2}(\theta_0)h| \leq B\} \to P\{\chi_k^2(b^2) \geq c_{k,1-\alpha}\} . \tag{13.93}$$

PROOF. First suppose $h_n \to h$ with h satisfying $|I^{1/2}(\theta)h| \geq b$. By the Continuous Mapping Theorem, under $\theta_0 + h_n n^{-1/2}$, Corollary 12.4.1 implies that

$$Z_n^T I^{-1}(\theta_0)Z_n \stackrel{d}{\to} \chi_k^2(|I^{1/2}(\theta_0)h|^2) .$$

Hence, the limiting power of ϕ_n^* against such a sequence is

$$P\{\chi_k^2(|I^{1/2}(\theta_0)h|^2) \geq c_{k,1-\alpha}\} \geq P\{\chi_k^2(b^2) \geq c_{k,1-\alpha}\} , \tag{13.94}$$

where the last inequality follows since the family of noncentral chi-squared distributions with fixed degrees of freedom and varying noncentrality parameter has monotone likelihood ratio. Now, if the result (13.93) were false, there would exist a sequence h_n satisfying $b \leq |I^{1/2}(\theta_0)h| \leq B$ and such that the limiting power of

ϕ_n^* under h_n is less than the right hand side of (13.94). But, h_n lies in a compact set, so we can extract a further subsequence h_{n_j} (if necessary) so that h_{n_j} converges. Applying the argument leading to (13.94) to such a subsequence results in a contraction. ■

We will later apply these results to obtain some asymptotically maximin tests of goodness of fit in Sections 14.3 and 14.4.

Note that the construction of asymptotically optimal tests in the multi-sided case depends on the existence of an optimal test for testing the mean vector $h = 0$ when $X \sim N(h, I^{-1}(\theta_0))$ and $I^{-1}(\theta_0)$ is a known nonsingular covariance matrix. For this problem, if the alternatives are specified by $|I^{1/2}(\theta_0)h| \geq b$, then the maximin test rejects for large values of $X^T \Sigma^{-1}(\theta_0) X$. But, the maximin optimality of this test need not hold if the alternative parameter space is specified differently; see Problem 8.30. Moreover, if C is any closed, convex set in $\mathbb{R}^k$, then the test that accepts if and only if $X \in C$ is admissible; see Problem 6.39. Thus, the optimality of the maximin test is not so compelling, particularly when $k > 1$.

13.6 Applications to Nonparametric Models

13.6.1 Nonparametric Mean

Let $X_1, \ldots, X_n$ be i.i.d. with c.d.f. F, mean $\mu(F)$ and variance $\sigma^2(F)$. Assume $F \in \tilde{\mathbf{F}}$, where $\tilde{\mathbf{F}}$ satisfies (11.77). We now would like to derive an optimality property of the t-test for the mean in a nonparametric setting. Theorem 11.4.5 implies that the power of the t-test is bounded away from α for distributions F whose standardized mean $n^{1/2}\mu(F)/\sigma(F)$ is bounded away from 0. It is then of interest to measure a test sequence by its maximin power over such alternatives, with the goal of finding the test that asymptotically maximizes the minimum power over such alternatives. Consider testing $\mu(F) = 0$ against the alternatives $\mu(F)/\sigma(F) \geq \delta/n^{1/2}$. By Theorem 11.4.5, the limiting minimum power of the t-test is $1 - \Phi(z_{1-\alpha} - \delta)$. We now show that this is indeed the optimal limiting maximin power in a nonparametric setting.

If the unknown family of distributions $\tilde{\mathbf{F}}$ contains the family $N(\theta, 1)$ for $\theta \geq 0$, then an optimality result is easy to obtain. Indeed, for any sequence of test functions $\phi_n = \phi_n(X_1, \ldots, X_n)$ which satisfies $E_F(\phi_n) \to \alpha$ for any $F \in \tilde{\mathbf{F}}$ with mean 0, we have

$$\limsup_n \inf_{\{F \in \tilde{\mathbf{F}},\ \mu(F)/\sigma(F) \geq \delta n^{-1/2}\}} E_F(\phi_n)$$

$$\leq \limsup_n E_{F=N(\delta n^{-1/2},1)}(\phi_n) = 1 - \Phi(z_{1-\alpha} - \delta) \ ,$$

since the right hand side is the optimal limiting power for testing $\theta = 0$ versus $\theta = \delta/n^{1/2}$ in the normal location model $N(\theta, 1)$. Hence, the t-test is asymptotically maximin since its limiting minimum power attains this bound.

If the family of distributions $\tilde{\mathbf{F}}$ does not contain the normal distributions, the above argument does not work. For example, suppose we consider distributions supported on $[-1, 1]$. Then, we can still obtain an optimality result for the t-test, as long as $\tilde{\mathbf{F}}$ satisfies (11.77). To this end, let $\mathbf{F}_0$ denote the family of all distributions on $[-1, 1]$. Let ϕ_n be any test sequence satisfying $E_F(\phi_n) \to \alpha$ if $F \in \mathbf{F}_0$

and $\mu(F) = 0$. Fix any such F with $\mu(F) = 0$ and $\sigma(F) > 0$. The smallest power over a large class of alternatives can always be bounded above by the smallest power over a smaller class. If the smaller class is chosen appropriately, the testing problem for the smaller model (which will be a parametric model that we have previously studied) will have relevance for the larger class (the nonparametric model we would like to study). So, introduce the parametric submodel with density

$$p_\theta(x) = \exp(\theta x - C(\theta)) \tag{13.95}$$

with respect to F. This is a one-parameter exponential family, and so the conditions of Theorem 13.3.2 are satisfied. Let

$$\mu_\theta = \int_{-1}^{1} x p_\theta(x) dF(x)$$

be the mean of p_θ and let σ_θ^2 be its variance. Since $\mu(F) = 0$, $\mu_0 = 0$. In addition, $\mu_\theta = C'(\theta)$ and $\sigma_\theta^2 = C''(\theta)$, so that $C'(0) = 0$ and $C''(0) = \sigma^2(F) > 0$. Then,

$$\frac{\mu_\theta}{\sigma_\theta} = \frac{C'(\theta)}{[C''(\theta)]^{1/2}} = \frac{\theta[C''(0)]^{1/2} + o(\theta)}{[C''(\theta)]^{1/2}} = \theta\sigma(F) + o(\theta)$$

as $\theta \to 0$. Also, for this model, $I(\theta) = C''(\theta)$, so that $I(0) = \sigma^2(F)$. It is also easy to check that the family (13.95) satisfies (11.77), at least for small enough θ (Problem 13.58).

With δ fixed, let θ_n be any fixed sequence such that $n^{1/2}\theta_n > \delta/\sigma(F)$ and $n^{1/2}\theta_n \to \delta/\sigma(F)$. Then,

$$n^{1/2}\mu_{\theta_n}/\sigma_{\theta_n} = n^{1/2}\theta_n/\sigma(F) + o(1)$$

as $\theta_n \to 0$. Thus, $n^{1/2}\mu_{\theta_n}/\sigma_{\theta_n} > \delta$ for all sufficiently large n. So, the problem of testing $\theta = 0$ versus $\theta = \theta_n$ is relevant to the nonparametric mean problem because $\theta = 0$ corresponds to a distribution in the null hypothesis parameter space while $\theta = \theta_n$ corresponds to a distribution in the alternative hypothesis parameter space (sequence). Hence, for any test sequence ϕ_n,

$$\limsup_n \inf_{F \in \mathbf{F}_0,\ n^{1/2}\mu(F)/\sigma(F) \geq \delta} E_F(\phi_n) \leq \limsup_n E_{\theta_n}(\phi_n) \ .$$

The right hand side is bounded above by the optimal limiting power for testing $\theta = 0$ versus $\theta = \theta_n$. The limiting value was obtained in Theorem 13.3.2 (with $h = \delta/\sigma(F)$) and is equal to

$$1 - \Phi(z_{1-\alpha} - h\sigma(F)) = 1 - \Phi(z_{1-\alpha} - \delta) \ .$$

Hence, we have shown that

$$\limsup_n \inf_{F \in \mathbf{F}_0,\ n^{1/2}\mu(F)/\sigma(F) \geq \delta} E_F(\phi_n) \leq 1 - \Phi(z_{1-\alpha} - \delta) \ .$$

But, the t-test attains the right hand side, and so is asymptotically maximin.

Of course, one can obtain a bound using other parametric submodels. The family p_θ chosen above certainly works in that it yields an optimality result for the t-test. To gain some insight into why this family works, let us consider the more general family of densities with densities

$$p_{T,\theta}(x) = \exp[\theta T(x) - C_T(\theta)]$$

with respect to F. This assumes that the function $T(x)$ is bounded on $[-1,1]$, or at least that $T(X)$ has a moment generating function if X has distribution F. Let

$$\mu_{T,\theta} = \int x p_{T,\theta}(x) dF(x)$$

and $\sigma^2_{T,\theta}$ be the variance of $p_{T,\theta}$. The functions $\mu_{T,\theta}$ and $\sigma_{T,\theta}$ are infinitely differentiable in θ. Then,

$$\frac{\partial \mu_{T,\theta}}{\partial \theta} = \int x[T(x) - C'_T(\theta)] p_{T,\theta}(x) ,$$

so that

$$\mu'_{T,0} = \int x[T(x) - C'_T(0)] dF(x) = Cov_F[X, T(X)] .$$

Then,

$$\mu_{T,\theta} = \theta Cov_F[X, T(X)] + o(\theta)$$

and

$$\sigma_{T,\theta} = \sigma(F) + o(\theta)$$

as $\theta \to 0$. Hence,

$$\frac{\mu_{T,\theta}}{\sigma_{T,\theta}} = \frac{\theta Cov_F[X, T(X)]}{\sigma(F)} + o(\theta)$$

as $\theta \to 0$. Assume $Cov_F[X, T(X)] \neq 0$, in which case we may assume without loss of generality that it is positive (or replace T with $-T$). Let θ_n be any fixed sequence with $n^{1/2}\theta_n > \delta\sigma(F)/Cov_F[X, T(X)]$ and

$$n^{1/2}\theta_n \to \delta\sigma(F)/Cov_F[X, T(X)] .$$

Then,

$$n^{1/2}\frac{\mu_{T,\theta_n}}{\sigma_{T,\theta_n}} = n^{1/2}\theta_n \frac{Cov_F[X, T(X)]}{\sigma(F)} + o(1) .$$

So, $n^{1/2}\mu_{T,\theta_n}/\sigma_{T,\theta_n} > \delta$ for all sufficiently large n. Thus, for any test sequence ϕ_n,

$$\limsup_n \inf_{F \in \mathbf{F}_0,\ n^{1/2}\mu(F)/\sigma(F) \geq \delta} E_F(\phi_n) \leq E_{T,\theta_n}(\phi_n) ,$$

where E_{T,θ_n} denotes expectation with respect to p_{T,θ_n}. Note that, for this model, the Information at $\theta = 0$ satisfies

$$I_T(0) = C''_T(0) = Var_F^{1/2}[T(X)] .$$

The best limiting power among asymptotically level α tests of $\theta = 0$ versus $\theta = \theta_n$ was obtained in Theorem 13.3.2 (with $h = \delta\sigma(F)/Cov_F[X, T(X)]$) as

$$1 - \Phi(z_{1-\alpha} - hI_T^{1/2}(0)) = 1 - \Phi(z_{1-\alpha} - \delta\sigma(F)I_T^{1/2}(0)/Cov_F[X, T(X)]) .$$

This reduces to the previous bound in the case $T(X) = X$. The sharpest possible result is obtained by choosing T to minimize the right hand side, which is equivalent to maximizing

$$\frac{Cov_F[X, T(X)]}{\{Var_F(X) Var_F[T(X)]\}^{1/2}} .$$

By the Cauchy-Schwarz inequality, this is bounded above by 1, and the resulting value of 1 is attained when $X = T(X)$.

Thus, in some sense, the model with $T(X) = X$ is least favorable in that it is the hardest parametric submodel to achieve high (limiting) power. The idea of using a parametric submodel to obtain efficiency results in nonparametric models dates back to Stein (1956b).

13.6.2 Nonparametric Testing of Functionals

Suppose $X_1, \ldots, X_n$ are i.i.d. $P \in \mathbf{P}$. In this section, the family $\mathbf{P}$ is a nonparametric family. Specifically, we would like to consider problems where we do not assume much or anything about P. Thus, $\mathbf{P}$ could be the family of *all* distributions on some sample space S, but it might be restricted by moment or smoothness conditions, in which case $\mathbf{P}$ is still quite large.

Let $\theta(\cdot)$ be a statistical functional; that is, $\theta(P)$ is a real-valued function of P, defined for $P \in \mathbf{P}$. For example, if P is a distribution on $\mathbb{R}$, $\theta(P)$ could be the mean of P, or the variance of P. In such cases, $\mathbf{P}$ could be the set of all distributions with finite variance. Or, if P is a distribution on $\mathbb{R}^2$, $\theta(P)$ might be the correlation of P, defined on the set $\mathbf{P}$ of all distributions whose marginals have a finite nonzero variance.

We wish to test the null hypothesis $\theta(P) \leq 0$ against $\theta(P) > 0$. Fix P with $\theta(P) = 0$. In order to assess the power of a test at some distribution Q near P, we will consider parametric submodels that contain P. The basic idea is that the power attainable in the full nonparametric model can be no greater than for any submodel.

Let $L^2(P)$ denote the space of (equivalence classes of) functions u which are square integrable with respect to P. The inner product is given by

$$\langle u, v \rangle_P = \int u(x) v(x) dP(x),$$

and $|u|_P^2 = \langle u, u \rangle_P$. Also, let $L_0^2(P)$ denote the subset of $u \in L^2(P)$ satisfying $\int u(x) dP(x) = 0$. By Problem 12.6, if $u \in L_0^2(P)$, we can construct a one-dimensional q.m.d. family $P_{u,t}$ indexed by t in some neighborhood of 0, such that $P_{u,0} = P$ and the score function at $t = 0$ is u. For example, if u is bounded and $|t| \leq [\sup_x |u(x)|]^{-1}$, then we can take $P_{u,t}$ to be the distribution with density with respect to P given by

$$\frac{dP_{u,t}}{dP}(x) = 1 + t u(x) . \tag{13.96}$$

(Note that $P_{u,t} \in \mathbf{P}$ if $\mathbf{P}$ is the set of all probabilities on S, but if there are restrictions on P, this construction may not work.)

In order to test $\theta(P_{u,t})$ along such a parametric submodel, we assume that $\theta(\cdot)$ is differentiable in the sense

$$\frac{\theta(P_{u,t}) - \theta(P)}{t} \to \langle u, \tilde{\theta}_P \rangle_P \quad \text{as } t \to 0 , \tag{13.97}$$

for some function $\tilde{\theta}_P \in L^2(P)$. Evidently, this condition implies that, as a real-valued function of the real variable t, $\theta(P_{u,t})$ is differentiable at $t = 0$.[4] Note that, if $\tilde{\theta}_P$ satisfies (13.97), then so does $\tilde{\theta}_P + c$ for any constant c; we will henceforth assume $\int \tilde{\theta}_P(x)dP(x) = 0$.

Example 13.6.1 (Linear Functionals) A statistical functional is *linear* if it can be represented as

$$\theta(P) = \int f(x)dP(x) \tag{13.98}$$

for some function $f \in L^2(P)$. In this case, if $P_{u,t}$ is given by (13.96), then

$$\frac{\theta(P_{u,t}) - \theta(P)}{t} = \langle u, f \rangle_P$$

with no error term; that is,

$$\tilde{\theta}_P(x) = f(x) - \int f(x)dP \ . \tag{13.99}$$

Even if $P_{u,t}$ is not specifically of the form (13.96), then it can be shown that $\theta(P)$ is differentiable in the sense of (13.97) with $\tilde{\theta}_P$ given by (13.99) if

$$\sup_{P \in \mathbf{P}} E_P[f^2(X)] < \infty \ ;$$

see Bickel et al. (1993, p.457-458). In particular, if f is a bounded function on a set S and $\mathbf{P}$ is the set of all probabilities on S, then $\theta(\cdot)$ is differentiable in the sense of (13.97). ■

Next, for testing $\theta(P) \leq 0$ against $\theta(P) > 0$, we obtain an upper bound for the limiting local power function along a one-dimensional q.m.d. submodel, among tests that are pointwise consistent in level. Note that, under (13.97),

$$\theta(P_{u,t}) = \theta(P) + t\langle \tilde{\theta}_P, u \rangle_P + o(t) \qquad \text{as } t \to 0 \ , \tag{13.100}$$

which implies $\theta(P_{u,t}) > 0$ for all small $t > 0$ if $\langle \tilde{\theta}_P, u \rangle_P > 0$.

By Lemma 13.3.1(ii), if $h > 0$ and $\langle \theta_P, u \rangle_P > 0$, then (Problem 13.59)

$$\limsup_n E_{P_{u,hn^{-1/2}}}(\phi_n) \leq 1 - \Phi(z_{1-\alpha} - h|u|_P) \ . \tag{13.101}$$

Fix $\delta > 0$ and let

$$h = h(u, \delta) = \frac{\delta}{\langle \tilde{\theta}_P, u \rangle_P} \ ;$$

then, $n^{1/2}\theta(P_{u,h(u,\delta)n^{-1/2}}) \to \delta$. The bound (13.101) at $h(u,\delta)n^{-1/2}$ becomes

$$\limsup_n E_{P_{u,h(u,\delta)n^{-1/2}}}(\phi_n) \leq 1 - \Phi(z_{1-\alpha} - \frac{\delta|u|_P}{\langle \tilde{\theta}_P, u \rangle_P}) \ . \tag{13.102}$$

---

[4]The condition (13.97) further asserts that, as a function of u, the limiting value on the right side of (13.97) is linear in u as u varies in $L_0^2(P)$. In fact, the Riesz representation theorem (see Theorem 6.4.1 of Dudley (1989)) asserts that any linear function of u must be of the form $\langle u, \tilde{\theta} \rangle_P$ for some $\tilde{\theta}$.

As u varies, the bound is smallest when $|u|_P / \langle \tilde{\theta}_P, u \rangle_P$ is minimized. But, by Cauchy-Schwarz,

$$\frac{|u|_P}{\langle \tilde{\theta}_P, u \rangle_P} \geq \frac{1}{|\tilde{\theta}_P|_P} ,$$

and equality occurs when $u = \tilde{\theta}_P$. Note that, when $u = \tilde{\theta}_P$, the bound (13.102) becomes

$$1 - \Phi(z_{1-\alpha} - \frac{\delta}{|\tilde{\theta}_P|_P}) . \tag{13.103}$$

Moreover, taking $u = \tilde{\theta}_P$ corresponds to the least favorable family (generalizing the results of the previous subsection for the mean). Actually, we will obtain a stronger result which will allow us to construct locally AUMP tests. First, we obtain an upper bound which is smaller than (13.101) and is generally attainable for all u.

Theorem 13.6.1 *Let $X_1, \ldots, X_n$ be i.i.d. $P \in \mathbf{P}$, where $\mathbf{P}$ is the set of all probabilities on space S (endowed with a σ-field). Assume $\theta(\cdot)$ is differentiable in the sense (13.97). Fix P with $\theta(P) = 0$, $u \in L_0^2(P)$, and let $\{P_{u,t}\}$ denote a q.m.d. submodel, defined for t in some neighborhood of 0 with $P_{u,0} = P$ and score function u. Let $\phi_n = \phi_n(X_1, \ldots, X_n)$ be a sequence of level α tests of $\theta(P) \leq 0$. If $\langle \tilde{\theta}_P, u \rangle_P > 0$ and $h > 0$, then,*

$$\limsup_n E_{P_{u,hn^{-1/2}}}(\phi_n) \leq 1 - \Phi\left(z_{1-\alpha} - h \frac{\langle \tilde{\theta}_P, u \rangle_P}{|\tilde{\theta}|_P} \right) . \tag{13.104}$$

Proof. Without loss of generality, assume $|u|_P^2 = 1$. Let $v = \tilde{\theta}_P - \langle \tilde{\theta}_P, u \rangle_P u$. Note that $v \in L_0^2(P)$, $\langle u, v \rangle_P = 0$, and

$$\langle v, v \rangle_P = |\tilde{\theta}_P|_P^2 - \langle \tilde{\theta}_P, u \rangle_P^2 .$$

Consider a two-dimensional parametric submodel P_{u,v,t_1,t_2} indexed by (t_1, t_2) in some neighborhood of the origin in $\mathbb{R}^2$ such that the score function at $(t_1, t_2) = (0, 0)$ is $(u, v)^T$. (See Problem 12.7 for a construction.) The experiments $P^n_{u,v,h_1 n^{-1/2}, h_2 n^{-1/2}}$ converge to a normal experiment where you observe $(Z_1, Z_2)^T$ with mean $E(Z_i) = h_i$, $Var(Z_1) = 1$, $Var(Z_2) = |v|_P^2$ and $Cov(Z_1, Z_2) = 0$ (since $\langle u, v \rangle_P = 0$).

Fix h_1 and h_2 and let $t_i = th_i$. Then, $h_1 u + h_2 v$ is the score function for the family $P_{u,v,h_1 t, h_2 t}$ indexed by t. Moreover,

$$\theta(P_{u,v,th_1,th_2}) - \theta(P) = t \langle h_1 u + h_2 v, \tilde{\theta}_P \rangle_P + o(t) .$$

So, if $\langle h_1 u + h_2 v, \tilde{\theta}_P \rangle_P < 0$, we have

$$\limsup E_{P_{u,v,h_1 n^{-1/2}, h_2 n^{-1/2}}}(\phi_n) \leq \alpha .$$

Therefore, by Theorem 13.4.1, the local limiting power of ϕ_n along any subsequence can be bounded above by the power of $\phi = \phi(Z_1, Z_2)$, where

$$E_{h_1,h_2}(\phi) \leq \alpha \quad \text{if } \langle h_1 u + h_2 v, \tilde{\theta}_P \rangle_P < 0 ,$$

and by continuity the result holds if $\langle h_1 u + h_2 v, \tilde{\theta}_P \rangle_P = 0$ as well. But, the UMP level α test for testing

$$h_1 \langle u, \tilde{\theta}_P \rangle_P + h_2 \langle v, \tilde{\theta}_P \rangle_P \le 0$$

rejects if

$$Z_1 \langle u, \tilde{\theta}_P \rangle_P + Z_2 \langle v, \tilde{\theta}_P \rangle_P > z_{1-\alpha} \sqrt{\langle u, \tilde{\theta}_P \rangle_P^2 + \langle v, \tilde{\theta}_P \rangle_P^2 |v|_P^2} = z_{1-\alpha} |\tilde{\theta}_P|_P \ ,$$

which has power with $h_1 = h$ and $h_2 = 0$ given by the right side of (13.104). ∎

Remark 13.6.1 The tests ϕ_n need not be exact level α. All that is required is that $\limsup_n E_{P_{u,hn^{-1/2}}}(\phi_n) \le \alpha$ if h has the opposite sign of $\langle \tilde{\theta}_P, u \rangle_P$. This must hold for u in the statement of (13.104) as well as any linear combination of u and $\tilde{\theta}_P$.

The result and the proof applies even if $\mathbf{P}$ is not the set of all probabilities on S. What is required is that the two-dimensional model P_{u,v,t_1,t_2} used in the proof also belongs to $\mathbf{P}$. Also, it only required that the differentiability condition need only hold for submodels $P_{u,v,h_1 t, h_2 t}$. For semiparametric models, the result needs to be modified, but a similar result holds; see Theorem 25.44 of van der Vaart (1998). ∎

Next, we consider tests whose power attains the bound (13.104).

Example 13.6.2 (Linear Functionals, continued) Let $\hat{P}_n$ be the empirical measure, i.e., $\hat{P}_n\{E\}$ is the proportion of observations that fall in E. Then, tests of $\theta(P)$ can be based on $\theta(\hat{P}_n) = n^{-1} \sum_i f(X_i)$. Under $\theta(P) = 0$,

$$n^{1/2} \theta(\hat{P}_n) \stackrel{d}{\to} N(0, |f|_P^2) \ .$$

Since $|f|_P$ is unknown, consider the test that rejects when $n^{1/2}\theta(\hat{P}_n)/S_n > z_{1-\alpha}$, where

$$S_n^2 = \frac{1}{n} \sum_{i=1}^n [f(X_i) - \theta(\hat{P}_n)]^2 \ .$$

Under P, $S_n^2 \stackrel{P}{\to} |f|_P^2$; by contiguity, this holds under $P^n_{u,hn^{-1/2}}$ as well. By Example 12.3.8, under $P^n_{u,hn^{-1/2}}$,

$$n^{1/2} \theta(\hat{P}_n) \stackrel{d}{\to} N(h \langle f, u \rangle_P, |f|_P^2) \ .$$

By Slutsky's Theorem, under $P_{u,hn^{-1/2}}$,

$$n^{1/2} \theta(\hat{P}_n)/S_n \stackrel{d}{\to} N(h \frac{\langle f, u \rangle_P}{|f|_P}, 1) \ .$$

Therefore, the limiting power of the above test against $P^n_{u,hn^{-1/2}}$ is the upper bound (13.104). Moreover, the convergence to the limiting power is uniform in h for $0 \le h \le c$ and any $c > 0$ (Problem 13.61). The resulting test is locally AUMP against all such alternatives. For example, the result applies to one-sided tests of $\theta(P) = P\{E\}$, and tests based on the empirical measure are asymptotically LAUMP. ∎

Example 13.6.3 (Variance Functional) Suppose P is a distribution on $\mathbb{R}$, and $\mathbf{P}$ is the set of all distributions with a uniformly bounded fourth moment. Let $\sigma^2(P)$ denote the variance of P, and $\mu(P)$ denote the mean of P. The problem is to test $\sigma^2(P) \le \sigma_0^2$. Let $\theta(P) = \sigma^2(P) - \sigma_0^2$. Then, the conditions of Theorem 13.6.1 hold with

$$\tilde{\theta}_P(x) = [x - \mu(P)]^2 - \theta(P) ,$$

and the test that rejects when $n^{1/2}[\theta(\hat{P}_n) - \sigma_0^2]/S_n > z_{1-\alpha}$ attains the bound (13.104), where S_n^2 is a consistent estimator of the variance of $[X_i - \mu(P)]^2$, such as

$$S_n^2 = n^{-1} \sum_i (X_i - \bar{X}_n)^4 - \sigma^4(\hat{P}_n) .$$

The details are left to Problem 13.63. ■

In general, consider tests of $\theta(P)$ based on $\theta(\hat{P}_n)$. This implicitly assumes $\theta(\cdot)$ is defined for empirical measures. Suppose $\theta(\hat{P}_n)$ is an asymptotically linear statistic in the sense that

$$n^{1/2}[\theta(\hat{P}_n) - \theta(P)] = \int \tilde{\theta}_P d(\hat{P}_n - P) + o_P(1) . \tag{13.105}$$

This can be verified directly in examples where $\theta(\hat{P}_n)$ is a smooth function of sample means, such as the previous example. Otherwise, θ must be differentiable in an appropriate sense, but such an approach is beyond the scope of the treatment here; see Serfling (1980, Chapter 10) or van der Vaart and Wellner (1996, Section 3.9). Note that (13.105) implies that, under P,

$$n^{1/2}[\theta(\hat{P}_n) - \theta(P)] \stackrel{d}{\to} N(0, |\tilde{\theta}_P|_P^2) .$$

In order to construct an optimal test, it is necessary to construct a consistent estimator of $|\tilde{\theta}_P|_P$. Assuming S_n is such a consistent estimator, the test that rejects for large $n^{1/2}\theta(\hat{P}_n)/S_n$ is asymptotically LAUMP, by the same argument used in Example 13.6.2. General approaches for constructing an estimator of the asymptotic variance of $n^{1/2}\theta(\hat{P}_n)$, as well as a means of estimating its sampling distribution, are provided by bootstrap resampling and subsampling, which will be discussed in Chapter 15.

13.7 Problems

Section 13.1

Problem 13.1 (i). Let P_i have density p_i with respect to a dominating measure μ. Show that $\|P_1 - P_0\|_1$ defined by $\int |p_1 - p_0| d\mu$ is independent of the choice of μ and is a metric.
(ii). Show the Hellinger distance defined in (13.12) is also independent of μ and is a metric.

Problem 13.2 Show that $\|P_1 - P_0\|_1$ can also be computed as

$$2 \sup_B |P_1(B) - P_0(B)| ,$$

where the supremum is over all measurable sets B. In addition, it may be computed as

$$\sup_{\{\phi : |\phi| \le 1\}} \left| \int \phi(x) dP_1(x) - \int \phi(x) dP_0(x) \right| ,$$

where the supremum is over all measurable functions ϕ such that $\sup_x |\phi(x)| \le 1$.

Problem 13.3 (i) Suppose X is a random variable taking values in a sample space S with probability law P. Let ω_0 and ω_1 be disjoint families of probability laws. Assume that, for every $Q \in \omega_1$ and any $\epsilon > 0$, there exists a subset A of S (which may depend on ϵ) such that $Q(A) \ge 1 - \epsilon$ and such that, if X has distribution Q, then the conditional distribution of X given $X \in A$ is a distribution in ω_0; call it P_ϵ. Show $\|Q - P_\epsilon\|_1 \to 0$ as $\epsilon \to 0$.
(ii) Based on data X with probability law P, consider the problem of testing the null hypothesis $P \in \omega_0$ versus $P \in \omega_1$. Suppose that, for every $Q \in \omega_1$, there exists a sequence $\{P_k\}$ with $P_k \in \omega_0$ such that $\|Q - P_k\|_1 \to 0$ as $k \to \infty$. Show that if a test ϕ is level α, then $E_Q[\phi(X)] \le \alpha$ for all $Q \in \omega_1$.
(iii) Suppose $X_1, \ldots, X_n$ are i.i.d. on the real line. Let ω_0 be distributions with a finite mean and ω_1 those without a finite mean. Apply (i) and (ii) to show that no level α test of ω_0 versus ω_1 has power $> \alpha$ against any $Q \in \omega_1$.

[Such nonexistence results data back to Bahadur and Savage (1956); see Lemma 11.4.4. This example in (iii) and others are treated in Romano (2004), which also contains many references on such problems.]

Problem 13.4 Let P_θ be uniform on $[0, \theta]$. Let $\theta_n = \theta_0 + h/n$. Calculate the limit of $nH^2(P_{\theta_0}, P_{\theta_0 + h/n})$. If $h > 0$, let ϕ_n be the UMP level α test which rejects when the maximum order statistic is too large. Evaluate the limit of the power of ϕ_n against the alternative θ_n.

Problem 13.5 Prove Lemma 13.1.1.

Problem 13.6 Consider testing $P_{\theta_0}^n$ versus $P_{\theta_n}^n$ and assume $nH^2(P_{\theta_0}, P_{\theta_n}) \to 0$. Let ϕ_n be any test sequence such that $\limsup E_{\theta_0}(\phi_n) \le \alpha$. Show that $\limsup E_{\theta_n}(\phi_n) \le \alpha$.

Problem 13.7 Let P_θ be $N(\theta, 1)$. Fix h and let $\theta_n = hn^{-1/2}$. Compute $S(P_0^n, P_{\theta_n}^n)$ and its limiting value. Compare your result with the upper bound obtained from Theorem 13.1.3.

Problem 13.8 If $I(\theta_0)$ is a positive definite Information matrix, show $h = 0$ if and only if $\langle h, I(\theta_0) h \rangle = 0$.

Problem 13.9 Let $X_1, \ldots, X_n$ be i.i.d. according to a model $\{P_\theta, \theta \in \Omega\}$, where θ is real-valued. Consider testing $\theta = \theta_0$ versus $\theta = \theta_n$ at level α (α fixed, $0 < \alpha < 1$). Show that it is possible to have $nH^2(P_{\theta_0}, P_{\theta_n}) \to c < \infty$ and still have a sequence of level α tests $\phi_n = \phi_n(X_1, \ldots, X_n)$ such that $E_{\theta_n}(\phi_n) \to 1$. *Hint:* Take P_θ uniform on $[0, \theta]$ and $\theta_n = \theta_0 - h/n$ for $h > 0$.

Problem 13.10 Suppose $\|P_n - Q_n\|_1 \to 0$. Show that P_n and Q_n are mutually contiguous. Furthermore, show that, for any sequence of test functions ϕ_n, $\int \phi_n dP_n - \int \phi_n dQ_n \to 0$.

Problem 13.11 For a q.m.d. family, show $nH^2(P_{\theta_0+hn^{-1/2}}, P_{\theta_0+h_n n^{-1/2}}) \to 0$ whenever $h_n \to h$. Then, show $P^n_{\theta_0+h_n n^{-1/2}}$ is contiguous to $P^n_{\theta_0}$ whenever $h_n \to h$.

Problem 13.12 Use Problem 13.11 to show that Theorem 12.2.3 (i) remains valid if h is replaced by h_n as long as h_n falls in a bounded subset of $\mathbb{R}^k$. Also, show part (ii) of Theorem 12.2.3 generalizes if h in the left hand side of the convergence (12.14) is replaced by $h_n \to h$ in $\mathbb{R}^k$. Further assume (12.84) and show that, for any $c > 0$, the supremum over h such that $|h| \le c$ of the left side of (12.13) tends to 0 in probability under θ_0.

Problem 13.13 Use problem 13.11 to prove Theorem 12.4.1 when $h_n \to h$.

Problem 13.14 Give an example where $\|Q_n - P_n\|_1 \to \delta > 0$ but P_n and Q_n are mutually contiguous.

Problem 13.15 Let P_n and Q_n be two sequences of probability measures defined on $(\Omega_n, \mathcal{F}_n)$. Assume they are contiguous. Assume further that both of them are product measures, i.e.

$$P_n = \prod_{i=1}^n P_{n,i} \quad \text{and} \quad Q_n = \prod_{i=1}^n Q_{n,i} \; .$$

Let $\|Q - P\|_1$ denote the total variation distance between P and Q. Show that

$$\sup_n \sum_{i=1}^n \|Q_{n,i} - P_{n,i}\|_1^2 < \infty \; .$$

Problem 13.16 Let $f(x)$ be the triangular density on $[-1, 1]$ defined by

$$f(x) = (1 - |x|)I\{x \in [-1, 1]\} \; .$$

Let P_θ be the distribution with density $f(x - \theta)$. Find the asymptotic behavior of $H(P_{\theta_0}, P_{\theta_0+h})$ as $h \to 0$, where H is the Hellinger distance. Compare your result with q.m.d. families.

Section 13.2

Problem 13.17 Under the assumptions of Theorem 13.2.1, suppose $\theta_k \to \theta_0$ and $\beta > \alpha > 0$. Show, for any $N < \infty$, there does not exist a test ϕ_k with $k \le N$ such that $\liminf_k E_{\theta_k}(\phi_k) \ge \beta$.

Problem 13.18 Under the assumptions of Example 13.2.1, show that the squared efficacy of the Wald test is $I(\theta_0)$.

Problem 13.19 Suppose $\Omega_0 = \{\theta_0\}$. In order to determine $c = c(n, \alpha)$ in (13.32), define $c(n, \alpha)$ to be

$$c(n, \alpha) = \inf\{d : \ P_{\theta_0}\{T_n > d\} \leq \alpha\} \ .$$

Argue that this choice of $c(n, \alpha)$ satisfies (13.32). What if $T_n > d$ is replaced by $T_n \geq d$?

Problem 13.20 For a double exponential location family, calculate the Pitman AREs among pairwise comparisons of the t-test, the Wilcoxon test, and the Sign test.

Problem 13.21 Prove the inequality (13.30). *Hint:* The quantity (13.29) is invariant with respect to scale. By taking $\sigma^2 = 1$, the problem reduces to choosing f to minimize $\int f^2$ subject to f being a mean 0 density with variance 1. Using the method of undetermined multipliers, it is sufficient to minimize

$$\int [f^2(x) + 2b(x^2 - a^2)f(x)]dx \ ,$$

where a and b are chosen so that f is a mean 0 density with variance 1.

Problem 13.22 Suppose $X_1, \ldots, X_n$ are i.i.d. Poisson with unknown mean θ. The problem is to test $\theta = \theta_0$ versus $\theta > \theta_0$. Consider the test that rejects for large $\bar{X}_n$ and the test that rejects for large

$$S_n^2 = \frac{1}{n-1} \sum_{i=1}^{n} (X_i - \bar{X}_n)^2 \ .$$

Compute the Pitman ARE.

Problem 13.23 Suppose $X_1, \ldots, X_n$ are i.i.d. $N(0, \sigma^2)$. Let $T_{n,1} = \bar{Y}_n = n^{-1} \sum_{i=1}^n Y_i$, where $Y_i = X_i^2$. Also, let $T_{n,2} = (2n)^{-1} \sum_{i=1}^n (Y_i - \bar{Y}_n)^2$. For testing $\sigma = 1$ versus $\sigma > 1$, does the Pitman asymptotic relative efficiency of $T_{n,1}$ with respect to $T_{n,2}$ exist? If so, find it.

Section 13.3

Problem 13.24 For testing $\theta = \theta_0$ versus $\theta > \theta_0$, define two test sequences ϕ_n and ψ_n to be *asymptotically equivalent under the null hypothesis* if $\phi_n - \psi_n \to 0$ in probability under θ_0. Does this imply that, if θ_0 is the true value, the probability the tests reach the same conclusion tends to 1? Show that, under q.m.d., asymptotic equivalence under the null hypothesis also implies that, under an alternative sequence $\theta_{n,h} = \theta_0 + hn^{-1/2}$,

$$E_{\theta_{n,h}}(\phi_n) - E_{\theta_{n,h}}(\psi_n) \to 0 \ .$$

Furthermore, assume at least one of the two, say ϕ_n is nonrandomized. Then, conclude the tests are *asymptotically equivalent* in the sense that the probability the tests reach the same conclusion tends to 1, both under θ_0 and a sequence $\theta_{n,h}$.

Problem 13.25 Under the q.m.d. assumptions of this section, show that $\phi_{n,h}$ given by (13.34) and $\tilde{\phi}_n$ given by (13.43) are asymptotically equivalent in the sense of Problem 13.24 for testing θ_0 against $\theta_0 + hn^{-1/2}$.

Problem 13.26 Let $X_1, \ldots, X_n$ be i.i.d. $N(\theta, 1)$. For testing $\theta = 0$ against $\theta > 0$, let ϕ_n be the UMP level α test. Let $\tilde{\phi}_n$ be the test which rejects if $\bar{X}_n \geq b_n/n^{1/2}$ or $\bar{X}_n \leq -a_n/n^{1/2}$, where $b_n = z_{1-\alpha} + n^{-1/4}$ and a_n is then determined to meet the level constraint. Are the tests asymptotically equivalent? Show that, for all $\theta \geq 0$,

$$\frac{1 - E_\theta(\phi_n)}{1 - E_\theta(\tilde{\phi}_n)} \to 0 \quad \text{as } n \to \infty \; .$$

How do you interpret this result? [Lehmann (1949)]

Problem 13.27 Prove Lemma 13.3.1 (iii). *Hint:* Problems 13.12-13.13.

Problem 13.28 Prove Theorem 13.3.1.

Problem 13.29 Prove the equivalence of Definition 13.3.2 and the definition in the statement immediately following Definition 13.3.2. What is an equivalent characterization for LAUMP tests?

Problem 13.30 For testing θ_0 versus θ_n, let ϕ_n^* be a test satisfying

$$\limsup_n E_{\theta_0}(\phi_n^*) = \alpha^* < \alpha$$

and $E_{\theta_n}(\phi_n^*) \to \beta^*$.
(i) Show there exists a test sequence ψ_n satisfying $\limsup_n E_{\theta_0}(\psi_n) = \alpha$ and a number β such that

$$\lim E_{\theta_n}(\psi_n) = \beta \geq \beta^* \; ,$$

and this last inequality is strict unless $\beta^* = 1$.
(ii) Hence, show that, under the conditions of Theorem 13.3.3, any LAUMP level α test sequence ϕ_n^* satisfies $E_{\theta_0}(\phi_n^*) \to \alpha$.

Problem 13.31 Suppose Z_n is any sequence of random variables such that $Var_{\theta_n}(Z_n) \leq 1$ while $E_{\theta_n}(Z_n) \to \infty$. Here, θ_n merely indicates the distribution of Z_n at time n. Show that, under θ_n, $Z_n \to \infty$ in probability.

Problem 13.32 In the double exponential location model of Example 13.3.2, show that a MLE estimator is a sample median $\hat{\theta}_n$. The test that rejects the null hypothesis if $n^{1/2}\hat{\theta}_n > z_{1-\alpha}$ is AUMP and is asymptotically equivalent to Rao's score test in the sense of Problem 13.24.

Problem 13.33 For the Cauchy location model of Example 13.3.3, consider the estimator $\hat{\theta}_n$ defined by (13.59). Show that the test that rejects when $n^{1/2}\hat{\theta}_n > 2^{1/2} z_{1-\alpha}$ is AUMP. Is the estimator location equivariant? Is the estimator $\hat{\theta}_n = \hat{\theta}_n(X_1, \ldots, X_n)$ monotone in the sense it is nondecreasing as any one component X_i increases?

Problem 13.34 Let $X_1, \ldots, X_n$ be i.i.d. according to a q.m.d. location model $f(x-\theta)$. Let $\hat{\theta}_n$ be any location equivariant estimator satisfying (13.58) (such as an efficient likelihood estimator). For testing $\theta \le 0$ against $\theta > 0$, show that the test that rejects when $n^{1/2}\hat{\theta}_n > I^{-1/2}(0) z_{1-\alpha}$ is AUMP.

Problem 13.35 Assume the conditions of Theorem 13.3.3. Assume ϕ_n is LAUMP level α. Suppose the power function of ϕ_n is nondecreasing in θ, for $\theta \ge \theta_0$. Show ϕ_n is also AUMP level α.

Problem 13.36 Assume the conditions of Example 13.3.1. Further assume f is strongly unimodal, i.e., $-\log(f)$ is convex. Show the test $\tilde{\phi}_n$ given by (13.43) is AUMP level α. *Hint:* Use Problem 13.35.

Problem 13.37 Suppose $X_1, \ldots X_n$ are i.i.d. Poisson(λ). Consider testing the null hypothesis $H_0 : \lambda = \lambda_0$ versus the alternative, $H_A : \lambda > \lambda_0$.
(i) Consider the test ϕ_n^1 with rejection region $n^{1/2}[\bar{X}_n - \lambda_0] > z_{1-\alpha}\lambda_0^{1/2}$, where $\Phi(z_\alpha) = \alpha$ and Φ is the cdf of a standard normal random variable. Find the limiting power of this test against $\lambda_0 + hn^{-1/2}$.
(ii) Alternatively, let g be a differentiable, monotone increasing function with $g'(\lambda_0) > 0$, and consider the test ϕ_n^g with rejection region

$$n^{1/2}[g(\bar{X}_n) - g(\lambda_0)] > z_{1-\alpha} g'(\lambda_0)\lambda_0^{1/2} .$$

Show that ϕ_n^1 and ϕ_n^g are equivalent in the sense that, for any $b > 0$,

$$\sup_{0 \le h \le b} E_{\lambda_0 + hn^{-1/2}} |\phi_n^1 - \phi_n^g| \to 0 .$$

(iii) Can we replace b by ∞?

Problem 13.38 Suppose $X_1, \ldots X_n$ are i.i.d. $N(\theta, 1+\theta^2)$. Consider testing $\theta = \theta_0$ versus $\theta > \theta_0$ and let ϕ_n be the test that rejects when $n^{1/2}[\bar{X}_n - \theta_0] > z_{1-\alpha}(1 + \theta_0^2)^{1/2}$.
(i) Compute the limiting power of this test against $\theta_0 + hn^{-1/2}$.
(ii) Is this test AUMP?

Problem 13.39 Define appropriate extensions of the definitions of LAUMP and AUMP to two-sided testing of a real parameter. Let $X_1, \ldots, X_n$ be i.i.d. $N(\theta, 1)$. Show that neither LAUMP nor AUMP tests exist for testing $\theta = 0$ against $\theta \ne 0$.

Section 13.4

Problem 13.40 Suppose $\{Q_{n,h}, h \in \mathbb{R}^k\}$ is asymptotically normal according to Definition 13.4.1, with Z_n and C satisfying (13.62). Show the matrix C is uniquely determined. Moreover, if $\tilde{Z}_n$ is any other sequence also satisfying (13.62), then $Z_n - \tilde{Z}_n \to 0$ in $Q_{n,h}$-probability for any h.

Problem 13.41 Suppose $\{Q_{n,h}, h \in \mathbb{R}^k\}$ is asymptotically normal. Show that Q_{n,h_1} and Q_{n,h_2} are mutually contiguous for any h_1 and h_2.

Problem 13.42 Assume $\{Q_{n,h}, h \in \mathbb{R}^k\}$ is asymptotically normal according to Definition 13.4.1, with Z_n and C satisfying (13.62). Show that, under $Q_{n,h}$, $Z_n \xrightarrow{d} N(Ch, C)$.

Problem 13.43 Let $dN(h, C)$ denote the density of the normal distribution with mean vector $h \in \mathbb{R}^k$ and positive definite covariance matrix C. Prove that $\exp(\langle h, x\rangle - \frac{1}{2}\langle h, Ch\rangle)dN(0, C)(x)$ is the density of $N(Ch, C)$ evaluated at x. *Hint:* Use characteristic functions.

Section 13.5

Problem 13.44 In the location scale model of Example 13.5.2, verify the expressions for the Information matrix. Deduce that the matrix is diagonal if f is an even function.

Problem 13.45 For the location scale model of Example 13.5.2 with $f(x) = C(\beta)\exp[-|x|^\beta]$, argue that the family is q.m.d. if $\beta > 1/2$.

Problem 13.46 For the location scale model in Problem 13.45, show that, for testing $\mu \leq 0$ versus $\mu > 0$, argue that the Wald test is LAUMP if $\beta \geq 1$. If $\hat{\sigma}_n$ is replaced by any consistent estimator of σ, does the LAUMP property continue to hold? If $1/2 < \beta < 1$, argue that the Rao test is LAUMP.

Problem 13.47 In Example 13.5.3, for testing $\rho \leq 0$ versus $\rho > 0$, find the optimal limiting power of the LAUMP against alternatives $hn^{-1/2}$. Compare with the case where the means and variances are known. Generalize to the case of testing $\rho \leq \rho_0$ against $\rho > \rho_0$.

Problem 13.48 Derive the inequality (13.74) under general conditions which assume the model is asymptotically normal.

Problem 13.49 Assume (13.75) and the setup described there. Show that the test that rejects when $g(\hat{\theta}_n) > z_{1-\alpha}\hat{\sigma}_n$ is pointwise level α and has a power function such that there is equality in (13.74).

Problem 13.50 Verify (13.76) as well as the form of the matrix $C(\theta_0)$.

Problem 13.51 Assume the conditions of Theorem 13.5.1, Consider the problem of testing $g(\theta) = 0$ against $g(\theta) \neq 0$. Restrict attention to tests ϕ_n that are asymptotically unbiased in the sense

$$\liminf_n \inf_{\{\theta:\ g(\theta)\neq 0\}} E_\theta(\phi_n) \geq \alpha ,$$

as well as (13.69). Prove a result analogous to Theorem 13.5.1. *Hint:* See Problem 5.10.

Problem 13.52 Consider the one-sample $N(\mu, 1)$ problem for testing $|\mu| \geq \Delta$ versus $|\mu| < \Delta$. Show that the level α test based on combining the two one-sided UMP level α tests has size strictly less than α.

Problem 13.53 Show that the size of the TOST test considered in Example 13.5.5 is α.

Problem 13.54 Let $C = C(\alpha, \delta, \sigma)$ be defined by (13.86). Show that $C > \delta - \sigma z_{1-\alpha}$. Use this to show that, in Example 13.5.6, the limiting power of ϕ_n^* always exceeds that of ϕ_n^{IUT}.

Problem 13.55 As in Example 13.5.7, consider testing $\theta = \theta_0$ versus $\theta \neq \theta_0$. Suppose ϕ_n is asymptotically level α and asymptotically unbiased in the sense

$$\liminf_n E_{\theta_0 + hn^{-1/2}}(\phi_n) \geq \alpha$$

for any $h \neq 0$. Argue that, among such tests ϕ_n, the two-sided Rao test $\phi_{n,2}$ is LAUMP.

Problem 13.56 Generalize Example 13.5.7 to the case of testing $\theta = \theta_0$ versus $\theta \neq \theta_0$ in the presence of nuisance parameters.

Problem 13.57 Under the conditions of Theorem 13.5.5 used to prove an asymptotic maximin result for Rao's test, derive analogous optimality results for both the Wald and likelihood ratio tests.

Section 13.6

Problem 13.58 Show that the family of densities (13.95) satisfies (11.77) for small enough θ.

Problem 13.59 Verify (13.101).

Problem 13.60 Compare the bounds (13.101) and (13.104). For what u is each attainable? Why is (13.101) generally not attainable for all u, even though there exists a test for the submodel $\{P_{u,t}\}$ for which the bound is attainable.

Problem 13.61 In Example 13.6.2, argue that the given test attains the optimal limiting power uniformly in h, for $0 \leq h \leq c$ and any $c > 0$.

Problem 13.62 In Theorem 13.6.1, compute the limiting power against $P_{u,hn^{-1/2}}$ where h is chosen so that $n^{1/2}\theta(P_{u,hn^{-1/2}}) \to \delta$. [The solution does not depend on u but only on the value of δ, which was noted by Pfanzagl and Wefelmeyer (1985).]

Problem 13.63 Provide the details for the optimality claimed in Example 13.6.3 for testing the variance in a nonparametric setting.

Problem 13.64 Let **P** be the set of all joint distributions in $\mathbb{R}^2$ on some compact set. Let $\theta(P)$ denote the correlation functional. For testing $\theta(P) \leq 0$, construct an asymptotically optimal test in a nonparametric setting.

Problem 13.65 Consider testing the difference of two population means $\mu(P_X) - \mu(P_Y) \leq 0$ in a nonparametric setting. Generalize Theorem 13.6.1 to obtain locally AUMP tests.

13.8 Notes

The Hellinger distance introduced in Section 13.1 was fundamental in Kakutani (1948) and does not seem to have been employed by Hellinger (Le Cam and Yang (2000), p. 48). The use of Hellinger distance to construct estimators and tests is developed in Beran (1977) and Simpson (1989).

The concept of Pitman asymptotic relative efficiency can be traced to an unpublished set of his lecture notes in (1949); Noether (1955) published a slightly more general result. The inequality (13.30) is due to Hodges and Lehmann (1956). Further results and references can be found in Serfling (1980) and Nikitin (1995). Some important alternative concepts of efficiency can be found in Bahadur (1960, 1965), Kallenberg (1982, 1983), and Inglot, Kallenberg and Ledwina (2000). Some numerical calculations are given in Groeneboom and Oosterhoff (1981). Higher order asymptotic comparisons can be approached through the concept of *deficiency*, introduced in Hodges and Lehmann (1970). Some general results for rank and permutation tests in the one-sample problem are obtained in Albers, Bickel and van Zwet (1976); analogous results for the two-sample problem are obtained in Bickel and van Zwet (1978). Pitman efficiencies of multivariate spatial sign and rank tests are considered in Peters and Randles (1991) and Möttönen, Oja and Tienari (1997). Asymptotic efficiency of rank tests is studied in Behnen and Neuhaus (1989) and Hájek, Sidák, and Sen (1999). Higher order efficiency is also considered in Bening (2000).

Our approach to large sample efficiency of tests is largely due to ideas in Wald (1939, 1941ab, 1943), though his assumptions were too strong. He focused on MLEs and the tests now known as Wald tests. Wald basically argued that one could construct optimal large sample tests based on the normal approximation to the MLE. A more formal approach was later provided by Le Cam's (1964, 1972) elegant notion of convergence of experiments, of which convergence to a normal experiment in the sense of Definition 13.4.1 is an important special case. This approach was used in Choi, Hall and Schick (1996). For references of (local) asymptotically normal experiments in time series models, see Hallin et al. (1999). Generalizations to limiting Poisson experiment and locally asymptotically quadratic experiments are discussed in Le Cam and Yang (2000). Roussas (1972) formulated and developed the concept of AUMP tests. The proof of Theorem 13.4.1 is based on Lemma 3.4.4 of Rieder (1994). The results in Section 13.5.2 are obtained in Romano (2005). Nonparametric tests of equivalence are studied in Janssen (2000b); also see Wellek (2003). The reduction of a nonparametric problem to a parametric one through the use of a least favorable family is due to Stein (1956b), and is prominent in the work of Koshevnik and Levit (1976), Pfanzagl (1982, 1985), Bickel et al (1993) and Janssen (1999), among others. The proof of Theorem 13.6.1 is based on the more general result Theorem 25.44 of van der Vaardt (1998). Efficiency of nonparametric confidence intervals is discussed in Low (1997) and Romano and Wolf (2000).

14

Testing Goodness of Fit

14.1 Introduction

So far, the principal framework of this book has been optimality (either exact or asymptotic) in situations where both the hypothesis and the class of alternatives were specified by parametric models. In the present chapter, we shall take up the crucial problem of testing the validity of such models, the hypothesis of *goodness of fit*. For example, we would like to know whether a set of measurements $X_1, \ldots, X_n$ is consonant with the assumption that the X's are an i.i.d. sample from a normal distribution.

A difficulty in testing such a hypothesis is that the class of alternatives typically is enormously large and can no longer be described by a parametric model. As a result, although some asymptotic optimality results are presented, they are isolated; no general asymptotic optimality theory seems to exist for this problem. In fact, there is growing evidence, such as the results of Janssen (2000a) (see Theorem 14.6.2), that any test can achieve high asymptotic power against local or contiguous alternatives for at most a finite-dimensional parametric family.

Because of the importance of the problem of testing goodness of fit, we shall nevertheless consider this problem here. However, the focus will no longer be on optimality. Instead, we shall present some of the principal methods that have been proposed and study their relative strengths and weaknesses.

For the sake of simplifying a very complicated problem we shall consider the case where $X_1, \ldots, X_n$ are i.i.d. according to some probability distribution P, and shall mostly assume that the null hypothesis $P = P_0$ completely specifies the distribution. While this assumption frequently is not fulfilled in applications, it makes it possible to cover some principal features of the problem which carry over to the more complex case of composite hypotheses.

In the case where the observations are real-valued, we index the unknown distribution by the underlying c.d.f. F and the problem is to test $F = F_0$. We will typically consider the case where F_0 is the uniform distribution on $(0, 1)$. This special case can be generalized to the problem of testing the simple null hypothesis H that $X_1, \ldots, X_n$ are i.i.d. from any fixed continuous c.d.f. F on the real line. To see how, define $Y_i = F(X_i)$, so that the Y_i are i.i.d. U(0,1) under H (Problem 3.22); then, test the hypothesis that $Y_1, \ldots, Y_n$ are i.i.d. uniform on $[0, 1]$.

Let $\hat{F}_n$ be the empirical c.d.f., which uniformly tends to F with probability one, by the Glivenko-Cantelli theorem. For testing the simple null hypothesis $F = F_0$, a natural starting point is to base a test statistic on some measure of discrepancy between $\hat{F}_n$ and F_0. In particular, if d is any metric on the space of distribution functions, then $d(\hat{F}_n, F_0)$ could serve as a test statistic. A classical choice is $d = d_K$, the Kolmogorov-Smirnov metric, which historically was the first test of goodness of fit that is (pointwise) consistent against any alternative. This test is studied in Section 14.2, but many other choices are possible; see 14.2.2. Two such choices are the Cramér-von Mises statistic and the Anderson-Darling statistic; in fact, these choices are often much more powerful than the Kolmogorov-Smirnov test.

In Section 14.3, the classical Chi-squared test is studied, and its asymptotic properties are derived. The class of Neyman smooth tests is considered in Section 14.4; it includes the Chi-squared test as a special case, and serves to motivate the class of weighted quadratic test statistics studied in Section 14.5. The difficulty of constructing goodness of fit tests with good power against broad alternatives is studied in Section 14.6.

14.2 The Kolmogorov-Smirnov Test

14.2.1 Simple Null Hypothesis

Suppose $X_1, \ldots, X_n$ are i.i.d. real-valued observations with c.d.f. F, and consider the problem of testing the simple null hypothesis that $F = F_0$ versus $F \neq F_0$. The classical Kolmogorov-Smirnov goodness of fit test statistic, introduced in Section 6.13 and Example 11.2.12, is

$$T_n \equiv \sup_{t \in \mathbb{R}} n^{1/2} |\hat{F}_n(t) - F_0(t)| = n^{1/2} d_K(\hat{F}_n, F_0) , \tag{14.1}$$

where d_K is the Kolmogorov-Smirnov distance

$$d_K(F, G) = \sup_t |F(t) - G(t)| .$$

Note that $d_K(F, G) = 0$ if and only if $F = G$.

The distribution of T_n under F_0 is the same for all continuous F_0 (Problem 11.57). Let $s_{n,1-\alpha}$ be the $1 - \alpha$ quantile of the distribution of T_n under any continuous F_0. The Kolmogorov-Smirnov test rejects the null hypothesis if $T_n > s_{n,1-\alpha}$. If F_0 is not continuous, using $s_{n,1-\alpha}$ results in a test that has level less than α (Problem 11.58), but in principle, one can determine (or simulate) a critical value that yields an exact level α test for this situation. Much of the remaining discussion in the section will focus on the case where the critical value $s_{n,1-\alpha}$ is used (but the arguments apply more generally). For references to tables of critical values and finite sample power calculations, see the references given in Example 11.2.12.

In order to study the limiting behavior of T_n, introduce the function

$$B_n(t) = n^{1/2}[\hat{F}_n(t) - F_0(t)] . \tag{14.2}$$

For each t, $B_n(t)$ is a real-valued random variable; in addition, $B_n(\cdot)$ can be viewed as a random function (or process) on $[0, 1]$, called the *empirical process.* By the multivariate Central Limit Theorem, if the null hypothesis is true, then for any $t_1, \ldots, t_k$,

$$[B_n(t_1), \ldots, B_n(t_k)] \stackrel{d}{\to} [B(t_1), \ldots, B(t_k)] , \tag{14.3}$$

where $[B(t_1), \ldots, B(t_k)]$ has the multivariate normal distribution with mean 0 and covariance matrix Σ, whose (i, j)th entry $\sigma_{i,j}$ is given by

$$\sigma_{i,j} = \begin{cases} F_0(t_i)(1 - F_0(t_i)) & \text{if } i = j \\ F_0(\min(t_i, t_j)) - F_0(t_i)F_0(t_j) & \text{otherwise.} \end{cases} \tag{14.4}$$

By the Continuous Mapping Theorem, it follows that, for any $t_1, \ldots, t_k$,

$$\max_{1,\ldots,k} n^{1/2}|\hat{F}_n(t_i) - F_0(t_i)| \stackrel{d}{\to} \max_{1,\ldots,k} |B(t_i)| . \tag{14.5}$$

In fact, $B(\cdot)$ itself can be represented as a random continuous process on $[0, 1]$, called the Brownian Bridge process. The study of random functions and empirical processes is beyond the scope of this book, but it is developed in Pollard (1984) and van der Vaart and Wellner (1996). However, the result (14.5) provides both insight and a basis for a rigorous treatment of the limiting behavior of T_n, which is the supremum over all t, and not just a finite set, of $|B_n(t)|$. It turns out that T_n has a limiting distribution which is continuous and strictly increasing on $(0, \infty)$. More specifically, Kolmogorov (1933) showed that if F_0 is continuous, then for any $d > 0$,

$$P\{T_n > d\} \to 2\sum_{k=1}^{\infty}(-1)^{k+1} \exp(-2k^2 d^2) .$$

The $1 - \alpha$ quantile of this distribution will be denoted by $s_{1-\alpha}$.

We now discuss some power properties of the Kolmogorov-Smirnov test.

Theorem 14.2.1 *The Kolmogorov-Smirnov test is pointwise consistent in power against any fixed $F \neq F_0$; that is,*

$$P_F\{T_n > s_{n,1-\alpha}\} \to 1$$

as $n \to \infty$.

PROOF. By the Glivenko-Cantelli theorem, under an alternative F,

$$\sup_t |\hat{F}_n(t) - F_0(t)| \to d_K(F, F_0) > 0$$

almost surely, and so $T_n \to \infty$ almost surely. Hence, by Slutsky's theorem,

$$P_F\{T_n > s_{n,1-\alpha}\} \to 1 ,$$

since $s_{n,1-\alpha} \to s_{1-\alpha} < \infty$. ■

For an alternative instructive proof of consistency (due to Massey (1950)), fix any F with $d_K(F, F_0) > 0$. Then, there exists some t with $F(t) \neq F_0(t)$. First, assume $F(t) > F_0(t)$. Then,

$$P_F\{T_n > s_{n,1-\alpha}\} \geq P_F\{\left|n^{1/2}[\hat{F}_n(t) - F_0(t)]\right| > s_{n,1-\alpha}\}$$

$$\geq P_F\{n^{1/2}[\hat{F}_n(t) - F(t)] \geq s_{n,1-\alpha} - n^{1/2}[F(t) - F_0(t)]\} , \tag{14.6}$$

which tends to 1 as $n \to \infty$ since the left side in the probability expression is bounded in probability while the right hand tends to $-\infty$. Hence, the limiting power is 1 against any F if there exists a t with $F(t) > F_0(t)$. By similar reasoning, the limiting power is 1 against F with $F(t) < F_0(t)$ for some t, and hence for any $F \neq F_0$.

We now show that the Kolmogorov-Smirnov test is uniformly consistent in power against alternatives F satisfying $n^{1/2} d_K(F, F_0) \geq \Delta_n$, as long as $\Delta_n \to \infty$.

Theorem 14.2.2 *Let $X_1, \ldots, X_n$ be i.i.d. random variables with c.d.f. F. For testing $F = F_0$ against $F \neq F_0$, the power of the Kolmogorov-Smirnov test tends to one uniformly over all alternatives F satisfying $n^{1/2} d_k(F, F_0) \geq \Delta_n$ if $\Delta_n \to \infty$ as $n \to \infty$; that is,*

$$\inf\left\{P_F\{T_n > s_{n,1-\alpha}\} : \ n^{1/2} d_K(F, F_0) \geq \Delta_n\right\} \to 1$$

if $\Delta_n \to \infty$.

PROOF. Let F_n be any sequence satisfying $n^{1/2} d_K(F_n, F_0) \geq \Delta_n$. By the triangle inequality,

$$d_K(F_n, F_0) \leq d_K(F_n, \hat{F}_n) + d_K(\hat{F}_n, F_0) ,$$

which implies

$$T_n \geq \Delta_n - n^{1/2} d_K(\hat{F}_n, F_n) .$$

Therefore,

$$P_{F_n}\{T_n > s_{n,1-\alpha}\} \geq P_{F_n}\{n^{1/2} d_K(\hat{F}_n, F_n) \leq \Delta_n - s_{n,1-\alpha}\} . \tag{14.7}$$

But, by Problem 11.60, under F_n, $n^{1/2} d_K(\hat{F}_n, F_n)$ is tight. Since $\Delta_n \to \infty$ and $s_{n,1-\alpha}$ has a finite limit, it follows that $\Delta_n - s_{n,1-\alpha} \to \infty$ and therefore

$$P_{F_n}\{T_n > s_{n,1-\alpha}\} \to 1 . \ ■$$

One can also obtain nonasymptotic lower bounds to the power of the Kolmogorov-Smirnov test by using (14.7). For example, application of the Dvoretzky Kiefer Wolfowitz inequality (Theorem 11.2.18) yields

$$P_{F_n}\{T_n > s_{n,1-\alpha}\} \geq 1 - 2\exp[-2(\Delta_n - s_{n,1-\alpha})^2] , \tag{14.8}$$

if $n^{1/2} d_K(F_n, F_0) \geq \Delta_n$ and $\Delta_n > s_{n,1-\alpha}$ (Problem 14.2).

It follows from Theorem 14.2.2 that the Kolmogorov-Smirnov test is uniformly consistent in power against alternatives F such that $d_K(F, F_0) \geq \Delta$, for any fixed $\Delta > 0$. Note, however, that for any fixed n and Δ, the rejection probability may be less than α; that is, the Kolmogorov-Smirnov test is biased, as shown by Massey (1950).

It also follows from Theorem 14.2.2 that the limiting power of the Kolmogorov-Smirnov test against a sequence of alternatives F_n is arbitrarily close to one for sequences F_n tending to F_0 sufficiently slowly. In the opposite direction, by the triangle inequality,

$$P_F\{T_n > s_{n,1-\alpha}\} \le P_F\{n^{1/2} d_K(\hat{F}_n, F) + n^{1/2} d_K(F, F_0) > s_{n,1-\alpha}\} , \quad (14.9)$$

which implies the power of the Kolmogorov-Smirnov test is poor against sequences of alternatives tending to F_0 sufficiently fast (Problem 14.4). More specifically, the following holds.

Theorem 14.2.3 *For testing $F = F_0$ at level α, the limiting power of the Kolmogorov-Smirnov test is no better than α against any sequence of alternatives F_n satisfying $n^{1/2} d_K(F_n, F_0) \to 0$; that is,*

$$\limsup_n P_{F_n}\{T_n > s_{n,1-\alpha}\} \le \alpha .$$

Thus, the Kolmogorov Smirnov test cannot distinguish sequences that are at a distance $o(n^{-1/2})$ from F_0, where distance refers to the metric d_K. In fact, *no* test can have good power against all sequences F_n satisfying $n^{1/2} d_K(F_n, F_0) \to 0$. To prove this statement, consider a smooth parametric model containing F_0, such as a one-parameter exponential family having density of the form

$$\exp(\theta T(x) - A(\theta)) dF_0(x) .$$

Let F_n denote the c.d.f. corresponding to this density with $\theta = h_n n^{-1/2}$. Note that $d_K(F_n, F_0) = O(h_n n^{-1/2})$ (Problem 14.5). Then, the AMP test sequence for testing $\theta = 0$ (corresponding to F_0) against $\theta_n = h_n n^{-1/2}$ has limiting power α if $h_n \to 0$.

One can also obtain an upper bound to the power against alternatives F_n satisfying

$$n^{1/2} d_K(F_n, F_0) \to \delta < s_{1-\alpha} .$$

By (14.9),

$$P_F\{T_n > s_{n,1-\alpha}\} \le P_F\{d_K(\hat{F}_n, F) > n^{-1/2} s_{n,1-\alpha} - d_K(F, F_0)\} .$$

Then, by the Dvoretzky, Kiefer and Wolfowitz Inequality (Theorem 11.2.18), the last expression is bounded above by

$$2 \exp\{-2n[s_{n,1-\alpha} n^{-1/2} - d_K(F, F_0)]^2\} .$$

Therefore, if F_n is a sequence satisfying

$$n^{1/2} d_K(F_n, F_0) \to \delta < s_{1-\alpha} ,$$

then the limiting power against F_n is bounded above by

$$2 \exp[-2(s_{1-\alpha} - \delta)^2] .$$

So far, we have obtained crude upper and lower bounds to the power of the Kolmogorov-Smirnov test, and it follows from Theorems 14.2.2 and 14.2.3

that, like the parametric situations considered earlier, it is against sequences of alternatives F_n with

$$n^{1/2} d_K(F_n, F_0) \to \delta \quad (0 < \delta < \infty)$$

that we expect the power of the test to tend to limits strictly between α and 1. Let us now sketch an approach to calculating the exact limiting power against a local sequence of alternatives F_n. Consider the normalized difference

$$d_n(t) = n^{1/2}[F_n(t) - F_0(t)] ,$$

and assume that for some function d

$$\sup_t |d_n(t) - d(t)| \to 0 .$$

Note the basic identity

$$n^{1/2}[\hat{F}_n(t) - F_0(t)] = n^{1/2}[\hat{F}_n(t) - F_n(t)] + d_n(t) . \tag{14.10}$$

Under F_n, $n^{1/2}[\hat{F}_n(t) - F_n(t)]$ has mean 0 and variance

$$F_n(t)[1 - F_n(t)] \to F_0(t)[1 - F_0(t)] .$$

For fixed t, the Lindeberg Central Limit Theorem (see Problem 11.13) implies that, under F_n,

$$n^{1/2}[\hat{F}_n(t) - F_n(t)] \xrightarrow{d} B(t) ,$$

where $B(t)$ has the same limiting normal distribution $N(0, F_0(t)[1 - F_0(t)])$ that arose when studying the limiting behavior (14.3) of the empirical process $B_n(t)$ (defined in 14.2) under F_0. Hence, under F_n, (14.10) implies that

$$n^{1/2}[\hat{F}_n(t) - F_0(t)] \xrightarrow{d} B(t) + d(t) \sim N\left(d(t), F_0(t)[1 - F_0(t)]\right) .$$

Similarly, for any fixed $t_1, \ldots, t_k$, under F_n,

$$n^{1/2}[\hat{F}_n(t_1) - F_0(t_1), \ldots, \hat{F}_n(t_k) - F_0(t_k)] \xrightarrow{d} [B(t_1) + d(t_1), \ldots, B(t_k) + d(t_k)] .$$

By the Continuous Mapping Theorem, it then follows that, under F_n

$$\max_{1,\ldots,k} n^{1/2}|\hat{F}_n(t_i) - F_0(t_i)| \xrightarrow{d} \max_{1,\ldots,k} |B(t_i) + d(t_i)| . \tag{14.11}$$

This result suggests that, under F_n,

$$\sup_t n^{1/2}|\hat{F}_n(t) - F_0(t)| \xrightarrow{d} \sup_t |B(t) + d(t)| ,$$

where $B(t)$ is the Brownian Bridge process which was introduced at the beginning of this section. This suggested result does in fact hold, and so the limiting power of the Kolmogorov-Smirnov test against F_n can be expressed as

$$P\{\sup_t |B(t) + d(t)| > s_{1-\alpha}\} . \tag{14.12}$$

The evaluation of this expression involves so-called general boundary-crossing probabilities and is beyond the present treatment; see Siegmund (1986) and the references given in Shorack and Wellner (1986), Section 4.2. Approximations to this limiting power are also obtained in Hájek, Sidák and Sen (1999), Section 7.4.

The results in this section show that the limiting power of the Kolmogorov-Smirnov test against alternatives F_n satisfying $n^{1/2} d_K(F_n, F_0) \to \delta$ is 0 or 1 unless δ is finite and positive. Moreover, the result (14.12) can be used to show that typically, the limiting power is strictly between α and 1. Surprisingly, and in distinction to the typical parametric situation, the limiting power can be α or 1 against a sequence of alternatives F_n satisfying $n^{1/2} d_K(F_n, F_0) \to \delta$ even if $0 < \delta < \infty$; for a construction, see Problem 14.6.

14.2.2 Extensions of the Kolmogorov-Smirnov Test

The basis of the Kolmogorov-Smirnov test is a measure of discrepancy between the hypothesized distribution function F_0 and the empirical (cumulative) distribution function $\hat{F}_n$. Any such statistic is called an EDF statistic. In particular, if d is a metric on the space of distribution functions, any statistic of the form $d(\hat{F}_n, F_0)$ is an EDF statistic. with the choice $d = d_K$ corresponding to the Kolmogorov-Smirnov statistic.

A second class of EDF statistics is given by the Cramér-von Mises family of statistics

$$V_n = n \int_{-\infty}^{\infty} [\hat{F}_n(x) - F_0(x)]^2 \psi(x) dF_0(x) .$$

Taking $\psi(x) = 1$ yields the Cramér-von Mises statistic, while

$$\psi(x) = \{F_0(x)[1 - F_0(x)]\}^{-1}$$

yields the Anderson-Darling statistic. Both choices will be studied in Section 14.5.

Tests based on EDF statistics can be used to test composite null hypothesis. For example, suppose it is desired to test whether the underlying c.d.f. is F_θ for some θ lying in a parameter space Θ_0, and that $\hat{\theta}_n$ is some reasonable estimator of θ. Then, an EDF test statistic is defined by some measure of discrepancy between $\hat{F}_n$ and $F_{\hat{\theta}_n}$. For example, for testing normality with unspecified mean μ and variance σ^2, a Kolmogorov-Smirnov test statistic is given by

$$\sup_x |\hat{F}_n(x) - \Phi\left(\frac{x - \bar{X}_n}{\hat{\sigma}_n}\right)| , \tag{14.13}$$

where $\Phi(\cdot)$ is the standard normal c.d.f. and $(\bar{X}_n, \hat{\sigma}_n)$ is the MLE for (μ, σ) under the normal model. It is easy to see that, under the null hypothesis, the distribution of (14.13) does not depend on (μ, σ) (Problem 14.9), and critical values can be approximated by simulation. Many other tests have been proposed to test for normality; see D'Agostino and Stephens (1986).

Unfortunately, for testing general parametric submodels indexed by θ, the asymptotic null distribution of an EDF statistic with estimated parameters depends on θ, which limits their use. For discussion and references to the literature of this problem, see D'Agostino and Stephens (1986) and De Wet and Randles (1987). An alternative approach based on the bootstrap is given in Beran (1986) and Romano (1988); see Example 15.6.5.

EDF tests can be extended to the case where the observations are not real-valued. Suppose $X_1, \ldots, X_n$ are i.i.d. P (on some arbitrary space). The natural extension of the empirical c.d.f. is the *empirical measure*, defined by

$$\hat{P}_n(E) = \frac{1}{n} \sum_{i=1}^{n} I\{X_i \in E\} .$$

Then, EDF test statistics can be constructed by some measure of discrepancy between $\hat{P}_n$ and a hypothesized P_0 (or $P_{\hat{\theta}_n}$ in the composite null hypothesis case). See Shorack and Wellner (1986), who also discuss the two-sample problem of comparing two samples by a measure of discrepancy between the empirical c.d.f.s of the samples.

14.3 Pearson's Chi-squared Statistic

14.3.1 Simple Null Hypothesis

In this section, we return to the simple goodness of fit problem for categorical data that was briefly considered in Example 12.4.6. As before, we are dealing with a sequence of n independent trials, each resulting in one of $k+1$ possible outcomes named $1, \ldots, k+1$. The jth outcome occurs with probability p_j on any given trial, so that $\sum_{j=1}^{k+1} p_j = 1$. Let Y_j be the number of trials resulting in outcome j. The joint distribution of $(Y_1, \ldots, Y_{k+1})$ is the multinomial distribution

$$P\{Y_1 = y_1, \ldots, Y_{k+1} = y_{k+1}\} = \frac{n!}{y_1! \cdots y_{k+1}!} p_1^{y_1} \cdots p_{k+1}^{y_{k+1}} , \tag{14.14}$$

with $\sum_{j=1}^{k+1} y_j = n$. The parameter space Ω is

$$\Omega = \{(p_1, \ldots, p_k) \in \mathbb{R}^k \; : \; p_i \geq 0, \; \sum_{j=1}^{k} p_j \leq 1\} \tag{14.15}$$

since $p_{k+1} = 1 - \sum_{j=1}^{k} p_j$.

Consider testing the simple null hypothesis $p_j = \pi_j$ for $j = 1, \ldots, k+1$ against the alternatives $p_j \neq \pi_j$ for some j. It will be assumed that $\pi_1, \ldots, \pi_k$ is an interior point of Ω.

A standard test, proposed by Pearson (1900), rejects for large values of Pearson's Chi-squared statistic, given by

$$Q_n = \sum_{j=1}^{k+1} \frac{(Y_j - n\pi_j)^2}{n\pi_j} . \tag{14.16}$$

This test was already introduced in Example 12.4.6 as an approximation to the likelihood ratio test, and it was shown that the limiting null distribution of Q_n as $n \to \infty$ is the Chi-squared distribution with k degrees of freedom. Below, we will give a direct argument of this result in Theorem 14.3.1. Thus, if $c_{k,1-\alpha}$ is the $1-\alpha$ quantile of χ_k^2, then the test that rejects when $Q_n > c_{k,1-\alpha}$ is asymptotically level α. The accuracy of the Chi-squared approximation to the exact null distribution of the test statistic is discussed for example by Radlow and Alf (1975); for more accurate approximations in this and related problems, see McCullagh (1985, 1986) and the literature cited there.

Consider next a fixed alternative

$$(p_1, \ldots, p_{k+1}) \neq (\pi_1, \ldots, \pi_{k+1}) .$$

If, for some j, $p_j \neq \pi_j$, then

$$Q_n \geq n(\frac{Y_j}{n} - \pi_j)^2 \xrightarrow{P} \infty$$

since $Y_j/n \xrightarrow{P} p_j$, by the law of large numbers. Hence, the power against such an alternative tends to one.

As in Example 11.2.5, a more discriminating result is obtained by considering local alternatives $p_j^{(n)}$ of the form

$$p_j^{(n)} = \pi_j + n^{-1/2} h_j ,$$

where $\sum_{j=1}^{k+1} h_j = 0$. We shall now show that, against such an alternative sequence, the limiting power is nondegenerate.

Theorem 14.3.1 *Assume the above multinomial setup.*
(i) Under the null hypothesis H: $p_j = \pi_j$ *for* $j = 1, \ldots, k+1$, $Q_n \xrightarrow{d} \chi_k^2$, *the Chi-squared distribution with* k *degrees of freedom.*
(ii) Under the alternative hypothesis (sequence) K: $p_j^{(n)} = \pi_j + n^{-1/2} h_j$ *where* $\sum_{j=1}^{k+1} h_j = 0$, $Q_n \xrightarrow{d} \chi_k^2(\lambda)$, *the noncentral Chi-squared distribution with* k *degrees of freedom and noncentrality parameter*

$$\lambda = \sum_{j=1}^{k+1} \frac{h_j^2}{\pi_j} . \tag{14.17}$$

(iii) The power of the χ^2 *test based on* Q_n *against the alternatives in (ii) with not all the* h_j *equal to 0 tends to a limit strictly greater than* α *and less than 1. This holds if the test is carried out using an exact level* α *critical value, or any critical value sequence tending to* $c_{k,1-\alpha}$ *in probability (such as* $c_{k,1-\alpha}$ *itself).*

PROOF. The proof of (i) is an application of the multivariate CLT followed by the continuous mapping theorem. Let V_n be the $k \times 1$ vector defined by

$$V_n^T = n^{1/2}(\frac{Y_1}{n} - \pi_1, \ldots, \frac{Y_k}{n} - \pi_k) . \tag{14.18}$$

By the multivariate CLT, $V_n \xrightarrow{d} N(0, \Sigma)$, where the $k \times k$ covariance matrix Σ has (i,j) entry (Problem 14.12 (i))

$$\sigma_{i,j} = \begin{cases} \pi_i(1-\pi_i) & \text{if } j = i \\ -\pi_i \pi_j & \text{otherwise.} \end{cases} \tag{14.19}$$

It can be checked that Σ has inverse $\Sigma^{-1} = A$, where A has (i,j) entry given by (Problem 14.12 (ii))

$$a_{i,j} = \begin{cases} \frac{1}{\pi_i} + \frac{1}{\pi_{k+1}} & \text{if } j = i \\ \frac{1}{\pi_{k+1}} & \text{otherwise.} \end{cases} \tag{14.20}$$

Hence, $A^{1/2} V_n \xrightarrow{d} N(0, I_k)$, where I_k is the $k \times k$ identity matrix. By the Continuous Mapping Theorem 11.2.13,

$$(A^{1/2} V_n)^T (A^{1/2} V_n) \xrightarrow{d} \chi_k^2 .$$

But, the left hand side is $V_n^T A V_n$, which in turn is equal to

$$n \sum_{j=1}^{k} \frac{1}{\pi_j} (\frac{Y_j}{n} - \pi_j)^2 + \frac{n}{\pi_{k+1}} \sum_{i=1}^{k} \sum_{j=1}^{k} (\frac{Y_i}{n} - \pi_i)(\frac{Y_j}{n} - \pi_j) .$$

The last term reduces to

$$n[\sum_{j=1}^{k} (\frac{Y_j}{n} - \pi_j)]^2 / \pi_{k+1} = n(\frac{Y_{k+1}}{n} - \pi_{k+1})^2 / \pi_{k+1} ,$$

where, in the last equality, we have used $\sum_{j=1}^{k} Y_j = n - Y_{k+1}$ and $\sum_{j=1}^{k} \pi_j = 1 - \pi_{k+1}$. Thus, $V_n^T A V_n = Q_n$.

The proof of (ii) is similar. First, note that

$$V_n^T = n^{1/2} (\frac{Y_1}{n} - p_1^{(n)}, \ldots, \frac{Y_k}{n} - p_k^{(n)}) + (h_1, \ldots, h_k) .$$

It follows from the Cramér-Wold device and the Berry-Esseen Theorem (Problem 14.13) that, under the alternative sequence,

$$V_n \stackrel{d}{\to} N(h, \Sigma) . \tag{14.21}$$

Therefore,

$$A^{1/2} V_n \stackrel{d}{\to} N(A^{1/2} h, I_k)$$

and so

$$(A^{1/2} V_n)^T (A^{1/2} V_n) \stackrel{d}{\to} \chi_k^2(\lambda) ,$$

where

$$\lambda = (A^{1/2} h)^T (A^{1/2} h) = h^T A h ;$$

simple algebra shows that $h^T A h$ agrees with the expression (14.17) for λ and the proof of (ii) follows.

The proof of (iii) is left as an exercise (Problem 14.15). ∎

We are now in a position to prove an optimality result for Pearson's Chi-squared test in the multinomial goodness of fit problem. The problem is to test the null hypothesis $p = \pi$, where π is the vector with jth component π_j. The goal is to show Pearson's Chi-squared test is asymptotically maximin over an appropriate (shrinking) set of alternatives p which tend to π at rate $n^{-1/2}$. First, note that the Information matrix $I(p)$ with (i, j) entry $a_{i,j}$ is given by

$$a_{i,j} = \begin{cases} \frac{1}{p_i} + \frac{1}{p_{k+1}} & \text{if } j = i \\ \frac{1}{p_{k+1}} & \text{otherwise.} \end{cases} \tag{14.22}$$

(Problem 14.14). Let $h^T = (h_1, \ldots, h_k)$ and set $h_{k+1} = -\sum_{i=1}^{k} h_i$ so that $\sum_{i=1}^{k+1} h_i = 0$. Then,

$$|I^{1/2}(\pi) h|^2 = \sum_{i=1}^{k+1} \frac{h_i^2}{\pi_i} .$$

Theorem 14.3.2 *Assume the above multinomial setup.*
(i) For any test sequence ϕ_n *such that* $E_\pi(\phi_n) \to \alpha$,

$$\limsup_{n \to \infty} \inf \{ E_{\pi + hn^{-1/2}}(\phi_n) : \sum_{i=1}^{k+1} \frac{h_i^2}{\pi_i} \ge b^2 , \ \pi + hn^{-1/2} \in \Omega \}$$

$$\le P\{ \chi_k^2(b^2) > c_{k,1-\alpha} \} \ . \tag{14.23}$$

(ii) Pearson's Chi-squared test ϕ_n^*, *which rejects when*

$$\sum_{i=1}^{k+1} \frac{(Y_i - n\pi_i)^2}{n\pi_i} > c_{k,1-\alpha} \ ,$$

is asymptotically maximin in the sense that the inequality in (14.23) is an equality when $\phi_n = \phi_n^*$. *Thus,* ϕ_n^* *maximizes*

$$\liminf_{n} \{ E_{\pi + hn^{-1/2}}(\phi_n) : \sum_{i=1}^{k+1} \frac{h_i^2}{\pi_i} \ge b^2 , \ \pi + hn^{-1/2} \in \Omega \}$$

among all tests with asymptotic level α.

Proof. Theorem 13.5.4 immediately implies (i). To prove (ii), assume the opposite. Let R denote the right side of (14.23). Then, there exists a sequence of alternatives $h^{(n)}$ (with ith component denoted $h_i^{(n)}$) satisfying

$$\sum_{i=1}^{k+1} \frac{[h_i^{(n)}]^2}{\pi_i} \ge b^2 , \quad \sum_{i=1}^{k+1} h_i^{(n)} = 0$$

such that

$$E_{\pi + h^{(n)} n^{-1/2}}(\phi_n^*) \to \ell \ ,$$

and ℓ is strictly less than R. Since

$$\sum_{i=1}^{k+1} \frac{[h_i^{(n)}]^2}{\pi_i} \ge b^2 ,$$

we cannot have $h_i^{(n)} \to 0$ for every i.

We also cannot have $[h_i^{(n)}]^2 \to \infty$ for any i, for then

$$E_{\pi + h^{(n)} n^{-1/2}}(\phi_n^*) \to 1 \ ,$$

which would be a contradiction since $R < 1$. To see why this expectation would tend to 1, suppose $h_i^{(n)} \to \infty$ (and a similar argument holds if $h_i^{(n)} \to -\infty$). Then,

$$E_{\pi + h^{(n)} n^{-1/2}}(\phi_n^*) \ge P_{\pi + h^{(n)} n^{-1/2}} \left\{ \frac{(Y_i - n\pi_i)^2}{n\pi_i} > c_{k,1-\alpha} \right\}$$

$$> P_{\pi + h^{(n)} n^{-1/2}} \left\{ n^{1/2} (\frac{Y_i}{n} - \pi_i) > c_{k,1-\alpha}^{1/2} \right\}$$

$$= P_{\pi + h^{(n)} n^{-1/2}} \left\{ n^{1/2} \left[\frac{Y_i}{n} - (\pi_i + h_i^{(n)} n^{-1/2}) \right] + h_i^{(n)} > c_{k,1-\alpha}^{1/2} \right\} \ . \tag{14.24}$$

But, by Chebyshev's inequality,

$$n^{1/2}\left[\frac{Y_i}{n} - (\pi_i + h_i^{(n)} n^{-1/2})\right]$$

is bounded in probability, since it has mean 0 and variance bounded by one. Hence, (14.24) tends to one and so

$$E_{\pi+h^{(n)}n^{-1/2}}(\phi_n^*) \to 1 .$$

The same conclusion holds along any subsequence n_k satisfying $h_i^{(n_k)} \to \infty$.

Thus, we must have $h_i^{(n)} \asymp 1$ for every i. By passing to subsequences which converge, assume

$$h_i^{(n)} \to h_i^{(\infty)} < \infty \ , \ \text{and} \ \lambda \equiv \sum_{i=1}^{k+1} \frac{[h_i^{(\infty)}]^2}{\pi_i} \geq b^2 .$$

The limiting power was obtained in Theorem (14.3.1) with $h_i^{(n)} = h_i$ fixed, but the argument applies with obvious modifications to sequences that converge; moreover, this limiting power is

$$P\{\chi_k^2(\lambda) > c_{k,1-\alpha}\} \geq P\{\chi_k^2(b^2) > c_{k,1-\alpha}\} ,$$

since the family of Chi-squared distributions has monotone likelihood ratio. This again yields a contradiction. The same conclusion holds for any subsequence, because we can apply the argument to further subsequences where $h_i^{(n)}$ converges along the subsubsequences. ■

The above result states that the Chi-squared test is asymptotically maximin for the multinomial goodness of fit problem. The same result holds for the likelihood ratio test (Problem 14.16). Moreover, the above argument shows that the worst case power over alternatives $\pi + hn^{-1/2}$ with

$$\sum_{i=1}^{k} h_i^2/\pi_i \geq b^2$$

occurs (asymptotically) when $\sum_{i=1}^{k} h_i^2/\pi_i = b^2$.

14.3.2 Chi-squared Test of Uniformity

So far, we have been concerned with testing the parameters of a multinomial model. Let us now return to the problem stated at the beginning of Section 14.2, where $X_1, \ldots, X_n$ are i.i.d. real-valued observations with c.d.f. F, and the problem is that of testing the null hypothesis H that $F = F_0$, where $F_0(t) = t$ is the uniform c.d.f. on $(0, 1)$. To reduce this problem of goodness of fit to that of testing a multinomial hypothesis, fix a positive integer k and divide the unit interval into $k + 1$ subintervals of length $1/(k + 1)$; for $j = 1, \ldots, k + 1$, let Y_j be the number of X_i observations that fall in the interval $I_{k,j}$ defined by

$$I_{k,j} = [(j-1)/(k+1), j/(k+1)) .$$

Under the null hypothesis, the joint distribution of $(Y_1, \ldots, Y_{k+1})$ is multinomial based on n trials and equal class probabilities of $1/(k + 1)$. So, one can test H

by using the Chi-squared test which rejects for large values of

$$\sum_{j=1}^{k+1} \frac{(Y_j - \frac{n}{k+1})^2}{\frac{n}{k+1}} .$$

It follows that, for fixed k, the Chi-squared test is consistent against any alternative distribution F which does not assign equal probability to all intervals $I_{k,j}$.

Next, consider a sequence of alternative densities f_n of the form

$$f_n(x) = 1 + b_n u(x) , \tag{14.25}$$

where u satisfies $\int_0^1 u(x)dx = 0$ and $\int u^2(x)dx < \infty$. Then, f_n assigns probability

$$\int_{I_{k,j}} [1 + b_n u(x)]dx = \frac{1}{k+1} + b_n \int_{I_{k,j}} u(x)dx$$

to $I_{k,j}$. By Theorem 14.3.1 (ii), with k fixed and $b_n = hn^{-1/2}$, the limiting power of the Chi-squared test is given by

$$P\{\chi_k^2(\lambda_k) > c_{k,1-\alpha}\} ,$$

where

$$\lambda_k = h^2(k+1) \sum_{j=1}^{k+1} \left[\int_{I_{k,j}} u(x)dx \right]^2 .$$

Note that, if

$$\int_{I_{k,j}} u(x)dx$$

is not zero for at least one j, then the noncentrality parameter λ_k is positive. Also, if u is continuous except at most a finite number of points, then

$$\lambda_k \to \lambda_\infty \equiv h^2 \int_0^1 u^2(x)dx \quad \text{as } k \to \infty . \tag{14.26}$$

Note that for any fixed k, λ_k can be 0 even if $\lambda_\infty > 0$. Indeed, the Chi-squared test has power equal to the size of the test against any distribution that has mass $1/(k+1)$ on each subintervals, and so for fixed k, the Chi-squared test is not consistent against all alternatives.

Therefore, it is tempting to allow $k = k_n$ to increase with n in order to obtain power against an even broader range of alternatives. On the other hand, if λ_k approaches λ_∞ quite fast, then it would be undesirable to let k_n increase too quickly. To illustrate this point, consider the following example. Let $u_0(x) = 1$ for $x \le 1/2$ and $u_0(x) = -1$ for $x > 1/2$. Then, $\lambda_k = \lambda_\infty = h^2$ for all k odd. If $k = 1$, then the limiting power of the Chi-squared test against f_n given by

$$f_n(x) = 1 + hn^{-1/2} u_0(x)$$

is

$$P\{\chi_1^2(h^2) > c_{1,1-\alpha}\} .$$

If instead, $k = 2j + 1$ with $j \ge 1$, the limiting power is exactly

$$P\{\chi_k^2(h^2) > c_{k,1-\alpha}\} .$$

Notice that the noncentrality parameter is the same for all odd k. But, for fixed h, this probability is decreasing in k its limiting value is α as $k \to \infty$, as shown by the following lemma.

Lemma 14.3.1 *Let $M(k,h)$ be defined as*

$$M(k,h) = P\{\chi_k^2(h^2) > c_{k,1-\alpha}\} , \tag{14.27}$$

where $\chi_k^2(h^2)$ denotes a noncentral Chi-squared variable with k degrees of freedom and noncentrality parameter h^2.
(i) For fixed h, $M(k,h)$ is nonincreasing in k, and is strictly decreasing if $h \neq 0$.
(ii) If $h_k \to h$ for some finite h, $M(k,h_k) \to \alpha$ as $k \to \infty$. In particular, $M(k,h) \to \alpha$ as $k \to \infty$.
(iii) If $(2k)^{-1/2} h_k^2 \to c$ as $k \to \infty$, then

$$M(k,h_k) \to 1 - \Phi(z_{1-\alpha} - c) .$$

PROOF. The proof of (i) is left as an exercise (Problem 14.17). To prove (ii), let $Z_1, Z_2, \ldots$ denote i.i.d. standard normal variables. By the Central Limit Theorem,

$$(2k)^{-1/2}\left(\sum_{i=1}^{k} Z_i^2 - k\right) \xrightarrow{d} N(0,1) , \tag{14.28}$$

which implies

$$(2k)^{-1/2}(c_{k,1-\alpha} - k) \to z_{1-\alpha} \tag{14.29}$$

as $k \to \infty$. Of course, the result (14.28) holds even if the $i = 1$ term is omitted from the sum. Hence,

$$M(k,h_k) = P\{(Z_1 + h_k)^2 + \sum_{i=2}^{k} Z_i^2 > c_{k,1-\alpha}\}$$

$$= P\{(2k)^{-1/2}(Z_1+h_k)^2 + (2k)^{-1/2}\left(\sum_{i=2}^{k} Z_i^2 - k\right) > (2k)^{-1/2}(c_{k,1-\alpha} - k)\} . \tag{14.30}$$

By (14.29), the right side of the last expression tends to $z_{1-\alpha}$. Also, as $k \to \infty$,

$$(2k)^{-1/2}(Z_1 + h_k)^2 \xrightarrow{P} 0 .$$

By Slutsky's Theorem, the left side of (14.30) tends in distribution to $N(0,1)$. The result (ii) follows by another application of Slutsky's Theorem. The proof of (iii) is similar. The only difference is that the term

$$(2k)^{-1/2}(Z_1 + h_k)^2 \xrightarrow{P} c$$

if $(2k)^{-1/2} h_k^2 \to c$. ■

Thus, the results in (i) and (ii) of Lemma 14.3.1 show that the choice $k = 1$ is optimal for the situation with $u = u_0$. The point is that increasing k too much decreases the limiting power. Furthermore, if k is quite large, the limiting power is approximately α. This latter conclusion applies to any alternative sequence of the form (14.25) with $b_n = n^{-1/2}$; also see Problem 14.19.

Mann and Wald (1942) considered the optimal choice of k_n. In particular, let

$$d_K(F, F_0) = \sup_t |F(t) - t| .$$

Mann and Wald (1942) determined an optimal rate for k_n which satisfies $k_n = O(n^{2/5})$, and show that with such an optimal rate the limiting power is $1/2 > \alpha$ against a sequence of alternatives F_n satisfying $n^{2/5} d_K(F_n, F_0) \to \infty$. This result on optimal rates is somewhat contradicted by the above analysis and other results that indicate that the best choice of k_n is rather small; see Stuart and Ord (1991, Chapter 30).)

It is interesting to compare the results of Mann and Wald with the fact that the Kolmogorov-Smirnov goodness of fit test has limiting power one if $n^{1/2} d_K(F_n, F_0) \to \infty$, as shown in Theorem 14.2.2. It follows that the Kolmogorov Smirnov test (and this is also true of Cramér von-Mises test) is asymptotically superior to the Chi-squared test in this case. However, it has been pointed out that this superiority is connected with the choice of distance with which one measures deviations from F_0. If one replaces the Kolmogorov-Smirnov distance with an L_2 distance based on the integral of the squared difference in densities (satisfying smoothness conditions), then the Chi-squared test can asymptotically outperform the Kolmogorov-Smirnov test; see Ingster (1993). We will later obtain further results, since Chi-squared tests can be viewed as a special case of the more general class of Neyman smooth tests that will be studied in Section 14.4.

14.3.3 Composite Null Hypothesis

Next, we consider the application of the Chi-squared test to composite hypotheses. First, suppose data $(Y_1, \ldots, Y_{k+1})$ has the multinomial distribution (14.14), where Y_j is the number of trials resulting in outcome j and p_j is the probability of the jth outcome for any given trial. The full model allows the p_j to vary freely, subject to their being nonnegative and summing to one.

Consider testing the null hypothesis that the p_j are of the form

$$p_j = f_j(\beta_1, \ldots, \beta_q) , \qquad j = 1, \ldots, k+1,$$

where the f_j are known functions of $\beta = (\beta_1, \ldots, \beta_q)$, and β varies in a subset of $\mathbb{R}^q$ for some $q < k$. For testing the simple null hypothesis that $p_j = f_j(\beta)$, $1 \le j \le k$, for a fixed value of β, the Chi-squared test is based on the statistic

$$Q_n(\beta) = \sum_{j=1}^{k+1} \frac{(Y_j - n f_j(\beta))^2}{n f_j(\beta)} . \tag{14.31}$$

If β is unspecified, Fisher (1928b) suggested the test statistic $Q_n(\hat{\beta}_n)$, where $\hat{\beta}_n$ is a MLE of β under the null hypothesis submodel (or any efficient estimator). Following Fisher, Neyman (1949) recommends $Q_n(\tilde{\beta}_n)$, where $\tilde{\beta}_n$ is chosen to minimize $Q_n(\beta)$ (in which case $\tilde{\beta}_n$ is called a minimum Chi-squared estimator). Not surprisingly, it is typically the case that, under the null hypothesis,

$$Q_n(\hat{\beta}_n) - Q_n(\tilde{\beta}_n) \xrightarrow{P} 0 .$$

Example 14.3.1 (Fisher linkage model) Fisher (1928b) postulated a genetics model with 4 possible types of offspring, whose probabilities are of the form

$$(p_1, p_2, p_3, p_4) = \frac{1}{4}(2 + \beta, 1 - \beta, 1 - \beta, \beta)$$

for some $\beta \in (0, 1)$. In the above notation, $f_1(\beta) = 2 + \beta$, $f_2(\beta) = f_3(\beta) = 1 - \beta$, and $f_4(\beta) = \beta$. (The parameter β depends on the linkage between the two genetic factors under consideration.) To test the validity of such a model, a Chi-squared test can be employed. To estimate β, it is easily checked (Problem 14.23) that the likelihood equation is

$$\frac{Y_1}{2+\beta} - \frac{(Y_2 + Y_3)}{1 - \beta} + \frac{Y_4}{\beta} = 0 , \tag{14.32}$$

which reduces to a quadratic equation, and the MLE $\hat{\beta}_n$ is the root of this equation that lies in $[0, 1]$. The resulting test statistic is then $Q_n(\hat{\beta}_n)$. ■

Just as in the case of simple null hypothesis, if the null hypothesis is true, then (Problem 14.20)

$$2\log(R_n) - Q_n(\hat{\beta}_n) \xrightarrow{P} 0 . \tag{14.33}$$

Thus, under the assumptions of Theorem 12.4.2 (iii), it follows that, under the null hypothesis,

$$Q_n(\hat{\beta}_n) \xrightarrow{d} \chi^2_{k-q} . \tag{14.34}$$

As in the case of a simple null hypothesis, the problem of testing a composite hypothesis of goodness of fit can be reduced to the multinomial case. Suppose $X_1, \ldots, X_n$ are i.i.d. according to a model $\{P_\theta,\ \theta \in \Omega\}$, where $\Omega \subset \mathbb{R}^k$. The null hypothesis specifies $\theta = f(\beta)$ for some fixed function f from $\mathbb{R}^q$ to $\mathbb{R}^k$. Now, partition the range of the X_i into $k+1$ sets $E_1, \ldots, E_{k+1}$, and let $P_\theta\{E_i\}$ be the probability of E_i under θ. Let Y_j denote the number of X_i falling in E_j and let

$$Q_n(\beta) = \sum_{j=1}^{k+1} \frac{(Y_j - nP_{f(\beta)}\{E_i\})^2}{nP_{f(\beta)}\{E_i\}} .$$

Then, a test can be based on $Q_n(\hat{\beta}_n)$, where $\hat{\beta}_n$ is an estimator of β assuming the null hypothesis submodel.

Just as in the case of a simple null hypothesis, the choice of k (and now also of the sets E_i) is complex; note the references in the previous subsection.[1] In addition, a further complication arises, which is the choice of estimator $\hat{\beta}_n$. If the estimator is an efficient likelihood estimator based on the likelihood of the categorized data $Y_1, \ldots, Y_{k+1}$, then we have returned to the setting of the multinomial case considered at the beginning of this section, and the limiting distribution of $Q_n(\hat{\beta}_n)$ is Chi-squared. On the other hand, one might also estimate β based on the likelihood of the original sample $X_1, \ldots, X_n$. In this case, Chernoff and

[1]For randomly chosen partitions, see Chapter 2 of Greenwood and Nikulin (1996) and Theorem 5.7.1 of Lehmann (1999). Data-based partitions occur, for example, when the number of observations falling in any set is small and one then combines such sets.

Lehmann (1954) showed that $Q_n(\hat{\beta}_n)$ need not be Chi-squared. For an example, see Problem 14.24.

14.4 Neyman's Smooth Tests

Suppose that $X_1, \ldots, X_n$ are i.i.d. according to a probability distribution P on some sample space S. Consider testing the simple null hypothesis $P = P_0$, where P_0 is some fixed probability distribution on S. When $S = \mathbb{R}$, one possible test is the Kolmogorov-Smirnov test, discussed in Section 14.2, which was seen to be consistent in power against any fixed alternative, and uniformly consistent against the large class of distributions F with $d_K(F, F_0) > \Delta$ for any small Δ. Even so, the Kolmogorov-Smirnov test can have poor power against local alternatives; see Problems 14.6 and 14.7. In fact, whenever the family of alternative distributions is large, it is unlikely that there will exist a single test that will perform uniformly well across against all of them, and certainly no UMP test will exist. For a q.m.d. family indexed by a real-valued parameter, one can construct AUMP tests, as discussed in Section 13.3. However, even if the family of alternatives is q.m.d. and indexed by a parameter in $\mathbb{R}^2$, there exists no test that is asymptotically uniformly optimal (Problem 14.25). Thus, one goal might be to construct tests that perform well across a fairly broad range of alternatives. In this spirit, Neyman (1937b) considered large parametric families of alternatives and derived tests that asymptotically maximize minimum (and average) power against these alternatives. Such tests will be described in this section.

Consider the parametric model of densities $p_\theta(x)$ with respect to P_0 given by

$$p_\theta(x) = C_k(\theta) \exp[\sum_{j=1}^{k} \theta_j T_j(x)] \ , \tag{14.35}$$

where k is some positive integer so that $\theta \in \mathbb{R}^k$. Setting $T_0(x) = 1$, the functions $T_1, \ldots, T_k$ are chosen so that $T_0, \ldots, T_k$ is a set of orthonormal functions on $L^2(P_0)$, the space of functions that are square integrable with respect to P_0; that is

$$Cov_0[T_i(X_1), T_j(X_1)] = \int_S T_i(x) T_j(x) dP_0(x) = \delta_{i,j} \ ,$$

where $\delta_{i,j} = 1$ if $i = j$ and $\delta_{i,j} = 0$ if $i \neq j$. This implies $E_0(T_j) = 0$ for $j = 1, \ldots, k$. The normalizing constant $C_k(\theta)$ is given by

$$C_k(\theta) = \{\int_S \exp[\sum_{j=1}^{k} \theta_j T_j(x)] dP_0(x)\}^{-1} \ . \tag{14.36}$$

Let Ω_k denote the set of θ where the integral in (14.36) is finite so that p_θ is a proper density. We will also assume 0 is an interior point of Ω_k, in which case the family of densities constitutes a k-parameter exponential family of full rank. The null hypothesis asserts $\theta = 0$.

Example 14.4.1 (Testing uniformity using Legendre polynomials) As a prototype, consider the goodness of fit problem of testing that $X_1, \ldots, X_n$ are i.i.d. from the uniform distribution on $[0, 1]$, so that $S = [0, 1]$ and P_0 is the uniform distribution on $[0, 1]$. For this problem, Neyman (1937b) chose $T_j(x)$ to be a polynomial of degree j. Specifically, set $T_0(x) = 1$, $T_1(x) = \sqrt{3}(2x - 1)$, $T_2(x) = \sqrt{5}(6x^2 - 6x + 1)$, $T_3(x) = \sqrt{7}(20x^3 - 30x^2 + 12x - 1)$, and so on, so that T_j is constructed to be a polynomial of degree j such that it is orthogonal to $T_0, \ldots T_{j-1}$, and its square integrates to one. The polynomials T_j are the so-called normalized Legendre polynomials. ■

Returning to the general case, we next derive Neyman's test as a special case of Rao's score test for testing $\theta = 0$ in the parametric model. The family of densities (14.35) is a k-parameter exponential family in natural form. By Example 12.2.6, this family is q.m.d. at $\theta = \theta_0 = 0$. By Theorem 12.2.2, the score vector at $\theta_0 = 0$ (12.73) is given by

$$Z_n^T = n^{-1/2}\left(\frac{\partial}{\partial \theta_1} \log L_n(\theta), \ldots, \frac{\partial}{\partial \theta_k} \log L_n(\theta)\right)\Bigg|_{\theta=0} ,$$

where $L_n(\theta)$ is the likelihood function

$$L_n(\theta) = C_k^n(\theta) \exp[\sum_{i=1}^n \sum_{j=1}^k \theta_j T_j(X_i)] .$$

Hence,

$$\frac{\partial}{\partial \theta_m} \log[L_n(\theta)] = n \frac{\partial}{\partial \theta_m} \log[C_k(\theta)] + \sum_{i=1}^n T_m(X_i) .$$

But, by Problem 2.16,

$$-\frac{\partial}{\partial \theta_m} \log[C_k(\theta)] = E_\theta[T_m(X_i)] ,$$

which is 0 when $\theta = 0$ (since we are assuming $T_0(x) = 1$ and T_m is orthogonal to T_0). Hence, the score vector at θ_0 reduces to

$$Z_n^T = n^{-1/2} \left(\sum_{i=1}^n T_1(X_i), \ldots, \sum_{i=1}^n T_k(X_i) \right) . \tag{14.37}$$

By the orthogonality of the T_i, we have $Cov[T_i(X_1), T_j(X_1)] = \delta_{i,j}$. Arguing directly, the Multivariate Central Limit Theorem implies that, under $\theta = 0$,

$$Z_n \stackrel{d}{\to} N(0, I_k) ,$$

where I_k is the $k \times k$ identity matrix. Moreover, the Fisher Information at $\theta = 0$ is $I(0) = I_k$. Therefore, Rao's score test rejects for large values of

$$Z_n^T I^{-1}(0) Z_n = Z_n^T Z_n = \sum_{j=1}^k Z_{n,j}^2 ,$$

where

$$Z_{n,j} = n^{-1/2} \sum_{i=1}^n T_j(X_i) . \tag{14.38}$$

Let $c_{k,1-\alpha}$ be the $1-\alpha$ quantile of the χ^2-distribution with k degrees of freedom. By the Continuous Mapping Theorem,

$$Z_n^T Z_n \xrightarrow{d} \chi_k^2 ,$$

and so the test ϕ_n^* which rejects when $Z_n^T Z_n > c_{k,1-\alpha}$ is asymptotically consistent in level. The test ϕ_n^* will be referred to as *Neyman's smooth test.* (Of course, one can always replace $c_{k,1-\alpha}$ by the exact $1-\alpha$ quantile of the finite sampling null distribution of $Z_n^T Z_n$, or the null distribution can be simulated.)

Example 14.4.2 (Continuation of Example 14.4.1) In this case,

$$Z_{n,1}^2 = [n^{-1/2} \sum_{i=1}^{n} \sqrt{3}(2X_i - 1)]^2 = 12[n(\bar{X}_n - \frac{1}{2})^2] . \tag{14.39}$$

Thus, $Z_{n,1}^2$ is large when the sample mean differs 1/2, from the hypothesized mean. Similarly, $Z_{n,j}^2$ is large when the first j sample moments differ greatly from those of $U(0,1)$. ■

Example 14.4.3 (The χ^2 test) As in Section 14.3, consider the goodness of fit problem for testing a multinomial distribution with $k+1$ categories. For concreteness, suppose $X_1, \ldots, X_n$ are i.i.d., each X_i taking the value e_j with probability p_j, where e_j is the vector with 1 in the jth component and 0 in the remaining k components. Then, the chi-squared statistic Q_n given by (14.16) can be viewed as a Neyman smooth test. Recall V_n given by (14.18) and the matrix A given by (14.20). Now, let Z_n be the vector $A^{1/2}V_n$, so that $Q_n = Z_n^T Z_n$. Furthermore, the probability mass function of X_i can be written in the form (14.35) with T_j satisfying $n^{-1/2}\sum_i T_j(X_i)$ equal to the jth component of Z_n (Problem 14.26). (Note, however, that unlike the Legendre polynomials of Example 14.4.1, the functions T_j depend on k, so that we really have a triangular array of orthonormal functions.) ■

14.4.1 Fixed k Asymptotics

Assuming the model (14.35) holds, we can apply Corollary 12.4.1 to conclude that, under $h/n^{1/2}$,

$$Z_n^T Z_n \xrightarrow{d} \chi_k^2(|h|^2) . \tag{14.40}$$

We now apply Theorems 13.5.4 and 13.5.5 in order to obtain an asymptotic maximin property for ϕ_n^*.

Theorem 14.4.1 *Assume the model (14.35) and assume $\theta = 0$ is an interior point of Ω_k. Consider the problem of testing $\theta = 0$.*
(i) For any sequence of tests ϕ_n such that $E_0(\phi_n) \to \alpha$ and any b and B satisfying $0 < b < B \le \infty$,

$$\limsup_{n\to\infty} \inf\{E_{hn^{-1/2}}(\phi_n) : \ b \le |h| \le B\} \le P\{\chi_k^2(b^2) > c_{k,1-\alpha}\} , \tag{14.41}$$

where $\chi_k^2(b^2)$ is noncentral Chi-squared with k degrees of freedom and noncentrality parameter b^2.

(ii) Neyman's smooth test ϕ_n^* is asymptotically maximin in the sense that, for any $0 < b < B < \infty$,

$$\inf\{E_{hn^{-1/2}}(\phi_n^*) :\ b \le |h| \le B\} \to P\{\chi_k^2(b^2) > c_{k,1-\alpha}\} \ . \tag{14.42}$$

Thus, for any $0 < b < B < \infty$, ϕ_n^* maximizes

$$\lim_n \inf\{E_{hn^{-1/2}}(\phi_n) :\ b \le |h| \le B\}$$

among all tests with asymptotic level α.

PROOF. Theorem 13.5.4 implies (14.41) and Theorem 13.5.5 implies (14.42). ■

The result (14.41) holds if $B = \infty$ (since the inf over a larger set is bounded above by the inf over a smaller set). In many cases, one can replace B by ∞ in (14.42) as well. For example, suppose $Var_\theta[T_j(X_1)]$ is a uniformly bounded function of θ. Then, (14.42) holds if $B = \infty$ (Problem 14.27). This condition is satisfied, for example, if the $T_j(x)$ are uniformly bounded functions of x, as they are in Neyman's choice of the Legendre polynomials.

Theorem 14.4.1 states an asymptotic maximin property over alternatives θ that are $O(n^{-1/2})$ from $\theta = 0$. Of course, Neyman's smooth test is also consistent in power against any fixed $\theta \ne 0$. Actually, it is consistent in power against a broad range of alternatives, not just alternatives in the parametric model (14.35).

To make this statement more precise, first consider Neyman's original construction with $k = 1$ for testing the hypothesis of uniformity, as described in Example 14.4.1. Then, the test statistic reduces to (14.39). The test statistic is designed to have power against distributions with mean not equal to 1/2 and it serves this purpose. For, under an alternative distribution P on $(0, 1)$ with mean $\mu(P) \ne 1/2$, the power of the test which rejects when $Z_{n,1}^2 > c_{1,1-\alpha}$ tends to 1. To see why, note that by the Weak Law of Large Numbers,

$$(\bar{X}_n - \frac{1}{2})^2 \xrightarrow{P} (\mu(P) - \frac{1}{2})^2 > 0 \ ,$$

and so

$$12n(\bar{X}_n - \frac{1}{2})^2 \xrightarrow{P} \infty \ .$$

Therefore, by Slutsky's Theorem,

$$P\{12n(\bar{X}_n - \frac{1}{2})^2 > c_{1,1-\alpha}\} \to 1 \ .$$

The point is that the test will be consistent against any alternative P with mean $\mu(P) \ne 1/2$, even if P is not a member of the parametric model (14.35).

Similarly, for $k > 1$, Neyman's test will be consistent against any distribution P, as long as the first k moments of P are not identical to the first k moments of the uniform distribution (Problem 14.28). Thus, Neyman's test for testing $P = P_0$ has good power across a broader range of distributions than just the original parametric model (14.35).

Example 14.4.4 (Limiting Power Against a Contiguous Sequence) Consider a sequence of alternative densities of the form

$$f_n(x) = 1 + b_n u(x) \ , \tag{14.43}$$

where $b_n \to 0$ and u satisfies

$$\int_0^1 u(x)dx = 0 \ .$$

Assume $\sup |u(x)| < \infty$, so that f_n is a density for b_n small enough. If we set $b_n = hn^{-1/2}$, we can calculate the limiting power of Neyman's smooth test against f_n as follows. The family of densities $1 + \theta u(x)$ is q.m.d. at $\theta = 0$ (Problem 12.6) with score function $n^{-1/2} \sum_i u(X_i)$. If P_n denotes the probability distribution with density f_n with $b_n = hn^{-1/2}$, then P_n^n is contiguous to P_0^n. Under $\theta = 0$, $(Z_n^T, n^{-1/2} u(X_i))$ is asymptotically multivariate normal. By the multivariate generalization of Corollary 12.3.2 obtained in Problem 12.33, under f_n with $b_n = hn^{-1/2}$,

$$Z_n^T \xrightarrow{d} N(c, I_k) \ ,$$

where c is the vector with jth component given by

$$c_j = Cov(Z_{n,j}, hn^{-1/2} \sum_i u(X_i)) = h\langle T_j, u \rangle \ ,$$

and

$$\langle T_j, u \rangle = \int_0^1 T_j(x) u(x) dx \ .$$

Hence, under f_n,

$$Z_n^T Z_n \xrightarrow{d} \chi_k^2(\delta^2) \ , \tag{14.44}$$

where

$$\delta^2 = h^2 \sum_{j=1}^k \langle T_j, u \rangle^2 \ .$$

Thus, the limiting power is $M(k, \delta^2)$, with $M(k, h)$ defined by (14.27). Note that if u is represented as $u(x) = \sum_{j=1}^k \gamma_j T_j(x)$, then by Parseval's identity (see A.7),

$$\sum_{j=1}^k \langle T_j, u \rangle^2 = \int_0^1 u^2(x) dx \ .$$

Thus, Neyman's test has limiting power exceeding α against alternatives of the form (14.43) with $b_n \asymp n^{-1/2}$ if u is in the span of $T_1, \ldots, T_k$. ■

14.4.2 Neyman's Smooth Tests With Large k

In the previous section, Neyman's smooth test was shown to be an asymptotically maximin procedure for the parametric model (14.35) with k fixed. Obviously, the larger the value of k, the greater the number of orthogonal directions used to construct the test statistic. For fixed k, consistency of Neyman's smooth test holds for a restricted class of alternatives. For example, Neyman's construction results in a test of uniformity that is consistent in power against any distribution that does not have the same first k moments as that of the uniform distribution. This suggests the possibility that, if we let k increase with n, we can obtain consistency against all distributions because on the unit interval, a distribution is uniquely determined by its moments; see Feller (1971), Section VII.3. To investigate this possibility, we now develop some basic properties of the test based on

$$S_{n,k_n} = \sum_{j=1}^{k_n} Z_{n,j}^2 \ , \tag{14.45}$$

where k_n is some fixed sequence satisfying $k_n \to \infty$.

For fixed k, we saw that, under H_0,

$$\sum_{j=1}^{k} Z_{n,j}^2 \xrightarrow{d} \chi_k^2 \ .$$

If k is large, the Chi-squared distribution with k degrees of freedom is approximately $N(k, 2k)$, and so it is reasonable to expect that, under H_0,

$$\frac{\sum_{j=1}^{k_n} Z_{n,j}^2 - k_n}{(2k_n)^{1/2}} \xrightarrow{d} N(0,1) \ .$$

In order to prove this convergence, we need the following lemma, due to Bentkus (2003), which can be viewed as a multivariate version of the Berry-Esseen Theorem. In the statement of the result, let $\mathcal{E}_k$ denote the class of Euclidean balls in $\mathbb{R}^k$; that is, the family of sets $\{y \in \mathbb{R}^k : \ |x-y| < r\}$ as $x \in \mathbb{R}^k$ and $r > 0$ vary. Also, let $\mathcal{C}_k$ denote the class of convex sets in $\mathbb{R}^k$.

Lemma 14.4.1 *Let $Y_1, Y_2, \ldots, Y_n$ be i.i.d. random vectors in $\mathbb{R}^k$ with mean vector 0 and $k \times k$ identity covariance matrix I_k. Let $\beta = E(|Y_i|^3)$, and let $Z^{(k)}$ denote a multivariate normal random vector with mean 0 and covariance matrix I_k. Then,*

$$\sup_{B \in \mathcal{C}_k} \left| P\{n^{-1/2} \sum_{i=1}^{n} Y_i \in B\} - P\{Z^{(k)} \in B\} \right| \le 400 k^{1/4} \beta n^{-1/2} \ .$$

If $\mathcal{C}_k$ is replaced by $\mathcal{E}_k$, then the right side can be replaced by the upper bound $C\beta n^{-1/2}$, where C is an absolute constant (independent of k). Hence,

$$\sup_{t \in \mathbb{R}} \left| P\{|n^{-1/2} \sum_{i=1}^{n} Y_i|^2 \le t\} - P\{|Z^{(k)}|^2 \le t\} \right| \le C\beta n^{-1/2} \ .$$

We now apply the lemma with

$$Y_i = (T_1(X_i), \ldots, T_k(X_i)) \tag{14.46}$$

so that

$$S_{n,k} = \left| n^{-1/2} \sum_{i=1}^{n} Y_i \right|^2 \ .$$

Note that

$$\beta = E\left([T_1^2(X_i) + \cdots + T_k^2(X_i)]^{3/2} \right) \ .$$

By Minkowski's Inequality (Problem 14.30),

$$\beta^{2/3} \le \sum_{j=1}^{k} E[|T_j(X_i)|^3]^{2/3} . \tag{14.47}$$

If

$$\sup_j E[|T_j(X_i)|^3] \le B < \infty ,$$

then, $\beta \le Bk^{3/2}$. Hence, the following is true.

Theorem 14.4.2 *Consider S_{n,k_n} given by (14.45), where*

$$Z_{n,j} = n^{-1/2} \sum_{i=1}^{n} T_j(X_i) ,$$

and let $T_0 = 1$, and $T_0, T_1, T_2, \ldots$ be an infinite sequence of orthonormal functions on $L_2(P_0)$. Assume

$$\sup_j E_{P_0}[|T_j(X_i)|^3] = B < \infty . \tag{14.48}$$

If $k_n \to \infty$ and $k_n^3/n \to 0$, then, under $P = P_0$,

$$\frac{S_{n,k_n} - k_n}{(2k_n)^{1/2}} \xrightarrow{d} N(0,1) .$$

PROOF. Apply the lemma with Y_i given by (14.46). Then,

$$\left| P\{S_{n,k_n} \le t(2k_n)^{1/2} + k_n\} - P\{|Z^{(k_n}|^2 \le t(2k_n)^{1/2} + k_n\} \right|$$

is bounded above by

$$(Bk_n)^{3/2} n^{-1/2} \to 0 .$$

But, by the Central Limit Theorem,

$$P\{|Z^{(k_n)}|^2 \le t(2k_n)^{1/2} + k_n\} \to \Phi(t) , \tag{14.49}$$

where Φ is the standard normal c.d.f., and the result follows. ∎

Under the assumptions of Theorem 14.4.2, the sequence of tests that rejects when

$$\frac{S_{n,k_n} - k_n}{(2k_n)^{1/2}} > z_{1-\alpha} \tag{14.50}$$

is asymptotically level α.

Example 14.4.5 Let

$$T_j(x) = \sqrt{2}\cos(\pi j x) .$$

Such a choice arises in the construction of the Cramér-von Mises test, which will be discussed further in Example 14.5.1. Under the null hypothesis $P = P_0 = U(0,1)$,

$$E_{P_0}[|T_j(X_i)|^3] \le \sqrt{2} E_{P_0}[T_j^2(X_i)] = \sqrt{2} .$$

Hence, the condition (14.48) is satisfied. ∎

Next, we consider the power of (14.50) (with $k_n \to \infty$) against a fixed alternative. As in Theorem 14.5.1, suppose P is any probability distribution such that

$$E_P[T_j(X_1)] \neq E_{P_0}[T_j(X_1)]$$

for some j. Then, for such a j,

$$\frac{Z_{n,j}^2}{n} = \left[\frac{1}{n}\sum_{i=1}^{n} T_j(X_i)\right]^2 \xrightarrow{P} \{E_P[T_j(X_1)]\}^2 > 0 ,$$

by the Weak Law of Large Numbers. Hence,

$$\frac{S_{n,k_n} - k_n}{(2k_n)^{1/2}} \geq \frac{Z_{n,j}^2 - k_n}{(2k_n)^{1/2}} = \frac{\frac{Z_{n,j}^2}{n} - \frac{k_n}{n}}{\frac{(2k_n)^{1/2}}{n}} \xrightarrow{P} \infty$$

if $k_n/n \to 0$. Hence, the test (14.50) (or the test that rejects if $S_{n,k_n} > c_{k_n,1-\alpha}$) satisfies

$$P\{\frac{S_{n,k_n} - k_n}{(2k_n)^{1/2}} > z_{1-\alpha}\} \geq P\{\frac{Z_{n,j}^2 - k_n}{(2k_n^{1/2}} > z_{1-\alpha}\} \to 1$$

and is therefore pointwise consistent in power against P.

Note that the condition $k_n/n \to 0$ is a sufficient condition to ensure the test statistic $[S_{n,k_n} - k_n]/(2k_n)^{1/2}$ tends to ∞ in probability under an alternative P. The stronger condition $k_n^3/n \to 0$ is sufficient to show asymptotic normality under the null hypothesis. These conditions can be weakened, but the message is that one can obtain consistency against a broad family of distributions by letting k increase with n.

Next, we discuss the limiting power of the test (14.50) against a local sequence of alternatives. Suppose we consider alternatives of the form (14.35) used in the construction of Neyman's smooth tests. Specifically, consider the family of densities indexed by $\theta_1 \in \mathbb{R}$ given by

$$p_{\theta_1}(x) = C_1(\theta_1)\exp[\theta_1 T_1(x)] \ .$$

Fix $h > 0$. For testing $\theta_1 = 0$ versus $\theta_1 = hn^{-1/2}$ at level α, the limiting power of an asymptotically most powerful test sequence is $1 - \Phi(z_{1-\alpha} - h)$, by Lemma 13.3.1. This optimal limiting power exceeds α for $h > 0$ and approaches 1 as $h \to \infty$.

Now, consider the limiting power of Neyman's smooth test with any fixed k against the same sequence of alternatives. By (14.40), if k is fixed, the limiting power against $hn^{-1/2}$ of the test that rejects when $S_{n,k} > c_{k,1-\alpha}$ is $M(k,h)$ given by (14.27). Lemma 14.3.1 implies that, for large k, the power of the test that rejects for large $S_{n,k}$ is nearly α, against the sequence of alternatives defined by $\theta_1 = hn^{-1/2}$. In other words, Neyman's smooth test has poor power against such a sequence of alternatives, even though this family of alternatives is included in the original parametric model (14.35) leading to the derivation of the Neyman smooth tests. Moreover, one can show (Problem 14.32) that, assuming the conditions of Theorem 14.4.2, under $\theta_1 = hn^{-1/2}$,

$$[S_{n,k_n} - k_n]/(2k_n)^{1/2} \xrightarrow{d} N(0,1) \tag{14.51}$$

as $n, k_n \to \infty$. Thus, the limiting distribution of the normalized S_{n,k_n} is the same under $\theta_1 = 0$ as under the sequence $\theta_1 = hn^{-1/2}$. Hence, the limiting power is α against either sequence.

In order for the limiting power to be nontrivial against local alternatives, it is necessary to consider alternatives that converge to H_0 at a rate slower than the usual parametric rate $n^{-1/2}$. For example, let f_n be defined as in (14.43), but with b_n not of the form $hn^{-1/2}$. By (14.44), if k is fixed, under f_n, $S_{n,k}$ is approximately distributed as $\chi_k^2(\delta_k^2)$, where

$$\delta_k^2 = nb_n^2 \sum_{j=1}^{k} \langle T_j, u \rangle^2 \ .$$

But,

$$\frac{\chi_k^2(\delta_k^2) - k}{(2k)^{1/2}} \stackrel{d}{\to} N(\mu, 1)$$

if $\delta_k^2/(2k)^{1/2} \to \mu$ as $k \to \infty$. Therefore, one might expect that, under f_n,

$$\frac{S_{n,k_n} - k_n}{(2k_n)^{1/2}} \stackrel{d}{\to} N(\mu, 1) \tag{14.52}$$

if

$$\frac{nb_n^2 \sum_{j=1}^{k_n} \langle T_j, u \rangle^2}{(2k_n)^{1/2}} \to \mu \ .$$

Now, if $T_0, T_1, T_2, \ldots$ form a complete orthonormal system for the space of square integrable functions on $(0, 1)$, then,

$$\sum_{j=1}^{k_n} \langle T_j, u \rangle^2 \to \int_0^1 u^2(x) dx \ .$$

Therefore, if we take $b_n = (2k_n)^{1/4}/n^{1/2}$, we expect that (14.52) holds, where

$$\mu = \int_0^1 u^2(x) dx \ .$$

In fact, such a result is proved in Eubank and LaRiccia (1992) in the case $T_j(x) = \sqrt{2}\cos(\pi j x)$ if $k_n^5/n^2 \to 0$. The conclusion is that Neyman's test with increasing order k_n has nonnegligible power against alternatives converging to the null at rate $k_n^{1/4}/n^{1/2}$. This result suggests that k_n should not increase too quickly.

Further theoretical results concerning Neyman's smooth tests, especially in regard to the choice of k, can be found in Eubank and LaRiccia (1992), Ledwina (1994), Kallenberg and Ledwina (1995), Fan (1996) and Inglot and Ledwina (1996). This growing literature includes simulation studies which show that Neyman's smooth tests perform well across a broad range of alternatives and are competitive with existing tests.

14.5 Weighted Quadratic Test Statistics

In the construction of Neyman's smooth tests based on k, equal weight was given to the first k *directions* determined by the orthonormal functions $T_1, T_2, \ldots$.

Instead, one might consider modifying the test statistic so that different weights are given to different directions; with such a modification, it becomes possible to consider an infinite number of directions. Such weighted quadratic test statistics are considered in this section.

Under the setup and notation of Section 14.4, consider the problem of testing the simple null hypothesis $H_0 : P = P_0$. Let $T_0 = 1$ and suppose $T_0, T_1, , T_2, \ldots$ is an infinite sequence of orthonormal functions on $L_2(P_0)$. Let $Z_{n,j}$ be defined by (14.38) and consider the test statistic

$$W_n = \sum_{j=1}^{\infty} a_j Z_{n,j}^2 , \tag{14.53}$$

where a_j is a sequence of nonnegative numbers. Typically, we would choose a_j to decrease with j, so that less weight is given to the jth component making up W_n. Note that W_n is only computable if only finitely many a_j are nonzero, or - as will be exemplified later - the infinite sum can be explicitly evaluated by an alternative computable formula.

Let F_{W_n} denote the c.d.f. of W_n under P_0, and set

$$w_{n,1-\alpha} = \inf\{x : F_{W_n}(x) \geq 1 - \alpha\} .$$

The following result summarizes some basic properties of W_n.

Theorem 14.5.1 *Assume $a_j \geq 0$ and $\sum_j a_j < \infty$.*
(i) Under H_0, W_n is a well-defined random variable; that is, $W_n < \infty$ with probability one.
(ii) Under H_0,

$$W_n \xrightarrow{d} W = \sum_{j=1}^{\infty} a_j Z_j^2 ,$$

where $Z_1, Z_2, \ldots$ are i.i.d. $N(0,1)$ random variables, and W has a continuous distribution function F_W which is strictly increasing on $(0, \infty)$.
(iii) Let $w_{1-\alpha}$ denote the $1 - \alpha$ quantile of the distribution of W, so that

$$F_W(w_{1-\alpha}) = 1 - \alpha .$$

Then, $w_{n,1-\alpha} \to w_{1-\alpha}$.
(iv) Assume a_j is such that $a_j > 0$. Suppose P is any probability distribution such that

$$E_P[T_j(X_1)] \neq E_{P_0}[T_j(X_1)] \tag{14.54}$$

(where the expectation on the left side is assumed to exist). Then, the limiting power of the test that rejects when $W_n > w_n(1 - \alpha)$ against the alternative P is one. Hence, if all the a_j satisfy $a_j > 0$, then the test is consistent in level against any P which satisfies (14.54) for some j.

PROOF. First, note that

$$0 \leq E_{\theta_0}(W_n) = \sum_{j=1}^{\infty} a_j Var_{\theta_0}(Z_{n,j}) \leq \sum_{j=1}^{\infty} a_j E_{\theta_0} T_j^2(X_1) = \sum_{j=1}^{\infty} a_j < \infty .$$

Part (i) follows, since a nonnegative random variable with a finite mean is finite with probability one. To prove (ii), first note that W is a well-defined random variable since

$$E(W) = \sum_{j=1}^{\infty} a_j < \infty .$$

Now, let

$$W^{(k)} = \sum_{j=1}^{k} a_j Z_j^2 .$$

Then, $W^{(k)} \stackrel{d}{\to} W$ as $k \to \infty$. Indeed,

$$0 \le W - W^{(k)} = \sum_{j=k+1}^{\infty} a_j Z_j^2 \stackrel{P}{\to} 0$$

since, by Markov's Inequality (Problem 11.26), for $\delta > 0$,

$$P\{W - W^{(k)} > \delta\} \le \frac{E(W - W^{(k)})}{\delta} = \frac{\sum_{j=k+1}^{\infty} a_j}{\delta} \to 0$$

as $k \to \infty$. Moreover, the distribution of W is continuous and strictly increasing (Problem 14.33). To show that W_n converges in distribution to W, write

$$W_n = W_n^{(k)} + R_n^{(k)} ,$$

where

$$W_n^{(k)} = \sum_{j=1}^{k} a_j Z_{n,j}^2 .$$

For any fixed k, the Multivariate Central Limit Theorem yields

$$(Z_{n,1}, \ldots, Z_{n,k}) \stackrel{d}{\to} (Z_1, \ldots, Z_k) .$$

By the Continuous Mapping Theorem,

$$P\{W_n \le t\} \le P\{W_n^{(k)} \le t\} \to P\{W^{(k)} \le t\} .$$

Therefore, for any k,

$$\limsup_n P\{W_n \le t\} \le P\{W^{(k)} \le t\}$$

and so

$$\limsup_n P\{W_n \le t\} \le \lim_{k \to \infty} P\{W^{(k)} \le t\} = P\{W \le t\} . \tag{14.55}$$

Similarly, for any $\delta > 0$,

$$P\{W_n \le t\} \ge P\{W_n \le t,\ \ R_n^{(k)} < \delta\} \ge P\{W_n^{(k)} \le t - \delta,\ \ R_n^{(k)} < \delta\} .$$

Using the general inequality $P(AB) \ge P(A) - P(AB^c)$ yields

$$P\{W_n \le t\} \ge P\{W_n^{(k)} \le t - \delta\} - P\{R_n^{(k)} \ge \delta\} .$$

But, by Markov's Inequality,

$$P\{R_n^{(k)} \geq \delta\} \leq \delta^{-1} E(R_n^{(k)}) \leq \delta^{-1} \sum_{j=k+1}^{\infty} a_j \ .$$

Hence, for any δ and k,

$$P\{W_n \leq t\} \geq P\{W_n^{(k)} \leq t - \delta\} - \delta^{-1} \sum_{j=k+1}^{\infty} a_j$$

and so

$$\liminf_n P\{W_n \leq t\} \geq P\{W^{(k)} \leq t - \delta\} - \delta^{-1} \sum_{j=k+1}^{\infty} a_j \ .$$

Now, let $k \to \infty$ to conclude

$$\liminf_n P\{W_n \leq t\} \geq P\{W \leq t - \delta\} \ .$$

Letting $\delta \to 0$ and using the continuity of the distribution of W, we conclude

$$\liminf_n P\{W_n \leq t\} \geq P\{W \leq t\} \tag{14.56}$$

Combining (14.55) and (14.56) yields (ii).

Part (iii) follows from Lemma 11.2.1. To prove (iv), suppose j is such that

$$E_P[T_j(X_1)] \neq E_{P_0}[T_j(X_1)] \ .$$

By the Law of Large Numbers,

$$\frac{1}{n} \sum_{i=1}^{n} T_j(X_i) \xrightarrow{P} E_P[T_j(X_1)]$$

and so

$$|Z_{n,j}| = |n^{1/2} \cdot \frac{1}{n} \sum_{i=1}^{n} \{T_j(X_i) - E_{P_0}[T_j(X_i)]\}| \xrightarrow{P} \infty \ .$$

Therefore,

$$P\{W_n > w_n(1-\alpha)\} \geq P\{a_j Z_{n,j}^2 > w(1-\alpha)\} \to 1 \ . \blacksquare$$

Note that the conclusion (iv) holds if the critical value of the test $w_{n,1-\alpha}$ is replaced by $w_{1-\alpha}$. Using either critical value results in a test that is asymptotically consistent in level. Of course, one can achieve exact level α if F_{W_n} is not continuous by rejecting H_0 if $W_n > w_{n,1-\alpha}$ and possibly randomizing if $W_n = w_{n,1-\alpha}$. But, the above result also implies $W_n = w_{n,1-\alpha}$ with probability tending to 0.

Thus, we can conclude that the test that rejects for large W_n is consistent in power against a broad family of alternatives. Indeed, for a given set of orthonormal functions $T_1, T_2, \ldots$, let Ω_k denote the family of densities (14.35) with k fixed. Let W_n be of the form (14.53) with positive, summable weights a_j. Then, the test that rejects for large W_n is consistent in power against any $P \neq P_0$ in $\bigcup_{k=1}^{\infty} \Omega_k$. Actually, letting Ω'_k denote the family of distributions P such that

$$E_P[T_k(X_1)] \neq E_{P_0}[T_k(X_1)] \ .$$

Then, the test is consistent in power against any P in $\bigcup_{k=1}^{\infty} \Omega'_k$. In contrast, Neyman's smooth tests are consistent in power against Ω_k and $\bigcup_{j=1}^{k} \Omega'_j$, where k is fixed.

For example, for testing uniformity using the normalized Legendre polynomials $T_1, T_2, \ldots$, the test that rejects for large W_n is consistent in power against any P that is not the uniform distribution, since P and the uniform distribution cannot have the same sequence of moments.

Example 14.5.1 (The Cramér-von Mises Test) Let $X_1, \ldots, X_n$ be i.i.d. real-valued random variables with c.d.f. F. For testing $F = F_0$, the Cramér-von Mises statistic is given by

$$C_n = n \int_{-\infty}^{\infty} [\hat{F}_n(x) - F_0(x)]^2 dF_0(x) \ , \tag{14.57}$$

where $\hat{F}_n(x)$ is the empirical c.d.f.

$$\hat{F}_n(x) = \frac{1}{n} \sum_{i=1}^{n} I\{X_i \le x\} \ .$$

The distribution of C_n under F_0 is the same for all F_0 which are continuous (Problem 14.34). Hence, we now assume that $F_0(x) = x$. Now, C_n can actually be represented as a weighted quadratic test statistic W_n with

$$T_j(x) = \sqrt{2} \cos(\pi j x) \ , \qquad j = 1, 2, \ldots$$

and $a_j = 1/(\pi^2 j^2)$. To see this, note that the functions $\sqrt{2}\sin(\pi j x)$, $j = 1, 2, \ldots$ form an orthonormal basis of the space $L_2[0,1]$, the (equivalence class of) functions that are square integrable on $[0,1]$ (see Section A.3). By Parseval's formula (A.7), it follows that

$$C_n = n \sum_{j=1}^{\infty} \{ \int_0^1 [\hat{F}_n(x) - x] \sqrt{2} \sin(\pi j x) dx \}^2 \ .$$

By integration by parts (Billingsley (1995), Theorem 18.4),

$$\int_0^1 [\hat{F}_n(x) - x] \sqrt{2} \sin(\pi j x) dx = \frac{-1}{\pi j} \int_0^1 \sqrt{2} \cos(\pi j x) d(\hat{F}_n(x) - x)$$

$$= \frac{-1}{\pi j} \int_0^1 \sqrt{2} \cos(\pi j x) d\hat{F}_n(x) = -\frac{1}{\pi j n} \sum_{i=1}^{n} T_j(X_i) = -\frac{Z_{n,j}}{\pi j n^{1/2}} \ .$$

Hence,

$$C_n = \sum_{j=1}^{\infty} \frac{1}{\pi^2 j^2} Z_{n,j}^2 \ ,$$

as required.

By Theorem 14.5.1, it follows that, under the null hypothesis,

$$C_n \xrightarrow{d} \sum_{j=1}^{\infty} \frac{1}{\pi^2 j^2} Z_j^2 \ ,$$

where $Z_1, Z_2, \ldots$ is a sequence of i.i.d. standard normal random variables. It also follows that the test is pointwise consistent in power against any alternative c.d.f. F for which

$$E_F[T_j(X_1)] = \int_0^1 \sqrt{2}\cos(\pi j x) dF(x) \neq \int_0^1 \sqrt{2}\cos(\pi j x) dF_0(x) = 0$$

for some j. But,

$$\int_0^1 \cos(\pi j x) dF(x) = 0 \quad \text{for all } j = 1, 2, \ldots$$

implies $F = F_0$ (Problem 14.36), and so the test is pointwise consistent in power against any $F \neq F_0$. ■

Example 14.5.2 (The Anderson-Darling Test) As in Example 14.5.1 for testing $F(x) = F_0(x) = x$, consider the Anderson-Darling statistic defined by

$$A_n = n \int_0^1 \frac{[\hat{F}_n(x) - x]^2}{x(1-x)} dx \quad . \tag{14.58}$$

It can be shown (Problem 14.37) that A_n has the form (14.38) of a weighted quadratic test statistic with

$$a_j = \frac{1}{j(j+1)}$$

and $T_j(x)$ the jth normalized Legendre polynomial on $[0, 1]$ (used in Neyman's original proposal of Neyman's smooth tests; see Section 14.4). Thus,

$$A_n = \sum_{j=1}^{\infty} \frac{1}{j(j+1)} Z_{n,j}^2 \quad , \tag{14.59}$$

(while Neyman's test corresponds to $\sum_{j=1}^{k} Z_{n,j}^2$). It then follows that, under $F = F_0$,

$$A_n \xrightarrow{d} \sum_{j=1}^{\infty} \frac{1}{j(j+1)} Z_j^2 \ .$$

In fact, many test statistics defined by an integral of the form

$$\int_0^1 U^2(x) dx$$

can be rewritten in the form of a weighted quadratic test statistic. A general treatment of such integral tests of fit can be found in Chapter 5 of Shorack and Wellner (1986); also, see van der Vaart and Wellner (1996).

Theorem 14.5.1 considered the behavior of a general weighted quadratic test under the null hypothesis $P = P_0$ and under a fixed alternative. Next, we would like to consider the behavior of W_n under a sequence of local alternatives P_n.

Suppose P_n has density p_n and P_0 has density p_0 with respect to some common measure μ. Consider the likelihood ratio based on n i.i.d. observations $X_1, \ldots, X_n$ given by

$$L_n = L_n(X_1, \ldots, X_n) = \frac{\prod_{i=1}^n p_n(X_i)}{\prod_{i=1}^n p_0(X_i)} \ .$$

Assume, under P_0,

$$\log(L_n) = n^{-1/2} \sum_{i=1}^{n} \tilde{\eta}(X_i) - \frac{\sigma^2}{2} + o_{P_0^n}(1) , \tag{14.60}$$

where

$$E_{P_0}[\tilde{\eta}(X_i)] = 0$$

and

$$0 < E_{P_0}[\tilde{\eta}^2(X_i)] = \sigma^2 < \infty .$$

Then, the Central Limit Theorem implies that, under P_0,

$$\log(L_n) \xrightarrow{d} N(-\frac{\sigma^2}{2}, \sigma^2)$$

and $\{P_n^n\}$ and $\{P_0^n\}$ are contiguous (by Corollary 12.3.1). Furthermore, under P_0,

$$Z_{n,j} = n^{-1/2} \sum_{i=1}^{n} T_j(X_i) \xrightarrow{d} N(0,1) .$$

By the bivariate Central Limit Theorem, under P_0, $(Z_{n,j}, \log(L_n))$ is asymptotically bivariate normal with asymptotic covariance

$$c_j = Cov_{P_0}[T_j(X_1), \tilde{\eta}(X_1)] . \tag{14.61}$$

It follows from Corollary 12.3.2 that, under P_n,

$$Z_{n,j} \xrightarrow{d} N(c_j, 1) .$$

Similarly, for any fixed integer k and constants $\alpha_1, \ldots, \alpha_k$, under P_0,

$$\sum_{j=1}^{k} \alpha_j Z_{n,j} = n^{-1/2} \sum_{i=1}^{n} \sum_{j=1}^{k} \alpha_j T_j(X_i) \xrightarrow{d} N(0, \sum_{j=1}^{k} \alpha_j^2)$$

and

$$(\sum_{j=1}^{k} \alpha_j Z_{n,j}, \log(L_n))$$

is asymptotically bivariate normal with covariance

$$Cov_{P_0}(\sum_{j=1}^{k} \alpha_j Z_{n,j}, \log(L_n)) = Cov_{P_0}(\sum_{j=1}^{k} \alpha_j T_j(X_i), \tilde{\eta}(X_i)) = \sum_{j=1}^{k} \alpha_j c_j .$$

Hence, under P_n,

$$\sum_{j=1}^{k} \alpha_j Z_{n,j} \xrightarrow{d} N(\sum_{j=1}^{k} \alpha_j c_j, 1) ,$$

again by Corollary 12.3.2. By the Cramér-Wold device, it follows that, under P_n,

$$(Z_{n,1}, \ldots, Z_{n,k}) \xrightarrow{d} (Z_1 + c_1, \ldots, Z_k + c_k) , \tag{14.62}$$

where $Z_1, \ldots, Z_k$ are i.i.d. $N(0,1)$. This suggests that, under P_n,

$$W_n \xrightarrow{d} \sum_{j=1}^{\infty} a_j (Z_j + c_j)^2 \ .$$

In fact, the following result is true.

Theorem 14.5.2 *Let W_n be defined by (14.38) with $a_j \geq 0$ and*

$$\sum_{j=1}^{\infty} a_j < \infty \ .$$

(i) Assume, based on n i.i.d. observations from P_n, for any k,

$$(Z_{n,1}, \ldots, Z_{n,k}) \xrightarrow{d} (Z_1 + c_1, \ldots, Z_k + c_k) \ , \tag{14.63}$$

where $Z_1, Z_2, \ldots$ are i.i.d. $N(0,1)$. If $\sum_{j=1}^{\infty} a_j c_j^2 < \infty$, then

$$W_n \xrightarrow{d} \sum_{j=1}^{\infty} a_j (Z_j + c_j)^2 \ . \tag{14.64}$$

(ii) If P_n is such that the loglikelihood ratio L_n satisfies (14.60), then, under P_n, (14.63) holds with c_j given by (14.61). Furthermore, $\sum_j a_j c_j^2 < \infty$ and so (14.64) holds as well.

Proof. The proof of (i) is a straightforward generalization of Theorem 14.5.1. (Note that it can be generalized further in that the $Z_{n,j}$ need not be a normalized average and the Z_j need not be normal nor independent.) To prove (ii), note that (14.63) holds by the discussion leading to (14.62). Moreover,

$$\sum_{j=1}^{\infty} a_j c_j^2 = \sum_{j=1}^{\infty} a_j Cov_{P_0}^2[T_j(X_i), \tilde{\eta}(X_i)]$$

$$\leq \sum_{j=1}^{\infty} a_j Var_{P_0}[T_j(X_i)] Var_{P_0}[\tilde{\eta}(X_i)] = Var_{P_0}[\tilde{\eta}(X_i)] \cdot \sum_{j=1}^{\infty} a_j < \infty \ .$$

Hence, the condition (14.63) in (i) holds. ∎

Example 14.5.3 (Limiting Power Calculation) As in Example 14.4.4, let $f_n(x)$ be given by (14.43) with $b_n = hn^{-1/2}$. As noted in Example 14.4.4, under f_n,

$$(Z_{n,1}, \ldots, Z_{n,k})^T \xrightarrow{d} N(c, I_k) \ ,$$

where c has jth component $c_j = h\langle T_j, u \rangle$. Note that

$$\sum_j a_j c_j^2 \leq h^2 \int_0^1 u^2(x) dx \sum_j a_j < \infty \ .$$

Therefore, by Theorem 14.5.2, (14.64) holds. ∎

Assume the hypothesis in Theorem 14.5.2 (ii). Let $w_{1-\alpha}$ be the $1-\alpha$ quantile of the limit distribution under the null hypothesis. Then, the limiting power against P_n is given by

$$P\{\sum_{j=1}^{\infty} a_j (Z_j + c_j)^2 > w_{1-\alpha}\} . \tag{14.65}$$

If there exists a nonzero c_j for which $a_j > 0$, then (14.65) exceeds α (Problem 14.41). For example, if $a_j > 0$ for all j, then the requirement is that there exists some j for which c_j is nonzero. But, this must be the case if $1, T_1, T_2, \ldots$ form an orthonormal basis for $L_2(P_0)$, because Parseval's identity implies

$$0 < Var_{P_0}[\tilde{\eta}(X_1)] = \sum_{j=1}^{\infty} c_j^2 .$$

It follows that not all c_j can be 0.

Thus, unlike Neyman's smooth test with $k_n \to \infty$, the limiting power for W_n is nontrivial against certain contiguous alternatives, and so it appears that tests based on W_n are better at detecting alternatives that are close to H_0. However, we now show that the limiting power of W_n can be α against a contiguous sequence of alternatives.

Example 14.5.4 (Another Local Power Calculation) Let

$$T_j(x) = \sqrt{2} \cos(\pi j x) .$$

Set $p_\theta(x) = C(\theta) \exp[\theta T_B(x)]$. If B is fixed and large, the limiting distribution of W_n against $\theta = hn^{-1/2}$ is given by the distribution of $a_B(Z_B + h)^2$. Since $a_B \to 0$ as $B \to \infty$, it follows that

$$a_B(Z_B + h)^2 \xrightarrow{P} 0$$

as $B \to \infty$. Therefore, the limiting power against such a sequence is small. In order to obtain a limiting value of α, let

$$f_n(x) = C_n(\theta) \exp[\theta T_n(x)] . \tag{14.66}$$

Then, if $\theta = hn^{-1/2}$, the limiting power of the test based on W_n against such a sequence is α, even though P_n^n is contiguous to P_0^n, where P_n is the distribution with density f_n when $\theta = hn^{-1/2}$ (Problem 14.39). ■

A difficulty in applying a weighted quadratic test statistic is the computation of critical values and power. Of course, one may resort to Monte Carlo simulation of the null distribution. Alternatively, the representation of the limiting distribution as that of

$$W = \sum_{j=1}^{\infty} a_j (Z_j + c_j)^2 \tag{14.67}$$

can be useful. For example, the null distribution (in the case $c_j = 0$) has characteristic function

$$\zeta_W(t) = \prod_{j=1}^{\infty} (1 - 2ia_j t)^{-1/2}$$

(Problem 14.40). In the special case of the Cramér-von Mises test, Smirnov inverted ζ_W (see Durbin (1973)) and obtained

$$P\{W > x\} = \frac{1}{\pi}\sum_{j=1}^{\infty}(-1)^{j+1}\int_{(2j-1)^2\pi^2}^{4j^2\pi^2}\frac{1}{y}\sqrt{\frac{-\sqrt{y}}{\sin(\sqrt{y})}}\exp(-\frac{xy}{2})dy .$$

Alternatively, one may truncate the series (14.67) to a finite sum and use numerical methods; see Durbin and Knott (1972). Another possibility is to match moments of W to a Pearson family of distributions, as done by Stephens (1976).

Some numerical power comparisons between competing goodness of fit tests can be found in Durbin and Knott (1972) and Stephens (1974), where both the Anderson-Darling and Cramér-von Mises statistics outperform the Kolmogorov-Smirnov test. A further comparison is presented in D'Agostino and Stephens (1986), Section 8.14. However, Example 14.5.4 shows that tests based on weighted quadratic statistics W_n can have poor power against higher frequency alternatives, such as (14.66). In the case of the Cramér-von Mises statistic and the Anderson-Darling statistic, this can be explained by the rapid downweighting of the a_j. Moreover, several simulation studies have demonstrated that Neyman's smooth tests can outperform tests based on W_n over a wide range of alternatives; see Miller and Quesenberry (1979), Rayner and Best (1989) and Eubank and LaRiccia (1992). In summary, both Neyman's smooth tests and weighted quadratic tests offer viable approaches to testing goodness of fit, but neither approach is asymptotically uniformly optimal. Unfortunately, we will see in the next section that no test can perform uniformly well against local or contiguous alternatives when the family of possible alternatives is large.

14.6 Global Behavior of Power Functions

For testing uniformity, the Kolmogorov-Smirnov and the weighted quadratic tests such as the Cramér-von Mises test are consistent in power against any alternative. Even the Chi-squared test with a finite number of partitions and the Neyman smooth tests with finite k are consistent in power against a broad range of alternatives. However, as we will see in this section, the power of any goodness of fit test is poor against a local sequence of (contiguous) alternatives, except possibly in a finite (bounded) number of directions, even with increasing sample size. Such a statement is not surprising for Neyman's smooth tests with k fixed, since then only a finite number of orthogonal directions are used. While a quadratic test statistic gives positive weight to infinitely many components, the weights a_j satisfy $\sum_j a_j < \infty$; this condition evidently entails

$$\sum_{j=k+1}^{\infty} a_j < \epsilon$$

for large enough k, so that the test essentially only uses a finite number of directions as well; roughly, the test behaves similar to the corresponding test obtained by summing over only the first k components. (For a rigorous statement, see Milbrodt and Strasser (1990, Remark 2.6) and Janssen (1995).) Thus, while consistency may hold against any fixed alternative as $n \to \infty$, there remains the possibility that, for any fixed sample size n, any test will perform poorly against a broad range of alternatives. Moreover, one cannot simply increase k to obtain power against a broader family of distributions. As we saw in the case of the Chi-squared test of uniformity with $k+1$ cells, while increasing k increases the set of consistent alternatives, it will decrease the limiting power against contiguous alternatives. Roughly speaking, we will see that one can only obtain reasonable power locally across a family of distributions of fixed bounded dimension.

In order to make this precise, first consider the following normal model, which arises as the limiting experiment for testing goodness of fit in Section 14.4. The argument leading to the optimality result (14.42) was based on the fact that, for the parametric model P_θ of densities p_θ given by (14.35), the experiment $\{P^N_{hn^{-1/2}}\}$ is (locally) asymptotically normal at $\theta_0 = 0$, where the limit experiment $\{Q_h\}$ consists of observing $Z^T = (Z_1, \ldots, Z_k)$ and the Z_i are independent with $Z_i \sim N(h_i, 1)$. In this model, for testing $h = 0$ against $|h| \geq b$, the maximin test rejects when $\sum_{i=1}^k Z_i^2 > c_{k,1-\alpha}$. The maximin power of this test over alternatives $|h| \geq b$ is given by the right side of (14.42), which is denoted by

$$M(k, b) = P\{\chi_k^2(b^2) > c_{k,1-\alpha}\} .$$

By Lemma 14.3.1, $M(k, b) \to \alpha$ as $k \to \infty$. Thus, in the limiting normal experiment with k large, one cannot test h against $|h| \geq b$ uniformly well in all directions. To put this another way, consider the r_k-dimensional subspace V_k of $\mathbb{R}^k$ which, without loss of generality, we take to be spanned by the first r_k axes of the original k-dimensional space. Then, the maximin power against alternatives in V_k with $\sum_{i=1}^k h_i^2 = b^2$ is attained by $h_1 = \cdots = h_{r_k} = b/r_k$ and $h_{r_k+1} = \cdots = h_k = 0$. The same argument used in Lemma 14.3.1 shows that the maximum power will tend to α if $r_k \to \infty$. Therefore, in order for the power to be bounded away from α as $k \to \infty$, we must require r_k bounded as $k \to \infty$. Thus, one cannot expect to construct tests with high power, except possibly in a finite-dimensional subspace. This point was made clear by Janssen (2000a), who provided more specific bounds on the dimension of the subspace. We now develop his results.

Lemma 14.6.1 *Suppose $Z_1, \ldots, Z_k$ are independent with Z_i distributed as $N(h_i, 1)$. Here, the parameter $(h_1, \ldots, h_k)$ varies in $\mathbb{R}^k$. Consider testing the null hypothesis that $h_i = 0$ for all i, against the alternative that not all the h_i are 0. Let $\phi = \phi(Z_1, \ldots, Z_k)$ be any test with $E_0(\phi) = \alpha$. Define e_i to be the unit vector in $\mathbb{R}^k$ with 1 in the ith component and 0 in the other components. Then, for each $H > 0$,*

$$\sum_{i=1}^k [\sup |E_{te_i}(\phi) - \alpha| : \ |t| \leq H]^2 \leq \alpha(1-\alpha)(\exp(H^2) - 1) . \tag{14.68}$$

PROOF. The function

$$g_i(t) = |E_{te_i}(\phi) - \alpha|$$

is continuous on $t \in [-H, H]$, and so it attains its maximum at some point t_i. Let

$$Y_i = \exp(t_i Z_i - \frac{t_i^2}{2}) - 1 .$$

Using the fact that $E[\exp(tZ)] = \exp(t^2/2)$ if Z is $N(0,1)$ yields $E_0(Y_i) = 0$ and

$$Var_0(Y_i) = \exp(t_i^2) - 1 \le \exp(H^2) - 1 \ .$$

Let φ denote the standard normal density. Then, the point of introducing the Y_i is that

$$E_{t_i e_i}[\phi(Z_1,\ldots,Z_k)] = \int \phi(z_1,\ldots,z_k)\varphi(z_i - t_i)\prod_{j\neq i}\varphi(z_j)\prod_{i=1}^k dz_i$$

$$= \int \phi(z_1,\ldots,z_k)\frac{\varphi(z_i - t_i)}{\varphi(z_i)}\prod_{i=1}^k \varphi(z_i)dz_i$$

$$= \int \phi(z_1,\ldots,z_k)\exp(t_i z_i - \frac{t_i^2}{2})\prod_{i=1}^k \varphi(z_i)dz_i = E_0[\phi(Z_1,\ldots,Z_k)Y_i] + \alpha$$

and so

$$E_{t_i e_i}[\phi(Z_1,\ldots,Z_k)] - \alpha = Cov_0(\phi, Y_i) \ .$$

Define

$$\beta_i = \begin{cases} \frac{Cov_0(\phi,Y_i)}{Var_0(Y_i)} & \text{if } Var_0(Y_i) > 0 \\ 0 & \text{otherwise.} \end{cases} \tag{14.69}$$

Note that, if $t_i \neq 0$, then $Var_0(Y_i) > 0$; if $t_i = 0$, then $Y_i = 0$ and $\beta_i = 0$. Define $\tilde{\phi}$ by the relation

$$\phi(Z_1,\ldots,Z_k) - \alpha = \sum_{i=1}^k \beta_i Y_i + \tilde{\phi} \ ,$$

so that

$$E_0(\tilde{\phi}) = 0 \ , \qquad E_0(\tilde{\phi}^2) < \infty$$

and

$$Cov_0(\tilde{\phi}, Y_i) = 0 \qquad i = 1,\ldots n \ .$$

This implies $\tilde{\phi}$ is uncorrelated with $\phi - \tilde{\phi}$, and so

$$Var_0(\phi) = Var_0(\tilde{\phi} + \phi - \tilde{\phi}) = Var_0(\tilde{\phi}) + Var_0(\phi - \tilde{\phi}) \ .$$

Therefore,

$$Var_0(\phi - \tilde{\phi}) \le Var_0(\phi) = E_0(\phi^2) - \alpha^2 \le \alpha(1-\alpha) \ .$$

Also,

$$\sum_{i=1}^k \beta_i^2 Var_0(Y_i) = Var_0(\sum_{i=1}^k \beta_i Y_i) = Var_0(\phi - \tilde{\phi}) \le \alpha(1-\alpha) \ . \tag{14.70}$$

But,

$$E_{t_i e_i}(\phi) - \alpha = \beta_i Var_0(Y_i)$$

implies

$$|E_{t_i e_i}(\phi) - \alpha|^2 \le \beta_i^2 Var_0(Y_i) \cdot Var_0(Y_i) \le \beta_i^2 Var_0(Y_i)(\exp(H^2) - 1) .$$

Summing over i and using the bound (14.70) yields the result. ■

Notice that the bound on the right side of (14.68) does not depend on k, the dimension of the parameter space. In fact, the same bound holds for tests based on an infinite sequence $Z_1, Z_2, \ldots$. In order to avoid certain technical aspects of likelihoods on infinite product spaces, we restrict attention to the case of k finite.

We now use the previous lemma to show that, for the normal testing problem studied in Lemma 14.6.1, the power of any level α test is poor, except possibly on a restricted range of alternatives. Thus, for fixed large k, it is impossible to construct a test that has high power in all directions (which certainly implies the same conclusion for any larger k or when $k = \infty$). The following notation will be used. For a set V in $\mathbb{R}^k$, let $V^\perp$ be defined as

$$V^\perp = \{x : \langle x, v\rangle = 0 \quad \text{for all} \quad v \in V\} .$$

Theorem 14.6.1 *Suppose $Z_1, \ldots, Z_k$ are independent, with Z_i normally distributed with mean h_i and variance one. The parameter $h = (h_1, \ldots, h_k)^T$ varies freely in $\mathbb{R}^k$. For testing $h = 0$ versus $h \ne 0$, let $\phi = \phi(Z_1, \ldots, Z_k)$ be any test with $E_0(\phi) = \alpha$. Fix any ϵ and any $H > 0$. Assume*

$$k > 1 + \epsilon^{-1}\alpha(1-\alpha)[\exp(H^2) - 1] . \tag{14.71}$$

Then, there exists a linear subspace V, whose dimension d is independent of k and ϕ, such that

$$\sup\{|E_h(\phi) - \alpha| : \ h \in V^\perp, \ |h| \le H\} \le \epsilon \tag{14.72}$$

and

$$d \le 1 + \epsilon^{-1}\alpha(1-\alpha)[\exp(H^2) - 1] . \tag{14.73}$$

In words, the power of ϕ is poor on $V^\perp \bigcap \{h : \ |h| \le H\}$.

PROOF. Let $V_0 = \{0\}$. We will inductively choose linear subspaces $V_n = \text{span}\{v_1, \ldots, v_n\}$ of $\mathbb{R}^k$ as follows. Given $v_1, \ldots, v_n$, let v_{n+1} be orthogonal to $v_1, \ldots, v_n$ and satisfy $|v_{n+1}| = 1$ and

$$\left[\sup |E_{tv}(\phi) - \alpha| : \ |t| \le H, \ v \in V_n^\perp, \ |v| = 1\right]^2 \le |E_{t_{n+1}v_{n+1}}(\phi) - \alpha|^2 + \frac{\epsilon}{2^{n+1}} .$$

Let $b_{n+1} = |E_{t_{n+1}v_{n+1}}(\phi) - \alpha|^2$. Choose m to be the smallest positive integer satisfying

$$b_m + \frac{\epsilon}{2^m} \le \epsilon . \tag{14.74}$$

To see that such an m exists and $m \le k$, note that Lemma 14.6.1 implies (possibly after an orthogonal transformation) that

$$\sum_{n+1}^{k} \left(b_n + \frac{\epsilon}{2^n}\right) \le \alpha(1-\alpha)[\exp(H^2) - 1] + \epsilon .$$

But, the assumption on k implies

$$\frac{\alpha(1-\alpha)[\exp(H^2)-1]}{\epsilon k}+\frac{1}{k}<1$$

which implies

$$\frac{1}{k}\sum_{n=1}^{k}\left(b_n+\frac{\epsilon}{2^n}\right)\leq\frac{\alpha(1-\alpha)[\exp(H^2)-1]}{k}+\frac{\epsilon}{k}<\epsilon\ .$$

Hence, there exists such an m with $m \leq k$. Let V in the statement of the theorem be V_{m-1}. Then (14.72) is satisfied because m satisfies (14.74). Moreover, since

$$b_j+\frac{\epsilon}{2^j}>\epsilon \quad for\ j=1,\ldots,m-1\ ,$$

we have

$$(m-1)\epsilon<\sum_{j=1}^{m-1}\left(b_j+\frac{\epsilon}{2^j}\right)\leq\alpha(1-\alpha)[\exp(H^2)-1]+\epsilon\ ,$$

where the last inequality follows from Lemma 14.6.1. Therefore,

$$m-1\leq 1+\epsilon^{-1}\alpha(1-\alpha)[\exp(H^2)-1]\ . \blacksquare$$

The point of Lemma 14.6.1 and Theorem 14.6.1 is that one cannot have high power uniformly in all orthonormal directions. This is not particularly surprising given that there are k observations and k parameters. Nevertheless, the statistician must then implicitly or explicitly construct a test so that the power is high in certain important directions.

We can obtain analogous results for the problem of testing $P = P_0$ based on n i.i.d. observations from P. Even with increasing n, the total amount of squared power greater than α of any test (sequence) is bounded.

Theorem 14.6.2 *Let $X_1,\ldots,X_n$ be i.i.d. P_θ, where P_θ has density p_θ given by (14.35) with $\theta\in\mathbb{R}^k$. For testing $\theta=0$ versus $\theta\neq 0$, let $\phi_n=\phi_n(X_1,\ldots,X_n)$ be any level α test. Fix $\epsilon>0$ and $H>0$, and assume k satisfies (14.71). Then, (i)*

$$\limsup_n\sum_{i=1}^{k}\left[\sup|E_{te_in^{-1/2}}(\phi_n)-\alpha|:\ |t|\leq H\right]^2 \tag{14.75}$$

$$\leq\alpha(1-\alpha)[\exp(H^2)-1]\ .$$

(ii) There exists a subspace V of $\mathbb{R}^k$ whose dimension d satisfies (14.73) (independent of k) such that

$$\limsup_n\sup\{|E_{hn^{-1/2}}(\phi_n)-\alpha|:\ h\in V^{\perp},\ |h|\leq H\}\leq\epsilon\ . \tag{14.76}$$

Proof. The sequence of models $P^n_{hn^{-1/2}}$ is asymptotically normal with identity covariance matrix I_k, in the sense of Definition 13.4.1. Indeed, the family is an exponential family and hence is quadratic mean differentiable. In fact, as previously pointed out, the score vector for this model is given by (14.37) and is asymptotically multivariate normal with mean 0 and identity covariance matrix. The proof then follows from Theorem 13.4.1, which compares the limiting power of any test sequence with that of a test for the normal model studied in Lemma 14.6.1. For the limiting normal experiment, an upper bound for the sum of squared powers is given in Lemma 14.6.1, and so this bound must hold asymptotically. Similarly, (ii) follows by Theorem 14.6.1. ■

Of course, the theorem has implications for testing $P = P_0$ against alternatives outside the parametric model (14.35). Indeed, since the right side of (14.75) does not depend on k, we may take $k = \infty$ on the left side and obtain the same result. That is, the squared infinite sum of deviations of power from α remains bounded. We have stated the result first for finite k since our proof then only requires convergence to a normal experiment in a finite dimensional space (as we have not considered infinite dimensional spaces).

In fact, Janssen (2000a) shows that this result holds for each n as well; that is, one can simply delete the limsup in (14.75). Thus, the power of any test sequence is essentially flat outside a space of dimension d, where d does not depend on n.

To explain the result a little further, fix $\theta \in \mathbb{R}^k$ and consider the one-dimensional model indexed by t with density $p_{t\theta}$ defined in (14.35). If we know that the actual distribution belongs to this one-dimensional exponential family submodel for some $t > 0$, then a UMP level α test sequence exists for testing $t = 0$ against $t > 0$, which we now denote by $\phi_\theta^* = \{\phi_{n,\theta}^*\}$; moreover,

$$\lim_n E_{t\theta n^{-1/2}}(\phi_{n,\theta}^*) = 1 - \Phi(z_{1-\alpha} - t|\theta|) \tag{14.77}$$

(Problem 14.42). We will now connect the performance of an arbitrary test sequence $\phi = \{\phi_n\}$ with the notion of asymptotic relative efficiency, as developed in Section 13.2. Let $N_\phi(t, \theta, \alpha, \beta)$ be the smallest sample size required to achieve power at least β if the true density is $p_{t\theta}$. In the case of ϕ_θ^*, it follows from (14.77) (or Theorem 13.2.1(iii)) that, if $|\theta| = 1$,

$$\lim_{t \to 0+} t^2 N_{\phi_\theta^*}(t, \theta, \alpha, \beta) = (z_\alpha - z_\beta)^2 . \tag{14.78}$$

With α and β fixed, choose any small $\delta > 0$, any ϵ satisfying $0 < \epsilon < \beta - \alpha$ and $H > 0$ large enough so that $(z_\alpha - z_\beta)^2/H^2 \le \delta$. For an arbitrary test ϕ, Theorem 14.6.2(ii) implies that there exists $V \subset \mathbb{R}^k$ of dimension d satisfying (14.73) such that, for all small t and $\theta \in V^\perp$ with $|\theta| = 1$, the power function at $t\theta$ is bounded above by $\alpha + \epsilon < \beta$, at least for t such that $tn^{1/2} \le H$. This in turn implies that n must satisfy $n^{1/2} t > H$ in order to achieve power β; thus,

$$\liminf_{t \to 0+} t^2 N_\phi(t, \theta, \alpha, \beta) \ge H^2 . \tag{14.79}$$

Combining (14.78) and (14.79) yields, for $\theta \in V^\perp$,

$$\limsup_{t \to 0+} \frac{N_{\phi_\theta^*}(t, \theta, \alpha, \beta)}{N_\phi(t, \theta, \alpha, \beta)} \le \frac{(z_\alpha - z_\beta)^2}{H^2} \le \delta . \tag{14.80}$$

If the limsup on the left side of (14.80) is replaced by a limit, which is shown to exist, the limiting value would be the Pitman ARE of ϕ with respect to ϕ_θ^* for the submodel $P_{t\theta}$. While we are not claiming such a limit exists, the interpretation of the result is the following. Except on a set of θ values of dimension d (independent of n and k), the test ϕ_θ^* requires approximately no more than a small proportion δ of the sample size required by ϕ to achieve power β. Therefore, it is not possible to simultaneously have high power along all "directions" θ, at least from this local point of view.

The possibility of high power for parameter values far from 0 (corresponding to $|t| > H$) remains however, and so this result does not contradict the uniform consistency result, Theorem 14.2.2, of the Kolmogorov-Smirnov test; there, the power tends to one against nonlocal alternatives. But, for testing goodness of fit against a broad nonparametric class of alternatives, Lemma 14.3.1 and Theorem 14.6.1 imply that any test (sequence) performs well locally only in some fixed finite dimensional subset of alternatives, even as n increases. To put it another way, any test has a preferred set of alternatives (of bounded dimension) for which its power is locally high. Unfortunately, it may be difficult to analyze the preferred alternatives for any particular test. For certain classes of tests, such as the integral tests of Cramér-von Mises or Anderson and Darling, there exist principle component decompositions of the test statistics, which lead to useful power calculations; see Shorack and Wellner (1986), Chapter 5. For the Kolmogorov-Smirnov test, it is known that it is roughly speaking more powerful to deviations of the median; see Milbrodt and Strasser (1990) and Janssen (1995) for a more careful statement. Since any given test sequence can only perform well for some finite dimensional set of alternatives, it seems natural to design tests that perform well on a given finite dimensional set, which is exactly the approach taken in the construction of Neyman's smooth tests. A general theory of efficiency of goodness of fit tests is developed in Nitikin (1995), who also compares distinct notions of efficiency; also see Janssen (2003). Unfortunately, different efficiency notions give rise to different tests. It appears that a proper choice of test must be based on some knowledge of the possible set of alternatives for a given experiment. By restricting attention to families of densities with different degrees of smoothness, asymptotically maximin results have been obtained; see Ingster and Suslina (2003).

14.7 Problems

Section 14.2

Problem 14.1 Verify (14.3).

Problem 14.2 (i) Let $X_1, \ldots, X_n$ be i.i.d. real-valued random variables with c.d.f. F. Consider testing $F = F_0$ against $F \neq F_0$ based on the Kolmogorov-Smirnov test. Fix F with $n^{1/2} d_K(F, F_0) > s_{n,1-\alpha}$. Show that

$$P_F\{T_n > s_{n,1-\alpha}\} \geq 1 - \frac{1}{4|n^{1/2} d_K(F, F_0) - s_{n,1-\alpha}|^2} .$$

Hint: Use (14.6) and Chebyshev's inequality.
(ii) Derive the alternative lower bound to the power of the Kolmogorov-Smirnov test given by (14.8). Compare the two lower bounds.

Problem 14.3 For testing $F = F_0$, where F_0 is the uniform (0,1) c.d.f., consider alternatives F_n to F_0 of the form

$$F_n(t) = (1 - \lambda_n)F_0(t) + \lambda_n G(t) ,$$

where $G \neq F_0$ is some fixed distribution. Show that, if $\lambda_n = \lambda n^{-1/2}$, then the limiting power of the Kolmogorov-Smirnov test is bounded away from α if λ is large enough.

Problem 14.4 Suppose F_n satisfies $n^{1/2} d_K(F_n, F_0) \to 0$. For testing $F = F_0$ at level α, show that the limiting power of the Kolmogorov-Smirnov test against F_n is no better than α. In the case that F_n is continuous for every n, show that the limiting power is equal to α.

Problem 14.5 (i) Suppose $\{P_\theta\}$ is q.m.d. at θ_0, where P_θ is a probability distribution on $\mathbb{R}$ with corresponding c.d.f. F_θ. Show that there exists $B = B_{\theta_0}(h) < \infty$ such that

$$\limsup_n n d_K^2(F_{\theta_0 + hn^{-1/2}}, \theta_0) \leq B_{\theta_0}(h)$$

and $B_{\theta_0}(h) \to 0$ as $h \to 0$.
(ii) Construct a sequence of probability distributions P_n on the real line with corresponding c.d.f.s F_n satisfying $d_K(F_n, F_0) \to 0$ but $H(P_n, P_0)$ is bounded away from 0, where H is the Hellinger metric. On the other hand, show that $H(P_n, P_0) \to 0$ implies $d_K(F_n, F_0) \to 0$.

Problem 14.6 Let F_0 be the uniform (0,1) c.d.f. and consider testing $F = F_0$ by the Kolmogorov Smirnov test.
(i) Construct a sequence of alternatives F_n to F_0 satisfying $n^{1/2} d_K(F_n, F_0) \to \delta$ with $0 < \delta < \infty$ such that the limiting power against F_n is α, even though there exist tests whose limiting power against F_n exceeds α.
(ii) Construct a sequence of alternatives F_n to F_0 satisfying $n^{1/2} d_K(F_n, F_0) \to \delta$ with $0 < \delta < \infty$ such that the limiting power against F_n is one.
[*Hint*: Fix $1 > \gamma_n > 0$ with $n^{1/2}\gamma_n \to \delta > 0$ and let $F_n(t)$ be defined by

$$F_n(t) = \begin{cases} 0 & \text{if } t < \gamma_n \\ t & \text{if } \gamma_n \leq t \leq 1. \end{cases} \tag{14.81}$$

Note that $d_K(F_n, F_0) = \gamma_n$ by construction. Let $U_1, \ldots, U_n$ be i.i.d. according to the uniform distribution on $(0, 1)$, and let $\hat{G}_n(t)$ denote the empirical c.d.f. of the U_i. Set

$$X_i = \begin{cases} U_i & \text{if } U_i \geq \gamma_n \\ \gamma_n & \text{if } U_i < \gamma_n, \end{cases} \tag{14.82}$$

so that $X_1, \ldots, X_n$ are i.i.d. with c.d.f. F_n. Let $\hat{F}_n(t)$ denote the empirical c.d.f. of the X_i. Argue that

$$\sup_t |\hat{F}_n(t) - t| \leq \max\left[\sup_t |\hat{G}_n(t) - t|, \gamma_n\right]$$

and

$$P_{F_n}\{T_n > s_{n,1-\alpha}\} \leq P\{n^{1/2} \sup_t |\hat{G}_n(t) - t| > s_{n,1-\alpha}\}$$

if $n^{1/2}\gamma_n < s_{n,1-\alpha}$. If $\delta < s_{1-\alpha}$, then this last condition will be satisfied for large enough n. Finally, the last displayed expression equals α.]

Problem 14.7 Let $\mathbf{F}$ be the family of distributions having density $F' = f$ on $(0,1)$ and let $F_0' = f_0$ be the uniform density. Consider testing the null hypothesis that $F = F_0$ based on the Kolmogorov Smirnov test. Show that, if $d_k(f, f_0)$ is the sup distance between densities and $0 < c < 1$, then, for every n,

$$\inf P_F\{T_n \geq s_{n,1-\alpha} : \ F \in \mathbf{F},\ d_K(F', f_0) \geq c\} \leq \alpha \ . \tag{14.83}$$

Argue that the result applies if d_K is replaced by the L^2 distance between densities. *Hint:* Consider densities of the form $f_\theta(t) = 1 + c\sin(2\pi\theta t)$. [Compare this result with Theorem 14.2.2. Ingster and Suslina (2003) argue that alternatives based on the sup distance between distribution functions are less natural than metrics between densities. This problem shows it is impossible for the Kolomogorv-Smirnov test to have power bounded away from α against such alternatives. In fact, this is true for *any* test; see Ingster (1993) and Section 14.6. However, by restricting the family of densities to have further smoothness properties, Ingster and Suslina (2003) have obtained positive results.]

Problem 14.8 Generalize Theorem 14.2.2 to any EDF test statistic of the form $n^{1/2}d(\hat{F}_n, F_0)$, if d is a metric weaker than the Kolmogorov-Smirnov metric d_k in the sense

$$d(F, G) \leq C d_K(F, G)$$

for some constant C. In particular, show the result applies to the Cramér-von Mises test.

Problem 14.9 For testing the null hypothesis that $X_1, \ldots, X_n$ are i.i.d. from a normal distribution with unknown mean μ and unknown variance σ^2, show that the null distribution of (14.13) does not depend on (μ, σ) (but it does depend on n). Describe a simulation method to approximate this null distribution. How can you construct a test that is exact level $\alpha = 0.05$ based on simulation? Generalize this problem to testing a general location-scale family.

Problem 14.10 Suppose $X_1, \ldots, X_n$ are i.i.d. with c.d.f F on the real line. The problem is to test the null hypothesis H_0 that the X_i are uniform on $(0, \theta]$ for some θ. Let $\hat{\theta}_n = \max(X_1, \ldots, X_n)$, and let $\hat{F}_n$ be the empirical distribution function. Let $d_K(F, G)$ be the Kolmogorov-Smirnov distance between F and G. Consider the test statistic

$$T_n = n^{1/2} d_K(\hat{F}_n, F_{\hat{\theta}_n}) \ ,$$

where F_θ is the uniform $(0, \theta)$ c.d.f. Under H_0, what is the limiting distribution of T_n?

Problem 14.11 Let $X_1, \cdots, X_n$ be a sample from the normal distribution with mean θ and variance 1, with cdf denoted by $F_\theta(\cdot)$. Let $\Phi(z)$ denote the standard normal cdf, so that $F_\theta(t) = \Phi(t - \theta)$. For any two cdfs F and G, let $\|F - G\|$ denote $\sup_t |F(t) - G(t)|$. Let $\hat{\theta}_n$ be the estimator of θ minimizing $\|\hat{F}_n - F_\theta\|$, where $\hat{F}_n(t) = n^{-1}\sum_{i=1}^n 1(X_i \le t)$ denotes the empirical cdf. In case you are worried about problems of existence or uniqueness, you may assume $\hat{\theta}_n$ is any estimator satisfying

$$\|\hat{F}_n - F_{\hat{\theta}_n}\| \le \inf_\theta \|\hat{F}_n - F_\theta\| + \epsilon_n ,$$

where ϵ_n is any sequence of positive constants tending to 0.
(i) Prove $\hat{\theta}_n$ is a consistent estimator of θ.
(ii) Suppose now the observations come from a cdf F, possibly nonnormal. The problem is to test the null hypothesis that F is normal with variance 1 against the alternative hypothesis that F is not. Consider the test statistic

$$T_n = \inf_\theta \|\hat{F}_n - F_\theta\| .$$

Argue, if F is $N(\theta, 1)$, then the distribution of T_n does not depend on θ.
(iii) If F is not normal with variance one, argue that T_n tends in probability to the constant $\gamma_F = \inf_\theta \|F - F_\theta\|$, and $\gamma_F > 0$.
(iv) Find a sequence of constants c_n so that the test that rejects iff $T_n \ge c_n$ has probability of a Type I error tending to 0, and has power tending to one for any fixed alternative F. *Hint:* Use the Dvoretzky, Kiefer, Wolfowitz Inequality.

Section 14.3

Problem 14.12 (i) Verify (14.19).
(ii) Verify (14.20).

Problem 14.13 Prove the convergence (14.21).

Problem 14.14 In the multinomial goodness of fit problem, calculate the Information matrix $I(p)$ given by (14.22).

Problem 14.15 Prove part (iii) of Theorem 14.3.1.

Problem 14.16 Show that the result Theorem 14.3.2 (ii) holds for the likelihood ratio test.

Problem 14.17 Prove Lemma 14.3.1(i).

Problem 14.18 Recall $M(k,h)$ defined by (14.27) and let F_k denote the c.d.f. of the central Chi-squared distribution with k degrees of freedom. Show that

$$M(k,h) = \alpha + \gamma_k \frac{h^2}{2} + o(h^2) \quad \text{as } h \to 0 ,$$

where

$$\gamma_k = F_k(c_{k,1-\alpha}) - F_{k+2}(c_{k,1-\alpha}) .$$

Problem 14.19 As in Section 14.3.2, consider the Chi-squared test for testing uniformity on $(0,1)$ based on $k+1$ cells; call if $\phi^*_{n,k}$. Fix any $B < \infty$ and $\epsilon > 0$. Let $\mathcal{U}_B$ be the set of u with $\int u = 0$ and $\int u^2 \le B$. For alternative sequences of the form (14.25) with $b_n = n^{-1/2}$, show that, if k is large enough (but fixed), then

$$\limsup_n \sup_{u:u \in \mathcal{U}_B} E_{f_n}(\phi^*_{n,k}) \le \alpha + \epsilon .$$

Problem 14.20 Verify (14.33).

Problem 14.21 Under the setup of Problem 12.61, determine a Chi-squared test statistic, as well as its limiting distribution under the null hypothesis. [For a discussion of the Chi-squared test for testing independence in a two-way table, see Diaconis and Efron (1985) and Loh (1989).]

Problem 14.22 The Hardy-Weinberg law says the following. If gene frequencies are in equilibrium, the genotypes AA, Aa, and aa occur in a population with frequencies θ^2, $2\theta(1-\theta)$, and $(1-\theta)^2$. In an i.i.d. sample of size n, with each outcome being an AA, Aa, or aa with the above probabilities, let X_1, X_2, and X_3 be the observed counts. For example, X_1 is the number of trials where the observation is AA. Note that $X_1 + X_2 + X_3 = n$. The joint distribution of (X_1, X_2, X_3) is a trinomial distribution. Hence,

$$P_\theta\{X_1 = x_1, X_2 = x_2, X_3 = x_3\} = \frac{n!}{x_1!x_2!x_3!}(\theta^2)^{x_1}[2\theta(1-\theta)]^{x_2}[(1-\theta)^2]^{x_3}$$

for any nonnegative integers x_1, x_2, and x_3 summing to n. Find the MLE and its limiting distribution (suitably normalized). Derive the likelihood ratio and chi-squared tests to test the Hardy-Weinberg law.

Problem 14.23 In Example 14.3.1, verify (14.32) and determine the MLE $\hat\beta_n$ for the linkage submodel being tested. Determine the limiting distribution of the Chi-squared statistic $Q_n(\hat\beta_n)$.

Problem 14.24 Consider the limit distribution of the Chi-squared goodness-of-fit statistic for testing normality if using the maximum likelihood estimators to estimate the unknown parameters. Specifically, suppose $X_1, \ldots, X_n$ are i.i.d. and the problem is to test whether the underlying distribution is $N(\theta, 1)$ for some θ. Group the observations into just 2 groups: positive observations and negative observations. Derive the limit distribution of the Chi-squared statistic using the sample mean to estimate θ and show it is not Chi-squared.

Section 14.4

Problem 14.25 Let $X_1, \ldots, X_n$ be i.i.d. F, and consider testing the null hypothesis that F is the uniform (0,1) c.d.f. For $\theta = (\theta_1, \theta_2) \in \mathbb{R}^2$, consider a family of alternative densities of the form

$$p_\theta(x) = C(\theta)\exp[\theta_1 T_1(x) + \theta_2 T_2(x)], \quad 0 < x < 1 .$$

Assume this two-parameter exponential family is well-defined for all small enough $|\theta|$, so that the family is a full rank exponential family which is q.m.d. at $\theta = 0$ with Information matrix at $\theta = 0$ denoted by $I(0)$. For the submodel with $\theta_2 = 0$, what is the optimal limiting power for testing $\theta_1 = 0$ against $\theta_1 = hn^{-1/2}$ at level α. Similarly, with $\theta_1 = 0$, what is the optimal limiting power for testing $\theta_2 = 0$ against $\theta_2 = hn^{-1/2}$. Prove that no level α test sequence exists whose limiting power simultaneously achieves these optimal values. *Hint:* If (Z_1, Z_2) is bivariate normal with (h_1, h_2), then no UMP test exists for testing $(h_1, h_2) = (0, 0)$.

Problem 14.26 In Example 14.4.3, show that the multinomal distribution can be written in the form (14.35) for the given orthogonal choice of functions T_j.

Problem 14.27 Show that (14.42) holds with $B = \infty$ if $Var_\theta[T_j(X_1)]$ is uniformly bounded in θ. *Hint:* Argue by contradiction. Suppose there exists h_n with $|h_n| \geq b$ such that

$$E_{h_n n^{-1/2}}(\phi_n^*) \to \ell ,$$

where ℓ is less than the right side of (14.42). This is a contradiction if

$$E_{h_n n^{-1/2}}(\phi_n^*) \to 1$$

if $|h_n| \to \infty$. By taking subsequences if necessary, assume the jth component $h_{n,j}$ of h_n satisfies $|h_{n,j}| \to \infty$. Then,

$$E_{h_n n^{-1/2}}(\phi_n^*) \geq P_{h_n n^{-1/2}}\{Z_{n,j}^2 > c_{k,1-\alpha}\} .$$

It now suffices to show $|Z_{n,j}| \to \infty$ in probability under $h_n n^{-1/2}$. But $|E_\theta[T_j(X_1)]|$ increases in θ (using properties of exponential families) while the variance of $Z_{n,j}$ remains bounded.

Problem 14.28 For testing $P = P_0$ in the model of densities (14.35) with T_j the normalized Legendre polynomials, show that Neyman's smooth test is consistent in power against any distribution P as long as the first k moments of P are not all identical to the first k moments of P_0.

Problem 14.29 Let $X_1, \ldots, X_n$ be i.i.d. random variables on [0,1] with unknown distribution P. The problem is to test $P = P_0$, the uniform distribution on $[0, 1]$. Assume a parametric model with densities of the form (14.35) for some fixed positive integer k. Set $T_0(x) = 1$ and assume the functions $T_1, \ldots, T_k$ are chosen so that $T_0, \ldots, T_k$ is a set of orthonormal functions on $L^2(P_0)$. Assume that

$$\sup_{x,j} |T_j(x)| < \infty ,$$

so that $C_k(\theta)$ is well-defined for all k-vectors θ. Let Λ_n be a probability distribution over values of θ and let $A(\phi_n, \Lambda_n)$ denote the average power of a test ϕ_n with respect to Λ_n; that is,

$$A(\phi_n, \Lambda_n) = \int_\theta E_\theta(\phi_n) d\Lambda_n(\theta) .$$

In particular, let Λ_n be the k-dimensional normal distribution with mean vector 0 and covariance matrix equal to n^{-1} times the identity matrix. Among tests ϕ_n such that $E_0(\phi_n) \to \alpha$, find one that maximizes

$$\lim_n A(\phi_n, \Lambda_n)$$

and find a simple expression for this limiting average power.

Problem 14.30 Use Minkowski's Inequality (Section A.3) to show (14.47).

Problem 14.31 Show (14.49).

Problem 14.32 Argue the validity of (14.51).

Section 14.5

Problem 14.33 In Theorem 14.5.1, show that W has a continuous, strictly increasing distribution function on $(0, \infty)$. *Hint:* Write $W = a_i Z_i^2 + R$ for some i with $a_i > 0$ and note that $a_i Z_i^2$ has a density.

Problem 14.34 Show that the distribution of the Cramér-von Mises test statistic (14.57) under F_0 is the same for all continuous distributions F_0.

Problem 14.35 Show that the Cramér-von Mises test statistic C_n given by (14.57) can be computed by

$$C_n = \frac{1}{12n} + \sum_{i=1}^{n} [X_{(i)} - \frac{2i-1}{2n}]^2 ,$$

where $X_{(1)} \leq \cdots \leq X_{(n)}$ denote the order statistics; see D'Agostino and Stephens (1986), p.101 for computing formulas for other test statistics based on the empirical distribution function.

Problem 14.36 Let F be a c.d.f. on $(0, 1)$. If

$$\int_0^1 \cos(\pi j x) dF(x) = 0$$

for all $j = 1, 2, \ldots$, then F must be the uniform distribution on $(0, 1)$. *Hint:* Integrate by parts and use the fact the functions $\sqrt{2} \sin(\pi j x)$ form a complete, orthonormal system for $L_2[0, 1]$.

Problem 14.37 Show that the Anderson-Darling statistic (14.58) can be rewritten in the form (14.59).

Problem 14.38 Consider W_n with $T_j(x) = \sqrt{2} \cos(\pi j x)$. Fix $\gamma_j \geq 0$ with $\gamma_j^2 < \infty$. Let

$$q_\theta(x) = C(\theta) \exp[\theta \sum_{j=1}^{\infty} \gamma_j T_j(x)] .$$

Show that, under $\theta = h n^{-1/2}$,

$$W_n \xrightarrow{d} \sum_j a_j (Z_j + h\gamma_j)^2 .$$

Problem 14.39 Verify the claims made in Example 14.5.4.

Problem 14.40 What is the characteristic function of the limiting random variable W of Theorem 14.5.1? As a special case, show that the characteristic function of the limiting null distribution of the Cramér-von Mises statistic is given by

$$\zeta(t) = \prod_{j=1}^{\infty} (1 - \frac{2t}{\pi j})^{-1/2} .$$

(Note this characteristic function was inverted by Smirnov; see Durbin (1973), p.32.)

Problem 14.41 Show that the expression (14.65) exceeds α if there exists a j for which $a_j > 0$ and $c_j \neq 0$. Also, show that (14.65) is an increasing function of $|c_j|$.

Section 14.6

Problem 14.42 Show why (14.77) is true.

Problem 14.43 Consider the setting of Problem 8.30 with $\delta = \delta_k \to 0$ as $k \to \infty$. At what rate should $\delta_k \to 0$ as $k \to \infty$ so that the limiting maximin power is strictly between α and 1?

14.8 Notes

Goodness of fit tests based on the empirical distribution function were introduced by Cramér (1928), von Mises (1931) and Kolmogorov (1933). A classical reference for the asymptotic theory of such tests is Durbin (1973); also see Kendall and Stuart (1979, Chapter 30), Neuhaus (1979) and Tallis (1983). Readable accounts of many goodness of fit tests can be found in D'Agostino and Stephens (1986) and Read and Cressie (1988). Methods particularly suitable for testing normality are discussed for example in Shapiro, Wilk, and Chen (1968), Hegazy and Green (1975), D'Agostino (1982), Hall and Welsh (1983), and Spiegelhalter (1983), and for testing exponentiality in Galambos (1982), Brain and Shapiro (1983), Spiegelhalter (1983), Deshpande (1983), Doksum and Yandell (1984), and Spurrier (1984). See also Kent and Quesenberry (1982). Modern treatments are provided by Shorack and Wellner (1986), van der Vaart and Wellner (1996) and Nikitin (1995). Some recent generalizations of the Kolmogorov-Smirnov test for testing goodness of fit are discussed in Beran and Millar (1986, 1988), Romano (1988), Khmaladze (1993), Cabaña and Cabaña (1997), Dümbgen (1998), and Polonik (1999).

The Chi-squared test was introduced by Pearson (1900). Cohen and Sackrowitz (1975) prove a finite sample local optimality property of the Chi-squared test in the case of testing a simple null hypothesis of equal cell probabilities. In the context of testing a multinomial, Hoeffding (1965) compares the Chi-squared and likelihood ratio tests while letting $\alpha \to 0$ as $n \to \infty$; he finds the likelihood ratio test superior if the number of cells is fixed, but notes the situation can be reversed otherwise. As mentioned in Section 14.3, the use of the Chi-squared test for testing goodness of fit for continuous observations is hampered by the apparent loss of information through data grouping and the choice of the number of groups. The choice of the number of groups is considered, among others, by Quine and Robinson (1985) and by Kallenberg, Oosterhoff, and Schriever (1985). A class of generalized Chi-squared tests is studied in Drost (1988, 1989), who uses the concept of Pitman asymptotic relative efficiency to study the effect of number of groups; a particular test, known as the Rao-Robson-Nikulin test, is advocated. In the case of nuisance parameters, Fisher (1924) argued that estimating nuisance parameters changes the limiting distribution of the Chi-squared statistic, contrary to early opinion. Chernoff and Lehmann (1954) showed that, when parameters are estimated by MLEs, the limiting distribution need not even be Chi-squared; also see de Wet and Randles (1987). For further discussion on the Chi-squared test, as well as its generalizations, see Kendall and Stuart (1979). A full account of the practical implementation of the Chi-squared test, including the accuracy of the Chi-squared approximation and choice of classes, as well as an extensive bibliography, are provided by Greenwood and Nikulin (1996).

Neyman's smooth tests were introduced in Neyman (1937b), which were seen to be a special case of the general score tests of Rao (1947). An elementary treatment is provided by Rayner and Best (1989), who also consider extensions to problems with nuisance parameters. The use of smooth tests for multinomial data with adaptive choice of order is advocated in Eubank (1997). For recent work on smooth tests for composite hypotheses, see Inglot, Kallenberg and Ledwina (1997), Pena (1998), and Fan and Lin (1998).

Goodness of fit tests based on the Kullback-Leibler divergence are studied in Barron (1989). Tests based on spacings are considered in Wells, Jammalamadaka and Tiwari (1993). Tests based on the likelihood ratio are given in Zhang (2002).

15
General Large Sample Methods

15.1 Introduction

In this chapter, we shall deal with situations where both the hypothesis and the class of alternatives may be nonparametric and where as a result it may be difficult even to construct tests (or confidence regions) that satisfactorily control the level (exactly or asymptotically). For such situations, we shall develop methods which achieve this modest goal under fairly general assumptions. A secondary aim will then be to obtain some idea of the power of the resulting tests.

In Section 15.2, we consider the class of randomization tests as a generalization of permutation tests. Under the randomization hypothesis (see Definition 15.2.1 below), the empirical distribution of the values of a given statistic recomputed over transformations of the data serves as a null distribution; this leads to exact control of the level in such models. When the randomization hypothesis holds, the construction applies broadly to any statistic. Efficiency properties ensue if the statistic is chosen appropriately.

In Section 15.3 we review some basic constructions of confidence regions and tests, which derive from the limiting distribution of an estimator or test sequence. This serves to motivate the bootstrap construction studied in Section 15.4; the bootstrap method offers a powerful approach to approximating the sampling distribution of a given statistic or estimator. The emphasis here is to find methods that control the level constraint, at least asymptotically. Like the randomization construction, the bootstrap approach will be asymptotically efficient if the given statistic is chosen appropriately; for example, see Theorem 15.4.2 and Corollary 15.4.1.

While the bootstrap is quite general, how does it compare in situations when other large sample approaches apply as well? In Section 15.5, we provide some support to the claim that the bootstrap approach can improve upon methods which rely on a normal approximation. The use of the bootstrap in the context of hypothesis testing is studied in Section 15.6.

While the bootstrap method is quite broadly applicable, in some situations, it can be inconsistent. A more general approach based on subsampling is presented in 15.7. Together, these approaches serve as valuable tools for inference without having to make strong assumptions about the underlying distribution.

15.2 Permutation and Randomization Tests

Permutation tests were introduced in Chapter 5 as a robust means of controlling the level of a test if the underlying parametric model only holds approximately. For example, the two-sample permutation t-test for testing equality of means studied in Section 11 of Chapter 5 has level α whenever the two populations have the same distribution under the null hypothesis (without the assumption of normality). In this section, we consider the large sample behavior of permutation tests and, more generally, randomization tests. The use of the term randomization here is distinct from its meaning in Sections 5.10. There, randomization was used as a device prior to collecting data, for example, by randomly assigning experimental units to treatment or control. Such a device allows for a meaningful comparison after the data has been observed, by considering the behavior of a statistic recomputed over permutations in the data. Thus, the term randomization referred to both the experimental design and the analysis of data by recomputing a statistic over permutations or randomizations (sometimes called rerandomizations) of the data. It is this latter use of randomization that we now generalize. Thus, the term randomization test will refer to tests obtained by recomputing a test statistic over transformations (not necessarily permutations) of the data.

A general test construction will be presented that yields an exact level α test for a fixed sample size, under a certain group invariance hypothesis. Then, two main questions will be addressed. First, we shall consider the robustness of the level. For example, in the two-sample problem just mentioned, the underlying populations may have the same mean under the null hypothesis, but differ in other ways, as in the classical Behrens-Fisher problem, where the underlying populations are normal but may not have the same variance. Then, the rejection probability under such populations is no longer α, and it becomes necessary to investigate the behavior of the rejection probability. In addition, we also consider the large sample power of permutation and randomization tests. In the two-sample problem when the underlying populations are normal with common variance, for example, we should like to know whether there is a significant loss in power when using a permutation test as compared to the UMPU t-test.

15.2.1 The Basic Construction

Based on data X taking values in a sample space $\mathcal{X}$, it is desired to test the null hypothesis H that the underlying probability law P generating X belongs to a certain family Ω_0 of distributions. Let $\mathbf{G}$ be a finite group of transformations g

of $\mathcal{X}$ onto itself. The following assumption, which we will call the *randomization hypothesis*, allows for a general test construction.

Definition 15.2.1 (Randomization Hypothesis) Under the null hypothesis, the distribution of X is invariant under the transformations in $\mathbf{G}$; that is, for every g in $\mathbf{G}$, gX and X have the same distribution whenever X has distribution P in Ω_0.

The randomization hypothesis asserts that the null hypothesis parameter space Ω_0 remains invariant under g in $\mathbf{G}$. However, here we specifically do not require the alternative hypothesis parameter space to remain invariant (unlike what was assumed in Chapter 6).

As an example, consider testing the equality of distributions based on two independent samples $(Y_1, \ldots, Y_m)$ and $(Z_1, \ldots, Z_n)$, which was previously considered in Sections 5.8-5.11. Under the null hypothesis that the samples are generated from the same probability law, the observations can be permuted or assigned at random to either of the two groups, and the distribution of the permuted samples is the same as the distribution of the original samples. (Note that a test that is invariant with respect to all permutations of the data would be useless here.)

To describe the general construction of a randomization test, let $T(X)$ be any real-valued test statistic for testing H. Suppose the group $\mathbf{G}$ has M elements. Given $X = x$, let

$$T^{(1)}(x) \leq T^{(2)}(x) \leq \cdots \leq T^{(M)}(x)$$

be the ordered values of $T(gx)$ as g varies in $\mathbf{G}$. Fix a nominal level α, $0 < \alpha < 1$, and let k be defined by

$$k = M - [M\alpha] , \tag{15.1}$$

where $[M\alpha]$ denotes the largest integer less than or equal to $M\alpha$. Let $M^+(x)$ and $M^0(x)$ be the number of values $T^{(j)}(x)$ $(j = 1, \ldots, M)$ which are greater than $T^{(k)}(x)$ and equal to $T^{(k)}(x)$, respectively. Set

$$a(x) = \frac{M\alpha - M^+(x)}{M^0(x)} .$$

Generalizing the construction presented in Section 5.8, define the randomization test function $\phi(x)$ to be equal to 1, $a(x)$, or 0 according to whether $T(x) > T^{(k)}(x)$, $T(x) = T^{(k)}(x)$, or $T(x) < T^{(k)}(x)$, respectively. By construction, for every x in $\mathcal{X}$,

$$\sum_{g \in \mathbf{G}} \phi(gx) = M^+(x) + a(x)M^0(x) = M\alpha . \tag{15.2}$$

The following theorem shows that the resulting test is level α, under the hypothesis that X and gX have the same distribution whenever the distribution of X is in Ω_0. Note that this result is true for *any* choice of test statistic T.

Theorem 15.2.1 *Suppose X has distribution P on $\mathcal{X}$ and the problem is to test the null hypothesis $P \in \Omega_0$. Let $\mathbf{G}$ be a finite group of transformations of $\mathcal{X}$ onto itself. Suppose the randomization hypothesis holds, so that, for every $g \in \mathbf{G}$, X and gX have the same distribution whenever X has a distribution P in Ω_0.*

Given a test statistic $T = T(X)$, let ϕ be the randomization test as described above. Then,

$$E_P[\phi(X)] = \alpha \quad \textit{for all} \quad P \in \Omega_0 \ . \tag{15.3}$$

PROOF. To prove (15.3), by (15.2),

$$M\alpha = E_P[\sum_g \phi(gX)] = \sum_g E_P[\phi(gX)] \ .$$

By hypothesis $E_P[\phi(gX)] = E_P[\phi(X)]$, so that

$$M\alpha = \sum_g E_P[\phi(X)] = ME_P[\phi(X)] \ ,$$

and the result follows. ■

To gain further insight as to why the construction works, for any $x \in \mathcal{X}$, let $\mathbf{G}^x$ denote the $\mathbf{G}$-orbit of x; that is,

$$\mathbf{G}^x = \{gx : \ g \in \mathbf{G}\} \ .$$

Recall from Section 6.2 that these orbits partition the sample space. The hypothesis in Theorem 15.2.1 implies that the conditional distribution of X given $X \in \mathbf{G}^x$ is uniform on $\mathbf{G}^x$, as will be seen in the next theorem. Since this conditional distribution is the same for all $P \in \Omega_0$, a test can be constructed to be level α conditionally, which is then level α unconditionally as well. Because the event $\{X \in \mathbf{G}^x\}$ typically has probability zero for all x, we need to be careful about how we state a result. As x varies, the sets $\mathbf{G}^x$ form a partition of the sample space. Let $\mathcal{G}$ be the σ-field generated by this partition.

Theorem 15.2.2 *Under the null hypothesis of Theorem 15.2.1, for any real-valued statistic $T = T(X)$, any $P \in \Omega_0$, and any Borel subset B of the real line,*

$$P\{T(X) \in B | X \in \mathcal{G}\} = M^{-1} \sum_g I\{T(gx) \in B\} \tag{15.4}$$

with probability one under P. In particular, if the M values of $T(gx)$ as g varies in $\mathbf{G}$ are all distinct, then the uniform distribution on these M values serves as a conditional distribution of $T(X)$ given that $X \in \mathbf{G}^x$.

PROOF. First, we claim that, for any $g \in \mathbf{G}$ and $E \in \mathcal{G}$, $gE = E$. To see why, assume $y \in E$. Then, $g^{-1}y \in E$, because $g^{-1}y$ is on the same orbit as y. Then, $gg^{-1}y \in gE$ or $y \in gE$. A similar argument shows that, if $y \in gE$, then $y \in E$, so that $gE = E$. Now, the right hand side of (15.4) is clearly $\mathcal{G}$-measurable, since the right hand side is constant on any orbit. We need to prove, for any $E \in \mathcal{G}$,

$$\int_E M^{-1} \sum_g I\{T(gx) \in B\} dP(x) = P\{T(X) \in B, \ X \in E\} \ .$$

But, the left hand side is

$$M^{-1} \sum_g \int_E I\{T(gx) \in B\} dP(x) = M^{-1} \sum_g P\{T(gX) \in B, \ X \in E\}$$

$$= M^{-1}\sum_g P\{T(gX) \in B,\ gX \in gE\} = M^{-1}\sum_g P\{T(gX) \in B,\ gX \in E\}\ ,$$

since $gE = E$. Hence, this last expression becomes (by the randomization hypothesis)

$$M^{-1}\sum_g P\{T(X) \in B,\ X \in E\} = P\{T(X) \in B,\ X \in E\}\ ,$$

as was to be shown. ■

Example 15.2.1 (One Sample Tests) Let $X = (X_1, \ldots, X_n)$, where the X_i are i.i.d. real-valued random variables. Suppose that, under the null hypothesis, the distribution of the X_i is symmetric about 0. This applies, for example, to the parametric normal location model when the null hypothesis specifies the mean is 0, but it also applies to the nonparametric model that consists of all distributions with the null hypothesis specifying the underlying distribution is symmetric about 0. For $i = 1, \ldots, n$, let ϵ_i take on either the value 1 or -1. Consider a transformation $g = (\epsilon_1, \ldots, \epsilon_n)$ of $\mathbb{R}^n$ that takes $x = (x_1, \ldots, x_n)$ to $(\epsilon_1 x_1, \ldots, \epsilon_n x_n)$. Finally, let $\mathbf{G}$ be the $M = 2^n$ collection of such transformations. Then, the randomization hypothesis holds, i.e., X and gX have the same distribution under the null hypothesis. ■

Example 15.2.2 (Two Sample Tests) Suppose $Y_1, \ldots, Y_m$ are i.i.d. observations from a distribution P_Y and, independently, $Z_1, \ldots, Z_n$ are i.i.d. observations from a distribution P_Z. Here, $X = (Y_1, \ldots, Y_m, Z_1, \ldots, Z_n)$. Suppose that, under the null hypothesis, $P_Y = P_Z$. This applies, for example, to the parametric normal two-sample problem for testing equality of means when the populations have a common (possibly unknown) variance. Alternatively, it also applies to the parametric normal two-sample problem where the null hypothesis is that the means and variances are the same, but under the alternative either the means or the variances may differ; this model was advocated by Fisher (1935a, p.122-124). Lastly, this setup also applies to the nonparametric model where P_Y and P_Z may vary freely, but the null hypothesis is that $P_Y = P_Z$. To describe an appropriate $\mathbf{G}$, let $N = m + n$. For $x = (x_1, \ldots, x_N) \in \mathbb{R}^N$, let $gx \in \mathbb{R}^N$ be defined by $(x_{\pi(1)}, \ldots, x_{\pi(N)})$, where $(\pi(1), \ldots, \pi(N))$ is a permutation of $\{1, \ldots, N\}$. Let $\mathbf{G}$ be the collection of all such g, so that $M = N!$. Whenever $P_Y = P_Z$, X and gX have the same distribution. In essence, each transformation g produces a new data set gx, of which the first m elements are used as the Y sample and the remaining n as the Z sample to recompute the test statistic. Note that, if a test statistic is chosen that is invariant under permutations within each of the Y and Z samples (which makes sense by sufficiency), it is enough to consider the $\binom{N}{m}$ transformed data sets obtained by taking m observations from all N as the Y observations and the remaining n as the Z observations (which, of course, is equivalent to using a subgroup $\mathbf{G}'$ of $\mathbf{G}$).

As a special case, suppose the observations are real-valued and the underlying distribution is assumed continuous. Suppose T is any statistic that is a function of the ranks of the combined observations, so that T is a *rank statistic* (previously studied in Sections 6.8 and 6.9). The randomization (or permutation) distribution can be obtained by recomputing T over all permutations of the ranks. In this sense, rank tests are special cases of permutation tests. ■

Example 15.2.3 (Tests of Independence) Suppose that X consists of i.i.d. random vectors $X = ((Y_1, Z_1), \ldots, (Y_n, Z_n))$ having common joint distribution P and marginal distributions P_Y and P_Z. Assume, under the null hypothesis, Y_i and Z_i are independent, so that P is the product of P_Y and P_Z. This applies to the parametric bivariate normal model when testing that the correlation is zero, but it also applies to the nonparametric model when the null hypothesis specifies Y_i and Z_i are independent with arbitrary marginal distributions. To describe an appropriate $\mathbf{G}$, let $(\pi(1), \ldots, \pi(n))$ be a permutation of $\{1, \ldots n\}$. Let g be the transformation that takes $((y_1, z_1), \ldots, (y_n, z_n))$ to the value $((y_1, z_{\pi(1)}), \ldots, (y_n, z_{\pi(n)}))$. Let $\mathbf{G}$ be the collection of such transformations, so that $M = n!$. Whenever Y_i and Z_i are independent, X and gX have the same distribution. ■

In general, one can define a p-value $\hat{p}$ of a randomization test by

$$\hat{p} = \frac{1}{M} \sum_g I\{T(gX) \geq T(X)\} . \tag{15.5}$$

It can be shown (Problem 15.2) that $\hat{p}$ satisfies, under the null hypothesis,

$$P\{\hat{p} \leq u\} \leq u \quad \text{for all } 0 \leq u \leq 1 . \tag{15.6}$$

Therefore, the nonrandomized test that rejects when $\hat{p} \leq \alpha$ is level α.

Because $\mathbf{G}$ may be large, one may resort to an approximation to construct the randomization test, for example, by randomly sampling transformations g from $\mathbf{G}$ with or without replacement. In the former case, for example, suppose $g_1, \ldots, g_{B-1}$ are i.i.d. and uniformly distributed on $\mathbf{G}$. Let

$$\tilde{p} = \frac{1}{B} \left[1 + \sum_{i=1}^{B-1} I\{T(g_i X) \geq T(X)\} \right] . \tag{15.7}$$

Then, it can be shown (15.3) that, under the null hypothesis,

$$P\{\tilde{p} \leq u\} \leq u \quad \text{for all } 0 \leq u \leq 1 , \tag{15.8}$$

where this probability reflects variation in both X and the sampling of the g_i. Note that (15.8) holds for any B, and so the test that rejects when $\tilde{p} \leq \alpha$ is level α even when a stochastic approximation is employed. Of course, the larger the value of B, the closer $\hat{p}$ and $\tilde{p}$ are to each other; in fact, $\hat{p} - \tilde{p} \to 0$ in probability as $B \to \infty$ (Problem 15.4). Approximations based on auxiliary randomization (such as the sampling of g_i) are known as stochastic approximations.

15.2.2 Asymptotic Results

We next study the limiting behavior of the randomization test in order to derive its large sample power properties. For example, for testing the mean of a normal distribution is zero with unspecified variance, one would use the optimal t-test. But if we use the randomization test based on the transformations in Example 15.2.1, we will find that the randomization test has the same limiting power as the t-test against contiguous alternatives, and so is LAUMP. Of course, for testing the mean, the randomization test can be used without the assumption of normality, and we will study its asymptotic properties both when the underlying distribution is symmetric so that the randomization hypothesis holds, and also when the randomization hypothesis fails.

Consider a sequence of situations with $X = X^n$, $P = P_n$, $\mathcal{X} = \mathcal{X}_n$, $\mathbf{G} = \mathbf{G}_n$, $T = T_n$, etc. defined for $n = 1, 2, \ldots$; notice we use a superscript for the data $X = X^n$. Typically, $X = X^n = (X_1, \ldots, X_n)$ consists of n i.i.d. observations and the goal is to consider the behavior of the randomization test sequence as $n \to \infty$.

Let $\hat{R}_n$ denote the *randomization distribution* of T_n defined by

$$\hat{R}_n(t) = M_n^{-1} \sum_{g \in \mathbf{G}_n} I\{T_n(gX^n) \le t\} . \tag{15.9}$$

We seek the limiting behavior of $\hat{R}_n(\cdot)$ and its $1 - \alpha$ quantile, which we now denote $\hat{r}_n(1-\alpha)$ (but in the previous subsection was denoted $T^{(k)}(X)$); thus,

$$\hat{r}_n(1-\alpha) = \hat{R}_n^{-1}(1-\alpha) = \inf\{t : \ \hat{R}_n(t) \ge 1-\alpha\} .$$

We will study the behavior of $\hat{R}_n$ under the null hypothesis and under a sequence of alternatives. First, observe that

$$E[\hat{R}_n(t)] = P\{T_n(G_n X^n) \le t\} ,$$

where G_n is a random variable that is uniform on $\mathbf{G}_n$. So, in the case the randomization hypothesis holds, $G_n X^n$ and X^n have the same distribution and so

$$E[\hat{R}_n(t)] = P\{T_n(X^n) \le t\} .$$

Then, if T_n converges in distribution to a c.d.f. $R(\cdot)$ which is continuous at t, it follows that

$$E[\hat{R}_n(t)] \to R(t) .$$

In order to deduce $\hat{R}_n(t) \stackrel{P}{\to} R(t)$ (i.e., the randomization distribution asymptotically approximates the unconditional distribution of T_n), it is then enough to show $Var[\hat{R}_n(t)] \to 0$. This approach for proving consistency of $\hat{R}_n(t)$ and $\hat{r}_n(1-\alpha)$ is used in the following result, due to Hoeffding (1952). Note that the randomization hypothesis is not assumed.

Theorem 15.2.3 *Suppose X^n has distribution P_n in $\mathcal{X}_n$, and $\mathbf{G}_n$ is a finite group of transformations from $\mathcal{X}_n$ to $\mathcal{X}_n$. Let G_n be a random variable that is uniform on $\mathbf{G}_n$. Also, let G'_n have the same distribution as G_n, with X^n, G_n, and G'_n mutually independent. Suppose, under P_n,*

$$(T_n(G_n X^n), T_n(G'_n X^n)) \stackrel{d}{\to} (T, T') , \tag{15.10}$$

where T and T' are independent, each with common c.d.f. $R(\cdot)$. Then, under P_n,

$$\hat{R}_n(t) \stackrel{P}{\to} R(t) \tag{15.11}$$

for every t which is a continuity point of $R(\cdot)$. Let

$$r(1-\alpha) = \inf\{t : \ R(t) \ge 1-\alpha\} .$$

Suppose $R(\cdot)$ is continuous and strictly increasing at $r(1-\alpha)$. Then, under P_n,

$$\hat{r}_n(1-\alpha) \xrightarrow{P} r(1-\alpha) .$$

PROOF. Let t be a continuity point of $R(\cdot)$. Then,

$$E_{P_n}[\hat{R}_n(t)] = P_n\{T_n(G_n X^n) \leq t\} \to R(t) ,$$

by the convergence hypothesis (15.10). It therefore suffices to show that $Var_{P_n}[\hat{R}_n(t)] \to 0$ or, equivalently, that

$$E_{P_n}[\hat{R}_n^2(t)] \to R^2(t) .$$

But,

$$E_{P_n}[\hat{R}_n^2(t)] = M_n^{-2} \sum_g \sum_{g'} P_n\{T_n(gX^n) \leq t,\ T_n(g'X^n) \leq t\}$$

$$= P_n\{T_n(G_n X^n) \leq t,\ T_n(G'_n X^n) \leq t\} \to R^2(t) ,$$

again by the convergence hypothesis (15.10). Hence, $\hat{R}_n(t) \to R(t)$ in P_n-probability. The convergence of $\hat{r}_n(1-\alpha)$ now follows by Lemma 11.2.1 (ii). ■

Note that, if the randomization hypothesis holds, then $T_n(X^n)$ and $T_n(G_n X^n)$ have the same distribution. The assumption (15.10) then implies the unconditional distribution of $T_n(X^n)$ under P_n converges to R in distribution. The conclusion is that the randomization distribution approximates this (unconditional) limit distribution in the sense that (15.11) holds.

Example 15.2.4 (One Sample Test, continuation of Example 15.2.1) In Example 15.2.1, first consider $T_n = n^{1/2}\bar{X}_n$. If P denotes the common distribution of the X_i, then $P_n = P^n$ is the joint distribution of the sample. Let P be any distribution with mean 0 and finite nonzero variance $\sigma^2(P)$ (not necessarily symmetric). We will verify (15.10) with $R(t) = \Phi(t/\sigma(P))$. Let $\epsilon_1, \ldots, \epsilon_n, \epsilon'_1, \ldots, \epsilon'_n$ be mutually independent random variables, each 1 or -1 with probability $\frac{1}{2}$ each. We must find the limiting distribution of

$$n^{-1/2} \sum_i (\epsilon_i X_i, \epsilon'_i X_i) .$$

But, the vectors $(\epsilon_i X_i, \epsilon'_i X_i)$, $1 \leq i \leq n$, are i.i.d. with

$$E_P(\epsilon_i X_i) = E_P(\epsilon'_i X_i) = E(\epsilon_i) E_P(X_i) = 0 ,$$

$$E_P[(\epsilon_i X_i)^2] = E(\epsilon_i^2) E_P(X_i^2) = \sigma^2(P) = E_P[(\epsilon'_i X_i)^2] ,$$

and

$$Cov_P(\epsilon_i X_i, \epsilon'_i X_i) = E_P(\epsilon_i \epsilon'_i X_i^2) = E(\epsilon_i) E(\epsilon'_i) E_P(X_i^2) = 0 .$$

By the bivariate Central Limit Theorem,

$$n^{-1/2} \sum_i (\epsilon_i X_i, \epsilon'_i X_i) \xrightarrow{d} (T, T') ,$$

where T and T' are independent, each distributed as $N(0, \sigma^2(P))$. Hence, by Theorem 15.2.3, we conclude

$$\hat{R}_n(t) \xrightarrow{P} \Phi(t/\sigma(P))$$

and

$$\hat{r}_n(1-\alpha) \xrightarrow{P} \sigma(P) z_{1-\alpha} .$$

Let ϕ_n be the randomization test which rejects when $T_n > \hat{r}_n(1-\alpha)$, accepts when $T_n < \hat{r}_n(1-\alpha)$ and possibly randomizes when $T_n = \hat{r}_n(1-\alpha)$. Since T_n is asymptotically normal, it follows by Slutsky's Theorem that

$$E_P(\phi_n) = P\{T_n > \hat{r}_n(1-\alpha)\} + o(1) \to P\{\sigma(P)Z > \sigma(P)z_{1-\alpha}\} = \alpha ,$$

where Z denotes a standard normal variable. In other words, we have deduced the following for the problem of testing the mean of P is zero versus the mean exceeds zero. By Theorem 15.2.1, ϕ_n is exact level α if the underlying distribution is symmetric about 0; otherwise, it is at least asymptotically pointwise level α as long as the variance is finite.

We now investigate the asymptotic power of ϕ_n against the sequence of alternatives that the observations are $N(hn^{-1/2}, \sigma^2)$. By the above, under $N(0, \sigma^2)$, $\hat{r}_n(1-\alpha) \to \sigma z_{1-\alpha}$ in probability. By contiguity, it follows that, under $N(hn^{-1/2}, \sigma^2)$, $\hat{r}_n(1-\alpha) \to \sigma z_{1-\alpha}$ in probability as well. Under $N(hn^{-1/2}, \sigma^2)$, T_n is $N(h, \sigma^2)$. Therefore, by Slutsky's Theorem, the limiting power of ϕ_n against $N(hn^{-1/2}, \sigma^2)$ is then

$$E_{P_n}(\phi_n) \to P\{\sigma Z + h > \sigma z_{1-\alpha}\} = 1 - \Phi(z_{1-\alpha} - \frac{h}{\sigma}) .$$

In fact, this is also the limiting power of the optimal t-test for this problem. Thus, there is asymptotically no loss in efficiency when using the randomization test as opposed to the optimal t-test, but the randomization test has the advantage that its size is α over all symmetric distributions. In the terminology of Section 13.2, the efficacy of the randomization test is $1/\sigma$ and its ARE with respect to the t-test is 1. In fact, the ARE is 1 whenever the underlying family is a q.m.d. location family with finite variance (Problem 15.6).

In fact, the randomization test that is based on T_n is identical to the randomization test that is based on the usual t-statistic t_n. To see why, first observe that the randomization test based on T_n is identical to the randomization test based on $\tilde{T}_n = T_n/(\sum_i X_i^2)^{1/2}$, simply because all "randomizations" of the data have the same value for the sum of squares. But, as was seen in Section 5.2, t_n is an increasing function of S_n for positive S_n. Hence, the one-sample t-test which rejects when t_n exceeds $t_{n-1,1-\alpha}$, the $1-\alpha$ quantile of the t-distribution with $n-1$ degrees of freedom, is equivalent to a randomization test based on the statistic t_n, except that $t_{n-1,1-\alpha}$ is replaced by the data-dependent value. Such an analogy was previously made for the two-sample test in Section 5.8.

The value of the randomization test is that one does not have to assume normality. On the other hand, the asymptotic results allow one to avoid the exact computation of the randomization distribution by approximating the critical value by the normal quantile $z_{1-\alpha}$ or even $t_{n-1,1-\alpha}$. The problem of whether to use $z_{1-\alpha}$ or $t_{n-1,1-\alpha}$ is solved in Diaconis and Holmes (1994), who also give algorithms for the exact evaluation of the randomization distribution. ■

In the previous example, it was seen that the randomization distribution approximates the (unconditional) null distribution of T_n in the sense that

$$\hat{R}_n(t) - P\{T_n \le t\} \xrightarrow{P} 0$$

if P has mean 0 and finite variance, since $P\{T_n \le t\} \to \Phi(t/\sigma(P))$. The following is a more general version of this result.

Theorem 15.2.4 *(i) Suppose $X_1, \ldots, X_n$ are i.i.d. real-valued random variables with distribution P, assumed symmetric about 0. Assume T_n is asymptotically linear in the sense that, for some function ψ_P,*

$$T_n = n^{-1/2} \sum_{i=1}^{n} \psi_P(X_i) + o_P(1) , \tag{15.12}$$

where $E_P[\psi_P(X_i)] = 0$ and $\tau_P^2 = Var_P[\psi_P(X_i)] < \infty$. Also, assume ψ_P is an odd function. Let $\hat{R}_n$ denote the randomization distribution based on T_n and the group of sign changes in Example 15.2.1. Then, the hypotheses of Theorem 15.2.3 hold with $P_n = P^n$ and $R(t) = \Phi(t/\tau(P))$, and so

$$\hat{R}_n(t) \xrightarrow{P} \Phi(t/\tau(P)) .$$

(ii) If P is not symmetric about 0, let F denote its c.d.f. and define a symmetrized version $\tilde{P}$ of P as the probability with c.d.f.

$$\tilde{F}(t) = \frac{1}{2}[F(t) + 1 - F(-t)] .$$

Assume T_n satisfies (15.12) under $\tilde{P}$. Then, under P,

$$\hat{R}_n(t) \xrightarrow{P} \Phi(t/\tau(\tilde{P})) \qquad \text{and} \qquad \hat{r}_n(1-\alpha) \xrightarrow{P} \tau(P) z_{1-\alpha} .$$

PROOF. Independent of $X^n = (X_1, \ldots, X_n)$ let $\epsilon_1, \ldots, \epsilon_n$ and $\epsilon_1', \ldots, \epsilon_n'$ be mutually independent, each ± 1 with probability $\frac{1}{2}$. Then, in the notation of Theorem 15.2.3, $G_n X^n = (\epsilon_1 X_1, \ldots, \epsilon_n X_n)$. Set $r_n(X_1, \ldots, X_n) = T_n - n^{-1/2} \sum \psi_P(X_i)$ so that $r_n(X_1, \ldots, X_n) \xrightarrow{P} 0$. Since $\epsilon_i X_i$ has the same distribution as X_i, it follows that $r_n(\epsilon_1 X_1, \ldots, \epsilon_n X_n) \xrightarrow{P} 0$, and the same is true with ϵ_i replaced by ϵ_i'. Then,

$$\left(T_n(G_n X^n), T_n(G_n' X^n)\right) = n^{-1/2} \sum_{i=1}^{n} \left(\psi_P(\epsilon_i X_i), \psi_P(\epsilon_i' X_i)\right) + o_P(1) .$$

But since ψ_P is odd, $\psi_P(\epsilon_i X_i) = \epsilon_i \psi_P(X_i)$. By the bivariate CLT,

$$n^{-1/2} \sum_{i=1}^{n} \left(\epsilon_i \psi_P(X_i), \epsilon_i' \psi_P(X_i)\right) \xrightarrow{d} (T, T') ,$$

where (T, T') is bivariate normal, each with mean 0 and variance τ_P^2, and

$$Cov(T, T') = Cov\left(\epsilon_i \psi_P(X_i), \epsilon_i' \psi_P(X_i)\right) = E(\epsilon_i) E(\epsilon_i') E_P[\psi_P^2(X_i)] = 0 ,$$

and so (i) follows.

To prove (ii), observe that, if X has distribution P and $\tilde{X}$ has distribution $\tilde{P}$, then $|X|$ and $|\tilde{X}|$ have the same distribution. But, the construction of the randomization distribution only depends on the values $|X_1|, \ldots, |X_n|$. Hence, the behavior of $\hat{R}_n$ under P and $\tilde{P}$ must be the same. But, the behavior of $\hat{R}_n$ under $\tilde{P}$ is given in (i). ∎

Example 15.2.5 (One-Sample Location Models) Suppose $X_1, \ldots, X_n$ are i.i.d. $f(x-\theta)$, where f is assumed symmetric about $\theta_0 = 0$. Assume the family is q.m.d. at θ_0 with score statistic Z_n. Thus, under θ_0, $Z_n \stackrel{d}{\to} N(0, I(\theta_0))$. Consider the randomization test based on $T_n = Z_n$ (and the group of sign changes). It is exact level α for all symmetric distributions. Moreover, $Z_n = n^{-1/2} \sum_i \tilde{\eta}(X_i, \theta_0)$, where $\tilde{\eta}$ can always be taken to be an odd function if f is even. So, the assumptions of Theorem 15.2.4 (i) hold. Hence, when $\theta_0 = 0$,

$$\hat{r}_n(1-\alpha) \to I^{1/2}(\theta_0) z_{1-\alpha} .$$

By contiguity, the same is true under $\theta_{n,h} = hn^{-1/2}$. By Theorem 13.2.1, the efficacy of the randomization test is $I^{1/2}(\theta_0)$. By Corollary 13.2.1, the ARE of the randomization test with respect to the Rao test that uses the critical value $z_{1-\alpha} I^{1/2}(\theta_0)$ (or even an exact critical value based on the true unconditional distribution of Z_n under θ_0) is 1. Indeed, the randomization test is AUMP. Therefore, there is no loss of efficiency in using the randomization test, and it has the advantage of being level α across symmetric distributions. ∎

Example 15.2.6 (Two-Sample Tests, Continuation of Example 15.2.2) Recall the setup of Example 15.2.2 where $Y_1, \ldots, Y_m$ are i.i.d. P_Y and, independently, $Z_1, \ldots, Z_n$ are i.i.d. P_Z, where P_Y and P_Z are now assumed to be distributions on the real line. Let $\mu(P)$ and $\sigma^2(P)$ denote the mean and variance, respectively, of a distribution P. Consider the test statistic

$$T_{m,n} = m^{1/2}(\bar{Y}_m - \bar{Z}_n) = m^{-1/2}[\sum_{i=1}^{m} Y_i - \frac{m}{n} \sum_{j=1}^{n} Z_j] . \tag{15.13}$$

Assume $m/n \to \lambda \in (0, \infty)$ as $m, n \to \infty$. If the variances of P_Y and P_Z are finite and nonzero and $\mu(P_Y) = \mu(P_Z)$, then

$$T_{m,n} \stackrel{d}{\to} N\left(0, \sigma^2(P_Y) + \lambda \sigma^2(P_Z)\right) .$$

We wish to study the limiting behavior of the randomization test based on the test statistic $T_{m,n}$. If the null hypothesis implies that $P_Y = P_Z$, then the randomization test is exact level α, though we may still require an approximation to its power. On the other hand, we may consider using the randomization test for testing the null hypothesis $\mu(P_Y) = \mu(P_Z)$, and the randomization test is no longer exact if the distributions differ.

Let $N = m + n$ and write

$$(X_1, \ldots, X_N) = (Y_1, \ldots, Y_m, Z_1, \ldots, Z_n) .$$

Independent of the Xs, let $(\pi(1), \ldots, \pi(N))$ and $(\pi'(1), \ldots, \pi'(N))$ be independent random permutations of $1, \ldots, N$. In order to verify the conditions for Theorem 15.2.3, we need to determine the joint limiting behavior of

$$(T_{m,n}, T'_{m,n}) = m^{-1/2}(\sum_{i=1}^{N} X_i W_i, \sum_{i=1}^{N} X_i W'_i) , \tag{15.14}$$

where $W_i = 1$ if $\pi(i) \le m$ and $W_i = -m/n$ otherwise; W_i' is defined with π replaced by π'. Note that $E(W_i) = E(X_i W_i) = 0$. Moreover, an easy calculation (Problem 15.8) gives

$$Var(T_{m,n}) = \frac{m}{n}\sigma^2(P_Y) + \sigma^2(P_Z) \tag{15.15}$$

and

$$Cov(T_{m,n}, T'_{m,n}) = m^{-1} \sum_{i=1}^{N} \sum_{j=1}^{N} E(X_i X_j W_i W_j') = 0 \ , \tag{15.16}$$

by the independence of the W_i and the W_i'. These calculations suggest the following result.

Theorem 15.2.5 *Assume the above setup with $m/n \to \lambda \in (0, \infty)$. If $\sigma^2(P_Y)$ and $\sigma^2(P_Z)$ are finite and nonzero and $\mu(P_Y) = \mu(P_Z)$, then (15.14) converges in law to a bivariate normal distribution with independent, identically distributed marginals having mean 0 and variance*

$$\tau^2 = \lambda\sigma^2(P_Y) + \sigma^2(P_Z) \ .$$

PROOF. Assume without loss of generality that $\mu(P_Y) = 0$. By the Cramér-Wold device (Theorem 11.2.3), it suffices to show, for any a and b,

$$m^{-1/2} \sum_{i=1}^{N} X_i (aW_i + bW_i') \xrightarrow{d} N\left(0, (a^2 + b^2)\tau^2\right) \ .$$

The argument follows by conditioning on the W_i and W_i' and writing the left side as

$$m^{-1/2} \sum_{i=1}^{m} Y_i (aW_i + bW_i') + m^{-1/2} \sum_{j=1}^{n} Z_j (aW_{m+j} + bW'_{m+j}) \ , \tag{15.17}$$

which becomes (conditionally) an independent sum of a linear combination of independent variables. It is not hard to check that $m^{-1} \sum_{i=1}^{m} (aW_i + bW_i')^2$ is bounded in probability (because its expectation is uniformly bounded) and

$$m^{-1} \max_i |aW_i + bW_i'|^2 \xrightarrow{P} 0 \ . \tag{15.18}$$

Thus, Lemma 11.3.3 can be applied (conditionally) to each term in (15.17) and the result follows. ■

Consider the problem of testing equality of means in the two-sample problem without imposing parametric assumptions on the underlying distributions, which can be viewed as a nonparametric version of the Behrens-Fisher problem. Theorem 15.2.3 and Theorem 15.2.5 imply that the randomization distribution is, in large samples, approximately a normal distribution with mean 0 and variance τ^2. Hence, the critical value of the randomization test that rejects for large values of $T_{m,n}$ converges in probability to $z_{1-\alpha}\tau$. On the other hand, the true sampling distribution of $T_{m,n}$ is approximately normal with mean 0 and variance

$$\sigma^2(P_Y) + \lambda\sigma^2(P_Z) \ ,$$

if $\mu(P_Y) = \mu(P_Z)$. These two distributions are identical if and only if $\lambda = 1$ or $\sigma^2(P_Y) = \sigma^2(P_Z)$. Therefore, for testing equality of means, the randomization test will be pointwise consistent in level even if P_Y and P_Z differ, as long as the variances of the populations are the same, or the sample sizes are roughly the same. In particular, when the underlying distributions have the same variance (as in the normal theory model assumed in Section 5.3 for which the two-sample t-test is UMPU), the two-sample t-test is asymptotically equivalent to the corresponding randomization test. This equivalence is not limited to the behavior under the null hypothesis; see Problem 15.10.

If the underlying variances differ and $\lambda \neq 1$, the permutation test based on $T_{m,n}$ given in (15.13) will have rejection probability that does not tend to α. However, if one replaces $T_{m,n}$ by the studentized version

$$\tilde{T}_{m,n} = T_{m,n} / \sqrt{S_Y^2 + \frac{m}{n} S_Z^2} \;, \tag{15.19}$$

where

$$S_Y^2 = (m-1)^{-1} \sum_{i=1}^{m} (Y_i - \bar{Y}_m)^2 \quad \text{and} \quad S_Z^2 = (n-1)^{-1} \sum_{j=1}^{n} (Z_j - \bar{Z}_n)^2 \;,$$

then the permutation test is pointwise consistent in level for testing equality of means, even when the underlying distributions have possibly different variances and the sample sizes differ (Problem 15.11).

Further results are given in Romano (1990). For example, two-sample permutations tests based on sample medians lead to tests that are not even pointwise consistent in level, unless the strict randomization hypothesis of equality of distributions holds. Thus, if testing equality of population medians based on the difference between sample medians, the asymptotic rejection probability of the randomization test need not be α even with the underlying populations have the same median.

15.3 Basic Large Sample Approximations

In the previous section, it was shown how permutation and randomization tests can be used in certain problems where the randomization hypothesis holds. Unfortunately, randomization tests only apply to a restricted class of problems. In this section, we discuss some generally used asymptotic approaches for constructing confidence regions or hypothesis tests based on data $X = X^n$. In what follows, $X^n = (X_1, \ldots, X_n)$ is typically a sample of n i.i.d. random variables taking values in a sample space S and having unknown probability distribution P, where P is assumed to belong to a certain collection $\mathbf{P}$ of distributions. Even outside the i.i.d. case, we think of the data X^n as coming from a model indexed by the unknown probability mechanism P. The collection $\mathbf{P}$ may be a parametric model indexed by a Euclidean parameter, but we will also consider nonparametric models.

We shall be interested in inferences concerning some parameter $\theta(P)$. By the usual duality between the construction of confidence regions and hypothesis tests, we can restrict the discussion to the construction of confidence regions. Let the range of θ be denoted Θ, so that

$$\Theta = \{\theta(P) : P \in \mathbf{P}\} .$$

Typically, Θ is a subset of the real line, but we also consider more general parameters. For example, the problem of estimating the entire cumulative distribution function (c.d.f.) of real-valued observations may be treated, so that Θ is an appropriate function space.

This leads to considering a *root* $R_n(X^n, \theta(P))$, a term first coined by Beran (1984), which is just some real-valued functional depending on both X^n and $\theta(P)$. The idea is that a confidence interval for $\theta(P)$ could be constructed if the distribution of the root were known. For example, an estimator $\hat{\theta}_n$ of a real-valued parameter $\theta(P)$ might be given so that a natural choice is $R_n(X^n, \theta(P)) = [\hat{\theta}_n - \theta(P)]$, or alternatively $R_n(X^n, \theta(P)) = [\hat{\theta}_n - \theta(P)]/s_n$, where s_n is some estimate of the standard deviation of $\hat{\theta}_n$.

When $\mathbf{P}$ is suitably large so that the problem is nonparametric in nature, a natural construction for an estimator $\hat{\theta}_n$ of $\theta(P)$ is the plug-in estimator $\hat{\theta}_n = \theta(\hat{P}_n)$, where $\hat{P}_n$ is the empirical distribution of the data, defined by

$$\hat{P}_n(E) = n^{-1} \sum_{i=1}^{n} I\{X_i \in E\} .$$

Of course, this construction implicitly assumes that $\theta(\cdot)$ is defined for empirical distributions so that $\theta(\hat{P}_n)$ is at least well-defined. Alternatively, in parametric problems for which $\mathbf{P}$ is indexed by a parameter ψ belonging to a subset Ψ of $\mathbb{R}^p$ so that $\mathbf{P} = \{P_\psi : \psi \in \Psi\}$, then $\theta(P)$ can be described as a functional $t(\psi)$. Hence, $\hat{\theta}_n$ is often taken to be $t(\hat{\psi}_n)$, where $\hat{\psi}_n$ is some desirable estimator of ψ, such as an efficient likelihood estimator.

Let $J_n(P)$ be the distribution of $R_n(X^n, \theta(P))$ under P, and let $J_n(\cdot, P)$ be the corresponding cumulative distribution function defined by

$$J_n(x, P) = P\{R_n(X^n, \theta(P)) \le x\}.$$

In order to construct a confidence region for $\theta(P)$ based on the root $R_n(X^n, \theta(P))$, the sampling distribution $J_n(P)$ or its appropriate quantiles must be known or estimated. Some standard methods, based on pivots and asymptotic approximations, are now briefly reviewed. Note that in many of the examples when the observations are real-valued, it is more convenient and customary to index the unknown family of distributions by the cumulative distribution function F rather than P. We will freely use both, depending on the situation.

15.3.1 Pivotal Method

In certain exceptional cases, the distribution $J_n(P)$ of $R_n(X^n, \theta(P))$ under P does not depend on P. In this case, the root $R_n(X^n, \theta(P))$ is called a *pivotal quantity* or a *pivot* for short. Such quantities were previously considered in Section 6.12. From a pivot, a level $1 - \alpha$ confidence region for $\theta(P)$ can be constructed by choosing constants c_1 and c_2 so that

$$P\{c_1 \le R_n(X^n, \theta(P)) \le c_2\} \ge 1 - \alpha . \tag{15.20}$$

Then, the confidence region

$$C_n = \{\theta \in \Theta : \; c_1 \leq R_n(X^n, \theta) \leq c_2\}$$

contains $\theta(P)$ with probability under P at least $1 - \alpha$. Of course, the coverage probability is exactly $1 - \alpha$ if one has equality in (15.20).

Classical examples where confidence regions may be formed from a pivot are the following.

Example 15.3.1 (Location and Scale Families) Suppose we are given an i.i.d. sample $X^n = (X_1, \ldots, X_n)$ of n real-valued random variables, each having a distribution function of the form $F[(x - \theta)/\sigma]$, where F is known, θ is a location parameter, and σ is a scale parameter. More generally, suppose $\hat{\theta}_n$ is location and scale equivariant in the sense that

$$\hat{\theta}_n(aX_1 + b, \ldots, aX_n + b) = a\hat{\theta}_n(X_1, \ldots, X_n) + b \; ;$$

also suppose $\hat{\sigma}_n$ is location invariant and scale equivariant in the sense that

$$\hat{\sigma}_n(aX_1 + b, \ldots, aX_n + b) = |a|\hat{\sigma}_n(X_1, \ldots, X_n) \; .$$

Then, the root $R_n(X^n, \theta(P)) = n^{1/2}[\hat{\theta}_n - \theta(P)]/\hat{\sigma}_n$ is a pivot (Problem 15.14). For example, in the case where F is the standard normal distribution function, $\hat{\theta}_n$ is the sample mean and $\hat{\sigma}_n^2$ is the usual unbiased estimate of variance, R_n has a t-distribution with $n-1$ degrees of freedom. For another example, if $\hat{\sigma}_n$ is location invariant and scale equivariant, then $\hat{\sigma}_n/\sigma$ is also a pivot, since its distribution will not depend on θ or σ, but will of course depend on F. When F is not normal, exact distribution theory may be difficult, but one may resort to Monte Carlo simulation of $J_n(P)$ (discussed below). This example can be generalized to a class of parametric problems where group invariance considerations apply, and pivotal quantities lead to equivariant confidence sets; see Section 6.12 and Problems 6.69-6.72. ∎

Example 15.3.2 (Kolmogorov-Smirnov Confidence Bands) Suppose that $X^n = (X_1, \cdots, X_n)$ be a sample of n real-valued random variables having a distribution function F. For a fixed value of x, a (pointwise) confidence interval for $F(x)$ can be based on the empirical distribution function $\hat{F}_n(x)$, by using the fact that $n\hat{F}_n(x)$ has a binomial distribution with parameters n and $F(x)$. The goal now is to construct a uniform or simultaneous confidence band for $\theta(F) = F$, so that it is required to find a set of distribution functions containing the true $F(x)$ for all x (or uniformly in x) with coverage probability $1 - \alpha$. Toward this end, consider the root

$$R_n(X^n, F) = n^{1/2} \sup_x |\hat{F}_n(x) - F(x)|.$$

Recall that, if F is continuous, then the distribution of $R_n(X^n, F)$ under F does not depend on F and so $R_n(X^n, F)$ is a pivot (Section 6.13 and Problem 11.57). As discussed in Section 6.13 and 14.2, the finite sample quantiles of this distribution have been tabled. Without the assumption that F is continuous, the distribution of $R_n(X^n, F)$ under F does depend on F, both in finite samples and asymptotically. ∎

In general, if $R_n(X^n, \theta(P))$ is a pivot, its distribution may not be explicitly computable or have a known tractable form. However, since there is only one distribution that needs to be known (and not an entire family indexed by P), the problem is much simpler than if the distribution depends on P. One can resort to Monte Carlo simulation to approximate this distribution to any desired level of accuracy, by simulating the distribution of $R_n(X^n, \theta(P))$ under P for any choice of P in $\mathbf{P}$. For further details, see Example 11.2.13.

15.3.2 Asymptotic Pivotal Method

In general, the above construction breaks down because $R_n(X^n, \theta(P))$ has a distribution $J_n(P)$ which depends on the unknown probability distribution P generating the data. However, it is then sometimes the case that $J_n(P)$ converges weakly to a limiting distribution J which is independent of P. In this case, the root (sequence) $R_n(X^n, \theta(P))$ is called an *asymptotic pivot*, and then the quantiles of J may be used to construct an asymptotic confidence region for $\theta(P)$.

Example 15.3.3 (Parametric Models) Suppose $X^n = (X_1, \ldots, X_n)$ is a sample from a model $\{P_\theta,\ \theta \in \Omega\}$, where Ω is a subset of $\mathbb{R}^k$. To construct a confidence region for θ, suppose $\hat{\theta}_n$ is an efficient likelihood estimator (as discussed in Section 12.4), satisfying

$$n^{1/2}(\hat{\theta}_n - \theta) \xrightarrow{d} N(0, I^{-1}(\theta)) ,$$

where $I(\theta)$ is the Fisher Information matrix, assumed continuous. Then, the root (expressed as a function of θ rather than P_θ)

$$R_n(X^n, \theta) = n(\hat{\theta}_n - \theta)^T I(\hat{\theta}_n)(\hat{\theta}_n - \theta)$$

is an asymptotic pivot. The limiting distribution is the χ^2_k, the Chi-squared distribution with k degrees of freedom, and the resulting confidence region is Wald's confidence ellipsoid introduced in Section 12.4.2. Alternatively, let

$$\tilde{R}_n(X^n, \theta) = \frac{\sup_{\beta \in \Omega} L_n(\beta)}{L_n(\theta)} ,$$

where $L_n(\theta)$ is the likelihood function (12.56). As discussed in Section 12.4.2, under regularity conditions, $2 \log \tilde{R}_n(X^n, \theta)$ is asymptotically χ^2_k, in which case $\tilde{R}_n(X^n, \theta)$ is an asymptotic pivot. ■

Example 15.3.4 (Nonparametric Mean) Suppose $X^n = (X_1, \ldots, X_n)$ is a sample of n real-valued random variables having distribution function F, and we wish to construct a confidence interval for $\theta(F) = E_F(X_i)$, the mean of the observations. Assume X_i has a finite nonzero variance $\sigma^2(F)$. Let the root R_n be the usual t-statistic defined by $R_n(X^n, \theta(F)) = n^{1/2}[\bar{X}_n - \theta(F)]/S_n$, where $\bar{X}_n$ is the sample mean and S_n^2 is the (unbiased version of the) sample variance. Then, $J_n(F)$ converges weakly to $J = N(0, 1)$, and so the t-statistic is an asymptotic pivot. ■

15.3.3 Asymptotic Approximation

The pivotal method assumes the root has a distribution $J_n(P)$ which does not depend on P, while the asymptotic pivotal method assumes the root has an asymptotic distribution $J(P)$ which does not depend on P. More generally, $J_n(P)$ converges to a limiting distribution $J(P)$ which depends on P, and we shall now consider this case. Suppose that this limiting distribution has a known form which depends on P, but only through some unknown parameters. For example, in the nonparametric mean example, the root $n^{1/2}[\bar{X}_n - \theta(F)]$ has the $N(0, \sigma^2(F))$ distribution, and so depends on F through the variance parameter $\sigma^2(F)$. An approximation of the asymptotic distribution is $J(\hat{P}_n)$, where $\hat{P}_n$ is some estimate of P. Typically, $J(P)$ is a normal distribution with mean zero and variance $\tau^2(P)$. The approximation then consists of a normal approximation based on an estimated variance $\tau^2(\hat{P}_n)$ which converges in probability to $\tau^2(P)$, and the quantiles of $J_n(P)$ may then be approximated by those of $J(\hat{P}_n)$. Of course, this approach depends very heavily on knowing the form of the asymptotic distribution as well as being able to construct consistent estimates of the unknown parameters upon which $J(P)$ depends. Moreover, the method essentially consists of a double approximation; first, the finite sampling distribution $J_n(P)$ is approximated by an asymptotic approximation $J(P)$, and then $J(P)$ is in turn approximated by $J(\hat{P}_n)$.

The most general situation occurs when the limiting distribution $J(P)$ has an unknown form, and methods to handle this case will be treated in the subsequent sections.

Example 15.3.5 (Nonparametric Mean, continued) In the previous example, consider instead the non-studentized root

$$R_n(X^n, \theta(F)) = n^{1/2}[\bar{X}_n - \theta(F)] .$$

In this case, $J_n(F)$ converges weakly to $J(F)$, the normal distribution with mean zero and variance $\sigma^2(F)$. The resulting approximation to $J_n(F)$ is the normal distribution with mean zero and variance S_n^2. Alternatively, one can estimate the variance by any consistent estimator, such as the sample variance $\sigma^2(\hat{F}_n)$, where $\hat{F}_n$ is the empirical distribution function. In effect, studentizing an asymptotically normal root converts it to an asymptotic pivot, and both methods lead to the same solution. (However, the bootstrap approach in the next section treats the roots differently.) ■

Example 15.3.6 (Binomial p) As in Example 11.2.7, Suppose X is binomial based on n trials and success probability p. Let $\hat{p}_n = X/n$. As in the previous example, the non-studentized root $n^{1/2}(\hat{p}_n - p)$ and the studentized root $n^{1/2}(\hat{p}_n - p)/[\hat{p}_n(1 - \hat{p}_n)]^{1/2}$ lead to the same approximate confidence interval given by (11.23). On the other hand, the Wilson interval (11.25) based on the root $n^{1/2}(\hat{p}_n - p)/[p(1 - p)]^{1/2}$ leads to a genuinely different solution which performs better in finite samples; see Brown, Cai and DasGupta (2001). ■

Example 15.3.7 (Trimmed mean) Suppose $X^n = (X_1, \dots, X_n)$ is a sample of n real-valued random variables with unknown distribution function F. Assume that F is symmetric about some unknown value $\theta(F)$. Let $\hat{\theta}_{n,\alpha}(X_1, \ldots, X_n)$ be the α-trimmed mean; specifically,

$$\hat{\theta}_{n,\alpha} = \frac{1}{n - 2[\alpha n]} \sum_{i=[\alpha n]+1}^{n-[\alpha n]} X_{(i)} \ ,$$

where $X_{(1)} \leq X_{(2)} \leq \cdots \leq X_{(n)}$ denote the order statistics and $k = [\alpha n]$ is the greatest integer less than or equal to αn. Consider the root $R_n(X^n, \theta(F)) = n^{1/2}[\hat{\theta}_{n,\alpha} - \theta(F)]$. Then, under reasonable smoothness conditions on F and assuming $0 \leq \alpha < 1/2$, it is known that $J_n(F)$ converges weakly to the normal distribution $J(F)$ with mean zero and variance $\sigma^2(\alpha, F)$, where

$$\sigma^2(\alpha, F) = \frac{1}{(1-2\alpha)^2} [\int_{F^{-1}(\alpha)}^{F^{-1}(1-\alpha)} (t - \theta(F))^2 dF(t) + 2\alpha(F^{-1}(\alpha) - \theta(F))^2]; \tag{15.21}$$

see Serfling (1980, p.236). Then, a very simple first-order approximation to $J(F)$ is $J(\hat{F}_n)$, where $\hat{F}_n$ is the empirical distribution. The resulting $J(\hat{F}_n)$ is just the normal distribution with mean zero and variance $\sigma^2(\alpha, \hat{F}_n)$. ■

The use of the normal approximation in the previous example hinged on the availability of a consistent estimate of the asymptotic variance. The simple expression (15.21) easily led to a simple estimator. However, a closed form expression for the asymptotic variance may not exist. A fairly general approach to estimating the variance of a statistic is provided by the *jackknife* estimator of variance, for which we refer the reader to Shao and Tu (1995, Chapter 2). However, the double approximation based on asymptotic normality and an estimate of the limiting variance may be poor. An alternative approach that more directly attempts to approximate the finite sample distribution will be presented in the next section.

15.4 Bootstrap Sampling Distributions

15.4.1 Introduction and Consistency

In this section, the bootstrap, due to Efron (1979), is introduced as a general method to approximate a sampling distribution of a statistic or a root (discussed in Section 15.3) in order to construct confidence regions for a parameter of interest. The use of the bootstrap to approximate a null distribution in the construction of hypothesis tests will be considered later as well.

The asymptotic approaches in the previous section are not always applicable, as when the limiting distribution does not have a tractable form. Even when a root has a known limiting distribution, the resulting approximation may be poor in finite samples. The bootstrap procedure discussed in this section is an alternative, more general, direct approach to approximate the sampling distribution $J_n(P)$. An important aspect of the problem of estimating $J_n(P)$ is that, unlike the usual problem of estimation of parameters, $J_n(P)$ depends on n.

The bootstrap method consists of directly estimating the exact finite sampling distribution $J_n(P)$ by $J_n(\hat{P}_n)$, where $\hat{P}_n$ is an estimate of P in $\mathbf{P}$. In this light, the bootstrap estimate $J_n(\hat{P}_n)$ is a simple *plug-in* estimate of $J_n(P)$.

In nonparametric problems, $\hat{P}_n$ is typically taken to be the empirical distribution of the data. In parametric problems where $\mathbf{P} = \{P_\psi : \psi \in \Psi\}$, $\hat{P}_n$ may be taken to be $P_{\hat{\psi}_n}$, where $\hat{\psi}_n$ is an estimate of ψ.

In general, $J_n(x, \hat{P}_n)$ need not be continuous and strictly increasing in x, so that unique and well-defined quantiles may not exist. To get around this and in analogy to (11.19), define

$$J_n^{-1}(1-\alpha, P) = \inf\{x : J_n(x, P) \geq 1-\alpha\} .$$

If $J_n(\cdot, P)$ has a unique quantile $J_n^{-1}(1-\alpha, P)$, then

$$P\{R_n(X^n, \theta(P)) \leq J_n^{-1}(1-\alpha, P)\} = 1-\alpha ;$$

in general, the probability on the left is at least $1-\alpha$. If $J_n^{-1}(1-\alpha, P)$ were known, then the region

$$\{\theta \in \Theta : \ R_n(X^n, \theta) \leq J_n^{-1}(1-\alpha, P)\}$$

would be a level $1-\alpha$ confidence region for $\theta(P)$. The bootstrap simply replaces $J_n^{-1}(1-\alpha, P)$ by $J_n^{-1}(1-\alpha, \hat{P}_n)$. The resulting bootstrap confidence region for $\theta(P)$ of nominal level $1-\alpha$ takes the form

$$B_n(1-\alpha, X^n) = \{\theta \in \Theta : R_n(X^n, \theta) \leq J_n^{-1}(1-\alpha, \hat{P}_n)\} . \tag{15.22}$$

Suppose the problem is to construct a confidence interval for a real-valued parameter $\theta(P)$ based on the root $|\hat{\theta}_n - \theta(P)|$ for some estimator $\hat{\theta}_n$. The interval (15.22) would then be symmetric about $\hat{\theta}_n$. An alternative equi-tailed interval can be based on the root $\hat{\theta}_n - \theta(P)$ and uses both tails of $J_n(\hat{P}_n)$; it is given by

$$\{\theta \in \Theta : J_n^{-1}(\frac{\alpha}{2}, \hat{P}_n) \leq R_n(X^n, \theta) \leq J_n^{-1}(1-\frac{\alpha}{2}, \hat{P}_n)\} .$$

A comparison of the two approaches will be made in Section 15.5.

Outside certain exceptional cases, the bootstrap approximation $J_n(x, \hat{P}_n)$ cannot be calculated exactly. Even in the relatively simple case when $\theta(P)$ is the mean of P, the root is $n^{1/2}[\bar{X}_n - \theta(P)]$, and $\hat{P}_n$ is the empirical distribution, the exact computation of the bootstrap distribution involves an n-fold convolution.[1] Typically, one resorts to a Monte Carlo approximation to $J_n(P)$, as introduced in Example 11.2.13. Specifically, conditional on the data X^n, for $j = 1, \ldots, B$, let $X_j^{n*} = (X_{1,j}^*, \ldots, X_{n,j}^*)$ be a sample of n i.i.d. observations from $\hat{P}_n$; X_j^{n*} is referred to as the jth bootstrap sample of size n. Of course, when $\hat{P}_n$ is the empirical distribution, this amounts to resampling the original observations with replacement. The bootstrap estimator $J_n(\hat{P}_n)$ is then approximated by the empirical distribution of the B values $R_n(X_j^{n*}, \hat{\theta}_n)$. Because B can be taken to be large (assuming enough computing power), the resulting approximation can be made arbitrarily close to $J_n(\hat{P}_n)$ (see Example 11.2.13), and so we will subsequently focus on the exact bootstrap estimator $J_n(\hat{P}_n)$ while keeping in mind it is usually only approximated by Monte Carlo simulation.

The bootstrap can then be viewed as a simple plug-in estimator of a distribution function. This simple idea, combined with Monte Carlo simulation, allows for quite a broad range of applications.

---

[1] Diaconis and Holmes (1994) show how the exact bootstrap distribution can be calculated in some examples.

We will now discuss the consistency of the bootstrap estimator $J_n(\hat{P}_n)$ of the true sampling distribution $J_n(P)$ of $R_n(X^n, \theta(P))$. Typically, one can show that $J_n(P)$ converges weakly to a nondegenerate limit law $J(P)$. Since the bootstrap replaces P by $\hat{P}_n$ in $J_n(\cdot)$, it is useful to study $J_n(P_n)$ under more general sequences $\{P_n\}$. In order to understand the behavior of the random sequence of distributions $J_n(\hat{P}_n)$, it will be easier to first understand how $J_n(P_n)$ behaves for certain fixed sequences $\{P_n\}$. For the bootstrap to be consistent, $J_n(P)$ must be smooth in P since we are replacing P by $\hat{P}_n$. Thus, we are led to studying the asymptotic behavior of $J_n(P_n)$ under fixed sequences of probabilities $\{P_n\}$ which are "converging" to P in a certain sense. Once it is understood how $J_n(P_n)$ behaves for fixed sequences $\{P_n\}$, it is easy to pass to random sequences $\{\hat{P}_n\}$.

In the theorem below, the existence of a continuous limiting distribution is assumed, though its exact form need not be explicit. Although the conditions of the theorem are strong, they can be verified in many interesting examples.

Theorem 15.4.1 *Let $\mathbf{C}_P$ be a set of sequences $\{P_n \in \mathbf{P}\}$ containing the sequence $\{P, P, \cdots\}$. Suppose that, for every sequence $\{P_n\}$ in $\mathbf{C}_P$, $J_n(P_n)$ converges weakly to a common continuous limit law $J(P)$ having distribution function $J(x, P)$. Let X^n be a sample of size n from P. Assume that $\hat{P}_n$ is an estimate of P based on X^n such that $\{\hat{P}_n\}$ falls in $\mathbf{C}_P$ with probability one. Then,*

$$\sup_x |J_n(x, P) - J_n(x, \hat{P}_n)| \to 0 \text{ with probability one.} \tag{15.23}$$

If $J(\cdot, P)$ is continuous and strictly increasing at $J^{-1}(1-\alpha, P)$, then

$$J_n^{-1}(1-\alpha, \hat{P}_n) \to J^{-1}(1-\alpha, P) \text{ with probability one.} \tag{15.24}$$

Also, the bootstrap confidence set $B_n(1-\alpha, X^n)$ given by equation (15.22) is pointwise consistent in level; that is,

$$P\{\theta(P) \in B_n(1-\alpha, X^n)\} \to 1-\alpha \ . \tag{15.25}$$

Proof. For the proof of part (15.23), note that the assumptions and Polya's Theorem (Theorem 11.2.9) imply that

$$\sup_x |J_n(x, P) - J_n(x, P_n)| \to 0$$

for any sequence $\{P_n\}$ in $\mathbf{C}_P$. Thus, since $\{\hat{P}_n\} \in \mathbf{C}_P$ with probability one, (15.23) follows. Lemma 11.2.1 implies $J_n^{-1}(1-\alpha, P_n) \to J^{-1}(1-\alpha, P)$ whenever $\{P_n\} \in \mathbf{C}_P$; so (15.24) follows. In order to deduce (15.25), the probability on the left side of (15.25) is equal to

$$P\{R_n(X^n, \theta(P)) \le J_n^{-1}(1-\alpha, \hat{P}_n)\} \ . \tag{15.26}$$

Under P, $R_n(X^n, \theta(P))$ has a limiting distribution $J(\cdot, P)$ and, by (15.24), $J_n^{-1}(1-\alpha, \hat{P}_n) \to J^{-1}(1-\alpha, P)$. Thus, by Slutsky's Theorem, (15.26) tends to $J(J^{-1}(1-\alpha, P), P) = 1-\alpha$. ■

Often, the set of sequences $\mathbf{C}_P$ can be described as the set of sequences $\{P_n\}$ such that $d(P_n, P) \to 0$, where d is an appropriate metric on the space of probabilities. Indeed, one should think of $\mathbf{C}_P$ as a set of sequences $\{P_n\}$ that are converging to P in an appropriate sense. Thus, the convergence of $J_n(P_n)$ to

$J(P)$ is locally uniform in the sense $d(P_n, P) \to 0$ implies $J_n(P_n)$ converges weakly to $J(P)$. Note, however, that the appropriate metric d will depend on the precise nature of the root.

When the convergences (15.23) and (15.24) hold with probability one, we say the bootstrap is strongly consistent. If these convergences hold in probability, we say the bootstrap is weakly consistent. In any case, (15.25) holds even if (15.23) and (15.24) only hold in probability; see Problem 15.16.

Example 15.4.1 (Parametric Bootstrap) Suppose $X^n = (X_1, \ldots, X_n)$ is a sample from a q.m.d. model $\{P_\theta,\ \theta \in \Omega\}$, where $\Omega \subset \mathbb{R}^k$. Suppose $\hat{\theta}_n$ is an efficient likelihood estimator in the sense that (12.62) holds. Suppose $g(\theta)$ is a differentiable map from Ω to $\mathbb{R}$ with nonzero gradient vector $\dot{g}(\theta)$. Consider the root $R_n(X^n, \theta) = n^{1/2}[g(\hat{\theta}_n) - g(\theta)]$, with distribution function $J_n(x, \theta)$. By Theorem 12.4.1, $J_n(x, \theta) \to J(x, \theta)$, where $J(x, \theta) = \Phi(x/\sigma_\theta)$ and

$$\sigma_\theta^2 = \dot{g}(\theta) I^{-1}(\theta) \dot{g}(\theta)^T .$$

One approach to estimating the distribution of $n^{1/2}[g(\hat{\theta}_n) - g(\theta)]$ is to use the normal approximation $N(0, \hat{\sigma}_n^2)$, where $\hat{\sigma}_n^2$ is a consistent estimator of σ_θ^2. For example, if $\dot{g}(\theta)$ and $I(\theta)$ are continuous in θ, then a weakly consistent estimator of σ_θ^2 is

$$\hat{\sigma}_n^2 = \dot{g}(\hat{\theta}_n) I^{-1}(\hat{\theta}_n) \dot{g}(\hat{\theta}_n)^T .$$

In order to calculate $\hat{\sigma}_n^2$, the forms of $\dot{g}(\cdot)$ and $I(\cdot)$ must be known. This approach of using a normal approximation with an estimator of the limiting variance is a special case of asymptotic approximation discussed in Subsection 15.3.3. Because it may be difficult to calculate a consistent estimator of the limiting variance, and because the resulting approximation may be poor, it is interesting to consider the bootstrap method. A discussion of higher order asymptotic comparisons will be discussed in Section 15.5. For now, we show the bootstrap approximation $J_n(x, \hat{\theta}_n)$ to $J(x, \theta)$ is weakly consistent.

Theorem 15.4.2 *Under the above setup, under θ,*

$$\sup_x |J_n(x, \theta) - J(x, \theta)| \to 0$$

and

$$\sup_x |J_n(x, \hat{\theta}_n) - J_n(x, \theta)| \to 0 \tag{15.27}$$

in probability; therefore, (15.25) holds.

Proof. By Theorem 12.4.1, for any sequence θ_n such that $n^{1/2}(\theta_n - \theta) \to h$, $J_n(x, \theta_n) \to J(x, \theta)$. In trying to apply the previous theorem, define $\mathbf{C}_\theta$ as the set of sequences $\{\theta_n\}$ satisfying $n^{1/2}(\theta_n - \theta) \to h$, for some finite h. (Rather than describe $\mathbf{C}_P$ as a set of sequences of distributions, we identify P_θ with θ and describe $\mathbf{C}_\theta$ as a set of sequences of parameter values.) Unfortunately, $\hat{\theta}_n$ does not fall in $\mathbf{C}_\theta$ with probability one because $n^{1/2}(\hat{\theta}_n - \theta)$ need not converge with probability one. However, we can modify the argument as follows. Since $n^{1/2}(\hat{\theta}_n - \theta)$ converges in distribution, we can apply the Almost Sure Representation Theorem (Theorem 11.2.19). Thus, there exist random variables $\tilde{\theta}_n$ and H defined on a common probability space such that $\hat{\theta}_n$ and $\tilde{\theta}_n$ have the same distribution and $n^{1/2}(\tilde{\theta}_n - \theta) \to H$ almost surely. Then, $\{\tilde{\theta}_n\} \in \mathbf{C}_\theta$ with probability one, and we can conclude

$$\sup_x |J_n(x, \tilde{\theta}_n) - J_n(x, \theta)| \to 0$$

almost surely. Since $\hat{\theta}_n$ and $\tilde{\theta}_n$ have the same distributional properties, so do $J_n(\hat{\theta}_n)$ and $J_n(\tilde{\theta}_n)$, and the result (15.27) follows. ■

A one-sided bootstrap lower confidence bound for $g(\theta)$ takes the form

$$g(\hat{\theta}_n) - n^{-1/2} J_n^{-1}(1 - \alpha, \hat{\theta}_n) .$$

The previous theorem implies, under θ,

$$J_n^{-1}(1 - \alpha, \hat{\theta}_n) \xrightarrow{P} \sigma_\theta z_{1-\alpha} .$$

Suppose now the problem is to test $g(\theta) = 0$ versus $g(\theta) > 0$. By the duality between tests and confidence regions, one possibility is to reject the null hypothesis if the lower confidence bound exceeds zero, or equivalently when $n^{1/2} g(\hat{\theta}_n) > J_n^{-1}(1 - \alpha, \hat{\theta}_n)$. This test is pointwise asymptotically level α because, by Slutsky's Theorem, $n^{1/2} g(\hat{\theta}_n)$ is asymptotically $N(0, \sigma_\theta^2)$ if $g(\theta) = 0$. The limiting power of this test against a contiguous sequence of alternatives is given in the following corollary.

Corollary 15.4.1 *Under the setup of Example 15.4.1 with θ satisfying $g(\theta) = 0$, the limiting power of the test that rejects when $n^{1/2} g(\hat{\theta}_n) > J_n^{-1}(1 - \alpha, \hat{\theta}_n)$ against the sequence $\theta_n = \theta + hn^{-1/2}$ satisfies*

$$P_{\theta_n}^n \{ n^{1/2} g(\hat{\theta}_n) > J_n^{-1}(1 - \alpha, \hat{\theta}_n) \} \to 1 - \Phi(z_{1-\alpha} - \sigma_\theta^{-1} \langle \dot{g}(\theta)^T, h \rangle) . \quad (15.28)$$

PROOF. The left hand side can be written as

$$P_{\theta_n}^n \{ n^{1/2} [g(\hat{\theta}_n) - g(\theta_n)] > J_n^{-1}(1 - \alpha, \hat{\theta}_n) - n^{1/2} g(\theta_n) \} . \quad (15.29)$$

Under P_θ^n, $J_n^{-1}(1-\alpha, \hat{\theta}_n)$ converges in probability to $\sigma_\theta z_{1-\alpha}$; by contiguity, under $P_{\theta_n}^n$, $J_n^{-1}(1 - \alpha, \hat{\theta}_n)$ converges to the same constant. Also, by differentiability of g and the fact that $g(\theta) = 0$

$$n^{1/2} g(\theta_n) \to \langle \dot{g}(\theta)^T, h \rangle .$$

By Theorem 12.4.1, the left hand side of (15.29) is asymptotically $N(0, \sigma_\theta^2)$. Letting Z denote a standard normal variable, by Slutsky's theorem, (15.29) converges to

$$P\{ \sigma_\theta Z > \sigma_\theta z_{1-\alpha} - \langle \dot{g}(\theta)^T, h \rangle \} ,$$

and the result follows. ■

In fact, it follows from Theorem 13.5.1 that this limiting power is optimal. The moral is that the bootstrap can produce an asymptotically optimal test, but only if the initial estimator or test statistic is optimally chosen. Otherwise, if the root is based on a suboptimal estimator, the bootstrap approach to approximating the sampling distribution of a root is so good that the bootstrap will not be optimal. For example, in a normal location model $N(\theta, 1)$, the bootstrap distribution based on the root $\bar{X}_n - \theta$ is exact as previously discussed (except possibly for simulation error), as is the bootstrap distribution for $T_n - \theta$, where T_n is any location equivariant estimator. But, taking T_n equal to the sample median would not lead to an AUMP test, since the bootstrap is approximating the distribution of the sample median, a suboptimal statistic in this case. Furthermore, this leads to the observation that the bootstrap can be used adaptively to approximate several distributions, and then inference can be based on the one with better properties; see Léger and Romano (1990a,b).

15.4.2 The Nonparametric Mean

In this section, we consider the case of Example 15.3.4, confidence intervals for the nonparametric mean. This example deserves special attention because many statistics can be approximated by linear statistics. We will examine this case in detail, since similar considerations apply to more complicated situations. Given a sample $X^n = (X_1, \ldots, X_n)$ from a distribution F on the real line, consider the problem of constructing a confidence interval for $\theta(F) = E_F(X_i)$. Let $\sigma^2(F)$ denote the variance of F. The conditions for Theorem 15.4.1 are verified in the following result.

Theorem 15.4.3 *Let F be a distribution on the line with finite, nonzero variance $\sigma^2(F)$. Let $J_n(F)$ be the distribution of the root $R_n(X^n, \theta(F)) = n^{1/2}[\bar{X}_n - \theta(F)]$.*

(i) Let $\mathbf{C}_F$ be the set of sequences $\{F_n\}$ such that F_n converges weakly to F, $\theta(F_n) \to \theta(F)$, and $\sigma^2(F_n) \to \sigma^2(F)$. If $\{F_n\} \in \mathbf{C}_F$, then $J_n(F_n)$ converges weakly to $J(F)$, where $J(F)$ is the normal distribution with mean zero and variance $\sigma^2(F)$.

(ii) Let $X_1, \ldots, X_n$ be i.i.d. F, and let $\hat{F}_n$ denote the empirical distribution function. Then, the bootstrap estimator $J_n(\hat{F}_n)$ is strongly consistent so that (15.23), (15.24), and (15.25) hold.

Proof of Theorem 15.4.3. For the purpose of proving (i), construct variables $X_{n,1}, \ldots, X_{n,n}$ which are independent with identical distribution F_n, and set $\bar{X}_n = \sum_i X_{n,i}/n$. We must show that the law of $n^{1/2}(\bar{X}_n - \mu(F_n))$ converges weakly to $J(F)$. It suffices to verify the Lindeberg Condition for $Y_{n,i}$, where $Y_{n,i} = X_{n,i} - \mu(F_n)$. This entails showing that, for each $\epsilon > 0$,

$$\lim_{n \to \infty} E[Y_{n,1}^2 1(Y_{n,1}^2 > n\epsilon^2)] = 0 . \tag{15.30}$$

Note that $Y_{n,1} \stackrel{d}{\to} Y$, where $Y = X - \mu(F)$ and X has distribution F, and $E(Y_{n,1}^2) \to E(Y^2)$. By the continuous mapping theorem (Theorem 11.2.13), $Y_{n,1}^2 \stackrel{d}{\to} Y^2$. Now, for any fixed $\beta > 0$ and all $n > \beta/\epsilon^2$,

$$E[Y_{n,1}^2 1(Y_{n,1}^2 > n\epsilon^2)] \le E[Y_{n,1}^2 1(Y_{n,1}^2 > \beta)] \to E[Y^2 1(Y^2 > \beta)] ,$$

where the last convergence holds if β is a continuity point of the distribution of Y^2, by (11.40). Since the set of continuity points of any distribution is dense and $E[Y^2 1(Y^2 > \beta)] \downarrow 0$ as $\beta \to \infty$, Lindeberg's Condition holds.

We now prove (ii) by applying Theorem 15.4.1; we must show that $\{\hat{F}_n\} \in \mathbf{C}_F$ with probability one. By the Glivenko-Cantelli theorem,

$$\sup_x |\hat{F}_n(x) - F(x)| \to 0 \quad \text{with probability one} .$$

Also, by the Strong Law of Large Numbers, $\theta(\hat{F}_n) \to \theta(F)$ with probability one and $\sigma^2(\hat{F}_n) \to \sigma^2(F)$ with probability one. Thus, bootstrap confidence intervals for the mean based on the root $R_n(X^n, \theta(F)) = n^{1/2}(\bar{X}_n - \theta(F))$ are asymptotically consistent in the sense of the theorem. ■

Remark 15.4.1 Let F and G be two distribution functions on the real line and define $d_p(F, G)$ to be the infimum of $\{E[|X - Y|^p]\}^{1/p}$ over all pairs of random variables X and Y such that X has distribution F and Y has distribution G. It can be shown that the infimum is attained and that d_p is a metric on the space of distributions having a pth moment. Further, if F has a finite variance $\sigma^2(F)$, then $d_2(F_n, F) \to 0$ is equivalent to F_n converging weakly to F and $\sigma^2(F_n) \to \sigma^2(F)$. Hence, Theorem 15.4.3 may be restated as follows. If F has a finite variance $\sigma^2(F)$ and $d_2(F_n, F) \to 0$, then $J_n(F_n)$ converges weakly to $J(F)$. The metric d_2 is known as the Mallow's metric. For details, see Bickel and Freedman (1981).

Continuing the example of the nonparametric mean, it is of interest to consider roots other than $n^{1/2}(\bar{X}_n - \theta(F))$. Specifically, consider the studentized root

$$R_n^s(X^n, \theta(F)) = n^{1/2}(\bar{X}_n - \theta(F))/\sigma(\hat{F}_n) , \tag{15.31}$$

where $\sigma^2(\hat{F}_n)$ is the usual bootstrap estimate of variance. To obtain consistency of the bootstrap method, called the bootstrap-t, we appeal to the following result.

Theorem 15.4.4 *Suppose F is a c.d.f. with finite nonzero variance $\sigma^2(F)$. Let $K_n(F)$ be the distribution of the root (15.31) based on a sample of size n from F.*

(i) Let $\mathbf{C}_F$ be defined as in Theorem 15.4.3. Then, for any sequence $\{F_n\} \in \mathbf{C}_F$, $K_n(F_n)$ converges weakly to the standard normal distribution.

(ii) Hence, the bootstrap sampling distribution $K_n(\hat{F}_n)$ is consistent in the sense that equations (15.23), (15.24), and (15.25) hold.

Before proving this theorem, we first need a weak law of large numbers for a triangular array that generalizes Theorem 11.2.10. The following lemma serves as a suitable version for our purposes.

Lemma 15.4.1 *Suppose $Y_{n,1}, \ldots, Y_{n,n}$ is a triangular array of independent random variables, the n-th row having c.d.f. G_n. Assume G_n converges in distribution to G and*

$$E[|Y_{n,1}|] \to E[|Y|] < \infty$$

as $n \to \infty$, where Y has c.d.f. G. Then,

$$\bar{Y}_n \equiv n^{-1} \sum_{i=1}^{n} Y_{n,i} \xrightarrow{P} E(Y)$$

as $n \to \infty$.

PROOF. Apply Lemma 11.4.2 and (11.40). ■

PROOF OF THEOREM 15.4.4. For the proof, let $X_{n,1}, \ldots, X_{n,n}$ be independent with distribution F_n. By Theorem 15.4.3 and Slutsky's Theorem, it is enough to show $\sigma^2(\hat{F}_n) \to \sigma^2(F)$ in probability under F_n. But,

$$\sigma^2(\hat{F}_n) = \frac{1}{n} \sum_i (X_{n,i} - \bar{X}_n)^2 \ .$$

Now, apply Lemma 15.4.1 on the Weak Law of Large Numbers for a triangular array with $Y_{n,i} = X_{n,i}$ and also with $Y_{n,i} = X_{n,i}^2$. The consistency of the bootstrap method based on the root (15.31) now follows easily. ■

It is interesting to consider how the bootstrap behaves when the underlying distribution has an infinite variance (but well-defined mean). The short answer is that the bootstrap procedure considered thus far will fail, in the sense that the convergence in expression (15.23) does not hold. The failure of the bootstrap for the mean in the infinite variance case was first noted by Babu (1984); further elucidation is given in Athreya (1987) and Knight (1989). In fact, a striking theorem due to Giné and Zinn (1989) asserts that the simple bootstrap studied thus far will work for the mean in the sense of strong consistency if and only if the variance is finite. For a nice exposition of related results, see Giné (1997).

Related results for the studentized bootstrap based on approximating the distribution of the root (15.31) were considered by Csörgö and Mason (1989) and Hall (1990). The conclusion is that the bootstrap is strongly or almost surely consistent if and only if the variance is finite; the bootstrap is weakly consistent if and only if X_i is in the domain of attraction of the normal distribution.

In fact, it was realized by Athreya (1985) that the bootstrap can be modified so that consistency ensues even with infinite variance. The modification consists of reducing the bootstrap sample size. Further results are given in Arcones and Giné (1989, 1991). In fact, In other instances where the simple bootstrap fails, consistency can often be recovered by reducing the bootstrap sample size. The benefit of reducing the bootstrap sample size was recognized first in Bretagnolle (1983). An even more general approach based on subsampling will be considered later in Section 15.7.

15.4.3 *Further Examples*

Example 15.4.2 (Multivariate Mean) Let $X^n = (X_1, \ldots, X_n)$ be a sample of n observations from F, where X_i takes values in $\mathbb{R}^k$. Let $\theta(F) = E_F(X_i)$ be equal to the mean vector, and let

$$S_n(X^n, \theta(F)) = n^{1/2}(\bar{X}_n - \theta(F)) \ , \tag{15.32}$$

where $\bar{X}_n = \sum_i X_i / n$ is the sample mean vector. Let

$$R_n(X^n, \theta(F)) = \|S_n(X^n, \theta(F))\| \ ,$$

where $\| \cdot \|$ is any norm on $\mathbb{R}^k$. The consistency of the bootstrap method based on the root R_n follows from the following theorem.

Theorem 15.4.5 *Let $L_n(F)$ be the distribution (in $\mathbb{R}^k$) of $S_n(X^n, \theta(F))$ under F, where S_n is defined in (15.32). Let $\Sigma(F)$ be the covariance matrix of S_n under F. Let $\mathbf{C}_F$ be the set of sequences $\{F_n\}$ such that F_n converges weakly to F and $\Sigma(F_n) \to \Sigma(F)$, so that each entry of the matrix $\Sigma(F_n)$ converges to the corresponding entry (assumed finite) of $\Sigma(F)$.*

(i) Then, $L_n(F_n)$ converges weakly to $L(F)$, the multivariate normal distribution with mean zero and covariance matrix $\Sigma(F)$.

(ii) Assume $\Sigma(F)$ contains at least one nonzero component. Let $\|\cdot\|$ be any norm on $\mathbb{R}^k$ and let $J_n(F)$ be the distribution of $R_n(X^n, \theta(F)) = \|S_n(X^n, \theta(F))\|$ under F. Then, $J_n(F_n)$ converges weakly to $J(F)$, which is the distribution of $\|Z\|$ when Z has distribution $L(F)$.

(iii) Suppose $X_1, \ldots, X_n$ are i.i.d. F with empirical distribution $\hat{F}_n$ (in $\mathbb{R}^k$). Then, the bootstrap approximation satisfies

$$\rho(J_n(F), J_n(\hat{F}_n)) \to 0 \text{ with probability one },$$

and bootstrap confidence regions based on the root R_n are consistent in the sense that the convergences (15.23) to (15.25) hold.

PROOF. The proof of (i) follows by the Cramer-Wold device (Theorem 11.2.3) and by Theorem 15.4.3 (i). To prove (ii), note that any norm $\|\cdot\|$ on $\mathbb{R}^k$ is continuous almost everywhere with respect to $L(F)$. A proof of this statement can be based on the fact that, for any norm $\|\cdot\|$, the set $\{x \in \mathbb{R}^k : \|x\| = c\}$ has Lebesgue measure zero because it is the boundary of a convex set. So, the continuous mapping theorem applies and so $J_n(F_n)$ converges weakly to $J(F)$.

Part (iii) follows because $\{\hat{F}_n\} \in \mathbf{C}_F$ with probability one, by the Glivenko-Cantelli theorem (on $\mathbb{R}^k$) and the strong law of large numbers. ■

Note the power of the bootstrap method. Analytical methods for approximating the distribution of the root $R_n = \|S_n\|$ would depend heavily on the choice of norm $\|\cdot\|$, but the bootstrap handles them all with equal ease.

Let $\hat{\Sigma}_n = \Sigma(\hat{F})$ be the sample covariance matrix. As in the univariate case, one can also bootstrap the root defined by

$$\tilde{R}_n(X^n, \theta(F)) = \|\hat{\Sigma}_n^{-1/2}(\bar{X}_n - \theta(F))\|, \tag{15.33}$$

provided $\Sigma(F)$ is assumed positive definite. In the case where $\|\cdot\|$ is the usual Euclidean norm, this root leads to confidence ellipsoid, i.e., a confidence set whose shape is an ellipsoid.

Example 15.4.3 (Smooth Functions of Means) Let $X_1, \ldots, X_n$ be i.i.d. S-valued random variables with distribution P. Suppose $\theta = \theta(P) = (\theta_1, \ldots, \theta_p)$, where $\theta_j = E_P[h_j(X_i)]$ and the h_j are real-valued functions defined on S. Interest focuses on θ or some function f of θ. Let $\hat{\theta}_n = (\hat{\theta}_{n,1}, \ldots, \hat{\theta}_{n,p})$, where $\hat{\theta}_{n,j} = \sum_{i=1}^n h_j(X_i)/n$. Assume moment conditions on the $h_j(X_i)$. Then, by the multivariate mean case, the bootstrap approximation to the distribution of $n^{1/2}(\hat{\theta}_n - \theta)$ is appropriately close in the sense

$$\rho\left(\mathcal{L}_P(n^{1/2}(\hat{\theta}_n - \theta)), \mathcal{L}_{\hat{P}_n^*}(n^{1/2}(\hat{\theta}_n^* - \hat{\theta}_n))\right) \to 0 \tag{15.34}$$

with probability one, where ρ is any metric metrizing weak convergence in $\mathbb{R}^p$ (such as the Bounded-Lipschitz metric introduced in Problem 11.23). Here, P_n^* refers to the distribution of the data resampled from the empirical distribution conditional on $X_1, \ldots X_n$. Moreover,

$$\rho\left(\mathcal{L}_P(n^{1/2}(\hat{\theta}_n - \theta)), \mathcal{L}(Z)\right) \to 0 , \tag{15.35}$$

where Z is multivariate normal with mean zero and covariance matrix Σ having (i,j)-th component

$$Cov(Z_i, Z_j) = Cov[h_i(X_1), h_j(X_1)].$$

To see why, define Y_i to be the vector in $\mathbb{R}^p$ with j-th component $h_j(X_i)$, so that we are exactly back in the multivariate mean case. Now, suppose f is an appropriately smooth function from $\mathbb{R}^p$ to $\mathbb{R}^q$, and interest now focuses on the parameter $\mu = f(\theta)$. Assume $f = (f_1, \ldots, f_q)^T$, where $f_i(y_1, \ldots, y_p)$ is a real-valued function from $\mathbb{R}^p$ having a nonzero differential at $(y_1, \cdots, y_p) = (\theta_1, \ldots, \theta_p)$. Let D be the $q \times p$ matrix with (i,j) entry $\partial f_i(y_1, \ldots, y_p)/\partial y_j$ evaluated at $(\theta_1, \ldots, \theta_p)$. Then, the following is true.

Theorem 15.4.6 *Suppose f is a function satisfying the above smoothness assumptions. If $E[h_j^2(X_i)] < \infty$, then equations (15.34) and (15.35) hold. Moreover,*

$$\rho\left(\mathcal{L}_P(n^{1/2}[f(\hat{\theta}_n) - f(\theta)]), \mathcal{L}_{P_n^*}(n^{1/2}[f(\hat{\theta}_n^*) - f(\hat{\theta}_n)])\right) \to 0$$

with probability one and

$$\sup_s \left| P\{\|f(\hat{\theta}_n) - f(\theta)\| \le s\} - P_n^*\{\|f(\hat{\theta}_n^*) - f(\hat{\theta}_n)\| \le s\}\right| \to 0$$

with probability one.

PROOF. The proof follows as equations (15.34) and (15.35) are immediate from the multivariate mean case, and the smoothness assumptions on f and the Delta Method imply that $n^{1/2}[f(\hat{\theta}_n) - f(\theta)]$ has a limiting multivariate normal distribution with mean 0 and covariance matrix $D\Sigma D^T$; see Theorem 11.2.14. ∎

Example 15.4.4 (Joint Confidence Rectangles) Under the assumptions of Theorem 15.4.6, a joint confidence set can be constructed for $(f_1(\theta), \ldots, f_q(\theta))$ with asymptotic coverage $1 - \alpha$. In the case where $\|x\| = \max|x_i|$, the set is a rectangle in $\mathbb{R}^q$. Such a set is easily described as

$$\{f(\theta) : \ |f_i(\hat{\theta}_n) - f_i(\theta)| \le \hat{b}_n(1-\alpha) \quad \text{for all } i\ \},$$

where $\hat{b}_n(1-\alpha)$ is the bootstrap approximation to the $1-\alpha$ quantile of the distribution of $\max_i |f_i(\hat{\theta}_n) - f_i(\theta)|$. Thus, a value for $f_i(\theta)$ is included in the region if and only if $f_i(\theta) \in f_i(\hat{\theta}_n) \pm \hat{b}_n(1-\alpha)$. Note, however, the intervals $f_i(\hat{\theta}_n) \pm \hat{b}_n(1-\alpha)$ may be unbalanced in the sense that the limiting coverage probability for each marginal parameter $f_i(\theta)$ may depend on i. To fix this, one could instead bootstrap the distribution of $\max_i |f_i(\hat{\theta}_n) - f_i(\theta)|/\hat{\sigma}_{n,i}$, where $\hat{\sigma}_{n,i}$ is some consistent estimate of the (i,i) entry of the asymptotic covariance matrix $D\Sigma D^T$ for $n^{1/2}f(\hat{\theta}_n)$. For further discussion, see Beran (1988a), who employs a transformation called prepivoting to achieve balance.

Example 15.4.5 (Uniform Confidence Bands for a c.d.f. F) Consider a sample $X^n = (X_1, \ldots, X_n)$ real-valued observations having c.d.f. F. The empirical c.d.f. $\hat{F}_n$ is then

$$\hat{F}_n(t) = n^{-1} \sum_{i=1}^{n} I\{X_i \leq t\} .$$

For two distribution functions F and G, define the Kolmogorov-Smirnov (or uniform) metric

$$d_K(F, G) = \sup_t |F(t) - G(t)| .$$

Now, consider the root

$$R_n(X^n, \theta(F)) = n^{1/2} d_K(\hat{F}_n, F) ,$$

whose distribution under F is denoted $J_n(F)$. As discussed in Example 11.2.12, $J_n(F)$ has a continuous limiting distribution. In fact, the following triangular array convergence holds. If $d_K(F_n, F) \to 0$, then $J_n(F_n) \xrightarrow{d} J(F)$; for a proof, see Politis, Romano, and Wolf (1999, p.20). Thus, we can define $\mathbf{C}_F$ to be the set of sequences $\{F_n\}$ satisfying $d_K(F_n, F) \to 0$. By the Glivenko-Cantelli Theorem, $d_K(\hat{F}_n, F) \to 0$ with probability one, and strong consistency of the bootstrap follows. The resulting uniform confidence bands for F are then consistent in the sense that (15.25) holds, and no assumption on continuity of F is needed (unlike the classical limit theory). This example has been generalized considerably, and the proof depends on the behavior of $n^{1/2}[\hat{F}_n(t) - F(t)]$, which can be viewed as a random function and is called the *empirical process.* The general theory of bootstrapping empirical processes is developed in van der Vaart and Wellner (1996) and in Chapter 2 of Giné (1997). In particular, the theory generalizes to quite general spaces S, so that the observations need not be real-valued. In the special case when S is k-dimensional Euclidean space, the k-dimensional empirical process was considered in Beran and Millar (1986). Confidence sets for a multivariate distribution based on the bootstrap can then be constructed which are pointwise consistent in level. ■

15.4.4 Stepdown Multiple Testing

Suppose data $X = X^n$ is generated from some unknown probability distribution P, where P belongs to a certain family of probability distributions Ω. For $j = 1, \ldots, s$, consider the problem of simultaneously testing hypotheses $H_j : P \in \omega_j$.

For any subset $K \subset \{1, \ldots, s\}$, let $H_K = \bigcap_{j \in K} H_j$ be the hypothesis that $P \in \bigcap_{j \in K} \omega_j$. Suppose that a test of the individual hypothesis H_j is based on a test statistic $T_{n,j}$, with large values indicating evidence against the H_j.

The goal is to construct a stepdown method that controls the familywise error rate (FWER). Recall that the FWER is the probability of rejecting at least one true null hypothesis. More specifically, if P is the true probability mechanism, let $I = I(P) \subset \{1, \ldots, s\}$ denote the indices of the set of true hypotheses; that is, $i \in I$ if and only $P \in \omega_i$. Then, FWER is the probability under P that any H_i with $i \in I$ is rejected. To show its dependence on P, we may write FWER $= \text{FWER}_P$. We require that any procedure satisfy that the FWER be no bigger than α (at least asymptotically).

Suppose H_i is specified a real-valued parameter $\beta_i(P) = 0$. Then, one approach to constructing a multiple test is to invert a simultaneous confidence region. Under the setup of Example 15.4.4, with $\beta_i(P) = f_i(\theta(P))$, any hypothesis H_i is rejected if $f_i(\hat{\theta}_n) > \hat{b}_n(1-\alpha)$. A procedure that uses a common critical value $\hat{b}_n(1-\alpha)$ for all the hypotheses is called a single-step method.

Another approach is to compute (or approximate) a p-value for each individual test, and then use Holm's method discussed in Section 9.1, However, Holm's method, which makes no assumptions about the dependence structure of the test statistics, can be improved by methods that implicitly or explicitly estimate this dependence structure. In this section, we consider a stepdown procedure that incorporates the dependence structure and thereby improves upon the two methods just described.

Let

$$T_{n,r_1} \geq T_{n,r_2} \geq \cdots \geq T_{n,r_s} \tag{15.36}$$

denote the observed ordered test statistics, and let H_{r_1}, $H_{r_2}, \ldots, H_{r_s}$ be the corresponding hypotheses.

Recall the stepdown method presented in Procedure 9.1.1. The problem now is how to construct the $\hat{c}_{n,K}(1-\alpha)$ so that the FWER is controlled, at least asymptotically. The following is an immediate consequence of Theorem 9.1.3, and reduces the multiple testing problem of asymptotically controlling the FWER to the single testing problem of asymptotically controlling the probability of a Type 1 error.

Corollary 15.4.2 *Let P denote the true distribution generating the data. Consider Procedure 9.1.1 based on critical values $\hat{c}_{n,K}(1-\alpha)$ which satisfy the monotonicity requirement: for any $K \supset I(P)$,*

$$\hat{c}_{n,K}(1-\alpha) \geq \hat{c}_{n,I(P)}(1-\alpha) \ . \tag{15.37}$$

If $\hat{c}_{n,I(P)}(1-\alpha)$ satisfies

$$\limsup_n P\{\max(T_{n,j} : \ j \in I(P)) > \hat{c}_{n,I(P)}(1-\alpha)\} \leq \alpha \ , \tag{15.38}$$

then $\limsup_n FWER_P \to \alpha$ *as* $n \to \infty$.

Under the monotonicity requirement (15.37), the multiplicity problem is effectively reduced to testing a single intersection hypothesis at a time. So, the problem now is to construct intersection tests whose critical values are monotone and asymptotically control the rejection probability.

We now specialize a bit and develop a concrete construction based on the bootstrap. Suppose hypothesis H_i is specified by $\{P : \ \theta_i(P) = 0\}$ for some real-valued parameter θ_i, and $\hat{\theta}_{n,i}$ is an estimate of θ_i. Also, let $T_{n,i} = \tau_n|\hat{\theta}_{n,i}|$ for some nonnegative (nonrandom) sequence $\tau_n \to \infty$; usually, $\tau_n = n^{1/2}$. The bootstrap method relies on its ability to approximate the joint distribution of $\{\tau_n[\hat{\theta}_{n,i} - \theta_i(P)] : \ i \in K\}$, whose distribution we denote by $J_{n,K}(P)$. Also, let $L_{n,K}(P)$ denote the distribution under P of $\max\{\tau_n|\hat{\theta}_{n,i} - \theta_i(P)| : \ i \in K\}$, with corresponding distribution function $L_{n,K}(x,P)$ and α-quantile

$$b_{n,K}(\alpha, P) = \inf\{x : L_{n,K}(x,P) \geq \alpha\} \ .$$

Let $\hat{Q}_n$ be some estimate of P. Then, a nominal $1-\alpha$ level bootstrap confidence region for the subset of parameters $\{\theta_i(P) : i \in K\}$ is given by

$$\{(\theta_i : i \in K) : \max_{i \in K} \tau_n |\hat{\theta}_{n,i} - \theta_i| \leq b_{n,K}(1-\alpha, \hat{Q}_n)\} .$$

So a value of 0 for $\theta_i(P)$ falls outside the region iff $T_{n,i} = \tau_n |\hat{\theta}_{n,i}| > b_{n,K}(1-\alpha, \hat{Q}_n)$. By the usual duality of confidence sets and hypothesis tests, this suggests the use of the critical value

$$\hat{c}_{n,K}(1-\alpha) = b_{n,K}(1-\alpha, \hat{Q}_n) , \tag{15.39}$$

at least if the bootstrap is a valid asymptotic approach for confidence region construction.

Note that, regardless of asymptotic behavior, the monotonicity assumption (15.37) is always satisfied for the choice (15.39). Indeed, for any Q and if $I \subset K$, $b_{n,I}(1-\alpha, Q)$ is the $1-\alpha$ quantile under Q of the maximum of $|I|$ variables, while $b_{n,K}(1-\alpha, Q)$ is the $1-\alpha$ quantile of these same $|I|$ variables together with $|K| - |I|$ variables.

Therefore, in order to apply Theorem 15.4.2 to conclude $\limsup_n \mathrm{FWER}_P \leq \alpha$, it is now only necessary to study the asymptotic behavior of $b_{n,K}(1-\alpha, \hat{Q}_n)$ in the case $K = I(P)$. For this, we assume the usual conditions for bootstrap consistency when testing the *single* hypothesis that $\theta_i(P) = 0$ for all $i \in I(P)$; that is, we assume the bootstrap consistently estimates the joint distribution of $\tau_n[\hat{\theta}_{n,i} - \theta_i(P)]$ for $i \in I(P)$. In particular, we assume

$$J_{n,I(P)}(P) \xrightarrow{d} J_{I(P)}(P) , \tag{15.40}$$

a nondegenerate limit law. Assumption (15.40) implies $L_{n,I(P)}(P)$ has a limiting distribution $L_{I(P)}(P)$, with c.d.f. denoted $L_{I(P)}(x, P)$. We will further assume $L_{I(P)}(P)$ is continuous and strictly increasing on its support. It follows that

$$b_{n,I(P)}(1-\alpha, P) \to b_{I(P)}(1-\alpha, P) , \tag{15.41}$$

where $b_{I(P)}(\alpha, P)$ is the α-quantile of the limiting distribution $L_{I(P)}(P)$.

Theorem 15.4.7 *Fix P and assume (15.40) and that $L_{I(P)}(P)$ is continuous and strictly increasing on its support. Let $\hat{Q}_n$ be an estimate of P satisfying: for any metric ρ metrizing weak convergence on $\mathbb{R}^{|I(P)|}$,*

$$\rho\left(J_{n,I(P)}(P), J_{n,I(P)}(\hat{Q}_n)\right) \xrightarrow{P} 0 . \tag{15.42}$$

Consider the generic stepdown method in Procedure 9.1.1 with $c_{n,K}(1-\alpha)$ equal to $b_{n,K}(1-\alpha, \hat{Q}_n)$. Then, $\limsup_n FWER_P \leq \alpha$.

PROOF. By the Continuous Mapping Theorem and a subsequence argument (Problem 15.28), the assumption (15.40) implies

$$\rho_1\left(L_{n,I(P)}(P), L_{n,I(P)}(\hat{Q}_n)\right) \xrightarrow{P} 0 , \tag{15.43}$$

where ρ_1 is any metric metrizing weak convergence on $\mathbb{R}$. It follows from Lemma 11.2.1(ii) that

$$b_{n,I(P)}(1-\alpha, \hat{Q}_n) \xrightarrow{P} b_{I(P)}(1-\alpha, P) .$$

By Slutsky's Theorem,

$$P\{\max(T_{n,j} : j \in I(P))\} > b_{n,I(P)}(1-\alpha, \hat{Q}_n)\} \to 1 - L_{I(P)}(b_{I(P)}(1-\alpha, P), P),$$

and the last expression is α. ■

Example 15.4.6 (Multivariate Mean) Assume $X_i = (X_{i,1}, \ldots, X_{i,s})$ are n i.i.d. random vectors with $E(|X_i|^2) < \infty$ and mean vector $\mu = (\mu_1, \ldots, \mu_s)$. Note that the vector X_i can have an arbitrary s-variate distribution, so that multivariate normality is not assumed as it was in Example 9.1.4. Suppose H_i specifies $\mu_i = 0$ and $T_{n,i} = n^{-1/2}|\sum_{j=1}^{n} X_{j,i}|$. Then, the conditions of Theorem 15.4.7 are satisfied by Example 15.4.2. Alternatively, one can also consider the studentized test statistic $t_{n,i} = T_{n,i}/S_{n,i}$, where $S_{n,i}^2$ is the sample variance of the ith components of the data (Problem 15.29). ■

Example 15.4.7 (Comparing Treatment Means) For $i = 1, \ldots, k$, suppose we observe k independent samples, and the ith sample consists of n_i i.i.d. observations $X_{i,1}, \ldots, X_{i,n_i}$ with mean μ_i and finite variance σ_i^2. Hypothesis $H_{i,j}$ specifies $\mu_i = \mu_j$, so that the problem is to compare all $s = \binom{k}{2}$ means. (Note that we are indexing hypotheses and test statistics now by 2 indices i and j.) Let $T_{n,i,j} = n^{1/2}|\bar{X}_{n,i} - \bar{X}_{n,j}|$, where $\bar{X}_{n,i} = \sum_{j=1}^{n} X_{i,j}/n_i$. Let $\hat{Q}_{n_i,i}$ be the empirical distribution of the ith sample. The bootstrap resampling scheme is to independently resample n_i observations from $\hat{Q}_{n,i}$, $i = 1, \ldots, k$. Then, Theorem 15.4.7 applies and it also applies to appropriately studentized statistics (Problem 15.30) The setup can easily accommodate comparisons of k treatments with a control group (Problem 15.31). ■

Example 15.4.8 (Testing Correlations) Suppose $X_1, \ldots, X_n$ are i.i.d. random vectors in $\mathbb{R}^k$, so that $X_i = (X_{i,1}, \ldots, X_{i,k})$. Assume $E|X_{i,j}|^2 < \infty$ and $Var(X_{i,j}) > 0$, so that the correlation between $X_{1,i}$ and $X_{1,j}$, namely $\rho_{i,j}$ is well-defined. Let $H_{i,j}$ denote the hypothesis that $\rho_{i,j} = 0$, so that the multiple testing problem consists in testing all $s = \binom{k}{2}$ pairwise correlations. Also let $T_{n,i,j}$ denote the ordinary sample correlation between variables i and j. (Note that we are indexing hypotheses and test statistics now by 2 indices i and j.) By Example 15.4.3, the conditions for the bootstrap hold because correlations are smooth functions of means. ■

15.5 Higher Order Asymptotic Comparisons

One of the main reasons the bootstrap approach is so valuable is that it can be applied to approximate the sampling distribution of an estimator in situations where the finite or large sample distribution theory is intractable, or depends on unknown parameters. However, even in relatively simple situations, we will see that there are advantages to using a bootstrap approach. For example, consider the problem of constructing a confidence interval for a mean. Under the assumption of a finite variance, the standard normal theory interval and the bootstrap-t are each pointwise consistent in level. In order to compare them, we must consider higher order asymptotic properties. More generally, suppose I_n is a nominal

$1 - \alpha$ level confidence interval for a parameter $\theta(P)$. Its coverage error under P is

$$P\{\theta(P) \in I_n\} - (1 - \alpha) \ ,$$

and we would like to examine the rate at which this tends to zero. In typical problems, this coverage error is a power of $n^{-1/2}$. It will be necessary to distinguish one-sided and two-sided confidence intervals because their orders of coverage error may differ.

Throughout this section, attention will focus on confidence intervals for the mean in a nonparametric setting. Specifically, we would like to compare some asymptotic methods based on the normal approximation and the bootstrap. Let $X^n = (X_1, \ldots, X_n)$ be i.i.d. with c.d.f. F, mean $\theta(F)$, and variance $\sigma^2(F)$. Also, let $\hat{F}_n$ denote the empirical c.d.f., and let $\hat{\sigma}_n = \sigma(\hat{F}_n)$.

Before addressing coverage error, we recall from Section 11.4.1 the Edgeworth expansions for the distributions of the roots

$$R_n(X^n, F) = n^{1/2}(\bar{X}_n - \theta(F))$$

and

$$R_n^s(X^n, F) = n^{1/2}(\bar{X}_n - \theta(F))/\hat{\sigma}_n \ ;$$

as in Section 15.4.2, their distribution functions under F are denoted $J_n(\cdot, F)$ and $K_n(\cdot, F)$, respectively. Let Φ and φ denote the standard normal c.d.f. and density, respectively.

Theorem 15.5.1 *Assume $E_F(X_i^4) < \infty$. Let ψ_F denote the characteristic function of F, and assume*

$$\limsup_{|s| \to \infty} |\psi_F(s)| < 1 \ . \tag{15.44}$$

Then,

$$J_n(t, F) = \Phi(t/\sigma(F)) - \frac{1}{6}\gamma(F)\varphi(t/\sigma(F))(\frac{t^2}{\sigma^2(F)} - 1)n^{-1/2} + O(n^{-1}) \ , \tag{15.45}$$

where

$$\gamma(F) = E_F[X_1 - \theta(F)]^3/\sigma^3(F)$$

is the skewness of F. Moreover, the expansion holds uniformly in t in the sense that

$$J_n(t, F) = [\Phi(t/\sigma(F)) - \frac{1}{6}\gamma(F)\varphi(t/\sigma(F))(\frac{t^2}{\sigma^2(F)} - 1)n^{-1/2}] + R_n(t, F) \ ,$$

where $|R_n(t, F)| \le C/n$ for all t and some $C = C_F$ which depends on F.

Theorem 15.5.2 *Assume $E_F(X_i^4) < \infty$ and that F is absolutely continuous. Then, uniformly in t,*

$$K_n(t, F) = \Phi(t) + \frac{1}{6}\gamma(F)\varphi(t)(2t^2 + 1)n^{-1/2} + O(n^{-1}) \ . \tag{15.46}$$

Note that the term of order $n^{-1/2}$ is zero if and only if the underlying skewness $\gamma(F)$ is zero, so that the dominant error in using a standard normal approximation to the distribution of the studentized statistic is due to skewness of the underlying distribution. We will use these expansions in order to derive some important properties of confidence intervals. Note, however, that the expansions are asymptotic results, and for finite n, including the correction term (i.e. the term of order $n^{-1/2}$) may worsen the approximation.

Expansions for the distribution of a root such as (15.45) and (15.46) imply corresponding expansions for their quantiles, which are known as *Cornish-Fisher Expansions.* For example, $K_n^{-1}(1-\alpha, F)$ is a value of t satisfying $K_n(t, F) = 1-\alpha$. Of course, $K_n^{-1}(1-\alpha, F) \to z_{1-\alpha}$. We would like to determine $c = c(\alpha, F)$ such that

$$K_n^{-1}(1-\alpha, F) = z_{1-\alpha} + cn^{-1/2} + O(n^{-1}) .$$

Set $1-\alpha$ equal to the right hand side of (15.46) with $t = z_{1-\alpha} + cn^{-1/2}$, which yields

$$\Phi(z_{1-\alpha} + cn^{-1/2}) + \frac{1}{6}\gamma(F)\varphi(z_{1-\alpha} + cn^{-1/2})(2z_{1-\alpha}^2 + 1)n^{-1/2} + O(n^{-1}) = 1-\alpha .$$

By expanding Φ and φ about $z_{1-\alpha}$, we find that

$$c = -\frac{1}{6}\gamma(F)(2z_{1-\alpha}^2 + 1) .$$

Thus,

$$K_n^{-1}(1-\alpha, F) = z_{1-\alpha} - \frac{1}{6}\gamma(F)(2z_{1-\alpha}^2 + 1)n^{-1/2} + O(n^{-1}) . \qquad (15.47)$$

In fact, under the assumptions of Theorem 15.5.2, the expansion (15.46) holds uniformly in t, and so the expansion (15.47) holds uniformly in $\alpha \in [\epsilon, 1-\epsilon]$, for any $\epsilon > 0$ (Problem 15.34). Similarly, one can show (Problem 15.35) that, under the assumptions of Theorem 15.5.1,

$$J_n^{-1}(1-\alpha, F) = \sigma(F)z_{1-\alpha} + \frac{1}{6}\sigma(F)\gamma(F)(z_{1-\alpha}^2 - 1)n^{-1/2} + O(n^{-1}) , \quad (15.48)$$

uniformly in $\alpha \in [\epsilon, 1-\epsilon]$.

Normal Theory Intervals. The most basic approximate upper one-sided confidence interval for the mean $\theta(F)$ is given by

$$\bar{X}_n + n^{-1/2}\hat{\sigma}_n z_{1-\alpha} , \qquad (15.49)$$

where $\hat{\sigma}_n^2 = \sigma^2(\hat{F}_n)$ is the (biased) sample variance. Its one-sided coverage error is given by

$$P_F\{\theta(F) \le \bar{X}_n + n^{-1/2}\hat{\sigma}_n z_{1-\alpha}\} - (1-\alpha)$$

$$= \alpha - P_F\{n^{1/2}(\bar{X}_n - \theta(F))/\hat{\sigma}_n < z_\alpha\} . \qquad (15.50)$$

By (15.46), the one-sided coverage error of this normal theory interval is

$$-\frac{1}{6}\gamma(F)\varphi(z_\alpha)(2z_\alpha^2 + 1)n^{-1/2} + O(n^{-1}) = O(n^{-1/2}) . \qquad (15.51)$$

Analogously, the coverage error of the two-sided confidence interval of nominal level $1 - 2\alpha$,

$$\bar{X}_n \pm n^{-1/2}\hat{\sigma}_n z_{1-\alpha} , \tag{15.52}$$

satisfies

$$P_F\{-z_{1-\alpha} \le n^{1/2}(\bar{X}_n - \theta(F))/\hat{\sigma}_n \le z_{1-\alpha}\} - (1 - 2\alpha)$$

$$= P\{n^{1/2}(\bar{X}_n - \theta(F))/\hat{\sigma}_n \le z_{1-\alpha}\} - P\{n^{1/2}(\bar{X}_n - \theta(F))\hat{\sigma}_n < -z_{1-\alpha}\} - (1 - 2\alpha) ,$$

which by (15.46) is equal to

$$[\Phi(z_{1-\alpha}) + \frac{1}{6}\gamma(F)\varphi(z_{1-\alpha})(2z_{1-\alpha}^2 + 1)n^{-1/2} + O(n^{-1})]$$

$$-[\Phi(-z_{1-\alpha}) + \frac{1}{6}\gamma(F)\varphi(-z_{1-\alpha})(2z_{1-\alpha}^2 + 1)n^{-1/2} + O(n^{-1})] - (1 - 2\alpha) = O(n^{-1}) ,$$

using the symmetry of the function φ. Thus, while the coverage error of the one-sided interval (15.49) is $O(n^{-1/2})$, the two-sided interval (15.52) has coverage error $O(n^{-1})$. The main reason the one-sided interval has coverage error $O(n^{-1/2})$ derives from the fact that a normal approximation is used for the distribution of $n^{1/2}(\bar{X}_n - \theta(F))/\hat{\sigma}_n$ and no correction is made for skewness of the underlying distribution. For example, if $\gamma(F) > 0$, the one-sided upper confidence bound (15.49) undercovers slightly while the one-sided lower confidence bound overcovers. The combination of overcoverage and undercoverage yields a net result of a reduction in the order of coverage error of two-sided intervals. Analytically, this fact derives from the key property that the $n^{-1/2}$ term in (15.46) is an even polynomial. (Note, however, that the one-sided coverage error is $O(n^{-1})$ if $\gamma(F) = 0$.) These results are in complete analogy with the corresponding results in Section 11.4.1 for error in rejection probability of tests of the mean based on the normal approximation.

Basic Bootstrap Intervals. Next, we consider bootstrap confidence intervals for $\theta(F)$ based on the root

$$R_n(X^n, \theta(F)) = n^{1/2}(\bar{X}_n - \theta(F)) . \tag{15.53}$$

It is plausible that the bootstrap approximation $J_n(t, \hat{F}_n)$ to $J_n(t, F)$ satisfies an expansion like (15.45) with F replaced by $\hat{F}_n$. In fact, it is the case that

$$J_n(t, \hat{F}_n) = \Phi(t/\hat{\sigma}_n) - \frac{1}{6}\gamma(\hat{F}_n)\varphi(t/\hat{\sigma}_n)(\frac{t^2}{\hat{\sigma}_n^2} - 1)n^{-1/2} + O_P(n^{-1}) . \tag{15.54}$$

Both sides of (15.54) are random and the remainder term is now of order n^{-1} in probability. Similarly, the bootstrap quantile function $J_n^{-1}(1 - \alpha, \hat{F}_n)$ has an analogous expansion to (15.48) and is given by

$$J_n^{-1}(1 - \alpha, \hat{F}_n) = \hat{\sigma}_n[z_{1-\alpha} + \frac{1}{6}\gamma(\hat{F}_n)(z_{1-\alpha}^2 - 1)n^{-1/2}] + O_P(n^{-1}) . \tag{15.55}$$

The validity of these expansions is quite technical and is proved in Hall (1992, Section 5.2), and a sufficient condition for them to hold is that F satisfies Cramér's condition and has infinitely many moments; such assumptions will remain in force for the remainder of this section. From (15.45) and (15.54), it follows that

$$J_n(t, \hat{F}_n) - J_n(t, F) = O_P(n^{-1/2})$$

because

$$\hat{\sigma}_n - \sigma(F) = O_P(n^{-1/2}) .$$

Thus, the bootstrap approximation $J_n(t, \hat{F}_n)$ to $J_n(t, F)$ has the same order of error as that provided by the normal approximation.

Turning now to coverage error, consider the one-sided coverage error of the nominal level $1-\alpha$ upper confidence bound $\bar{X}_n - n^{-1/2} J_n^{-1}(\alpha, \hat{F}_n)$, given by

$$P_F\{\theta(F) \le \bar{X}_n - n^{-1/2} J_n^{-1}(\alpha, \hat{F}_n)\} - (1-\alpha)$$

$$= \alpha - P_F\{n^{1/2}(\bar{X}_n - \theta(F)) < J_n^{-1}(\alpha, \hat{F}_n)\}$$

$$= \alpha - P_F\{n^{1/2}(\bar{X}_n - \theta(F))/\hat{\sigma}_n < z_\alpha + \frac{1}{6}\gamma(F)(z_\alpha^2 - 1)n^{-1/2} + O_P(n^{-1})\}$$

$$= \alpha - P_F\{n^{1/2}(\bar{X}_n - \theta(F))/\hat{\sigma}_n < z_\alpha + \frac{1}{6}\gamma(F)(z_\alpha^2 - 1)n^{-1/2}\} + O(n^{-1}) .$$

The last equality, though plausible, requires a rigorous argument, but follows from Problem 15.36. The last expression, by (15.46) and a Taylor expansion, becomes

$$-\frac{1}{2}\gamma(F)\varphi(z_\alpha)z_\alpha^2 n^{-1/2} + O(n^{-1}) ,$$

so that the one-sided coverage error is of the same order as that provided by the basic normal approximation. Moreover, by similar reasoning, the two-sided bootstrap interval of nominal level $1-2\alpha$, given by

$$[\bar{X}_n - n^{-1/2} J_n^{-1}(1-\alpha, \hat{F}_n), \bar{X}_n - n^{-1/2} J_n^{-1}(\alpha, \hat{F}_n)] , \tag{15.56}$$

has coverage error $O(n^{-1})$. Although these basic bootstrap intervals have the same orders of coverage error as those based on the normal approximation, there is evidence that the bootstrap does provide some improvement (in terms of the size of the constants); see Liu and Singh (1987).

Bootstrap-t Confidence Intervals. Next, we consider bootstrap confidence intervals for $\theta(F)$ based on the studentized root

$$R_n^s(X^n, \theta(F)) = n^{1/2}(\bar{X}_n - \theta(F))/\hat{\sigma}_n , \tag{15.57}$$

whose distribution under F is denoted $K_n(\cdot, F)$. The bootstrap versions of the expansions (15.46) and (15.47) are

$$K_n(t, \hat{F}_n) = \Phi(t) + \frac{1}{6}\gamma(\hat{F}_n)\varphi(t)(2t^2+1)n^{-1/2} + O_P(n^{-1}) \tag{15.58}$$

and

$$K_n^{-1}(1-\alpha, \hat{F}_n) = z_{1-\alpha} - \frac{1}{6}\gamma(\hat{F}_n)(2z_{1-\alpha}^2 + 1)n^{-1/2} + O_P(n^{-1}) . \tag{15.59}$$

Again, these results are obtained rigorously in Hall (1992), and a sufficient condition for their validity is that F is absolutely continuous with infinitely many moments. By comparing (15.46) and (15.58), it follows that

$$K_n(t, \hat{F}_n) - K_n(t, F) = O_P(n^{-1}) , \tag{15.60}$$

since $\gamma(\hat{F}_n) - \gamma(F) = O_P(n^{-1/2})$. Similarly,

$$K_n^{-1}(1-\alpha, \hat{F}_n) - K_n^{-1}(1-\alpha, F) = O_P(n^{-1}) . \tag{15.61}$$

Thus, the bootstrap is more successful at estimating the distribution or quantiles of the studentized root than its nonstudentized version.

Now, consider the nominal level $1-\alpha$ upper confidence bound $\bar{X}_n - n^{-1/2}\hat{\sigma}_n K_n^{-1}(\alpha, \hat{F}_n)$. Its coverage error is given by

$$P_F\{\theta(F) \le \bar{X}_n - n^{-1/2}\hat{\sigma}_n K_n^{-1}(\alpha, \hat{F}_n)\} - (1-\alpha)$$

$$= \alpha - P_F\{n^{1/2}(\bar{X}_n - \theta(F))/\hat{\sigma}_n < K_n^{-1}(\alpha, \hat{F}_n)\}$$

$$= \alpha - P_F\{n^{1/2}(\bar{X}_n - \theta(F))/\hat{\sigma}_n < z_\alpha - \frac{1}{6}\gamma(F)(2z_\alpha^2+1)n^{-1/2} + O_P(n^{-1})\} ,$$

since (15.59) implies the same expansion for $K_n^{-1}(\alpha, \hat{F}_n)$ with $\gamma(\hat{F}_n)$ replaced by $\gamma(F)$ (again using the fact that $\gamma(\hat{F}_n) - \gamma(F) = O_P(n^{-1/2})$). By Problem 15.36, this last expression becomes

$$\alpha - P_F\{n^{1/2}(\bar{X}_n - \theta(F))/\hat{\sigma}_n < z_\alpha - \frac{1}{6}\gamma(F)(2z_\alpha^2+1)n^{-1/2}\} + O(n^{-1}) .$$

Let

$$t_n = t_n(\alpha, F) = z_\alpha - \frac{1}{6}\gamma(F)(2z_\alpha^2+1)n^{-1/2} ,$$

so that $(t_n - z_\alpha) = O(n^{-1/2})$. Then, the coverage error becomes

$$\alpha - [\Phi(t_n) + \frac{1}{6}\gamma(F)\varphi(t_n)(2t_n^2+1)n^{-1/2} + O(n^{-1})] .$$

By expanding Φ and φ about z_α and combining terms that are $O(n^{-1})$, the last expression becomes

$$\alpha - \Phi(z_\alpha) - (t_n - z_\alpha)\varphi(z_\alpha) + O(n^{-1})$$

$$-\frac{1}{6}\gamma(F)[\varphi(z_\alpha) + (t_n - z_\alpha)\varphi'(z_\alpha) + O(n^{-1})](2z_\alpha^2+1)n^{-1/2} + O(n^{-1}) = O(n^{-1}) .$$

Thus, the one-sided coverage error of the bootstrap-t interval is $O(n^{-1})$ and is of smaller order than that provided by the normal approximation or the bootstrap based on a nonstudentized root. Intervals with one-sided coverage error of order $O(n^{-1})$ are said to be *second-order accurate*, while intervals with one-sided coverage error of order $O(n^{-1/2})$ are only *first-order accurate*.

A heuristic reason why the bootstrap based on the root (15.57) outperforms the bootstrap based on the root (15.53) is as follows. In the case of (15.53), the bootstrap is estimating a distribution that has mean 0 and unknown variance $\sigma^2(F)$. The main contribution to the estimation error is the implicit estimation of $\sigma^2(F)$ by $\sigma^2(\hat{F}_n)$. On the other hand, the root (15.57) has a distribution that is nearly independent of F since it is an asymptotic pivot.

The two-sided interval of nominal level $1-2\alpha$,

$$[\bar{X}_n - n^{-1/2}\hat{\sigma}_n K_n^{-1}(1-\alpha, \hat{F}_n), \bar{X}_n - n^{-1/2}\hat{\sigma}_n K_n^{-1}(\alpha, \hat{F}_n)] , \tag{15.62}$$

also has coverage error $O(n^{-1})$ (Problem 15.38). This interval was formed by combining two one-sided intervals. Instead, consider the absolute studentized root

$$R_n^t(X^n, \theta(F)) = |n^{1/2}(\bar{X}_n - \theta(F))|/\hat{\sigma}_n ,$$

whose distribution and quantile functions under F are denoted $L_n(t, F)$ and $L_n^{-1}(1-\alpha, F)$, respectively. An alternative two-sided bootstrap confidence interval for $\theta(F)$ of nominal level $1 - \alpha$ is given by

$$\bar{X}_n \pm n^{-1/2}\hat{\sigma}_n L_n^{-1}(1 - \alpha, \hat{F}_n) .$$

Note that this interval is symmetric about $\bar{X}_n$. Its coverage error is actually $O(n^{-2})$. The arguments for this claim are similar to the previous claims about coverage error, but more terms are required in expansions like (15.46).

Bootstrap Calibration. By considering a studentized statistic, the bootstrap-t yields one-sided confidence intervals with coverage error smaller than the non-studentized case. However, except in some simple problems, it may be difficult to standardize or studentize a statistic because an explicit estimate of the asymptotic variance may not be available. An alternative approach to improving coverage error is based on the following calibration idea of Loh (1987). Let $I_n = I_n(1-\alpha)$ be any interval with nominal level $1 - \alpha$, such as one given by the bootstrap, or a simple normal approximation. Its coverage is defined to be

$$C_n(1 - \alpha, F) = P_F\{\theta(F) \in I_n(1 - \alpha)\} .$$

We can estimate $C_n(1 - \alpha, F)$ by its bootstrap counterpart $C_n(1 - \alpha, \hat{F}_n)$. Then, determine $\hat{\alpha}_n$ to satisfy

$$C_n(1 - \hat{\alpha}_n, \hat{F}_n) = 1 - \alpha ,$$

so that $\hat{\alpha}_n$ is the value that results in the estimated coverage to be the nominal level. The calibrated interval then is defined to be $I_n(1 - \hat{\alpha}_n)$.

To fix ideas, suppose $I_n(1 - \alpha)$ is the one-sided normal theory interval $(-\infty, \bar{X}_n + n^{-1/2}\hat{\sigma}_n z_{1-\alpha}]$. We argued its coverage error is $O(n^{-1/2})$. More specifically,

$$C_n(1 - \alpha, F) = P_F\{n^{1/2}(\bar{X}_n - \theta(F))/\hat{\sigma}_n < z_\alpha\}$$

$$= 1 - \alpha + \frac{1}{6}\varphi(z_\alpha)(2z_\alpha^2 + 1)n^{-1/2} + O(n^{-1}) .$$

Under smoothness and moment assumptions, the bootstrap estimated coverage satisfies

$$C_n(1 - \alpha, \hat{F}_n) = 1 - \alpha + \frac{1}{6}\varphi(z_\alpha)\gamma(\hat{F}_n)(2z_\alpha^2 + 1)n^{-1/2} + O_P(n^{-1}) ,$$

and the value of $\hat{\alpha}_n$ is obtained by setting the estimated coverage equal to $1 - \alpha$. One can then show that

$$\hat{\alpha}_n - \alpha = -\frac{1}{6}\varphi(z_\alpha)\gamma(F)(2z_\alpha^2 + 1)n^{-1/2} + O_P(n^{-1}) . \qquad (15.63)$$

By using this expansion and (15.46), it can be shown that the interval $I_n(1 - \hat{\alpha}_n)$ has coverage $1 - \alpha + O(n^{-1})$, and hence is second-order accurate (Problem 15.39). Thus, calibration reduces the order of coverage error.

Other Bootstrap Methods. There are now many variations on the basic bootstrap idea that yield confidence regions that are second-order accurate, assuming the validity of Edgeworth Expansions like the ones used in this section. The calibration method described above is due to Loh (1987, 1991) and is essentially equivalent to Beran's (1987, 1988) method of prepivoting (Problem 15.43). Given an interval $I_n(1-\alpha)$ of nominal level $1-\alpha$, calibration produces a new interval, say $I_n^1(1-\alpha) = I_n(1-\hat{\alpha}_n)$, where $\hat{\alpha}_n$ is chosen by calibration. It is tempting to iterate this idea to further reduce coverage error. That is, now calibrate I_n^1 to yield a new interval I_n^2, and so on. Further reduction in coverage error is indeed possible (at the expense of increased computational effort). For further details on these and other methods such as Efron's BC_a method, see Hall and Martin (1988), Hall (1992) and Efron and Tibshirani (1993).

The analysis of this section was limited to methods for constructing confidence intervals for a mean, assuming the underlying distribution is smooth and has sufficiently many moments. But, many of the conclusions extend to smooth functions of means studied in Example 15.4.3. In particular, in order to reduce coverage error, it is desirable to use a root that is at least asymptotically pivotal, such as a studentized root that is asymptotically standard normal. Otherwise, the basic bootstrap interval (15.22) has the same order of coverage error as one based on approximating the asymptotic distribution. However, whether or not the root is asymptotically pivotal, bootstrap calibration reduces the order of coverage error. Of course, some qualifications are necessary. For one, even in the context of the mean, Cramér's condition may not hold, as in the context of a binomial proportion. Edgeworth expansions for such discrete distributions supported on a lattice are studied in Chapter 5 of Bhattacharya and Rao (1976) and Kolassa and McCullagh (1990); also see Brown, Cai and DasGupta (2001), who study the binomial case. In other problems where smoothness is assumed, such as inference for a density or quantiles, Edgeworth expansions for appropriate statistics behave somewhat differently than they do for a mean. Such problems are treated in Hall (1992).

15.6 Hypothesis Testing

In this section, we consider the use of the bootstrap for the construction of hypothesis tests. Assume the data X^n is generated from some unknown law P. The null hypothesis H asserts that P belongs to a certain family of distributions $\mathbf{P_0}$, while the alternative hypothesis K asserts that P belongs to a family $\mathbf{P_1}$. Of course, we assume the intersection of $\mathbf{P_0}$ and $\mathbf{P_1}$ is the empty set, and the unknown law P belongs to $\mathbf{P}$, the union of $\mathbf{P_0}$ and $\mathbf{P_1}$.

There are several approaches one can take to construct a hypothesis test. First, consider the case when the null hypothesis can be expressed as a hypothesis about a real- or vector-valued parameter $\theta(P)$. Then, one can exploit the familiar duality between confidence regions and hypothesis tests to test hypotheses about $\theta(P)$. Thus, a consistent in level test of the null hypothesis that $\theta(P) = \theta_0$ can be constructed by a consistent in level confidence region for $\theta(P)$ by the rule: accept the null hypothesis if and only if the confidence region includes θ_0. Therefore, all the methods we have thus far discussed for constructing confidence regions may be utilized: methods based on a pivot, an asymptotic pivot, an asymptotic approximation, or the bootstrap. Indeed, this was the bootstrap approach already considered in Corollary 15.4.1, and it was also the basis for the multiple test construction in Section 15.4.4.

However, not all hypothesis testing problems fit nicely into the framework of testing parameters. For example, consider the problem of testing whether the data come from a certain parametric submodel (such as the family of normal distributions) of a nonparametric model, the so-called goodness of fit problem. Or, when X_i is vector-valued, consider the problem of testing whether X_i has a distribution that is spherically symmetric.

Given a test statistic T_n, its distribution must be known, estimated, or approximated (at least under the null hypothesis), in order to construct a critical value. The approach taken in this section is to estimate the null distribution of T_n by resampling from a distribution obeying the constraints of the null hypothesis.

To be explicit, assume we wish to construct a test based on a real-valued test statistic $T_n = T_n(X^n)$ which is consistent in level and power. Large values of T_n reject the null hypothesis. Thus, having picked a suitable test statistic T_n, our goal is to construct a critical value, say $c_n(1-\alpha)$, so that the test which rejects if and only if T_n exceeds $c_n(1-\alpha)$ satisfies

$$P\{T_n(X^n) > c_n(1-\alpha)\} \to \alpha \text{ as } n \to \infty$$

when $P \in \mathbf{P_0}$. Furthermore, we require this rejection probability to tend to one when $P \in \mathbf{P_1}$. Unlike the classical case, the critical value will be constructed to be data-dependent (as in the case of a permutation test). To see how the bootstrap can be used to determine a critical value, let the distribution of T_n under P be denoted by

$$G_n(x, P) = P\{T_n(X^n) \le x\} .$$

Note that we have introduced $G_n(\cdot, P)$ instead of utilizing $J_n(\cdot, P)$ to distinguish from the case of confidence intervals where $J_n(\cdot, P)$ represents the distribution of a root which may depend both on the data and on P. In the hypothesis testing context, $G_n(\cdot, P)$ represents the distribution of a statistic (and not a root) under P. Let

$$g_n(1-\alpha, P) = \inf\{x : G_n(x, P) \ge 1-\alpha\} .$$

Typically, $G_n(\cdot, P)$ will converge in distribution to a limit law $G(\cdot, P)$, whose $1-\alpha$ quantile is denoted $g(1-\alpha, P)$.

The bootstrap approach is to estimate the null sampling distribution by $G_n(\cdot, \hat{Q}_n)$, where $\hat{Q}_n$ is an estimate of P in $\mathbf{P_0}$ so that $\hat{Q}_n$ satisfies the constraints of the null hypothesis, since critical values should be determined as if the null hypothesis were true. A bootstrap critical value can then be defined by $g_n(1-\alpha, \hat{Q}_n)$. The resulting nominal level α bootstrap test rejects H if and only if $T_n > g_n(1-\alpha, \hat{Q}_n)$.

Notice that we would not want to replace a $\hat{Q}_n$ satisfying the null hypothesis constraints by the empirical distribution function $\hat{P}_n$, the usual resampling mechanism of resampling the data with replacement. One might say that the bootstrap is so adept at estimating the distribution of a statistic that $G_n(\cdot, \hat{P}_n)$ is a good estimate of $G_n(\cdot, P)$ whether or not P satisfies the null hypothesis constraints. Hence, the test that rejects when T_n exceeds $g_n(1-\alpha, \hat{P}_n)$ will (under suitable conditions) behave asymptotically like the test that rejects when T_n exceeds $g_n(1-\alpha, P)$, and this test has an asymptotic probability of α of rejecting the null hypothesis, even if $P \in \mathbf{P_1}$. But, when $P \in \mathbf{P_1}$, we would want the test to reject with probability that is approaching one.

Thus, the choice of resampling distribution $\hat{Q}_n$ should satisfy the following. If $P \in \mathbf{P_0}$, $\hat{Q}_n$ should be near P so that $G_n(\cdot, P) \approx G_n(\cdot, \hat{Q}_n)$; then, $g_n(1-\alpha, P) \approx g_n(1-\alpha, \hat{Q}_n)$ and the asymptotic rejection probability approaches α. If, on the other hand, $P \in \mathbf{P_1}$, $\hat{Q}_n$ should not approach P, but some P_0 in $\mathbf{P_0}$. In this way, the critical value should satisfy

$$g_n(1-\alpha, \hat{Q}_n) \approx g_n(1-\alpha, P_0) \to g(1-\alpha, P_0) < \infty$$

as $n \to \infty$. Then, assuming the test statistic is constructed so that $T_n \to \infty$ under P when $P \in \mathbf{P_1}$, we will have

$$P\{T_n > g_n(1-\alpha, \hat{Q}_n)\} \approx P\{T_n > g(1-\alpha, P_0)\} \to 1$$

as $n \to \infty$, by Slutsky's Theorem.

As in the construction of confidence intervals, $G_n(\cdot, P)$ must be smooth in P in order for the bootstrap to succeed. In the theorem below, rather than specifying a set of sequences $\mathbf{C}_P$ as was done in Theorem 15.4.1, smoothness is described in terms of a metric d, but either approach could be used. The proof is analogous to the proof of Theorem 15.4.1.

Theorem 15.6.1 *Let X^n be generated from a probability law $P \in \mathbf{P_0}$. Assume the following triangular array convergence: $d(P_n, P) \to 0$ and $P \in \mathbf{P_0}$ implies $G_n(\cdot, P_n)$ converges weakly to $G(\cdot, P)$ with $G(\cdot, P)$ continuous. Moreover, assume $\hat{Q}_n$ is an estimator of P based on X^n which satisfies $d(\hat{Q}_n, P) \to 0$ in probability whenever $P \in \mathbf{P_0}$. Then,*

$$P\{T_n > g_n(1-\alpha, \hat{Q}_n)\} \to \alpha \qquad \text{as } n \to \infty .$$

Example 15.6.1 (Normal Correlation) Suppose (Y_i, Z_i), $i = 1, \ldots, n$ are i.i.d. bivariate normal with unknown means, variances, and correlation ρ. The null hypothesis specifies $\rho = \rho_0$ versus $\rho > \rho_0$. Let $T_n = n^{1/2}\hat{\rho}_n$, where $\hat{\rho}_n$ is the usual sample correlation. Under the null hypothesis, the distribution of T_n doesn't depend on any unknown parameters. So, if $\hat{Q}_n$ is any bivariate normal distribution with $\rho = \rho_0$, the bootstrap sampling distribution $G_n(\cdot, \hat{Q}_n)$ is exactly equal to the true null sampling distribution. Note, however, that inverting a parametric bootstrap confidence bound using the root $n^{1/2}(\hat{\rho}_n - \rho)$ would not be exact. ■

Example 15.6.2 (Likelihood Ratio Tests) Suppose $X_1, \ldots, X_n$ are i.i.d. according to a model $\{P_\theta,\ \theta \in \Omega\}$, where Ω is an open subset of $\mathbb{R}^k$. Assume θ is partitioned as (ξ, μ), where ξ is a vector of length p and μ is a vector of length $k-p$. The null hypothesis parameter space Ω_0 specifies $\xi = \xi_0$. Under the conditions of Theorem 12.4.2, the likelihood ratio statistic $T_n = 2\log(R_n)$ is asymptotically χ^2_p under the null hypothesis. Suppose $(\xi_0, \hat{\mu}_{n,0})$ is an efficient likelihood estimator of θ for the model Ω_0. Rather than using the critical value obtained from χ^2_p, one could bootstrap T_n. So, let $G_n(x, \theta)$ denote the distribution of T_n under θ. An appropriate parametric bootstrap test obeying the null hypothesis constraints is to reject the null when T_n exceeds the $1-\alpha$ quantile of $G_n(x, (\xi_0, \hat{\mu}_{n,0}))$. Beran and Ducharme (1991) argue that, under regularity conditions, the bootstrap test has error in rejection probability equal to $O(n^{-2})$, while the usual likelihood ratio test has error $O(n^{-1})$. Moreover, the bootstrap test can be viewed as an analytical approximation to a Bartlett-corrected likelihood ratio test (see Section 12.4.4). In essence, the bootstrap automatically captures the Bartlett correction and avoids the need for analytical calculation. As an example, recall Example 12.4.7, where it was observed the Bartlett-corrected likelihood ratio test has error $O(n^{-2})$. Here, the bootstrap test is exact (Problem 15.45). ■

Example 15.6.3 (Behrens-Fisher Problem Revisited) For $j = 1, 2$, let $X_{i,j}$, $i = 1, \ldots, n_j$ be independent with $X_{i,j}$ distributed as $N(\mu_j, \sigma_j^2)$. All four parameters are unknown and vary independently. The null hypothesis asserts $\mu_1 = \mu_2$ and the alternative is $\mu_1 > \mu_2$. Let $n = n_1 + n_2$, and for simplicity assume n_1 to be the integer part of λn for some $0 < \lambda < 1$. Let $(\bar{X}_{n,j}, S_{n,j}^2)$ be the usual unbiased estimators of (μ_j, σ_j^2) based on the jth sample. Consider the test statistic

$$T_n = (\bar{X}_1 - \bar{X}_2) / \sqrt{\frac{S_{n,1}^2}{n_1} + \frac{S_{n,2}^2}{n_2}} \ .$$

By Example 13.5.4, the test that rejects the null hypothesis when $T_n > z_{1-\alpha}$ is efficient. However, we now study its actual rejection probability.

The null distribution of T_n depends only on $\sigma^2 = (\sigma_1^2, \sigma_2^2)$ through the ratio σ_1/σ_2, and we denote this distribution by $G_n(\cdot, \sigma^2)$. Let $S_n^2 = (S_{n,1}^2, S_{n,2}^2)$. Like the method used in Problem 11.89, by conditioning on S_n^2, we can write

$$G_n(x, \sigma^2) = E[a(S_n^2, \sigma^2, x)] \ ,$$

where

$$a(S_n^2, \sigma^2, x) = \Phi[(1+\delta)^{1/2} x]$$

and

$$\delta = \sum_{j=1}^{2} n_j^{-1}(S_{n,j}^2 - \sigma_j^2) / \sum_{j=1}^{2} n_j^{-1} \sigma_j^2 \ .$$

By Taylor expansion and the moments of S_n^2, it follows that (Problem 15.46)

$$G_n(x, \sigma^2) = \Phi(x) + \frac{1}{n} b_n(x, \sigma^2) + O(n^{-2}) \ , \tag{15.64}$$

where

$$\frac{1}{n} b_n(x, \sigma^2) = -(x + x^3)\phi(x)\rho_n^2/4$$

is $O(n^{-1})$ and

$$\rho_n^2 = \sum_{j=1}^{2} (n_j - 1)^{-1} n_j^{-2} \sigma_j^4 / (\sum_{j=1}^{2} n_j^{-1} \sigma_j^2)^2 \ .$$

Correspondingly, the quantile function satisfies

$$G_n^{-1}(1-\alpha, \sigma^2) = z_{1-\alpha} + (z_{1-\alpha} + z_{1-\alpha}^3)\rho_n^2/4 + O(n^{-2}) \ . \tag{15.65}$$

It follows that the rejection probability of the asymptotic test that rejects when $T_n > z_{1-\alpha}$ is $\alpha + O(n^{-1})$.

Consider next the (parametric) bootstrap-t, which rejects when $T_n > G_n^{-1}(1-\alpha, S_n^2)$. Its rejection probability can be expressed as

$$1 - E[a(S_n^2, \sigma^2, G_n^{-1}(1-\alpha, S_n^2))] \ .$$

By Taylor expansion, it can be shown that the rejection probability of the test is $\alpha + O(n^{-2})$ (Problem 15.47). Thus, the bootstrap-t improves upon the asymptotic expansion. In fact, bootstrap calibration (or the use of prepivoting) further reduces the error in rejection probability to $O(n^{-3})$. Details are in Beran (1988), who further argues that the Welch method described in Section 11.3.1 behaves like the bootstrap-t method. Although the Welch approximation is based on elegant mathematics, the bootstrap approach essentially reproduces the analytical approximation automatically. ■

Example 15.6.4 (Nonparametric Mean) Let $X_1, \ldots, X_n$ be i.i.d. observations on the real line with probability law P, mean $\mu(P)$ and finite variance $\sigma^2(P)$. The problem is to test $\mu(P) = 0$ against either a one-sided or two-sided alternative. So, $\mathbf{P_0}$ is the set of distributions with mean zero and finite variance. In the one-sided case, consider the test statistic $T_n = n^{1/2}\bar{X}_n$, where $\bar{X}_n$ is the sample mean, since test statistics based on $\bar{X}_n$ were seen in Section 11.4 to possess a certain optimality property. We will also consider the studentized statistic $T_n' = n^{1/2}\bar{X}_n/S_n$, where we shall take S_n^2 to be the unbiased estimate of variance. To apply Theorem 15.6.1, let $\hat{Q}_n$ be the empirical distribution $\hat{P}_n$ shifted by $\bar{X}_n$ so it has mean 0. Then, the error in rejection probability will be $O(n^{-1/2})$ for T_n, and will be $O(n^{-1})$ for T_n', at least under the assumptions that F is smooth and has infinitely many moments; these statements follow from the results in Section 15.5 (Problem 15.49).

While shifting the empirical distribution works in this example, it is not easy to generalize when testing other parameters. Therefore, we consider the following alternative approach. The idea is to choose the distribution in $\mathbf{P_0}$ that is in some sense closest to the empirical distribution $\hat{P}_n$. One way to describe closeness is the following. For distributions P and Q on the real line, let $\delta_{KL}(P, Q)$ be the (forward) Kullback-Leibler divergence between P and Q (studied in Example 11.2.4), defined by

$$\delta_{KL}(P, Q) = \int log(\frac{dP}{dQ})dP \ . \tag{15.66}$$

Note that $\delta_{KL}(P, Q)$ may be ∞, δ_{KL} is not a metric, and it is not even symmetric in its arguments. Let $\hat{Q}_n$ be the Q that minimizes $\delta_{KL}(\hat{P}_n, Q)$ over Q in $\mathbf{P_0}$. This choice for $\hat{Q}_n$ can be shown to be well-defined and corresponds to finding the nonparametric maximum likelihood estimator of P assuming P is constrained to have mean zero. (Another possibility is to minimize the (backward) Kullback-Leibler divergence $\delta_{KL}(Q, \hat{P}_n)$.) By Efron (1981) (Problem 15.50), $\hat{Q}_n$ assigns mass w_i to X_i, where w_i satisfies

$$w_i \propto \frac{(1 + tX_i)^{-1}}{\sum_{j=1}^n (1 + tX_j)^{-1}}$$

and t is chosen so that $\sum_{i=1}^n w_i X_i = 0$. Now, one could bootstrap either T_n or T'_n from $\hat{Q}_n$.

In fact, this approach suggests an alternative test statistic given by $T''_n = n\delta_{KL}(\hat{P}_n, \hat{Q}_n)$, where $\hat{Q}_n$ is the Q minimizing the Kullback-Leibler divergence $\delta_{KL}(\hat{P}_n, Q)$ over Q in $\mathbf{P_0}$. This is equivalent to the test statistic used by Owen (1988, 2001) in his construction of empirical likelihood, who shows the limiting distribution of $2T''_n$ under the null hypothesis is Chi-squared with 1 degree of freedom. The wide scope of empirical likelihood is presented in Owen (2001). ■

Example 15.6.5 (Goodness of fit) The problem is to test whether the underlying probability distribution P belongs to a parametric family of distributions $\mathbf{P_0} = \{P_\theta, \theta \in \Theta_0\}$, where Θ_0 is an open subset of k-dimensional Euclidean space. Let $\hat{P}_n$ be the empirical measure based on $X_1, \ldots, X_n$. Let $\hat{\theta}_n \in \Theta_0$ be an estimator of θ. Consider the test statistic

$$T_n = n^{1/2} \delta(\hat{P}_n, P_{\hat{\theta}_n}) ,$$

where δ is some measure (typically a metric) between $\hat{P}_n$ and $P_{\hat{\theta}_n}$. (In fact, δ need not even be symmetric, which is useful sometimes: for example, consider the Cramér–von Mises statistic.) Beran (1986) considers the case where $\hat{\theta}_n$ is a minimum distance estimator, while Romano (1988) assumes that $\hat{\theta}_n$ is some asymptotically linear estimator (like an efficient likelihood estimator). For the resampling mechanism, take $\hat{Q}_n = P_{\hat{\theta}_n}$. Both Beran (1986) and Romano (1988) give different sets of conditions so that the above theorem is applicable, both requiring the machinery of empirical processes. ■

15.7 Subsampling

In this section, a general theory for the construction of approximate confidence sets or hypothesis tests is presented, so the goal is the same as that of the bootstrap. The basic idea is to approximate the sampling distribution of a statistic based on the values of the statistic computed over smaller subsets of the data. For example, in the case where the data are n observations which are independent and identically distributed, a statistic $\hat{\theta}_n$ is computed based on the entire data set and is recomputed over all $\binom{n}{b}$ data sets of size b. Implicit is the notion of a statistic sequence, so that the statistic is defined for samples of size n and b. These recomputed values of the statistic are suitably normalized to approximate the true sampling distribution.

This approach based on subsamples is perhaps the most general one for approximating a sampling distribution, in the sense that consistency holds under extremely weak conditions. That is, it will be seen that, under very weak assumptions on b, the method is consistent whenever the original statistic, suitably normalized, has a limit distribution under the true model. The bootstrap, on the other hand, requires that the distribution of the statistic is somehow locally smooth as a function of the unknown model. In contrast, no such assumption is required in the theory for subsampling. Indeed, the method here is applicable even in the several known situations which represent counterexamples to the bootstrap. However, when both subsampling and the bootstrap are consistent, the bootstrap is typically more accurate.

To appreciate why subsampling behaves well under such weak assumptions, note that each subset of size b (taken without replacement from the original data) is indeed a sample of size b from the true model. If b is small compared to n (meaning $b/n \to 0$), then there are many (namely $\binom{n}{b}$) subsamples of size b available. Hence, it should be intuitively clear that one can at least approximate the sampling distribution of the (normalized) statistic $\hat{\theta}_b$ by recomputing the values of the statistic over all these subsamples. But, under the weak convergence hypothesis, the sampling distributions based on samples of size b and n should be close. The bootstrap, on the other hand, is based on recomputing a statistic over a sample of size n from some estimated model which is hopefully close to the true model.

The use of subsample values to approximate the variance of a statistic is well-known. The Quenouille-Tukey jackknife estimates of bias and variance based on computing a statistic over all subsamples of size $n-1$ has been well-studied and is closely related to the mean and variance of our estimated sampling distribution with $b = n-1$. For further history of subsampling methods, see Politis, Romano, and Wolf (1999).

15.7.1 The Basic Theorem in the I.I.D. Case

Suppose $X_1, \ldots, X_n$ is a sample of n i.i.d. random variables taking values in an arbitrary sample space S. The common probability measure generating the observations is denoted P. The goal is to construct a confidence region for some parameter $\theta(P)$. For now, assume θ is real-valued, but this can and will be generalized to allow for the construction of confidence regions for multivariate parameters or confidence bands for functions.

Let $\hat{\theta}_n = \hat{\theta}_n(X_1, \ldots, X_n)$ be an estimator of $\theta(P)$. It is desired to estimate the true sampling distribution of $\hat{\theta}_n$ in order to make inferences about $\theta(P)$. Nothing is assumed about the form of the estimator.

As in previous sections, let $J_n(P)$ be the sampling distribution of the root $\tau_n(\hat{\theta}_n - \theta(P))$ based on a sample of size n from P, where τ_n is a normalizing constant. Here, τ_n is assumed known and does not depend on P. Also define the corresponding cumulative distribution function:

$$J_n(x, P) = P\{\tau_n[\hat{\theta}_n(X_1, \ldots, X_n) - \theta(P)] \le x\} .$$

Essentially, the only assumption that we will need to construct asymptotically valid confidence intervals for $\theta(P)$ is the following.

Assumption 15.7.1 There exists a limiting distribution $J(P)$ such that $J_n(P)$ converges weakly to $J(P)$ as $n \to \infty$.

This assumption will be required to hold for some sequence τ_n. The most informative case occurs when τ_n is such that the limit law $J(P)$ is nondegenerate.

To describe the subsampling method, consider the $N_n = \binom{n}{b}$ subsets of size b of the data $\{X_1, \ldots, X_n\}$; call them $Y_1, \ldots, Y_{N_n}$, ordered in any fashion. Thus, each

Y_i constitutes a sample of size b from P. Of course, the Y_i depend on b and n, but this notation has been suppressed. Only a very weak assumption on b will be required. In the consistency results that follow, it will be assumed that $b/n \to 0$ and $b \to \infty$ as $n \to \infty$. Now, let $\hat{\theta}_{n,b,i}$ be equal to the statistic $\hat{\theta}_b$ evaluated at the data set Y_i. The approximation to $J_n(x, P)$ we study is defined by

$$L_{n,b}(x) = N_n^{-1} \sum_{i=1}^{N_n} I\{\tau_b(\hat{\theta}_{n,b,i} - \hat{\theta}_n) \le x\} . \tag{15.67}$$

The motivation behind the method is the following. For any i, Y_i is actually a random sample of b i.i.d. observations from P. Hence, the *exact* distribution of $\tau_b(\hat{\theta}_{n,b,i} - \theta(P))$ is $J_b(P)$. The empirical distribution of the N_n values of $\tau_b(\hat{\theta}_{n,b,i} - \theta(P))$ should then serve as a good approximation to $J_n(P)$. Of course, $\theta(P)$ is unknown, so we replace $\theta(P)$ by $\hat{\theta}_n$, which is asymptotically permissible because $\tau_b(\hat{\theta}_n - \theta(P))$ is of order τ_b/τ_n, and τ_b/τ_n will be assumed to tend to zero.

Theorem 15.7.1 *Suppose Assumption 15.7.1 holds. Also, assume $\tau_b/\tau_n \to 0$, $b \to \infty$, and $b/n \to 0$ as $n \to \infty$.*

(i) If x is a continuity point of $J(\cdot, P)$, then $L_{n,b}(x) \to J(x, P)$ in probability.

(ii) If $J(\cdot, P)$ is continuous, then

$$\sup_x |L_{n,b}(x) - J_n(x, P)| \to 0 \text{ in probability} . \tag{15.68}$$

(iii) Let

$$c_{n,b}(1-\alpha) = \inf\{x : L_{n,b}(x) \ge 1 - \alpha\} .$$

and

$$c(1-\alpha, P) = \inf\{x : J(x, P) \ge 1 - \alpha\} .$$

If $J(\cdot, P)$ is continuous at $c(1-\alpha, P)$, then

$$P\{\tau_n[\hat{\theta}_n - \theta(P)] \le c_{n,b}(1-\alpha)\} \to 1 - \alpha \text{ as } n \to \infty . \tag{15.69}$$

Therefore, the asymptotic coverage probability under P of the confidence interval $[\hat{\theta}_n - \tau_n^{-1} c_{n,b}(1-\alpha), \infty)$ is the nominal level $1-\alpha$.

Proof. Let

$$U_n(x) = U_{n,b}(x, P) = N_n^{-1} \sum_{i=1}^{N_n} I\{\tau_b[\hat{\theta}_{n,b,i} - \theta(P)] \le x\} . \tag{15.70}$$

Note that the dependence of $U_n(x)$ on b and P will now be suppressed for notational convenience. To prove (i), it suffices to show $U_n(x)$ converges in probability to $J(x, P)$ for every continuity point x of $J(x, P)$. To see why, note that

$$L_{n,b}(x) = N_n^{-1} \sum_i I\{\tau_b[\hat{\theta}_{n,b,i} - \theta(P)] + \tau_b[\theta(P) - \hat{\theta}_n] \le x\} ,$$

so that for every $\epsilon > 0$,

$$U_n(x - \epsilon) I\{E_n\} \le L_{n,b}(x) I\{E_n\} \le U_n(x + \epsilon) I\{E_n\} ,$$

where $I\{E_n\}$ is the indicator of the event $E_n \equiv \{\tau_b|\theta(P) - \hat{\theta}_n| \leq \epsilon\}$. But, the event E_n has probability tending to one. So, with probability tending to one,

$$U_n(x - \epsilon) \leq L_{n,b}(x) \leq U_n(x + \epsilon)$$

for any $\epsilon > 0$. Hence, if $x + \epsilon$ and $x - \epsilon$ are continuity points of $J(\cdot, P)$, then $U_n(x \pm \epsilon) \to J(x \pm \epsilon, P)$ in probability implies

$$J(x - \epsilon, P) - \epsilon \leq L_{n,b}(x) \leq J(x + \epsilon, P) + \epsilon$$

with probability tending to one. Now, let $\epsilon \to 0$ so that $x \pm \epsilon$ are continuity points of $J(\cdot, P)$. Then, it suffices to show $U_n(x) \to J(x, P)$ in probability for all continuity points x of $J(\cdot, P)$. But, $0 \leq U_n(x) \leq 1$ and $E[U_n(x)] = J_b(x, P)$. Since $J_b(x, P) \to J(x, P)$, it suffices to show $Var[U_n(x)] \to 0$. To this end, suppose k is the greatest integer less than or equal to n/b. For $j = 1, \ldots, k$, let $R_{n,b,j}$ be equal to the statistic $\hat{\theta}_b$ evaluated at the data set $\hat{\theta}_b(X_{b(j-1)+1}, X_{b(j-1)+2}, \ldots, X_{b(j-1)+b})$ and set

$$\bar{U}_n(x) = k^{-1} \sum_{j=1}^{k} I\{\tau_b[R_{n,b,j} - \theta(P)] \leq x\} .$$

Clearly, $\bar{U}_n(x)$ and $U_n(x)$ have the same expectation. But, since $\bar{U}_n(x)$ is the average of k i.i.d. variables (each of which is bounded between 0 and 1), it follows that

$$Var[\bar{U}_n(x)] \leq \frac{1}{4k} \to 0$$

as $n \to \infty$. Intuitively, $U_n(x)$ should have a smaller variance than $\bar{U}_n(x)$, because $\bar{U}_n(x)$ uses the ordering in the sample in an arbitrary way. Formally, we can write

$$U_n(x) = E[\bar{U}_n(x)|\mathbf{X_n}] ,$$

where $\mathbf{X_n}$ is the information containing the original sample but without regard to their order. Applying the inequality $[E(Y)]^2 \leq E(Y^2)$ (conditionally) yields

$$E[U_n^2(x)] = E\{E[\bar{U}_n(x)|\mathbf{X_n}]\}^2 \leq \{E[\bar{U}_n^2(x)|\mathbf{X_n}]\} = E[\bar{U}_n^2(x)] .$$

Thus, $Var[U_n(x)] \to 0$ and (i) follows.

To prove (ii), given any subsequence $\{n_k\}$, one can extract a further subsequence $\{n_{k_j}\}$ so that $L_{n_{k_j}}(x) \to J(x, P)$ almost surely. Therefore, $L_{n_{k_j}}(x) \to J(x, P)$ almost surely for all x in some countable dense set of the real line. So, $L_{n_{k_j}}$ tends weakly to $J(x, P)$ and this convergence is uniform by Polya's Theorem. Hence, the result (ii) holds.

To prove (iii), $c_{n,b}(1 - \alpha) \overset{P}{\to} c(1 - \alpha, P)$ by Lemma 11.2.1 (ii). The limiting coverage probability now follows from Slutsky's Theorem. ■

The assumptions $b/n \to 0$ and $b \to \infty$ need not imply $\tau_b/\tau_n \to 0$. For example, in the unusual case $\tau_n = \log(n)$, if $b = n^\gamma$ and $\gamma > 0$, the assumption $\tau_b/\tau_n \to 0$ is not satisfied. In fact, a slight modification of the method is consistent without assuming $\tau_b/\tau_n \to 0$; see Politis, Romano, and Wolf (1999), Corollary 2.2.1. In regular cases, $\tau_n = n^{1/2}$, and the assumptions on b simplify to $b/n \to 0$ and $b \to \infty$.

The assumptions on b are as weak as possible under the weak assumptions of the theorem. However, in some cases, the choice $b = O(n)$ yields similar results;

this occurs in Wu (1990), where the statistic is approximately linear with an asymptotic normal distribution and $\tau_n = n^{1/2}$. This choice will not work in general; see Example 15.7.2.

Assumption 15.7.1 is satisfied in numerous examples, including all previous examples considered by the bootstrap.

15.7.2 Comparison with the Bootstrap

The usual bootstrap approximation to $J_n(x, P)$ is $J_n(x, \hat{Q}_n)$, where $\hat{Q}_n$ is some estimate of P. In many nonparametric i.i.d. situations, $\hat{Q}_n$ is taken to be the empirical distribution of the sample $X_1, \ldots, X_n$. In Section 15.4, we proved results to (15.68) and (15.69) with $L_{n,b}(x)$ replaced by $J_n(x, \hat{Q}_n)$. While the consistency of the bootstrap requires arguments specific to the problem at hand, the consistency of subsampling holds quite generally.

To elaborate a little further, we proved bootstrap limit results in the following manner. For some choice of metric (or pseudo-metric) d on the space of probability measures, it must be known that $d(P_n, P) \to 0$ implies $J_n(P_n)$ converges weakly to $J(P)$. That is, Assumption 15.7.1 must be strengthened so that the convergence of $J_n(P)$ to $J(P)$ is suitably locally uniform in P. In addition, the estimator $\hat{Q}_n$ must then be known to satisfy $d(\hat{Q}_n, P) \to 0$ almost surely or in probability under P. In contrast, no such strengthening of Assumption 15.7.1 is required in Theorem 15.7.1. In the known counterexamples to the bootstrap, it is precisely a certain lack of uniformity in convergence which leads to failure of the bootstrap.

In some special cases, it has been realized that a sample size trick can often remedy the inconsistency of the bootstrap. To describe how, focus on the case where $\hat{Q}_n$ is the empirical measure, denoted by $\hat{P}_n$. Rather than approximating $J_n(P)$ by $J_n(\hat{P}_n)$, the suggestion is to approximate $J_n(P)$ by $J_b(\hat{P}_n)$ for some b which usually satisfies $b/n \to 0$ and $b \to \infty$. The resulting estimator $J_b(x, \hat{P}_n)$ is obviously quite similar to our $L_{n,b}(x)$ given in (2.1). In words, $J_b(x, \hat{P}_n)$ is the bootstrap approximation defined by the distribution (conditional on the data) of $\tau_b[\hat{\theta}_b(X_1^*, \ldots, X_b^*) - \hat{\theta}_n]$, where $X_1^*, \ldots, X_b^*$ are chosen with replacement from $X_1, \ldots, X_n$. In contrast, $L_{n,b}(x)$ is the distribution (conditional on the data) of $\tau_b[\hat{\theta}_b(Y_1^*, \ldots, Y_b^*) - \hat{\theta}_n)]$, where $Y_1^*, \ldots, Y_b^*$ are chosen *without* replacement from $X_1, \ldots, X_n$. Clearly, these two approaches must be similar if b is so small that sampling with and without replacement are essentially the same. Indeed, if one resamples b numbers (or indices) from the set $\{1, \ldots, n\}$, then the chance that none of the indices is duplicated is $\Pi_{i=1}^{b-1}(1 - \frac{i}{n})$. This probability tends to 0 if $b^2/n \to 0$. (To see why, take logs and do a Taylor expansion analysis.) Hence, the following is true.

Corollary 15.7.1 *Under the further assumption that $b^2/n \to 0$, parts (i)–(iii) of Theorem 15.7.1 remain valid if $L_{n,b}(x)$ is replaced by the bootstrap approximation $J_b(x, \hat{P}_n)$.*

The bootstrap approximation with smaller resample size, $J_b(\hat{P}_n)$, is further studied in Bickel, Götze, and van Zwet (1997). In spite of the Corollary, we point out that $L_{n,b}$ is more generally valid. Indeed, without the assumption $b^2/n \to 0$, $J_b(x, \hat{P}_n)$ can be inconsistent. To see why, let P be any distribution on the real line with a density (with respect to Lebesgue measure). Consider any statistic $\hat{\theta}_n$,

τ_n, and $\theta(P)$ satisfying Assumption 15.7.1. Even the sample mean will work here. Now, modify $\hat{\theta}_n$ to $\tilde{\theta}_n$ so that the statistic $\tilde{\theta}_n(X_1, \ldots, X_n)$ completely misbehaves if any pair of the observations $X_1, \ldots, X_n$ are identical. The bootstrap approximation to the distribution of $\tilde{\theta}_n$ must then misbehave as well unless $b^2/n \to 0$, while the consistency of $L_{n,b}$ remains intact.

The above example, though artificial, was designed to illustrate a point. We now consider some further examples.

Example 15.7.1 (U-statistics of Degree 2) Let $X_1, \ldots, X_n$ be i.i.d. on the line with c.d.f. F. Denote by $\hat{F}_n$ the empirical distribution of the data. Let

$$\theta(F) = \int\int \omega(x, y) dF(x) dF(y)$$

and assume $\omega(x, y) = \omega(y, x)$. Assume $\int \omega^2(x, y) dF(x) dF(y) < \infty$. Set $\tau_n = n^{1/2}$ and $\hat{\theta}_n = \sum_{i<j} \omega(X_i, X_j)/\binom{n}{2}$. Then, it is well known that $J_n(F)$ converges weakly to $J(F)$, the normal distribution with mean 0 and variance given by

$$v^2(F) = 4\left\{\int [\omega(x, y) dF(y)]^2 dF(x) - \theta^2(F)\right\}.$$

Hence, assumption 15.7.1 holds. However, in order for the bootstrap to succeed, the additional condition $\int \omega^2(x, x) dF(x) < \infty$ is required. Bickel and Freedman (1981) give a counterexample to show the inconsistency of the bootstrap without this additional condition.

Interestingly, the bootstrap may fail even if $\int \omega^2(x, x) dF(x) < \infty$, stemming from the possibility that $v^2(F) = 0$. (Otherwise, Bickel and Freedman's argument justifies the bootstrap.) As an example, let $w(x, y) = xy$. In this case, $\theta(\hat{F}_n) = \bar{X}_n^2 - S_n^2/n$, where S_n^2 is the usual unbiased sample variance. If $\theta(F) = 0$, then $v(F) = 0$. Then, $n[\theta(\hat{F}_n) - \theta(F)]$ converges weakly to $\sigma^2(F)(Z^2 - 1)$, where Z denotes a standard normal random variable and $\sigma^2(F)$ denotes the variance of F. However, it is easy to see that the bootstrap approximation to the distribution of $n[\theta(\hat{F}_n) - \theta(F)]$ has a representation $\sigma^2(F)Z^2 + 2Z\sigma(F)n^{1/2}\bar{X}_n$. Thus, failure of the bootstrap follows.

In the context of U-statistics, the possibility of using a reduced sample size in the resampling has been considered in Bretagnolle (1983); an alternative correction is given by Arcones (1991). ∎

Example 15.7.2 (Extreme Order Statistic) The following counterexample is taken from Bickel and Freedman (1981). If $X_1, \ldots, X_n$ are i.i.d. according to a uniform distribution on $(0, \theta)$, let $X_{(n)}$ be the maximum order statistic. Then, $n[X_{(n)} - \theta]$ has a limit distribution given by the distribution of $-\theta X$, where X is exponential with mean one. Hence, Assumption 15.7.1 is satisfied here. However, the usual bootstrap fails. To see why, let $X_1^*, \ldots, X_n^*$ be n observations sampled from the data with replacement, and let $X_{(n)}^*$ be the maximum of the bootstrap sample. The bootstrap approximation to the distribution of $n[X_{(n)} - \theta]$ is the distribution of $n[X_{(n)}^* - X_{(n)}]$, conditional on $X_1, \ldots, X_n$. But, the probability mass at 0 for this bootstrap distribution is the probability that $X_{(n)}^* = X_{(n)}$, which occurs with probability $1 - (1 - \frac{1}{n})^n \to 1 - \exp(1)$. However, the true limiting distribution is continuous. Note in Theorem 15.7.1 that the conditions on b (with $\tau_n = n$) reduce to $b/n \to 0$ and $b \to \infty$. In this example, at least, it is clear that we cannot assume $b/n \to c$, where $c > 0$. Indeed, $L_{n,b}(x)$ places mass b/n at 0. Thus, while it is sometimes true that, under further conditions such as Wu (1990) assumes, we can take b to be of the same order as n, this example makes it clear that we cannot in general weaken our assumptions on b without imposing further structure. ■

Example 15.7.3 (Superefficient Estimator) Assume $X_1, \ldots, X_n$ are i.i.d. according the normal distribution with mean $\theta(P)$ and variance one. Fix $c > 0$. Let $\hat{\theta}_n = c\bar{X}_n$ if $|\bar{X}_n| \leq n^{-1/4}$ and $\hat{\theta}_n = \bar{X}_n$ otherwise. The resulting estimator is known as Hodges' superefficient estimator; see Lehmann and Casella (1998), p.440 and Problem 12.66. It is easily checked that $n^{1/2}(\hat{\theta}_n - \theta(P))$ has a limit distribution for every θ, so the conditions for our Theorem 15.7.1 remain applicable. However, Beran (1984) showed that the distribution of $n^{1/2}(\hat{\theta}_n - \theta(P))$ cannot be bootstrapped, even if one is willing to apply a parametric bootstrap! ■

We have claimed that subsampling is superior to the bootstrap in a first order asymptotic sense, since it is more generally valid. However, in many typical situations, the bootstrap is far superior and has some compelling second-order asymptotic properties. Some of these were studied in Section 15.5; also see Hall (1992). In nice situations, such as when the statistic or root is a smooth function of sample means, a bootstrap approach is often very satisfactory. In other situations, especially those where it is not known that the bootstrap works even in a first-order asymptotic sense, subsampling is preferable. Still, in other situations (such as the mean in the infinite variance case), the bootstrap may work, but only with a reduced sample size. The issue becomes whether to sample with or without replacement (as well as the choice of resample size). Although this question is not yet answered unequivocally, some preliminary evidence in Bickel et al. (1997) suggests that the bootstrap approximation $J_b(x, \hat{P}_n)$ might be more accurate; more details on the issue of higher-order accuracy of the subsampling approximation $L_{n,b}(x)$ are given in Chapter 10 of Politis, Romano, and Wolf (1999).

Because $\binom{n}{b}$ can be large, $L_{n,b}$ may be difficult to compute. Instead, an approximation may be employed. For example, let $I_1, \ldots I_B$ be chosen randomly with or without replacement from $\{1, 2, \ldots, N_n\}$. Then, $L_{n,b}(x)$ may be approximated by

$$\hat{L}_{n,b}(x) = \frac{1}{B} \sum_{i=1}^{B} I\{\tau_b(\hat{\theta}_{n,b,I_i} - \hat{\theta}_n) \leq x\}. \tag{15.71}$$

Corollary 15.7.2 *Under the assumptions of Theorem 15.7.1 and the assumption $B \to \infty$ as $n \to \infty$, the results of Theorem 15.7.1 are valid if $L_{n,b}(x)$ is replaced by $\hat{L}_{n,b}(x)$.*

PROOF. If the I_i are sampled with replacement, $\sup_x |\hat{L}_{n,b}(x) - L_{n,b}(x)| \to 0$ in probability by the Dvoretzky, Kiefer, Wolfowitz inequality. This result is also true in the case the I_i are sampled without replacement; apply Proposition 4.1 of Romano (1989b). ■

An alternative approach, which also requires fewer computations, is the following. Rather than employing all $\binom{n}{b}$ subsamples of size b from $X_1, \ldots, X_n$, just use the $n - b + 1$ subsamples of size b of the form $\{X_i, X_{i+1}, \dots, X_{i+b-1}\}$. Notice that the ordering of the data is fixed and retained in the subsamples. Indeed, this is the approach that is applied for time series data; see Chapter 3 of Politis, Romano and Wolf (1999), where consistency results in data-dependent situations are given. Even when the i.i.d. assumption seems reasonable, this approach may be desirable to ensure robustness against possible serial correlation. Most inferential procedures based on i.i.d. models are simply not valid (i.e., not even first order accurate) if the independence assumption is violated, so it seems worthwhile to account for possible dependencies in the data if we do not sacrifice too much in efficiency.

15.7.3 Hypothesis Testing

In this section, we consider the use of subsampling for the construction of hypothesis tests. As before, $X_1, \dots, X_n$ is a sample of n independent and identically distributed observations taking values in a sample space S. The common unknown distribution generating the data is denoted by P. This unknown law P is assumed to belong to a certain class of laws $\mathbf{P}$. The null hypothesis H asserts $P \in \mathbf{P_0}$, and the alternative hypothesis K is $P \in \mathbf{P_1}$, where $\mathbf{P_i} \subset \mathbf{P}$ and $\mathbf{P_0} \bigcup \mathbf{P_1} = \mathbf{P}$.

The goal is to construct an asymptotically valid test based on a given test statistic,

$$T_n = \tau_n t_n(X_1, \dots, X_n) ,$$

where, as before, τ_n is a fixed nonrandom normalizing sequence. Let

$$G_n(x, P) = P\{\tau_n t_n(X_1, \dots, X_n) \le x\} .$$

We will be assuming that $G_n(\cdot, P)$ converges in distribution, at least for $P \in \mathbf{P_0}$. Of course, this would imply (as long as $\tau_n \to \infty$) that $t_n(X_1, \dots, X_n) \to 0$ in probability for $P \in \mathbf{P_0}$. Naturally, t_n should somehow be designed to distinguish between the competing hypotheses. The theorem we will present will assume t_n is constructed to satisfy the following: $t_n(X_1, \dots, X_n) \to t(P)$ in probability, where $t(P)$ is a constant which satisfies $t(P) = 0$ if $P \in \mathbf{P_0}$ and $t(P) > 0$ if $P \in \mathbf{P_1}$. This assumption easily holds in typical examples.

To describe the test construction, as in Subsection 15.7.1, let $Y_1, \dots, Y_{N_n}$ be equal to the $N_n = \binom{n}{b}$ subsets of $\{X_1, \dots, X_n\}$, ordered in any fashion. Let $t_{n,b,i}$ be equal to the statistic t_b evaluated at the data set Y_i. The sampling distribution of T_n is then approximated by

$$\hat{G}_{n,b}(x) = N_n^{-1} \sum_{i=1}^{N_n} I\{\tau_b t_{n,b,i} \le x\} . \tag{15.72}$$

Using this estimated sampling distribution, the critical value for the test is obtained as the $1 - \alpha$ quantile of $\hat{G}_{n,b}(\cdot)$; specifically, define

$$g_{n,b}(1 - \alpha) = \inf\{x : \hat{G}_{n,b}(x) \ge 1 - \alpha\} . \tag{15.73}$$

Finally, the nominal level α test rejects H if and only if $T_n > g_{n,b}(1 - \alpha)$.

The following theorem gives the asymptotic behavior of this procedure, showing the test is pointwise consistent in level and pointwise consistent in power.

In addition, an expression for the limiting power of the test is obtained under a sequence of alternatives contiguous to a distribution in the null hypothesis.

Theorem 15.7.2 *Assume $b/n \to 0$ and $b \to \infty$ as $n \to \infty$.*

(i) Assume, for $P \in \mathbf{P_0}$, $G_n(P)$ converges weakly to a continuous limit law $G(P)$, whose corresponding cumulative distribution function is $G(\cdot, P)$ and whose $1-\alpha$ quantile is $g(1-\alpha, P)$. If $G(\cdot, P)$ is continuous at $g(1-\alpha, P)$ and $P \in \mathbf{P_0}$, then

$$g_{n,b}(1-\alpha) \to g(1-\alpha, P) \text{ in probability}$$

and

$$P\{T_n > g_{n,b}(1-\alpha)\} \to \alpha \text{ as } n \to \infty.$$

(ii) Assume the test statistic is constructed so that $t_n(X_1, \ldots, X_n) \to t(P)$ in probability, where $t(P)$ is a constant which satisfies $t(P) = 0$ if $P \in \mathbf{P_0}$ and $t(P) > 0$ if $P \in \mathbf{P_1}$. Assume $\liminf_n(\tau_n/\tau_b) > 1$. Then, if $P \in \mathbf{P_1}$, the rejection probability satisfies

$$P\{T_n > g_{n,b}(1-\alpha)\} \to 1 \text{ as } n \to \infty.$$

(iii) Suppose P_n is a sequence of alternatives such that, for some $P_0 \in \mathbf{P_0}$, $\{P_n^n\}$ is contiguous to $\{P_0^n\}$. Then,

$$g_{n,b}(1-\alpha) \to g(1-\alpha, P_0) \text{ in } P_n^n\text{-probability}.$$

Hence, if T_n converges in distribution to T under P_n and $G(\cdot, P_0)$ is continuous at $g(1-\alpha, P_0)$, then

$$P_n^n\{T_n > g_{n,b}(1-\alpha)\} \to Prob\{T > g(1-\alpha, P_0)\}.$$

The proof is similar to that of Theorem 15.7.1 (Problem 15.52).

Example 15.7.4 Consider the special case of testing a real-valued parameter. Specifically, suppose $\theta(\cdot)$ is a real-valued function from $\mathbf{P}$ to the real line. The null hypothesis is specified by $\mathbf{P_0} = \{P : \theta(P) = \theta_0\}$. Assume the alternative is one-sided and is specified by $\{P : \theta(P) > \theta_0\}$. Suppose we simply take

$$t_n(X_1, \ldots, X_n) = \hat{\theta}_n(X_1, \ldots, X_n) - \theta_0 .$$

If $\hat{\theta}_n$ is a consistent estimator of $\theta(P)$, then the hypothesis on t_n in part (ii) of the theorem is satisfied (just take the absolute value of t_n for a two-sided alternative). Thus, the hypothesis on t_n in part (ii) of the theorem boils down to verifying a consistency property and is rather weak, though this assumption can in fact be weakened further. The convergence hypothesis of part (i) is satisfied by typical test statistics; in regular situations, $\tau_n = n^{1/2}$. ■

The interpretation of part (iii) of the theorem is the following. Suppose, instead of using the subsampling construction, one could use the test that rejects when $T_n > g_n(1-\alpha, P)$, where $g_n(1-\alpha, P)$ is the exact $1-\alpha$ quantile of the true sampling distribution $G_n(\cdot, P)$. Of course, this test is not available in general because P is unknown and so is $g_n(1-\alpha, P)$. Then, the asymptotic power of the subsampling test against a sequence of contiguous alternatives $\{P_n\}$ to P with

P in $\mathbf{P_0}$ is the same as the asymptotic power of this fictitious test against the same sequence of alternatives. Hence, to the order considered, there is no loss in efficiency in terms of power.

15.8 Problems

Section 15.2

Problem 15.1 Generalize Theorem 15.2.1 to the case where $\mathbf{G}$ is an infinite group.

Problem 15.2 With $\hat{p}$ defined in (15.5), show that (15.6) holds.

Problem 15.3 (i) Suppose $Y_1, \ldots, Y_B$ are exchangeable real-valued random variables; that is, their joint distribution is invariant under permutations. Let $\tilde{q}$ be defined by

$$\tilde{q} = \frac{1}{B}\left[1 + \sum_{i=1}^{B-1} I\{Y_i \geq Y_B\}\right] .$$

Show, $P\{\tilde{q} \leq u\} \leq u$ for all $0 \leq u \leq 1$. *Hint:* Condition on the order statistics.
(ii) With $\tilde{p}$ defined in (15.7), show that (15.8) holds.
(iii) How would you construct a p-value based on sampling without replacement from $\mathbf{G}$?

Problem 15.4 With $\hat{p}$ and $\tilde{p}$ defined in (15.5) and (15.7), respectively, show that $\hat{p} - \tilde{p} \to 0$ in probability.

Problem 15.5 As an approximation to (15.9), let $g_1, \ldots, g_{B-1}$ be i.i.d. and uniform on $\mathbf{G}$. Also, set g_B to be the identity. Define

$$\tilde{R}_{n,B}(t) = \frac{1}{B}\sum_{i=1}^{B} I\{T_n(g_i X) \leq t\} .$$

Show, conditional on X,

$$\sup_t |\tilde{R}_{n,B}(t) - \hat{R}_n(t)| \to 0$$

in probability as $B \to \infty$, and so

$$\sup_t |\tilde{R}_{n,B}(t) - \hat{R}_n(t)| \to 0$$

in probability (unconditionally) as well. Do these results hold only under the null hypothesis? *Hint:* Apply Theorem 11.2.18. For a similar result based on sampling without replacement, see Romano (1989b).

Problem 15.6 Suppose $X_1, \ldots, X_n$ are i.i.d. according to a q.m.d. location model with finite variance. Show the ARE of the one-sample t-test with respect to the randomization t-test (based on sign changes) is 1 (even if the underlying density is not normal).

Problem 15.7 In Theorem 15.2.4, show the conclusion may fail if ψ_P is not an odd function.

Problem 15.8 Verify (15.15) and (15.16). *Hint:* Let S be the number of positive integers $i \le m$ with $W_i = 1$, and condition on S.

Problem 15.9 Provide the remaining details for the proof of Theorem 15.2.5.

Problem 15.10 In the two-sample problem of Example 15.2.6, suppose the underlying distributions are normal with common variance. For testing $\mu(P_Y) = \mu(P_Z)$ against $\mu(P_Y) > \mu(P_Z)$ compute the limiting power of the randomization test based on the test statistic $T_{m,n}$ given in (15.13) against contiguous alternatives of the form $\mu(P_Y) = \mu(P_Z) + hn^{-1/2}$. Show this is the same as the optimal two-sample t-test. Argue that the two tests are asymptotically equivalent in the sense of Problem 13.24.

Problem 15.11 Using Theorem 15.2.3, prove a result analogous to Theorem 15.2.5 with $T_{m,n}$ replaced by $\tilde{T}_{m,n}$ defined in (15.19). Deduce that the two-sample permutation test is consistent in level for testing equality of population means, as long as the underlying populations have a finite variance. [This result was proved in Janssen (1997) by an alternative method.]

Problem 15.12 Under the setting of Problem 11.52 for testing equality of Poisson means λ_i based on the test statistic T, show how to construct a randomization test based on T. Examine the limiting behavior of the randomization distribution under the null hypothesis and contiguous alternatives.

Problem 15.13 Suppose $(X_1, Y_1), \ldots (X_n, Y_n)$ are i.i.d. bivariate observations in the plane, and let ρ denote the correlation between X_1 and Y_1. Let $\hat{\rho}_n$ be the sample correlation

$$\hat{\rho}_n = \frac{\sum (X_i - \bar{X}_n)(Y_i - \bar{Y}_n)}{[\sum_i (X_i - \bar{X}_n)^2 \sum_j (X_j - \bar{Y}_n)^2]^2} .$$

(i) For testing independence of X_i and Y_i, construct a randomization test based on the test statistic $T_n = n^{1/2} |\hat{\rho}_n|$.
(ii) For testing $\rho = 0$ versus $\rho > 0$ based on the test statistic $\hat{\rho}_n$, determine the limit behavior of the randomization distribution when the underlying population is bivariate Gaussian with correlation $\rho = 0$. Determine the limiting power of the randomization test under local alternatives $\rho = hn^{-1/2}$. Argue that the randomization test and the optimal UMPU test (5.75) are asymptotically equivalent in the sense of Problem 13.24.
(iii) Investigate what happens if the underlying distribution has correlation 0, but X_i and Y_i are dependent.

Section 15.3

Problem 15.14 Assume $X_1, \ldots, X_n$ are i.i.d. according to a location scale model with distribution of the form $F[(x - \theta)/\sigma]$, where F is known, θ is a location parameter, and σ is a scale parameter. Suppose $\hat{\theta}_n$ is a location and scale equivariant estimator and $\hat{\sigma}_n$ is a location invariant, scale equivariant estimator. Then, show that the roots $[\hat{\theta}_n - \theta]/\hat{\sigma}_n$ and $\hat{\sigma}_n/\sigma$ are pivots.

Problem 15.15 Let $X = (X_1, \ldots, X_n)^T$ and consider the linear model

$$X_i = \sum_{j=1}^{s} a_{i,j}\beta_j + \sigma\epsilon_i \ ,$$

where the ϵ_i are i.i.d. F, where F has mean 0 and variance 1. Here, the $a_{i,j}$ are known, $\beta = (\beta_1, \ldots, \beta_s)^T$ and σ are unknown. Let A be the $n \times s$ matrix with (i, j) entry $a_{i,j}$ and assume A has rank s. As in Section 11.3.3, let $\hat{\beta}_n = (A^T A)^{-1} A^T X$ be the least squares estimate of β. Consider the test statistic

$$T_n = \frac{(n-s)(\hat{\beta}_n - \beta)(A^T A)(\hat{\beta}_n - \beta)}{sS_n^2} \ ,$$

where $S_n^2 = (X - A\hat{\beta}_n)^T (X - A\hat{\beta}_n)/(n - s)$. Is T_n a pivot when F is known?

Section 15.4

Problem 15.16 Suppose the convergences (15.23) and (15.24) only hold in probability. Show that (15.25) still holds.

Problem 15.17 In Theorem 15.4.1, one cannot deduce the uniform convergence result (15.23) without the assumption that the limit law $J(P)$ is continuous. Show that, without the continuity assumption for $J(P)$,

$$\rho_L(J_n(\hat{P}_n), J_n(P)) \to 0$$

with probability one, where ρ_L is the Lévy metric defined in Definition 11.2.3.

Problem 15.18 In Theorem 15.4.3 (i), show that the assumption that $\theta(F_n) \to \theta(F)$ actually follows from the other assumptions.

Problem 15.19 Reprove Theorem 15.4.3 under the assumption $E(|X_i|^3) < \infty$ by using the Berry-Esseen Theorem.

Problem 15.20 Prove the following extension of Theorem 15.4.3 holds. Let $\mathbf{D_F}$ be the set of sequences $\{F_n\}$ such that F_n converges weakly to a distribution G and $\sigma^2(F_n) \to \sigma^2(G) = \sigma^2(F)$. Then, Theorem (15.4.3) holds with $\mathbf{C}_F$ replaced by $\mathbf{D_F}$. (Actually, one really only needs to define $\mathbf{D_F}$ so that and sequence $\{F_n\}$ is tight and any weakly convergent subsequence of $\{F_n\}$ has the above property.) Thus, the possible choices for the resampling distribution are quite large in the sense that the bootstrap approximation $J_n(\hat{G}_n)$ can be consistent even if $\hat{G}_n$ is not at all close to F. For example, the choice where $\hat{G}_n$ is normal with mean $\bar{X}_n$ and variance equal to a consistent estimate of the sample variance results in consistency. Therefore, the normal approximation can in fact be viewed as a bootstrap procedure with a perverse choice of resampling distribution. Show the bootstrap can be inconsistent if $\sigma^2(G) \neq \sigma^2(F)$.

Problem 15.21 In the case that $\theta(P)$ is real-valued, Efron initially proposed the following construction, called the bootstrap *percentile* method. Let $\hat{\theta}_n$ be an estimator of $\theta(P)$, and let $\tilde{J}_n(P)$ be the distribution of $\hat{\theta}_n$ under P. Then, Efron's two-sided percentile interval of nominal level $1-\alpha$ takes the form

$$[\tilde{J}_n^{-1}(\frac{\alpha}{2}, \hat{P}_n), \tilde{J}_n^{-1}(1-\frac{\alpha}{2}, \hat{P}_n)] \ . \tag{15.74}$$

Also, consider the root $R_n(X^n, \theta(P)) = n^{1/2}(\hat{\theta}_n - \theta(P))$, with distribution $J_n(P)$. Write (15.74) as a function of $\hat{\theta}_n$ and the quantiles of $J_n(\hat{P}_n)$. Suppose Theorem 15.4.1 holds for the root R_n, so that $J_n(P)$ converges weakly to $J(P)$. What must be assumed about $J(P)$ so that $P\{\theta(P) \in I_n\} \to 1-\alpha$?

Problem 15.22 Let $\hat{\theta}_n$ be an estimate of a real-valued parameter $\theta(P)$. Suppose there exists an increasing transformation g such that

$$g(\hat{\theta}_n) - g(\theta(P))$$

is a pivot, so that its distribution does not depend on P. Also, assume this distribution is continuous, strictly increasing and symmetric about zero.
(i) Show that Efron's percentile interval (15.74), which may be constructed without knowledge of g, has exact coverage $1-\alpha$.
(ii) Show that the percentile interval is transformation equivariant. That is, if $\phi = m(\theta)$ is a monotone transformation of θ, then the percentile interval for ϕ is the percentile interval for θ transformed by m, at least if $\hat{\phi}_n$ is taken to be $m(\hat{\theta})_n$. This holds true for the theoretical percentile interval as well as its approximation due to simulation.
(iii) If the parameter θ only takes values in an interval I and $\hat{\theta}_n$ does as well, then the percentile interval is range-preserving in the sense that the interval is always a subset of I.

Problem 15.23 Suppose $\hat{\theta}_n$ is an estimate of some real-valued parameter $\theta(P)$. Let $H_n(x, \theta)$ denote the c.d.f. of $\hat{\theta}_n$ under θ, with inverse $H_n^{-1}(1-\alpha, \theta)$. The percentile interval lower confidence bound of level $1-\alpha$ is then $H_n^{-1}(\alpha, \hat{\theta}_n)$. Suppose that, for some increasing transformation g, and constants z (called the *bias correction*) and a (called the *acceleration constant*),

$$P\{\frac{g(\hat{\theta}_n) - g(\theta)}{1 + ag(\theta)} + z_0 \le x\} = \Phi(x) \ , \tag{15.75}$$

where Φ is the standard normal c.d.f.
(i) Letting $\hat{\phi}_n = g(\hat{\theta}_n)$, show that $\hat{\theta}_{n,L}$ given by

$$\hat{\theta}_{n,L} = g^{-1}\left\{\hat{\phi}_n + (z_\alpha + z)(1 + a\hat{\phi}_n)/[1 - a(z_\alpha + z_0)]\right\}$$

is an exact $1-\alpha$ lower confidence bound for θ.
(ii) Because $\hat{\theta}_{n,L}$ requires knowledge of g, let

$$\hat{\theta}_{n,BC_a} = H_n^{-1}(\beta, \hat{\theta}_n) \ ,$$

where

$$\beta = \Phi(z + (z_\alpha + z)/[1 - a(z_\alpha + z)] \ .$$

Show that $\hat{\theta}_{n,BC_a} = \hat{\theta}_{n,L}$. [The lower bound $\hat{\theta}_{n,BC_a}$ is called the BC_a lower bound and Efron shows one may take $z = \Phi^{-1}(G_n(\hat{\theta}_n, \hat{\theta}_n))$ and gives methods to estimate a; see Efron and Tibshirani (1993, Chapter 14).]

Problem 15.24 Assume the setup of Problem 15.23 and condition (15.75). Let θ_0 be any value of θ and let $\theta_1 = G_n^{-1}(1-\alpha, \theta_0)$. Let

$$\hat{\theta}_{n,AP} = G_n^{-1}(\beta', \hat{\theta}_n) ,$$

where

$$\beta' = G_n(\theta_0, \theta_1) .$$

Show that $\hat{\theta}_{n,AP}$ is an exact level $1-\alpha$ lower confidence bound for θ. [This is called the *automatic percentile* lower bound of DiCiccio and Romano (1989), and may be computed without knowledge of g, a or z. Its exactness holds under assumptions even weaker than (15.75).]

Problem 15.25 Let $X_1, \ldots, X_{n_X}$ be i.i.d. with distribution F_X, and let $Y_1, \ldots, Y_{n_Y}$ be i.i.d. with distribution F_Y. The two samples are independent. Let $\mu(F)$ denote the mean of a distribution F, and let $\sigma^2(F)$ denote the variance of F. Assume $\sigma^2(F_X)$ and $\sigma^2(F_Y)$ are finite. Suppose we are interested in $\theta = \theta(F_X, F_Y) = \mu(F_X) - \mu(F_Y)$. Construct a bootstrap confidence interval for θ of nominal level $1-\alpha$, and prove that it asymptotically has the correct coverage probability.

Problem 15.26 Let $X_1, \cdots, X_n$ be i.i.d. Bernoulli trials with success probability θ.
(i). As explicitly as possible, find a uniformly most accurate upper confidence bound for θ of nominal level $1-\alpha$. State the bound explicitly in the case $X_i = 0$ for every i.
(ii). Describe a bootstrap procedure to obtain an upper confidence bound for θ of nominal level $1-\alpha$. What does it reduce to for the previous data set?
(iii). Let $\hat{B}_{1-\alpha}$ denote your upper bootstrap confidence bound for θ. Then, $P_\theta(\theta \le \hat{B}_{1-\alpha}) \to 1-\alpha$ as $n \to \infty$. Prove the following.

$$\sup_\theta |P_\theta(\theta \le \hat{B}_{1-\alpha}) - (1-\alpha)|$$

does not tend to 0 as $n \to \infty$.

Problem 15.27 Let $X_1, \ldots, X_n$ be i.i.d. with c.d.f. F, mean $\mu(F)$ and finite variance $\sigma^2(F)$. Consider the root $R_n = n^{1/2}(\bar{X}_n^2 - \mu^2(F))$ and the bootstrap approximation to its distribution $J_n(\hat{F}_n)$, where $\hat{F}_n$ is the empirical c.d.f. Determine the asymptotic behavior of $J_n(\hat{F}_n)$. *Hint:* Distinguish the cases $\mu(F) = 0$ and $\mu(F) \neq 0$.

Problem 15.28 Show why (15.43) is true.

Problem 15.29 (i) Under the setup of Example 15.4.6, prove that Theorem 15.4.7 applies if studentized statistics are used.

(ii) In addition to the $X_1, \ldots, X_n$, suppose i.i.d. $Y_1, \ldots, Y_{n'}$ are observed, with $Y_i = (Y_{i,1}, \ldots, Y_{i,s})$. The distribution of Y_i need not be that of X_i. Suppose the mean of Y_i is $(\mu'_1, \ldots, \mu'_s)$. Generalize Example 15.4.6 to simultaneously test $H_i : \mu_i = \mu'_i$. Distinguish between two cases, first where the X_is are independent of the Y_js, and next where (X_i, Y_i) are paired (so $n = n'$) and X_i need not be independent of Y_i.

Problem 15.30 Under the setup of Example 15.4.7, provide the details to show that the FWER is asymptotically controlled.

Problem 15.31 Under the setup of Example 15.4.7, suppose that there is also an i.i.d. control sample $X_{0,1}, \ldots, X_{0,n_0}$, independent of the other Xs. Let μ_0 denote the mean of the controls. Now consider testing $H_i : \ \mu_i = \mu_0$. Describe a method that asymptotically controls the FWER.

Problem 15.32 Under the setup of Example 15.4.7, let F_i denote the distribution of the ith sample. Now, consider $H'_{i,j} : F_i = F_j$ based on the same test statistics. Describe a randomization test that has exact control of the FWER. [*Hint:* Recall Theorem 9.1.3(ii).]

Problem 15.33 Let $\epsilon_1, \epsilon_2, \ldots$ be i.i.d. $N(0,1)$. Let $X_i = \mu + \epsilon_i + \beta\epsilon_{i+1}$ with β a fixed nonzero constant. The X_i form a moving average process studied in Section 11.3.1.
(i) Examine the behavior of the nonparametric bootstrap method for estimating the mean using the root $n^{1/2}(\bar{X}_n - \mu)$ and resampling from the empirical distribution. Show that the coverage probability does not tend to the nominal level under such a moving average process.
(ii) Suppose $n = bk$ for integers b and k. Consider the following *moving blocks bootstrap* resampling scheme. Let $L_{i,b} = (X_i, X_{i+1}, \ldots, X_{i+b-1})$ be the block of b observations beginning at "time" i. Let $X_1^*, \ldots, X_n^*$ be obtained by randomly choosing with replacement k of the $n - b + 1$ blocks $L_{i,b}$; that is, $X_1^*, \ldots, X_b^*$ are the observations in the first sampled block, $X_{b+1}^*, \ldots, X_{2b}^*$ are the observations from the second sampled block, etc. Then, the distribution of $n^{1/2}[\bar{X}_n - \mu]$ is approximated by the *moving blocks bootstrap* distribution given by the distribution of $n^{1/2}[\bar{X}_n^* - \bar{X}_n]$, where $\bar{X}_n^* = \sum_{i=1}^n X_i^*/n$. If b is fixed, determine the mean and variance of this distribution as $n \to \infty$. Now let $b \to \infty$ as $n \to \infty$. At what rate should $b \to \infty$ so that the mean and variance of the moving blocks distribution tends to the same limiting values as the true mean and variance, at least in probability? [The moving blocks bootstrap was independently discovered by Künsch (1989) and Liu and Singh (1992). The stationary bootstrap of Politis and Romano (1994a) and other methods designed for dependent data are studied in Lahiri (2003).]

Section 15.5

Problem 15.34 Under the assumptions of Theorem 15.5.2, show that, for any $\epsilon > 0$, the expansion (15.47) holds uniformly in $\alpha \in [\epsilon, 1 - \epsilon]$.

Problem 15.35 Under the assumptions of Theorem 15.5.1, show that, for any $\epsilon > 0$, the expansion (15.48) holds uniformly in $\alpha \in [\epsilon, 1-\epsilon]$.

Problem 15.36 Suppose Y_n is a sequence of random variables satisfying

$$P\{Y_n \le t\} = g_0(t) + g_1(t)n^{-1/2} + O(n^{-1}) ,$$

uniformly in t, where g_0 and g_1 have uniformly bounded derivatives. If $T_n = O_P(n^{-1})$, then show, for any fixed (nonrandom) sequence t_n,

$$P\{Y_n \le t_n + T_n\} = g_0(t_n) + g_1(t_n)n^{-1/2} + O(n^{-1}) .$$

Problem 15.37 Assuming the expansions in the section hold, show that the two-sided bootstrap interval (15.56) has coverage error of order n^{-1}.

Problem 15.38 Assuming the expansions in the section hold, show that the two-sided bootstrap-t interval (15.62) has coverage error of order n^{-1}.

Problem 15.39 Verify the expansion (15.63) and argue that the resulting interval $I_n(1-\hat{\alpha}_n)$ has coverage error $O(n^{-1})$.

Problem 15.40 In the nonparametric mean setting, determine the one- and two-sided coverage errors of Efron's percentile method described in (15.74).

Problem 15.41 Assume F has infinitely many moments and is absolutely continuous. Under the notation of this section, argue that $n^{1/2}[J_n(t,\hat{F}_n) - J_n(t,F)]$ has an asymptotically normal limiting distribution, as does $n[K_n(t,\hat{F}_n) - K_n(t,F)]$.

Problem 15.42 (i) In a normal location model $N(\mu, \sigma^2)$, consider the root $R_n = n^{1/2}(\bar{X}_n - \mu)$, which is not a pivot. Show that bootstrap calibration, by parametric resampling, produces an exact interval.

(ii) Next, consider the root $n^{1/2}(S_n^2 - \sigma^2)$, where S_n^2 is the usual unbiased estimate of variance. Show that bootstrap calibration, by parametric resampling, produces an exact interval.

Problem 15.43 (i) Show the bootstrap interval (15.22) can be written as

$$\{\theta \in \Theta : \ J_n(R_n(X^n, \theta), \hat{P}_n) \le 1 - \alpha\} \tag{15.76}$$

if, for the purposes of this problem, $J_n(x,P)$ is defined as the left continuous c.d.f.

$$J_n(x,P) = P\{R_n(X^n, \theta(P)) < x\}$$

and $J_n^{-1}(1-\alpha, P)$ is now defined as

$$J_n^{-1}(1-\alpha, P) = \sup\{x : \ J_n(x,P) \le 1 - \alpha\} .$$

[*Hint:* If a random variable Y has left continuous c.d.f. $F(x) = P\{Y < x\}$ and $F^{-1}(1-\alpha)$ is the largest $1-\alpha$ quantile of F, then the event $\{X \le F^{-1}(1-\alpha)\}$ is identical to $\{F(X) \le 1-\alpha\}$ for any random variable X (which need not have distribution F). Why?]

(ii) The bootstrap interval (15.76) pretends that

$$R_{n,1}(X^n, \theta(P)) \equiv J_n(R_n(X^n, \theta(P)), \hat{P}_n)$$

has the uniform distribution on $(0, 1)$. Let $J_{n,1}(P)$ be the actual distribution of $R_{n,1}(X^n, \theta(P))$ under P, with left continuous c.d.f. denoted $J_{n,1}(x, P)$. This results in a new interval with R_n and J_n replaced by $R_{n,1}$ and $J_{n,1}$ in (15.76). Show that the resulting interval is equivalent to bootstrap calibration of the initial interval. [The mapping of R_n into $R_{n,1}$ by estimated c.d.f. of the former is called *prepivoting*. Beran (1987, 1988b) argues that the interval based on $R_{n,1}$ has better coverage properties than the interval based on R_n.]

Section 15.6

Problem 15.44 In Example 15.6.1, rather than exact evaluation of $G_n(\cdot, \hat{Q}_n)$, describe a simulation test of H that has exact level α.

Problem 15.45 In Example 15.6.2, why is the parametric bootstrap test exact for the special case of Example 12.4.7?

Problem 15.46 In the Behrens-Fisher problem, show that (15.64) and (15.65) hold.

Problem 15.47 In the Behrens-Fisher problem, verify the bootstrap-t has rejection probability equal to $\alpha + O(n^{-2})$.

Problem 15.48 In the Behrens-Fisher problem, what is the order of error in rejection probability for the likelihood ratio test? What is the order of error in rejection probability if you bootstrap the non-studentized statistic $n^{1/2}(\bar{X}_{n,1} - \bar{X}_{n,2})$.

Problem 15.49 In Example 15.6.4, with resampling from the empirical distribution shifted to have mean 0, what are the errors in rejection for the tests based on T_n and T'_n? How do these tests differ from the corresponding tests obtained through inverting bootstrap confidence bounds?

Problem 15.50 Let $X_1, \ldots, X_n$ be i.i.d. with a distribution P on the real line, and let $\hat{P}_n$ be the empirical distribution function. Find Q that minimizes, $\delta_{KL}(\hat{P}_n, Q)$, where δ_{KL} is the Kullback-Leibler divergence defined by (15.66).

Problem 15.51 Suppose $X_1, \ldots, X_n$ are i.i.d. real-valued with c.d.f. F. The problem is to test the null hypothesis that F is $N(\mu, \sigma^2)$ for some (μ, σ^2). Consider the test statistic

$$T_n = n^{1/2} \sup_t |\hat{F}_n(t) - \Phi((t - \bar{X}_n)/\hat{\sigma}_n)| ,$$

where $\hat{F}_n$ is the empirical c.d.f. and $(\bar{X}_n, \hat{\sigma}_n^2)$ is the MLE for (μ, σ^2) assuming normality. Argue that the distribution of T_n does not depend on (μ, σ^2) and describe an exact bootstrap test construction. [Such problems are studied in Romano (1988)].

Section 15.7

Problem 15.52 Prove Theorem 15.7.2. [*Hint*: For (ii), rather than considering $\hat{G}_{n,b}(x)$, just look at the empirical distribution of the values of $t_{n,b,i}$ (not scaled by τ_b) and show $\hat{G}^0_{n,b}(\cdot)$ converges in distribution to a point mass at $t(P)$.]

Problem 15.53 Prove a result for subsampling analogous to Theorem 15.4.7, but that does not require assumption (15.42). [Theorem 15.4.7 applies to testing real-valued parameters; a more general multiple testing procedure based on subsampling is given by Theorem 4.4 of Romano and Wolf (2004).]

Problem 15.54 To see how subsampling extends to a dependent time series model, assume $X_1, \ldots, X_n$ are sampled from a stationary time series model that is m-dependent. [Stationarity means the distribution of the $X_1, X_2, \ldots$ is the same as that of $X_t, X_{t+1}, \ldots$ for any t. The process is m-dependent if, for any t and m, $(X_1, \ldots, X_t)$ and $(X_{t+m+1}, X_{t+m+2}, \ldots)$ are independent; that is, observations separated in time by more than m units are independent.] Suppose the sum in the definition (15.67) of $L_{n,b}$ extends only over the $n - b + 1$ subsamples of size b of the form $(X_i, X_{i+1}, \ldots, X_{i+b-1})$; call the resulting estimate $\tilde{L}_{n,b}$. Under the assumption of stationarity and m-dependence, prove a theorem analogous to Theorem 15.7.1. [The theorem can be extended to much weaker types of dependence; see Politis, Romano, and Wolf (1999).]

15.9 Notes

Early references to permutations tests were provided at the end of Chapter 5. An elementary account is provided by Good (1994), who provides an extensive bibliography, and Edgington (1995). Multivariate permutation tests are developed in Pesarin (2001). The present large sample approach is due to Hoeffding (1952). Applications to block experiments is discussed in Robinson (1973). Expansions for the power of rank and permutation tests in the one- and two-sample problems are obtained in Albers, Bickel and van Zwet (1976) and Bickel and van Zwet (1978), respectively. A full account of the large sample theory of rank statistics is given in Hájek, Sidák, and Sen (1999). Robust two-sample permutation tests are obtained in Lambert (1985).

The bootstrap was discovered by Efron (1979), who coined the name. Much of the theoretical foundations of the bootstrap are laid out in Bickel and Freedman (1981) and Singh (1981). The development in Section 15.4 is based on Beran (1984). The use of Edgeworth expansions to study the bootstrap was initiated in Singh (1981) and Babu and Singh (1983), and is used prominently in Hall (1992). There have since been hundreds of papers on the bootstrap, as well as several book length treatments, including Hall (1992), Efron and Tibshirani (1993), Shao and Tu (1995), Davison and Hinkley (1997) and Lahiri (2003). Comparisons of bootstrap and randomization tests are made in Romano (1989b) and Janssen and Pauls (2003). Westfall and Young (1993) and van der Lann, Dudoit and Pollard (2004) apply resampling to multiple testing problems. Theorem 15.4.7 is based on Romano and Wolf (2004).

The method of empirical likelihood referred to in Example 15.6.4 is fully treated in Owen (2001). Similar to parametric models, the method of empirical likelihood can be improved through a Bartlett correction, yielding two-sided tests with error in rejection probability of $O(n^{-2})$; see DiCiccio, Hall and Romano (1991). Alternatively, rather than using the asymptotic Chi-squared distribution to get critical values, a direct bootstrap approach resamples from $\hat{Q}_n$. Higher order properties of such procedures are considered in DiCiccio and Romano (1990).

The roots of subsampling can be traced to Quenouille's (1949) and Tukey's (1958a) jackknife. Hartigan (1969) and Wu (1990) used subsamples to construct confidence intervals, but in a very limited setting. A general theory for using subsampling to approximate a sampling distribution is presented in Politis and Romano (1994b), including i.i.d. and data-dependent settings. A full treatment with numerous references is given by Politis, Romano, and Wolf (1999).

Appendix A
Auxiliary Results

A.1 Equivalence Relations; Groups

A relation: $x \sim y$ among the points of a space $\mathcal{X}$ is an equivalence relation if it is reflexive, symmetric, and transitive, that is, if

(i) $x \sim x$ for all $x \in \mathcal{X}$;

(ii) $x \sim y$ implies $y \sim x$;

(iii) $x \sim y$, $y \sim z$ implies $x \sim z$.

Example A.1.1 Consider a class of statistical decision procedures as a space, of which the individual procedures are the points. Then the relation defined by $\delta \sim \delta'$ if the procedures δ and δ' have the same risk function is an equivalence relation. As another example consider all real-valued functions defined over the real line as points of a space. Then $f \sim g$ if $f(x) = g(x)$ a.e. is an equivalence relation.

Given an equivalence relation, let D_x denote the set of points of the space that are equivalent to x. Then $D_x = D_y$ if $x \sim y$, and $D_x \cap D_y = 0$ otherwise. Since by (i) each point of the space lies in at least one of the sets D_x, it follows that these sets, the *equivalence classes* defined by the relation $\sim$, constitute a partition of the space.

A set G of elements is called a *group* if it satisfies the following conditions.

(i) There is defined an operation, group multiplication, which with any two elements $a, b \in G$ associates an element c of G. The element c is called the product of a and b and is denoted by ab.

(ii) Group multiplication obeys the associative law

$$(ab)c = a(bc).$$

(iii) There exists an element $e \in G$, called the *identity*, such that

$$ae = ea = a \qquad \text{for all} \quad a \in G.$$

(iv) For each element $a \in G$, there exists an element $a^{-1} \in G$, its *inverse*, such that

$$aa^{-1} = a^{-1}a = e.$$

Both the identity element and the inverse a^{-1} of any element a can be shown to be unique.

Example A.1.2 The set of all $n \times n$ orthogonal matrices constitutes a group if matrix multiplication and inverse are taken as group multiplication and inverse respectively, and if the identity matrix is taken as the identity element of the group. With the same specification of the group operations, the class of all non-singular $n \times n$ matrices also forms a group. On the other hand, the class of all $n \times n$ matrices fails to satisfy condition (iv).

If the elements of G are transformations of some space onto itself, with the group product ba defined as the result of applying first transformation a and following it by b, then G is called a transformation group. Assumption (ii) is then satisfied automatically. For any transformation group defined over a space $\mathcal{X}$ the relation between points of X given by

$$x \sim y \quad \text{if} \quad \text{there exists } a \in G \text{ such that } y = ax$$

is an equivalence relation. That it satisfies conditions (i), (ii), and (iii) required of an equivalence follows respectively from the defining properties (iii), (iv), and (i) of a group.

Let $\mathbb{C}$ be any class of $1:1$ transformations of a space, and let G be the class of all finite products $a_1^{\pm 1} a_2^{\pm 1} \ldots a_m^{\pm 1}$, with $a_1, \ldots, a_m \in \mathbb{C}$, $m = 1, 2, \ldots$, where each of the exponents can be $+1$ or -1 and where the elements $a_1, a_2, \ldots$ need not be distinct. Then it is easily checked that G is a group, and is in fact the smallest group containing $\mathbb{C}$.

A.2 Convergence of Functions; Metric Spaces

When studying convergence properties of functions it is frequently convenient to consider a class of functions as a realization of an abstract space $\mathcal{F}$ of points f in which convergence of a sequence f_n to a limit f, denoted by $f_n \to f$, has been defined.

Example A.2.1 Let μ be a measure over a measurable space $(\mathcal{X}, \mathcal{A})$.

(i) Let $\mathcal{F}$ be the class of integrable functions. Then f_n converges to f *in the mean* if[1]

$$\int |f_n - f| \, d_\mu \to 0. \tag{A.1}$$

(ii) Let $\mathcal{F}$ be a uniformly bounded class of measurable functions. The sequence is said to converge to f *weakly* if

$$\int f_n p \, d\mu \to \int f p \, d\mu \tag{A.2}$$

for all functions p that are integrable μ.

(iii) Let $\mathcal{F}$ be the class of measurable functions. Then f_n converges to f *pointwise* if

$$f_n(x) \to f(x) \qquad a.e. \ \mu. \tag{A.3}$$

A subset of $\mathcal{F}_0$ is *dense* in $\mathcal{F}$ if, given any $f \in \mathcal{F}$, there exists a sequence in $\mathcal{F}_0$ having f as its limit point. A space $\mathcal{F}$ is *separable* if there exists a countable dense subset of $\mathcal{F}$. A space $\mathcal{F}$ such that every sequence has a convergent subsequence whose limit point is in $\mathcal{F}$ is compact.[2] A space $\mathcal{F}$ is a *metric space* if for every pair of points f, g in $\mathcal{F}$ there is defined a metric (or distance) $d(f,g) \geq 0$ such that

(i) $d(f,g) = 0$ if and only if $f = g$;

(ii) $d(f,g) = d(g,f)$;

(iii) $d(f,g) + d(g,h) \geq d(f,h)$ for all f, g, h.

The space is a *pseudometric* space if (i) is replaced by

(i′) $d(f,f) = 0$ for all $f \in \mathcal{F}$.

A pseudometric space can be converted into a metric space by introducing the equivalence relation $f \sim g$ if $d(f,g) = 0$. The equivalence classes F, G, ... then constitute a metric space with respect to the metric $D(F,G) = d(f,g)$ where $f \in F$, $g \in G$.

In any pseudometric space a natural convergence definition is obtained by putting $f_n \to f$ if $d(f_n, f) \to 0$.

Example A.2.2 The space of integrable functions of Example A.2.1(i) becomes a pseudometric space if we put

$$d(f,g) = \int |f - g| \, d\mu$$

and the induced convergence definition is that given by (1).

[1] Here and in the examples that follow, the limit f is not unique. More specifically, if $f_n \to f$, then $f_n \to g$ if and only if $f = g$ (a.e. μ). Putting $f \sim g$ when $f = g$ (a.e. μ), uniqueness can be obtained by working with the resulting equivalence classes of functions rather than with the functions themselves.

[2] The term *compactness* is more commonly used for an alternative concept. which coincides with the one given here in metric spares. The distinguishing term *sequential compactness* is then sometimes given to the notion defined here.

Example A.2.3 Let $\mathcal{P}$ be a family of probability distributions over $(\mathcal{X}, \mathcal{A})$. Then $\mathcal{P}$ is a metric space with respect to the metric

$$d(P, Q) = \sup_{A \in \mathcal{A}} |P(A) - Q(A)|. \tag{A.4}$$

Lemma A.2.1 *If $\mathcal{F}$ is a separable pseudometric space, then every subset of $\mathcal{F}$ is also separable.*

PROOF. By assumption there exists a dense countable subset $\{f_n\}$ of $\mathcal{F}$. Let

$$S_{m,n} = \left\{ f : d(f, f_n) < \frac{1}{m} \right\},$$

and let A be any subset of $\mathcal{F}$. Select one element from each of the intersections $A \cap S_{m,n}$ that is nonempty, and denote this countable collection of elements by A_0. If a is any element of A and m any positive integer, there exists an element f_{n_m} such that $d(a, f_{n_m}) < 1/m$. Therefore a belongs to S_{m,n_m}, the intersection $A \cap S_{m,n_m}$ is nonempty, and there exists therefore an element of A_0 whose distance to a is $< 2/m$. This shows that A_0 is dense in A, and hence that A is separable. ■

Lemma A.2.2 *A sequence f_n of integrable functions converges to f in the mean if and only if*

$$\int_A f_n \, d\mu \to \int_A f \, d\mu \qquad \textit{uniformly for} \quad A \in \mathcal{A}. \tag{A.5}$$

PROOF. That (1) implies (5) is obvious, since for all $A \in \mathcal{A}$

$$\left| \int_A f_n \, d\mu - \int_A f \, d\mu \right| \leq \int |f_n - f| \, d\mu.$$

Conversely, suppose that (5) holds, and denote by A_n and A'_n the set of points x for which $f_n(x) > f(x)$ and $f_n(x) < f(x)$ respectively. Then

$$\int |f_n - f| \, d\mu = \int_{A_n} (f_n - f) \, d\mu - \int_{A'_n} (f_n - f) \, d\mu \to 0 \,. \; ■$$

Lemma A.2.3 *A sequence f_n of uniformly bounded functions converges to a bounded function f weakly if and only if*

$$\int_A f_n \, d\mu \to \int_A f \, d\mu \qquad \textit{for all } A \textit{ with } \mu(A) < \infty. \tag{A.6}$$

PROOF. That weak convergence implies (6) is seen by taking for p in (2) the indicator function of a set A, which is integrable if $\mu(A) < \infty$. Conversely (6) implies that (2) holds if p is any simple function $s = \sum a_i I_{A_i}$ with all the $\mu(A_i) < \infty$. Given any integrable function p, there exists, by the definition of the integral, such a simple function s for which $\int |p - s| \, d\mu < \epsilon/3M$, where M is a bound on the $|f|$'s. We then have

$$\left| \int (f_n - f) p \, d\mu \right| \leq \left| \int f_n (p - s) \, d\mu \right| + \left| \int f(s - p) \, d\mu \right| + \left| \int (f_n - f) s \, d\mu \right|.$$

The first two terms on the right-hand side are $< \epsilon/3$, and the third term tends to zero as n tends to infinity. Thus the left-hand side is $< \epsilon$ for n sufficiently large, as was to be proved. ■

Lemma A.2.4[3] *Let f and f_n, $n = 1, 2, \ldots$, be nonnegative integrable functions with*

$$\int f\, d\mu = \int f_n\, d\mu = 1.$$

Then pointwise convergence of f_n to f implies that $f_n \to f$ in the mean.

PROOF. If $g_n = f_n - f$, then $g \geq -f$, and the negative part $g_n^- = \max(-g_n, 0)$ satisfies $|g_n^-| \leq f$. Since $g_n(x) \to 0$ (a.e. μ), it follows from Theorem 2.2.2(ii) of Chapter 2 that $\int g_n^-\, d\mu \to 0$, and $\int g_n^+\, d\mu$ then also tends to zero, since $\int g_n\, d\mu = 0$. Therefore $\int |g_n|\, d\mu = \int (g_n^+ + g_n^-)\, d\mu \to 0$, as was to be proved.

Let P and P_n, $n = 1, 2, \ldots$ be probability distributions over $(\mathcal{X}, \mathcal{A})$ with densities p_n and p with respect to μ. Consider the convergence definitions

(a) $p_n \to p$ (a.e. μ);

(b) $\int |p_n - p|\, d\mu \to 0$;

(c) $\int g p_n\, d\mu \to \int g p\, d\mu$ for all bounded measurable g;

and

(b′) $P_n(A) \to P(A)$ uniformly for all $A \in \mathcal{A}$;

(c′) $P_n(A) \to P(A)$ for all $A \in \mathcal{A}$.

Then Lemmas A.2.2 and A.2.4 together with a slight modification of Lemma A.2.3 show that (a) implies (b) and (b) implies (c), and that (b) is equivalent to (b′) and (c) to (c′). It can further be shown that neither (a) and (b) nor (b) and (c) are equivalent.[4] ■

A.3 Banach and Hilbert Spaces

A set V is called a *vector space* (or linear space) over the reals if there exists a function $+$ on $V \times V$ to V and a function $\cdot$ on $\mathbf{R} \times V$ to V which satisfy for $x, y, z \in V$,

(i) $x + y = y + x$.
(ii) $(x + y) + z = z + (y + z)$.
(iii) There is a vector $\underline{0} \in V$: $x + \underline{0} = x$ for all $x \in V$.
(iv) $\lambda(x + y) = \lambda x + \lambda y$ for any $\lambda \in \mathbf{R}$.
(v) $(\lambda_1 + \lambda_2)x = \lambda_1 x + \lambda_2 x$ for $\lambda_i \in \mathbf{R}$.
(vi) $\lambda_1(\lambda_2 x) = (\lambda_1\lambda_2)x$ for $\lambda_i \in \mathbf{R}$.
(vii) $0 \cdot x = \underline{0}$, $1 \cdot x = x$.

---

[3] Scheffé (1947).
[4] Robbins (1948).

The operation + is called addition by scalars and $\cdot$ is multiplication by scalars. A nonnegative real-valued function $\|\ \|$ defined on a vector space is called a norm if

(i) $\|x\| = 0$ if and only if $x = \underline{0}$.
(ii) $\|x + y\| \le \|x\| + \|y\|$.
(iii) $\|\lambda x\| = |\lambda|\|x\|$.

A vector space with norm $\|\ \|$ is a then a metric space if we define the metric d to be $d(x, y) = \|x - y\|$.

A sequence $\{x_n\}$ of elements in a normed vector space V is called a Cauchy sequence if, given $\epsilon > 0$, there is an N such that for all $m, n \ge N$, we have $\|x_n - x_m\| < \epsilon$. A Banach space is a normed vector space that is complete in the sense that every Cauchy sequence $\{x_n\}$ satisfies $\|x_n - x\| \to 0$ for some $x \in V$.

Example A.3.1 (L^p spaces.) Let μ be a measure over a measurable space $(\mathcal{X}, \mathcal{A})$. Fix $p > 0$ and $L^p[\mathcal{X}, \mu]$ denote the measurable functions f such that $\int |f|^p d\mu < \infty$. If we identify equivalence classes of functions that are equal almost everywhere μ, then, for $p \ge 1$, this vector space becomes a normed vector space by defining

$$\|f\| = \|f\|_p = \left[\int |f|^p d\mu\right]^{1/p} .$$

In this case, the triangle inequality

$$\|f + g\|_p \le \|f\|_p + \|g\|_p$$

is known as *Minkowski's inequality.* Moreover, this space is a Banach space.[5]

A Hilbert space H is a Banach space for which there is defined a function $\langle x, y\rangle$ on $H \times H$ to $\mathbf{R}$, called the inner product of x and y, satisfying, for x_i, $y \in H$, $\lambda_i \in \mathbf{R}$,

(i) $\langle \lambda_1 x_1 + \lambda_2 x_2, y\rangle = \lambda_1\langle x_1, y\rangle + \lambda_2\langle x_2, y\rangle$.
(ii) $\langle x, y\rangle = \langle y, x\rangle$.
(iii) $\langle x, x\rangle = \|x\|^2$.

Two vectors x and y of H are called orthogonal if $\langle x, y\rangle = 0$. A collection $H_0 \subset H$ of vectors is called an orthogonal system if any two elements in H_0 are orthogonal. An orthogonal system is orthonormal if each vector in it has norm 1. An orthonormal system H_0 is called complete if $\langle x, h\rangle = 0$ for all $h \in H_0$ implies $x = \underline{0}$. In a separable Hilbert space, every orthonormal system is countable and there exists a complete orthonormal system. Letting $\{h_1, h_2, \ldots\}$ denote a complete orthonormal system, Parseval's identity says that, for any $x \in H$,

$$\|x\|^2 = \sum_{j=1}^{\infty}[\langle x, h_j\rangle]^2 . \tag{A.7}$$

Example A.3.2 (L^2 spaces.) In example A.3.1 with $p = 2$, the equivalence classes of square integrable functions is a Hilbert space with inner product given

[5] For proofs of the results in this section, see Chapter 5 of Dudley (1989).

by

$$\langle f_1, f_2 \rangle = \int f_1 f_2 d\mu \ .$$

If $\mathcal{X}$ is $[0,1]$ and μ is Lebesgue measure, then a complete orthonormal system is given by the functions $f_j(u) = \sqrt{2}\sin(\pi j u)$, $j = 1, 2, \ldots$. Therefore, for any square integrable function f, Parseval's identity yields

$$\int_0^1 f^2(u)du = 2\sum_{j=1}^{\infty}\left[\int_0^1 f(u)\sin(\pi j u)du\right]^2 .$$

A.4 Dominated Families of Distributions

Let $\mathcal{M}$ be a family of measures defined over a measurable space $(\mathcal{X}, \mathcal{A})$. Then $\mathcal{M}$ is said to be *dominated* by a σ-finite measure μ defined over $(\mathcal{X}, \mathcal{A})$ if each member of $\mathcal{M}$ is absolutely continuous with respect to μ. The family $\mathcal{M}$ is said to be *dominated* if there exists a σ-finite measure dominating it. Actually, if $\mathcal{M}$ is dominated there always exists a finite dominating measure. For suppose that $\mathcal{M}$ is dominated by μ and that $\mathcal{X} = \cup A_i$, with $\mu(A_i)$ finite for all i. If the sets A_i are taken to be mutually exclusive, the measure $\nu(A) = \sum \mu(A \cap A_i)/2^i \mu(A_i)$ also dominates $\mathcal{M}$ and is finite.

Theorem A.4.1[6] *A family $\mathcal{P}$ of probability measures over a Euclidean space $(\mathcal{X}, \mathcal{A})$ is dominated if and only if it is separable with respect to the metric* (4) *or equivalently with respect to the convergence definition*

$$P_n \to P \quad \textit{if} \quad P_n(A) \to P(A) \quad \textit{uniformly for} \quad A \in \mathcal{A}.$$

PROOF. Suppose first that $\mathcal{P}$ is separable and that the sequence $\{P_n\}$ is dense in $\mathcal{P}$, and let $\mu = \sum P_n/2^n$. Then $\mu(A) = 0$ implies $P_n(A) = 0$ for all n, and hence $P(A) = 0$ for all $P \in \mathcal{P}$. Conversely suppose that $\mathcal{P}$ is dominated by a measure μ, which without loss of generality can be assumed to be finite. Then we must show that the set of integrable functions $dP/d\mu$ is separable with respect to the convergence definition (5) or, because of Lemma A.2.2, with respect to convergence in the mean. It follows from Lemma A.2.1 that it suffices to prove this separability for the class $\mathcal{F}$ of all functions f that are integrable μ. Since by the definition of the integral every integrable function can be approximated in the mean by simple functions, it is enough to prove this for the case that $\mathcal{F}$ is the class of all simple integrable functions. Any simple function can be approximated in the mean by simple functions taking on only rational values, so that it is sufficient to prove separability of the class of functions $\sum r_i I_{A_i}$ where the r's are rational and the A's are Borel sets, with finite μ-measure since the f's are integrable. It is therefore finally enough to take for $\mathcal{F}$ the class of functions I_A, which are indicator functions of Borel sets with finite measure. However, any such set can be approximated by finite unions of disjoint rectangles with rational end

[6]Berger (1951b).

points. The class of all such unions is denumerable, and the associated indicator functions will therefore serve as the required countable dense subset of $\mathcal{F}$. ∎

An examination of the proof shows that the Euclidean nature of the space $(\mathcal{X}, \mathcal{A})$ was used only to establish the existence of a countable number of sets $A_i \in \mathcal{A}$ such that for any $A \in \mathcal{A}$ with finite measure there exists a subsequence A_i with $\mu(A_i) \to \mu(A)$. This property holds quite generally for any σ-field $\mathcal{A}$ which has a *countable number of generators*, that is, for which there exists a countable number of sets B_i such that $\mathcal{A}$ is the smallest σ-field containing the B_i.[7] It follows that Theorem A.4.1 holds for any σ-field with this property. Statistical applications of such σ-fields occur in sequential analysis, where the sample space $\mathcal{X}$ is the union $\mathcal{X} = \cup_i \mathcal{X}_i$ of Borel subsets $\mathcal{X}_i$ of i-dimensional Euclidean space. In these problems, $\mathcal{X}_i$ is the set of points $(x_1, \ldots, x_i)$ for which exactly i observations are taken. If $\mathcal{A}_i$ is the σ-field of Borel subsets of $\mathcal{X}_i$, one can take for $\mathcal{A}$, the σ-field generated by the $\mathcal{A}_i$, and since each $\mathcal{A}_i$ possesses a countable number of generators, so does $\mathcal{A}$.

If $\mathcal{A}$ does not possess a countable number of generators, a somewhat weaker conclusion can be asserted. Two families of measures $\mathcal{M}$ and $\mathcal{N}$ are *equivalent* if $\mu(A) = 0$ for all $\mu \in \mathcal{M}$ implies $\nu(A) = 0$ for all $\nu \in \mathcal{N}$ and vice versa.

Theorem A.4.2[8] *A family $\mathcal{P}$ of probability measures is dominated by a σ-finite measure if and only if $\mathcal{P}$ has a countable equivalent subset.*

PROOF. Suppose first that $\mathcal{P}$ has a countable equivalent subset $\{P_1, P_2, \ldots\}$. Then $\mathcal{P}$ is dominated by $\mu = \sum P_n/2^n$. Conversely, let $\mathcal{P}$ be dominated by a σ-finite measure μ, which without loss of generality can be assumed to be finite. Let $\mathcal{Q}$ be the class of all probability measures Q of the form $\sum c_i P_i$, where $P_i \in \mathcal{P}$, the c's are positive, and $\sum c_i = 1$. The class $\mathcal{Q}$ is also dominated by μ, and we denote by q a fixed version of the density $dQ/d\mu$. We shall prove the fact, equivalent to the theorem, that there exists Q_0 in $\mathcal{Q}$ such that $Q_0(A) = 0$ implies $Q(\mathcal{A}) = 0$ for all $Q \in \mathcal{Q}$.

Consider the class $\mathbb{C}$ of sets C in $\mathcal{A}$ for which there exists $Q \in \mathcal{Q}$ such that $q(x) > 0$ a.e. μ on C and $Q(C) > 0$. Let $\mu(C_i)$ tend to $\sup_{\mathbb{C}} \mu(C)$, let $q_i(x) > 0$ a.e. on C_i, and denote the union of the C_i by C_0. Then $q_0^*(x) \sum c_i q_i(x)$ agrees a.e. with the density of $Q_0 = \sum c_i Q_i$ and is positive a.e. on C_0, so that $C_0 \in \mathbb{C}$. Suppose now that $Q_0(A) = 0$, let Q be any other member of $\mathcal{Q}$, and let $C = \{x : q(x) > 0\}$. Then $Q_0(A \cap C_0) = 0$, and therefore $\mu(A \cap C_0) = 0$ and $Q(A \cap C_0) = 0$. Also $Q(A \cap \tilde{C}_0 \cap \tilde{C}) = 0$. Finally, $Q(A \cap \tilde{C}_0 \cap C) > 0$ would lead to $\mu(C_0 \cup [A \cap \tilde{C}_0 \cap C]) > \mu(C_0)$ and hence to a contradiction of the relation $\mu(C_0) = \sup_{\mathbb{C}} \mu(C)$, since $A \cap \tilde{C}_0 \cap C$ and therefore $C_0 \cup [A \cap \tilde{C}_0 \cap C]$ belongs to $\mathbb{C}$. ∎

[7] A proof of this is given for example by Halmos (1974, Theorem B of Section 40).

[8] Halmos and Savage (1949).

A.5 The Weak Compactness Theorem

The following theorem forms the basis for proving the existence of most powerful tests, most stringent tests, and so on.

Theorem A.5.1[9] (Weak compactness theorem). *Let μ be a σ-finite measure over a Euclidean space, or more generally over any measurable space $(\mathcal{X}\mathcal{A})$ for which $\mathcal{A}$ has a countable number of generators. Then the set of measurable functions ϕ with $0 \le \phi \le 1$ is compact with respect to the weak convergence (2).*

Proof. Given any sequence $\{\phi_n\}$, we must prove the existence of a subsequence $\{\phi_{n_j}\}$ and a function ϕ such that

$$\lim \int \phi_{n_i} p \, d\mu = \int \phi p \, d\mu$$

for all integrable p. If μ^* is a finite measure equivalent to μ, then p^* is integrable μ^* if and only if $p = (d\mu^*/d\mu)p^*$ is integrable μ, and $\int \phi p \, d\mu = \int \phi p^* \, d\mu^*$ for all ϕ. We may therefore assume without loss of generality that μ is finite. Let $\{p_n\}$ be a sequence of p's which is dense in the p's with respect to convergence in the mean. The existence of such a sequence is guaranteed by Theorem A.4.1 and the remark following it. If

$$\Phi_n(p) = \int \phi_n p \, d\mu,$$

the sequence $\Phi_n(p)$ is bounded for each p. A subsequence Φ_{n_k} can be extracted such that $\Phi_{n_k}(p_m)$ converges for each p_m by the following diagonal process. Consider first the sequence of numbers $\{\Phi_n(p_1)\}$ which possesses a convergent subsequence $\Phi_{n_1'}(p_1), \Phi_{n_2'}(p_1), \ldots$. Next the sequence $\Phi_{n_1'}(p_2), \Phi'_{n_2}(p_2), \ldots$ has a convergent subsequence $\Phi_{n_1''}(p_2), \Phi_{n_2''}(p_2), \ldots$. Continuing in this way, let $n_1 = n_1'$, $n_2 = n_2''$, $n_3''', \ldots$. Then $n_1 < n_2 < \ldots$, and the sequence $\{\Phi_{n_i}\}$ converges for each p_m. It follows from the inequality

$$\left| \int (\phi_{n_j} - \phi_{n_i}) p \, d\mu \right| \le \left| \int (\phi_{n_j} - \phi_{n_i}) p_m \, d\mu \right| + 2 \int |p - p_m| \, d\mu$$

that $\Phi_{n_i}(p)$ converges for all p. Denote its limit by $\Phi(p)$, and define a set function Φ^* over $\mathcal{A}$ by putting

$$\Phi^*(A) = \Phi(I_A).$$

Then Φ^* is nonnegative and bounded, since for all A, $\Phi^*(A) \le \mu(A)$. To see that it is also countably additive let $A = \cup A_k$, where the A_k are disjoint. Then $\Phi^*(A) = \lim \Phi^*_{n_i}(\cup A_k)$ and

$$\left| \int_{\cup A_k} \phi_{n_i} \, d\mu - \sum \Phi^*(A_k) \right| \quad \le \quad \left| \int_{\cup_{k=1}^m A_k} \phi_{n_i} \, d\mu - \sum_{k=1}^m \Phi^*(A_k) \right|$$

[9] Banach (1932). The theorem is valid even without the assumption of a countable number of generators; see Nölle and Plachky (1967) and Aloaglu's theorem, given for example in Royden (1988).

$$+ \left| \int_{\cup_{k=m+1}^{\infty} A_k} \phi_{n_i} \, d\mu - \sum_{k=m+1}^{\infty} \Phi^*(A_k) \right| .$$

Here the second term is to be taken as zero in the case of a finite sum $A = \cup_{k=1}^{m} A_k$, and otherwise does not exceed $2\mu(\cup_{k=m+1}^{\infty} A_k)$, which can be made arbitrarily small by taking m sufficiently large. For any fixed m the first term tends to zero as i tends to infinity. Thus Φ^* is a finite measure over $(\mathcal{X}, \mathcal{A})$. It is furthermore absolutely continuous with respect to μ, since $\mu(A) = 0$ implies $\Phi_{n_i}(I_A) = 0$ for all i, and therefore $\Phi(I_A) = \Phi^*(A) = 0$ We can now apply the Radon–Nikodym theorem to get

$$\Phi^*(A) = \int_A \phi \, d\mu \quad \text{for all } A,$$

with $0 \le \phi \le 1$. We then have

$$\int_A \phi_{n_i} \, d\mu \to \int_A \phi \, d\mu \quad \text{for all } A,$$

and weak convergence of the ϕ_{n_i} to ϕ follows from Lemma A.2.3. ■

References

Agresti, A. (1992). A survey of exact inference for contingency tables (with discussion). *Statistical Science* **7**, 131–177.

Agresti, A. (2002). *Categorical Data Analysis*, 2nd edition. John Wiley, New York.

Agresti, A. and Coull, B. (1998). Approximate is better than "exact" for interval estimation of binomial proportions. *American Statistician* **52**, 119–126.

Aiyar, R. J., Guillier, C. L., and Albers, W. (1979). Asymptotic relative efficiencies of rank tests for trend alternatives. *Journal of the American Statistical Association* **74**, 226–231.

Akritas, M., Arnold, S. and Brunner, E. (1997). Nonparametric hypotheses and rank statistics for unbalanced factorial designs. *Journal of the American Statistical Association* **92**, 258–265.

Albers, W. (1978). Testing the mean of a normal population under dependence. *Annals of Statistics* **6**, 1337–1344.

Albers, W., Bickel, P. and van Zwet, W. (1976). Asymptotic expansion for the power of distribution free tests in the one-sample problem. *Annals of Statistics* **4**, 108–156.

Albert, A. (1976). When is a sum of squares an analysis of variance? *Annals of Statistics* **4**, 775–778.

Anderson, T. W. (1967). Confidence limits for the expected value of an arbitrary bounded random variable with a continuous distribution function. *Bull. ISI* **43**, 249–251.

Anderson, T. W. (2003). *An Introduction to Multivariate Statistical Analysis*, 3rd edition. John Wiley, Hoboken, NJ. [Problem 6.19.]

Andersson, S. (1982). Distributions of maximal invariants using quotient measures. *Annals of Statistics* **10**, 955–961.

Anscombe, F. (1948). Transformations of Poisson, binomial and negative binomial data. *Biometrika* **35**, 246–254.

Antille, A., Kersting, G., and Zucchini, W. (1982). Testing symmetry. *Journal of the American Statistical Association* **77**, 639–651.

Arbuthnot, J. (1710). An argument for Divine Providence, taken from the constant regularity observ'd in the births of both sexes. *Phil. Trans.* **27**, 186–190.

Arcones, M. (1991). On the asymptotic theory of the bootstrap. Ph.D. thesis, The City University of New York.

Arcones, M. and Giné, E. (1989). The bootstrap of the mean with arbitrary bootstrap sample size. *Annals of the Institute Henri Poincaré* **25**, 457–481.

Arcones, M. and Giné, E. (1991). Additions and correction to "the bootstrap of the mean with arbitrary bootstrap sample size". *Annals of the Institute Henri Poincaré* **27**, 583–595.

Armsen, P. (1955). Tables for significance tests of 2×2 contingency tables. *Biometrika* **42**, 494–511.

Arnold, S. (1981). *The Theory of Linear Models and Multivariate Analysis.* John Wiley, New York.

Arnold, S. (1984). Pivotal quantities and invariant confidence regions. *Statistics and Decisions* **2**, 257–280.

Arrow, K. (1960). Decision theory and the choice of a level of significance for the t-test. In *Contributions to Probability and Statistics* (Olkin et al., eds.) Stanford University Press, Stanford, California.

Arvesen, J. N. and Layard, M. W. J. (1975). Asymptotically robust tests in unbalanced variance component models. *Annals of Statistics* **3**, 1122–1134.

Athreya, K. (1985). Bootstrap of the mean in the infinite variance case, II. Technical Report 86-21, Department of Statistics, Iowa State University.

Athreya, K. (1987). Bootstrap of the mean in the infinite variance case. *Annals of Statistics* **15**, 724–731.

Atkinson, A. and Donev, A. (1992). *Optimum Experimental Design.* Clarendon Press, Oxford.

Atkinson, A. and Riani, M. (2000). *Robust Regression Analysis.* Springer-Verlag, New York.

Babu, G. (1984). Bootstrapping statistics with linear combinations of chi-squares as weak limit. Sankhya Series A **56**, 85–93.

Babu, G. and Singh, K. (1983). Inference on means using the bootstrap. *Annals of Statistics* **11**, 999–1003.

Bahadur, R. (1955). A characterization of sufficiency. *Annals of Mathematical Statistics* **26**, 286–293.

Bahadur, R. (1960). Stochastic comparison of tests. *Annals of Mathematical Statistics* **31**, 279–295.

Bahadur, R. (1965). An optimal property of the likelihood ratio statistic. In *Proc. 5th Berkeley Symposium on Probab. Theory and Math. Statist.* **1**, Le Cam, L. and Neyman, J. (eds.), University of California Press, 13–26.

Bahadur, R. (1979). A note on UMV estimates and ancillary statistics. In *Contributions to Statistics*, J. Hájek Memorial Volume, Edited by Jureckova, Academia, Prague.

Bahadur, R. and Lehmann, E. L. (1955). Two comments on 'sufficiency and statistical decision functions'. *Annals of Mathematical Statistics* **26**, 139–142. [Problem 2.5.]

Bahadur, R. and Savage, L. J. (1956). The nonexistence of certain statistical procedures in nonparametric problems. *Annals of Mathematical Statistics* **27**, 1115–1122.

Bain, L. J. and Engelhardt, M. E. (1975). A two-moment chi-square approximation for the statistic $\log(\bar{X}/\tilde{X})$. *Journal of the American Statistical Association* **70**, 948–950.

Baker, R. (1995). Two permutation tests of equality of variances. *Statistics and Computing* **5**, 351–361.

Banach, S. (1932). *Théorie des Operations Linéaires.* Funduszu Kultury Narodowej, Warszawa.

Bar-Lev, S. and Plachky, D. (1989). Boundedly complete families which are not complete. *Metrika* **36**, 331–336.

Bar-Lev, S. and Reiser, B. (1982). An exponential subfamily which admits UMPU tests based on a single test statistic. *Annals of Statistics* **10**, 979–989.

Barankin, E. W. and Maitra, A. P. (1963). Generalizations of the Fisher–Darmois–Koopman–Pitman theorem on sufficient statistics. *Sankhyā Series A* **25**, 217–244.

Barlow, R. E., Bartholomew, D. J., Bremner, J. M., and Brunk, H. D. (1972). *Statistical Inference under Order Restrictions*, John Wiley, New York.

Barnard, G. A. (1976). Conditional inference is not inefficient. *Scandinavian Journal of Statistics* **3**, 132–134. [Problem 10.27.]

Barnard, G. A. (1995). Pivotal models and the fiducial argument. *International Statistical Review* **63**, 309–323.

Barnard, G. A. (1996). Rejoinder, Pivotal models and structural models. *International Statistical Review* **64**, 235–236.

Barndorff-Nielsen, O. (1978). *Information and Exponential Families in Statistical Theory.* John Wiley, New York. [Provides a systematic discussion of various concepts of ancillarity with many examples.]

Barndorff-Nielsen, O. (1983). On a formula for the distribution of the maximum likelihood estimator. *Biometrika* **70**, 343–365.

Barndorff-Nielsen, O., Cox, D. and Reid, N. (1986). Differential geometry in statistical theory. *International Statistics Review* **54**, 83–96.

Barndorff-Nielsen, O. and Hall, P. (1988). On the level-error after Bartlett adjustment of the likelihood ratio statistic. *Biometrika* **75**, 374–378.

Barndorff-Nielsen, O. and Pedersen, K. (1968). Sufficient data reduction and exponential families. *Math. Scand.* **2**, 197–202.

Barnett, V. (1999). *Comparative Statistical Inference*, 3rd edition. John Wiley, New York.

Barron, A. (1989). Uniformly powerful goodness of fit tests. *Annals of Statistics* **17**, 107–124.

Bartlett, M. S. (1937). Properties of sufficiency and statistical tests. *Proc. Roy. Sec. London, Ser. A* **160**, 268–282. [Points out that *exact* (that is, similar) tests can be obtained by combining the conditional tests given the different values of a sufficient statistic. Applications.]

Bartlett, M. S. (1957). A comment on D. V. Lindley's statistical paradox. *Biometrika* **44**, 533–534.

Basu, D. (1955). On statistics independent of a complete sufficient statistic. *Sankhyā* **15**, 377–380.

Basu, D. (1958). On statistics independent of a sufficient statistic. *Sankhyā* **20**, 223–226.

Basu, D. (1959). The family of ancillary statistics. *Sankhyā (A)* **21**. 247–256. [Problem 10.7.]

Basu, D. (1964). Recovery of ancillary information. *Sankhyā (A)* **26**, 3–16. [Problems 10.9, 10.11.]

Basu, D. (1978). On partial sufficiency: A review. *Journal of Statistical Planning and Inference* **2**, 1–13.

Basu, D. (1982). Basu theorems. In *Encycl. Statisti. Sci* **1**, 193-196.

Basu, S. (1999). Conservatism of the z confidence interval under symmetric and asymmetric departures from normality. *Annals of the Institute of Statistical Mathematics* **51**, 217–230.

Basu, S. and DasGupta, A. (1995). Robustness of standard confidence intervals for location parameters under departure from normality. *Annals of Statistics* **23**, 1433–1442.

Bayarri, M. and Berger, J. (2000). P-values for composite null hypotheses. *Journal of the American Statistical Association* **95**, 1127–1142.

Bayarri, M. and Berger, J. (2004). The interplay of Bayesian and frequentist analysis. *Statistical Science* **19**, 58–80.

Becker, B. (1997). Combination of p-values. *Encycl. Statist. update* **1**, 448–453.

Becker, N. and Gordon, I. (1983). On Cox's criterion for discriminating between alternative ancillary statistics. *International Statistical Review* **51**. 89–92.

Bednarski, T. (1984). Minimax testing between Prohorov neighbourhoods. *Statistics and Decisions* **2**, 281–292.

Behnen, K. and Neuhaus, G. (1989). *Rank Tests With Estimated Scores and Their Applications.* (Teubner Skripten zur Mathematischen Stochastik) B. G. Teubner, Stuttgart.

Bell, C. B., Blackwell, D. and Breiman, L. (1960). On the completeness of order statistics. *Annals of Mathematical Statistics* **31**, 794–797.

Bell, C. B. (1964). A characterization of multisample distribution-free statistics. *Annals of Mathematical Statistics* **35**, 735–738.

Bell, C. D. and Sen, P. (1984). Randomization procedures. In *Handbook of Statistics 4* (Krishnaiah and Sen, eds.), Elsevier.

Benichou, J., Fears, T. and Gail, M. (1996). A reminder of the fallibility of the Wald statistic. *American Statistician* **50**, 226–227.

Bening, V. (2000). *Asymptotic Theory of Testing Statistical Hypotheses: Efficient Statistics, Optimality, Power Loss, and Deficiency.* VSP Publishing, The Netherlands.

Benjamini, Y. (1983). Is the t-test really conservative when the parent distribution is long-tailed? *Journal of the American Statistical Association* **78**, 645–654.

Benjamini, Y. and Hochberg, Y. (1995). Controlling the false discovery rate: a practical and powerful approach to multiple testing, *Journal of the Royal Statistical Society Series B* **57**, 289–300.

Benjamini, Y. and Yekutieli, D. (2001). The control of the false discovery rate in multiple testing under dependency. *Annals of Statistics* **29**, 1165–1189.

Bennett, B. (1957). On the performance characteristic of certain methods of determining confidence limits. *Sankhyā* **18**, 1–12.

Bentkus, V. (2003). On the dependence of the Berry-Esseen bound on dimension. *Journal of Statistical Planning and Inference* **113**, 385–402.

Beran, R. (1974). Asymptotically efficient adaptive rank estimates in location models. *Annals of Statistics* **2**, 63–74.

Beran, R. (1977). Minimum Hellinger distance estimates for parametric models. *Annals of Statistics* **5**, 445–463.

Beran, R. (1984). Bootstrap methods in statistics. *Jahresberichte des Deutschen Mathematischen Vereins* **86**, 14–30.

Beran, R. (1986). Simulated power function. *Annals of Statistics* **14**, 151–173.

Beran, R. (1987). Prepivoting to reduce level error of confidence sets. *Biometrika* **74**, 151–173.

Beran, R. (1988a). Balanced simultaneous confidence sets. *Journal of the American Statistical Association* **83**, 679–686.

Beran, J. (1988b). Prepivoting test statistics: a bootstrap view of asymptotic refinements. *Journal of the American Statistical Association* **83**, 687–697.

Beran, J. (1999). Hajék-Inagaki convolution theorem. In *Encyclopedia of Statistical Sciences*, Update **3**, 294–297. John Wiley, New York.

Beran, J. and Ducharme, G. (1991). *Asymptotic Theory for Bootstrap Methods in Statistics.* Centre de recherches mathématiques, University of Montreal, Quebec.

Beran, R. and Millar, W. (1986). Confidence sets for a multivariate distribution. *Annals of Statistics* **14**, 431–443.

Beran, R. and Millar, W. (1988). A stochastic minimum distance test for multivariate parametric models. *Annals of Statistics* **17**, 125–140.

Beran, R. and Srivastava, M. S. (1985). Bootstrap tests and confidence regions for functions of a covariance matrix. *Annals of Statistics* **13**, 95–115.

Berger, A. (1951a). On uniformly consistent tests. *Annals of Mathematical Statistics* **22**, 289–293.

Berger, A. (1951b). Remark on separable spaces of probability measures. *Annals of Mathematical Statistics* **22**, 119–120.

Berger, J. (1985a). *Statistical Decision Theory and Bayesian Analysis*, 2nd edition. Springer, New York.

Berger, J. (1985b). The frequentist viewpoint of conditioning. In *Proc. Berkeley Conf. in Honor of J. Neyman and J. Kiefer* (Le Cam and Olshen, eds.), Wadsworth, Belmont, Calif.

Berger, J. (2003). Could Fisher, Jeffreys and Neyman have ageed on testing? (with discussion). *Statistical Science* **18**, 1–32.

Berger, J., Boukai, B. and Wang, Y. (1997). United frequentist and Bayesian testing of a precise hypothesis (with discussion). *Statistical Science* **12**, 133–160.

Berger, J., Brown, L. D. and Wolpert, R. (1994). A unified conditional frequentist and Bayesian test for fixed and sequential simple hypothesis testing. *Annals of Statistics* **22**, 1787–1807.

Berger, J., Liseo, B. and Wolpert, R. (1999). Integrated likelihood methods for eliminating nuisance parameters (with discussion). *Statistical Science* **14**, 1–28.

Berger, J. and Sellke, T. (1987). Testing a point null-hypothesis: The irreconcilability of significance levels and evidence. *Journal of the American Statistical Association* **82**, 112–122.

Berger, J. and Wolpert, R. (1988). *The Likelihood Principle*, 2nd edition, IMS Lecture Notes–Monograph Series, Hayward, CA.

Berger, R. (1982). Multiparameter hypothesis testing and acceptance sampling. *Technometrics*, **24**, 295–300.

Berger, R. and Boos, D. (1994). p-values maximized over a confidence set for the nuisance parameter. *Journal of the American Statistical Association* **89**, 1012–1016.

Berger, R. and Hsu, J. (1996). Bioequivalence trials, intersection-union tests and equivalence confidence sets (with discussion). *Statistical Science* **11**, 283–319.

Berk, R. (1970). A remark on almost invariance. *Annals of Mathematical Statistics* **41**, 733–735.

Berk, R. and Bickel, P. (1968). On invariance and almost invariance. *Annals of Mathematical Statistics* **39**, 1573–1576.

Berk, R. and Cohen, A. (1979). Asymptotically optimal methods of combining tests. *Journal of the American Statistical Association* **74**, 812–814.

Berk, R., Nogales, A. and Oyola, J. (1996). Some counterexamples concerning sufficiency and invariance. *Annals of Statistics* **24**, 902–905.

Bernardo, J. and Smith, A. (1994). *Bayesian Theory.* New York, John Wiley.

Bernoulli, D. (1734). Quelle est la cause physique de l'inclinaison des plans des orbites des planetes par rapport au plan de l'équateur de la revolution du soleil autour de son axe; Et d'oú vient que les inclinaisons de ces orbites sont differentes entre elles. *Recueil des Pièces qui ont Remporté le Prix de l'Académie Royale des Sciences* **3**, 93–122.

Bhat, U. and Miller, G. (2002). *Elements of Applied Stochastic Processes*, 3rd edition, John Wiley, New York.

Bhattacharya, P. K., Gastwirth, J. L., and Wright, A. L. (1982). Two modified Wilcoxon tests for symmetry about an unknown location parameter. *Biometrika* **69**, 377–382.

Bhattacharya, R. and Ghosh, J. (1978). On the validity of the formal Edgeworth expansion. *Annals of Statistics* **6**, 434–451.

Bhattacharya, R. and Rao, R. (1976). *Normal Approximation and Asymptotic Expansions*. John Wiley, New York.

Bickel, P. (1974). Edgeworth expansions in nonparametric statistics. *Annals of Statistics* **2**, 1–20.

Bickel, P. (1982). On adaptive estimation. *Annals of Statistics* **10**, 647–671.

Bickel, P. (1984). Parametric robustness: small biases can be worthwhile. *Annals of Statistics* **12**, 864–879.

Bickel, P. and Doksum, K. A. (1981). An analysis of transformations revisited. *Journal of the American Statistical Association* **76**, 296–311.

Bickel, P. and Doksum, K. A. (2001). *Mathematical Statistics*, volume I, 2nd edition. Prentice Hall, Upper Saddle River, New Jersey.

Bickel, P. and Freedman, D. (1981). Some asymptotic theory for the bootstrap. *Annals of Statistics* **9**, 1196–1217.

Bickel, P. and Ghosh, J. (1990). A decomposition for the likelihood ratio statistic and the Bartlett correction – a Bayesian argument. *Annals of Statistics* **18**, 1070–1090.

Bickel, P., Götze, F. and van Zwet, W. R. (1997). Resampling fewer than n observations: Gains, losses, and remedies for losses. *Statistica Sinica* **7**, 1–31.

Bickel, P., Klaassen, C., Ritov, Y., and Wellner, J. (1993). *Efficient and Adaptive Estimation for Semiparametric Models.* The John Hopkins University Press, Baltimore, MD.

Bickel, P. and Van Zwet, W. R. (1978). Asymptotic expansions for the power of distribution free tests in the two-sample problem. *Annals of Statistics* **6**, 937–1004.

Billingsley, P. (1961). Statistical methods in Markov chains. *Annals of Mathematical Statistics* **32**, 12–40.

Billingsley, P. (1968). *Convergence of Probability Measures.* John Wiley, New York.

Billingsley, P. (1995). *Probability and Measure*, 3rd edition. John Wiley, New York.

Birch, M. W. (1964). The detection of partial association, I The 2×2 case. *Journal of the Royal Statistical Society Series B,* **26**, 313–324.

Birnbaum, A. (1954a). Statistical methods for Poisson processes and exponential populations. *Journal of the American Statistical Association* **49**, 254–266.

Birnbaum, A. (1954b). Admissible test for the mean of a rectangular distribution. *Annals of Mathematical Statistics* **25** 157–161.

Birnbaum, A. (1955). Characterization of complete classes of tests of some multiparameter hypotheses, with applications to likelihood ratio tests. *Annals of Mathematical Statistics* **26**, 21–36.

Birnbaum, A. (1962). On the foundations of statistical inference (with discussion). *Journal of the American Statistical Association* **57**, 269–326.

Birnbaum Z. W. (1952). Numerical tabulation of the distribution of Kolmogorov's statistic for finite sample size. *Journal of the American Statistical Association* **47**, 431.

Birnbaum, Z. W. and Chapman, D. G. (1950). On optimum selections from multinormal populations. *Annals of Mathematical Statistics* **21**, 433–447. [Problem 3.46]

Bishop, Y. M. M., Fienberg, S. E., and Holland, P. W. (1975). *Discrete Multivariate Analysis: Theory and Practice*, MIT. Press, Cambridge, Mass.

Blackwell, D. (1951). On a theorem of Lyapunov. *Annals of Mathematical Statistics* **22**, 112–114.

Blackwell, D. and Dubins, L. E. (1975). On existence and non-existence of proper, regular conditional distributions. *Annals Probability* **3**, 741–752.

Blackwell, D. and Girshick, M. A. (1954). *Theory of Games and Statistical Decisions.* John Wiley, New York.

Blackwell, D. and Ramamoorthi, R. V. (1982). A Bayes but not classically sufficient statistic. *Annals of Statistics* **10**, 1025–1026.

Blair, R. C. and Higgins, J. J. (1980). A comparison of the power of Wilcoxon's rank-sum statistic to that of Student's t-statistic under various nonnormal distributions. *Journal of Educational Statistics* **5**, 309–335.

Blyth, C. R. (1970). On the inference and decision models of statistics (with discussion). *Annals of Statistics* **41**, 1034–1058.

Blyth, C. R. (1984). Approximate binomial confidence limits. Queen's Math. Preprint 1984–6, Queens' Univ., Kingston, Ontario.

Blyth, C. R. and Hutchinson, D. W. (1960). Tables of Neyman—shortest confidence intervals for the binomial parameter. *Biometrika* **47**, 481–491.

Blyth, C.R. and Staudte, R. (1995). Estimating statistical hypotheses. *Statistics and Probability Letters* **23**, 45–52.

Blyth, C.R. and Staudte, R. (1997). Hypothesis estimates and acceptability profiles for 2 × 2 contingency tables. *Journal of the American Statistical Association* **92**, 694–699.

Blyth, C. R. and Still, H. A. (1983). Binomial confidence intervals. *Journal of the American Statistical Association* **78**, 108–116.

Bohrer, R. (1973). An optimality property of Scheffé bounds. *Annals of Statistics* **1**, 766–772.

Bondar, J. V. (1977). A conditional confidence principle. *Annals of Mathematical Statistics* **5**, 881–891.

Bondar, J. V. and Milnes, P. (1981). Amenability: A survey for statistical applications of Hunt–Stein and related conditions on groups. *Zeitschrift für Wahrscheinlichkeitstheorie und verwandte Gebiete* **57**, 103–128.

Bondar, J. V. and Milnes, P. (1982). A converse to the Hunt-Stein theorem. Unpublished.

Bondessen, L. (1983). Equivariant estimators. in *Encyclopedia of Statistical Sciences*, Vol. 2. John Wiley, New York.

Boos, D. (1982). A test for asymmetry associated with the Hodges–Lehmann estimator. *Journal of the American Statistical Association* **77**, 647–651.

Boos, D. and Brownie, C. (1989). Bootstrap methods for testing homogeneity of variances. *Technometrics* **31**, 69–82.

Boos, D. and Hughes-Oliver, J. (1998). Applications of Basu's theorem. *American Statistician* **52**, 218–221.

Boschloo, R. D. (1970). Raised conditional level of significance for the 2×2 table when testing the equality of two probabilities. *Statistica Neerlandica* **24**, 1–35.

Bowker, A. H. (1948). A test for symmetry in contingency tables. *Journal of the American Statistical Association* **43**, 572–574.

Box, G. E. P. (1953). Non-normality and tests for variances. *Biometrika* **40**, 318–335.

Box, G. E. P. and Andersen, S. L. (1955). Permutation theory in the derivation of robust criteria and the study of departures from assumptions. *Journal of the Royal Statistical Society Series B* **17**, 1–34.

Box, G. E. P. and Cox, D. R. (1964). An analysis of transformations. *Journal of the Royal Statistical Society Series B* **26**, 211–252.

Box, G. E. P. and Cox, D. R. (1982). An analysis of transformations revisited, rebutted. *Journal of the American Statistical Association* **77**, 209–210.

Box, G. E. P., Hunter, W. G., and Hunter, J. S. (1978). *Statistics for Experimenters.* John Wiley, New York.

Box, G. E. P. and Tiao, G. C. (1964). A note on criterion robustness and inference robustness. *Biometrika* **51**, 169–173.

Box, G. E. P. and Tiao, G. C. (1973). *Bayesian Inference in Statistical Analysis.* Addison–Wesley, Reading, Mass.

Box, J. F. (1978). *R. A. Fisher: The Life of a Scientist.* John Wiley, New York.

Brain, C. W. and Shapiro, S. S. (1983). A regression test for exponentiality: Censored and complete samples. *Technometrics* **25**, 69–76.

Braun, H. (Ed.) (1994). *The collected works of John W. Tukey: Vo. VIII Multiple comparisons: 1948–1983.* Chapman & Hall, New York.

Bretagnolle, J. (1983). Limites du bootstrap de ceraines fonctionnelles. *Annals of the Institute Henri Poincaré* **3**, 281–296.

Brockwell, P. J. and Davis, R. A. (1991). *Time Series: Theory and Models*, 2nd edition. Springer, New York.

Broemeling, L. D. (1985). *Bayesian Analysis of Linear Models*. Marcel Dekker, New York.

Bross, I. D. J. and Kasten, E. L. (1957). Rapid analysis of 2×2 tables. *Journal of the American Statistical Association* **52**, 18–28.

Brown, K. G. (1984). On analysis of variance in the mixed model. *Annals of Statistics* **12**, 1488–1499.

Brown, L. D. (1964). Sufficient statistics in the case of independent random variables. *Annals of Mathematical Statistics* **35**, 1456–1474.

Brown, L. D. (1966). On the admissibility of invariant estimators of one or more location parameters. *Annals of Mathematical Statistics* **37**, 1087–1136.

Brown, L. D. (1967). The conditional level of Student's t-test. *Annals of Mathematical Statistics* **38**, 1068–1071.

Brown, L. D. (1978). An extension of Kiefer's theory of conditional confidence procedures. *Annals of Statistics* **6**, 59–71.

Brown, L. D. (1986). *Fundamentals of Statistical Exponential Families (With Application to Statistical Decision Theory)*. Institute of Statistical Mathematics Lecture Notes Monograph Series, **9**, Hayward, CA.

Brown, L. D. (1990). An ancillarity paradox which appears in muliple linear regression (with discussion). *Annals of Statistics* **18**, 471–538.

Brown, L. D. (1994). Minimaxity, more or less. In *Statistical Decision Theory and Related Topics V*, Gupta and Berger (eds.), 1–18. Springer-Verlag, New York.

Brown, L. D. (2000). Statistical decision theory. *Journal of the American Statistical Association* **95**, 1277–1281.

Brown, L. D., Cai, T. and DasGupta, A. (2001). Interval estimation for a binomial proportion. *Statistical Science* **16**, 101–133.

Brown, L. D., Cai, T. and DasGupta, A. (2002). Confidence intervals for a binomial proportion and asymptotic expansions. *Annals of Statistics* **30**, 160–201.

Brown, L. D., Casella, G. and Hwang, J. (1995). Optimal confidence sets, bioequivalence, and the limacon of Pascal. *Journal of the American Statistical Association* **90**, 880–889.

Brown, L. D., Cohen, A., and Strawderman, W. E. (1976). A complete class theorem for strict monotone likelihood ratio with applications. *Annals of Statistics* **4**, 712–722.

Brown, L. D., Johnstone, I. M. and MacGibbon, K. G. (1981). Variation diminishing transformations: A direct approach to total positivity and its statistical applications. *Journal of the American Statistical Association* **76**, 824–832.

Brown, L. D., Hwang, J. and Munk, A. (1997). An unbiased test for the bioequivalence problem. *Annals of Statistics* **25**, 2345–2367.

Brown, L. D. and Marden, J. (1989). Complete class results for hypothesis testing problems with simple null hypotheses. *Annals of Statistics* **17**, 209–235.

Brown, L. D. and Sackrowitz, H. (1984). An alternative to Student's t-test for problems with indifference zones. *Annals of Statistics* **12**, 451–469.

Brown, M. B. and Forsythe, A. (1974a). The small sample behavior of some statistics which test the equality of several means. *Technometrics* **16**, 129–132.

Brown, M. B. and Forsythe, A. (1974b). Robust tests for the equality of variances. *Journal of the American Statistical Association* **69**, 364–367.

Brownie, C. and Kiefer, J. (1977). The ideas of conditional confidence in the simplest setting. *Comm. Statist.* **A6**(10.8), 691–751.

Buehler, R. (1959). Some validity criteria for statistical inferences. *Annals of Mathematical Statistics* **30**, 845–863. [The first systematic treatment of relevant subsets, including Example 10.4.1.]

Buehler, R. (1982). Some ancillary statistics and their properties. *Journal of the American Statistical Association* **77**, 581–589. [A review of the principal examples of ancillaries.]

Buehler, R. (1983). Fiducial inference. In *Encyclopedia of Statistical Sciences*, Vol. 3, John Wiley, New York

Buehler, R. and Feddersen, A. P. (1963). Note on a conditional property of Student's t. *Annals of Mathematical Statistics* **34**. 1098–1100.

Burkholder, D. L. (1961). Sufficiency in the undominated case. *Annals of Mathematical Statistics* **32**, 1191–1200.

Cabaña, A. and Cabaña, E. (1997). Transformed empirical processes and modified Kolmogorov-Smirnov tests for multivariate distributions. *Annals of Statistics* **25**, 2388–2409.

Casella, G. (1987). Conditionally acceptable recentered set estimators. *Annals of Statistics* **15**, 1363–1371.

Casella, G. (1988). Conditionally acceptable frequentist solutions (with discussion). In *Statistical Decision Theory and Related Topis IV* **1**, 73–117.

Castillo, J. and Puig, P. (1999). The best test of exponentiality against singly truncated normal alternatives. *Journal of the American Statistical Association* **94**, 529–532.

Chambers, E. A. and Cox, D. R. (1967). Discrimination between alternative binary response models. *Biometrika* **54**, 573–578.

Chatterjee, S., Hadi, A. and Price, B. (2000). *Regression Analysis By Example*, 3rd edition. John Wiley, New York.

Chebychev, P. (1890). Sur deux théoremes relatifs aux probabilitiés. *Acta. Math.* **14**, 305-315.

Chen, L. (1995). Testing the mean of skewed distributions. *Journal of the American Statistical Association* **90**, 767–772.

Chernoff, H. (1949). Asymptotic studentization in testing of hypotheses. *Annals of Mathematical Statistics* **20**, 268–278.

Chernoff, H. (1954). On the distribution of the likelihood ratio statistic. *Annals of Mathematical Statistics* **25**, 579–586.

Chernoff, H. and Lehmann, E. L. (1954). The use of maximum likelihood estimates in χ^2 goodness of fit. *Annals of Mathematical Statistics* **25**, 579–586.

Chhikara, R. S. (1975). Optimum tests for the comparison of two inverse Gaussian distribution means. *Australian Journal of Statistics* **17**, 77–83.

Chhikara, R. S. and Folks, J. L. (1976). Optimum test procedures for the mean of first passage time distribution in Brownian motion with positive drift. *Technometrics* **18**, 189–193.

Chmielewski, M. A. (1981). Elliptically symmetric distributions: A review and bibliography. *International Statistical Review* **49**, 67–74.

Choi, K. and Marden, J. (1997). An approach to multivariate rank tests in multivariate analysis of variance. *Journal of the American Statistical Association* **92**, 1581–1590.

Choi, S., Hall, W. and Schick, A. (1996). Asymptotically uniformly most powerful tests in parametric and semiparametric models. *Annals of Statistics* **24**, 841–861.

Chou, Y. M., Arthur, K. H., Rosenstein, R. B., and Owen, D. B. (1984). New representations of the noncentral chi-square density and cumulative. *Communications in Statistics – Theory and Methods* **13**, 2673–2678.

Christensen, R. (1989). Lack-of-fit tests based on near or exact replicates. *Annals of Statistics* **17**, 673–683.

Christensen, R. (2000). Linear and loglinear models. *Journal of the American Statistical Association* **95**, 1290–1293.

Cima, J. A. and Hochberg, Y. (1976). On optimality criteria in simultaneous interval estimation. *Communications in Statistics – Theory and Methods* **A5**, 875–882.

Clinch, J. C. and Kesselman, H. J. (1982). Parametric alternatives to the analysis of variance. *Journal of Educational Statistics* **7**, 207–214.

Cochran, W. G. (1968). Errors of measurement in statistics. *Technometrics* **10**, 637–666.

Cohen, A. (1972). Improved confidence intervals for the variance of a normal distribution. *Journal of the American Statistical Association* **67**, 382–387.

Cohen, A., Gatsonis, C., and Marden, J. (1983). Hypothesis tests and optimality properties in discrete multivariate analysis. In *Studies in Econometrics, Time Series, and Multivariate Statistics* (Karlin et al., eds.), 379–405. Academic Press, New York.

Cohen, A., Kemperman, J. and Sackrowitz, H. (1994). Unbiased testing in exponential family regression. *Annals of Statistics* **22**, 1931–1946.

Cohen, A. and Marden, J. (1989). On the admissibility and consistency of tests for homogeneity of variances. *Annals of Statistics* **17**, 236–251.

Cohen, A. and Miller, J. (1976). Some remarks on Scheffés two-way mixed model. *American Statistician* **30**, 36–37.

Cohen, A. and Sackrowitz, H. (1975). Unbiasedness of the chi-square, likelihood ratio and other goodness of fit tests for the equal cell case. *Annals of Statistics* **3**, 959–964.

Cohen, A. and Sackrowitz, H. (1992). Improved tests for comparing treatments against a control and other one-sided problems. *Journal of the American Statistical Association* **87**, 1137–1144.

Cohen, A. and Strawderman, W. E. (1973). Admissibility implications for different criteria in confidence estimation. *Annals of Statistics* **1**, 363–366.

Cohen, J. (1962). The statistical power of abnormal-social psychological research: A review. *J. Abnormal and Soc. Psychology* **65**, 145–153.

Cohen, J. (1977). *Statistical Power Analysis for the Behavioral Sciences*, revised edition. Academic Press, New York. [Advocates the consideration of power attainable against the alternatives of interest, and provides the tables needed for this purpose for some of the most common tests.]

Cohen, L. (1958). On mixed single sample experiments. *Annals of Mathematical Statistics* **29**, 947–971.

Conover, W. J., Johnson, M. E. and Johnson, M. M. (1981). A comparative study of tests for homogeneity of variances, with applications to the outer continental shelf bidding data. *Technometrics* **23**, 351–361.

Cox, D. R. (1958). Some problems connected with statistical inference. *Annals of Mathematical Statistics* **29**, 357–372.

Cox, D. R. (1959). *Planning of Experiments.* John Wiley, New York.

Cox, D. R. (1961). Tests of separate families of hypotheses. In *Proc. 4th Berkeley Symp.*, Vol. 1, 105–123.

Cox, D. R. (1962). Further results on tests of separate families of hypotheses. *Journal of the Royal Statistical Society Series B* **24**, 406–423.

Cox, D. R. (1966). A simple example of a comparison involving quantal data. *Biometrika* **53**, 215–220.

Cox, D. R. (1970). *The Analysis of Binary Data*, Methuen, London. [An introduction to the problems treated in Sections 4.6-4.7 and some of their extensions.]

Cox, D. R. (1971). The choice between ancillary statistics. *Journal of the Royal Statistical Society* (*B*) **33**, 251–255.

Cox, D. R. (1977). The role of significance tests. *Scandinavian Journal of Statistics* **4**, 49–62.

Cramér, H. (1928). On the composition of elementary errors. *Skand. Aktuarietidskr.* **11**, 13-74, 141-186.

Cramér, H. (1937). *Random Variables and Probability Distributions.* Cambridge University Press, Cambridge.

Cramér, H. (1946). *Mathematical Methods of Statistics.* Princeton University Press.

Cressie, N. (1980). Relaxing assumptions in the one-sample t-test. *Australian Journal of Statistics* **22**, 143–153.

Csörgö, S. and Mason, D. (1989). Bootstrap empirical functions. *Annals of Statistics* **17**, 1447–1471.

Cvitanic, J. and Karatzas, I. (2001). Generalized Neyman-Pearson lemma via convex duality. *Bernoulli* **7**, 79–97.

Cyr, J. L. and Manoukian, E. B. (1982). Approximate critical values with error bounds for Bartlett's test of homogeneity of variances for unequal sample sizes. *Communications in Statistics – Theory and Methods* **11**, 1671–1680.

D'Agostino, R. (1982). Departures from normality, tests for. In *Encycl. Statist. Sci.* Vol. 2. John Wiley, New York.

D'Agostino, R. and Stephens, M. A. (1986). *Goodness-of-Fit Techniques*. Marcel Dekker, New York.

Dantzig, G. B. and Wald, A. (1951). On the fundamental lemma of Neyman and Pearson. *Annals of Mathematical Statistics* **22**, 87–93.

Darmois, G. (1935). Sur les lois de probabilite a estimation exhaustive. *C.R. Acad. Sci. Paris* **260**, 1265–1266.

DasGupta, A. (1991). Diameter and volume minimizing confidence sets in Bayes and classical problems. *Annals of Statistics* **19**, 1225–1243.

Davenport, J. M. and Webster, J. T. (1975). The Behrens-Fisher problem. An old solution revisited. *Metrika* **22**, 47–54.

David, H. A. (1981). *Order Statistics*, 2nd edition. John Wiley, New York.

Davison, A. and Hinkley, D. (1997). *Bootstrap Methods and their Application.* Cambridge University Press, Cambridge.

Dawid, A. P. (1975). On the concepts of sufficiency and ancillarity in the presence of nuisance parameters. *Journal of the Royal Statistical Society Series B* **37**, 248–258.

Dawid, A. P. (1977). Discussion of Wilkinson: On resolving the controversy in statistical inference. *Journal of the Royal Statistical Society* **39**, 151–152. [Problem 10.12.]

Dayton, C. (2003). Information criteria for pairwise comparisons. *Psychological Methods* **8**, 61–71.

de Leeuw, J. (1992). Introduction to Akaike's (1973) paper "Information theory and an extension of the maximum likelihood principle". Appeared in *Breakthroughs in Statistics*, volume I, Kotz, S. and Johnson, N. L. eds., Springer-Verlag, New York.

de Moivre, A. (1733). *The Doctrine of Chances*, 3rd edition (1756) has been reprinted by Chelsea, New York (1967).

Dempster, A. P. (1958). A high dimensional two-sample significance test. *Annals of Mathematical Statistics* **29**, 995–1010.

Deshpande, J. V. (1983). A class of tests for exponentiality against increasing failure rate average alternatives. *Biometrika* **70**, 514–518.

Deuchler, G. (1914). Ueber die Methoden der Korrelationsrechnung in der Paedagogik und Psychologic. *Z. Pädag. Psychol.* **15**, 114–131, 145–159, 229–242.

Devroye, L. (1986). *Non-Uniform Random Variate Generation.* Springer-Verlag, New York.

de Wet, T. and Randles, R. (1987). On the effect of substituting parameter estimators in limiting χ^2 U and V statistics. *Annals of Statistics* **15**, 398–412.

Diaconis, P. (1988). *Group representations in probability and statistics.* IMS Lecture Notes, **11**, Institute of Statistical Mathematics, Hayward, CA.

Diaconis, P. and Efron, B. (1985). Testing for independence in a two-way table. New interpretations of the chi-square statistic (with discussion). *Annals of Statistics* **13**, 845–913.

Diaconis, P. and Holmes, S. (1994). Gray codes for randomization procedures. *Statistics and Computing* **4**, 287-302.

DiCiccio, T., Hall, P., and Romano, J. P (1991). Empirical likelihood is Bartlett-correctable. *Annals of Statistics* **19**, 1053–1061.

DiCiccio, T. and Romano, J. P. (1989). The automatic percentile method: accurate confidence limits in parametric models. *Canadian Journal of Statistics* **17**, 155–169.

DiCiccio, T. and Romano, J. (1990). Nonparametric confidence limits by resampling and least favorable distributions. *International Statistical Review* **58**, 59–76.

DiCiccio, T. and Stern, S. (1994). Frequentist and Bayesian Bartlett correction of test statistics based on adjusted profile likelihoods. *Journal of the Royal Statistical Society Series B* **56**, 397–408.

Dobson, A. (1990). *An Introduction to Generalized Linear Models.* Chapman & Hall, London.

Doksum, K. A. and Yandell, B. S. (1984). Tests for exponentiality. In *Handbook of Statistics* (Krishnaiah and Sen, editors), Vol. 4, 579–611.

Donoghue, J. (2004). Implementing Shaffer's multiple comparison procedure for a large number of groups. To appear in *Recent Developments in Multiple Comparison Procedures*, IMS Lecture Notes Monograph Series.

Donoho, D. (1988). One-sided inference about functionals of a density. *Annals of Statistics* **16**, 1390–1420.

Draper, D. (1981). *Rank-Based Robust Analysis of Linear Models*, Ph.D. Thesis, Dept. of Statistics, University of California. Berkeley.

Draper, D. (1983). *Rank-Based Robust Analysis of Linear Models. I. Exposition and Background*, Tech. Report No. 17, Dept. of Statistics, University of California, Berkeley.

Drost, F. (1988). *Asymptotics for Generalized Chi-Square Goodness-of-Fit Tests.* Centrum voor Wiskunde en Informatica **48**, Amsterdam.

Drost, F. (1989). Generalized chi-square goodness-of-fit tests for location-scale models when the number of classes tends to infinity. *Annals of Statistics* **17**, 1285–1300.

Dudley, R. (1989). *Real Analysis and Probability.* Wadsworth, Belmont.

Dudoit, S., Shaffer, J. P. and Boldrick, J. (2003). Multiple hypothesis testing in microarray experiments. *Statistical Science* **18**, 71–103.

Dümbgen, L. (1998). New goodness-of-fit tests and their application to nonparametric confidence sets. *Annals of Statistics* **26**, 288–314.

Duncan, D. B. (1955). Multiple range and multiple F-tests. *Biometrics* **11**, 1–42.

Durbin, J. (1970). On Bimbaum's theorem on the relation between sufficiency, conditionality, and likelihood. *Journal of the American Statistical Association* **65**, 395–398.

Durbin, J. (1973). *Distribution theory for tests based on the sample distribution function.* SIAM Philadelphia, PA.

Durbin, J. and Knott, M. (1972). Components of Cramér-von Mises statistics. Part I. *Journal of the Royal Statistical Society B* **34**, 290–307.

Dvoretzky, A., Kiefer, J. and Wolfowitz, J. (1953). Sequential decision problems for processes with continuous time parameter. Testing hypotheses. *Annals of Mathematical Statistics* **24**, 254–264.

Dvoretzky, A., Kiefer, J. and Wolfowitz. J. (1956). Asymptotic minimax character of the sample distribution function and the classical multinomial estimator. *Annals of Mathematical Statistics* **27**, 642–669.

Dvoretzky, A., Wald, A. and Wolfowitz, J. (1951). Elimination of randomization in certain statistical decision procedures and zero-sum two-person games. *Annals of Mathematical Statistics* **22**, 1–21.

Eaton, M. (1983). *Multivariate Statistics.* John Wiley, New York.

Eaton, M. (1989). *Group Invariance Applications in Statistics.* Institute of Statistical Mathematics, Hayward, CA.

Edelman, D. (1990). An inequality of optimal order for the tail probabilities of the T statistic under symmetry. *Journal of the American Statistical Association* **85**, 120–122.

Edgeworth, F. Y. (1885). *Methods of Statistics*, Jubilee volume of the Statist. Soc., E. Stanford, London.

Edgeworth, F. Y. (1905). The law of error. *Proc. Camb. Philos. Soc.* **20**, 36–45.

Edgeworth F. Y. (1908–09). On the probable errors of frequency constants. *J. Roy. Statist. Soc.* **71**, 381–397, 499–512, 651–678; **72**, 81–90. [Edgeworth's work on maximum-likelihood estimation and its relation to the results of Fisher in the same area is reviewed by Pratt (1976). Stigler (1978) provides a systematic account of Edgeworth's many other important contributions to statistics.]

Edgington, E. S. (1995). *Randomization Tests*, 3rd edition. Marcel Dekker, New York.

Edwards, A. W. F. (1963). The measure of association in a 2×2 table. *Journal of the Royal Statistical Society Series B* **126**,109–114.

Edwards, A. W. F. (1983). Fiducial distributions. In *Encycl. of Statist. Sci.*, Vol. 3. John Wiley, New York.

Efron, B. (1969). Student's t-test under symmetry conditions. *Journal of the American Statistical Association* **64**, 1278–1302.

Efron, B. (1979). Bootstrap methods: Another look at the jackknife. *Annals of Statistics* **7**, 1–26.

Efron, B. (1981). Nonparametric standard errors and confidence intervals (with discussion). *Canadian Journal of Statistics* **9**, 139–172.

Efron, B. (1982). *The Jackknife, the Bootstrap and Other Resampling Plans.* SIAM, Philadelphia.

Efron, B. and Tibshirani, R. (1993). *An Introduction to the Bootstrap.* Chapman & Hall, New York.

Elfving, G. (1952). Sufficiency and completeness. *Ann. Acad. Sci. Fennicae (A)*, No. 135.

Engelhardt, M. and Bain, L. J. (1977). Uniformly most powerful unbiased tests on the scale parameter of a gamma distribution with a nuisance shape parameter. *Technometrics* **19**, 77–81.

Engelhardt, M. and Bain, L. J. (1978). Construction of optimal unbiased inference proceures for the parameters of the gamma distribution. *Technometrics* **20**, 485–489.

Eubank, R. (1997). Testing goodness of fit with multinomial data. *Journal of the American Statistical Association* **92**, 1084–1093.

Eubank, R. and LaRiccia, V. (1992). Asymptotic comparison of Cramér-von Mises and nonparametric function estimation techniques for testing goodness-of-fit. *Annals of Statistics* **20**, 2071-2086.

Falk, M. and Kohne, W. (1984). A robustification of the sign test under mixing conditions. *Annals of Statistics* **12**, 716–729.

Fan, J. (1996). Test of significance based on wavelet thresholding and Neyman's truncation. *Journal of the American Statistical Association* **91**, 674–688.

Fan, J. and Lin, S. (1998). Test of significance when data are curves. *Journal of the American Statistical Association* **93**, 1007–1021.

Fan, J., Zhang, C. and Zhang, J. (2001). Generalized likelihood ratio statistics and Wilks phenomenon. *Annals of Statistics* **29**, 153–193.

Faraway, J. and Sun, J. (1995). Simultaneous confidence bands for linear regression with heteroscedastic errors. *Journal of the American Statistical Association* **90**, 1094–1098.

Farrell, R. (1985a). *Multivariate Calculation: Use of the Continuous Groups.* Springer, Berlin.

Farrell, R. (1985b). *Techniques of Multivariate Calculation.* Springer, Berlin.

Fears, T., Benichou, J. and Gail, M. (1996). A reminder of the fallibility of the Wald statistic. *American Statistician* **50**, 226–227.

Feller, W. (1948). On the Kolmogorov–Smirnov limit theorems for empirical distributions. *Annals of Statistics* **19**, 177–189.

Feller, W. (1968). *An Introduction to Probability Theory and its Applications*, 3rd edition, Vol. 1. John Wiley, New York.

Feller, W. (1971). *An Introduction to Probability Theory and its Applications*, Vol. 2, 2nd edition. John Wiley, New York.

Fenstad, G. U. (1983). A comparison between the U and V tests in the Behrens-Fisher problem. *Biometrika* **70**, 300–302.

Ferguson, T. S. (1967). *Mathematical Statistics: A Decision Theoretic Approach.* Academic Press, New York.

Ferguson, T. S. (1996). *A Course in Large Sample Theory.* Chapman & Hall, New York.

Fienberg, S. (1980). *The Analysis of Cross-Classified Categorical Data*, 2nd edition. MIT Press, Cambridge, Massachusetts.

Fienberg, S. and Tanur, J. (1996). Reconsidering the fundamental contributions of Fisher and Neyman in experimentation and sampling. *International Statistical Review* **64**, 237-253.

Finch, P. D. (1979). Description and analogy in the practice of statistics (with discussion). *Biometrika* **66**, 195–208.

Finner, H. (1994). Two-sided tests and one-sided confidence bounds. *Annals of Statistics* **22**, 1502–1516.

Finner, H. (1999). Stepwise multiple test procedures and control of directional errors. *Annals of Statistics* **27**, 274–289.

Finner, H. and Roters, M. (1998). Asymptotic comparison of step-down and step-up multiple test procedures based on exchangeable test statistics. *Annals of Statistics* **26**, 505–524.

Finner, H. and Roters, M. (2001). On the false discovery rate and expected type I errors, *Biometric Journal* **43**, 995-1005.

Finney, D. J. (1948). The Fisher–Yates test of significance in 2×2 contingency tables. *Biometrika* **35**, 145–156.

Finney, D. J., Latscha, R., Bennett, B., Hsu, P. and Horst, C. (1963, 1966). *Tables for Testing Significance in a* 2×2 *Contingency Table*, Cambridge U.P.

Fisher, R. A. (1922). On the mathematical foundations of theoretical statistics. *Phil. Trans. Roy. Soc. London Series A* **222**, 309–368.

Fisher, R. A. (1924). The conditions under which chi square measures the discrepancy between observation and hypothesis. *Journal of the Royal Statistical Society* **87**, 442–450.

Fisher, R.A. (1925a). Theory of statistical estimation. *Proc. Cambridge Phil. Soc.* **22**, 700–725. [These papers develop a theory of point estimation (based on the maximum likelihood principle) and the concept of sufficiency. The factorization theorem is given in a form which is formally weaker but essentially equivalent to (1.20). First use of term ancillary.]

Fisher, R. A. (1925b). *Statistical Methods for Research Workers*, 1st edition (14th edition, 1970), Oliver and Boyd, Edinburgh.

Fisher, R. A. (1928a). The general sampling distribution of the multiple correlation coefficient. *Proc. Roy. Soc. Series A* **121**, 654–673. [Derives the noncentral χ^2- and noncentral beta-distributions and the distribution of the sample multiple correlation coefficient for arbitrary values of the population multiple correlation coefficient.]

Fisher R. A. (1928b). On a property connecting the χ^2 measure of discrepancy with the method of maximum likelihood. Atti de Congresso Internazionale dei Mathematici, Bologna **6**, 94–100.

Fisher, R. A. (1930). Inverse probability. *Proc. Cambridge Philos. Soc.* **26**, 528–535.

Fisher, R. A. (1934a). *Statistical Methods for Research Workers*, 5th and subsequent eds., Oliver and Boyd, Edinburgh, Section 21.02. [Proposes the conditional tests for the hypothesis of independence in a 2 × 2 table.]

Fisher, R. A. (1934b). Two new properties of mathematical likelihood. *Proc. Roy. Soc. (A)* **144**, 285–307. [Introduces the idea of conditioning on ancillary statistics and applies it to the estimation of location parameters.]

Fisher R. A. (1935a). *The Design of Experiments*, 1st edition (8th edition, 1966). Oliver and Boyd, Edinburgh. [Contains the basic ideas concerning permutation tests. In particular, points out how randomization provides a basis for inference and proposes the permutation version of the t-test as not requiring the assumption of normality.]

Fisher R. A. (1935b). The logic of inductive inference (with discussion). *Journal of the Royal Statistical Society* **98**, 39–82.

Fisher R. A. (1936). Uncertain inference. *Proc. Amer. Acad. Arts and Sci.* **71**, 245–258.

Fisher, R. A. (1956a). On a test of significance in Pearson's Biometrika tables (No. 11). *Journal of the Royal Statistical Society (B)* **18**, 56–60. (See also the discussion of this paper by Neyman, Bartlett, and Welch in the same volume, pp. 288–302.) [Exhibits a negatively biased relevant subset for the Welch–Aspin solution of the Behren–Fisher problem.]

Fisher, R. A. (1956b, 1959, 1973). *Statistical Methods and Scientific Inference.* Oliver and Boyd, Edinburgh (1956, 1959); Hafner, New York (1973). [In Chapter IV the author gives his views on hypothesis testing and in particular discusses his ideas on the Behrens-Fisher problem. Contains Fisher's last comprehensive statement of his views on many topics, including ancillarity and the Behrens–Fisher problem.]

Fisher, R. A. (1971–1973). *Collected Papers* (J. H. Bennett, ed.), University of Adelaide.

Fisher, R. A. (1973). *Statistical Methods and Scientific Inference*, 3rd edition, Hafner, New York.

Folks, J. L. and Chhikara, R. S. (1978). The inverse Gaussian distribution and its statistical applications—a review (with discussion). *Journal of the Royal Statistical Society Series B* **40**, 263–289.

Forsythe, A. and Hartigan, J. A. (1970). Efficiency of confidence intervals generated by repeated subsample calculations. *Biometrika* **57**, 629–639.

Fourier, J. B. J. (1826). *Recherches Statistiques sur la Ville de Paris el le Départment de la Seine, Vol.* 3.

Franck, W. E. (1981). The most powerful invariant test of normal versus Cauchy with applications to stable alternatives. *Journal of the American Statistical Association* **76**, 1002–1005.

Fraser, D. A. S. (1953). *Canadian Journal of Mathematics* **6**, 42–45.

Fraser, D. A. S. (1956). Sufficient statistics with nuisance parameters. *Annals of Mathematical Statistics* **27**, 838–842.

Fraser, D. (1996). Comment on "Pivotal inference and the fiducial argument." *International Statistical Review* **64**, 231-235.

Freedman, D. and Lane, D. (1982). Significance testing in a nonstochastic setting. In *Festschrift for Erich L. Lehmann* (Bickel. Doksum, and Hodges, eds.), Wadsworth, Belmont, Calif.

Freeman, M. F. and Tukey, J. W. (1950). Transformations related to the angular and the square root. *Annals of Mathematical Statistics* **21**, 607–611.

Freiman, J. A., Chalmers, T. C., Smith, H. and Kuebler, R. R. (1978). The importance of beta, the type II error and sample size in the design and interpretation of the randomized control trial. *New England Journal of Medicine* **299**, 690–694.

Frisén, M. (1980). Consequences of the use of conditional inference in the analysis of a correlated contingency table. *Biometrika* **67**, 23–30.

Fuller, W. (1996). *Introduction to Statistical Time Series*, 2nd Edition, John Wiley, New York.

Gabriel, K. R. (1964). A procedure for testing the homogeneity of all sets of means in analysis of variance. *Biometrics* **20**, 459–477.

Gabriel, K. R. and Hall W. J. (1983). Rerandomization inference on regression and shift effects: Computationally feasible methods. *Journal of the American Statistical Association* **78**, 827–836.

Gabriel, K. R. and Hsu, C. F. (1983). Evaluation of the power of rerandomization tests, with application to weather modification experiments. *Journal of the American Statistical Association* **78**, 766–775.

Galambos, J. (1982). Exponential distribution. In *Encycl. Statist. Sci.*, Vol. 2, John Wiley, New York.

Gan, L. and Jiang, J. (1999). A test for global maximum. *Journal of the American Statistical Association* **94**, 847–854.

Garside, G. R. and Mack, C. (1976). Actual type 1 error probabilities for various tests in the homogeneity case of the 2×2 contingency table. *American Statistician* **30**, 18–21.

Gart, J. J. (1970). Point and interval estimation of the common odds ratio in the combination of 2×2 tables with fixed marginals. *Biometrika* **57**, 471–475.

Garthwaite, P. (1996). Confidence intervals from randomization tests. *Biometrics* **52**, 1387–1393.

Gastwirth, J. L. and Rubin, H. (1971). Effect of dependence on the level of some one-sample tests. *Journal of the American Statistical Association* **66**, 816–820.

Gauss, C. F. (1809). *Theoria motus corporum coelestium in sectionibus conicis solem ambientium.* Hamburg.

Gauss, C. F. (1816). Bestimmung der Genauigkeit der Beobachtungen. *Z. Astron. and Verw. Wiss* **1**. (Reprinted in Gauss' collected works, Vol. 4, pp. 109–119.)

Gavarret, J. (1840). *Principes Gènèraux de Statistique Mèdicale*, Paris.

George, E. I. and Casella, G. (1994). An empirical Bayes confidence report. *Statistica Sinica* **4**, 617–638.

Ghosh, J. (1961). On the relation among shortest confidence intervals of different types. *Calcutta Statist. Assoc. Bull.* 147-152.

Ghosh, J., Morimoto, H. and Yamada, S. (1981). Neyman factorization and minimality of pairwise sufficient subfields. *Annals of Statistics* **9**, 514–530.

Ghosh, M. (1948). On the problem of similar regions. *Sankhyā* **8**, 329–338.

Gibbons, J. (1986). Ranking procedures. In *Encycl. Statist. Sci.* **7**, 588–592.

Gibbons, J. (1988). Selection procedures. In *Encycl. Statist. Sci.* **8**, 337-345.

Gibbons, J. and Chakraborti, S. (1992). *Nonparametric statistical inference*, 3rd edition. Marcel Dekker, New York.

Giesbrecht, F. and Gumpertz, M. (2004). *Planning, Construction, and Statistical Analysis of Comparative Experiments.* John Wiley, New York.

Giné, E. (1997). *Lectures on Some Aspects of the Bootstrap.* École d'Été de Calcul de Probabilités de Saint-Flour.

Giné, E. and Zinn, J. (1989). Necessary conditions for the bootstrap of the mean. *Annals of Statistics* **17**, 684–691.

Giri, N. and Kiefer, J. (1964). Local and asymptotic minimax properties of multivariate tests. *Annals of Mathematical Statistics* **35**, 21–35.

Giri, N., Kiefer, J. and Stein, C. M. (1963). Minimax character of Hotelling's T^2 test in the simplest case. *Annals of Mathematical Statistics* **34**, 1524–1535.

Girshick, M. A., Mosteller, F. and Savage, L. J. (1946). Unbiased estimates for certain binomial sampling problems with applications. *Annals of Mathematical Statistics* **17**, 13–23. [Problem 4.12.]

Glaser, R. E. (1976). The ratio of the geometric mean to the arithmetic mean for a random sample from a gamma distribution. *Journal of the American Statistical Association* **71**, 481–487.

Glaser, R. E. (1982). Bartlett's test of homogeneity of variances. *Encycl. Statist. Sci.* **1**, 189–191.

Gleser, L. J. (1985). Exact power of goodness-of-fit tests of Kolmogorov type for discontinuous distributions. *Journal of the American Statistical Association* **80**, 954–958.

Gleser, L. J. and Hwang, J. (1987). The nonexistence of $100(1-\alpha)\%$ confidence sets of finite expected diameter in errors-in-variables and related models. *Annals of Statistics* **15**, 1351–1362.

Gokhale, D. V. and Johnson, N. S. (1978). A class of alternatives to independence in contingency tables. *Journal of the American Statistical Association* **73**, 800–804.

Good, P. (1994). *Permutation Tests, A Practical Guide to Resampling Methods for Testing Hypotheses*. Springer-Verlag, New York.

Goodman, L. A. and Kruskal, W. (1954, 1959). Measures of association for cross classification. *Journal of the American Statistical Association* **49**, 732–764; **54**, 123–163.

Goutis, C. and Casella, G. (1991). Improved invariant confidence intervals for a normal variance. *Annals of Statistics* **19**, 2015–2031.

Goutis, C. and Casella, G. (1992). Increasing the confidence in student's t interval. *Annals of Statistics* **20**, 1501–1513.

Graybill, F. A. (1976). *Theory and Application of the Linear Model.* Duxbury Press, North Scituate, Mass.

Green, B. F. (1977). A practical interactive program for randomization tests of location. *American Statistician* **31**, 37–39.

Greenwood, P. and Nikulin, M. (1996). *A Guide to Chi-Squared Testing.* John Wiley, New York.

Grenander, U. (1981). *Abstract Inference.* John Wiley, New York.

Groeneboom, P. (1980). *Large Deviations and Asymptotic Efficiencies.* Mathematisch Centrum, Amsterdam, The Netherlands.

Groeneboom, P. and Oosterhoof, J. (1981). Bahadur efficiency and small sample efficiency. *International Statistical Review* **49**, 127–141.

Guenther, W. C. (1978). Some remarks on the runs tests and the use of the hypergeometric distribution. *American Statistician* **32**, 71–73.

Gupta, A. and Vermeire, L. (1986). Locally optimal tests for multiparameter hypotheses. *Journal of the American Statistical Association* **81**, 819–825.

Haberman, S. J. (1974). *The Analysis of Frequency Data.* University of Chicago Press.

Haberman, S. J. (1982). Association, Measures of In *Encycl. Statist. Sci.*, Vol. 1. John Wiley, New York, 130–136.

Hájek, J. (1962). Asymptotically most powerful rank order tests. *Annals of Mathematical Statistics* **33**, 1124–1147.

Hájek, J. (1967). On basic concepts of statistics, In *Proc. Fifth Berkeley Symp. Math. Statist. and Probab.*, Univ. of Calif. Press, Berkeley.

Hájek, J. (1970). A characterization of limiting distributions of regular estimates. *Zeitschrift für Wahrscheinlichkeitstheorie und verwandte Gebiete* **14**, 323–330.

Hájek, J. (1972). Local asymptotic minimax and admissibility in estimation. *Proceedings of the Sixth Berkeley Symposium on Mathematical Statistics and Probability* **1**, 175–194.

Hájek, J. and Sidák, Z. (1967). *Theory of Rank Tests.* C.S.A.V. Prague and Academic Press.

Hájek, J. Sidák, Z. and Sen, P. (1999). *Theory of Rank Tests*, 2nd edition. Academic Press, San Diego.

Hald, A. (1990). *A History of Probability and Statistics (and Their Applications Before 1750).* John Wiley, New York.

Hald, A. (1998). *A History of Mathematical Statistics (from 1750 to 1930)*. John Wiley, New York.

Hall, P. (1982). Improving the normal approximation when constructing one-sided confidence intervals for binomial or Poisson parameters. *Biometrika* **69**, 647–652.

Hall, P. (1986). On the bootstrap and confidence intervals. *Annals of Statistics* **14**, 1431–1452.

Hall, P. (1990). Asymptotic properties of the bootstrap for heavy-tailed distributions. *Annals of Probability* **18**, 1342–1360.

Hall, P. (1992). *The Bootstrap and Edgeworth Expansion.* Springer, New York.

Hall, P. and Jing, B. (1995). Uniform coverage bounds for confidence intervals and Berry-Esseen theorems for Edgeworth expansions. *Annals of Statistics* **23**, 363–375.

Hall, P. and Martin, M. (1988). On bootstrap resampling and iteration. *Biometrika* **75**, 661-671.

Hall, P. and Padmanabhan, A. (1997). Adaptive inference and the two sample scale problem. *Technometrics* **39**, 412–422.

Hall, P. and Welsh, A. H. (1983). A test for normality based on the empirical characteristic function. *Biometrika* **70**, 485–489.

Hall, W. and Mathiason, D. (1990). On large-sample estimation and testing in parametric models. *International Statistical Review* **58**, 77–97.

Hall, W., Wijsman, R. and Ghosh, J. (1965). The relationship between sufficiency and invariance with applications in sequential analysis. *Annals of Mathematical Statistics* **36**, 575–614.

Hallin, M., Taniguchi, M., Serroukh, A. and Choy, K. (1999). Local asymptotic normality for regression models with long memory disturbance. *Annals of Statistics* **27**, 2054–2080.

Halmos, P. R. (1974). *Measure Theory.* Springer, New York.

Halmos, P. R. and Savage, L. J. (1949). Application of the Radon–Nikodym theorem to the theory of sufficient statistics. *Annals of Mathematical Statistics* **20**, 225–241. [First abstract treatment of sufficient statistics; the factorization theorem. Problem 10.]

Hamilton, J. D. (1994). *Time Series Analysis.* Princeton University Press, Princeton.

Hartigan, J. A. (1969). Using subsample values as typical values. *Journal of the American Statistical Association* **64**, 1303–1317.

Harville, D. A. (1978). Alternative formulations and procedures for the two-way mixed model. *Biometrics* **34**, 441–454.

Hastie, T. and Tibshirani, R. (1990). *Generalized Additive Models.* Chapman & Hall, London.

Hastie, T. and Tibshirani, R. (1997). Generalized additive models. In *Encycl. Statist. Sci. update* **1**, 261–269.

Haytner, A. and Hsu, J. (1994). On the relationship between stepwise decision procedures and confidence sets. *Journal of the American Statistical Association* **89**, 128–136.

Hedges, L. and Olkin, I. (1985). *Statistical methods for meta-analysis.* Academic Press, Orlando.

Hegazy, Y. A. S. and Green, J. R. (1975). Some new goodness-of-fit tests using order statistics. *Applied Statistics* **24**, 299–308.

Hegemann, V. and Johnson, D. E. (1976). The power of two tests for nonadditivity. *Journal of the American Statistical Association* **71**, 945–948.

Heritier, S. and Ronchetti, E. (1994). Robust bounded-influence tests in general parametric models. *Journal of the American Statistical Association* **89**, 897–904.

Hettmansperger, T. P. (1984). *Statistical Inference Based on Ranks.* John Wiley, New York.

Hettmansperger, T. and McKean, J. (1998). *Robust Nonparametric Statistical Methods.* Arnold, London.

Hettmansperger, T., McKean, J. and Sheather, S. (2000). Robust nonparametric methods. *Journal of the American Statistical Association* **95**, 1308–1312.

Hettmansperger, T., Möttönen, J. and Oja, H. (1997). Affine-invariant multivariate one-sample signed-rank tests. *Journal of the American Statistical Association* **92**, 1591–1600.

Hinkley, D. (1977). Conditional inference about a normal mean with known coefficient of variation. *Biometrika*, **64**, 105–108.

Hinkley, D. and Runger, G. (1984). The analysis of transformed data. (with discussion). *Journal of the American Statistical Association* **79**, 302–320.

Hipp, C. (1974). Sufficient statistics and exponential families. *Annals of Statistics* **2**, 1283–1292.

Hobson, E. W. (1927). *Theory of Functions of a Real Variable*, 3rd edition, Vol. 1. Cambridge University Press, p. 194.

Hochberg, Y. and Tamhane, A. (1987). *Multiple Comparison Procedures.* John Wiley, New York.

Hocking, R. R. (1973). A discussion of the two-way mixed model. *American Statistician* **27**, 148–152.

Hocking, R. R. (2003). *Methods and Applications of Linear Models*, 2nd edition. John Wiley, New York.

Hocking, R. R. and Speed, F. M. (1975). A full rank analysis of some linear model problems. *Journal of the American Statistical Association* **70**, 706–712.

Hodges, J. L., Jr. (1957). The significance probability of the Smirnov two-sample test. *Arkiv für Matematik* **3**, 469–486.

Hodges, J. L., Jr. and Lehmann, E. L. (1954). Testing the approximate validity of statistical hypotheses. *Journal of the Royal Statistical Society Series B* **16**, 261–268.

Hodges, J. L., Jr. and Lehmann, E. L. (1956). The efficiency of some non-parametric competitors of the t-test. *Annals of Mathematical Statistics* **27**, 324–335.

Hodges, J. L., Jr. and Lehmann, E.L. (1970). Deficiency. *Annals of Mathematical Statistics* **41**, 783–801.

Hoeffding W. (1951). 'Optimum' nonparametric tests. in *Proc. 2nd Berkeley Symposium on Mathematical Statistics and Probability*, Univ. of Calif. Press., Berkeley, 83–92.

Hoeffding, W. (1952). The large-sample power of tests based on permutations of observations. *Annals of Mathematical Statistics* **23**, 169–192.

Hoeffding, W. (1956). The role of assumptions in statistical decisions. In *Proc. Third Berkeley Symposium on Mathematical Statistics and Probability*, edited by Neyman, University of California Press, Berkeley, CA.

Hoeffding, W. (1965). Asymptotically optimal tests for multinomial distributions (with discussion). *Annals of Mathematical Statistics* **36**, 369–408.

Hoeffding, W. (1977). Some incomplete and boundedly complete families of distributions. *Annals of Statistics* **5**, 278–291.

Hoel, P. G. (1948). On the uniqueness of similar regions. *Annals of Mathematical Statistics* **19**, 66–71. [Theorem 4.3.1 under regularity assumptions.]

Hogg, R. V. (1972). More light on the kurtosis and related statistics. *Journal of the American Statistical Association* **67**, 422–424.

Holm, S. (1979). A simple sequentially rejective multiple test procedure. *Scandinavian Journal of Statistics* **6**, 65–70.

Holm, S. (1999). Multiple confidence sets based on stagewise tests. *Journal of the American Statistical Association* **94**, 489–495.

Hooper, P. M. (1982a). Sufficiency and invariance in confidence set estimation. *Annals of Statistics* **10**, 549–555.

Hooper, P. M. (1982b). Invariant confidence sets with smallest expected measure. *Annals of Statistics* **10**, 1283–1294.

Hotelling, H. (1931). The generalization of Student's ratio. *Annals of Mathematical Statistics* **2**, 360–378.

Hotelling, H. (1936). Relations between two sets of variates. *Biometrika* **28**, 321–377. [One of the early papers making explicit use of invariance considerations.]

Hotelling, H. (1953). New light on the correlation coefficient and its transforms. *Journal of the Royal Statistical Society Series B* **15**, 193–224.

Hotelling, H. (1961). The behavior of some standard statistical tests under non-standard conditions. *Proceedings of the Fourth Berkeley Symposium of Mathematical Statistics* Prob. **1**, 319–360.

Hsu, C. T. (1940). On samples from a normal bivariate population. *Annals of Mathematical Statistics* **11**, 410–426.

Hsu, J. (1996). *Multiple Comparisons: Theory and Methods.* Chapman & Hall, London.

Hsu, P. (1941). Analysis of variance from the power function stand-point. *Biometrika* **32**, 62–69. [Shows that the test (7.7) is UMP among all tests whose power function depends only on the noncentrality parameter.]

Hsu, P. (1945). On the former function of the E^2-test and the T^2-test. *Annals of Mathematical Statistics* **16**, 278–286. [Obtains a result on best average power for the T^2-test analogous to that of Chapter 7, Problem 7.5.]

Huang, J. S. and Ghosh, M. (1982). A note on strong unimodality of order statistics. *Journal of the American Statistical Association* **77**, 929–930.

Huber, P. J. (1965). A robust version of the probability ratio test. *Annals of Mathematical Statistics* **36**, 1753–1758.

Huber, P. J. (1973). Robust regression: Asymptotics, conjectures and Monte Carlo. *Annals of Statistics* **1**, 799–821. [Obtains the robustness conditions (11.55) and (11.57); related results are given by Eicker (1963).]

Huber, P. J. (1981). *Robust Statistics.* John Wiley, New York.

Huber, P. J. and Strassen, V. (1973, 1974). Minimax tests and the Neyman–Pearson lemma for capacities. *Annals of Statistics* **1**, 251–263; **2**, 223–224.

Hunt, G. and Stein, C. M. (1946). Most stringent tests of statistical hypotheses. [In this paper. which unfortunately was never published, a general theory of invariance is developed for hypothesis testing.]

Hwang, J. and Brown, L. D. (1991). Estimated confidence under the validity constraint. *Annals of Statistics* **19**, 1964–1977.

Hwang, J. and Casella, G. (1982). Minimax confidence sets for the mean of a multivariate normal distribution. *Annals of Statistics* **10**, 868–881.

Hwang, J., Casella, G., Robert,C., Wells, M. and Farrell, R. (1992). Estimation of accuracy in testing. *Annals of Statistics* **20**, 490–509.

Ibragimov, I. and Has'minskii, R. (1981). *Statistical Estimation.* Springer-Verlag, New York.

Ibragimov, J. A. (1956). On the composition of unimodal distributions (Russian). *Teoriya Veroyatnostey* **1**, 283–288; Eng]. transl., *Theor. Probab. Appl.* **1** (1956), 255–260.

Inglot, T., Kallenberg, W. and Ledwina, T. (1997). Data driven smooth tests for composite hypotheses. *Annals of Statistics* **25**, 1222-1250.

Inglot, T., Kallenberg, W. and Ledwina, T. (2000). Vanishing shortcoming and asymptotic relative efficiency. *Annals of Statistics* **28**, 215-238.

Inglot, T. and Ledwina, T. (1996). Asymptotic optimality of data-driven Neyman's tests for uniformity. *Annals of Statistics* **24**, 1982–2019.

Ingster, Y. (1993). Asymptotically minimax hypothesis tests for nonparametric alternatives I, II, III. *Math. Methods Statist.* **2**, 85–114, 171–189, 249–268.

Ingster, Y. and Suslina, I. (2003). *Nonparametric Goodness-of-Fit Testing Under Gaussian Models.* Springer Lecture Notes in Statistics **169**, Springer-Verlag, New York.

Isaacson, S. L. (1951). On the theory of unbiased tests of simple statistical hypotheses specifying the values of two or more parameters. *Annals of Mathematical Statistics* **22**, 217–234. [Introduces type D and E tests.]

Jagers, P. (1980). Invariance in the linear model—an argument for χ^2 and F in nonnormal situations. *Statistics* **11**, 455–464.

James, A. T. (1954). Normal multivariate analysis and the orthogonal group. *Annals of Mathematical Statistics* **25**, 40–75.

James, G. S. (1951). The comparison of several groups of observations when the ratios of the population variances are unknown. *Biometrika* **38**, 324–329.

James, G. S. (1954). Tests of linear hypotheses in univariate and multivariate analysis when the ratios of the population variances are unknown. *Biometrika* **41**, 19–43.

Janssen, A. (1995). Principal component decomposition of non-parametric tests. *Probability Theory and Related Fields* **101**, 193–209.

Janssen, A. (1997). Studentized permutation tests for non-i.i.d. hypotheses and the generalized Behrens-Fisher problem. *Statistics and Probability Letters* **36**, 9–21.

Janssen, A. (1999). Testing nonparametric statistical functionals with applications to rank tests. *Journal of Statistical Planning and Inference* **81**, 71–93, Erratum **92**,

Janssen, A. (2000a). Global power functions of goodness of fit tests. *Annals of Statistics* **28**, 239–253.

Janssen, A. (2000b). Nonparametric bioequivalence for tests for statistical functionals and their efficient power functions. *Statistics and Decisions* **18**, 49–78.

Janssen, A. (2003). Which power of goodness of fit tests can really be expected: intermediate versus contiguous alternatives. *Statistics and Decisions* **21**, 301–325.

Janssen, A. and Pauls, T. (2003). How do bootstrap and permutation tests work? *Annals of Statistics* **31**, 768–806.

Jensen, J. (1993). A historical sketch and some new results on the improved log likelihood ratio statistic. *Scandinavian Journal of Statistics* **20**, 1–15.

Jockel, K. (1986). Finite sample properties and asymptotic efficiency of Monte Carlo tests. *Annals of Statistics* **14**, 336–347.

Johansen, S. (1979). *Introduction to the Theory of Regular Exponential Families*, Lecture Notes, No. 3, Inst. of Math. Statist., University of Copenhagen.

Johansen, S. (1980). The Welch–James approximation to the distribution of the residual sum of squares in a weighted linear regression. *Biometrika* **67**, 85–92.

John, R. D. and Robinson, J. (1983a). Edgeworth expansions for the power of permutation tests. *Annals of Statistics* **11**, 625–631.

John, R. D. and Robinson, J. (1983b). Significance levels and confidence intervals for permutation tests. *Journal of Statistical Computation and Simulation* **16**, 161–173.

Johnson, N. L. and Kotz, S. (1969). *Distributions in Statistics: Discrete Distributions.* Houghton Mifflin, New York.

Johnson, N. L. and Kotz, S. (1970). *Distributions in Statistics: Continuous Univariate Distributions* (2 vols.). Houghton Mifflin,New York.

Johnson, N. L., Kotz, S. and Balakrishnan, N. (1995). *Continuous Univariate Distributions* **2**, 2nd edition. John Wiley, New York.

Johnson, N. L., Kotz, S. and Kemp, A. (1992). *Univariate Discrete Distributions*, 2nd edition. John Wiley, New York.

Joshi, V. (1982). Admissibility. In *Encycl. Statist. Sci.* **1**, 25–29.

Kabe, D. G. and Laurent, A. G. (1981). On some nuisance parameter free uniformly most powerful tests. *Biometrics Journal* **23**, 245–250.

Kakutani, S. (1948). On the equivalence of infinite product measures. *Annals of Mathematical Statistics* **49**, 214–224.

Kalbfleisch, J. D. (1975). Sufficiency and conditionality (with discussion). *Biometrika* **62**, 251–259.

Kallenberg, W. (1982). Chernoff efficiency and deficiency. *Annals of Statistics* **10**, 583–594.

Kallenberg, W. (1983). Intermediate efficiency, theory and examples. *Annals of Statistics* **11**, 170-1-82.

Kallenberg, W. C. M. et al. (1984). *Testing Statistical Hypotheses: Worked Solutions*, CWI Syllabus No. 3, Centrum voor Wiskunde en Informatien, Amsterdam.

Kallenberg, W. and Ledwina, T. (1995). Consistency and Monte Carlo simulation of a data driven version of smooth goodness-of-fit tests. *Annals of Statistics* **23**, 1594–1608.

Kallenberg, W. and Ledwina, T. (1999). Data-driven rank tests for independence. *Journal of the American Statistical Association* **94**, 285–301.

Kallenberg, W. C. M., Oosterhoff J. and Schriever B. F. (1985). The number of classes in chi-squared goodness-of-fit tests. *Journal of the American Statistical Association* **80**, 959–968.

Kanoh, S. and Kusunoki, U. (1984). One sided simultaneous bounds in linear regression. *Journal of the American Statistical Association* **79**, 715–719.

Kappenman, R. F, (1975). Conditional confidence intervals for the double exponential distribution parameters. *Technometries* **17**, 233–235.

Kariya, T. (1981). Robustness of multivariate tests. *Annals of Statistics* **9**, 1267–1275.

Kariya, T. (1985). *Testing in the Multivariate Linear Model.* Kinokuniya, Tokyo.

Kariya, T. and Sinha, B. (1985). Nonnull and optimality robustness of some tests. *Annals of Statistics* **13**, 1182–1197.

Karlin, S. (1957). Pòlya type distributions. II. *Annals of Mathematical Statistics* **28**, 281–308.

Karlin, S. (1968). *Total Positivity*, Vol. I, Stanford U.P. Stanford, Calif. [Properties of TP distributions, including Problems 3.50–3.53.]

Karlin, S. and Rubin, H. (1956). The theory of decision procedures for distributions with monotone likelihood ratio. *Annals of Mathematical Statistics* **27**. 272–299. [General theory of families with monotone likelihood ratio, including Theorem 3.4.2. For further developments of this theory, see Brown, Cohen, and Strawderman (1976).]

Karlin, S. and Taylor, H. (1975). *A First Course in Stochastic Processes*, 2nd ed., Academic Press, San Diego, CA.

Kempthorne, O. (1955). The randomization theory of experimental inference. *Journal of the American Statistical Association* **50**, 946–967.

Kempthorne, P. (1988). Controlling risks under different loss functions: The compromise decision problem. *Annals of Statistics* **16**, 1594-1608.

Kendall, M. G. (1970). *Rank Correlation Methods*, 4th edition. Griffin, London.

Kendall M. G. and Stuart, A. (1979). *The Advanced Theory of Statistics*, 4th edition, Vol. 2. MacMillan, New York.

Kent, J. and Quesenberry, C. P. (1982). Selecting among probability distributions used in reliability. *Technometrics* **24**, 59–65.

Khmaladze, E. (1993). Goodness of fit problem and scanning innovation martingales. *Annals of Statistics.* **21**, 798–829.

Kiefer, J. (1958). On the nonrandomized optimality and randomized nonoptimality of symmetrical designs. *Annals of Mathematical Statistics* **29**, 675–699. [Problem 8.6(ii).]

Kiefer, J. (1977a). Conditional confidence statements and confidence estimators (with discussion). *Journal of the American Statistical Association* **72**, 789–827. [The key paper in Kiefer's proposed conditional confidence approach.]

Kiefer, J. (1977b). Conditional confidence and estimated confidence in multidecision problems (with applications to selections and ranking). *Multivariate Analysis* **IV**, 143–158.

Kiefer, J. and Schwartz, R. (1965). Admissible Bayes character of T^2-, R^2-, and other fully invariant tests for classical multivariate normal problems. *Annals of Mathematical Statistics* **36**, 747–770.

King, M. L. and Hillier, G. H. (1985). Locally best invariance tests of the error covariance matrix of the linear regression model. *Journal of the Royal Statistical Society Series B* **47**, 98–102.

Knight, K. (1989). On the bootstrap of the sample mean in the infinite variance case. *Annals of Statistics* **17**, 1168–1175.

Koehn, U. and Thomas, D. L. (1975). On statistics independent of a sufficient statistic: Basu's Lemma. *American Statistician* **29**, 40–41.

Kolassa, J. and McCullagh, P. (1990). Edgeworth series for lattice distributions. *Annals of Statistics* **18**, 981–985.

Kolmogorov, A. (1933). Sulla adeterminazione empirica di una legge di distribuzione. *Giorn. Inst. Ital. Attuari* **4**, 83–91.

Kolmogorov, A. (1942). Sur l'estimation statistique des paramètres de la loi de Gauss. *Bull. Acad. Sci. URSS Ser. Math.* **6**, 3–32. (Russian–French summary.) [Definition of sufficiency in terms of distributions for the parameters.]

Kolodziejczyk, S. (1935). An important class of statistical hypotheses. *Biometrika* **37**, 161–190. [Discussion of the general linear univariate hypothesis from the likelihood-ratio point of view.]

Koopman, B. (1936). On distributions admitting a sufficient statistic. *Trans. Amer. Math. Soc.* **39**, 399–409.

Korn, E., Troendle, J., McShane, L. and Simon, R. (2004). Controlling the number of false discoveries: Applications to high-dimensional genomic data. *Journal of Statistical Planning and Inference* **124**, 379–398.

Koschat, M. (1987). A characterization of the Fieller solution. *Annals of Statistics* **15**, 462–468.

Koshevnik, Y. and Levit, B. (1976). On a non-parametric analogue of the information matrix. *Theory of Probability and its Applications* **21**, 738–753.

Kotz, S., Wang, Q., and Hung, K. (1990). Interrelations among various definitions of bivariate positive dependence. In *Topics in Statistical Dependence.* Block, Sampson and Savits, eds. (1990). IMS Lecture Notes **16**, Hayward, CA.

Kowalski, J. (1995). Complete classes of tests for regularly varying distributions. *Annals of the Institute of Statistical Mathematics* **47**, 321–350.

Koziol, J. A. (1983). Tests for symmetry about an unknown value based on the empirical distribution function. *Communications in Statistics* **12**, 2823–2846.

Krafft, O. and Witting, H. (1967). Optimale tests under ungünstigsten Verteilungen. *Zeitschrift für Wahrscheinlichkeitstheorie und verwandte Gebiete* **7**, 289–302.

Kraft, C. (1955). Some conditions for consistency and uniform consistency of statistical procedures. *Univ. of Calif. Publ. in Statist.* **2**, 125–142.

Kruskal, W. (1954). The monotonicity of the ratio of two non-central t density functions. *Annals of Mathematical Statistics* **25**, 162–165.

Kruskal, W. H. (1957). Historical notes on the Wilcoxon unpaired two-sample test. *Journal of the American Statistical Association* **52**, 356–360.

Kruskal, W. H. (1978). Significance, Tests of. In *International Encyclopedia of Statistics*, Free Press and Macmillan, New York and London.

Künsch, H. R. (1989). The jackknife and the bootstrap for general stationary observations. *Annals of Statistics* **17**, 1217–1241.

Lahiri, S. N. (2003). *Resampling Methods for Dependent Data.* Springer, New York.

Lambert, D. (1985). Robust two-sample permutation tests. *Annals of Statistics* **13**, 606–625.

Lambert, D. and Hall, W. (1982). Asymptotic lognormality of p-values. *Annals of Statistics* **10**, 44–64.

Laplace, P. S. (1773). Mèmoire sur l'inclinaison moyenne des orbites des comètes. *Mem. Acad. Roy. Sci. Paris* **7** (1776). 503–524.

Laplace, P. S. (1812). *Thèorie Analytique des Probabilitès*, Paris. (The 3rd edition of 1820 is reprinted as Vol. 7 of Laplace's collected works.)

Lawless, J. F. (1972). Conditional confidence interval procedures for the location and scale parameters of the Cauchy and logistic distributions. *Biometrika* **59**, 377–386.

Lawless, J. F. (1973). Conditional versus unconditional confidence intervals for the parameters of the Weibull distribution. *Journal of the American Statistical Association* **68**, 655–669.

Lawless, J. F. (1978). Confidence interval estimation for the Weibull and extreme value distributions. *Technometries* **20**, 355–368.

Le Cam, L. (1953). On some asymptotic properties of maximum likelihood estimates and related Bayes estimates. In *Univ. Calif. Publs. Statistics*, Vol. 1, pp. 277–329, Univ. of California Press, Berkeley and Los Angeles.

Le Cam, L. (1956). On the asymptotic theory of estimation and testing hypotheses. *Proc. 3rd Berkeley Symposium I*, 129–156.

Le Cam, L. (1958). Les propiétés asymptotiques des solutions de Bayes. *Publ. Inst. Statist. Univ. Paris.* **VII** (3-4, 17–35.

Le Cam, L. (1960). Locally asymptotically normal families of distributions. *Univ. California Publ. Statist.* **3**, 37–98.

Le Cam, L. (1964). Sufficiency and approximate sufficiency. *Annals of Mathematical Statistics* **35**, 1419–1455.

Le Cam, L. (1969). *Theorie Asymptotique de la Decision Statistique.* Presses de l'Université de Montreal.

Le Cam, L. (1970). On the assumptions used to prove asymptotic normality of maximum likelihood estimators. *Annals of Mathematical Statistics* **41**, 802–828.

Le Cam, L. (1972). Limits of experiments. *Proc. 6th Berkeley Symp. on Math. Stat. and Prob. I*, 245–261.

Le Cam, L. (1979). On a theorem of J. Hájek. In *Contributions to Statistics: J. Hájek Memorial Volume* (Jureckova, ed.), Academia, Prague.
[Rigorous and very general treatment of the large-sample theory of maximum-likelihood estimates, with a survey of the large previous literature on the subject.]

Le Cam, L. (1986). *Asymptotic Methods in Statistical Decision Theory.* Springer-Verlag, New York.

Le Cam, L. (1990). On the standard asymptotic confidence ellipsoids of Wald. *International Statistical Review.* **58**, 129–152.

Le Cam, L. and Yang, G. (2000). *Asymptotics in Statistics, Some Basic Concepts*, 2nd edition. Springer-Verlag, New York.

Ledwina, T. (1994). Data-driven version of Neyman's smooth test of fit. *Journal of the American Statistical Association* **89**, 1000–1005.

Léger, C. and Romano, J. P. (1990a). Bootstrap adaptive estimation: the trimmed-mean example. *Canadian Journal of Statistics* **18**, 297–314.

Léger, C. and Romano, J. P. (1990b). Bootstrap choice of tuning parameters. *Annals of the Institute of Statistical Mathematics* **42**, 709–735.

Lehmann, E. L. (1949). Some comments on large sample tests. In *Proceedings of the Berkeley Symposium on Mathematical Statistics and Probability*, University of CA Press, Berkeley. [Problem 13.26.]

Lehmann, E. L. (1950). Some principles of the theory of testing hypotheses. *Annals of Mathematical Statistics* **21**, 1–26.

Lehmann, E. L. (1951a). A general concept of unbiasedness. *Annals of Mathematical Statistics* **22**, 587–597. [Definition (1.8); Problems 1.2, 1.3, 1.4, 1.6, 1.7, and 1.14.]

Lehmann, E. L. (1951b). Consistency and unbiasedness of certain nonparametric tests. *Annals of Mathematical Statistics* **22**, 165–179.

Lehmann, E. L. (1952a). Testing multiparameter hypotheses. *Annals of Mathematical Statistics* **23**, 541–552.

Lehmann, E. L. (1952b). On the existence of least favorable distributions. *Annals of Mathematical Statistics* **23**, 408–416.

Lehmann, E. L. (1955). Ordered families of distributions. *Annals of Mathematical Statistics* **26**, 399–419. [Lemma 8.2.1; Problems 8.2, 8.9 (This problem is a corrected version of Theorem 8.5.1 of the paper in question. Thanks to R. Blumenthal for pointing out an error in the statement of this theorem in the paper.) and 8.10.]

Lehmann, E. L. (1958). Significance level and power. *Annals of Mathematical Statistics* **29**, 1167–1176.

Lehmann, E. L. (1961). Some model I problems of selection. *Annals of Mathematical Statistics* **32**, 990–1012.

Lehmann, E. L. (1980). An interpretation of completeness and Basu's theorem. *Journal of the American Statistical Association* **76**, 335–340. [Problem 5.70.]

Lehmann, E. L. (1985a). The Neyman-Pearson theory after 50 years. In *Proc. Neyman–Kiefer Conference* (LeCam and Olshen. eds.), Wadsworth, Belmont, CA.

Lehmann, E. L. (1985b). The Neyman-Pearson Lemma. In *Encycl. Stat. Sci.* **6**, 224–230.

Lehmann, E. L. (1993). The Fisher, Neyman-Pearson theories of testing hypotheses: one theory or two? *Journal of the American Statistical Association* **78**, 1242–1249.

Lehmann, E. L. (1997). Testing statistical hypotheses: the story of a book. *Statistical Science* **12**, 48–52.

Lehmann, E. L. (1998). *Nonparametrics: Statistical Methods Based on Ranks*, revised first edition. Prentice Hall, Upper Saddle River, New Jersey. [Previous edition by Holden-Day (1975).]

Lehmann, E. L. (1999). *Elements of Large-Sample Theory.* Springer-Verlag, New York.

Lehmann, E. L. and Casella, G. (1998). *Theory of Point Estimation*, Second Edition, Springer-Verlag, New York.

Lehmann, E. L. and Loh, W-Y. (1990). Pointwise versus uniform robustness in some large-sample tests and confidence intervals. *Scandinavian Journal of Statistics* **17**, 177–187.

Lehmann, E. L. and Rojo, J. (1992). Invariant directional orderings. *Annals of Statistics* **20**, 2100–2110.

Lehmann, E. L. and Romano, J. P. (2005). Generalizations of the familywise error rate. Technical Report 2003-37, Department of Statistics, Stanford University, to appear in *Annals of Statistics*.

Lehmann, E. L., Romano, J. P., and Shaffer, J. P. (2003). On optimality of stepdown and stepup procedures. Technical Report 2003-12, Department of Statistics, Stanford University.

Lehmann, E. L. and Scheffé, H. (1950, 1955). Completeness, similar regions, and unbiased estimation. *Sankhyā* **10**, 305–340; **15**, 219–236. [Introduces the concept of completeness. Theorem 4.4.1 and applications.]

Lehmann, E. L. and Shaffer, J. P. (1979). Optimal significance levels for multistage comparison procedures. *Annals of Statistics* **7**, 27–45.

Lehmann, E. L. and Stein, C. M. (1948). Most powerful tests of composite hypotheses. *Annals of Mathematical Statistics* **19**, 495–516. [Theorem 3.8.1 and applications.]

Lehmann, E. L. and Stein, C. M. (1949). On the theory of some non-parameteric hypotheses. *Annals of Mathematical Statistics* **20**, 28–45. [Develops the theory of optimum permutation tests, Problem 8.33.]

Lehmann, E. L. and Stein, C. M. (1953). The admissibility of certain invariant statistical tests involving a translation parameter. *Annals of Mathematical Statistics* **24**, 473–479.

Lentner, M. M. and Buehler, R. (1963). Some inferences about gamma parameters with an application to a reliability problem. *Journal of the American Statistical Association* **58**, 670–677.

Levy, K. J. and Narula, S. C. (1974). Shortest confidence intervals for the ratio of two normal variances. *Canadian Journal of Statistics* **2**, 83–87.

Lexis, W. (1875). *Einleitung in die Theorie der Bevölkerungsstatistik*, Strassburg.

Lexis, W. (1877). *Zur Theorie der Massenerscheinungen in der Menschlichen Gesellschaft*, Freiburg.

Liang, K. Y. (1984). The asymptotic efficiency of conditional likelihood methods. *Biometrika* **71**, 305–313.

Liang, K.Y. and Self, S. G. (1985). Tests for homogeneity of odds ratio when the data are sparse. *Biometrika* **72**, 353–358.

Lieberman, G. J. and Owen, D. B. (1961). *Tables of the Hypergeometric Probability Distribution*, Stanford University Press.

Lindley, D. V. (1957). A statistical paradox. *Biometrika* **44**, 187–192.

Linnik, Y. V., Pliss, V. A. and Salaevskii, O. V. (1968). On the theory of Hotelling's test (Russian). Dok. AN SSSR **168**, 743–746.

Littell, R. C. and Louv, W. C. (1981). Confidence regions based on methods of combining test statistics. *Journal of the American Statistical Association* **76**, 125–130.

Liu, H. and Berger, R. (1995). Uniformly more powerful one-sided tests for hypotheses about linear inequalities. *Annals of Statistics* **23**, 55-72.

Liu, R. Y. and Singh,K. (1987). On a partial correction by the bootstrap. *Annals of Statistics* **15**, 1713–1718.

Liu, R. Y. and Singh, K. (1992). Moving blocks jackknife and bootstrap capture weak dependence. In *Exploring the Limits of Bootstrap*, 225–248. Edited by LePage, R. and Billard, L., John Wiley, New York.

Liu, R. Y. and Singh, K. (1997). Notions of limiting P-values based on data depth and bootstrap. *Journal of the American Statistical Association* **92**, 266–277.

Loh, W.-Y. (1984a). Strong unimodality and scale mixtures. *Annals of the Institute of Statistical Mathematics* **36**, 441–450.

Loh, W.-Y. (1984b). Bounds on ARE's for restricted classes of distributions defined via tail-orderings. *Annals of Statistics* **12**, 685–701.

Loh, W.-Y. (1985). A new method for testing separate families of hypotheses. *Journal of the American Statistical Association* **80**, 362–368.

Loh, W.-Y. (1987). Calibrating confidence coefficients. *Journal of the American Statistical Association* **82**, 155–162.

Loh, W.-Y. (1989). Bounds on the size of the χ^2-test of independence in a contingency table. *Annals of Statistics* **17**, 1709–1722.

Loh, W.-Y. (1991). Bootstrap calibration for confidence interval construction and selection. *Statistica Sinica* **1**, 479–495.

Loomis, L. H. (1953). *An Introduction to Abstract Harmonic Analysis*. Van Nostrand, New York.

Lorenzen, T. J. (1984). Randomization and blocking in the design of experiments. *Communications in Statistics – Theory and Methods* **13**, 2601–2623.

Lou, W. (1996). On runs and longest run tests: a method of finite Markov chain embedding. *Journal of the American Statistical Association* **91**, 1595–1601.

Low, M. (1997). On nonparametric confidence intervals. *Annals of Statistics* **25**, 2547–2554.

Lyapounov, A. M. (1940). Sur les fonctions-vecteurs complètement additives, *Izv. Akad. Nauk SSSR Ser. Mat.* **4**, 465–478.

Maatta, J. and Casella, G. (1987). Conditional properties of interval estimators of the normal variance. *Annals of Statistics* **15**, 1372–1388.

Mack, G. A. and Skillings, J. H. (1980). A Friedman type rank test for main effects in a two-factor ANOVA. *Journal of the American Statistical Association* **75**, 947–951.

Madansky, A. (1962). More on length of confidence intervals. *Journal of the American Statistical Association* **57**, 586–599.

Mandelbaum, A. and Rüschendorf, L. (1987). Complete and symmetrically complete families of distributions. *Annals of Statistics* **15**, 1229–1244.

Mann, H. and Wald, A. (1942). On the choice of the number of intervals in the application of the chi-square test. *Annals of Mathematical Statistics* **13**, 306–317.

Mantel, N. (1987). Understanding Wald's test for exponential families. *American Statistician* **41**, 147–149.

Marasinghe, M. C. and Johnson, D. E. (1981). Testing subhypotheses in the multiplicative interaction model. *Technometrics* **23**, 385–393.

Marcus, R., Peritz, E. and Gabriel, K. R. (1976). On closed testing procedures with special reference to ordered analysis of variance. *Biometrika* **63**, 655–660.

Marden, J. (1982a). Minimal complete classes of tests of hypotheses with multivariate one-sided alternatives. *Annals of Statistics* **10**, 962–970.

Marden, J. (1982b). Combining independent noncentral chi-squared or F-tests. *Annals of Statistics* **10**, 266–270.

Marden, J. (1985). Combining independent one-sided noncentral t or normal mean tests. *Annals of Statistics* **13**, 1535–1553.

Marden, J. (1991). Sensitive and sturdy p-values. *Annals of Statistics* **19**, 918–934.

Marden, J. (2000). Hypothesis testing: from p-values to Bayes factors. *Journal of the American Statistical Association* **95**, 1316–1320.

Marden, J. and Muyot, M. (1995). Rank tests for main and interaction effects in analysis of variance. *Journal of the American Statistical Association* **90**, 1388–1398.

Marden, J. and Perlman, M. (1980). Invariant tests for means with covariates. *Annals of Statistics* **8**, 25–63.

Mardia, K. V. and Zemroch, P. J. (1978). *Tables of the F- and Related Distributions with Algorithms.* Academic Press, London. [Extensive tables of critical values for the central F- and related distributions.]

Maritz, J. S. (1979). A note on exact robust confidence intervals for location. *Biometrika* **66**, 163–166. [Problem 5.46(ii).]

Marshall, A. W. and Olkin, I. (1979). *Inequalities: Theory of Majorization and Its Applications.* Academic Press, New York.

Martín, A. and Tapia, J. (1998). On determining the p-value in 2×2 multinomial trials. *Journal of Statistical Planning and Inference* **69**, 33–49.

Massart, P. (1990). The tight constant in the Dvoretsky-Kiefer-Wolfowitz inequality. *Annals of Probability* **18**, 1269–1283.

Massey, F. J. (1950). A note on the power of a non-parametric test. *Annals of Mathematical Statistics* **21**, 440–443.

Mathew, M. and Sinha, B. (1988a). Optimum tests for fixed effects and variance components in balanced models. *Journal of the American Statistical Association* **83**, 133–135.

Mathew, M. and Sinha, B. (1988b). Optimum tests in unbalanced two-way models without interaction. *Annals of Statistics* **16**, 1727–1740.

Mattner, L. (1993). Some incomplete but boundedly complete location families. *Annals of Statistics* **21**, 2158–2162.

Mattner, L. (1996). Complete order statistics in parametric models. *Annals of Statistics* **24**, 1265–1282.

McCullagh, P. (1985). On the asymptotic distribution of Pearson's statistic in linear exponential-family models. *International Statistical Review* **53**, 61–67.

McCullagh, P. (1986). The conditional distribution of goodness-of-fit statistics for discrete data. *Journal of the American Statistical Association* **81**, 104–107.

McCullagh, P. and Nelder, J. (1989). *Generalized Linear Models*, 2nd edition. Chapman & Hall, London.

McCulloch, C. and Searle, S. (2001). *Generalized, Linear, and Mixed Models*. John Wiley, New York.

McDonald, L. L., Davis, B. M. and Milliken, G. A. (1977). A nonrandomized unconditional test for comparing two proportions in 2×2 contingency tables. *Technometrics* **19**, 145–158.

McKean, J. and Schrader, R. M. (1982). The use and interpretation of robust analysis of variance. In *Modern Data Analysis* (Launer and Siegel. eds.). Academic Press, New York.

Mee, R. (1990). Confidence intervals for probabilities and tolerance regions based on a generalization of the Mann-Whitney statistic. *Journal of the American Statistical Association* **85**, 793–800.

Meeks, S. L. and D'Agostino, R. (1983). A note on the use of confidence limits following rejection of a null hypothesis. *American Statistician* **37**, 134–136.

Meng, X. (1994). Posterior predictive p-values. *Annals of Statistics* **22**, 1142–1160.

Michel, R. (1979). On the asymptotic efficiency of conditional tests for exponential families. *Annals of Statistics* **7**, 1256–1263.

Milbrodt, H. and Strasser, H. (1990). On the asymptotic power of the two-sided Kolmogorov-Smirnov test. *Journal of Statistical Planning and Inference* **26**, 1–23.

Millar, W. (1983). The minimax principle in asymptotic statistical theory. In *Ecole d'Eté de Probabilités de Saint Flour XI 1981* (P.L. Hennequin, ed.), 75–266. Lecture Notes in Mathematics **976**, Springer-Verlag, Berlin.

Millar, W. (1985). Nonparametric applications of an infinite dimensional convolution theorem. *Zeitschrift für Wahrscheinlichkeitstheorie und verwandte Gebiete* **68**, 545–556.

Miller, F. R., Neill, J. and Sherfey, B. (1998). Maximin clusters for near-replicate regression lack of fit tests. *Annals of Statistics* **26**, 1411–1433.

Miller, F. L. and Quesenberry, C. (1979). Power studies of tests for uniformity II. *Communications in Statist. Simulation. Comput.* **8**, 271–290.

Miller, J. (1977a). Asymptotic properties of maximum likelihood estimates in the mixed model of the analysis of variance. *Annals of Statistics* **5**, 746–762.

Miller, R. G. (1977b). Developments in multiple comparisons 1966–1976. *Journal of the American Statistical Association* **72**, 779–788.

Miller, R. G. (1981). *Simultaneous Statistical Inference*, 2nd edition. Springer, New York.

Miller, R. G. (1986). *Beyond Anova.* John Wiley, New York.

Miwa, T. and Hayter, T. (1999). Combining the advantages of one-sided and two-sided test procedures for comparing several treatment effects. *Journal of the American Statistical Association* **94**, 302–307.

Montgomery, D. (2001). *Design and Analysis of Experiments*, 5th edition. John Wiley, New York.

Morgan, W. A. (1939). A test for the significance of the difference between the two variances in a sample from a normal bivariate population. *Biometrika* **31**, 13–19.

Morgenstem, D. (1956). Einfache Beispiele zweidimensionaler Verteilungen, *Mitteil. Math. Statistik* **8**, 234–235.

Mosteller, F. and Tukey, J. W. (1977). *Data Analysis and Regression: A Second Course in Statistics*, Addison-Wesley, MA.

Möttönen, J., Oja, H. and Tienari, J. (1997). On the efficiency of multivariate spatial sign and rank tests. *Annals of Statistics* **25**, 542–552.

Mudholkar, G. S. (1983). Fisher's z-transformation. *Encyclopedia of Statistical Science* **3**, 130-135. (S. Kotz, N.L. Johnson, C.B. Read, eds.)

Müller, C. (1998). Optimum robust testing in linear models. *Annals of Statistics* **26**, 1126–1146.

Murphy, S. and van der Vaart, A. (1997). Semiparametric likelihood ratio inference. *Annals of Statistics* **25**, 1471–1509.

Nachbin, L. (1965). *The Haar Integral.* Van Nostrand, New York.

Naiman, D. Q. (1984a). Average width optimality of simultaneous confidence bounds. *Annals of Statistics* **12**, 1199–1214.

Naiman, D. Q. (1984b). Optimal simultaneous confidence bounds. *Annals of Statistics* **12**, 702–715.

Nandi, H. K. (1951). On type B_1 and type B regions. *Sankhyā* **11**, 13–22. [One of the cases of Theorem 4.4.1, under regularity assumptions.]

Neuhaus, G. (1979). Asymptotic theory of goodness of fit tests when parameters are present: A survey. *Statistics* **10**, 479–494.

Neyman, J. (1923). On the application of probability theory to agriculture experiments. Essay on Principles. Section 9. Translated and edited by D. Dabrowska and T. Speed in (1990), *Statistical Science* **5**, 465–480, with comments by D. Rubin. The Polish original appeared in Roczniki Nauk Rolniczych Tom X (1923), 1–51 (*Annals of Agricultural Sciences*).

Neyman, J. (1935a). Sur un teorema concernente le cosidette statistiche sufficienti. *Giorn. Ist. Ital. Att.* **6**. 320–334.

Neyman, J. (1935b). Sur la vérification des hypothèses statistiques composées. *Bull. Soc. Math. France* **63**, 246–266. [Defines, and shows how to derive, tests of type B, that is, tests which are LMP among locally unbiased tests in the presence of nuisance parameters.]

Neyman, J. (1937a). Outline of a theory of statistical estimation based on the classical theory of probability. *Phil. Trans. Roy. Soc. Ser. A.* **236**, 333–380.

Neyman, J. (1937b). Smooth test for goodness of fit. *Skand. Aktuarietidskr.* **20**, 150–199.

Neyman, J. (1938a). L'estimation statistique traitée comme un problème classique de probabilité. *Actualités Sci. et Ind.* **739**, 25–57.

Neyman, J. (1938b). *Lectures and Conferences on Mathematical Statistics and Probability*, 1st edition (2nd edition, 1952), Graduate School, U.S. Dept. of Agriculture, Washington.

Neyman, J. (1939). On statistics the distribution of which is independent of the parameters involved in the original probability law of the observed variables, *Statist. Res. Mem.* **2**, 59–89. [Essentially Theorem 5.1.2 under regularity assumptions.]

Neyman, J. (1941a). On a statistical problem arising in routine analyses and in sampling inspection of mass distributions. *Annals of Mathematical Statistics* **12**, 46–76. [Theory of tests of composite hypotheses that are locally unbiased and locally most powerful.]

Neyman, J. (1941b). Fiducial argument and the theory of confidence intervals. *Biometrika* **32**, 128–150.

Neyman, J. (1949). Contribution to the theory of the χ^2 test. In *Proc. Berkeley Symposium on Mathematical Statistics and Probability*, Univ. of California Press, Berkeley, 239–273. [Gives a theory of χ^2 tests with restricted alternatives.]

Neyman, J. (1952). *Lectures and Conferences on Mathematical Statistics*, 2nd edition Washington Graduat School, U.S. Dept. of Agriculture, 43–66. [An account of various approaches to the problem of hypothesis testing.]

Neyman, J. (1967). *A Selection of Early Statistical Papers of J. Neyman*, Univ. of California Press, Berkeley. [Puts forth the point of view that statistics is primarily concerned with how to behave under uncertainty rather than with determining the values of unknown parameters, with inductive behavior rather than with inductive inference.]

Neyman, J. and Pearson, E. S. (1928). On the use and interpretation of certain test criteria for purposes of statistical inference. *Biometrika* **20A**, 175–240, 263–295.

Neyman, J. and Pearson, E. S. (1933a). On the testing of statistical hypotheses in relation to probability a priori. *Proc. Cambridge Phil. Soc.* **29**, 492–510.

Neyman, J. and Pearson, E. S. (1933b). On the problem of the most efficient tests of statistical hypotheses. *Phil. Trans Roy. Soc. Ser. A* **231**, 289–337.

Neyman, J. and Pearson, E. S. (1936a). Contributions to the theory of testing statistical hypotheses. I. Unbiased critical regions of type A and type A_1. *Statist. Res. Mem.* **1**. 1–37.

Neyman, J. and Pearson, E. S. (1936b). Sufficient statistics and uniformly most powerful tests of statistical hypotheses. *Statist. Res. Mem.* **1**, 113–137. [Problem 3.4(ii).]

Neyman, J. and Pearson, E. S. (1936, 1938). Contributions to the theory of testing statistical hypotheses. *Statist. Res. Mem.* **1**, 1–37; **2**, 25–57. [Defines unbiasedness and determines both locally and UMP unbiased tests of certain classes of simple hypotheses. Discusses tests of types A, that is, tests which are LMP among locally unbiased tests when no nuisance parameters are present.]

Neyman, J. and Pearson, E. S. (1967). *Joint Statistical Papers of J. Neyman and E. S. Pearson*, Univ. of California Press, Berkeley. [In connection with the problem of hypothesis testing, suggests assigning weights for the various possible wrong decisions and the use of the minimax principle.]

Nicolaou, A. (1993). Bayesian intervals with good frequentist behaviour in the presence of nuisance parameters. *Journal of the Royal Statistical Society Series B* **55**, 377–390.

Niederhausen, H. (1981). Scheffer polynomials for computing exact Kolmogorov–Smirnov and Renyi type distributions. *Annals of Statistics* **9**, 923–944.

Nikitin, Y. (1995). *Asymptotic Efficiency of Nonparametric Tests.* Cambridge University Press.

Noether, G. (1955). On a theorem of Pitman. *Annals of Mathematical Statistics* **26**, 64-68.

Nogales, A. and Oyola, J. (1996). Some remarks on sufficiency, invariance and conditional independence. *Annals of Statistics* **24**, 906–909.

Nogales, A., Oyola, J. and Pérez, P. (2000). Invariance, almost invariance and sufficiency. *Statistica*, LX, 277–286.

Nölle, G. and Plachky, D. (1967). Zur schwachen Folgenkompaktheit von Testfunktionen. *Zeitschrift für Wahrscheinlichkeitstheorie und verwandte Gebiete* **8**, 182–184.

Odén, A. and Wedel, H. (1975). Arguments for Fisher's permutation test. *Annals of Statistics* **3**, 518–520.

Olshen, R. A. (1973). The conditional level of the F-test. *Journal of the American Statistical Association* **68**, 692–698.

Oosterhoff, J. and van Zwet, W. (1979). A note on contiguity and Hellinger distance. *Contributions to Statistics*, Reidel, Dordrecht-Boston, Mass. London, 157–166.

Owen, A. (1988). Empirical likelihood ratio confidence intervals for a single functional. *Biometrika* **72**, 45–58.

Owen, A. (1995). Nonparametric likelihood confidence bounds for a distribution function. *Journal of the American Statistical Association* **90**, 516–521.

Owen, A. (2001). *Empirical Likelihood.* Chapman & Hall, New York.

Owen, D. B. (1985). Noncentral t-distribution. *Encycl. Statist. Sci.* **6**, 286–290.

Pace, L. and Salvan, A. (1990). Best conditional tests for separate families of hypotheses. *Journal of the Royal Statistical Society Series B* **52**, 125–134.

Pachares, J. (1961). Tables for unbiased tests on the variance of a normal population. *Annals of Mathematical Statistics* **32**, 84–87.

Patel, J. K. and Read, C. B. (1982). *Handbook of the Normal Distribution.* Marcel Dekker, New York.

Paulson, E. (1941). On certain likelihood ratio tests associated with the exponential distribution. *Annals of Mathematical Statistics* **12**, 301–306. [Discusses the power of the tests of Problem 5.15.]

Pawitan, Y. (2000). A reminder of the fallibility of the Wald statistic: likelihood explanation. *American Statistician* **54**, 54–56.

Pearson, E. S. (1929). Some notes on sampling tests with two variables. *Biometrika* **21**. 337–360.

Pearson, E. S. (1966). The Neyman–Pearson story: 1926–1934. In *Research Papers in Statistics: Festschrift for J. Neyman* (F. N. David, ed.), John Wiley, New York.

Pearson, E. S. and Hartley, H. O. (1972). *Biometrika Tables for Statisticians.* Cambridge University Press, Cambridge.

Pearson, K. (1900). On the criterion that a given system of deviations from the probable in the case of a correlated system of variables is such that it can be reasonably supposed to have arisen from random sampling. *Philosophical Magazine, Series 5* **50**, 157-175. (Reprinted in: *Karl Pearson's Early Statistical Papers*, Cambridge University Press, 1956). [The χ^2-test is proposed for testing a simple multinomial hypothesis, and the limiting distribution of the test criterion is obtained under the hypothesis. The test is extended to composite hypotheses but contains an error in the degrees of freedom of the limiting distribution; a correct solution for the general case was found by Fisher (1924a). Applications.]

Peisakoff, M. (1951). *Transformation of Parameters*, unpublished thesis. Princeton Univ. [Extends the Hunt-Stein theory of invariance to more general classes of decision problems; see Problem 1.11(ii). The theory is generalized further in Kiefer (1957, 1966) and Kudo (1955).]

Pena, E. (1998). Smooth goodness-of-fit tests for composite hypothesis in hazard based models. *Annals of Statistics* **26**, 1935–1971.

Pereira, B. (1977). Discriminating among separate models: A bibliography. *International Statistical Review* **45**, 163–172.

Peritz, E. (1965). On inferring order relations in analysis of variance. *Biometrics* **21**, 337–344.

Perlman, M. (1969). One-sided testing problems in multivariate analysis. *Annals of Mathematical Statistics* **40**, 549–567. [Correction: *Annals of Mathematical Statistics* **42** (1971), 1777.]

Perlman, M. (1972). On the strong consistency of approximate maximum likelihood estimators. *Proceedings of the Sixth Berkeley Symposium in Mathematical Statistics* **1**, University of California Press, 263–281.

Perlman, M. and Wu, L. (1999). The Emperor's new tests. *Statistical Science* **14**, 355–381.

Pesarin, F. (2001). *Multivariate Permutation Tests With Applications in Biostatistics.* John Wiley, Chichester, England.

Peters, D. and Randles, R. (1991). A bivariate signed rank test for the two-sample location problem. *Journal of the Royal Statistical Society Series B* **53**, 493–504.

Pfanzagl, J. (1967). A technical lemma for monotone likelihood ratio families. *Annals of Mathematical Statistics* **38**, 611–613.

Pfanzagl, J. (1968). A characterization of the one parameter exponential family by existence of uniformly most powerful tests. *Sankhyā Series A* **30**, 147–156.

Pfanzagl, J. (1974). On the Behrens-Fisher problem. *Biometrika* **61**, 39–47.

Pfanzagl, J. (1979). On optimal median unbiased estimators in the presence of nuisance parameters. *Annals of Statistics* **7**, 187–193.

Pfanzagl, J. (with the assistance of W. Wefelmeyer) (1982). *Contributions to a General Asymptotic Theory.* Springer-Verlag, New York.

Pfanzagl, J. (with the assistance of W. Wefelmeyer) (1985). *Asymptotic Expansions for General Statistical Models.* Springer-Verlag, New York, NY.

Piegorsch, W. W. (1985a). Admissible and optimal confidence bounds in simple linear regression. *Annals of Statistics* **13**, 801–817.

Piegorsch, W. W. (1985b). Average width optimality for confidence bands in simple linear regression. *Journal of the American Statistical Association* **80**, 692–697.

Pierce, D. A. (1973). On some difficulties in a frequency theory of inference. *Annals of Statistics* **1**, 241–250.

Pitman, E. J. G. (1937, 1938a). Significance tests which may be applied to samples from any population, *J. Roy. Statist. Soc. Suppl.* **4**, 119–130, 225–232; Biometrika **29**, 322–335.

Pitman, E. J. G. (1938b). The estimation of the location and scale parameters of a continuous population of any given form. *Biometrika* **30**, 391–421.

Pitman, E. J. G. (1939a). A note on normal correlation. *Biometrika* **31**, 9–12. [Problem 5.39(i).]

Pitman, E. J. G. (1939b). Tests of hypotheses concerning location and scale parameters. *Biometrika* **31**, 200–215. [In these papers the restriction to invariant procedures is introduced for estimation and testing problems involving location and scale parameters.]

Pitman, E. J. G. (1949). Lecture notes on nonparametric statistical inference, unpublished. [Develops the concept of relative asymptotic efficiency and applies it to several examples including the Wilcoxon test.]

Plackett, R. L. (1977). The marginal totals of a 2×2 table. *Biometrika* **64**. 37–42. [Discusses the fact that the marginals of a 2×2 table supply some, but only little, information concerning the odds ratio. See also Barndorff-Nielsen (1978), Example 10.8.]

Plackett, R. L. (1981). *The Analysis of Categorical Data*, 2nd edition. MacMillan, New York.

Politis, D. N. and Romano, J. P. (1994a). The stationary bootstrap. *Journal of the American Statistical Association* **89**, 1303–1313.

Politis, D. N. and Romano, J. P. (1994b). Large sample confidence regions based on subsamples under minimal assumptions. *Annals of Statistics* **22**, 2031–2050.

Politis, D. N., Romano, J. P. and Wolf, M. (1999). *Subsampling.* Springer, New York.

Pollard, D. (1984). *Convergence of Stochastic Processes.* Springer-Verlag, New York.

Pollard, D. (1997). Another look at differentiability in quadratic mean. In *Festschrift for Lucien Le Cam*, 305–314. Springer-Verlag, New York.

Polonik, W. (1999). Concentration and goodness-of-fit in higher dimensions: (Asymptotically) distribution-free methods. *Annals of Statistics* **27**, 1210–1229.

Posten, H. O., Yeh, H. C. and Owen, D. B. (1982). Robustness of the two-sample t-test under violations of the homogeneity of variance assumption. *Communications in Statistics* **11**, 109–126.

Pratt, J. W. (1958). Admissible one-sided tests for the mean of a rectangular distribution. *Annals of Mathematical Statistics* **29**, 1268–1271.

Pratt, J. W. (1961a). Length of confidence intervals. *Journal of the American Statistical Association* **56**, 549–567.

Pratt, J. W. (1961b). Review of Testing Statistical Hypotheses by E L. Lehmann. *Journal of the American Statistical Association* **56**. 163–167. [Problems 10.27, 10.28.]

Pratt, J. W. (1962). A note on unbiased tests. *Annals of Mathematical Statistics* **33**, 292–294.

Pratt, J. W. (1964). Robustness of some procedures for the two-sample location problem. *Journal of the American Statistical Association* **59**, 665–680. [Proposes and illustrates approach (ii) of Section 1.]

Prescott, P. (1975). A simple alternative to Student's t. *Applied Statistics* **24**, 210–217.

Przyborowski, J. and Wilenski, H. (1939). Homogeneity of results in testing samples from Poisson series. *Biometrika* **31**, 313–323. [Derives the UMP similar test for the equality of two Poisson parameters.]

Pukelsheim, F. (1993). *Optimal Design of Experiments.* John Wiley, New York.

Quenouille, M. (1949). Approximate tests of correlation in time series. *Journal of the Royal Statististical Society Series B* **11**, 68–84.

Quesenberry, C. P. and Starbuck, R. R. (1976). On optimal tests for separate hypotheses and conditional probability integral transformations. *Communications in Statistics (A)* **1**, 507–524.

Quine, M. P. and Robinson, J. (1985). Efficiencies of chi-square and likelihood ratio goodness-of-fit tests. *Annals of Statistics* 13, 727–742.

Radlow, R. and Alf, E. (1975). An alternative multinomial assessment of the accuracy of the chi-squared test of goodness of fit. *Journal of the American Statistical Association* **70**, 811–813.

Ramachandran, K. V. (1958). A test of variances. *Journal of the American Statistical Association* **53**, 741–747.

Ramsey, P. H. (1980). *Journal of Educational Statistics* **5**, 337–349.

Randles, R. and Wolfe, D. A. (1979). *Introduction to the Theory of Nonparametric Statistics.* John Wiley, New York.

Rao, C. R. (1947). Large sample tests of statistical hypotheses concerning several parameters with applications to problems of estimation. *Proc. Camb. Phil. Soc.* **44**, 50-57.

Rao, C. R. (1963). Criteria of estimation in large samples. *Sankhyā* **25**, 189–206.

Rao, C. R. and Wu, Y. (2001). On model selection (with discussion). In *Model Selection*, Lahiri, P. ed., IMS Lecture Notes–Monograph Series, volume 38.

Rao, P. (1968). Estimation of the location of the cusp of a continuous density. *Annals of Mathematical Statistics* **39**, 76–87.

Rayner, J. and Best, D. (1989). *Smooth Tests of Goodness of Fit.* Oxford University Press, Oxford.

Read, T. and Cressie, N. (1988). *Goodness-of-Fit Statistics for Discrete Multivariate Data.* Springer-Verlag, New York.

Reid, C. (1982). *Neyman from Life.* Springer, New York.

Reinhardt, H. E. (1961). The use of least favorable distributions in testing composite hypotheses. *Annals of Mathematical Statistics* **32**, 1034–1041.

Richmond, J. (1982). A general method for constructing simultaneous confidence intervals. *Journal of the American Statistical Association* **77**, 455–460.

Rieder, H. (1977). Least favorable pairs for special capacities. *Annals of Statistics* **5**, 909–921.

Rieder, H. (1994). *Robust Asymptotic Statistics.* Springer-Verlag, New York.

Ripley, B. (1987). *Stochastic Simulation.* Wiley, New York.

Robbins, H. (1948). Convergence of distributions. *Annals of Mathematical Statistics* **19**, 72–76.

Robert, C. (1993). A note on Jeffreys-Lindley paradox. *Statistica Sinica* **3**, 603–605.

Robert, C. (1994). *The Bayesian Choice.* Springer-Verlag, New York.

Robertson, T., Wright, F. and Dykstra, R. (1988). *Order Restricted Statistical Inference.* John Wiley, New York.

Robins, J., van der Vaart, A. and Ventura, V. (2000). Asymptotic distribution of *p*-values in composite null models. *Journal of the American Statistical Association* **95**, 1143-1172.

Robinson, G. (1976). Properties of Student's t and of the Behrens–Fisher solution to the two means problem. *Annals of Statistics* **4**, 963–971. [Correction (1982). *Ann. Statist.* **10**, 321.]

Robinson, G. (1979a). Conditional properties of statistical procedures. *Annals of Statistics* **7**, 742–755.

Robinson, G. (1979b). Conditional properties of statistical procedures for location and scale parameters. *Annals of Statistics* **7**, 756–771. [Basic results concerning the existence of relevant and semirelevant subsets for location and scale parameters, including Example 10.4.1.]

Robinson, G. (1982). Behrens-Fisher problem. In *Encycl. Statist. Sci.* **1**, 205-209.

Robinson, J. (1973). The large-sample power of permutation tests for randomization models. *Annals of Statistics* **1**, 291–296.

Robinson, J. (1983). Approximations to some test statistics for permutation tests in a completely randomized design. *Australian Journal of Statistics* **25**, 358–369. [Discusses the asymptotic performance of the permutation version of the F-test in randomized block experiments.]

Rojo, J. (1983). *On Lehmann's General Concept of Unbiasedness and Some of Its Applications*, Ph.D. Thesis. University of California, Berkeley.

Romano, J. P. (1988). A bootstrap revival of some nonparametric distance tests. *Journal of the American Statistical Association* **83**, 698–708.

Romano, J. P. (1989a). Do boostrap confidence procedures behave well uniformly in P? *Canadian Journal of Statistics* **17**, 75–80.

Romano, J. P. (1989b). Bootstrap and randomization tests of some nonparametric hypotheses. *Annals of Statistics* **17**, 141–159.

Romano, J. P. (1990). On the behavior of randomization tests without a group invariance assumption. *Journal of the American Statistical Association* **85**, 686–692.

Romano, J. P. (2004). On nonparametric testing, the uniform behavior of the t-test, and related problems. *Scandinavian Journal of Statistics*, to appear.

Romano, J. P. (2005). Optimal testing of equivalence hypothesis. *Annals of Statistics*, to appear.

Romano, J. P. and Shaikh, A. M. (2004). On control of the false discovery proportion. Department of Statistics Technical Report 2004-31, Stanford University, Stanford, CA.

Romano, J. P. and Siegel, A.F. (1986). *Counterexamples in Probability and Statistics*. Wadsworth, Belmont.

Romano, J. P. and Thombs, L. A. (1996). Inference for autocorrelations under weak assumptions. *Journal of the American Statistical Association* **91**, 590–600.

Romano, J. P. and Wolf, M. (2000). Finite sample nonparametric inference and large sample efficiency. *Annals of Statistics* **28**, 756–778.

Romano, J. P. and Wolf, M. (2004). Exact and approximate stepdown methods for multiple testing. *Journal of the American Statistical Association*, to appear.

Ronchetti, E. (1982). Robust alternatives to the F-test for the linear model. In *Probability and Statistical Inference* (Grossman, Pflug, and Wertz, eds.), D. Reidel, Dordrecht.

Rosenthal, R. and Rubin, D. B. (1985). Statistical analysis: summarizing evidence versus establishing facts. *Psych. Bull* **97**, 527–529,

Ross, S. (1996). *Stochastic Processes*, 2nd edition. John Wiley, New York.

Rothenberg, T. J. (1984). Hypothesis testing in linear models when the error covariance matrix is nonscalar. *Econometrica* **52**, 827–842.

Roussas, G. (1972) *Contiguous Probability Measures: Some Applications in Statistics*, Cambridge University Press.

Roy, K. K. and Ramamoorthi, R. V. (1979). Relationship between Bayes, classical and decision theoretic sufficiency. *Sankhyā* **41**, 48–58.

Roy, S. N. and Bose, R. C. (1953). Simultaneous confidence interval estimation. *Annals of Mathematical Statistics* **24**, 513–536.

Royden, H. L. (1988). *Real Analysis.* 3rd ed., Macmillan, New York.

Ruist, E. (1954). Comparison of tests for non-parametric hypotheses. Arkiv Mat. **3**, 133–136. [Problem 8.7.]

Rukhin, A. (1993). Bahadur efficiency of tests of separate hypotheses and adaptive test statistics. *Journal of the American Statistical Association* **88**, 161–165.

Runger, G. and Eaton, M. (1992). Most powerful invariant permutation tests. *Journal of Multivariate Analysis* **42**, 202–209.

Ruppert, D., Wand, M. P. and Carroll, R. J. (2003). *Semiparametric Regression.* Cambridge University Press.

Sackrowitz, H. and Samuel-Cahn, E. (1999). P-values as random variables – expected p-values. *American Statistician* **53**, 326–331.

Sahai, H. and Khurshid, A. (1995). *Statistics in Epidemiology: Methods, Techniques and Applications.* CRC Press, Boca Raton, Florida.

Sahai, H. and Ojeda, M. (2004). *Analysis of Variance for Random Models.* Birkhäuser, Boston.

Salaevskii, Y. (1971). *Essay in Investigations in Classical Problems of Probability Theory and Mathematical Statistics* (V. M. Kalinin and O. V. Salaevskii, eds.) (Russian), Leningrad Seminars in Math., Vol. 13, Steklov Math. Inst.; Engl. transl., Consultants Bureau, New York.

Sanathanan, L. (1974). Critical power function and decision making. *Journal of the American Statistical Association* **69**, 398–402.

Sarkar, S. K. (2002). Some results on false discovery rate in stepwise multiple testing procedures, *Annals of Statistics* **30**, 239–257.

Savage, L. J. (1962). *The Foundations of Statistical Inference.* Methuen, London.

Savage, L. J. (1972). *The Foundations of Statistics*, 2nd edition. Dover, New York.

Savage, L. J. (1976). On rereading R. A. Fisher (with discussion). *Annals of Statistics* **4**, 441–500.

Schafer, G. (1982). Lindley's paradox (with discussion). *Journal of the American Statistical Association* **77**, 325–351.

Schafer, G. (1988). Sharp null hypotheses. In *Encycl. Statist. Sci.* **8**, 433–436.

Scheffé, H. (1942). On the ratio of the variances of two normal populations. *Annals of Mathematical Statistics* **13**, 371–388.

Scheffé, H. (1943). On a measure problem arising in the theory of non-parametric tests. *Annals of Mathematical Statistics* **14**, 227–233.
[Proves the completeness of order statistics.]

Scheffé, H. (1947). A useful convergence theorem for probability distribution functions. *Annals of Mathematical Statistics* **18**, 434–438.

Scheffé, H. (1956). A 'mixed model' for the analysis of variance. *Annals of Mathematical Statistics* **27**, 23–36 and 251–271.

Scheffé, H. (1959). *Analysis of Variance.* John Wiley, New York.

Scheffé, H. (1970). Practical solutions of the Behrens-Fisher problem. *Journal of the American Statistical Association* **65**, 1501–1504. [Introduces the idea of logarithmically shortest confidence intervals for ratios of scale parameters.]

Scheffé, H. (1977). A note on a reformulation of the S-method of multiple comparison (with discussion). *Journal of the American Statistical Association* **72**, 143–146. [Problem 7.18.]

Schervish, M. (1995). *Theory of Statistics.* Springer-Verlag, New York.

Schoenberg, I. J. (1951). On Pólya frequency functions. I. *J. Analyse Math.* **1**, 331–374. [Example 8.2.1.]

Scholz, F. W. (1982). Combining independent P-values. In *A Festschrift for Erich L. Lehmann* (Bickel, Doksum, and Hodges, eds.), Wadsworth, Belmont. Calif.

Schuirmann, D. (1981). On hypothesis testing to determine if the mean of a normal distribution is contained in a known interval. *Biometrics* **37**, 617.

Schwartz, R. E. (1967a). Locally minimax tests. *Annals of Mathematical Statistics* **38**, 340–360.

Schwartz, R. (1967b). Admissible tests in multivariate analysis of variance. *Annals of Mathematical Statistics* **38**, 698–710.

Schwartz, R. (1969). Invariant proper Bayes tests for exponential families. *Amer. Math. Statist.* **40**, 270–283.

Schweder, T. (1988). A significance version of the basic Neyman-Pearson theory for scientific hypothesis testing. *Scandinavian Journal of Statistics* **15**, 225–242.

Schweder, T. and Spjøtvoll, E. (1982). Plots of P-values to evaluate many tests simultaneously. *Biometrika* **69**, 493–502.

Seal, H. L. (1967). Studies in the history of probability and statistics XV. The historical development of the Gauss linear model. *Biometrika* **54**, 1–24.

Searle, S. (1987). *Linear Models and Unbalanced Data.* John Wiley, New York.

Seber, G. A. F. (1977). *Linear Regression Analysis.* John Wiley, New York.

Seber, G. A. F. (1984). *Multivariate Observations.* John Wiley, New York.

Seidenfeld, T. (1992). R. A. Fisher's fiducial argument and Bayes' theorem. *Statistical Science* **7**, 358–368.

Sellke, T., Bayarri, J. and Berger, J. (2001). Calibration of p-values for testing precise null hypotheses. *American Statistician* **55**, 62–71.

Serfling R. H. (1980). *Approximation Theorems of Mathematical Statistics.* John Wiley, New York.

Severini, T. (1993). Bayesian interval estimates which are also confidence intervals. *Journal of the Royal Statistical Society Series B* **55**, 533–540.

Shaffer, J. P. (1973). Defining and testing hypotheses in multi-dimensional contingency tables. *Psych. Bull.* **79**, 127–141.

Shaffer, J. P. (1977a). Multiple comparisons emphasizing selected contrasts: An extension and generalization of Dunnett's procedure. *Biometrics* **33**, 293–303.

Shaffer, J. P. (1977b). Reorganization of variables in analysis of variance and multidimensional contingency tables. *Psych. Bull.* **84**, 220–228.

Shaffer, J. P. (1980). Control of directional errors with stagewise multiple test procedures. *Annals of Statistics* **8**, 1342–1347.

Shaffer, J. P. (1981). Complexity: an interpretability criterion for multiple comparisons. *Journal of the American Statistical Association* **76**, 395–401.

Shaffer, J. P. (1984). Issues arising in multiple comparisons among populations. In *Proc. Seventh Conference on Probab. Theory* (Iosifescu, ed.). Edit. Acad. Republ. Soc. Romania. Bucharest.

Shaffer, J. P. (1986). Modified sequentially rejective multiple test procedures. *Journal of the American Statistical Association* **81**, 826–831.

Shaffer, J. P. (1995). Multiple hypothesis testing: A review. *Annual Review of Psychology* **46**, 561–584.

Shaffer, J. P. (2002). Optimality results in multiple hypothesis testing. In *The First Erich L. Lehmann Symposium – Optimality*, Rojo and Pérez-Abren (eds.), IMS Lecture Notes **44**, Beachwood, Ohio.

Shao, J. (1999). *Mathematical Statistics.* Springer, New York.

Shao, J. and Tu, D. (1995). *The Jackknife and the Bootstrap.* Springer, New York.

Shapiro, S. S., Wilk M. B. and Chen H. J. (1968). A comparative study of various tests of normality. *Journal of the American Statistical Association* **63**, 1343–1372.

Shewhart, W. and Winters, F. (1928). Small samples – new experimental results. *Journal of the American Statistical Association* **23**, 144–153.

Shorack, G. (1972). The best test of exponentiality against gamma alternatives. *Journal of the American Statistical Association* **67**, 213–214.

Shorack, G. and Wellner, J. (1986). *Empirical Processes with Applications to Statistics.* John Wiley, New York.

Shorrock, G. (1990). Improved confidence intervals for a normal variance. *Annals of Statistics* **18**, 972–980.

Shuster, J. (1968). On the inverse Gaussian distribution function. *Journal of the American Statistical Association* **63**, 1514–1516.

Siegmund, D. (1985). *Sequential Analysis: Tests and Confidence Intervals.* Springer-Verlag, New York.

Siegmund, D. (1986). Boundary crossing probabilities and statistical applications. *Annals of Statistics* **14**, 361–404.

Sierpinski, W. (1920). Sur les fonctions convexes measurables. *Fundamenta Math.* **1**, 125–129.

Silvapulle, M. and Silvapulle, P. (1995). A score test against one-sided alternatives. *Journal of the American Statistical Association* **90**, 342–349.

Silvey, S. D. (1980). *Optimal Design: An Introduction to the Theory of Parameter Estimation*, Chapman & Hall, London.

Simpson, D. (1989). Hellinger deviance tests: efficiency, breakdown points, and examples. *Journal of the American Statistical Association* **84**, 107–113.

Singh, K. (1981). On the asymptotic accuracy of Efron's bootstrap. *Annals of Statistics* **9**, 1187–1195.

Small, C., Wang, J. and Yang, Z. (2000). Eliminating multiple root problems in estimation (with discussion). *Statistical Science* **15**, 313–341.

Smirnov, N. V. (1948). Tables for estimating the goodness of fit of empirical distributions. *Annals of Mathematical Statistics* **19**, 279–281.

Smith, D. W. and Murray, L. W. (1984). An alternative to Eisenhart's Model II and mixed model in the case of negative variance estimates. *Journal of the American Statistical Association* **79**, 145–151.

Sophister (G. Story) (1928). Discussion of small samples drawn from an infinite skew population. *Biometrika* 20A, 389–423.

Speed, F. M., Hocking, R. R. and Hackney, O. P. (1979). Methods of analysis of linear models with unbalanced data. *Journal of the American Statistical Association* **73**, 105–112.

Speed, T. (1987). What is an analysis of variance? (with discussion). *Annals of Statistics* **15**, 885–941.

Speed, T. (1990). Introductory remarks on Neyman (1923). *Statistical Science* **5**, 463–464.

Spiegelhalter, D. J. (1983). Diagnostic tests of distributional shape. *Biometrika* **70**, 401–409.

Spjøtvoll, E. (1967). Optimum invariant tests in unbalanced variance components models. *Annals of Mathematical Statistics* **38**, 422–428.

Spjøtvoll, E. (1972). On the optimality of some multiple comparison procedures. *Annals of Mathematical Statistics* **43**, 398–411.

Spjøtvoll, E. (1974). Multiple testing in analysis of variance. *Scandinavian Journal of Statistics* **1**, 97–114,

Sprott, D. A. (1975). Marginal and conditional sufficiency. *Biometrika* **62**, 599–605.

Spurrier, J. D. (1984). An overview of tests for exponentiality. *Communications in Statistics – Theory and Methods* **13**, 1635–1654.

Spurrier, J. D. (1999). Exact confidence bounds for all contrasts of three or more regression lines. *Journal of the American Statistical Association* **94**, 483–488.

Stein, C. M. (1951). A property of some tests of composite hypotheses. *Annals of Mathematical Statistics* **22**, 475–476. [Problem 3.58.]

Stein, C. M. (1956a). The admissibility of Hotelling's T^2-test. *Annals of Mathematical Statistics* **27**, 616–623.

Stein, C. M. (1956b). Efficient nonparametric testing and estimation. in *Proc. 3rd Berkeley Symp. Math. Statist. and Probab.* Univ. of Calif. Press, Berkeley.

Stein, C. M. (1962). Confidence sets for the mean of a multivariate normal distribution. *Journal of the Royal Statistical Society Series B* **24**, 265-296.

Stein, C. M. (1981). Estimation of the mean of a multivariate normal distribution. *Annals of Statistics* **9**, 1135–1151.

Stephens, M. (1974). EDF Statistics for goodness-of-fit and some comparisons. *Journal of the American Statistical Association* **69**, 730–737.

Stephens, M. (1976). Asymptotic results for goodness-of-fit statistics with unknown parameters. *Annals of Statistics* **4**, 357–369.

Stigler, S. M. (1977). Eight centuries of sampling inspection: The trial of the Pyx. *Journal of the American Statistical Association* **72**, 493–500.

Stigler, S. M. (1978). Francis Ysidro Edgeworth, Statistician (with discussion). *Journal of the Royal Statistical Society Series A* **141**, 287–322.

Stigler, S. M. (1986). Laplace's 1774 memoir on inverse probability. *Statistical Science* **1**, 359-378.

Stone, C. J. (1975). Adaptive maximum likelihood estimators of a location parameter. *Annals of Statistics* **3**, 267–294.

Stone, C. J. (1981). Admissible selection of an accurate and parsimonious normal linear regression model. *Annals of Statistics* **9**, 475–485.

Stone, M. and von Randow, R. (1968). Statistically inspired conditions on the group structure of invariant experiments and their relationships with other conditions on locally compact topological groups. *Zeitschrift für Wahrscheinlichkeitstheorie und verwandte Gebiete* **10**, 70–78.

Strasser, H. (1985). *Mathematical Theory of Statistics.* Walter de Gruyter, Berlin.

Stuart, A. and Ord, J. (1987). *Kendall's Advanced Theory of Statistics*, Vol. 1, 5th edition. Oxford University Press, New York.

Stuart, A. and Ord, J. (1991). *Kendall's Advanced Theory of Statistics*, Vol. 2, 5th edition. Oxford University Press, New York.

Stuart, A., Ord., J. and Arnold, S. (1999). *Kendall's Advanced Theory of Statistics*, Vol. 2A, 6th edition. Oxford University Press, New York..

Student (W.S. Gosset) (1908). On the probable error of the mean. *Biometrika* **6**, 1–25.

Student (W.S. Gosset) (1927). Errors of routine analysis. *Biometrika* **19**, 151–164.

Sugiura, N. (1965). An example of the two-sided Wilcoxon test which is not unbiased. *Annals of the Institute of Statistical Mathematics* **17**, 261–263

Sutton, C. (1993). Computer-intensive methods for tests about the mean of an asymmetrical distribution. *Journal of the American Statistical Association* **88**, 802–810.

Sverdrup, E. (1953). Similarity. unbiasedness, minimaxibility and admissibility of statistical test procedures. *Skand. Aktuar. Tidskrift* **36**, 64–86. [Theorem 4.3.1 and results of the type of Theorem 4.4.1. Applications including the 2×2 table.]

Swed, F. S. and Eisenhart, C. (1943). Tables for testing randomness of grouping in a sequence of alternatives. *Annals of Mathematical Statistics* **14**, 66–87.

Takeuchi, K. (1969). A note on the test for the location parameter of an exponential distribution. *Annals of Mathematical Statistics* **40**, 1838–1839.

Tallis, G. M. (1983). Goodness of fit. In *Encycl. Statist. Sci.*, Vol. 3. John Wiley, New York.

Tan, W. Y. (1982). Sampling distributions and robustness of t, F and variance-ratio in two samples and ANOVA models with respect to departure from normality. *Communications in Statistics – Theory and Methods* **11**, 2485–2511.

Tang, D. (1994). Uniformly more powerful tests in a one-sided multivariate problem. *Journal of the American Statistical Association* **89**, 1006–1011.

Tate, R. F. and Klett, G. W. (1959). Optimal confidence intervals for the variance of a normal distribution. *Journal of the American Statistical Association* **54**, 674–682.

Taylor, H. and Karlin, S. (1998). *An Introduction to Stochastic Modeling*, 3rd edition. Academic Press, San Diego, CA.

Thompson, W. R. (1936). On confidence ranges for the median and other expectation distributions for populations of unknown distribution form. *Annals of Mathematical Statistics* **7**, 122–128. [Problem 3.57.]

Tiku, M. L. (1967). Tables of the power of the F-test. *Journal of the American Statistical Association* **62**, 525–539.

Tiku, M. L. (1972). More tables of the power of the F-test. *Journal of the American Statistical Association* **67**, 709–710.

Tiku, M. L. (1985a). Noncentral chi-square distribution. *Encycl. Statist. Sci.* **6**, 276–280.

Tiku, M. L. (1985b). Noncentral F-distribution. *Encycl. Statist. Sci.* **6**, 280–294.

Tiku, M. L. and Balakrishnan, N. (1984). Testing equality of population variances the robust way. *Communications in Statistics – Theory and Methods* **13**, 2143–2159.

Tiku, M. L. and Singh, M. (1981). Robust test for means when population variances are unequal. *Communications in Statistics – Theory and Methods* **A10**, 2057–2071.

Tocher, K. D. (1950). Extension of Neyman–Pearson theory of tests to discontinuous variates. *Biometrika* **37**, 130–144. [Proves the optimum property of Fisher's exact test.]

Tong, Y. L. (1980). *Probability Inequalities in Multivariate Distributions.* Academic Press, New York.

Tritchler, D. (1984). On inverting permutation tests. *Journal of the American Statistical Association* **79**, 200–207.

Troendle, J. (1995). A stepwise resampling method of multiple testing. *Journal of the American Statistical Association* **90**, 370–378.

Tseng, Y. and Brown, L. D. (1997). Good exact confidence sets and minimax estimators for the mean vector of a multivariate normal distribution. *Annals of Statistics* **25**, 2228–2258.

Tukey, J. W. (1949a). One degree of freedom for non-additivity. *Biometrics* **5**, 232–242.

Tukey, J. W. (1949b). Standard confidence points. Unpublished Report 16, Statist, Res. Group, Princeton Univ. (To be published in Tukey's *Collected Works*, Wadsworth, Belmont, Calif.)

Tukey, J. W. (1953). The problem of multiple comparisons. Published in *The Collected Works of John W. Tukey: Multiple Comparisons, Volume VIII.* (1999). Edited by H. Braun, CRC Press, Boca Raton, Florida. [This MS, unpublished until 1999, was widely distributed and exerted a strong influence on the development and acceptance of multiple comparison procedures. It pioneered many of the basic ideas, including the T-method and a first version of Lemma 9.3.1.]

Tukey, J. W. (1958a). Bias and confidence in not quite large samples (abstract). *Annals of Mathematical Statistics* **29**, 614.

Tukey, J. W. (1958b). A smooth invertibility theorem. *Annals of Mathematical Statistics* **29**, 581–584.

Tukey, J. W. (1960). A survey of sampling from contaminated distributions. In *Contributions to Probability and Statistics* (Olkin, ed.), Stanford University Press.

Tukey, J. W. (1991). The philosophy of multiple comparisons. *Statistical Science* **6**, 100-116.

Tukey, J. W. and McLaughlin, D. H. (1963). Less vulnerable confidence and significance procedures for location based on a single sample: Trimming/Winsorization 1. *Sankhyā* **25**, 331–352.

Turnbull, H. (1952). *Theory of Equations*, 5th ed., Oliver and Boyd, Edinburgh.

Tweedie, M. C. K. (1957). Statistical properties of inverse Gaussian distributions I, II. *Annals of Mathematical Statistics* **28**, 362–377, 696–705.

Unni, K. (1978). *The Theory of Estimation in Algebraic and Analytic Exponential Families with Applications to Variance Components Models*, unpublished Ph.D. Thesis, Indian Statistical Institute.

Uthoff, V. A. (1970). An optimum test property of two well-known statistics. *Journal of the American Statistical Association* **65**, 1597–1600.

Uthoff, V. A. (1973). The most powerful scale and location invariant test of normal versus double exponential. *Annals of Statistics* **1**, 170–174.

Vadiveloo, J. (1983). On the theory of modified randomization tests for nonparametric hypotheses. *Communications in Statistics* **A12**, 1581–1596.

Vaeth, M. (1985). On the use of Wald's test in exponential families. *International Statistical Review* **53**, 199–214.

van Beek, P. (1972). An application of Fourier methods to the problem of sharpening the Berry-Esseen inequality. *Zeitschrift für Wahrscheinlichkeitstheorie und verwandte Gebiete* **23**, 187-196.

van der Laan, M., Dudoit, S. and Pollard, K. (2004). Multiple testing. Part II. Step-down procedures for control of the familywise error rate. *Statistical Applications in Genetics and Molecular Biology* **3**, Article 14.

van der Vaart, A. (1988). *Statistical Estimation in Large Parameter Spaces.* C.W.I. Tract 44, Amsterdam.

van der Vaart, A. (1998). *Asymptotic Statistics.* Cambridge University Press.

van der Vaart, A. and Wellner, J. (1996). *Weak Convergence and Empirical Processes.* Springer, New York.

Venable, T. C. and Bhapkar, V. P. (1978). Gart's test of interaction in a $2 \times 2 \times 2$ contingency table for small samples. *Biometrika* **65**, 669–672.

von Mises, R. (1931). *Wahrscheinlichkeitsrechnung.* Franz Deuticke, Leipzig, Germany.

Vu, H. and Zhou, S. (1997). Generalization of likelihood ratio tests under nonstandard conditions. *Annals of Statistics* **25**, 897–916.

Wacholder, S. and Weinberg, C. R. (1982). Paired versus two-sample design for a clinical trial of treatments with dichotomous outcome: Power considerations. *Biometrics* **38**, 801–812.

Wald, A. (1939). Contributions to the theory of statistical estimation and testing hypotheses. *Annals of Mathematical Statistics* **10**, 299–326. [A general formulation of statistical problems containing estimation and testing problems as special cases. Discussion of Bayes and minimax procedures.]

Wald, A. (1941a). Asymptotically most powerful tests of statistical hypotheses. *Annals of Mathematical Statistics* **12**, 1–19.

Wald, A. (1941b). Some examples of asymptotically most powerful tests. *Annals of Mathematical Statistics* **12**, 396–408.

Wald, A. (1942). On the power function of the analysis of variance test. *Annals of Mathematical Statistics* **13**, 434–439. [Problem 7.5. This problem is also treated by Hsu, "On the power function of the E^2-test and the T^2-test", *Annals of Mathematical Statistics* **16** (1945), 278–286.]

Wald, A. (1943). Tests of statistical hypotheses concerning several parameters when the number of observations is large. *Trans. Amer. Math. Soc.* **54**, 426–482. [General asymptotic distribution and optimum theory of likelihood ratio (and asymptotically equivalent) tests.]

Wald, A. (1949). Note on the consistency of the maximum likelihood estimate. *Annals of Mathematical Statistics* **20**, 595–601.

Wald, A. (1950). *Statistical Decision Functions.* John Wiley, New York. [Definition of most stringent tests.]

Wald, A. (1958). *Selected Papers in Statistics and Probability by Abraham Wald.* Stanford Univ. Press. [Defines and characterizes complete classes of decision procedures for general decision problems. The ideas of this and the preceding paper were developed further in a series of papers culminating in Wald's book (1950).]

Wallace, D. (1958). Asymptotic approximations to distributions. *Annals of Mathematical Statistics* **29**, 635–654.

Wallace, D. (1980). The Behrens–Fisher and Fieller–Creasy problems. In *R. A. Fisher: An Appreciation* (Fienberg and Hinkley. eds.) Springer. New York, pp. 119–147.

Walsh, J. E. (1949). Some significance tests for the median which are valid under very general conditions. *Annals of Mathematical Statistics* **20**, 64–81. [Lemma 6.7.1; proposes the Wilcoxon one-sample test in the form given in Problem 6.48. The equivalence of the two tests was shown by Tukey in an unpublished mimeographed report dated 1949. Contains a result related to Problem 4.13.]

Wang, H. (1999). Brown's paradox in the estimated confidence approach. *Annals of Statistics* **27**, 610–626.

Wang, Y. Y. (1971). Probabilities of the type I errors of the Welch tests for the Behrens-Fisher problem. *Journal of the American Statistical Association* **66**, 605–608.

Weisberg, S. (1985). *Applied Linear Regression*, 2nd edition. John Wiley, New York.

Welch, B. L. (1939). On confidence limits and sufficiency with particular reference to parameters of location. *Annals of Mathematical Statistics* **10**, 58–69.

Welch, B. L. (1951). On the comparison of several mean values: An alternative approach. *Biometrika* **38**, 330–336.

Welch, W. (1990). Construction of permutation tests. *Journal of the American Statistical Association* **85**, 693–698.

Wellek, S. (2003). *Testing Statistical Hypotheses of Equivalence.* Chapman & Hall/CRC.

Wells, M., Jammalamadaka, S. and Tiwari, R. (1993). Large sample theory of spacings statistics for tests of fit for the composite hypothesis. *Journal of the Royal Statistical Society Series B* **55**, 189–203.

Westfall, P. H. (1989). Power comparisons for invariant variance ratio tests in mixed ANOVA models. *Annals of Statistics* **17**, 318–326.

Westfall, P. H. (1997). Multiple testing of general contrasts using logical constraints and correlations. *Journal of the American Statistical Association* **92**, 299–306.

Westfall, P. H. and Young, S. (1993). *Resampling-Based Multiple Testing: Examples and Methods for P-Value Adjustment.* John Wiley, New York.

Westlake, W. (1981). Response to T. B. L. Kirkwood: bioequivalence testing – a need to rethink. *Biometrics* **37**, 589–594.

Wijsman, R. (1979). Constructing all smallest simultaneous confidence sets in a given class, with applications to MANOVA. *Annals of Statistics* **7**, 1003–1018.

Wijsman, R. (1980). Smallest simultaneous confidence sets with applications in multivariate analysis. *Journal of Multivariate Analysis* **V**, 483–498.

Wijsman, R. (1990). *Invariant Measures on Groups and Their Use in Statistics.* IMS Lecture Notes. Institute of Mathematical Statistics, Hayward, CA.

Wilcoxon, F. (1945). Individual comparisons by ranking methods. *Biometries* **1**, 90–83. [Proposes the two tests bearing his name. (See also Deuchler, 1914.)]

Wilk, M. B. and Kempthorne, O. (1955). Fixed, mixed, and random models. *Journal of the American Statistical Association* **50**, 1144–1167.

Wilks, S. S. (1938). The large-sample distribution of the likelihood ratio for testing composite hypotheses. *Annals of Mathematical Statistics* **9**, 60–62. [Derives the asymptotic distribution of the likelihood ratio when the hypothesis is true.]

Williams, D. (1991). *Probability With Martingales.* Cambridge University Press, Cambridge, England.

Wilson, E. B. (1927). Probable inference, the law of succession, and statistical inference. *Journal of the American Statistical Association* **22**, 209–212.

Wolfowitz, J. (1949). The power of the classical tests associated with the normal distribution. *Annals of Mathematical Statistics* **20**, 540–551.
[Proves Lemma 6.5.1 for a number of special cases. Proves that the standard tests of the univariate linear hypothesis and for testing the absence of multiple correlation are most stringent among all similar tests and possess certain related optimum properties.]

Wolfowitz, J. (1950). Minimax estimates of the mean of a normal distribution with known variance. *Annals of Mathematical Statistics* **21**, 218–230.

Working, H. and Hotelling, H. (1929). Application of the theory of error to the interpretation of trends. *Journal of the American Statistical Association* **24**, Mar. Suppl., 73–85.

Wu, C. F. (1990). On the asymptotic properties of the jackknife histogram. *Annals of Statistics* **18**, 1438–1452.

Wu, C. F. and Hamada, M. (2000). *Experiments: Planning, Analysis and Parameter Design.* John Wiley, New York.

Wynn, H. P. (1984). An exact confidence band for one-dimensional polynomial regression. *Biometrika* **71**, 375–379.

Wynn, H. P. and Bloomfield, P. (1971). Simultaneous confidence bands in regression analysis (with discussion). *Journal of the Royal Statistical Society Series B* **33**, 202–217.

Yamada, S. and Morimoto, H. (1992). Sufficiency. In *Current Issues in Statistical Inference: Essays in Honor of D. Basu.* Gosh and Pathak (eds.), IMS Lecture Notes **17**, Hayward, CA.

Yanagimoto, T. (1990). Dependence ordering in statistical models and other notions. In *Topics in Statistical Dependence*, Block, Sampson and Savits (eds.) (1990), IMS Lecture Notes **16**, Hayward, CA.

Yuen, K. K. (1974). The two-sample trimmed t for unequal population variances. *Biometrika* **61**, 165–170.

Zabell, S. (1992). R. A. Fisher and the fiducial argument. *Statistical Science* **7**, 369–387.

Zhang, J. (2002). Powerful goodness-of-fit tests based on the likelihood ratio. *Journal of the Royal Statistical Society Series B* **64**, 281–294.

Zhang, J. and Boos, D. (1992). Bootstrap critical values for testing homogeneity of covariance matrices. *Journal of the American Statistical Association* **87**, 425–429.

Author Index

Subject Index

Springer Texts in Statistics *(continued from page ii)*

Lehmann: Testing Statistical Hypotheses, Second Edition
Lehmann and Casella: Theory of Point Estimation, Second Edition
Lindman: Analysis of Variance in Experimental Design
Lindsey: Applying Generalized Linear Models
Madansky: Prescriptions for Working Statisticians
McPherson: Applying and Interpreting Statistics: A Comprehensive Guide, Second Edition
Mueller: Basic Principles of Structural Equation Modeling: An Introduction to LISREL and EQS
Nguyen and Rogers: Fundamentals of Mathematical Statistics: Volume I: Probability for Statistics
Nguyen and Rogers: Fundamentals of Mathematical Statistics: Volume II: Statistical Inference
Noether: Introduction to Statistics: The Nonparametric Way
Nolan and Speed: Stat Labs: Mathematical Statistics Through Applications
Peters: Counting for Something: Statistical Principles and Personalities
Pfeiffer: Probability for Applications
Pitman: Probability
Rawlings, Pantula and Dickey: Applied Regression Analysis
Robert: The Bayesian Choice: From Decision-Theoretic Foundations to Computational Implementation, Second Edition
Robert and Casella: Monte Carlo Statistical Methods, Second Edition
Rose and Smith: Mathematical Statistics with *Mathematica*
Ruppert: Statistics and Finance: An Introduction
Santner and Duffy: The Statistical Analysis of Discrete Data
Saville and Wood: Statistical Methods: The Geometric Approach
Sen and Srivastava: Regression Analysis: Theory, Methods, and Applications
Shao: Mathematical Statistics, Second Edition
Shorack: Probability for Statisticians
Shumway and Stoffer: Time Series Analysis and Its Applications
Simonoff: Analyzing Categorical Data
Terrell: Mathematical Statistics: A Unified Introduction
Timm: Applied Multivariate Analysis
Toutenburg: Statistical Analysis of Designed Experiments, Second Edition
Wasserman: All of Statistics: A Concise Course in Statistical Inference
Whittle: Probability via Expectation, Fourth Edition
Zacks: Introduction to Reliability Analysis: Probability Models and Statistical Methods